Welding and Metal Fabrication

Larry Jeffus

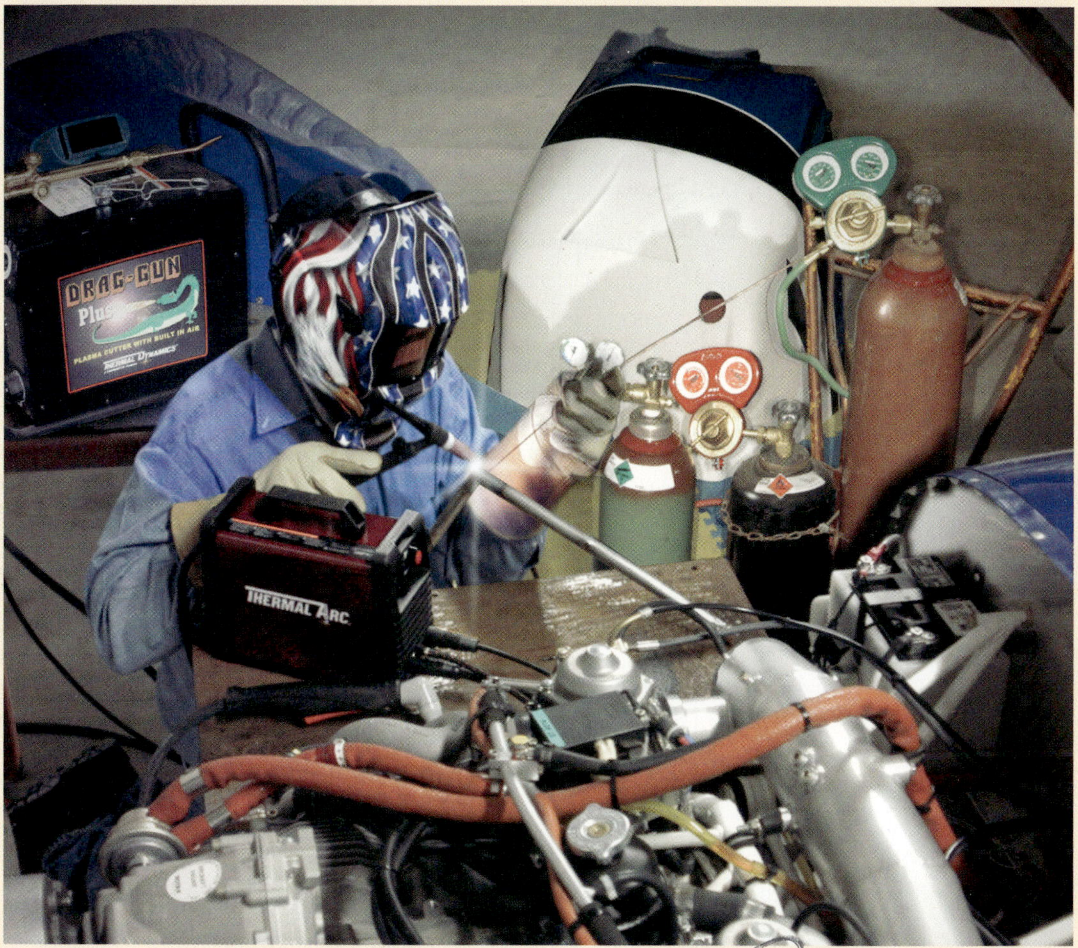

DELMAR
CENGAGE Learning™

Australia • Brazil • Japan • Korea • Mexico • Singapore • Spain • United Kingdom • United States

DELMAR
CENGAGE Learning™

Welding and Metal Fabrication

Larry Jeffus

Vice President, Editorial: **Dave Garza**

Director of Learning Solutions: **Sandy Clark**

Executive Editor: **Dave Boelio**

Managing Editor: **Larry Main**

Senior Product Manager: **Sharon Chambliss**

Editorial Assistant: **Jillian Borden**

Vice President, Marketing: **Jennifer Baker**

Executive Marketing Manager:
Deborah S. Yarnell

Marketing Specialist: **Mark Pierro**

Production Director: **Wendy Troeger**

Production Manager: **Mark Bernard**

Content Project Manager: **Joan Conlon**

Art Director: **Joy Kocsis**

Technology Project Manager:
Christopher Catalina

Library of Congress Control Number: 2010941439

ISBN-13: 978-1-4180-1374-5

ISBN-10: 1-4180-1374-9

Delmar
5 Maxwell Drive
Clifton Park, NY 12065-2919
USA

Cengage Learning products are represented in Canada by Nelson Education, Ltd.

For your lifelong learning solutions, visit **delmar.cengage.com**

Visit our corporate website at **cengage.com**

Printed in the USA
1 2 3 4 5 6 7 XX 13 12 11

Contents

Projects

Preface

For many welding jobs it can be as important to be able to lay out and fit up the weldment as it is to be able to weld it. This textbook, therefore, combines the skills of measuring, cutting, shaping, fitting, welding, and finishing.

The cover photo of *Welding and Metal Fabrication* was taken of me in my hangar as I was making a GTA weld on tubing. Aircraft tubing is used to make a number of parts for an aircraft, including the engine mount. Today, many light-sport aircraft, like this KitFox, can be built and flown by individuals.

All of the welding equipment required to build a homebuilt light-sport aircraft is shown in the cover photo. With some hand tools, a few power tools, and the welding and fabrication skills taught in this book, you too could build your own aircraft. There is nothing like the feeling of freedom one experiences as the pilot of your own plane.

Whether you build your own aircraft or you build anything else, there is a sense of accomplishment and a great feeling of pride in being able to point at something and say, "I made that." Over 30 years ago, I owned a welding company that made agricultural equipment, and I still look at farm equipment as I drive through the countryside to see if one of my units is out there in the field. Recently, my niece was in a volleyball playoff. It was being held in an empty warehouse that had been converted into a gym. I proudly told my wife, "I made the ramp for cars and light trucks at the loading dock on this building." I had made it over 20 years ago, and it was still there and being used to drive into the warehouse.

Not only have I fitted and welded things on my jobs, I have made parts for airplanes, barbeque grills, step and tow truck bumpers, truck racks, farm gates, wood stoves, compost bins, car jack stands, bases for power tools, toys, furniture, tools, car trailers, boat trailers, utility trailers, and hundreds of other big and small welded fabrications.

Welding and Metal Fabrication is designed to help you develop all of the skills to become a highly paid versatile welder. In addition, it is designed to make the process of learning to weld interesting and rewarding by having everything you weld on become something you can take home and use. The projects within each of the fabrication and welding chapters are designed to be functional even though the welds may be your first attempts at welding. So at first do not be overly critical of your welding skills, they will improve as you advance through the textbook. In that way, both your fabrication skills and your welding skills will improve together.

This textbook is the result of my more than 45 years of welding and fabrication experience. In addition to my personal experiences, I have drawn valuable welding and fitting information from many friends, colleagues, and former students. I know that not everyone who learns welding and fitting will use it to earn their primary paycheck; for some it will be a hobby or part-time job. For me, welding has not only been a lifelong, very profitable career, it has been my hobby too. This is what I had in mind when I wrote this textbook.

In the welding field, the ability to lay out, cut out, and assemble a welded part can be as important as good welding skills. This textbook is designed to give you all of the skills needed to be successful in welding. The chapters fall into four general categories:

- **General and Background** chapters cover important information that will help you work safely, be able to read drawings, and be a better all-round welding employee.
- **Theory and Application** chapters cover the equipment, materials, and procedures related to a single welding or cutting process.
- **Fabrication and Welding** chapters cover how to use each of the processes to produce a finished project and also cover related fabrication techniques.
- **Supplemental and Technical** chapters cover material that will help you solve welding problems you may encounter on the job by giving you information about other processes, metal identification, filler metal selection, and testing and certification.

As an example of the importance in learning proper welding skills, I offer the following true story regarding overwelding. It is a common problem that often results when welders believe that if a little weld is good, then a bigger weld is probably better. Overwelding is so common that it has its own term—"gorilla welding." Gorilla welds often are referred to as strong and ugly. I once subscribed to this myth. It is easy to argue that they are ugly, but are they really strong?

When I attended Hiwassee College in Madisonville, Tennessee, in the 1960s, I worked as a welder in a local shop that specialized in farm equipment repair welding. Like many young welders, I thought I was the world's best welder. My welds never cracked. I even convinced the shop owner to offer this warranty on my welds: "If our welds crack, we fix them for free."

To ensure that my welds did not fail, I made the biggest gorilla welds you have ever seen. Everyone knew that if I welded it, my welds would not fail. Although my welds never cracked, the base metal alongside my weld often did.

Cracks beside my welds meant my warrantee did not apply, so I could bill the farmer for my new welds. Although I stayed busy rewelding parts with cracks alongside my welds, my customers were happy with my work because they also thought a bigger weld was better.

By the time a local farmer got rid of his dump trailer that I had kept "fixing" for him, there was a 3-inch wide series of welds on the hinge point. Today, I realize that my welded repairs failed because of the size of the welds.

Often I made large welds on thin sheet metal parts that were subjected to vibration as the equipment was used in the fields. Each time a crack appeared next to one of my previous welds, I would just add another weld. Not all overwelding today is as blatant as mine was, but it still is a problem.

In addition to being costly, overwelding can produce a welded joint that cannot withstand the designed forces or vibration. Overwelded joints are not as flexible, and the resulting joint stresses are focused alongside the weld, which is why cracks always appeared just alongside my welds.

A good rule of thumb on weld sizes is that the weld size should not be much more than the thickness of the metal being joined, and the weld should have a smooth contour with the base metal.

The material in this textbook is designed to give you the skills to fabricate and weld projects so you never create the same problems for your customers that I once did.

SUPPLEMENTS

Accompanying the text is a carefully prepared supplements package, which includes an Instructor's Guide, an Instructor Resources CD, and a Workbook.

The Instructor's Guide contains chapter objectives, answers to the end of chapter review questions and answers to the questions in the Workbook.

The Instructor Resources CD contains PowerPoint lecture slides that present the highlights of each chapter, an ExamView computerized test bank, an electronic version of the Instructor's Guide, and an Image Library that includes images from the text.

In the Workbook, each chapter includes a variety of review questions that correspond with the chapter objectives to provide a comprehensive, in-depth review of material covered in the chapter. Questions include sentence completion, multiple choice, and figure-labeling exercises.

The *Welding Principles and Practices on DVD* series explains the concepts and shows the procedures students need to understand to become proficient and professional welders. Four DVDs cover Shielded Metal Arc Welding, Gas Metal Arc Welding, Flux Cored Arc Welding, and Oxyacetylene Welding in detail. The main subject areas are further broken down into subsections on each DVD for easy comprehension. The DVD set offers instructors and students the best welding multimedia learning tool at the fingertips.

Features of the Text

Objectives are a brief list of the most important topics to study stated at the beginning of each chapter.

Key Terms are the most important technical words you will learn in the chapter. These are listed at the beginning of each chapter following the objectives and appear in bold print where they are first used. These terms are also defined in the glossary at the end of the book.

Cautions summarize critical safety rules. They alert you to operations that could hurt you or others. Safety rules are covered in detail in Chapter 2 and also mentioned throughout the text wherever they apply to specific discussions, practice exercises, or experiments.

Practices are hands-on exercises designed to build your welding skills. Each practice describes in detail what skill you will learn and what equipment, supplies, and tools you will need to complete the exercise.

Experiments are designed to allow you to see what effect changes in the process settings, operation, or techniques have on the type of weld produced. Many are group activities and will help you work as a team.

Projects incorporate print reading, layout, cutting, fabricating, welding, and finishing techniques to make learning to weld more interesting while expanding the skill base that a welding fabricator needs. The projects are small, requiring minimum materials, while maximizing learning opportunities. The welds made on projects within the chapters are more decorative than structural, so even novice welders can make them.

Paperwork at the conclusion of each of the welding projects is designed to help students learn some of the skills needed for the business side of welding. The student is asked to fill out three forms: a "Time Sheet" where they list the hours spent on each task performed; a "Parts List", which includes a complete list of parts; and in some instances, a "Bill of Materials" where students will calculate the total project cost including material, labor, and markup.

Math problems similar to the types of math a welder normally would use are included. The math problems are laid out sequentially so they are easier to follow.

Spark Your Imagination features are designed to encourage creative thinking when problem solving. It is often important to look at other ways to solve challenges that you may encounter.

Shop Talk are hints and suggestions about good work habits and shortcuts that can make you a better worker.

Notes are designed to point out important information or give you related information that will help you understand the material or broaden your knowledge.

Illustrations include clear, colorful detailed photos, line art, graphs, and tables designed to help you visualize concepts and guide you as you learn to weld.

Summaries review the important points in the chapter and serve as a useful study tool. In the chapters that contain projects, an additional, larger, and more complex project is included in the summary. This larger project combines the fabrication and welding skills learned while building the smaller chapter projects.

Review Questions help measure the skills and knowledge you learned in the chapter. Each question is designed to help you apply and understand the information in the chapter.

Glossary definitions are provided for commonly used welding and fabrication terms. Additional line art in the glossary will also help you gain a greater understanding of challenging terms. Also, Spanish translations are provided for the glossary terms.

Appendix has Time Sheet, Parts List, and Bill of Materials forms, and material specification tables to aid in calculating the bills of material and invoices for the welding and fabrication projects in the chapters.

Acknowledgments

To bring a book of this size to publication requires the assistance of many individuals, and the author and publisher would like to thank the following for their unique contributions to this edition.

Special thanks are due to the following companies for their contributions to the text: American Welding Society, Inc.; SkillsUSA; Frommelt; Speedglas Inc.; Mine Safety Appliances Company; Lincoln Electric Company; Kedman Co.; Huntsman Product Division; NASA; Timely Products Co., Inc.; J. S. Staedtler, Inc.; Woodworker's Supply; Arcon Welding, LLC; ESAB Welding & Cutting Products; TWECO®, a Thermadyne® Company; Hobart Brothers Company; Dynatorque; Silver Dollar City; American Torch Tip; Controls Corporation of America; Western Enterprises; Crouse-Hinds Company; Prince & Izant Co.; Reynolds Metals Company; Victor Equipment, a Thermadyne Company; Arcair®, a Thermadyne® Company; Ingersoll-Rand; Caterpillar, Inc.; Laser Machining, Inc.; Praxair, Inc.; Messer Eutectic; M Engineering, Inc.; ARC Machines, Inc.; Fanuc Robotics North America, Inc.; Merrick Engineering, Inc.; Applied Robotics, Inc.; George VanderVoort; Buehler, Ltd., Lake Bluff, Illinois, Tempil Division; Big Three Industries, Inc.; Tinius Olsen Testing Machine Co., Inc.; Magnaflux Corporation; Newage Testing Instruments, Inc.; Garland Welding Supply Co., Inc.; Atlas Copco, City of Garland, Texas; and Garland Power and Light.

The following individuals who reviewed the book and/or gave the author ideas and suggestions that have been incorporated into the text: Ted Alberts, Bryan Baker, M.Ed., Wanda Sue Benton, Allen Bellamy, Douglas A. Desrochers, Ruben Felix, Drew Fontenot, Ross Gandy, Tom Gingras, Mike Jaeger, Chester Kemp, Wayne Knuppel, Danny Lott, Donald Montgomery, Jack E. Paige, Patrick Pico, Paul H. Plourde, Frank Rose, Winford E. Sartin, Dan Swearengen, Jim Barnett, Craig Berendzen, Samuel Butler, Ashley Black, Kevin Gratton, Steve Hrabal, Clive Lugmayer, James Owens, Tim Strouse, Gus Perez, Dewayne Roy, John Paul Gable, Randy Squibb, Robert L. Cross, Matt Siddens, Frank Wilkins, Joe Smith, Randy Zajic, and Traywick Duffie.

The author also would like to express his deepest appreciation to Thermal Dynamics and Jay Jones for all the props and help they provided for the preparation of this text. In addition he would like to thank Garland Welding Supply Co., Inc., for the loan of material and supplies for photo shoots.

My daughters, Wendy and Amy, and Marilyn's sons Ben and Sam for all of the general office help they provided.

About the Author

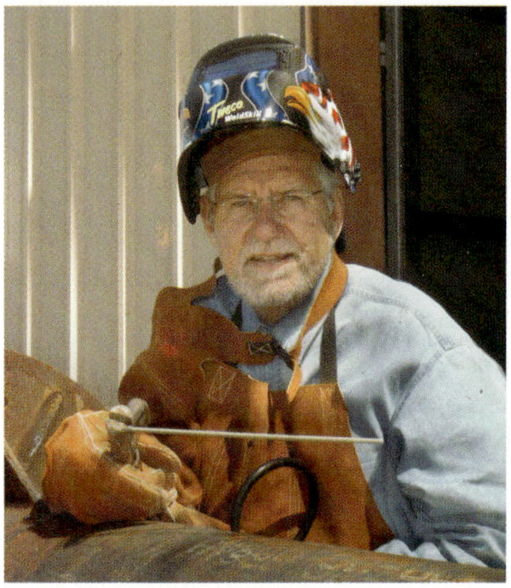

During my junior year of high school, I learned to weld in the metal shop. There, I was taught basic welding and fabrication, and I was able to build a number of projects in the shop using oxyacetylene welding, shielded metal arc welding, and brazing.

The practice welds helped me develop welding skills, and building the projects allowed me to start developing my fabrication skills. By the end of my junior year, I had become a fairly skilled welder.

In high school, I joined the Vocational Industrial Clubs of America (VICA), now SkillsUSA-VICA. SkillsUSA brings together educators, administrators, the corporate world, labor organizations, trade associations, and the government in a coordinated effort to address the need for a globally competitive, skilled workforce. The mission of SkillsUSA is to help students become world-class workers and responsible citizens. Through my involvement in SkillsUSA, I learned a great deal about industry and business and also the value of integrity, responsibility, citizenship, service, and respect. In addition, I developed leadership skills, established goals, and learned the value of performing quality work. These are all things that I still use in my life today.

During my senior year at New Bern High School, I was given an opportunity to join Mr. Z. T. Koonce's first class in a new program called Industrial Cooperative Training (ICT). ICT was a cooperative work experience program that coordinated school experiences with real jobs. The ICT program allowed me to attend high school in the morning to complete my academic courses for graduation and work as a welder in the afternoons. Mr. Koonce helped me get an afternoon job working as a welder in a local shipyard. Today, there are programs similar to ICT that are available in many areas. The programs may be called internship, Earn-While-You-Learn, apprenticeship, and so on.

Welding in the shipyard helped me improve my welding skills. As my skills improved, my supervisor allowed me to work on more difficult jobs. I welded on barges, military landing craft, tugboats, PT boats, small tankers, and marine vessels. I still remember being assigned my first welding fabrication job that I was to do all by myself. It was to fabricate a section of the retaining wall that went around the edge of the deck on a utility barge.

After graduation I went off to college. With my welding and fabrication skills, I was able to get a job in a small welding shop in Madisonville, Tennessee, while I attended Hiwassee College. After graduating from Hiwassee, I found other welding jobs that allowed me to continue my education at the University of Tennessee, where I earned a bachelor's degree. After four years, I had both a college degree and four years of industrial experience, which together qualified me for my job as a vocational teacher.

During my career as a welder, I have welded on military tanks, pressure vessels, oil well drilling equipment, farm equipment, buildings, race cars, aircraft, and more. As a career education teacher, I have taught in high schools, schools for special education, schools for the deaf, and three colleges. I also work as a consultant to schools and the welding industry.

Larry Jeffus is a recognized welding instructor with many years of experience teaching welding technology at the community college level. He has been actively involved in the American Welding Society, having served on the General Education Committee and as the chairman of the North Texas Section of the American Welding Society. He serves on a number of boards and advisory committees where he is a spokesman for career and technical education.

Chapter 1

Introduction

OBJECTIVES

After completing this chapter, the student should be able to:

- Discuss the role welding plays in the manufacture of modern products today.
- Explain the primary steps used in welding fabrication.
- Describe the most popular welding and cutting processes.
- Discuss the importance of careful and accurate part assembly for welding fabrication.
- List the types of jobs available in the welding industry.
- Convert from standard units to metric (SI) units and from SI units to standard units.

KEY TERMS

American Welding Society (AWS)
certification
coalescence
flux cored arc welding (FCAW)
forge welding (FOW)
gas metal arc welding (GMAW)
gas tungsten arc welding (GTAW)
oxyfuel gas (OF)
oxyfuel gas cutting (OFC)
oxyfuel gas welding (OFW)
plasma arc cutting (PAC)
semiautomatic process
shielded metal arc welding (SMAW)
torch or oxyfuel brazing (TB)
welding

INTRODUCTION

The ability to put things together to build a useful tool has been important since the dawn of humanity. Early civilizations used vines or rope to tie stones to sticks to make tools such as axes. Later, glues or cements were used to hold parts together. **Forge welding (FOW)** was used to join smaller pieces of

FIGURE 1-1 Example of forge welding done around 1850, in Baltimore, Maryland. Larry Jeffus

metal that could be heated in a forge and hammered together, Figure 1-1. At the dawn of the Iron Age, rivets were used to fabricate large metal structures like bridges, boilers, trains, and ships, Figure 1-2. But with the advent of modern welding, cutting, and brazing, civilization began advancing more rapidly. In fact, modern civilization could not exist without welding. Today, everything we touch was manufactured using some welding process or was made on equipment that was welded.

The skills of welding, cutting, and brazing are an essential part of metal fabrication.

- *Metal fabrication* is the building, shaping, and assembling of a product, equipment, or machine from raw metal stock. Metal fabrication can be done using rivets, bolts, welding, and so forth.
- A *welded metal fabrication* is primarily assembled using one or more of the following processes: welding, thermal cutting, or brazing.
- A *weldment* is an assembly in which its component parts are all joined by welding.

In some cases, a welded fabricated part may require some postweld finishing such as grinding, drilling, machining, or painting to complete the fabrication.

Welding Applications

Modern welding techniques are employed in the construction of numerous products. Ships, buildings, bridges, and recreational rides are examples of welded fabrications, Figure 1-3.

The exploration of space would not be possible without modern welding techniques. From the very beginning of early rockets to today's aerospace industry, welding has played an important role. Many of aerospace welding advancements have helped improve our daily lives.

Many experiments aboard the Space Station have involved welding and metal joining. The International Space Station was constructed using many advanced welding techniques. Someday, welders will be required to build even larger structures in the vacuum of space.

Welding is used extensively in the manufacture of automobiles, farm equipment, home appliances, computer components, mining equipment, and construction equipment. Railway equipment, furnaces, boilers, air-conditioning units, and hundreds of other

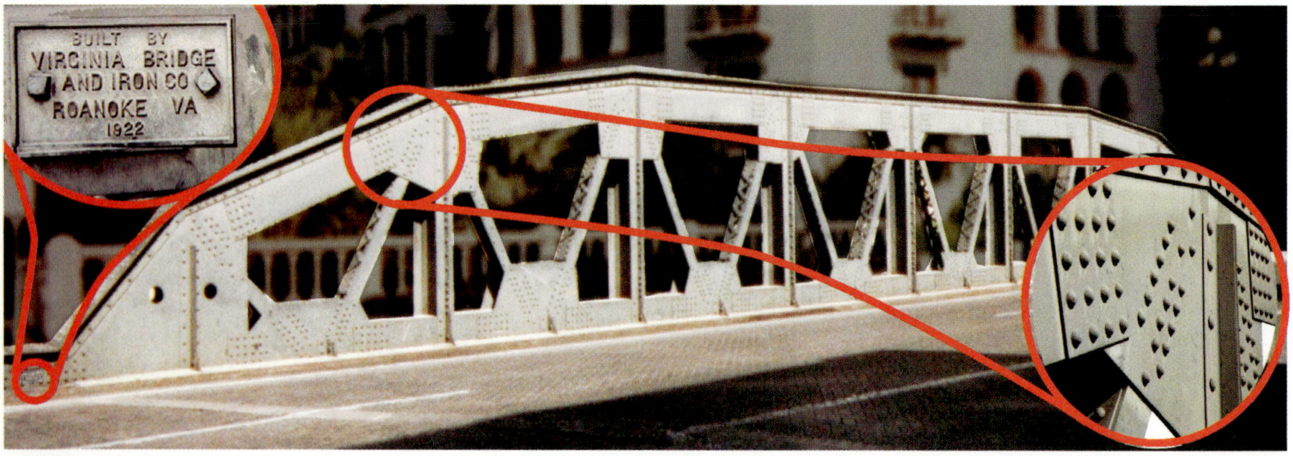

FIGURE 1-2A This all-riveted bridge was built in 1922, and is still in use today in San Antonio, Texas. Larry Jeffus

FIGURE 1-2B Riveted boiler used in a gold mine in Eagle Nest, New Mexico. Larry Jeffus

FIGURE 1-2C Riveted narrow-gauge logging train, once used along the coast of the Olympic Peninsula in Washington State. Larry Jeffus

products we use in our daily lives are also joined together by some type of welding process.

Fabrication Steps

The process of metal fabrication can be divided into several, often distinct steps. Following are the primary steps for fabrication:

- Layout—the process of drawing lines on the raw metal stock according to the parts drawings and specifications, Figure 1-4.
- Cut out—the process of removing all of the unwanted material around the laid-out part or sometimes just cutting material to the desired length. Some of the most common methods of cutting out the parts are flame cutting, plasma cutting, sawing, and punching, Figure 1-5.
- Assembling—the process of placing all the parts together in the correct location and orientation with each other. The parts may be held in place with small welds called tack welds or by some type of clamp, Figure 1-6.
- Welding—the process of permanently attaching the parts together to form the finished part, Figure 1-7.
- Finishing—can be accomplished by any number of different processes such as grinding, polishing, drilling, machining, painting, etc., Figure 1-8.

Not all metal fabrication includes all of the steps, and the difficulty of each step varies with the complexity of the fabrication. In addition, sometimes the order in which each step is done may change. For example,

FIGURE 1-3 Welded car ferry used on the Cherry Branch–Minnesott Beach crossing of the Neuse River in North Carolina. Larry Jeffus

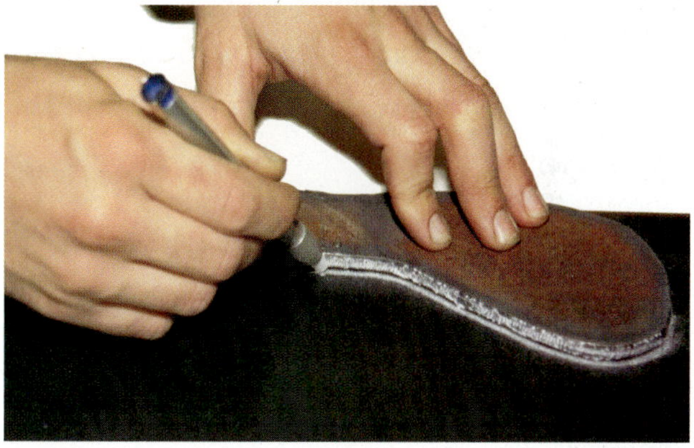

FIGURE 1-4 One way to lay out parts, is to trace them. Larry Jeffus

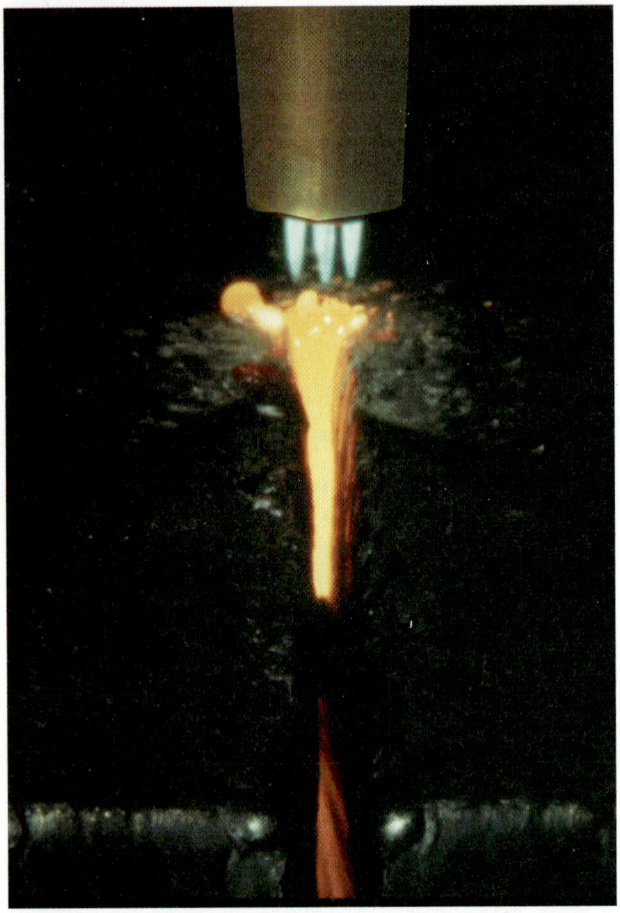

FIGURE 1-5 Oxyacetylene torch cutting a 2-in. (50.8-mm) thick steel plate. Larry Jeffus

FIGURE 1-7 Gas tungsten arc welding was used to join this stainless steel flange to pipe. Larry Jeffus

FIGURE 1-8 Angle grinding a weld to prepare it for finish painting. Larry Jeffus

FIGURE 1-6 Magnetic alignment clamp used to hold pipe in the correct placement for welding. Larry Jeffus

it may be necessary to wait until part of the assembly has been welded before laying out the location of an additional part; or a part may be trimmed to fit once other parts have been welded in place.

WELDING DEFINED

Most people think of **welding** as either a gas torch or electric arc welding process. They also think of it as just melting metal together. In the earlier history of welding, that was true, but welding is a lot more than that today. For example, welds can be made without an arc or flame with the induction welding (IW) process; without heat using the pressure welding (PW) process; or with an explosive using the explosion welding (EW) process. In fact, welding today is much more than the basics; it can be a very sophisticated process.

The **American Welding Society's (AWS)** definition of welding is very technical to reflect the differences in the welding processes used today. The AWS definition of *welding* states that welding is "a localized coalescence of metals or nonmetals produced either by heating the materials to the required welding temperatures, with or without the application of pressure, or by the application of pressure alone, and with or without the use of filler materials." The term **coalescence** means the fusion or growing together of the grain structure of the materials being welded. The definition includes the terms *metals or nonmetals* because materials such as plastics ceramics, and so forth, are not metals and they can be welded. The phrase *with or without the application of pressure* is important because without the application of significant pressure, some of the processes would not work, such as electric resistance welding (ERW) and friction welding (FW). In some welding processes only pressure is used to cause localized coalescence such as the PW and EW processes. And the last part of the definition says *with or without the use of filler materials,* meaning welded joints can be made by using only the base material.

A nontechnical definition of welding would be that welding is the joining together of the surface(s) of a material by the application of heat only, pressure only, or with heat and pressure together so that the surfaces fuse together. A filler material may or may not be added to the joint.

Weld Quality

We would like to think that every weld is made perfectly, with not even a slight flaw or imperfection allowed. However, that is not possible. The higher the welding standard, the higher the cost to produce a weld to that standard. Therefore, we often say that a weld must be fit for service. *Fit for service* means that there is a reasonable expectation that the weld will never fail as long as the weldment is used as it was designed to be used. So the quality of welding required for weldments differs depending on the intended service of the weldment. For example, a weld that is made on a high-pressure oil refinery vessel must be of an extremely high quality, Figure 1-9. A weld failure on such a vessel would be catastrophic, causing great property damage and possible loss of life. However, if a weld made on a driveway gate failed, it might be inconvenient but not likely to cause a significant loss of life like a high-pressure vessel failing, Figure 1-10.

To further illustrate this point, Figure 1-11 shows two welds made on two different vehicles. The weld shown in Figure 1-11A was made in 1945 by my grandfather on the family farm. He made the weld using bare metal electrodes. This farm trailer made

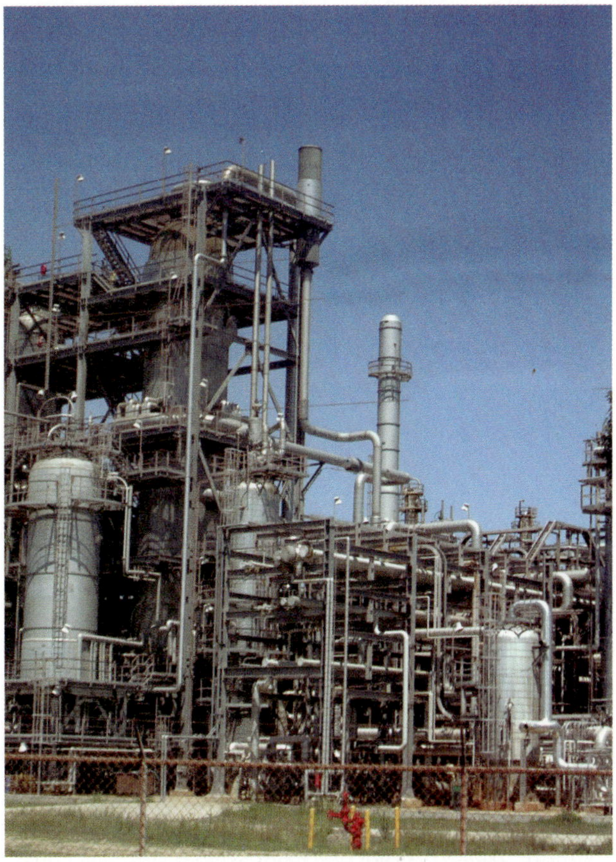

FIGURE 1-9 An oil refinery is an example of a structure that requires critical welds to ensure its safe operation.
Larry Jeffus

FIGURE 1-10 A farm gate is an example of a structure containing noncritical welds. Larry Jeffus

FIGURE 1-11 (A) Welded in 1945, using bare metal electrodes. (B) Welded in 2008 with GTA welding. Larry Jeffus

FIGURE 1-12 Farm trailer, 3 miles per hour. Larry Jeffus

FIGURE 1-13 Formula 1 race car, 300 miles per hour. Larry Jeffus

from a Model A Ford car axle is still being used, Figure 1-12. The weld shown in Figure 1-11B was made in 2008 to hold the front suspension on a Formula 1 race car, Figure 1-13. The weld on the trailer tongue has lasted for more than half a century traveling around the farm at 3 or 4 miles per hour. But it would last only a few seconds traveling around a racetrack at 300 miles per hour on a race car.

Good welders always try to make perfect welds no matter what code or standard is specified.

WELDING AND CUTTING PROCESSES

Welding processes differ greatly in the manner in which heat, pressure, or both heat and pressure are applied and in the type of equipment used. Table 1-1 lists various welding and allied processes. Some 67 welding processes are listed, requiring hammering, pressing, or rolling to effect the coalescence in the weld joint. Other methods bring the metal to a fluid state, and the edges flow together.

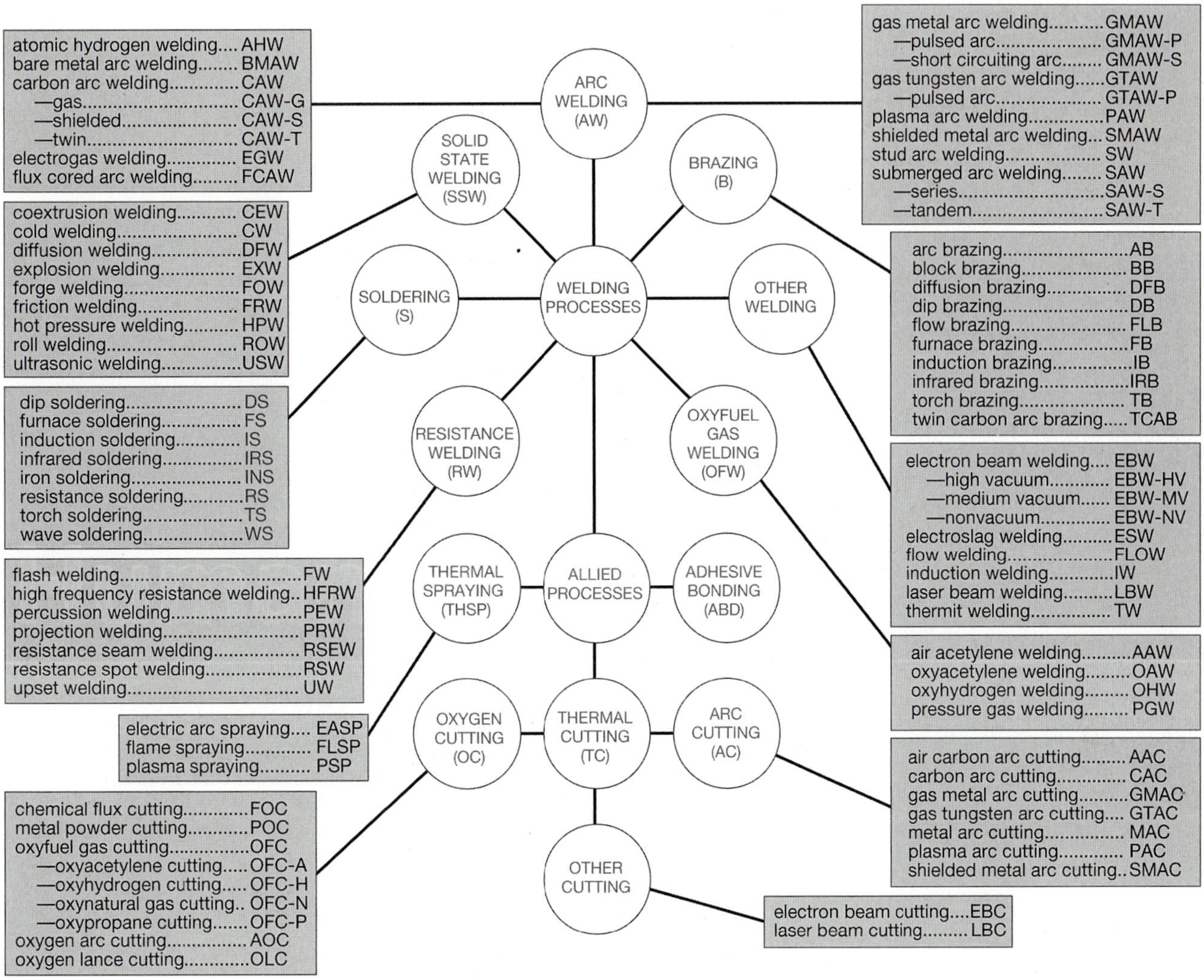

Table 1-1 Master Chart of Welding and Allied Processes *(American Welding Society)*

The most popular welding processes are gas metal arc welding(GMAW) , flux cored arc welding (FCAW), shielded metal arc welding (SMAW), gas tungsten arc welding (GTAW), oxyacetylene welding (OAW), and torch or oxyfuel brazing (TB). The two most popular thermal cutting processes are oxyacetylene cutting (OAW) and plasma arc cutting (PAC).

Welders, like many professionals, have developed *jargon*, nonstandard terms for many of the welding processes. For example, the oxyacetylene welding process is a part of the larger group of processes known as **oxyfuel gas welding (OFW).** Sometimes it is referred to as *gas welding* and *torch welding*. Shielded metal arc welding is sometimes referred to as *stick welding, rod welding,* or just *arc welding.* As you begin your work career, you will learn the various names used in your area, but you should always keep in mind and use the more formal terms whenever possible.

Gas Metal Arc Welding

Gas metal arc welding (GMAW) uses a solid electrode wire that is continuously fed from a spool, through the welding cable assembly, and out through the gun. A shielding gas flows through a separate tube in the cable assembly, out of the welding gun nozzle, and around the electrode wire. The welding power flows through a cable in the cable assembly and is transferred to the electrode wire at the welding gun. The GMA weld is produced as the arc melts the end of the continuously fed filler electrode wire and the surface of the base metal. The molten electrode metal transfers across the arc and becomes part of the weld. The gas shield flows out of the welding gun nozzle to protect the molten weld from atmospheric contamination.

GMA welding is extremely fast and economical because it can produce long welds rapidly that require very little postweld cleanup. This process can be used to weld metal ranging in thickness from thin-gauge sheet metal to heavy plate metal by making only a few changes in the welding setup.

Flux Cored Arc Welding

Flux cored arc welding (FCAW) uses a flux core electrode wire that is continuously fed from a spool, through the welding cable assembly, and out through the gun. The welding power also flows through the cable assembly. Some welding electrode wire types must be used with a shielding gas, as in GMA welding, but others have enough shielding, which is produced as the flux core vaporizes. The welding current melts both the filler wire and the base metal. When some of the flux vaporizes, it forms a gaseous cloud that protects the surface of the weld. Some of the flux that melts travels across the arc with the molten filler metal where it enters the molten weld pool. Inside the molten weld metal, the flux gathers up impurities and floats them to the surface where it forms a slag covering on the weld as it cools.

Although slag must be cleaned from the FCA welds after completion, the advantages of this process, including high quality, versatility, and welding speed offset this requirement.

Gas metal arc welding and flux cored arc welding are very different welding processes, but they use very similar welding equipment, Figure 1-14. Both GMA and FCA welding are classified as **semiautomatic processes** because the filler metal is automatically fed into the welding arc, and the welder manually moves the welding gun along the joint being welded. GMA and FCA welding are the first choice for many welding fabricators because these processes are cost effective, produce high-quality welds, and are flexible and versatile. In addition to welding supply stores, many others stores such as hardware stores, building supply stores, automotive supply stores, and others carry GMA/FCA welding equipment and filler metals.

Shielded Metal Arc Welding

Shielded metal arc welding (SMAW) uses a 14-in.-(350-mm) long consumable stick electrode that both conducts the welding current from the electrode holder to the work, and as the arc melts the end of the electrode away, it becomes part of the weld metal. The welding arc vaporizes the solid flux that covers the electrode so that it forms an expanding gaseous cloud to protect the molten weld metal. In addition to protecting molten weld metal, fluxes also perform a number of beneficial functions for the weld, depending on the type of electrode being used.

SMA welding equipment can be very basic compared to that used in other welding processes. It

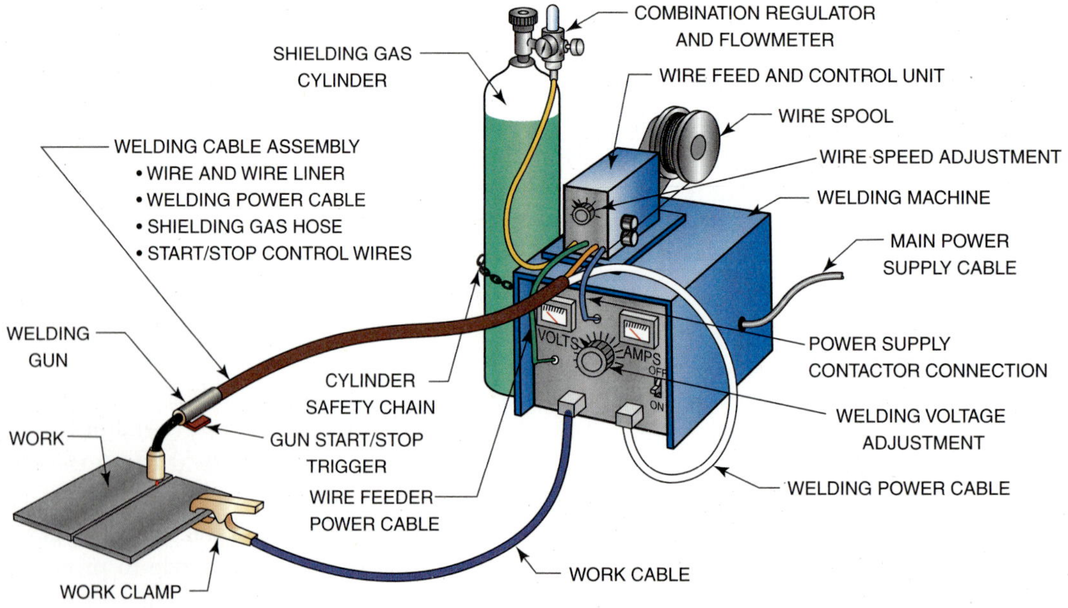

FIGURE 1-14 Gas metal arc (GMA) and/or flux cored arc (FCA) welding equipment.

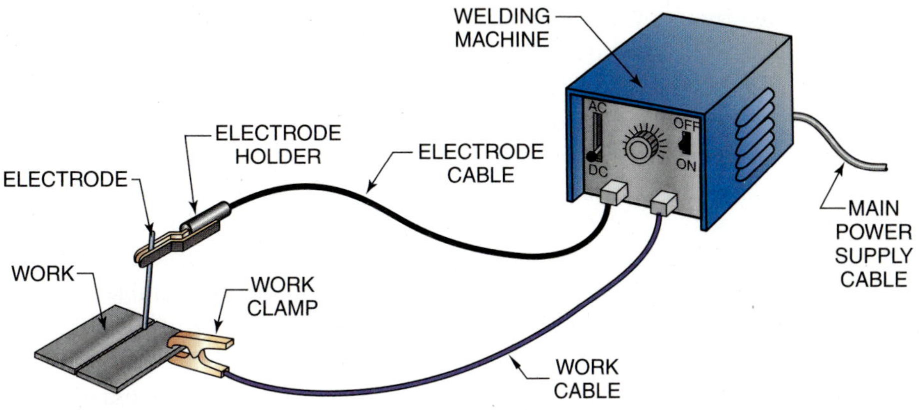

FIGURE 1-15 Shielded metal arc (SMA) welding equipment. © Cengage Learning 2012

can consist of a welding transformer and two welding cables with a work clamp and electrode holder, Figure 1-15. There are more types and sizes of SMA welding electrodes than there are filler metal types and sizes for any other welding process. This wide selection of filler metal allows welders to select the best electrode type and size to fit their specific welding job requirements. So, a wide variety of metal types and metal thicknesses can be joined with one machine.

Gas Tungsten Arc Welding

Gas tungsten arc welding (GTAW) uses a nonconsumable electrode made of tungsten. In GTA welding the arc between the electrode and the base metal melts the base metal and the end of the filler metal as it is manually dipped into the molten weld pool. A shielding gas flowing from the gun nozzle protects the molten weld metal from atmospheric contamination. A foot or thumb remote-control switch may be added to the basic GTA welding setup to allow the welder better control, Figure 1-16. This remote-control switch is often used to start and stop the welding current as well as make adjustments in the power level.

GTA welding is the cleanest of all of the manual welding processes. But because there is no flux used to clean the weld in GTA welding, all surface contamination such as oxides, oil, and dirt must be cleaned from the part being welded and the filler metal so it does

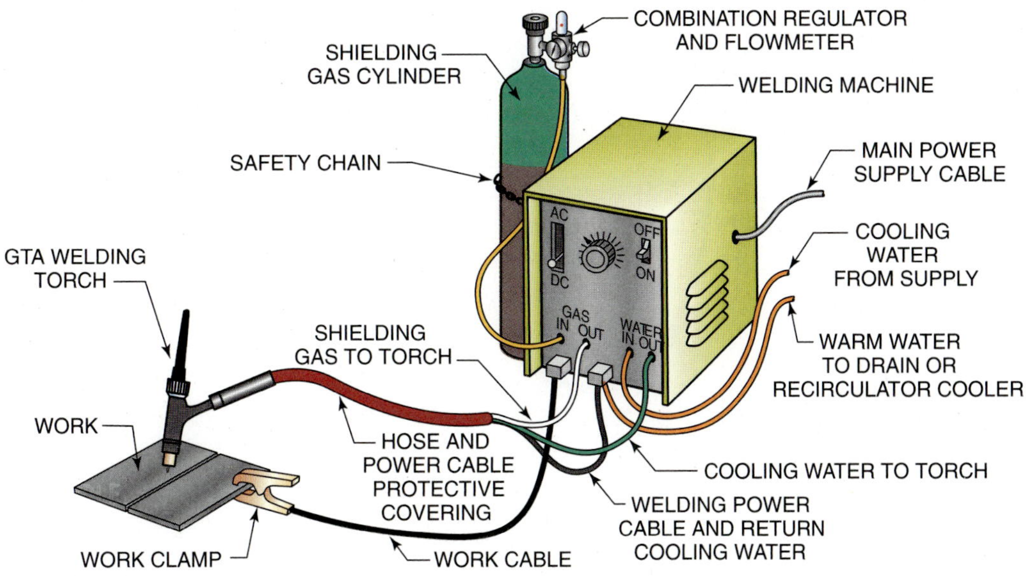

FIGURE 1-16 Gas tungsten arc (GTA) welding equipment. © Cengage Learning 2012

not contaminate the weld. Even though GTA welding is slower and requires a higher skill level as compared to other manual welding processes, it is still in demand because it can be used to make extremely high-quality welds in applications in which weld integrity is critical. In addition, there are metal alloys that can be joined only with the GTA welding process.

Oxyacetylene Welding, Brazing, and Cutting

Oxyacetylene welding (OAW) and **torch or oxyfuel brazing (TB)** can be done with the same equipment, and oxyfuel gas cutting (OFC) uses very similar equipment, Figure 1-17.

In OA welding and TB, a high-temperature flame is produced at the torch tip by burning oxygen and a fuel gas. The most common fuel gas is acetylene; however, other combinations of oxygen and fuel gases **(oxyfuel gas [OF])** can be used for welding such as hydrogen, Mapp®, or propane. In OF welding the base metal is melted, and a filler metal may be added to reinforce the weld. No flux is required to make an OF weld.

In TB, the metal is heated to a sufficient temperature but below its melting point so that a brazing alloy can be melted and bond to the hot base metal. A flux may be used to help the brazing alloy bond to the base metal. Both OF welding and TB are used primarily on smaller, thinner-gauge metals.

THERMAL CUTTING PROCESSES

There are a number of thermal cutting processes such as oxyfuel gas cutting and plasma arc cutting (PAC). They are the most commonly used in most welding shops. Air carbon arc cutting (AAC) is also frequently used, and many larger fabrication shops have started using laser beam cutting (LBC).

Oxyfuel Gas Cutting

Oxyfuel gas cutting (OFC) uses the high-temperature flame to heat the surface of a piece of steel to a point where a forceful stream of oxygen flowing out a center hole in the tip causes the hot steel to burn away, leaving a gap or cut. Because OF cutting relies on the rapid oxidation of the base metal at elevated temperatures to make a cut, the types of metals and alloys that it can be used on are limited. OF cutting can be used on steel from a fraction of an inch thick to several feet, depending on the capacity of the torch and tip being used.

Plasma Arc Cutting

Plasma arc cutting (PAC) uses a stiff, highly ionized, extremely hot column of gas to almost instantly vaporize the metal being cut. Most ionized plasma is formed as high-pressure air is forced through a very

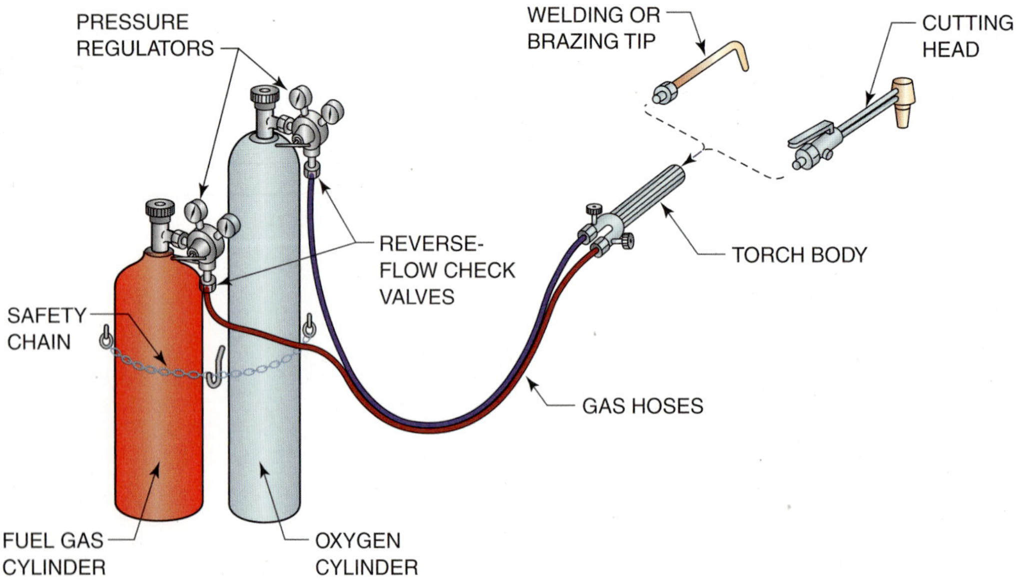

FIGURE 1-17 Oxyfuel gas welding (OFW) and oxyfuel cutting (OFC) equipment.

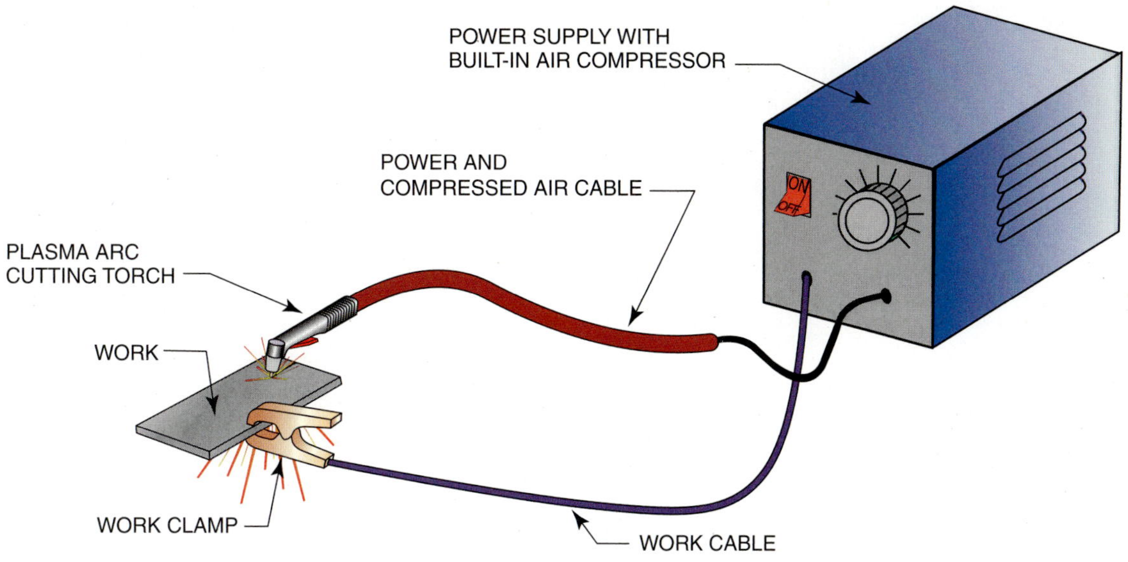

FIGURE 1-18 Plasma arc cutting (PAC) equipment. © Cengage Learning 2012

small opening between a tungsten electrode and the torch tip, Figure 1-18. As the air is ionized, it heats up, expands, and exits the torch tip at supersonic speeds. PAC does not rely on rapid oxidation of the metal being cut like OFC, so almost any metal or alloy can be cut.

PA cutting equipment consists of a transformer power supply, plasma torch and cable, work clamp and cable, and an air supply. Some PA cutting equipment has self-contained air compressors. Because the PA cutting process can be performed at some very high travel speeds, it is often used on automated cutting machines. The high travel speeds and very low heat input help to reduce or eliminate part distortion, a common problem with some OF cutting.

DEMONSTRATIONS, PRACTICES, AND PROJECTS

The welding *Demonstrations* in the textbook are designed to show you how something works, reacts to heating or welding, or how you might be able to perform a task. They may be done individually or as a group.

The welding *Practices* in this textbook are designed for you to develop a specific welding skill. Welding is a combination of technical knowledge and skill. You can develop the technical knowledge through reading and studying the text, and you can develop the skill and art of welding by performing the welds laid out in each practice and project. Learning to weld requires practicing each weld. Often, you have to make the weld several times before you develop the eye–hand coordination. The more time you spend practicing welding, the better your skills will become.

The welding *Projects* in this textbook are designed to help you both improve your welding skills and develop your fabricating skills. The beginning welding projects are designed so that you can make them even though you may not have developed all of your welding skills yet. That is not to say that you should not always try to make high-quality welds every time you weld. Making perfect welds is every welder's desire; however, a high skill level comes with practice. If you follow the project drawings and specifications, every project should result in a usable product being produced.

Assembling the Parts

The assembly process for a weldment can be as simple as holding a part in place with one hand as you make a tack weld using your other hand. But most of the time, it is much more complicated, requiring clamps, jigs, or fixtures to hold the parts in place for tack welding or finish welding, Figure 1-19. A variety of hand tools such as squares, magnetic angle blocks, clamps, and locking pliers are commonly used to align and hold the parts for welding.

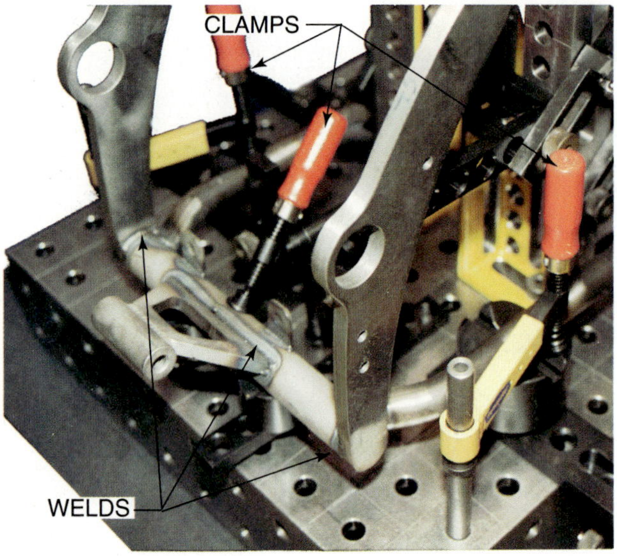

FIGURE 1-19 **Motorcycle frame clamped in a welding jig.** Larry Jeffus

Getting the parts of an assembly properly positioned may take more time than it takes to do the welding. Time spent accurately positioning the part is not wasted, because welding parts in the wrong place can result in time being wasted removing the welds and repositioning a part.

Selection of the Joining Process

Many different welding processes can be used to tack weld or finish weld. Some of the factors to consider are whether this is a single weldment or whether a large number of welds will be required. When a large number of welds are needed, then a slightly faster welding process would be worth using. Another factor to consider is the metal thickness and joint design. You have more choices when welds are made in a welding shop rather than in the field. For example, when it is too windy, FCA and GMA welding cannot be used outside, so SMA welding may need to be used for these field welds.

OCCUPATIONAL OPPORTUNITIES IN WELDING

The American welding industry has contributed to the widespread growth of the welding and allied processes. Without welding, much of what we use on a daily basis could not be manufactured. The list of these products grows every day, thus increasing the number of jobs for people with welding skills. The need to fill these well-paying jobs is not concentrated in major metropolitan areas but exists throughout the country and the world. Because of the diverse nature of the welding industry, the exact job duties of each skill area vary. The following are general descriptions of the job classifications used in our profession; specific tasks may vary from one location to another.

Welders perform the actual welding. They are the skilled craftspeople who, through their own labor, produce the welds on a variety of complex products, Figure 1-20.

Tack welders, also skilled workers, often help the welder by making small welds to hold parts in place. The tack weld must be correctly applied so that it is strong enough to hold the assembly and still not interfere with the finished welding.

Welding operators, often skilled welders, operate machines or automatic equipment used to make welds.

Welders' helpers are employed in some welding shops to clean slag from the welds and help move and position weldments for the welder.

Welder assemblers or welder fitters, position all the parts in their proper places and make these ready for the tack welders. These skilled workers must be able to interpret blueprints and welding procedures.

FIGURE 1-20 **Sometimes welding must be done in confined spaces such as this pumping station.** Larry Jeffus

They also must have knowledge of the effects of contraction and expansion of the various types of metals.

Welding inspectors are often required to hold a special **certification** such as the one supervised by the American Welding Society known as Certified Welding Inspector (CWI). To become a CWI, candidates must pass a test covering the welding process, blueprint reading, weld symbols, metallurgy, codes and standards, and inspection techniques. Vision screening is also required on a regular basis once the technical skills have been demonstrated.

Welding shop supervisors may or may not weld on a regular basis, depending on the size of the shop. In addition to their welding skills, they must demonstrate good management skills by effectively planning jobs and assigning workers.

Welding salespersons may be employed by supply houses or equipment manufacturers. These jobs require a broad understanding of the welding process as well as good marketing skills. Good salespersons are able to provide technical information about their products to convince customers to make a purchase.

Welding shop owners are often welders who have a high degree of skill and knowledge of small-business management and prefer to operate their own businesses. These individuals may specialize in one field, such as hardfacing, repair, and maintenance, or specialty fabrications, or they may operate as subcontractors of manufactured items. A welding business can be as small as one individual, one truck, and one portable welder or as large as a multimillion-dollar operation employing hundreds of workers.

Welding engineers design, specify, and oversee the construction of complex weldments. The welding engineer may work with other engineers in areas such as mechanics, electronics, chemicals, or civil engineering in the process of bringing a new building, ship, aircraft, or product into existence. The welding engineer is required to know all of the welding processes and metallurgy as well as have good math, reading, communication, and design skills. This person usually has an advanced college degree and possesses a professional certification.

In many industries, the welder, welding operator, and tack welder must be able to pass a performance test to a specific code or standard.

The highest paid welders are those who have the education and skills to read blueprints and perform the required work to produce a weldment to strict specifications. Large industrial concerns employ workers who serve as support for the welders. These engineers and technicians must have knowledge of chemistry, physics, metallurgy, electricity, and mathematics. Engineers are responsible for the research, design, development, and fabrication of a project. Technicians work as part of the engineering staff. These individuals may oversee the actual work for the engineer by providing the engineer with progress reports as well as chemical, physical, and mechanical test results. Technicians may also require engineers to build prototypes for testing and evaluation.

Another group of workers employed by the industry does layouts or makes templates. These individuals have had drafting experience and have a knowledge of operations such as punching, cutting, shearing, twisting, and forming, among others. The layout is generally done directly on the material. A template is used for repetitive layouts and is made from sheet metal or other suitable materials.

The flame-cutting process is closely related to welding. Some operators use handheld torches, and others are skilled operators of oxyfuel cutting machines. These machines range from simple mechanical devices to highly sophisticated, computer-controlled, multiple-head machines that are operated by specialists.

Metric Units

Both standard and metric (SI) units are given in this text. The SI units are in parentheses ( ) following the standard unit. When nonspecific values are used—for example, "set the gauge at 2 psig" where 2 is an

1/4 in. = 6 mm
1/2 in. = 13 mm
3/4 in. = 18 mm
1 in. = 25 mm
2 in. = 50 mm
1/2 gal = 2 L
1 gal = 4 L
1 lb = 1/2 K
2 lb = 1 K
1 psig = 7 kPa
1°F = 2°C

Table 1-2 Metric Conversion Approximations

By using an approximation for converting standard units to metric, it is possible to quickly have an idea of how large or heavy an object is in the other units. For estimating, it is not necessary to be concerned with the exact conversions.

TEMPERATURE

Units

°F (each 1° change)	=	0.555°C (change)
°C (each 1° change)	=	1.8°F (change)
32°F (ice freezing)	=	0°Celsius
212°F (boiling water)	=	100°Celsius
–460°F (absolute zero)	=	0°Rankine
–273°C (absolute zero)	=	0°Kelvin

Conversions

°F to °C ______ °F – 32 = ______ × .555 = ______ °C

°C to °F ______ °C × 1.8= ______ + 32 = ______ °F

LINEAR MEASUREMENT

Units

1 inch	=	25.4 millimeters
1 inch	=	2.54 centimeters
1 millimeter	=	0.0394 inch
1 centimeter	=	0.3937 inch
12 inches	=	1 foot
3 feet	=	1 yard
5280 feet	=	1 mile
10 millimeters	=	1 centimeter
10 centimeters	=	1 decimeter
10 decimeters	=	1 meter
1000 meters	=	1 kilometer

Conversions

in. to mm	______ in.	×	25.4	=	______	mm
in. to cm	______ in.	×	2.54	=	______	cm
ft to mm	______ ft	×	304.8	=	______	mm
ft to m	______ ft	×	0.3048	=	______	m
mm to in.	______ mm	×	0.0394	=	______	in.
cm to in.	______ cm	×	0.3937	=	______	in.
mm to ft	______ mm	×	0.00328	=	______	ft
m to ft	______ m	×	3.28	=	______	ft

AREA MEASUREMENT

Units

1 sq in.	=	0.0069 sq ft
1 sq ft	=	144 sq in.
1 sq ft	=	0.111 sq yd
1 sq yd	=	9 sq ft
1 sq in.	=	645.16 sq mm
1 sq mm	=	0.00155 sq in.
1 sq cm	=	100 sq mm
1 sq m	=	1000 sq cm

Conversions

sq in. to sq mm ______ sq in. × 645.16 = ______ sq mm

sq mm to sq in. ______ sq mm × 0.00155 = ______ sq in.

VOLUME MEASUREMENT

Units

1 cu in.	=	0.000578 cu ft
1 cu ft	=	1728 cu in.
1 cu ft	=	0.03704 cu yd
1 cu ft	=	28.32 L
1 cu ft	=	7.48 gal (U.S.)
1 gal (U.S.)	=	3.737 L
1 cu yd	=	27 cu ft
1 gal	=	0.1336 cu ft
1 cu in.	=	16.39 cu cm
1 L	=	1000 cu cm
1 L	=	61.02 cu in.
1 L	=	0.03531 cu ft
1 L	=	0.2642 gal (U.S.)
1 cu yd	=	0.769 cu m
1 cu m	=	1.3 cu yd

Conversions

cu in. to L	______ cu in.	×	0.01638	=	______	L
L to cu in.	______ L	×	61.02	=	______	cu in.
cu ft to L	______ cu ft	×	28.32	=	______	L
L to cu ft	______ L	×	0.03531	=	______	cu ft
L to gal	______ L	×	0.2642	=	______	gal
gal to L	______ gal	×	3.737	=	______	L

WEIGHT (MASS) MEASUREMENT

Units

1 oz	=	0.0625 lb
1 lb	=	16 oz
1 oz	=	28.35 g
1 g	=	0.03527 oz
1 lb	=	0.0005 ton
1 ton	=	2000 lb
1 oz	=	0.283 kg
1 lb	=	0.4535 kg
1 kg	=	35.27 oz
1 kg	=	2.205 lb
1 kg	=	1000 g

Conversions

lb to kg	______ lb	×	0.4535	=	______	kg
kg to lb	______ kg	×	2.205	=	______	lb
oz to g	______ oz	×	0.03527	=	______	g
g to oz	______ g	×	28.35	=	______	oz

PRESSURE AND FORCE MEASUREMENTS

Units

1 psig	=	6.8948 kPa
1 kPa	=	0.145 psig
1 psig	=	0.000703 kg/sq mm
1 kg/sq mm	=	6894 psig
1 lb (force)	=	4.448 N
1 N (force)	=	0.2248 lb

Conversions

psig to kPa	______ psig	×	6.8948	=	______	kPa
kPa to psig	______ kPa	×	0.145	=	______	psig
lb to N	______ lb	×	4.448	=	______	N
N to lb	______ N	×	0.2248	=	______	psig

VELOCITY MEASUREMENTS

Units

1 in./sec	=	0.0833 ft/sec
1 ft/sec	=	12 in./sec
1 ft/min	=	720 in./sec
1 in./sec	=	0.4233 mm/sec
1 mm/sec	=	2.362 in./sec
1 cfm	=	0.4719 L/min
1 L/min	=	2.119 cfm

Conversions

ft/min to in./sec ____ ft/min × 720 = ____ in./sec

in./min to mm/sec ____ in./min × .4233 = ____ mm/sec

mm/sec. to in./min ____ mm/sec × 2.362 = ____ in./min

cfm to L/min ____ cfm × 0.4719 = ____ L/min

L/min to cfm ____ L/min × 2.119 = ____ cfm

TABLE 1-3 Table of Conversions: U.S. Customary (Standard) Units and Metric Units (SI)

U.S. Customer (Standard) Units						
°F	=	degrees Fahrenheit				
°R	=	degrees Rankine				
	=	degrees absolute F				
lb	=	pound				
psi	=	pounds per square inch				
	=	lb per sq in.				
psia	=	pounds per square inch absolute				
	=	psi + atmospheric pressure				
in.	=	inches	=	in.	=	"
ft	=	foot or feet	=	ft	=	'
sq in.	=	square inch	=	in.		
sq ft	=	square foot	=	ft		
cu in.	=	cubic inch	=	in.		
cu ft	=	cubic foot	=	ft		
ft-lb	=	foot-pound				
ton	=	ton of refrigeration effect				
qt	=	quart				
Metric Units (SI)						
°C	=	degrees Celsius				
°K	=	Kelvin				
mm	=	millimeter				
cm	=	centimeter				
cm^2	=	centimeter squared				
cm^3	=	centimeter cubed				
dm	=	decimeter				
dm^2	=	decimeter squared				
dm^3	=	decimeter cubed				
m	=	meter				
m^2	=	meter squared				
m^3	=	meter cubed				
L	=	liter				
g	=	gram				
kg	=	kilogram				
J	=	joule				
kJ	=	kilojoule				
N	=	newton				
Pa	=	pascal				
kPa	=	kilopascal				
W	=	watt				
kW	=	kilowatt				
MW	=	megawatt				
Miscellaneous Abbreviations						
P	=	pressure	sec	=	seconds	
h	=	hours	r	=	radius of circle	
D	=	diameter	π	=	3.1416 (a constant used in determining the area of a circle)	
A	=	area				
V	=	volume				
∞	=	infinity				

Table 1-4 Abbreviations and Symbols

approximate value—the SI units have been rounded off to the nearest whole number. Round off occurs in these cases, to agree with the standard value and because whole numbers are easier to work with. SI units are not rounded off only when the standard unit is an exact measurement.

Often students have difficulty understanding metric units because exact conversions are used even when the standard measurement was an approximation. Rounding off the metric units makes understanding the metric system much easier, Table 1-2, page 13. By using this approximation method, you can make most standard-to-metric conversions in your head without using a calculator.

Once you have learned to use approximations for metric, you will find it easier to make exact conversions whenever necessary. Conversions must be exact in the shop when a part is dimensioned with one system's units and the other system must be used to fabricate the part. For that reason you must be able to make those conversions. Table 1-3 and Table 1-4 are set up to be used with or without the aid of a calculator. Many calculators today have built-in standard–metric conversions. Of course, it is a good idea to know how to make these conversions with and without these aids. Practice making such conversions whenever the opportunity arises.

Welding Video Series

Delmar/Cengage Learning, in cooperation with the author, has produced a series of DVDs. Each of the four DVD sets covers specific equipment setup and operation for welding, cutting, soldering, or brazing.

REVIEW QUESTIONS

1. Define *metal fabrication.*
2. Define welded *metal fabrication.*
3. What is a weldment?
4. List 20 items that are manufactured using welding or thermal cutting processes.
5. List the five steps that might be followed to fabricate a project.
6. What is a nontechnical definition of *welding?*
7. Explain the term *fit for service.*
8. List the six most popular welding processes.

9. Name the two most popular thermal cutting processes.
10. List the major parts of a GMA welding setup.
11. List the major parts of an FCA welding setup.
12. List the major parts of an SMA welding setup.
13. List the major parts of a GTA welding setup.
14. List the major parts of an OF cutting setup.
15. List the major parts of a PA cutting setup.
16. Why can more types of metal be cut with PAC than with OFC?
17. List some ways that parts can be held in place for tack welding or finish welding.
18. List some factors that should be considered when selecting a welding process.
19. List some of the types of jobs that are available in the welding industry.
20. Using Table 1-3, convert the following standard units to SI units.
    - **a.** 212°F to ________ °C
    - **b.** 2 in. to ________ mm
    - **c.** 4 lb to ________ k
    - **d.** 10 psig to ________ kPa
21. Using Table 1-3, convert the following SI units to standard units.
    - **a.** 100 °C to ________ °F
    - **b.** 400 mm to ________ in.
    - **c.** 10 k to ________ lb
    - **d.** 100 kPa to ________ psig

Chapter 2

Welding Safety

OBJECTIVES

After completing this chapter, the student should be able to:

- Explain how to work safely.
- Identify each degree of burn and describe how to provide first aid.
- List the types of protective clothing a welder should wear.
- Explain the importance of proper ventilation and respiratory protection.
- Describe how to safely lift, climb, and handle materials.
- Demonstrate electrical safety.

KEY TERMS

acetone
acetylene
earmuffs
earplugs
electrical ground
electrical resistance
exhaust pickups
flash burn
flash glasses
forced ventilation
full face shield
goggles
ground-fault circuit interpreter (GFCI)
infrared light
material specification data sheet (MSDS)
natural ventilation
safety glasses
type A fire extinguisher
type B fire extinguisher
type C fire extinguisher
type D fire extinguisher
ultraviolet light
valve protection cap
ventilation
visible light
warning label
welding helmet

INTRODUCTION

Accident prevention is the main intent of this chapter. The safety information included in this text is intended as a guide. There is no substitute for caution and common sense. A safe job is no accident; it takes work to make the job safe. Each person working must do their own part to make the job safe.

Welding fabrication is a very large and diverse industry. This chapter concentrates on only that portion of welding fabrication safety related to the areas of light metal. You must read, learn, and follow all safety rules, regulations, and procedures for those areas.

Light welding fabrication, like all other areas of welding work, has a number of potential safety hazards. These hazards need not result in anyone being injured. Learning to work safely is as important as learning to be a skilled welding fabrication worker.

You must approach new jobs with your safety in mind. Your safety is your own responsibility, and you must shoulder that responsibility. It is not possible to anticipate all of the possible dangers in every job. This text may not cover some dangers. You can get specific safety information from welding equipment manufacturers and their local suppliers, your local college and university, and the World Wide Web.

If an accident does occur on a welding site, it can have consequences far beyond just the person being injured. Serious accidents can result in local, state, or national investigations. For example, if the federal office of the Occupational Safety and Health Administration (OSHA) becomes involved, the jobsite may be closed for hours, days, weeks, months, or even permanently. While the jobsite is closed for the investigation, you may be off without pay. If it is determined that your intentional actions contributed to the accident, you may lose your job, be fined, or worse. Always follow the rules; never engage in horseplay or play practical jokes while at work.

BURNS

Burns are one of the most common and painful injuries that occur in welding fabrication. Burns can be caused by ultraviolet light rays as well as by contact with hot welding material. The chance of infection is high with burns because of the dead tissue that results. It is important that all burns receive proper medical treatment to reduce the chance of infection. Burns are divided into three classifications, depending upon the degree of severity. The three classifications include first-degree, second-degree, and third-degree burns.

First-Degree Burns

First-degree burns have occurred when the surface of the skin is reddish in color, tender, and painful; they do not involve any broken skin, Figure 2-1. The first step in treating a first-degree burn is to immediately put the burned area under cold water (not iced) or apply cold water compresses (clean towel, washcloth, or handkerchief soaked in cold water) until the pain decreases. Then cover the area with sterile bandages or a clean cloth. Do not apply butter or grease, or any other home remedies or medications, without a doctor's recommendation.

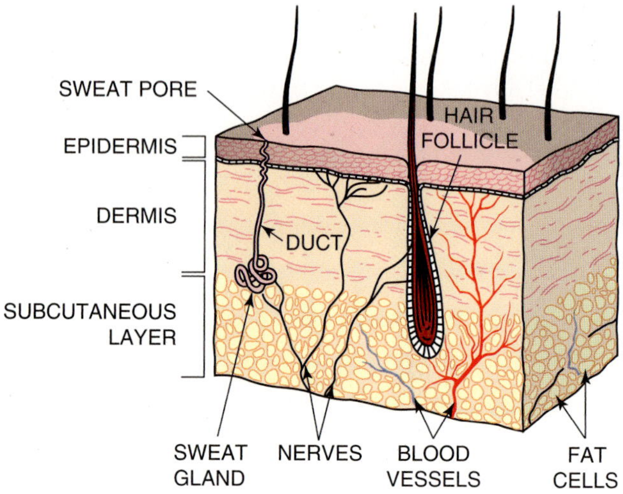

FIGURE 2-1 **First-degree burn—only the skin surface (epidermis) is affected.** © Cengage Learning 2012

Second-Degree Burns

Second-degree burns have occurred when the surface of the skin is severely damaged, resulting in the formation of blisters and possible breaks in the skin, Figure 2-2. Again, the most important first step in treating a second-degree burn is to put the area under

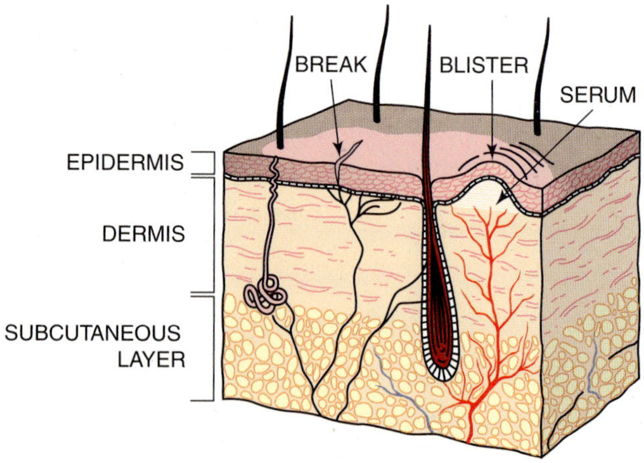

FIGURE 2-2 **Second-degree burn—the epidermal layer is damaged, forming blisters or shallow breaks.** © Cengage Learning 2012

cold water (not iced) or apply cold water compresses until the pain decreases. Gently pat the area dry with a clean towel, and cover the area with a sterile bandage or clean cloth to prevent infection. Seek medical attention. If the burns are around the mouth or nose, or involve singed nasal hair, breathing problems may develop. Do not apply ointments, sprays, antiseptics, or home remedies. Note: In an emergency, any cold liquid you drink, for example, water, cold tea, soft drinks, or milk shake, can be poured on a burn. The purpose is to lower the skin temperature as quickly as possible to reduce tissue damage.

Third-Degree Burns

Third-degree burns have occurred when the surface of the skin and possibly the tissue below the skin appear white or charred. There may be cracks or breaks in the skin, Figure 2-3. Initially, little pain is present because the nerve endings have been destroyed. Do not remove any clothes that are stuck to the burn. Do not put ice water or ice on the burns; this could intensify the shock reaction. Do not apply ointments, sprays, antiseptics, or home remedies. If the victim is on fire, smother the flames with a blanket, rug, or jacket. Breathing difficulties are common with burns around the face, neck, and mouth; be sure that the victim is breathing. Place a cold cloth or cool water on burns of the face, hands, or feet to cool the burned areas. Cover the burned area with thick, sterile, nonfluffy dressings. Call for an ambulance immediately if needed; people with even small third-degree burns need to consult a doctor.

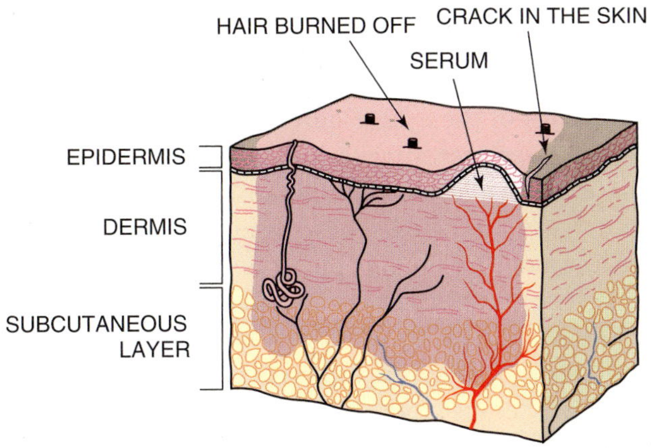

FIGURE 2-3 Third-degree burn—the epidermis, dermis, and the subcutaneous layers of tissue are destroyed.
© Cengage Learning 2012

Burns Caused by Light

Some types of light can cause burns. There are three types of light—ultraviolet, infrared, and visible. Ultraviolet and infrared are not visible to the unaided human eye but can cause burns. During welding, one or more of the three types of light may be present. Arc welding and arc cutting produce all three kinds of light, but gas welding produces only the less hazardous visible and infrared lights.

The light from the welding process can be reflected from walls, ceilings, floors, or any other large surface. This reflected light is as dangerous as the direct welding light. To reduce the danger from reflected light, welding shops, if possible, should be painted flat black. Flat black reduces the reflected light by absorbing more of it than any other color. If the welding cannot be moved away from other workers, screen off the welding arc with welding curtains that will absorb the welding light, Figure 2-4. These special portable welding curtains may be either transparent or opaque. Transparent welding curtains are made of a special high-temperature, flame-resistant plastic that prevents the harmful light from passing through.

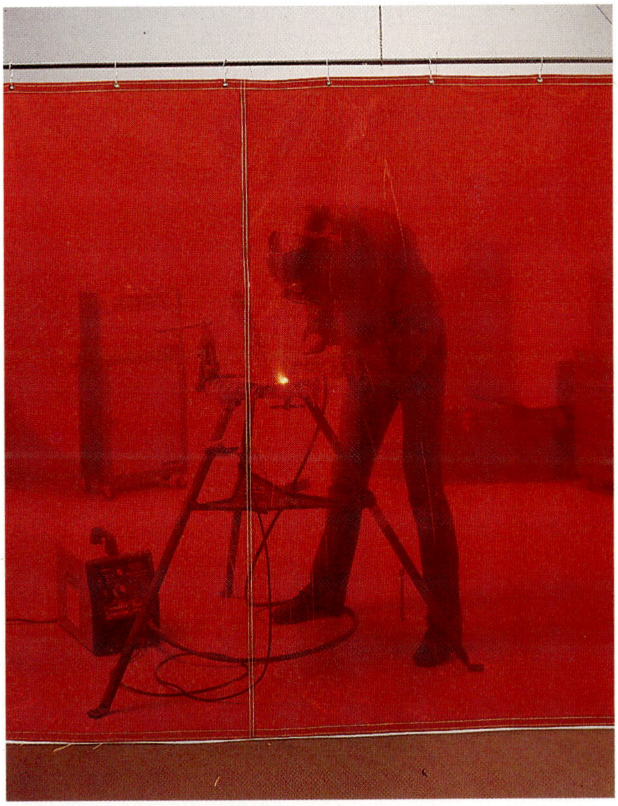

FIGURE 2-4 Portable welding curtains.
Frommelt Safety Products

CAUTION

Welding curtains must always be used to protect other workers in an area that might be exposed to the welding light.

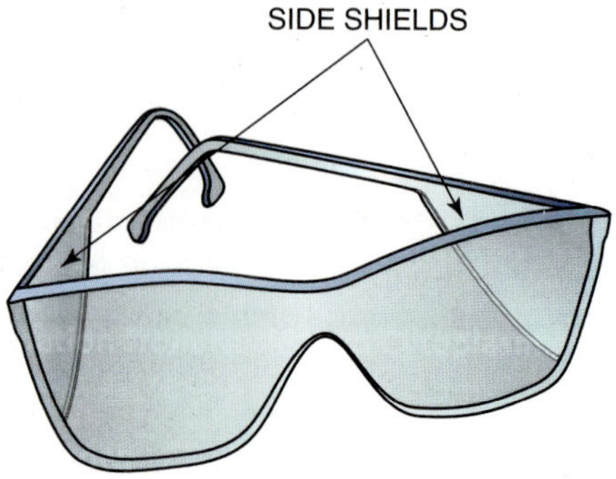

FIGURE 2-5 Safety glasses with side shields.

Ultraviolet Light (UV)

Ultraviolet light waves are the most dangerous. They can cause first-degree and second-degree burns to the eyes or to any exposed skin. Because you cannot see or feel ultraviolet light while being exposed to it, you must stay protected when in the area of any arc welding processes. The closer a person is to the arc and the higher the current, the quicker a burn may occur. The ultraviolet light is so intense during some welding processes that eyes can receive a **flash burn** within seconds, and the skin can be burned within minutes. Ultraviolet light can pass through loosely woven clothing, thin clothing, light-colored clothing, and a damaged or poorly maintained arc **welding helmet.**

Infrared Light

Infrared light is the light wave that is felt as heat. Although infrared light can cause burns, a person will immediately feel this type of light. Therefore, burns can easily be avoided.

Visible Light

Visible light is the light that we see. It is produced in varying quantities and colors during welding. Too much visible light may cause temporary night blindness (poor eyesight under low light levels). Too little visible light may cause eyestrain, but visible light is not hazardous.

Whether burns are caused by ultraviolet light or hot material, they can be avoided if proper clothing and other protection are worn.

EYE AND EAR PROTECTION

Face and Eye Protection

Eye protection must be worn in the shop at all times. Eye protection can be **safety glasses** with side shields, Figure 2-5; **goggles;** or a **full face shield.** For better protection when working in brightly lit areas or outdoors, some welders wear **flash glasses,** which are special, lightly tinted safety glasses. These safety glasses provide protection from both flying debris and reflected light.

Suitable eye protection is important because you cannot immediately detect excessive exposure to arc light. Welding light damage occurs often without warning, like a sunburn's effect that is felt the following day. Therefore, welders must take appropriate precautions in selecting filters or goggles that are suitable for the process being used, Table 2-1. Selecting the correct shade lens is also important because both extremes of too light or too dark can cause eyestrain. New welders often select a lens that is too dark, assuming it will give them better protection, but this results in eyestrain in the same manner as if they were trying to read in a poorly lit room. In reality, any approved arc welding lens filters out the harmful ultraviolet light. Select a lens that lets you see comfortably. At the very least, the welder's eyes must not be strained by excessive glare from the arc.

Ultraviolet light can burn the eye in two ways. It can injure the retina, which is the back of the eye. Burns on the retina are not painful but may cause some loss of eyesight. Ultraviolet light can also burn the whites of the eyes, Figure 2-6. The whites of the eyes are very sensitive, and burns are very painful. The eyes are easily infected because, as with any burn, many cells are killed. These dead cells in the moist environment of the eyes promote the growth of bacteria that cause infection. When the eye is burned, it feels as though there is something in the eye, but without a professional examination, it is impossible to know. Because there may be something in the eye and because of the high risk of infection, home remedies or other medicines should never be used for eye burns. Any time you receive an eye injury, you should see a doctor.

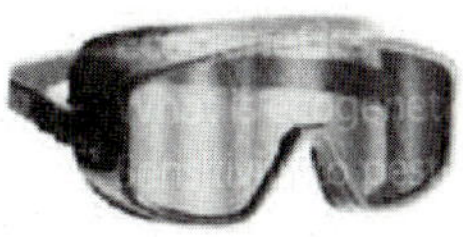

1

Goggles, flexible fitting, regular ventilation

2

Goggles, flexible fitting, hooded ventilation

3

Goggles, cushioned fitting, rigid body

4

Spectacles

5

Spectacles, eyecup type eye shields

6

Spectacles, semi-flat-fold side shields

7

Welding goggles, eyecup type, tinted lenses

7A

Chipping goggles, eyecup type, tinted lenses

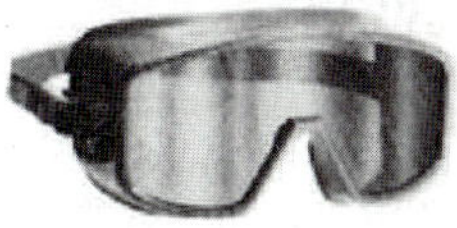

8

Welding goggles, cover-spec type, tinted lenses

8A

Chipping goggles, cover-spec type, clear safety lenses

9

Welding goggles, cover-spec type, tinted plate lens

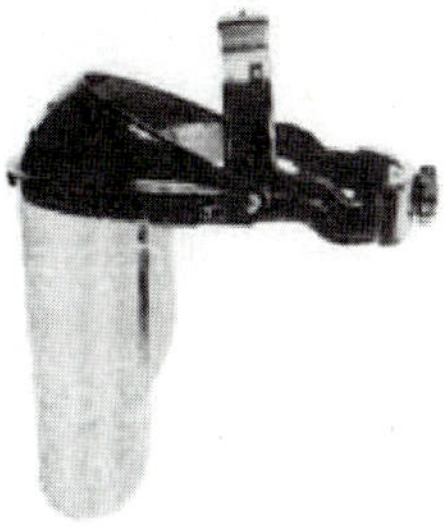

10

Face shield, plastic or mesh window (see caution note)

11

Welding helmet

**Non-side-shield spectacles are available for limited hazard use requiring only frontal protection.*

Applications

Operation	Hazards	Protectors
Acetylene-Burning Acetylene-Cutting Acetylene-Welding	Sparks, Harmful Rays, Molten Metal, Flying Particles	7,8,9
Chemical Handling	Splash, Acid Burns, Fumes	2 (for severe exposure add 10)
Chipping	Flying Particles	1,2,4,5,6,7A,8A
Electric (Arc) Welding	Sparks, Intense Rays, Molten Metal	11 (in combination with 4,5,6 in tinted lenses advisable)
Furnace Operations	Glare, Heat, Molten Metal	7,8,9 (for severe exposure add 10)
Grinding-Light	Flying Particles	1,3,5,6 (for severe exposure add 10)
Grinding-Heavy	Flying Particles	1,3,7A,8A (for severe exposure add 10)
Laboratory	Chemical Splash, Glass Breakage	2 (10 when in combination with 5,6)
Machining	Flying Particles	1,3,5,6 (for severe exposure add 10)
Molten Metals	Heat, Glare, Sparks, Splash	7,8 (10 in combination with 5,6 in tinted lenses)
Spot Welding	Flying Particles, Sparks	1,3,4,5,6 (tinted lenses advisable, for severe exposure add 10)

CAUTION:
Face shields alone do not provide adequate protection. Plastic lenses are advised for protection against molten metal splash.
Contact lenses, in and of themselves, do not provide eye protection in the industrial sense and shall not be worn in a hazardous environment without appropriate covering safety eyewear.

Table 2-1 Huntsman Selector Chart *(Kedman Co., Huntsman Product Division)*

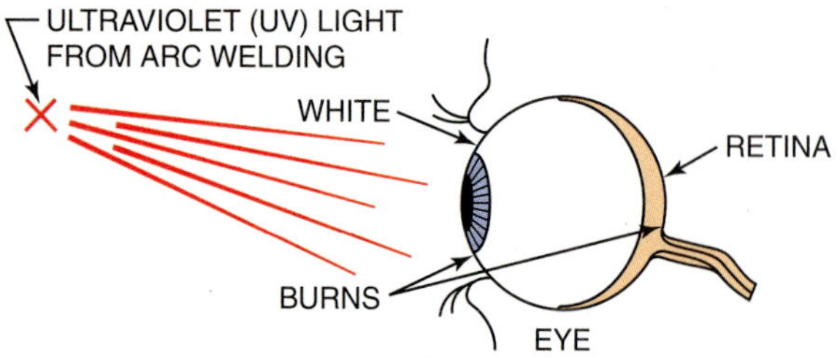

FIGURE 2-6 The eye can be burned on the white or on the retina by ultraviolet light.
© Cengage Learning 2012

Even with quality welding helmets, like the one shown in Figure 2-7, the welder must check for potential problems that may occur from accidents or daily use. Small, undetectable leaks of ultraviolet light in an arc welding helmet can cause a welder's eyes to itch or feel sore after a day of welding. To prevent these leaks, make sure that the lens gasket is installed correctly, Figure 2-8. The outer and inner clear lens must be plastic. As shown in Figure 2-9, the lens can be checked for cracks by twisting it between your fingers. Worn or cracked spots on a helmet must be repaired. Tape can be used as a temporary repair until the helmet can be replaced or permanently repaired.

Safety glasses with side shields are adequate for general use, but if you are doing heavy grinding, chipping, or overhead work, you should wear goggles or a full face shield in addition to safety glasses, Figure 2-10. Safety glasses are best for general protection because they must be worn under an arc welding helmet at all times.

FIGURE 2-7 Typical arc welding helmets used to provide eye and face protection during welding.
Larry Jeffus

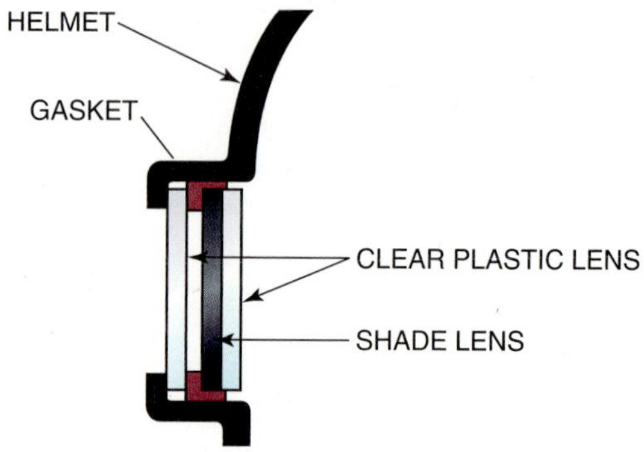

FIGURE 2-8 The correct placement of the gasket around the shade lens is important because it can stop ultraviolet light from bouncing around the lens assembly.
© Cengage Learning 2012

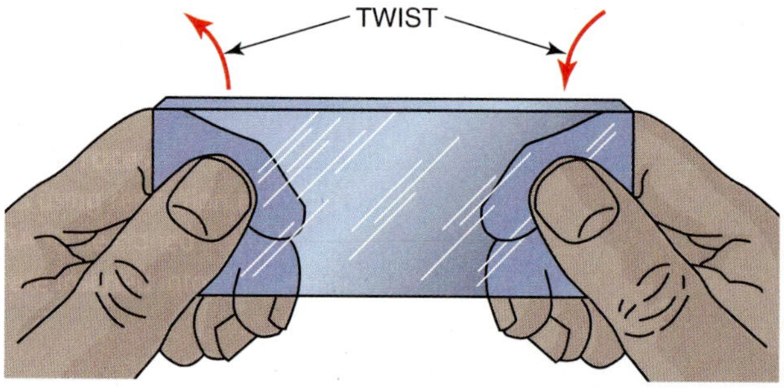

FIGURE 2-9 To check the shade lens for possible cracks, gently twist it. © Cengage Learning 2012

FIGURE 2-10 Full face shield.
© Cengage Learning 2012

OSHA PERMISSIBLE EXPOSURE

Average Daily Sound Level	Time of Noise Exposure
90	8 Hours
95	4 Hours
100	2 Hours
105	1 Hour
110	30 Minutes
111	15 Minutes

SOUND LEVEL dBA

Sound Level	SOUND SOURCE
0	Threshold of Hearing
20	Buzzing Insect
40	Transformer Welder
60	Speech
95	Hand Grinder
120	Jet Plane and Threshold of Pain

Table 2-2 OSHA Permissible Sound Exposure Levels

Ear Protection

The welding environment can be very noisy. The sound level is at times high enough to cause pain and some loss of hearing if the welder's ears are unprotected, Table 2-2. Hot sparks can also drop into an open ear, causing severe burns.

Ear protection is available in several forms. One form of protection is **earmuffs** that cover the outer ear completely, Figure 2-11. Another form of protection is **earplugs** that fit into the ear canal, Figure 2-12. Both of these protect a person's hearing, but only the earmuffs protect the outer ear from burns.

FIGURE 2-11 Earmuffs provide complete ear protection and can be worn under a welding helmet. MSA

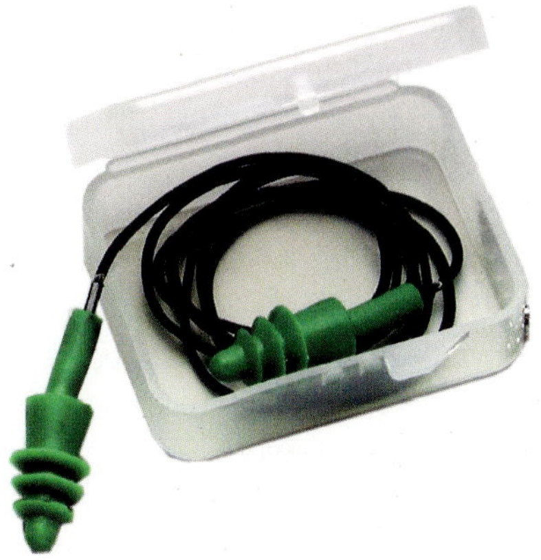

FIGURE 2-12 Earplugs used as protection from noise only. MSA

CAUTION

Damage to your hearing caused by high sound levels may not be detected until later in life, and the resulting loss in hearing is not recoverable. Your hearing will not improve with time, and each exposure to high levels of sound will further damage your hearing.

RESPIRATORY PROTECTION

All welding and cutting processes produce undesirable by-products, such as harmful dusts, fumes, mists, gases, smokes, sprays, or vapors. For your safety and the safety of others, your primary objective is to prevent these contaminants from forming and collecting in the area's atmosphere. This can be accomplished as much as possible by thorough cleaning of surface contaminants before starting work, and confinement of the operation to outdoor or open spaces.

Production of welding fumes and vapor by-products cannot be avoided. They are created when the temperature of metals and fluxes is raised above the temperatures at which they boil or decompose. Most of the by-products are recondensed in the weld. However, some do escape into the atmosphere, producing the haze that occurs in improperly ventilated welding shops. Some fluxes used in welding electrodes produce fumes that may irritate the welder's nose, throat, and lungs.

When welders must work in an area where effective controls to remove airborne welding by-products are not feasible, employers must provide equipment such as respirators when necessary to protect workers' health. The respirators must be applicable and suitable for the purpose intended. The welding shop will establish and implement a written respiratory protection program with work site-specific procedures indicating where respirators are necessary to protect welders' health or whenever respirators are required by the shop. Welders are responsible for following the welding shop's established written respiratory protection program. Guidelines for the respiratory protection program are available from the OSHA office in Washington, DC.

Training must be a part of the welding shop's respiratory protection program. This training should include instruction in the following procedures:

- Proper use of respirators, including techniques for putting them on and removing them.
- Schedules for cleaning, disinfecting, storing, inspecting, repairing, discarding, and performing other aspects of maintenance of the respiratory protection equipment.
- Selection of the proper respirators for use in the workplace, and any respiratory equipment limitations.
- Procedures for testing for tight-fitting respirators.
- Proper use of respirators in both routine and reasonably foreseeable emergency situations.
- Regular evaluation of the effectiveness of the program.

All respiratory protection equipment used in a welding shop should be certified by the National Institute for Occupational Safety and Health (NIOSH). Some of the types of respiratory protection equipment that may be used include the following:

- Air-purifying respirators have an air-purifying filter, cartridge, or canister that removes specific air contaminants by passing ambient air through the air-purifying element.
- Atmosphere-supplying respirators supply breathing air from a source independent of the ambient atmosphere; this includes both the supplied-air respirators and self-contained breathing apparatus type units.
- Demand respirators are atmosphere-supplying respirators that admit breathing air to the facepiece only when a negative pressure is created inside the facepiece by inhalation.

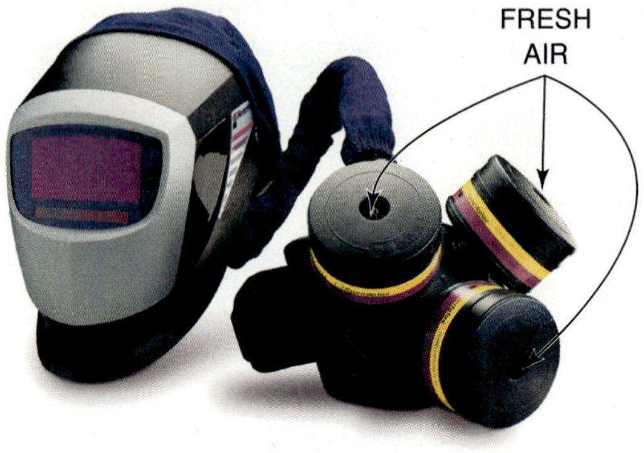

FIGURE 2-13 Filtered fresh air is forced into the welder's breathing area.

- Positive pressure respirators are respirators in which the pressure inside the respiratory inlet covering exceeds the ambient air pressure outside the respirator.
- Powered air-purifying respirators (PAPRs) are air-purifying respirators that use a blower to force the ambient air through air-purifying elements to the inlet covering, Figure 2-13.
- Self-contained breathing apparatuses (SCBAs) are atmosphere-supplying respirators for which the breathing air source is designed to be carried by the user.
- Supplied-air respirators (SARs), or airline respirators, are atmosphere-supplying respirators that have air piped in through a flexible hose from a large central air supply.

Respiratory protection equipment used in many welding applications is of the filtering facepiece (dust mask) type, Figure 2-14. These mask types use the negative pressure as you inhale to draw air through a filter, which is an integral part of the facepiece. In areas of severe contamination, you may use a hood-type respirator, which covers your head and neck and may also cover portions of your shoulders and torso.

CAUTION

Welding or cutting must never be performed on drums, barrels, tanks, vessels, or other containers until they have been emptied and cleaned thoroughly, eliminating all flammable materials and all substances (such as detergents, solvents, greases, tars, or acids) that might produce flammable, toxic, or explosive vapors when heated.

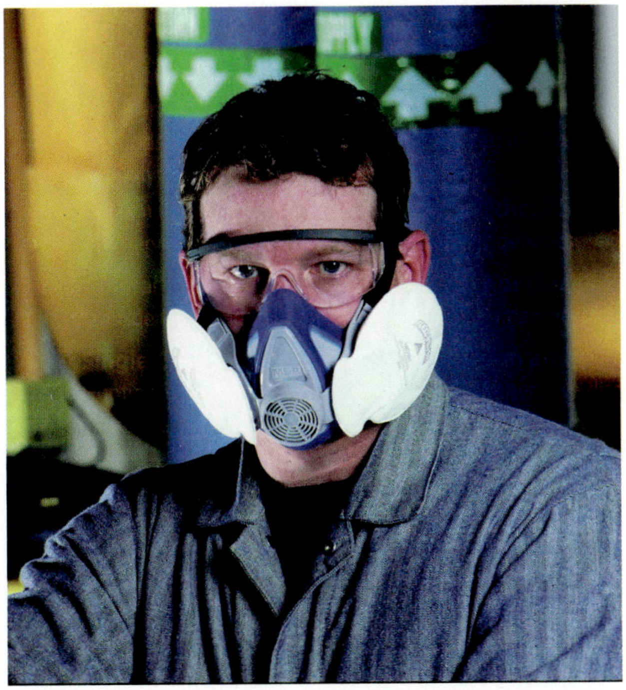

FIGURE 2-14 Typical respirator for contaminated environments. The filters can be selected for specific types of contaminant. MSA

Some materials used as paint, coating, or plating on metals to prevent rust or corrosion can cause respiratory problems. Other potentially hazardous materials include alloys used in metals to give them special properties.

Before it is welded or cut, any metal that has been painted or has any grease, oil, or chemicals on its surface must be thoroughly cleaned. This cleaning may be done by grinding, sandblasting, or applying an approved solvent.

CAUTION

Never weld, cut, or braze on any metal that has any chemical residue. Even small amounts trapped in crevices, such as those in the hopper or between parts, must be thoroughly removed before you can safely perform any welding, cutting, or brazing. All of these chemicals can release deadly gases when heated. Some are explosive.

Most paints containing lead have been removed from the market. But some industries, such as marine or ship applications, still use these lead-based paints. Often, old machinery surfaces may have lead-based paint coatings. Old solder often contains lead alloys. The welding and cutting of lead-bearing alloys or metals whose surfaces have been painted with lead-based paint can generate lead oxide fumes. Inhalation and ingestion of lead oxide fumes and other lead compounds cause lead poisoning. Symptoms include a metallic taste in the mouth, a loss of appetite, nausea, abdominal cramps, and insomnia. In time, anemia and a general weakness, chiefly in the muscles of the wrists, develop.

Both cadmium and zinc are plating materials used to prevent iron or steel from rusting. Cadmium is often used on bolts, nuts, hinges, and other hardware items, and it gives the surface a yellowish-gold appearance. Acute exposure to high concentrations of cadmium fumes can produce severe lung irritation. Long-term exposure to low levels of cadmium in air can result in emphysema (a disease affecting the lung's ability to absorb oxygen) and can damage the kidneys.

Zinc galvanizing may be found on pipes, sheet metal, water tanks, bolts, nuts, and many other types of hardware. Zinc plating that is thin may appear as a shiny, metallic patchwork or crystal pattern; thicker, hot-dipped zinc appears rough and may look dull. Zinc is used in large quantities in the manufacture of brass and is found in brazing rods. Inhalation of zinc oxide fumes can occur when you are welding or cutting on these materials. Exposure to these fumes is known to cause metal fume fever; its symptoms are very similar to those of common influenza.

Concern has been expressed about the possibility of lung cancer being caused by some of the chromium compounds that are produced when stainless steels are welded.

CAUTION

Take extreme caution to avoid the fumes produced when welding is done on dirty or used metal. Any chemicals that are on the metal can become mixed with the welding fumes, a combination that can be extremely hazardous. All metal must be cleaned before welding to avoid this potential problem.

Rather than take chances, welders should recognize that fumes of any type, regardless of their source, should not be inhaled. The best way to avoid problems is to provide adequate **ventilation.** If this is not possible, breathing protection should be used.

Potentially dangerous gases can also be present in a welding shop. Proper ventilation or respirators are necessary when welding in confined spaces, regardless of the welding process being used. Ozone is a gas that is produced by the ultraviolet radiation in the air in the vicinity of arc welding and cutting operations. Ozone is very irritating to all mucous membranes, with excessive exposure producing pulmonary edema. Other effects of exposure to ozone include headache, chest pain, and dryness in the respiratory tract.

Phosgene is formed when ultraviolet radiation decomposes chlorinated hydrocarbon. Fumes from chlorinated hydrocarbons can come from solvents such as those used for degreasing metals and from refrigerants from air-conditioning systems. They decompose in the arc to produce a potentially dangerous chlorine acid compound. This compound reacts with the moisture in the lungs to produce hydrogen chloride, which in turn destroys lung tissue. For this reason, any use of chlorinated solvents should be well away from welding operations in which ultraviolet radiation or intense heat is generated. Any welding or cutting on refrigeration or air-conditioning piping must be done only after the refrigerant has been completely removed in accordance with Environmental Protection Agency (EPA) regulations.

Care also must be taken to avoid the infiltration of any fumes or gases, including argon or carbon dioxide, into a confined working space, such as when welding in tanks or cylinders. The collection of some fumes and gases in a work area can go unnoticed by the welders. Concentrated fumes or gases can cause a fire or explosion if they are flammable, asphyxiation if they replace the oxygen in the air, or death if they are toxic.

Despite these fumes and other potential hazards in welding shops, welders have been found to be as healthy as workers employed in other industrial occupations.

VENTILATION

The actual welding area should be outside or well-ventilated. Excessive fumes, ozone, or smoke may collect in the welding area; ventilation should be provided for their removal. **Natural ventilation** is best, but forced ventilation may be required. Areas that have 10,000 cu ft (283 m^3) or more per welder, or that have ceilings 16 ft (4.9 m) high or higher, Figure 2-15, may not require forced ventilation unless fumes or smoke begins to collect.

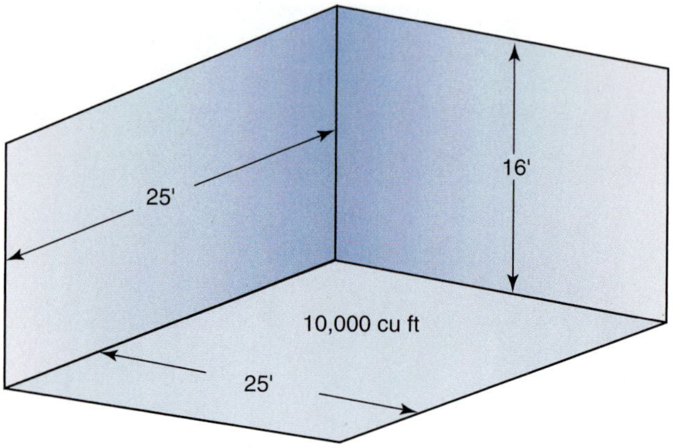

FIGURE 2-15 A room with a ceiling 16 ft (4.9 m) high may not require forced ventilation for one welder. Natural ventilation is best for welding areas.
Larry Jeffus

Forced Ventilation

Small shops or shops with large numbers of welders require **forced ventilation.** Forced ventilation can be general or localized using fixed or flexible **exhaust pickups,** Figure 2-16. General room ventilation must be at a rate of 2000 cu ft (56 m^3) or more per person welding. Localized exhaust pickups must have a suction strong enough to pull welding fumes away from the welder at a velocity of 100 linear feet (30.5 m) per minute, Figure 2-17. Local, state, or federal regulations may require that welding fumes be treated to remove hazardous components before they are released into the atmosphere.

Any system of ventilation should draw the fumes or smoke away before rising past the level of the welder's face.

FIGURE 2-16 Flexible exhaust pickup. Larry Jeffus

FIGURE 2-17 An exhaust pickup. Larry Jeffus

Forced ventilation is always required when welding on metals that contain zinc, lead, beryllium, cadmium, mercury, copper, austenitic manganese, or other materials that give off dangerous fumes.

MATERIAL SPECIFICATION DATA SHEET (MSDS)

All manufacturers of potentially hazardous materials must provide to the users of their products detailed information regarding possible hazards resulting from the use of their products. These **material specification data sheets** are often called **MSDSs.** They must be provided to anyone using the product or anyone working in the area where the products are in use. Often companies will post these sheets on a bulletin board or put them in a convenient place near the work area. Some states have right-to-know laws that require specific training of all employees who handle or work in areas with hazardous materials.

WASTE MATERIAL DISPOSAL

Welding shops generate waste materials. Much of the waste is scrap metal. All scrap metal, including electrode stubs, can be easily recycled. Recycling metal is good for the environment and can be a source of revenue for the welding shop.

Some of the other waste, such as burned flux, cleaning solvents, and dust collected in shop air filtration systems, may be considered hazardous materials. Check with the material manufacturer or an environmental consultant to determine if any waste material is considered hazardous. Throwing hazardous waste material into the trash, pouring it on the ground, or dumping it down the drain is illegal. Before you dispose of any welding shop waste that is considered hazardous, you must first consult local, state, and federal regulations. Protecting our environment from pollution is everyone's responsibility.

LADDER SAFETY

Falls are a major cause of injury and death. Improper use of ladders is often a factor in these falls. Always keep in mind when erecting a ladder that even short step stools can pose a potential fall hazard. Never approach a climb assuming that because it is not high, it cannot be that dangerous. All ladder usage poses a danger to your safety. Some students think that if a ladder starts to fall they will just "jump clear." You cannot jump clear if the ladder under you has given way, because there is nothing solid under your feet for you to jump from. When a ladder falls, you fall. Keep the area around the base of the ladder clear so if you do fall, it will not be onto debris or equipment.

Types of Ladders

Both stepladders and straight ladders are used extensively in welding fabrication. Straight ladders may be single section or extension-type ladders. Most ladders are made from wood, aluminum, or fiberglass, and each type has its advantages and disadvantages, Table 2-3. All ladders used in welding fabrication should be listed with the American National Standards Institute (ANSI) or Underwriters Laboratories (UL) to ensure they are constructed to a standard of safety.

Ladder Inspection

Over time, ladders can become worn or damaged and should be inspected each time they are used. Look for loose or damaged steps, rungs, rails, braces, and safety feet. Check to see that all hardware is tight, including hinges, locks, nuts, bolts, screws, and rivets. Wooden ladders must be checked for cracks, rot, or wood decay. Never use a defective ladder. Make any

Material	Advantages	Disadvantages
Wood	Electrically nonconductive	Long-term exposure to weather will cause rotting
Aluminum	Lightweight Weather resistant	Electrically conductive Shakier than wood or fiberglass
Fiberglass	Electrically nonconductive Weather resistant	Heavier than aluminum and wood Fiberglass splinters

Table 2-3 Major Advantages and Disadvantages of Typical Ladder Materials

necessary repairs before it is used or, if it cannot be repaired, replace it.

Rules for Ladder Use

Read the entire ladder manufacturer's list of safety rules before using the ladder for the first time. Stepladders must be locked in the full opened position with the spreaders. Straight or extension ladders must be used at the proper angle; either too steep or too flat is dangerous, Figure 2-18.

The following are general safety and usage rules for ladders:

- Follow all recommended practices for safe use and storage.
- Do not exceed the manufacturer's recommended maximum weight limit for the ladder.

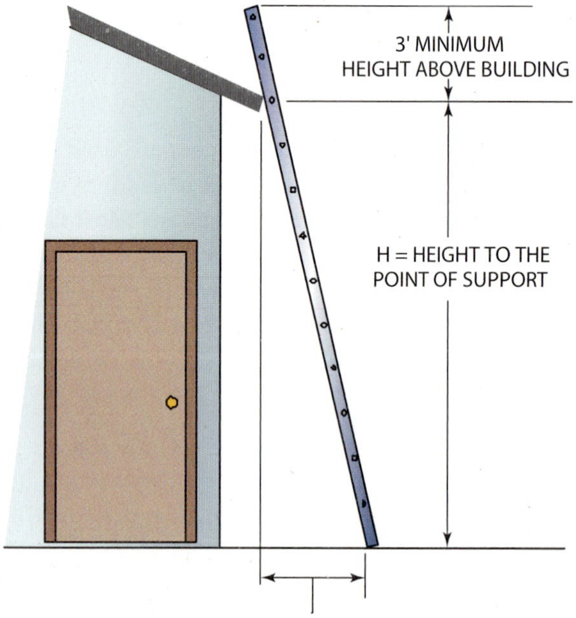
3' MINIMUM
HEIGHT ABOVE BUILDING
H = HEIGHT TO THE
POINT OF SUPPORT
THE BASE OF A LADDER SHOULD BE SET OUT
A DISTANCE EQUAL TO 1/4 OF THE HEIGHT TO
THE POINT OF SUPPORT (H/4)

FIGURE 2-18 Make sure the ladder is leaning at the proper angle. © Cengage Learning 2012

- Before setting up a ladder, make certain that it will be erected on a level, solid surface.
- Never use a ladder in a wet or muddy area where water or mud will be tracked up the ladder's steps or rungs. Only climb or descend ladders with clean, dry shoes.
- Tie the ladder securely in place.
- Climb and descend the ladder cautiously.
- Do not carry tools and supplies in your hand as you climb or descend a ladder. Use a rope to raise or lower the items once you are safely in place.
- Never use ladders around live electrical wires.
- Never use a ladder that is too short for the job, requiring you to reach or stand on the top step.
- Wear well-fitted shoes or boots.

CAUTION

Turn off the power to any wires that will be near your ladder when you are working. Never use ladders near live electrical wires. Never move a ladder in the upright position near live electrical wires.

ELECTRICAL SAFETY

Electric shock can cause injuries and even death unless proper precautions are taken. Most welding and cutting operations involve electrical equipment in addition to the arc welding power supplies. Grinders, electric motors on automatic cutting machines, and drills are examples. Most electrical equipment in a welding shop is powered by alternating-current (AC) sources having input voltages ranging from 115 to 460 volts. However, fatalities have occurred when working with equipment operating at less than 80 volts. Most electric shocks in the welding industry do not occur from contact with welding electrode holders but as a result of accidental contact with bare or poorly insulated conductors. **Electrical resistance**

Welding Safety Checklist

Hazard	Factors to Consider	Precaution Summary
Electric shock can kill 	• Wetness • Welder in or on workpiece • Confined space • Electrode holder and cable insulation	• Insulate welder from workpiece and ground using *dry* insulation. Rubber mat or dry wood. • Wear *dry, hole-free* gloves. (Change as necessary to keep dry.) • Do not touch electrically "hot" parts or electrode with bare skin or wet clothing. • If wet area and welder cannot be insulated from workpiece with dry insulation, use a semiautomatic, constant-voltage welder or stick welder with voltage reducing device. • Keep electrode holder and cable insulation in good condition. Do not use if insulation is damaged or missing.
Fumes and gases can be dangerous 	• Confined area • Positioning of welder's head • Lack of general ventilation • Electrode types, i.e., manganese, chromium, etc. See MSDS • Base metal coatings, galvanize, paint	• Use ventilation or exhaust to keep air breathing zone clear, comfortable. • Use helmet and positioning of head to minimize fume in breathing zone. • Read warnings on electrode container and material safety data sheet (MSDS) for electrode. • Provide additional ventilation/exhaust where special ventilation requirements exist. • Use special care when welding in a confined area. • Do not weld unless ventilation is adequate.
Welding sparks can cause fire or explosion 	• Containers which have held combustibles • Flammable materials	• Do not weld on containers which have held combustible materials (unless strict AWS F4.1 procedures are followed). Check before welding. • Remove flammable materials from welding area or shield from sparks, heat. • Keep a fire watch in the area during and after welding. • Keep a fire extinguisher in the welding area. • Wear fire-retardant clothing and hat. Use earplugs when welding overhead.
Arc rays can burn eyes and skin 	• Process: gas-shielded arc most severe	• Select a filter lens which is comfortable for you while welding. • Always use a helmet when welding. • Provide nonflammable shielding to protect others. • Wear clothing which protects skin while welding.
Confined space 	• Metal enclosure • Wetness • Restricted entry • Heavier than air gas • Welder inside or on workpiece	• Carefully evaluate adequacy of ventilation, especially where electrode requires special ventilation or where gas may displace breathing air. • If basic electric shock precautions cannot be followed to insulate welder from work and electrode, use semiautomatic, constant-voltage equipment with cold electrode or stick welder with a voltage-reducing device. • Provide welder helper and method of welder retrieval from outside enclosure.
General work area hazards	• Cluttered area	• Keep cables, materials, tools neatly organized.
	• Indirect work (welding ground) connection	• Connect work cable as close as possible to area where welding is being performed. Do *not* allow alternate circuits through scaffold cables, hoist chains, ground leads.
	• Electrical equipment	• Use only double insulated or properly grounded equipment. • Always disconnect power to equipment before servicing.
	• Engine-driven equipment	• Use in only open, well-ventilated areas. • Keep enclosure complete and guards in place. • See Lincoln service shop if guards are missing. • Refuel with engine off. • If using auxiliary power, OSHA may require GFI protection or assured grounding program (or isolated windings if less than 5 KW).
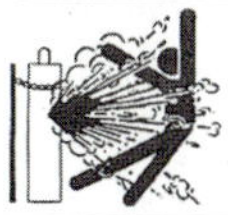	• Gas cylinders	• Never touch cylinder with the electrode. • Never lift a machine with cylinder attached. • Keep cylinder upright and chained to support.

FIGURE 2-19 Note the warning information contained on this typical label, which may be attached to welding equipment or in the equipment owner's manual.
The Lincoln Electric Company

is lowered in the presence of water or moisture, so welders must take special precautions when working under damp or wet conditions, including perspiration. Figure 2-19 shows a typical **warning label** shipped with the welding equipment.

CAUTION

Welding cables must never be spliced within 10 ft (3 m) of the electrode holder.

Cables must be checked periodically to be sure that they have not become frayed, and if they have, they must be replaced immediately.

Never allow the metal parts of electrodes or electrode holders to touch the skin or wet coverings on the body. Dry gloves in good condition must always be worn. Rubber-soled shoes are advisable. Precautions against accidental contact with bare-conducting surfaces must be taken when the welder is required to work in cramped kneeling, sitting, or prone positions. Insulated mats or dry wooden boards are desirable protection from the earth.

Welding circuits must be turned off when the workstation is left unattended. The main power supply must be turned off and locked or tagged to prevent electrocution when you are working on the welder, welding leads, electrode holder, torches, wire feeder, guns, or other parts. Since the electrode holder is energized when coated electrodes are changed, the welder must wear dry gloves.

The workpiece being welded, and the frame or chassis of all electrically powered machines, must be connected to a good **electrical ground.** The work lead from the welding power supply is not an electrical ground and is not sufficient. A separate lead is required to ground the workpiece and power source.

Electrical connections must be tight. Terminals for welding leads and power cables must be shielded from accidental contact by personnel or by metal objects. Cables must be used within their current-carrying and duty-cycle capacities; otherwise, they will overheat and break down the insulation rapidly. Cable connectors for lengthening leads must be insulated.

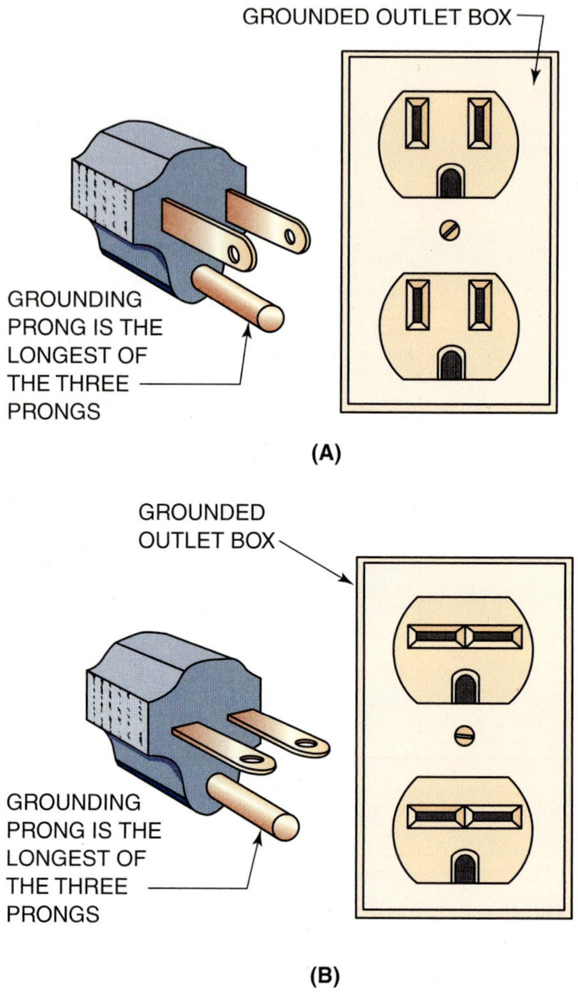

FIGURE 2-20 **(A) A three-prong grounding plug for use with up to 150-volt tools and (B) a grounding plug for use with 150- to 250-volt tools.** © Cengage Learning 2012

Electrical Safety Systems

For protection from electrical shock, the standard portable tool is built with either of two equally safe systems: external grounding or double insulation.

A tool with external grounding has a wire that runs from the housing through the power cord to a third prong on the power plug. When this third prong is connected to a grounded, three-hole electrical outlet, the grounding wire will carry any current that leaks past the electrical insulation of the tool away from the user and into the ground. In most electrical systems, the three-prong plug fits into a three-prong, grounded receptacle. If the tool is operated at less than 150 volts, it has a plug like that shown in Figure 2-20A. If it is used at 150 to 250 volts, it has a plug like that shown in Figure 2-20B. In either type, the green (or green and yellow) conductor in the tool cord is the grounding wire. Never connect the grounding wire to a power terminal.

All electrical plugs must be grounded, and it is against many local, state, and national electrical safety codes to remove or disable the ground plug on a power cord. However, some older buildings may have old ungrounded, two-prong electrical outlets, so an adapter must be used. If there is any uncertainty about whether a receptacle is properly grounded, have it checked by a qualified electrician.

A double-insulated power tool has an extra layer of electrical insulation that eliminates the need for a three-prong and grounded outlet. Double-insulated tools do not require grounding and therefore have a two-prong plug. In addition, double-insulated tools are always labeled as such on their nameplate or case, Figure 2-21.

FIGURE 2-21 Typical portable power tool nameplate.

VOLTAGE WARNINGS

Before connecting a tool to a power supply, be sure that the voltage supplied is the same as that specified on the nameplate of the tool. A power source with a voltage greater than that specified for the tool can lead to serious injury to the user as well as damage to the tool. Using a power source with a voltage lower than the rating on the nameplate is harmful to the motor.

Tool nameplates also bear a figure with the abbreviation *amps* (for amperes, a measure of electric current). This refers to the current-drawing requirement of the tool. The higher the input current, the more powerful the motor.

Extension Cords

When there is some distance from the power source to the work area or if the portable tool is equipped with a stub power cord, an extension cord must be used. When using extension cords on portable power tools, the size of the conductors must be large enough to prevent an excessive drop in voltage. A voltage drop is the lowering of the voltage at the power tool from that of the voltage at the supply. This occurs because of resistance flow in the wire. A voltage drop causes loss of power, overheating, and possible motor damage. Table 2-4 shows the correct size extension cord to use based on cord length and nameplate amperage rating. If in doubt, use the next larger size. The smaller the gauge number of an extension cord, the larger the cord.

Only three-wire, grounded extension cords connected to properly grounded, three-wire receptacles should be used. Two-wire extension cords with two-prong plugs should not be used. Current specifications require outdoor receptacles to be protected with **ground-fault circuit interpreter (GFCI)** devices. These safety devices are often referred to as GFIs.

Nameplate Amperes	Cord Length in Feet							
	25	50	75	100	125	150	175	200
1	16	16	16	16	16	16	16	16
2	16	16	16	16	16	16	16	16
3	16	16	16	16	16	16	14	14
4	16	16	16	16	16	14	14	12
5	16	16	16	16	14	14	12	12
6	16	16	16	14	14	12	12	12
7	16	16	14	14	12	12	12	10
8	14	14	14	14	12	12	10	10
9	14	14	14	12	12	10	10	10
10	14	14	14	12	12	10	10	10
11	12	12	12	12	10	10	10	8
12	12	12	12	12	10	10	8	8

Table 2-4 Recommended Extension Cord Sizes for Use with Portable Electric Tools

When using extension cords, keep the following safety tips in mind:

- Always connect the cord of a portable electric power tool into the extension cord before the extension cord is connected to the outlet.
- Always unplug the extension cord from the receptacle before unplugging the cord of the portable power tool from the extension cord.
- Extension cords should be long enough to make connections without being pulled taut, creating unnecessary strain or wear.
- Be sure that the extension cord does not come in contact with sharp objects or hot surfaces. The cords should not be allowed to kink, nor should they be dipped in or splattered with oil, grease, or chemicals.
- Before using a cord, inspect it for loose or exposed wires and damaged insulation. If a cord is damaged, replace it. This also applies to the tool's power cord.
- Extension cords should be checked frequently while in use to detect unusual heating. Any cable that feels more than slightly warm to a bare hand placed outside the insulation should be checked immediately for overloading.
- See that the extension cord is positioned so that no one trips or stumbles over it.
- To prevent the accidental separation of a tool cord from an extension cord during operation, make a knot as shown in Figure 2-22A or use a cord connector as shown in Figure 2-22B.

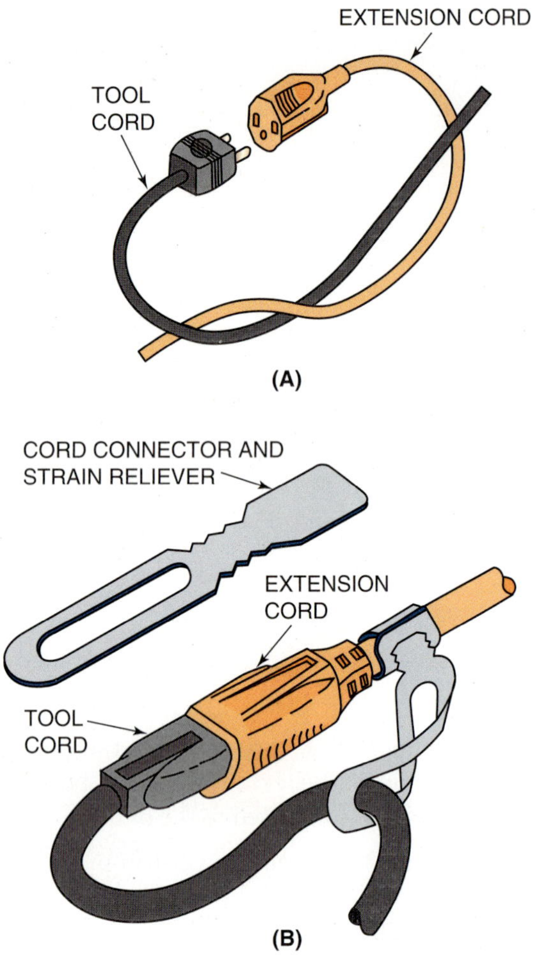

FIGURE 2-22 (A) A knot will prevent the extension cord from accidentally pulling apart from the tool cord during operation. (B) A cord connector will serve the same purpose.

- Use an extension cord that is long enough for the job but not excessively long.
- Extension cords that go through dirt and mud must be cleaned before storing.

Safety Rules for Portable Electric Tools

In all tool operation, safety is simply the removal of any element of chance. Following are a few safety precautions that should be observed. These are general rules that apply to all power tools. They should be strictly obeyed to avoid injury to the operator and damage to the power tool.

- Know the tool. Learn the tool's applications and limitations as well as its specific potential hazards by reading the manufacturer's literature.
- Ground the portable power tool unless it is double insulated. If the tool is equipped with a three-prong plug, it must be plugged into a three-hole electrical receptacle. If an adapter is used to accommodate a two-pronged receptacle, the adapter wire must be attached to a known ground. Never remove the third prong.
- Do not expose the power tool to rain. Do not use a power tool in wet locations.
- Keep the work area well-lighted. Avoid chemical or corrosive environments.
- Because electric tools spark, portable electric tools should never be started or operated in the presence of propane, natural gas, gasoline, paint thinner, **acetylene,** or other flammable vapors that could cause a fire or explosion.
- Do not force a tool. It will do the job better and more safely if operated at the rate for which it was designed.
- Use the right tool for the job. Never use a tool for any purpose other than that for which it was designed.
- Wear eye protectors. Safety glasses or goggles protect the eyes while you operate power tools.
- Wear a face or dust mask if the operation creates dust.
- Take care of the power cord. Never carry a tool by its cord or yank it to disconnect it from the receptacle.
- Secure your work with clamps. It is safer than using your hands, and it frees both hands to operate the tool.
- Do not overreach when operating a power tool. Keep proper footing and balance at all times.
- Maintain power tools. Follow the manufacturer's instructions for lubricating and changing accessories. Replace all worn, broken, or lost parts immediately.
- Disconnect the tools from the power source when they are not in use.
- Form the habit of checking to see that any keys or wrenches are removed from the tool before turning it on.
- Avoid accidental starting. Do not carry a plugged-in tool with your finger on the switch. Be sure the switch is off when plugging in the tool.
- Be sure accessories and cutting bits are attached securely to the tool.
- Do not use tools with cracked or damaged housings.
- When operating a portable power tool, give it your full and undivided attention; avoid dangerous distractions.

GENERAL WORK CLOTHING

Because of the amount and temperature of hot sparks, metal, and slag produced during welding, cutting, or brazing; and the fact that special protective clothing cannot be worn at all times, it is important to choose general work clothing that minimizes the possibility of getting burned.

Wool clothing (100% wool) is the best choice but difficult to find. All-cotton (100% cotton) clothing is a good second choice, and it is the most popular material used. Synthetic materials, including nylon, rayon, and polyester, must be avoided because they are easily melted, they produce a hot, sticky ash (because it sticks, burns can be more severe), and some produce poisonous gases. The clothing must also stop ultraviolet light from passing through it. This is accomplished if the material chosen is a dark color, thick, and tightly woven.

The following are some guidelines for selecting work clothing:

- Shirts must be long-sleeved to protect the arms, have a high-buttoned collar to protect the neck, be long enough to tuck into the pants to protect the waist, and have flaps on the pockets to keep sparks out (or have no pockets), Figure 2-23.
- Pants must have legs long enough to cover the tops of the boots and must be without cuffs that would catch sparks.
- Boots must have high tops to keep out sparks; have steel toes to prevent crushed toes, Figure 2-24; and have smooth tops to prevent sparks from being trapped in seams.
- Caps should be thick enough to prevent sparks from burning the top of a welder's head.

FIGURE 2-23 The top button of the shirt worn by the welder should always be buttoned to prevent severe burns to the neck. Larry Jeffus

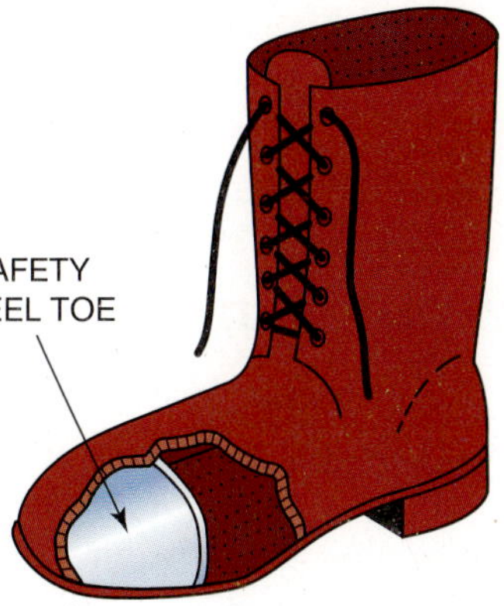

FIGURE 2-24 Safety boots with steel toes are required by many welding shops. © Cengage Learning 2012

All clothing must be free of frayed edges and holes. The clothing must be relatively tight-fitting to prevent excessive folds or wrinkles that might trap sparks.

Butane lighters and matches may catch fire or explode if they are subjected to welding heat or sparks. There is no safe place to carry these items when welding. They must always be removed from the welder's pockets and placed a safe distance away before any work is started.

CAUTION

There is no safe place to carry butane lighters or matches while welding or cutting. They can catch fire or explode if subjected to welding heat or sparks. Butane lighters may explode with the force of 1/4 stick of dynamite. Matches can erupt into a ball of fire. Both butane lighters and matches must always be removed from the welder's pockets and placed a safe distance away before any work is started.

SPECIAL PROTECTIVE CLOTHING

Each person must wear general work clothing in the shop. In addition to this clothing, extra protection is needed for each person who is in direct contact with hot materials. Leather is often the best material to use, as it is lightweight, flexible, resists burning, and is readily available. Synthetic insulating materials are also available. Ready-to-wear leather protection includes capes, jackets, aprons, sleeves, gloves, caps, pants, kneepads, and spats, among other items.

Hand Protection

All-leather, gauntlet-type gloves should be worn when doing any welding, Figure 2-25. Gauntlet gloves that have a cloth liner for insulation are best for hot work. Noninsulated gloves give greater flexibility for fine work. Some leather gloves are available with a canvas gauntlet top, which should be used for light work only.

When a great deal of manual dexterity is required for gas tungsten arc welding, brazing, soldering, oxyfuel gas welding, and other delicate processes, soft leather gloves may be used, Figure 2-26. All-cotton gloves are sometimes used when doing very light welding.

FIGURE 2-25 All-leather, gauntlet-type welding gloves. Larry Jeffus

FIGURE 2-26 For welding that requires a great deal of manual dexterity, soft leather gloves can be worn. Larry Jeffus

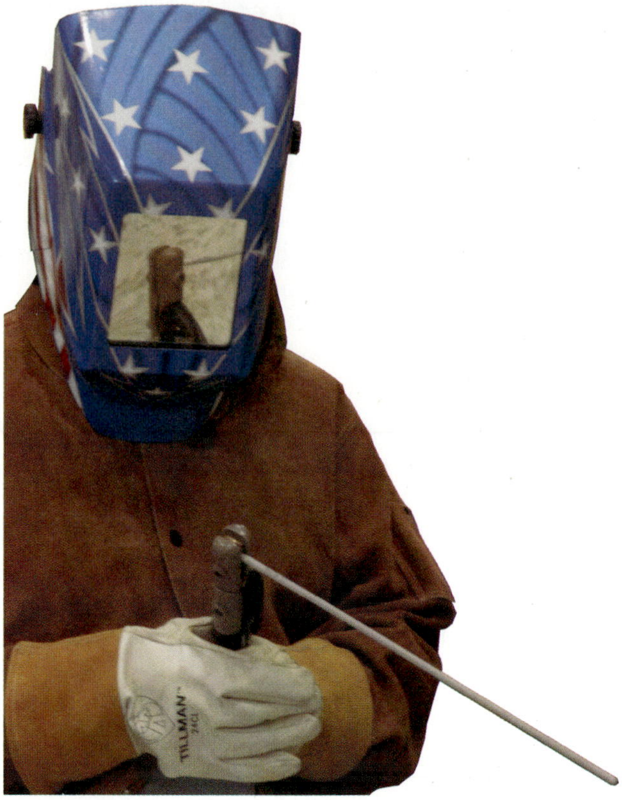

FIGURE 2-27 Full leather jacket. Larry Jeffus

Body Protection

Full leather jackets and capes protect a welder's shoulders, arms, and chest, Figure 2-27. A jacket, unlike the cape, protects a welder's back and complete chest. A cape is open and much cooler but offers less protection. The cape can be used with a bib apron to provide some additional protection while leaving the back cooler. Either the full jacket or the cape with a bib apron should be worn for any out-of-position work.

Waist and Lap Protection

Bib aprons or full aprons protect a welder's lap. Welders especially need to protect their laps if they squat or sit while working and when they bend over or lean against a table.

Arm Protection

For some vertical welding, a full- or half-sleeve can protect a person's arm, Figure 2-28. The sleeves work best if the work level is not above the welder's chest. Work levels higher than this usually require a jacket or cape to keep sparks off the welder's shoulders.

FIGURE 2-28 Full leather sleeve. Larry Jeffus

Leg and Foot Protection

When heavy cutting or welding is being done and a large number of sparks are falling, leather pants and spats should be used to protect the welder's legs and feet. If the weather is hot and full leather pants are uncomfortable, leather aprons with leggings are available. Leggings can be strapped to the legs, leaving the back open. Spats prevent sparks from burning through the front of lace-up boots.

HANDLING AND STORING CYLINDERS

Oxygen and fuel gas cylinders or other flammable materials must be stored separately. The storage areas must be separated by 20 ft (6.1 m), or by a wall 5 ft (1.5 m) high with at least a 1/2-hour (hr) burn rating, Figure 2-29. The purpose of the distance or wall is to keep the heat of a small fire from causing the oxygen cylinder safety valve to release. If the safety valve were to release the oxygen, a small fire would become a raging inferno.

Inert gas cylinders may be stored separately or with oxygen cylinders. Empty cylinders must be stored separately from full cylinders, although they may be stored in the same room or area. All cylinders must be stored vertically and have the protective caps screwed on firmly.

Securing Gas Cylinders

Cylinders must be secured with a chain or other device so that they cannot be knocked over accidentally.

Storage Areas

Cylinder storage areas must be located away from halls, stairwells, and exits so that in case of an emergency they will not block an escape route. Storage areas should also be located away from heat, radiators, furnaces, and welding sparks. Storage areas should be situated so that unauthorized people cannot tamper with the cylinders. A warning sign that reads "Danger—No Smoking, Matches, or Open Lights," or similar wording, should be posted in the storage area, Figure 2-30.

Cylinders with Valve Protection Caps

Cylinders equipped with a **valve protection cap** must have the cap in place unless the cylinder is in use. The protection cap prevents the valve from being broken off if the cylinder is knocked over. If the valve of a full high-pressure cylinder (argon, oxygen, $C0_2$, and mixed gases) is broken off, the cylinder can fly around the shop like a missile if it has not been secured properly.

Never lift a cylinder by the safety cap or the valve. The valve can easily break off or be damaged.

When you are moving cylinders, the valve protection cap must be on, especially if the cylinders are mounted on a truck or trailer for out-of-shop work. The cylinders must never be dropped or handled roughly.

General Precautions

Use warm water (not boiling) to loosen cylinders that are frozen to the ground. Any cylinder that leaks, has a bad valve, or gas-damaged threads must be identified and reported to the supplier. A piece of soapstone is used to write the problem on the cylinder. If the leak

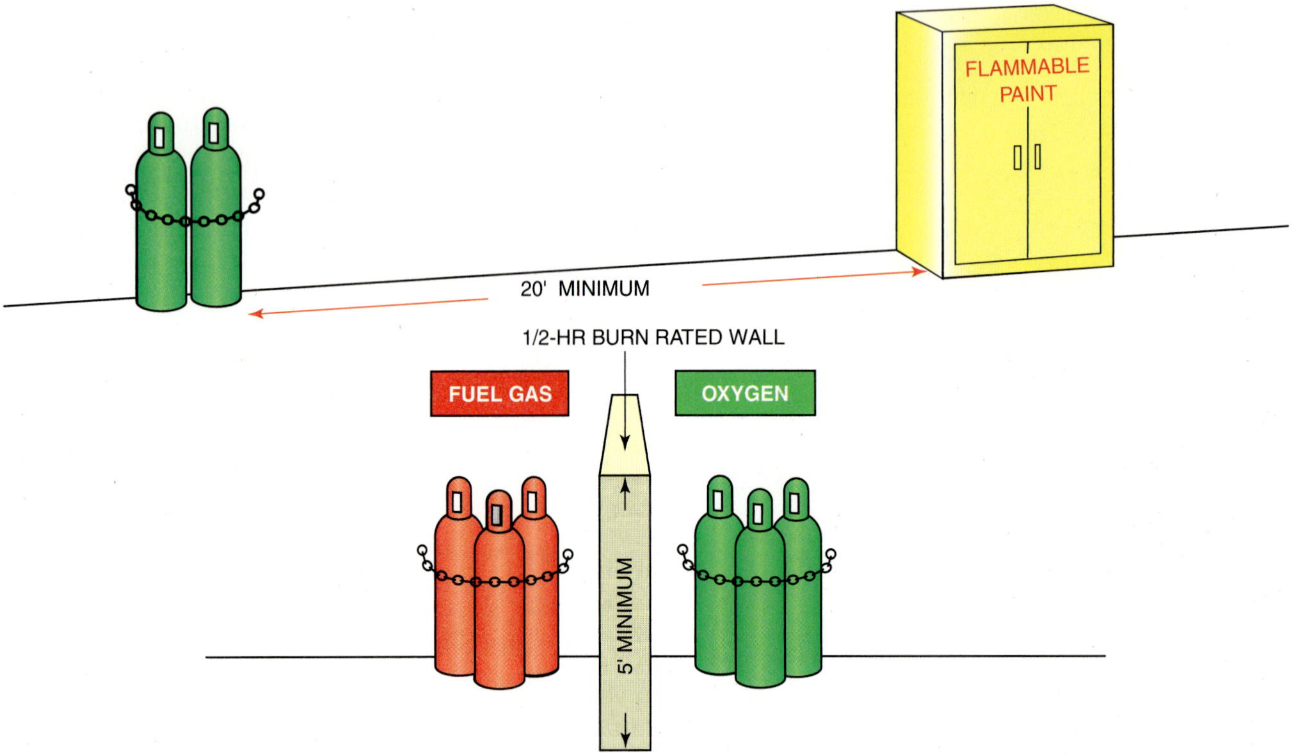

FIGURE 2-29 The minimum safe distance between stored fuel gas cylinders and any flammable material is 20 ft (6.1 m) or a wall 5 ft (1.5 m) high. © Cengage Learning 2012

cannot be stopped by closing the cylinder valve, the cylinder should be moved to a vacant lot or an open area. The pressure should then be slowly released after a warning sign is posted, Figure 2-31.

Acetylene cylinders that have been lying on their sides must stand upright for four hours or more before they are used. The acetylene is absorbed in **acetone,** and the acetone is absorbed in a filler material. The filler does not allow the liquid to settle back away from

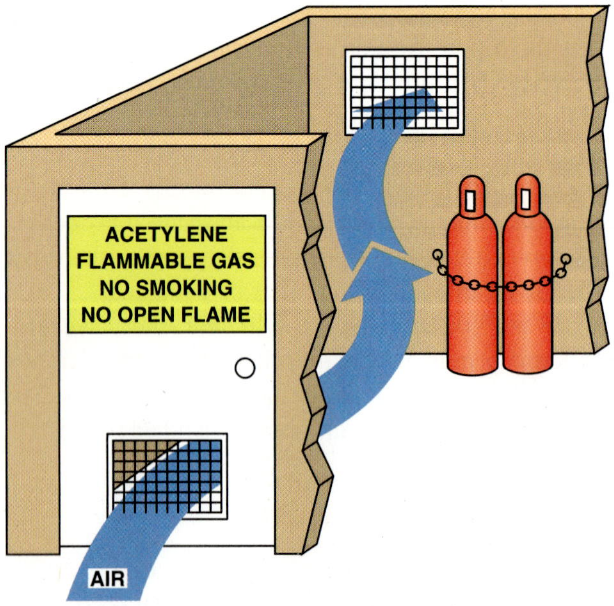

FIGURE 2-30 A separate room used to store acetylene must have good ventilation and should have a warning sign posted on the door. © Cengage Learning 2012

FIGURE 2-31 Move a leaking fuel gas cylinder out of the building or any work area. The pressure should slowly be released after a warning of the danger is posted.
© Cengage Learning 2012

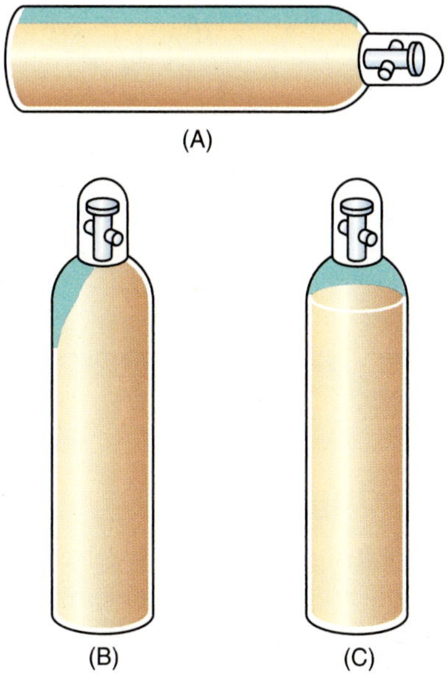

FIGURE 2-32 The acetone in an acetylene cylinder must have time to settle before the cylinder can be used safely. © Cengage Learning 2012

the valve very quickly, Figure 2-32. If the cylinder has been in a horizontal position, using it too soon after it is placed in a vertical position may draw acetone out of the cylinder. Acetone lowers the flame temperature and can damage the regulator or torch valve settings.

FIRE PROTECTION

Fire is a constant danger to the welder. The possibilities of fire cannot always be removed, but they should be minimized. Highly combustible material like paint, fuel, oil, and any large quantity of material that might start more than a small fire must be moved 35 ft (10.7 m) from any welding or cutting.

However, in welding fabrication it is not always possible to weld or cut 35 ft (10.7 m) or more away from some combustible materials. When welding outside or at a remote jobsite, there may be dry grass, wood chips, or numerous other items that will burn around your work areas. If you must work around combustible materials, wet the area first and then keep a bucket of water and a fire extinguisher handy and use a fire watch, Figure 2-33A.

Fire Watch

A fire watch can be provided by any person who knows how to sound the alarm and use a fire extinguisher. The fire extinguisher must be the type required to put out a fire of the type of combustible materials near the welding. Combustible materials that cannot be removed from the welding area should be soaked with water or covered with sand or noncombustible insulating blankets, whichever is available.

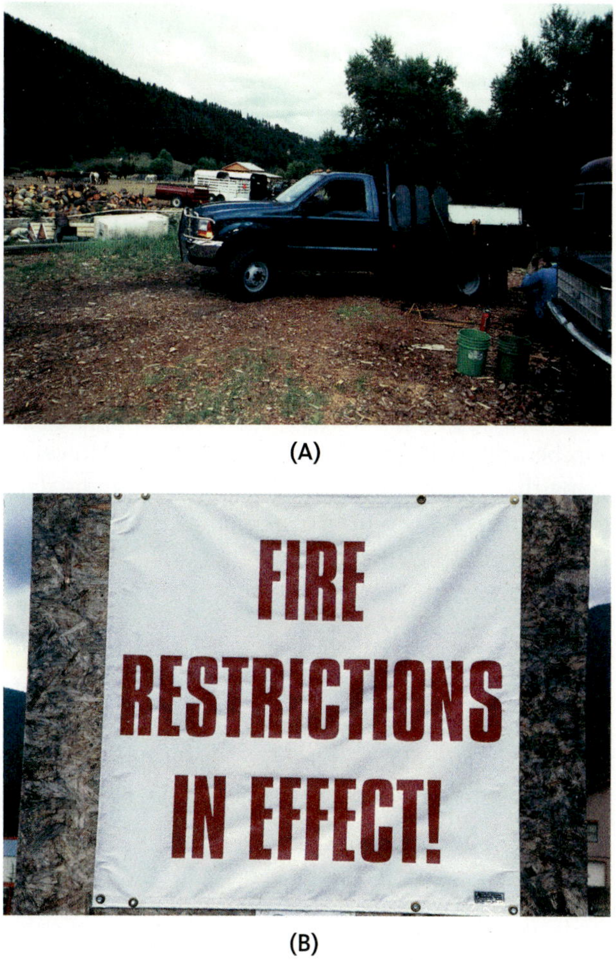

FIGURE 2-33 (A) Two buckets of water and a fire extinguisher are ready if needed. (B) You probably should not weld outdoors when a sign like this is posted. Larry Jeffus

> **CAUTION**
>
> Never weld or cut in a dry area or any area that has been posted by the county or state as fire restricted, Figure 2-33B. You could be held legally responsible if you ignore the warning and a fire starts.

Fire Extinguishers

The four kinds of fire extinguishers are type A, type B, type C, and type D. Each is designed to put out fires

FIGURE 2-34 Type A fire extinguisher symbol.
© Cengage Learning 2012

FIGURE 2-36 Type C fire extinguisher symbol.
© Cengage Learning 2012

on certain types of materials. Some fire extinguishers can be used on more than one type of fire. However, using the wrong kind of fire extinguisher can be dangerous, causing the fire to spread, causing electrical shock, or causing an explosion.

Type A Extinguishers

Type A fire extinguishers are used for combustible solids (articles that burn), such as paper, wood, and cloth. The symbol for a type A extinguisher is a green triangle with the letter *A* in the center, Figure 2-34.

Type B Extinguishers

Type B fire extinguishers are used for combustible liquids, such as oil, gas, and paint thinner. The symbol for a type B extinguisher is a red square with the letter *B* in the center, Figure 2-35.

Type C Extinguishers

Type C fire extinguishers are used for electrical fires. For example, they are used on fires involving motors, fuse boxes, and welding machines. The symbol for a type C extinguisher is a blue circle with the letter *C* in the center, Figure 2-36.

FIGURE 2-35 Type B fire extinguisher symbol.
© Cengage Learning 2012

Type D Extinguishers

Type D fire extinguishers are used on fires involving combustible metals, such as zinc, magnesium, and titanium. The symbol for a type D extinguisher is a yellow star with the letter *D* in the center, Figure 2-37.

Location of Fire Extinguishers

Fire extinguishers should be of a type that can be used on the combustible materials located nearby, Figure 2-38. The extinguishers should be placed so that they can be easily removed without reaching over

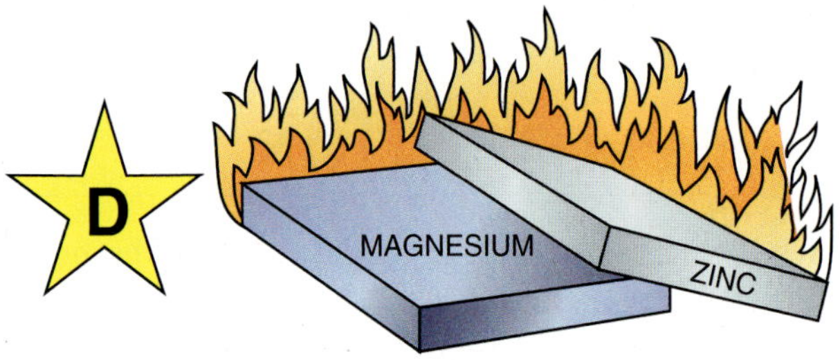

FIGURE 2-37 Type D fire extinguisher symbol.
© Cengage Learning 2012

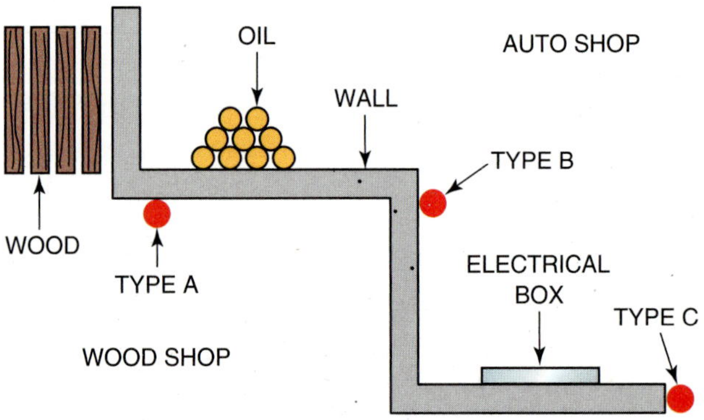

FIGURE 2-38 The type of fire extinguisher provided should be appropriate for the materials being used in the surrounding area. © Cengage Learning 2012

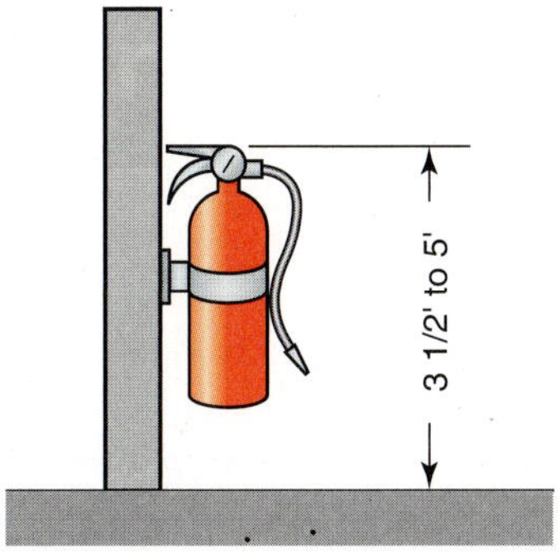

FIGURE 2-39 Mount the fire extinguisher so that it can be lifted easily in an emergency.
© Cengage Learning 2012

FIGURE 2-41 Point the extinguisher at the material burning, not at the flames. © Cengage Learning 2012

combustible material. They should also be placed at a level low enough to be easily lifted off the mounting, Figure 2-39. The location of fire extinguishers should be marked with red paint and signs, high enough so that their location can be seen from a distance over people and equipment. The extinguishers should also be marked near the floor so they can be found even if a room is full of smoke, Figure 2-40.

Use

A fire extinguisher works by breaking the fire triangle of heat, fuel, and oxygen. Most extinguishers both cool

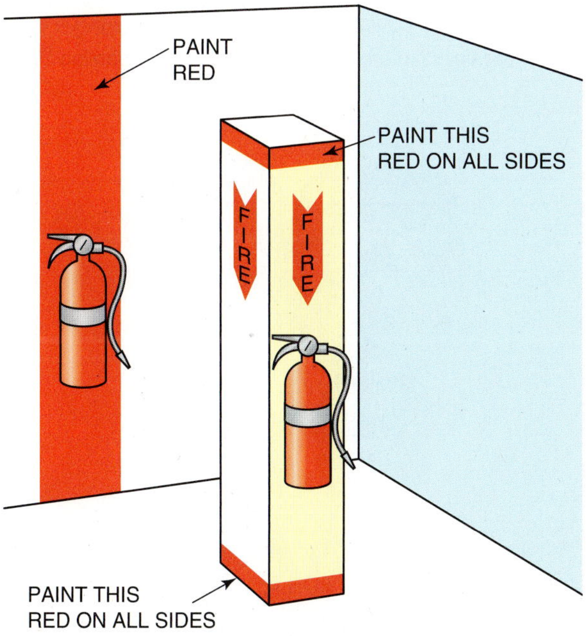

FIGURE 2-40 The location of fire extinguishers should be marked so they can be located easily in an emergency.
© Cengage Learning 2012

the fire and remove the oxygen. They use a variety of materials to extinguish the fire. The majority of fire extinguishers found in welding shops use foam, carbon dioxide, a soda-acid gas cartridge, a pump tank, or dry chemicals.

When using a foam extinguisher, do not spray the stream directly into the burning liquid. Allow the foam to fall lightly on the base of the fire.

When using a carbon dioxide extinguisher, direct the discharge as close to the fire as possible, first at the edge of the flames and gradually to the center.

When using a soda-acid gas cartridge extinguisher, place your foot on the footrest and direct the stream at the base of the flames.

When using a dry chemical extinguisher, direct the extinguisher at the base of the flames. In the case of type A fires, follow up by directing the dry chemicals at the remaining material still burning. Therefore, the extinguisher must be directed at the base of the fire where the fuel is located, Figure 2-41.

EQUIPMENT MAINTENANCE

A routine schedule of equipment maintenance aids in detecting potential problems such as leaking coolant, loose wires, poor grounds, frayed insulation, or split hoses. Small problems, if fixed in time, can prevent the loss of valuable time due to equipment breakdown or injury.

Any maintenance beyond routine external maintenance should be referred to a trained service technician.

In most areas, it is against the law for anyone but a licensed electrician to work on arc welders and anyone but a factory-trained repair technician to work on regulators. Electrical shock and exploding regulators can cause serious injury or death.

Hoses

Hoses must be used only for the gas or liquid for which they were designed. Green hoses are to be used only for oxygen, and red hoses are to be used only for acetylene or other fuel gases. Using unnecessarily long lengths of hoses should be avoided. Never use oil, grease, lead, or other pipe-fitting compounds for any joints. Hoses should also be kept out of the direct line of sparks. Any leaking or bad joints in gas hoses must be repaired.

WORK AREA

The work area should be kept picked up and swept clean. Collections of steel, welding electrode stubs, wire, hoses, and cables are difficult to work around and easy to trip over. An electrode caddy can be used to hold the electrodes and stubs, Figure 2-42. Hooks can be made to hold hoses and cables, and scrap steel should be thrown into scrap bins.

If a piece of hot metal is going to be left unattended, write the word *hot* on it before leaving. This procedure can also be used to warn people of hot tables, vises, firebricks, and tools.

FIGURE 2-42 **An easy-to-build electrode caddy can be used to hold both electrodes and stubs.** Larry Jeffus

HAND TOOLS

The welder uses hand tools to do necessary assembly and disassembly of parts for welding as well as to perform routine equipment maintenance.

The adjustable wrench is the most popular tool used by the welder. During use, this wrench should be adjusted tightly on the nut and pushed so that most of the force is on the fixed jaw, Figure 2-43. When a wrench is being used on a tight bolt or nut, the wrench should be pushed with the palm of an open hand or pulled to prevent injuring the hand. If a nut or bolt is too tight to be loosened with a wrench, obtain a longer wrench. A cheater bar should not be used. The fewer points a box end wrench or socket has, the stronger it is and the less likely it is to slip or damage the nut or bolt, Figure 2-44.

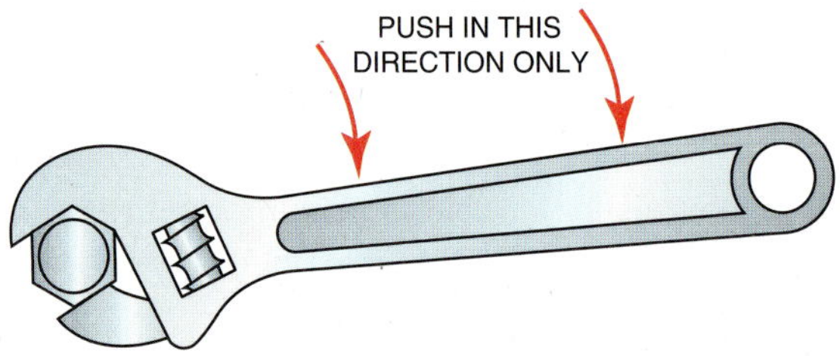

FIGURE 2-43 **The adjustable wrench is stronger when used in the direction indicated.** © Cengage Learning 2012

FIGURE 2-44 **The fewer the points, the less likely the wrench is to slip.** Larry Jeffus

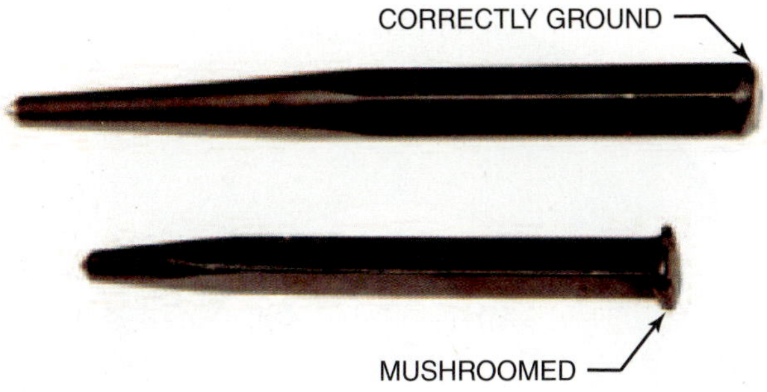

FIGURE 2-45 **Any mushroomed heads must be ground off.** Larry Jeffus

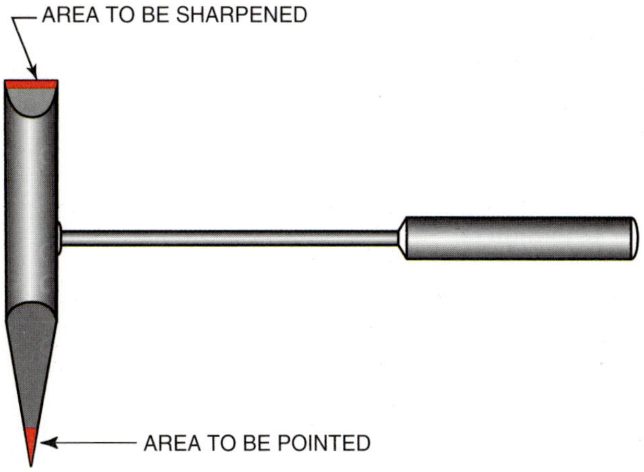

FIGURE 2-47 **Welder's chipping hammer; grind the head periodically.** © Cengage Learning 2012

Striking a hammer directly against a hard surface such as another hammer face or anvil may result in chips flying off and causing injury

The mushroomed heads of chisels, punches, and the faces of hammers should be ground off, Figure 2-45.

Chisels and punches that are going to be hit harder than a slight tap should be held in a chisel holder or with pliers to eliminate the danger of injuring your hand. A handle should be placed on the tang of a file to avoid injuring your hand, Figure 2-46. A file can be kept free of chips by rubbing a piece of soapstone on it before it is used.

It is important to remember to use the correct tool for the job. Do not try to force a tool to do a job that it was not designed to do.

Hand Tool Safety

Hand tools used in welding fabrication should be treated properly and not abused. Many accidents can be avoided by using the right tool for the job. For instance, use a tool that is the correct size for the work rather than one that is too large or too small.

Keep hand tools clean to protect them from corrosion. Wipe off any accumulated dirt, mud, and grease. Occasionally dip the tools in cleaning fluids or solvents and wipe them clean. Lubricate adjustable and other moving parts to prevent wear and misalignment.

Make sure that hand tools are sharp, Figure 2-47. Sharp tools make work easier, improve the accuracy of the work, save time, and are safer than dull tools. When sharpening, redressing, or repairing tools; shape, grind, hone, file, fit, and set them properly using other tools suited to each purpose. For sharpening tools, either an oilstone or a grindstone is preferred. If grinding on an abrasive wheel is required, grind only a small amount at a time. Hold the tool lightly against the wheel to prevent overheating, and frequently dip the part being ground in water to keep it cool. This will protect the hardness of the metal and help to retain the sharpness of the cutting edge. Be sure to wear safety goggles when sharpening or redressing tools.

When carrying tools, protect the cutting edges and carry the tools in such a way that you will not endanger yourself or others. Carry pointed or sharp-edged tools in pouches or holsters.

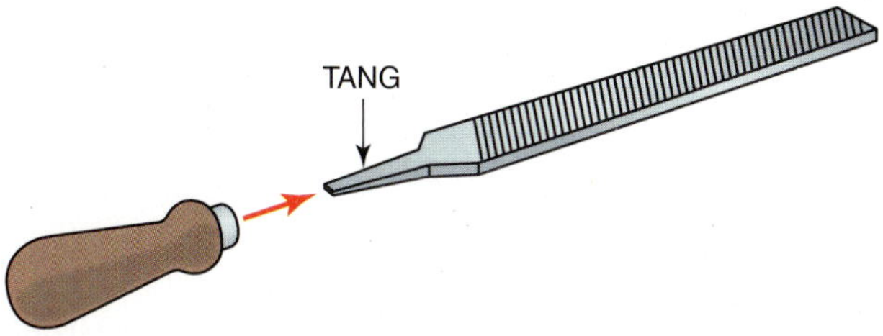

FIGURE 2-46 **To protect yourself from the sharp tang of a file, always use a handle with a file.** © Cengage Learning 2012

Hammer Safety

Keep hammer handles secure and safe. Check wedges and handles frequently. Be sure heads are wedged tightly on handles. Keep handles smooth and free of rough or jagged edges. Do not rely on friction tape to secure split handles or to prevent handles from splitting. Replace handles that are split or chipped or that cannot be refitted securely.

When swinging a hammer, be absolutely certain that no one is within range or can come within range of the swing or be struck by flying material. Always allow plenty of room for arm and body movements when hammering on anything.

The following safety precautions generally apply to all hammers:

- Check to see that the handle is tight before using any hammer. Never use a hammer with a loose or damaged handle.
- Always use a hammer of suitable size and weight for the job.
- Discard or repair any tool if the face shows excessive wear, dents, chips, mushrooming, or improper redressing.
- Rest the face of the hammer on the work before striking to get the feel or aim; then, grasp the handle firmly with the hand near the end of the handle. Move the fingers out of the way before striking with force.
- A hammer blow should always be struck squarely, with the hammer face parallel to the surface being struck. Always avoid glancing blows and over-and-under strikes.
- For striking another tool (cold chisel, punch, wedge, etc.), the face of the hammer should be proportionately larger than the head of the tool. For example, a 1/2-in. (13-mm) cold chisel requires at least a 1-in. (24-mm) hammer face.
- Never use one hammer to strike another hammer.
- Do not use the end of the handle of any tool for tamping or prying; it might split.

POWER TOOLS

All power tools must be properly grounded to prevent accidental electrical shock. If even a slight tingle is felt while using a power tool, stop and have the tool checked by an electrical technician. Power tools should never be used with force or allowed to overheat from excessive or incorrect use. If an extension cord is used, it should have a large enough current rating to carry the load, Table 2-4. An extension cord that is too small will cause the tool to overheat.

Safety glasses must be worn at all times when using any power tools.

FIGURE 2-48 **Always check to be sure that the grinding stone and the grinder are compatible before installing a stone.** Larry Jeffus

Grinders

Grinding using a pedestal grinder or a portable grinder is required to do many welding jobs correctly. Often it is necessary to grind a groove, remove rust, or smooth a weld. Grinding stones have the maximum revolutions per minute (RPM) listed on the paper blotter, Figure 2-48. They must never be used on a machine with a higher-rated RPM. If grinding stones are turned too fast, they can explode.

Grinding Stone

Before a grinding stone is put on the machine, the stone should be tested for cracks. You can do this by tapping the stone in four places and listening for a sharp ring, which indicates it is good, Figure 2-49. A dull sound indicates that the grinding stone is cracked and should not be used. Once a stone has been installed and has been used, it may need to be trued and balanced by using a special tool designed for that purpose, Figure 2-50. Truing keeps the stone face flat and sharp for better results.

Types of Grinding Stones: Each grinding stone is made for grinding specific types of metal. Most stones are for ferrous metals, meaning iron, cast iron, steel, and stainless steel, among others. Some

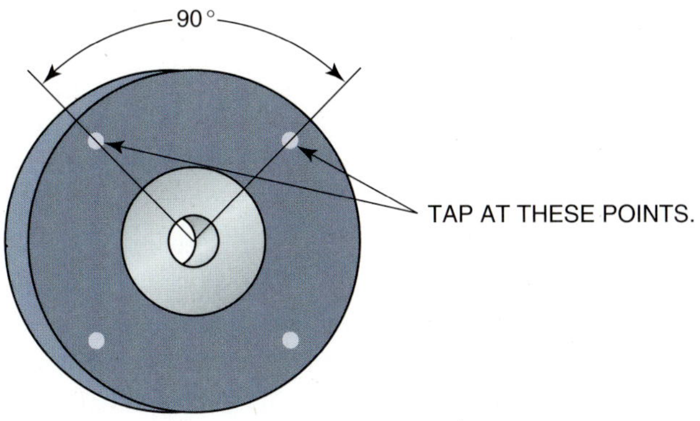

FIGURE 2-49 **Grinding stones should be checked for cracks before they are installed.** © Cengage Learning 2012

FIGURE 2-50 **Use a grinding stone redressing tool as needed to keep the stone in balance.** Larry Jeffus

stones are made for nonferrous metals such as aluminum, copper, and brass. If a ferrous stone is used to grind nonferrous metal, the stone becomes glazed (the surface clogs with metal) and may explode due to frictional heat building up on the surface. If a nonferrous stone is used to grind ferrous metal, the stone is quickly worn away.

When the stone wears down, keep the tool rest adjusted to within 1/16 in. (2 mm), Figure 2-51, so that the metal being ground cannot be pulled between the tool rest and the stone surface. Stones should not be used when they are worn down to the size of the paper blotter. If small parts become hot from grinding, pliers can be used to hold them. Gloves should never be worn when grinding. If a glove gets caught in a stone, the whole hand may be drawn in.

The sparks from grinding should be directed down and away from other welders or equipment.

Drills

Before starting to drill, secure the workpiece as necessary and fasten it in a vise or clamp. Holding a small

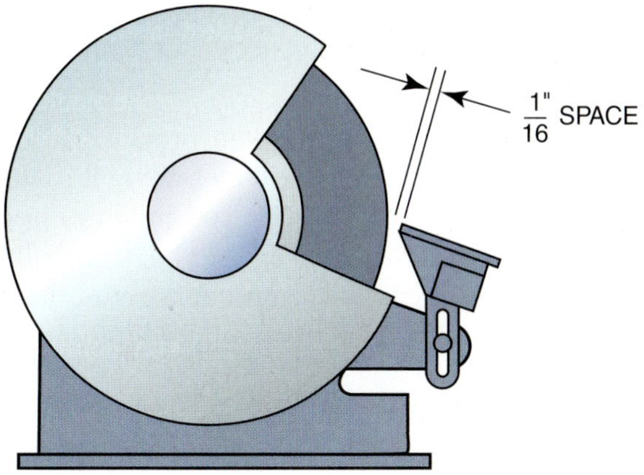

FIGURE 2-51 **Keep the tool rest adjusted.** © Cengage Learning 2012

item in your hand can cause injury if it is suddenly seized by the bit and whirled from your grip. This is most likely to happen just as the bit breaks through the hole at the back side of the work. All sheet metal tends to cause the bit to grab as it goes through. This can be controlled by reducing the pressure on the drill just as the bit starts to go through the workpiece.

Carefully center the drill bit in the jaws of the chunk and securely tighten it. Do not insert the bit off-center because it will wobble and probably break when it is used. Drill bits that are 1/4 in. (6 mm) may be hand tightened in the drill chuck to prevent them from snapping if they are accidentally grabbed. Hand tightening the small bits allows them to spin in the chuck if necessary, thus reducing bit breakage. This technique does not always work because some chunks cannot hold the bits securely enough to prevent them from spinning during normal use. In these cases, the chunk must be tightened securely with a chuck key.

When possible, center-punch the workpiece before drilling to prevent the drill bit from moving across the surface as the drilling begins. After centering the drill bit tip on the exact point at which the hole is to be drilled, start the motor by pulling the trigger switch. Never apply a spinning drill bit to the work. With a variable-speed drill, run it at a very low speed until the cut has begun. Then gradually increase to the optimum drill speed.

Except when it is desirable to drill a hole at an angle, hold the drill perpendicular to the face of the work. Align the drill bit and the axis of the drill in the direction that the hole is to go and apply pressure only along this line, with no sideways or bending pressure. Changing the direction of pressure will distort the dimensions of the hole and might snap a small drill bit.

Use just enough steady, even pressure to keep the drill cutting. Guide the drill by leading it slightly, if needed, but do not force it. Too much pressure can cause the bit to break or overheat. Too little pressure will keep the bit from cutting and dull its edges due to the friction created by sliding over the surface.

If the drill becomes jammed in the hole, release the trigger immediately, remove the drill bit from the work, and determine the cause of the stalling or jamming. Do not squeeze the trigger on or off in an attempt to free a stalled or jammed drill. When using a reversing-type model, the direction of the rotation may be reversed to help free a jammed bit. Be sure the direction of the rotation is reset before trying to continue the drilling.

Reduce the pressure on the drill just before the bit cuts through the work to avoid stalling in metal. When the bit has completely penetrated the work and is spinning freely, withdraw it from the work while the motor is still running, and then turn off the drill.

METAL CUTTING MACHINES

Many types of metal cutting machines are used in the welding shop—for example, shears, punches, cut-off machines, and band saws. Their advantages include little or no postcutting cleanup, a wide variety of metals that can be cut, and the fact that the metal is not heated.

CAUTION

Before operating any power equipment for the first time, you must read the manufacturer's safety and operating instructions and should be assisted by someone who has experience with the equipment. Be sure your hands are clear of the machine before the equipment is started. Always turn off the power and lock it off before working on any part of the equipment.

Shears and Punches

Welders frequently use shears and punches in the fabrication of metal for welding. These machines can be operated either by hand or by powerful motors. Hand-operated equipment is usually limited to thin sheet stock or small bar stock. Powered equipment can be used on material an inch or more in thickness and several feet wide, depending on the equipment's rating. Their power can be a potential danger if these machines are not used correctly. Both shears and punches are rated by the thickness, width, and type of metal that can be used to work with safely. Failure to follow these limitations can result in damage to the equipment, damage to the metal being worked, and injury to the operator.

Shears work like powerful scissors. The metal being cut should be placed as close to the pivoting pin as possible, Figure 2-52. The metal being sheared must be securely held in place by the clamp on the shear before it is cut. If you are cutting a long piece of metal that is not being supported by the shear table, then portable supports must be used. As the metal is being cut, it may suddenly move or bounce around; if you are holding on to it, this can cause a serious injury.

Power punches are usually either hydraulic or flywheel operated. Both types move quickly, but only the hydraulic type can usually be stopped midstroke. Once

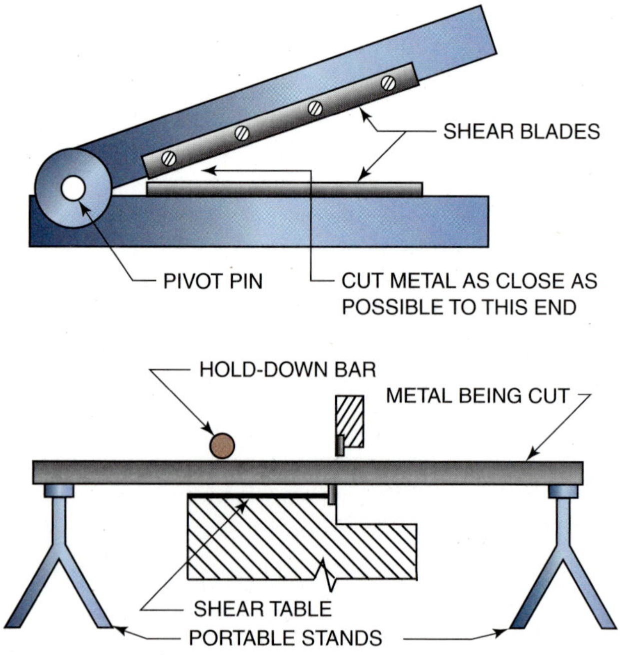

FIGURE 2-52 Power shear. © Cengage Learning 2012

the flywheel-type punch has been engaged, in contrast, it will make a complete cycle before it stops. Because punches move quickly or may not be stopped, it is very important that the operator's two hands be clear of the machine and that the metal be held firmly in place by the machine clamps before starting the punching operation.

Cut-Off Machines

Cut-off machines may use abrasive wheels or special saw blades to make their cuts. Most abrasive cut-off wheels spin at high speeds (high RPMs) and are used dry (without a coolant). Most saws operate much more slowly and with a liquid coolant. Both types of machines produce quality cuts in a variety of bar- or structural-shaped metals. The cuts require little or no postcut cleanup. Always wear eye protection when operating these machines. Before a cut is started, the metal must be clamped securely in the machine vise. Even the slightest movement of the metal can bind or break the wheel or blade. If the machine has a manual feed, the cutting force must be applied at a smooth and steady rate. Apply only enough force to make the cut without dogging down the motor. Use only reinforced abrasive cut-off wheels that have an RPM rating equal to or higher than the machine-rated speed.

Band Saws

Band saws can be purchased as vertical or horizontal, and some can be used in either position. Some band saws can be operated with a cooling liquid and are called wet saws; but most operate dry. The blade guides must be adjusted as closely as possible to the metal being cut. The cutting speed and cutting pressure must be low enough to prevent the blade from overheating. When using a vertical band saw with a manual feed, you must keep your hands away from the front of the blade so that if your hand slips, it will not strike the moving blade. If the blade breaks, sticks, or comes off the track, turn off the power, lock it off, and wait for the band saw drive wheels to come to a complete stop before touching the blade. Be careful of hot flying chips.

Portable band saws can be used to make cuts through pipes, angles, and other similarly shaped materials up to approximately 6 in. (152 mm) in size. These portable saws can make quick work of cutting small-diameter material; however, the cut may not be as square as those made with a fixed band saw. In most cases, the slight out-of-square cut is made well within the tolerance needed for general fabrication work.

MATERIAL HANDLING

Proper lifting, moving, and handling of large, heavy-welded assemblies is important to the safety of the workers and the weldment. Improper work habits can cause serious personal injury as well as cause damage to equipment and materials.

Lifting

When you are lifting a heavy object, the weight of the object should be distributed evenly between both hands, and your legs (not your back) should be used to lift, Figure 2-53. Do not try to lift a large or bulky object without help if the object is heavier than you can lift with one hand.

Hoists or Cranes

Hoists or cranes can be overloaded with welded assemblies. The capacity of the equipment should be checked before trying to lift a load. Keep any load as close to the ground as possible while it is being moved. Pushing a load on a crane is better than pulling a load. It is advisable to stand to one side of ropes, chains, and cables that are being used to move or lift

FIGURE 2-53 Lift with your legs, not your back.

FIGURE 2-54 Never stand in-line with a rope, chain, or cable that is being used to move or lift a load. © Cengage Learning 2012

a load, Figure 2-54. If they break and snap back, they will miss hitting you. If it is necessary to pull a load, use a rope, Figure 2-55.

Hauling

Much of the fabrication work may be done away from the buildings or shops. This means that the equipment used for welding, such as torches, grinders, gas cylinders, and portable welders, has to be hauled out to the job in the work area. Keeping everything together makes it safer and easier to haul. Depending on the operation's needs, a truck or trailer can be set up for welding.

To secure the safety of oxyacetylene cylinders, a special cylinder holder can be built in the vehicle. It is important that fuel and oil for the engine generator welder not be stored in the toolboxes. If fuel or oil gets on the oxygen and acetylene regulators, they can explode when connected to the cylinders

FIGURE 2-55 When moving a load overhead, stay out of the way of the load in case it falls. © Cengage Learning 2012

for use. Even a small amount of liquid or vapors from fuel and oil can ruin welding electrodes. Electrodes contaminated with fuel or oil produce porous, weak welds.

CAUTION

It is dangerous to store fuel and oil in the same toolbox as regulators and electrodes. The regulators can explode and the electrodes may not work correctly. Securely store fuel and oil containers in the open vehicle bed, Figure 2-56.

Often the tie-downs in a truck bed are connected high along the bed rail, which makes it difficult to secure small items like fuel and oil containers. Magnetic strap connectors, like Magna-Ties™, make it easier to safely secure small items in the bed of welding vehicles.

FIGURE 2-56 Secure oxygen and fuel cylinders on vehicles. Remove regulators and replace cylinder valve safety caps any time the vehicle is in motion. Larry Jeffus

SUMMARY

Safety in all aspects of welding fabrication is of utmost importance. When done correctly, welding and fabrication are safe. You are responsible for your own safety. If you do not think a job can be done safely, do not start the job. If you have safety concerns about a job, get help.

You must read and follow all of the manufacturer's operation and safety literature on any equipment. Do not assume that any new equipment has the exact same operating instructions or safety rules. As equipment changes, even slightly, manufacturers update the literature; you must read it to know that you are working in the safest manner.

Periodically performing maintenance, servicing, and doing a safety check of equipment make doing every job safer. In addition to maintaining safety, keeping equipment well-maintained reduces operating costs. Equipment that is in good working order makes the job go better, faster, and safer.

Further safety information is available in *Safety for Welders* by Larry F. Jeffus, published by Delmar/Cengage Learning and the U.S. Department of Agriculture.

REVIEW QUESTIONS

1. What are the most common and painful injuries that occur in welding fabrication?
2. What three classifications are burns divided into?
3. What burns are reddish in color, tender, and painful and do not involve any broken skin?
4. Why is it important to cool the burned area?
5. Why should you not put ice or ice water on a burn?
6. What are the two types of light that can cause burns?
7. What problems can reflected welding light cause?
8. What should be done if welding cannot be moved away from other workers in the area?
9. How long might it take to burn your eyes with UV light?

10. What problem might visible light cause?
11. When must eye protection be worn?
12. Why can an injured (burned) eye become infected?
13. What type of eye and face protection should be worn for heavy grinding?
14. According to Table 2-2, what is the maximum number of hours a welder can grind without wearing ear protection?
15. Which welding and cutting processes may produce undesirable respiratory by-products that might require some type of protection?
16. List some of the items that should be included in a respirator-protection training program.
17. Which type of respirator uses a blower to force the ambient air through air-purifying elements to the inlet covering?
18. What is a possible way of telling if a bolt, nut, or hinge is plated with cadmium?
19. How is ozone produced by welding?
20. Why must welding be done outside or in a well-ventilated area?
21. List some of the metals that may give off dangerous fumes and always require forced ventilation when being welded.
22. Ladders used for welding fabrication should be listed with what organizations?
23. List the general safety rules for ladder use.
24. Why must welding cables be checked periodically?
25. What is a double-insulated power tool?
26. Refer to Table 2-4 to find the wire gauge needed for an extension cord if it is going to be 100 ft (30.5 m) long and supply 9 amps to the power tool.
27. What is the correct term for a device sometimes referred to as a GFI?
28. Clothing made from what materials must be avoided in the welding shop?
29. Why must butane lighters and matches not be carried in a welding shop?
30. List the types of ready-to-wear leather-protective clothing for welding.
31. Why must fuel gas cylinders or other flammable materials be stored separately from oxygen cylinders?
32. When must the valve protection cap be on a cylinder?
33. Why must an acetylene cylinder stand upright for four hours after its has been stored on its side?
34. Who can stand fire watch for a welder?
35. List the four types of fire extinguishers and tell what type of fire each can be used to fight.
36. Why should a wrench be pushed with the palm of an open hand or pulled?
37. How can hand tools be cleaned?
38. How should a hammer blow be made?
39. How can you tell what the maximum RPMs a grinding stone can be used at?
40. Why might small drill bits be hand tightened?
41. What should you do before operating any power equipment for the first time?
42. What can happen if fuel or oil gets on oxygen and acetylene regulators?

Chapter 3

Shop Math

OBJECTIVES

After completing this chapter, the student should be able to:

- Solve basic welding fabrication math problems.
- Round numbers.
- Convert mixed units, fractions, and decimal fractions.
- Reduce fractions and decimal fractions.

KEY TERMS

common denominator
converting or conversion
decimal fractions
denominator
dimensioning
equation
formula
fractions
mixed units
numerator
rounding
sequence of mathematical operations
tolerance
whole numbers

INTRODUCTION

The most common use of math in welding fabrication is for **dimensioning.** Most welding shops use standard or English dimensioning with feet, inches, and **fractions.** A few shops use metric dimensioning, and even fewer use feet, inches, and decimal fractions, Figure 3-1. The math functions of addition, subtraction, multiplication, and division of dimensions are easier in the metric and the feet, inches, and decimal fraction systems than they are in the feet, inches, and fraction systems because calculators can be used.

Almost all welding shop dimensioning uses **mixed units.** An example of a mixed unit is feet and inches, with inches based on 12 inches to the foot. So the largest number of inches is 11 because when you add 1 more inch it becomes 12 inches, which is expressed as 1 foot. Mixed units present unique problems for addition, subtraction, multiplication, and division because each

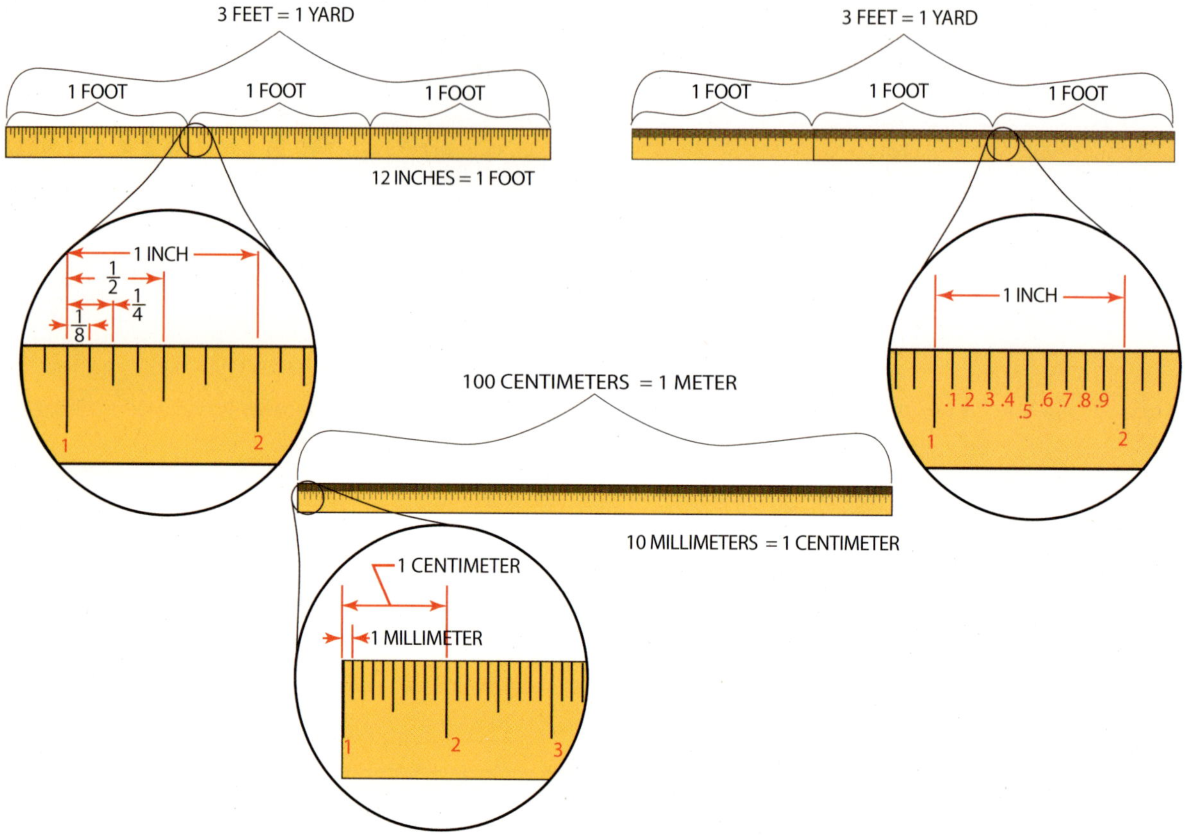

FIGURE 3-1 Three types of measuring rules. © Cengage Learning 2012

type of unit must be worked separately. For example, you cannot add feet to inches without first converting the feet to inches.

Most welding shops encourage the use of calculators because they eliminate arithmetic errors. There are a few calculators that can add, subtract, multiply, and divide standard fractional dimensions; and some can even work mixed units of feet and inches. However, most math using fractions must be worked manually.

This chapter concentrates on basic math and reviews the use of mixed units, decimal fractions, and fractions.

TYPES OF NUMBERS

Welding fabricators use only a few types of numbers on a regular basis. They most commonly use the following types of numbers:

- Whole numbers—**Whole numbers** are numbers used to express units in increments of 1, so they can be divided evenly by the number 1. Examples of whole numbers are 1, 2, 3, 4, 5, 6, 7, 8, 9, 10. They are the easiest type of numbers to use with all types of mathematical functions.
- Decimal fractions—A **decimal fraction** is a number that uses a decimal point to denote a unit that is smaller than 1. Examples of decimal fractions are 0.1, 0.35, 0.9518, 1.1, 5.4, 3.14, and 125.1234. Decimal fractions are expressed in units that are 10 times, 100 times, 1000 times, and so on, smaller than 1, Figure 3-2. They can be added and subtracted as easily as whole numbers. Decimal fractions can be used with whole numbers.
- Mixed units—Mixed units are measurements containing numbers that are expressed in two or more different units. An example of a mixed unit is a linear measurement such as 2 ft 6 in., with part of the measurement expressed in feet (2) and the other part in inches (6). Other examples of mixed units are angular measurements such as 45° 0', weight measurements such as 8 lb 4 oz, and time measurements such as 3 hrs 20 min 15 sec. The most common types of mixed units used in welding fabrication are linear dimensions, angular dimensions, weight, and time. Linear dimensions use units of feet and

1,000,000.0 millions
100,000.0 hundred thousand
10,000.0 ten thousand
1000.0 thousand
100.0 hundreds
10.0 tens
1.0 ones
tenths 0.1
hundredths 0.01
thousandths 0.001
ten thousandths 0.0001
hundred thousandths 0.00001
millionths 0.000001

FIGURE 3-2 Number and decimal terminology.

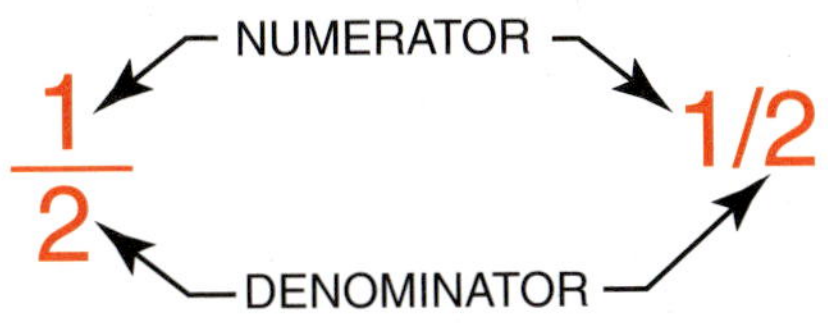

FIGURE 3-3 Fraction identification.

inches; angular dimensions use degrees, minutes, and seconds; weight uses pounds and ounces; and time uses hours, minutes, and seconds. The different units may be separated with a space ( ) or dash (–). It is important to keep the different number units straight when performing mathematical functions.

- Fractions—A fraction is two or more numbers used to express a unit smaller than one. Examples of fractions are 1/2, 3/4, 5/16, 2 3/8, and 9 1/2. A dash (–) or slash (/) is used to separate the top and bottom numbers of a fraction. The **denominator** is the bottom number of a fraction, and the **numerator** is the top number, Figure 3-3. Fractions can be the hardest type of number to work with because they often cannot be added, subtracted, multiplied, or divided without first having to make some type of **conversion.** Fractions can include whole numbers, such as the "9" in 9 1/2.

GENERAL MATH RULES

All of the math in this textbook will be set up and worked in the same manner.

1st Step The equation or formula will be stated on the first line.

2nd Step Explain the meaning of any variables.

3rd Step State the problem's known values and what answer is needed.

4th Step Write the equation or formula.

5th Step Write the known values in place of the variables.

6th Step One mathematical calculation will be performed per line.

7th Step Add as many lines as necessary to complete the problem's calculations.

8th Step Give the answer, including units.

9th Step Explain the answer in a written statement.

The working of the math problems in this very structured way is intended to make it easier for you to look back at the examples in this textbook and work new problems you might encounter in a welding fabrication shop. It can be very frustrating to know the equation or formula but not remember the sequence of mathematical functions.

EQUATIONS AND FORMULAS

An **equation** is a mathematical statement in which both sides are equal to each other; for example, 2X = 1Y. In this equation, the value of X is always going to be 1/2 of the value of Y. An example of an equation used in metal fabrication would be: the number of hours worked (hrs) × pay per hour ($) = total labor bill (T), or hrs × $ = T. If either the hours or the pay rate goes up, the total bill goes up too and vice versa.

To find the labor cost, use the following formula:

$$\text{hrs} \times \$ = \text{T}$$

Where:
hrs = hours worked
$ = hourly rate
T = total labor bill in dollars

Find the total labor bill for 4 hours of work at $15 per hour, Figure 3-4.

$$\text{hrs} \times \$ = \text{T}$$
$$4 \times 15 = 60$$

The total pay for the 4 hours of work at $15 per hour would be $60.

A **formula** is a mathematical statement of the relationship of items. It also defines how one cell

JOB CARD		
Job Weld out Go Cart Frame	Date 7/10/10	Welder lfj
Starting Time	Ending Time	Total Time
7:00 am	11:00 am	4 hrs
	Total Hours	4 hrs
	Hourly Rate	x $15/hr
	Total Labor	$60.00

FIGURE 3-4 Bill of materials. © Cengage Learning 2012

of data relates to another cell of data; for example, wt = [(l" × w" × t") ÷ 1728] × wt/ft. In this formula you must first find the volume of material by multiplying length (l) times width (w) times thickness (t) before dividing that number by the number of cubic inches in a cubic foot (1728), then multiply that number by the material's weight per cubic foot (wt/ft^3).

The **sequence of mathematical operations** is important when working formulas and equations. For example, $6 \times 4 \div 2 \times 5 = 60$, but $(6 \times 4) \div (2 \times 5) = 2.4$. When a formula has more than one mathematical operation, the operations must be performed in the following order:

1st Perform all operations within parentheses.

2nd Resolve any exponents.

3rd Do all multiplication and division working from left to right.

4th Do all addition and subtraction working from left to right.

To find the total weight of a piece of material, use the following formula:

$$\text{wt} = [(\text{l}'' \times \text{w}'' \times \text{t}'') \div 1728] \times \text{wt/ft}$$

Where:

wt = total weight
l" = length in inches
w" = length in inches
t" = thickness in inches
1728 = number of cubic inches in a cubic foot
wt/ft = weight of material in pounds per cubic foot

Find the total weight of a piece of steel that is 144 in. long, 12 in. wide, and 2 in. thick if the steel weighs 490 lb per cubic foot:

$$\text{wt} = [(\text{l}'' \times \text{w}'' \times \text{t}'') \div 1728] \times \text{wt/ft}$$
$$\text{wt} = [(144'' \times 12'' \times 2'') \div 1728] \times 490$$
$$\text{wt} = [(1728 \times 2'') \div 1728] \times 490$$
$$\text{wt} = (3456 \times 1728) \times 490$$
$$\text{wt} = 2 \times 490$$
$$\text{wt} = 980$$

The total weight of this piece of steel is 980 lb.

WHOLE NUMBERS

Adding and Subtracting Whole Numbers

Most fabrication drawings contain as many dimensions as possible. Sometimes the welder may have to do some basic math to find a dimension that is not given on the drawing in order to complete the project. A common math problem that welders may encounter relates to how much metal stock is needed for a project. This information may be included in the bill of materials, but when it is omitted, you will have to do the math yourself. For example, if you want to know how many feet of pipe you need and the dimensions are whole numbers such as 10' + 5', all you do is add them together using the following formula:

Add $L_t = L_1 + L_2$

Where:

L_t = total length of pipe needed
L_1 = length of part 1
L_2 = length of part 2

Add
$$L_t = L_1 + L_2$$
$$L_t = 10' + 5'$$
$$L_t = 15'$$

The total feet of pipe required is 15 ft.

Most pipe comes in 20-ft sections. To find out how much scrap pipe you will have once you cut out the 15 ft needed in the previous problem, you subtract the whole dimensions 15 ft from 20 ft

Subtract $L_t = L_1 - L_2$

Where:

L_t = total length of scrap
L_1 = length of part 1
L_2 = length of part 2

Subtract
$$L_t = L_1 - L_2$$
$$L_t = 20' - 15'$$
$$L_t = 5'$$

The length of scrap pipe is 5 ft, Figure 3-5.

The same thing would be done when adding or subtracting dimensions if they were all in inches units, such as 5" + 3" or 9" – 4".

Add
$$L_t = L_1 + L_2$$
$$L_t = 5'' + 3''$$
$$L_t = 8''$$

or

Subtract
$$L_t = L_1 - L_2$$
$$L_t = 9'' - 4''$$
$$L_t = 5''$$

Multiply and Divide Whole Numbers

Multiplication can be used to calculate the area, determine the weight of a weldment, estimate the total time required to build a weldment, and so on. Multiplying is the same thing as adding the same number to itself over and over; for example, $2 \times 4 = 8$ is the same thing as $2 + 2 + 2 + 2 = 8$. An example of a multiplication problem would be if you want to cut off four pieces of bar stock. Each piece of bar will be 5 ft long, Figure 3-6. To find out how many total feet of bar stock you will need, you would multiply 5 ft times 4 pieces using the following:

Multiply
$$5' \times 4$$
$$5' \times 4 = 20'$$

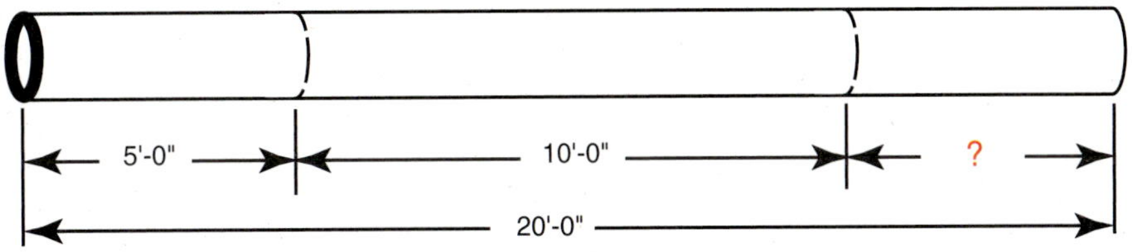

FIGURE 3-5 Pipe layout adding and subtracting whole numbers. © Cengage Learning 2012

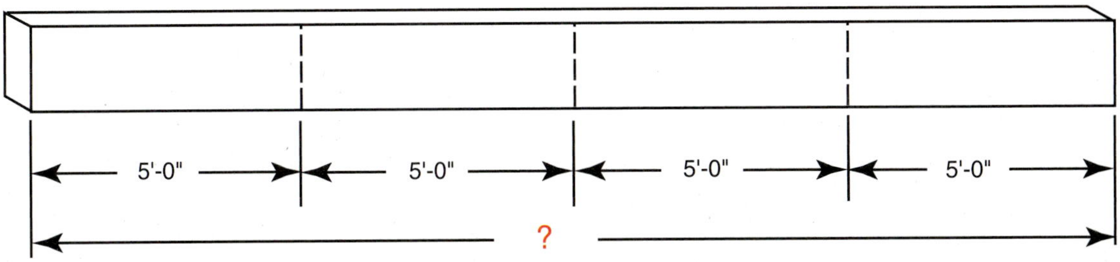

FIGURE 3-6 Bar layout multiplying whole numbers. © Cengage Learning 2012

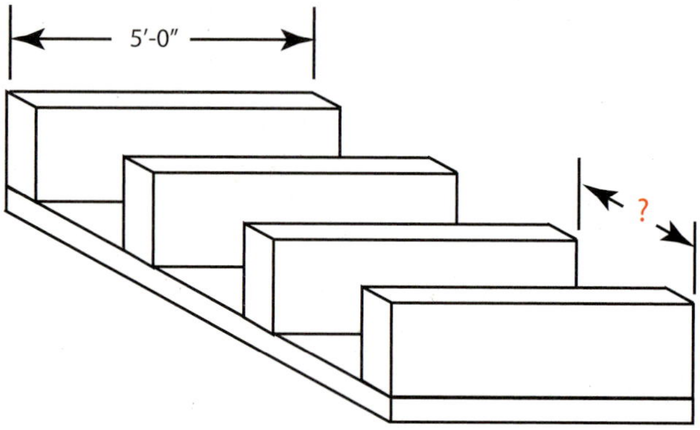

FIGURE 3-7 Part layout dividing whole numbers.

Division can be used to locate the center of an object, determine the even spacing of parts, find the number of parts that can be cut from a piece of metal, and so on. Division is the process of subtracting the same number over and over again; for example, 20 ÷ 4 = 5 is the same thing as repeatedly subtracting 4 from 20: 20 − 4 = 16 (one time), then 16 − 4 = 12 (two times), then 12 − 4 = 8 (three times), then 8 − 4 = 4 (four times), now 4 − 4 = 0 (five times). So, 4 goes into 20 five times. An example of a division problem would be to find the spacing distance between the four pieces of bar stock you just cut off, assuming you want to space them evenly on a 20-in.-(508-mm) wide 5-ft-(1.524-m) long plate, Figure 3-7. Use the following to determine the center spacing:

Divide 20" ÷ 4

20" ÷ 4 = 5"

DECIMAL FRACTIONS

Adding and Subtracting Decimal Fractions

Decimal fractions are added and subtracted just like whole numbers as long as the decimal points are kept in a vertical line. An example of a welding fabrication decimal fraction problem would be to find the total length of two pieces of pipe when one is 5.5 ft and the other is 2.25 ft.

Add 5.5' + 2.25'

$$\begin{array}{r} 5.5 \\ +\,2.25 \\ \hline 7.75 \end{array}$$

When adding decimal fractions you carry the number forward to the left column just like adding whole numbers. So, 0.9 + 0.1 would become 1.0.

Add 8.2 + 4.6 + 9.4 + 11.25

$$\begin{array}{r} 8.2 \\ 4.6 \\ 9.4 \\ +\,11.25 \\ \hline 33.45 \end{array}$$

Subtract 9.8 − 3.2

$$\begin{array}{r} 9.8 \\ -\,3.2 \\ \hline 6.6 \end{array}$$

When subtracting decimal fractions, you can borrow from the left column just like subtracting whole numbers. So, 1.0 − 0.9 would become 0.1.

Subtract 11.3 − 7.5

$$\begin{array}{r} 11.3 \\ -\,7.5 \\ \hline 3.8 \end{array}$$

Multiply and Divide Decimal Fractions

Multiplying Decimal Fractions

When multiplying decimal fractions, you need to ensure that the decimal point is placed in the correct position in the answer. The rule is that you will have the same number of digits to the right of the decimal point in the answer as there were to the right of the decimal points in both of the numbers being multiplied; for example, if the two numbers being multiplied have a total of five digits to the right of the decimal points, then the answer would have five digits to the right of the decimal point.

To multiply 5.4 times 2.3, use the following steps:

1st Step
$$\begin{array}{r} 5.4 \\ \times\,2.3 \\ \hline \end{array}$$

2nd Step
$$\begin{array}{r} 5.4 \\ \times\,2.3 \\ \hline 12 \end{array}$$

3rd Step
$$\begin{array}{r} 5.4 \\ \times\,2.3 \\ \hline 12 \\ 15 \end{array}$$

4th Step
```
  5.4
× 2.3
   12
  15
  8
```

5th Step
```
  5.4
× 2.3
   12
  15
  8
 10
```

6th Step
```
   5.4
 × 2.3
    12
   15
   8
+ 10
 1242
```

7th Step
```
   5.4} 1 PLACE
 × 2.3} 1 PLACE
    12
   15
   8
+ 10
 12.42} 2 PLACES
```

Dividing Decimal Fractions

When dividing decimal fractions you must keep the decimal point in the correct location. The decimal point in the answer of a division problem is placed before any calculations are done, Figure 3-8.

1st Step Move the decimal point in the divisor to make it a whole number, and then move the decimal point in the dividend the same direction and the same number of places as it was moved in the divisor. Add zeros to the dividend if needed to have enough places to move the decimal point.

2nd Step Mark the decimal point in the answer directly above the decimal point you just placed in the divisor.

3rd Step Divide the divisor into the dividend as many times as it will go completely. Subtract that number from the dividend.

4th Step Repeat the third step as many times as needed until the remainder is zero or you start repeating the decimal figures.

WRITE THE PROBLEM	10 ⟌ 1.	0.1 ⟌ 1	1.1 ⟌ 1.1
LOCATE THE DECIMAL POINT	10 ⟌ 1.0	0.1. ⟌ 10 .	1.1. ⟌ 1.1.0
PLACE THE DECIMAL POINT	10 ⟌ 1.0	0 1. ⟌ 10 .	1 1. ⟌ 11.0
DIVIDE	.1 / 10 ⟌ 1.0 / -1 0 / 0	10. / 1. ⟌ 10. / -1 / 00 / 0 / 0	1. / 1 1. ⟌ 11. / -11 / 0
	(A)	(B)	(C)

FIGURE 3-8 Dividing decimal fractions. © Cengage Learning 2012

To find out how many 1.5-ft pieces you could cut out of a 7.8-ft long piece of metal, you would do the following:

Divide 7.8 by 1.5

1st Step 1.5 ⟌ 7.8

2nd Step 15. ⟌ 78.0

3rd Step
```
        5.
15. ⟌ 78.0
     -75.
      3.0
```

4th Step
```
        5.2
15. ⟌ 78.0
     -75.
      3.0
     -3.0
        0
```

Rounding Numbers

When dividing numbers, we often get a whole number followed by a long decimal fraction. When we divide 10 by 3, for example, we get 3.3333333. For all practical purposes, we need not lay out weldments to an accuracy greater than the second decimal place. We would, therefore, round off this number to 3.33, a dimension that would be easier to work with in the welding shop.

RULE: When rounding off a number, look at the number to the right of the last significant place to be used. If this number is less than 5, drop it and leave the remaining number unchanged. If this number is 5 or greater, increase the last significant number by 1 and record the new number.

- Round off 15.6549 to the second decimal place. Because the number in the third place to the right of the decimal point is less than 5, the number would be rounded to 15.65.
- Round off 8.2764 to the second decimal place. Because the number in the third place to the right of the decimal point is 5 or more, the number would be rounded to 8.28.
- Round off 0.8539 to the third decimal place. Because the number in the fourth place to the right of the decimal point is 5 or more, the number would be rounded to 0.854.
- Round off 156.8244 to the first decimal place. Because the number in the second place to the right of the decimal point is less than 5, the number would be rounded to 156.8.

MIXED UNITS

Adding and Subtracting Mixed Units

When adding mixed units such as feet and inches, you have to add each type of number together first. For example, you would add the inches to inches and feet to feet. An example of a mixed unit problem that you might find in welding would be to see how many feet of metal stock you need if one piece is 10 ft 6 in. long and the other is 3 ft 5 in. long, Figure 3-9. The first step would be to write the numbers in columns with feet over feet and inches over inches. The second step would be to add the inches to the inches and feet to the feet.

Add 10' 6" + 3' 5"

	Feet Column	Inch Column
1st Step	10'	6"
2nd Step	+ 3'	5"
	13'	11"

or

Add 10' 6" + 3' 5"

1st Step 10 + 3' = 13'
2nd Step 6" + 5" = 11"
3rd Step 13' + 11" = 13' 11"

When subtracting mixed units, use the same steps as those used for adding. For example, to see how many feet of scrap pipe you have left from a 7 ft 8 in. piece when 5 ft 5 in. will be cut off, Figure 3-10, you would do the following:

Subtract 7' 8" – 5' 5"

	Feet Column	Inch Column
1st Step	7'	8"
2nd Step	– 5'	5"
	2'	3"

or

Subtract 7' 8" – 5' 5"

1st Step 7' – 5' = 2'
2nd Step 8" – 5" = 3"
3rd Step 2' + 3" = 2' 3"

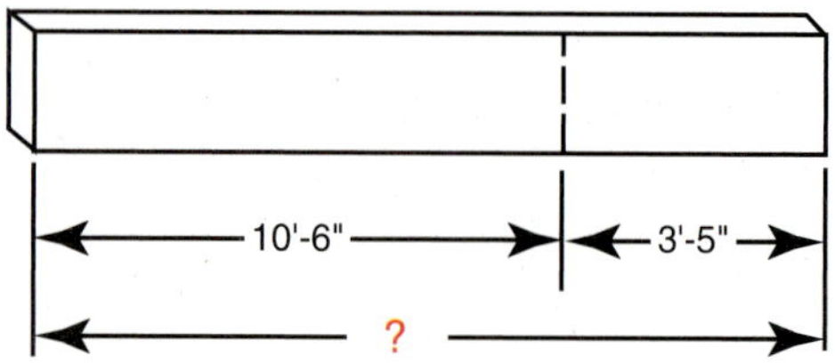

FIGURE 3-9 Adding mixed numbers. © Cengage Learning 2012

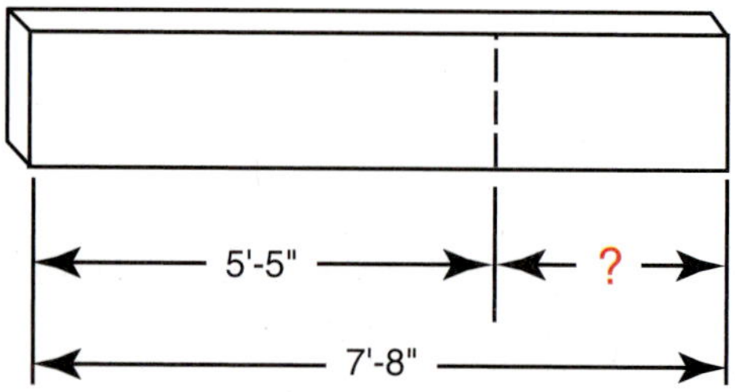

FIGURE 3-10 Subtracting mixed numbers. © Cengage Learning 2012

Feet and Inches	
Ft Fraction	Inches
1/12	1
2/12	2
3/12	3
4/12	4
5/12	5
6/12	6
7/12	7
8/12	8
9/12	9
10/12	10
11/12	11
12/12	12

Pounds and Ounces	
Lb Fraction	Ounces
1/16	1
2/16	2
3/16	3
4/16	4
5/16	5
6/16	6
7/16	7
8/16	8
9/16	9
10/16	10
11/16	11
12/16	12
13/16	13
14/16	14
15/16	15
16/16	16

Table 3-1 Feet and Inches; Pounds and Ounces.

When some mixed units are added, the sum can be reduced. For example, if you add 7 in. to 7 in., the total would be 14 in., which is the same as 1 ft 2 in. To reduce any inches equal to or larger than 12 in. to feet and inches you divide the inches by 12. For example, how long will a piece of steel bar need to be if you are going to cut both a 7 ft 8 in. piece and a 6 ft 6 in. piece from it? Reduce the answer in inches to feet and inches using Table 3-1. Pounds and ounces are also mixed units, which can be reduced by dividing the ounces by 16.

Add 7' 8" + 6' 6"

	Feet Column	Inch Column
1st Step	7'	8"
2nd Step	+ 6'	6"
	13'	14"
Reduce and Add		
3rd Step	13'	14 ÷ 12
4th Step	13 + 1	2
5th Step	14'	2"

or

Add 7' 8" + 6' 6"

1st Step 7' + 6' = 13'
2nd Step 8" + 6" = 14"
3rd Step 13' + 14" = 13' 14"

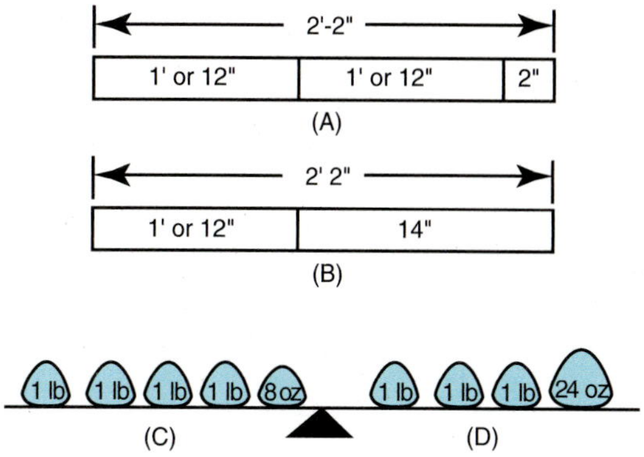

FIGURE 3-11 Mixed units. © Cengage Learning 2012

Reduce 14" to feet and inches

4th Step 14 ÷ 12 = 1' and 2/12"
5th Step 2/12 = 2"

Re-add 13' 0" + 1' 2"

6th Step 13' + 1' = 14'
7th Step 0" + 2" = 2"
8th Step 14' + 2" = 14' 2"

When subtracting one mixed unit from another and the small unit being subtracted is larger than the small unit it is being subtracted from—for example, 1 ft 4 in. from 2 ft 2 in.—you must make that unit larger. You can increase the small unit in a mixed number by subtracting one whole large unit by dividing it into the smaller units. For example, 2 ft 2 in., Figure 3-11A, is the same dimension as 1 ft 14 in., Figure 3-11B; and 4 lb 8 oz, Figure 3-11C, is the same weight as 3 lb 24 oz, Figure 3-11D.

Convert 2' 2"

1st Step 2' = 1' 12"
2nd Step 1' 12" + 2" = 1' 14"

Convert 4 lb 8 oz

1st Step 4 lb = 3 lb 16 oz
2nd Step 3 lb 16 oz + 8 oz = 3 lb 24 oz

With 2 ft 2 in. converted to 1 ft 14 in., you can now subtract 1 ft 4 in. the same way that you subtracted mixed numbers before.

Subtract 1' 14" – 1' 4"

1st Step 1' – 1' = 0'
2nd Step 14" – 4" = 10"
3rd Step 0' + 10" = 10"

To subtract 2 lb 10 oz from 4 lb 8 oz, you have to convert 1 lb to 16 oz and add that to the 8 oz as shown in Figure 3-11C and D. Now 4 lb 8 oz has become 3 lb 24 oz.

Subtract	3 lb 24 oz – 2 lb 10 oz
1st Step	3 lb – 2 lb = 1 lb
2nd Step	24 oz – 10 oz = 14 oz
3rd Step	1 lb + 12 oz = 1 lb 14 oz

When adding multiple mixed numbers, add all of the inches first and then all of the feet. Then reduce the inches to feet and add the new mixed number to get the final answer. Follow the same process when adding pounds and ounces. Find the total length of angle iron needed if the following pieces are to be cut out.

Add 2' 6" + 5' 6" + 7' 3" + 8' 2" + 1' 1" + 3' 9"

	Feet Column	Inch Column
1st Step		
	2'	6"
	5'	6"
	7'	3"
	8'	2"
	1'	1"
2nd Step	+ 3'	9"
	26'	27"

Reduce and Add

3rd Step	26'	27 ÷ 12
4th Step	26 + 2	3
5th Step	28'	3"

or

Add 2' 6" + 5' 6" + 7' 3" + 8' 2" + 1' 1" + 3' 9"

1st Step	2' + 5' + 7' + 8' + 1' + 3' = 26'
2nd Step	6" + 6" + 3 + 2" + 1" + 9" = 27"

Reduce 27" to feet and inches

3rd Step	27 ÷ 12" = 2' 3/12"
4th Step	3/12" = 3"

Add 26' 0" + 2' 3"

5th Step	26' + 2' = 28'
6th Step	0" + 3" = 3"
7th Step	28' + 3" = 28' 3"

Multiply and Divide Mixed Numbers

The easiest way to multiply and divide mixed numbers is to convert the large units to the smallest units. Multiply or divide the small units as if they were whole numbers. Then convert the answer back to a mixed number. You might need to multiply mixed numbers to find the area of a piece of metal or to find out how much material you need to cover the floor of a trailer. Find the area of a small lawn tractor bed that is 4 ft long and 3 ft 6 in. wide.

1st Step Convert the feet to inches by multiplying by 12.

2nd Step Add the converted inches to the already existing inches.

3rd Step Multiply the inches in the multiplicand by the multiplier to get the square inches of the material needed.

4th Step Divide the answer by 144 (number of square inches in a square foot) to find the square feet.

Multiply 4' 0" × 3' 6"

1st Step
4'
× 12
48"
3'
× 12
36"

2nd Step
48" + 0 = 48"
36" + 6 = 42"

3rd Step
48"
× 42"
96
+ 192
2016

4th Step
$$\begin{array}{r} 14\text{ ft}^2 \\ 144\,\overline{)\,2016} \\ \underline{144} \\ 576 \\ \underline{576} \\ 0 \end{array}$$

The small lawn tractor trailer will need 14 sq ft of metal to cover the bed. To divide mixed numbers convert the larger units to the smallest units then divide the whole number. Once the answer is derived, it may be necessary to convert the answer back to feet and inches; pounds and ounces; or hours, minutes, and seconds. This conversion can cause some problems because it is not often exactly even. The easiest way to make these conversions is with a conversion table or refer to the section on conversions that appears later in this chapter.

FRACTIONS

Fractions are commonly used in welding fabrication dimensioning. They are how we express parts (fractions) of inches when measuring. Figure 3-12 shows the common inch fractions most often used in welding fabrication. Fractions smaller than 1/16 of an inch, such as 1/32 and 1/64 of an inch, are not commonly used in fabrication. When they are used, it is most often with some form of automated or machine welding or cutting process.

Adding and Subtracting Fractions

The fractional dimensions found on welding fabrication drawings have a denominator of 2, 4, 8, or 16. If the denominators are not the same, they can easily be made the same, Table 3-2. When all of the denominators of the fractions are the same, adding and subtracting can be accomplished quickly by following these simple rules.

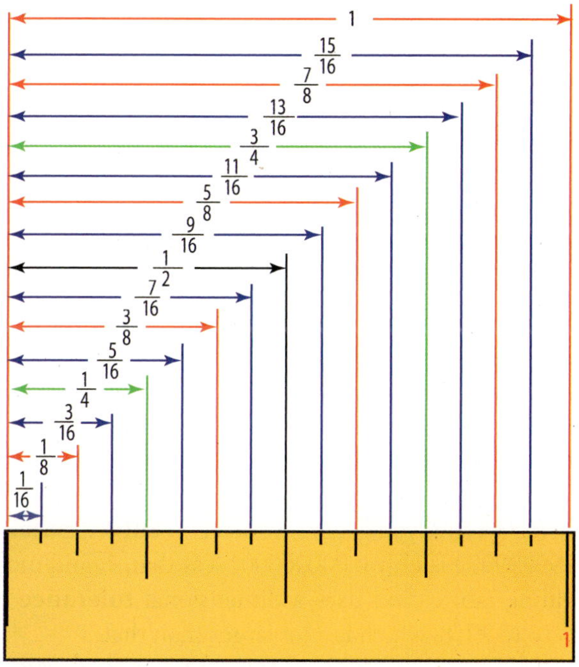

FIGURE 3-12 Fractions of an inch. © Cengage Learning 2012

Inch	Half	Fourth	Eighths	Sixteenths
				1/16
			1/8	2/16
				3/16
		1/4	2/8	4/16
				5/16
			3/8	6/16
				7/16
	1/2	2/4	4/8	8/16
				9/16
			5/8	10/16
				11/16
		3/4	6/8	12/16
				13/16
			7/8	14/16
				15/16
1	2/2	4/4	8/8	16/16

Table 3-2 Converting Fractions to a Common Denominator

1. Add the numerators.
2. Reduce the fraction to the lowest **common denominator.**

Add 1/8 and 3/8. Both denominators are the same, so all you have to do is add the numerators.

$$\text{Add } \frac{1}{8}+\frac{3}{8}$$

$$\frac{1+3}{8}$$

$$\frac{4}{8}$$

Subtract 5/8 from 7/8. Both denominators are the same, so all you have to do is subtract the numerators.

$$\text{Subtract } \frac{7}{8}-\frac{5}{8}$$

$$\frac{7-5}{8}$$

$$\frac{2}{8}$$

Finding the Fraction's Common Denominator

When the denominators of two fractions to be added or subtracted are different, one or both must be converted so that they are the same. To convert a denominator, multiply both the numerator and the denominator of the fraction by the same number. For example, to convert 1/4 to 16ths, you would multiply both the numerator and denominator by 4: 4 × 1 = 4 and 4 × 4 = 16, which is 4/16. There are conversion tables available to convert the denominators and numerators, Table 3-2. To add 1/2 and 1/16 where both denominators are different, you must first find the common denominator. In this case it is 16.

$$\text{Convert the Fraction } \frac{1}{2} = \frac{8}{16}$$

$$\text{Add } \frac{8}{16} + \frac{1}{16}$$

$$\frac{8+1}{16}$$

$$\frac{9}{16}$$

Reducing Fractions

Some fractions can be reduced to a higher denominator. For example, 4/8 is the same as 1/2, although when you are working with a rule or tape measure on the job, you can locate either measurement. Therefore, such reductions are necessary only when you are working with several different dimensions or various fractional units. Also, you would not want to ask for a 4/8-in. thick piece of steel.

The normal way to reduce a fraction is to find the largest number that can be divided into both the denominator and numerator. If you are good at math, that can be easily done, but if you are not good at doing math in your head, there is an alternate way of reducing fractions. For reducing fractions in a welding fabrication shop, it is often easiest to divide both the numerator and denominator by 2. This method will simplify the reduction because all the fractional units found on shop rules and tapes are divisible by 2, for example, halves, fourths, eighths, sixteenths, and thirty seconds. Using this method may require more than one reduction, but the simplicity of dividing by 2 offsets the time needed to repeat the reduction. For example, both the denominator and numerator of 4/8 can be divided by 2, so 4 ÷ 2 = 2 and 8 ÷ 2 = 4. That would make the new fraction 2/4, which can be reduced again by dividing the denominator and numerator one more time by 2. This last division results in 2/4 being reduced to 1/2. Reduction of fractions becomes easier with practice. Reduce 14/16 and 4/16 to their lowest common denominators.

$$\text{Reduce } \frac{14}{16}$$

$$\frac{14 \div 2 = 7}{16 \div 2 = 8}$$

$$\frac{7}{8}$$

The new fraction for 14/16 is 7/8, which is the lowest form.

$$1^{st} \text{ Reduction } \frac{4}{16}$$

$$\frac{4 \div 2 = 2}{16 \div 2 = 8}$$

$$\frac{7}{8}$$

$$2^{nd} \text{ Reduction } \frac{2}{8}$$

$$\frac{2 \div 2 = 1}{8 \div 2 = 4}$$

$$\frac{1}{4}$$

The new fraction for 4/16 is 1/4, which is the lowest form.

Multiply and Divide Fractions

Although there is a way to multiply and divide fractions, the easiest way is to convert the fraction to a decimal fraction and, if needed, convert it back to a fraction once you have an answer. By converting the fraction to a decimal fraction, you can easily use a calculator to do the math. One slight problem with doing that is that some fractions when converted to decimal fractions and then divided result in recurring decimals; for example, 1/3 becomes 0.333333333. If you add 0.333333333 + 0.333333333 + 0.333333333, you get 0.999999999, which is not equal to 1, but if you add 1/3 + 1/3 + 1/3, you get 3/3, which is 1. But realistically, 0.999999999 is only 0.000000001 less than 1, and most welding fabrication uses a dimensional **tolerance** of ±1/16 to ±1/8, which is a lot larger than that.

Table 3-3 lists the conversions for most fractions to decimal fractions. Also, converting fractions to

INCH FRACTION	INCH DECIMAL	mm
1/16	1.5875	0.062500
1/8	3.1750	0.125000
3/16	4.7625	0.187500
1/4	6.3500	0.250000
5/16	7.9375	0.312500
3/8	9.5250	0.375000
7/16	11.1125	0.437500
1/2	12.7000	0.500000
9/16	14.2875	0.562500
5/8	15.8750	0.625000
11/16	17.4625	0.687500
3/4	19.0500	0.750000
13/16	20.6375	0.812500
7/8	22.2250	0.875000
15/16	23.8125	0.937500
1	25.4000	1.000000

Table 3-3 Conversion of Fractions, Decimals, and Millimeters

decimal fractions and decimal fractions back to fractions is covered later in this chapter.

Multiplying Fractions

To multiply fractions without converting, follow these steps:

1st Multiply the two numerators to get the new numerator.

2nd Multiply the two denominators to get the new denominator.

3rd Simplify the resulting fraction, if possible.

To find the area of a piece of metal that is 3/4 in. long by 7/8 in. wide, you would multiply the length by the width as follows:

$$A = L \times W$$

Where:

A = area
L = length
W = width

Multiply

$$A = L \times W$$
$$A = \frac{3}{4} \times \frac{7}{8}$$
$$A = \frac{3 \times 7 = 21}{4 \times 8 = 32}$$
$$A = \frac{21}{32}$$

Dividing Fractions

To divide fractions, invert the divisor, then multiply using steps 2 and 3 in the previous problem.

To find out how many 1/4-in. pieces could be cut out of a 3/4-in. piece of metal, you would do the following:

$$\text{Divide Number} = \frac{3}{4} \div \frac{1}{4}$$
$$\text{Number} = \frac{3}{4} \times \frac{4}{1}$$
$$\text{Number} = \frac{3 \times 4 = 12}{4 \times 1 = 4}$$
$$\text{Number} = \frac{12}{4}$$

The fraction 12/4 can be reduced to 3, so you could cut three pieces from the 3/4-in. piece of metal.

CONVERTING NUMBERS

Dimensions on a drawing are usually given in a consistent format and unit type. Very seldom will you find that you must make conversions to interpret a drawing. However, you may be asked to install or mount a unit such as a winch on a truck bumper or motor on a piece of equipment. Things like this may have a different unit of measurement. If the part is available, you can simply measure it using a tape or rule of the same type as your drawing. Sometimes you will be working from the manufacturer's drawing, and then you may have to make conversions of measurements.

Conversion of measurements is also often used to make it easier to lay out the part. Conversions can also be used to reduce confusion with your coworkers on a job. It is much easier to understand 1 ft 1 1/8 in. rather than trying to find 105/8 in. on a scale, even though they are the same.

Converting Fractions to Decimals

From time to time it may be necessary to convert fractional numbers to decimal numbers. A fraction-to-decimal conversion is needed before most calculators can be used to solve problems containing fractions. There are some calculators that will allow the inputting of fractions without converting them to decimals.

RULE: To convert a fraction to a decimal, divide the numerator (top number in the fraction) by the denominator (bottom number in the fraction).

To convert 3/4 to a decimal:

$$3 \div 4 = 0.75$$

To convert 7/8 to a decimal:

$$7 \div 8 = 0.875$$

Tolerances

All measuring, whether on a part or on the drawing, is in essence an estimate because no matter how accurately the measurement was made, there will always be a more accurate way of making it. The more accurate the measurement, the more time it takes. To save time while still making an acceptable part, dimensioning tolerances have been established. Most drawings usually state a dimensioning tolerance, the amount by which the part can be larger or smaller than the stated dimensions and still be acceptable. Tolerances are usually expressed as plus (+) and minus (–). If the tolerance is the same for both the plus and the minus, it can be written using the symbol ±, Table 3-4. In addition to the tolerance for a part, there may be an overall tolerance for the completed weldment. This dimension ensures that, if all the parts are either too large or too small, their cumulative effect will not make the completed weldment too large or too small. Most weldments use a tolerance of ±1/16 in. or ±1/8 in., Figure 3-13.

		Acceptable Dimensions	
Dimension	Tolerance	Minimum	Maximum
12"	±1/8"	11 7/8"	12 1/8"
2' 8"	±1/4"	2' 7 3/4"	2' 8 1/4"
10'	±1/8"	9' 11 7/8"	10' 1/8"
11"	±0.125	10.875"	11.125"
6'	±0.25	5' 11.75"	6' 0.25"
250 mm	±5 mm	245 mm	255 mm
300 mm	+5 mm–0 mm	300 mm	305 mm
175 cm	±10 mm	174 cm	176 cm

Table 3-4 Examples of Tolerances

Converting Decimals to Fractions

This process is less exact than the conversion of fractions to decimals. Except for specific decimals, the conversion will leave a remainder unless a small enough fraction is selected. For example, the decimal 0.765 is very close to the decimal 0.75, which easily converts to the fraction 3/4. The difference between 0.765 and 0.75 is 0.015 (0.765 – 0.75 = 0.015), which is within the acceptable tolerance for most welding applications. If you are working to a ±1/8-in. tolerance that has up to a 1/4-in. difference from the minimum to maximum dimensions, a measurement of 3/4 is acceptable. More accurately, 0.765 can be converted to 49/64 in., a dimension that would be hard to lay out and impossible to cut using a hand torch.

RULE: To convert a decimal to a fraction, multiply the decimal by the denominator of the fractional units desired; that is, for 8ths (1/8) use 8, for 4ths (1/4) use 4, and so on. Place the whole number (dropping or rounding off the decimal remainder) over the fractional denominator used.

To convert 0.75 to 4ths:

$$0.75 \times 4 = 3.0 \text{ or } 3/4$$

To convert 0.75 to 8ths:

$$0.75 \times 8 = 6.0 \text{ or } 6/8, \text{ which will reduce to } 3/4$$

To convert 0.51 to 4ths:

$$0.51 \times 4 = 2.04 \text{ or } 2/4, \text{ which will reduce to } 1/2$$

Conversion Charts

Occasionally, a welder must convert the units used on the drawing to the type of units used on the layout rule or tape. Fortunately, charts that can be easily used to convert between fractions, decimals, and metric units are available. To use these charts, Table 3-3, locate the original dimension and then look at the dimension in the adjacent column(s) of the new units needed.

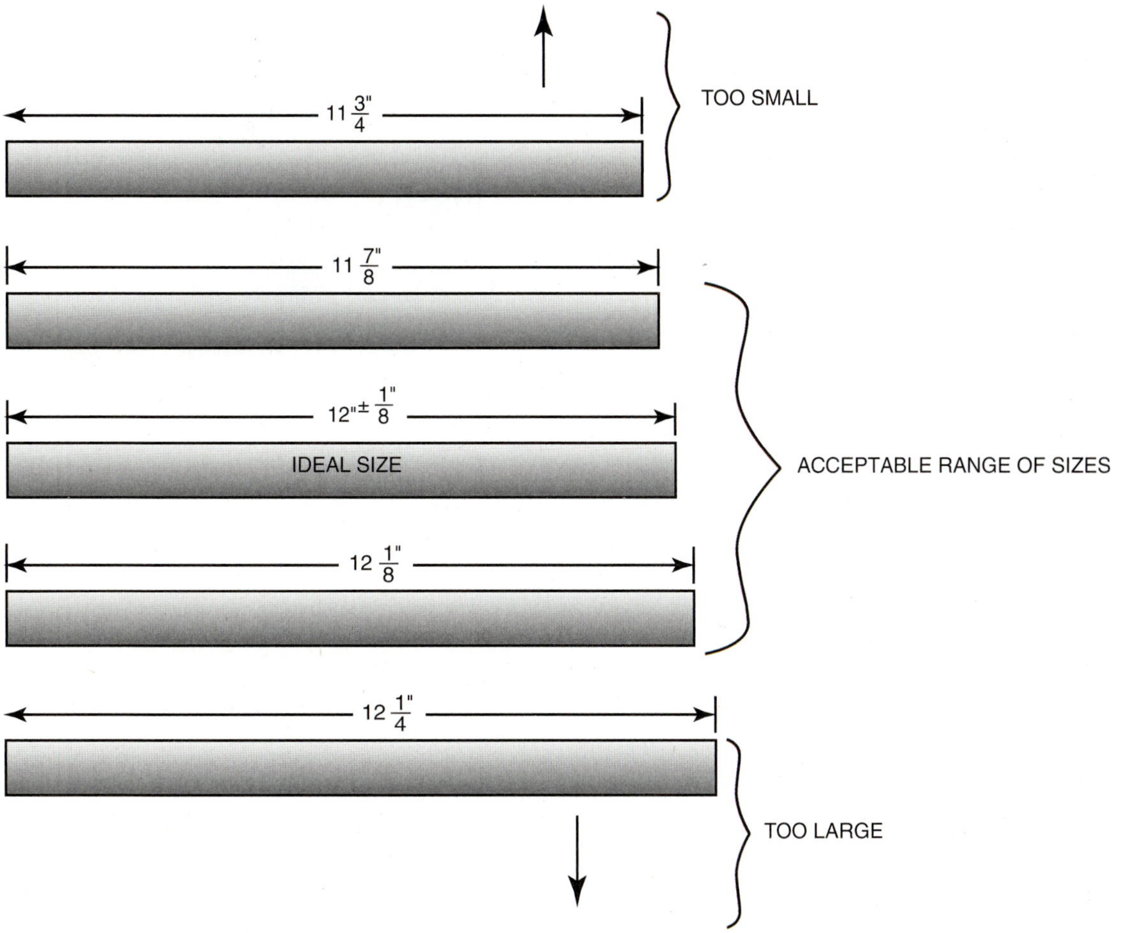

FIGURE 3-13 Tolerances. © Cengage Learning 2012

To convert 1/16 in. to millimeters:

1/16 in. = 1.5875 mm

To convert 0.5 in. to a fraction:

0.5 in. = 1/2 in.

To convert 0.375 in. to millimeters:

0.375 in. = 9.525 mm

To convert 25 mm to a decimal inch:

25 mm = 0.98425 in.

To convert 19 mm to a fractional inch:

19 mm = 3/4 in. (approximately)

Both metric-to-standard conversions and standard-to-metric conversions result in answers that often contain long strings of decimal numbers. Often this new converted number, because of the decimals now attached to it, cannot be easily located on the rule or tape. In addition, most of the layout and fabrication work welders perform do not require such levels of accuracy. These small decimal fractions, in inches or millimeter scales, represent such a small difference that they cannot be laid out with a steel rule or tape. Such small differences can be important to some weldments, but in these cases some machining is required to obtain that level of accuracy. Because these small units are not normally included in a layout, they can be rounded off. Round off millimeter units to the nearest whole number; for example, 19.050 mm would be 19 mm and 1.5875 mm would be 2 mm, and so on. Round off decimal inch units to the nearest 1/16-in. fractional unit; 0.47244 in. would become 0.5 in. (1/2 in.), and 0.23622 in. would become 0.25 in. (1/4 in.). In both cases of rounding, the whole number obtained is well within most welding layout and fabrication drawing tolerances.

Using the rounding-off method of conversions with the conversion chart makes the following converted units easier to locate on rules and tapes.

To convert 1/2 in. to millimeters:

1/2 in. = 13 mm

To convert 0.625 in. to millimeters:

0.625 in. = 16 mm

To convert 2 3/4 in. to millimeters:

$$2 \times 25.4 = 50.8$$
$$3/4 = 19.0$$
$$50.8 + 19 = 69.8$$
69.8 rounded to 70 mm

To convert 5.5 in. to millimeters:

$$5 \times 25.4 = 127.0$$
$$0.5 = 12.7$$
$$127 + 12.7 = 139.7$$
139.7 rounded to 140 mm

To convert 10 mm to a fraction inch:

10 mm = 3/8 in.

To convert 14 mm to a decimal inch:

14 mm = 0.5625 in.

To convert 300 mm to a fraction inch:

300 ÷ 25.4 = 11.81 in. rounded to 11 13/16 in.

To convert 240 mm to a decimal inch:

240 ÷ 25.4 = 9.44 in. rounded to 9 7/16 in.

FIGURE 3-14 **Measuring tape.** © Cengage Learning 2012

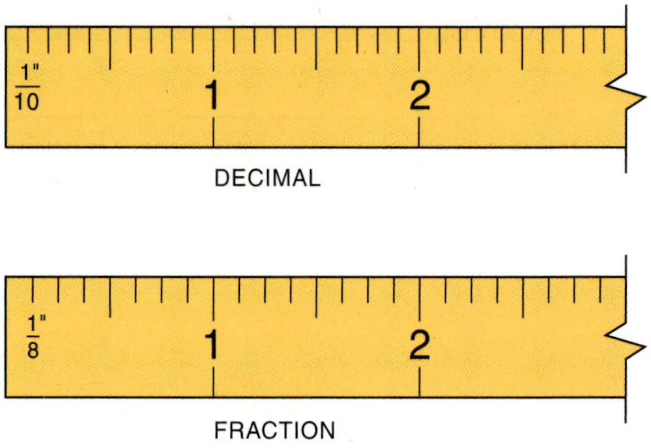

FIGURE 3-15 **Decimal and fraction rules.** © Cengage Learning 2012

Measuring

Measuring for most welded fabrications does not require accuracies greater than what can be obtained with a steel rule or a steel tape, Figure 3-14. Both steel rules and steel tapes are available in standard and metric units. Standard unit rules and tapes are available in fractional and decimal units, Figure 3-15.

SUMMARY

Math is a very important part of welding metal fabrication. You will often use your math skills as a welder fabricator to locate parts, lay out parts, calculate material needs, and determine cost. Math is a skill, and like the skill of welding, it requires practice to become an expert. As you weld on projects, take every opportunity to practice your math. As you practice math, you will find shortcuts to some math problem solving.

Once you have learned the mathematical processes, a calculator can be a big help. A word of caution about using a calculator: Do a quick estimate of the answer you expect from a problem. That way, when you get an answer from the calculator, you can compare the two. Sometimes we hit the wrong key on a calculator accidentally, so the answer is not even close to being correct. If it differs greatly, recheck your work.

REVIEW QUESTIONS

1. Give an example of a dimension containing mixed units.
2. List three examples of a whole number.
3. List three examples of decimal fractions.
4. List three examples of fractions.
5. When a formula has more than one mathematical operation, in what order must the operations be performed?
6. If you need two pieces of pipe—one must be 15 ft and the other 10 ft—what is the total amount of pipe needed?
7. A job requires 500 pieces of bar stock, each 11 in. long. What is the total length of bar stock required in inches?
8. A 21-ft long piece of new steel must be sheared into 7-ft long pieces. How many pieces can be obtained?
9. Find the total length of two pieces of pipe when one is 5.25 ft and the other is 3.5 ft.
10. How much scrap pipe will you have once you cut out 5.5 ft from a 20-ft length of pipe?
11. A welder uses 3.1875 cu ft of acetylene gas to cut one angle iron. How much acetylene gas would be needed to cut 20 angle irons?
12. When dividing decimal fractions, what must be done before starting any calculations?
13. A piece of frame material is 50 ft long. If we cut it into 6.25-ft long pieces, how many pieces will we have?
14. Round 75.235 to the second decimal place.
15. How many total feet of metal stock would you need if one piece is 12 ft 5 in. long and the other is 7 ft 3 in. long?
16. How many feet of scrap pipe will you have left from a 9 ft 6 in. piece when 4 ft 2 in. is cut off?
17. How many total feet of metal stock would you need if one piece is 11 ft 9 in. long and the other is 6 ft 5 in. long?
18. List in order the fractional dimensions of 1 in. in 16ths.
19. How thick will the finished part be if two pieces of metal are welded together if one is 3/4 in. and the other is 5/16 in.?
20. How much metal is left if 1/8 in. is ground off a 5/16-in. thick plate?
21. What is the minimum and maximum length a part can be if it is shown as needing to be 5 7/8 in. ± 1/8?

Chapter 4
Reading Technical Drawings

OBJECTIVES

After completing this chapter, the student should be able to:

- Explain the purpose of a set of drawings and what information is contained on them.
- Identify 10 types of lines used on mechanical drawings.
- Describe what mechanical and pictorial drawings are.
- Name the various special views that can be shown on drawings.
- Explain dimensioning on drawings.
- Discuss how a drawing can be scaled.
- Compare sketches and mechanical drawings.
- Demonstrate the ability to make a sketched mechanical drawing.
- Illustrate how to use graph paper to make a scaled mechanical drawing.
- List the advantages of using computer-aided drafting software to make mechanical drawings.

KEY TERMS

alphabet of lines
architectural scale
bill of materials
break line
cavalier drawings
center line
computer drafting programs
computer drawing programs
cutting plane line
cut-a-ways
detail views
dimension line
drawing scale
engineering scale
extension line
hidden line
isometric drawings
leaders and arrows
mechanical drawings
object line
orthographic projections
phantom lines
pictorial drawings
project routing information
raster lines
rotated views
section line
section view
set of drawings
sketching
specifications
title box
vector lines

INTRODUCTION

Drawings are the tools that let us accurately communicate with each other in a very technical way. It is said that "a picture is worth a thousand words, but a mechanical drawing is worth millions." How many words would you have to use to tell someone how to build a jet airliner, a tanker ship, or even something as simple as a door hinge? Try it. How would you describe the thickness, diameter, and angle of the countersink of the screw holes and their locations? What would be the radius of the hinge pin? And the list of things you need to describe about the making of a hinge goes on and on. We can do all of that and more with easily understood mechanical drawings.

As you look at a basic mechanical drawing, you can see the object's shape, size, and location of its parts. However, as the drawing becomes more and more complex, it can be more difficult to see how everything relates. This chapter helps you see the basic layout of mechanical drawing and how the various parts of a drawing relate.

MECHANICAL DRAWINGS

Mechanical drawings have been used for centuries. Leonardo da Vinci (1452–1519) made mechanical drawings extensively for his inventive works. Mechanical drawings are produced in a similar format worldwide. There are a few differences in how the views are laid out, but even with that they are understandable. The two most common methods are third-angle projection and first-angle projection, Figure 4-1.

SHOP TALK

Did you know that mechanical drawings are often called the universal language? Many of Leonardo da Vinci's drawings still exist today and are as easily understood now as when they were drawn. For that reason mechanical drawings have been called the universal language. There are many different cultures with a wide variety of beliefs among the many nations of the world. Some of their beliefs and practices may seem strange to us, but much of what we do may seem strange to them too. Notwithstanding these differences and the many different languages and measuring systems used around the globe, the basic shape of an object and location of components can be determined from any good drawing. Drawings are so universal that they were placed on the Voyager 1 and 2 space probes that were sent to travel out of our solar system in 1977. The drawings depict a little about us and where the satellite came from so that if someday it is found by another planet's civilization, they will know a little something about us and our world.

A group of drawings, known as a **set of drawings,** should contain enough information to enable a welder to produce the weldment. The set of drawings may contain various pages showing different aspects of the project to aid in its fabrication. The pages may include the following: title page, pictorial drawing, assembly drawing, detailed drawing, and exploded view, Figure 4-2.

In addition to the actual shape as described by the various lines, a set of drawings may contain additional information such as the title box, specifications, project routing information, and bill of materials. The **title box,** which will appear in one corner of the drawing, may contain the name of the part, company name, scale of the drawing, date of the drawing, who

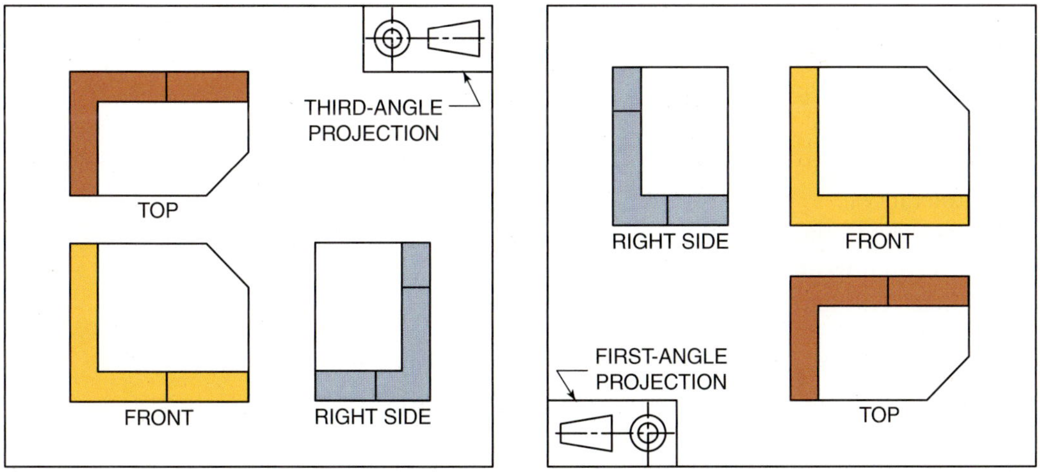

FIGURE 4-1 Two different layout methods used to show drawing views.

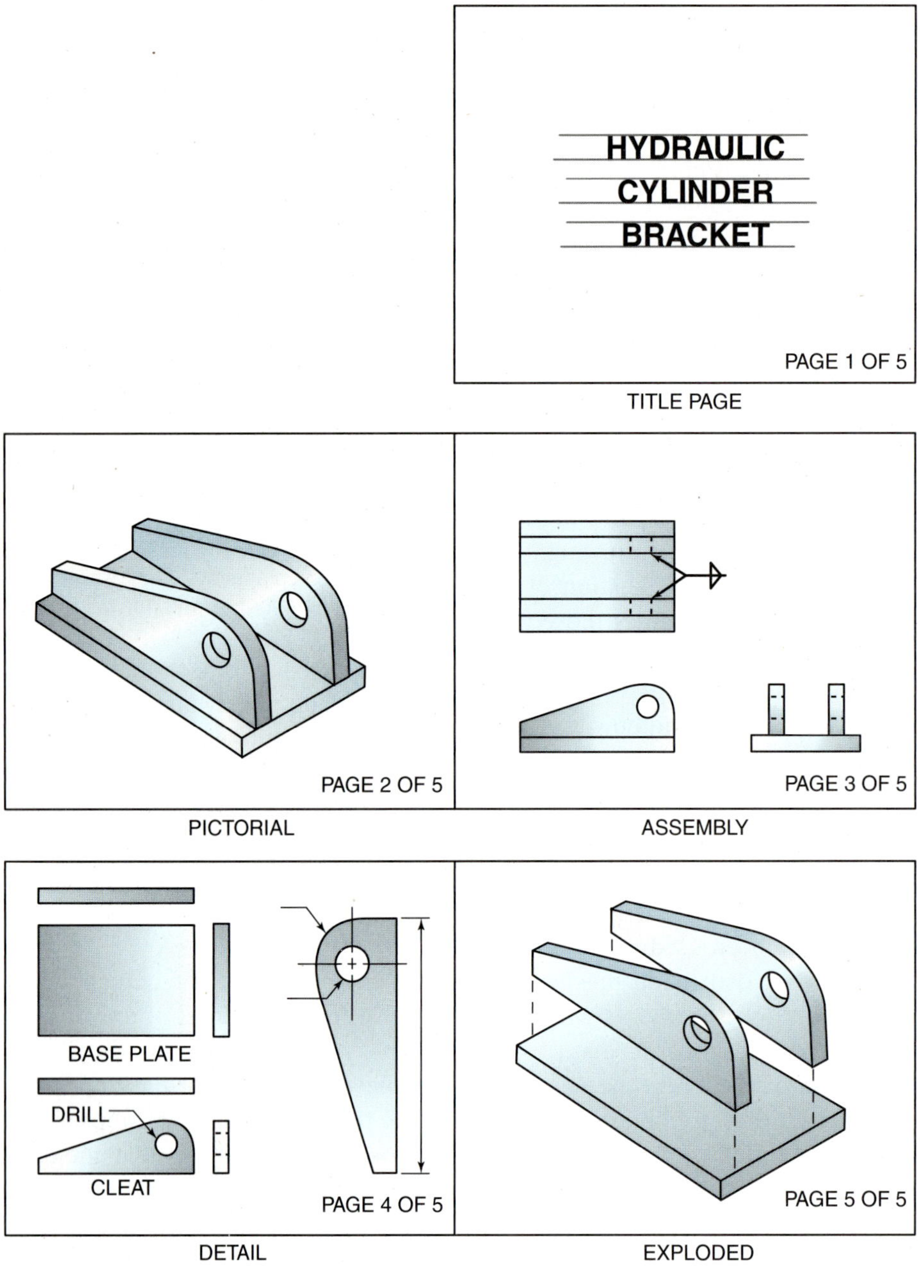

FIGURE 4-2 Example of a set of drawings. © Cengage Learning 2012

made the drawing, drawing number, number of drawings in this set, and tolerances.

The **specifications** detail the type and grade of material to be used, including base metal, consumables such as filler metal, and hardware such as nuts and bolts.

The **project routing information** is used in large shops that employ an assembly line process. This information lets you know where the parts should be sent once you have completed your part of the assembly process.

A **bill of materials** can also be included in the set of drawings. This is a list of the various items that are needed to build the weldment, Figure 4-3.

Lines

Different types of lines are used to represent various parts of the object being illustrated. The various line types are collectively known as the **alphabet of lines,** Table 4-1 and Figure 4-4.

BILL OF MATERIALS				
Part	Number Required	Type of Material	Size	
			Standard Units	SI Units

FIGURE 4-3 Typical bill of materials form.

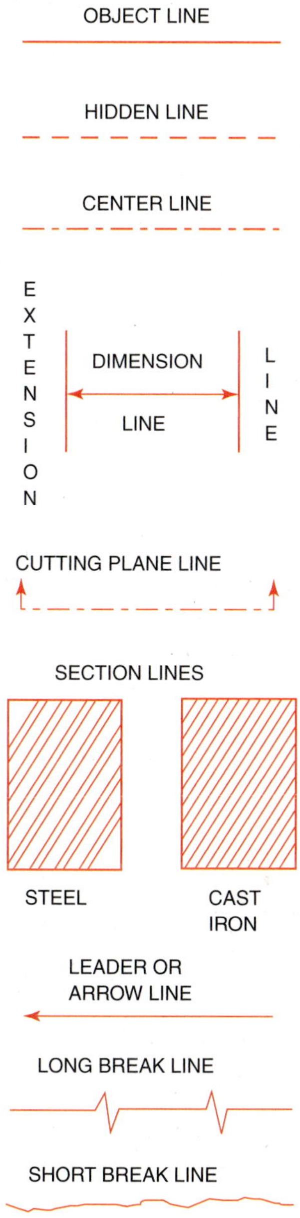

Table 4-1 Alphabet of Lines Used for Mechanical Drawings

Following are the types of lines used on drawings:

- **Object line**—Object lines show the edge of an object; the intersection of surfaces that form corners or edges; and the extent of a curved surface, such as the sides of a cylinder, Figure 4-4A.
- **Hidden lines**—Hidden lines show the same features as object lines except that the corners, edges, and curved surfaces cannot be seen because they are hidden behind the surface of the object, Figure 4-4B.
- **Center lines**—Center lines show the center point of circles and arcs and round or symmetrical objects. They also locate the center point for holes, irregular curves, and bolts, Figure 4-4C.
- **Extension lines**—Extension lines are the lines that extend from an object and locate the points being dimensioned, Figure 4-4D.
- **Dimension lines**—Dimension lines are drawn so that their ends touch the object being measured, or they may touch the extension line extending from the object being measured. Numbers in the dimension line or next to it, give the size or length of an object, Figure 4-4E.
- **Cutting plane lines**—Cutting plane lines represent an imaginary cut through the object. They are used to expose the details of internal parts that would not be shown clearly with hidden lines, Figure 4-4F.
- **Section lines**—Section lines show the surface that has been imaginarily cut away with a cutting plane line to show internal details. Different patterns of section lines are used to show different types of materials. The evenly spaced diagonal lines for cast iron are often used as the default pattern, Figure 4-4G.

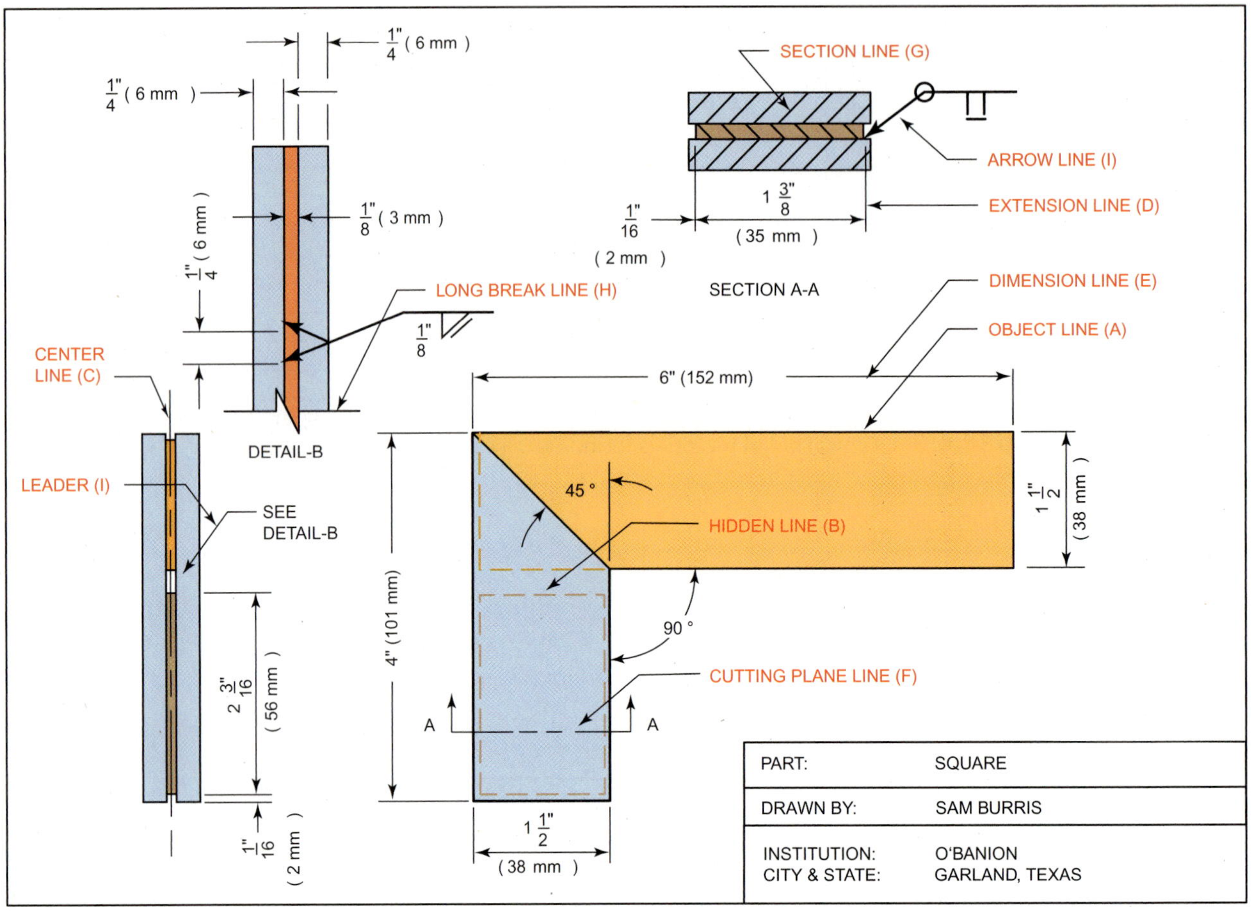

FIGURE 4-4 Drawing with the alphabet of lines identified. © Cengage Learning 2012

- **Break lines**—There are two types of break lines: long break lines and short break lines. Both show that part of an object has been removed. This is often done when a long uniform object needs to be shortened to fit the drawing page, Figure 4-4H.
- **Leaders and arrows**—Leaders (the straight part) and arrows (the pointed end) point to a part to identify it, show the location, and/or are the basis of a welding symbol, Figure 4-4I.
- **Phantom lines**—Phantom lines show an alternate position of a moving part or the extent of motion such as the on/off position of a light switch. They can also be used as a placeholder for a part that will be added later.

SPARK YOUR IMAGINATION

How would you show the front view of the cylinder in Figure 4-5? How would the side view look, and would it be much different from the front view? What are some other shapes that might have similar appearances in more than one view?

Types of Drawings

Drawings used for most welding projects can be divided into two categories—**orthographic projections** (mechanical drawings) and pictorial. The projection

FIGURE 4-5 How would the side view of this cylinder look? © Cengage Learning 2012

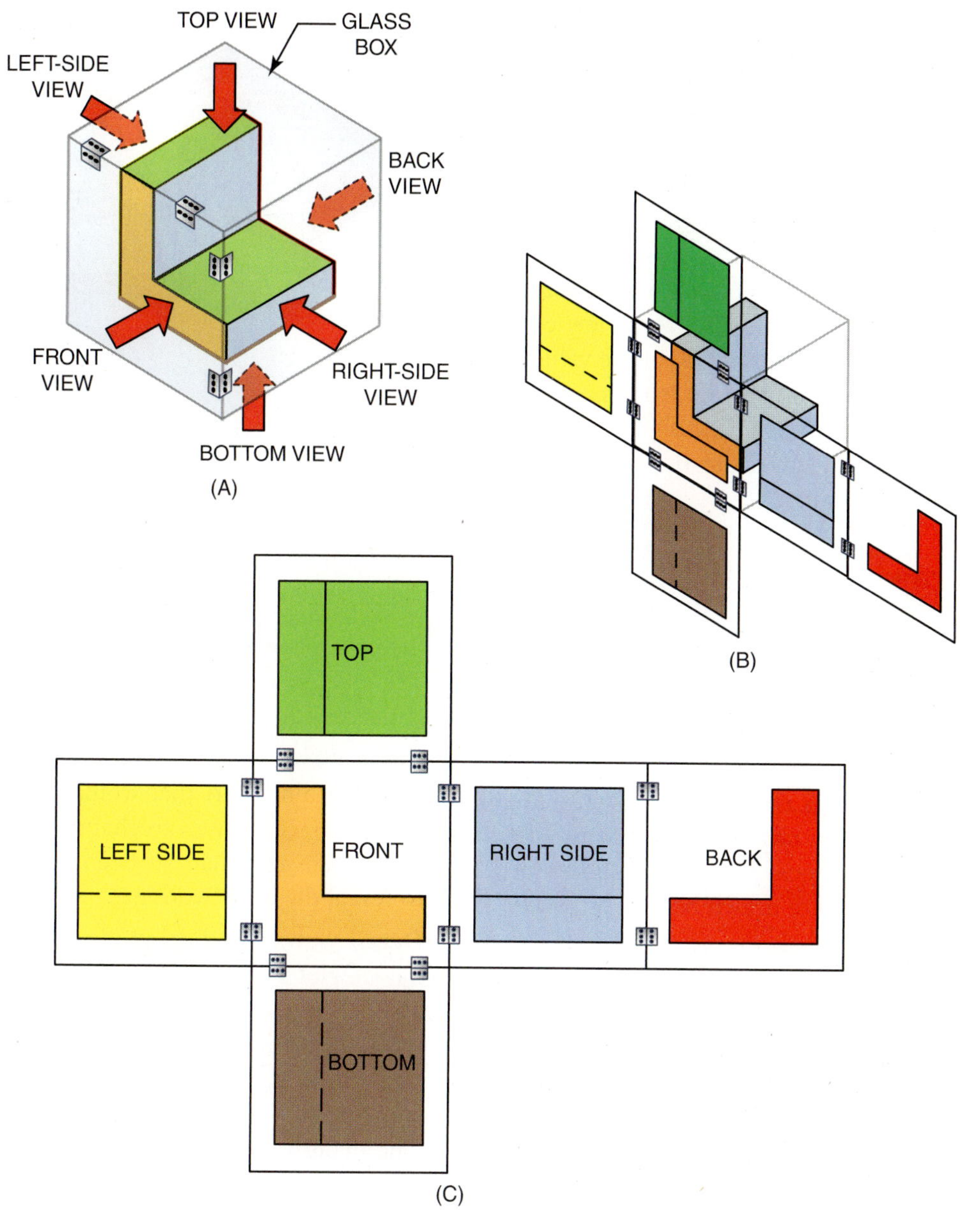

FIGURE 4-6 The arrangement of views of an object as if it were viewed in a glass box and then the box unfolded. © Cengage Learning 2012

drawings are made as if you were looking through the sides of a glass box at the object and tracing its shape on the glass, Figure 4-6A. If all of the sides of the object were traced and the box unfolded and laid out flat, Figure 4-6B, there would be six basic views shown, Figure 4-6C.

Pictorial drawings present the object in a more realistic or understandable form. These drawings usually appear as one of two types, isometric or cavalier, Figure 4-7. The more realistic perspective drawing form is seldom used for welding projects.

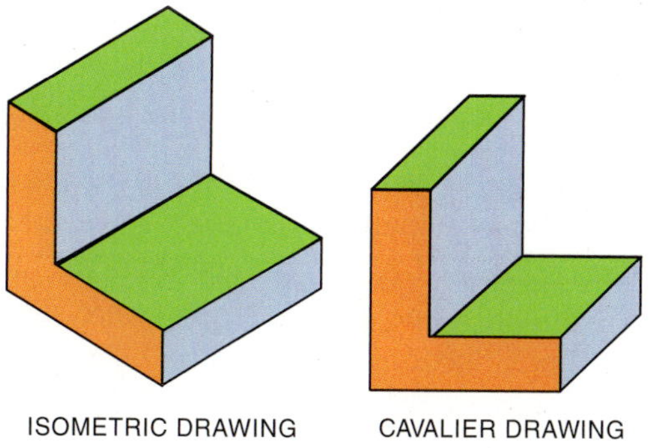

FIGURE 4-7 Two types of pictorial drawings. © Cengage Learning 2012

Mechanical Drawings

Usually not all of the six views are required to build the weldment. Only those needed are normally provided, usually only the front, right side, and top views. Sometimes only one or two of these views are needed.

The front view is not necessarily the front of the object. The front view is selected because when the object is viewed from this direction, its overall shape is best described. As an example, the front view of a car or truck would probably be the side of the vehicle because viewing the vehicle from its front may not show enough detail to let you know whether it is a car, light truck, SUV, or van. From the front, most vehicles may look very similar.

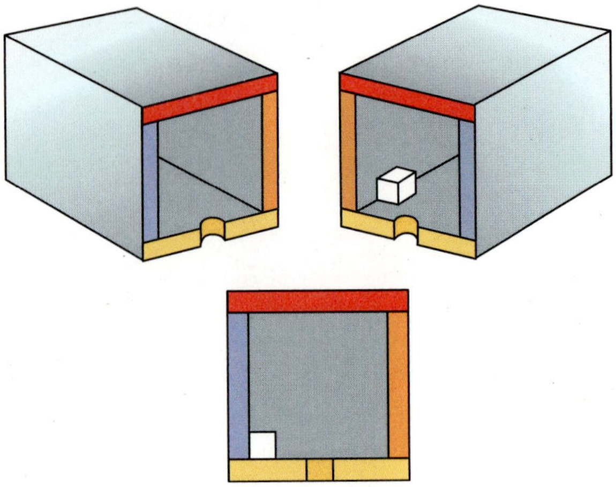

FIGURE 4-8 **Sectioning of parts can be used to show internal details more clearly.** © Cengage Learning 2012

Pictorial Drawings

Of the two types of pictorial drawings, the isometric drawing is more picturelike. **Isometric drawings** are drawn at a 30° angle so it appears as if you are looking at one corner. As with all drawings, all of the lines that are parallel on the object appear parallel on the drawing.

On **cavalier drawings** one surface, usually the front, is drawn flat to the page. It appears just like the front view of a mechanical drawing. The lines for the top and side surfaces are drawn back at an angle, usually 30°, 45°, or 60°.

SPARK YOUR IMAGINATION

Both types of pictorial drawings are used in this book. Look through this chapter and make a list of all of the figure numbers that show pictorial drawings. Look around your shop and classroom and see how many posters, signs, or drawings have pictorial drawings. Make a list of the different items you find.

Special Views

Special views may be included on a drawing to help describe the object so it can be made accurately. Special views on some drawings may include the following:

- **Section view**—The section view is made as if you were to saw away part of the object to reveal internal details, Figure 4-8. This is done when the internal details would not be as clear if they were shown as hidden lines. Sections can either be fully across the object or just partially across it. The imaginary cut surface is set off from other noncut surfaces by section lines drawn at an angle on the cut surfaces. On some drawings the type of section line used illustrates the type of material that the part was made with. The location at which this imaginary cut takes place is shown using a cutting plane line, Figure 4-9.
- **Cut-a-ways**—The cut-a-way view is used to show detail within a part that would be obscured by the part's surface. Often a freehand break line is used to outline the area that has been imaginarily removed to reveal the inner workings.
- **Detail views**—The detail view is usually an external view of a specific area on the part. Detail views are used to show small details of an area on a part without having to draw the entire part larger. Sometimes only a small portion of a view has significance, and this area can be shown in a detail view. The detail view can be drawn at the same scale or larger if needed. By showing only what is needed within the detail, the part drawn can be clearer and not require such a large page.
- **Rotated views**—A rotated view can be used to show a surface of the part that would not normally be drawn square to any one of the six normal view planes. If a surface is not square to the viewing angle, then lines may be distorted. For example, a circle when viewed at an angle looks like an ellipse, Figure 4-10.

Dimensioning

Often it is necessary to look at other views to locate all of the dimensions required to build the object. By knowing how the views are arranged, it becomes easier to locate dimensions. Length dimensions can be found on the front and top views. Height dimensions

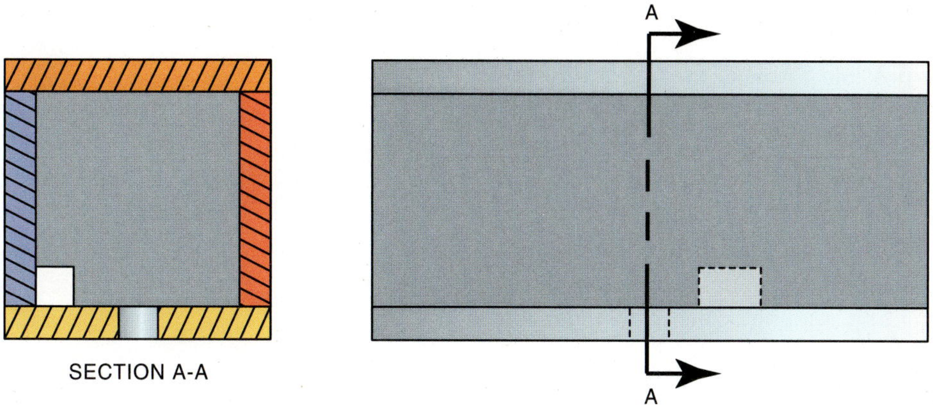

FIGURE 4-9 Using a cutting plane line. © Cengage Learning 2012

can be found on the front and right-side views. Width dimensions can be found on the top and right-side views, Figure 4-11. Locating dimensions on these views is consistent with both the first-angle perspective or third-angle perspective layouts.

If the needed dimensions cannot be found on the drawings, do not try to obtain them by measuring the drawing itself. Even if the original drawing was made very accurately, the paper it is on changes size with changes in humidity. Copies of the original drawing are never the exact same size. The most acceptable way of determining missing dimensions is to contact the person who made the drawing.

Keep the drawing clean and well away from any welding. Avoid writing or doing calculations on the drawing unless you are noting changes. Often there is a need to make a change in the part as it is being fabricated. When these changes are added to the "as drawn" drawing, the drawing is referred to as the "as built" drawing. It is important to keep these "as built" drawings so they will be filed following the project for use at a later date. The better care you take with the drawings, the easier it will be for someone else to use them.

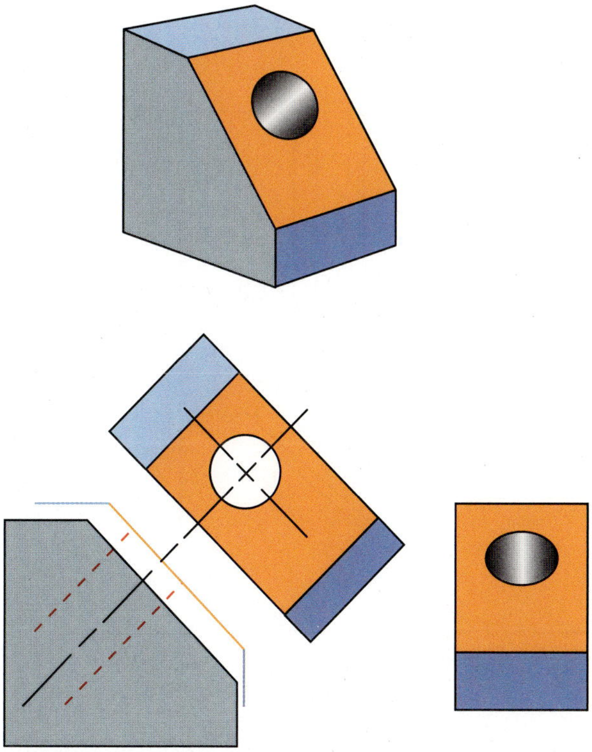

FIGURE 4-10 Note that the round hole looks misshapen or elliptical in the right-side view but looks round in the special view. © Cengage Learning 2012

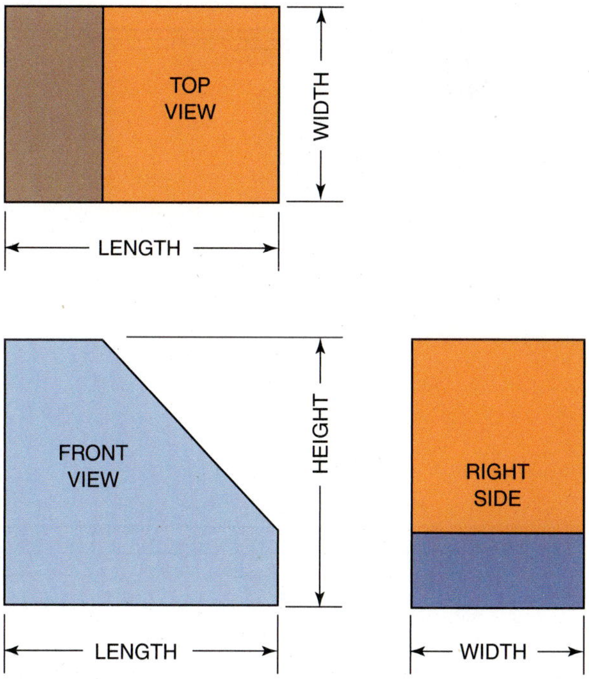

FIGURE 4-11 Drawing dimension locations.
© Cengage Learning 2012

Drawing Scale

It would not be possible to make every drawing the same size as the parts being made; for that reason you must change the scale. When we use a **drawing scale,** we are saying that the part being drawn is drawn smaller or larger than it really is. An easy scale to use is 1 in. equals 1 ft, so that if we draw a line that is 10 in. long, it represents an actual distance of 10 ft. To aid in making and reading scaled drawings, you can use a drafting tool called a *scale,* Figure 4-12. A scale is a special type of ruler that is marked with different units. There are two commonly used scales: the **architectural scale** and the **engineering scale.** The architectural scale is the one that is most often used for mechanical welding drawings.

Architectural scales are divided into fractions of inches. Some of the common units are: 1/8, 3/32, 1/4, 3/16, 3/8, and 3/4. These scales can be used to represent different lengths and units of measure. For example, the 1/4 scale can be used with inches so that 1/4 in. equals 1 in. or to represent feet as in 1/4 in. equals 1 ft. It could be used with yards, meters, miles, or any other standard unit.

Engineering scales are divided into decimals of inches. Some of the common units are 10th, 20th, 30th, 40th, 50th, and 60th. As with the architectural scale, these units can be used to represent any number of distances.

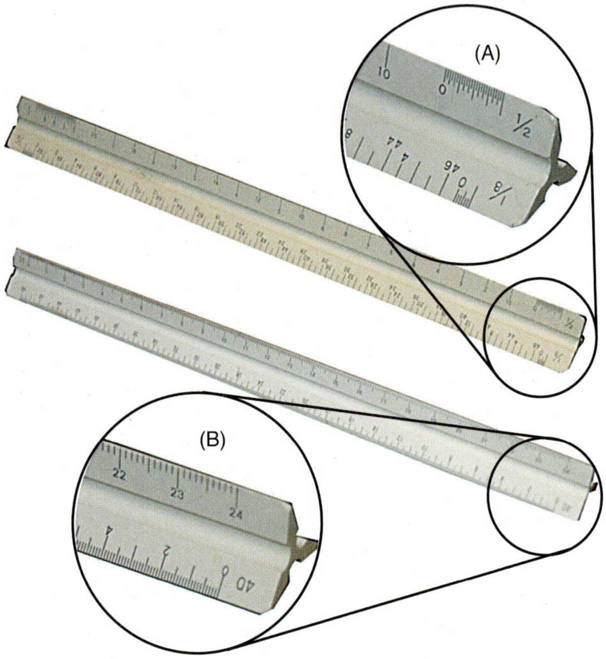

FIGURE 4-12 Drawing scales: (A) fraction, (B) decimal.

Not all drawings are scaled down; some are made larger, Figure 4-13A. Most often, details are drawn at a larger scale so that the important parts can be seen more clearly, Figure 4-13B.

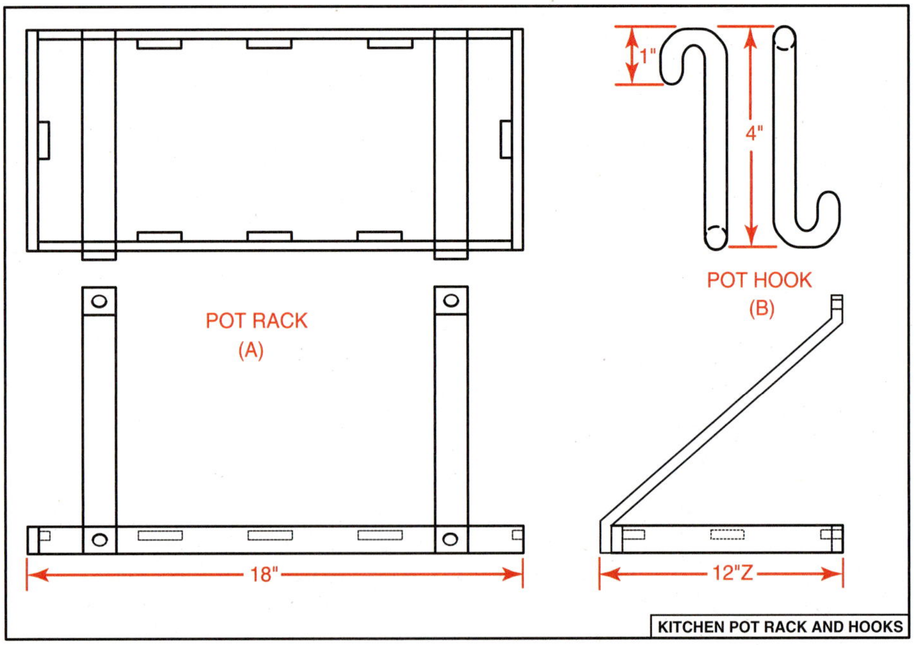

FIGURE 4-13 Examples of scaling down and scaling up on a drawing.

MECHANICAL DRAWING VIEW COLOR CODES			
VIEW NAME	VIEW COLOR	VIEW NAME	VIEW COLOR
FRONT	ORANGE	LEFT SIDE	YELLOW
TOP	GREEN	BACK	RED
RIGHT SIDE	BLUE	BOTTOM	BROWN

Table 4-2 Color-Coding Used in This Chapter to Help Identify Views

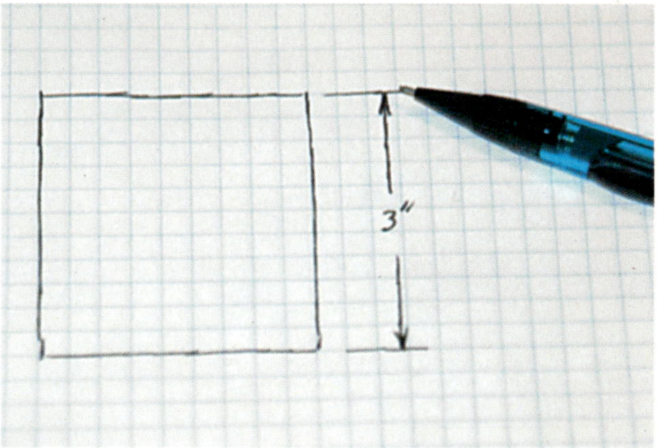

FIGURE 4-14 Sketching using draft paper. © Cengage Learning 2012

READING MECHANICAL DRAWINGS

The pictorial drawing of the block in the glass cube shown in Figure 4-6A is color-coded. The same color coding has been used for all of the multiview drawings in this chapter to help you identify the views. Table 4-2 lists the colors and views used. The front surface in all the views is colored orange, the top view is green, the right side is blue, the left side is yellow, and the bottom is brown. Mechanical drawings you work with in the field are not color-coded.

In addition, the lines that represent the hidden surfaces of the block in the glass cube have been color-coded. For example, you can see the red line for the back, which is visible on the top and right-side views.

The color-coding is intended to help you learn how to read a mechanical drawing. The important element to remember is that the front view is the main view. The standard views include the front, top, and right-side views. Not all drawings use all of the standard views, and drawings may or may not include additional views. For that reason, you need to know where each of the views are and how they relate to each other.

SKETCHING

Sketching is a quick and easy way of producing a drawing that can be used in the welding shop. Sketches and mechanical drawings have some similarities and some differences. They are similar in that both contain the necessary information to produce a welded project. The main way that they are different is that a sketch is a quick way of drawing an object and sketches may not be drawn to scale. Mechanical drawings take more time and are drawn to scale. They should both contain all the necessary information to build the desired weldment.

NOTE: When graph paper is used, sketches can be easily drawn to scale, Figure 4-14.

A sketch can be any drawing that is made without the extensive use of drafting instruments or computer-aided design (CAD). Sketches are generally drawings that are made in the shop by the welder, shop supervisor, or customer. A sketched drawing can be made with or without the use of drawing tools such as scales, straight edges, curves, and circle templates, Figure 4-15. When drawing tools are used, the

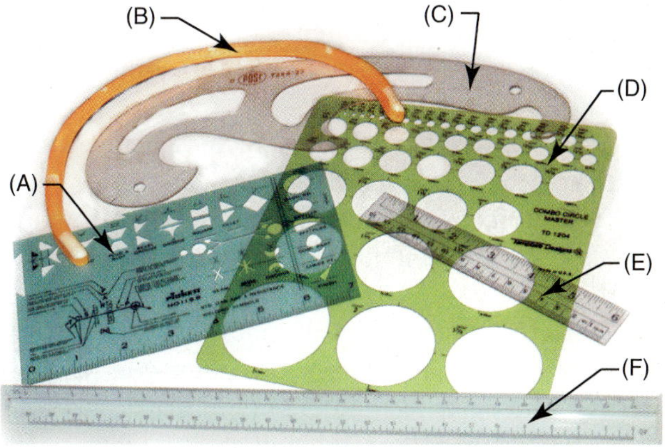

FIGURE 4-15 Drafting tools: (A) welding symbol template, (B) flexible curve, (C) French curve, (D) circle template, (E) rule, (F) scale. Larry Jeffus

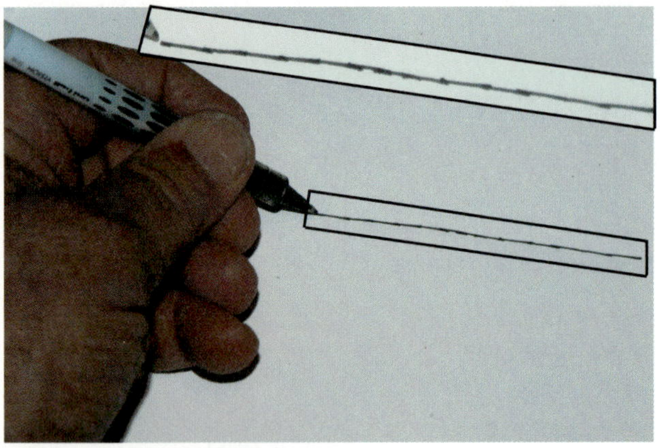

FIGURE 4-16 When sketching, use short strokes.
Larry Jeffus

FIGURE 4-17 Overlap each sketched stroke so they form a solid line. © Cengage Learning 2012

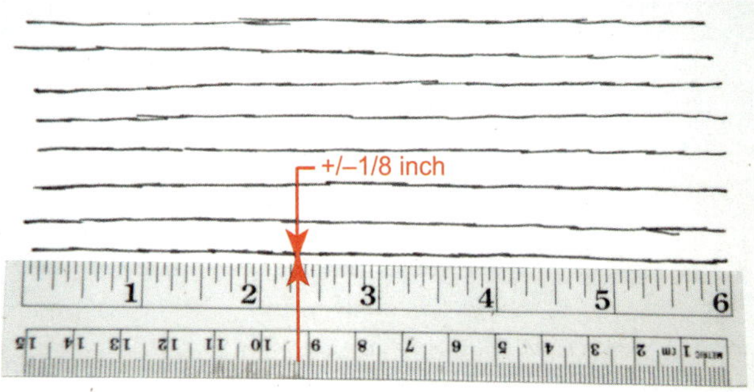

FIGURE 4-18 Practice sketching until you can keep your line within ±1/8 in. (3 mm) of being straight.
Larry Jeffus

drawing may have straighter lines but take longer to produce. Straight lines are not always necessary to the actual production of the parts being welded as long as the drawing is clear.

Sketching is the process of making a line on a drawing by making a series of quick, short strokes with the pencil or pen, Figure 4-16. The technique of sketching a line may seem slow and awkward in the beginning, compared to just holding the pencil on the paper and drawing a line. But a sketched line can be much straighter and faster once you have developed the skill. Each stroke may not be straight, but the line produced can be straight.

PRACTICE 4-1

Sketching Straight Lines

Using a pencil and unlined paper, you are going to sketch a series of 6-in. (152-mm) long straight lines.

Practice sketching from right to left and left to right. You may find that it is easier for you to go in one direction, rather than the other. The direction you sketch is not as important as your ability to make straight lines.

Start by making a small mark or dot approximately 6 in. (152 mm) from the point you plan on starting your sketched line. Do not measure; just estimate the distance. Part of being able to make quick sketches is the ability to judge lengths. The mark will give you an aiming point. Look at the point as you start the series of short sketched marks. Make each sketch mark about 1/2 in. (13 mm) to 3/4 in. (19 mm) long and make them in a quick, smooth series. Overlap each mark so they form a solid line, Figure 4-17. You may want to make the sketched line very light initially and go back over it to make it darker. It may be easier to make the sketch marks in the direction of travel or in the opposite direction—try both.

Once you have completed six or eight lines, lay a straight edge next to the lines and see how straight you were able to make them. Keep practicing sketching straight lines until you are able to make 6-in. (152-mm) long lines that are within ±1/8 in. (3 mm) of being straight, Figure 4-18.◆

PRACTICE 4-2

Sketching Circles and Arcs

Using a pencil and unlined paper, you are going to sketch a series of circles.

Start by sketching two light construction lines that cross at right angles. Construction lines are often used in drawing as guides to the finished line. Construction lines should be very light so they can be left on the drawing or easily erased.

Make two marks on each of the construction lines about 1/2 in. (13 mm) from the center point. These

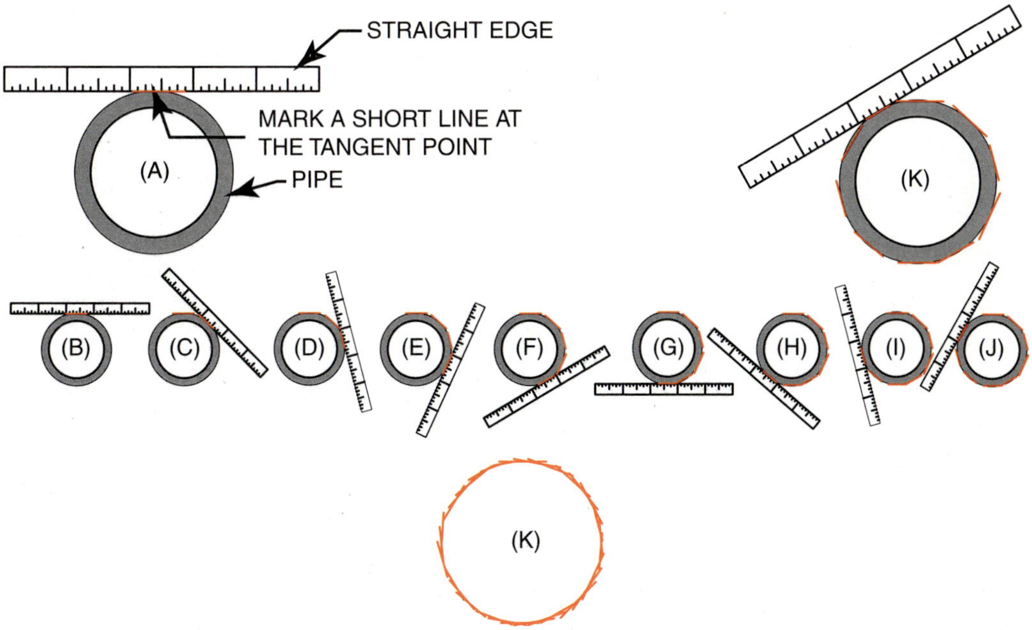

FIGURE 4-19 A circle can be sketched as a series of straight lines.

points will serve as your aiming points as you sketch the circle. The circle you sketch will be tangent to these points. A tangent straight line is one that meets a circle at a point where the circular line and straight line are going in the same direction. Much like placing a 12-in. (25-mm) ruler on a round pipe, where the ruler and pipe meet is the tangent point. If you were to make a short, straight line at the tangent point and keep doing this all the way around the pipe, you would wind up with a circle drawn from a series of short straight lines, Figure 4-19.

Sketch a tangent line starting at the top mark, keep sketching and gradually turn the line toward the mark on the next construction line. Once you have completed the first quarter of the circle, you may find it easier to continue if you turn the paper. Repeat the sketching process until you have completed sketching the circle.

Repeat this process making several different size circles. On larger circles it may be helpful to make more construction lines. Using a circle template, check your circles for accuracy. Continue making sketched circles until you can draw them in several sizes within ±1/8 in. (3 mm) of round.

NOTE: Sometimes it is easier to draw small circles in a square box. You can do this by first drawing a box using construction lines then drawing the circle inside the box.◆

PRACTICE 4-3

Sketching a Block

Using a pencil and unlined paper, you are going to sketch a mechanical drawing showing three views of a block as shown in Figure 4-20.

NOTE: In the mechanical drawing type called *orthographic projection*, the views must be arranged properly. The top view is drawn straight above the front view, and the right-side view is aligned to the right. This arrangement makes it possible for you to see distinguishing lines in one view and locate the corresponding lines in another view. This is how you find all of the location dimensions or even how you can identify material. For example, in one view angle, iron might be shown as parallel lines while in another view you would see the distinctive "L" shape. The same would apply to a cylinder that would appear as parallel lines in one view but as a circle in a different view.

Start by sketching construction lines as shown in Figure 4-21A. These lines will form the boxes for the front, top, and right-side views. Darken the lines that make up the object's lines so it is easier to see, Figure 4-21B.

Repeat this practice using the shaped objects shown in Figure 4-22.◆

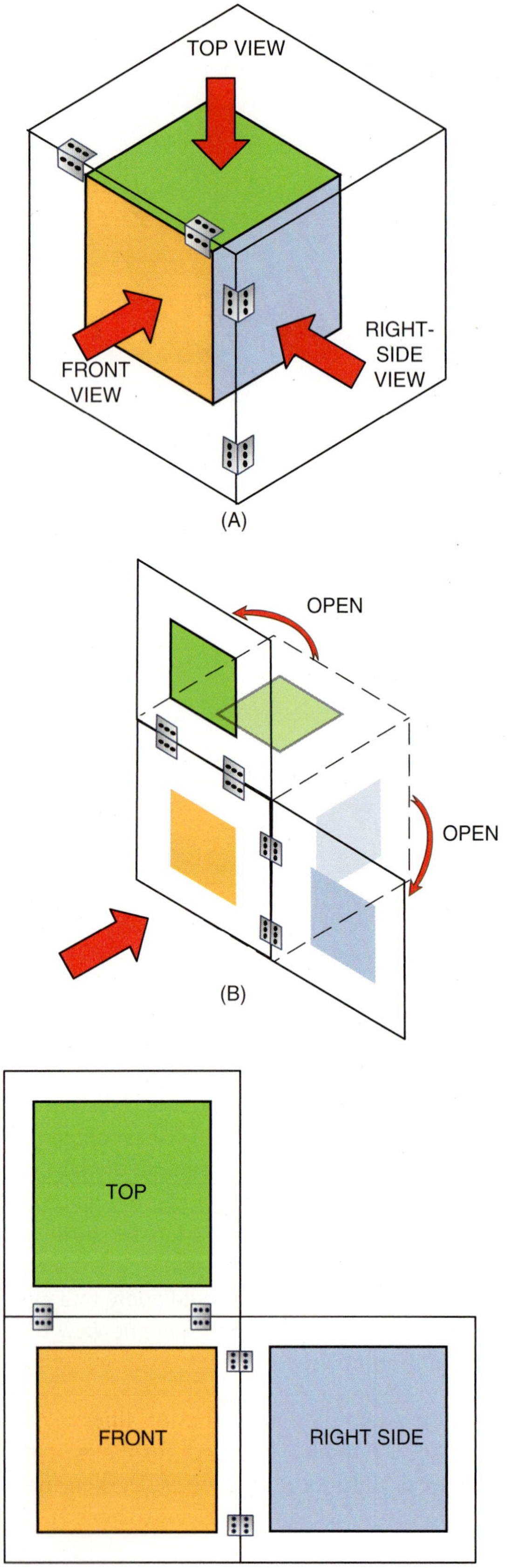

FIGURE 4-20 Practice 4-3.

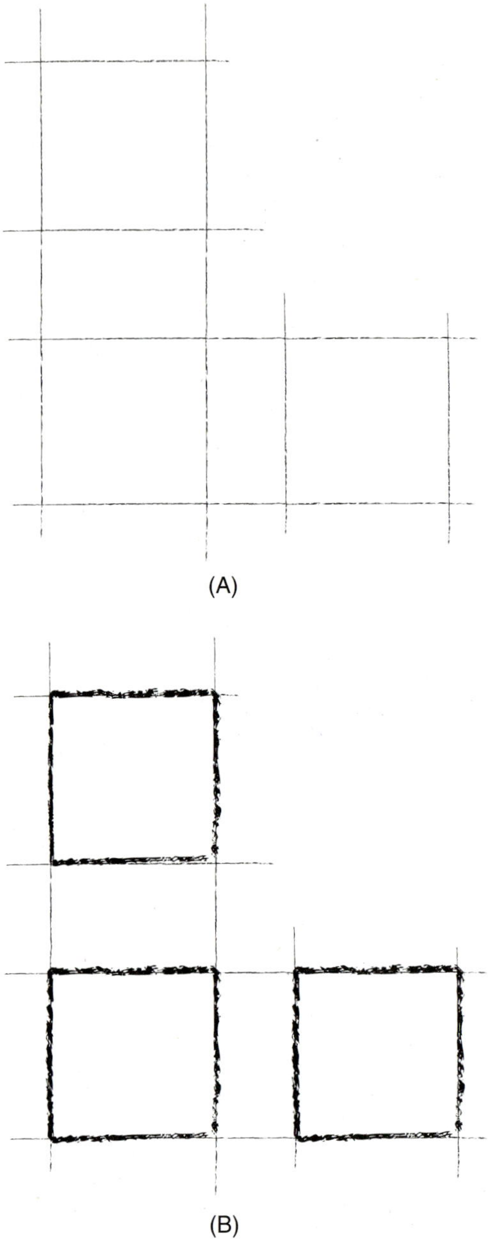

FIGURE 4-21 **(A) First, lightly sketch the lines and then (B) darken the object lines.**

SPARK YOUR IMAGINATION

Figure 4-23 shows some differently shaped front views that all have the same shapes for the top and right-side views. There are lots of possibilities for other shapes of front views that can be used with these same top and right-side views. Sketch as many differently shaped front views as you can for the top and side views shown. Hint: Curved surfaces do not show lines on other views unless they are tangent to the view, and solid object lines hide hidden lines.

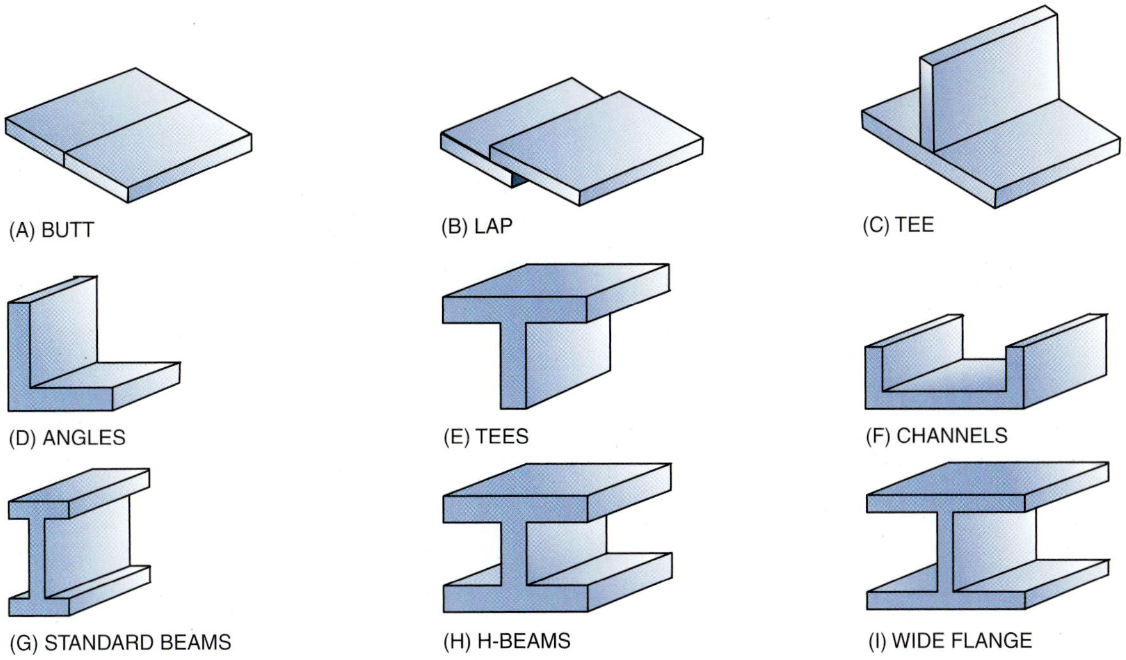

FIGURE 4-22 Additional shapes to be sketched for Practice 4-3. © Cengage Learning 2012

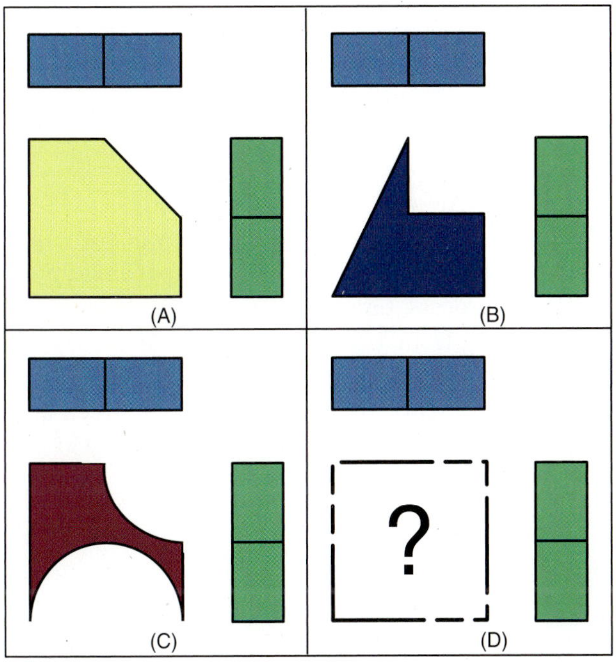

FIGURE 4-23 What other shapes could the front view of this object be? © Cengage Learning 2012

PRACTICE 4-4

Sketch a Candlestick Holder

Using a pencil and unlined paper, you are going to sketch a three-view mechanical drawing of the candlestick holder shown in Figure 4-24.

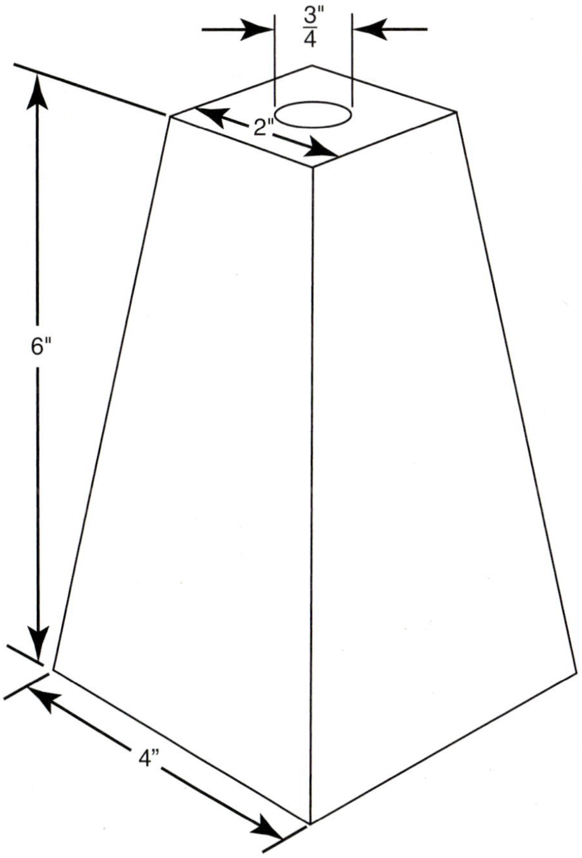

FIGURE 4-24 Practice 4-4. © Cengage Learning 2012

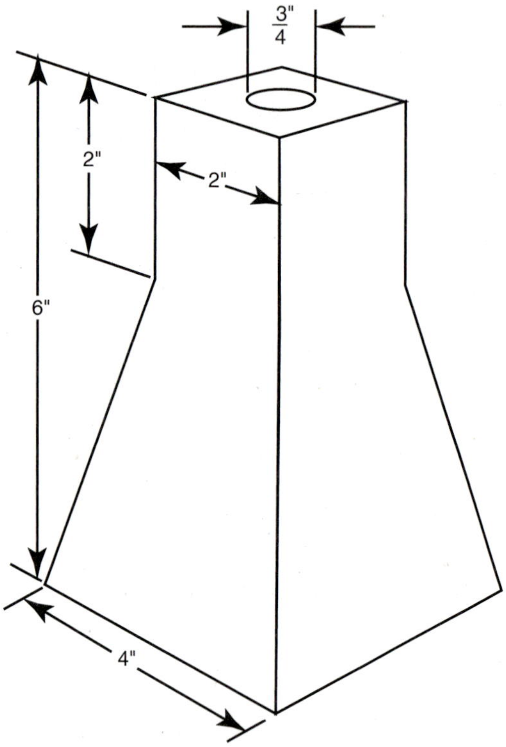

FIGURE 4-25 Additional object to be sketched for Practice 4-4. © Cengage Learning 2012

Repeat the process using the candlestick holder shown in Figure 4-25.

NOTE: All you have to do to lay out the angle for the candlestick holder shown in Figure 4-24 is measure down the correct distance from the top and measure out the correct distance at that point from the center line. Connecting the points will automatically give the angle. But most important, remember that sketches do not have to be exactly to scale as long as all of the needed dimensions are shown.

Erasers and Erasing

Most pencil erasers have an abrasive action on the paper as they are used to erase pencil marks. This can sometimes cause the top surface of the paper to be roughed up or rubbed off. If you are not careful in erasing with a pencil eraser, you can damage the paper's surface, making it hard to redraw over that area.

Plastic erasers are usually white. These erasers do not have an abrasive and will not damage the paper's surface like pencil erasers. Plastic erasers are very effective in removing unwanted pencil lines.

Sometimes you need to erase a small part of a line without removing or smudging a nearby line. There are thin metal tools called "eraser shields" that are used in drafting to protect the neighboring line from erasure, Figure 4-26. An easy substitute for this tool is any scrap piece of paper. Simply cover the line you do not want to erase, and rub the eraser away from the edge so the edge is not wrinkled.

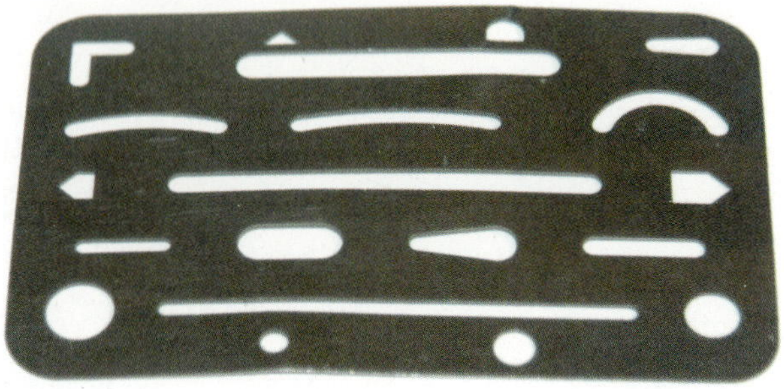

FIGURE 4-26 Eraser shield. © Cengage Learning 2012

NOTE: A Post-it Note can be used to make a great temporary eraser shield.

Both correction tape and fluid do not erase errors but cover them up so corrections can be made. They are easy to use and work well; however, if you are drawing with a pencil, they may be more difficult to draw over than an erased line.

Graph Paper

Making a sketch on graph paper is a way of both making your drawing more accurate and speeding up the sketching process. Graph paper is available with grid sizes ranging from 1/8-in. (3-mm) to 1/4-in. (6-mm) squares, Figure 4-27. Other sizes are available. Many copiers do not copy the light blue or light green lines on graph paper, so if you need the lines later, you may

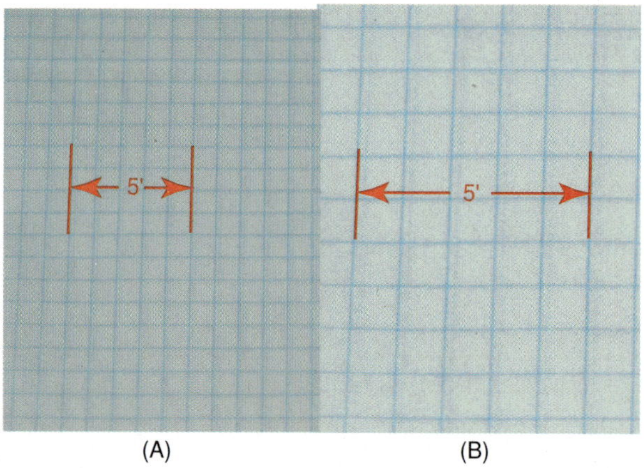

FIGURE 4-27 Graph paper. © Cengage Learning 2012

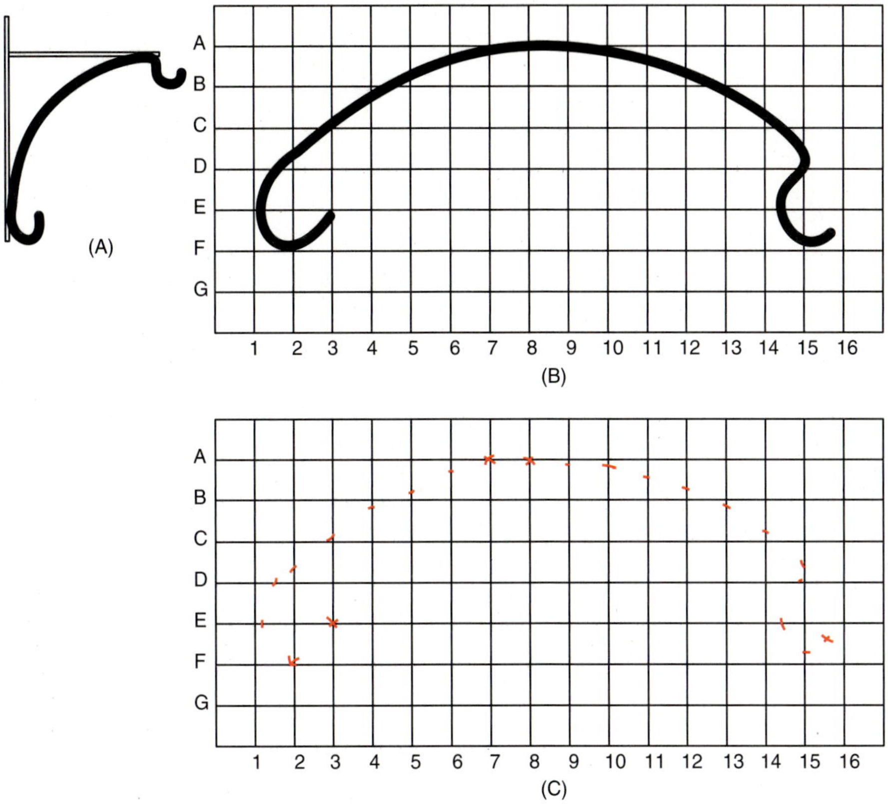

FIGURE 4-28 Practice 4-5. © Cengage Learning 2012

need to make the copy machine copy darker or use another color of lined paper. Usually, the inch lines are a little darker on the graph paper, which makes it easier to count when measuring.

Even though you have lines to follow on graph paper, you may find that you can make a better-looking drawing by sketching over the grid lines rather than trying to just follow the grid line with your pencil. Graph paper does lend itself to the use of straight edges and other drafting tools, but with practice sketching is faster and works well.

PRACTICE 4-5

Sketching Curves and Irregular Shapes

Using a pencil and graph paper, you are going to sketch a front view of the plant hanger shown in Figure 4-28.

Curves and irregular shapes can be easily drawn using a grid such as that found on graph paper. The first thing you need to do is locate a series of points on the graph paper that coincide with points on the curve you are copying. Start with the easy points where the lines on the paper cross at a point on the object. For example, one end of the curve starts at the intersection of lines E-3, so put a dot there. Next, the curve is tangent to lines F-2, so put a dot there. Follow the curve around, putting additional dots at the other intersecting points.

Once all of the easy dots are located, you are going to have to make some estimates for the next series of dots. For example, the curve almost touches the 1 line as it crosses the E line. Put a dot there and at similar points where the curve crosses other lines.

After you have located all of the points for the curve, sketch a line through all of the points, Figure 4-29. Refer to the figure to see how the line

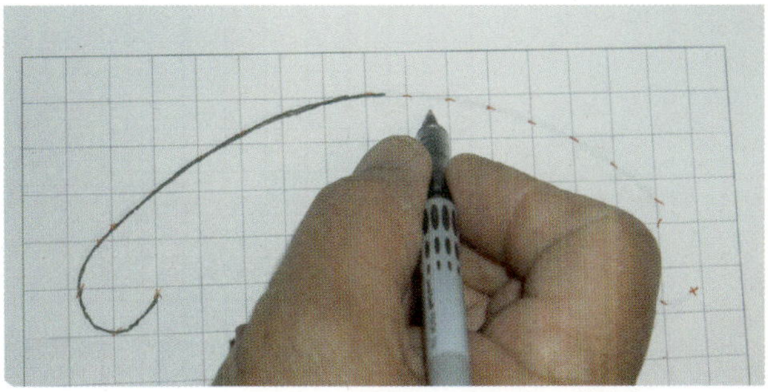

FIGURE 4-29 Sketching a curved line. Larry Jeffus

should pass through the point. If you are not sure, you can always add additional points that are not on lines to help guide your sketching.

SHOP TALK

Did you know you can draw graph lines on a project to help with the layout?

If you are laying out a curved object to be cut out, it is often easy to draw grid lines on the metal as a way of aiding in the layout. You can use a chalk line or straight edge and marker to create your grid. Also, you do not have to draw the full grid; you may be able to just draw the grid around the area where your curved line is to be drawn.

NOTE: The curve you made is called *free form,* and as the name implies, it is not an exacting process. If you are working on an exacting curve, there will often be center points for you to follow, Figure 4-30. You would use a compass to lay out this curve.

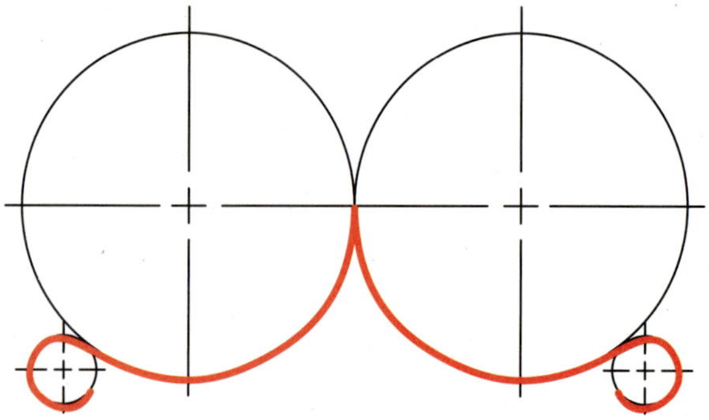

FIGURE 4-30 Sketching curves using a center point. © Cengage Learning 2012

Repeat this practice and make a three-view drawing of the candlestick holder shown in Figure 4-31.◆

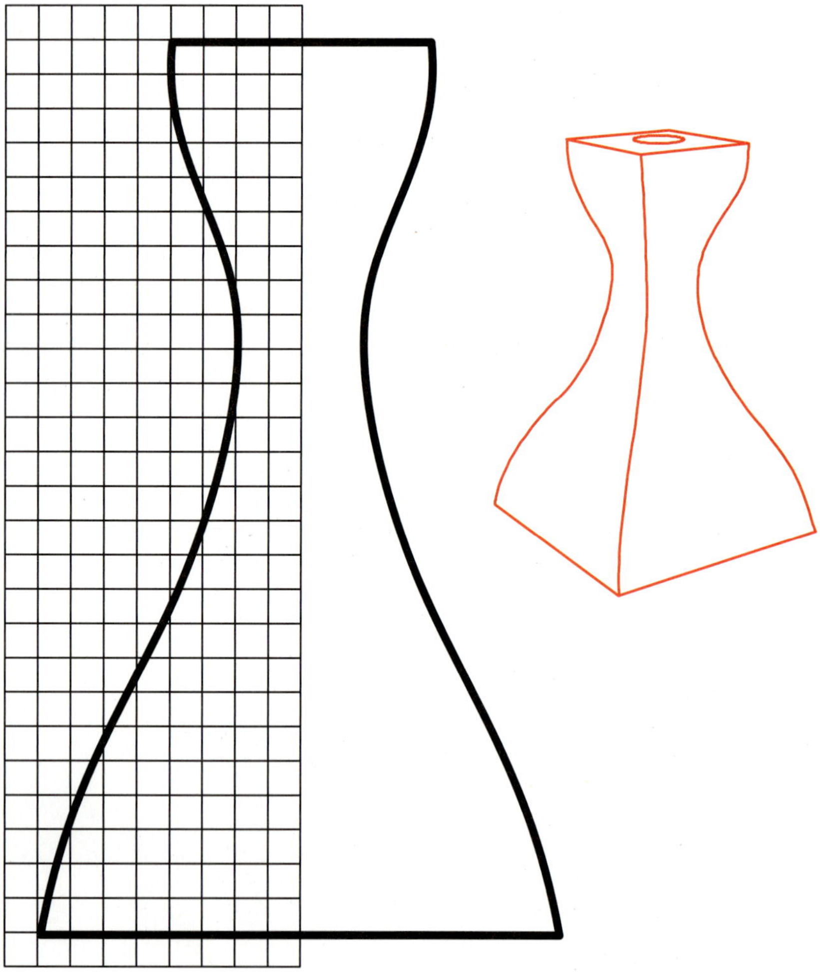

FIGURE 4-31 Practice sketching the curved object. © Cengage Learning 2012

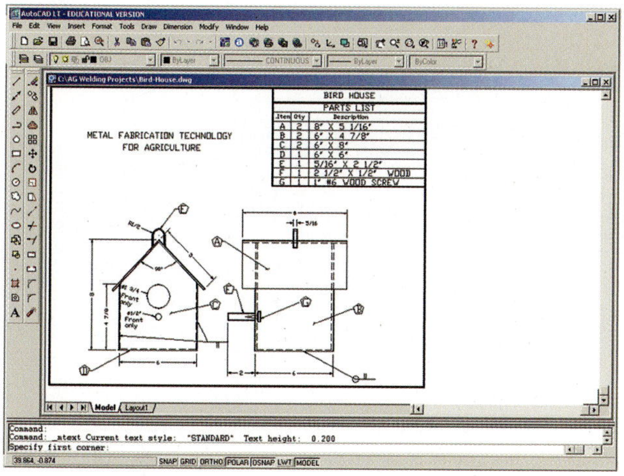

FIGURE 4-32 AutoCAD LT® drafting program used to draw a birdhouse. Larry Jeffus

COMPUTERS AND DRAWINGS

Computers have made it much easier to draw plans for projects. The welded birdhouse project shown in Figure 4-32 was drawn on AutoCAD LT®. There are a number of **computer drawing programs** like AutoCAD LT® available. **Computer drafting programs** use **vector lines,** which is different from most drawing programs, which use raster art. Computers see vector lines as lines, and they see **raster lines** as if they were part of a picture. Because vector drawings are seen by the computer as lines, you can zoom in and out, measure, resize, reshape, or rotate the drawing, and the lines stay crisp and sharp. For example, as the lines on the roof of the vector-drawn birdhouse in Figure 4-33 are magnified 300 and 500 times, they stay sharp.

Computers see raster images as a series of small squares called pixels. Raster drawings are commonly known as bitmap drawings because the computer maps the location of every little bit (pixel) of the drawing. When these pixels are very small, your eye sees them as a line; but as you zoom in, they start looking like a bunch of colored squares. For example, as the lines on the roof of the bitmap-drawn birdhouse in Figure 4-34 are magnified 300 and 500 times, they look like a group of colored squares not even recognizable as lines. Bitmap lines have a softer appearance, and they work best in art and photographic programs. The sharp crisp lines of vector drawings work best for mechanical drawings.

Two-dimensional drafting programs, abbreviated 2D, like AutoCAD LT®, allow you to make mechanical drawings accurately for projects. Vector

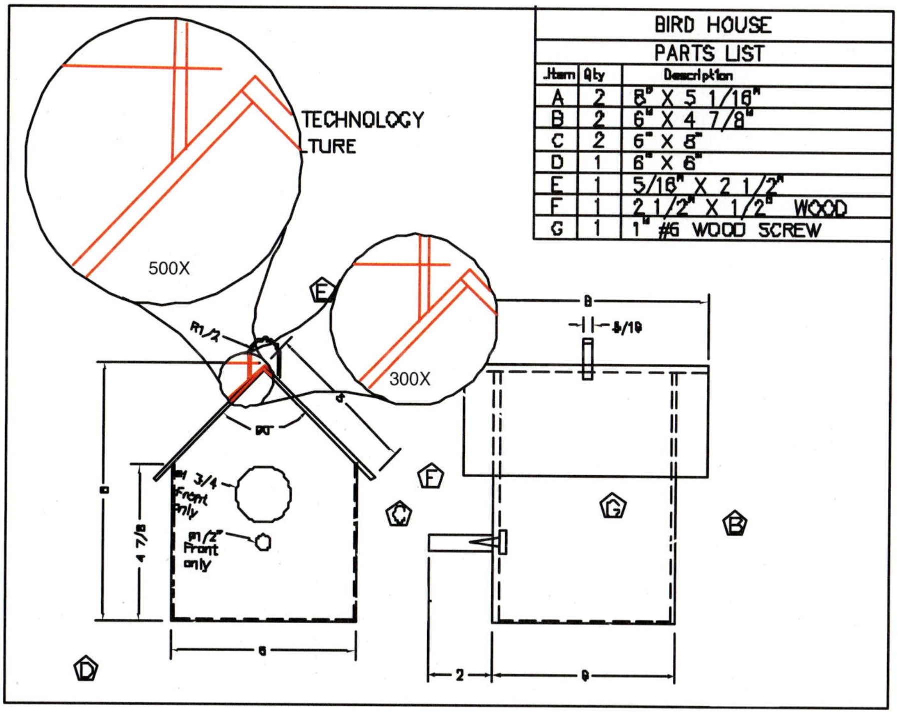

FIGURE 4-33 Vector line art drawing. © Cengage Learning 2012

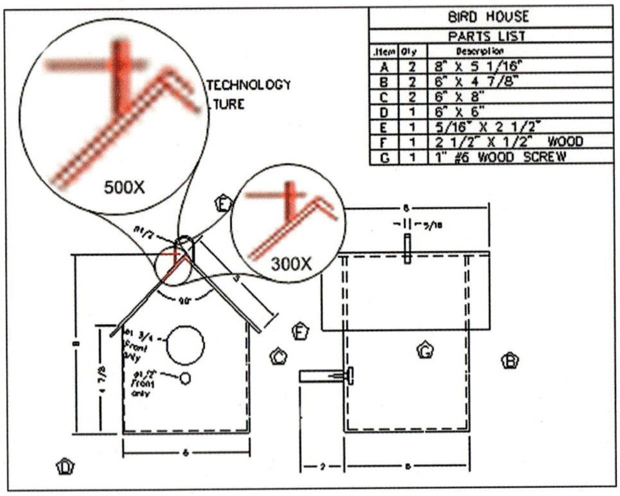

FIGURE 4-34 **Bitmap line art drawing.** © Cengage Learning 2012

FIGURE 4-36 **Well-designed trailer for hauling hay or equipment.** Larry Jeffus

drafting programs allow you to draw trailers, barns, or other large projects more accurately than they can be built. Using the pull-down and dimensioning options, you can set the precision for this birdhouse drawing at 1/256 in. (0.01 mm), Figure 4-35. Accuracy is important when parts must fit together and move without interfering with each other. It is also helpful when planning projects like the trailer in Figure 4-36. The back of the trailer has ramps built on it that can be flipped down for easier loading of equipment. The ramps are hinged to the back of the trailer so that when they are in the up position, the dropped back end of the bed is level. AutoCAD LT® makes designing the trailer easier, and leveling the bed makes it easier to haul hay, Figure 4-37.

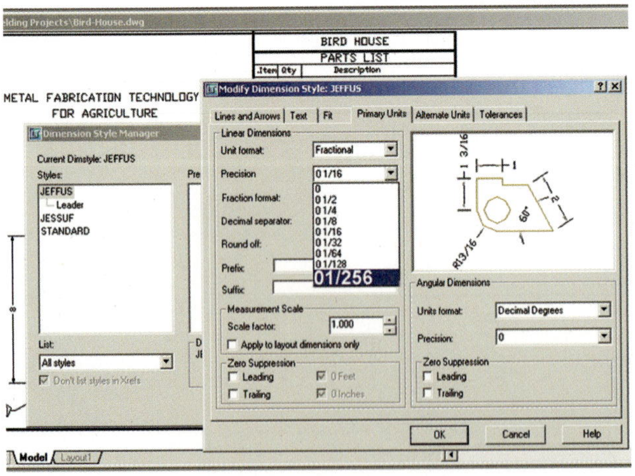

FIGURE 4-35 **The drawing's precision can be set on an AutoCad LT® drawing by using the dimension style manager.** Larry Jeffus

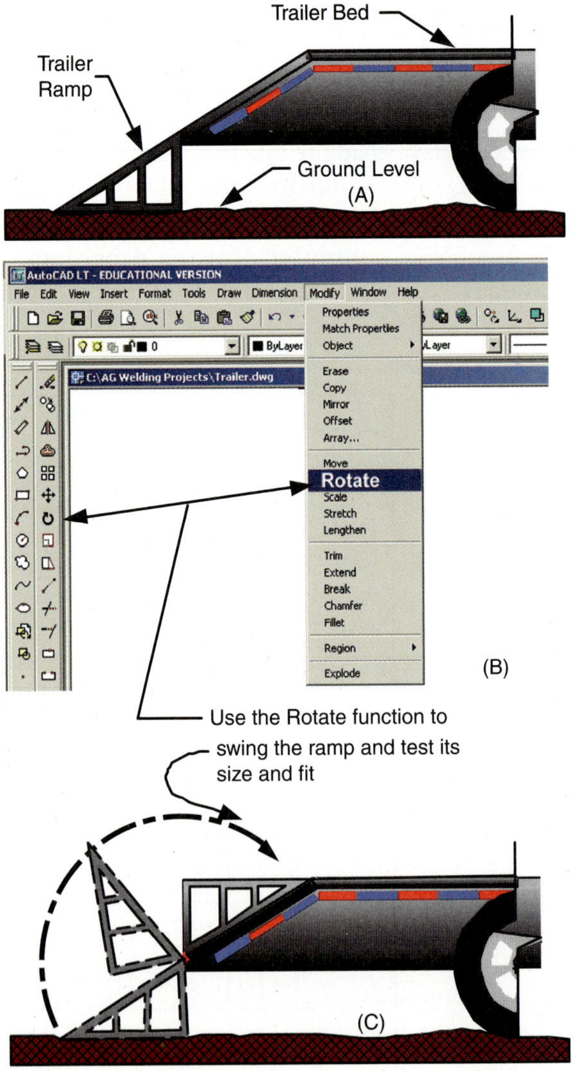

FIGURE 4-37 **The AutoCAD LT® rotate feature can be used to check art alignment and fit-up.** Larry Jeffus

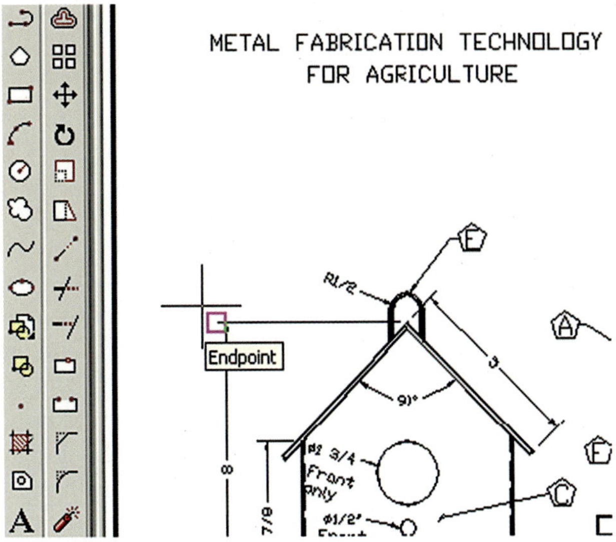

FIGURE 4-38 AutoCAD LT® makes it easy to connect lines using its endpoint connecting feature. Larry Jeffus

Lines on a mechanical drawing connect to other lines on the drawing. It is that connection between lines that allows the object being drawn to take shape. Very rarely would a mechanical drawing line just be drawn by itself. Computerized drafting programs make it easy to make lines connect. As the curser nears the end of a line in AutoCAD LT®, as with many vector drafting programs, the end of the line changes colors, Figure 4-38. If you move closer, the line will "jump" to join the end of the first line drawn. This feature of "joining lines," like other features, can be adjusted or turned off as needed. This ability to customize the settings on a drafting program makes the program much easier and faster to use.

The readability and ease of understanding a drawing is dependent on the drawing's layout and location of dimensions and notes. Drafting programs allow you to move things around to make them clearer and easier to understand. This feature is important because on pencil drawings erasing and redrawing can result in a messy drawing, which can make it harder to understand.

On-screen help with computer drafting is available in several general ways. One method of accessing on-screen help is to hover the curser over a function button and a one- or two-word description of the button will appear, Figure 4-39. A more complete listing of help is available on screen by clicking the "?" button on the tool bar at the top of the screen, Figure 4-40. Some programs, like AutoCAD LT®, provide an Active Assistance function that pops up the first time a function button is clicked, Figure 4-41. The Active Assistance tool provides step-by-step instructions on how to use that function.

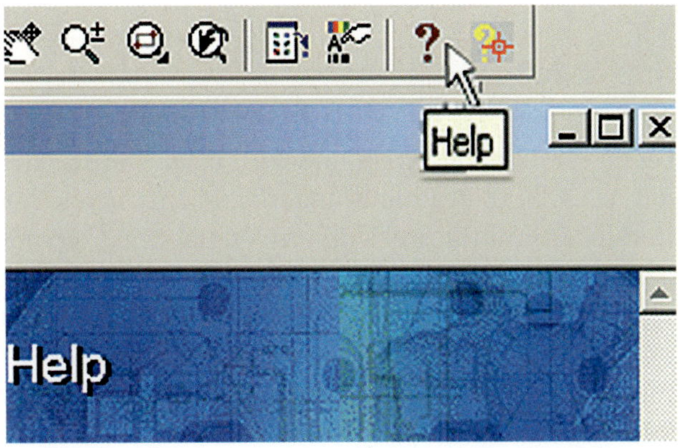

FIGURE 4-39 Hovering above an icon will generate a one- or two-word identification for that icon. Larry Jeffus

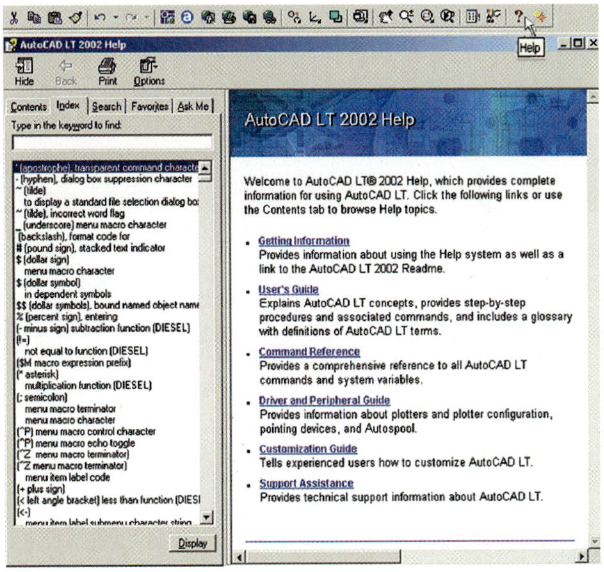

FIGURE 4-40 On-screen help will guide you through many of the program's features. Larry Jeffus

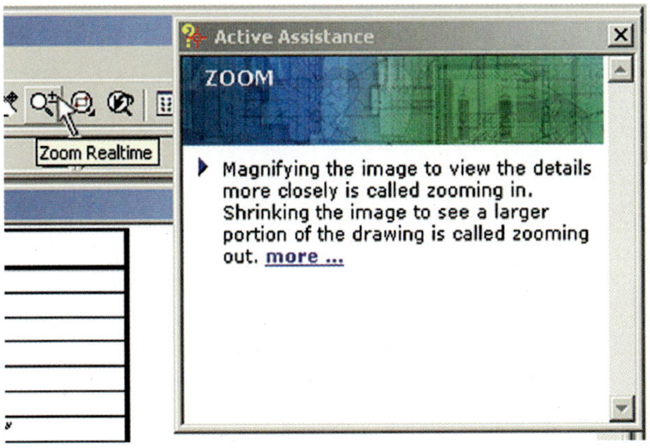

FIGURE 4-41 Active Assistance details specific program features. Larry Jeffus

SUMMARY

Technical drawings are a universal language, so once you are able to read and understand them, you can literally communicate with any engineer or technician anywhere in the world. Of course, you would not be able to read the notes, but everything else—part location, method of assembly, surface, finish—is all important and easily understood. In addition to being able to read drawings, you should be able to make accurate and easily read sketches so that you can communicate your idea clearly.

Before starting to work on a project, you should look at the entire set of drawings. It may be advantageous for you to even make a quick sketch of how you understand that the part should be assembled. This can help you avoid misreading or misunderstanding details that might be contained within the drawing. A part may be referenced on more than one page; therefore, on large complex drawing sets that may contain 10, 20, 30, or more pages, you will have to make sure that the references are all consistent. For example, the part may be shown to be located 3 in. from the end on one page, and several pages back it may be shown to be located 3 1/2 in. These differences are the result of revisions to the part and drawings during the design process or updating. Discrepancies in the drawings do not often occur, but they can be costly if not caught.

Practice reading drawings any time you can because with practice comes greater speed and accuracy.

REVIEW QUESTIONS

1. What information may be included in a set of drawings to aid in the fabrication of the project?
2. What information is provided in the title box on a set of drawings?
3. What is a bill of materials?
4. Sketch an object that contains all of the necessary elements so that you can use the following line types: object line, hidden line, center line, extension line, and dimension line. Label each line type using leaders and arrows.
5. What are the two types of break lines and why are they used?
6. What type of line would be used on a drawing to show an alternate position of a moving part?
7. What is the difference between orthographic projections (mechanical drawings) and pictorial drawings?
8. Which three views are usually provided on a mechanical drawing?
9. Describe how to tell the difference between an isometric drawing and a cavalier drawing.
10. What kind of information does a section view on a drawing provide?
11. What view might you draw to show small details of an area on a part without having to draw the entire part larger?
12. Which of the standard views would you look at to find the height of an object?
13. What is the difference between "as drawn" and "as built" drawings?
14. What is the most commonly used scale for mechanical drawings?
15. What is the difference between architectural scales and engineering scales?
16. What is the difference between a sketch and a mechanical drawing?
17. What is the advantage of vector over raster in computer-aided mechanical drawing?

Chapter 5

Welding Joint Design, Welding Symbols

OBJECTIVES

After completing this chapter, the student should be able to:

- Sketch the five basic welding joints.
- Explain the factors that must be considered when choosing a weld joint design.
- List and explain five ways that forces cause stress in welds.
- Discuss the factors to consider when selecting a weld joint design.
- List and explain the information that can be included on a welding symbol.
- Describe the various types of welds.
- Sketch a welding symbol and identify the components.

KEY TERMS

bevel
chamfering
edge preparation
faying
fillet weld
flange weld
gouging
groove welds
intermittent
J-groove
nondestructive
out-of-position
outside corner joint
plug weld
radius
reinforcement
root
specifications
spot weld
U-groove
V-groove
welding symbols

WELD JOINT DESIGN

The term *weld joint design* refers to the way pieces of metal are put together or aligned with each other. The five basic joint designs are butt joints, lap joints, tee joints, outside corner joints, and edge joints. Figure 5-1 illustrates the way the joint members come together.

- Butt joint—In a butt joint the edges of the metal meet so that the thickness of the joint is approximately equal to the thickness of the metal. The metal surfaces are usually parallel with each other, although there can be some difference in thickness or misalignment of the plates. Butt joints can be welded from one side or both sides with some form of groove weld.
- Lap joint—In a lap joint the edges of the metal overlap so that the thickness of the joint is approximately equal to the combined thickness of both pieces of metal. The distance the surfaces overlap each other may vary from a fraction of an inch to several inches or even feet. Lap welds are usually joined by making a **fillet weld** along the edge of one plate, joining it to the surface of the other. There are several alternate ways of welding lap joints where the weld is made through one or both pieces of metal joining the lap in the center of the overlap.

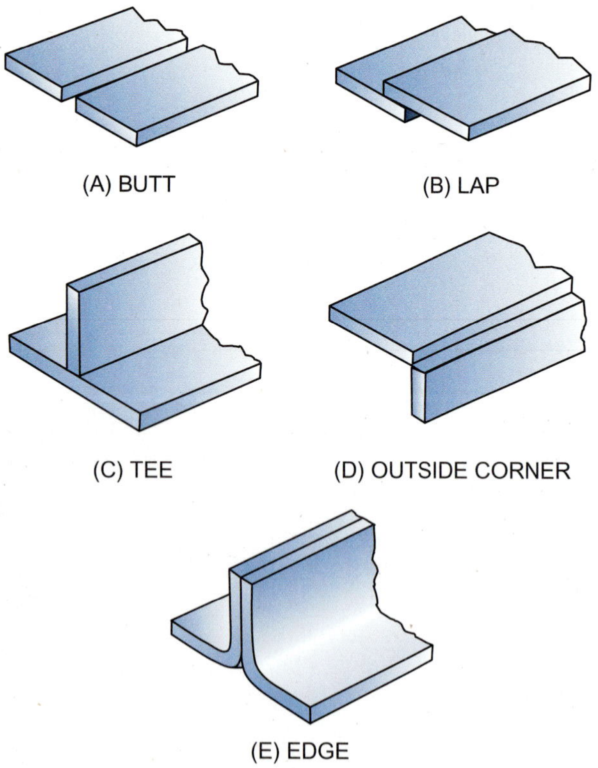

FIGURE 5-1 Types of joints. © Cengage Learning 2012

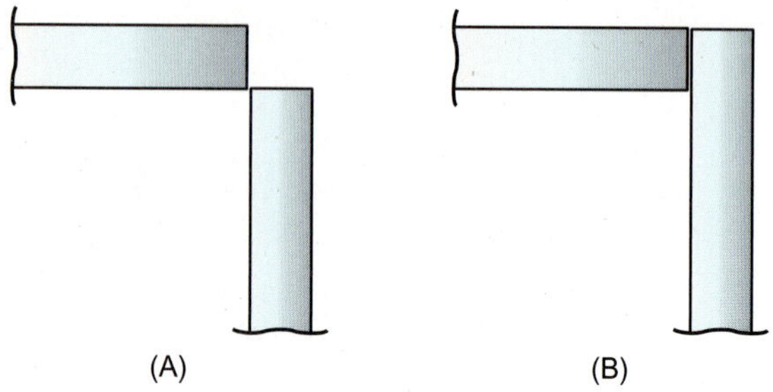

FIGURE 5-2 Two ways of fitting up an outside corner joint. © Cengage Learning 2012

Some examples of this would be plug welds, seam welds, and stir welds. Welds can be made on one side or both sides of the joint.

- Tee joint—In a tee joint the edge of a piece of metal is placed on the surface of another piece of metal. Usually the parts are placed at a 90° angle with each other. Tee joints can be welded with a fillet weld applied to the surfaces, or a weld can be made in a precut groove in the edge of the joining plate. In a few cases, a fillet weld can be made on the top of a groove weld on a tee joint. Welds can be made on one side or both sides of the joint.
- Outside corner joint—In an **outside corner joint,** the edges of the metal are brought together at an angle, usually around 90° to each other. The edges can meet at the corner evenly or they can overlap, Figure 5-2. The outside corner joint can be welded on both sides, with the outside being made as a groove weld and the inside as a fillet weld.
- Edge joint—In an edge joint the metal surfaces are placed together so the edges are even. One or both plates may be formed by bending them at an angle, Figure 5-3. Edge joints are usually welded only on one side.

Welding drawings and **specifications** usually tell you exactly which joint design will be used for all of the welds to be made. Often a welding engineer or designer has determined the best type of joint to be

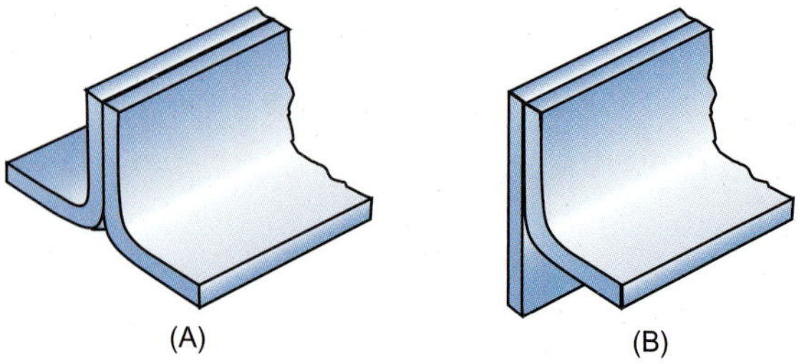

FIGURE 5-3 Edge joints. © Cengage Learning 2012

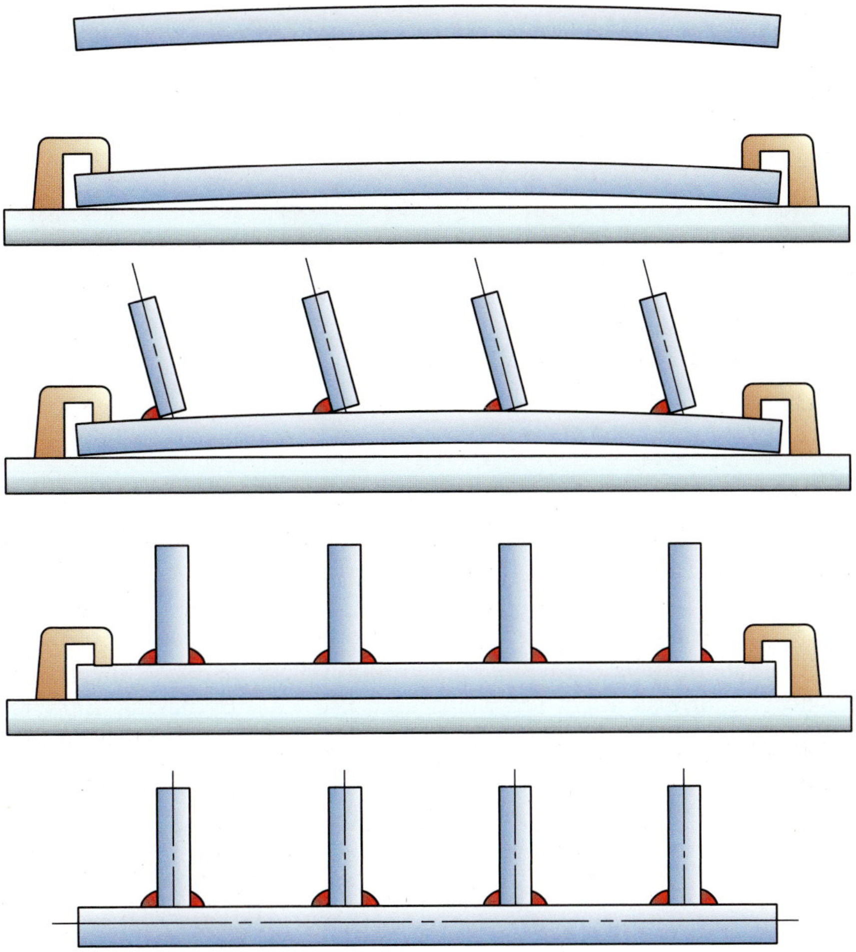

FIGURE 5-4 A method of controlling distortion.

used. However, on small projects or on some repair welding jobs, you will be making the decision as to the joint design to be used. The way the pieces of metal fit together may determine the joint design that must be used. For example, the part shown in Figure 5-4 can be made using only tee joints. Joints on every weldment are not as easily determined.

If you are the one choosing the weld joint design, you must consider a number of factors. Some of the factors involve the type and thickness of metal being welded, the welding position, welding process, finished weld properties, and any code requirements. The selection of the best joint design for a specific weldment requires that you carefully consider all of the various factors. Each factor, if considered alone, could result in a part that might not be able to be fabricated or meet the strength requirements. For example, a narrower joint angle requires less filler metal, and that results in lower welding cost. But if the angle is too small for the welding process being used, the weld cannot be made strong enough. A large weld may be stronger, but it may result in the part being distorted so badly that it becomes useless.

The purpose of a welded joint is to join parts together so that the completed weldment can withstand the stresses. The forces acting on a weld cause stresses. Forces cause stresses in five ways: tensile, compression, bending, torsion, and shear, Figure 5-5. If the stresses are excessive, the part can fail. The ability of a welded joint to withstand these forces depends both upon the joint design and the weld integrity. Some joints can withstand some types of forces better than others.

Some of the factors that affect the selection of a specific weld joint design include welding process, edge preparation, joint dimensions, metal thickness, metal type, welding position, codes or standards, and cost.

Welding process. The welding process to be used has a major effect on the selection of the joint design. Each welding process has characteristics that affect its performance. Some processes are easily used in

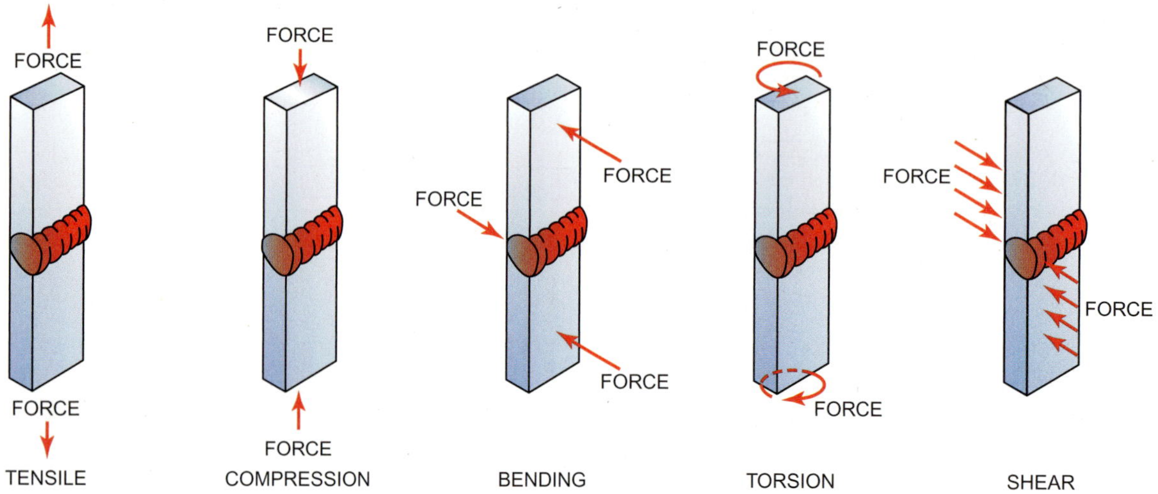

FIGURE 5-5 Forces on a weld. © Cengage Learning 2012

any position; others may be restricted to one or more positions. The rate of travel, penetration, deposition rate, and heat input also affect the welds used on some joint designs. For example, a square butt joint can be made in very thick plates using either electroslag or electrogas welding, but not many other processes can be used on such a joint design.

Edge preparation. The area of the metal's surface that is melted during the welding process is called the **faying** surface. The faying surface can be shaped before welding to increase the weld's strength; this is called **edge preparation.** The edge preparation may be the same on both members of the joint, or each side can be shaped differently, Figure 5-6. Reasons for preparing the faying surfaces for welding include the following:

- Codes and standards—Some codes and standards require specific edge preparations.
- Metals—Some metals must be grooved to successfully weld them, such as thick magnesium, which must be U-grooved; or cast iron cracks, which must be drill-stopped and grooved, Figure 5-7.
- Deeper weld penetration—With the metal removed by grooving or beveling the metal's edge, it is easier for the molten weld metal to completely fuse through the joint. In some cases, it is possible to make a through-thickness weld from one side.
- Smooth appearance—The weld's surface can be ground smooth with the base metal so that the weld "disappears." This can be done for appearance or so that the weld does not interfere with the sliding or moving of parts along the surface.
- Increased strength—A weld should be as strong as or stronger than the base metal being joined. By having 100% joint fusion and an appropriate amount of weld **reinforcement,** the weld can meet its strength requirement.

SPARK YOUR IMAGINATION

Make a list of all the different types of welded joints and weldments that you find as you walk around your welding shop, see on construction equipment, or find on the way to your school or job.

Joint dimensions. In some cases the exact size, shape, and angle can be specified for a groove, Figure 5-8. If exact dimensions are not given, you may make the groove any size that you feel necessary; but remember, the wider the groove, the more welding it will require to complete, Figure 5-9.

Metal thickness. As the metal becomes thicker, you must change the joint design to ensure a sound weld. On thin sections it is often possible to make full penetration welds using a square butt joint. Square butt joints take less preparation time and less welding time. But with thicker plates or pipe, the edge must be prepared with a groove on one or both sides. The edge may be shaped with either a **bevel, V-groove, J-groove,** or **U-groove.**

When welding on thick plate or pipe, it is often impossible for the welder to get 100% penetration without some type of groove being used. The groove may be cut into just one of the plates or pipes or both. On

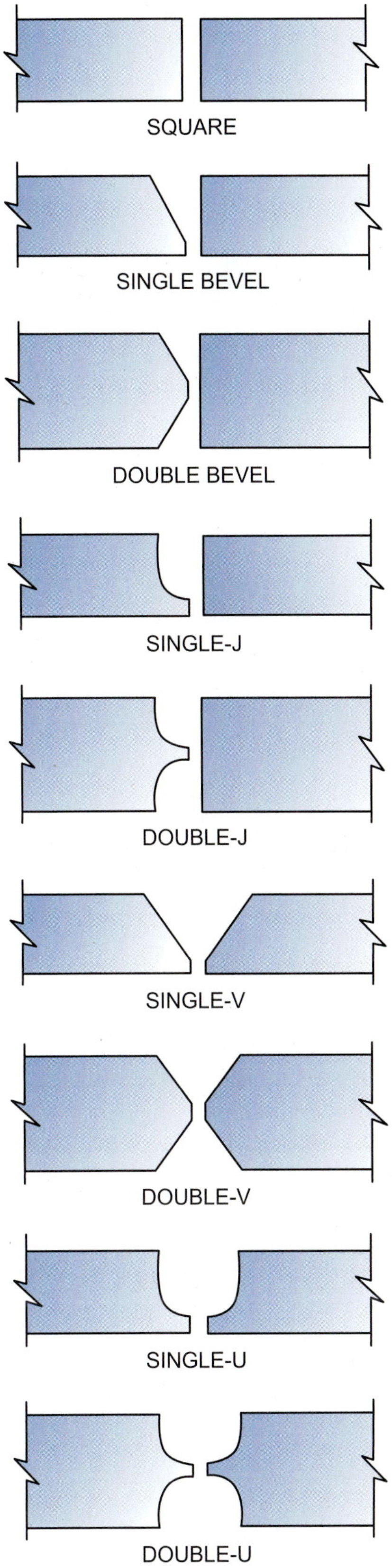

FIGURE 5-6 Joint edge preparation. American Welding Society

FIGURE 5-7A Drill each end of the crack. Larry Jeffus

FIGURE 5-7B The drill hole will stop the crack from lengthening as it is being repaired. Larry Jeffus

FIGURE 5-7C Grind a U-groove all the way along the crack. (*continued*) Larry Jeffus

FIGURE 5-7D continued Grind the weld before finishing. Larry Jeffus

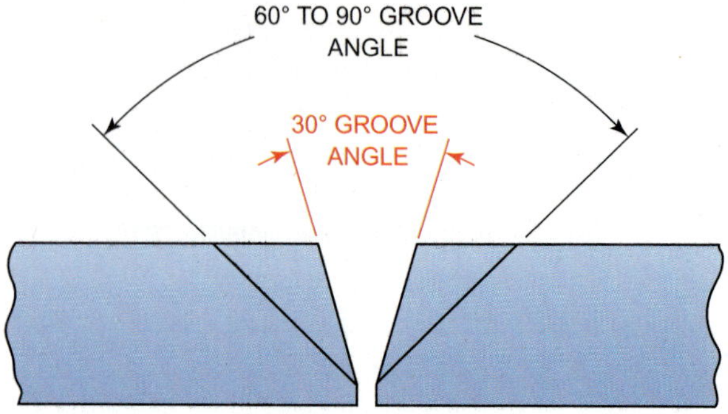

FIGURE 5-9 A smaller groove angle can reduce both weld time and weld metal required to make the joint. © Cengage Learning 2012

some plates it can be cut both inside and outside of the joint. The groove may be ground, flame cut, gouged, sawed, or machined on the edge of the plate before or after the assembly. Bevels and V-grooves are best if they are cut before the parts are assembled. J-grooves and U-grooves can be cut either before or after assembly. The lap joint is seldom prepared with a groove because little or no strength can be gained by grooving this joint.

For most welding processes, plates that are thicker than 3/8 in. (10 mm) may be grooved on both the inside and outside of the joint. Plate in the flat position is usually grooved on only one side unless it can be repositioned or it is required to be welded on both sides. Tee joints in a thick plate are easier to weld and will have less distortion if they are grooved on both sides.

Sometimes plates are either grooved and welded or just welded on one side and then back-gouged and welded, Figure 5-10. Back **gouging** is a process of cutting a groove in the back side of a joint that has been

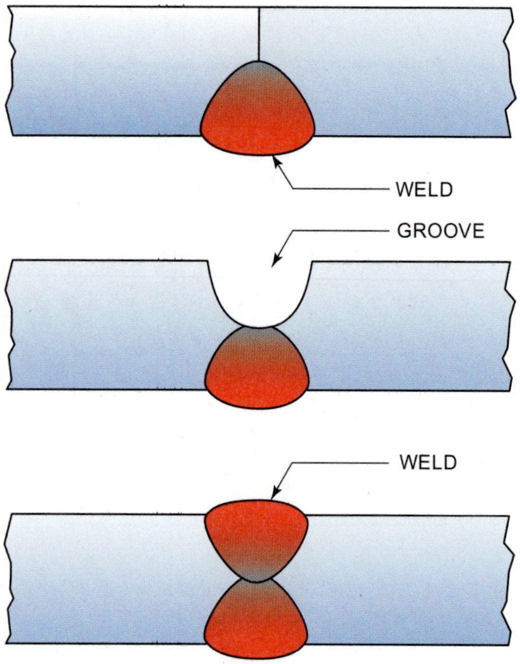

FIGURE 5-10 Back gouging a weld joint to ensure 100% joint penetration. © Cengage Learning 2012

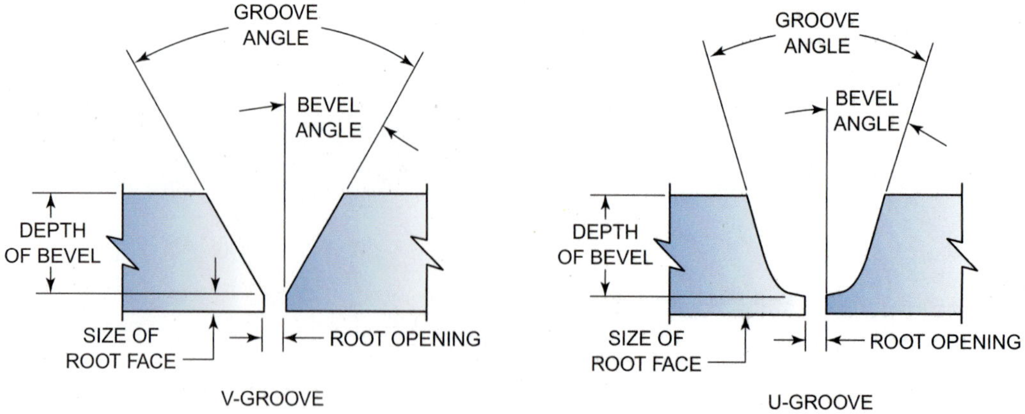

FIGURE 5-8 U-groove terminology. American Welding Society

welded. Back gouging can ensure 100% joint fusion at the **root** and remove discontinuities of the root pass.

Metal type. Because some metals have specific problems with thermal expansion, crack sensitivity, or distortion, the joint design selected must help control these problems. For example, magnesium is very susceptible to postweld stresses, and the U-groove works best for thick sections.

Welding position. The most ideal welding position for most joints is the flat position because it allows for larger molten weld pools to be controlled. Usually the larger that a weld pool can be, the faster the joint can be completed. When welds are made in any position other than the flat position, they are referred to as being done **out-of-position**. Some types of grooves work better in out-of-position welding than others; for example, the bevel joint is often the best choice for horizontal butt welding, Figure 5-11.

- Plate Welding Positions—The American Welding Society has divided plate welding into four basic positions for grooves (G) and fillet (F) welds as follows:
  - Flat 1G or 1F—When welding is performed from the upper side of the joint, and the face of the weld is approximately horizontal, Figure 5-12A and B.
  - Horizontal 2G or 2F—The axis of the weld is approximately horizontal, but the type of weld dictates the complete definition. For a fillet weld, welding is performed on the upper side of an approximately vertical surface. For a groove weld, the face of the weld lies in an approximately vertical plane, Figure 5-12C and D.

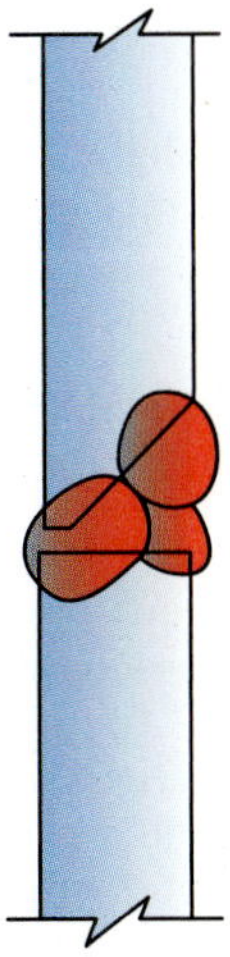

FIGURE 5-11 Weld position for a horizontal joint.
© Cengage Learning 2012

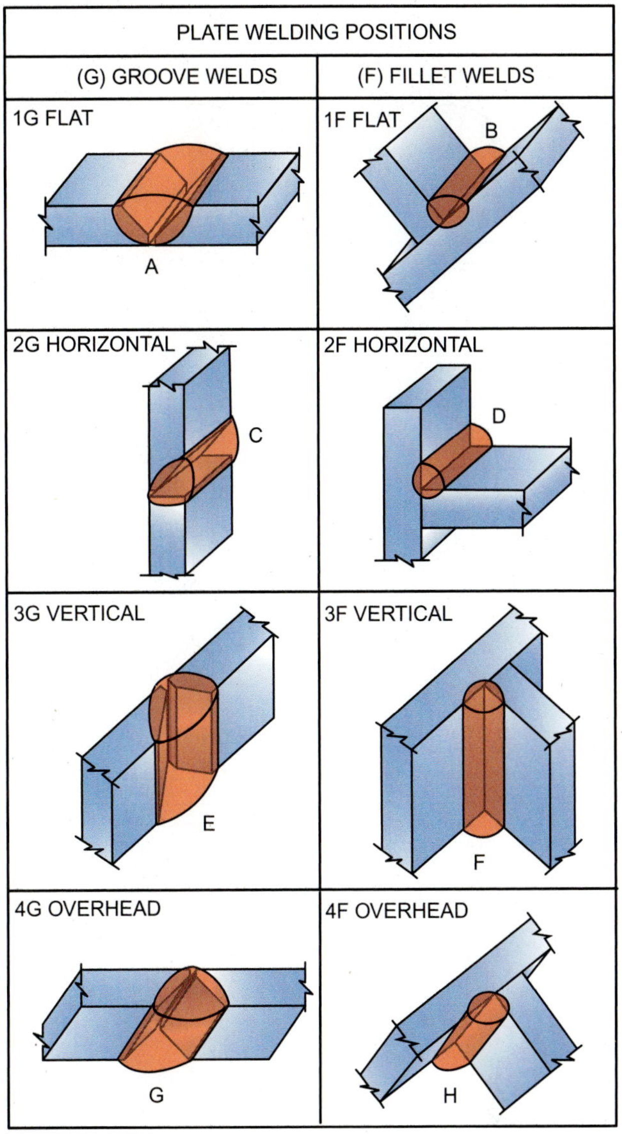

FIGURE 5-12 Plate welding positions.
American Welding Society

  - Vertical 3G or 3F—The axis of the weld is approximately vertical, Figure 5-12E and F.
  - Overhead 4G or 4F—When welding is performed from the underside of the joint, Figure 5-12G and H.
- Pipe Welding Positions—The American Welding Society has divided pipe welding into five basic positions:
  - Horizontal rolled 1G—When the pipe is rolled either continuously or intermittently so that the weld is performed within 0° to 15° of the top of the pipe, Figure 5-13A.
  - Horizontal fixed 5G—When the pipe is parallel to the horizon, and the weld is made vertically around the pipe, Figure 5-13B.

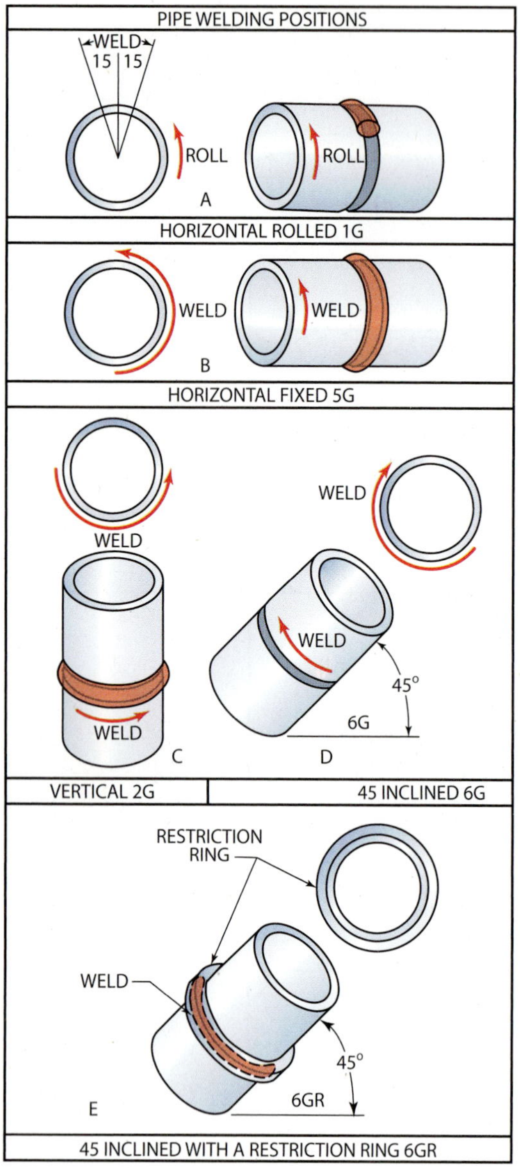

FIGURE 5-13 **Pipe welding positions.**
American Welding Society

- Vertical 2G—The pipe is vertical to the horizon, and the weld is made horizontally around the pipe, Figure 5-13C.
- Inclined 6G—The pipe is fixed in a 45°-inclined angle, and the weld is made around the pipe, Figure 5-13D.
- Inclined with a restriction ring 6GR—The pipe is fixed in a 45° inclined angle, and there is a restricting ring placed around the pipe below the weld groove, Figure 5-13E.

Code or standards requirements. The type, depth, angle, and location of the groove is usually determined by a code or standard that has been qualified for the specific job. Organizations such as the American Welding Society (AWS), American Society of Mechanical Engineers (ASME), and the American Bureau of Ships are a few of the agencies that issue such codes and specifications. The most common codes or standards are the AWS D1.1 and the ASME Boiler and Pressure Vessel (BPV) Section IX. The joint design for a specific set of specifications is often known as prequalified. These joints have been tested and found to be reliable for the weldments for specific applications. The joint design can be modified, but the cost to have the new design accepted under the standard being used is often prohibitive.

Cost. Almost any weld can be made in any material in any position if cost is not a factor. A number of items affect the cost of producing a weld. Joint design can be a major way to control welding cost. Changes in the design can reduce cost while still meeting the weldment's strength requirements. Making the groove angle smaller can help, Figure 5-14, reducing the welding filler metal required to complete the weld as well as reducing the time required to fill the larger groove opening. Good joint design must be a consideration for any project to be competitive and cost effective.

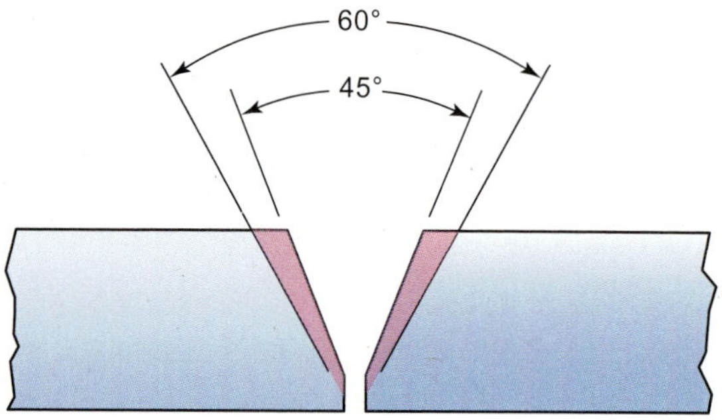

FIGURE 5-14 **Even a slight change in groove angle can save time and money.** © Cengage Learning 2012

SPARK YOUR IMAGINATION

We have all heard the expression "Time is money." On your next welding practice, record the time it takes to make the weld. How many inches of weld can you produce in a minute? To find out, you can divide the time you took into the length of the weld to determine how many inches per minute you weld. If you know how many inches of weld you can make in a minute, then how would you find out how many inches of the same weld you could make in an hour if you did not stop for breaks?

Do not be discouraged if you are making only a fraction of an inch of weld per minute. With some welding processes and material thicknesses, even a highly skilled professional welder may not do much better. You are still learning!

WELDING SYMBOLS

The use of **welding symbols** enables a designer to indicate clearly to the welder, important detailed information regarding the weld. The information in the welding symbol can include details for the weld such as length, depth of penetration, height of reinforcement, groove type, groove dimensions, location, process, filler metal, strength, number of welds, weld shape, and surface finishing. All of this information would normally be included on the welding assembly drawings.

Welding symbols are a shorthand language for the welder. They save time and money and serve to ensure understanding and accuracy. The American Welding Society has standardized welding symbols. Some of the more common symbols for welding are reproduced in this chapter. If more information is desired about symbols or how they apply to all forms of manual and automatic machine welding, these symbols can be found in the complete manual *Standard Symbols for Welding, Brazing, and Nondestructive Examination*, ANSI/AWS A2.4, published as an American National Standard by the American Welding Society.

Figure 5-15 shows the basic components of welding symbols, consisting of a reference line with an arrow on one end. Other information relating to various features of the weld are shown by symbols, abbreviations, and figures located around the reference line. A tail is added to the basic symbol as necessary for the placement of specific information.

INDICATING TYPES OF WELDS

Welds are classified as follows: fillets, grooves, flange, plug or slot, spot or protecting, seam, back or backing, and surfacing. Each type of weld has a specific symbol that is used on drawings to indicate the weld. A fillet weld, for example, is designated by a right triangle.

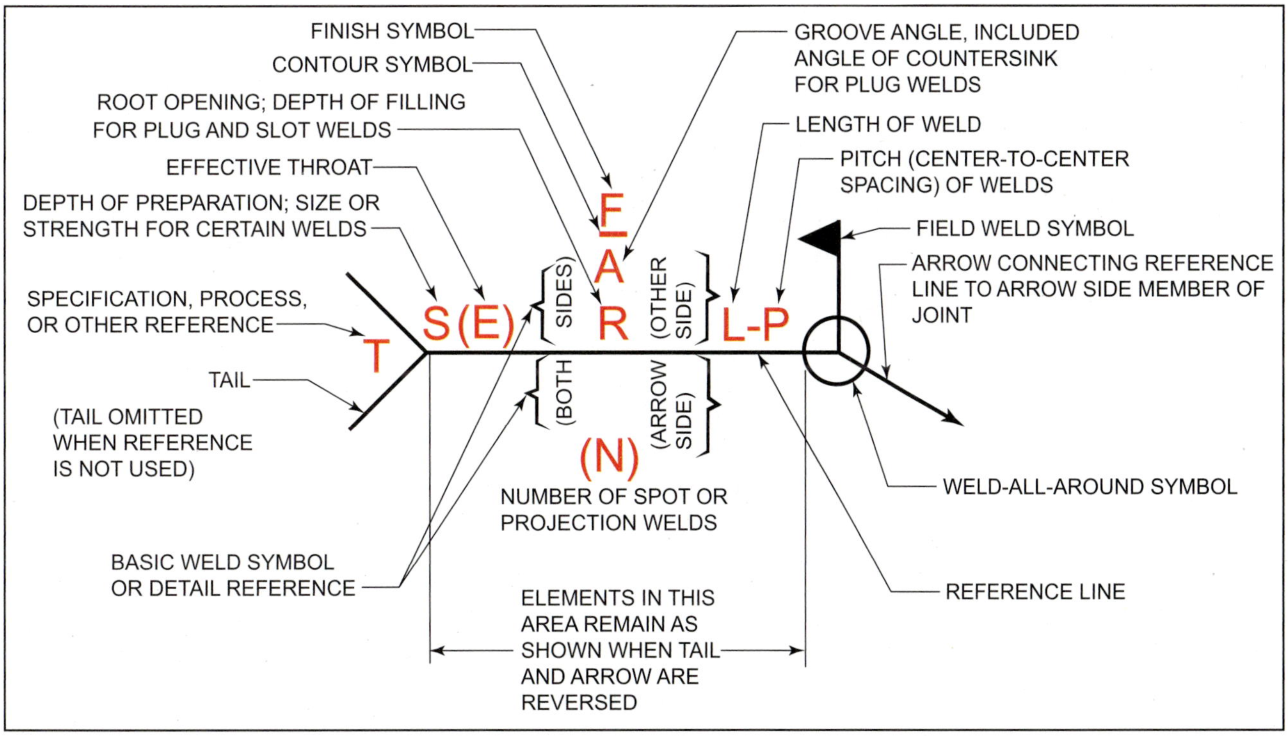

FIGURE 5-15 Standard locations of elements of a welding symbol. American Welding Society

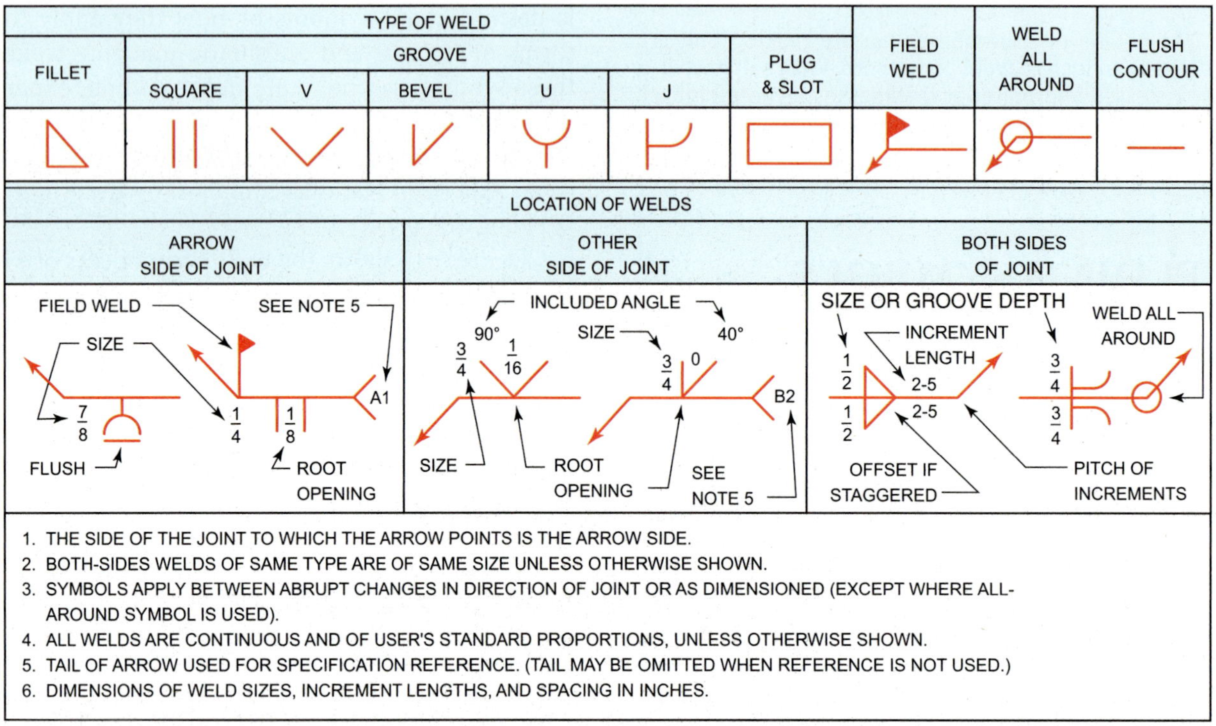

FIGURE 5-16 **Welding symbols for different types of welds.** American Welding Society

A plug weld is indicated by a rectangle. All of the basic symbols are shown in Figure 5-16.

WELD LOCATION

Welding symbols are applied to the reference line at the base. All reference lines have an arrow side (near side) and other side (far side). Accordingly, the terms *arrow side, other side,* and *both sides* are used to locate the weld with respect to the joint. The reference line is always drawn horizontally. An arrow line is drawn from one end or both ends of a reference line to the location of the weld. The arrow line can point to either side of the joint and extend either upward or downward.

If the weld is to be deposited on the arrow side of the joint (near side), the desired weld symbol is placed below the reference line, Figure 5-17A.

If the weld is to be deposited on the other side of the joint (far side), the weld symbol is placed above the reference line, Figure 5-17B. When welds are to be deposited on both sides of the same joint, the same weld symbol appears above and below the reference line, Figure 5-17C and D.

The tail is added to the basic welding symbol when it is necessary to designate the welding specifications, procedures, or other supplementary information needed to make the weld, Figure 5-18. The notation placed in the tail of the symbol may indicate the welding process to be used, the type of filler metal needed,

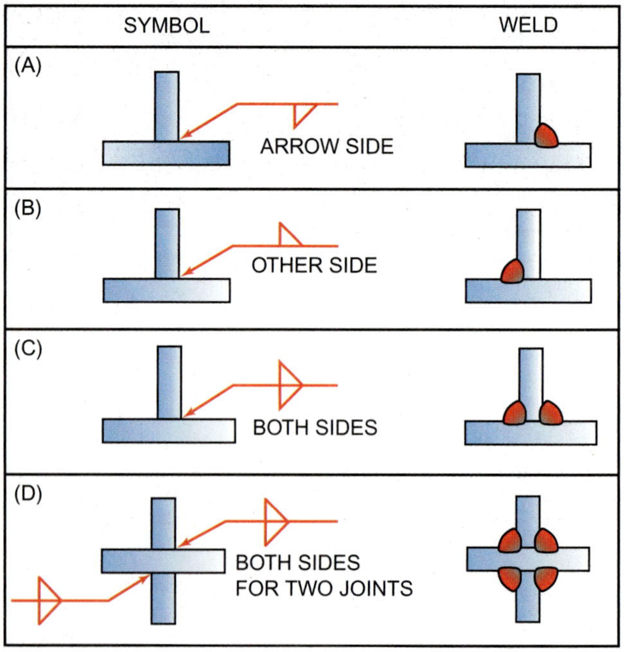

FIGURE 5-17 **Designating weld location.**
American Welding Society

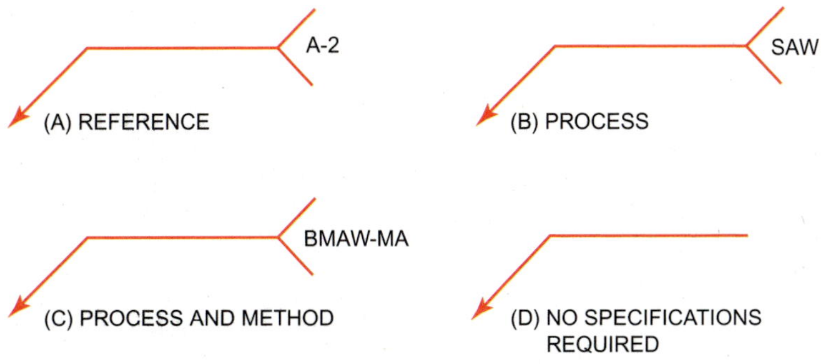

FIGURE 5-18 Location of specification, process, and other references on weld symbols. American Welding Society

whether or not peeling or root chipping is required, and other information pertaining to the weld. If notations are not used, the tail of the symbol is omitted.

For joints that are to have more than one weld, a symbol is shown for each weld.

LOCATION SIGNIFICANCE OF ARROW

In the case of fillet and groove welding symbols, the arrow connects the welding symbol reference line to one side of the joint. The surface of the joint that the arrow point actually touches is considered to be the arrow side of the joint. The side opposite the arrow side of the joint is considered to be the other (far) side of the joint.

On a drawing, when a joint is illustrated by a single line and the arrow of a welding symbol is directed to the line, the arrow side of the joint is considered to be the near side of the joint.

For welds designated by the plug, slot, spot, seam, resistance, flesh, upset, or projection welding symbols, the arrow connects the welding symbol reference line to the outer surface of one of the members of the joint at the center line of the desired weld. The member to which the arrow points is considered to be the arrow side member. The remaining member of the joint is considered to be the other side member.

FILLET WELDS

NOTE: A fillet weld is approximately triangular in shape. It is used to join lap joints, tee joints, or corner joints where the joint is at an approximate right angle.

Dimensions of fillet welds are shown on the same side of the reference line as the weld symbol and are shown to the left of the symbol, Figure 5-19A. When both sides of a joint have the same size fillet welds, one or both may be dimensioned as shown in

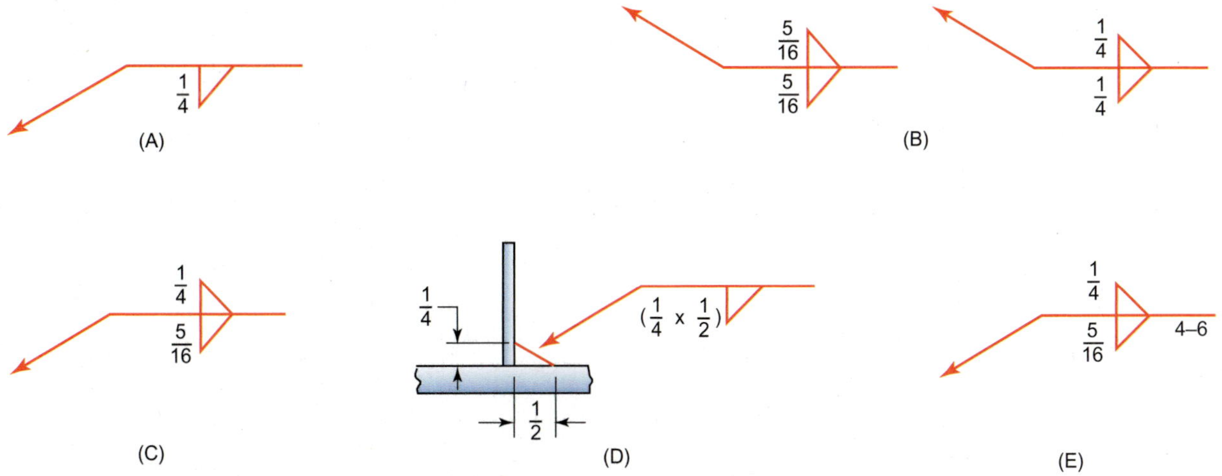

FIGURE 5-19 Dimensioning fillet weld symbols. American Welding Society

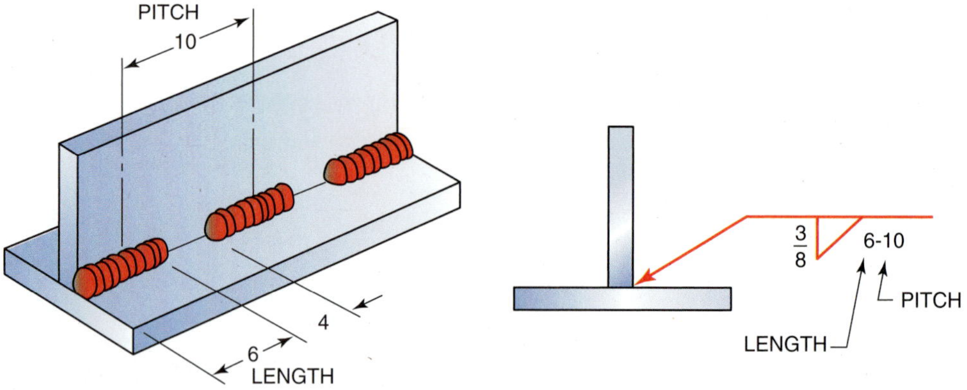

FIGURE 5-20 Dimensioning intermittent fillet welds. American Welding Society

Figure 5-19B. When both sides of a joint have different size fillet welds, both are dimensioned, Figure 5-19C. When the dimensions of one or both welds differ from the dimensions given in the general notes, both welds are dimensioned. The size of a fillet weld with unequal legs is shown in parentheses to the left of the weld symbol, Figure 5-19D. The length of a fillet weld, when indicated on the welding symbol, is shown to the right of the weld symbol, Figure 5-19E. In **intermittent** fillet welds, the length and pitch increments are placed to the right of the weld symbol, Figure 5-20. Intermittent welds are often used in sheet metal to both reduce the heat input to the joint and to stop cracking from continuing through the joint. Each individual weld serves as a crack-stopper. If a crack starts in a single intermittent weld, it has to restart in the next weld before it can continue on down the joint. For these reasons you must make the intermittent welds as designed, even though making a continuous weld might be faster. The first number represents the length of the weld, and the second number represents the pitch or the distance between the center of two welds.

NOTE: Unequal legged fillet welds are used when one piece of metal is much thinner than the other. The short leg of the fillet allows less heat input to the thinner metal and reduces the chance of burnthrough.

PRACTICE 5-1

Referring to the weld symbols shown in Figure 5-19 and using a pencil and paper, sketch a cross section of each of the welds. Include the dimensions for the fillet weld size for the six welding symbols shown in the figure. See Figure 5-21 for an example of how to do this practice.◆

PLUG WELDS

NOTE: A plug weld is made by welding through a round hole in the top plate to fuse the bottom plate. The hole in the top plate may or may not be filled completely by the weld. Plug welds are used to make lap joints.

Holes in the arrow side member of a joint for plug welding are indicated by placing the weld symbol below the reference line. Holes in the other side member of a joint for plug welding are indicated by placing the weld symbol above the reference line.

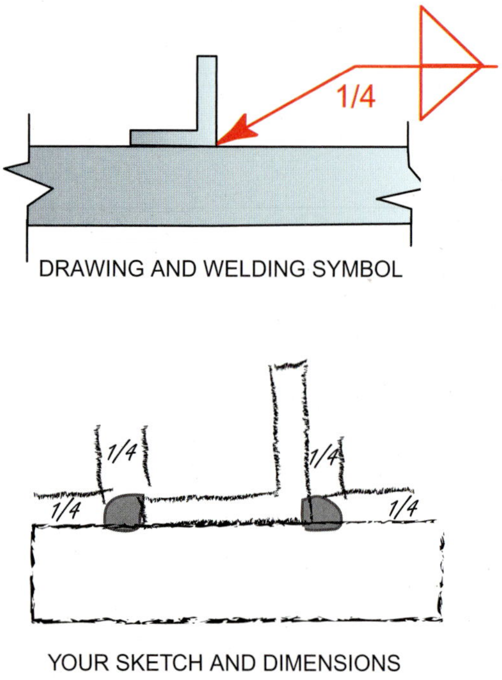

FIGURE 5-21 Practice 5-1. © Cengage Learning 2012

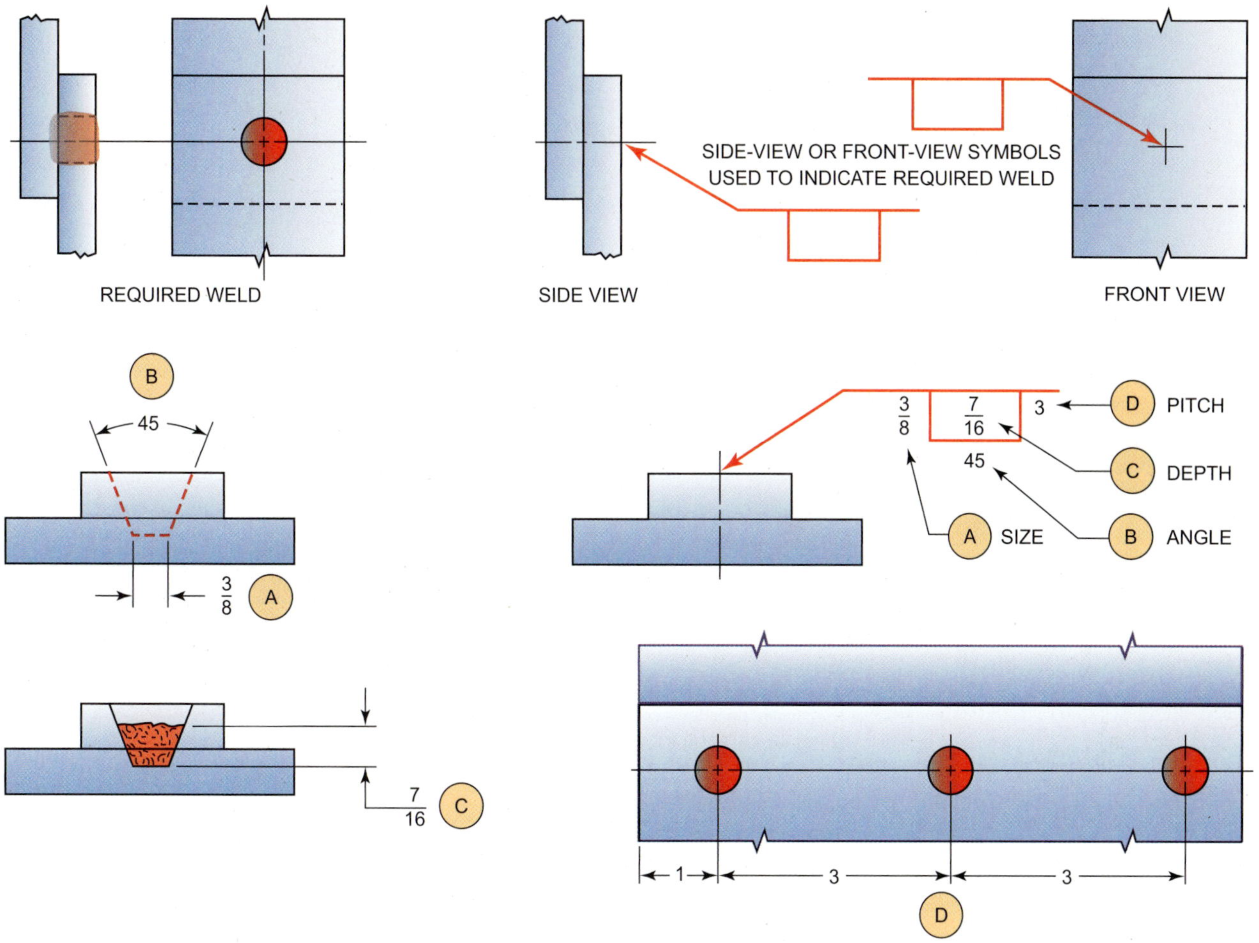

FIGURE 5-22 Applying dimensions to plug welds. American Welding Society

Refer to Figure 5-22 for the location of the dimensions used on plug welds. The diameter or size is located to the left of the symbol (A). The angle of the sides of the hole, if not square, is given above the symbol (B). The depth of buildup, if not completely flush with the surface, is given in the symbol (C). The center-to-center dimensioning or pitch is located on the right of the symbol (D).

SPOT WELDS

NOTE: A spot weld is approximately round and is created between the two overlapping surfaces being joined.

Dimensions of **spot welds** are indicated on the same side of the reference line as the weld symbol, Figure 5-23. Such welds are dimensioned either by size or strength. The size is designated as the diameter of the weld expressed in fractions or in decimal hundredths of an inch. The size is shown with or without inch marks to the left of the weld symbol. The center-to-center spacing (pitch) is shown to the right of the symbol.

The strength of spot welds is shown as the minimum shear strength in pounds (Newton's) per spot and is shown to the left of the symbol, Figure 5-24A. When a definite number of spot welds are desired in a certain joint, the quantity is placed above or below the weld symbol in parentheses, Figure 5-24B.

PRACTICE 5-2

Referring to the weld symbols shown in Figure 5-24, and using a pencil and paper, sketch a cross section of each of the welds. Include the dimensions for the spot weld size for the five welding symbols shown in the figure. See Figure 5-21 for an example of how to do this practice.◆

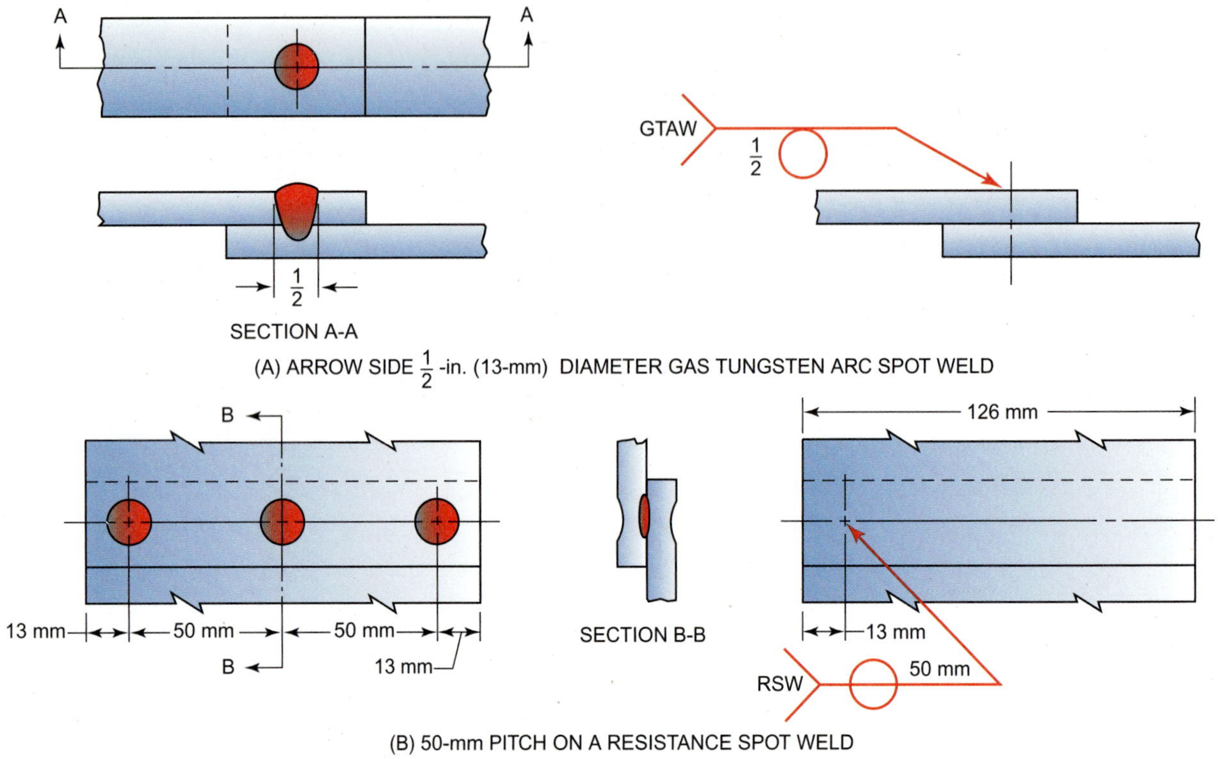

FIGURE 5-23 **Spot welding symbols.** American Welding Society

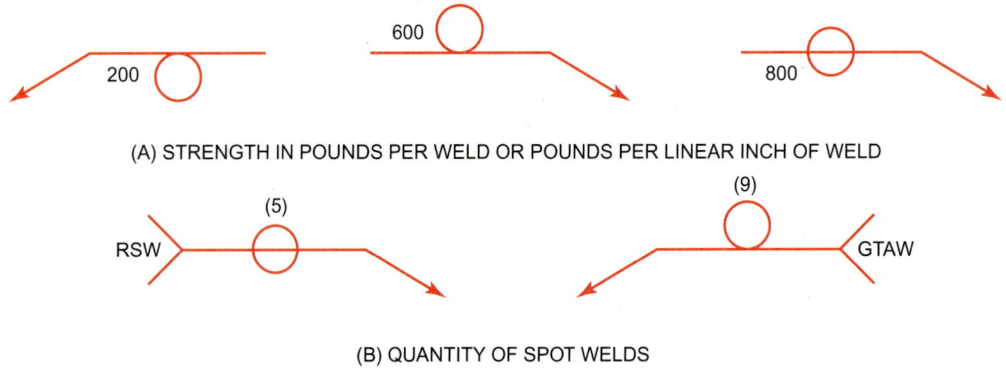

FIGURE 5-24 **Designating strength, number, and process of spot welds.** American Welding Society

SEAM WELDS

NOTE: **Seam welds are continuous along the overlapping surfaces. They can be made by producing a series of overlapping spot welds or be one continuous resistance weld.**

Dimensions of seam welds are shown on the same side of the reference line as the weld symbol. Dimensions relate to either size or strength. The size of seam welds is designated as the width of the weld expressed in fractions or decimal hundredths of an inch. The size is shown with or without the inch marks to the left of the weld symbol, Figure 5-25A. When the length of a seam weld is indicated on the symbol, it is shown to the right of the symbol, Figure 5-25B. When seam welding extends for the full distance between abrupt changes in the direction of welding, a length dimension is not required on the welding symbol.

The strength of seam welds is designated as the minimum acceptable shear strength in pounds per linear inch. The strength value is placed to the left of the weld symbol, Figure 5-26.

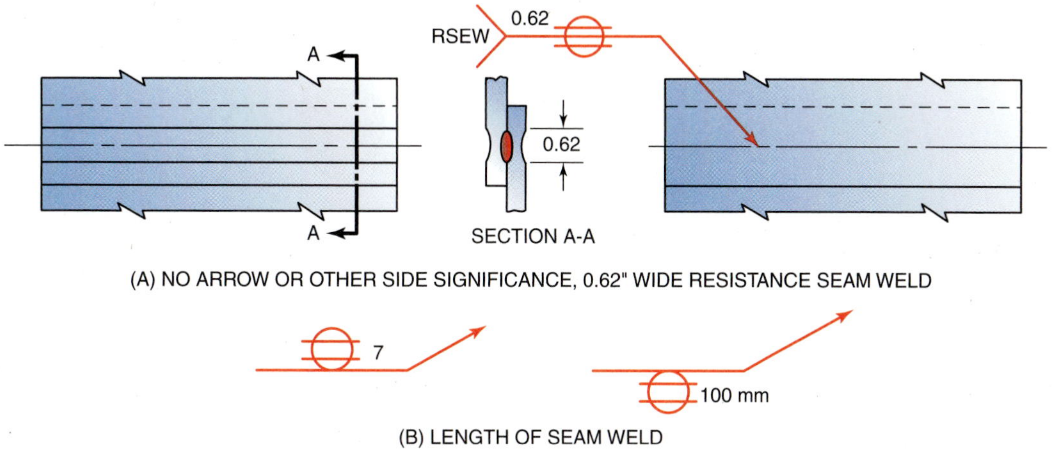

FIGURE 5-25 Designating the size of seam welds. American Welding Society

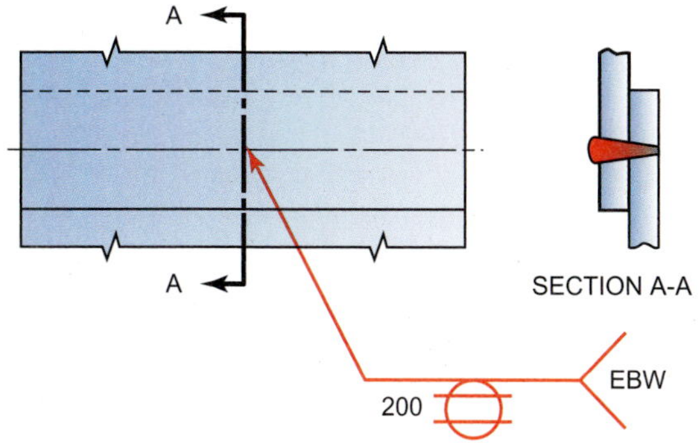

FIGURE 5-26 Strength of a seam weld. American Welding Society

GROOVE WELDS

NOTE: A groove weld is made in the space cut into the joint between two pieces being joined.

Joint strengths can be improved by making some type of groove preparation before the joint is welded. There are seven types of grooves. The groove can be made in one or both plates or on one or both sides. By cutting the groove in the plate, the weld can penetrate deeper into the joint, helping to increase the joint strength without restricting flexibility.

The grooves can be cut in base metal in a number of different ways. The groove can be oxyacetylene cut, air carbon arc cut, plasma arc cut, machined, sawed, and so forth. The various features of groove welds are as follows:

- Single-groove and symmetrical double-groove welds that extend completely through the members being joined. No size is included on the weld symbol, Figure 5-27A and B.
- Groove welds that extend only partway through the parts being joined. The size as measured from the top of the surface to the bottom (not including reinforcement) is included to the left of the welding symbol, Figure 5-27C.

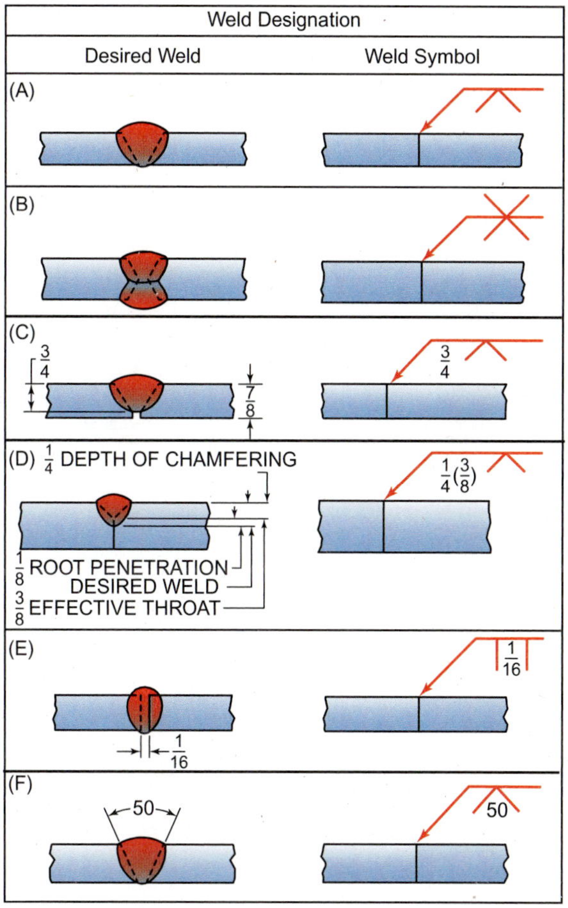

FIGURE 5-27 Designating groove weld location, size, and root penetration. American Welding Society

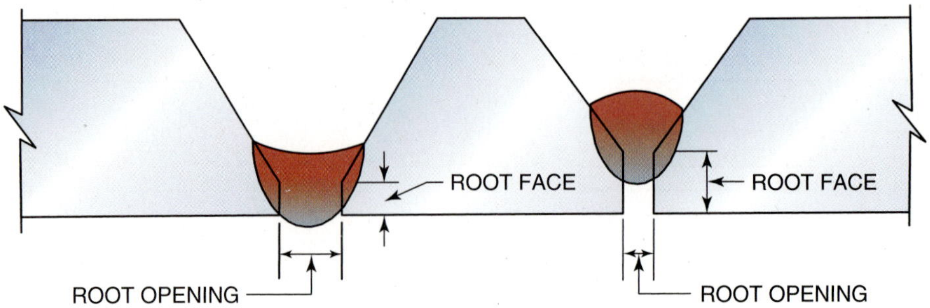

FIGURE 5-28 **Effect that root dimensioning can have on groove weld penetration.** American Welding Society

- The size of groove welds with a specified *effective throat* is indicated by showing the depth of groove preparation with the effective throat appearing in parentheses and placed to the left of the weld symbol, Figure 5-27D. The size of square groove welds is indicated by showing the root penetration. The depth of **chamfering** and the root penetration is read in that order from left to right along the reference line.
- The root opening of groove welds is the user's standard unless otherwise indicated. The root opening of groove welds, when not the user's standard, is shown inside the weld symbol, Figure 5-27E and F.
- The root face's main purpose is to minimize the burnthrough that can occur with a feather edge. The size of the root face is important to ensure good root fusion, Figure 5-28.
- The size of flare groove welds is considered to extend only to the tangent points of the members, Figure 5-29.

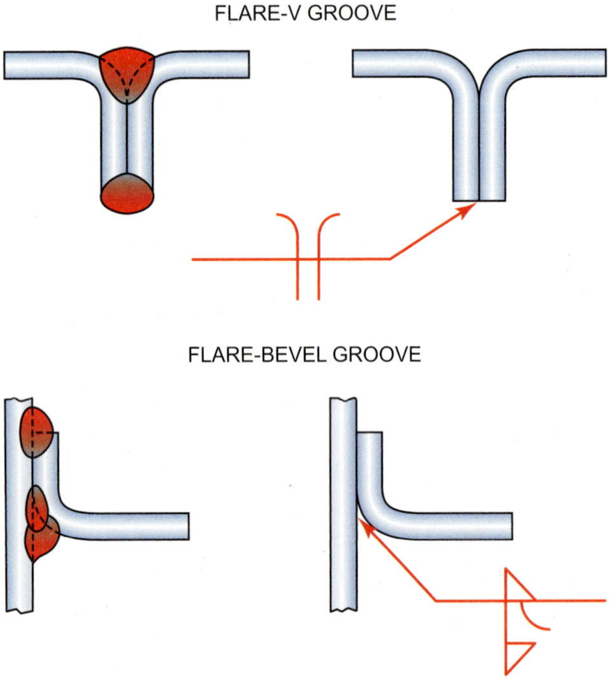

FIGURE 5-29 **Designating flare-V and flare-bevel groove welds.** American Welding Society

BACKING

A backing (strip) is a piece of metal that is placed on the back side of a weld joint to prevent the molten metal from dripping through the open root. It helps to ensure that 100% of the base metal's thickness is fused by the weld. The backing must be thick enough to withstand the heat of the root pass as it is burned in. A backing strip may be used on butt joints, tee joints, and outside corner joints, Figure 5-30.

The backing may be either left on the finished weld or removed following welding. If the backing is to be removed, the letter *R* is placed in the backing

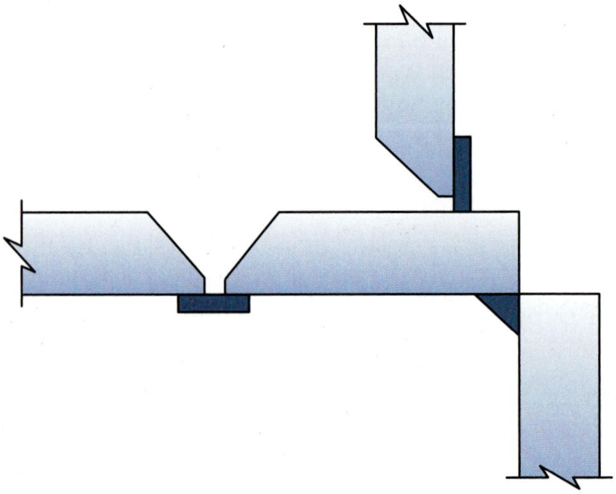

FIGURE 5-30 **Backing strips.** © Cengage Learning 2012

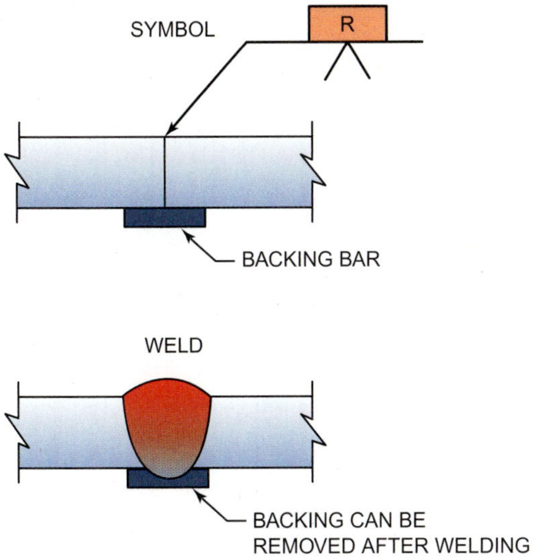

FIGURE 5-31 Backing strip weld symbol.
American Welding Society

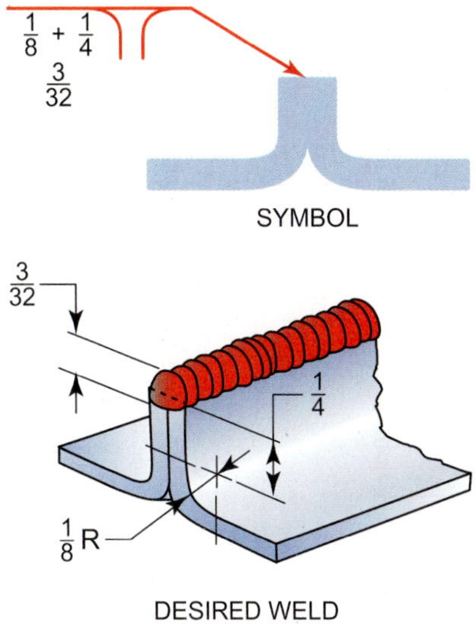

FIGURE 5-32 Applying dimensioning to flare welds.
American Welding Society

symbol, Figure 5-31. The backing is often removed from a finished weld because it can be a source of stress concentration and a crevice to promote rusting.

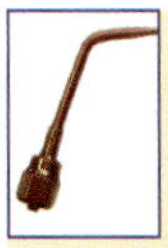

SHOP TALK

Why would you want to remove a backing strip from a weld?

There are several reasons to remove a backing strip from a weld. The strip can collect moisture and cause the part to rust. A backing strip can keep a weldment from bending evenly under a heavy load, and if it cannot bend uniformly, it will break.

PRACTICE 5-3

Referring to the weld joints shown in Figure 5-30 and using a pencil and paper, sketch a cross section of each, showing a weld and the appropriate welding symbol.◆

FLANGE WELDS

NOTE: Flange welds are used on thin metal as a way of stiffening the edge so there is less distortion. It can also make it easier to make a weld on these thin sections without excessive burnthrough. A flange weld is made along the edge of the metal that has been bent upward, forming a ridge or flange.

The following welding symbols are used for light-gauge metal joints where the edges to be joined are bent to form a flange or flare.

- Edge flange welds are shown by the edge flange weld symbol.
- Corner flange welds are indicated by the corner flange weld symbol.
- Dimensions of flange welds are shown on the same side of the reference line as the weld symbol and are placed to the left of the symbol, Figure 5-32. The **radius** and height above the point of tangency are indicated by showing both the radius and the height separated by a plus sign.
- The size of the flange weld is shown by a dimension placed outward from the flanged dimensions.

PRACTICE 5-4

Referring to the AWS weld test shown in Figure 5-33 and using a pencil and paper, sketch a cross section of each of the welds. Include the dimensions for the groove, if any, and the finished weld size for the seven welding symbols used on the drawing. See Figure 5-21 for an example of how to do this practice.◆

NONDESTRUCTIVE TESTING SYMBOLS

The increased use of **nondestructive** testing (NDT) as a means of quality assurance has resulted in the development of standardized symbols. The designer or engineer uses them to indicate the area to be tested and the type of test to be used. The inspection symbol

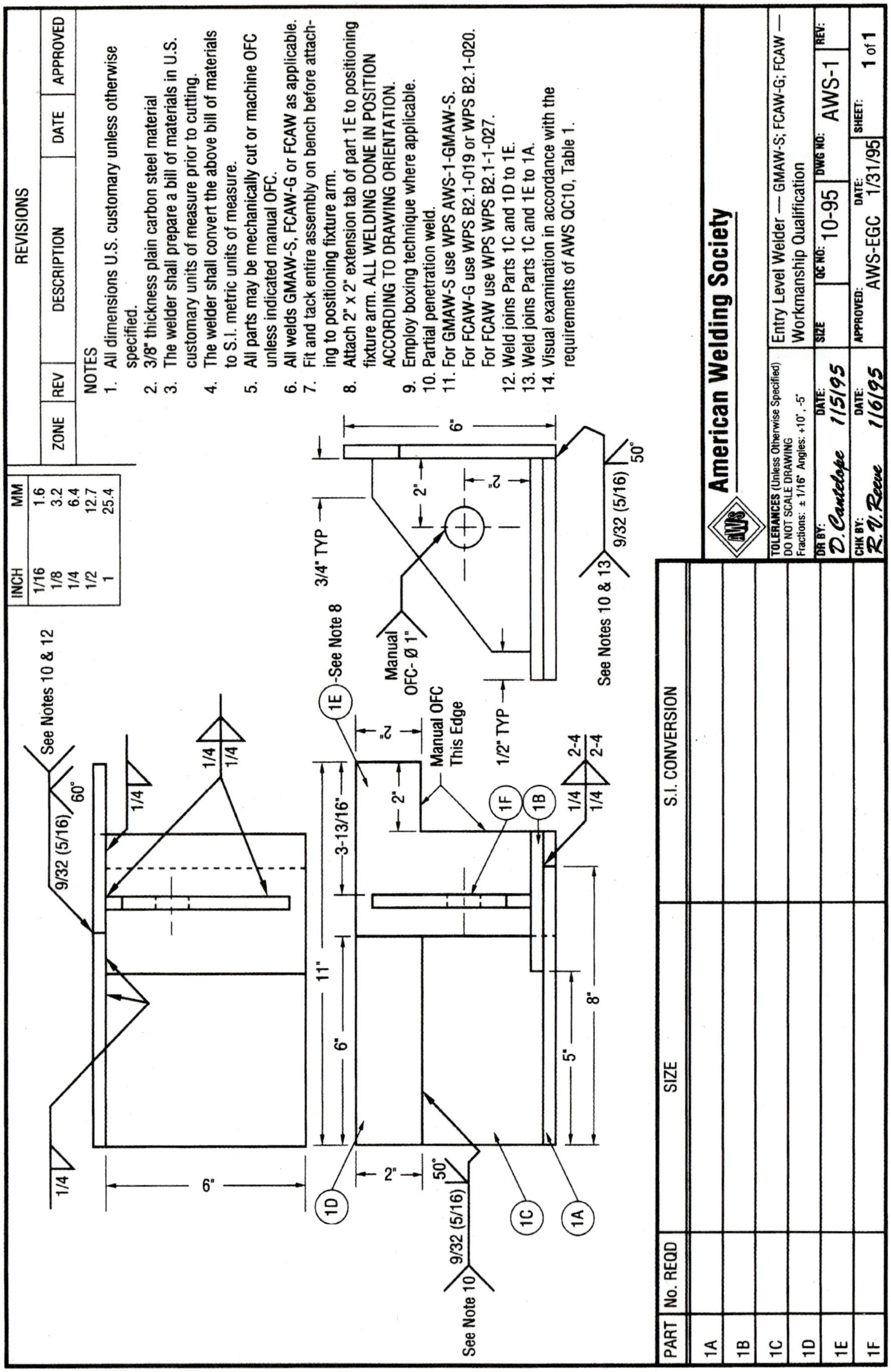

FIGURE 5-33 American Welding Society Workmanship Qualification Test. American Welding Society

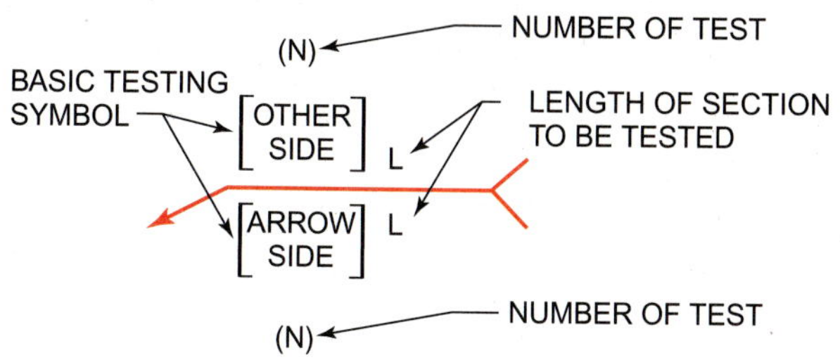

FIGURE 5-34 Basic nondestructive testing symbol.
American Welding Society

uses the same basic reference line and arrow as the welding symbol, Figure 5-34.

The symbol for the type of nondestructive test to be used, Table 5-1, is shown with a reference line. The location above, below, or on the line has the same significance as it does with a welding symbol. Symbols above the line indicate other side, symbols below the line indicate arrow side, and symbols on the line indicate no preference for the side to be tested, Figure 5-35. Some tests may be performed on both sides; therefore, the symbol appears on both sides of the reference line.

Type of Nondestructive Test	Symbol
Visual	VT
Penetrant	PT
Dye penetrant	DPT
Fluorescent penetrant	FPT
Magnetic particle	MT
Eddy current	ET
Ultrasonic	UT
Acoustic emission	AET
Leak	LT
Proof	PRT
Radiographic	RT
Neutron radiographic	NRT

Table 5-1 Standard Nondestructive Testing Symbols.
(American Welding Society)

Two or more tests may be required for the same section of weld. Figure 5-36 shows methods of combining testing symbols to indicate more than one type of test to be performed.

The length of weld to be tested or the number of tests to be made can be noted on the symbol. The length may either be given to the right of the test symbol, usually in inches, or can be shown by the arrow line, Figure 5-37. The number of tests to be

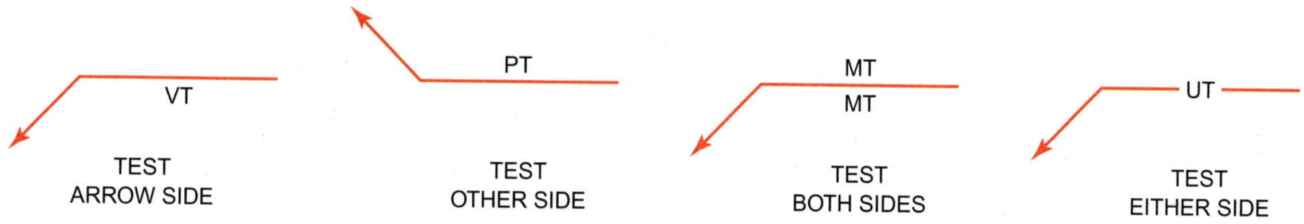

FIGURE 5-35 Testing symbol used to indicate the side that is to be tested. American Welding Society

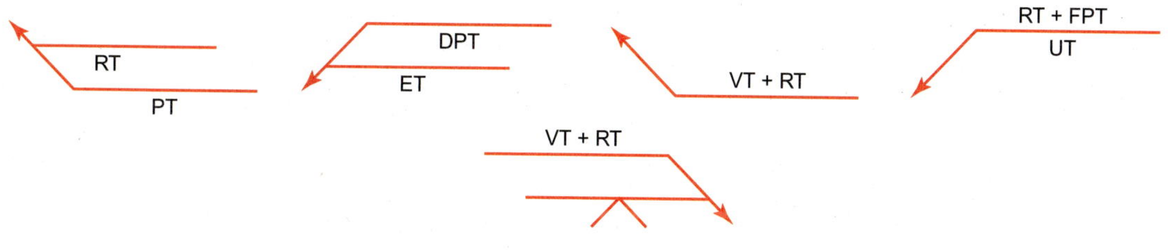

FIGURE 5-36 Method of combining testing symbols. American Welding Society

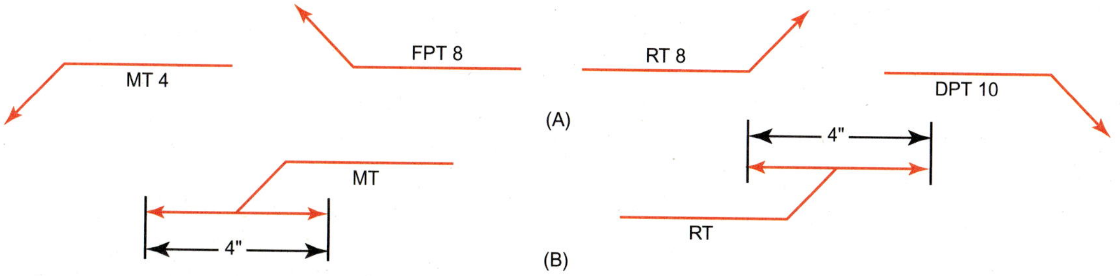

FIGURE 5-37 Two methods of designating the length of the weld to be tested.
American Welding Society

FIGURE 5-38 Method of specifying the number of tests to be made. American Welding Society

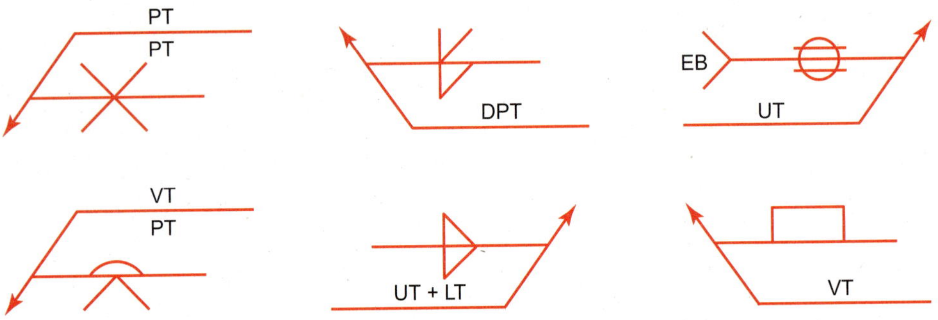

FIGURE 5-39 Combination of weld and nondestructive testing symbols. American Welding Society

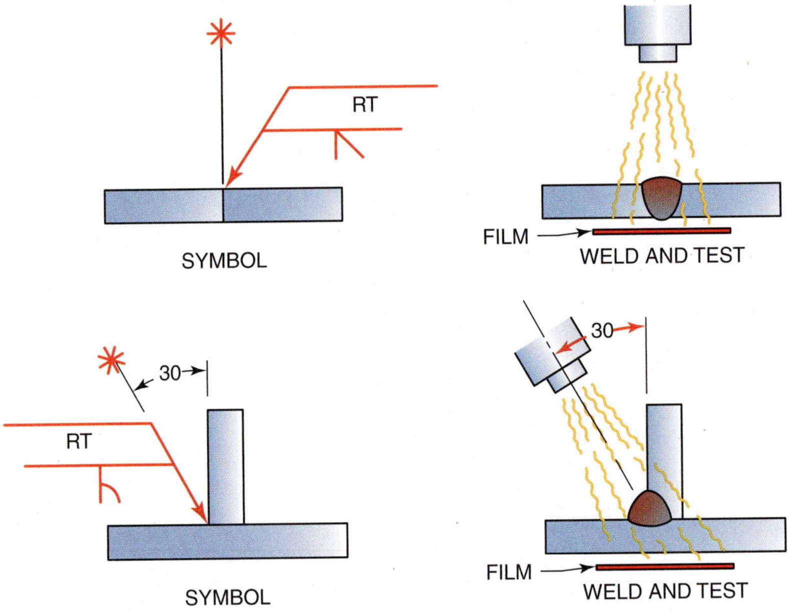

FIGURE 5-40 Combination symbol for weld and radiation source location for testing. American Welding Society

made is given in parentheses ( ) above or below the test symbol, Figure 5-38.

The welding symbols and nondestructive testing symbols can be combined into one symbol, Figure 5-39. The combination symbol may help both the welder and inspector to identify welds that need special attention. A special symbol can be used to show the direction of radiation used in a radiographic test, Figure 5-40.

PRACTICE 5-5

Referring to the weld test shown in Figure 5-39 and using a pencil and paper, sketch a cross section of each of the welds shown by the six welding symbols. Write a sentence for each, describing the testing that is specified in each welding symbol. Refer to Figure 5-21 for an example of how to do this practice.◆

SUMMARY

Understanding the mechanics of joint design is essential so that you can recognize and anticipate the various forces that will be applied to a weldment in the field. For example, a one-sided fillet weld on a tee joint would withstand much more force on the welded side than it would on the back side of the weld. So it is important that you recognize the limitations of weld joint configurations to select the best design to withstand these forces. Engineers use all of the appropriate codes and standards when designing a welded structure. They also have taken into consideration all of the forces that can be anticipated. For that reason it is important that you follow their weld specifications closely.

From time to time, you will be asked to make changes in structures as part of a modification or repair. In the field, you will be expected to understand the types of forces being applied to the weldment, and to determine the best joint design to prevent these forces from causing a structural failure.

REVIEW QUESTIONS

1. List and describe the five basic joint designs.
2. What factors must you consider when choosing a weld joint design?
3. What are the five ways that forces cause stress on a weld?
4. What is a faying surface?
5. Why should you prepare the faying surfaces for welding?
6. What effect does metal thickness have on the joint design?
7. What are some methods of cutting a groove?
8. What is back gouging and what advantages does it offer?
9. What is the most ideal welding position for most joints and why?
10. When welds are made in any position other than a flat position, what are they called?
11. List and identify each of the four plate welding positions.
12. List and identify each of the five pipe welding positions.
13. Name three agencies that issue codes and/or specifications.
14. What does it mean when a joint design has been prequalified?
15. What can affect the cost of producing a weld?
16. What detailed information can the welding symbol include?
17. What are the basic components of a welding symbol?
18. List the types of welds.
19. If the desired weld symbol is placed below the reference line, on which side of the joint should the weld be?
20. If the desired weld symbol is placed above the reference line, on which side of the joint should the weld be?
21. If the tail is added to the basic welding symbol, what information can it provide?
22. List two reasons an intermittent weld is used.
23. How is the strength of seam welds designated?
24. How can cutting a groove in a plate improve a joint's strength?
25. Describe several features of groove welds.
26. What is the purpose of placing a backing on the back side of a weld joint?

Chapter 6

Fabricating Techniques and Practices

OBJECTIVES

After completing this chapter, the student should be able to:

- Explain the various safety issues related to fabrication.
- List the advantages of using preformed parts for fabrication.
- List the advantages of using custom fabrication parts.
- Demonstrate an understanding of the proper placement of tack welds.
- Demonstrate the use of location and alignment points when assembling a project.
- Explain how to adjust parts to meet the tolerance.
- Describe how to control weld distortion.
- Lay out and trace parts.
- Identify common sizes and shapes of metals used in weldments.
- Describe how to assemble and fit up parts for welding.

KEY TERMS

angles
assembly
chalk line
clamps
contour marker
custom fabrication
fabrication
fitting
fixtures
kerf
layout
nesting
scribe/punch
tack welds
template
thermal conductivity
thermal expansion
tolerance
tubing
warp
weld distortion
weldment

INTRODUCTION

The first step in almost every welding operation is the **assembly** of the parts to be joined by welding. At the very basic level, this assembly can be just placing two pieces of metal flat on a table and tack welding them together for practice welding. At a higher level is the assembly of complex equipment, buildings, ships, or other large welded structures. The important thing to remember, however, is that no matter how large or complicated the welded structure, it is assembled one piece at a time. That is true for a simple project you build as part of your welding shop learning in school or for the ship, rocket engine, or building you might construct one day.

FABRICATION

The difference between a weldment and a fabrication is that a weldment is an assembly whose parts are all welded together, but a **fabrication** is an assembly whose parts may be joined by a combination of methods including welds, bolts, screws, adhesives, and so on. All weldments are fabrications, but not all fabrications are weldments.

In addition to straight welding, welders are often required to assemble parts together to form a **weldment.** The weldment may form a completed project or may only be part of a larger structure. Some weldments are composed of two or three parts; others may have hundreds or even thousands of individual parts, Figure 6-1. Even the largest weldments start by placing two parts together.

The number and type of steps required to take a plan and create a completed project will vary depending on the complexity and size of the finished weldment. All welding projects start with a plan. This plan can range from a simple one that may exist only in the mind of the welder, or it can be complex and composed of a set of drawings. As a beginning welder, you must learn how to follow a set of drawings to produce a finished weldment.

We are now fabricating large structures in space such as the International Space Station, Figure 6-2. The station is being assembled in space from large sections that were built here on earth. Most of the assemblies require some type of welding. Someday we expect to be welding in space. Research for welding in space dates back to the 1960s with experiments that were done on board the U.S. Skylab. Today, that research continues with experiments in the Space

FIGURE 6-1 Petrol chemical refinery near Point Comfort, Texas, along the South Texas coast. Larry Jeffus

FIGURE 6-2A The neutral buoyancy tank allows divers to work in spacesuits to simulate the microgravity of space. NASA

FIGURE 6-2B International Space Station. NASA

Shuttle program and in conjunction with the International Space Station project.

Safety

As with any welding, safety is of primary concern for fabrication of weldments. Fabrication may present some potential safety problems not normally encountered in straight shop welding. Unlike most practice welding, much of the larger fabrication work may need to be performed outside an enclosed welding booth. Additionally, several welders may be working on a structure at the same time. You must let the other workers in the area know that you are going to be welding so they can protect themselves from the arc light, sparks, and other possible hazards. Tell them about the hazards because you cannot assume that they know

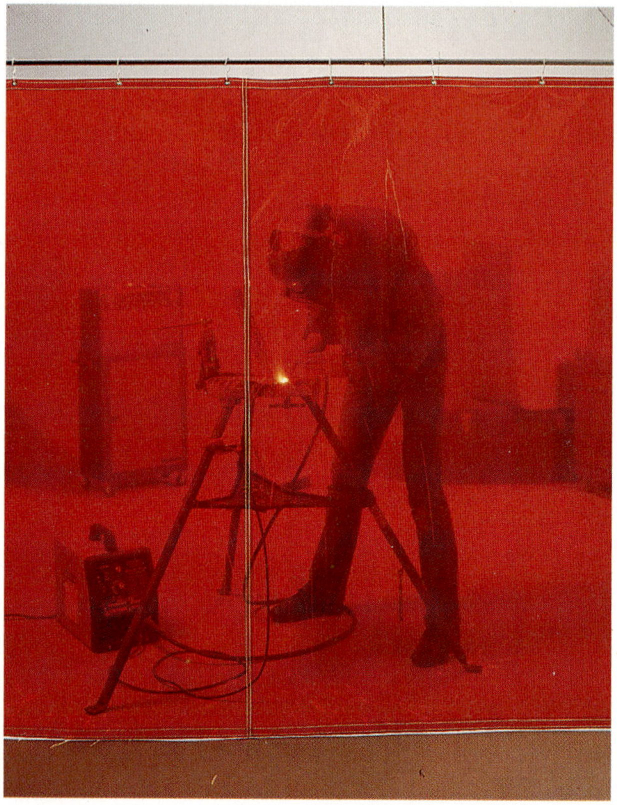

FIGURE 6-3 Portable welding curtain. Frommelt Safety Products

about the hazards of welding. Extra care must be taken to ensure that burns do not occur on you or the other welders from the arc or hot sparks. When possible, you should erect portable welding curtains, Figure 6-3.

> **SHOP TALK**
>
> Did you know that conflicts can arise because you are trying to do your job?
>
> Sometimes conflicts occur between a welder working outside of a welding shop and others working in the same area. Sometimes these conflicts can occur between different trade groups or with individuals who feel you are in their space. Keep the lines of communication open by asking them what their needs are for time, access, space, and other pertinent requirements. It may be necessary to have a group meeting among all the parties affected to resolve any conflict.

Ventilation is also important because the normal shop ventilation may not extend to the fabrication area. A portable fan may be needed to help blow the welding fumes away from the work area. Be sure the fan blows the fumes away from you and others.

Often you will be working in an area that has welding cables, torch hoses, extension cords, and other trip hazards lying on the floor. These must be flat on the floor and should be covered if they are in a walkway to prevent accidental tripping. Keep all of the scrap metal and other debris picked up; a neat work area is a safe work area.

As the fabrication grows in size, it will become heavier. Make sure it is stable and not likely to fall. A weldment that starts out stable and well-supported can become unstable and likely to fall as it grows in size. Keeping it well-supported is important especially if you have to crawl under it to work on the bottom side welds. Check with your supervisor or shop safety officer before working under any weldment.

These and other safety concerns are covered in Chapter 2, "Welding Safety." You should also read any safety booklets supplied with the equipment before starting any project.

PRACTICE 6-1

Conflict Resolution

Form a small group and divide up into two or more factions representing differing viewpoints that might be in conflict on a jobsite. One faction may be the electricians, carpenters, plumbers, or other skilled trades; another might be the owner or supervisor over the area; and the last faction would be the welder.

Using the role-playing technique and discuss various situations that might cause conflicts and how these may be resolved so that everyone feels there is a win-win result. Ask your instructor to help if you have difficulty developing the possible conflict scenarios. Have each person in the group write a short paragraph describing how the conflict was resolved.◆

PARTS AND PIECES

Welded fabrications can be made from precut and preformed parts, or they can be made from parts cut and formed by hand. In most cases, weldments are made using a variety of precut and preformed parts along with handmade pieces.

Preshaped pieces may be precut, bent, machined, or otherwise prepared before you receive them. This is a common practice in large shops and on large-run projects. Large shops may have an entire department dedicated to material and part preparation. When a large number of the same items are made in a large-run shop, the shop may outsource some of the parts to shops that specialize in mass producing items. When making an assembly with precut and formed parts, little or no on-the-job **fitting** may be required. That, of course, depends on how accurately the parts were prepared.

The opposite end of the spectrum from preformed parts is **custom fabrication,** in which all or most of the assembly is handmade. This might include cutting, bending, grinding, drilling, or other similar processes. Almost all weldments were produced by hand until the introduction of automated equipment for cutting, bending, and machining. Today, many items, including some large machines and almost all repair work, are still custom-fabricated.

Advantages of using preformed parts include the following:

- Cost—Shops that specialize in cutting out mass numbers of similar parts can do it less expensively than a shop that makes the same parts one-by-one by hand.
- Speed—High-speed cutting and forming machines can produce a large number of items quickly.
- Accuracy—Automated equipment can make parts more accurately than they can be made by hand.
- Less waste—The wise use of materials is important to both control cost and to conserve natural resources.

Advantages of custom-fabricating parts include the following:

- Originals—It is not practical to set up an automated process when there will be only one of a kind or a limited number of an item produced.
- Prototypes—Often, even if there are going to be thousands or even tens of thousands of a weldment produced, the first one, the prototype, must be made by hand to be sure that everything works as it was planned.
- Repairs—Seldom would it be necessary to make a large number of the same part or piece when making a repair on a damaged or worn item.
- Custom jobs—Sometimes people want to have something special or unique built or modified just for them.

TACK WELDS

Tack welds are the welds, usually small in size, that are made during the assembly to hold all of the parts of a weldment together so they can be finish-welded.

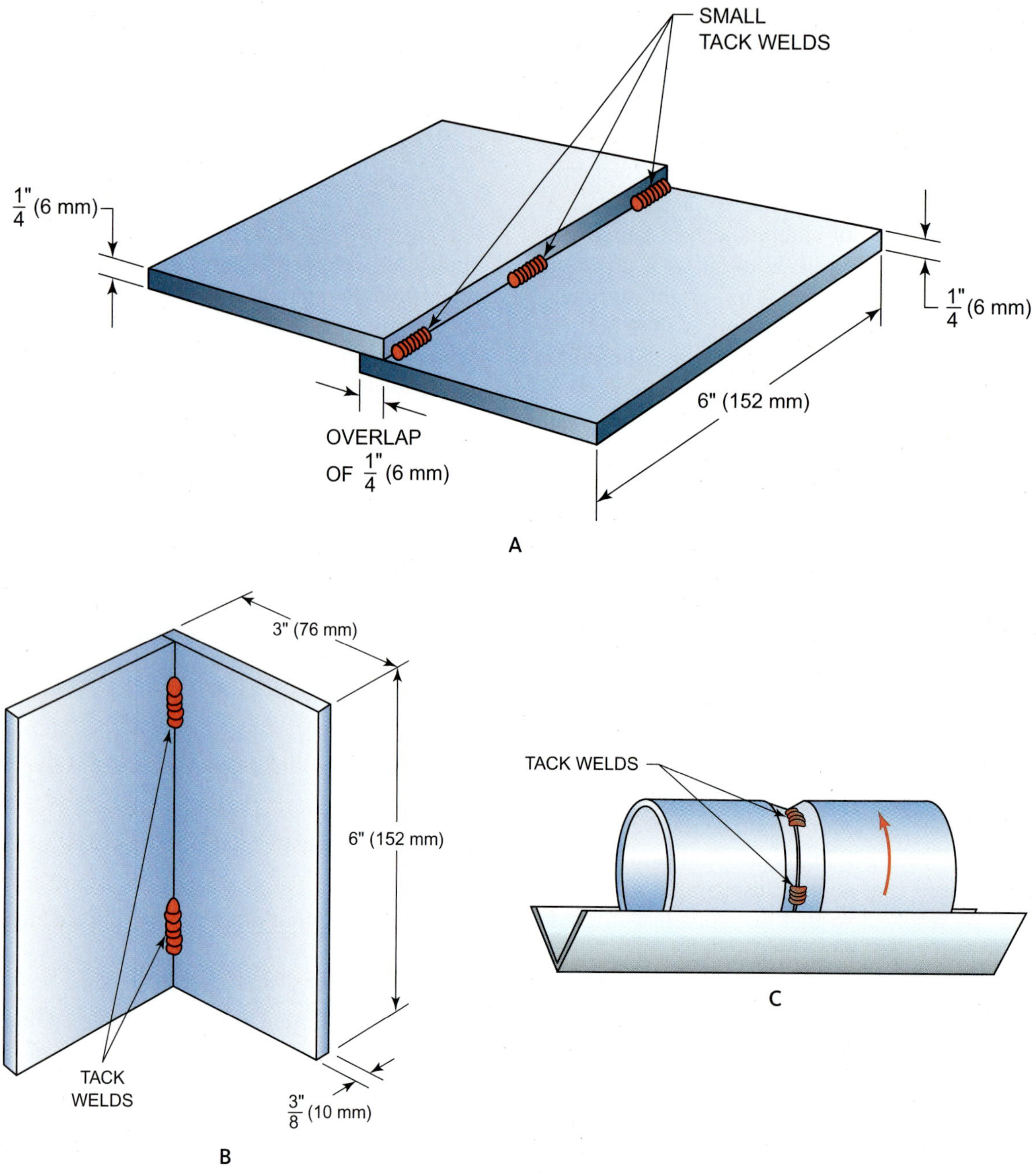

FIGURE 6-4 Tack welds. © Cengage Learning 2012

Making good tack welds is one of the keys to assembly work. Tack welds must also be small enough to be incorporated into the final weld without causing a discontinuity in its size or shape, Figure 6-4. They must be strong enough to hold the parts in place for welding but small enough so that they become an unseen part of the finished weld. Deciding on the number, size, and location of tack welds takes some planning. Some of the factors to consider regarding the number of tack welds include the following:

- Thickness of the metal—A large number of very small tack welds should be used on thin metal sections, while a few large tack welds may be used for thicker metal parts.
- Length and shape of the joint—Obviously, short joints take fewer welds, but some long, straight joints may have very few tack welds compared to a shorter joint that is very curvy.
- Welding stresses—All welds create stress in the surrounding metal as they cool and shrink. Larger welds

produce greater stresses that might pull tack welds loose from an adjoining part if the tack welds are not strong enough to withstand the welding stresses.

- Tolerances—The more exacting the tolerance for the finished weldment, the more tack welds are required.
- Fit-up—When custom bending parts during the fit-up process, it may be necessary to use a large number of small tack welds to keep the parts in alignment and make the bends more uniform.

Tack welds must be made in accordance with any welding procedure with an appropriate filler metal. They must be located well within the joint so that they can be completely remelted into the finished weld. Posttack welding cleanup is required to remove any slag or impurities that may cause finished weld flaws. Sometimes the ends of a tack weld must be ground down to a taper to improve their tie-in to the finished weld metal.

Make sure that your tack welds are not going to be weld defects in the completed weldment. A good tack weld is one that does its job by holding parts in place, yet is undetectable in the finished weld.

On harder metals like steel, you can often hear a tack weld break; however, for some soft metals like aluminum, the tack weld may separate quietly. Depending on the type of metal and its size and thickness, a breaking tack weld can make a small, sharp snap or a deep, resounding thump. You might hear one break while you are welding or sometime afterward. Do not assume that a broken tack weld has no effect and continue to weld. A broken tack can allow parts to shift well out of tolerance. If you continue to weld, it may be impossible to pull the loose part back into position, which could result in the weldment not meeting its specifications. Sometimes this is referred to as "making scrap metal" and not a weldment.

LOCATION AND ALIGNMENT POINTS

Locating parts is easier when the parts being assembled are lined up on an edge starting at a corner, Figure 6-5. This makes it fairly easy for the assembler to put the pieces in their proper positions. However, if the parts being assembled are to be fitted in the middle of a surface or edge, placing them accurately becomes more difficult. Sometimes there are alignment slots, notches, or points made onto the parts to aid during the assembly, Figure 6-6.

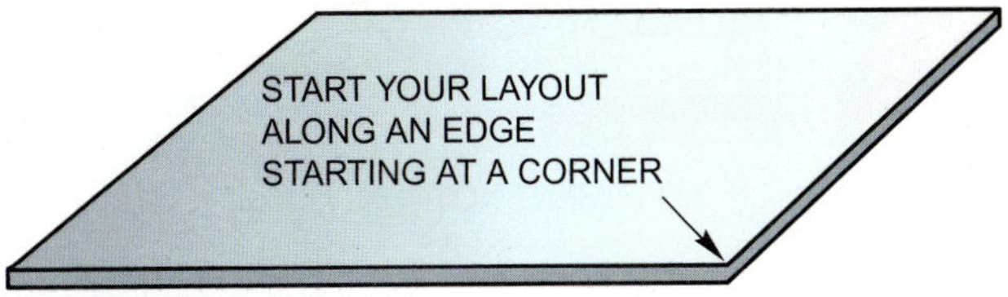

FIGURE 6-5 Laying out parts squarely is easier when they are placed along an edge or at the corner of the plate.

Locating parts along an edge or at the corner is the easiest way to determine where they are to be aligned. You must look at the drawing to see how the edges are fitted. This is much more important on thicker materials than it is with thin stock, Figure 6-7A. Of course, even on thin material it can be important if the part is to be built to a very tight tolerance. In that case, if the joint should be assembled as shown in Figure 6-7A but is assembled as shown in Figure 6-7B; the overall length in one direction decreases by the thickness of the material, and in the other direction the dimension increases by the thickness of the material.

On thicker materials it is easy to see how the overall dimensions of a weldment could change if the parts are not properly aligned during fit-up. But in addition to not being the correct size, sometimes the weld itself will not be as strong if the joint is not aligned properly. The reason the weld might not be as strong as it is designed to be is because the tensile strength of thick metal plate differs depending on the direction in which the load is placed on the plate as compared to the rolling direction of the plate, Figure 6-8A. Metal plate, much like wood, will break in one direction easier than in another. In this aspect, steel plate is similar to a wooden board in that the direction of the rolled grain of a plate and the direction of the wood grain in a board affect their strength. Many common materials have a grain; for example, when you tear a newspaper down the page, it tears fairly easily and straight. However, when you try to tear it across the page, the tear is much more jagged, Figure 6-8B.

OVERALL TOLERANCE

When fabricating a weldment that is made up of a number of parts all welded together, there is a potential problem that the overall size of the weldment can be wrong. As discussed in Chapter 3, every part

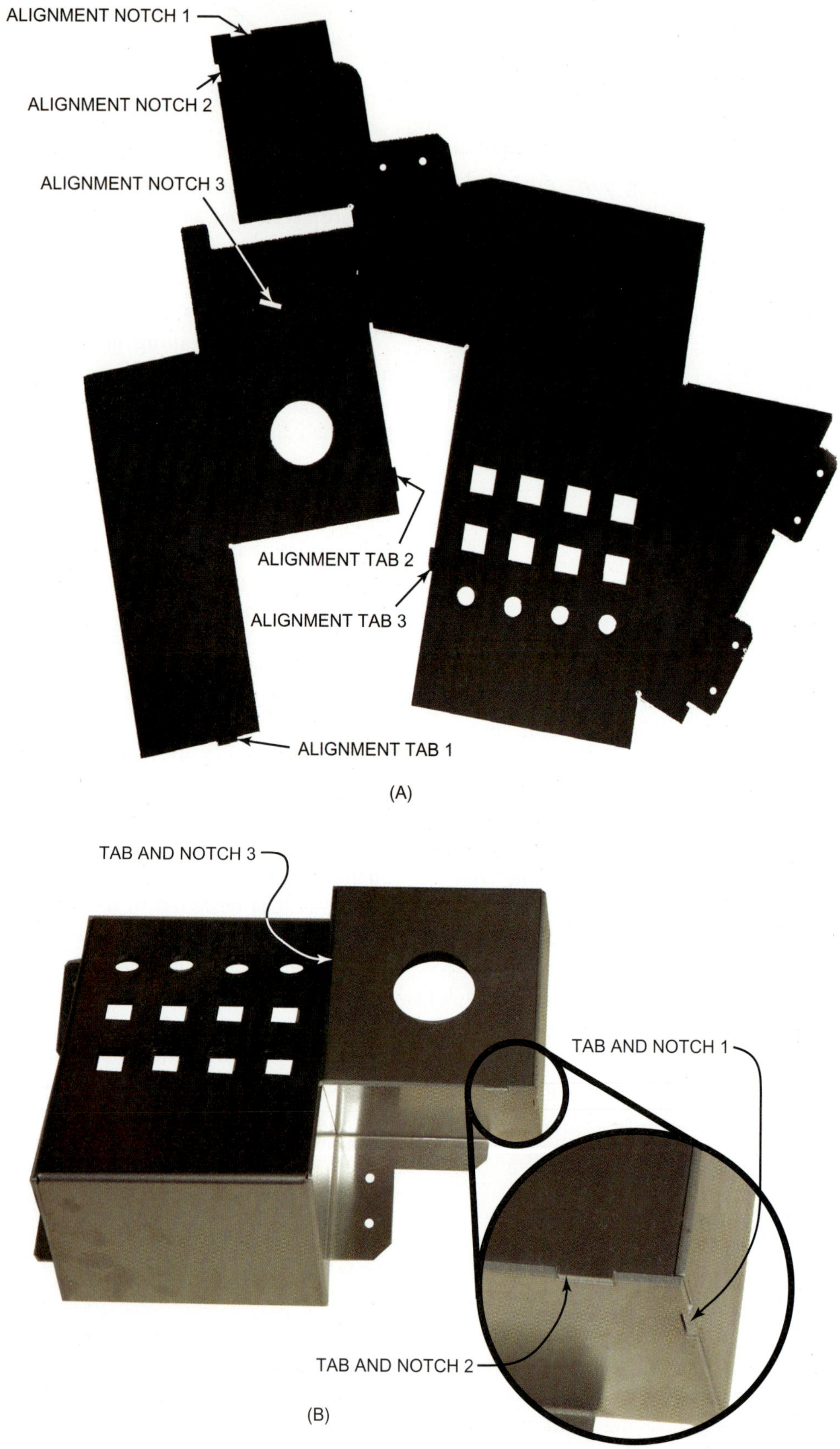

FIGURE 6-6 Complex parts that are to be formed may use alignment slots, notches, or points. Larry Jeffus

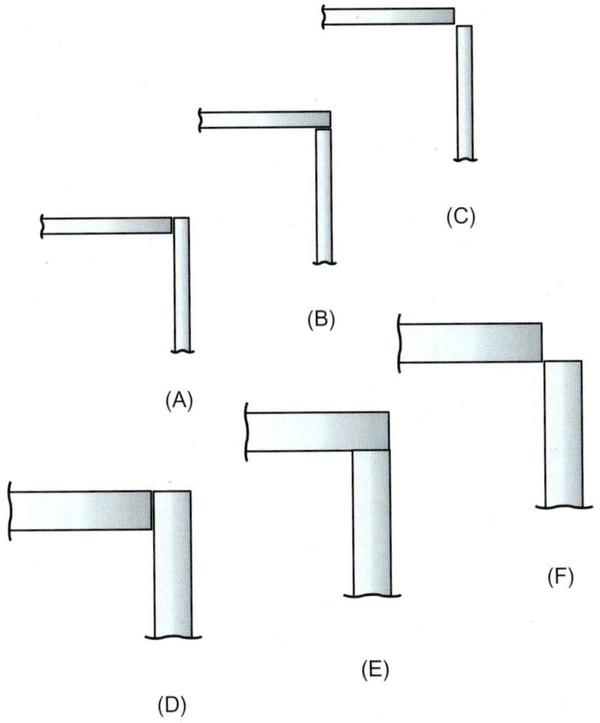

FIGURE 6-7 **Part placement during assembly can affect the overall dimension, especially when the metal being assembled is thick.** © Cengage Learning 2012

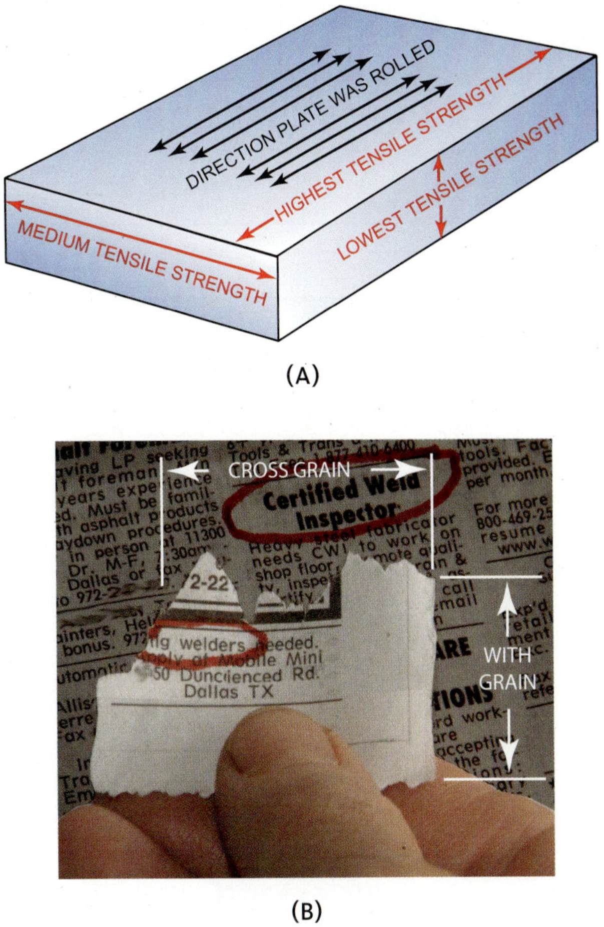

FIGURE 6-8 **(A) A metal's strength is affected by the direction it is rolled because of its grain structure. (B) The effect of grain structure can be seen on a newspaper as it is torn.** Larry Jeffus

manufactured has a **tolerance.** A part's tolerance is the amount that a part can be bigger or smaller than it should be and still be acceptable. The more exact a part's tolerance, the more time it takes to make and, therefore, the more it costs. In most cases, the welding engineer has calculated the effect of these slight variations in size when designing a weldment. As the weldment fabricator, you must take these tolerances into consideration as you make the assembly to ensure that the overall size of the weldment is within its tolerance.

As the number of parts that make up a weldment increases, the problem of compounding the errors increases. For example, if there are 8 parts and each part is 1/8 in. (3 mm) larger than its ideal size but within its ±1/8 in. (3 mm) tolerance, the overall length of the finished weldment could be 8/8ths or 1 in. (25 mm) too long. Likewise, if each of the parts was 1/8 in. (3 mm) shorter, the overall length would be 1 in. (25 mm) too short, Figure 6-9. Therefore, an assembler must be mindful of both the size of each part and the overall size of the assembly. In addition to tolerances for size, parts also have angle tolerances. For example, each of the 10 pieces that make up the star in Figure 6-10 are off by only 1°. But as you can see, when they are assembled, the last corner does not fit, making the weldment unacceptable.

Ideally, all the parts for a weldment fit up perfectly; however, in reality that does not always happen. You cannot just throw out all of the parts that do not fit in order to find the ones that would make the perfect star. That is especially true if the parts are made within the correct tolerance. When parts like the ones in this star are made on the shorter side of the tolerance, you might be able to "loosen up" the joint tolerance to make the overall star work. Welded joints, like parts, have tolerances. By slightly adjusting the alignment of each of the 10 pieces, the star can be made within its overall acceptable tolerance. In this case, by making sure that all the joints stay within tolerance, the complete star can be made without having to recut any of the parts. If you can make this assembly by adjusting the joint tolerance, you can assemble

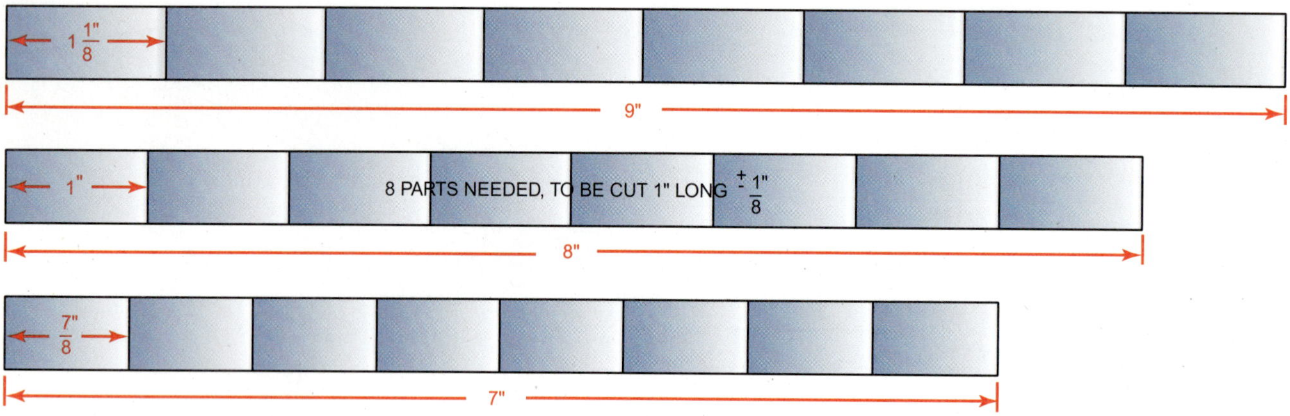

FIGURE 6-9 Lay out the parts to avoid compounding dimensioning errors.

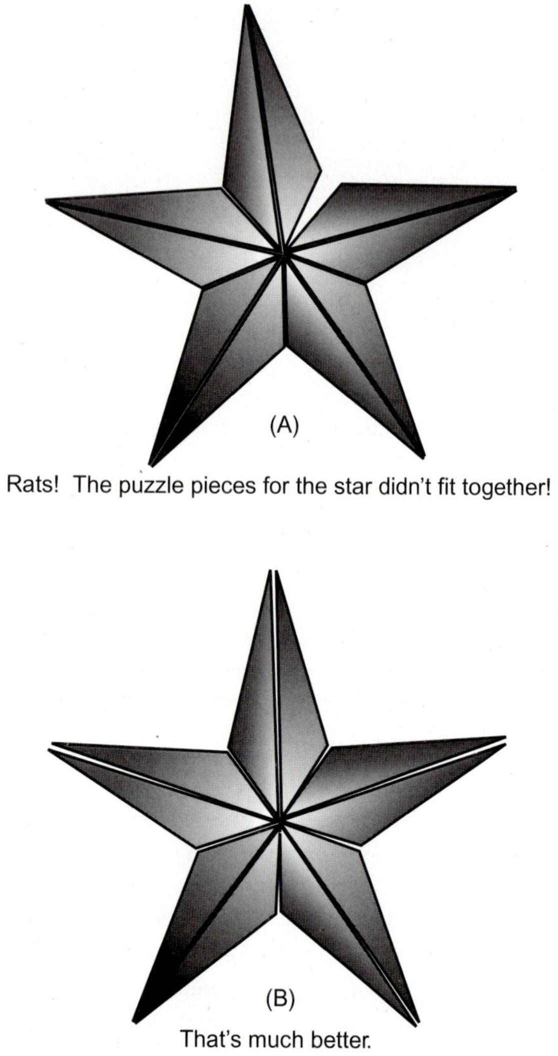

FIGURE 6-10 Small errors on lots of parts can become a big error on the finished assembly.

it faster, and because each of the parts is exactly the same, it will look perfect, Figure 6-10.

What else could be done to make this star fit up? Well, if the height tolerance allows some adjustment, then the fit-up can be made even better. As the height is reduced, the gap at the inside edges will close. So, by slightly flattening the star, its fit-up can be improved.

Whenever possible, try to get the parts to fit without having to recut or grind them; but remember, the finished weldment must be within tolerance. You want to avoid recutting and grinding because both will add time and cost to the finished project. However, you must remember that in some cases the only way that the weldment can be assembled within overall tolerance is to recut or grind some or all of the parts. If the finished weldment is not within tolerance, it may be unusable. If you must grind a part to fit, try to do as little grinding as possible to get the parts to fit up. Hand-grinding is a time-consuming operation, and as Benjamin Franklin once said, "Remember that time is money."

Where you recut or grind a part can sometimes greatly affect the time required. For example, if the pieces used in the star need to be ground to fit, you might want to grind along the short side, which would be faster, Figure 6-11A. In addition, it might be possible to grind only part of the edge to get the parts to fit up within tolerance. In the case of the parts shown in Figure 6-11B, the required root opening tolerance of 1/4 in. (6 mm) ±1/8 in. (3 mm) will allow the part's edge to be ground unevenly as long as the root opening tolerance is maintained, Figure 6-11C. Note how the root opening varies but stays within the acceptable tolerance. The root opening is 1/4 in. (6 mm) at

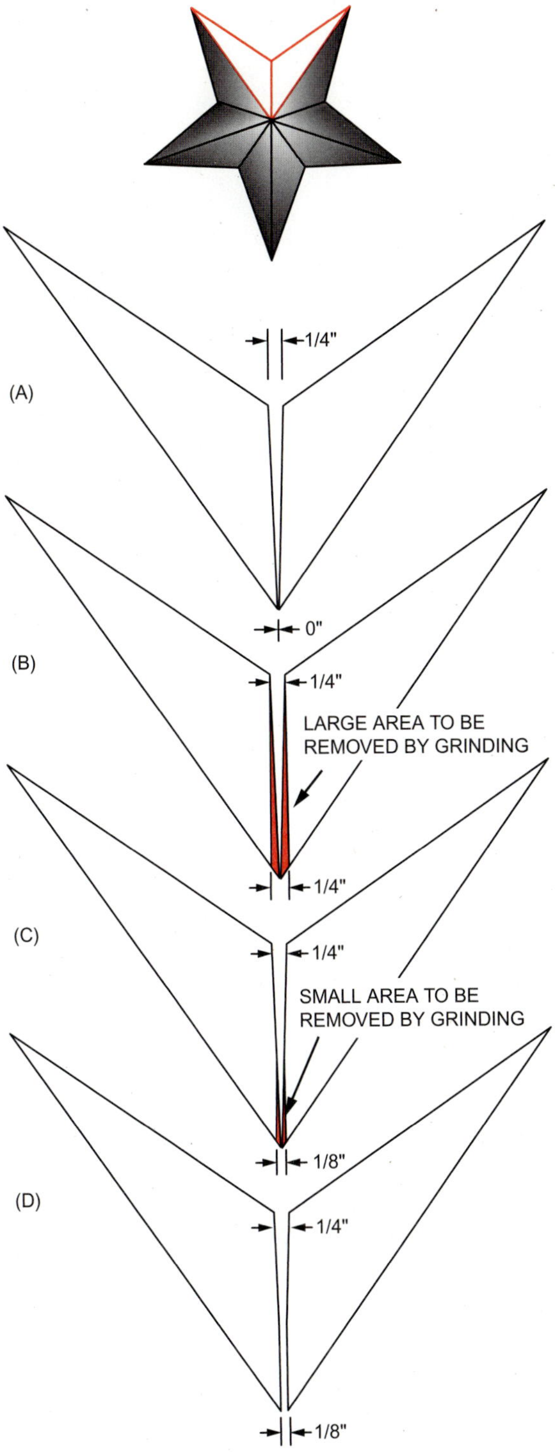

FIGURE 6-11 Trimming parts efficiently can save time.

one end, which is acceptable; but at the point where it becomes too close (less than 1/8 in. [3 mm]), begin grinding. The result as shown in Figure 6-11D meets the part's fit-up specifications and requires a minimum amount of grinding.

SHOP TALK

Do you know why it is so important to make the root opening exactly as it is specified on the drawings?

If the root opening is too narrow, you may not be able to get 100% root penetration with the weld as required. In some cases, a welder may be able to push the weld through the narrower root opening and get the penetration. To get the penetration, some aspect of the welding procedure specification would have to be changed. It may require that a higher-than-normal current be used or a different electrode size be used than is specified.

If the root opening is too wide, you may be able to make the weld, but it would be too large. Larger welds result in more filler metal being added and more heat input to the base metal. Larger welds may cause greater **weld distortion** and have larger heat-affected zones. Both can result in a weld that will not withstand the part's designed strength specifications.

Even a quick visual inspection by a welding inspector would reveal that the weld is unacceptable and must be repaired. You cannot deviate from the root opening specified in the welding procedure; to do so is wrong.

WELD DISTORTION

To make it easier to understand what is happening to the metal during welding, we must first define the terms used. Most dictionaries define the terms *distortion* and **warp** and the terms *distorted* and *warped* very similarly. However, in this textbook the terms *distortion* and *warp* will be used as active and temporary events such as "the part warps during welding" or "weld distortion affects the weldment." In these cases, it should be understood that once the welding is over, the metal will return to nearly its prewelded shape. The terms distort*ed* and warp*ed* as in "the weld distorted the plate" or "the weld warped the weldment" are past tense and refer to the fact that the postwelded metal, after it has cooled, has been significantly misshapen as a result of the welding process.

All metals distort by expansion when heated and distort by contraction when cooled. Parts return to their original shape when cooled if the heating is uniform and their shapes are symmetrical. However, if the heating and/or the parts' shapes are not symmetrical, the parts to some degree will be permanently distorted as the result of the heat/cooling cycle. Almost

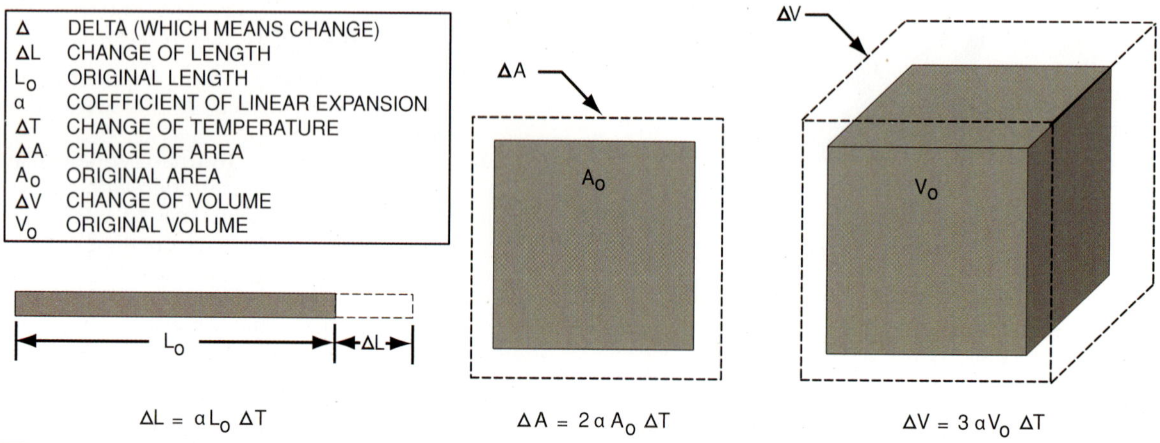

PROPERTIES OF METALS

TYPE OF METAL	COEFFICIENT OF LINEAR EXPANSION X 10^{-6} PER DEGREE		RELATIVE THERMAL CONDUCTIVITY COPPER = 1
	in/°F	mm/°K	
ALUMINUM	13.0	24.0	0.52
BRASS	11.0	19.0	0.28
BRONZE	16.6	29.9	0.15
COPPER	9.4	17.0	1.00
GOLD	8.2	14.2	0.76
GRAY CAST IRON	6.0	10.8	0.12
INCONEL	7.0	12.6	0.04
LEAD	15.1	28.0	0.08
MAGNESIUM	14.3	25.7	0.40
MONEL	7.5	13.5	0.07
NICKEL	7.4	13.3	0.16
SILVER	10.0	18.0	1.07
STEEL, LOW CARBON	6.7	12.1	0.17
STEEL, MEDIUM CARBON	6.7	12.1	0.17
STEEL, HIGH CARBON	7.2	13.0	0.17
STEEL, STAINLESS			
AUSTENITIC	9.6	17.3	0.12
MATENSITIC	9.5	17.1	0.17
FERRITIC	5.5	9.9	0.17
TANTALUM	3.6	6.5	0.13
TIN	13.0	23.4	0.15
TITANIUM	4.0	7.2	0.04
TUNGSTEN	2.4	4.3	0.42
ZINC	22.1	39.8	0.27

Table 6-1 Thermal Expansion Properties of Metals

every welding process involves some heat cycling. Most welding heat cycling is not symmetrical; so as a result, the weldment will be distorted to some degree.

The two factors that affect the degree to which a metal will distort and possibly remain distorted are its rate of **thermal expansion** and its rate of **thermal conductivity,** Table 6-1. Basically, the higher the coefficiency of thermal expansion, the greater the metal distorts. From Table 6-1, we see that tungsten has the smallest coefficiency of expansion, and zinc has the largest coefficiency of expansion. What this means is that if the same size pieces of tungsten and zinc are heated to the same temperature, the zinc will expand a lot more. Using the following linear expansion formula:

$$DL = \alpha L_O \Delta T$$

Where:

ΔL = the change in length
α = the coefficiency of linear expansion
L_O = the original length
ΔT = the change in temperature, which is the difference between the beginning temperature and the ending temperature ($T_1 - T_2$). (A change in temperature is always expressed as a positive number whether the temperature was increased or decreased.)

Using the formula just given, calculate the change in length of a 100-in. (2540-mm) long tungsten bar and the change in length of a 100-in. (2540-mm) long zinc bar. The beginning temperature for both bars is 70°F, and the ending temperature is 570°F. Use Table 6-1 to determine the coefficiency of thermal expansion for each metal.

$$\Delta L = \alpha L_O \Delta T$$

From Table 6-1: α (alpha) for tungsten is $2.4^{*10^{-6}}$ or 0.0000024 and for zinc is $22.1^{*10^{-6}}$ or 0.0000221.

Tungsten	**Zinc**
$\Delta L = \alpha L_O \Delta T$	$\Delta L = \alpha L_O \Delta T$
$\Delta L = 0.0000024 \times 100 \times 500°F$	$\Delta L = 0.0000221 \times 100 \times 500°F$
$\Delta L = 0.12$ in. (3 mm)	$\Delta L = 1.105$ in. (28 mm)

As you can see, the α for zinc at 22.1 is about 10 times greater than the α for tungsten at 2.4; and the expansion in length for the zinc bar was about the same—10 times greater than the tungsten bar. The length of the heated tungsten bar would be 100.12 in. (2543 mm), and the length of the heated zinc bar would be 101.105 in. (2568 mm).

Some metals have high values of coefficients of expansion and others have much lower values. The metals with the higher values tend to distort more during welding and those with lower values distort less.

SPARK YOUR IMAGINATION

All metals expand when heated and contract when cooled. An example of how we use thermal expansion in our daily life is when we hold a sealed jar lid under the hot water faucet to loosen the seal for easier opening.

List some other examples of thermal expansion or contraction that we might observe in our daily life. Hint: All materials, not only metals, expand and contract with temperature changes.

EXPERIMENT 6-1

Thermal Expansion

In this experiment, you will observe the effect that heating has on metals having different coefficiencies of thermal expansion. You will need three bars of metal of about the same size and length (2 ft [0.61 m] to 3 ft [0.914] long), one each of aluminum, steel, and stainless steel; a vise; a measuring tape; an oxyfuel torch; required safety equipment; and pencil and paper. Working in a small group of three or four students, select a leader and have the leader assign tasks to each group member. Someone will need to work the torch for heating the metal bars, one or two people will need to do the measuring, and someone will need to record the observations.

Wearing all required safety equipment and following all shop and equipment manufacturer safety procedures, clamp all three metal bars horizontally in a vise so that you can measure the movement at the far end of the bars as they are heated and cooled, Figure 6-12. Measure and record the distance from the bar tips to the floor or workbench top. Next, with the metal bars clamped in place and the oxyfuel torch set up, lit, and adjusted according to all manufacturer and shop safety requirements, begin heating all three bars at the same time by moving the torch flame back and forth across all of them about 2 in. (51 mm) from the vise. When the metal bars have been heated a little, remeasure the tip distances and record the distance next to the first measurements. Continue heating the metal and take two more measurements. Be careful not to overheat and melt the aluminum; it melts at a lower temperature than the other metals.

Let the bars cool, then remeasure the tip distance. You may even use a fine mist of water from a spray bottle to speed the process. If you use water, be very careful because when the water hits the hot metal, it may turn into steam. Steam can easily burn you.

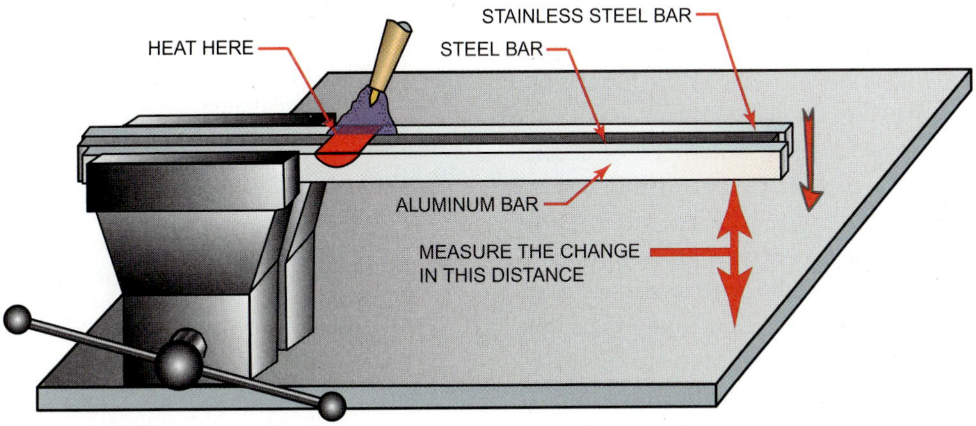

FIGURE 6-12 Experiment 6-1: Thermal Expansion. © Cengage Learning 2012

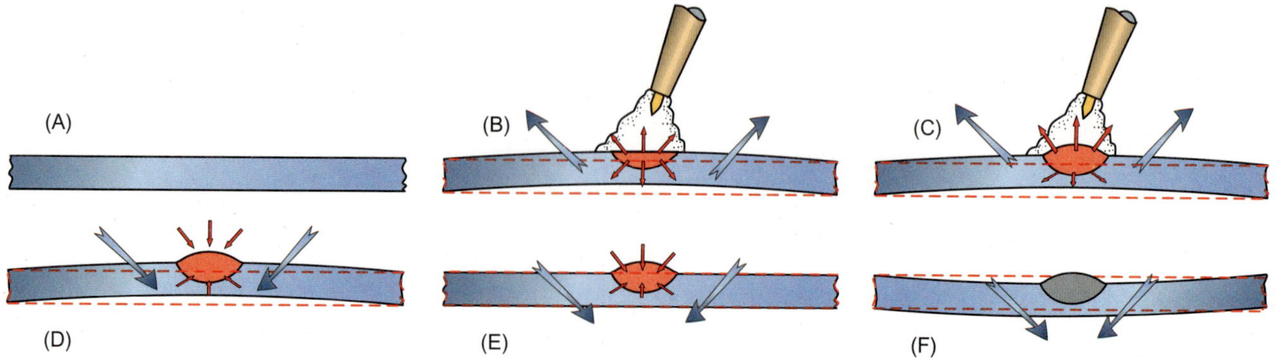

FIGURE 6-13 Heating or welding on metal causes it to distort. © Cengage Learning 2012

As a group, discuss the following questions and write your answers to the questions on a sheet of paper to be turned into your instructor.

1. Did all of the bars move the same distance?
2. If not, which bar moved farthest? Which bar moved the least?
3. Did all of the bars move at the same rate?
4. When the bars cooled, did they return to their original location?
5. If not, how did their final cool position compare to their original position?
6. How did what you observed in this experiment compare to the metals' coefficiencies of expansion as shown in Table 6-1?

Thermal conductivity is another factor that affects the degree of distortion experienced during welding. As you remember, the more uniform a part is heated, the more uniform the expansion and contraction, which can result in the metal being less distorted following the heat–cooling cycle. As a way of showing the comparison of how a metal conducts heat, other metals are related to copper. Copper is given the value of 1.0. Most metals conduct heat a little less quickly than copper. Silver at 1.07 conducts heat a little faster than copper. So, when looking at Table 6-1, you see that all the metals listed have values less than 1.0 and conduct heat slower than copper. Metals with low thermal conductivity like austenitic stainless steel at 0.12 and a higher coefficiency of expansion of 9.6 in./°F (17.3 mm/°K) tend to be more distorted during welding.

You cannot usually change from one metal that is known for its problem with distortion to one that is less likely to be distorted. If the specifications call for the part to be made out of stainless steel or any other metal that tends to be distorted from welding, that is the metal you have to use. But there are ways of controlling distortion to keep the distorted part within tolerance. Weld shrinkage is the primary reason welds cause metal to be distorted.

If you take a flat plate and begin to heat it in the center to make a weld, it begins to expand away from the heat, Figure 6-13A. Initially, the expansion is uniform in all directions and the plate is bent away from the heat, Figure 6-13B. But as the metal gets even hotter, it becomes softer and it reaches the molten state where it becomes fluid. It no longer is strong enough to bend the plate, so it begins to expand outward from the plate, Figure 6-13C. The result is that the hot metal has actually expanded outward more than it has expanded inward toward the cooler surrounding metal. At this time the bending of the plate stops. Once the heat is removed, the metal begins to cool and shrink uniformly, Figure 6-13D. This uniform shrinkage causes the plate to bend toward the weld, Figure 6-13E. Because the heat caused the metal to become a little thicker where it was heated, there is more metal there to shrink. As the plate cools, it bends toward the heated side, resulting in a slight bow in the plate, Figure 6-13F. To recap the process, as metal is heated, it initially bends away from the heat, but when it is allowed to cool, it bends even farther back toward the heated spot.

To learn how to control distortion, you need to know the factors in addition to conduction and expansion that affect the degree that distortion occurs during welding. Following are some of the factors that affect the degree at which distortion occurs:

- Heat input—The lower the heat input to the part, the less the distortion.
- Metal shrinkage—As weld metal cools, it contracts or pulls together, causing distortion. Small welds produce less weld shrinkage.

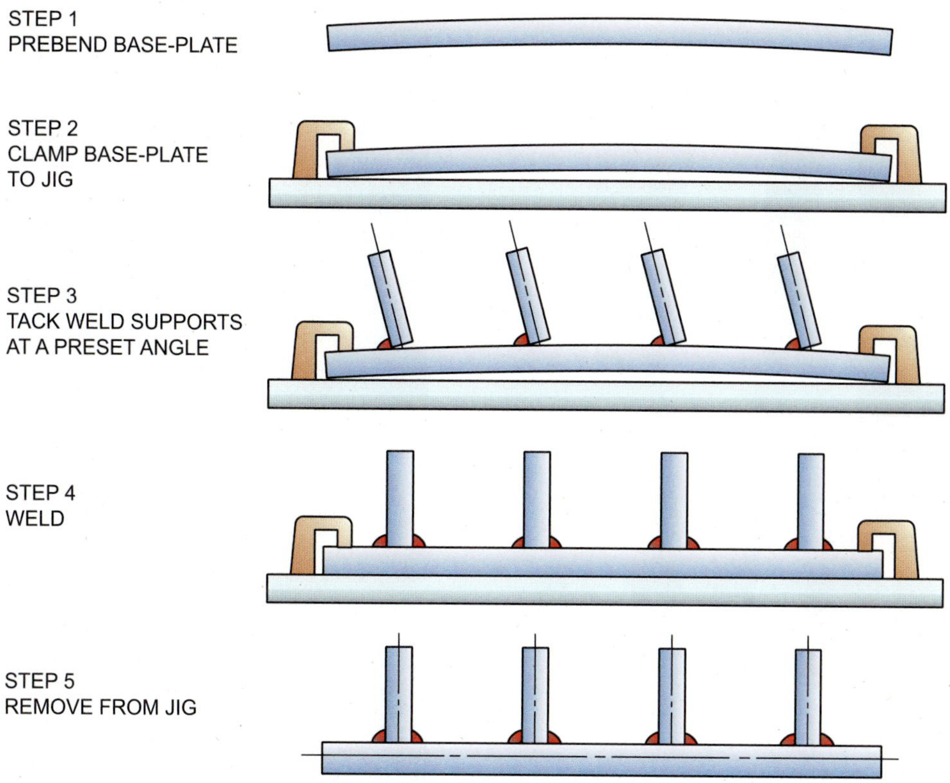

FIGURE 6-14 **Jigs and fixtures can be used to minimize weld distortion.** © Cengage Learning 2012

- Weld uniformity—Welds that are more uniform in shape or symmetrical like U-groove welds cause the metal to expand uniformly and contract uniformly, resulting in less distortion.
- Small welds—If a large weld is required, it can be built up from a series of small welds. Each of the small welds reduces the stress caused by the previous weld, thus reducing the total degree of distortion caused by making a large weld.
- Offset the parts before welding—In Figure 6-14, the plates are slightly offset by a slight angle. Without this offset, the weld would bend past the 90° desired angle as it cooled.◆

LAYOUT

Parts for fabrication may require that the welder lay out lines and locate points for cutting, bending, drilling, and assembling. Lines may be marked with a soapstone, metal marker, a **chalk line,** scratched with a metal scribe, or punched with a center punch. Soapstone and welding metal markers, Figure 6-15, are made to withstand most welding and cutting temperatures without vanishing like a line from a felt-tip marker will. A potential problem when using markers not specifically manufactured for welding is that they may contain elements like sulfur that can contaminate welds. Use only approved materials for marking metal for welding or cutting.

When marking a straight edge on a metal part, a small gap can result if the soapstone point is not sharp enough to fit tightly up against the straight edge, which can cause the parts to not fit together properly. The end of the soapstone should be sharpened to increase accuracy. You can use a grinder to sharpen it or just rub it on some scrap metal or rough concrete, Figure 6-16.

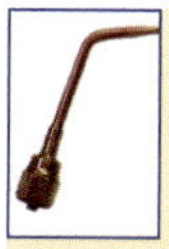

SHOP TALK

Do you know you should keep your work area neat?

Soapstone dust from sharpening can leave a mess. If you rub it on a rough surface, make sure it is not going to be something that will look unsightly once you are finished. Do not make a mess.

A chalk line will make a long, straight line on metal and is best used on large jobs with long, straight lines,

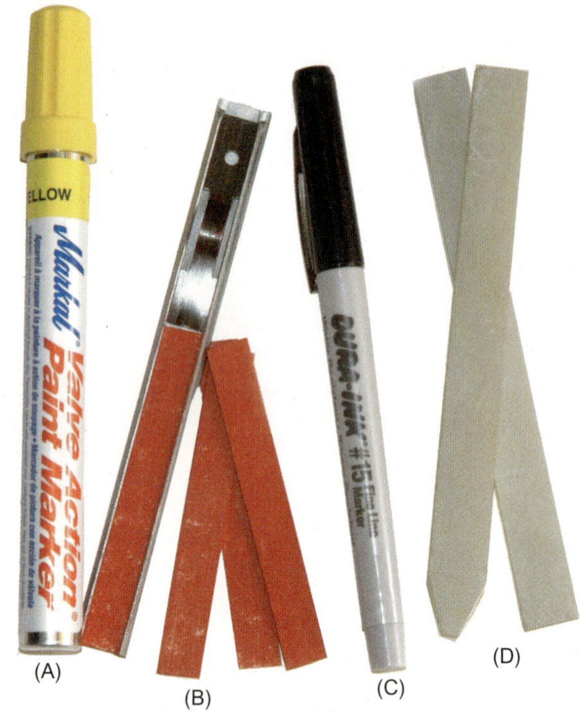

FIGURE 6-15 (A) Paint marker, (B) grease marker, (C) felt-tip marker, (D) soapstone. Larry Jeffus

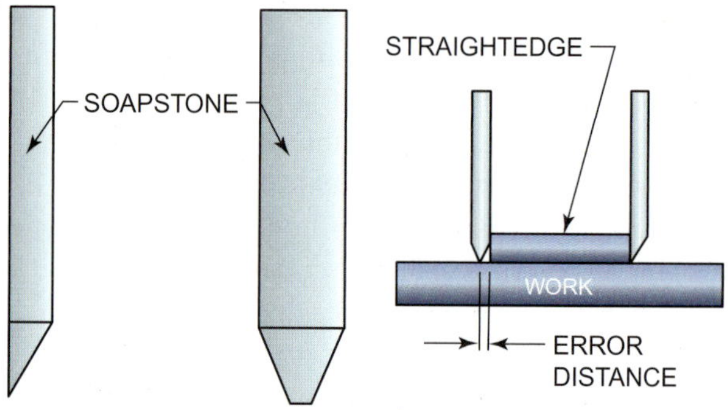

FIGURE 6-16 Proper method of sharpening a soapstone. © Cengage Learning 2012

Figure 6-17. Keep the tip of the chalk line reel pointed up to prevent an excess amount of chalk in the chalk-line reel from coming out as the string is pulled out. This will keep your work area cleaner and reduce the need to refill the reel's chalk so often.

NOTE: Keeping the chalk dry is important. The powdered chalk used in a chalk-line reel will become useless if it gets too damp or wet. It is very hard to clean out the gooey chalk mess if you let your reel get rained on or drop it in a puddle.

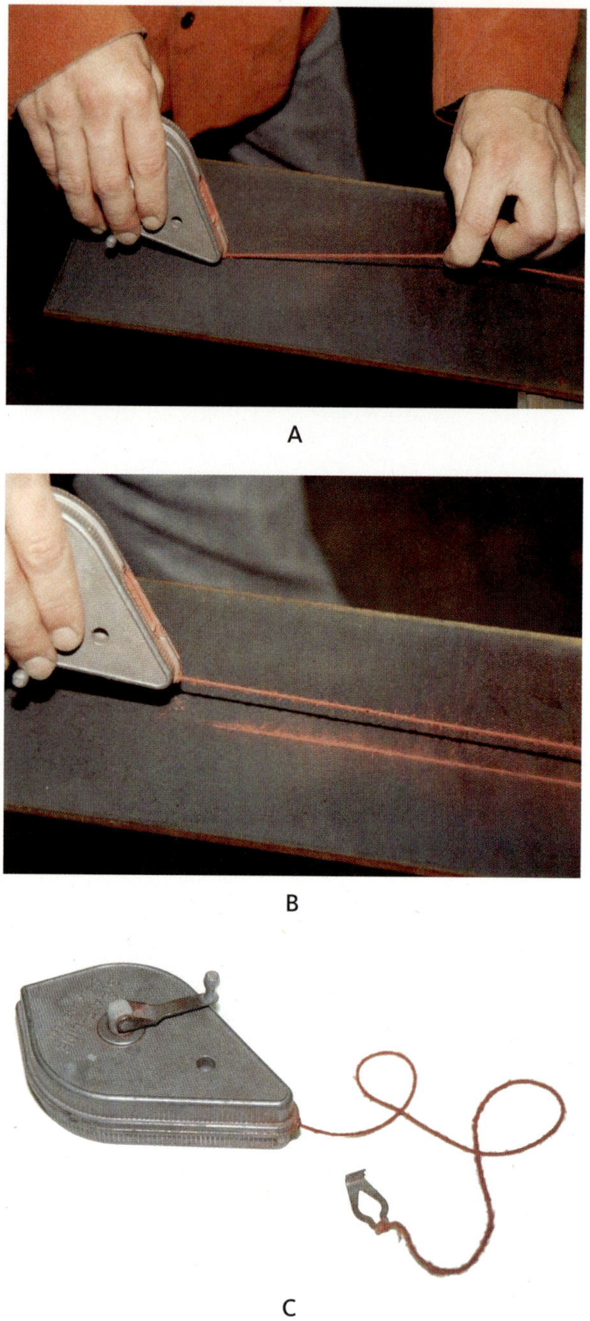

FIGURE 6-17 (A) Pull the chalk line tight and then snap the line. (B) Check to see that the line is dark enough to be easily seen. (C) Chalk-line reel; powdered chalks are available in several different colors. Larry Jeffus

Either a **scribe** or a **punch** can be used to lay out an accurate line, but the punched line can be easier to see when cutting. A punch can be held as shown in Figure 6-18, with the tip just above the surface of the metal. When the punch is struck with a lightweight hammer, it will make a mark. If you move your hand along the line and rapidly strike the punch, it will leave a series of punch marks for the cut to follow.

FIGURE 6-18 Holding the punch slightly above the surface allows it to be struck repeatedly, making punched marks to form a punched line that is easily seen for cutting. © Cengage Learning 2012

Starting your **layout** along an existing edge or from the corner can both help you make a more accurate layout and conserve materials. An existing edge is easier to use to hook the end of your tape measure or line up a square. In addition, the corners and edges of new plates and sheets are straight and square. So, by starting in a corner or along the edge, you can both take advantage of the preexisting cut as well as reduce wasted material.

It is often easy to mistakenly cut the wrong line. With your visibility somewhat limited by the cutting goggles or shield, it is not always possible to see far enough ahead to know when to stop cutting. Stopping to look can result in a problem with restarting the cut; therefore, you want to make sure that the lines are clearly marked. This can become more of a problem if one person lays out the parts and another makes the cuts. Even if you are doing both jobs, it is easy to cut the wrong line. Sometimes there may be layout lines used to help locate parts or there may be bend lines, both of which must not be mistakenly cut. To avoid making a cutting mistake, lines should be identified as to whether they are being used for cutting, locating bends, drill centers, or assembly locations. The lines that are not to be cut may be marked with an *X*, or they may be identified by writing directly on the part, Figure 6-19. Mark the *X* at the beginning and end of the line; and if it is a long line, you may want to make several more *X*s along the full length of the line. You should also mark the side of the line that is scrap so that the kerf is removed from that side, which will leave the part the proper size, Figure 6-19. When possible, erase all unneeded lines before starting the cut.

Some shops will have their own shorthand methods for identifying layout lines, or you may develop your own system. Failure to develop and use a system

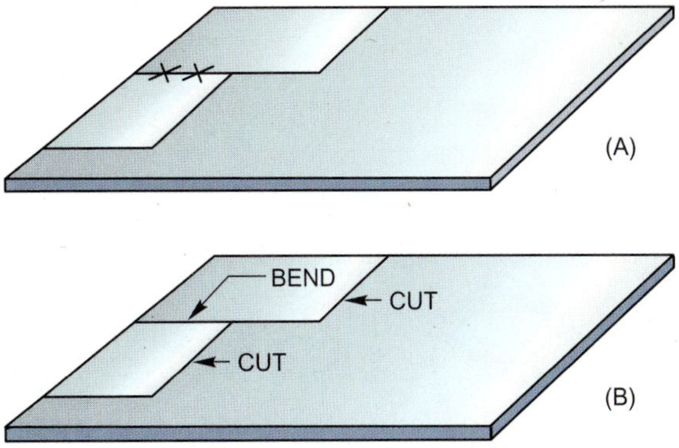

FIGURE 6-19 Identifying layout lines to avoid mistakes during cutting. © Cengage Learning 2012

for identifying lines will ultimately result in a mistakenly made cut. In a welding shop, there are only those who have made the wrong cut and those who will make the wrong cut. When it happens, check with the welding shop supervisor to see what corrective steps may be taken. One advantage for most welding assemblies is that many errors in cutting can be repaired by welding. There are often prequalified procedures established for just such an event, so check before deciding to scrap the part.

The process of laying out a part may be affected by the following factors:

- Material shape—Figure 6-20 lists the most common shapes of metal used for fabrication. Flat stock such as sheets and plates are the easiest to lay out, and the most difficult shapes to work with are pipes and round tubing.
- Part shape—Parts with square and straight cuts are easier to lay out than parts with angles, circles, curves, and irregular shapes.
- Tolerance—The smaller or tighter the tolerance that must be maintained, the more difficult the layout.
- Nesting—The placement of parts together in a manner that will minimize the waste created is called **nesting.**

SPARK YOUR IMAGINATION

Many differently shaped materials have been used for building items found in your welding shop. Make a list of the items in your school's welding shop. Include on the list all the different types of materials used in their construction. Use Figure 6-20 to help identify as many of the different types of materials used as possible.

Parts with square or straight edges are the easiest to lay out. Simply measure the distance and use

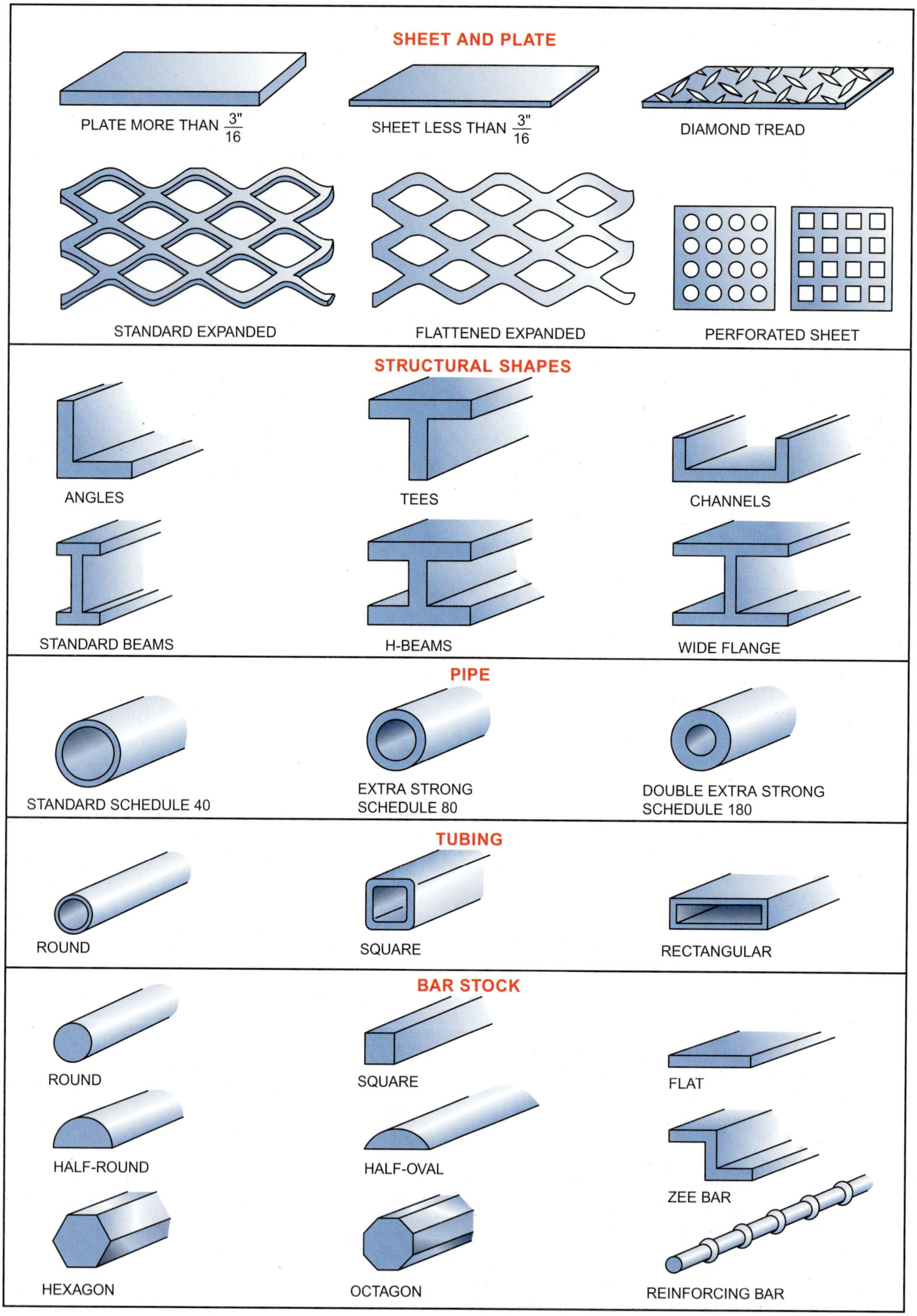

FIGURE 6-20 Standard metal shapes; most are available with different surface finishes, such as hot-rolled, cold-rolled, or galvanized.

FIGURE 6-21 Using a square to draw a straight line.
Larry Jeffus

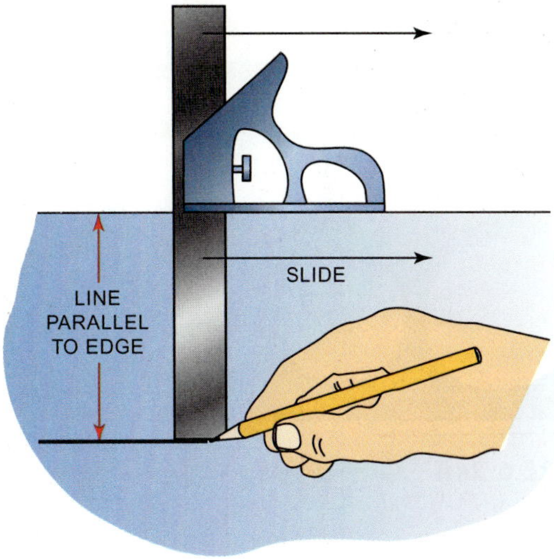

FIGURE 6-22 Using a combination square to lay out a line parallel to the edge of a plate.

a square or straight edge to lay out the line to be cut, Figure 6-21. Straight cuts that are to be made parallel to an edge can be drawn by using a combination square and a piece of soapstone. Set the combination square to the correct dimension, and drag it along the edge of the plate while holding the soapstone at the end of the combination square's blade, Figure 6-22.

PRACTICE 6-2

Laying Out Square, Rectangular, and Triangular Parts

Using a piece of metal or paper, soapstone or pencil, tape measure, and square, you are going to lay out the parts shown in Figure 6-23. The parts must be laid out within ±1/16 in. (2 mm) of the dimensions. Convert the dimensions into SI metric units of measure, and write the SI units next to the standard dimensions for the parts.

NOTE: Check your work and see how you did. If you have laid out item C correctly, you can cut it out, use it as a pattern, and cut out nine more just like it to make a star.

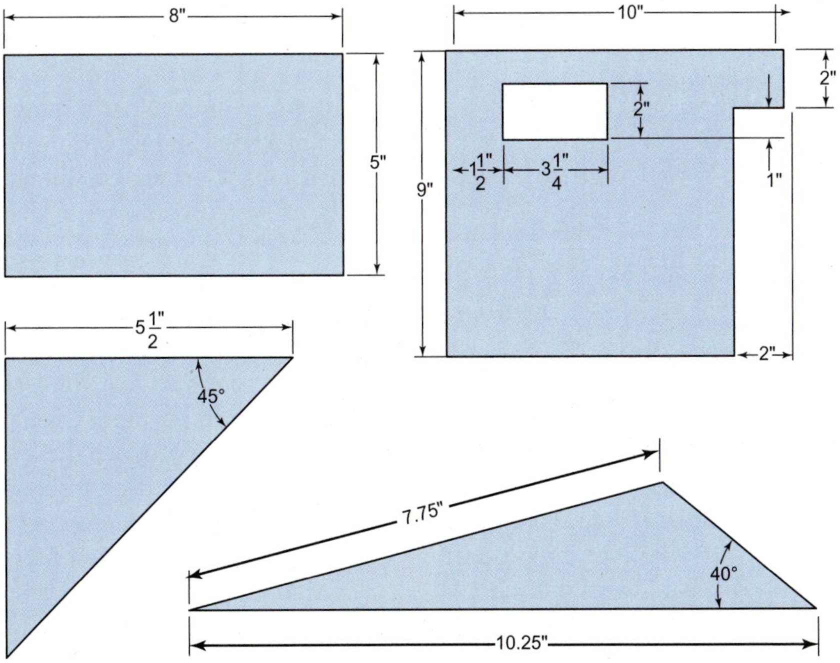

FIGURE 6-23 Practice 6-2.

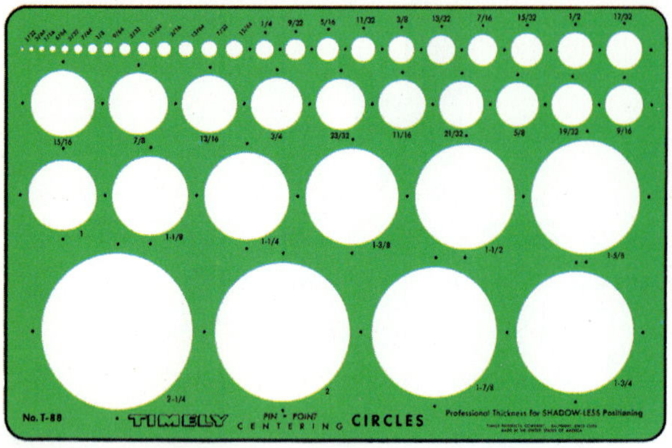

FIGURE 6-24A Circle template. Timely Products Co., Inc.

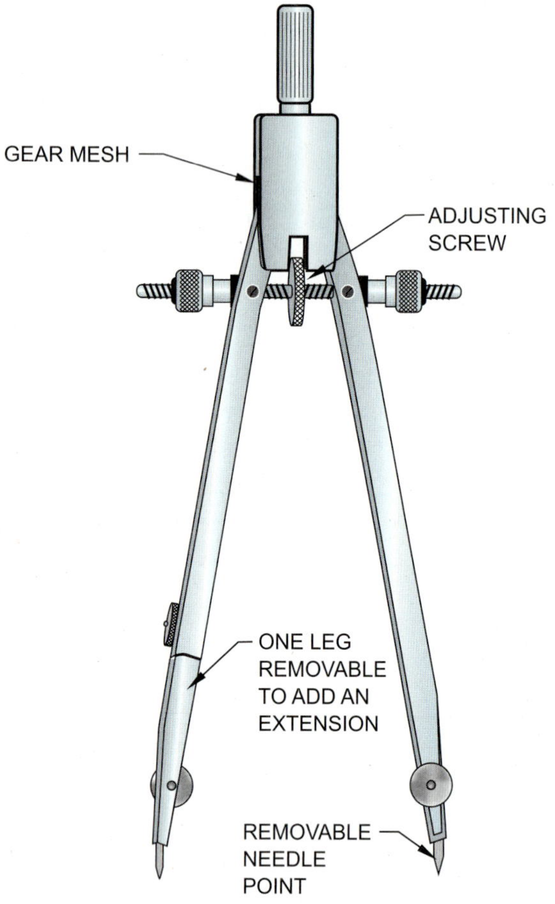

FIGURE 6-24B Compass. Larry Jeffus

Circles, arcs, and curves can be laid out by either using a compass or circle **template,** Figure 6-24. The diameter is usually given for a hole or round part, and the radius is usually given for arcs and curves, Figure 6-25. The center of the circle, arc, or curve may be located using dimension lines and center lines. Curves and arcs that are to be made tangent to another line may be dimensioned with only their radiuses, Figure 6-25.◆

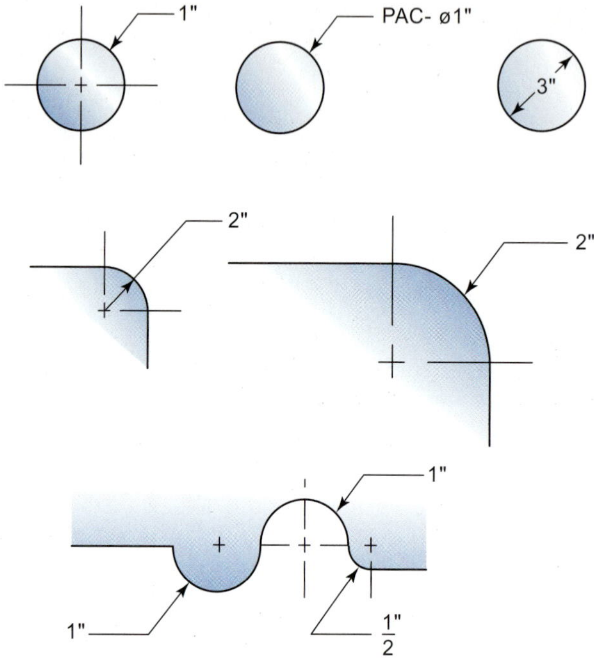

FIGURE 6-25 Methods of dimensioning arcs, curves, radii, and circles. © Cengage Learning 2012

PRACTICE 6-3

Laying Out Circles, Arcs, and Curves

Using a piece of metal or paper that is approximately 8 1/2 by 11 in., soapstone or pencil, tape measure, compass or circle template, and square, you are going to lay out the parts shown in Figure 6-26. The parts must be laid out within ±1/16 in. (2 mm) of the dimensions. Convert the dimensions into SI metric units of measure and write the SI units next to the standard dimensions for the parts.◆

Nesting

Laying out parts so that the least amount of scrap is produced is important. Odd-shaped and unusual-sized parts often produce the largest amount of scrap. Computers can be used to lay out nested parts with the least scrap. Some computerized cutting machines can be programmed to nest parts.

Manually nesting of parts may require several tries at laying out the parts to achieve the lowest possible scrap.

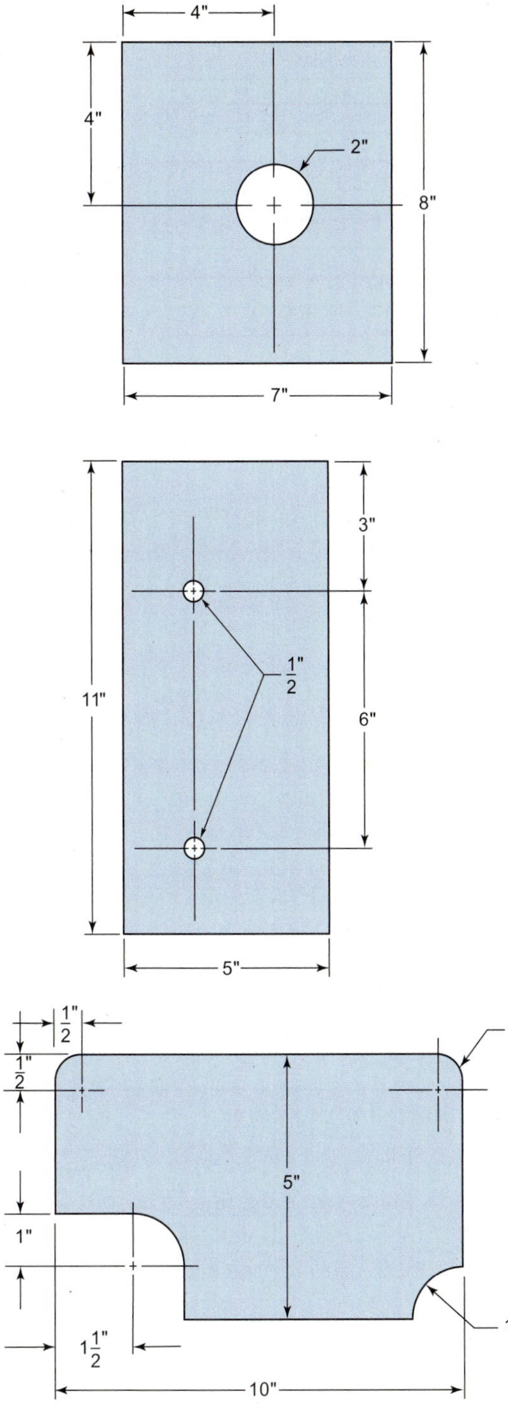

FIGURE 6-26 Practice 6-3. © Cengage Learning 2012

PRACTICE 6-4

Nesting Layout

Using metal or paper that is approximately 8 1/2 by 11 in., soapstone or pencil, tape measure, and square, you are going to lay out the parts shown in Figure 6-27, in a manner that will result in the least scrap, assuming a

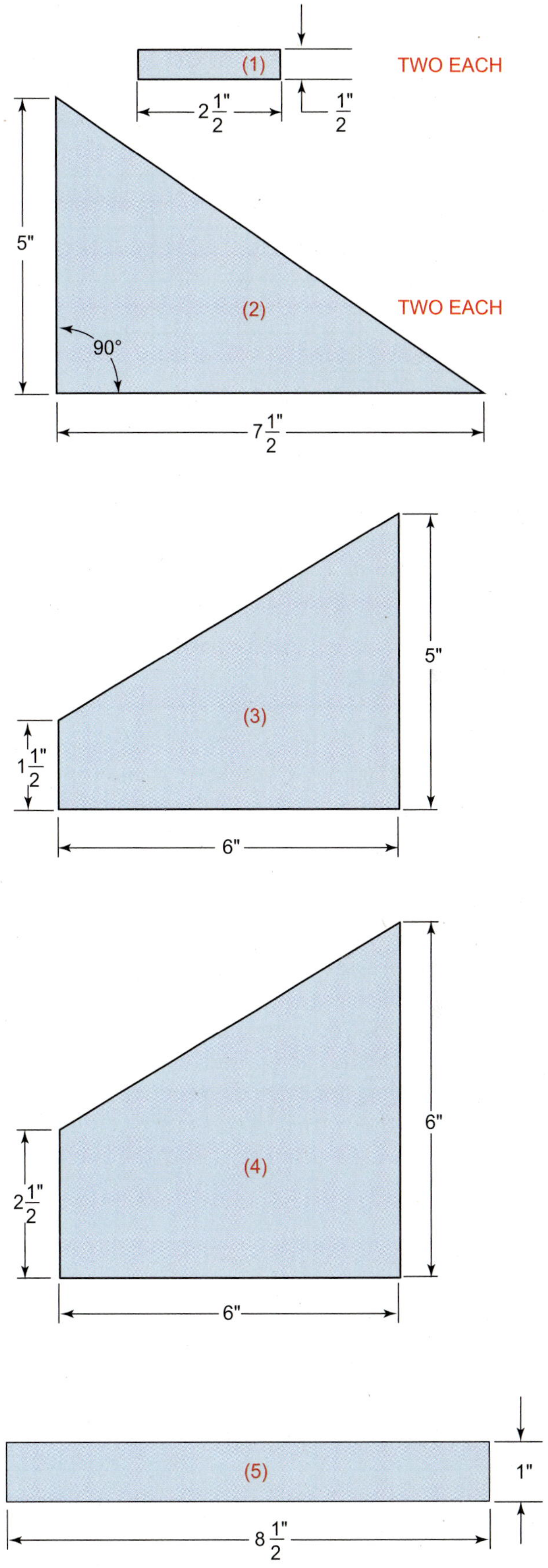

FIGURE 6-27 Practice 6-4. © Cengage Learning 2012

BILL OF MATERIALS				
Part	Number Required	Type of Material	Size	
			Standard Units	SI Units

Table 6-2 Bill of Materials Form

kerf width of zero. Use as many 8 1/2 by 11 in. pieces of stock as would be necessary to produce the parts that you are using in your layout. The parts must be laid out within ±1/16 in. (2 mm) of the dimensions. Convert the dimensions into SI metric units of measure.◆

PRACTICE 6-5

Bill of Materials

Using the parts laid out in Practice 6-4 and paper and pencil, you are going to fill out the Bill of Materials Form shown in Table 6-2.◆

Kerf Space

All cutting processes except shearing remove some metal, leaving a small gap or space. This gap or space is called the **kerf.** You must allow for the kerf width when parts are being laid out side by side. The width of material being removed by the cut's kerf varies depending on the cutting process used. Of the cutting processes used in most shops, the metal saw produces one of the smallest kerfs while the handheld oxyfuel cutting torch can produce one of the widest. Saw kerfs often range around 1/16 in. (2 mm) wide while a very good oxyfuel cut's kerf will be around 1/8 in. (3 cm) wide.

When only one or two parts are being cut, the kerf width may not need to be added to the part dimension. This space may be taken up during assembly by the root gap required for a joint. However, if a large number of parts are being cut out of a single piece of stock, the kerf width can add up. This combining of kerf width from each cut can increase the stock required for cutting out the parts, Figure 6-28.

PRACTICE 6-6

Allowing Space for the Kerf

Using a pencil, paper 8 1/2 by 11 in., measuring tape or rule, and square, you are going to lay out four rectangles 2 1/2 by 5 1/4 in. down one side of the paper, leaving 3/32 in. (2 mm) for the kerf.

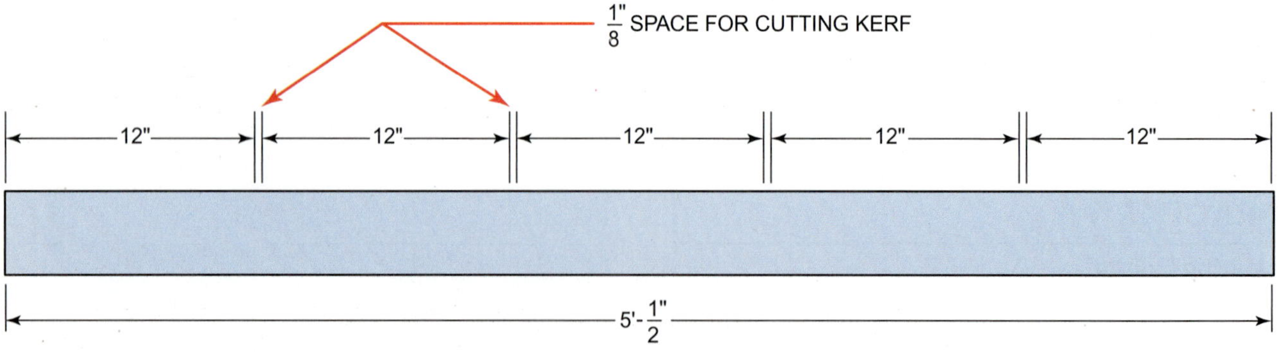

FIGURE 6-28 Because of the kerf, an additional 1/2 in. (13 mm) of stock would be required to make these five 1-ft (0.30-m) long pieces.

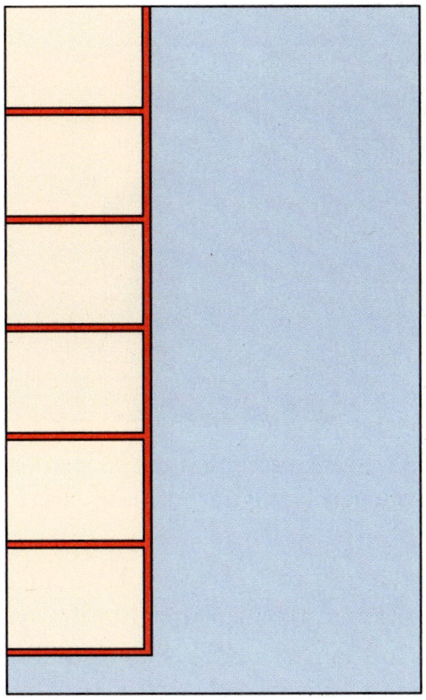

FIGURE 6-29 Kerf is marked so it will be cut between the lines. © Cengage Learning 2012

There are two common ways to provide for kerf spacing. One method is to draw a double line and make your cut between the lines, Figure 6-29. Another way is to lay out a single line and place an *X* on the side of the line that the cut is to be made on, Figure 6-30. You do not have to leave a kerf space along the sides drawn next to the edge of the paper or next to the scrap. What is the total size (length and width) of material needed to lay out these four parts?◆

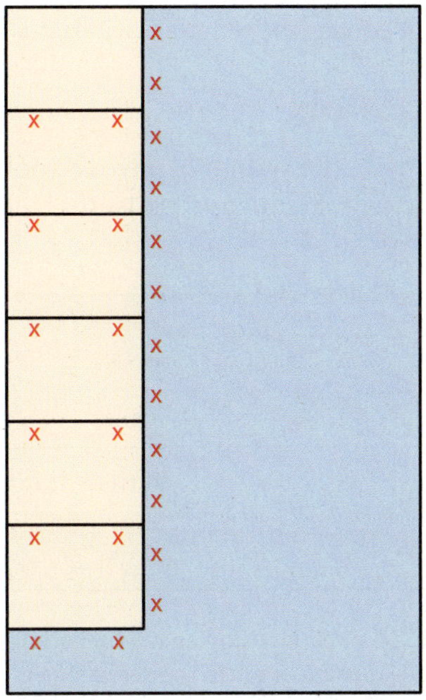

FIGURE 6-30 *X*s mark which side of the line is to be cut. © Cengage Learning 2012

PRACTICE 6-7

Lay Out Nested Parts

Using metal or paper that is approximately 18 by 24 in., soapstone or pencil, tape measure, and square, you are going to lay out the parts shown in Figure 6-31, in a manner that will result in the least scrap, assuming

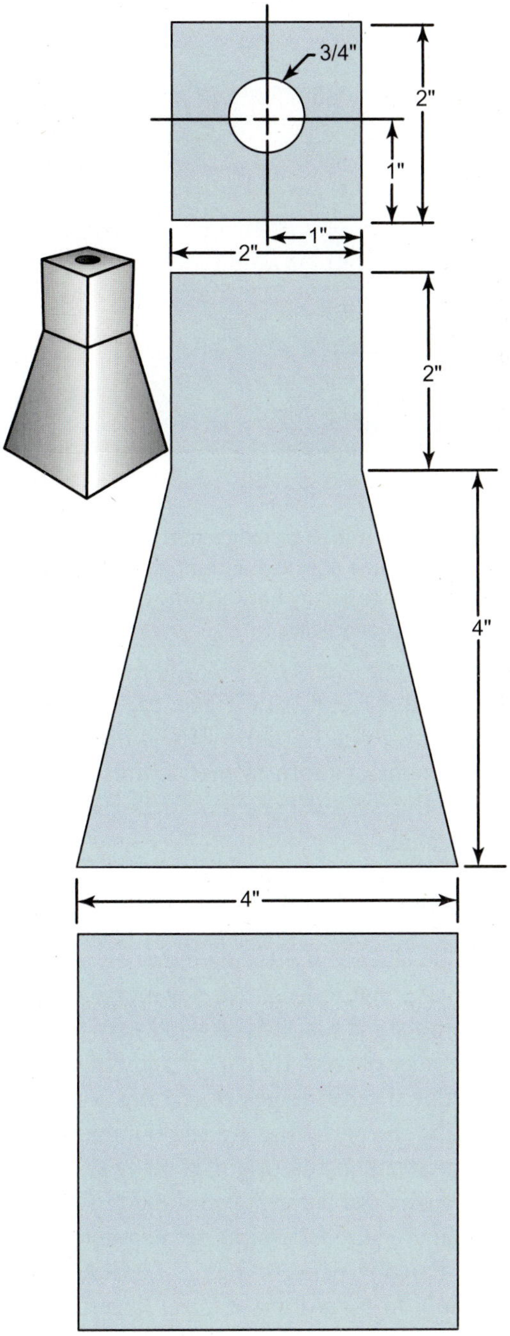

FIGURE 6-31 Practice 6-7. © Cengage Learning 2012

(A)

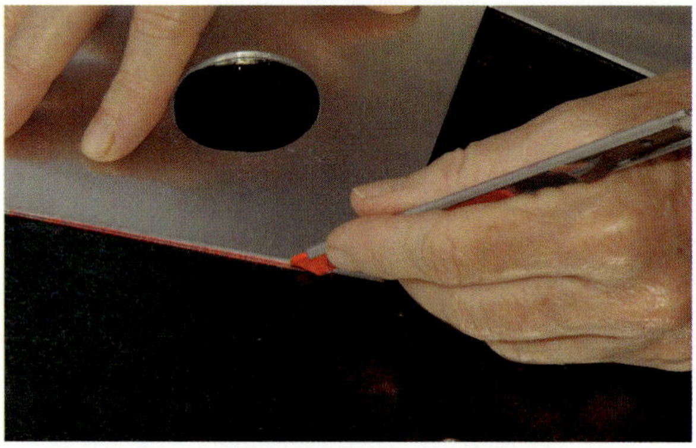

(B)

FIGURE 6-32 When tracing a part as a template, always use the same part as the template each time. This reduces the chance of any new errors getting transferred to other parts. Larry Jeffus

a 1/8-in. (3-mm) kerf width. The parts must be laid out within ±1/16 in. (2 mm) of the dimensions. Convert the dimensions into SI metric units of measure. Make note that you will need four side pieces to make this candlestick.

Parts Tracing

Parts can be laid out by tracing either an existing part or a template, Figure 6-32. An advantage to tracing a part or template is that it is fast and easy to mark out a large number of parts that are exactly the same. One of the major disadvantages of tracing parts or templates is that the parts that are laid out will not be any more accurate than the originals being traced. If you are using a part as a pattern, you should mark that part so that you are sure to use the same one for all of the layouts. Changing parts or using one that was traced earlier can compound any errors. It can also result in each part getting ever so slightly larger, until they

FIGURE 6-33 Make sure that the soapstone is held tightly into the part being traced. Larry Jeffus

are completely out of tolerance. To reduce the error caused by tracing, be sure that the line you draw is made tight to the part's edge, Figure 6-33. The inside edge of the line is the exact size of the part. Make the cut on the line or to the outside so that the part will be the correct size once it is cleaned up, Figure 6-34.

Sometimes a template may be made for a part. Templates can be useful if the part is complex and needs to fit into an existing weldment. They are also helpful when a large number of the same part is to be made or when the part is only occasionally used. The advantages of using templates are that once the detailed layout work is completed, exact replicas can be made any time they are needed. Templates can be made out of heavy paper, cardboard, wood, sheet metal, or other appropriate material. The sturdier the material, the longer the template will last.◆

Layout Tools

In addition to the standard layout tools such as squares, straight edges, tape measures, and markers,

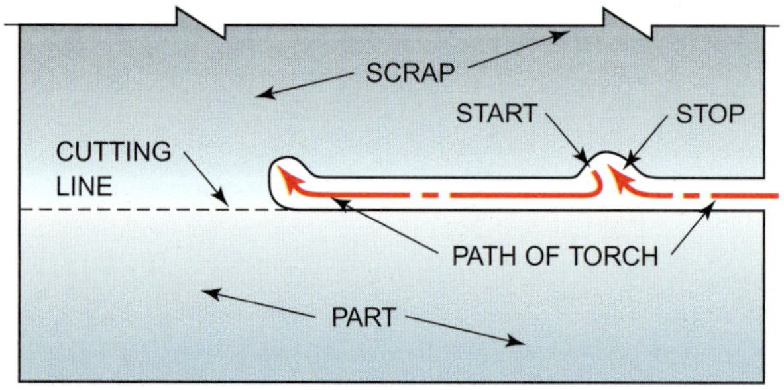

FIGURE 6-34 When using an oxyfuel or plasma torch to make a cut, turning out into the scrap side of the line when you have to stop will make restarting smoother. © Cengage Learning 2012

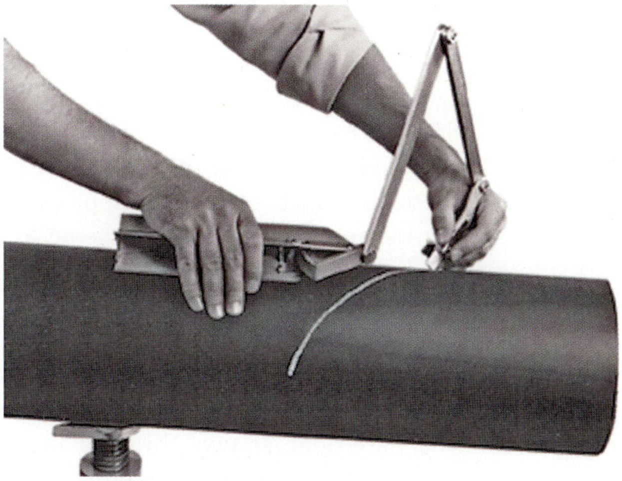

FIGURE 6-35 Pipe being laid out using a contour marker. Larry Jeffus

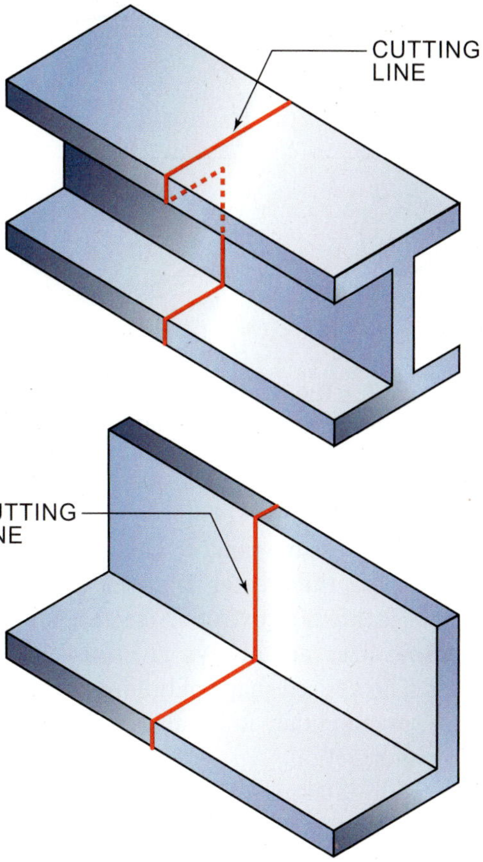

FIGURE 6-36 Laying out structural shapes. © Cengage Learning 2012

a number of special tools have been developed to aid in laying out parts. One such tool is the **contour marker,** Figure 6-35. These markers are highly accurate when properly used. They require a certain amount of practice to gain experience. Once familiar with this tool, the user can lay out an almost infinite variety of joints within the limits of the tool being used. An advantage of tools like the contour marker is that all sides of a cut in structural shapes and pipe can be laid out from one side without relocating the tool, Figure 6-36.

MATERIAL SHAPES

Metal stock can be purchased in a wide variety of shapes, sizes, and materials. Weldments may be constructed from a combination of sizes and shapes of metal. Usually, only a single type of metal is used in most weldments unless a special property such as corrosion-resistance is needed. In those cases, dissimilar metals may be joined into the fabrication at such locations as needed. The most common metal used is carbon steel, and the most common shapes used are plate, sheet, pipe, tubing, and angles. For that reason, most of the fabrication covered in this chapter concentrates on carbon steel in those commonly used shapes. Transferring the fabrication skills learned in this chapter to the other metals and shapes should require only a little practice time.

Plate is usually 3/16 in. (4.7 mm) or thicker and measured in inches and fractions of inches. Plates are available in widths ranging from 12 in. (305 mm) up to 96 in. (2438 mm) and lengths from 8 ft (2.4 m) to 20 ft (6 m). Plate thickness can range up to 12 in. (305 mm).

Sheets are usually 3/16 in. (4.7 mm) or less and measured in gauge or decimals of an inch. There are several different gauge standards that are used. The two most common are the Manufacturer's Standard Gauge for Sheet Steel; used for carbon steels; and the American Wire Gauge, which is used for most nonferrous metals such as aluminum and brass.

Pipe is dimensioned by its diameter and schedule or strength. Pipe that is smaller than 12 in. (305 mm) is dimensioned by its inside diameter, and the outside diameter is given for pipe that is 12 in. (305 mm) in diameter and larger, Figure 6-37. The strength of pipe is given as a schedule. Schedules 10 through 180 are available; schedule 40 is often considered a standard strength. The wall thickness for pipe is determined by

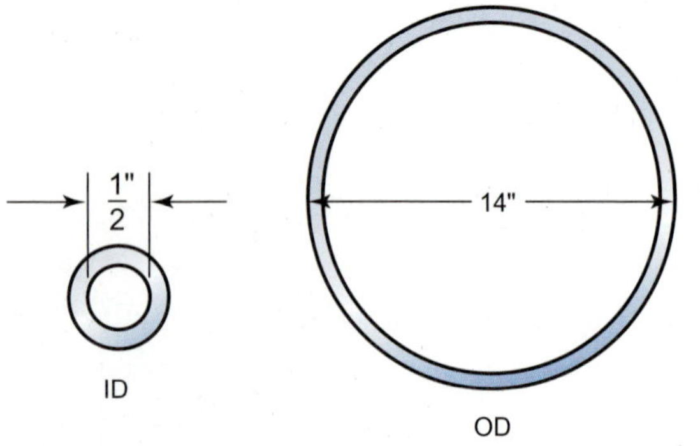

FIGURE 6-37 The inside diameter is abbreviated ID, and the outside diameter is abbreviated OD. © Cengage Learning 2012

its schedule (pressure range). The larger the diameter of the pipe, the greater its area. Pipe is available as welded (seamed) or extruded (seamless).

Tubing sizes are always given as the outside diameter. The desired shape of tubing, such as square, round, or rectangular, must also be listed with the ordering information.

The wall thickness of tubing is measured in inches (millimeters) or as Manufacturer's Standard Gauge for Sheet Metal. Tubing should also be specified as rigid or flexible. The strength of tubing may also be specified as the ability of tubing to withstand compression, bending, or twisting loads.

Angles, sometimes referred to as angle iron, are dimensioned by giving the length of the legs of the angle and their thickness, Figure 6-38. Stock lengths of angles are 20 ft, 30 ft, 40 ft, and 60 ft (6 m, 9.1 m, 12.2 m, and 18.3 m).

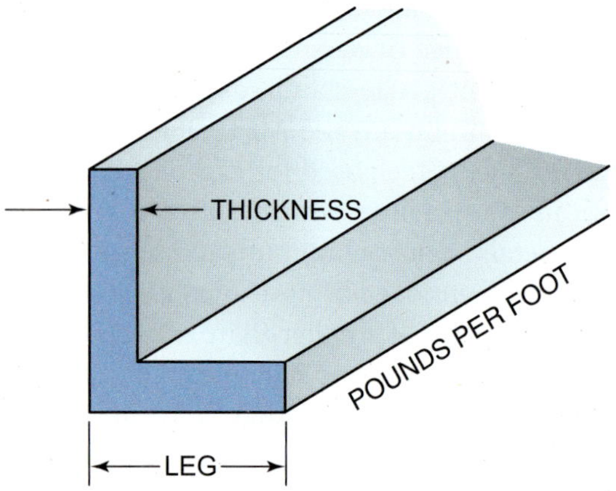

FIGURE 6-38 Standard specifications used for sizing angle iron. © Cengage Learning 2012

ASSEMBLY

The assembling process, bringing together all of the parts of the weldment, requires a proficiency in several areas, Figure 6-39. You must be able to read the drawing and interpret the information provided to properly locate each part. An assembly drawing has the necessary information, both graphically and dimensionally, to allow the various parts to be properly located as part of the weldment. If the assembly drawings include either pictorial or exploded views, this process is much easier for the beginning assembler; however, most assembly drawings are done as two, three, or more orthographic views, Figure 6-40. Orthographic views are more difficult to interpret until you have developed an understanding of their various elements.

On very large projects such as buildings or ships, a corner or center line is established as a base line. This is the point where all measurements for all part location begins. When working with smaller weldments, a single part may be selected as such a starting point. Often, selection of the part to serve as the base is easy because all the other parts are to be joined to this central part. However, on other weldments the selection is strictly up to the assembler.

To start the assembly, make a selection of the largest or most central part to be the base for your assembly. Once this is done, all other parts will be aligned to this one part. Using a base also helps prevent location and dimension errors. A slight misalignment of

FIGURE 6-39 Welded assembly fabricated at SkillsUSA competition. Larry Jeffus

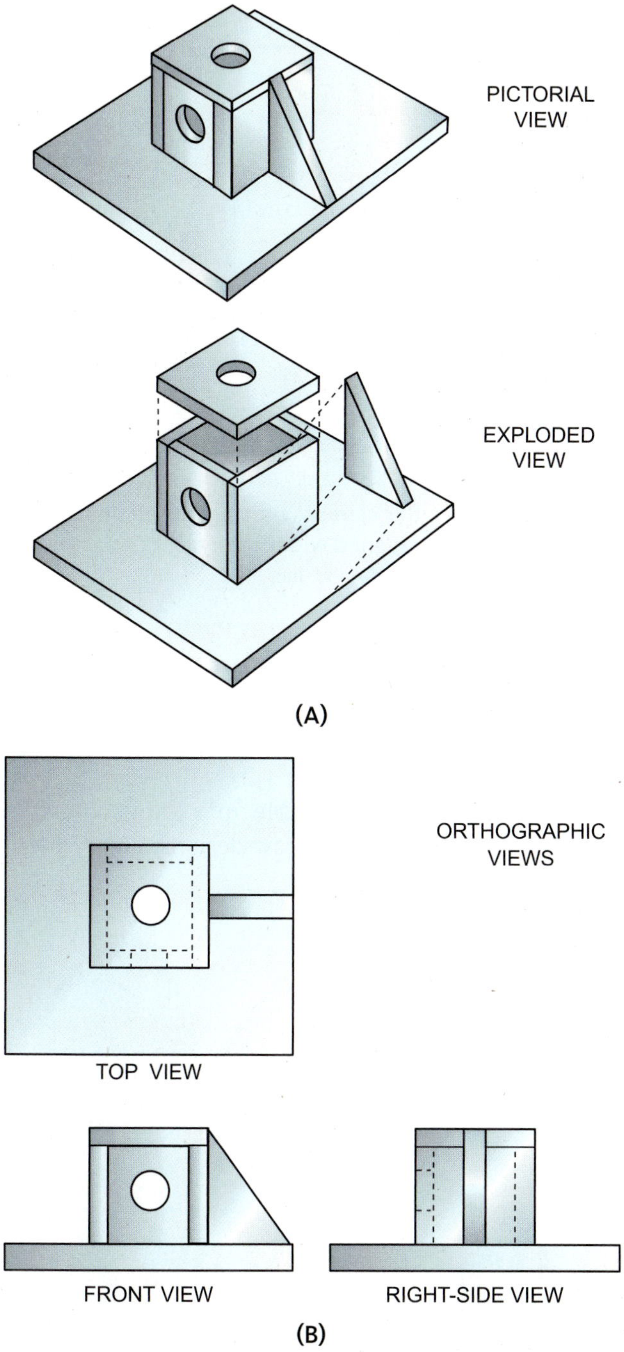

FIGURE 6-40 **Types of drawings that can be used to show a weldment assembly.** © Cengage Learning 2012

one part, even within tolerances, can be compounded with other misalignments, resulting in an unacceptable weldment. Using a base line or base part results in a more accurate assembly.

Identify each part of the assembly, and mark each piece for future reference. If needed, you can hold the parts together and compare their orientation to the drawing. Locate points on the parts that can be easily identified on the drawing such as holes, notches, and so on, Figure 6-41. Now mark the parts as to their top, front, or other such orientation, which will help you locate them during the assembly.

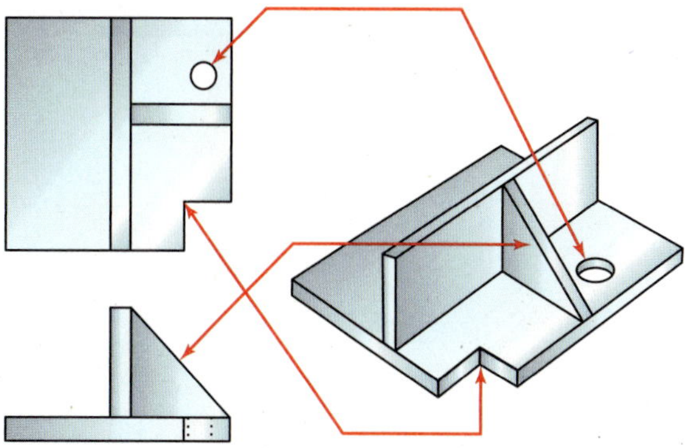

FIGURE 6-41 **Identify unique points or shapes on parts to aid in assembly.** © Cengage Learning 2012

Layout lines and other markings can be made on the base to locate other parts. Using a consistent method of marking helps prevent mistakes. One method is to draw parallel lines on both parts where they meet, Figure 6-42.

After the parts have been identified and marked, they can be either held or clamped into place. Holding the parts in alignment by hand for tack welding is fast but often leads to errors and is not recommended for beginning assemblers. Experienced assemblers recognize that clamping the parts

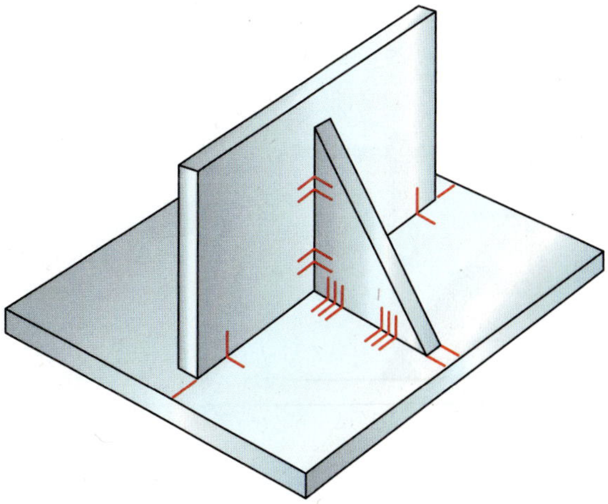

FIGURE 6-42 **Layout marks on parts can help locate the parts for tack welding.** © Cengage Learning 2012

FIGURE 6-43 C-clamp being used to hold plates for tack welding. Larry Jeffus

in place before tack welding is a much more accurate method, Figure 6-43.

ASSEMBLY TOOLS

Clamps

A variety of types of **clamps** can be used to temporarily hold parts in place so that they can be tack welded.

- C-clamps are one of the most commonly used clamps. They come in a variety of sizes, Figure 6-44. Some C-clamps have been specially designed for welding. These clamps may have a spatter cover over the screw, and others have their screws made of spatter-resistant materials such as copper alloys.
- Bar clamps are useful for clamping larger parts. Bar clamps have a sliding lower jaw that can be snugged up against the part before tightening the screw-clamping end, Figure 6-45. They are available in a variety of lengths.
- Pipe clamps are very similar to bar clamps. The advantage of pipe clamps is that the ends can be attached to a section of standard 1/2-in. (13-mm) pipe. This allows for greater flexibility in length, and the pipe can easily be changed if it becomes damaged.
- Locking pliers are available in a range of sizes with a number of various jaw designs, Figure 6-46.

FIGURE 6-44 C-clamps come in a variety of sizes. Larry Jeffus

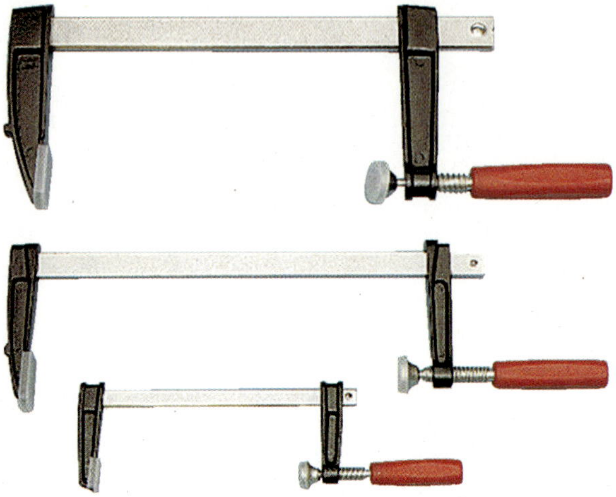

FIGURE 6-45 Bar clamps, which are used on larger weldments, can be opened wider than most C-clamps. Strong Hand Tools & Good Hand, Inc.

FIGURE 6-46 Three common jaw types on locking pliers. Strong Hand Tools & Good Hand, Inc.

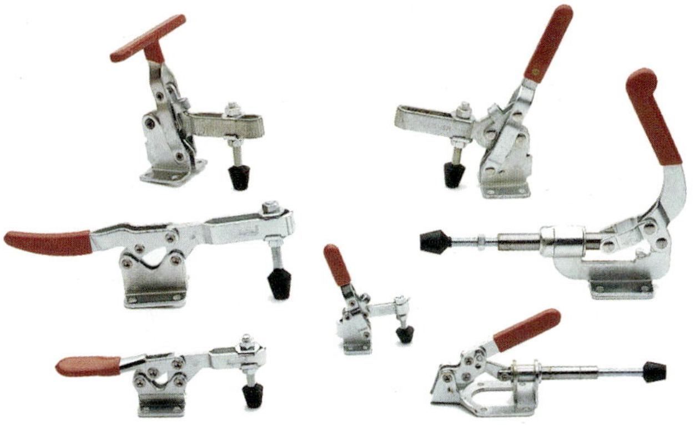

FIGURE 6-47 **Toggle-type clamps.** Strong Hand Tools & Good Hand, Inc.

The versatility and gripping strength make locking pliers very useful. Some locking pliers have a self-adjusting feature that allows them to be moved between different thicknesses without the need to readjust them.

- Cam-lock clamps are specialty clamps that are often used in conjunction with a jig or fixture. They can be preset, which allows for faster work, Figure 6-47.
- Specialty clamps such as these for pipe welding, Figure 6-48, are available for many different types of jobs. Such specialty clamps make it possible to do faster and more accurate assembling.

Fixtures

Fixtures are devices that are made to aid in assemblies and fabrication of weldments. Fixtures may align, position, or support parts. They may have clamping devices permanently attached to speed up their use, Figure 6-49. Each fixture is unique to the weldment being fabricated; therefore, they are used when a number of similar parts are to be made. They can increase speed and accuracy in the assembly of parts. They must be strong enough to support the weight of the parts, able to withstand the rigors of repeated assemblies, and remain in tolerance. Often, locating pins or other devices are used to ensure proper part location. A well-designed fixture allows adequate room for the welder to make the necessary tack welds. Some parts are left in the fixture through the entire welding process to reduce distortion. Making fixtures for every job is cost prohibitive and not necessary for a skilled assembler.

(A)

(B)

(C)

FIGURE 6-48 **Specialty pipe clamps and alignment tools.** Larry Jeffus

A

B

FIGURE 6-49 A fixture for holding a motorcycle frame for welding. Larry Jeffus

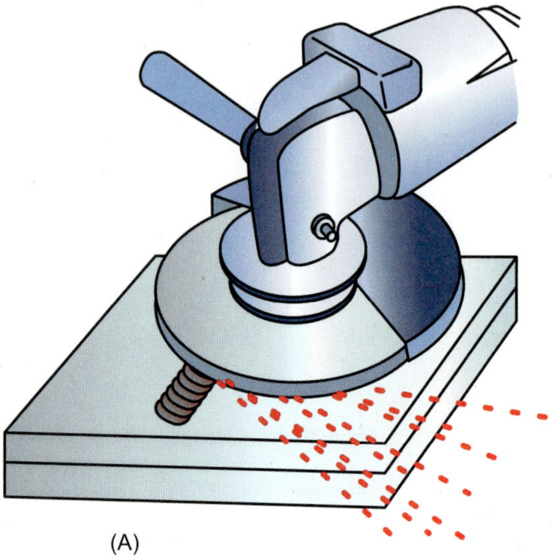

(A)

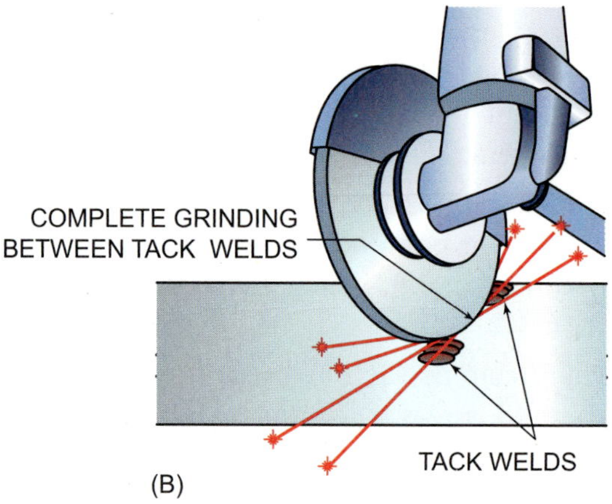

(B)

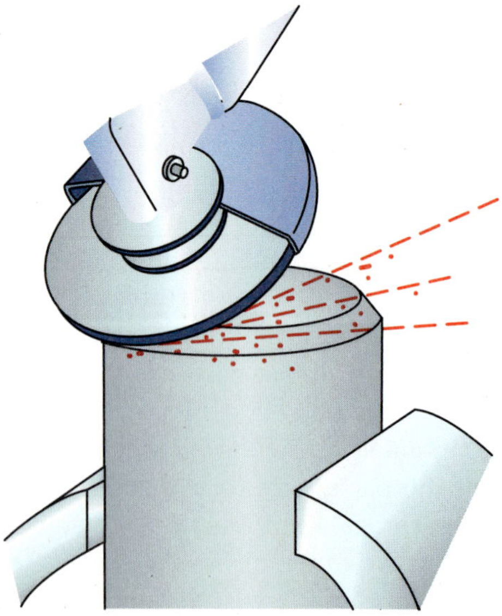

(C)

FIGURE 6-50 Abrasive grinding disk can be used to (A) remove excessive weld metal, (B) cut a groove, or (C) prepare a bevel for welding. © Cengage Learning 2012

FITTING

Not all parts fit exactly as they were designed. There may be slight imperfections in cutting or distortion of parts due to welding, heating, or mechanical damage. Some problems can be solved by grinding away the problem area. Hand-grinders are most effective for this type of problem, Figure 6-50. Other situations may require that the parts be forced into alignment.

A simple way of correcting slight alignment problems is to make a small tack weld in the joint and then use a hammer and possibly an anvil to pound the part into place, Figure 6-51. Small tacks applied in this manner will become part of the finished weld. Be sure not to strike the part in a location that damages the surface, which could render the finished part unsightly or unusable.

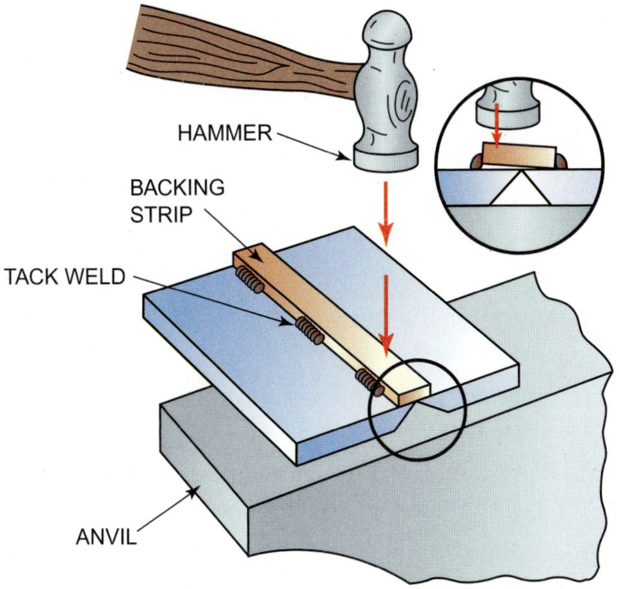

FIGURE 6-51 **A hammer can be used to bend metal into alignment after a tack weld has been made.** © Cengage Learning 2012

More aligning force can be applied using cleats or dogs with wedges or jacks. Cleats or dogs are pieces of metal that are temporarily attached to the weldment's parts to enable them to be forced into place. Jacks will do a better job if the parts must be moved more than about 1/2 in. (13 mm), Figure 6-52. Any time that cleats or dogs are used, they must be removed and the area ground smooth.

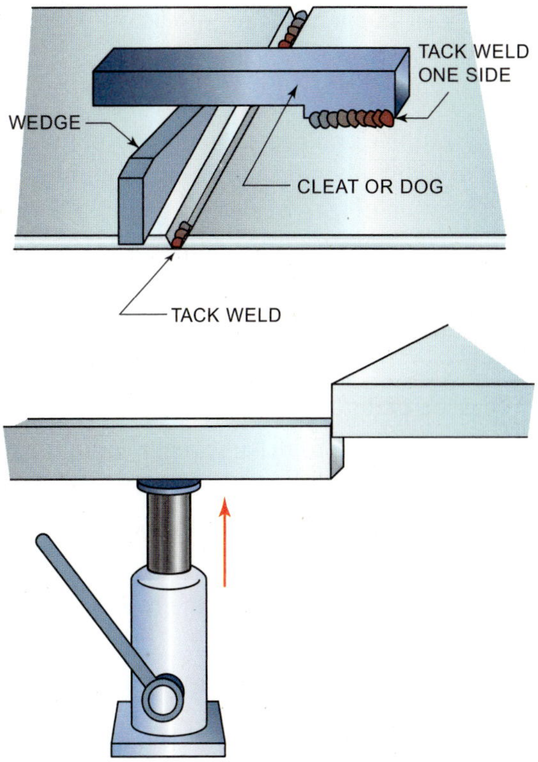

FIGURE 6-52 **(A) Cleat and wedge used to align parts. (B) Hydraulic jack used to realign a part.** © Cengage Learning 2012

Some codes and standards will not allow cleats or dogs to be welded to the base metal. In these cases, more expensive and time-consuming fixtures must be constructed to help align the parts if needed.

SUMMARY

One of the greatest experiences as a welder/fabricator is completing work on a piece of equipment, building, trailer, or other structure. You can proudly point to it and say, "I helped to make that." Learning layout and fabrication techniques will let you someday be able to experience the sense of pride when you are then are able to say, "I built that all by myself." Welded structures are an enduring monument to your skill as a craftsman, so it is important that each and every time you build a project, that you do it as if it were going to be on display, because it is.

REVIEW QUESTIONS

1. What precautions must you take when welding around other workers?
2. How can you minimize the potential for conflict when welding around other workers?
3. What safety hazard can occur as a fabrication grows in size?
4. What are four advantages for using preformed parts in a fabrication?

5. What are four advantages of custom fabricating parts?
6. What is the purpose of tack welds?
7. What are five considerations that you must take into account when deciding on the number, size, and location of tack welds?
8. Does it matter if one of your tack welds breaks during welding, and why?
9. How does the tensile strength of a metal affect a weld?
10. What is a part's tolerance?
11. What can you do to make parts fit without having to recut or regrind them?
12. Why should you avoid recutting or grinding parts whenever possible?
13. What problem can result if the root opening of a weld is too narrow?
14. What problem can result if the root opening of a weld is too wide?
15. How will a metal distort when heated? How will a metal distort when cooled?
16. What two factors affect the degree to which a metal will distort and possibly remain distorted?
17. Using the formula for linear expansion, $\Delta L = \alpha L_O \Delta T$, calculate the new lengths of 200-in. long tungsten and zinc bars that have a temperature change of 800°F.
18. Do metals having higher values of coefficients of expansion tend to distort more or less than those with lower values?
19. What is kerf space?
20. Which of the cutting processes will leave the smallest kerf?
21. What are two common ways to provide for the kerf spacing?
22. Why should you use the same part as your pattern every time you trace a new part?
23. What is the most common metal used and the most common shapes?
24. What is the assembly process?
25. What is a base line?
26. Describe three types of clamps used to temporarily hold parts in place so that they can be tack welded.
27. What is a simple way of correcting slight alignment problems?

Chapter 7

Welding Shop Practices

OBJECTIVES

After completing this chapter, the student should be able to:

- Discuss job skills that will help ensure that a welder will be a more valuable employee.
- Explain why materials should be used efficiently.
- Give examples of ways to conserve metal.
- Tell how to conserve electrical energy in a welding shop.
- Explain why welding shops should recycle scrap metal.
- Tell what can be done to ensure the safe operation of equipment in the shop.
- Recognize hand signals used to communicate with crane operators.

KEY TERMS

colleagues
employees
employer
employment
flaw
foreman
horseplay
interpersonal
owner
profit
salary
workplace

INTRODUCTION

Once you have a job, there are lots of things to learn in a welding shop other than just welding. As an employee, you become part of a team of workers who together can produce high-quality products efficiently. How well you work with others is as important as how well you can weld. You will be judged not only on your welding skills but your efficiency and productivity. As a new employee, you will have little time to develop a working knowledge of the shop and its organization. This can be a time for you to ask questions while you watch and learn.

Sometimes something as simple as learning other workers' names can help you fit into the shop environment faster.

JOB-RELATED SKILLS

Often, the success of welders on the job may not only be based on their welding and fabricating skills. They are often judged by their ability to use their time wisely, work with others, and think ahead. These are qualities that are often referred to as a welder's character and can be the basis for your reputation. These are life skills and attitudes that apply to everything you do. They should be practiced all the time, not just on the job.

Time Management

Often, in a welding shop there are periods of time when you can not continue welding on your part because you are waiting for it to cool down, be machine sandblasted, ground, and so on. Making good use of this time is important. Often, you can use the time to prep the next part that you will be working on—perhaps grind, position, mark, or record data in the part's log and so forth. If there is nothing else to be done, you can clean up your welding area, straighten up your tools, or pick up parts and supplies that are needed or will be needed from the parts room or welding supply. Check with your supervisor or boss to see if there is another job you might work on or if there is another welder you might help.

The main purpose that you are hired for by any company is for your productivity. Your productivity is your ability to get jobs done and make the company profitable. If you are not working, you are not making the company profitable. Do not simply find a comfortable spot to sit down and wait. Nothing brings unwanted attention to yourself more than being the one who is always "just sitting around." Wasting time is, indeed, wasting your company's money.

Teamwork

Often a job will go faster if you work with one or more people. For example, in large layouts it is faster to have someone on the far end of a plate who can move and tape the chalk line so you do not have to move from one end to the other. During assembly, having a coworker hold a part in place as you tack weld it can be faster than having to clamp the part securely in place so that you can do the tack welding.

Developing a friendly working relationship with coworkers, which will allow you to more easily ask for help, is important. Some of your coworkers will be more willing than others to lend a hand. Often, it is the way you ask for help that will make a difference in others being willing to help you. For example, some **employees** will respond favorably if you simply come up and say, "I need a hand. Can you come?" while others may want to be given the option. For example, "If you have time, could you come and help?" Role-playing in class will help you develop these **interpersonal** skills so that once you are on the job, you can more easily get the assistance you need.

It is important to note that you should not be offended if someone says, "No, I can't help you." They may have a project that is on a time line. In the same vein, you may need to turn someone down when you feel that taking time out of your schedule to solve or work with them on a problem or give assistance will put you behind schedule on your project.

Planning and Thinking Ahead

At the beginning of the day, you may want to spend a few minutes talking with your boss and **colleagues** about how to best approach a project so that it will move through the shop most efficiently. During a long weld, you may want to spend time thinking about your next task and think it through so that once the weld is complete, you can move on seamlessly and begin the next project.

Labor costs, the money your **employer** spends on your **salary**, or the money you earn as a result of your work in your own business is a major cost factor in the finished product. It may also be the most flexible part of the project's expense. For example, a higher-speed welding process such as gas metal arc welding (GMAW) or flux cored arc welding (FCAW) may take a little longer to set up but may ultimately save time to complete the project. Conversely, you might find that shielded metal arc welding would be the most efficient way when there are a number of small welds to be made on a larger structure. For example, when working on a utility trailer, it would be much easier to move around with a shielded metal arc electrode holder than to try and manipulate a gas metal arc (GMA) welding gun.

Sometimes the total time spent on a project can be reduced by building a jig or fixture that can help align parts and tack them in place for welding. In some cases, something can be simply attached to a welding table to hold two angle iron pieces at a right angle

FIGURE 7-1 Welding jig used to hold square tubing in alignment for welding. Larry Jeffus

when building a frame, Figure 7-1. Jigs and fixtures can often be fabricated from scrap material. In some cases, they might be constructed from materials salvaged from earlier jigs or fixtures.

It is important to note that a considerable amount of time can be wasted in building jigs and fixtures when a very limited production run is anticipated. Some degree of judgment must be exercised to avoid this time delay.

Work Ethics

Honesty, attendance, and punctuality in the **workplace** are critical. The **owner** or shop **foreman** schedules work based on the assumption that employees will be on the job ready and willing to work. When you are late or absent, the job you were scheduled to do for the day may not get done. This can cause undue delays in production. It is not just a matter of *if you don't come in, you don't get paid;* it is more important than that. If you do not come in, the shop can not get the product out in time, and employees may not get paid. That can result in your being terminated and/or the shop losing its ability to be competitive and having to close. A sick day is only for when you are truly sick and not to be used simply as a vacation day. Because of employee abuse, some employers now require a note from a doctor any time that an employee uses a sick day. Being on time is one of the many factors that employers consider when deciding about pay raises, job advancements, and in some cases, continued **employment**.

NOTE: Your boss is your boss and not your friend. Your boss is not as interested in what you did last night or what you are going to do tonight as he is interested in your being there on time and giving the job 100% of your attention and effort during the day. You need to be on time or early and ready and willing to start work the moment your shift starts.

It is important to have a friendly and professional work environment, but **horseplay** and practical jokes have no place on the job. They are disruptive and can be extremely dangerous.

Take care of personal business on your personal time. When you use your cell phone to make personal phone calls or to text a friend, you are in essence stealing time from your employer. The welding shop environment, with hot sparks, dust, and dirt, is not the best environment for your cell phone; leave it with your personal belongings. If you need to check it for messages, do so during your break or at lunch and not while you are supposed to be working.

Sometimes workers feel that it is OK to take a few screws, bolts, welding rods, or other materials from the job for a personal project they are working on at home. Honesty is important; never take any materials home from your shop. This can be considered stealing even if the material was in the scrap bin. Getting caught taking something home is bad, but rumors springing up about your honesty can be even worse.

CONSERVING MATERIALS AND SUPPLIES

A company's **profit** margin and ability to compete when bidding on jobs is affected by a number of factors. One of the most important can be the efficient use of materials and supplies. Also, being conservative with materials and supplies is good for the environment.

Metal Conservation

The cost of welding materials is a major part of any project's total price. One way to reduce material cost is to look for ways to reuse leftover material from a previous job. You might be able to weld together two or more smaller pieces of stock such as pipe, angle iron, T bar, or other similar materials. This can make these smaller pieces into one single piece long enough to fit. A little welding and grinding can even make it look almost new. There are a few cases in which this may not work, such as on critical structures or when the customer has specified all new stock.

Another important way to save on material costs is to make sure that you are being as conservative as possible with materials when you are laying out the parts on new stock. Figure 7-2 shows three ways to lay out three rings that are 4 ft in diameter and 2 in. wide. In layout Figure 7-2A, the rings are laid out on two sheets of a 4 × 8-ft plate. This layout leaves a 4 × 4-ft piece of scrap. In layout Figure 7-2B, all three rings can fit on a single 4 × 8-ft piece of plate. This would require six 2-in. long welds to assemble the rings. The most material-conserving layout is Figure 7-2C. In this layout, a total of twelve 2-in. long welds are required to make the three rings. This layout leaves a 3-ft 3-in. × 4-ft piece of scrap metal, which is almost the same size as was left from the second sheet in layout Figure 7-2A.

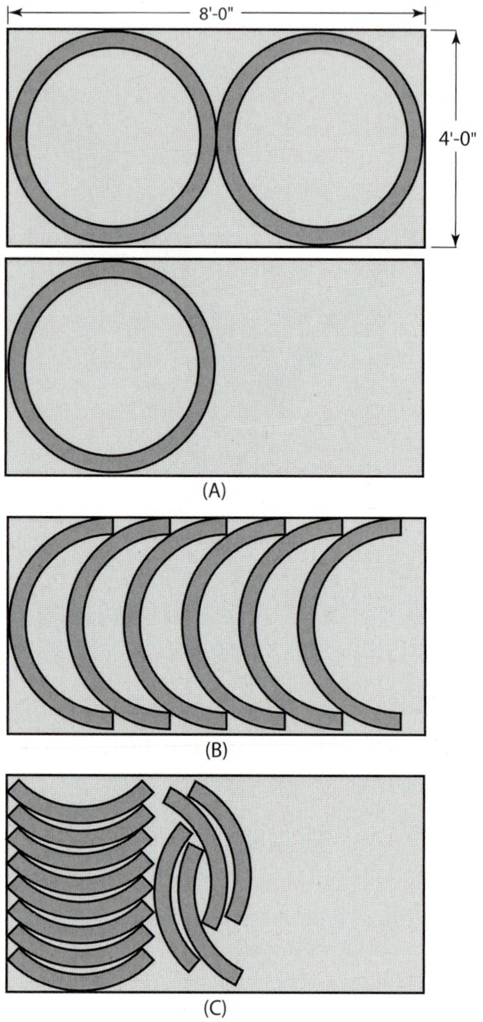

FIGURE 7-2 (A) Three rings cut out of two plates. (B) Three rings cut out of one plate. (C) Three rings cut out of approximately 1/2 a plate. © Cengage Learning 2012

Electrode Conservation

Most SMA welders try to leave a welding electrode stub that is no more than a 1 1/2 in. (38 mm) to 2 in. (51 mm) long, Figure 7-3. Even with a short stub, approximately 35% of the total electrode weight purchased is lost due to losses of spatter, slag, fumes, and the stub. If longer stubs are left, the percentage of wasted electrode can increase dramatically. Sometimes a weld joint ends before the electrode is completely used down to a short stub. These longer, partially consumed electrode stubs can be reused to make tack welds.

NOTE: Sometimes it is difficult to reuse some mineral-based low-hydrogen electrodes because it can be difficult to restart an arc. To make this easier, practice restarting these electrodes.

Sometimes partially used electrodes may cause a small **flaw** in the weld as they are restarted. For this reason, they may not be used when very stringent code requirements will not allow this type of flaw.

Making welds of the correct size and not excessively large, saves weld material, can save postweld grinding, and reduces the electric and labor costs. Flat position welds can be made at a higher rate of travel than most other weld positions. Positioning the work so that as much as possible of the welding can be done in the flat position makes it easier on the welder, more comfortable for the welder, and increases the welding productivity.

Never continue to weld when you observe excessive spatter being produced. Excessive spatter wastes both filler metal and requires longer postweld cleanup, Figure 7-4. Excessive SMA welding spatter

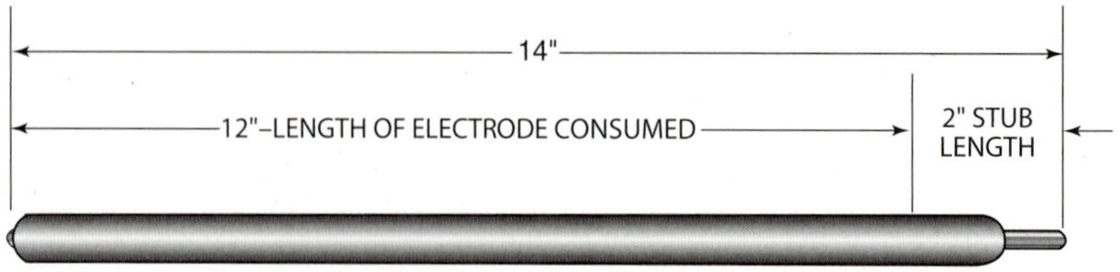

FIGURE 7-3 SMA welding electrodes should be used down to within 2 in. (51 mm) of the end. © Cengage Learning 2012

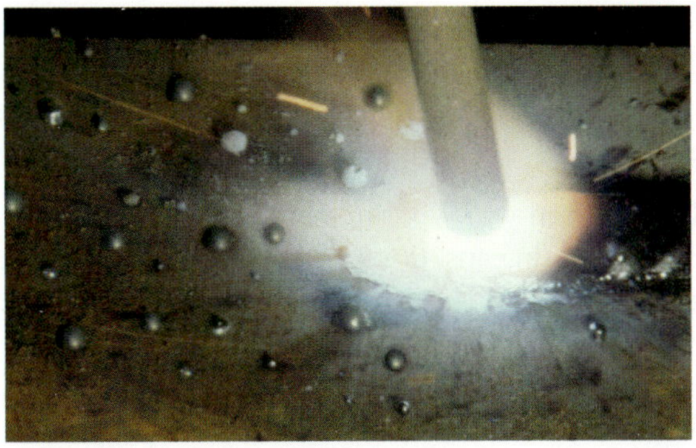

FIGURE 7-4 Excessive weld spatter caused by a high wind blowing shielding away. Larry Jeffus

can be caused by arc blow, wrong welding current or polarity, too long an arc length, or damp electrodes. Sometimes high winds can blow enough of the gaseous cloud away to cause excessive spatter.

Both GMA and FCA can produce excessive spatter when the welding voltage and amperage are not set within the electrode manufacturer's specifications. In some cases, the wrong shielding gas mixture for the electrode type can also cause spatter.

Gas Conservation

Shielding gases are used for all GMA and gas tungsten arc (GTA) welding and for some FCA welding processes. These shielding gases are used to protect the molten weld metal from atmospheric contamination. Several factors affect the shielding gas flow rate range required to provide adequate protection. These factors include the nozzle diameter, the nozzle-to-work distance, welding current, pre- and postpurge gas flow times, and draft or wind speed.

When possible, set the shielding gas flow rate on GMA, GTA, and FCA welding processes at the lower end of the flow rate range to reduce the shielding gas cost. Refer to the manufacturer's data sheet for the recommended flow rate range. Table 7-1 shows examples of the shielding gas flow rates for GTA welding. From Table 7-1 you can see that for metal of 16-gauge thickness, the gas flow rate is from 20 to 30 cubic feet per hour (CFH) when using a nozzle that is from 1/4 in. (6 mm) to 3/8 in. (10 mm) in diameter. Setting the flow rate at 20 CFH would be a one-third savings in shielding gas used as compared to the higher setting of 30 CFH.

Metal Thickness	Oxygen		Acetylene	
	PSI	CFH	PSI	CFH
1/4 in.	20-25	38-45	5-10	8-12
3/8 in.	20-28	55-60	5-10	8-12
1/2 in.	30-35	70-75	5-10	8-12

Table 7-2 Example of Oxyacetylene Pressures and Flow Rates for Cutting

Setting the gas pressures as low as possible within the manufacturer's recommended pressures for oxyfuel cutting reduces gas cost. Both too low a gas pressure and too high a gas pressure result in poor-quality cuts. So, setting the pressure correctly saves both gas costs and improves the quality of the cuts being made.

Table 7-2 shows examples of pressures and flow rates for oxyacetylene cutting torch tips used for various thicknesses of metal. There can be an approximate 10% savings in oxygen if the lower oxygen gas setting for 3/8-in. (10-mm) thick metal is used.

NOTE: Cutting torch tips must be clean to ensure that you can get clean cuts at the lower gas pressure settings.

Some high-pressure gas cylinder valves, such as those used on oxygen, argon, and mixtures of argon and oxygen, have small leaks around the cylinder valve stem, Figure 7-5. Although most of these leaks are very small, over a long period of time they can result in a significant loss of gas. Turning off the cylinder valve any time that the gas is not needed both reduces that possible leak and is required by safety rules.

Energy Conservation

All welding and cutting processes except for the few oxyfuel processes use large quantities of electrical power.

		GAS FLOW				
METAL THICKNESS	FILLER METAL DIA. OF ER4043	RATES CFM	PURGING TIMES		NOZZLE SIZE IN.	AMPERAGE MIN. MAX.
			PREPURGING	POSTPURGING		
18 ga.	3/32"	20 to 30	10 to 15 sec.	10 to 25 sec.	1/4 to 3/8	40 to 60
17 ga.	3/32"	20 to 30	10 to 15 sec.	10 to 25 sec.	1/4 to 3/8	50 to 70
16 ga.	3/32"	20 to 30	10 to 15 sec.	10 to 25 sec.	1/4 to 3/8	60 to 75

Table 7-1 Example of GTA Welding Shielding Gas Flows

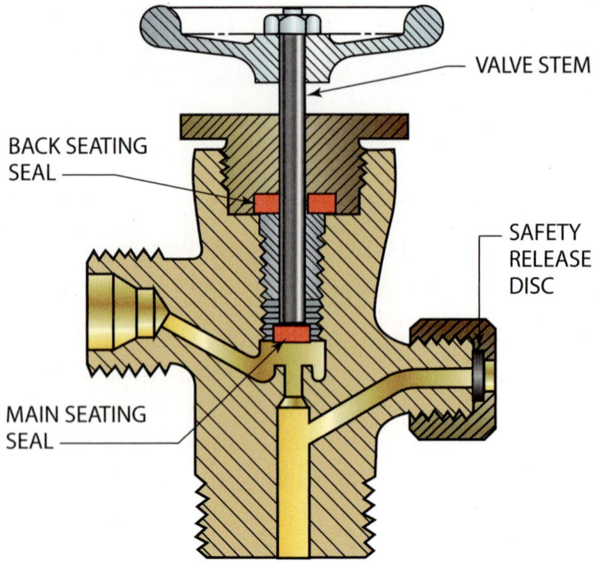

FIGURE 7-5 **High-pressure gas cylinder valves have two valve seals: a main seal and one back seal to prevent gas leaks.**

Equipment manufacturers have improved the operating efficiencies of newer models of their product lines. Using more efficient pieces of equipment reduces your shop's electrical costs. New equipment can be more efficient; however, it is not always cost effective for a welding shop to purchase new equipment just to save electrical energy.

The best way of saving energy when using any welding equipment is to turn off the machines when they are not being used. Check with the equipment manufacturer to see if turning off the power might cause any problems. A problem with turning off equipment immediately after a long period of welding might be that the cooling fan may not have removed enough internal heat from the welder.

Some equipment may have a small quantity of background power consumption even if the main equipment power switch is off. In these cases, the main power disconnect or the main power breaker must be turned off. The main power disconnect or main power breaker is not usually located on the equipment but is on the wall near the equipment.

Recycling

Recycling weld shop materials is important because it both helps the environment and can be a source of revenue for the shop as the scrap is sold to a recycler. Mixed metal recycling has a lower scrap value, so it is important, if possible, to keep different types of metal segregated in the waste stream in the shop. For example, keeping the aluminum separated from steel is important, but if you have a large amount of a particular grade of aluminum, it should also be segregated. The same goes for the many varieties of stainless steel.

NOTE: Often you will find that individuals will drop scrap paper, tin cans, drink cans, and other trash in the recycle container. This can result in a lower value of the scrap when it is sent to the recycler.

SPARK YOUR IMAGINATION

It is important to protect our environment. In a small group, create an action plan to improve some aspect of your shop's environment and its impact on your community's environment.

GENERAL SHOP PRACTICES

Operating Equipment

Lots of pieces of equipment in a welding shop are not typically found in most school welding shops. For example, production shops often have overhead hoists or cranes, fork trucks, punch presses, shears, and other similar production equipment. It is important to read and follow all manufacturers' operating and safety instructions for each piece of equipment. It is also important to have someone demonstrate the proper and safe operating techniques and possibly observe you using the equipment before you begin operating it by yourself. In some cases, such as the operation of overhead cranes, significant damage and danger can be caused when the equipment is overstressed by attempting to pick up a part when its weight is greater than the rated capacity of the hoist.

Hand Signals

Welding shops tend to be noisy places at times. This background noise can make it difficult to use two-way radios to communicate from the weld shop floor to crane operators. At other times a two-way radio may not be available, so there are standard hand signals used by most industries to communicate with the crane operator. Hand signals can provide the necessary instruction for the crane operator to safely and accurately place a component.

Figure 7-6 shows the basic hand signals most often used by crane operators and the related movement

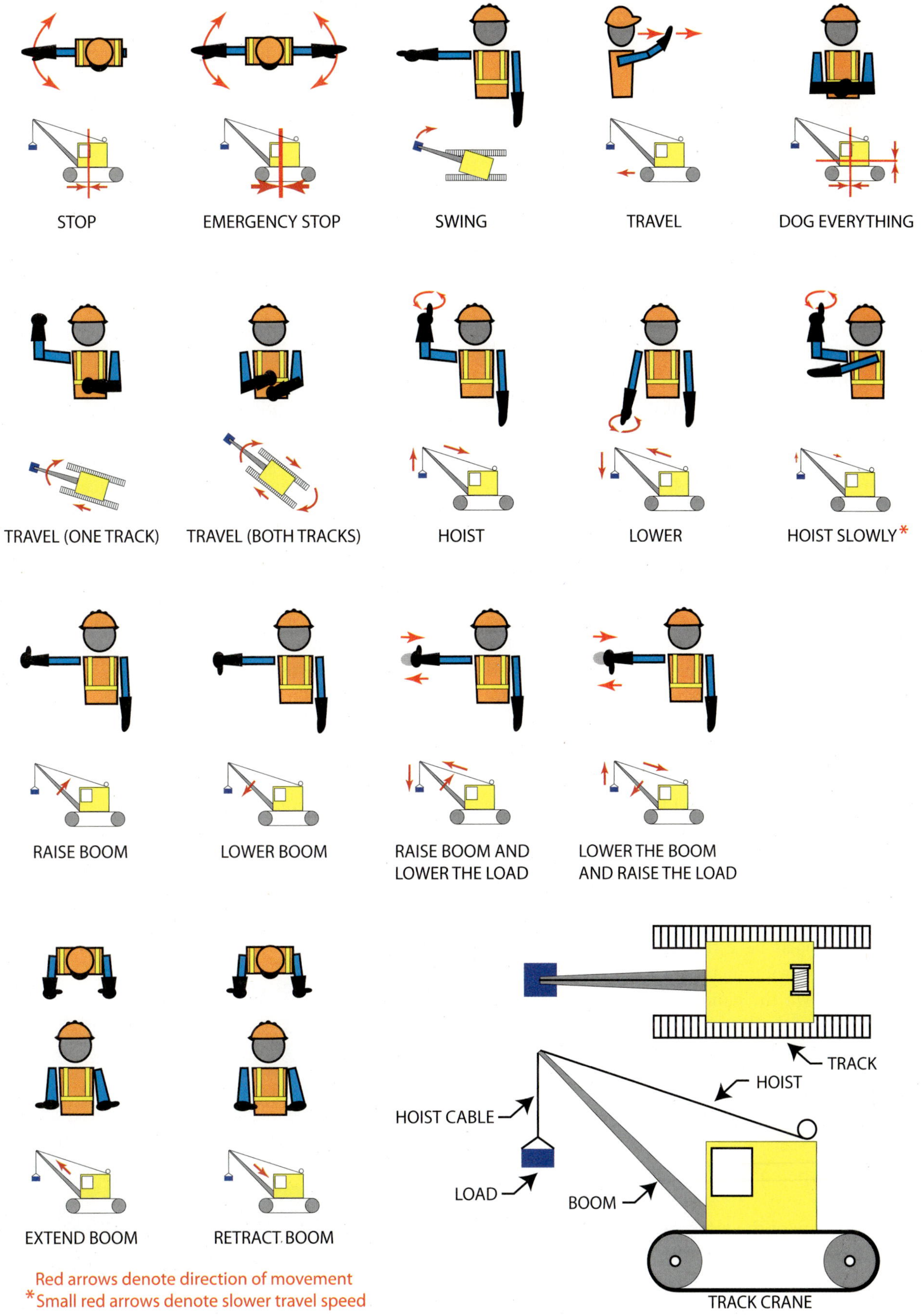

FIGURE 7-6 Crane hand signals. © Cengage Learning 2012

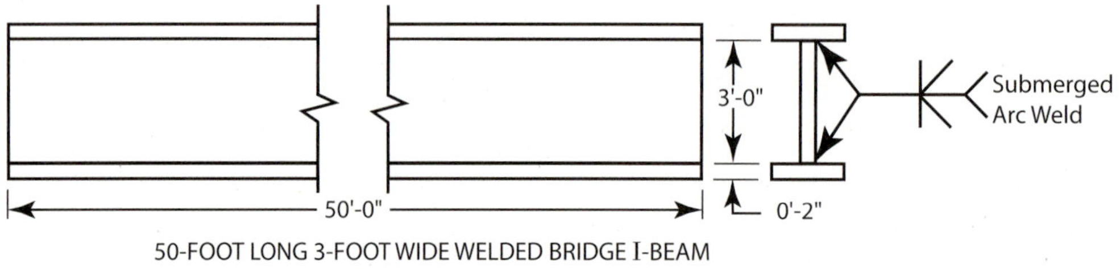

FIGURE 7-7 Welded 50-ft (15.2-m) bridge beam. © Cengage Learning 2012

of a crane. A basic track crane is depicted in the illustration, but the same hand signals can be used to communicate to any crane operator using any other type of equipment. The red arrows shown around the illustration's hand for "HOIST" indicate that this motion continues as long as the crane operator is lifting a load. Sometimes, the faster you make the circular motion with your hand, the faster the operator will lift the load. However, speed and safety are not equal; safety is the most important aspect of any operation.

As the boom is lowered, the load will both travel away from the crane and become lower. The reverse is true for raising the boom, where the load gets closer to the crane and becomes higher. To move a load horizontally toward or away from the crane, the operator must both raise the boom and lower the load or lower the boom and raise the load. These hand signals are performed by opening and closing your fisted hand with your thumb pointed up or down.

PRACTICE

In a small group, each person will make a list of five crane hand signals. Take turns giving the hand signals to the other group members. Each team member should write down the hand signals as they interpret them. Turn all of the lists in to your instructor after each group member has had a turn.

Outsourcing

Occasionally, when a large project is undertaken in the shop, there may be considerable advantages to outsourcing some of the work. Some of the reasons a shop might outsource are because the project requires equipment your shop does not have or another shop can produce it more cost effectively. For example, in the construction of large beams where welds are being made on a 2-in. (51-mm) thick plate forming an I-beam for a bridge, Figure 7-7; outsourcing that weld to a shop that has a large submerged arc welding machine might be the only way of producing it.

Another example would be when a large number of welds are to be made in a high-production run that would lend its production to a robot that your company does not own. Robots can produce these welds faster and more accurately than they could be produced manually in a shop.

NOTE: Often, welders want to do everything themselves; however, making a wise business decision relative to outsourcing can ultimately increase the shop's productivity and increase your earnings.

An example of cost-effectively outsourcing a lot of small parts might be to use a machine shop to punch them out rather than using your shop's flame-cutting equipment, Figure 7-8.

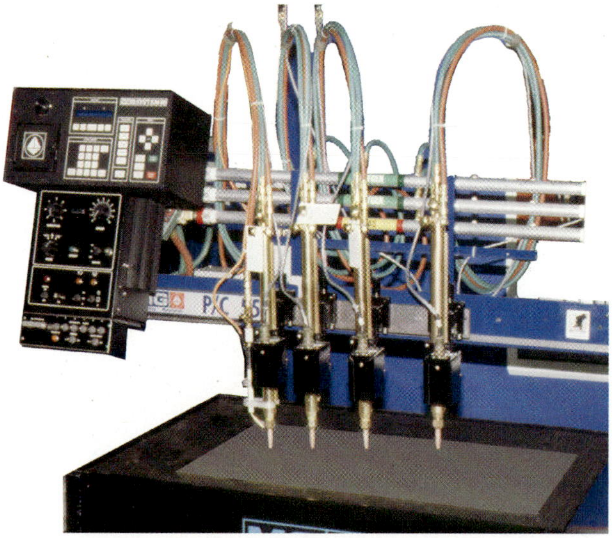

FIGURE 7-8 Four cutting head automatic flame-cutting machine. Larry Jeffus

SUMMARY

Once you get a job, your learning should not stop. You should familiarize yourself with everything you can about the shop, the equipment, and how the shop's systems work together. The more you learn, the more valuable you become to your company. It is that value along with your welding skills that can make you very successful.

REVIEW QUESTIONS

1. What should you do while you are waiting for your part to cool down?
2. Discuss how to develop a good working relationship with coworkers.
3. When would it be time-efficient to build a jig or fixture to help align parts?
4. How can employee tardiness and absences adversely affect a business?
5. What are some ways that leftover metal from a previous job can be used?
6. Draw an example of how parts might be laid out to conserve as much of the stock as possible.
7. What length of electrode stub should SMA welders leave?
8. Why is approximately 35% of every SMA welding electrode wasted?
9. In what situation should partially used electrodes not be used?
10. Why is it important to make correctly sized welds?
11. Why should a welder stop welding when excessive spatter is being produced?
12. What can cause excessive SMA welding spatter?
13. Why are shielding gases used for all GMA and GTA welding and some FCA welding processes?
14. Using Table 7-1, find the range of shielding gas flow rates for GTA welding on 17-gauge thickness.
15. Why is it important to set the oxygen pressure as low as possible when making an oxyfuel cut?
16. What is the best way of saving on electrical costs when using welding equipment?
17. Why is it important to recycle scrap metal?
18. What steps can be taken to ensure the safe operation of equipment in the welding shop?
19. How might hand signals be used in a welding shop?
20. When should a welding shop consider outsourcing a project to another shop?

Chapter 8

Shielded Metal Arc Equipment, Setup, and Operation

OBJECTIVES

After completing this chapter, the student should be able to:

- Describe the shielded metal arc welding process.
- List the three units used to describe an electric current and tell how they affect SMA welding.
- List the three types of welding current used in SMA welding.
- Describe how open-circuit and closed-circuit voltage affect SMA welding.
- Describe the force that causes arc blow and explain how it can be controlled.
- Explain how each type of welding power source produces the welding current.
- Determine the duty cycle for any given welder and amperage setting.
- Demonstrate how to set up a welding workstation.

KEY TERMS

amperage
anode
cathode
duty cycle
electrons
inverter
magnetic lines of flux
open circuit voltage
operating voltage
output
rectifier
step-down transformer
voltage
wattage
welding cables
welding leads

INTRODUCTION

Shielded metal arc welding (SMAW) is a welding process that uses a flux-covered metal electrode to carry an electrical current, Figure 8-1. The current forms an arc across the gap between the end of the electrode and the work.

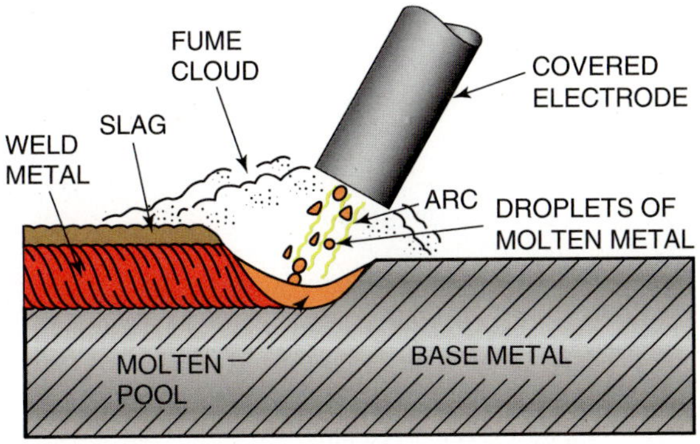

FIGURE 8-1 Shielded metal arc welding.
American Welding Society

The electric arc creates sufficient heat and temperature to melt both the end of the electrode and the base metal being welded. Molten metal from the end of the electrode travels across the arc to the molten pool on the metal being welded. The metal from the electrode and the molten base metal are mixed together to form the weld. The high temperature at the electrode end causes the flux covering around the electrode to burn or vaporize into a gaseous cloud. This gaseous cloud surrounds, purifies, and protects at the end of the electrode and molten pool of base metal. Some of the electrode flux forms a molten protective slag on top of the molten weld pool. As the arc moves away, the weld metal cools, forming one solid piece of metal.

SMAW is the most widely used welding process for metal fabrication because of its low cost, flexibility, portability, and versatility. The welding machine itself can be as simple as a 110-volt, step-down transformer that can be plugged into a normal electrical outlet. The electrodes are available from a large number of manufacturers in packages ranging from 1 lb (0.5 kg) to 50 lb (22 kg).

The SMAW process is very versatile because the same SMA welding machine can be used to make a wide variety of weld joint designs in a wide variety of metal types and thicknesses, and in all positions:

- Joint designs—In addition to the standard butt, lap, tee, and outside corner joints, at SMAW has been certified to be used to weld every possible joint design.
- Metal types—Although mild steel is the most common SMA-welded metal; stainless steel, aluminum, and cast iron are easily SMA welded.
- Metal thickness—Metal as thin as 16 gauge and approximately 1/16 in. (2 mm) thick up to several feet thick can be SMA welded.
- All positions—The flat welding position is the easiest and most productive because large welds can be made fast using SMA welding, but the process can be used to make welds in any position.

SMAW is a very portable process because it is easy to move the equipment from the shop to the job site. Engine-driven generator-type SMA welders are available that can be used almost anywhere. The limited amount of equipment required for the process makes moving easy.

WELDING CURRENT

Welding current is the term used to describe the electricity that jumps across the arc gap between the end of the electrode and the metal being welded. An electric current is the flow of **electrons.** The resistance to the flow of electrons (electricity) produces heat. The greater the electrical resistance, the greater the heat and temperature that the arc will produce. Air has a high resistance to current flow, so there is a lot of heat and temperature produced by the SMA welding arc. Electrons flow from negative (–) to positive (+), Figure 8-2.

Electrical Measurement

Three units are used to describe any electrical current. The three units are voltage (V), amperage (A), and wattage (W).

- **Voltage,** or volts (V), is the measurement of electrical pressure in the same way that pounds per square inch is a measurement of water pressure. Voltage controls the maximum gap that the electrons can jump to form the arc. A higher voltage can jump a larger gap. Welding voltage is associated with the welding temperature.
- **Amperage,** or amps (A), is the measurement of the total number of electrons flowing, in the same way that gallons are a measurement of the amount of water flowing. Amperage controls the size of the arc. Amperage is associated with the welding heat.
- **Wattage,** or watts (W), is a measurement of the amount of electrical energy or power in the arc. Watts are calculated by multiplying voltage (V) times amperes (A), Figure 8-3. Watts are associated with welding power or how much heat and temperature an arc produces, Figure 8-4.

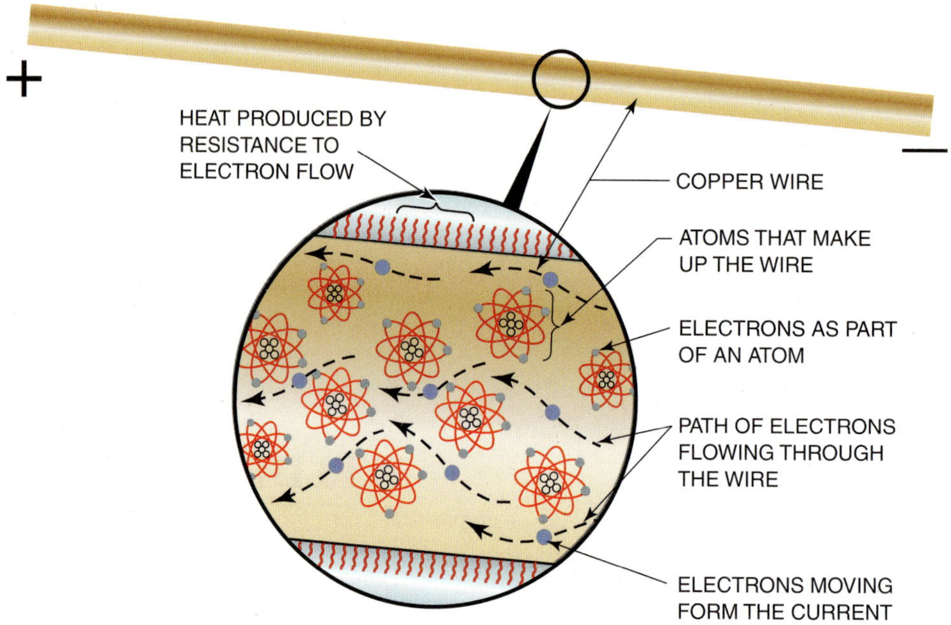

FIGURE 8-2 Electrons traveling along a conductor. © Cengage Learning 2012

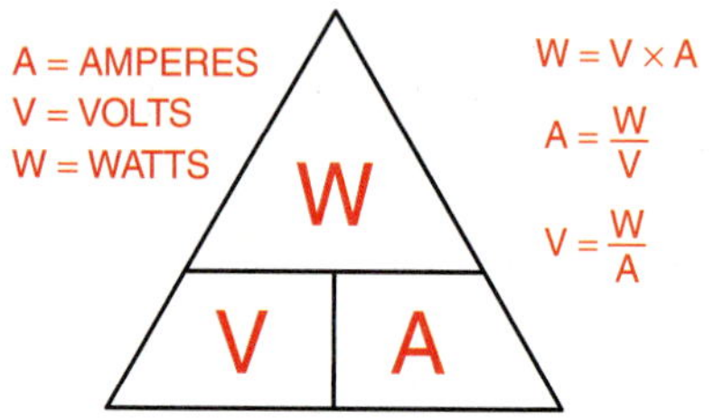

FIGURE 8-3 Ohm's Law. © Cengage Learning 2012

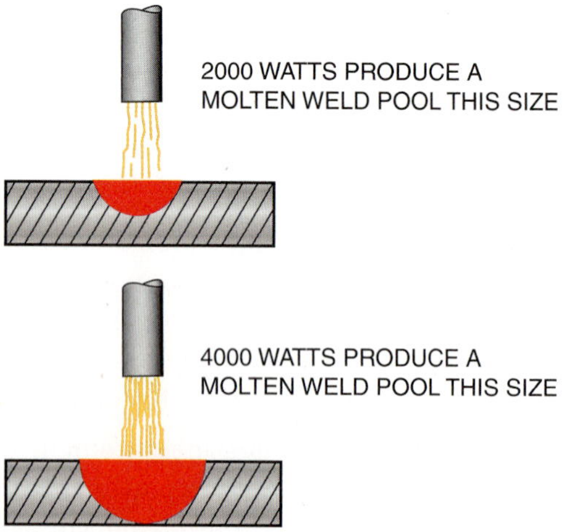

FIGURE 8-4 The molten weld pool size depends upon the energy (watts), the metal mass, and thermal conductivity. © Cengage Learning 2012

SMA Welding Arc Temperature and Heat

The terms *temperature* and *heat* are explained in Chapter 1.

The temperature of a welding arc is dependent on the voltage, arc length, and atmosphere. The arc temperature can range from around 5500°F to above 36,000°F, but most SMA welding arcs have effective temperatures around 11,000°F. The voltage and arc length are closely related. The shorter the arc, the lower the arc voltage and the lower the temperature produced, and as the arc lengthens, the resistance increases, thus causing a rise in the arc voltage and temperature.

Most shielded metal arc welding electrodes have chemicals added to their coverings to stabilize the arc. These arc stabilizers form conductive ions that make the arc more stable and reduce the arc resistance. This makes it easier to hold an arc. By lowering the resistance, the arc stabilizers also lower the arc temperature. Other chemicals within the gaseous cloud around the arc may raise or lower the resistance.

The amount of heat produced by the arc is determined by the amperage. The higher the amperage setting, the higher the heat produced by the welding arc, and the lower the amperage setting, the lower the heat produced. Each diameter of electrode has a recommended minimum and maximum amperage range and

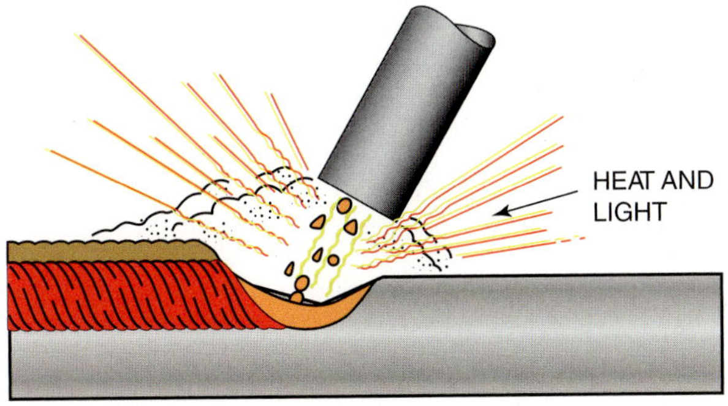

FIGURE 8-5 Energy is lost from the weld in the forms of radiation and convection.
© Cengage Learning 2012

therefore a recommended heat range. If you were to try to put too many amps through a small diameter electrode, it would overheat and could even melt. If the amperage setting is too low for an electrode diameter, the end of the electrode may not melt evenly, if at all.

Not all of the heat produced by an arc reaches the weld. Some of the heat is radiated away in the form of light and heat waves, Figure 8-5. Some additional heat is carried away with the hot gases formed by the electrode covering. Heat is also lost through conduction in the work. In total, about 50% of all heat produced by an arc is missing from the weld.

The 50% of the remaining heat produced by the arc is not distributed evenly between both ends of the arc. This distribution depends on the composition of the electrode's coating and type of welding current.

TYPES OF CURRENTS

The three different types of currents used for welding are direct-current electrode negative (DCEN), direct-current electrode positive (DCEP), and alternating current (AC).

DCEN: In direct-current electrode negative, the electrode is negative, and the work is positive, Figure 8-6. The electrons are leaving the electrode and traveling across the arc to the surface of the metal being welded. This results in approximately one-third of the welding heat on the electrode and two-thirds on the metal being welded. The former term for DCEN was DCSP, which meant direct-current straight polarity. DCEN welding current produces a high electrode melting rate.

DCEP: In direct-current electrode positive, the electrode is positive, and the work is negative, Figure 8-7.

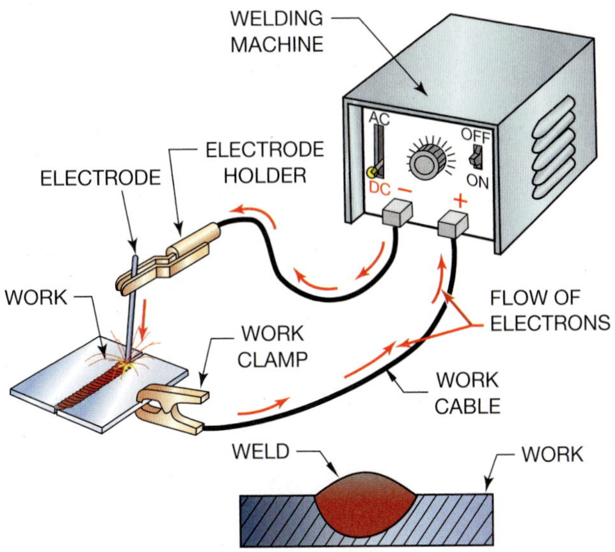

FIGURE 8-6 Electrode negative (DCEN), straight polarity (DCSP). © Cengage Learning 2012

The electrons are leaving the surface of the metal being welded and traveling across the arc to the electrode. This results in approximately two-thirds of the welding heat on the electrode and one-third on the metal being welded. The former term for DCEP was DCRP, which meant direct-current reverse polarity. DCEP current produces the best welding arc characteristics.

AC: In alternating current, the electrons change direction every 1/120 of a second so that the electrode and work alternate from **anode** to **cathode**, Figure 8-8. The rapid reversal of the current flow causes the

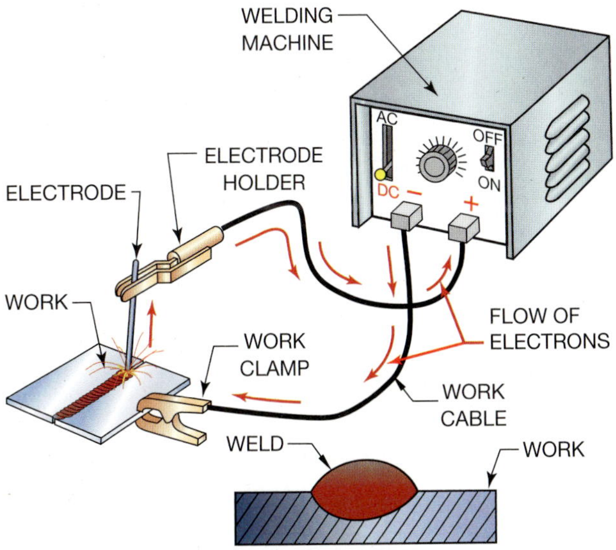

FIGURE 8-7 Electrode positive (DCEP), reverse polarity (DCRP). © Cengage Learning 2012

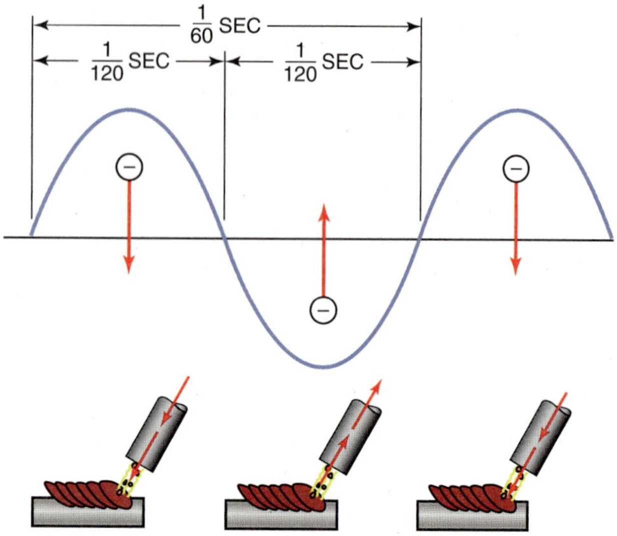

FIGURE 8-8 Alternating current (AC). © Cengage Learning 2012

welding heat to be evenly distributed on both the work and the electrode, that is, half on the work and half on the electrode, Figure 8-9. The even heating gives the weld bead a balance between penetration and buildup.

The amount of heat on the electrode and work are a factor in the weld buildup and penetration. However, the actual or effective heat input to the metal being welded is affected by the type of electrode being used. In some cases, the higher heat on the electrode end causes it to produce a more forceful arc that digs into the base metal, resulting in a deeper weld. So, the electrode type can have as much to do with the weld shape and weld penetration as the type of current selected.

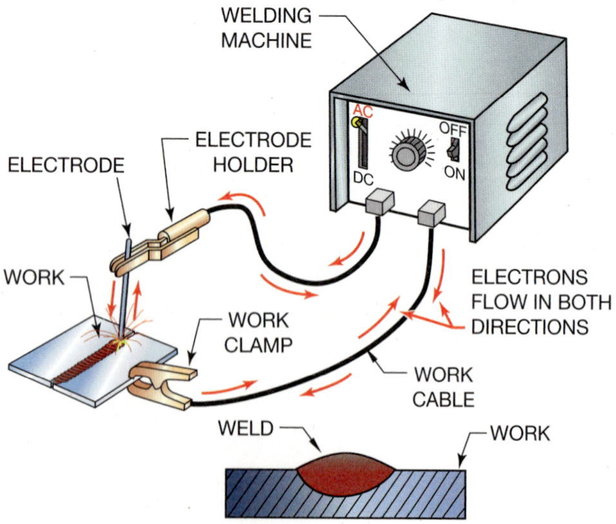

FIGURE 8-9 In an alternating current, electrons flow back and forth. © Cengage Learning 2012

WELDING POWER

The shielded metal arc welding process (SMAW) requires a constant current arc voltage characteristic, illustrated by the red line in Figure 8-10. Gas tungsten arc welding (GTAW) also uses this same type of welding power, but gas metal arc welding (GMAW) and flux cored arc welding (FCAW) both use a different type of welding power called constant voltage.

The SMA welding machines' voltage **output** decreases as current increases. This output power supply provides a reasonably high open circuit voltage before the arc is struck. The high open circuit voltage quickly stabilizes the arc. The arc voltage rapidly drops to the lower closed circuit level after the arc is struck. Following this short starting surge, the power (watts) remains almost constant despite the changes in arc length. With a constant voltage output, small changes in arc length would cause the power (watts) to make large swings. The welder would lose control of the weld.

OPEN CIRCUIT VOLTAGE

Open circuit voltage is the voltage at the electrode before striking an arc (with no current being drawn). The open circuit voltage is much like the higher surge

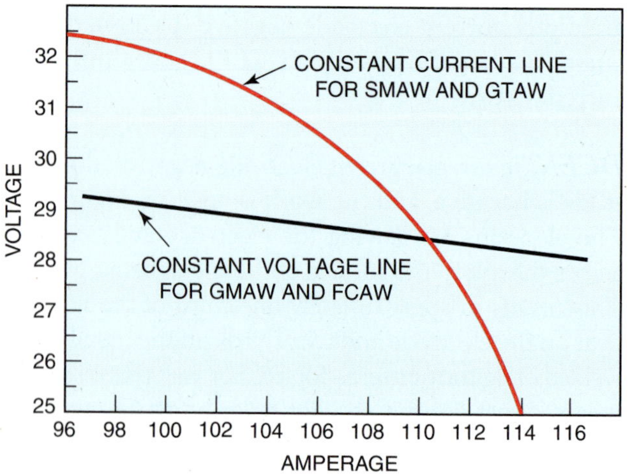

FIGURE 8-10 Constant voltage (CV) and constant current (CC). © Cengage Learning 2012

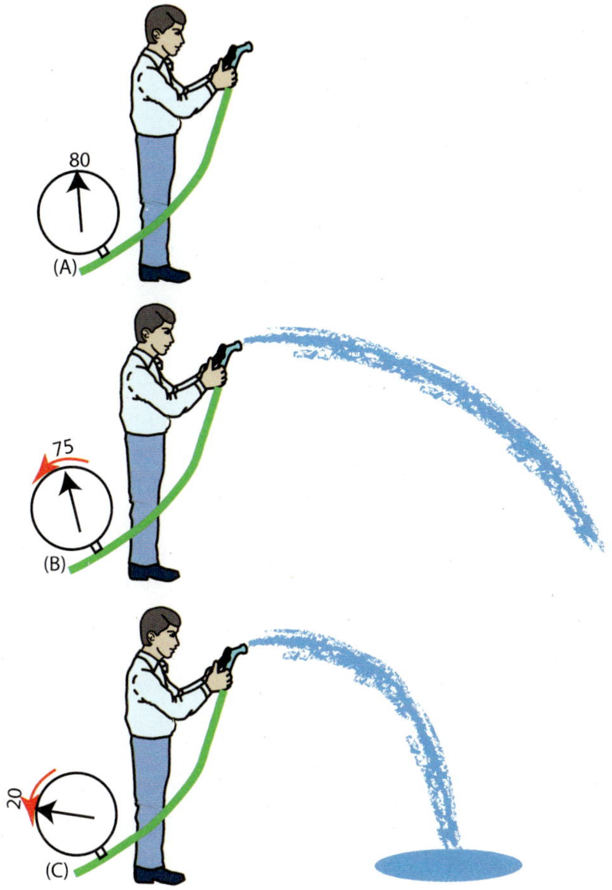

FIGURE 8-11 Electricity can have an initial surge much like the surge of water when a garden hose nozzle is first opened. © Cengage Learning 2012

of pressure you might observe when a water hose nozzle is first opened, Figure 8-11A and B. It is easy to see that the initial pressure from the garden hose was higher than the pressure of the continuous flow of water. The open circuit voltage is usually between 50 and 80 V. The higher the open circuit voltage, the easier it is to strike an arc because of the initial higher voltage pressure.

CAUTION

The maximum safe open circuit voltage for welders is 80 V. Higher voltages increase the chance of electrical shock.

OPERATING VOLTAGE

Operating, welding, or closed circuit voltage, is the voltage at the arc during welding. **Operating voltage** is much like the water pressure observed as the water hose is being used, Figure 8-11C. The operating voltage will vary with arc length, type of electrode being used, type of current, and polarity. The welding voltage will be between 17 and 40 V.

ARC BLOW

When electrons flow, they create lines of magnetic force that circle around the path of flow, Figure 8-12. These lines of magnetic force are referred to as **magnetic lines of flux.** They space themselves evenly along a current-carrying wire. If the wire is bent, the flux lines on one side are compressed together, and those on the other side are stretched out, Figure 8-13. The unevenly spaced flux lines try to straighten the wire so that the lines can be evenly spaced once again. The force that they place on the wire is usually small, so the wire does not move. However, when welding with very high amperages, 600 amperes or more, the force may actually cause the wire to move.

The welding current flowing through a plate or any residual magnetic fields in the plate results in unevenly spaced flux lines. These uneven flux lines can, in turn, cause the arc between the electrode and the work to move during welding. The term *arc blow*

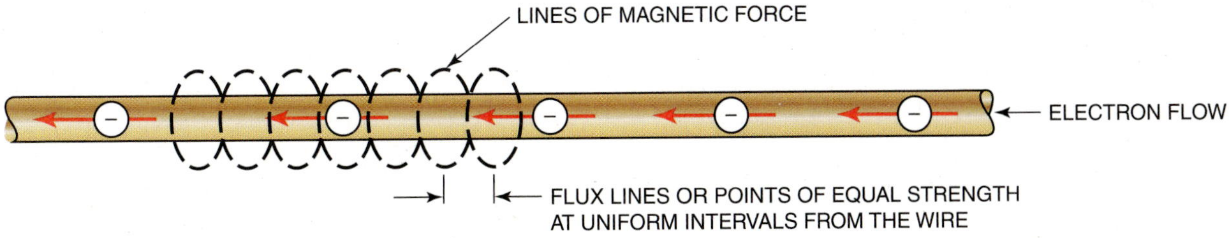

FIGURE 8-12 Magnetic force around a wire. © Cengage Learning 2012

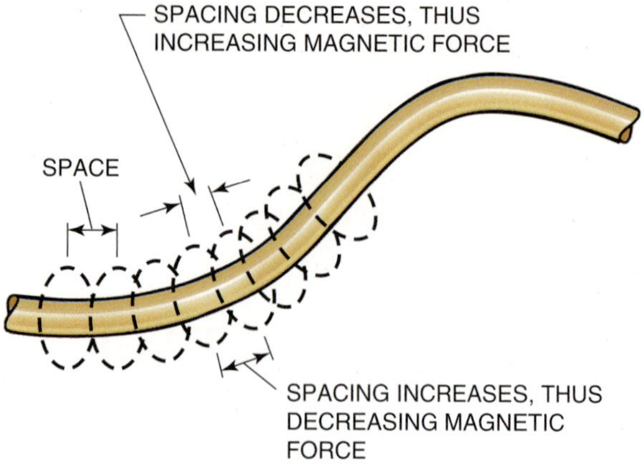

FIGURE 8-13 Magnetic forces concentrate around bends in wires. © Cengage Learning 2012

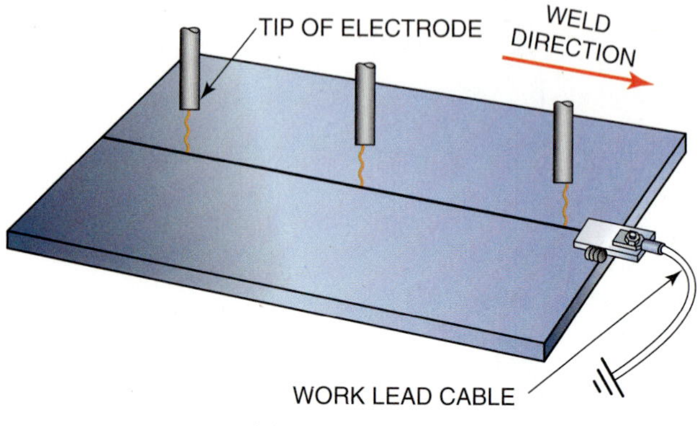

FIGURE 8-15 Correct current connections to control arc blow. © Cengage Learning 2012

refers to this movement of the arc. Arc blow makes the arc drift like a string would drift in the wind. Arc blow can be more of a problem when the magnetic fields are the most uneven such as when they are concentrated in corners, at the ends of plates, and when the work lead is connected to only one side of a plate, Figure 8-14.

The more complex a weldment becomes, the more likely arc blow will become a problem. Complex weldments can distort the magnetic lines of flux in unexpected ways. If you encounter severe arc blow during a weld, stop welding and take corrective measures to control or reduce the arc blow.

Arc blow can be controlled or reduced by connecting the work lead to the end of the weld joint and then welding away from the work lead, Figure 8-15. Another way of controlling arc blow is to use two work leads, one on each side of the weld. The best way to eliminate arc blow is to use an alternating current.

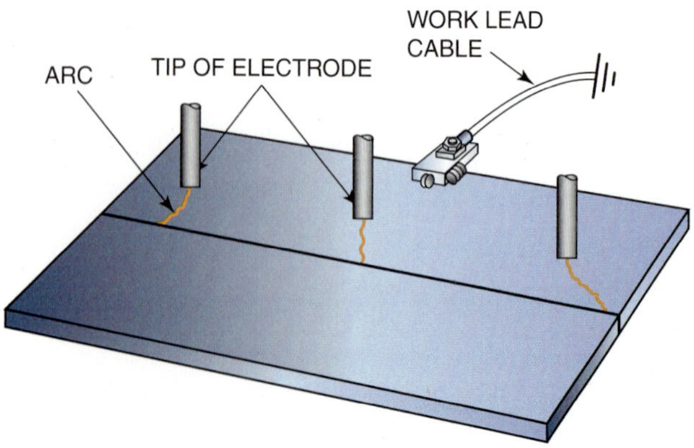

FIGURE 8-14 Arc blow. © Cengage Learning 2012

Because an alternating current changes directions, the flux lines do not become strong enough to bend the arc before the current changes direction. If it is impossible to move the work connection or to change to AC, a very short arc length can help control the arc blow. A large tack weld or a change in the electrode angle can also help control arc blow.

Arc blow may not be a problem as you are learning to weld in the shop because most welding tables are all steel. However, if you are using a pipe stand to hold your welding practice plates, arc blow can become a problem. Try reclamping your practice plates.

TYPES OF POWER SOURCES

Two types of electrical devices can be used to produce the low-voltage, high-amperage current combination that arc welding requires. One type uses step-down transformers. Because transformer-type welding machines are quieter, more energy efficient, require less maintenance, and are less expensive, they are now the industry standard. The other type uses an engine to drive an alternator or generator. Engine-powered generators are widely used for portable welding.

Transformer-Type Welding Machines

A transformer welding machine is connected to a high-voltage alternating current (AC) that is supplied to the welding shop by the utility company. It converts it to the low-voltage welding current. The heart of these

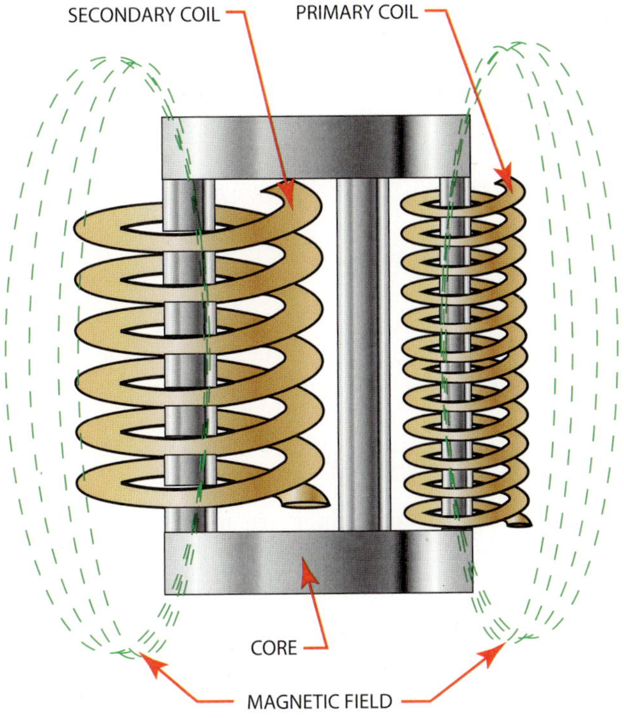

FIGURE 8-16 Parts of a step-down transformer.

welders is the **step-down transformer.** All transformers have the following three major components:

- Primary coil—The winding that is attached to the incoming electrical power.
- Secondary coil—The winding that has the electrical current induced and is connected to the welding lead and work leads.
- Core—Made of laminated sheets of steel and is used to concentrate the magnetic field produced in the primary winding into the secondary winding, Figure 8-16.

As electrons flow through a wire that is wound into the primary coil, each wire's weak magnetic field is combined to produce a much stronger central magnetic field. The magnetic field is constantly building and collapsing as the alternating current cycles back and forth 60 times a second. As the magnetic field passes back and forth over the secondary coil, an electric current is induced in the wires of this winding. The iron core in the center of these coils increases the concentration of the magnetic field, Figure 8-17.

A welding transformer has more turns of wire in the primary winding than in the secondary winding and is known as a step-down transformer. A step-down transformer takes a high-voltage, low-amperage current and changes it into a low-voltage, high-amperage current. Except for some power lost by heat within a transformer, the power (watts) into a transformer equals the power (watts) out because the volts and amperes are mutually increased and decreased.

A transformer welder is a step-down transformer. It takes the high line voltage (110 V, 220 V, 440 V, etc.) and low-amperage current (30 A, 50 A, 60 A, etc.) and changes it into 17 to 45 V at 190 to 590 A.

Welding machines can be classified by the method by which they control or adjust the welding current. The major classifications are multiple coil, called taps; movable coil or movable core, Figure 8-18; and inverter type.

Multiple Coil Welders

The multiple-coil machine, or tap-type machine, allows the selection of different current settings by tapping into the secondary coil at a different turn value. The greater the number of turns, the higher the

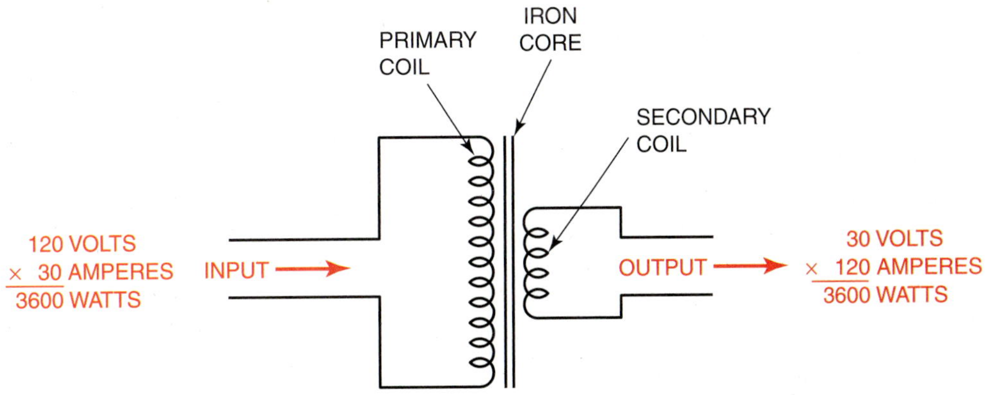

FIGURE 8-17 Diagram of a step-down transformer.

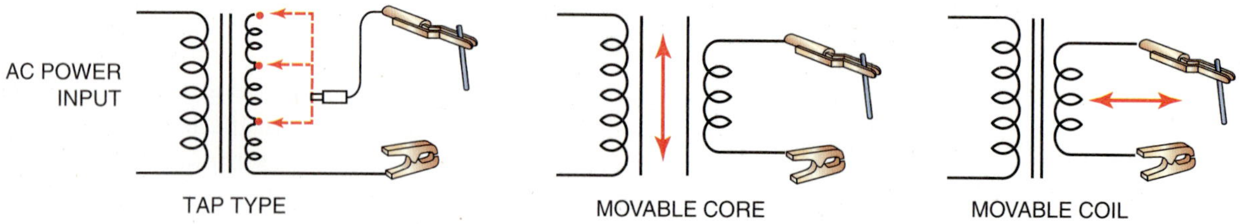

FIGURE 8-18 Three basic types of transformer welding machines. © Cengage Learning 2012

amperage induced in the turns. These machines may have a large number of fixed amperes, Figure 8-19, or they may have two or more amperages that can be adjusted further with a fine adjusting knob. The fine adjusting knob may be marked in amperes, or it may be marked in tenths, hundredths, or in any other unit.

Movable Coil or Core Welders

Movable coil or movable core machines are adjusted by turning a hand wheel that moves the internal parts closer together or farther apart. The adjustment may also be made by moving a lever, Figure 8-20. These machines may have a high and low range, but they do not have a fine adjusting knob. The closer the primary and secondary coils are, the greater the induced current; the greater the distance between the coils, the smaller the induced current, Figure 8-21. Moving the core in, concentrates more of the magnetic force on the secondary coil, thus increasing the current. Moving the core out, allows the field to disperse, and the current is reduced, Figure 8-22.

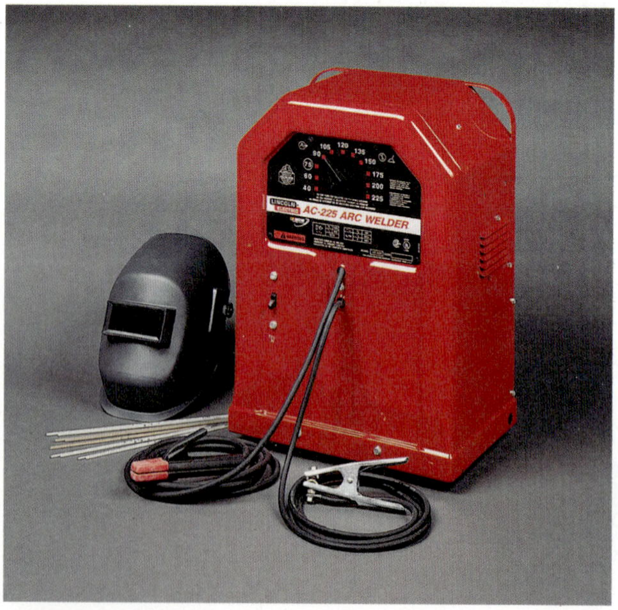

FIGURE 8-19 Tap-type transformer welding machine.
The Lincoln Electric Company

Inverter Welders

Inverter welding machines are much smaller than other types of machines of the same amperage range. This smaller size makes the welder much more portable as well as increases the energy efficiency. In a standard welding transformer, the iron core used to

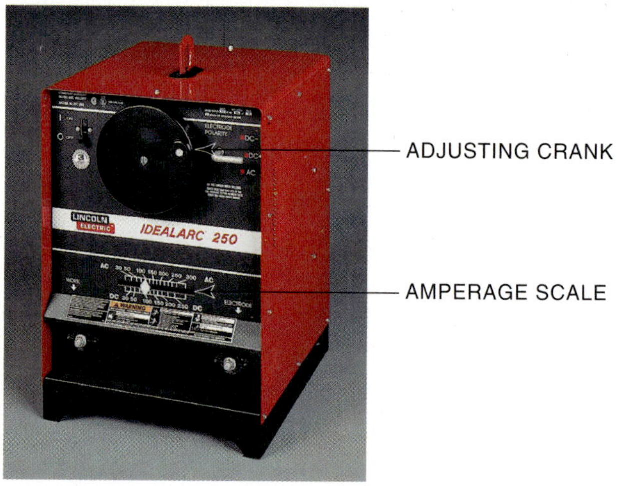

FIGURE 8-20 A movable core–type welding machine.
The Lincoln Electric Company

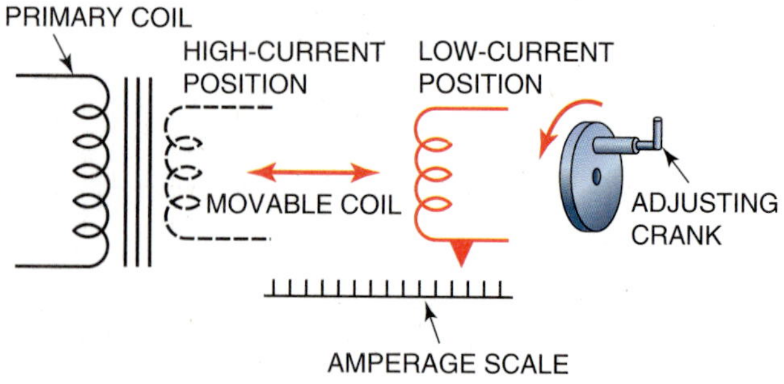

FIGURE 8-21 Movable coil. © Cengage Learning 2012

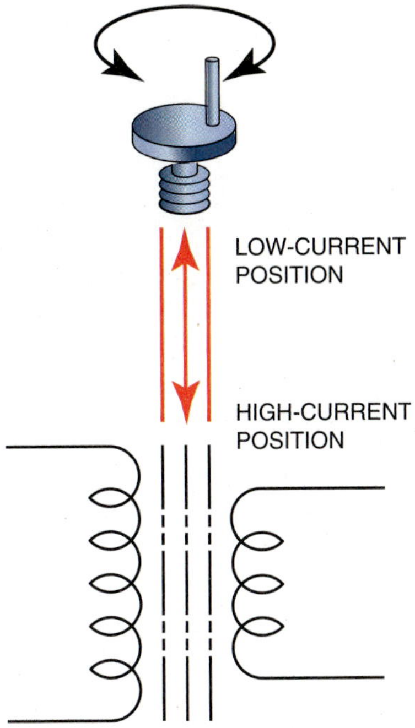

FIGURE 8-22 Moving the core out allows the field to disperse, and the current is reduced.

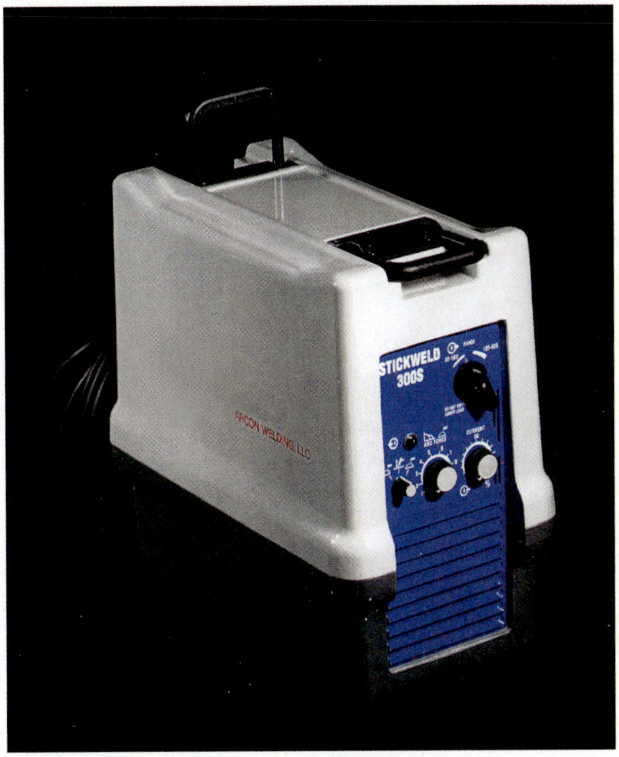

FIGURE 8-23 Typical 300-ampere inverter-type power supply weighing only 70 lb. Arcon Welding, LLC

concentrate the magnetic field in the coils must be a specific size. The size of the iron core is determined by the length of time it takes for the magnetic field to build and collapse. By using solid-state electronic parts, the incoming power in an inverter welder is changed from 60 cycles a second to several thousand cycles a second. This higher frequency allows a transformer that may be as light as 7 lb to do the work of a standard transformer weighing 100 lb. Additional electronic parts remove the high frequency for the output welding power.

The use of electronics in the inverter-type welder allows it to produce any desired type of welding power. Before the invention of this machine, each type of welding required a separate machine. Now a single welding machine can produce the specific type of current needed for shielded metal arc welding, gas tungsten arc welding, gas metal arc welding, and plasma arc cutting. Because the machine can be light enough to be carried closer to work, shorter welding cables can be used. The welder does not have to walk as far to adjust the machine. Welding machine power wire is cheaper than welding cables. Some manufacturers produce machines that can be stacked so that when you need a larger machine, all you have to do is add another unit to your existing welder, Figure 8-23.

GENERATOR AND ALTERNATOR WELDERS

Generators and alternators both produce welding electricity from a mechanical power source. Both devices have an armature that rotates and a stator that is stationary. As a wire moves through a magnetic force field, electrons in the wire are made to move, producing electricity.

In an alternator, magnetic lines of force rotate inside a coil of wire, Figure 8-24. An alternator can produce AC only. In a generator, a coil of wire rotates inside a magnetic force. A generator can produce AC or DC. It is possible for alternators and generators to both use diodes to change the AC to DC for welding.

In generators, the welding current is produced on the armature and is picked up with brushes, Figure 8-25. In alternators, the welding current is produced on the stator, and only the small current for the electromagnetic force field goes across the brushes. Therefore, the brushes in an alternator are smaller and last longer. Alternators can be smaller in size and lighter in weight than generators and still produce the same amount of power.

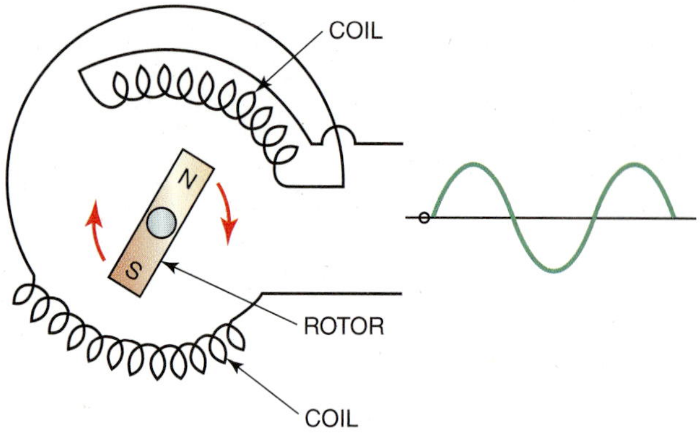

FIGURE 8-24 Schematic diagram of an alternator. © Cengage Learning 2012

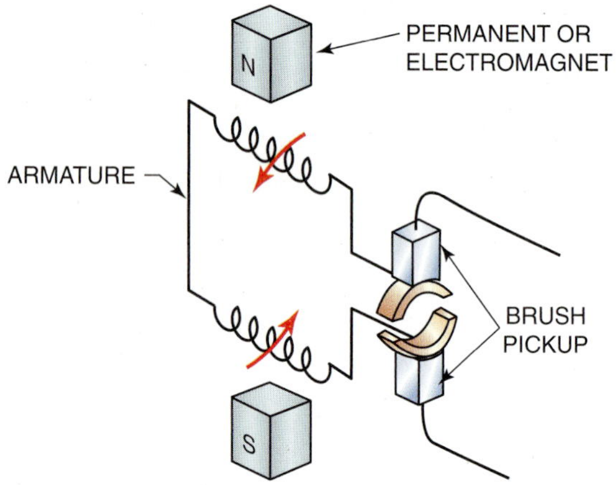

FIGURE 8-25 Diagram of a generator. © Cengage Learning 2012

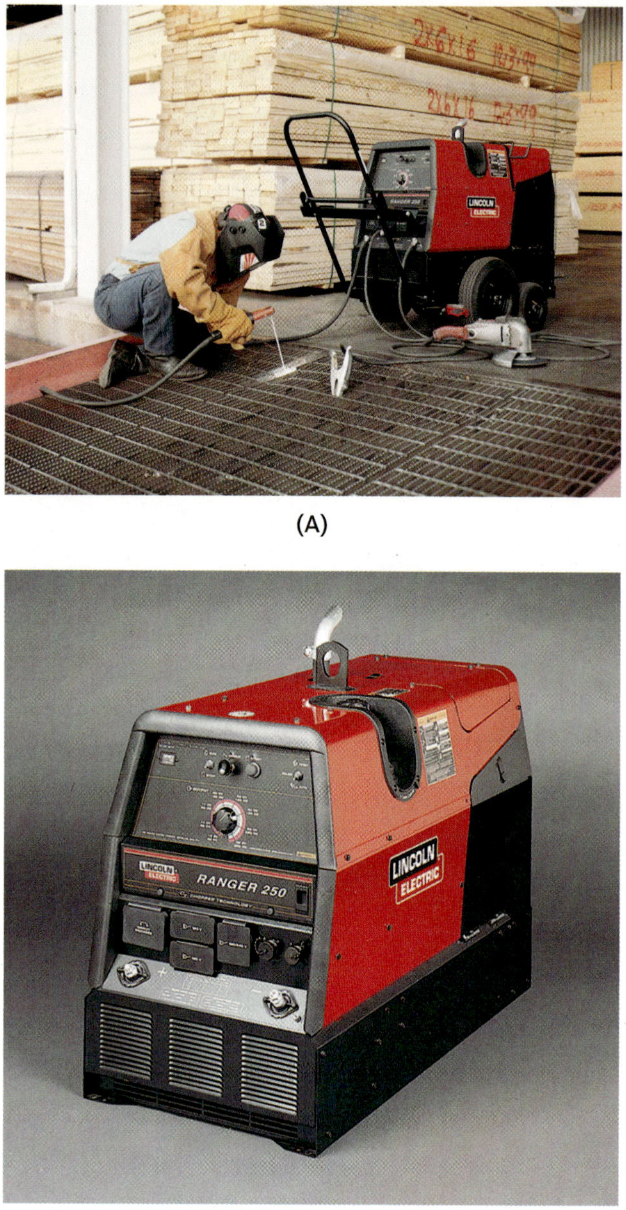

(A)

(B)

FIGURE 8-26 A portable engine generator welder. The Lincoln Electric Company

Older engine-driven generators and alternators may run at the welding speed all the time. Newer engine-driven machines have a controller that reduces their speed to an idle when welding stops. This controller saves fuel and reduces wear on the welding machine. To strike an arc when using this type of welder, stick the electrode to the work for a second. When you hear the welding machine (welder) pick up speed, remove the electrode from the work, and strike an arc. In general, the voltage and amperage are too low to start a weld, so shorting the electrode to the work should not cause the electrode to stick. A timer can be set to control the length of time that the welder maintains speed after the arc is broken. The time should be set long enough to change electrodes without losing speed.

Portable welders often have 110- or 220-volt plug outlets, which can be used to run grinders, drills, lights, and other equipment. The power provided may be AC or DC. If DC is provided, only equipment with brush-type motors or tungsten light bulbs can be used. If the plug is not specifically labeled 110 volts AC, check the owner's manual before using it for such devices as radios or other electronic equipment. A typical portable welder is shown in Figure 8-26.

It is recommended that a routine maintenance schedule for portable welders be set up and followed. By checking the oil, coolant, battery, filters, fuel, and other parts, the life of the equipment can be extended. A checklist can be posted on the welder, Table 8-1.

Check Each Day before Starting
Oil level
Water level
Fuel level
Check Each Monday
Battery level
Cables
Fuel line filter
Check at Beginning of Month
Air filter
Belts and hoses
Change oil and filter
Check Each Fall
Antifreeze
Test battery
Pack wheel bearings
Change gas filter

The owner's manual should be checked for any additional items that might be needed.

Table 8-1 Portable Welder Checklist
(ESAB Welding & Cutting Products)

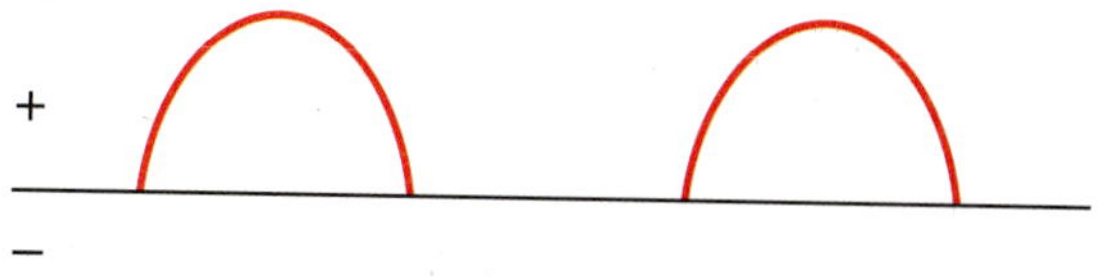

FIGURE 8-28 One rectifier in a welding power supply results in pulsing power. © Cengage Learning 2012

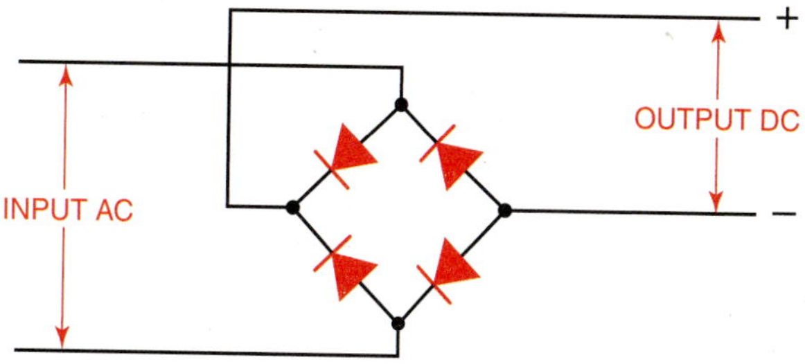

FIGURE 8-29 Bridge rectifier. © Cengage Learning 2012

CONVERTING AC TO DC

An alternating welding current can be converted to direct current by using a series of rectifiers. A **rectifier** allows a current to flow in one direction only, Figure 8-27. If one rectifier is added, the welding power appears as shown in Figure 8-28. It would be difficult to weld with pulsating power such as this. A series of rectifiers, known as a bridge rectifier, can modify the alternating current so that it appears as shown in Figure 8-29.

Rectifiers become hot as they change AC to DC. They must be attached to a heat sink and cooled by having air blown over them. The heat produced by a rectifier reduces the power efficiency of the welding machine. Figure 8-30 shows the amperage dial of a typical machine. Notice that at the same dial settings for AC and DC, the DC is at a lower amperage. The difference in amperage (power) is due to heat lost in the rectifiers. The loss in power makes operation with AC more efficient and less expensive compared to DC.

A DC adapter for small AC machines is available from manufacturers. For some types of welding, AC does not work properly.

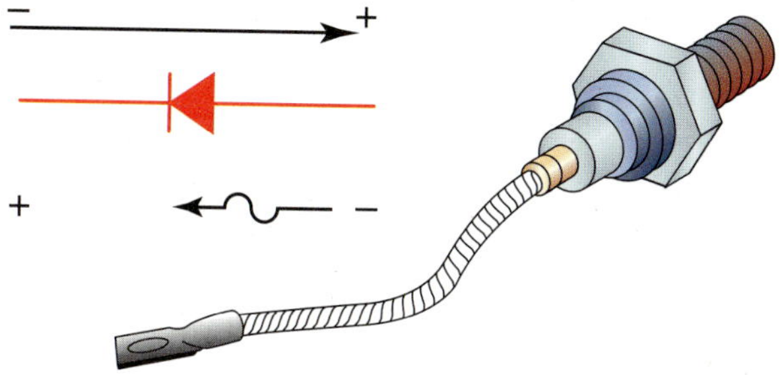

FIGURE 8-27 Rectifier. © Cengage Learning 2012

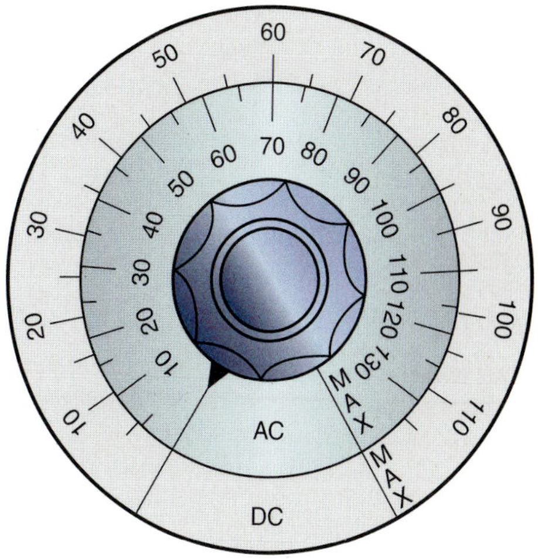

FIGURE 8-30 Typical dial on an AC-DC transformer rectifier welder. © Cengage Learning 2012

DUTY CYCLE

The **duty cycle** is the percentage of time a welding machine can be used continuously. Most SMA welding machines cannot be used 100% of the time because they produce some internal heat at the same time that they produce the welding current. SMA welders are rarely used every minute for long periods of time. The welder must take time to change electrodes, change positions, or change parts.

The duty cycle of a welding machine increases as the amperage is reduced and decreases as the amperage is raised, Figure 8-31. Most SMA welding machines weld at a 60% rate or less. Therefore, most manufacturers list the amperage rating for a 60% duty cycle on the nameplate that is attached to the machine. Other duty cycles are given on a graph in the owner's manual. A 60% duty cycle means that out of every 10 minutes, the machine can be used for 6 minutes at the maximum rated current. When providing power at this level, it must be cooled off for 4 minutes out of every 10 minutes.

The manufacturing cost of power supplies increases in proportion to their rated output and duty cycle. To reduce their price, it is necessary to reduce either their rating or their duty cycle. For this reason, some home-hobby welding machines may have duty cycles as low as 20% even at a low welding setting of 90 to 100 amperes. The duty cycle on these machines should never be exceeded because a buildup of the internal temperature can cause the transformer insulation to break down, damaging the power supply.

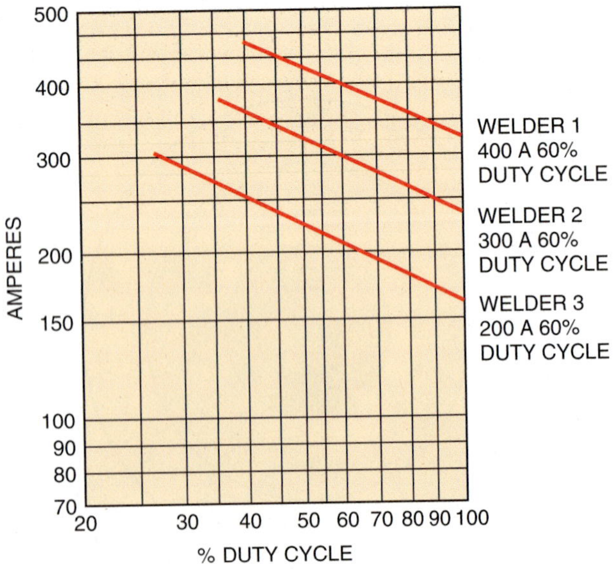

FIGURE 8-31 Duty cycle of a typical shielded metal arc welding machine. © Cengage Learning 2012

WELDER ACCESSORIES

A number of items must be used with a welding machine to complete the setup. The major items are the welding cables, the electrode holders, and the work clamps.

Welding Cables

The terms **welding cables** or **welding leads** are used to mean the same thing. Cables used for welding must be flexible, well-insulated, and the correct size for the job. Most welding cables are made from stranded copper wire. Some manufacturers sell a newer type of cable made from aluminum wires. The aluminum wires are lighter and less expensive than copper. Because aluminum as a conductor is not as good as copper for a given wire size, the aluminum wire should be one size larger than would be required for copper.

The insulation on welding cables will be exposed to hot sparks, flames, grease, oils, sharp edges, impact, and other types of wear. To withstand such wear, only specially manufactured insulation should be used for welding cable. Several new types of insulation are available that will give longer service against these adverse conditions.

As electricity flows through a cable, the resistance to the flow causes the cable to heat up and increase the voltage drop. To minimize the loss of power and prevent overheating, the electrode cable and work cable must be the correct size. Table 8-2 lists the minimum size cable that is required for each amperage and length. Large welding lead sizes make electrode manipulation difficult. Smaller cable can be spliced to the electrode end of a large cable to make it more flexible. This whip-end cable must not be over 10 ft (3 m) long.

CAUTION

A splice in a cable should not be within 10 ft (3 m) of the electrode because of the possibility of electrical shock.

Length of Cable (ft)	Length of Cable (m)	Copper Welding Lead Sizes								
Amperes		100	150	200	250	300	350	400	450	500
50	15	2	2	2	2	1	1/0	1/0	2/0	2/0
75	23	2	2	1	1/0	2/0	2/0	3/0	3/0	4/0
100	30	2	1	1/0	2/0	3/0	4/0	4/0		
125	38	2	1/0	2/0	3/0	4/0				
150	46	1	2/0	3/0	4/0					
175	53	1/0	3/0	4/0						
200	61	1/0	3/0	4/0						
250	76	2/0	4/0							
300	91	3/0								
350	107	3/0								
400	122	4/0								

Length of Cable (ft)	Length of Cable (m)	Aluminum Welding Lead Sizes								
Amperes		100	150	200	250	300	350	400	450	500
50	15	2	2	1/0	2/0	2/0	3/0	4/0		
75	23	2	1/0	2/0	3/0	4/0				
100	30	1/0	2/0	4/0						
125	38	2/0	3/0							
150	46	2/0	3/0							
175	53	3/0								
200	61	4/0								
225	69	4/0								

Table 8-2 Welding Lead Sizes

Splices and end lugs are available from suppliers. Be sure that a good electrical connection is made whenever splices or lugs are used. A poor electrical connection will result in heat buildup, voltage drop, and poor service from the cable. Splices and end lugs must be well-insulated against possible electrical shorting, Figure 8-32.

Electrode Holders

The electrode holder should be of the proper amperage rating and in good repair for safe welding. Electrode holders are designed to be used at their maximum amperage rating or less. Higher amperage values cause the holder to overheat and burn up. If the holder is too large for the amperage range being used, manipulation is hard, and operator fatigue increases. Make sure that the correct size holder is chosen, Figure 8-33.

CAUTION

Never dip a hot electrode holder in water to cool it off. The problem causing the holder to overheat should be repaired.

A properly sized electrode holder can overheat if the jaws are dirty or too loose, or if the cable is loose. If the holder heats up, welding power is being lost. In addition, a hot electrode holder is uncomfortable to work with.

FIGURE 8-32 Power lug protection is provided by insulators. ESAB Welding & Cutting Products

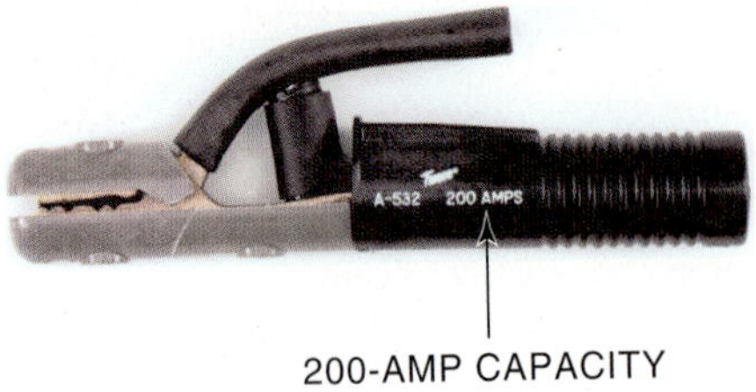

FIGURE 8-33 The amperage capacity of an electrode holder is often marked on its side. Thermadyne Industries, Inc.

Replacement springs, jaws, insulators, handles, screws, and other parts are available to keep the holder in good working order, Figure 8-34. To prevent excessive damage to the holder, welding electrodes should not be burned too short. A 2-in. (51-mm) electrode stub is short enough to minimize electrode waste and save the holder.

Work Clamps

The work clamp must be the correct size for the current being used, and it must clamp tightly to the material. Heat can build up in the work clamp, reducing welding efficiency, just as was previously described for the electrode holder. Power losses in the work clamp are often overlooked. The clamp should be touched occasionally to find out if it is getting hot.

In addition to power losses due to poor work lead clamping, a loose clamp may cause arcing that can damage a part. If the part is to be moved during welding, a swivel-type work clamp may be needed, Figure 8-35. It may be necessary to weld a tab to thick parts so that the work lead can be clamped to the tab, Figure 8-36.

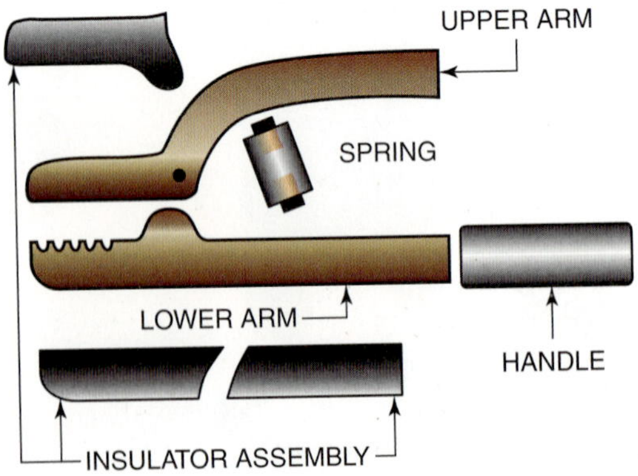

FIGURE 8-34 Replaceable parts of an electrode holder. © Cengage Learning 2012

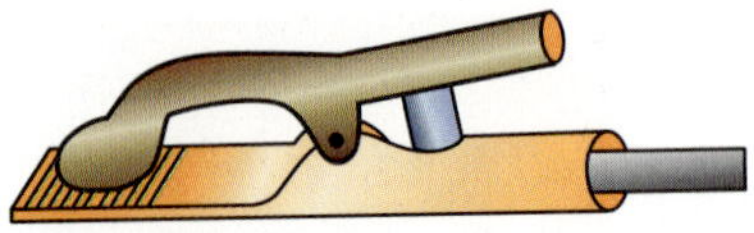

FIGURE 8-35 A work clamp may be attached to the workpiece. © Cengage Learning 2012

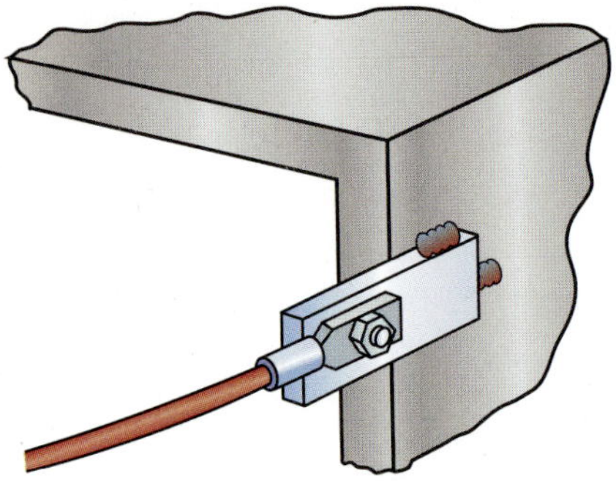

FIGURE 8-36 Tack welded ground to part. © Cengage Learning 2012

EQUIPMENT SETUP

Arc welding machines should be located near the welding site but far enough away so that they are not covered with spark showers. The machines may be stacked to save space, but there must be room enough between the machines for proper air to circulate to keep the machines from overheating. The air that is circulated through the machine should be as free as possible of dust, oil, and metal filings. Even in a good location, the power should be turned off periodically and the machine blown out with compressed air, Figure 8-37.

FIGURE 8-37 Slag, chips from grinding, and dust must be blown out occasionally so that they will not start a fire or cause a short-out or other type of machine failure. © Cengage Learning 2012

The welding machine should be located away from cleaning tanks and any other sources of corrosive fumes that could be blown through it. Water leaks must be fixed and puddles cleaned up before a machine is used.

Power to the machine must be fused, and a power shutoff switch provided. The switch must be located so that it can be reached in an emergency without touching either the machine or the welding station. The machine case or frame must be grounded.

The welding cables should be sufficiently long to reach the workstation but not so long that they must always be coiled. Cables must not be placed on the floor in aisles or walkways. If workers must cross a walkway, the cable must be installed overhead, or it must be protected by a ramp, Figure 8-38. The welding machine and its main power switch should be off while a person is installing or working on the cables.

The workstation must be free of combustible materials. Screens should be provided to protect other workers from the arc light.

The welding cable should never be wrapped around arms, shoulders, waist, or any other part of the body. If the cable was caught by any moving equipment, such as a forklift, crane, or dolly, a welder could be pulled off balance or more seriously injured. If it is necessary to hold the weight off the cable so that the welding can be done more easily, a free hand can be used. The cable should be held so that if it is pulled, it can be easily released.

FIGURE 8-38 To prevent people from tripping when cables must be placed in walkways, lay two blocks of wood beside the cables. © Cengage Learning 2012

CAUTION

The cable should never be tied to scaffolding or ladders. If the cable is caught by moving equipment, the scaffolding or ladder may be upset, causing serious personal injury.

Check the surroundings before starting to weld. If heavy materials are being moved in the area around you, there should be a safety watch. A safety watch can warn a person of danger while that person is welding.

SUMMARY

It is important that you understand basic principles of the SMA welding process. This background will help you better understand problems that might arise as you are welding or to more quickly understand the operation of a new piece of equipment. For example, failure to control arc blow can result in weld failures. Also, understanding electricity will help you interpret information given on manufacturers' tables, charts, and equipment specifications.

Before starting any new job or welding operation, be sure to check with the equipment manufacturer's safety guidelines for proper operation and maintenance. Follow all recommended guidelines.

Keeping your work area clean and orderly shows a sense of pride and craftsmanship as well as helps to prevent accidents.

REVIEW QUESTIONS

1. What are some advantages of SMA welding?
2. Explain the term *welding current.*
3. What three units are used to describe any electrical current?
4. Explain the relationship between voltage and arc length.
5. Explain the relationship between the amperage setting and the amount of heat produced by the arc.

6. How is heat produced by an arc lost before it reaches the weld?
7. What are the three different types of current used for welding?
8. Compare DCEN to DCEP.
9. How do the electrons flow in AC?
10. What kind of current is required for the SMAW process?
11. What is open circuit voltage?
12. What is closed circuit voltage?
13. What is arc blow?
14. What should you do if you encounter severe arc blow during a weld?
15. List three ways to control or reduce arc blow.
16. What are the two types of electrical devices used to produce a low-voltage, high-amperage current combination?
17. What are the advantages of transformer-type welding machines?
18. Describe the three major components of a transformer.
19. How is the current adjusted on a movable coil or core?
20. What are the advantages of inverter welding machines?
21. What are the differences between an alternator and a generator?
22. How can an alternating welding current be converted to direct current?
23. Why can't most SMA welding machines be used continuously?
24. What does a 60% duty cycle mean?
25. Why should the electrode cable and work cable be the correct size?
26. What can cause an electrode holder to overheat?
27. What problems can a loose work clamp cause?
28. Make a list of items to consider when properly setting up an arc welding machine.

Chapter 9

Shielded Metal Arc Welding Plate

OBJECTIVES

After completing this chapter, the student should be able to:

- Demonstrate the safe way to set up a welding station.
- Explain the differences among F2, F3, and F4 electrodes.
- Demonstrate welding fabrication skills.
- Demonstrate welding skills by making square butt, outside corner, lap, and tee joints.

KEY TERMS

amperage range
arc length
cellulose-based fluxes
chill plate
electrode angle
electrode manipulation
lap joint
leading electrode angle
mineral-based fluxes
outside corner joint
rutile-based fluxes
spatter
square butt joint
striking the arc
stringer bead
tack welds
tee joint
trailing electrode angle
weave pattern

INTRODUCTION

Shielded metal arc welding (SMAW) or stick welding is the most commonly used welding process. Stick welding is popular for a number of reasons. It can be used to make strong durable welds in a wide range of metal thicknesses and types. Transformer-type welders can sit for years without being used and still work when needed, and the welding rods have an almost unlimited storage life, as long as they are kept dry. These factors mean that the equipment is there when you need it to make a repair or build a project.

In addition to the standard rods that can be used to make welds in steel 1/8 in. (3 mm) and thicker, there are a wide variety of specialty electrodes. These specialty rods allow you to use the same welding machine to weld on cast iron, stainless steel, and aluminum. There are rods for cutting that will cut cast iron and stainless steel; these are metals that cannot be cut with oxygen and acetylene. Wear-resistant and buildup rods can be used to repair scraper and front-end loader blades and buckets.

The basic stick-welding skills you will learn in this chapter are used for almost every welding fabrication project. Welding is a skill that takes practice to perfect. The repetition in the chapter practices and projects is designed to give you the opportunity to develop your skills. The more time you spend welding, the better your welding skills will become.

ELECTRODES

Arc welding electrodes used for practice welds are grouped into three filler metal classes (F number) according to their major welding characteristics. The groups are E6010 and E6011, E6012 and E6013, and E7016 and E7018.

F3 E6010 and E6011 Electrodes

Both of these electrodes have **cellulose-based fluxes.** As a result, these electrodes have a forceful arc with little slag left on the weld bead. E6010 and E6011 are the most utilitarian welding electrodes for welding fabrication. They can be used on metal that has a little rust, oil, or dirt without seriously affecting the weld's strength. The E6010 electrodes can weld only with direct current (DC) welding machines. Because E6011 electrodes can be used with alternating current (AC), smaller transformer-type welders that put out only AC welding current can be used.

F2 E6012 and E6013 Electrodes

These electrodes have rutile-based fluxes, giving a smooth, easy arc with a thick slag left on the weld bead. Both E6012 and E6013 are easy electrodes to use. They do not have forceful arcs, so they can be used on thinner metals such as some thicker sheet metal gauges that are used as guards on equipment.

F4 E7016 and E7018 Electrodes

Both of these electrodes have a mineral-based flux. The resulting arc is smooth and easy, with a very heavy slag left on the weld bead. Of these two electrodes, E7018 is the one used most often to make high-strength welds on equipment. Store these electrodes in a dry place. If they get wet, they will still weld, but the welds will not be as strong.

The cellulose- and rutile-based groups F3 and F2 of electrodes have characteristics that make them the best electrodes for starting specific welds. The electrodes with the cellulose-based fluxes do not have heavy slags that may interfere with the welder's view of the weld. This feature is an advantage for flat tee and lap joints. Electrodes with the **rutile-based fluxes** (giving an easy arc with low **spatter**) are easier to control and are used for fillet, stringer beads, and butt joints.

Unless a specific electrode is required for a job, welders can select what they consider to be the best electrode for a specific weld. Welders often have favorite electrodes to use on specific jobs.

Electrodes with **mineral-based fluxes,** group F4, should be the last choice. Welds with a good appearance are more easily made with these electrodes, but strong welds are hard to obtain. Without special care being taken during the start of the weld, porosity will be formed in the weld.

Figure 9-1 shows a general comparison of the weld penetration and buildup each electrode will produce with approximately the same amperage. More information on electrode selection can be found in Chapter 26.

Effect of Too High or Too Low Current Settings

Each welding electrode must be operated in a particular current (amperage) range, Table 9-1. Welding with the current set too low results in poor fusion

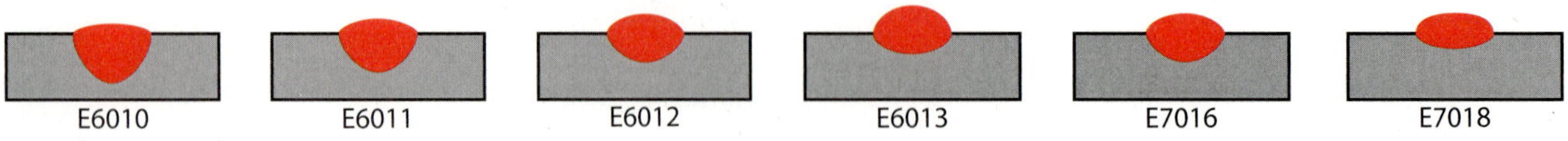

FIGURE 9-1 Comparison of electrode type and weld bead shapes. © Cengage Learning 2012

AWS CLASSIFICATION	ELECTRODE DIAMETER AND AMPERAGE RANGE		
	3"/32	1"/8	5"/32
E6010	40–80	70–130	110–165
E6011	50–70	85–125	130–160
E6012	40–90	75–130	120–200
E6013	40–85	70–120	130–160
E7016	75–105	100–150	140–190
E7018	70–110	90–165	125–220

Table 9-1 Welding Amperage Range for Common Electrode Types and Sizes

and poor arc stability, Figure 9-2. The weld may have slag or gas inclusions because the molten weld pool was not fluid long enough for the flux to react. Little or no penetration of the weld into the baseplate may also be evident. With the current set too low, the arc length is very short. A very short arc length results in frequent shortening and sticking of the electrode.

The core wire of the welding electrode is limited in the amount of current it can carry. As the current is increased, the wire heats up because of electrical resistance. This preheating of the wire causes some of the chemicals in the covering to be burned out too early, Figure 9-3. The loss of the proper balance of elements causes poor arc stability. This condition leads to spatter, porosity, and slag inclusions.

An increase in the amount of spatter is also caused by a longer arc. The weld bead made at a high amperage setting is wide and flat with deep penetration. The spatter is excessive and is mostly hard. The spatter is called hard because it fuses to the baseplate and is difficult to remove, Figure 9-4. The electrode covering is discolored more than 1/8 in. (3 mm) to

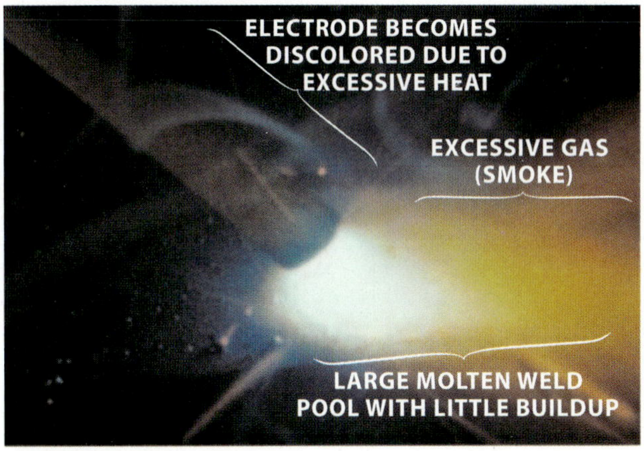

FIGURE 9-3 Welding with too high an amperage.
Larry Jeffus

1/4 in. (6 mm) from the end of the electrode. Extremely high settings may also cause the electrode to discolor, crack, glow red, or burn.

Electrode Size and Heat

The selection of the correct size of welding electrode for a weld is determined by the skill of the welder, the thickness of the metal to be welded, and the size of the metal. The 1/8-in. (3-mm) electrode is the most commonly used size for metal fabrication. It can be used to make welds on thin metal up to thick plates. Using small diameter electrodes requires less skill than using large diameter electrodes. The deposition rate, or the rate that weld metal is added to the weld, is slower when small diameter electrodes are used. Small diameter electrodes will make acceptable welds on a thick plate, but more time is required to make the weld.

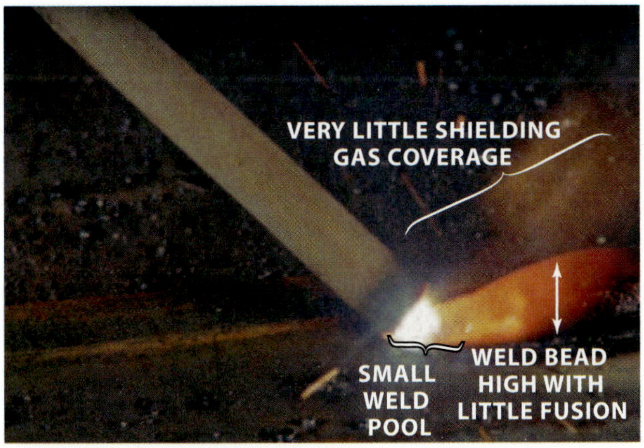

FIGURE 9-2 Welding with the amperage set too low.
Larry Jeffus

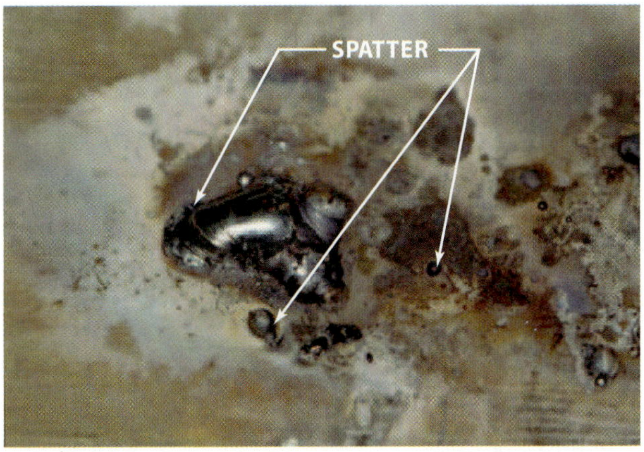

FIGURE 9-4 Hard weld spatter is fused to base metal and is difficult to remove. Larry Jeffus

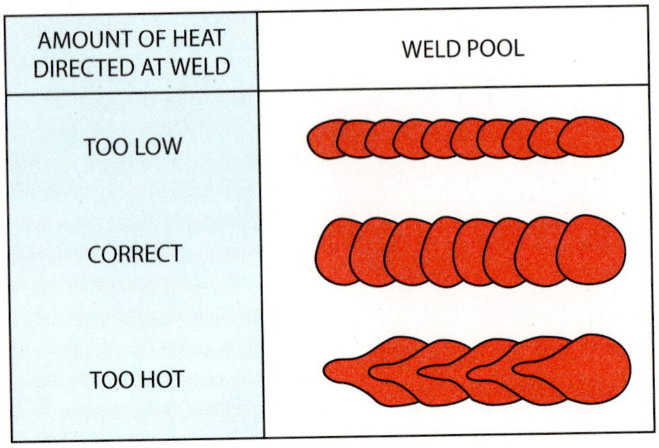

FIGURE 9-5 The effect on the shape of the molten weld pool caused by the heat input.

FIGURE 9-6 Welding with too short an arc length.
Larry Jeffus

Large diameter electrodes may overheat the metal if they are used with thin or small pieces of metal. To determine if a weld is too hot, watch the shape of the trailing edge of the molten weld pool, Figure 9-5. Rounded ripples indicate that the weld is cooling uniformly and that the heat is not excessive. If the ripples are pointed, the weld is cooling too slowly because of excessive heat. Extreme overheating can cause a burn-through, which is hard to repair.

To correct an overheating problem, a welder can turn down the amperage, use a shorter arc, travel at a faster rate, use a **chill plate** (a large piece of metal used to absorb excessive heat), or use a smaller electrode at a lower current setting.

Arc Length

The **arc length** is the distance that the arc must jump from the end of the electrode to the plate or weld pool surface. As the weld progresses, the electrode becomes shorter as it is consumed. To maintain a constant arc length, the electrode must be lowered continuously. Maintaining a constant arc length is important, as too great a change in the arc length will adversely affect the weld.

As the arc length is shortened, metal transferring across the gap may short out the electrode, causing it to stick to the plate. The weld that results is narrow and has a high buildup, Figure 9-6.

Long arc lengths produce more spatter because the metal being transferred may drop outside of the molten weld pool. The weld is wider and has little buildup, Figure 9-7.

There is a narrow range for the arc length in which it is stable, metal transfer is smooth, and the bead shape is controlled. Factors affecting the length are the class of electrode, joint design, metal thickness, and current setting.

Some welding electrodes, such as E7024, have a thick flux covering. The rate at which the covering melts is slow enough to permit the electrode coating to be rested against the plate. The arc burns back inside the covering as the electrode is dragged along, touching the joint, Figure 9-8. For this class of welding electrode, the arc length is maintained by the electrode covering. E7024 electrodes require very little welding skill to use. Because of the size of the molten weld pool, they are usually used only in the flat position on thick metal.

An arc will jump to the closest metal conductor. On joints that are deep or narrow, the arc is pulled to one side and not to the root, Figure 9-9. As a result, the root fusion is reduced or may be nonexistent,

FIGURE 9-7 Welding with too long an arc length.
Larry Jeffus

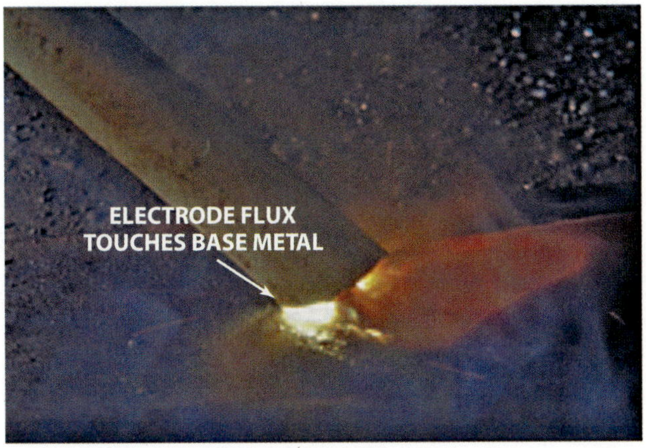

FIGURE 9-8 Welding with a drag technique.
Larry Jeffus

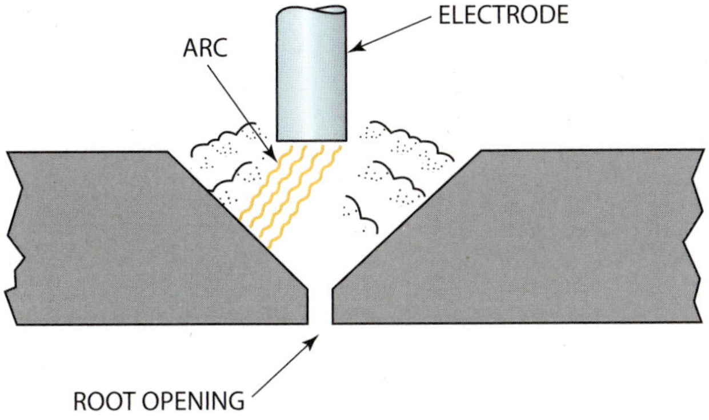

FIGURE 9-9 The arc may jump to the closest metal, reducing root penetration. © Cengage Learning 2012

thus causing a poor weld. If a very short arc is used, the arc is forced into the root for better fusion.

Because shorter arcs produce less heat and penetration, they are best suited for use on thin metal or thin-to-thick metal joints. Using this technique, metal as thin as 16 gauge can be arc welded easily. Higher amperage settings are required to maintain a short arc that gives good fusion with a minimum of slag inclusions. The higher settings, however, must be within the **amperage range** for the specific electrode.

Finding the correct arc length often requires some trial and adjustment. Most welding jobs require an arc length of 1/8 in. (3 mm) to 3/8 in. (10 mm), but this distance varies. It may be necessary to change the arc length when welding to adjust for varying welding conditions.

Electrode Angle

The **electrode angle** is measured from the electrode to the surface of the metal. The term used to identify the electrode angle is affected by the direction of travel, generally leading or trailing, Figure 9-10. The relative angle is important because there is a jetting force blowing the metal and flux from the end of the electrode to the plate.

Leading Electrode Angle

A **leading electrode angle** pushes molten metal and slag ahead of the weld, Figure 9-11. When welding in the flat position, caution must be taken to prevent cold lap and slag inclusions. The solid metal ahead of the weld cools and solidifies the molten filler metal and slag before they can melt the solid metal. This rapid cooling prevents the metals from fusing together, Figure 9-12. As the weld passes over this area, heat from the arc may not melt it. As a result, some cold lap and slag inclusions are left.

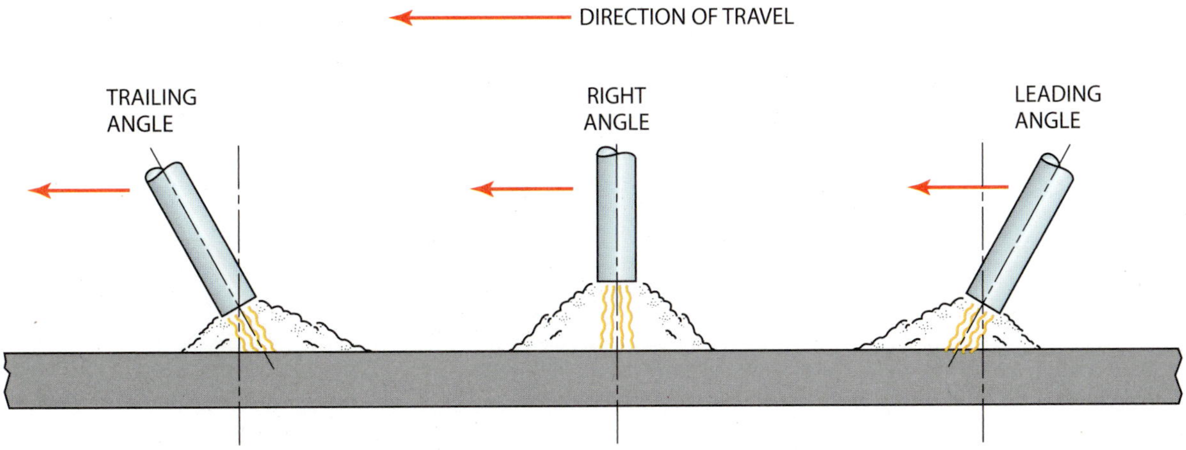

FIGURE 9-10 Direction of travel and electrode angle. © Cengage Learning 2012

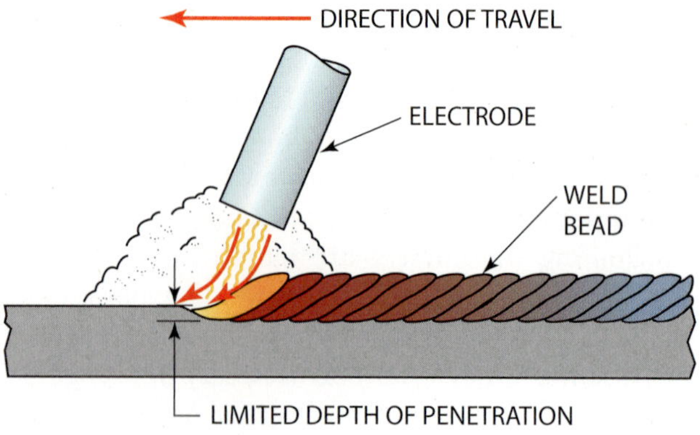

FIGURE 9-11 Leading, lag, or pushing electrode angle. © Cengage Learning 2012

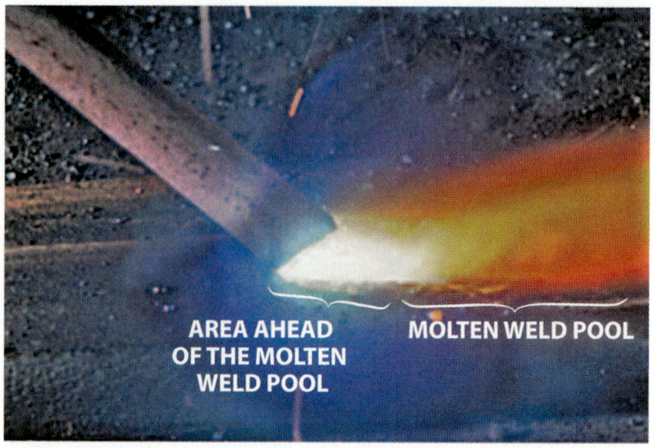

FIGURE 9-13 Metal being melted ahead of the molten weld pool helps to ensure good weld fusion. Larry Jeffus

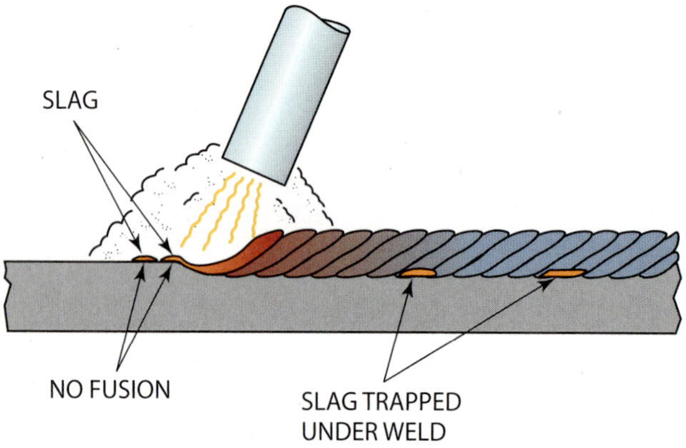

FIGURE 9-12 Some electrodes, such as E7018, may not remove the deposits ahead of the molten weld pool, resulting in discontinuities within the weld. © Cengage Learning 2012

The following are suggestions for preventing cold lap and slag inclusions:

- Use as little leading electrode angle as possible.
- Ensure that the arc melts the base metal completely, Figure 9-13.
- Use a penetrating-class electrode that causes little buildup.
- Move the arc back and forth across the molten weld pool to fuse both edges.

A leading angle can be used to minimize penetration or to help hold molten metal in place for vertical welds, Figure 9-14.

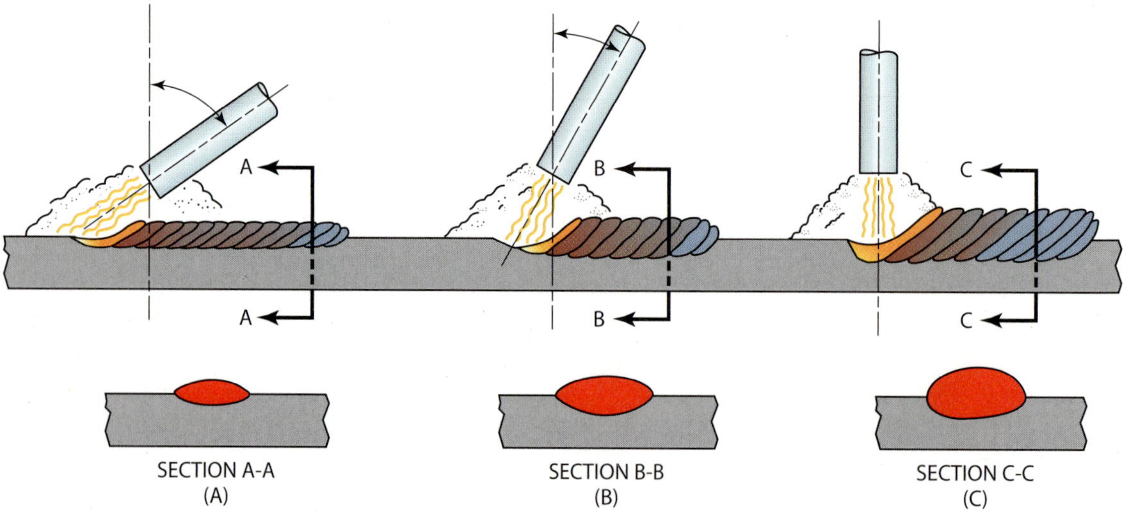

FIGURE 9-14 Effect of a leading angle on weld bead buildup, width, and penetration. As the angle increases toward the vertical position (C), penetration increases. © Cengage Learning 2012

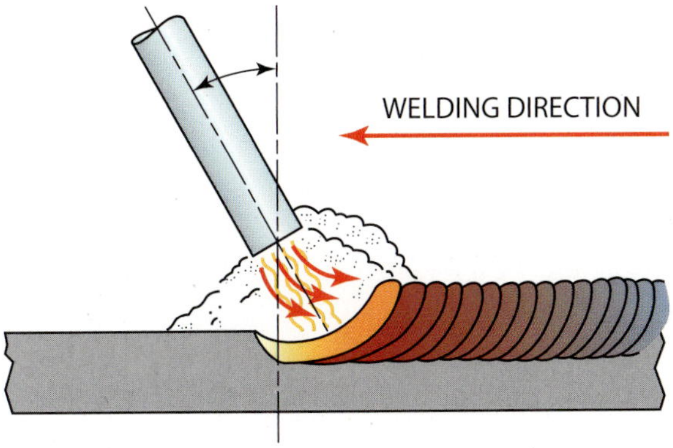

FIGURE 9-15 Trailing electrode angle. © Cengage Learning 2012

Trailing Electrode Angle

A **trailing electrode angle** pushes the molten metal away from the leading edge of the molten weld pool toward the back where it solidifies, Figure 9-15. As the molten metal is forced away from the bottom of the weld, the arc melts more of the base metal, which results in deeper penetration. The molten metal pushed to the back of the weld solidifies and forms reinforcement for the weld, Figure 9-16.

Electrode Manipulation

The movement or weaving of the welding electrode, called **electrode manipulation**, can control the following characteristics of the weld bead: penetration, buildup, width, porosity, undercut, overlap, and slag inclusions. The exact **weave pattern** for each weld is often the personal choice of the welder. However, some patterns are especially helpful for specific welding situations. The pattern selected for a flat (1G) butt joint is not as critical as is the pattern selection for other joints and other positions.

Many weave patterns are available for the welder to use. Figure 9-17 shows 10 different patterns that can be used for most welding conditions.

The circular pattern is often used for flat position welds on butt, tee, and outside corner joints, and for buildup or surfacing applications. The circle can be made wider or longer to change the bead width or penetration, Figure 9-18.

The "C" and square patterns are both good for most 1G (flat) welds but can also be used for vertical (3G) positions. These patterns can also be used if there is a large gap to be filled when both pieces of metal are nearly the same size and thickness.

The "J" pattern works well on flat (1F) lap joints, all vertical (3G) joints, and horizontal (2G) butt and lap (2F) welds. This pattern allows the heat to be concentrated on the thicker plate, Figure 9-19. It also allows the reinforcement to be built up on the metal deposited during the first part of the pattern. As a result, a uniform bead contour is maintained during out-of-position welds.

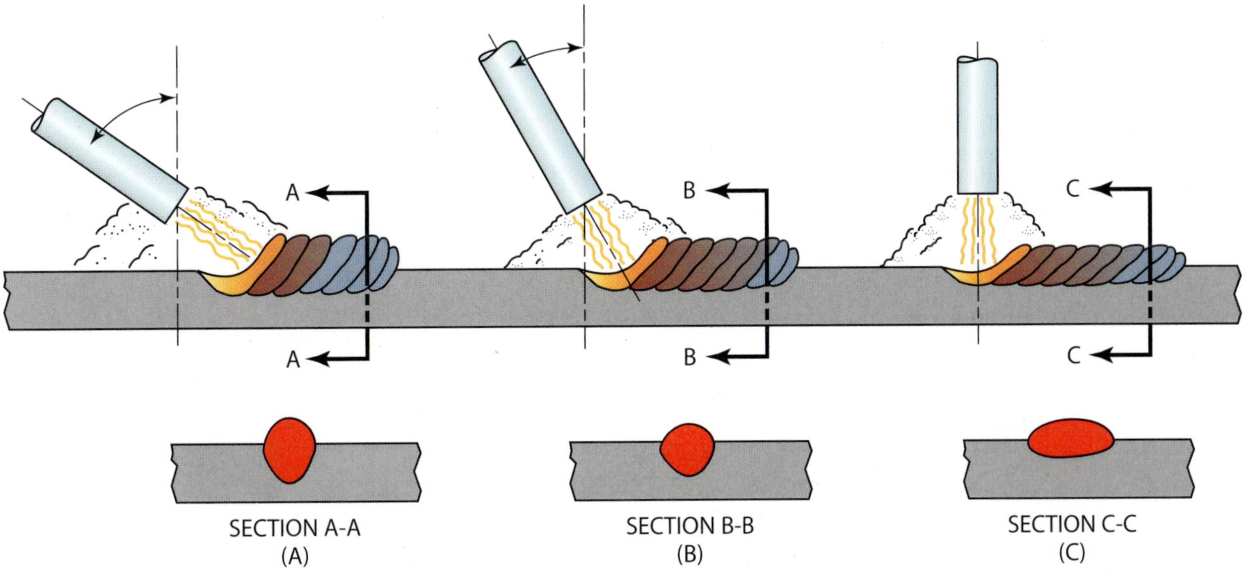

FIGURE 9-16 Effect of a trailing angle on weld bead buildup, width, and penetration. Section A-A shows more weld buildup due to a greater angle of the electrode. © Cengage Learning 2012

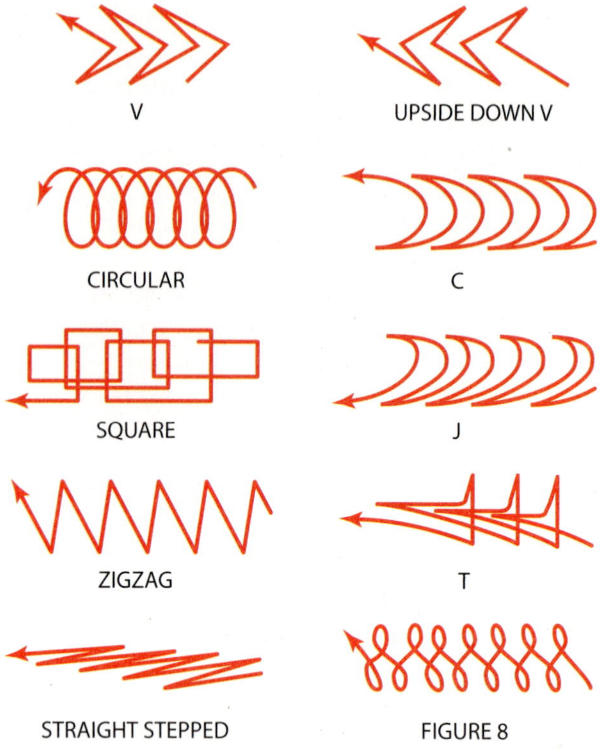
V
UPSIDE DOWN V
CIRCULAR
C
SQUARE
J
ZIGZAG
T
STRAIGHT STEPPED
FIGURE 8

FIGURE 9-17 Weave patterns.

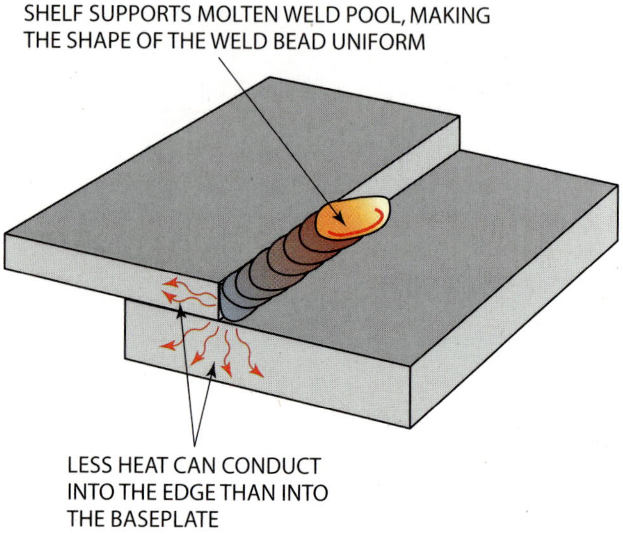
SHELF SUPPORTS MOLTEN WELD POOL, MAKING THE SHAPE OF THE WELD BEAD UNIFORM
LESS HEAT CAN CONDUCT INTO THE EDGE THAN INTO THE BASEPLATE

FIGURE 9-19 The "J" pattern allows the heat to be concentrated on the thicker plate.

The "T" pattern works well with fillet welds in the vertical (3F) and overhead (4F) positions, Figure 9-20. It can also be used for deep groove welds for the hot pass. The top of the "T" can be used to fill in the toe of the weld to prevent undercutting.

The straight step pattern can be used for stringer beads, root pass welds, and multiple pass welds in all positions. For this pattern, the smallest quantity of metal is molten at one time as compared to other patterns. Therefore, the weld is more easily controlled. At the same time that the electrode is stepped forward, the arc length is increased so that no metal is deposited ahead of the molten weld pool, Figure 9-21 and Figure 9-22. This action allows the molten weld pool to cool to a controllable size. In addition, the arc burns off any paint, oil, or dirt from the metal before it can contaminate the weld.

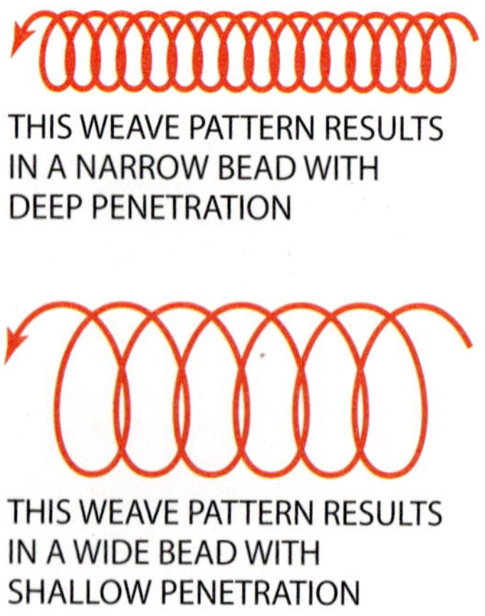
THIS WEAVE PATTERN RESULTS IN A NARROW BEAD WITH DEEP PENETRATION
THIS WEAVE PATTERN RESULTS IN A WIDE BEAD WITH SHALLOW PENETRATION

FIGURE 9-18 Changing the weave pattern width to change the weld bead characteristics.

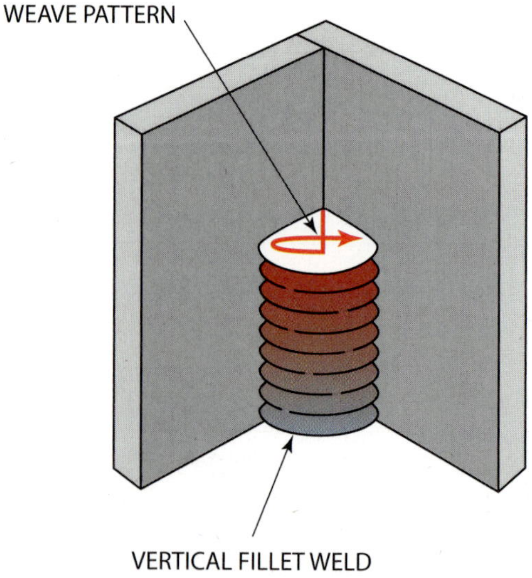
WEAVE PATTERN
VERTICAL FILLET WELD

FIGURE 9-20 "T" pattern.

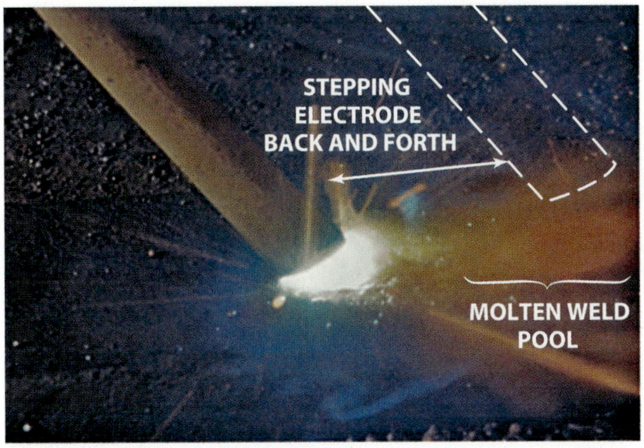

FIGURE 9-21 The electrode is moved slightly forward and then returned to the weld pool. Larry Jeffus

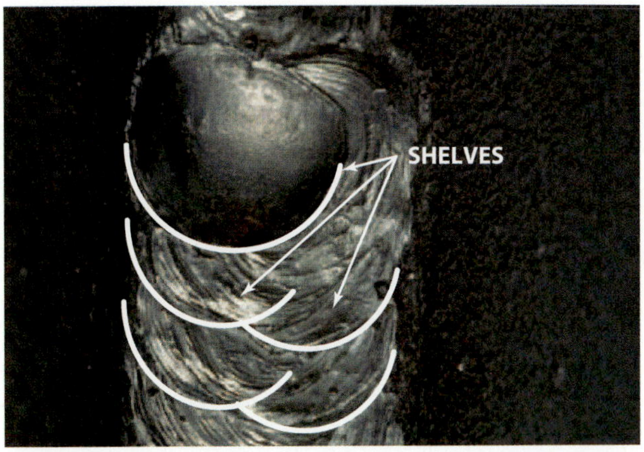

FIGURE 9-23 Using the shelf to support the molten pool for vertical welds. Larry Jeffus

The figure 8 pattern and the zigzag pattern are used as cover passes in the flat and vertical positions. Do not weave more than two-and-a-half times the width of the electrode. These patterns deposit a large quantity of metal at one time. A shelf can be used to support the molten weld pool when making vertical welds using either of these patterns, Figure 9-23.

Practice Welds

Practice welds are grouped according to the type of joint and the class of welding electrode. The welder or instructor should select the order in which the welds are made. The stringer beads should be practiced first in each position before the welder tries the different joints in each position. Some time can be saved by starting with the stringer beads. If this is done, it is not necessary to cut or tack the plate together, and a number of beads can be made on the same plate.

FIGURE 9-22 The electrode does not deposit metal or melt the base metal. Larry Jeffus

Students will find it easier to start with butt joints. The lap, tee, and outside corner joints are all about the same level of difficulty.

Starting with the flat position allows the welder to build skills slowly so that out-of-position welds become easier to do. The horizontal tee and lap welds are almost as easy to make as the fillet welds. Overhead welds are as simple to make as vertical welds, but they are harder to position. Horizontal butt welds are more difficult to perform than most other welds.

Positioning of the Welder and the Weld Plate

The welder should be in a relaxed, comfortable position before starting to weld. A good position is important for both the comfort of the welder and the quality of the welds. Welding in an awkward position can cause welder fatigue, which leads to poor welder coordination and poor-quality welds. Welders must have enough freedom of movement so that they do not need to change position during a weld. Body position changes should be made only during electrode changes.

When the welding helmet is down, the welder is blind to the surroundings. Due to the arc, the field of vision of the welder is also very limited. These factors often cause the welder to sway. To stop this swaying, the welder should lean against or hold onto a stable object. When welding, even if a welder is seated, touching a stable object will make the welder more stable and will make welding more relaxing.

(A)

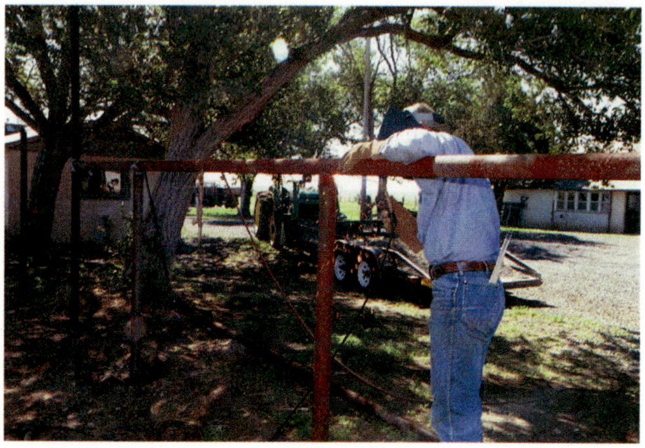

(B)

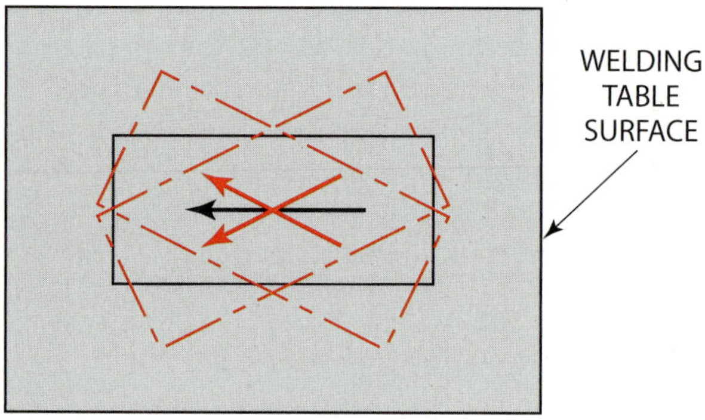

(C)

FIGURE 9-24 Not all field welding can be done in a comfortable position. Larry Jeffus

Welding is easier if the welder can find the most comfortable angle. The welder should be in either a seated or a standing position in front of the welding table. The welding machine should be turned off. With an electrode in place in the electrode holder, the welder can draw a straight line along the plate to be welded. By turning the plate to several different angles, the welder should be able to determine which angle is most comfortable for welding, Figure 9-24A, B, and C.

PRACTICE 9-1

Shielded Metal Arc Welding Safety

Skill to be learned: The safe setup of a welding station and the use of proper personal protective equipment (PPE).

Using a welding workstation, welding machine, welding electrodes, welding helmet, eye and ear protection, welding gloves, proper work clothing, and any special protective clothing that may be required, demonstrate to your instructor and other students the safe way to prepare yourself and the welding workstation for welding. Include in your demonstration appropriate references to burn protection, eye and ear protection, material specification data sheets, ventilation, electrical safety, general work clothing, special protective clothing, and area cleanup. ◆

STRIKING AN ARC

All welds start with an arc strike. It is the process of establishing a stable arc between the end of the electrode and the work. At first, striking an arc can be difficult because it may seem that the end of the electrode wants to stick to the plate. With practice, you will be able to strike an arc and establish a weld bead without much thought. One important thing to remember is that on most code welding jobs, an arc strike outside of the weld zone may be considered a defect. Start now building a habit of always **striking the arc** in the weld joint just ahead of where you are going to be welding. That way the arc strike will become part of the finished weld.

PROJECT 9-1

Striking the Arc to Build a Hot Plate

Skill to be learned: The ability to start and hold an arc to produce a weld bead.

Project Description

Hot plates serve two main purposes—to protect the surface of a kitchen countertop or table from damage caused by a hot skillet or other cooking utensil and to prevent a cold surface from cooling a dish before the food can be served. The hot plate you are going to fabricate serves both of these purposes. The cork or felt pads placed on the four corners of the bottom of the hot plate serve as insulators and surface protectors. They keep any heat picked up by the hot plate from being transferred to the surface and keep the metal hot plate from scratching the surface.

The short welds on the surface of the hot plate limit the contact area between the hot plate and the hot dish. This reduces the heat transfer from the dish to the hot plate.

Project Materials and Tools

The following items are needed to fabricate the hot plate.

Materials

- 5 to 10 1/8-in. diameter E6011 electrodes
- 1 6-in. square piece of 1/4-in. thick mild steel plate
- 4 cork or felt pads
- Paint
- Clear finish

Tools

- Arc welder
- PPE
- Angle grinder
- Wire brush
- Square
- Soapstone

Layout

Using a square, 12-in. rule and soapstone, lay out the plate as shown in the project drawing, Figure 9-25.

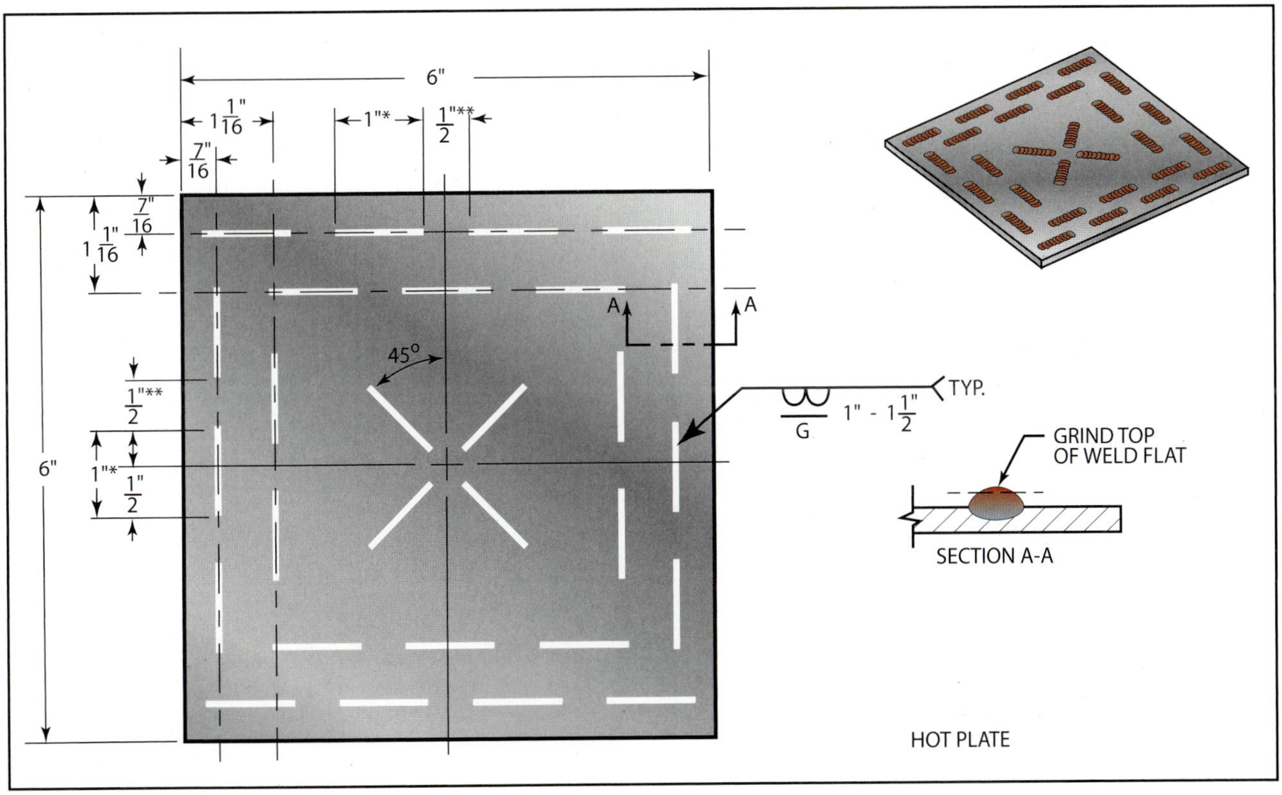

FIGURE 9-25 Project 9-1. © Cengage Learning 2012

Welding

Using a properly set up and adjusted arc welding machine and the proper safety protection as demonstrated in Practice 9-1, you will make a series of weld beads on the hot plate surface following the soapstone layout.

NOTE: Before starting on any project, test the setup of the welder by making a few test welds on a piece of scrap metal. That way you can make any adjustments in the welding amperage settings before starting on the project.

With the electrode held over the plate, lower your helmet. Scratch the electrode across the plate (like striking a large match), Figure 9-26. As the arc is established, slightly raise the electrode to the desired arc length. Hold the arc in one place until the molten weld pool builds to the desired size. Slowly lower the electrode as it burns off, and slowly move it forward to start the weld bead.

NOTE: If the electrode sticks to the plate, quickly squeeze the electrode holder lever to release the electrode. Break the electrode free by bending it back and forth a few times. Do not touch the electrode without gloves because it will still be hot. If the flux breaks away from the end of the electrode, throw out the electrode because restarting the arc will be very difficult, Figure 9-27.

Break the arc by rapidly raising the electrode after completing a 1-in. long weld bead. You will be making another weld on the next marked spot. Move the electrode end over the starting point for the next weld. Restart the arc as you did before, and make another short weld. Repeat this process until you have made all of the welds.

A more accurate way of striking the arc is to hold the electrode steady by resting it on your free hand like a pool cue over the desired starting point. After lowering your helmet, swiftly bounce the electrode against the plate, Figure 9-28. A lot of practice is required to develop the speed and skill needed to prevent the electrode from sticking to the plate. But striking an arc in an incorrect spot may cause damage to the base metal.

NOTE: Sometimes the welding fumes may partially cover the marks next to the weld you just made. If that happens, do not wipe the fumes off because you may erase the lines. Instead, use the soapstone to retrace

FIGURE 9-27 If the flux is broken off the end completely or on one side, the arc can be erratic or forced to the side. Larry Jeffus

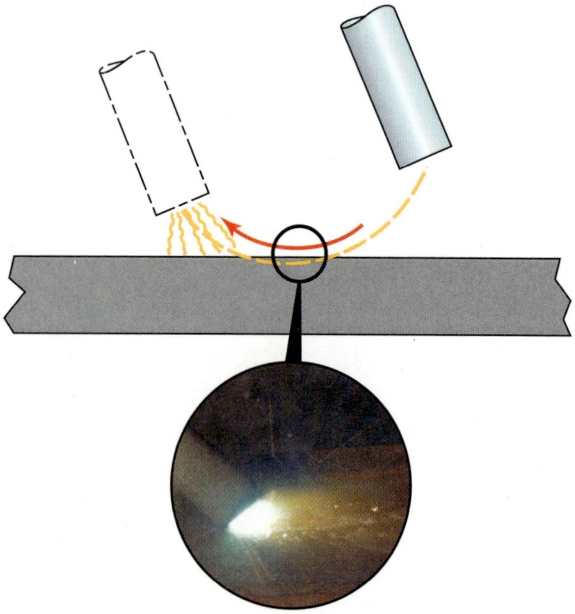

FIGURE 9-26 Scratch striking an arc. © Cengage Learning 2012

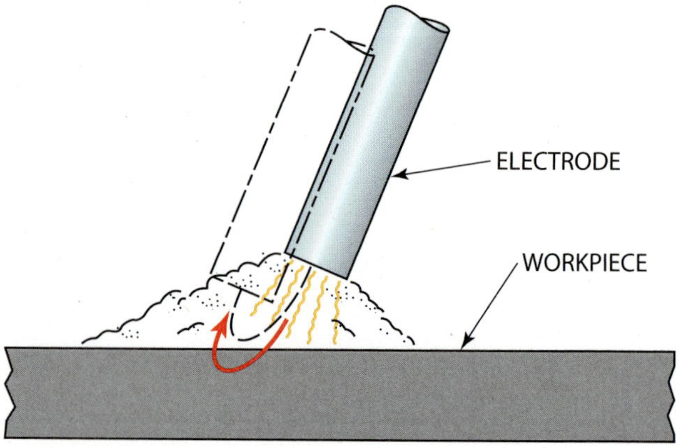

FIGURE 9-28 Bounce striking an arc to start it on a spot. © Cengage Learning 2012

the line. You can usually do this by hand without the need for a straightedge.

As you make more and more of the welds, you will become more efficient at starting the arc. When you have completed all of the welds, turn off the welding machine and clean up your work area.

Finishing

Cool off the metal, chip the slag, and wire brush the surface. Sometimes the metal may warp from the heat and welds. If necessary, you can flatten your hot plate enough for it to be used by using a hammer and anvil.

Paint the top and side surfaces with a latex paint. Once the paint is dry, grind the tops off of the welds as shown in the drawing. Do not grind too hard because that could cause excessive heat, which could damage the paint. Check the surface to see that it is flat enough to prevent a pot or dish from rocking. Do any additional grinding to make the tops of the welds flat. Paint the hot plate with a coat of clear latex finish to keep it from rusting.

Paperwork

Complete a copy of the time sheet in Appendix I, the bill of materials in Appendix III, or use forms provided by your instructor.◆

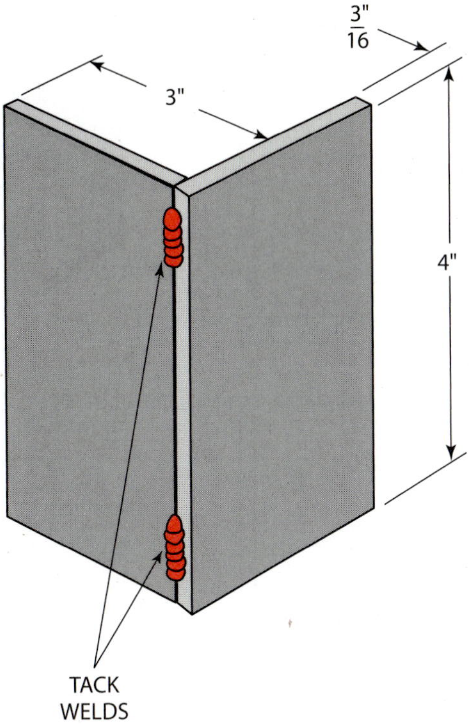

FIGURE 9-29 Tack welding outside corner joint.

TACK WELDS

Tack welds are a temporary method of holding parts in place until they can be completely welded. Usually, all of the parts of a weldment should be assembled before any finishing welding is started. This will help reduce distortion. Tack welds must be strong enough to withstand any forces applied during assembly and any force caused by weld distortion during final welding. They must also be small enough to be incorporated into the final weld without causing a discontinuity in its size or shape, Figure 9-29.

PROJECT 9-2

Tack Welding Assembly of a Pencil Holder

Skill to be learned: **The tack welding of parts in place so they can be finished welded.**

Project Description

The pencil holder will be assembled and tack welded in preparation for welding. The same skills you learned by making short welds as you fabricated the hot plate will be used to make the pencil holder. In this application short stringer welds will be used as tack welds to hold the parts in place so they can be welded.

Project Materials and Tools

The following items are needed to fabricate the pencil holder.

Materials

- 1 to 2 1/8-in. diameter E6012 or E6013 electrodes
- 4 × 4-in. rectangular pieces of 3/16-in. thick mild steel plate
- 1 3-in. square piece of 3/16-in. thick mild steel plate

Tools

- Arc welder
- Plasma cutting torch, oxyfuel cutting torch, or shear
- PPE
- Wire brush
- Chipping hammer
- Square
- C-clamp or magnetic clamp
- Soapstone

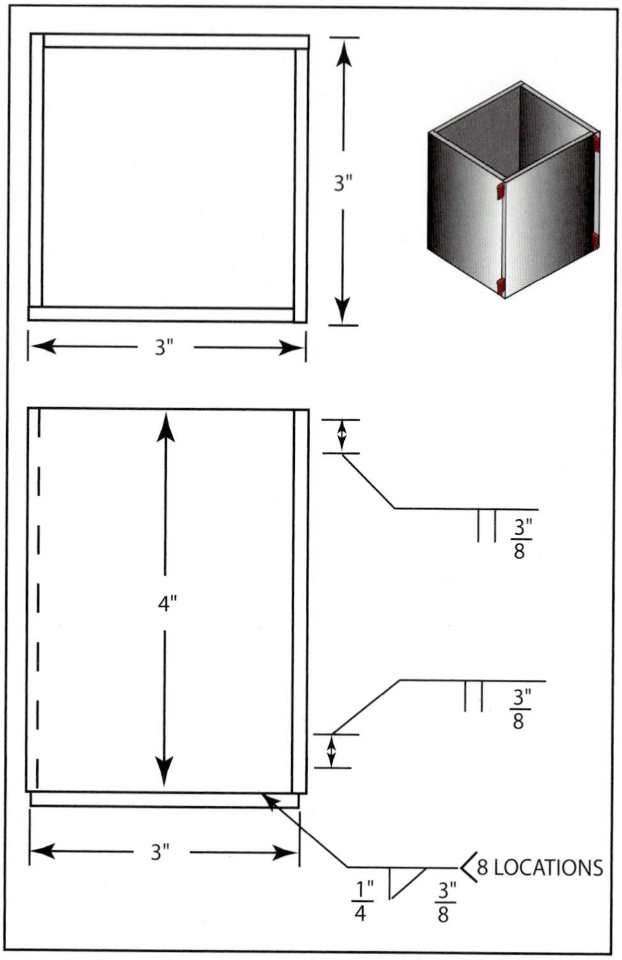

FIGURE 9-30 Project 9-2. © Cengage Learning 2012

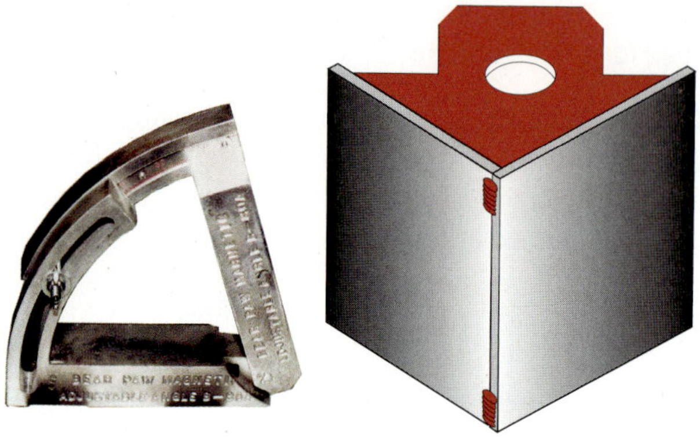

(A) (B)

FIGURE 9-31 Magnetic clamps. Larry Jeffus

Layout

Using a square, 12-in. rule and soapstone, lay out the plate as shown in the project drawing, Figure 9-30.

Cutting Out

If flat 3/16-in. bar stock is available, all you will have to do is shear the metal to length. However, if you are using sheet stock, you will have to thermally cut the parts using either a flame or plasma torch. (See Chapters 21 and 22 for details on flame and plasma cutting.) If the parts are thermally cut, clean off any slag before you assemble the parts.

Fabrication

The parts will need to be held square as they are tack welded. A C-clamp or magnetic clamp can be used to hold the parts in place for tack welding, Figure 9-31. Sometimes you can hold the parts in place with your gloved hand and make a small tack weld. If you use your square, keep it as far away as possible from the tack weld so it will not be damaged.

Make the tack welds as shown in the project drawing, Figure 9-30. Keep the tack welds small so they will not affect the finished weld.

Paperwork

Complete a copy of the time sheet in Appendix I, the bill of materials in Appendix III, or use forms provided by your instructor.

STRINGER BEADS

A straight weld bead on the surface of a plate with little or no side-to-side electrode movement is known as a **stringer bead.** Stringer beads are used by students to practice maintaining arc length, weave patterns, and electrode angle so that their welds will be straight, uniform, and free from defects. They are also used by experienced welders to set the welding machine amperage.

An example of an application for stringer beads is using them to build up a worn surface or apply a chemical- or mechanical-resistant weld metal to the surface. They may also be used to add an effect to the surface of a piece of art.

The stringer bead should be straight. A beginning welder needs time to develop the skill of viewing the entire welding area. At first, the welder sees only the arc, Figure 9-32. With practice, the welder begins to see parts of the molten weld pool. After much practice, the welder will see the molten weld pool (front, back, and both sides), slag, buildup, and

FIGURE 9-32 New welders frequently see only the arc and sparks from the electrode. Larry Jeffus

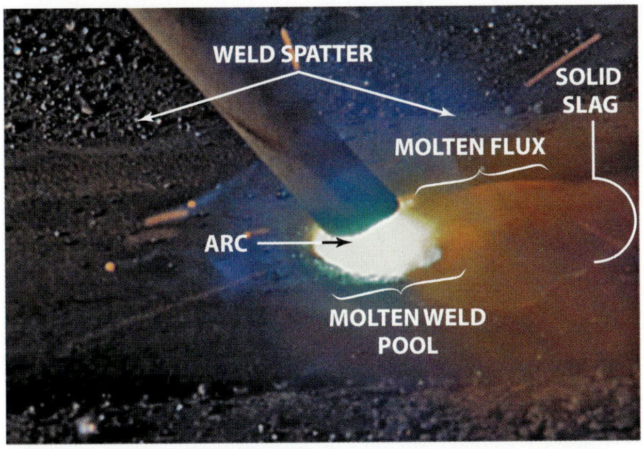

FIGURE 9-33 More experienced welders can see the molten pool, metal being transferred across the arc, and penetration into the base metal. Larry Jeffus

the surrounding plate, Figure 9-33. Often at this skill level, the welder may not even notice the arc.

A straight weld is easily made once the welder develops the ability to view the entire welding zone. The welder will occasionally glance around to ensure that the weld is straight. In addition, it can be noted if the weld is uniform and free from defects. The ability of the welder to view the entire weld area is demonstrated by making consistently straight and uniform stringer beads.

PROJECT 9-3

Stringer Bead to Surface a Pencil Holder

Skill to be learned: The control of the welding electrode to produce uniform weld beads in the flat position.

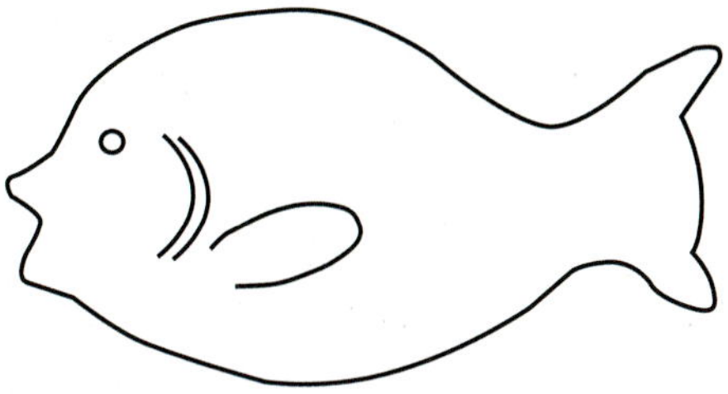

FIGURE 9-34 Another arc striking project. © Cengage Learning 2012

Project Description

The surface welds applied to the pencil holder are for artistic decoration only. They add depth and individual interest to the finished project. Other projects with a similar application of welds might include a fish on which the welds appear to be scales or birds on which they appear to be feathers, Figure 9-34.

Project Materials and Tools

The following items are needed to finish the pencil holder.

Materials

- 5 to 10 1/8-in. diameter E6012 or E6013 electrodes
- 1 assembled pencil holder
- 4 cork or felt pads
- Paint or clear finish

Tools

- Arc welder
- PPE
- Square
- Steel rule
- Angle grinder
- Wire brush
- Soapstone

Welding

Using a properly set up and adjusted arc welding machine and the proper safety protection as demonstrated in Practice 9-1, you will make a series of stringer weld beads on the surface of the pencil holder that was assembled in Project 9-2.

The first stringer weld bead will actually be a square butt joint weld. It will be used to prevent the plates from excessively distorting as the remainder of the stringer welds are made. Start the weld at the edge

FIGURE 9-35 Making the weld narrow with a higher buildup by using a short arc length can add dimension to the finished project's surface appearance. Larry Jeffus

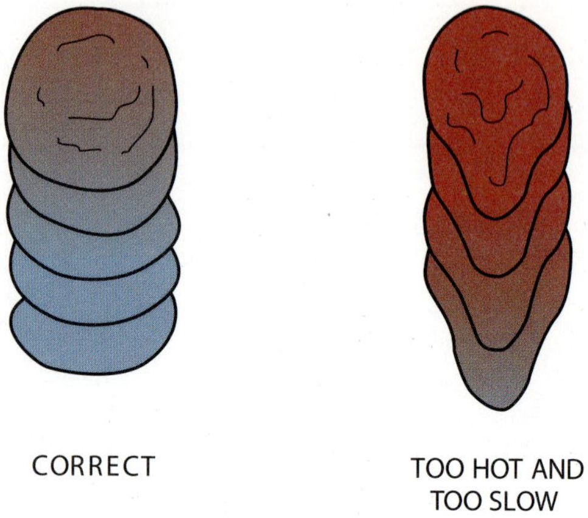

FIGURE 9-37 Comparison of the appearance of weld beads with correct temperature and overheated base metal. © Cengage Learning 2012

of the plate and weld all the way to the opposite end. Try to keep the weld width and travel speed consistent all the way down the plate, Figure 9-35. When the first weld is completed, roll the pencil holder over so that the next weld can be made to close the next butt joint. Repeat this until all four corners are welded.

Cool the part and look at the welds. See how consistent you were in your travel speed and electrode movement. You can tell when you were welding faster because the weld will be thinner with less buildup, Figure 9-36. The weld will be wider with more buildup when you travel slower. The width of the weld, spacing of the weld bead ripples, and smoothness of the sides of the weld are all indicators of how constant you were in your side-to-side manipulation.

Alternate both the direction of the weld and the side that you are welding on. The alternating of the direction will make the finished welds look neater, and

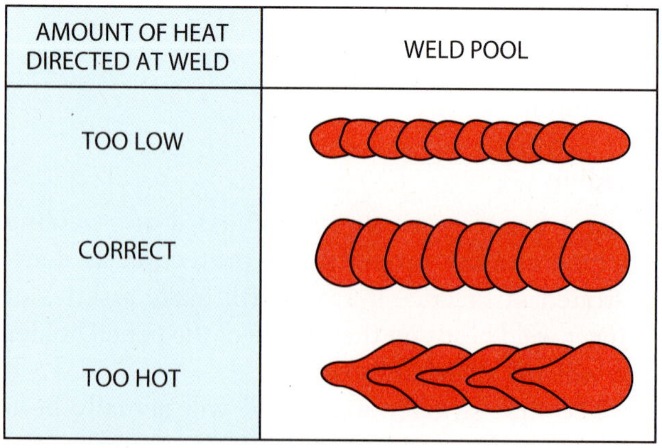

FIGURE 9-36 Comparison of weld beads and heat input. © Cengage Learning 2012

by switching sides you will not be as likely to overheat the metal. Watch the back side of the weld pool to see if the weld is getting too hot, Figure 9-37. Chip and wire brush the welds after each pass. Look at the weld bead for uniformity in width, height, and spacing of ripples.

NOTE: If the project is becoming overheated, you can cool it by quenching it in water. Normally, welds are not quenched in water because that might cause cracks and it makes them brittle. But the weld's strength is not a factor in the usability of this artistic project.

When all of the sides are completely covered with weld beads, cool the project. Next, you will make the last series of welds, numbers 3 and 4, as shown in the drawing, Figure 9-38.

Finishing

Cool off the metal, chip the slag, and wire brush the surface. A power wire brush can be used to give the welds a polished, shiny appearance. The project can be painted with a clear coat or colored coat of latex paint. Stick the felt or cork pads on the corners of the bottom to protect the surface of the furniture.

Paperwork

Complete a copy of the time sheet in Appendix I, the bill of materials in Appendix III, or use forms provided by your instructor.◆

Square Butt Joint

The **square butt joint** is made by tack welding two flat pieces of plate together, Figure 9-39. The space between the plates is called the root opening or root

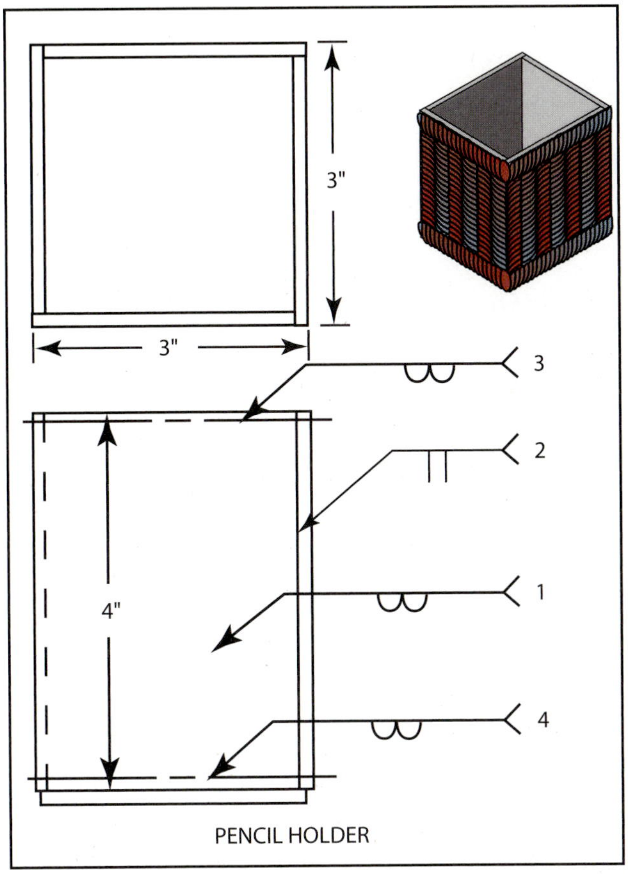

FIGURE 9-38 Project 9-3. © Cengage Learning 2012

FIGURE 9-39 The tack weld should be small and uniform to minimize its effect on the final weld. Larry Jeffus

gap. Changes in the root opening affect penetration. As the space increases, the weld penetration also increases. The root opening for most butt welds varies from 0 to 1/8 in. (3 mm). Excessively large openings can cause burnthrough or a cold lap at the weld root, Figure 9-40.

Square butt joints are used to join plate and structural steel shapes up to 1/4-in. thick. The advantage of using the square butt joint is that often the edges of the metal can be used just as they are without having to do additional edge beveling. Not having to bevel the edge of the metal saves time and money for the fabricator, Figure 9-41.

PROJECT 9-4

Butt Welds to Build a Smoke Box

Skill to be learned: The control of the welding electrode to produce uniform butt weld joints in the flat position.

Project Description

A smoke box is a small flat box that wood chips are placed in so that food cooked on a gas barbecue grill will have a wood fire flavor. The smoke box is used by placing several hickory, mesquite, or other aromatic wood chips in the box and covering them with the box lid. The lid prevents the wood from burning so it only smokes.

With the box closed, place it as close to the grill burners as possible. If your barbeque grill has lava stones, move enough of them aside so you can place the smoke box just above the burner flames. If your barbeque grill uses ceramic plates, one can be moved to make room for the smoke box, or just set it on top of the ceramic plate. The only thing that is important is that the box be placed so it will get hot enough for the wood chips inside to smoke.

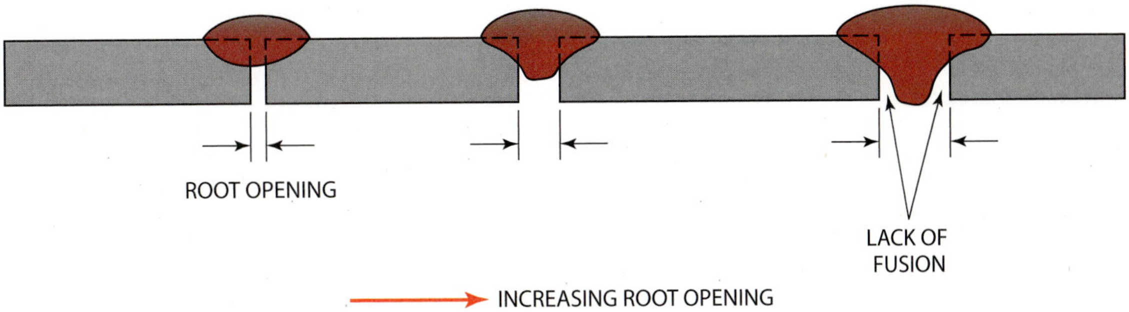

FIGURE 9-40 Effect of root opening on weld penetration. © Cengage Learning 2012

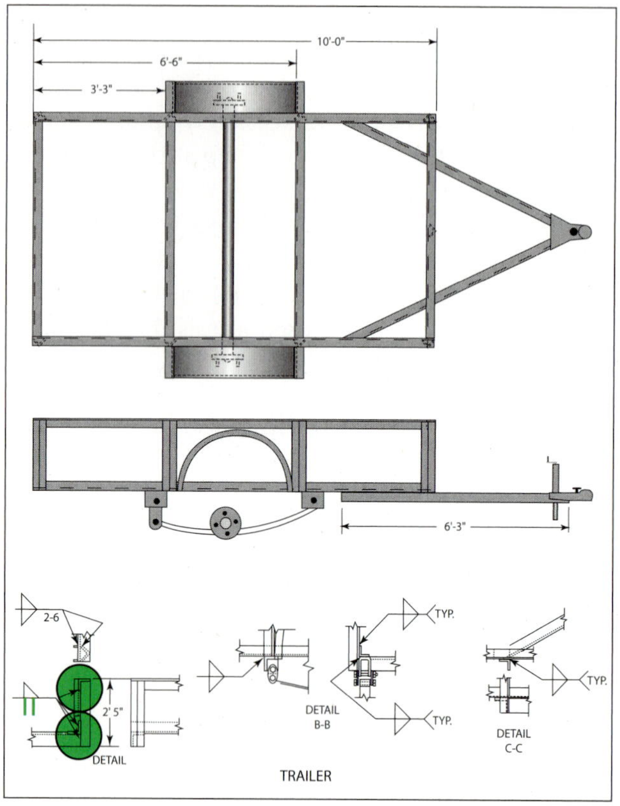

FIGURE 9-41 Square butt joints on utility trailer.

Project Materials and Tools

The following items are needed to fabricate the smoke box.

Materials

- 24 to 30 1/8-in. diameter F2 or F3 class electrodes
- 8 1 × 6-in. pieces of 3/16-in. to 1/4-in. thick mild steel plate
- 7 1 × 5 1/2-in. pieces of 3/16-in. to 1/4-in. thick mild steel plate
- 2 1 × 7-in. pieces of 3/16-in. to 1/4-in. thick mild steel plate

Tools

- Arc welder
- Plasma cutting torch, oxyfuel cutting torch, or shear
- PPE
- Wire brush
- Angle grinder
- Chipping hammer
- Square
- Tape measure
- C-clamp or magnetic clamp
- Soapstone

Layout

Using a square, 12-in. rule and soapstone, lay out the bar stock as shown in the project drawing, Figure 9-42.

Cutting Out

The first step is to cut the mild steel bar stock to length with a shear or thermally.

Fabrication

This project will be fabricated in steps.

1st Step—Assemble the top

- Assemble and tack weld together six of the 5 1/2-in. long bars into a square flat plate with no root space, Figure 9-42.

2nd Step—Assemble the smoke box

- Assemble and tack weld together six of the 6-in. long bars into a square flat plate with a 1/8-in. root space, Figure 9-42.
- Tack weld the two 1 × 7-in. and two of the 1 × 6-in. long bars on the sides of the first plate, tack welded together to form a shallow box.

3rd Step—Forming the handle

- The smoke box handle will be made from one of the 5 1/2-in. long pieces of metal bar. Using an oxyfuel torch, heat across one end of the bar 1/2 in. from the end. Place the hot end in a vise, and bend it at a right angle. Repeat this process on the other end to make the long U-shaped handle.

Welding

Using a properly set up and adjusted arc welding machine and the proper safety protection as demonstrated in Practice 9-1, you will make a series of welds on the joints on the top of the smoke box.

These welds can be made using the skills that you developed while doing the stringer beads on the pencil holder. The difference in making these welds and the stringer beads is that these welds must join two pieces of metal together. To do this you must watch the weld bead to see that it is flowing together and make the weld straight down the joint. The molten weld metal formed on the two sides of the joint will not just flow together forming a single weld bead joining the plates. You must make sure you are traveling

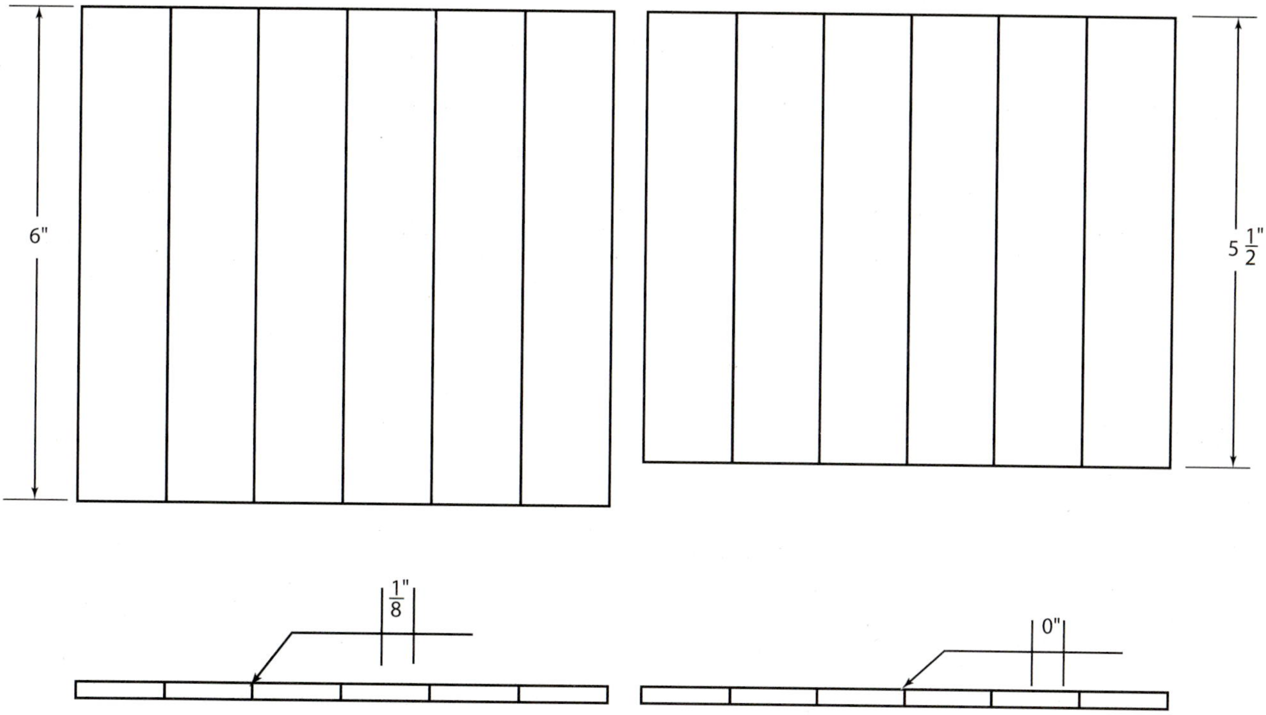

FIGURE 9-42 Project 9-4. © Cengage Learning 2012

slowly enough so that the weld side is adequate to allow the molten weld pools to flow together and not just form on the two edges of the weld, Figure 9-43.

The welds must be straight so that the weld beads are evenly spaced on both sides of the weld joints, Figure 9-44. You must watch the leading edge of the weld pool to make sure your weld bead is centered on the weld joint. If the weld does not follow along the center of the weld joint, the joint will not be as strong as it should be. Weak weld joints can be serious.

Cool, chip, and inspect each weld joint after you have finished the weld. Look at the weld to see if it has uniform width and buildup and follows the weld joint. A lack of uniform width might indicate you were not using a consistent weave pattern. A lack in uniformity in buildup might indicate that your rate of travel down the joint was not steady.

Finish welding the top by making all the welds along both sides of the weld joint on the top plate

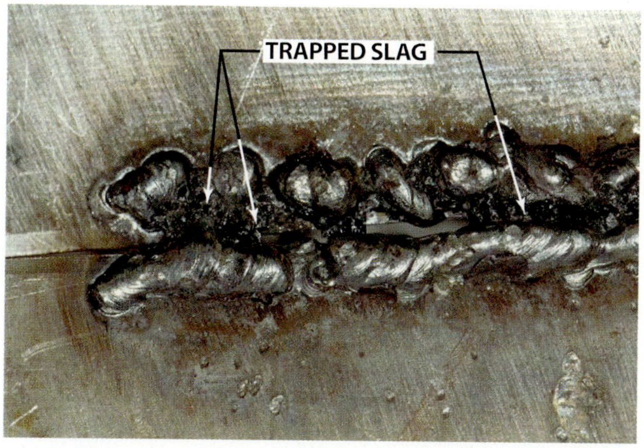

FIGURE 9-43 Moving the electrode from side to side too quickly can result in slag being trapped between the plates. Larry Jeffus

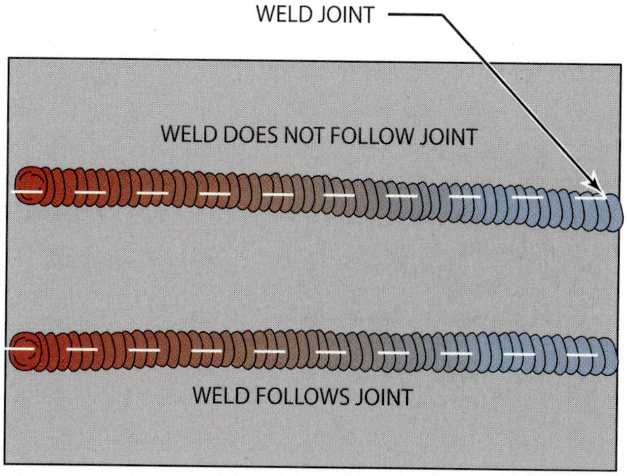

FIGURE 9-44 The weld bead must track down the center of the joint so it is equally distributed on both sides. © Cengage Learning 2012

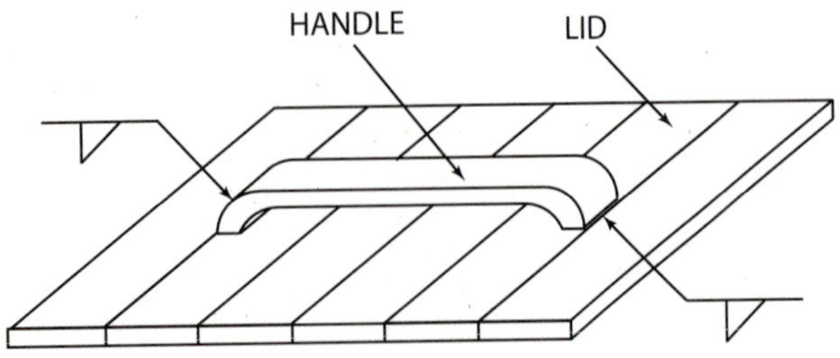

FIGURE 9-45 Smoke box handle.
© Cengage Learning 2012

until you have completed all 12 welds. Remember to cool and chip the welds between each one so you can see what changes in technique you need to make. This also ensures that the weldment does not become overheated.

Measure the length of the box handle and center it from both sides and the ends of the box top. Tack weld both ends before welding it in place with short welds across the outside of both ends of the handle, Figure 9-45.

The root space on the box may make it easier to follow the weld joint, but it makes it harder to get the weld metal to flow together. Too fast a travel speed along this joint will result in a lack of fusion between the two plates. To ensure that this does not happen, you must watch the back edge of the molten weld bead. If it is smooth and rounded, the molten weld metal is joining. Sometimes this can be difficult to see because of the flux covering. The lighter flux produced by F3 class electrodes does not cause this problem as much as the class F2 and F4 electrodes do. One indicator that the weld metal is not joining is a slight difference in the shade of red at the trailing edge of the weld. When there is a gap in the weld metal, the flux will have a slightly lighter red color.

NOTE: There can be enough weld stresses produced to break a tack weld. When tack welds break, they often produce a sharp pinging sound. If you hear this sound while you are welding, stop and locate the broken tack. Use clamps, if necessary, to force the joint back into position before tack welding it into place again.

Weld the joint on the outside of the box first. As before, cool, chip, and inspect each of the welds after they are completed. Make the butt welds around the side of the plate so that the sides are secured to the bottom plate. Use the same techniques for these welds as you did for the other butt joint welds.

The butt welds inside the box may have some degree of arc blow. If arc blow occurs, refer back to that section in this chapter to see how you might solve the problem. Try changing from a DC to AC welding current, and try changing the direction in which you are welding in addition to the other suggested arc blow controls. Observe how each arc blow control worked in this application. Under different conditions the results you observed here, other than changing the class of welding current, may differ.

Limited visibility may be another problem you might have when making the inside welds. The welds must be started all the way at the end, next to the side. This will require a trailing electrode angle, but you cannot end the weld with the same angle. There are several ways of solving this problem. One way is to gradually change from a trailing angle at the beginning to a leading angle at the end of the weld, Figure 9-46. The second way to ensure that the weld is made all the way across the joint is to stop the weld in the center of the joint. Clean and chip the weld end crater and complete the weld by starting on the opposite end and welding back to the center, Figure 9-47. Try both methods to see which works best for you.

Remember to cool, chip, and inspect the welds after each pass to see what changes you might need to make in your technique.

Finishing

Cool off the metal, chip the slag, and wire brush the surface. Grind any welds that interfere with the

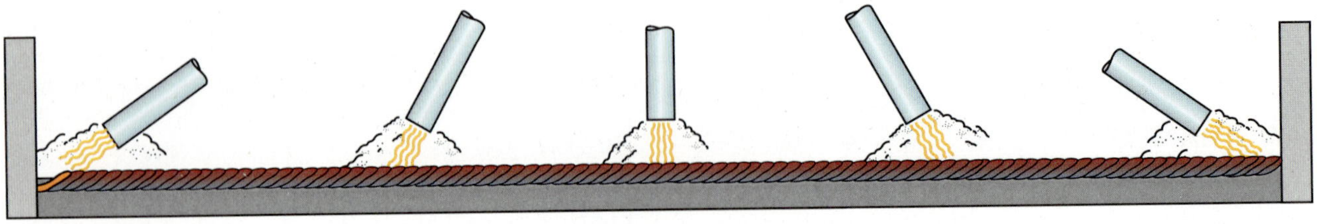

FIGURE 9-46 Rotating the electrode to improve visibility helps when welding in a confined space. © Cengage Learning 2012

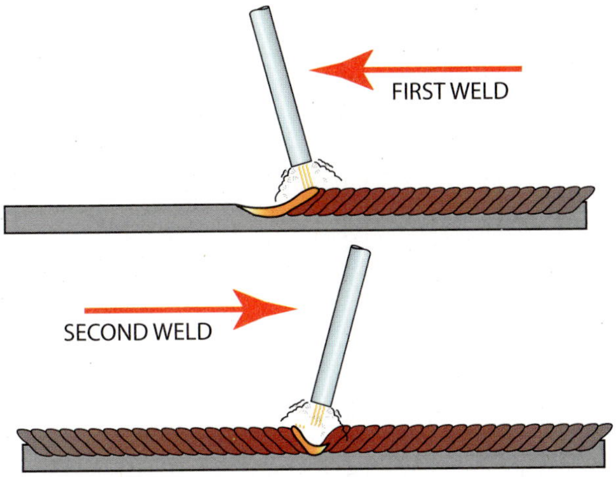

FIGURE 9-47 Starting the weld at the edge and welding to the center. © Cengage Learning 2012

top fitting into the box. No additional finishing is required.

Paperwork

Complete a copy of the time sheet in Appendix I, the bill of materials in Appendix III, or use forms provided by your instructor.◆

Outside Corner Joint

An **outside corner joint** is made by placing the plates at an angle to each other, with the edges forming a V-groove, Figure 9-48. The angle between the plates may range from a very slight 15° angle to almost flat at a 165° angle. There may or may not be a slight root opening left between the plate edges. Small tack welds should be made approximately 1/2 in. (13 mm) from each end of the joint. The weld bead should completely fill the V-groove formed by the plates and may have a slightly convex surface buildup.

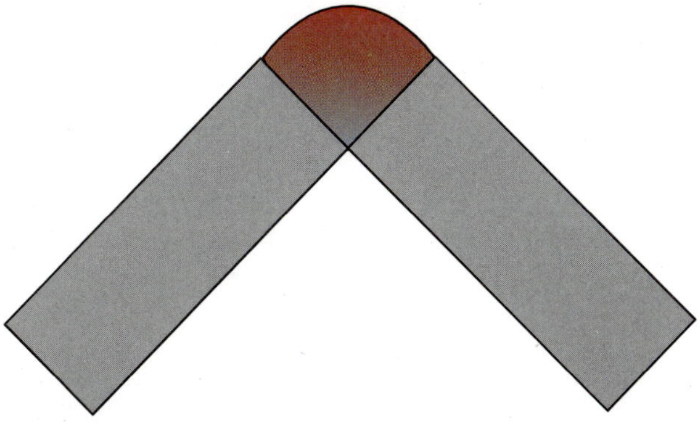

FIGURE 9-48 V formed by an outside corner joint. © Cengage Learning 2012

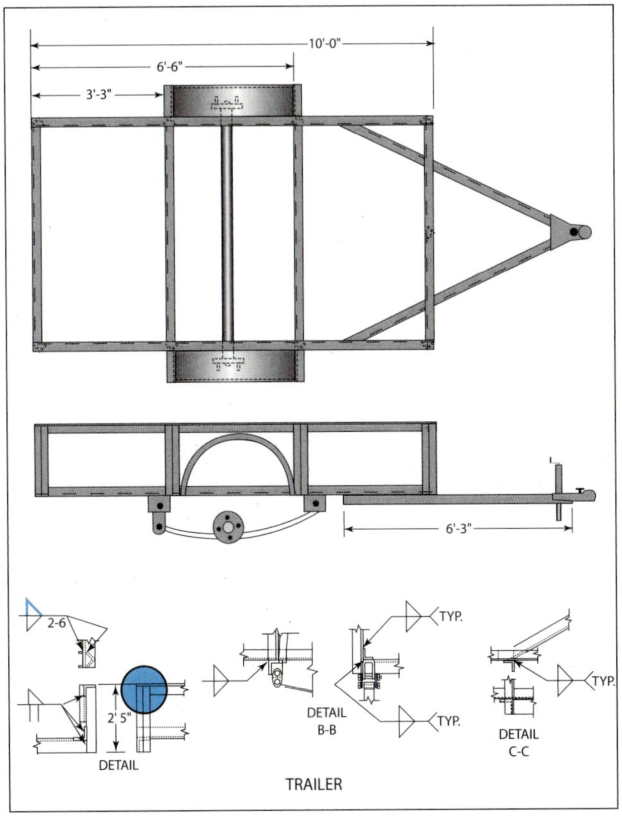

FIGURE 9-49 Outside corner joints on utility trailer. © Cengage Learning 2012

Outside corner joints are used at the corners of tanks, boxes, and ships. Four plates can be made into a square tube; three can be made into a triangular tube, Figure 9-49.

PROJECT 9-5

Outside Corner Welds to Build a Candlestick

Skill to be learned: The control of the welding electrode to produce uniform outside corner welds in the flat position.

Project Description

Candles come in a variety of sizes and lengths. It is common to display them on some type of stand. The stand can serve to raise a short candle above other decorations, or several different heights of candlesticks may be used to display candles in an arrangement. Most candles today are made from "drip-free" wax to avoid wax damaging furniture or fabric such as tablecloths. Some people prefer the nostalgic appearance of candle wax running down the side of a candle. You can build your candlestick holder with a top plate large enough to catch this dripping wax or small enough to allow the wax to run down the holder too.

Project Materials and Tools

The following items are needed to fabricate the pair of four-sided candlestick holders.

Materials

- 10 to 15 1/8-in. diameter F2 or F3 class electrodes
- 8 1 1/2 × 8-in. pieces of 3/16-in.- to 1/4-in. thick mild steel plate
- 2 2-in. square pieces of 3/16-in. to 1/4-in. thick mild steel plate
- 2 3-in. square pieces of 3/16-in. to 1/4-in. thick mild steel plate
- 8 cork or felt pads

Tools

- Arc welder
- Plasma cutting torch, oxyfuel cutting torch, or shear
- PPE
- Wire brush
- Angle grinder
- Chipping hammer
- Square
- Steel rule
- C-clamp or magnetic clamp
- Soapstone

Layout

Using a square, 12-in. rule and soapstone, lay out the bar stock as shown in the project drawing, Figure 9-50.

Cutting Out

The first step is to cut the mild steel bar stock to length with a shear or thermally.

Fabrication

Tack weld the 1 1/2-in. metal bars together so they make a square tube. The corners of the plates should meet so they form a 90° V, Figure 9-51. Place the tack welds about 1 in. from the ends of the joints so they do not interfere with the starting or stopping of the welds. The top and bottom plate will be assembled following the welding of the outside corner joints.

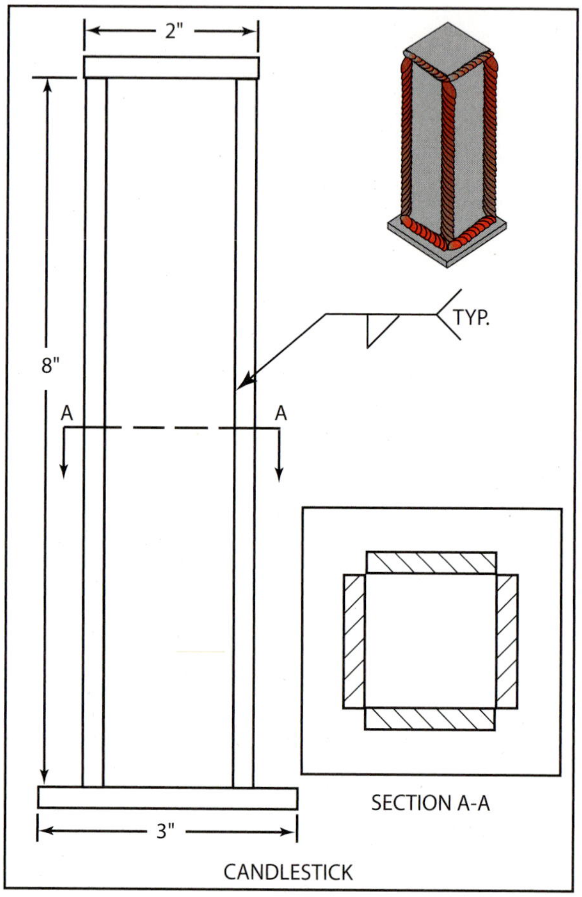

FIGURE 9-50 Project 9-5. © Cengage Learning 2012

Welding

Make sure to strike the arc in the V-groove of the outside corner joint. Although F3 class electrodes may produce a lot of weld spatter that can stick to the sides of the plates, it does not have the same effect that

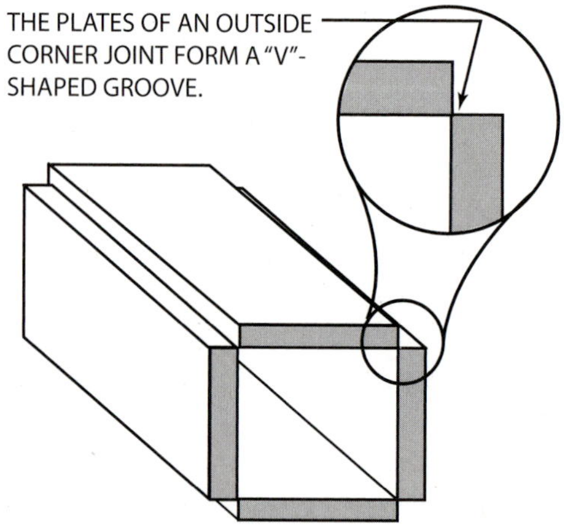

FIGURE 9-51 Outside corner joint layout. © Cengage Learning 2012

FIGURE 9-52 Effect of arc strike outside of the weld bead. Larry Jeffus

striking the arc has on the plate surface, Figure 9-52. The weld spatter can be cleaned off or left to add texture to the finished candlestick.

There may be some arc blow at the ends of the weld. The appearance of these welds is part of the finished part's appearance. So, you cannot start at the ends and weld to the center of the weld as you may have done when welding out the smoke box. Because the welds' appearance is important, if arc blow does occur, you cannot just stop. You will need to control the arc and weld pool to keep them uniform. Try holding a very short arc length as a way of controlling the arc blow if it becomes a problem.

Cool, chip, and inspect each weld joint after you have finished the weld. Look at the weld to see if it has uniform width and buildup and follows the weld joint. A lack of uniform width might indicate you were not using a consistent weave pattern. A lack in uniformity in buildup might indicate that your rate of travel down the joint was not steady.

When all of the welds are completed on both candlesticks, lay out the center of the top and bottom plates. Center the square tube on the bottom plate and square it to the baseplate before making a single small tack weld on one corner. Stand the candlestick up next to the square, Figure 9-53. Measure the distance from each of the sides of the top of the tube to see that it is standing up vertically on the base. If the tube is not vertical within 1/8 in., the tack weld can be bent slightly to make the tube vertical. If the tack weld cannot bend enough, it can be broken off and a new tack made.

NOTE: A magnetic level can be used to check if the tube is vertical before tack welding, Figure 9-54.

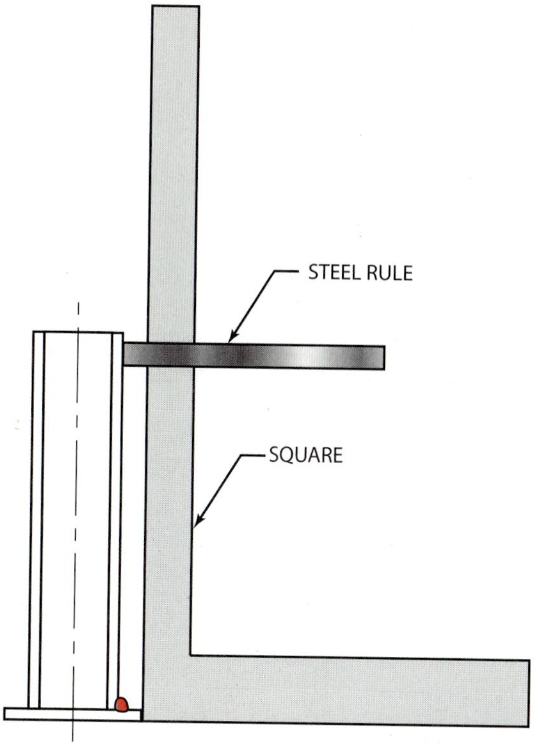

FIGURE 9-53 Checking squareness with a square. © Cengage Learning 2012

Once the tube is vertical on the baseplate, make two or three more tack welds to hold it in place. Recheck the tube to see that it is still vertical within tolerance.

The top plate must be centered on the tube and square to the tube and to the baseplate. One way to accomplish this is to stand the tube on the top plate and use a square and measuring tape to align the plate before tack welding it in place, Figure 9-55.

Make a fillet weld around the top and bottom plates to hold the tube in place. Refer to the following tee joint welding instruction, if necessary, to make these welds.

The height of the second candlestick must be within ±1/8 in. of the first one. You may need to grind the end of the tube or increase the joint spacing to ensure that both candlesticks are the same height.

Finishing

Cool and chip the welds. There are a number of ways to finish this project in addition to painting it a color or a clear finish. Following are some possibilities:

- Leave the welds and spatter as they are.
- Grind all the welds and surfaces smooth.
- Write a greeting, holiday saying, your name, draw a heart or other symbol with weld beads, or cut them in with a torch.

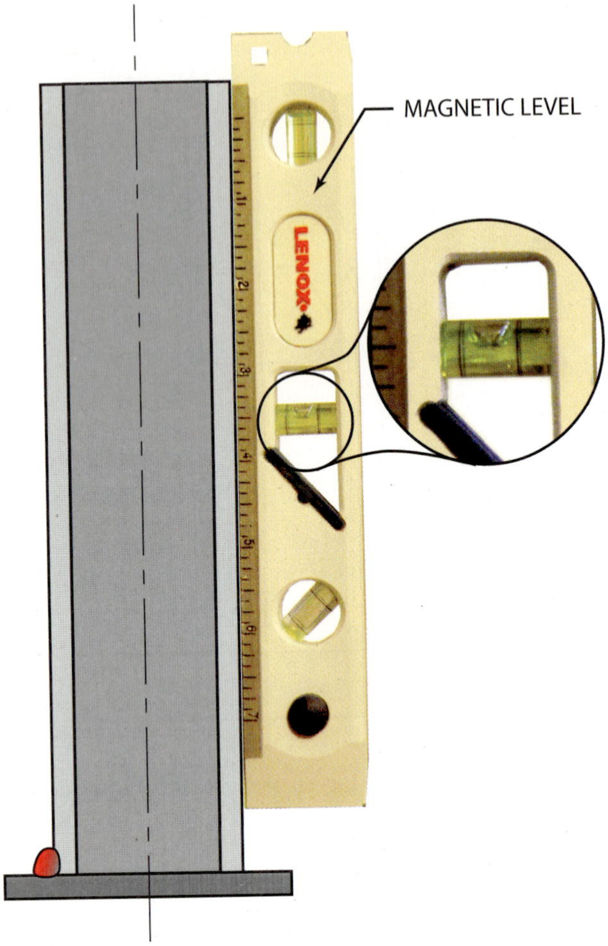

FIGURE 9-54 Checking squareness with a level.
© Cengage Learning 2012

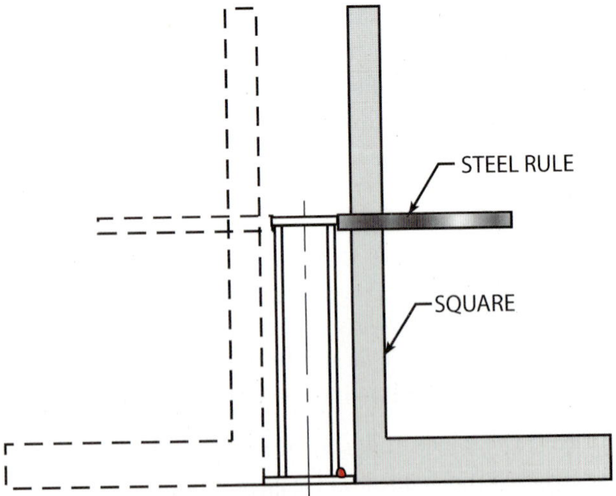

FIGURE 9-55 Centering the top plate.
© Cengage Learning 2012

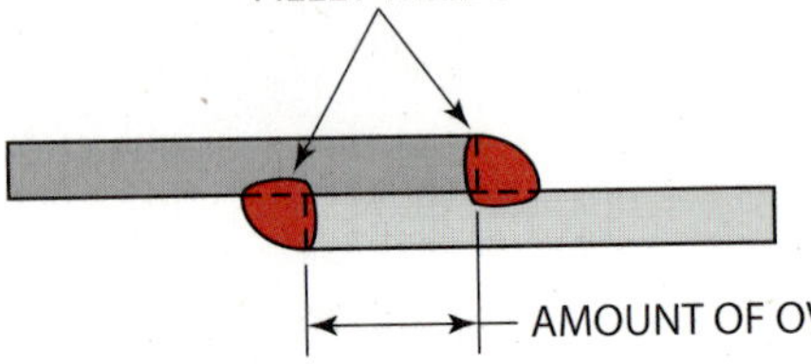

FIGURE 9-56 Lap joint. © Cengage Learning 2012

Paperwork

Complete a copy of the time sheet in Appendix I, the bill of materials in Appendix III, or use forms provided by your instructor.◆

Lap Joint

A **lap joint** is made by overlapping the edges of the two plates, Figure 9-56. The joint can be welded on one side or both sides with a fillet weld. In most cases, both sides of the joint should be welded. When just one side is welded, the joint is not as strong, and water can cause rust to form in the joint space, Figure 9-57.

As the fillet weld is made on the lap joint, the buildup should equal the thickness of the plate, Figure 9-58.

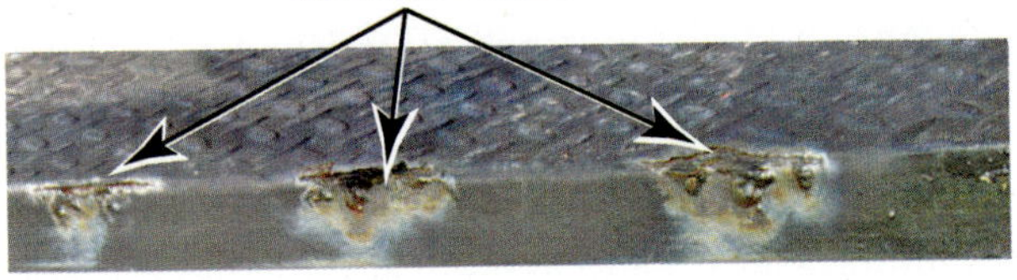

FIGURE 9-57 Crevice corrosion caused by water trapped in the space between metal plates at a lap joint.
Larry Jeffus

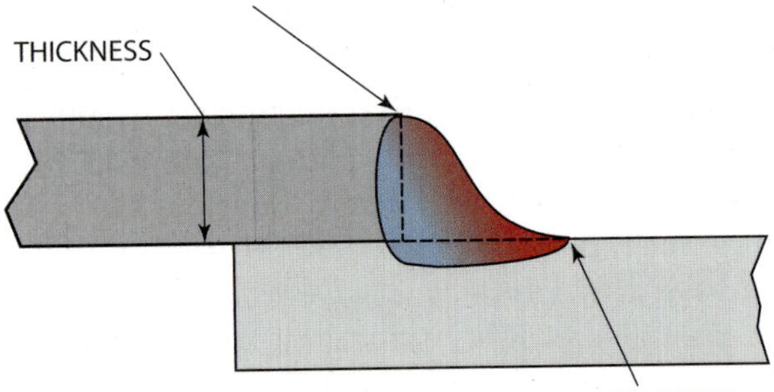

FIGURE 9-58 Correct lap joint bead shape.
© Cengage Learning 2012

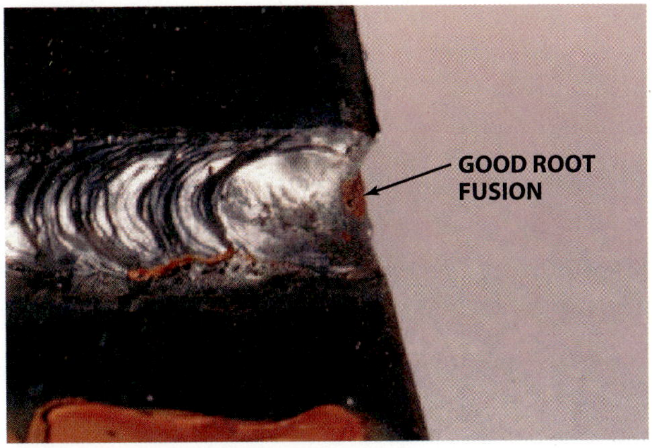

FIGURE 9-59 Correct crater fill at the end of a lap joint.
Larry Jeffus

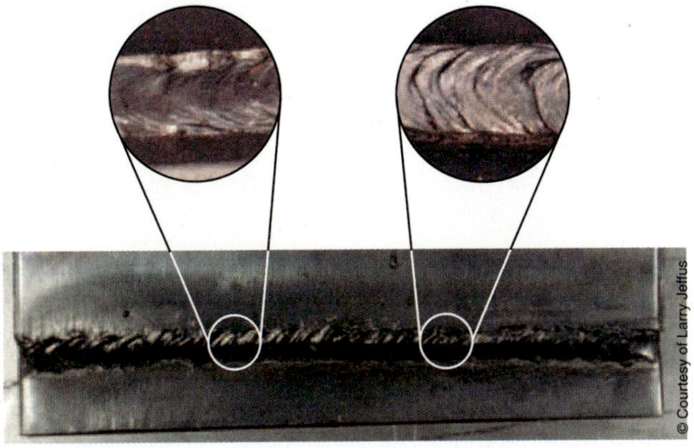

FIGURE 9-60 Lap joint weld. Larry Jeffus

A good weld will have a smooth transition from the plate surface to the weld. If this transition is abrupt, it can cause stresses that will weaken the joint.

Penetration for lap joints does not improve their strength; complete fusion is required. The root of fillet welds must be melted to ensure a completely fused joint, Figure 9-59. But if the molten weld pool shows a notch during the weld, this is an indication that the root is not being fused together. The weave pattern will help prevent this problem, Figure 9-60.

Lap joints are often used when joining plates or when plates are stiffened with an angle iron or another structural shape. They are also used when joining structural steel, Figure 9-61.

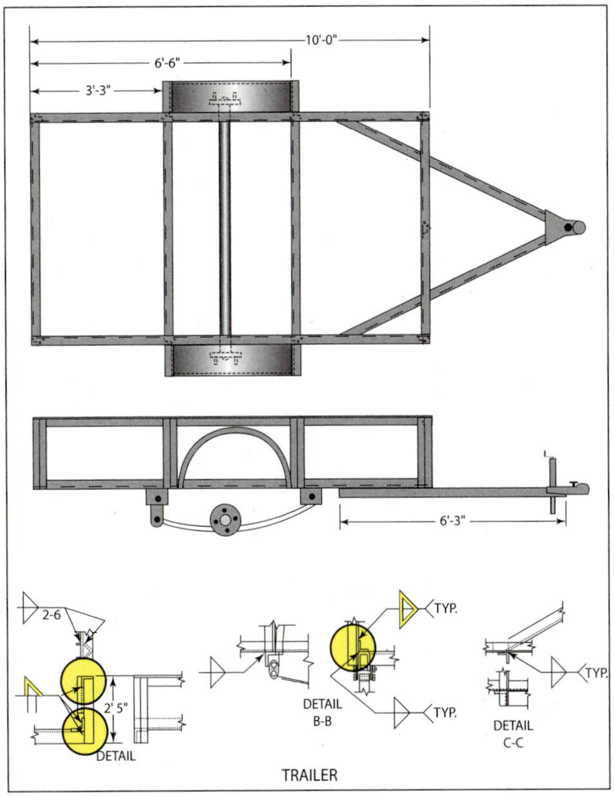

FIGURE 9-61 Lap joints on utility trailer.
© Cengage Learning 2012

PROJECT 9-6

Fillet Welds in Lap Joints to Build a Birdhouse Roof

Skill to be learned: The control of the welding electrode to produce uniform fillet welds in lap joints in the flat position. In addition, the square butt and outside corner welding skills learned in the previous projects will be reinforced. Measuring, fitting, and assembling fabrication skills and techniques will be developed.

Project Description

The roof of a house is usually covered with shingles that are lapped one on top of the other. The lap joints on the roof of your birdhouse are designed to give it the look of shingles.

This project will be fabricated in three parts.

Part 1—The roof of the birdhouse will be fabricated using lap joint welds.

Part 2—The walls and floor will be fabricated using butt joint welds.

Part 3—The corners of the walls, floor, and roof ridge will be joined using outside corner welds.

Part 1 Project Materials and Tools

The following items are needed to fabricate the birdhouse roof.

Materials

- 24 to 30 1/8-in. diameter F2 or F3 class electrodes
- 12 1 1/2 × 8-in. pieces of 3/16-in. to 1/4-in. thick mild steel plate

Tools

- Arc welder
- Plasma cutting torch, oxyfuel cutting torch, or shear
- PPE
- Wire brush
- Angle grinder
- Chipping hammer
- Square
- Steel rule
- C-clamp or magnetic clamp
- Soapstone

Layout

Using a square, 12-in. rule and soapstone, lay out the bar stock as shown in the project drawing, Figure 9-62.

Cutting Out

The first step is to cut the mild steel bar stock to length with a shear or thermally.

Fabrication

The two roof panels will be constructed with five 1 1/2 × 8-in. pieces of bar stock. The roof must be 5 1/2 in. wide, and the overlaps of each joint must be equal. To determine the amount of the overlap required, you must find the total width of all five of the 1 1/2-in. wide bars.

Total width = number of bars × width of each bar
Total width = 5 × 1 1/2
Total width = 7 1/2 in.

To find the overlap dimension for each of the four overlaps, you must first find the total overlap. To get this you must subtract the roof width of 5 1/2 in. from the total bar width of 7 1/2 in.

Total overlap = total width − roof width
Total overlap = 7 1/2 − 5 1/2
Total overlap = 2 in.

The five bars make a total of four overlap joints. So the dimension of each overlap will be 1/4 of the total overlap.

Overlap = total overlap ÷ number of overlaps
Overlap = 2 ÷ 4
Overlap = 1/2 in.

Lay out the bars so that they overlap 1/2 in. and have a total width of 5 1/2 in. Holding the bars in place, make a small tack weld on the end of each overlap, Figure 9-63. Placing the tack welds on the ends will keep them out of the way of the lap welds.

Once both lap weld panels are tack welded together, recheck to see that the overall panel width

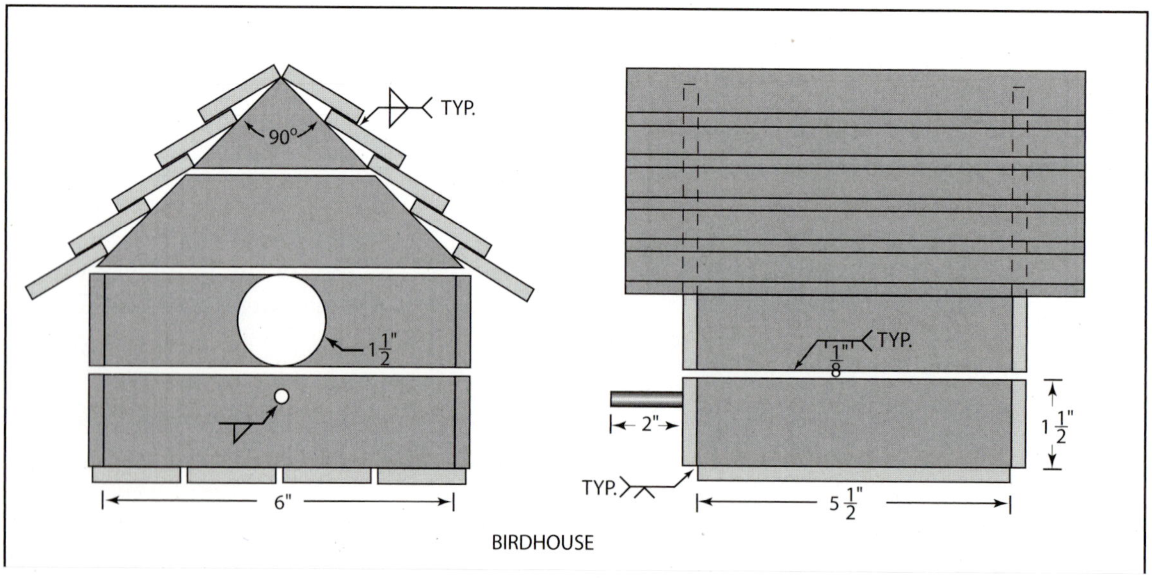

FIGURE 9-62 Project 9-6. © Cengage Learning 2012

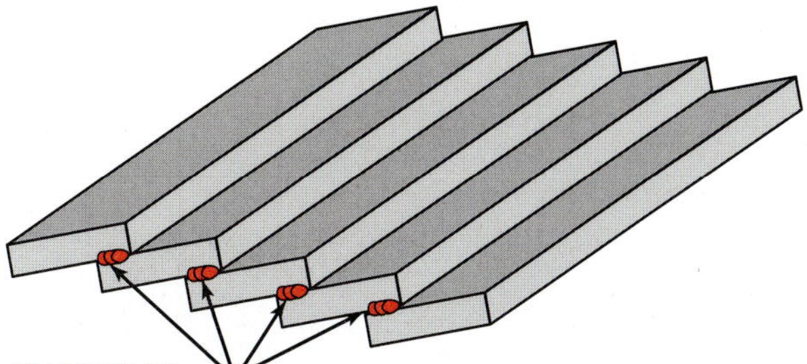

FIGURE 9-63 Tack welding lap joints.

of 5 1/2 in. is accurate. If necessary, grind off a tack weld and adjust the overlap so that the finished panel width is correct within ±1/8 in.

Welding

The lap joint panel will need to be braced at an approximate 45° angle so that the face of the weld will be flat. Also, the welds will have to be alternated from side to side to minimize the distortion of the roof panel.

The welding heat is not evenly distributed between the two plates of a lap joint. The edge of the top plate is easily overheated as compared to the surface of the bottom plate, Figure 9-64. To keep the top plate's edge from melting back, direct most of the arc's heat on the surface of the bottom plate. Using the "J" weave pattern with the long top of the J on the bottom plate surface, and the short bottom of the J just touching the top edge of the top plate should keep the weld away from the top plate edge.

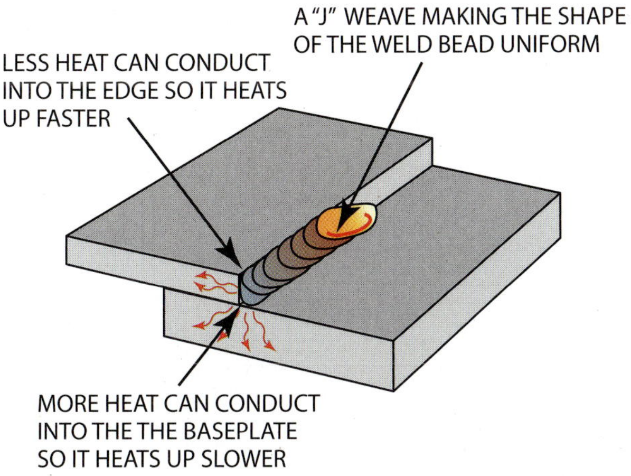

FIGURE 9-64 Effect of different plate thicknesses on joint heating.

One problem that occurs with all fillet welds is slag entrapment at the root of the weld. This occurs as a result of not getting enough of the arc's heat down in the root of the joint. More heat can be directed into the joint root by dipping the tip of the electrode into the root as the J-weld pattern passes over the weld root.

Cool, chip, and inspect each weld joint after you have finished the weld. Look at the weld to see if it has uniform width and buildup and follows the weld joint. A lack of uniform width might indicate you were not using a consistent weave pattern. A lack in uniformity in buildup might indicate that your rate of travel down the joint was not steady. Match your welding speed so that each weld is made with only one electrode.

Remember to alternate the welds from side to side to reduce weld distortion. When the welding is completed on the first panel, weld the second roof panel lap joints.

Part 2 Project Materials and Tools

The following items are needed to fabricate the walls and floor of the birdhouse.

Materials

- 24 to 30 1/8-in. diameter F2 or F3 class electrodes
- 8 1 1/2 × 6-in. pieces of 3/16-in. to 1/4-in. thick mild steel plate
- 8 1 1/2-by 5 1/2-in. pieces of 3/16-in. to 1/4-in. thick mild steel plate
- 1 3/8-in. diameter 2-in. long piece of round bar stock or reinforcement rod

Tools

- Arc welder
- Plasma cutting torch, oxyfuel cutting torch, or shear
- PPE
- Wire brush
- Angle grinder
- Chipping hammer
- Square
- Steel rule
- C-clamp or magnetic clamp
- Soapstone

Layout

Using a square, 12-in. rule and soapstone, lay out the bar stock as shown in the project drawing, Figure 9-62.

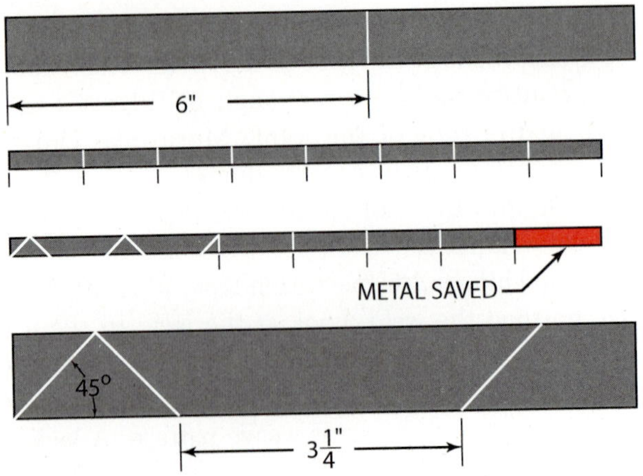

FIGURE 9-65 Laying out the birdhouse.
© Cengage Learning 2012

The front and back wall butt panels may be laid out in one of two ways. They can be laid out as a square, or they can be laid out and precut to shape, Figure 9-65. The square layout is easiest, and the precut saves metal. Both methods are acceptable.

Cutting Out

The first step is to cut the mild steel bar stock to length with a shear or thermally.

Fabrication

This project will be fabricated in steps.

1st Step—Assemble the bottom

- Assemble and tack weld together four of the 5 1/2-in. long bars into a square flat plate with a 1/8-in. root space, Figure 9-66.

2nd Step—Assemble both sides

- Assemble and tack weld together two panels of the 5 1/2-in. long bars into rectangular flat plates with a 1/8-in. root space.

3rd Step—Assemble both the front and back

- Assemble and tack weld together two panels of the 6-in. long bars into square flat plates with a 1/8-in. root space.
- If the end pieces are not cut to shape, you must make the tack welds inside the area that will remain after the ends are cut to shape. That way, the parts will stay together after the ends are cut.

4th Step—Assemble the sides, bottom, front, and back panels

- Assemble both sides and both end panels on the bottom, Figure 9-67. Put the tack welds to the inside so they will not affect the welds. The corners will form outside corner joints and can be tack welded either on the inside or outside of the birdhouse. The welding stresses may be strong as the butt joints are welded, so make several tack welds along the joints.

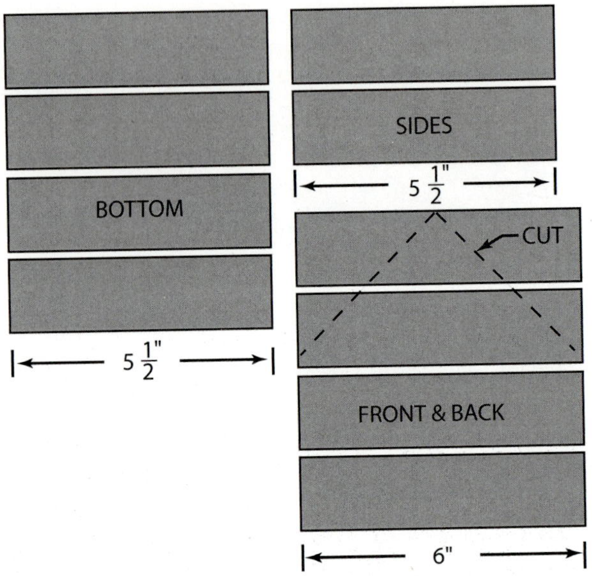

FIGURE 9-66 Birdhouse panel layout.
© Cengage Learning 2012

Welding

Using a properly set up and adjusted arc welding machine and the proper safety protection as demonstrated in Practice 9-1, you will make a series of welds on the joints on the top of the birdhouse.

The welds must be straight so that the weld beads are evenly spaced on both sides of the weld joints, Figure 9-44. Alternate the welds made from panel to panel so you do not overheat the weldment and to reduce weld stresses.

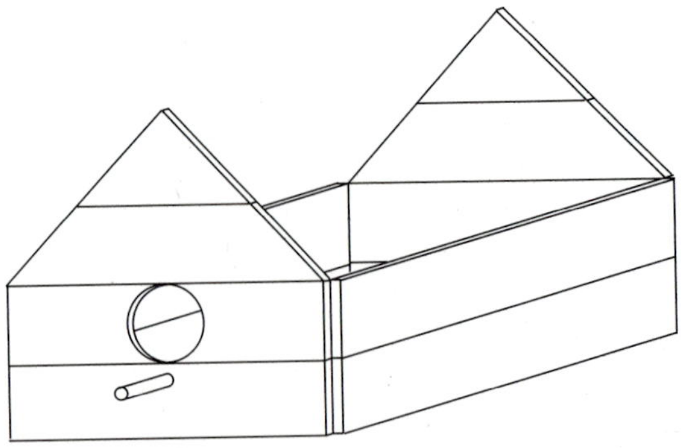

FIGURE 9-67 Birdhouse panel assembly.
© Cengage Learning 2012

Cool, chip, and inspect each weld joint after you have finished the weld. Look at the weld to see if it has uniform width and buildup and follows the weld joint. A lack of uniform width might indicate you were not using a consistent weave pattern. A lack in uniformity in buildup might indicate that your rate of travel down the joint was not steady.

Finish all the butt joint welds before starting on the outside corner welds. If necessary, grind off any of the ends of the butt welds if they extend into the outside corner joint groove. This will make it easier to keep the outside corner welds even. Make the welds that join the bottom to the sides and the bottom to the front and back panels first so that the birdhouse corner welds will be made last and will be on top of those welds. This will give the birdhouse a more finished look.

Set the two roof panels on top of the assembled birdhouse. You may have to shift them around to find the best fit. The ridge of the panels should be as straight and uniform as possible. Tack weld the panels together. Make a small fast outside corner weld across the ridge of the roof panels. Making the weld fast and small will minimize the weld distortion. Do not weld the roof to the base of the birdhouse. Leaving it loose will allow you to clean out the house after each season.

The 3/8-in. diameter 2-in. long piece of round bar stock or reinforcement rod is welded in place last.

Finishing

Lay out the 1 1/2-in. hole on the front side and thermally cut it out.

Cool and chip the welds and any slag left from cutting the hole. A power wire brush can be used to give the welds a polished, shiny appearance. Fit the roof onto the base. It should fit solidly. If it has distorted from the weld, a hammer and anvil can be used to reshape it as necessary.

The project can be painted with a clear coat or colored coat of latex paint.

Paperwork

Complete a copy of the time sheet in Appendix I, the bill of materials in Appendix III, or use forms provided by your instructor.◆

Tee Joint

The **tee joint** is made by tack welding one piece of metal on another piece of metal at a right angle, Figure 9-68. After the joint is tack welded together, the slag is chipped from the tack welds. If the slag is not removed, it will cause a slag inclusion in the final weld.

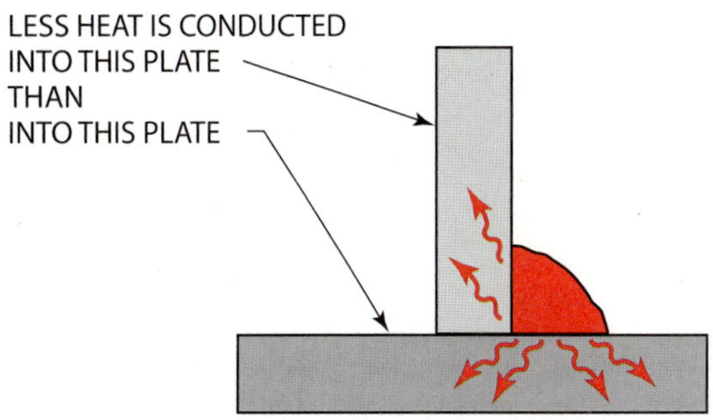

FIGURE 9-68 Tee joint. © Cengage Learning 2012

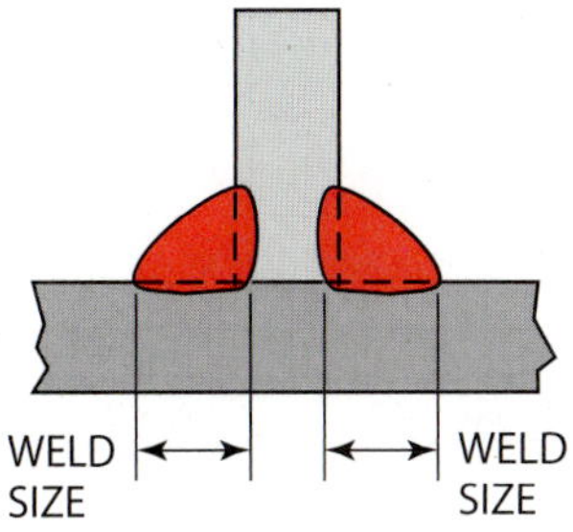

FIGURE 9-69 Tee joint welded on both sides. © Cengage Learning 2012

The heat is not distributed uniformly between both plates during a tee weld. The plate that forms the stem of the tee can conduct heat away faster than the baseplate. Heat escapes into the baseplate in two directions. When using a weave pattern, most of the heat should be directed to the baseplate to keep the weld size more uniform and to help prevent an undercut.

A welded tee joint can be strong if it is welded on both sides, even without having deep penetration, Figure 9-69. The weld will be as strong as the base plate if the size of the two welds equals the total thickness of the baseplate. The weld bead should have a flat or slightly concave appearance to ensure the greatest strength and efficiency, Figure 9-70.

Tee joints can be used to join two pieces of metal at a right angle. The joining of a ship's bulkhead to the hull forms a tee joint. Joists in buildings form a tee joint with the header, Figure 9-71.

PROJECT 9-7

Fillet Welds on Tee Joints to Make a C-Clamp

Skill to be learned: The control of the welding electrode to produce uniform fillet welds in tee joints in the flat position. Additional welding skills learned

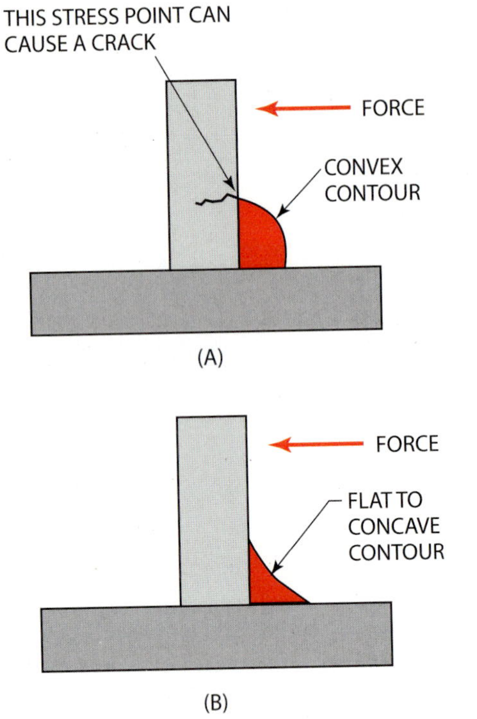

FIGURE 9-70 The stresses are distributed more uniformly through a flat or concave fillet weld.
© Cengage Learning 2012

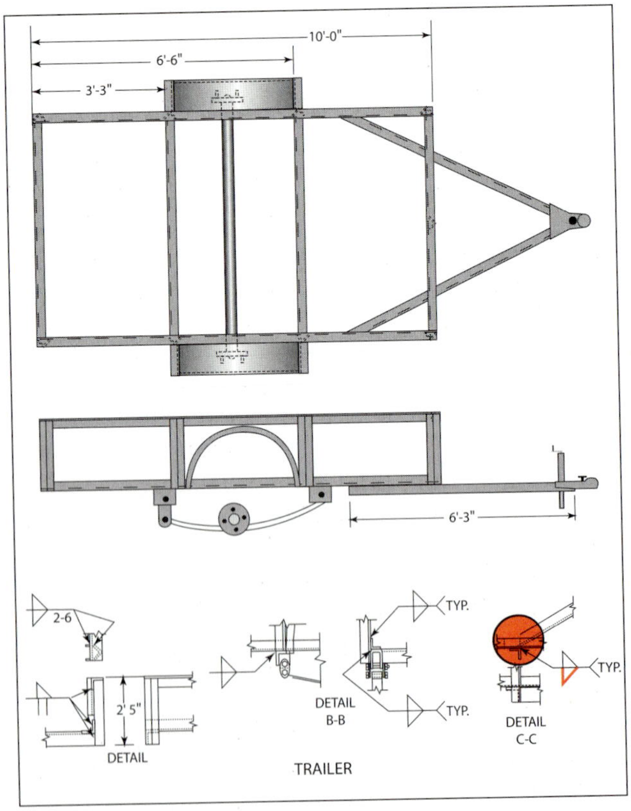

FIGURE 9-71 Tee joints on a utility trailer.
© Cengage Learning 2012

in the previous projects will be reinforced. Detailed fabrication skills and techniques required to align and square parts will be developed.

Project Description

C-clamps are one of the most commonly used fabrication hand tools. They are used to both hold parts in place until tack welds can be made and to force parts together. Sometimes enough force can be applied with one or more C-clamps to bend one part so it fits properly to another part.

As the bolt is turned, forcing parts together, it is important that the upper jaw turn on the bolt so that no twisting force is applied to the parts. If the top jaw twists, it can cause both the top and bottom C-clamp jaws to slip. This C-clamp has an upper jaw assembly of a nut and three different washers that allow the upper jaw to turn freely as the clamping pressure is applied with the C-clamp. A little light oil can be used so the jaw will turn easier.

Project Materials and Tools

The following items are needed to fabricate the C-clamp.

Materials

- 5 to 10 1/8-in. diameter F2 and/or F3 electrodes
- 2 1 × 12-in. pieces of 3/16-in. to 1/4-in. thick mild steel plate to form the back beam
- 2 1 × 5-in. pieces of 3/16-in. to 1/4-in. thick mild steel plate to form the bottom beam
- 2 1 × 6-in. pieces of 3/16-in. to 1/4-in. thick mild steel plate to form the top beam
- 3/8-in. diameter 4-in. long rod handle
- 3/4-10NC 14-in. bolt
- 3/4-10NC connector
- 3/4-10NC nut upper jaw part
- 3/4-in. flat washer upper jaw part
- 1-in. flat washer upper jaw part
- 3/8-in. washer upper jaw part
- 1-in. diameter 3/16-in. to 1/4-in. thick lower jaw

Tools

- Arc welder
- PPE
- Angle grinder
- Wire brush
- Square

- Soapstone
- Arc welder
- Plasma cutting torch, oxyfuel cutting torch, or shear
- PPE
- Wire brush
- Chipping hammer
- Square
- Steel rule
- C-clamp or magnetic clamp

Layout

Using a square, 12-in. rule and soapstone, lay out the plate as shown in the project drawing, Figure 9-72.

Cutting Out

The first step is to cut the mild steel bar stock to length with a shear or thermally.

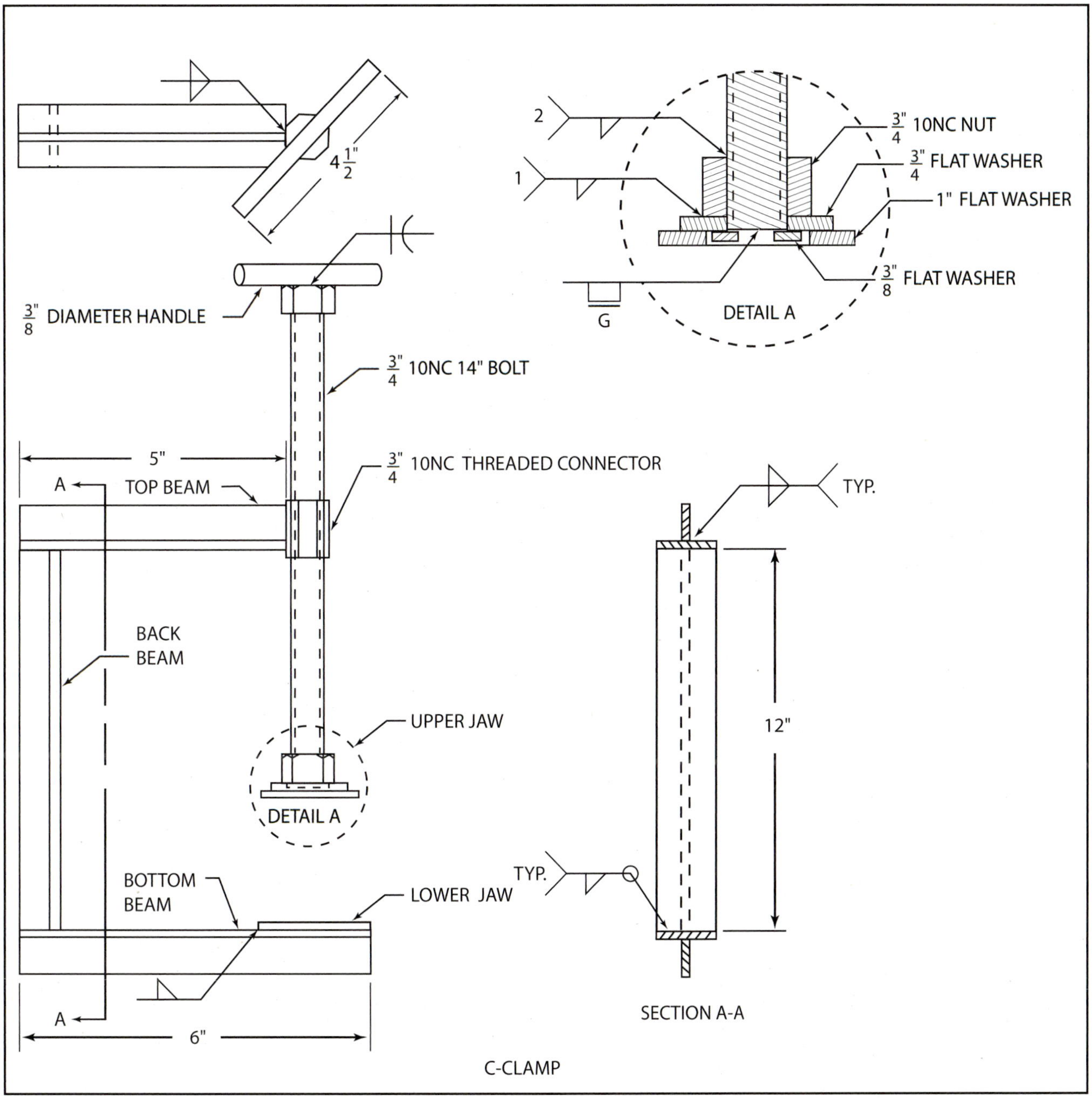

FIGURE 9-72 Project 9-7. © Cengage Learning 2012

Fabrication

This project will be fabricated in steps.

1st Step—Assemble the top, back, and bottom tee beams

- Center one 1-in. bar in the center of the matching length bar.
- Make tack welds on opposite sides of the vertical tee joint bar to keep it square.
- The tack welds should be approximately 1/2 in. long and about 1 in. from the ends of the joint.
- Make an additional set of tack welds in the center of the 12-in. long back beam.

2nd Step—Assembling C-clamp beams

- Once the three tee beams are welded, they can be assembled according to the layout shown in Figure 9-72.
- Use the square or the magnetic clamp to hold the beams squarely together.
- Make tack welds on the beam joints to hold them in alignment for welding.

3rd Step—Assemble the bolt and top jaw assembly

- Screw the 3/4-10NC 14-in. connector onto the 3/4-10NC 14-in. bolt
- Place the 3/4-in. flat washer on top of and in the center of the 1-in. flat washer and use the C-clamp to hold them tightly together on the welding table. Use the steel rule to check the center.
- Make four small fillet welds as shown on the number 1 welding symbol in Detail A, Figure 9-72. Cool and chip the welds.
- Screw the 3/4-10NC nut just far enough onto the bolt so that the assembled washers will fit as shown in Detail A, Figure 9-72.
- Place the 3/8-in. washer on the end of the bolt and hold it in place with your gloved finger.
- Rest the electrode on your gloved hand, lower your helmet, and fill the hole in the 3/8-in. washer by making a small plug weld.
- Screw the nut down on the bolt far enough so that the top of the plug weld can be ground flat.
- Screw the nut back down to the end so that the washer assembly is close but not tight to the 3/8-in. washer. The washer assembly needs to be loose enough to turn but not so loose that the 3/8-in. washer will extend past the washer assembly as the C-clamp is tightened, Figure 9-73.

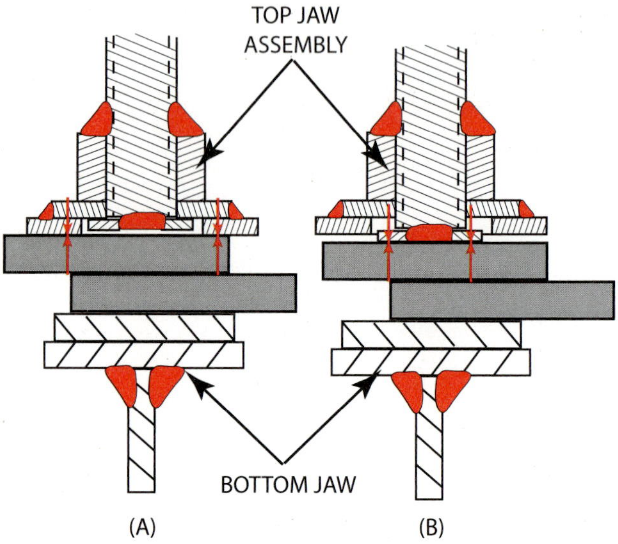

FIGURE 9-73 C-clamp screw assembly.

NOTE: If the center washer touches when the C-clamp is tightened on a surface, it may not stay in place. If this happens, you can add another 1-in. washer to the one already on the top jaw assembly by tack welding the edges of the washers together.

- Protect the bolt threads from weld spatter, and make four small fillet welds to hold the nut in place as shown on the number 2 welding symbol in Detail A, Figure 9-72. Cool and chip the welds.
- Center the 3/8-in. rod on top of the head of the bolt.
- Protect the bolt threads from weld spatter, and weld the rod to the bolt head.

4th Step—Mounting the screw assembly to the C-clamp beams

- Run the connector up the threads of the bolt so that the upper C-clamp jaw is touching the bottom jaw and the connector is even with the top beam.
- Align a flat side of the connector so it is centered on the top beam. Grind off any excessive weld bead material that might interfere with these parts fitting tightly together.
- Protect the bolt threads from weld spatter, and tack weld the connector to the top beam.
- Run the bolt in and out to be sure it is lined up with the lower jaw and moves freely.

Welding

Using a properly set up and adjusted arc welding machine and the proper safety protection as demonstrated in Practice 9-1, you will make a series of weld beads on the C-clamp surface following the soapstone layout.

Set the tee joint so that the weld surface will be flat. Strike an arc directly in the bottom of the joint about 1/2 in. away from the end. Keep a slightly longer-than-normal arc length as you bring the tip of the electrode to the end of the tee joint. Lower the tip of the electrode and establish the welding pool. Make sure the welding pool extends across the joint so both sides of the tee are being fused together.

NOTE: Allow these welds to cool slowly. Unlike the other projects in this chapter, the strength of these welds is very important. You want to pay special attention to the welds that join the top, back, and bottom beams together.

Once all of the beams have been welded, they can be assembled as outlined in the second step above. It is important that the welds you make to hold the beams together are strong. If you feel that any of these welds is not as strong as it should be, you can use the angle grinder to remove the weld. Grinding out a weld to reweld a joint is common practice in high-strength welding applications. When you are grinding out the weld, try to avoid grinding away any of the base metal.

Once the beams are welded, check to see that weld distortion did not cause them to twist out of shape by placing the C-clamp frame on a flat surface. The frame should lie flat and not "rock." If it is not flat, you can use a vise and twist it back into shape before attaching the screw.

Weld spatter can be a problem if it gets on the screw threads. F3 electrodes produce a lot more hard spatter than do F2 or F4 class electrodes. *Hard spatter* is spatter that fuses to the metal surface and cannot just be wiped off. For that reason you may want to use one of these electrodes to assemble and attach the C-clamp screw. But you will have to protect the screw threads from spatter regardless of the class of electrode you use.

The screw threads can be protected by using one of the commercially available anti-spatter products. You can also wrap the threads with a damp shop towel. Do not use a wet towel because it could quench the welds, making them brittle and weak.

Finishing

Cool off the metal, chip the slag, and wire brush the surface. As long as the C-clamp is kept dry, it will not rust. If you are using it in a damp area, you may want to spray it with a light oil to prevent rusting.

One advantage of building your own C-clamp is that if the screw threads get damaged by weld spatter, you can cut off the end and remove and replace the bolt.

Paperwork

Complete a copy of the time sheet in Appendix I, the bill of materials in Appendix III, or use forms provided by your instructor.◆

SUMMARY

In this chapter you have learned enough layout, cutting, fabrication, and welding skills to build the utility trailer shown in Figure 9-74. As you have built projects in this chapter, you have been learning the welding skills needed to build this trailer. The welds on the trailer are similar to the ones you have been practicing. Examples of similar welds have been highlighted in figures in this chapter. This utility trailer is small and designed to be lightweight and easily built using the skills that you have already learned. The small size of this utility trailer will allow it to be positioned so that all of the welds can be made in the flat position. The axel, springs, trailer hitch, and lighting required to make it fully functional and street legal are available commercially or over the Internet.

SMAW is often the process of choice for building one-of-a-kind or special projects like the utility trailer. In a factory, this trailer might be made on a production line with the FCA or GMA welded processes, but SMAW would probably be the best process if only one or two units are needed.

At one time almost everything welded was welded with the shielded metal arc welding process. But today, more and more welds are being done with gas metal arc welding and flux cored arc welding processes. However, no one believes SMAW is a dying process.

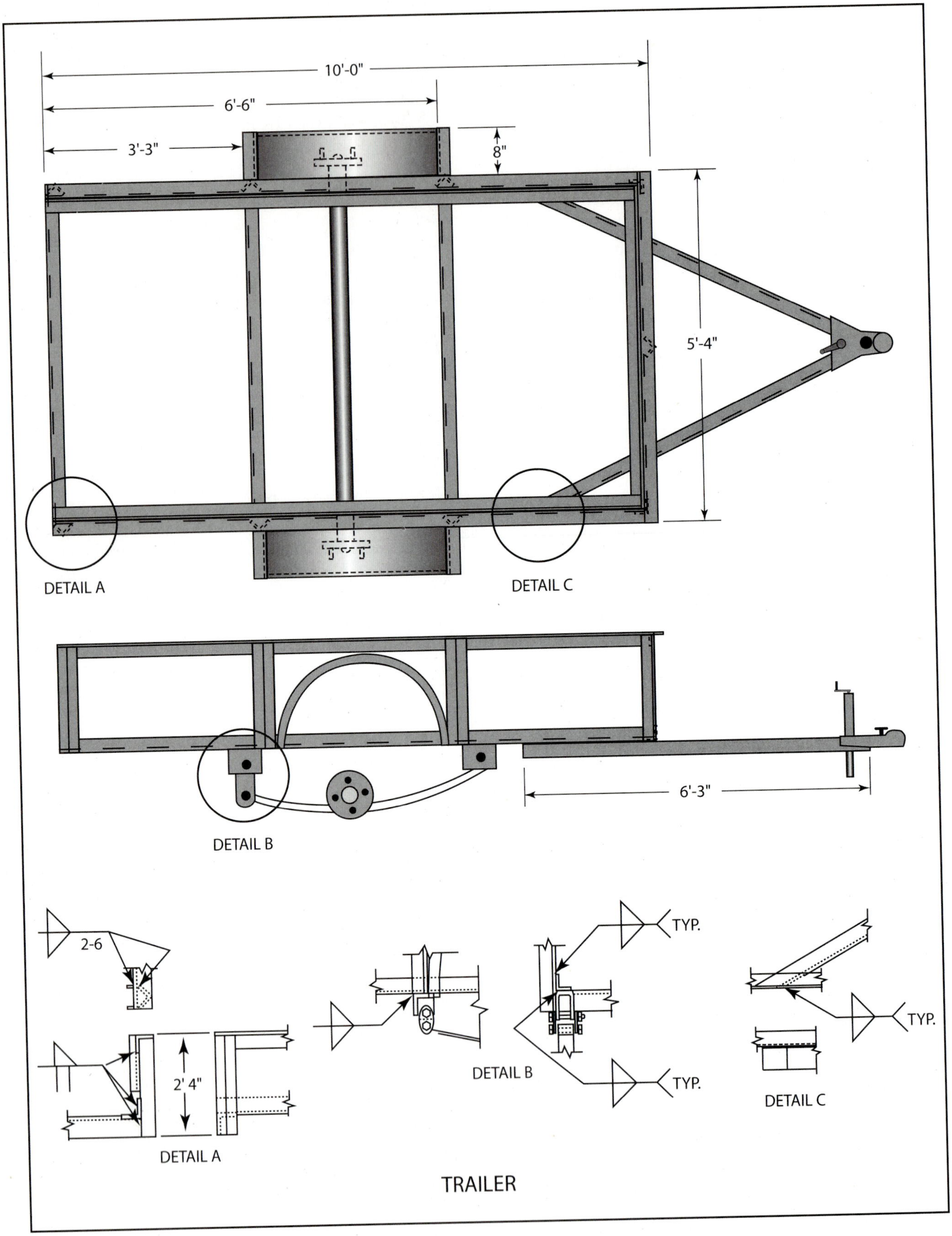

FIGURE 9-74 Utility trailer drawings. © Cengage Learning 2012

It will be a vital skill for many years to come, primarily because it is very flexible and versatile compared to the other processes. Some of its flexibility comes from the fact that it is much easier to change the class or size of an SMAW electrode than to change the electrode wire in a GMAW or FCAW welder. Some of its versatility comes from the fact that it can be used under less-than-ideal conditions. SMAW can be used when it is wet or windy or on rusty or dirty metal and still make acceptable welds.

REVIEW QUESTIONS

1. What electrode types are found in the following classes: F3, F2, and F4?
2. What class of electrodes is the most utilitarian for welding fabrication?
3. What class of electrodes is best for welding on thin metals?
4. What class of electrodes has a smooth easy arc and produces welds with very heavy slag?
5. According to Figure 9-1, which electrode would have the deepest penetration and least buildup?
6. According to Table 9-1, what would the amperage range be for a 1/8-in. E7018 electrode?
7. If an electrode's core wire overheats, what effect can this have on the weld?
8. How can you determine if a weld is too hot?
9. What can happen if the arc length is too short?
10. Which electrode angle may push metal and slag ahead of the weld?
11. What welding electrode manipulation weave pattern is good for flat lap joints?
12. What can cause welder fatigue that can lead to poor welds?
13. List some of the PPE required to weld safely.
14. Why should you always practice striking the arc in the weld groove?
15. What should you do before starting to weld on any project?
16. What should you do if the electrode gets stuck to the metal?
17. What is the purpose of a tack weld?
18. Why are tack welds kept small?
19. Why do welders not normally quench welds in water?
20. What is the normal range for the root opening for a square butt weld?
21. Up to what thickness of plate and structural steel are square butt joints used?
22. What sound might you hear if weld stresses break a tack weld?
23. What effect would changing from a DC to AC welding current have on arc blow?
24. What can a magnetic level be used for?
25. Why is the J weave pattern used for lap joint welds?
26. Why is most of the heat directed to the baseplate of a tee joint?

Chapter 10
Shielded Metal Arc Welding Pipe

OBJECTIVES

After completing this chapter, the student should be able to:

- Explain the difference between how pipe and tubing are used.
- Explain the difference between pipe used for piping systems versus pipe used for structural applications.
- Demonstrate welding fabrication skills.
- Demonstrate welding skills by making square butt and grooved pipe welds in the 1G, 2G, and 5G positions.

KEY TERMS

concave root surface
horizontal fixed pipe position
horizontal rolled pipe position
land
pipe
root face
root gap
root suck back
tubing
vertical fixed pipe position
welding uphill or downhill

INTRODUCTION

Pipe welding is used to both fabricate piping systems to carry any number of fluids, such as water, petroleum, and air; and to build structural items such as gates, truck racks, and go-cart frames. This chapter concentrates more on building structures. However, with few exceptions, the skills you will learn can be used for piping systems. For example, most piping systems use forged fittings while structural items use shop-fabricated fittings. The difference is that forged fitting types are limited and a lot more expensive, while fabricated fittings can be made to fit any angle and cost only a little time and effort.

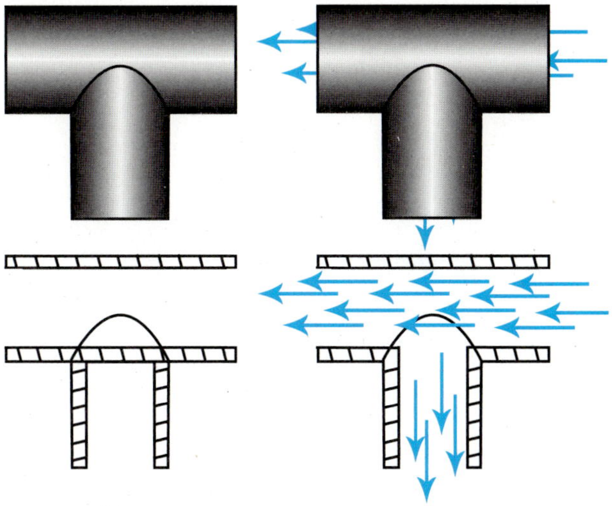

FIGURE 10-1 The difference between structural pipe (A) and a piping system (B). © Cengage Learning 2012

Another difference is that in most cases there is no reason to cut out the center of a structural fitting because nothing will be flowing through the pipes, Figure 10-1. Both pipe tee joints in Figure 10-1A and B look the same in the drawings above. But as you can see, only the Figure 10-1B drawing pipes are connected internally. Either pipe tee would be structurally as strong, but only the pipe in Figure 10-1B could be part of a piping system.

A pipe's strength works well as a structural component. It can withstand the usual forces that are applied to any welded structure, Figure 10-2. Pipe is especially good at withstanding a twisting force. The major drawback to using pipe as a structural component is

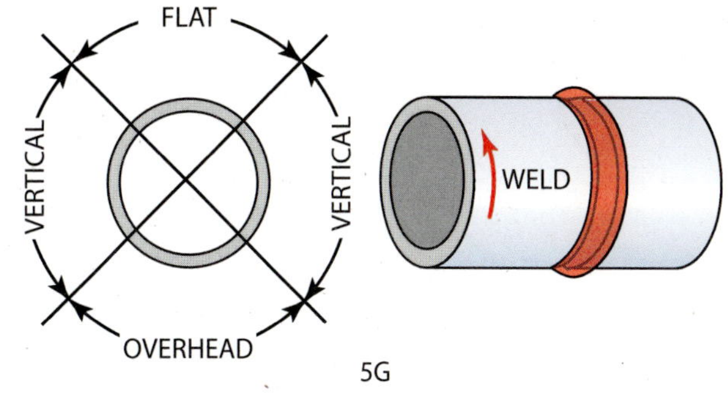

FIGURE 10-3 Welding positions on a fixed horizontal positioned pipe. © Cengage Learning 2012

the difficulty in fitting the parts together. In all pipes except the butt joint, a higher degree of fitting skill is required as compared to most structural shapes. The other problem with using pipe is that the weld position is constantly changing. For example, to make a weld around a pipe in the horizontal fixed position, the weld position can change from flat on top to vertical on the sides and overhead on the bottom, Figure 10-3. In this chapter, you will learn the basics needed to make pipe fittings and develop the welding skills required to weld the fittings.

PIPE AND TUBING

Although pipe and tubing may look alike, they have different types of specifications and uses. Sometimes they are interchangeable, but sometimes they are not. **Pipe** is primarily used to carry fluids, but it is also used for structural applications. **Tubing** can be used for both but it is primarily used for structural applications.

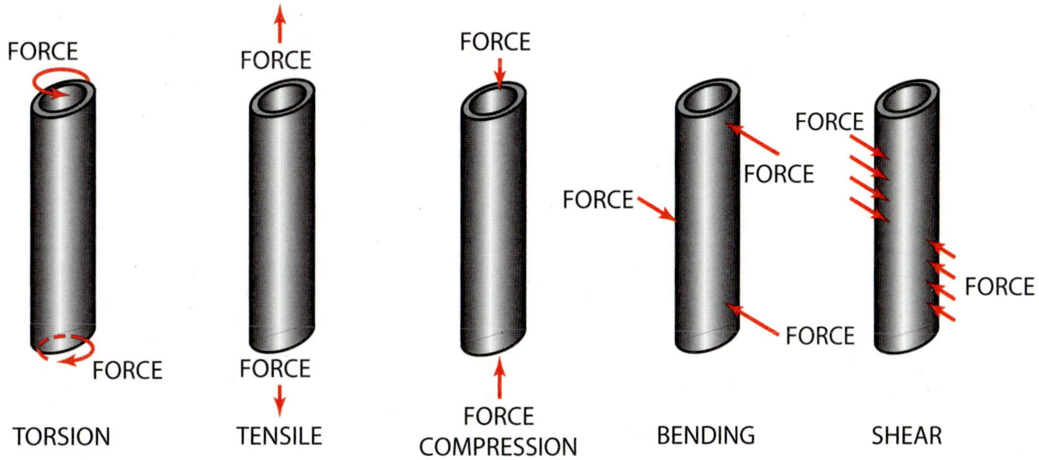

FIGURE 10-2 External mechanical stresses on pipe. © Cengage Learning 2012

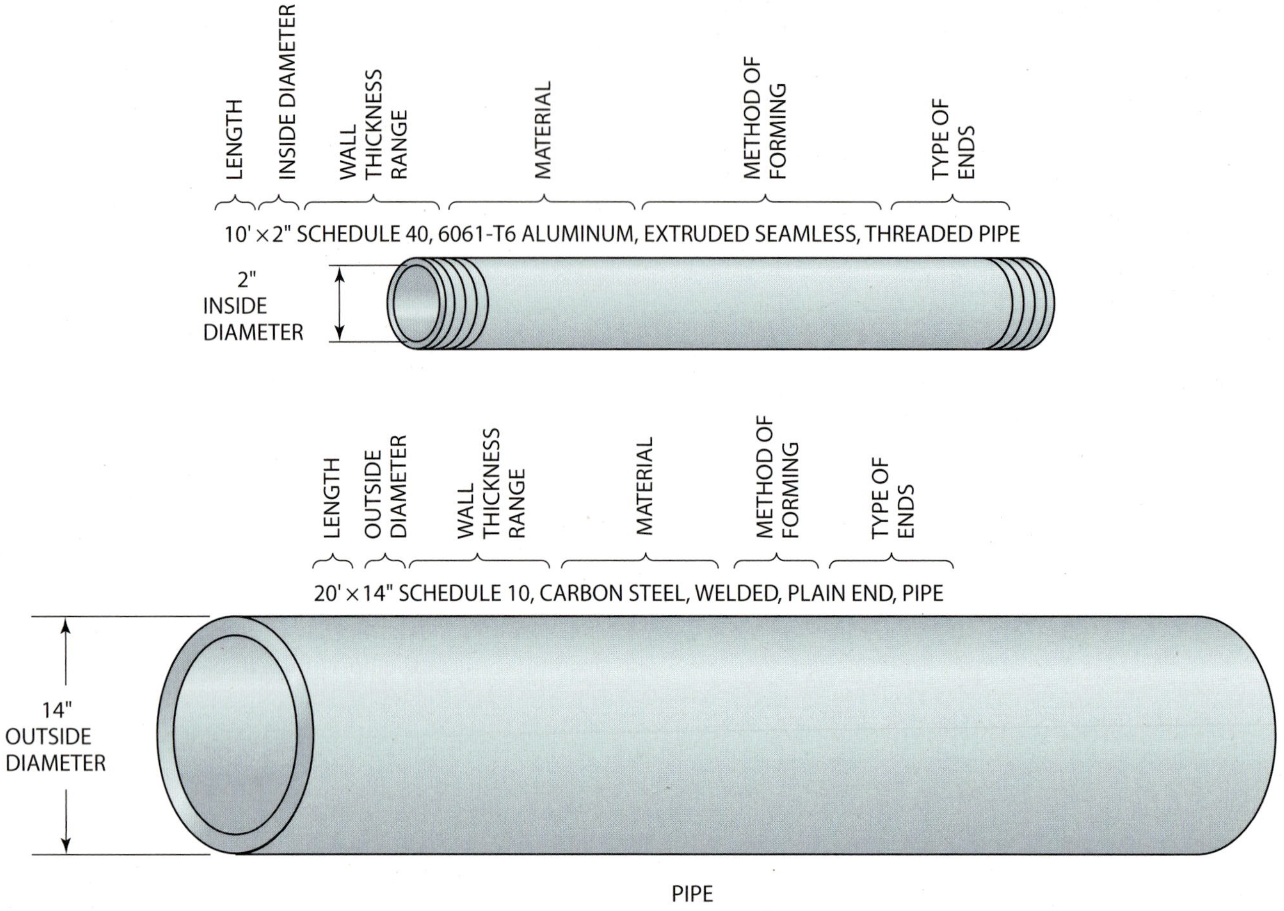

FIGURE 10-4 Typical specifications used when ordering pipe.

The specifications for pipe sizes are given as the inside diameter (ID) for pipe 12 in. (305 mm) in diameter or smaller and as the outside diameter (OD) for pipe larger than 12 in. (305 mm) in diameter. The wall thickness for pipe is determined by its schedule. Schedules 10 through 180 are available; schedule 40 is often considered a standard strength, Figure 10-4.

Tubing sizes are always given as the outside diameter. The wall thickness of tubing is measured in inches or as U.S. standard sheet metal gauge thickness. The desired shape of tubing, such as square, round, or rectangular, must also be listed with the ordering information. Tubing strength is the ability of tubing to withstand compression, bending, or twisting loads. Tubing should also be specified as rigid or flexible, Figure 10-5.

Pipe and tubing are both available as welded (seamed) or extruded (seamless). In this chapter, the term *pipe* will refer to pipe and/or tubing. The welding sequence, procedures, and skills can be used on round tubing.

Preparation and Fit-Up

The end of a pipe may be cut square, or it may be beveled. Often on small diameter pipe that is being used for structural applications, the ends of the pipe are welded using a square butt joint. Larger diameter pipe and that being used in a piping system are usually beveled for maximum penetration and high joint strength, Figure 10-6. Square butt joints are faster to make and can be just as strong as beveled joints if the weld thickness is equal to the metal thickness, Figure 10-7. Beveled pipe joints eliminate the unfused area often found on square butt joints. This unfused root face can trap water and cause rust in piping systems.

The end can be beveled by flame cutting, machining, grinding, or a combination process. It is important that the bevel be at the correct angle, about 37 1/2° depending on specifications, and that the end meet squarely with the mating pipe. The sharp or feathered inner edge of the bevel should be ground flat forming a chamfer. This area is called a **root face** or **land.** Final shaping should be done with a grinder so that the **root gap** is uniform.

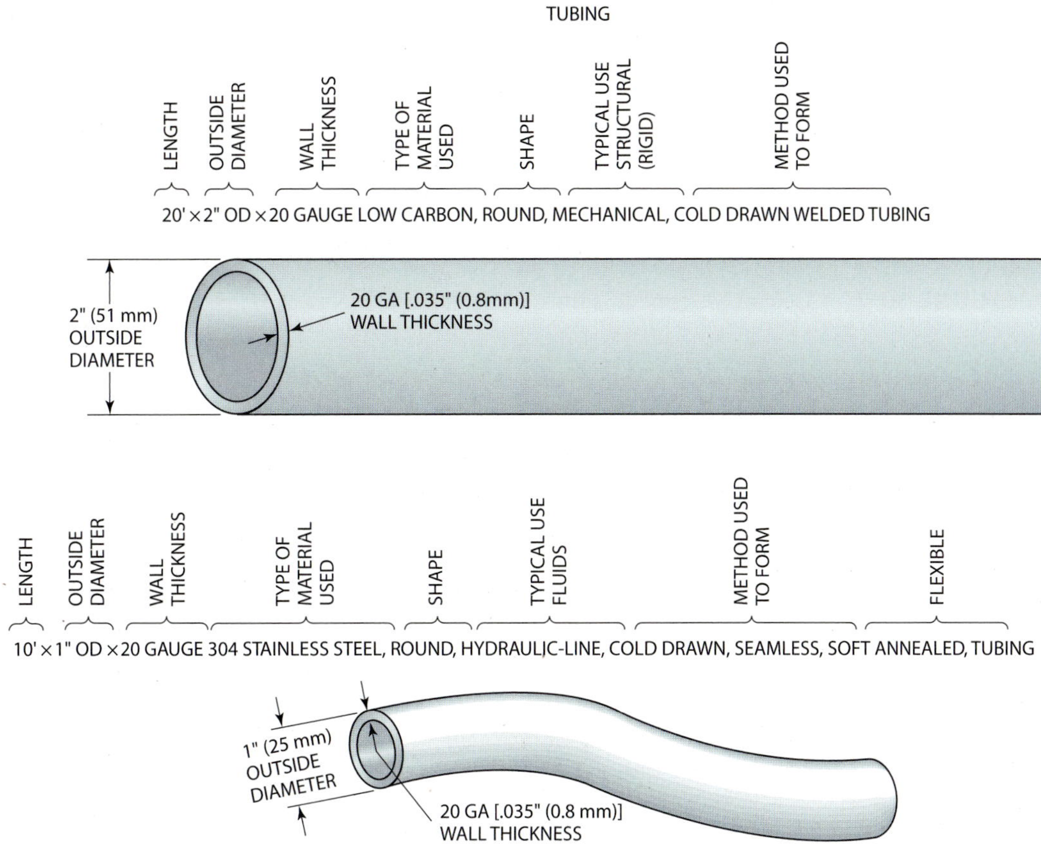

FIGURE 10-5 Typical specifications used when ordering tubing.

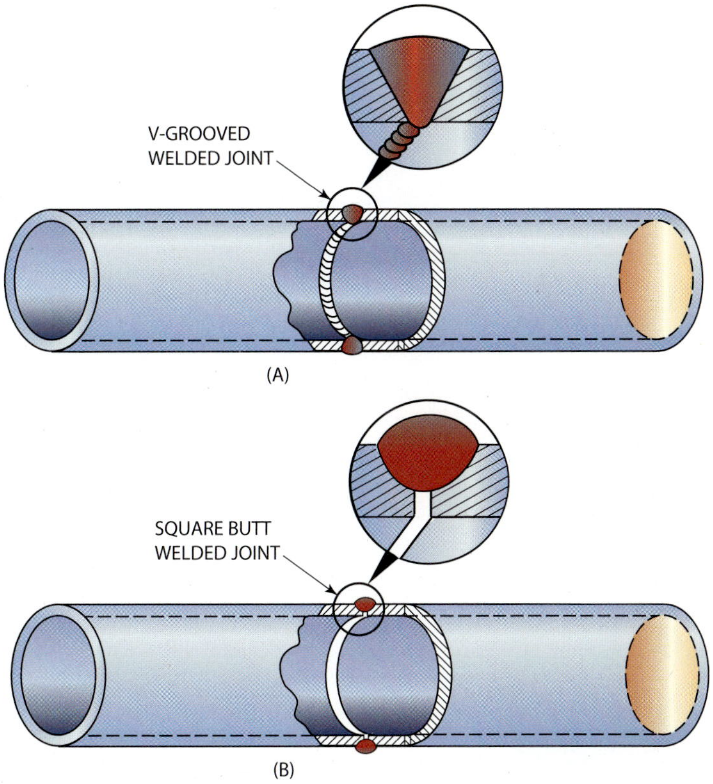

FIGURE 10-6 Comparison of pipe weld root penetration.

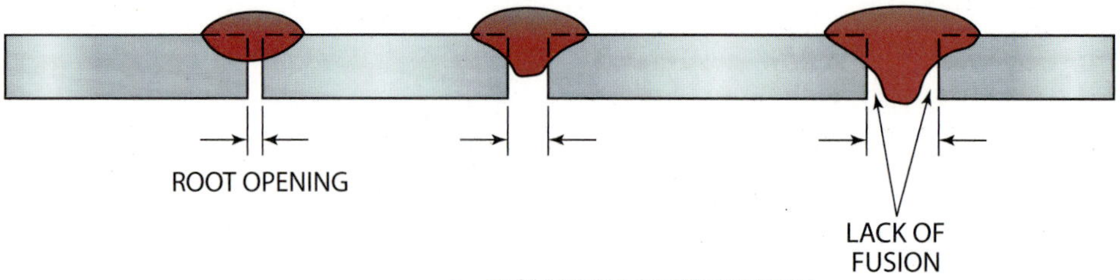

FIGURE 10-7 **Effect of root opening on joint penetration.** © Cengage Learning 2012

The bevel on the end of the pipe can be machine cut using a portable pipe beveling machine, Figure 10-8, or a handheld torch. Chapter 17 describes how to set up and operate flame-cutting equipment. A turntable, similar to the one shown in Figure 10-9, can be made in the school shop and used for beveling short pieces of pipe. The turntable can be used vertically or horizontally. By turning the table slowly with pipe held between the clamps, the welder can use a hand torch to produce smooth pipe bevels. Large-production welding shops may also use machines designed specifically for beveling pipe. These machines will accurately cut a 37 1/2° angle on the pipe.

The 37 1/2° angle allows easy access for the electrode with a minimum amount of filler metal required to fill the groove, Figure 10-10.

FIGURE 10-8 **Pipe beveling machine.** Larry Jeffus

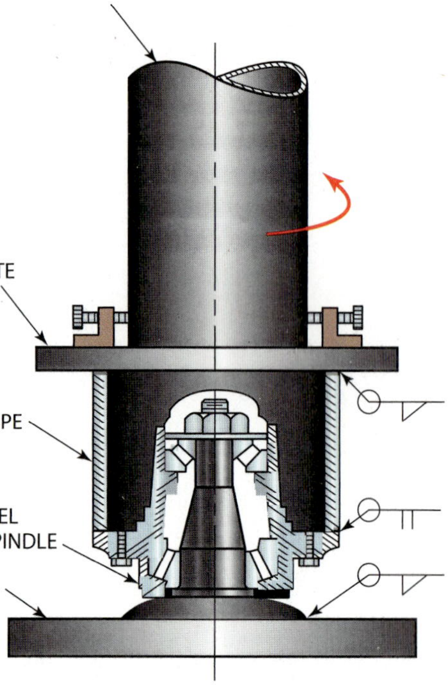

FIGURE 10-9 **Turntable built from a front wheel assembly.** © Cengage Learning 2012

The root face helps a welder control both penetration and **root suck back.** Root suck back is caused by the surface tension of the molten metal trying to pull itself into a ball, forming a **concave root surface,**

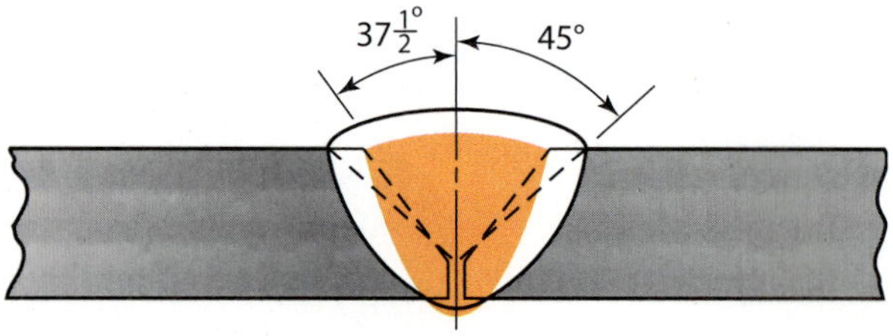

FIGURE 10-10 **The 37 1/2° angled joint may use nearly 50% less filler metal, time, and heat compared to the 45° angled joint.** © Cengage Learning 2012

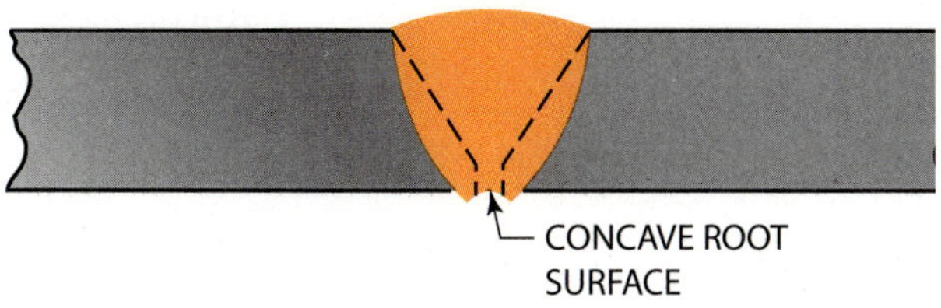

FIGURE 10-11 Root surface concavity. © Cengage Learning 2012

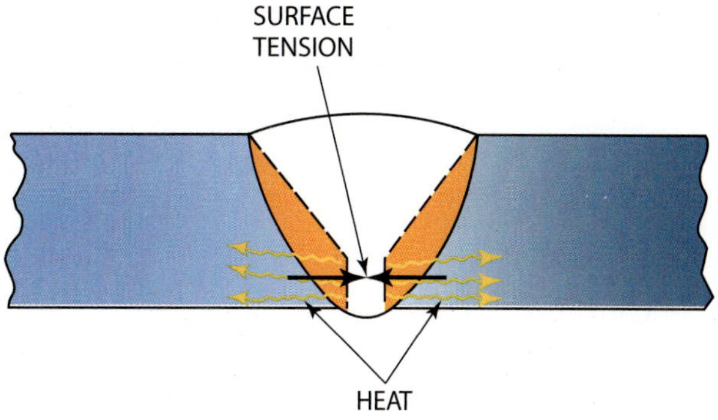

FIGURE 10-12 Heat is drawn out of the molten weld pool, and surface tension holds the pool in place. © Cengage Learning 2012

Figure 10-11. The root face allows a larger molten weld pool to be controlled. Penetration control is improved because there is more metal near the edge to absorb excessive arc heat. This makes machine adjustments less critical by allowing the molten weld pool to be quickly cooled between each electrode movement, Figure 10-12.

Fitting pipe together and holding it in place for welding become more difficult as the diameter of the pipe gets larger. Devices for clamping and holding pipe in place are available, or a series of wedges and dogs can be used, Figure 10-13.

Pipe Welding

The constant transitioning from one weld position to another weld is a problem that must be overcome when welding horizontal fixed position pipe. The rate of change in welding position is slower with large diameter pipes, but the large diameter pipes require more time to weld. When a welder first starts welding, a large diameter pipe should be used to make learning this transition easier. As welders develop the skill and technique of pipe welding, they can change to the small diameter pipe sizes.

Pipe used for piping systems is often welded using either E6010 or E6011 electrodes for the complete weld. Both of these electrodes work well for out-of-position welding. Sometimes these electrodes are used for the root pass, and E7018 electrodes are used to complete the joint. E7018 makes a stronger weld, and it can also be done using the E7018 electrode for the entire weld, Figure 10-14. Pipe used for structural applications may be welded the same way, or it may be welded with E6012 or E6013 if a smooth weld appearance is more important than strength.

PIPE WELDING PASSES

Grooved pipe welds are made up of several separate weld passes. These passes are needed to fill the groove, Figure 10-15. The passes include the following.

Root Weld

A root weld is the first weld in a joint, Figure 10-16. It is part of a series of welds that make up a multiple

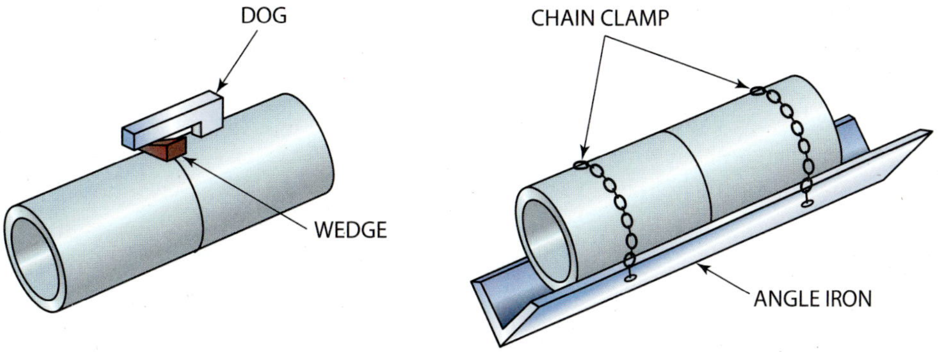

FIGURE 10-13 Shop fabrications used to align pipe joints. © Cengage Learning 2012

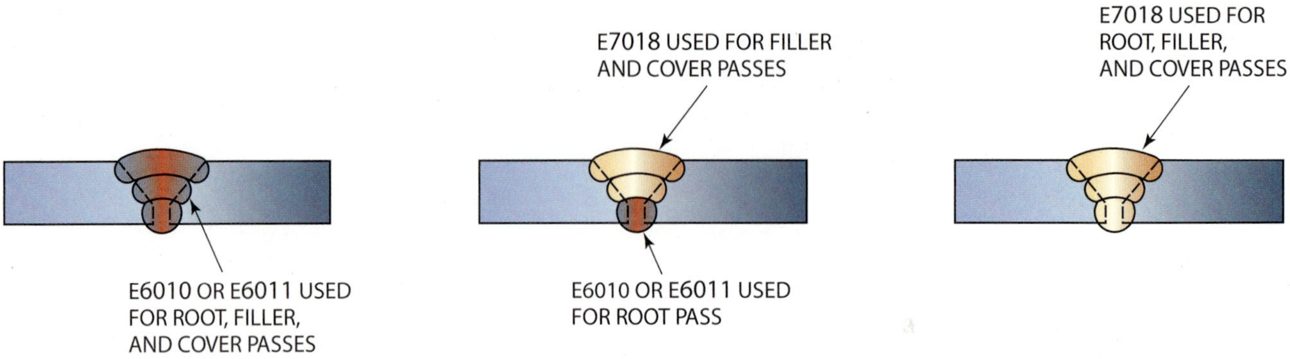

FIGURE 10-14 **Single or multiple types of electrodes may be used when producing a pipe weld. The electrode selected is most often controlled by a code or specification.** © Cengage Learning 2012

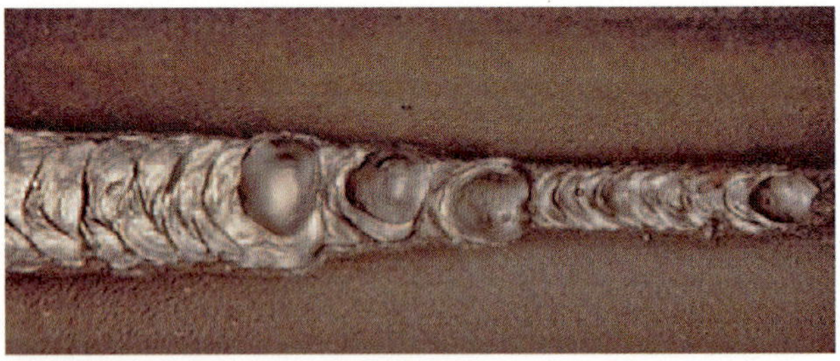

FIGURE 10-15 **Uniformity in each pass shows a high degree of welder skill and increases the probability that the weld will pass testing.** Larry Jeffus

pass weld. The root weld is used to establish the contour and depth of penetration. The most important part of a root weld is the internal root face or, in the case of pipe, the inside surface. The face, or outside shape, or contour of the root weld is not so important.

For high-strength piping systems, the root face of the weld is important. It should be clean, smooth, and uniform, Figure 10-16. If the face of the weld has excessive buildup, a grinder can be used to smooth and reshape the face of the root pass. Grinding removes slag along the sides of the weld bead and makes it easier to add the next pass.

Not all root passes are ground. Pipe that is to be used in low- and medium-pressure systems is not usually ground. Grinding each root pass takes extra time and does not give the welder the experience of using a hot pass. Most slag must be completely removed by chipping and wire brushing before the hot pass is used.

Hot Pass

The hot pass is used to quickly burn out small amounts of slag trapped along the edge of the root pass. This is slag that cannot be removed easily by chipping or wire brushing. A fast travel speed will form a concave weld bead that is easy to clean for the welds that will follow.

Filler Pass

After thoroughly removing slag from the weld groove by chipping, wire brushing, or grinding, it is time to fill the groove. The filler pass(es) may be either a series of stringer beads, Figure 10-17, or a weave bead,

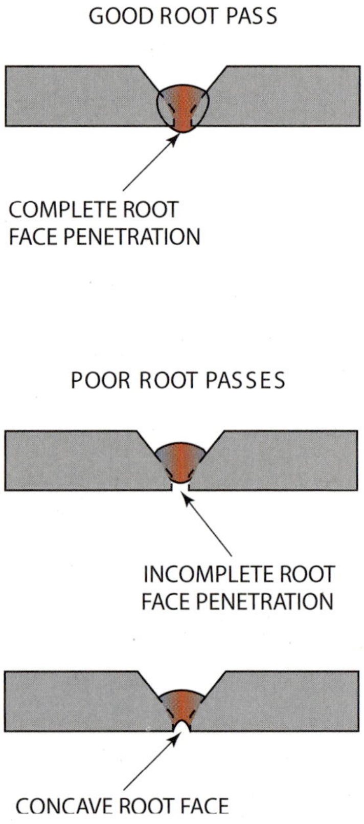

FIGURE 10-16 **Root passes.** © Cengage Learning 2012

FIGURE 10-17 Filler pass using stringer beads. Larry Jeffus

FIGURE 10-18 Filler pass using weave bead. The bead width should not be more than two times the rod diameter. Larry Jeffus

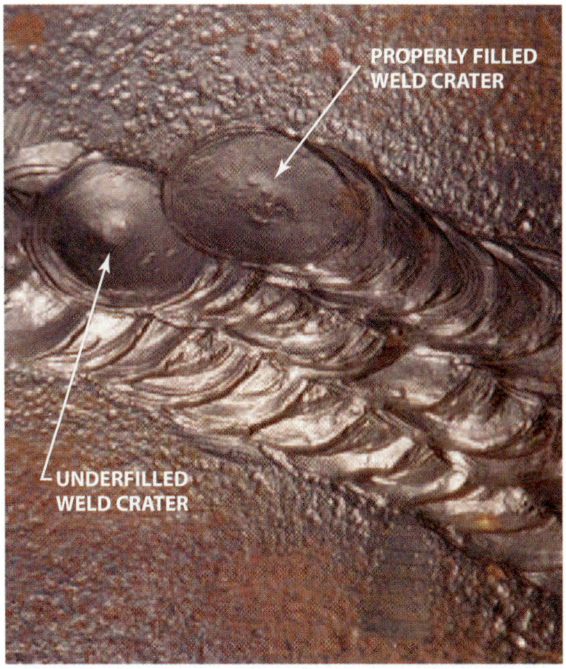

FIGURE 10-19 The weld crater should be filled to prevent cracking, and cleaned of slag before the arc is restarted. Larry Jeffus

Figure 10-18. Stringer beads require less welder skill because of the small amount of metal that is molten at one time. The weld bead crater must be cleaned before the next electrode is started, Figure 10-19. When the bead has gone completely around the pipe, it should continue past the starting point so that good fusion is ensured, Figure 10-20. The locations of

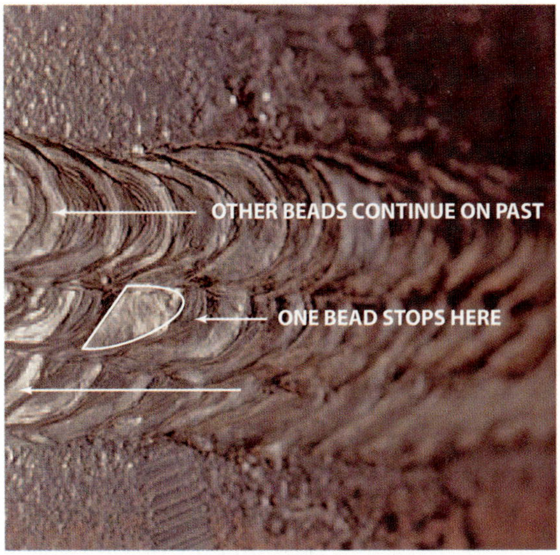

FIGURE 10-20 Avoid starting and stopping all weld passes in the same area. Larry Jeffus

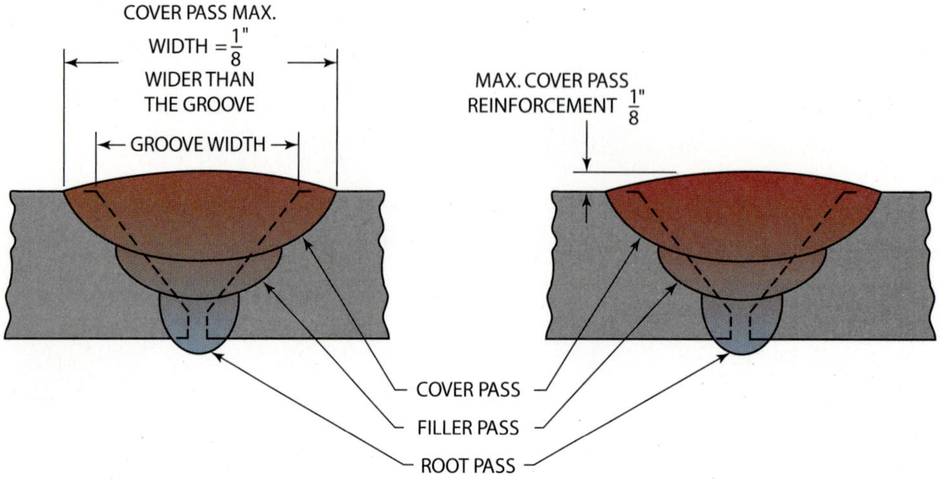

FIGURE 10-21 Excessively wide or builtup welds restrict pipe expansion at the joint, which may cause premature failure. Check the appropriate code or standard for exact specifications. © Cengage Learning 2012

starting and stopping spots for each weld pass must be staggered. Starting and stopping spots may not be as strong as the weld itself, so staggering them eliminates creating a possible weak place in the finished weld. The weld groove should be filled level with these beads so that it is ready for the cover pass.

Cover Pass

The final covering on a grooved weld is referred to as the cover pass or cap. It may be a weave or stringer bead. The cover pass should not be too wide or have too much reinforcement, Figure 10-21. Cover passes that are excessively large will reduce the pipe's strength, not increase it. A large cover pass will cause the stresses in the pipe to be concentrated at the sides of the weld. An oversized weld will not allow the pipe to expand and contract uniformly along its length. This concentration is similar to the restriction that a rubber band would have on an inflated balloon if it were put around its center.

1G Horizontal Rolled Pipe Position

The **horizontal rolled pipe position** is commonly used in fabrication shops where structures or small systems can be positioned for the convenience of the welder, Figure 10-22. The consistent high quality and quantity of welds produced in this position make it very desirable for both the welder and the company.

The penetration and buildup of the weld are controlled more easily with the pipe in this position. Weld visibility and welder comfort are improved so that welder fatigue is less of a problem. The pipe can be rolled continuously with some types of positioners, and the weld can be made in one continuous bead.

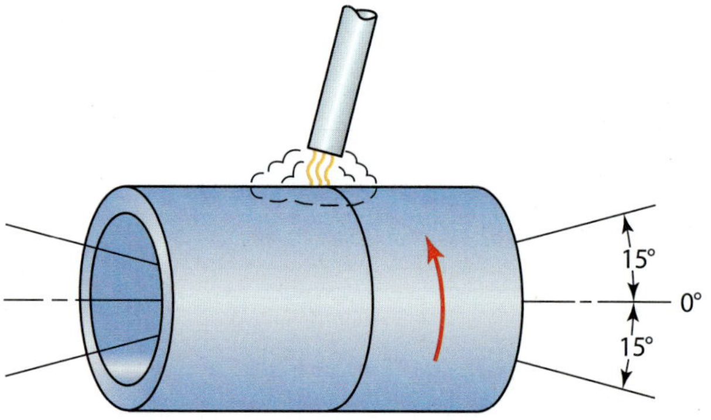

FIGURE 10-22 1G position. The pipe is rolled horizontally. © Cengage Learning 2012

Because of the ease in welding and the level of skill required, welders who are certified in this position may not be qualified to make welds in other positions.

PROJECT 10-1

1G Stringer Beads on a Bookend

Skill to be learned: The ability to control the weld bead on a horizontal rolled pipe.

Project Description

This decorative bookend has the appearance of being wrapped by a rope. The weld bead spirals around the pipe just like a rope around a pipe. Try to make each start and stop as smooth and uniform as possible so the weld bead looks continuous.

Project Materials and Tools

The following items are needed to fabricate the rope bookend.

Materials

- 5 to 10 1/8-in. diameter E6011 and/or E6012 electrodes
- 1 4-in. long piece of 2-in. diameter steel pipe
- 1 4 × 3-in. piece of 1/4-in. thick mild steel plate
- 4 cork or felt pads
- Clear finish or paint

Tools

- Arc welder
- Plasma cutting torch, oxyfuel cutting torch, or shear
- PPE
- Wire brush
- Chipping hammer
- Square and compass
- C-clamp or magnetic clamp
- Soapstone

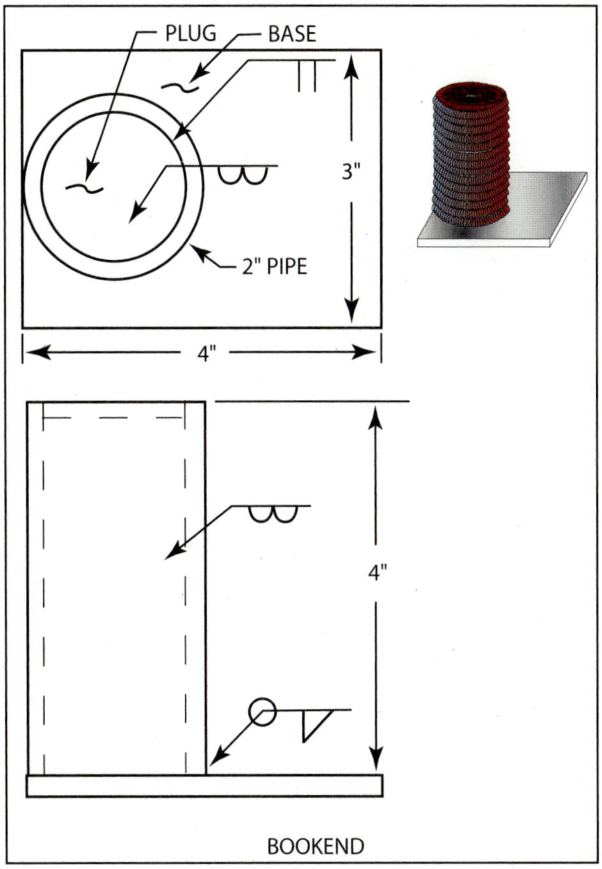

FIGURE 10-23 Project 10-1. © Cengage Learning 2012

Layout

Using a square, 12-in. rule and soapstone, lay out the plate as shown in the project drawing, Figure 10-23. If you are using 3-in. bar stock, just mark the length of the baseplate otherwise you will have to lay out the width and length. Measure the inside diameter of the pipe, and use the compass to lay out the circle for the plug. Mark the pipe to length; if it is going to be cut using a hand thermal torch, a line must be drawn all the way around the pipe, Figure 10-24.

Cutting Out

The pipe can be cut using a band saw, abrasive disk, pipe cutter, thermal torch, or other available process. The baseplate can be cut with a shear, or thermally. The circle for the plug can be thermally cut.

Fabrication

Grind the circular plug, if necessary, so it will fit into the end of the pipe. Make three small tack welds to hold the plug in place.

NOTE: The pipe end plug can be held in place several other ways. For example, one way would be to bend three short pieces of wire so they fit under the plug and rest on top of the pipe. Another way would be to tack weld short metal tabs to the plug so that they rest on the pipe holding the plug in place, Figure 10-25. A third way would be to use a magnetic clamp. A fourth would be to grind a slight taper on the side of the plug so it can be tapped (hammered) in place. Make the tack welds only on one side of the tab. After the tack welds are made to hold the plug in place, the wires and/or tabs can be removed with a grinder.

The pipe will be attached to the base after it has been welded. This will make both welding the stringer beads easier and will prevent the baseplate from getting

FIGURE 10-24 Pipe layout. © Cengage Learning 2012

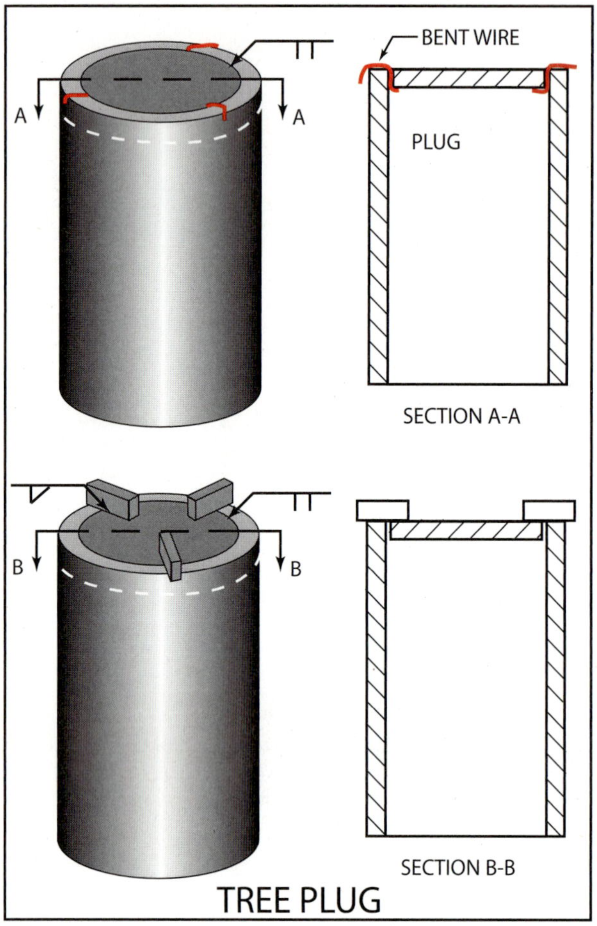

FIGURE 10-25 Two ways to locate an end plate. © Cengage Learning 2012

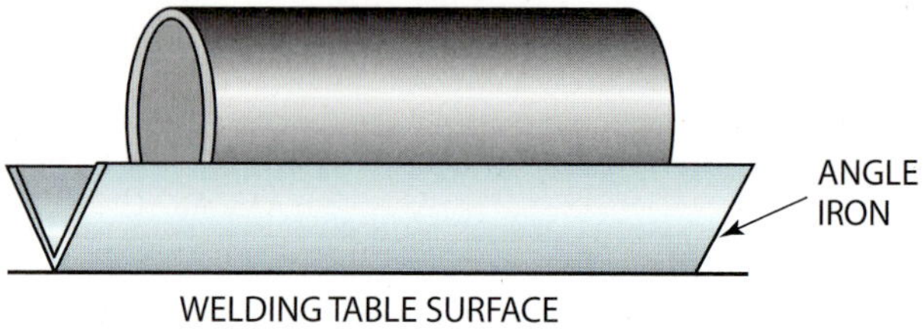

FIGURE 10-26 Angle iron can be used to hold pipe in alignment. © Cengage Learning 2012

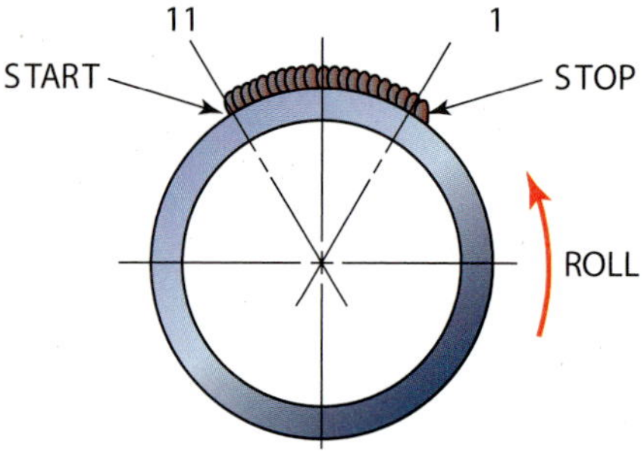

FIGURE 10-27 1G pipe welding area. © Cengage Learning 2012

covered with weld spatter. Use the square and C-clamp or a magnetic clamp to hold the pipe square to the baseplate. Make three tack welds to hold it in place.

Welding

Place the pipe horizontally on the welding table in a vee block made of angle iron, Figure 10-26. The vee block will hold the pipe steady and allow it to be moved easily between each bead. Strike an arc on the pipe at the 11 o'clock position. Make a stringer bead over the 12 o'clock position, stopping at the 1 o'clock position, Figure 10-27. Roll the pipe until the end of the weld is at the 11 o'clock position. Clean the weld crater by chipping and wire brushing.

Strike the arc again and establish a molten weld pool at the leading edge of the weld crater. With the molten weld pool reestablished, move the electrode back on the weld bead just short of the last full ripple, Figure 10-28. This action will both reestablish good fusion and keep the weld bead size uniform. Now that the new weld bead is tied into the old weld, continue

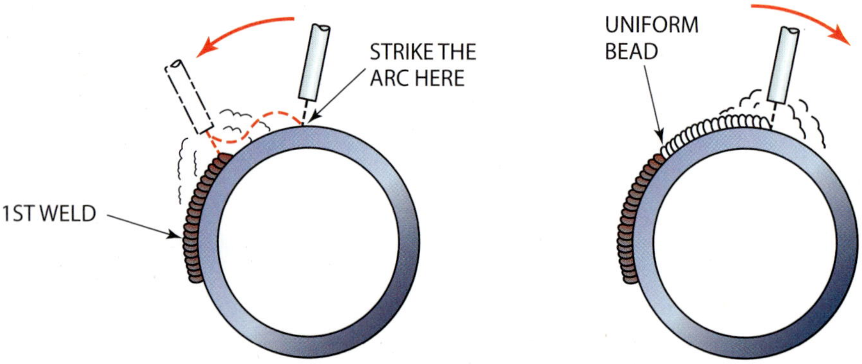

FIGURE 10-28 Keeping the weld uniform is important when restarting the arc. © Cengage Learning 2012

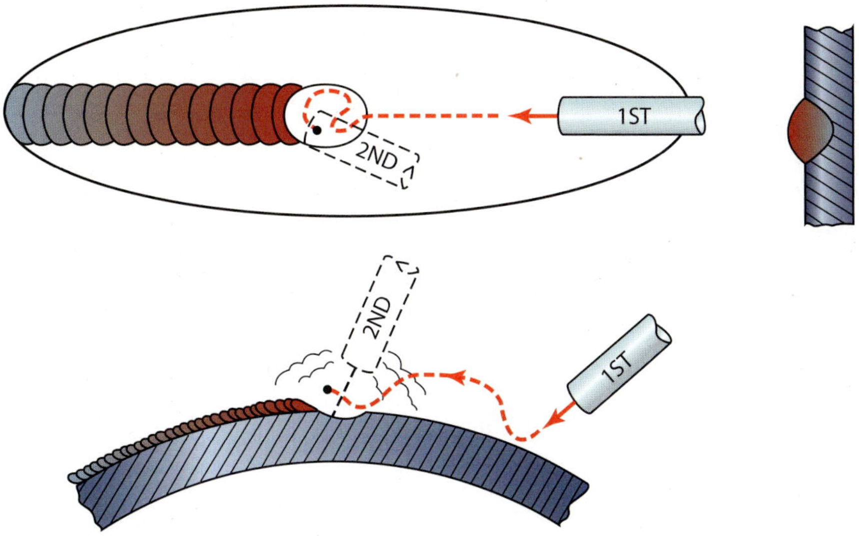

FIGURE 10-29 Restarting the weld. © Cengage Learning 2012

welding to the 1 o'clock position again. Stop welding, roll the pipe, clean the crater, and resume welding. Keep repeating this procedure until the weld is completely around the pipe. Before the last weld is started, clean the end of the first weld so that the end and beginning beads can be tied together smoothly. When you reach the beginning bead, swing your electrode around on both sides of the weld bead. A poor beginning of a weld bead is always high and narrow and has little penetration, Figure 10-29. By swinging the weave pattern (the "C" pattern is best) on both sides of the bead, you can make the bead correctly so the width is uniform. The added heat will give deeper penetration at the starting point. Hold the arc in the crater for a moment until it is built up, but do not overfill the crater.

Cool and chip the welds once the entire outside of the pipe is covered with stringer beads. Follow the tack welding procedure to assemble the pipe to the baseplate. Make a weld around the base of the pipe to hold it to the baseplate.

Finishing

Cool off the metal, chip the slag, and wire brush the surface. A power wire brush can be used to give the welds a polished, shiny appearance. The project can be painted with a clear coat or colored coat of latex paint.

Paperwork

Complete a copy of the time sheet in Appendix I, the bill of materials in Appendix III, or use forms provided by your instructor.◆

5G Horizontal Fixed Pipe Position

The 5G **horizontal fixed pipe position** is the pipe welding position used most often. A weld being produced on a horizontal pipe is being made in flat, vertical up or vertical down, and overhead positions. It is important to keep each of these parts of the weld positions uniform in appearance and of high quality. Pipe in the 5G position must be within 15° of horizontal, Figure 10-30.

When practicing these welds, mark the top of the pipe for future reference. Moving the pipe will make welding easier, but the same side must stay on the top at all times.

The weld can be performed by **welding uphill or downhill.** Practice making the welds both ways. When making the root pass on a pipe joint, the welding direction is usually made based on the code or standard. If the weld direction is not specified, then

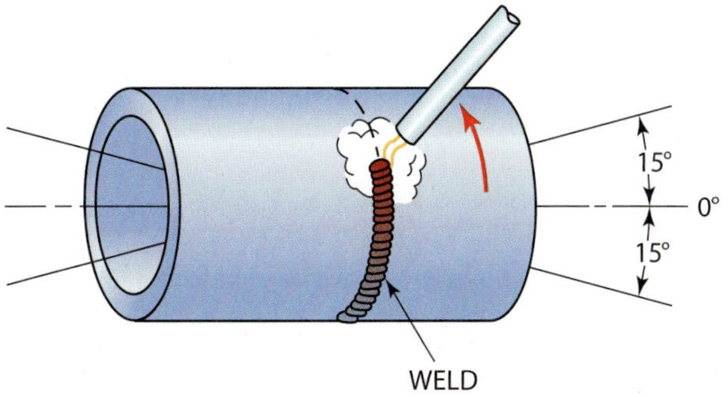

FIGURE 10-30 5G horizontal fixed position.
© Cengage Learning 2012

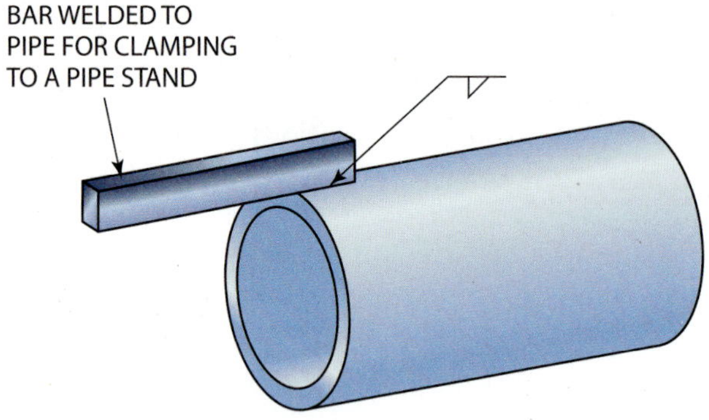

FIGURE 10-31 Holding the pipe in place by welding a piece of flat stock to the pipe and then clamping the flat stock to a pipe stand. © Cengage Learning 2012

the quality of the fit-up influences that choice of root pass welding direction. A close parallel root opening can be welded uphill or downhill. A root opening that is wide or uneven must be welded uphill.

The pipe may be removed from the welding position for chipping, wire brushing, or grinding. The pipe can be held in place by welding a piece of flat stock to it and clamping the flat stock to a pipe stand, Figure 10-31.

The electrode angle should always be upward, Figure 10-32. Changing the angle toward the top and bottom helps control the bead shape. The bead, if welded downhill, should start before the 12 o'clock position and continue past the 6 o'clock position to ensure good fusion and tie-in of the welds. The arc must always be struck inside the joint preparation groove.

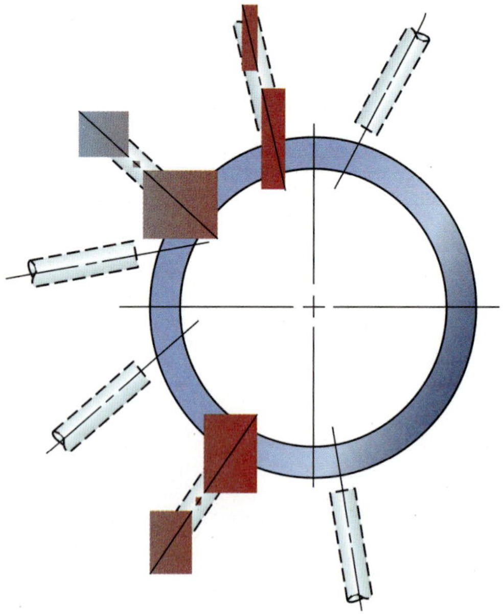

FIGURE 10-32 Electrode angle.
© Cengage Learning 2012

PROJECT 10-2

5G Stringer Beads on a Bookend

Skill to be learned: The ability to control the weld bead on a horizontal fixed position pipe.

Project Description

This is the mating piece for the bookend built in Project 10-1. It will have the same decorative appearance of being wrapped by a rope.

Project Materials and Tools

The following items are needed to fabricate the rope bookend.

Materials

- 5 to 10 1/8-in. diameter E6011 and/or E6012 electrodes
- 1 4-in. long piece of 2-in. diameter steel pipe
- 1 4 × 3-in. piece of 1/4-in. thick mild steel plate
- 4 cork or felt pads
- Clear finish or paint

Tools

- Arc welder
- Plasma cutting torch, oxyfuel cutting torch, or shear
- PPE
- Wire brush
- Chipping hammer
- Square and compass
- C-clamp or magnetic clamp
- Soapstone

Layout

Using a square, 12-in. rule and soapstone, lay out the plate as shown in the project drawing, Figure 10-23. If you are using 3-in. bar stock, just mark the length of the baseplate; otherwise, you will have to lay out the width and length. Measure the inside diameter of the pipe and use the compass to lay out the circle for the plug. Mark the pipe to length; if it is going to be cut using a hand thermal torch, a line must be drawn all the way around the pipe.

Cutting Out

The pipe can be cut using a band saw, abrasive disk, pipe cutter, thermal torch, or other available process. The baseplate can be cut with a shear, or thermally. The circle for the plug can be thermally cut.

Fabrication

Grind the circular plug, if necessary, so it will fit into the end of the pipe. Make three small tack welds to hold the plug in place.

The pipe will be attached to the base after it has been welded. This will make both welding the stringer beads easier and will prevent the baseplate from getting covered with weld spatter. Use the square and C-clamp or a magnetic clamp to hold the pipe square to the baseplate. Make three tack welds to hold it in place.

Clamp the pipe horizontally between waist and chest level. Starting at the 11 o'clock position, make a downhill straight stringer bead through the 12 o'clock and 6 o'clock positions. Stop at the 7 o'clock position, Figure 10-33. Using a new electrode, start at the 5 o'clock position and make an uphill straight stringer bead through the 6 o'clock and 12 o'clock positions. Stop at the 1 o'clock position. Change the electrode angle to control the molten weld pool.

Cool and chip the welds once the entire outside of the pipe is covered with stringer beads. Follow the tack welding procedure to assemble the pipe to the baseplate. Make a weld around the base of the pipe to hold it to the baseplate.

Finishing

Cool off the metal, chip the slag, and wire brush the surface. A power wire brush can be used to give the welds a polished, shiny appearance. The project can be painted with a clear coat or colored coat of latex paint.

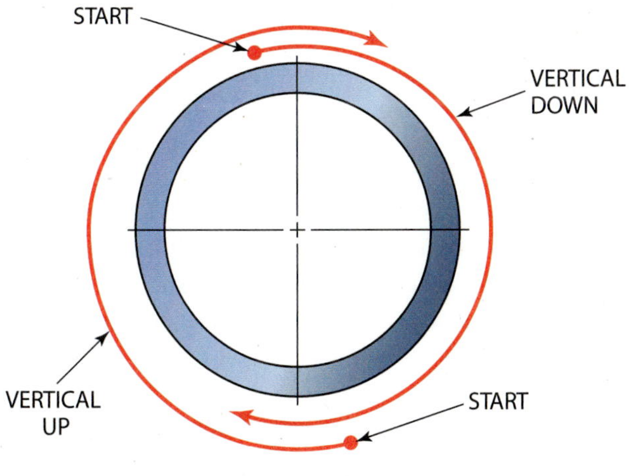

FIGURE 10-33 Stop at the 7 o'clock position.
© Cengage Learning 2012

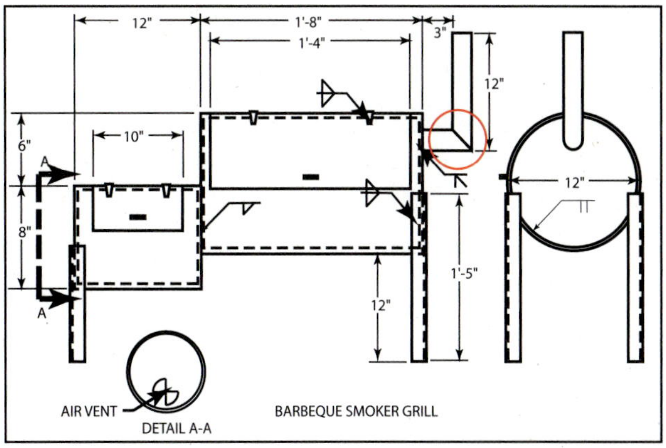

FIGURE 10-34 Horizontal fixed pipe joint on barbeque smoker grill. © Cengage Learning 2012

Paperwork

Complete a copy of the time sheet in Appendix I, the bill of materials in Appendix III, or use forms provided by your instructor.◆

Horizontal Pipe Welds

Rolling horizontal pipe is the easiest and fastest way of making horizontal pipe welds. However, not all horizontal pipe assemblies can be rolled, so it is important that you develop the skills needed to make welds in the horizontal fixed position. The skills that you will be learning when building this project are the same skills you will need to build the barbeque smoker grill shown in Figure 10-34.

PROJECT 10-3

1G and 5G V-Grooves on a Birdhouse

Skill to be learned: The ability to control the weld beads on a multiple pass horizontal V-grooved pipe.

Project Description

This pipe-ringed birdhouse will be fabricated all at one time and welded in two different positions. The first two welds will be made in the 1G horizontal rolled position and the last two welds will be in the 5G horizontal fixed position.

This birdhouse can be used to attract a number of different types of birds depending on the hole size and floor height, Table 10-1.

BIRD SPECIES	ENTRANCE SPECIFICATION	
	ENTRANCE DIAMETER IN INCHES	DISTANCE ABOVE FLOOR
BLUEBIRD	1 1/2	6
CAROLINE WREN	1 1/2	4–6
CHICKADEE	1 1/8	6–8
CRESTED FLYCATCHER	2	6–8
DOWNY WOODPECKER	1 1/4	6–8
HOUSE WREN	1 1/4	4–6
NUTHATCH	1 1/4	6–8
PURPLE MARTIN	2 1/2	2
TITMOUSE	1 1/4	6–8
TREE SWALLOW	1 1/2	1–5

Table 10-1 Birdhouse Specifications

Project Materials and Tools

The following items are needed to fabricate the birdhouse.

Materials

- 8 to 25 1/8-in. diameter F3 group (E6010 and/or E6011) electrodes
- 10 to 30 1/8-in. diameter E7018 electrodes
- 5 2-in. long pieces of 8-in. diameter steel pipe
- 1 10 × 10-in. piece of 3/16-in. to 1/4-in. thick mild steel plate
- 1 7 1/2-in. diameter piece of 3/16-in. to 1/4-in. thick mild steel plate
- 1 1 1/2 × 8-in. piece of 3/16-in. to 1/4-in. thick mild steel plate
- 4 1 1/2 × 1-in. pieces of 3/16-in. to 1/4-in. thick mild steel plate
- Clear finish or paint

Tools

- Arc welder
- Plasma cutting torch, oxyfuel cutting torch, or shear
- PPE
- Wire brush
- Chipping hammer
- Square and compass
- C-clamp or magnetic clamp
- Soapstone

Layout

Using a square, 12-in. rule and soapstone, lay out the plate as shown in the project drawing, Figure 10-35. Lay out the 10-in. square piece, and use a hand or automated torch to cut out the piece. Remember to account for the cut's kerf so that the part is not under- or oversized, Figure 10-36.

Mark the pipe to length; if it is going to be cut using a hand thermal torch, a line must be drawn all the way around the pipe. In this case, you do not have to allow for the kerf because the pipe will be beveled and the joint will have a 1/8-in. root gap. The root gap will be very close in size to the kerf.

Measure the inside diameter of the pipe and use the compass to lay out the circle for the bottom plug.

Cutting Out

The best way to cut the pipe is with a 37 1/2° bevel, Figure 10-37. If you do not have a pipe beveling cutter, you may use an angle grinder to bevel the pipe end. The baseplate can be cut with a shear, or thermally. The circle for the floor can be thermally cut. Make four 1-in. diameter semicircular cuts on the floor to allow for drainage and ventilation.

NOTE: **Do not let the pipe sections fall to the floor as they are cut off. This can cause them to bend out of round, making fit-up more difficult.**

Drill four 1/4-in. holes in the end of the mounting strap. These will be used to screw the birdhouse to a tree or post. The holes should be evenly spaced near the four corners of the strap.

Fabrication

Tack weld two pieces of pipe together as shown in Figure 10-38. Place the pipe horizontally in a vee block on the welding table. Sometimes pipe can be slightly out of round and will need to be reshaped. Make four small tack welds evenly spaced around the pipe. Making the tack welds somewhere around the 1:30, 4:30, 7:30, and 10:30 clock positions will keep them out of the way of your root pass starts and stops. Make sure you do not strike the arc outside of the weld groove.

NOTE: **Assemble all five pieces of pipe before welding. Weld distortion may make it difficult to fit them after the first two 1G welds are completed.**

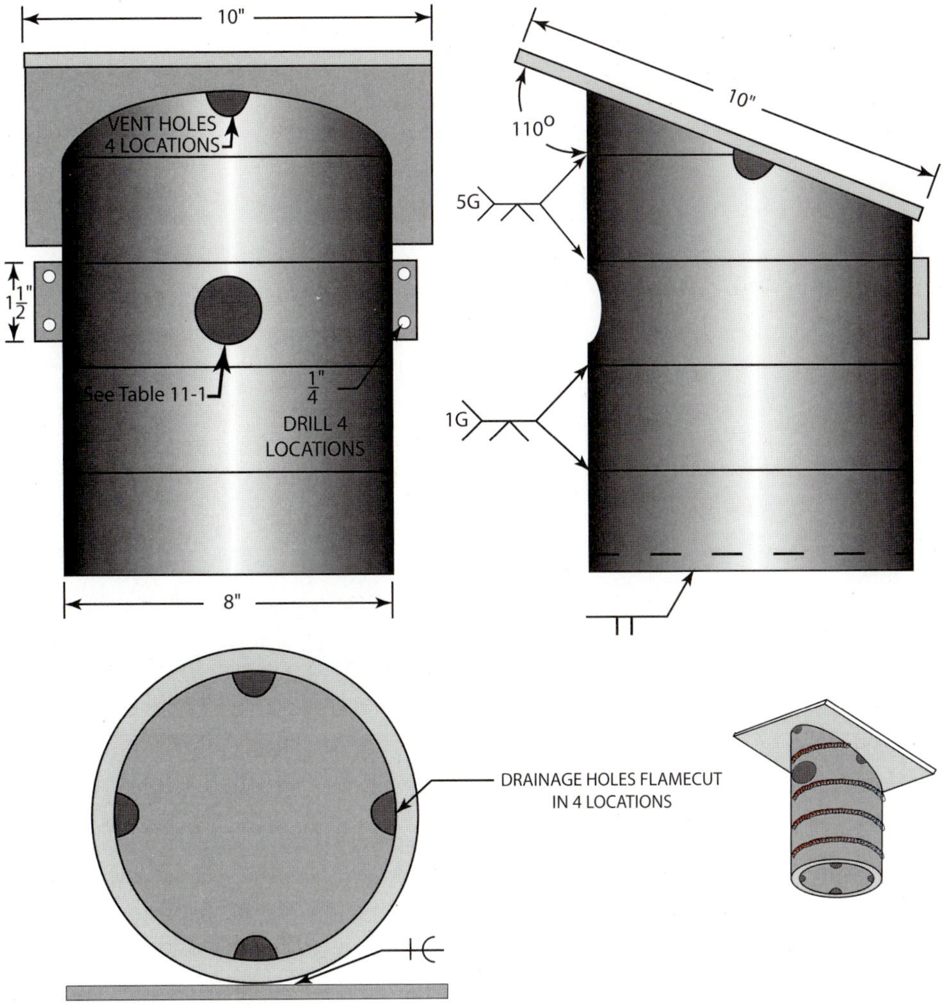

FIGURE 10-35 Project 10-3. © Cengage Learning 2012

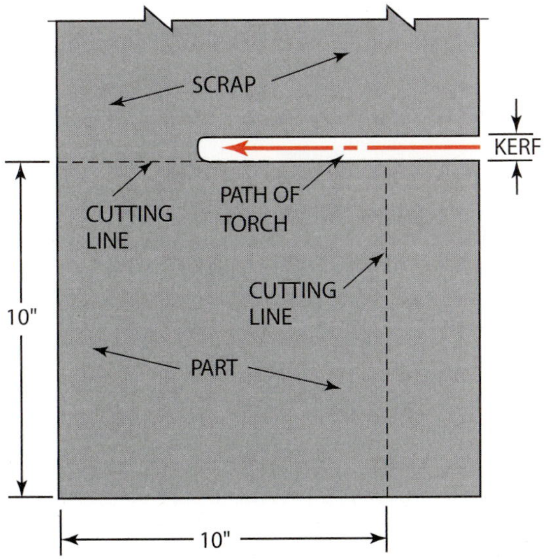

FIGURE 10-36 Making the cut so that the kerf does not affect the finished part's size. © Cengage Learning 2012

1G Welding

Start the root pass at the 11 o'clock position. Using a very short arc and high-current setting, weld toward the 1 o'clock position. Stop and roll the pipe, chip the slag, and repeat the weld until you have completed the root pass.

Clean the root pass by chipping and wire brushing. The root pass should not be ground this time. Replace the pipe in the vee block on the table so that the hot pass can be done. Turn up the machine amperage enough to remelt the root weld surface, for the hot pass. Use a stepped electrode pattern, moving forward each time the molten weld pool washes out the slag, and returning each time the molten weld pool is nearly all solid, Figure 10-39. Weld from the 11 o'clock position to the 1 o'clock position before stopping, rolling, and chipping the weld. Repeat this procedure until the hot pass is complete.

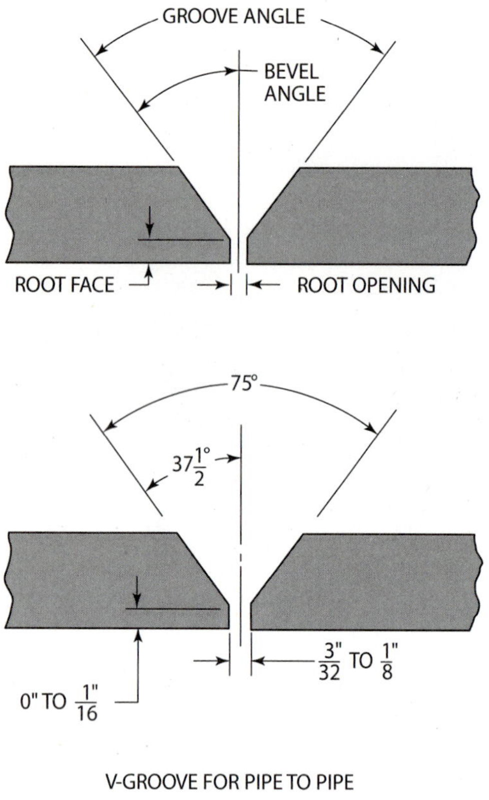

FIGURE 10-37 V-groove angle and terminology. © Cengage Learning 2012

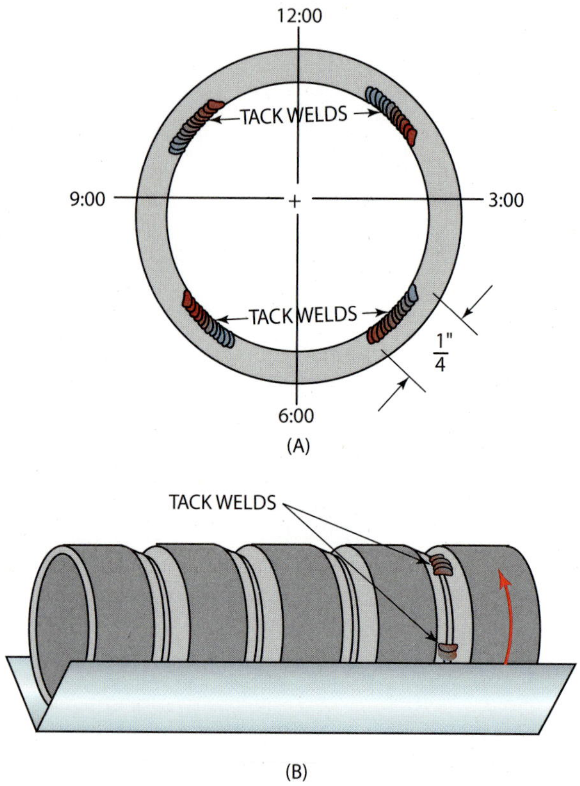

FIGURE 10-38 Angle iron being used to hold pipe joints in alignment. © Cengage Learning 2012

The filler pass and cover pass may be the same pass on this joint, Figure 10-40. Turn down the machine amperage. Use a "T," "J," "C," or zigzag pattern for this weld. Start the weld at the 10 o'clock position, and stop at the 12 o'clock position. Sweep the electrode so that the molten weld pool melts out any slag trapped by the hot pass. Watch the back edge of the bead to see that the molten weld pool is filling the groove completely. Turn, chip, and continue the bead until the weld is complete.

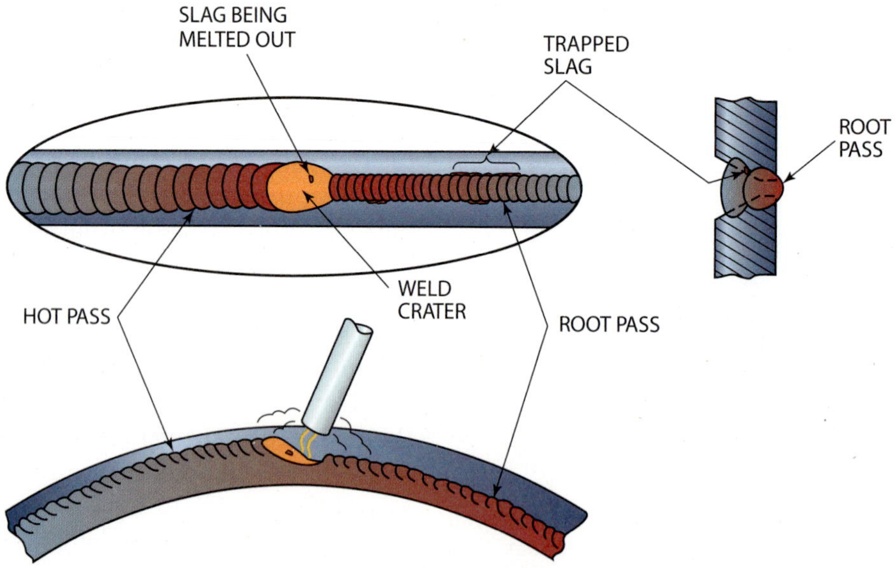

FIGURE 10-39 Hot pass. © Cengage Learning 2012

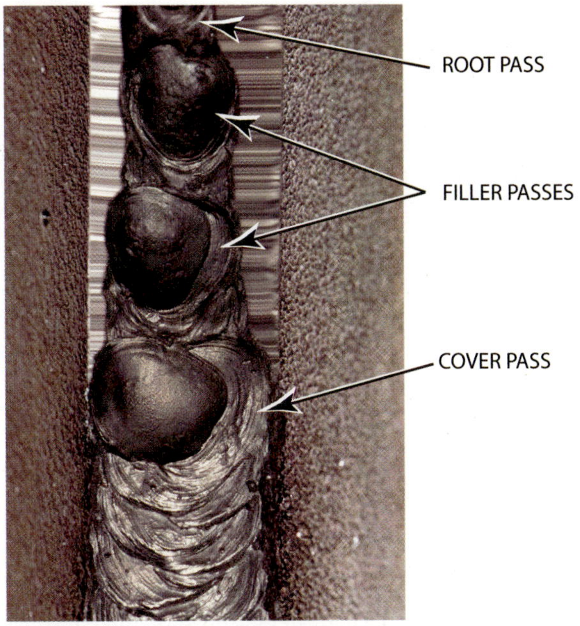

FIGURE 10-40 Multiple weld pass identification and locations. Larry Jeffus

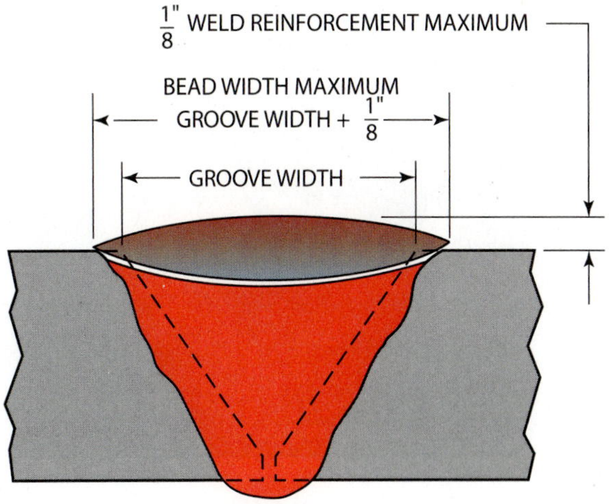

FIGURE 10-41 Cover pass identification and specifications. © Cengage Learning 2012

5G Welding

Mark the top of the pipe and mount it horizontally between waist and chest level. The mark will let you remove and replace the pipe for cleaning without losing the top. Often in a pre-job weld test, you may not be allowed to move the pipe for chipping or cleaning.

Weld the root pass uphill or downhill using E6010 or E6011 electrodes. Use a thin grinding disk to clean up the root pass. Be careful not to grind the groove wider as you try to get out trapped slag. Making the groove wider can make the finished weld too wide, Figure 10-41.

Use E7018 electrodes for the filler and cover passes with stringer or weave patterns. Any time an E7018 low-hydrogen-type electrode is to be used, the weave pattern, if used, must not be any larger than two-and-a-half times the diameter of the electrode. This weave cannot be any larger than 5/16 in. wide. Start welding at one end of the plate and weld to the other end. The weld bead should be about 5/16 in., wide having no more than 1/8 in. of uniform buildup, Figure 10-41. The weld buildup should have a smooth transition at the toe, with the plate and the face being somewhat convex. Undercut at the toe and a concave or excessively builtup face are the most common problems. Watch the sides of the bead for undercut. When undercut occurs, keep the electrode just ahead of the spot until it is filled in. There should be a smooth transition between the weld and the plate. The shape of the bead face can be controlled by watching the trailing edge of the molten weld pool. That trailing edge is the same as the finished bead.

Deep penetration is not required. After the weld is completely cooled, chip and inspect it for uniformity and defects.

Square Butt Weld

The floor of the birdhouse is welded in place with a square butt joint weld. The drainage/ventilation reliefs on the bottom can be used to hold the floor in place for welding. The other option is to place the floor in place on the first pipe section and tack weld it on the inside before adding the additional pipe sections.

Make the square butt weld in the flat position, starting and stopping at each of the reliefs.

Roof Cut

The roof of the birdhouse has a 20° slope to the back. This angle can be laid out several ways. The easiest way is to use a contour marker, Figure 10-42. This tool can be set to the angle, and it can be used to draw a line all the way around the pipe. A pipe "Wrap-A-Round" is a flexible tool that can be used to lay out pipe, Figure 10-43. Instructions on its use are supplied by the tool manufacturer.

A third way is to lay out the pipe diameter on a sheet of paper and measure the offset. This is not as accurate as calculating the offset using geometry, but for this project it is close enough, Figure 10-44.

Use a torch and cut the top of the birdhouse off so that it is evenly sloped. Make four 1-in. diameter

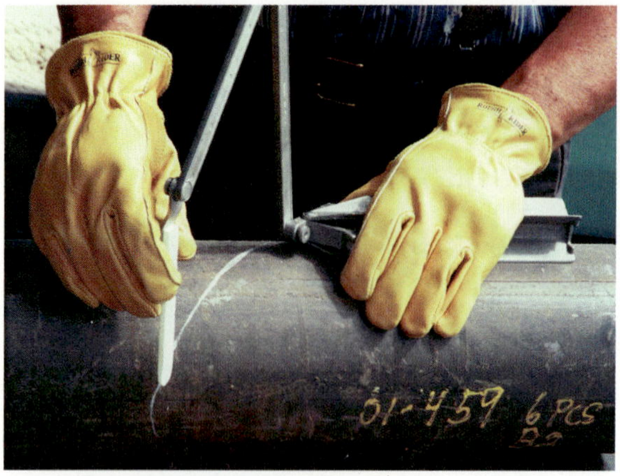

FIGURE 10-42 Laying out a 45° angle on a pipe using a Curv-O-Mark® tool. Larry Jeffus

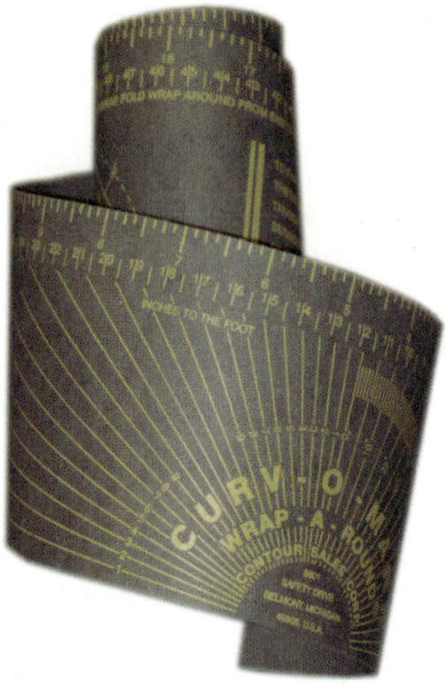

FIGURE 10-43 Wrap-A-Round used to lay out pipe. Larry Jeffus

semicircular cuts on the top of the pipe to allow for ventilation.

Flare-Bevel Weld

The strap used to hold the birdhouse to a tree or post is welded to the back of the pipe section with a flare-bevel weld.

Roof Layout and Assembly

The roof is not welded to the pipe but is kept in place by four small metal tabs. The tabs are made from 1 1/2-in. bar stock and are welded to the top so they stick inside the pipe, keeping it from sliding off. To locate the tabs, place the pipe on the roof and draw a line around it, Figure 10-45. Measure back 3/4 in. from the line and tack weld the tabs as shown in the

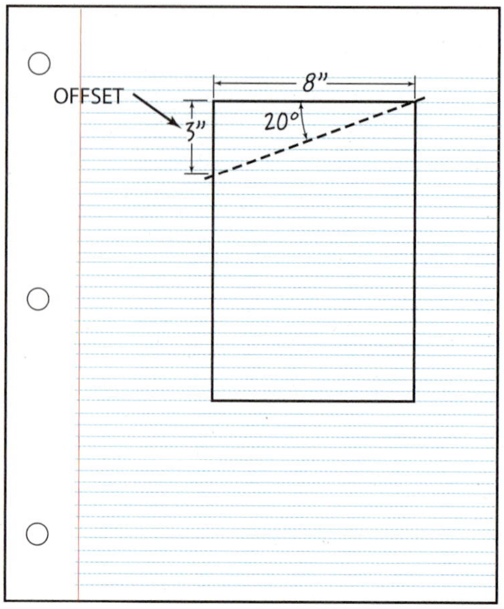

FIGURE 10-44 Laying out pipe by measuring the offset. © Cengage Learning 2012

figure. Check to see that the roof will fit onto the pipe and make any adjustments to the tabs as needed. Weld the tabs in place with small fillet welds on both sides of the tabs. Check the fit and make adjustments with a grinder to make it looser, or a hammer to make it tighter.

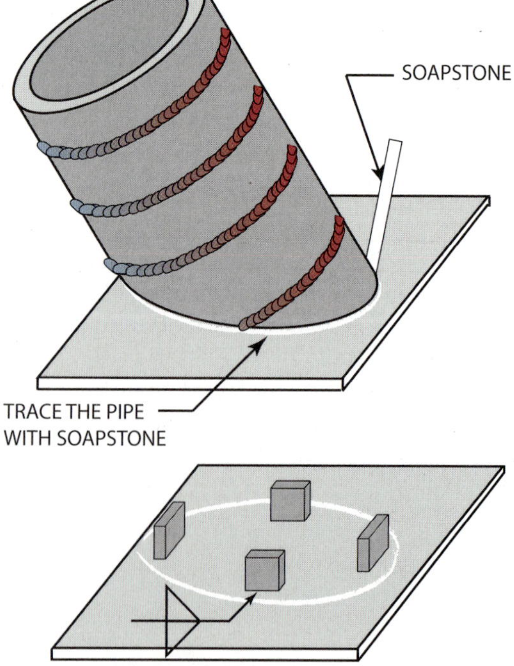

FIGURE 10-45 Locating tabs for the birdhouse top. © Cengage Learning 2012

Cutting the Hole

The size of the hole and the height that it is above the floor may affect the type of bird that chooses to nest in your birdhouse.

Finishing

Cool off the metal, chip the slag, and wire brush the surface. A power wire brush can be used to give the welds a polished, shiny appearance. The project can be painted with a clear coat or colored coat of latex paint.

Paperwork

Complete a copy of the time sheet in Appendix I, the bill of materials in Appendix III, or use forms provided by your instructor.◆

2G Vertical Fixed Pipe Position

In the 2G **vertical fixed pipe position,** the pipe is vertical and the weld is horizontal, Figure 10-46. With these welds, the welder does not need to change welding positions constantly. The major problem that welders face when welding pipe in this position is that the area to be welded is often located in corners. Because of this location, reaching the back side of the weld is difficult. In the projects that follow, you may turn the pipe between welds. As a welder gains more experience, welds in tight places will become easier.

The welds must be completed in the correct sequence, Figure 10-47. The root pass goes in as with other joints. To reduce the sagging of the bottom of the weld, increase the electrode to the work angle.

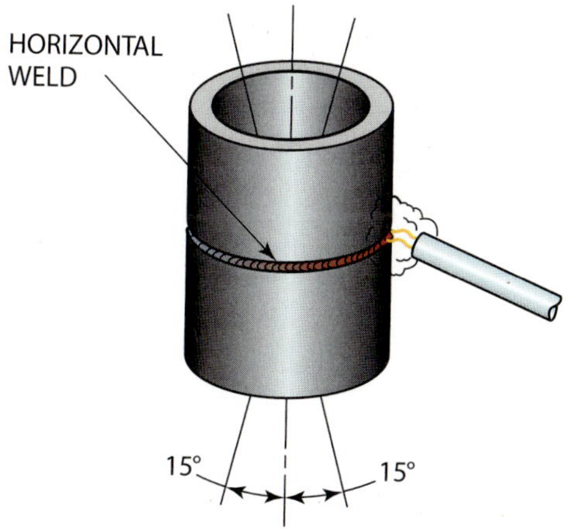

FIGURE 10-46 2G position. The pipe is fixed vertically, and the weld is made horizontally around it.

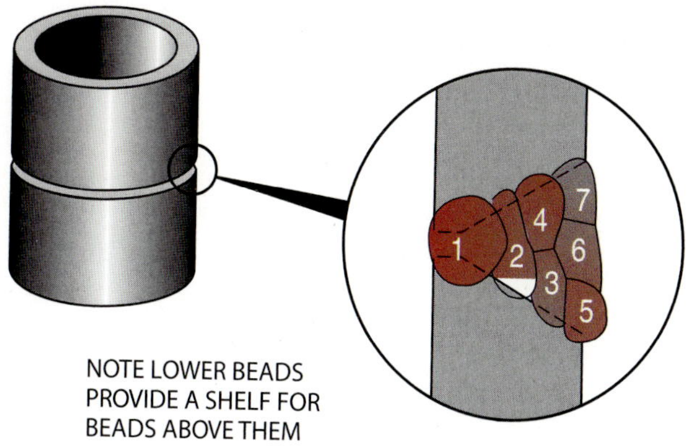

FIGURE 10-47 2G pipe welding position.

As long as the weld is burned in well and does not have cold lap on the bottom, the weld is correct. Each of the filler and cover welds that will follow must be supported by the previous weld bead.

Horizontal Welds on Vertical Pipe

Gravity causes the greatest problem with horizontal welds. The cooled, solidified weld metal is directly below the molten weld pool on vertical welds to hold it in place. On horizontal welds, the weld bead has to be made so that some support of the molten weld pool exists because the weld bead has to be uniform in shape, Figure 10-48. The skills that you will be learning when building this project are the same skills you will need to build the barbecue smoker grill shown in Figure 10-49.

PROJECT 10-4

2G Stringer Beads on a Boat Anchor

Skill to be learned: The ability to control the weld bead around a vertical pipe.

Project Description

This mushroom-type boat anchor is designed to be both functional and disposable. Too often boaters lose expensive anchors because they become snared on the bottom in rocks, dead trees, or other debris. This anchor is easily built and works great for holding your boat over a favorite fishing spot or in place as you take a swim. If it gets snagged, you can just cut it loose and make another. Keep the expensive, pretty anchor for those times when you know you are anchoring over a smooth, clean, snag-free bottom.

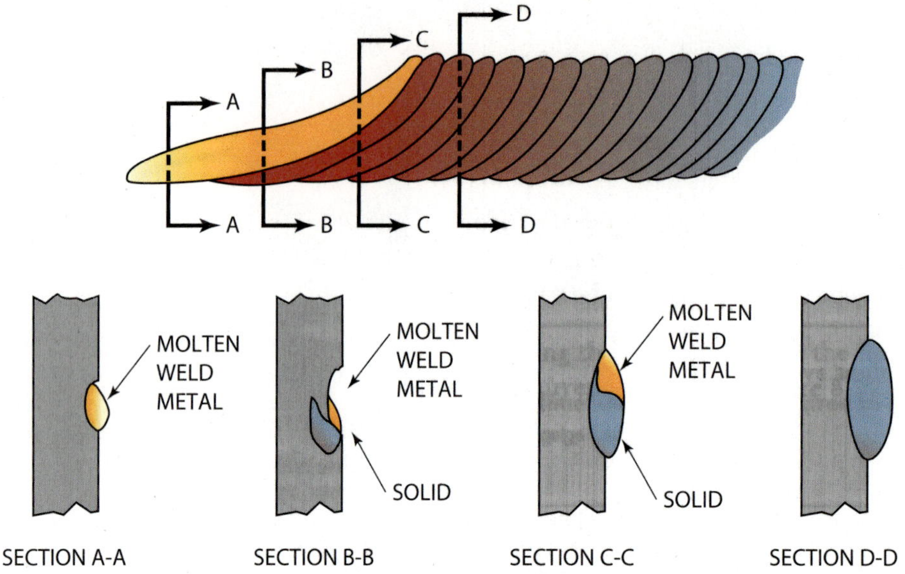

FIGURE 10-48 Electrode movement for a horizontal weld bead.
© Cengage Learning 2012

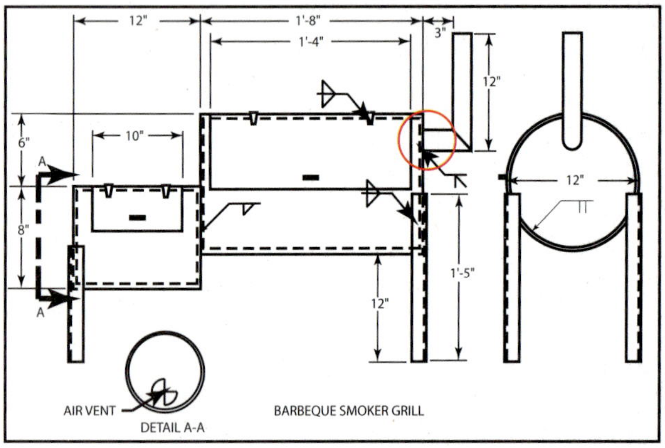

FIGURE 10-49 Horizontal weld on the barbeque smoker grill. © Cengage Learning 2012

The size of this anchor can be changed to meet your boat's needs. Smaller ones can be used on rowboats, and larger ones can be used on small commercial boats.

Project Materials and Tools

The following items are needed to fabricate the boat anchor.

Materials

- 5 to 10 1/8-in. diameter E6011 and/or E6012 electrodes
- 1 2-in. long piece of 8-in. diameter steel pipe
- 1 8-in. diameter piece of 3/16-in. to 1/4-in. thick mild steel plate
- 1 12-in. long piece of 3/4-in. diameter steel pipe
- 1 1-in. long piece of 1-in. diameter steel pipe
- Clear finish or paint

Tools

- Arc welder
- Plasma cutting torch, oxyfuel cutting torch, or shear
- PPE
- Wire brush
- Chipping hammer
- Square and compass
- C-clamp or magnetic clamp
- Soapstone

Layout

Using a square, 12-in. rule and soapstone, lay out the plate as shown in the project drawing, Figure 10-50. Mark the pipes to length; if the cuts are going to be made using a hand thermal torch, a line must be drawn all the way around the pipe. Measure the outside diameter of the pipe and use the compass to lay out the circle for the plug. If the pipe is cut, you could lay it on top of the plate and draw a chalk line around it to mark your circle.

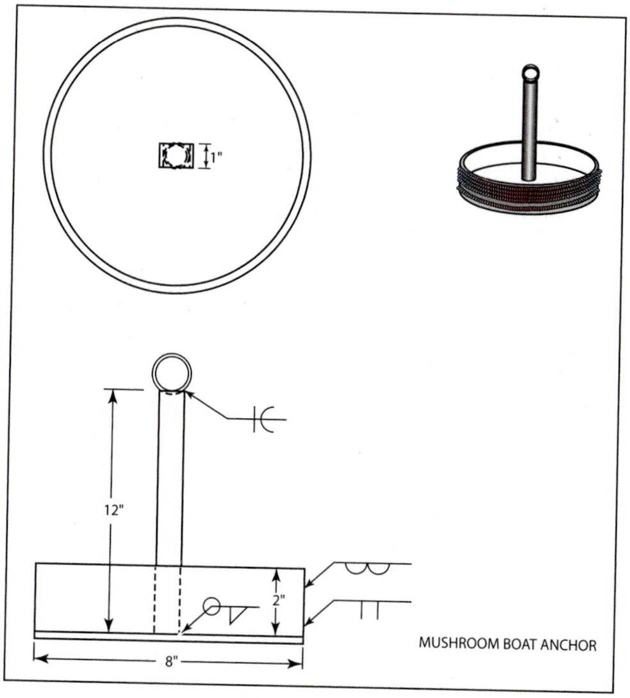

FIGURE 10-50 Project 10-4. © Cengage Learning 2012

Draw an *X* across the circle for the plug to locate the center. It is easier to locate the center before the plug is welded into the pipe. The 3/4-in. pipe that will be used for the anchor rope will be welded in the center of this plate.

Cutting Out

The pipes can be cut using a band saw, abrasive disk, pipe cutter, thermal torch, or other available processes. The baseplate can be cut with a shear, or thermally. The circle for the plug can be thermally cut.

Fabrication

Use the square or magnetic clamp to hold the 3/4-in. pipe vertically in the center of the circular plug. Make one small tack weld to hold the pipe in place, and move the square or magnetic clamp 90° around the pipe and check for square. If necessary, adjust the pipe and make another tack weld. Remove the square or magnetic clamp, and make two more tack welds to hold the pipe in place until it is welded.

Center the plug on the pipe, and make three or four small tack welds to hold the plate to the pipe.

Place the 1-in. piece of 1-in. diameter pipe on top of the vertical pipe. Make one tack weld on both sides of the pipe to hold it in place.

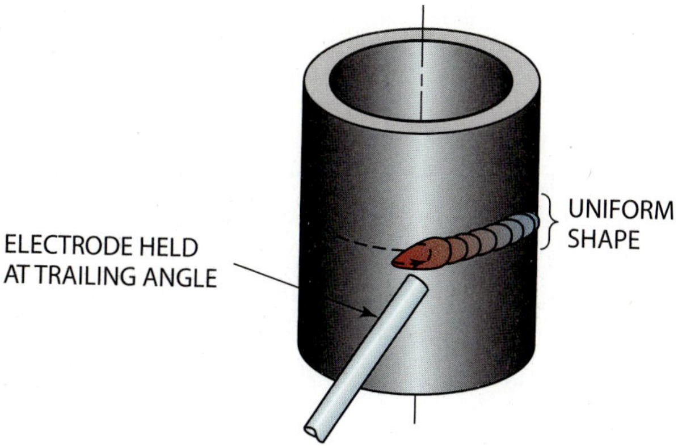

FIGURE 10-51 Electrode position for horizontal weld. © Cengage Learning 2012

Welding

You are going to make a 1-in. wide series of stringer weld beads around the top edge of the pipe. These welds are to add a little weight and decoration to the anchor. Start by placing the pipe in a vertical position at a comfortable height.

Strike an arc and use the "J"-weave pattern so that the molten weld pool will be supported by the lower edge of the solidified metal, Figure 10-51. Keep the electrode at an upward and trailing angle so that the arc force will help to keep the weld in place. When you have either used up all of the electrodes or have welded as far around the pipe as you can comfortably reach, break the arc. It is always a good practice to break the arc back over the weld crater.

Chip and clean the weld. Examine the weld for uniformity and any weld defects. An undercut along the top edge and overlap along the bottom edge of the weld are the most common problems with horizontal welds. If you have either or both of these, try adjusting your "J"-weave pattern and travel speed. Sometimes traveling a little faster will make a smaller weld bead that is less likely to have these problems.

Continue making weld beads around the pipe until you have made approximately a 1-in. wide weld. Cool and clean the welds.

The plug can be welded by putting a piece of steel under the anchor so that it is raised slightly off of the welding table. This will make it a little easier to make the horizontal square butt weld and keep you from accidentally welding the anchor to the welding table. Treat this weld as if it were a square butt pipe joint, and make the weld just as you made the horizontal stringer beads.

The 3/4-in. pipe will be welded to the plug with a small fillet weld. Chip any slag from the tack welds. Strike an arc in the weld zone and make a fillet weld as far around the pipe as possible. When you have to stop, clean the slag from the weld before restarting. This weld must be watertight to prevent rusting inside the pipe.

The small piece of the 1-in. pipe at the top of the pipe forms a loop for an anchor chain or rope to be attached. Weld all the way round the top of the 3/4-in. pipe so that this area is made watertight too.

Finishing

Cool off the metal, chip the slag, and wire brush the surface. The inside of the rope loop should be deburred or reamed to remove any sharp edges that might damage the anchor rope. If a cleat is to be used, this is not necessary. The project can be painted with a clear coat or colored coat of latex paint.

NOTE: If the anchor rope is not tied to the anchor but is threaded through the loop, it can easily be removed if the anchor gets snagged. All you have to do is release one end of the rope from the boat and pull it through the loop, leaving the anchor on the bottom of the lake.

Paperwork

Complete a copy of the time sheet in Appendix I, the bill of materials in Appendix III, or use forms provided by your instructor.◆

PROJECT 10-5

2G V-Grooved Weld on a Chimney Charcoal Starter

Skill to be learned: The ability to control the weld beads on a multiple pass vertical V-grooved pipe.

Project Description

Some people feel there is nothing like the taste of food cooked over a charcoal fire. However, getting the charcoal started can often be a problem. Because of air quality concerns, some cities and communities do not allow the use of charcoal lighter fluid. Some people prefer not to use lighter fluid because it takes so long to start the charcoal, and it takes some time for the odor to leave even after the charcoal is lit.

A chimney-type charcoal starter is a great answer to these problems. It takes only a few pieces of wadded-up paper to start the charcoal, and within a few minutes, all of the briquettes are evenly lit. The chimney works by inducing a convection draft of air. The paper starts the process by both heating the bottom charcoal briquettes and drawing in air. As the briquettes at the bottom begin to glow, more air is drawn in and spreads the heat upward to the rest of the briquettes. It is almost like having a small fan blowing air through the charcoal.

Once all of the charcoal is burning in this heavy-duty chimney starter, the briquettes are dumped out of the starter, glowing hot and ready to start cooking.

Project Materials and Tools

The following items are needed to fabricate the chimney charcoal starter.

Materials

- 10 to 20 1/8-in. diameter F3 group (E6010 and/or E6011) electrodes
- 10 to 30 1/8-in. diameter E7018 electrodes
- 5 2-in. long pieces of 8-in. diameter steel pipe
- 2 1 × 3 7/8-in. pieces of 3/16-in. to 1/4-in. thick mild steel plate
- 1 4 × 5-in. piece of 16-gauge sheet metal
- 1 3/4-in. wooden dowel rod
- 1 6-in. long 1/4-20NC all-thread with two nuts and flat washers
- 7 1/8-in. diameter 36-in. long ER316 bare metal electrodes
- Heat-resistant paint

Tools

- Arc welder
- Plasma cutting torch, oxyfuel cutting torch, or shear
- PPE
- Wire brush
- Chipping hammer
- Ball-peen hammer
- Hacksaw
- Electric drill and bits
- Wire-cutting pliers
- Square and compass
- C-clamp or magnetic clamp
- Soapstone

Layout

Using a square, 12-in. rule and soapstone, lay out the project as shown in the project drawing, Figure 10-52.

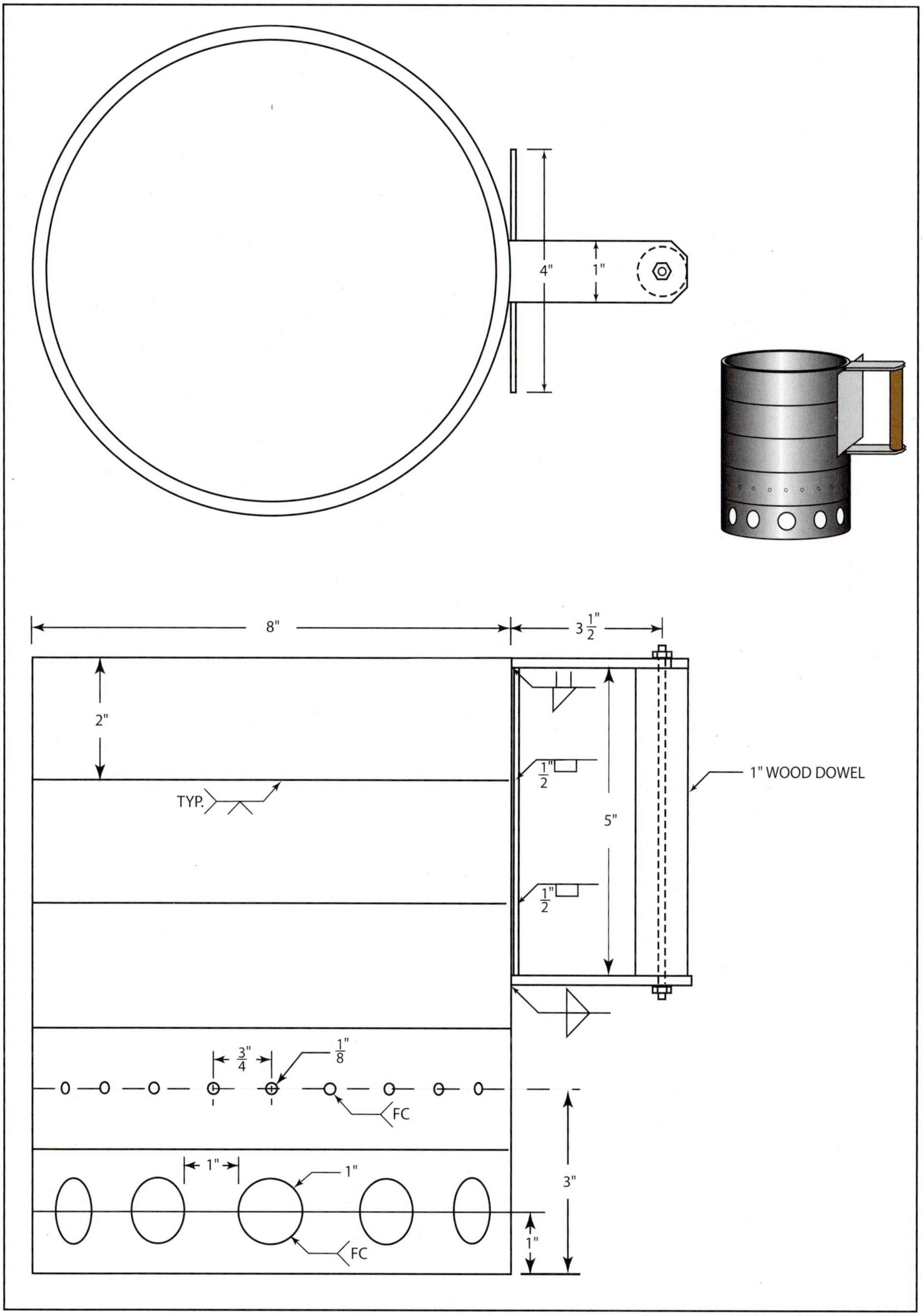

FIGURE 10-52 Project 10-5. © Cengage Learning 2012

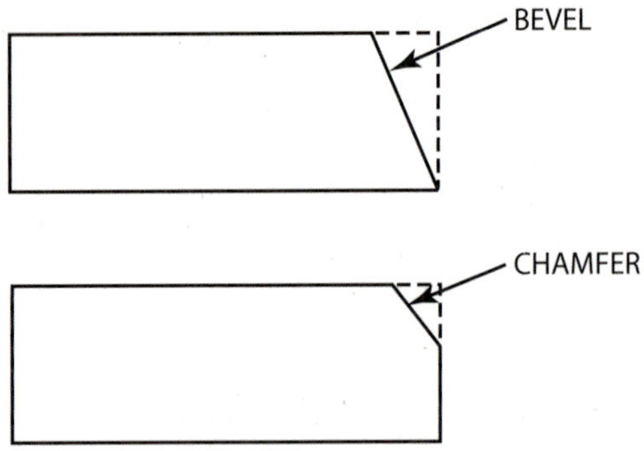

FIGURE 10-53 Comparing a bevel to a chamfer.

Mark the pipe to length; if it is going to be cut using a hand thermal torch, a line must be drawn all the way around the pipe. In this case, you do not have to allow for the kerf because the pipe will be beveled and the joint will have a 1/8-in. root gap. The root gap will be very close in size to the kerf.

Measure and mark two 1 × 4-in. pieces of bar stock with soapstone. Mark a 45° chamfer on both corners of one end.

NOTE: A bevel extends all the way across the part's surface. A chamfer removes only a corner of the surface, Figure 10-53. The term *bevel* is commonly misused. For example, rather than saying "Bevel the corner of that plate," the correct statement would be "Chamfer the corner of that plate."

Cutting Out

The best way of cutting the pipe is with a 37 1/2° bevel, Figure 10-37. If you do not have a pipe beveling cutter, you may use an angle grinder to bevel the pipe end.

NOTE: Do not let the pipe sections fall to the floor as they are cut off. This can cause them to bend out of round, making fit-up more difficult.

Drill one 1/4-in. hole in the center of the handle supports at the chamfered end 3/8 in. from the end. This is so that the wooden dowel rod handle can be held in place with the all-thread rod.

Fabrication

Tack weld the pieces of pipe together. Place the pipe horizontally in a vee block on the welding table. Sometimes pipe can be slightly out of round and will need to be reshaped to fit. Make four small tack welds evenly spaced around the pipe. Make sure you do not strike the arc outside of the weld groove.

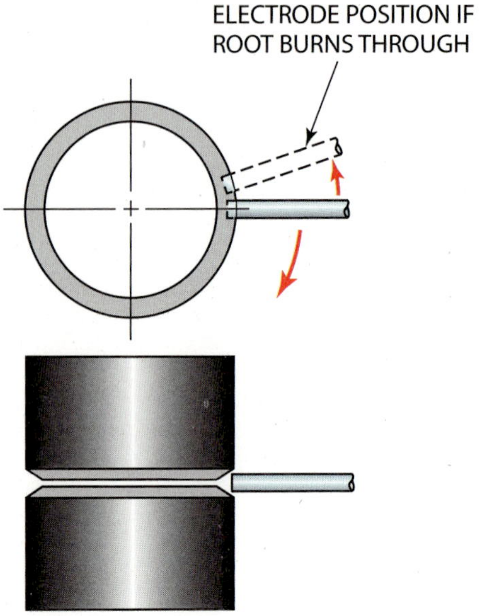

FIGURE 10-54 Electrode position for root pass.

Root Weld

Place the pipe vertically on the welding table. Hold the electrode at a 90° angle to the pipe axis and with a slight trailing angle, Figure 10-54. The electrode should be held tightly into the joint. If a burnthrough occurs, quickly push the electrode back over the burnthrough while increasing the trailing angle. This action forces the weld metal back into the opening.

If the root gap is uniform, the weld can be made with little or no electrode movement. However, if the root gap is uneven, then you may need to use a step pattern. At the end of each weld bead, stop, chip, and clean the weld crater before starting the next weld. When the root pass is complete, chip the surface slag, and then clean out the trapped slag by grinding or chipping, or use a hot pass.

Hot Pass

The inside root surface contour is more important than the weld face to the overall quality of the weld. That is because the weld face can be cleaned up and easily repaired if needed, but the root surface is much more difficult to fix. One way of cleaning up trapped slag or other similar root weld face problems is to use a hot pass. The hot pass need only burn the root pass clean, Figure 10-55. Undercut on the top pipe is

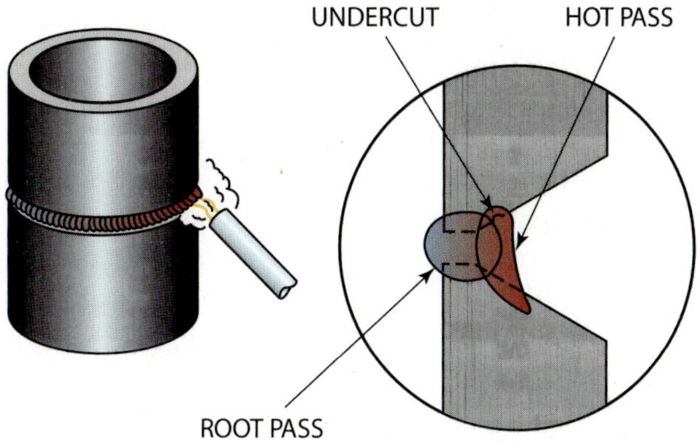

FIGURE 10-55 Electrode position for hot pass.

acceptable. The hot pass is as the name implies—a weld that is made with higher-than-normal current settings. It is usually made with the same electrode as the root pass with a higher-than-normal travel rate because little weld metal needs to be deposited.

The root can be ground with an angle grinder and a pipe disk. You should try both methods of cleaning the root pass as you work on these horizontal welds.

Filler Pass

Use E7018 electrodes for the filler and cover passes with a stringer pattern. The weave beads are not recommended with this electrode and position because they tend to undercut the top and overlap the bottom edge.

Strike an arc, and if necessary, use the "J"-weave pattern so that the molten weld pool will be supported by the lower edge of the solidified metal, Figure 10-56. Keep the electrode at an upward and trailing angle so that the arc force will help to keep the weld in place. Keep the molten weld pool's size small because it will be easier to control. Large welds tend to have more undercut along the top edge and overlap along the lower edge.

When you have either used up all of the electrode or have welded as far around the pipe as you can comfortably reach, break the arc. It is always a good practice to break the arc back over the weld crater.

Chip and clean the weld. Examine the weld for uniformity and any weld defects. An undercut along the top edge and overlap along the bottom edge of the weld are the most common problems with horizontal welds. If you have either or both of these, try adjusting your "J"-weave pattern and travel speed. Sometimes traveling a little faster will make a smaller weld bead that is less likely to have these problems.

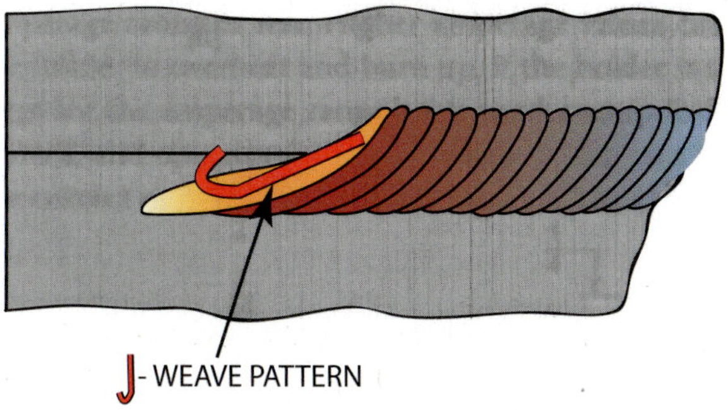

FIGURE 10-56 "J"-weave pattern.

Cover Pass

The cover pass is very important to the weld. It is the surface that everyone sees, and it is the one that they use to first judge the weld quality. Small stringer beads are the best type of welds to use for the cover pass. You are less likely to have too much reinforcement (buildup) with stringer beads. Most weld reinforcement on pipe welds should not exceed 1/8 in.

When you have to stop the weld and restart, clean the weld crater. Start the next weld's arc at a slight distance ahead of the weld crater. Bring the electrode back to the weld crater and when the molten weld pool has increased to the size of the weld crater, start moving the weld forward. This will ensure that the weld size is consistent. A good welder can sometimes even make the restarting point difficult to impossible to find because it is so consistent with the original weld.

Layout Holes

There are two sets of holes around the base of the chimney charcoal starter. The larger 1-in. holes are for ventilation. The second set of smaller 1/8-in. holes are for the stainless steel welding rods. These rods will support the charcoal briquettes inside the chimney so that they can get lots of fresh air. It is the fresh air that gets the charcoal burning faster. Lay out both sets of holes as shown in Figure 10-52.

Cutting the Holes

Use an oxyfuel or plasma torch to cut the 1-in. holes. Refer to Chapters 21 or 22 for detailed instruction on using these torches. The 1/8-in. holes can be drilled or cut using either type of torch. If they are going to be torch cut, test the size of the hole by putting a 1/8-in. welding rod into each hole once they are cut to check the size.

Assembling the Heat Shield

The heat shield is made out of 16-gauge sheet metal and is designed to keep the wooden handle from getting too hot while you are starting the charcoal. The 4 × 5-in. piece of sheet metal can be sheared, sawed, or thermally cut to size.

The heat shield will be attached to the pipe chimney with two plug welds. The plug welds will be made on the top two groove welds. The slight gap that this will create between the heat shield and the chimney pipe will make the heat shield work more efficiently.

Locate the tops of the groove welds and mark them on the heat shield. Drill or torch cut two 1/2-in. holes in the heat shield. Lay the chimney pipe on its side and place the heat shield on it. Hold the heat shield in place and strike an arc in the center of the hole on top of the groove weld bead. Hold the arc in the center and let the molten weld pool grow in size until it flows out to the heat shield. If you try to move the arc over to the sheet metal, it may melt back away from the heat, making an ever increasingly larger hole. When the first weld is completed, the weld shrinkage may pull the second plug weld up and away from the pipe. You can use a clamp or tap around the first plug weld with a hammer to get the heat shield back in place. Make the last plug weld the same way as the first.

Assembling the Handle

The first thing to do is to mark and cut a 5-in. long piece off of the wooden dowel rod. The hacksaw can be used to cut this piece of wood. Mark the center of the dowel rod and clamp it in a vise. Use the drill and a 5/16-in. to 3/8-in. drill bit to drill a hole through the dowel rod.

Grind, shear, saw, or torch cut the chamfer corners of the metal bars for the top and bottom of the handle. Cool and clean the bars if necessary. Mark the center of the chamfered end of the 1-in. bar. Use the same drill and bit to drill one hole in both bars.

Put the end of the all-thread rod through the metal washers, bars, and wood dowel and tighten the assembly using the nuts. With the assembly tightly held together with the all-thread rod, cut off the excessive length of the all-thread rod using the hacksaw. Grind the ends of the all-thread rod smooth after the saw cut if necessary.

Hold the handle in place on the chimney pipe and make two tack welds, one on the top and one on the bottom bars, to hold it in place for welding. If necessary, spread the bars so that they touch the pipe above and below the heat shield. Weld the handle support bars in place.

Finishing

Cool off the metal, chip the slag, and wire brush the surface. The stainless steel welding rods are inserted through the small holes. Bend the ends so that the rods are held in place. The rods may be installed several different ways as long as the grid opening that they make is small enough to keep the charcoal briquettes from falling through, Figure 10-57.

The project can be painted with high-temperature paint.

Paperwork

Complete a copy of the time sheet in Appendix I, the bill of materials in Appendix III, or use forms provided by your instructor.◆

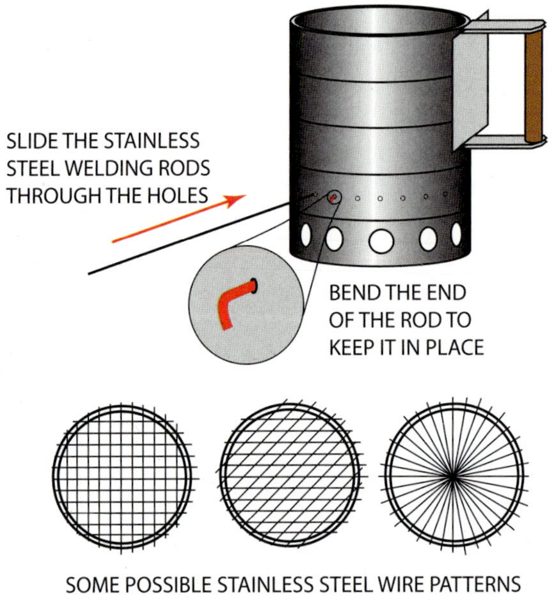

FIGURE 10-57 Bottom grid for charcoal starter.

SUMMARY

In this chapter, you have learned enough layout, cutting, fabrication, and welding skills to build the barbeque smoker grill shown in Figure 10-58. As you have built projects in this chapter, you have been learning the welding skills needed to build this barbeque smoker grill. The welds on the barbeque smoker grill are similar to the ones you have been practicing. Examples of similar welds have been highlighted in the figures in this chapter. This small barbeque smoker grill is designed to be used on a table or stand. Often these types of barbeque smoker grills are larger. The larger grills work great but are too large for a family outing or if you are having only a few friends over. Storing the larger grills can also be a problem. The firebox opens from the side, making it easy to load charcoal and wood chips. You can look up recipes and accessories for your barbeque smoker grill on the World Wide Web.

SMAW is often the process used to join piping systems. It is not as fast as some other processes such as FCAW or GMAW. But SMA welding has proved to be such a highly reliable process that it is still widely used. A good SMAW pipe welder is a highly sought-after artisan.

Because pipe welding presents definite challenges to the welder due to the constantly changing weld position, it is often considered the pinnacle of the welding trade. Practicing will help you learn to control the molten weld pool so that you can recognize the subtle changes as it transitions through the various weld positions. Over time, you will begin to make the welding technique changes as second nature.

A key to making a successful pipe weld is as much in the preparation of the pipe joint as it is in your welding skills. Properly setting up and having the weld joint accurately aligned before welding is essential in producing quality welded pipe joints.

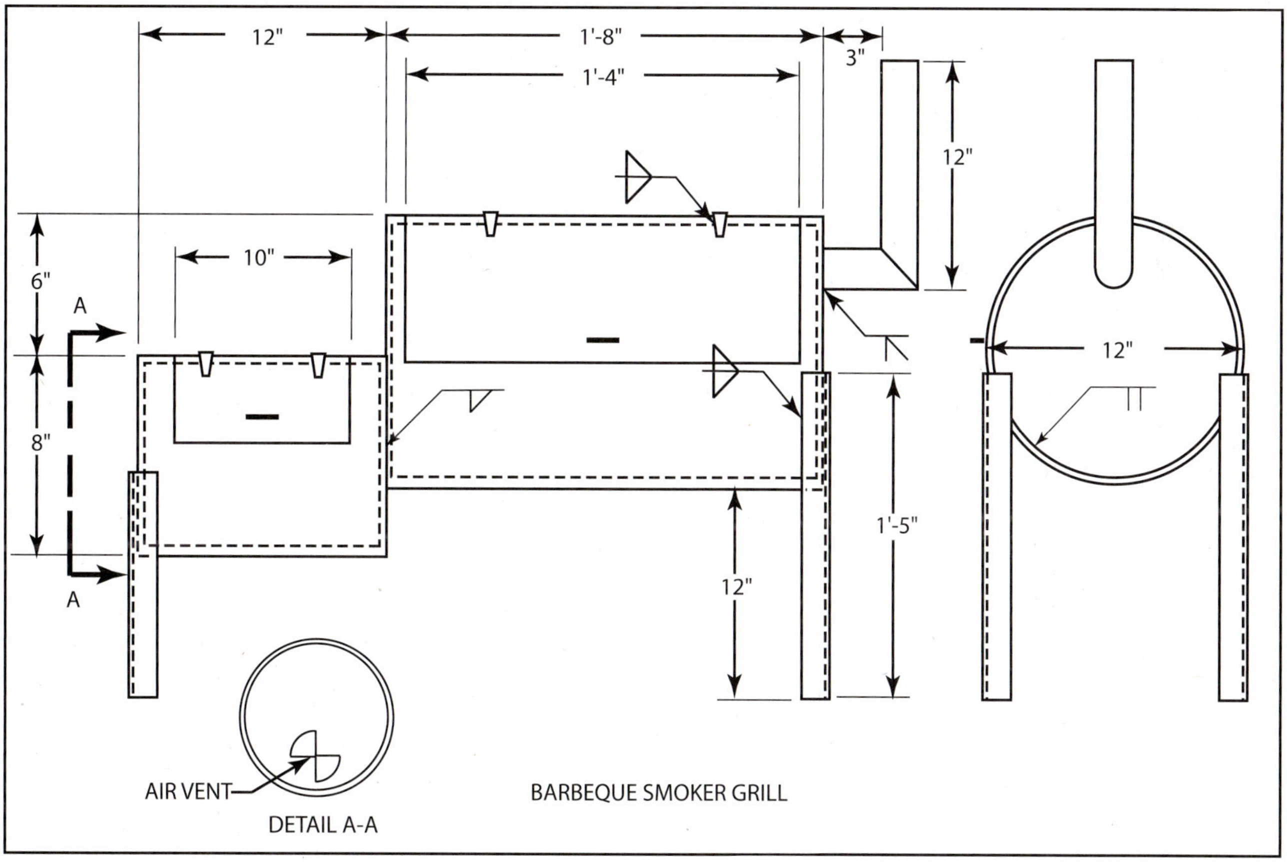

FIGURE 10-58 Barbeque smoker grill plans. © Cengage Learning 2012

REVIEW QUESTIONS

1. Why is it not necessary to cut the center out of structural pipe?
2. List the forces that are applied to welded structural pipe.
3. What is a major drawback to using pipe as structural components?
4. What is the difference between how a 3-in. (76-mm) pipe and a 3-in. (76-mm) tube would be measured?
5. What is the advantage of welding square butt pipe joints?
6. What is the advantage of the welded beveled pipe joint?
7. What are the terms that can be used to describe the flat area on the inside edge of a beveled pipe joint?
8. What pipe welding position requires the welder to transition from one welding position to another welding position?
9. List the four different weld passes that might be required to make a complete grooved pipe weld.
10. How can the problem of slag trapped on the sides of the face of a root weld be removed?
11. Why are the starting and stopping spots of the fillet welds staggered?
12. Sketch, identify, and dimension a cover pass.
13. What makes the 1G welding position desirable for both the welder and the company?
14. List four ways a pipe end plug can be held in place.
15. As expressed in clock position, how far across the top of a 1G pipe would be considered the flat position?
16. List the welding positions that make up a 5G weld.
17. How is the root pass welding direction influenced by weld fit-up?
18. Why must you account for the cut's kerf?
19. At what clock positions should the tack welds be made on a 5G pipe to keep them out of the way of the root pass starts and stops?
20. How can sagging along the bottom of a 2G weld bead be reduced?
21. What should the angle of the electrode be when making a root pass on 2G pipe?

Chapter 11

Gas Metal Arc Welding Equipment and Materials

OBJECTIVES

After completing this chapter, the student should be able to:

- Explain the gas metal arc (GMA) welding process and discuss its advantages.
- Explain the purpose of a shielding gas and how it is delivered to the weld.
- Identify the various components that make up a GMA welding station.
- Define the common electrical terms associated with a welding power supply.
- Describe the path that the electrical current takes in the welding process.
- Compare the four major types of wire feed systems.
- List the parts of a GMAW gun and describe how it works.
- State the most commonly used shielding gases and gas blends and what factors should be considered when choosing one.
- Choose the correct gas flow rate using welding guides.
- Describe the four modes of metal transfer and what factors should be considered when selecting one.
- Discuss the various features of GMAW electrodes, including sizes, coatings, cast, and helix, and the proper handling of the electrodes.
- Explain the meaning of the letters and numbers in the American Welding Society (AWS) GMAW electrode classification code.
- Demonstrate how to properly set up a GWA welding installation.
- Demonstrate how to properly thread an electrode wire through a welding installation.

KEY TERMS

argon (Ar)
argon-CO_2
argon-oxygen
axial spray metal transfer
bird nesting
carbon dioxide (CO_2)
cast
conduit liner
constant current (CC)

constant voltage (CV)
contact tube
electrode extension
feed rollers
flow rate
globular transfer
GMAW
GMAW-P
GMAW-S
helix
pinch effect
pull-type
pulsed-arc metal transfer
push-pull-type
push-type
short-circuit transfer
spray transfer
triggering
wire feed speed

INTRODUCTION

GMA welding uses a solid welding wire that is fed automatically at a constant speed as an electrode. An arc is generated between the wire and the base metal, and the resulting heat from the arc melts the welding wire and base metal to join the parts together, Figure 11-1. This is a semiautomatic arc welding process because wire is fed automatically at a constant rate and you provide the gun movement. During the welding process, a shielding gas protects the weld from the atmosphere and prevents oxidation of the base metal. The type of shielding gas used depends on the base material to be welded.

This process is often referred to by its original name, MIG, which stands for metal inert gas welding. The early GMA welding process used only inert gases for shielding, so the name MIG welding applied. Today, there are many different gases used for GMA welding—some are inert, nonreactive under all conditions, and others are reactive and can combine under some conditions. To reflect the addition of reactive gases, the term *MAG* (metal active gas) has been added to the lexicon of welding terms. The American Welding Society's standard term is *gas metal arc welding* (GMAW). However, in some industries the terms *MIG* and *MAG* are the most commonly used. The process has had other names through the years such as wire welding, but whatever the name, the process is the same.

The advantages of GMA welding over conventional electrode-type arc (stick) welding are numerous.

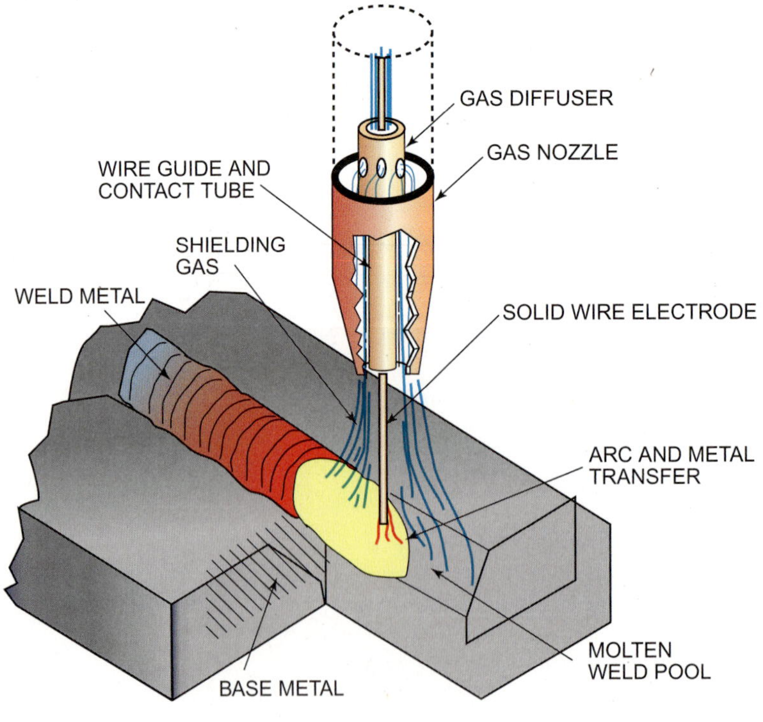

FIGURE 11-1 Gas metal arc welding. American Welding Society

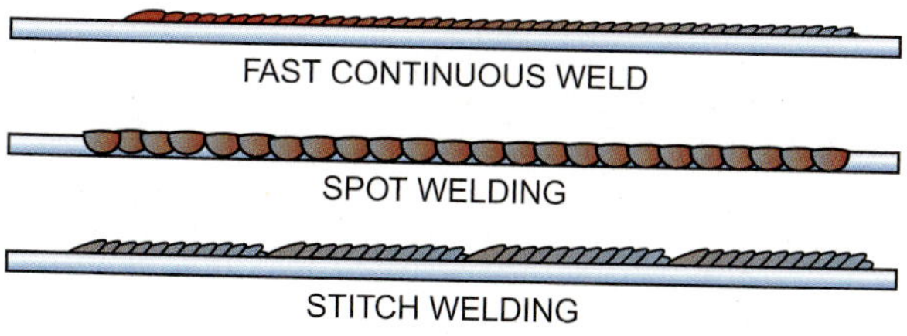

FIGURE 11-2 Ways of controlling distortion.

Some of the advantages of GMA welding are as follows:

- Easy to learn—The typical welder can learn to use GMAW equipment with just a few hours of instruction and practice. More time may be required to master the adjusting of the equipment.
- Speed and quality—GMA welding can produce higher-quality welds faster and more consistently than conventional stick electrode welds.
- Flexibility—The same size filler metal and type of shielding gas can be used to make welds on thin or thick metal by simply adjusting the welding current.
- Low distortion—The faster welding speeds and low currents reduce heat damage to adjacent areas that can cause strength loss and warping. Because there is no flux to remove between weld beads, other techniques, such as producing a series of spot welds or stitch welds, can be used to control distortion, Figure 11-2.
- Weld pool control—The small molten weld pool is easily controlled. This allows GMA welding to tolerate larger gaps and misfits. Gaps can be welded by making several short welds on top of each other.
- Out of position—GMA welding can easily be used in all positions because the weld pool is small and the metal is molten for a very short time.
- Tack welding—GMA welds are easily started in the correct spot because the wire is not energized until the gun trigger is depressed, so arc strikes outside of the weld groove can be avoided.
- Efficient—Some GMA welding procedures have a 98% efficiency of weld metal deposited as compared to shielded metal arc welding (SMAW), which is only 65% efficient under the most ideal conditions.

GMA welding is flexible and can be used for both new construction and repair work. Almost any welding that would be done with either an arc or gas welder can be done faster with GMA welding. In addition, it is possible to weld high-strength steel (HSS), high-strength low-alloy (HSLA), aluminum, stainless steel, and many other metals.

GMA WELDING EQUIPMENT

Although every manufacturer's GMA welding equipment is designed differently, it is all set up in a similar manner. Figure 11-3 identifies the various components that make up a GMA welding station.

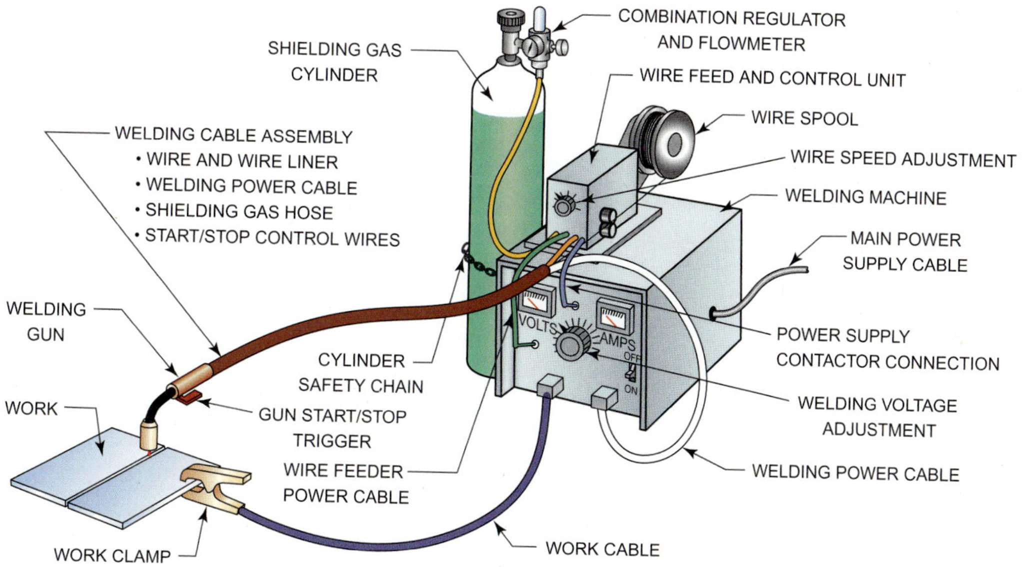

FIGURE 11-3 GMA welding equipment.

FIGURE 11-4 A 200-ampere constant-voltage power supply for multipurpose GMAW applications.
ESAB Welding & Cutting Products

Power Supply

The main piece of equipment is the power source, which is often called the welder. It produces direct current (DC) welding power. Most GMA welding is performed with the electrode positive, direct current electrode positive (DCEP). Depending upon the machine's capacity, its power can range from 40 to 600 amperes with 10 to 40 volts. Typical power supplies are shown in Figure 11-4.

Because of the long periods of continuous use, GMA welding machines have a 100% duty cycle. This allows the machine to be run for long periods of time or continuously without damage.

To better understand the terms used to describe the different welding power supplies you need to know the following electrical terms:

- Voltage or volts (V) is a measurement of electrical pressure, in the same way that pounds per square inch is a measurement of water pressure.
- Electrical potential means the same thing as voltage and is usually expressed by using the term *potential* (P or E). The terms *voltage, volts,* and *potential* can all be interchanged when referring to electrical pressure.
- Amperage or amps (A) is the measurement of the total number of electrons flowing, in the same way that gallons is a measurement of the amount of water flowing.
- Electrical current means the same thing as amperage and is usually expressed by using the term *current* (C or I). The terms *amperage, amps,* and *current* can all be interchanged when referring to electrical flow.

GMA welding power supplies produce a constant arc voltage. This type of machine is referred to as either **constant voltage (CV)** or constant potential (CP). Both CV and CP have the same meaning. This is in contrast to SMAW "stick" welding power supplies, which are **constant current (CC)**-type machines. It is impossible to make acceptable welds using the wrong type of power supply. GMA welding machines are available as the type that plug into the building power (transformer-rectifiers) or as engine-generators (portable welders), Figure 11-5. Some machines can produce both CV and CC welding currents. This enables them to be used for both GMA and SMA welding by simply flipping a switch.

Current Path

The path of the welding current from the welding machine to the arc and back through the work cable to the welding machine is as follows. The current leaves the welding power source positive terminal. It then flows through a heavy flexible copper cable to the welding power terminal on the wire feed unit. This terminal is located on or close to the end of the welding gun assembly. The current passes through the gun assembly connector and is transferred through a stranded copper cable inside of the welding gun cable to the welding gun handle. In the welding handle, the stranded copper cable is clamped into one end of a connector block. In the other end of the connector block, the conductor tube is connected. The current passes through the conductor tube and is transferred to the gas diffuser. The gas diffuser transfers the current to the **contact tube.** The contact tube transfers the current to the welding wire. This means that the welding wire has to carry the full welding current only a short distance to the arc. Because of the very small diameter of the welding wire, it would be impossible for it to carry the full welding current for any distance

(A)

(B)

FIGURE 11-5 **(A) Combination GMA and SMA transformer welding power supply.** Courtesy of ESAB Welding & Cutting Products **(B) Combination GMA and SMA engine generator welding power supply.** Thermadyne Industries, Inc.

without overheating. The current travels across the arc to the work, and it conducts the current back to the work clamp. The work clamp is connected to a welding cable, which returns the current back to the negative terminal on the welder, Figure 11-6.

The welding filler metal used in GMA welding is so small it is incapable of carrying the full welding current for more than a fraction of an inch. It is for this reason that the welding current is transferred through a number of components from the welding machine to the gun assembly, where it is finally transferred to the filler metal near the point of the actual welding arc. It is important to note that the welding current does travel back through the filler metal to

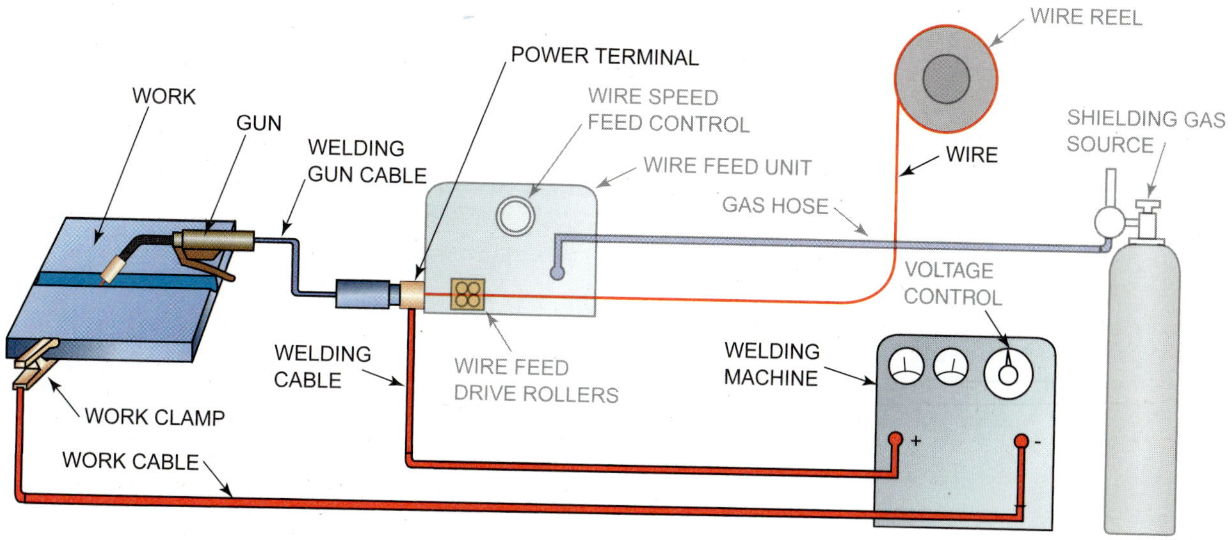

FIGURE 11-6 **Schematic layout of the welding circuit for a typical GMA welder.** Larry Jeffus

the spool on the wire feed unit. This means that the wire feed coil is actually connected to the electrode side of the welding circuit and would arc out if it accidentally comes into contact with the work side. Such arcing can more easily occur when metal-framed coils of filler metal are being used.

Wire Feed Unit

The purpose of the wire feeder is to provide a steady, uniform, and reliable supply of wire to the weld. It is very important that the wire be fed through the gun assembly smoothly and evenly because any change in the **wire feed speed** will result in a change in the weld produced. Even the slightest change or wire chatter can result in a weld bead defect or may even cause burn back. Burn back results when the wire momentarily stops feeding, and the arc backs up to fuse the welding contact tip, resulting in damage to the equipment, Figure 11-7.

The speed of the motor used in wire feed units can be continuously adjusted over the desired wire feed speed range. There are four major types of wire feed systems. They are push-type, pull-type, push-pull-type, and spool gun-type. The push, pull, and push-pull wire feed system names are descriptive based on how the wire is moved from the coil through the conduit to the gun.

FIGURE 11-7 Excessive heat can damage the contact tube. Larry Jeffus

Push-Type Wire Feed System

The drive rollers are made of hardened steel and have a groove cut into the surface. The groove size must match the welding wire size. The rollers are interchangeable to accommodate different sizes of welding wires. Depending on the type of wire feed unit being used, there may be one or two sets of rollers. Within the set of rollers, one or both rollers may be driven or power rollers. These are the rollers that actually move the wire through the system. When only one roller is a drive roller, the other roller is referred to as an idler. The advantage of having two drive rollers in a roller set is that the wire can be pushed through a longer gun cable.

The wire **feed rollers** are clamped securely against the wire to provide the necessary friction to push the wire through the conduit to the gun. The clamping pressure applied on the wire can be adjusted. One or both of the rollers will be grooved. The groove in the roller aids in wire alignment and lessens the chance of slippage. The wire must be held in alignment with the out feed guide. Most manufacturers provide rollers with smooth or knurled U-shaped or V-shaped grooves, Figure 11-8. Knurled rollers (knurling is a series of ridges cut into the groove) help grip larger diameter wires. They are mainly used when additional roller traction is needed to provide a smooth filler wire feed. Soft wires, such as aluminum and copper, are easy to damage if knurled rollers are used. U-grooved rollers should be used when feeding soft wires. Even V-grooved rollers

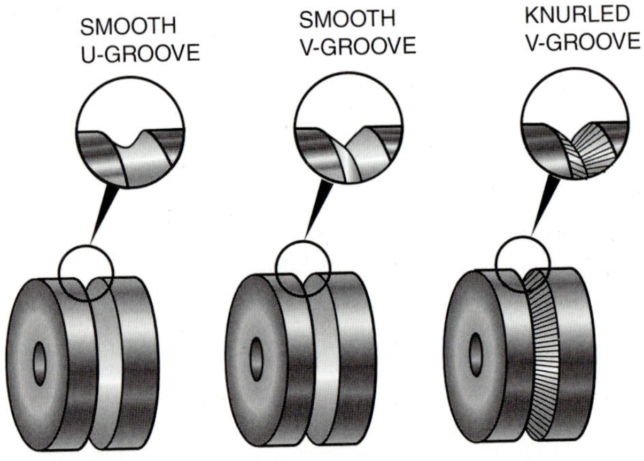

FIGURE 11-8 Feed rollers. © Cengage Learning 2012

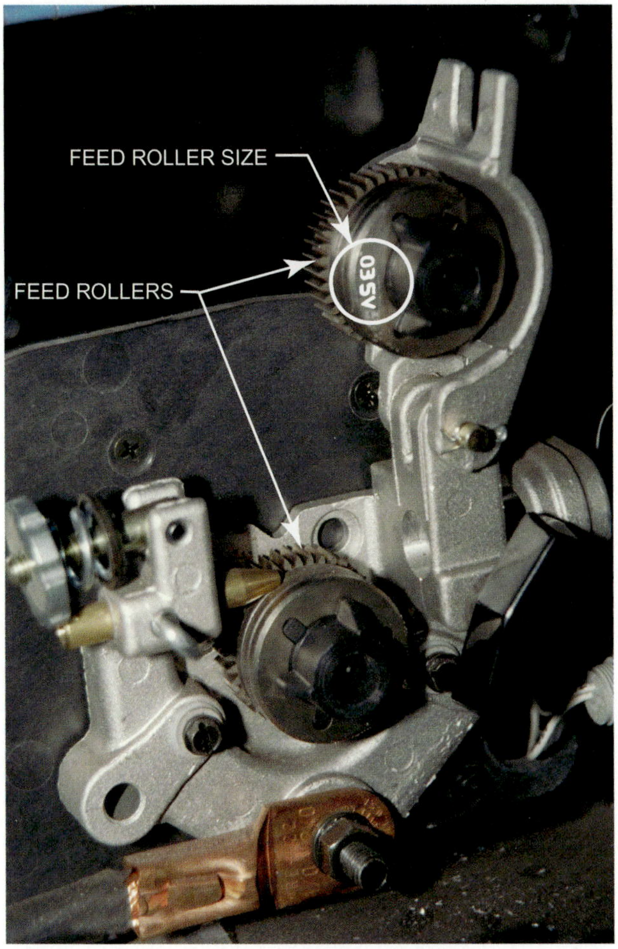

FIGURE 11-9 Check to be certain that the feed rollers are the correct size for the wire being used. Larry Jeffus

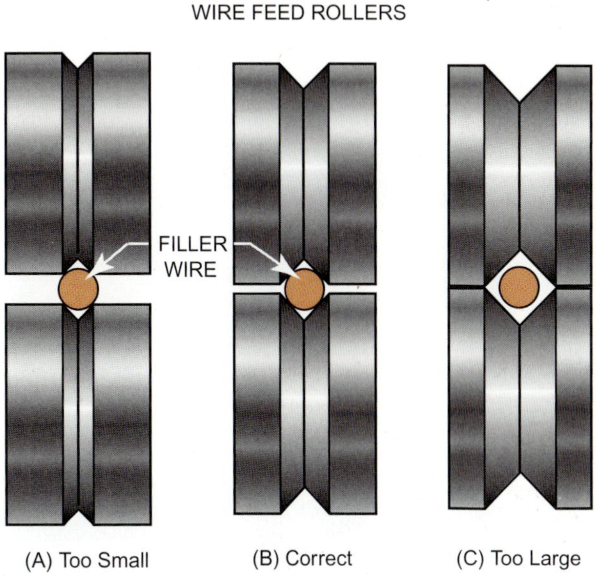

FIGURE 11-10 If the wire feed rollers are too small, the welding wire could be damaged. If the wire feed rollers are too large, the rollers will not grip the wire. © Cengage Learning 2012

can distort the surface of soft wire, causing wire feed problems. V-grooved rollers are best-suited for hard wires, such as mild steel and stainless steel.

The groove in the feed roller must be properly sized to fit the filler wire diameter being fed. The size of the filler wire to be used with a roller is usually stamped on the side of the roller, Figure 11-9. If the groove on the roller is too small for the wire being fed, the wire may wander out of the groove and may not be fed in proper alignment to the out feed guide. If the groove in the roller is too large, the wire may not make firm enough contact with the roller to be fed smoothly, Figure 11-10.

In the **push-type** system, the small diameter electrode wire must have enough stiffness to be pushed through the conduit without kinking. Mild steel and stainless steel can be readily pushed 15 ft (5 m) to 25 ft (8 m), but aluminum is much harder to push over 10 ft (3 m).

Pull-Type Wire Feed System

In **pull-type** systems, a small motor is located in the gun to pull the wire through the conduit. Using this system, it is possible to move even soft wire over great distances. A disadvantage is that the gun is slightly heavier and a may be a little more difficult to use.

Push-Pull-Type Wire Feed System

The **push-pull-type** wire feed system has two motors, one located at each end. One motor is located at the wire feed end and is pushing the wire, while the other motor is located at the gun end pulling the wire. These motors work together so that the wire is being both pushed and pulled at the same time. This type of wire feed system provides very smooth and reliable wire feeding, which reduces welding defects and burn back, which are common problems with some push-feed systems.

Spool Gun-Type

A spool gun is a compact, self-contained wire feed unit consisting of a small drive motor and a wire supply, Figure 11-11. This system allows you to move more freely around a job with only a power lead and

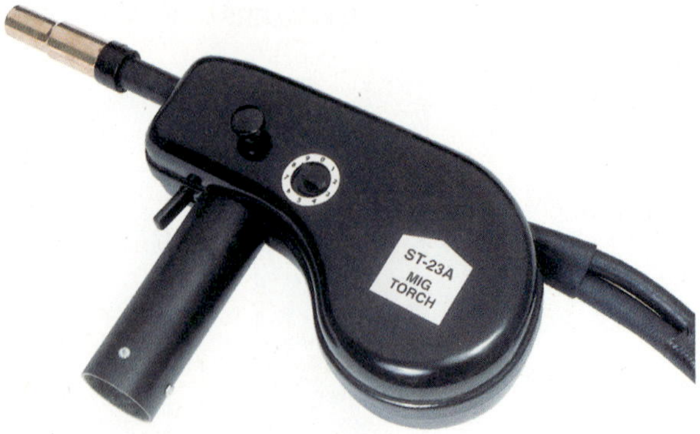

(A)

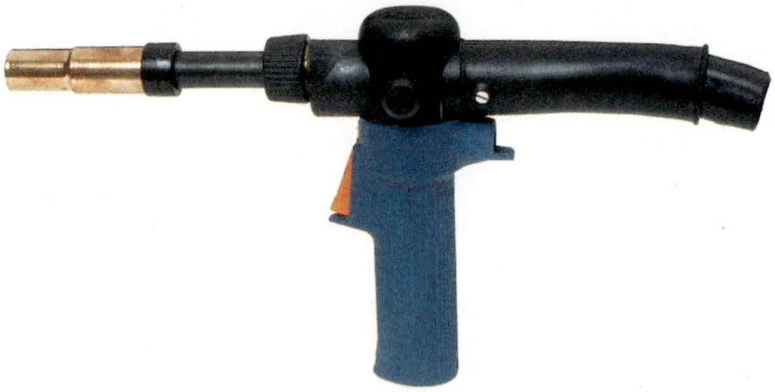

(B)

FIGURE 11-11 Feeder/gun for GMA welding.
ESAB Welding & Cutting Products

shielding gas hose connecting it to the power supply. The major control system is usually mounted on the welder. The feed rollers and motor are found in the gun just behind the nozzle and contact tube. Because of the short, straight distance that the wire must be moved, very soft wires such as aluminum and copper can be used. A small spool of welding wire is located just behind the feed rollers.

Feed Roller Tension

The feed rollers clamp onto the filler wire, and as the rollers spin, they push or pull the wire from the reel to the welding gun. For the feed rollers to properly grip and feed the filler metal, they must have proper tension applied. This tension is normally adjustable. It can usually be adjusted by turning a screw, bolt, or hand wheel, Figure 11-12. If the tension is too light, the wire feed can be erratic and may actually stop momentarily as you manipulate the welding gun. Too tight a wire feed pressure can result in bird-nesting of the filler wire if the wire becomes jammed at the tip

FIGURE 11-12 Adjust the wire feed tensioner.
© Cengage Learning 2012

or in the liner, Figure 11-13. To set the proper adjustment, press the gun's trigger and allow the wire to feed. The wire is electrically charged with the welding current and it can short out, so be careful and do not allow the wire to come in contact with the work or other grounded metal objects. As the wire is being fed, increase the roller tension until the wire feeds smoothly. To be sure the wire is not tensioned too tight, gently apply pressure to the wire spool with your hand. The wire feed roller is tensioned properly when the wire spool can be stopped without excessive pressure. A secondary advantage of not having excessive wire feed roller pressure is that if a welding cable, shielding gas hose, or other such item becomes tangled in the filler metal spool, the spool will be stopped before damage might occur to the wire, hose, or other tangled objects.

FIGURE 11-13 "Bird's nest" in the filler wire at the feed rollers. © Cengage Learning 2012

Reel Tension

The spindle that the reel or coil is mounted on has a friction adjustment. This adjustment allows the wire spool to be stopped quickly and not to coast once the welding trigger is released. Spool coasting after the trigger is released can result in two potential problems. One problem is that the loose wire can become tangled as welding is resumed. A more serious problem that occurs virtually each time that the slack is taken up is a momentary stoppage of the wire feed. This stoppage can result in burn back of the contact tube. To prevent coasting, the tension on the spindle must be properly set. It is set properly when the spool stops almost immediately each time that the welding gun trigger is released. Excessively high drag tension can result in wire feed problems also. To set the tension, use the gun trigger to start and stop the wire feed. Increase or decrease the wire feed tension while starting and stopping the feed until a smooth stop and restart occur without wire feed hesitation.

Wire Feed Guide

The out feed guide is a small pointed tube that guides the wire from the feed rollers into the beginning of the wire liner.

The steel, copper, brass, or plastic wire guides are such a simple component of the feeder system that it is easy to forget about them. They allow the wire to smoothly enter and exit the drive rolls.

The out feed guide should be aligned so that it is set as close as possible to the feed roller and in a straight line with the groove, Figure 11-14. The farther away the guide is from the feed roller, the more likely you are to have misfeed problems. Any misalignment of the out feed guide increases drag on the filler metal, which can lead to erratic wire feeding. On some copper-clad or softer filler wires, small amounts of copper cladding or slivers of the soft metal may be scraped away by the misaligned guide. These small particles are then carried into the wire conduit liner where they can accumulate, resulting in wire feed problems that can be remedied only by replacing the conduit liner itself.

Conduit Wire Liner

The **conduit liner** is a flexible hollow tube through which the wire is fed from the wire feed unit to the gun. It is positioned through the body of the gun and into the gun conductor tube. Most conduit liners are springlike in appearance because they are made from a hardened steel wire that is wound in a continuous spiral. Liners used with aluminum and other soft alloys often have a Teflon® liner to reduce the wire friction to allow the wire to feed more smoothly.

Replacement conduit liners can be ordered to fit individual guns, or they may be cut to length to fit some gun lengths from the feed unit to the GMAW gun.

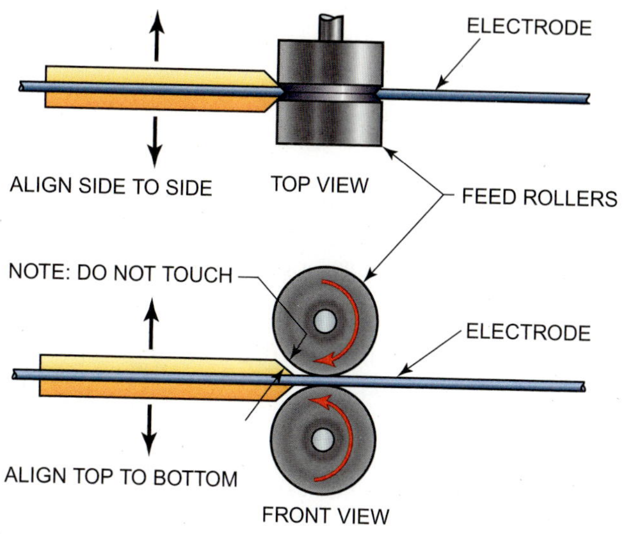

FIGURE 11-14 Feed roller and conduit alignment.
© Cengage Learning 2012

GMAW Guns

The main part of the GMA gun is called the gun body. The gun body is made of a heat-resistant plastic. It is connected to the end of the welding cable. Its major parts are the gun trigger, gun body, conductor tube, conduit, gas diffuser, contact tube, and gas nozzle, Figure 11-15.

The gun trigger is a toggle switch (off/on switch) and is attached to the gun body. When the trigger is pressed, electrical contact is made, activating the wire feeder. This simultaneously starts the wire feed, weld current, and shielding gas so that the welding arc can begin. As long as the trigger is not depressed, the arc will not be struck. So, you can hold the gun with the wire touching the metal being welded without an arc starting until the trigger is pulled. This feature of GMA welding helps you to always start the weld exactly where you need it.

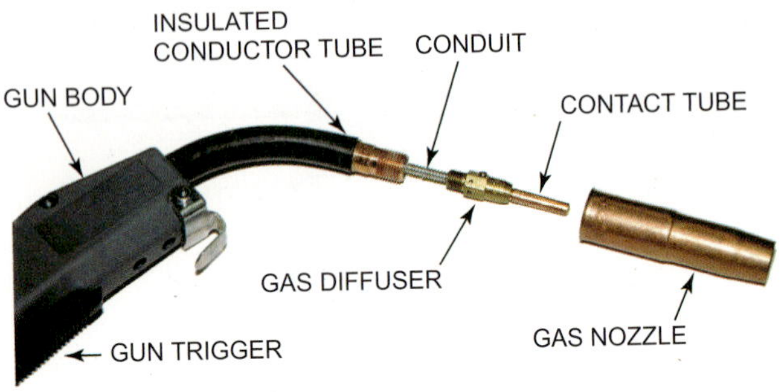

FIGURE 11-15 Typical parts of a GMA welding gun.
© Cengage Learning 2012

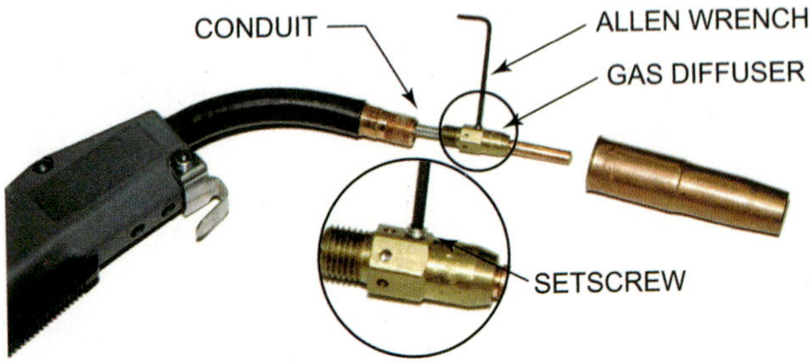

FIGURE 11-17 GMA welding gun assembly.
© Cengage Learning 2012

NOTE: On some small home/hobby-type GMA welding machines, the gun trigger starts and stops only the electrode feed motor. It may not start and stop the welding current, so the electrode is electrically charged all the time. Therefore, any time you touch the electrode to the work an arc will occur.

The insulated conductor tube, sometimes called the gun neck, is attached to the body of the gun. Conductor tubes are available as straight, 45°, 90°, and/or flexible, Figure 11-16. The most commonly used conductor tubes have a 45° angle. The choice of the angle of the conductor tube on the job is usually yours to make. Sometimes you may find that one angle of conductor tube gives you better access to the weld joint, or you may find one type of conductor tube more comfortable to use than the others. The connector tube is hollow so both the electrode liner and the shielding gas can pass through it. The conductor tube connects to the welding power cable in the handle, and it conducts the welding current to the end of the gun assembly; therefore, it is heavily insulated to prevent accidental arcing.

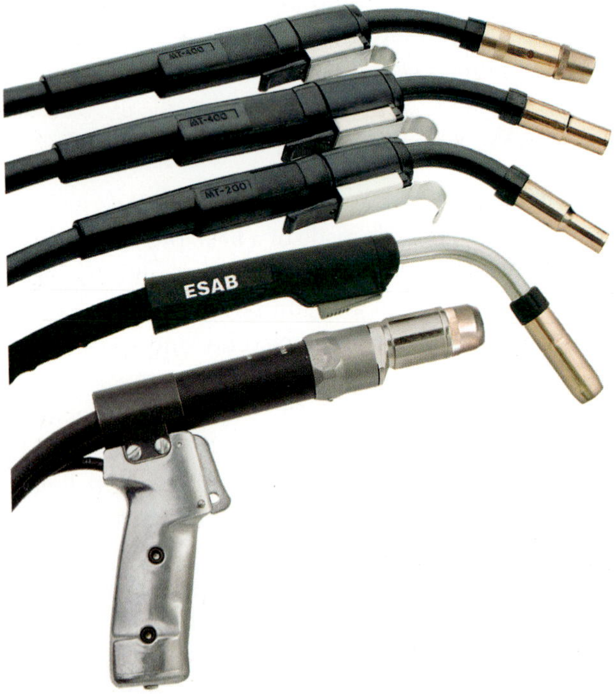

FIGURE 11-16 GMA welding guns are available in a variety of sizes and shapes. ESAB Welding & Cutting Products

The gas diffuser screws into the conductor tube and is electrically hot when the trigger is depressed. The gas diffuser allows the shielding gas to be dispersed or diffused around the contact tube, enveloping the entire weld area for greater protection, Figure 11-17.

The contact tube is screwed into the gas diffuser. The contact tube is a short replaceable copper conductor that transfers the electricity from the gun to the electrode wire. Over time the hole in the contact tube wears out and spatter builds up on the outside, resulting in the need for the contact tube to be replaced. Contact tubes come in a variety of sizes to fit the various diameter electrode wires. A contact tube with the proper inside diameter must be selected. It is important that the contact tube be sized correctly for the filler wire being used. Too large a contact tube does not provide adequate electrical conductivity from the contact tube to the wire, which can result in the contact tube overheating and the wire becoming stuck in the tube. Too small a contact tube creates excessive drag on the wire, which may result in the wire sticking in the tube. If the wire sticks for any reason, a burn back will occur, destroying the contact tube.

The contact tube may get as hot as 300°F (150°C) when you are welding with an air-cooled GMAW torch. Larger diameter, heavy-duty contact tubes are

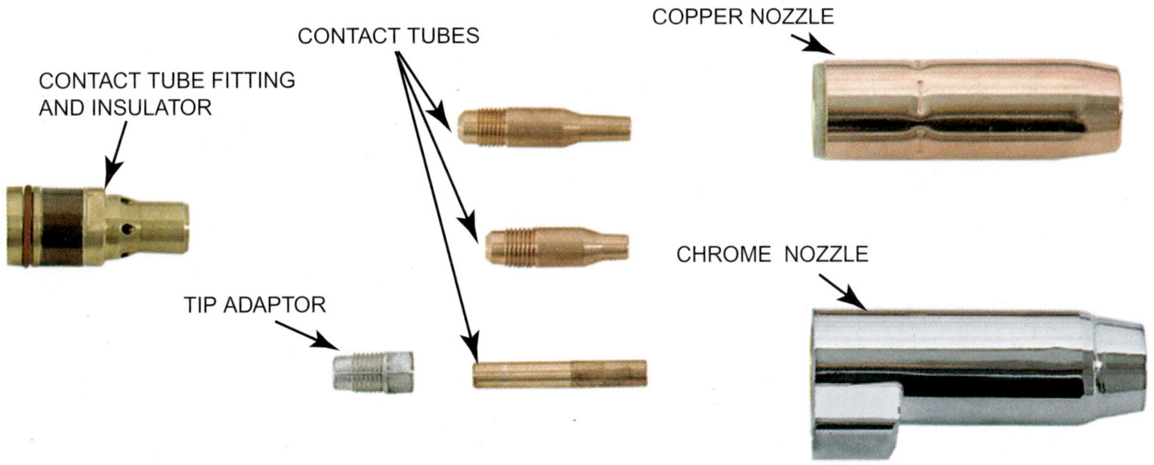

FIGURE 11-18 Typical replaceable parts of a GMA welding gun. ESAB Welding & Cutting Products

available from some manufacturers. You may want to consider using these heavier-duty tubes when welding with high-amperage settings because they can withstand the higher heat input better than the standard contact tubes.

When welding aluminum, it is very important to properly size the contact tube. You may want to try several different sized tubes to find the one that works best for your application. On any filler wire, but particularly with aluminum wire, the contact tube can be removed and cleaned with a set of oxyacetylene torch tip cleaners to extend the useful life of the contact tube.

To keep the electrically hot components from accidentally arcing, an insulating device called the gun nozzle insulator is used, Figure 11-18. The insulator is screwed onto the gas diffuser. On some types of GMA welding guns, the insulator ring can accidentally be screwed on backward. If the insulator is screwed on backward, the nozzle will not slide on far enough, the **electrode extension** will be excessively long, and poor welding will result.

The nozzle is the hollow metal tube located at the end of the gun assembly through which the welding wire passes and the shielding gas flows. Most nozzles are made of copper, although other materials such as brass and chrome plating are also used. Copper nozzles resist welding spatter and can withstand the welding heat. Chrome-plated nozzles are significantly more expensive but do resist weld spatter buildup much better than copper nozzles.

Nozzles come in a variety of diameters. The larger diameter nozzles provide better gas shielding for the weld but restrict the visibility significantly more than the smaller diameter nozzles. The nozzle size also affects the shielding gas flow rate. Obviously, the larger diameter nozzles require a higher shielding gas flow rate than the smaller diameter ones. In most cases, you can pick the size nozzle you prefer using; however, some welding procedures may specify the nozzle size and shielding gas flow rate that must be used.

Figure 11-19 shows an exploded view of a GMAW gun so that you can identify the various components of the guns. Although there are many brands of welding guns, the names used for the parts are generally the same.

Work Lead

It is important that the work clamp be securely connected to the work. GMA welding is very sensitive to changes in arc voltage. A loose or poor connection results in increased circuit resistance and a decrease in the arc's voltage. Voltage changes affect GMA weld quality. A more significant welding problem occurs when this resistance varies during the course of a weld. Such variations can dramatically adversely affect your ability to maintain weld bead control. GMA welding is more significantly affected by changes in the arc voltage than is SMAW (stick) welding. To be sure that you have a good work connection, remove any paint, dirt, rust, oil, or other surface contamination at the point where the work clamp is connected to the weldment.

Another problem that can cause resistance at the work clamp is when the clamp's internal spring

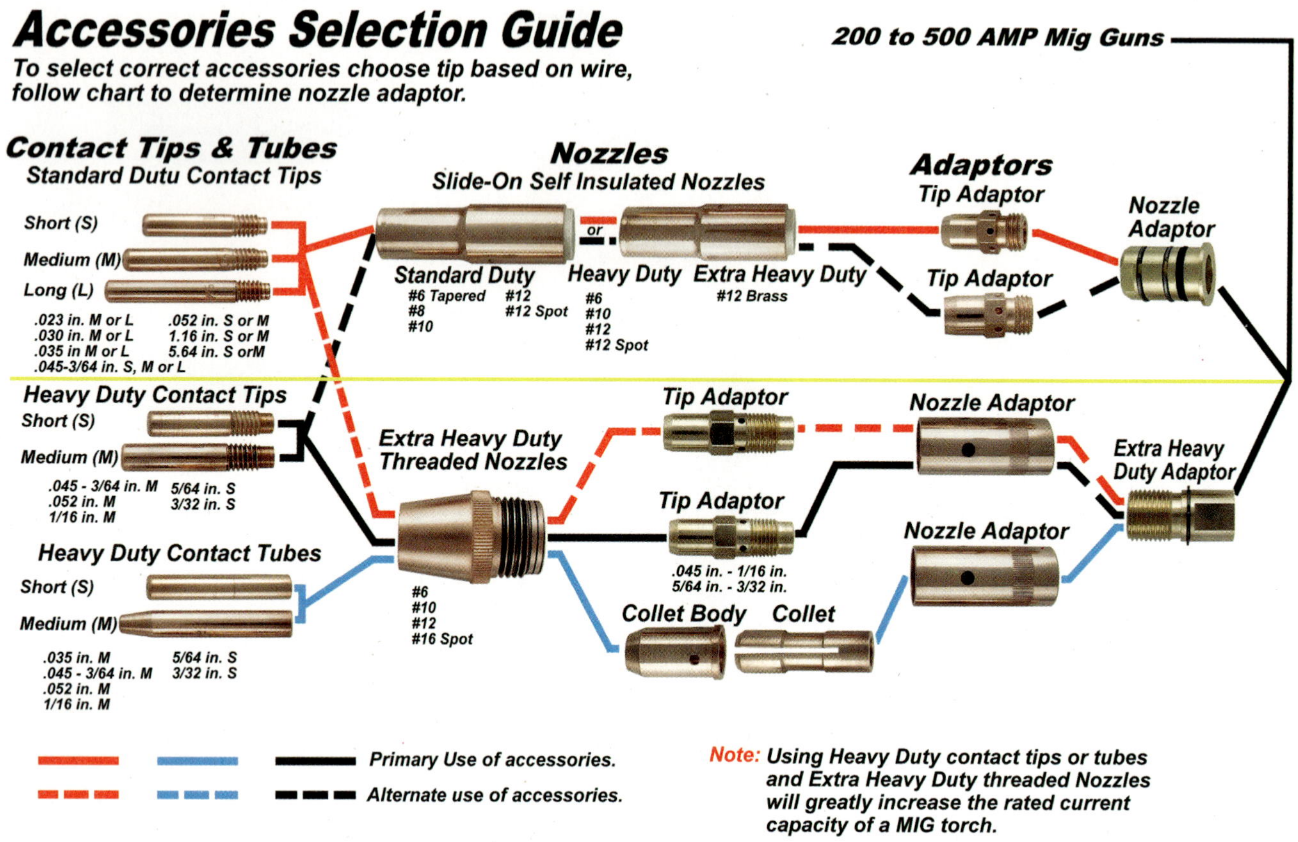

FIGURE 11-19 Accessories and parts selection guide for a GMA welding gun. ESAB Welding & Cutting Products

weakens. The spring provides the clamping pressure that holds the clamp onto the work. As this spring becomes weaker over time, it may not provide good electrical contact. The spring itself is not usually replaceable, so the entire work clamp may need to be replaced if it does not grip tightly enough to the work.

NOTE: The work clamp may become hot because it is connected to the part being welded. However, one way of determining that the work clamp is not clamping tightly enough is when it becomes hot during welding and it is not connected to a hot piece of metal. In this case, the clamp is becoming hot because of electrical resistance due to its poor connection to the work. Getting hot like this is a good indication that it needs to be repaired or replaced.

Shielding Gas Flowmeter

The shielding gas flowmeter measures the **flow rate** in cubic feet per hour (CFH) or in metric measure as liters per minute (L/min). The flow is controlled by opening a small valve at the base of the flowmeter. Some flowmeters indicate the flow rate by comparing the height that a small ball floats on top of the gas stream to a vertical scale on the transparent tube on the meter. Others flowmeters use a dial or digital display.

On ball-type flowmeters, if the tube is not vertical, the reading is not accurate. Less dense gases, such as helium and hydrogen, will not support the ball on as high a column with the same flow rate as a denser gas, like argon. So, to get accurate readings, be sure that the gas being used is read on the proper flow scale.

Welder Connections

Many smaller GMA welding units and some larger commercial units have the wire feed mechanism built into the welder so that no connections are required, Figure 11-20. Some types of GMA welding machines and wire feed units come as separate devices, Figure 11-21. These units require some connections before the welder can be used.

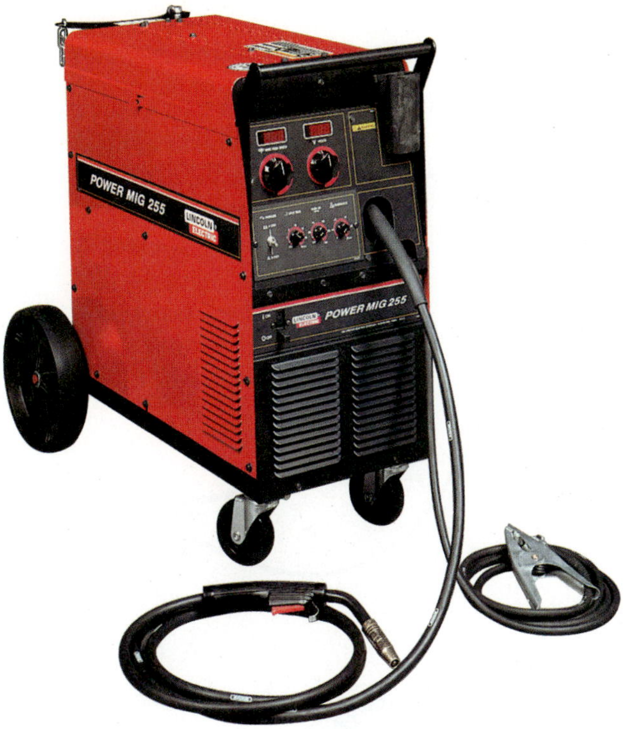

FIGURE 11-20 **Semiautomatic GMA welding setup.**
The Lincoln Electric Company

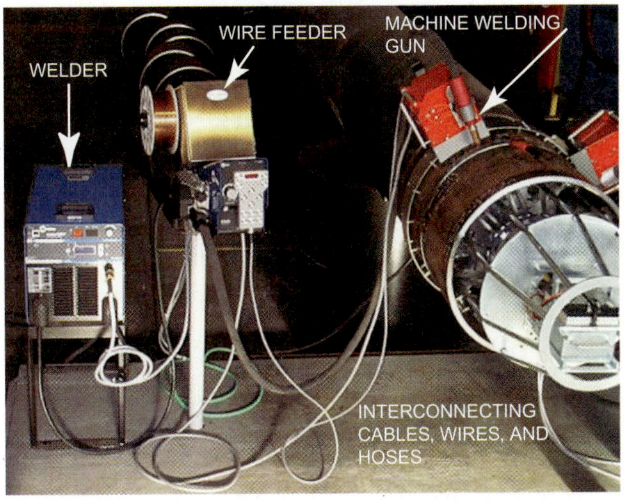

FIGURE 11-21 **Typical interconnecting cables and wires for an automatic GMA welding station.**
Dynatorque Technologies, Inc.

Having the wire feed assembly separate from the welding machine has several advantages:

- Portability—The wire feed unit can be moved some distance from the welding machine, allowing you to use a shorter welding gun assembly cable, thus reducing wire feed problems. There are fewer problems to having longer interconnecting power and control cables between the welding machine and wire feed units as compared to the limitations on the wire feed cable.
- Flexibility—Having the wire feed unit separate allows you to change the wire feed assembly without the necessity of purchasing another complete welding machine. This makes it easier to use different types of wire feed units for special applications. Wire feed units such as a standard wire feed system, spool gun, and machine welding assembly can all be easily interchanged with the same welding power supply when they are separate units.
- Space savings—Welding machines can be located so that they are out of the way such as under a welding table, some distance from the welding area, or even overhead. The wire feed unit can be suspended at the end of a specially designed boom, Figure 11-22. This keeps both the welder and the wire feed unit out of the way of the operation and clear of sparks and hot metal.
- Equipment replacement—By having separate components it is easier to upgrade either the welding power supply or wire feed unit when such changes become necessary.

GMAW SHIELDING GASES

The GMA shielding gas can be provided from a compressed gas cylinder or from a central gas piped manifold system. Individual cylinders provide the greatest portability, while the manifold system can offer the greatest potential savings by reducing the number of cylinders a shop must rent.

The most commonly used gases are carbon dioxide (CO_2), argon (Ar), and helium (He). Sometimes a blend of these gases is used to obtain the best possible welding performance. And occasionally a trace of oxygen (O) can be added to the mixture for making welds on some ferrous (steel) alloys.

Often codes, standards, and weld specifications give a range of gas flows and choices of gases or blends to be used for a weld. So, you will need to make some shielding gas choices even when there are welding specifications.

The most commonly used gases for ferrous metals are CO_2, argon with 2 to 5% oxygen added, and argon with 25% CO_2 added. Most nonferrous metals are

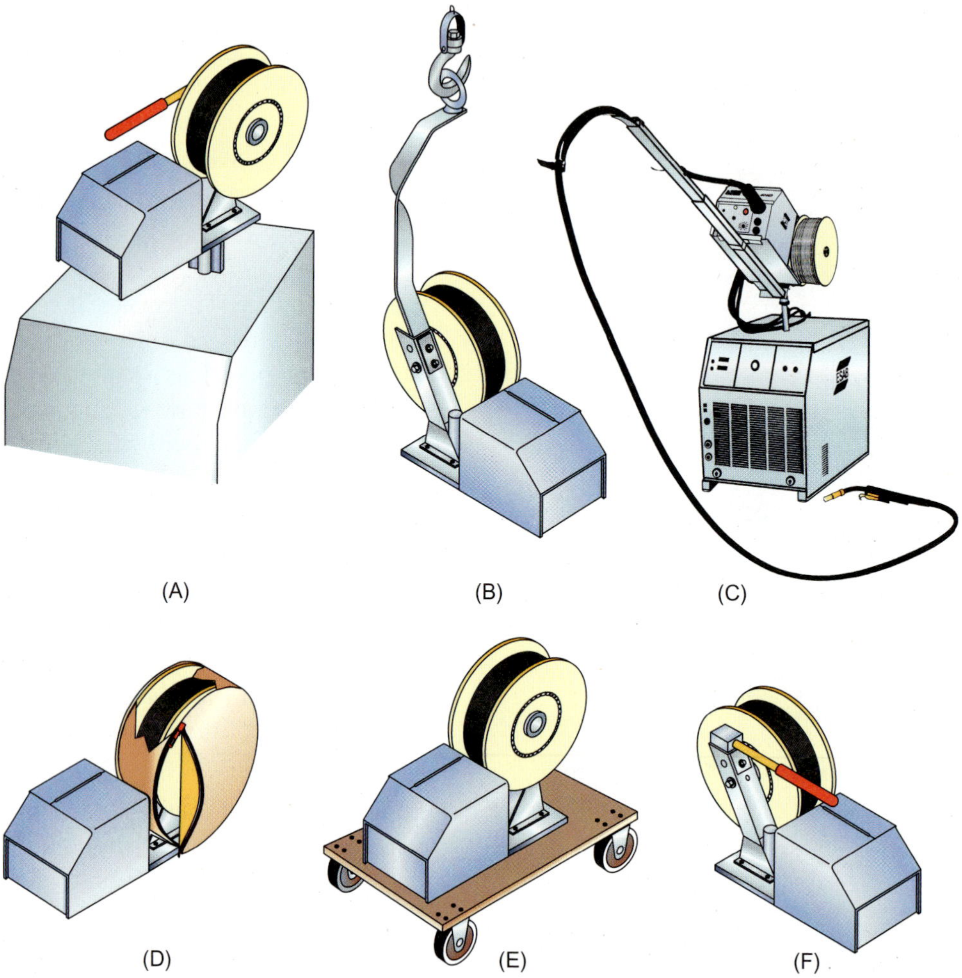

FIGURE 11-22 A variety of accessories is available for most electrode feed systems: (A) swivel post, (B) boom hanging bracket, (C) counterbalance mini-boom, (D) spool cover, (E) wire feeder wheel cart, (F) carrying handle. ESAB Welding & Cutting Products

welded using inert gases such as argon, helium, or a blend of argon and helium. To better make a selection of which shielding gas or shielding gas blends to use, you must consider many factors. Table 11-1 lists common GMA welded metals and some of the shielding gases and gas blends that can be used. It also lists the commonly used GMA welding gases and gas blends as they relate to metals and welding processes. The specific shielding gas or gas blend used affects the GMA weld being produced. Factors to be considered when selecting a shielding gas or blend may include the following:

- Metal transfer method
- Weld bead shape, penetration, and width of fusion zone
- Welding speed
- Weld discontinuities
- Weld spatter
- Metal transfer efficiency
- Type of base and filler metals
- Welding position
- Cost of the gas
- Total welding cost

Argon (Ar)

The atomic symbol for **argon** is *Ar*, and it is an inert gas. *Inert gases* do not react with any other substance and are insoluble in molten metal. One hundred percent argon is used on nonferrous metals such as aluminum, copper, magnesium, nickel, and their alloys,

GMAW Metals, Shielding Gases, and Gas Blends														
Metals	Gases		Blends of Two Gases									Blends of Three Gases		
			Argon + Oxygen			Argon + Carbon Dioxide			Argon + Helium					
	Argon (Ar)	CO_2	Ar + 1% O_2	Ar + 2% O_2	Ar + 5% O_2	Ar + 5% CO_2	Ar + 10% CO_2	Ar + 25% CO_2	Ar + 25% He	Ar + 50% He	Ar + 75% He	Ar + CO_2 + O_2	Ar + CO_2 + Nitrogen	Ar + CO_2 + Helium
Aluminum	X								X	X	X			X
Copper Alloys	X								X	X	X			
Stainless Steel		X	X	X	X			X			X	X	X	X
Steel		X	X	X	X	X	X	X	X			X		
Magnesium	X							X	X	X	X			
Nickel Alloys	X								X	X	X			

(A)

GMAW Shielding Gases, Gas Blends, Metals, and Welding Process							
Gases/Blend	Gas Reaction	Application	Remarks	Gas/Blend	Gas Reaction	Application	Remarks
Argon (Ar)	Inert	Nonferrous metals	Provides spray transfer	CO_2	Oxidizing	Mild, low alloy steels and stainless steel	Least expensive gas, deep penetration with short-circuiting or globular transfer
Helium (Me)	Inert	Aluminum and magnesium	Very hot arc for welds on thick sections, usually used in gas blends to increase the arc temperature and penetration	Nitrogen	Almost inert	Copper and copper alloys	Has high heat input with globular transfer
Ar + 1% O_2	Oxidizing	Stainless steel	Oxygen provides arc stability	Ar + 25% He	Inert	Al, Mg, copper, nickel, and their alloys	Higher heat input than Ar, for thicker metal
Ar + 2% O_2	Oxidizing	Stainless steel	Oxygen provides arc stability	Ar + 50% He	Inert	Al, Mg, copper, nickel, and their alloys	Higher heat in arc use on heavier thickness with spray transfer
Ar + 5% O_2	Oxidizing	Mild and low alloy steel	Provides spray transfer	Ar + 75% He	Inert	Copper, nickel, and their alloys	Highest heat input
Ar + 5% CO_2	Oxidizing	Low alloy steel	Pulse spray and short-circuit transfer in out-of-position welds	Ar + CO_2 + O_2	Oxidizing	Low alloy steel and some stainless steel	All metal transfer for automatic and robotic applications
Ar + 10% CO_2	Oxidizing	Low alloy steel	Same as above with a wider, more fluid weld pool	Ar + CO_2 + N	Almost insert	Stainless steel	All metal transfer, excellent for thin gauge material
Ar + 25% CO_2	Oxidizing	Mild, low alloy steels and stainless steel	Smooth weld surface, reduces penetration with short-circuiting transfer	Ar + 7.5% Ar + 2.5% CO_2	Almost insert	Stainless steel and some low alloy steels	Excellent toughness, excellent arc stability, wetting characteristics, and bead contour; little spatter with short-circuiting transfer

(B)

Table 11-1 (A) Metals Compared to GMAW Shielding Gases and Gas Blends (B) GMAW Shielding Gases and Gas Blends Compared with Metals

but 100% argon is not normally used for making welds on ferrous metals.

Because argon is denser than air, it effectively shields welds by pushing the lighter air away. Argon is relatively easy to ionize. Easily ionized gases can carry long arcs at lower voltages, making them less sensitive to changes in arc length.

Argon gas is naturally found in all air and is collected in air separation plants. There are two methods of separating air to extract nitrogen, oxygen, and argon. In the cryogenic process air is supercooled to temperatures that cause it to liquefy and the gases are separated. In the noncryogenic process, molecular sieves (strainers with very small holes) separate the various gases much like using a screen to separate sand from gravel.

Argon Gas Blends

Oxygen, carbon dioxide, helium, and nitrogen can be blended with argon to change argon's welding characteristics. Adding reactive gases (oxidizing) such as oxygen or carbon dioxide to argon tends to stabilize the arc, promote favorable metal transfer, and minimize spatter. As a result, the penetration pattern is improved, and undercutting is reduced or eliminated. Adding helium or nitrogen gases (nonreactive or inert) increases the arc heat for deeper penetration.

The amount of the reactive gases, oxygen or carbon dioxide, required to produce the desired effects is quite small. As little as a half percent change in the amount of oxygen produces a noticeable effect on the weld. Most of the time, blends containing 1 to 5% of oxygen are used. Carbon dioxide may be added to argon in the range of 20 to 30%. Blends of argon with less than 10% carbon dioxide may not have enough arc voltage to give the desired results. The most commonly used argon-CO_2 blend is 25% CO_2.

When you are using oxidizing shielding gases with oxygen or carbon dioxide added, a suitable filler wire containing deoxidizers should be used to prevent porosity in the weld. The presence of oxygen in the shielding gas can also cause some loss of certain alloying elements, such as chromium, vanadium, aluminum, titanium, manganese, and silicon.

Helium (He)

The atomic symbol for helium is *He*, and it is an inert gas that is a product of the natural gas industry. It is removed from natural gas as the gas undergoes separation (fractionation) for purification or refinement.

Helium is lighter than air, thus its flow rates must be about twice as high as argon's for acceptable stiffness in the gas stream to be able to push air away from the weld. Proper protection is difficult in drafts unless high flow rates are used. It requires a higher voltage to ionize, which produces a much hotter arc. There is a noticeable increase in both the heat and temperature of a helium arc. This hotter arc makes it easer to make welds on thick sections of aluminum and magnesium.

Small quantities of helium are blended with other heavier gases. These blends take advantage of the heat produced by the lightweight helium and weld coverage by the other heavier gas. Thus, each gas is contributing its primary advantage to the blended gas.

Carbon Dioxide (CO_2)

Carbon dioxide is a compound made up of one carbon atom (C) and two oxygen atoms (O_2), and its molecular formula is CO_2. One hundred percent carbon dioxide is widely used as a shielding gas for GMA welding of steels. It allows higher welding speed, better penetration, and good mechanical properties and costs less than the inert gases. The chief drawback in the use of carbon dioxide is the less-steady-arc characteristics and a considerable increase in weld spatter. The spatter can be kept at a minimum by maintaining a very short, uniform arc length. CO_2 can produce sound welds provided a filler wire having the proper deoxidizing additives is used.

Nitrogen (N)

The atomic symbol for nitrogen is *N*. It is not an inert gas but is relatively nonreactive to the molten weld pool. It is often used in blended gases to increase the arc's heat and temperature. One hundred percent nitrogen can be used to weld copper and copper alloys.

GAS FLOW RATE

The correct flow rate can be set by checking welding guides that are available, Table 11-2. These welding guides list the gas flow required for various nozzle sizes and welding amperage settings. Some welders

ELECTRODE		WELDING POWER			SHIELDING GAS			BASE METAL	
Type	Size	Amps	Wire Feed Speed IPM (cm/min)	Volts	Type	Flow	Nozzle	Type	Thickness
E70S-1	0.030	60 to 140	150 to 350	14 to 16	CO_2 or Ar + CO_2	30 to 50	1/2 in	Low-Carbon Steel	16 ga to 1/8 in
E70S-1	0.035	90 to 160	180 to 300	15 to 19	CO_2 or Ar + CO_2	30 to 50	1/2 in	Low-Carbon Steel	16 ga to 1/8 in
E70S-3	0.035	180 to 230	400 to 550	25 to 27	Ar + 2% to 5% O_2	30 to 50	5/8 in	Low-Carbon Steel	1/4 in to 1/2 in
E70S-3	0.045	260 to 340	300 to 500	25 to 30	Ar + 2% to 5% O_2	30 to 50	5/8 in	Low-Carbon Steel	1/4 in to 1/2 in
E70S-3	0.035	90 to 120	180 to 300	15 to 19	CO_2 or 75%Ar/$CO_2$25%	30 to 50	5/8 in	Low-Carbon Steel	1/4 in to 1/2 in
E70S-3	0.045	130 to 200	125 to 200	17 to 20	CO_2 or 75%Ar/$CO_2$25%	30 to 50	5/8 in	Low-Carbon Steel	1/4 in to 1/2 in

Table 11-2 GMA Welding Guide

feel that a higher gas flow provides better weld coverage, but that is not always the case. High gas flow rates both waste shielding gases and may lead to contamination. The contamination comes from turbulence in the gas at high flow rates. Air is drawn into the gas envelope by the venturi effect around the edge of the nozzle. Also, the air can be drawn in under the nozzle if the torch is held at too sharp an angle to the metal, Figure 11-23.

NOTE: If you need more shielding gas coverage in a windy or drafty area, use both a larger diameter gas nozzle and a higher gas flow rate. The larger the nozzle size, the higher the permissible flow rate without causing turbulence. Larger nozzle sizes may restrict your visibility of the weld. You might also consider setting up a wind barrier to protect your welding from the wind, Figure 11-24.

Shielding Gas Cost

The price of the gas is not the only factor that must be considered when selecting a shielding gas for GMA welding. For example, using CO_2 will produce the most spatter and the average efficiency will be about 93%. Using a 75% argon and 25% CO_2 gas mixture results in somewhat less spatter and an efficiency of approximately 96%. A 98% argon and 2% oxygen mixture will produce even less spatter, and the average efficiency will be about 98%. The loss of filler metal in the weld

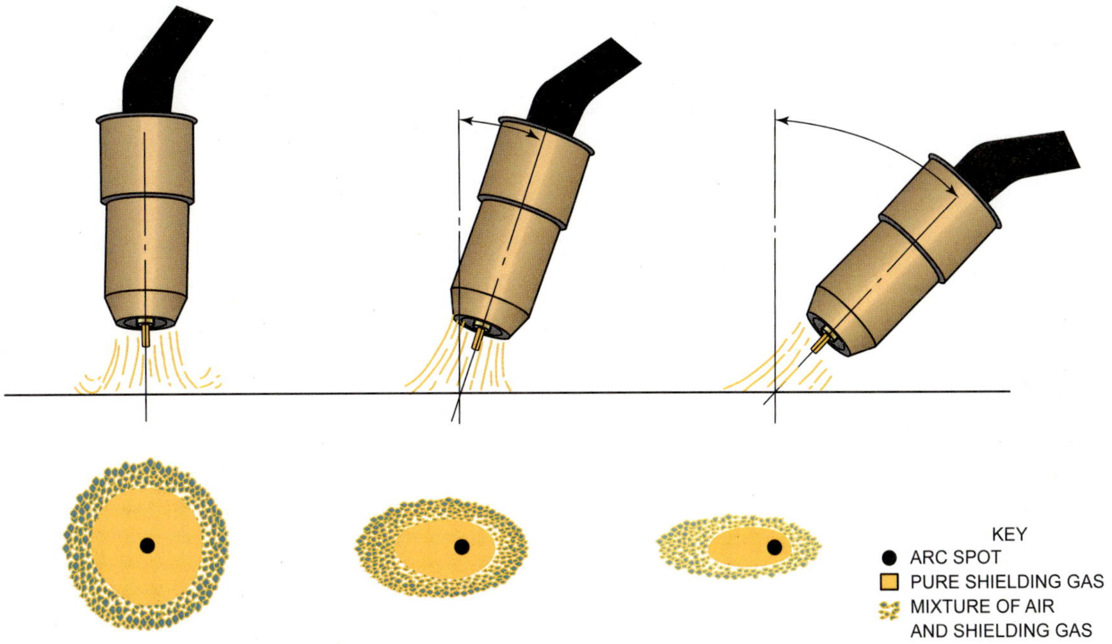

FIGURE 11-23 The welding gun angle affects the shielding gas coverage for the molten weld pool.

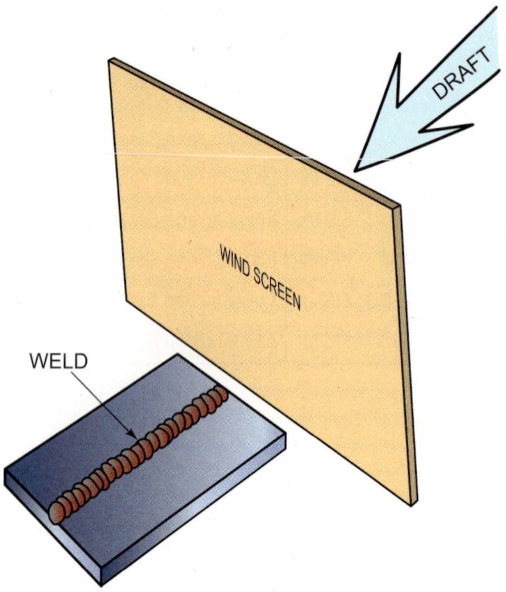

FIGURE 11-24 A wind screen can help prevent the shielding gas from being blown away.
© Cengage Learning 2012

among the various shielding gases is not significant. The significance is that the metal lost becomes mostly weld spatter, and if the welded part has to be cleaned of spatter, that cost can be significant.

Another cost factor is weld speed, and it is affected by the shielding gas. Both **argon-oxygen** and **argon-CO_2** blends can be used for the faster process of axial spray arc metal transfer, whereas straight CO_2 cannot. Axial spray arc provides the highest metal transfer rate and that means higher productivity. But that is only a benefit if you are welding on thick sections that can withstand the higher heat of this transfer method. In other words, there may not be a savings using the mixed gases if you are not welding on thick stock.

WELD METAL TRANSFER METHODS

The **GMAW** process is unique in that there are several modes of transferring the filler metal from the wire to the weld. Each mode of metal transfer has its own unique characteristics. The mode of metal transfer is the mechanism by which the molten filler metal is transferred across the arc to the base metal. The modes of metal transfer are short-circuit transfer (**GMAW-S**), axial spray transfer, globular transfer, and pulsed-arc transfer (**GMAW-P**). To change from one metal transfer mode to another, all you have to do is make the necessary changes in the voltage and amperage settings. In other words, it is the equipment setup and not the equipment or filler metal that determines the GMAW mode of metal transfer. Therefore, it is possible to make welds with each of the metal transfer modes by simply changing the voltage and amperage without even having to change the type of filler wire. In some cases, it may be necessary to change the shielding gas; for example, CO_2 cannot be used to make welds in all three transfer modes, but $Ar + O_2$ can.

Selecting the mode of metal transfer used depends on the welding power source, the wire electrode size, the type and thickness of material, the type of shielding gas used, and the best welding position for the task.

Short-Circuit Transfer—GMAW-S

In short-circuit metal transfer, the electrode actually momentarily comes in contact with the base metal surface, and the arc is momentarily shorted out. That is why it is called short-circuit metal transfer because at the moment that the electrode comes in contact with the surface, it electrically shorts out the arc and that is when a small drop of molten metal is transferred to the weld.

NOTE: **The terms *short* or *short circuit* is often confused with the terms for an open or broken circuit. When someone flips on a light switch and the light does not come on, it is incorrect to say that it must be shorted. If the light did not come on, you have an open circuit. An open circuit is one that does not have any current flow. However, in a short circuit all the possible power available tries to flow. When two wires bump together creating a shower of sparks, you have a short. So, if you switch on the light and a shower of sparks flies from the light fixture, you would be correct in saying you have a short. When you turn a switch on and nothing happens, you have an open circuit; if sparks fly, you have a short circuit; and if your light comes on, you have a closed circuit.**

The following description explains the short-circuit metal transfer process in detail. As the gun trigger is depressed, the wire feed unit is energized. An arc is established between the electrode and the base metal surface, forming a small molten weld pool. The wire electrode is fed out faster than the arc can melt the end away. This results in the electrode advancing and bridging the gap between the contact tube and the

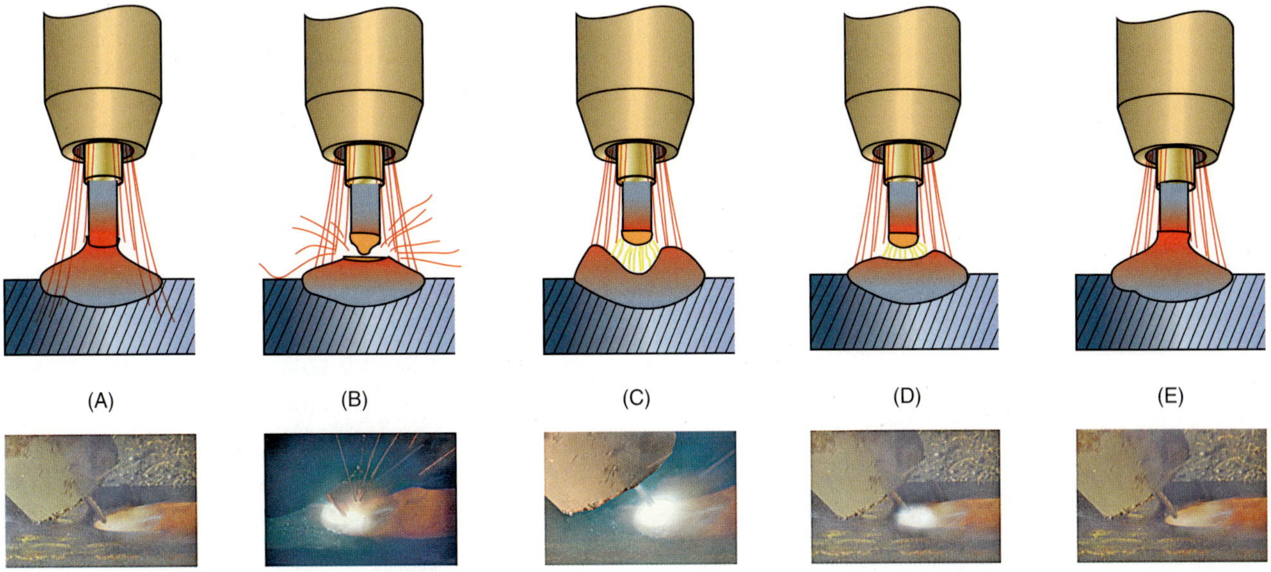

FIGURE 11-25 Schematic of short-circuit transfer. © Cengage Learning 2012

molten spot on the base metal. When the electrode touches the base metal, it shorts out, and the arc is extinguished. The short circuit becomes so hot that the electrode melts in two, and the arc is reestablished. The small piece of the electrode that shorted out into the molten weld pool is left in the molten weld pool as the arc reforms. The surge of current when the short is cleared causes the arc to melt the end of the electrode back up toward the contact tube, forming a small ball on the end of the wire. Once the momentary surge of current dissipates, the welding current cannot melt the wire fast enough, and the wire touches the molten weld pool and again shorts out. This short-circuit process is repeated approximately 20 to 200 times per second, each time transferring the small molten ball and the end of the wire to the base metal, Figure 11-25.

A properly functioning short-circuit weld makes a smooth frying sound.

The short-circuit mode of transfer is the most common process used with GMAW. It is used on thin or properly prepared thick sections of material, and when joining thick to thin materials. GMAW-S can be used with a wide range of electrode diameters and shielding gases.

The 0.023-, 0.030-, 0.035-, and 0.045-diameter (0.6-, 0.8-, 0.9- and 1.2-mm) wire electrodes are commonly used for the manual short-circuit welding mode. The common shielding gases used on carbon steel are 100% carbon dioxide, a blend of 25% carbon dioxide and 75% argon, or a blend of argon with a trace of oxygen.

All welding positions can be used for short-circuit transfer. It is the most widely used of the metal transfers for general repairs to mild steel. The amperage range may be from as low as 35 for materials of 24 gauge, to amperages of 225 for materials up to 1/8 in. (3 mm) in thickness on square groove weld joints. Thicker base metals can be welded if edges are properly prepared and cut at an angle (beveled) to provide a complete joint weld penetration.

SPARK YOUR IMAGINATION

The speed at which the short-circuit metal transfer method occurs is so fast that it cannot be seen. It literally happens in less time than the blink of your eye. List some other things that happen so fast that they may look like they are standing still or you cannot see a single part.

Globular Transfer

Globular transfer is generally used on thin materials and at a very low current range. In globular transfer, the arc melts the end of the electrode, forming a molten ball of metal. When the ball of metal becomes large enough, its surface tension cannot hold it onto the end of the wire. It falls across the arc, landing in the molten weld pool. Because there is little control over where the glob of metal lands and the weld pool tends to be a small landing target, this process is rarely used by itself. It is used more commonly in combination with pulsed spray transfer. With this

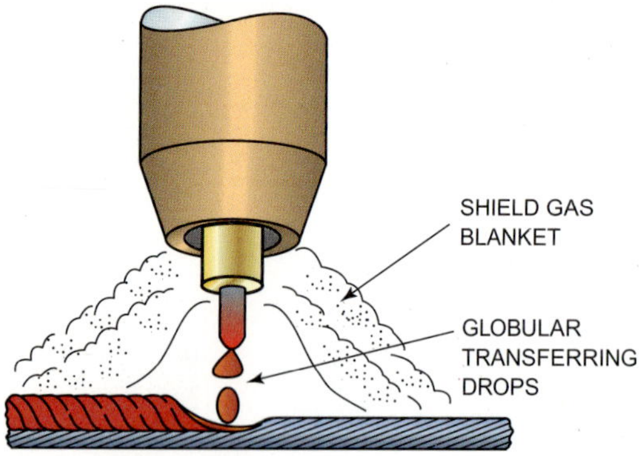

(A)

(B)

FIGURE 11-26 Globular transfer.

combination of processes, the molten weld pool is larger and the glob is more likely to land in the molten pool, Figure 11-26.

Axial Spray Transfer

The GMA **axial spray metal transfer** mode uses the highest voltage and amperage settings compared to other processes. Often this process is referred to simply as **spray transfer.** The current setting is high enough so that the end of the electrode is melted away rapidly and only small molten droplets are formed, Figure 11-27. These small droplets are transferred down the center axis of the arc to the molten weld pool. Axial spray transfer is a popular process used in manufacturing where high-deposition rates are required and deep weld joint penetration is desired. On ferrous metals it can use a blend of 95 to 98% argon and 2 to 5% oxygen. The added percentage of oxygen allows greater weld penetration. On nonferrous metals, 100% argon or blends of argon with helium or nitrogen are used.

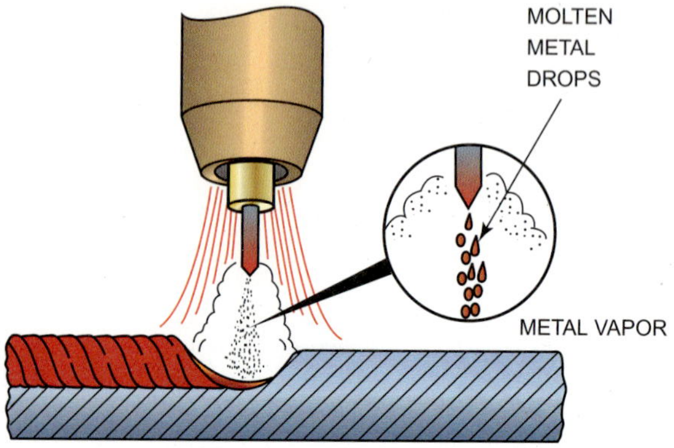

(A)

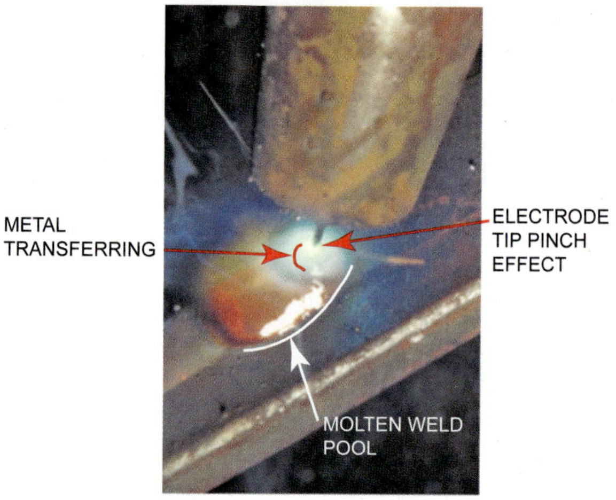

(B)

FIGURE 11-27 Axial spray metal transfer. Note the pinch effect of filler wire and the symmetrical metal transfer column.

This process generally uses larger diameter wire electrodes, so it requires a higher amperage range. The higher the amperage range, the faster the weld bead progresses and the joint filled. Increased current flow combines with the high percentage of argon, causing a **pinch effect** on the molten ball of wire electrode much like the effect that pinching has on a rubber water hose. When you pinch the end of the rubber water hose, the water exits the hose in a spray pattern. That is what happens to the end of the electrode in axial spray transfer.

The axial spray transfer process is very hot and virtually free of any spatter. The sound produced by axial spray transfer is a quiet, hissing sound, unlike the short-circuit process that makes a raspy, frying sound. A disadvantage of axial spray transfer is that it produces a very fluid weld pool that may be difficult to control in out-of-position welds. The high welding current produces a great deal of UV light, so

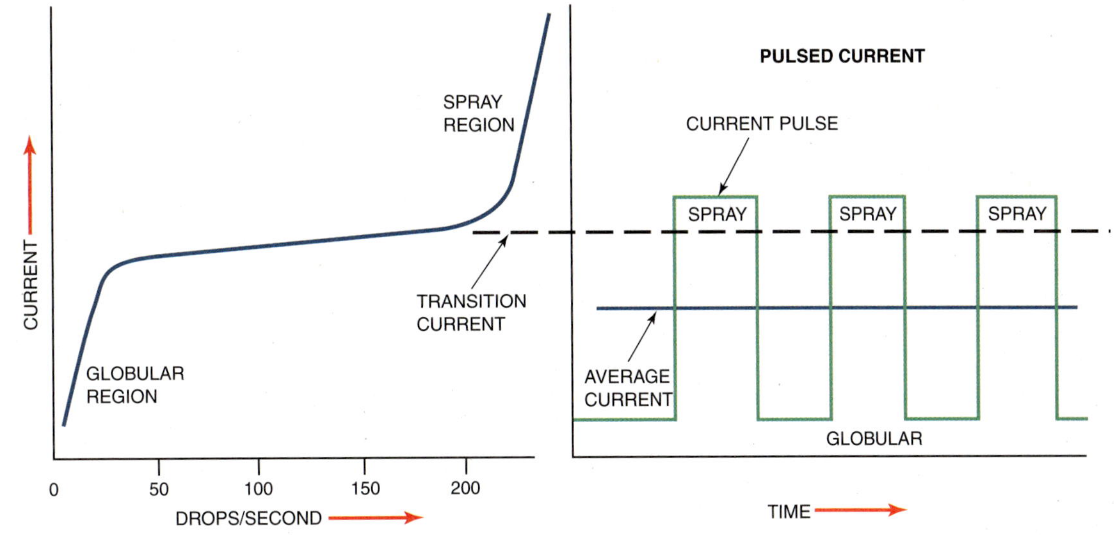

FIGURE 11-28 **Mechanism of pulsed-arc spray transfer at a low average current.**

you need extra burn protection for your eyes, hands, and arms.

> **SPARK YOUR IMAGINATION**
>
> The end of the GMA welding electrode turns into a stream of liquid droplets in the axial spray transfer method. When something forms into tiny liquid droplets, it is referred to as having been atomized. What are some other atomized liquids we see or use? What does the word *vaporized* mean and how does it differ from atomized?

> **CAUTION**
>
> A darker filter lens is required when using the spray transfer mode.

Pulsed-Arc Transfer—GMAW-P

The current produced by the **pulsed-arc transfer** mode (GMAW-P) is a dual-pulsed current. One pulse of high current is for the axial spray transfer mode, and the other lower pulse of current is for the globular transfer mode. This pulsing of the current levels permits the use of the high-amperage axial spray transfer mode and the low-amperage globular transfer mode to be used together so that the total heat input to the weld is lower. The high current of the spray mode produces good penetration and fusion, and the low current of the globular transfer allows the weld pool to cool and contract slightly, so that it is easier to control.

The lower total current and heat input provided by the pulsed-arc transfer permits its use on thinner-gauge base metals. It is also easier to use for out-of-position welding than the straight axial spray process. Pulsed arc produces less spatter and higher weld speeds than does the short-circuit metal transfer. The graph in Figure 11-28 shows the relationship between welding current, deposition, and the time for the pulses of a typical pulsed-arc transfer.

GMAW ELECTRODES

The electrode used with GMAW is a very long coil of solid wire. It is a filler wire electrode and is continuously fed off a reel attached to the wire feed unit. A series of drive rolls, guide tubes, a conduit liner, and a contact tube make the delivery of the electrically charged electrode seem effortless. When the electrode makes contact with the base metal, an arc is established, and the end of the electrode melts to produce the weld bead.

Although there is no flux or slag associated with GMA welding, you will occasionally notice an intermittent brown, tan, or black glass-like substance on the surface of the solidified weld bead. These are impurities that have been floated to the weld surface as they combine with the deoxidizers in the electrode. Deoxidizers, such as manganese and silicone, can be added to the electrode wire in small amounts to help remove impurities in or on the metal being welded.

The glass-like substance can pop off the weld bead as it cools and does not cause any damage to the weld surface.

CAUTION

Always wear your safety glasses when looking at the surface of the cooling weld bead.

Electrode Diameters

Wire electrodes are produced with diameters of 0.023, 0.030, 0.035, 0.045, and 0.062 (0.6, 0.8, 0.9, 1.2, and 1.6 mm). Other larger diameters are available for production work and can include sizes such as 5/64, and 7/64 in. (2.0 and 2.8 mm).

Some wire electrodes have a thin copper coating. This coating provides some protection to the electrode, preventing it from rusting; and improves the electrical contact between the wire electrode and the contact tube. Electrodes with this coating may look like copper wire because of the very thin copper cladding. The amount of copper is so thin that it either burns off or is diluted into the weld pool with no significant reaction.

Electrode Cast and Helix

The electrode wire is wound (rolled) on spools, reels, or coils made of formed metal, wood, pressed fiber, or plastic, Figure 11-29. All of them are collectively called the electrode package. Sometimes drums are used, with the wire coiled inside of the drum.

The **cast** is the diameter of the circle formed by an unrestricted coil of wire. The **helix** is the twist in the wire. To see the cast and helix of a GMA welding wire, snip off several feet of electrode. Let it fall to the floor and observe how it forms a circle. One complete circle or the diameter across the circle is known as the cast of the wire. You will notice that the wire electrode does not lie flat and one end will be slightly higher than the other. This is caused by the twist or helix of the wire. The helix is measured in height and is the maximum height from the flat surface to the highest point of the wire from the flat surface.

The manufacturer purposely puts the cast and helix in the wire as part of the manufacturing process. The slight bend of the cast in the electrode wire ensures a positive contact with the energized contact tube. The helix twists the wire, causing a rubbing effect on the inside of the contact tube. If the wire as not twisted, the contact tube would quickly wear a groove where the cast caused it to rub, but by twisting the wire, a more uniform wear pattern occurs inside the contact tube, Figure 11-30.

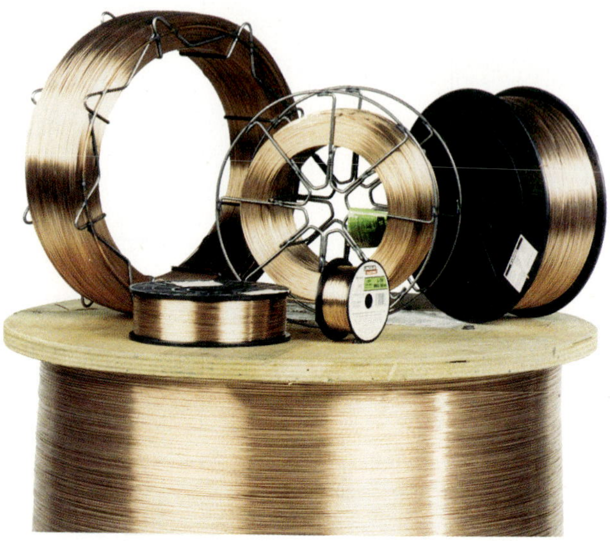

FIGURE 11-29 Filler metals are available in a variety of package types, sizes, and weights.
The Lincoln Electric Company

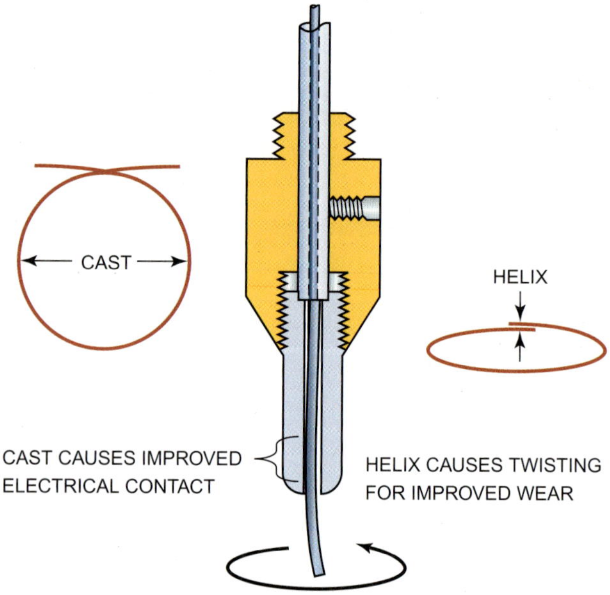

FIGURE 11-30 The cast of the welding wire causes it to rub firmly inside of the contact tube for good electrical contact. The helix causes the electrode to twist inside of the contact tube so that the tube is worn uniformly.
© Cengage Learning 2012

Electrode Handling

Wire electrodes may be wrapped in plastic sealed bags, wrapped in a special paper, or sealed in cans or cardboard boxes to protect the electrode from the elements. A small bag of a moisture-absorbing material, crystal desiccant, is sometimes placed in the shipping containers. This material is included in the packing to protect the wire electrode from rust-causing moisture.

Mishandling of the electrode wire can result in wire feed problems and in some cases cause weld contamination. The wire can become tangled if the electrode package is thrown, dropped, or mishandled. Oxidation or even rusting may occur if the electrode is stored in a damp location. Always keep the electrode wire dry, and handle it carefully.

GMAW ELECTRODE CLASSIFICATION

The American Welding Society (AWS) has a standardized method of identifying GMA welding electrodes. This standard uses a series of letters and numbers to group filler metals into specific classifications. The AWS specification is for the chemical and physical properties of the weld produced by the filler metal and not the specific composition of the wire. This allows manufacturers to make slight changes in the electrode composition as long as the weld produced with the electrode meets the group specifications.

Most manufacturers of filler metals have trade names unique to their products. A comparison chart is available from each manufacturer that lists its product names as they relate to specific AWS-numbered electrodes. These charts are helpful when it is necessary to make sure that a particular wire meets a code or standard or when you are changing from one supplier to another.

Carbon Steel and Low-Alloy Wire Electrodes

ER70S-2 is an example of the American Welding Society's classification system for mild steel GMAW filler metals. The *E* in the prefix shows that the wire is classified as an electrode. The *R* indicates that this particular filler metal could also be provided as a cut length or rod for GTA or OF welding.

The 70 that follows the *ER* designates the minimum tensile strength of the deposited weld metal in thousands. For example, ER70 indicates that the weld metal, as deposited, will have a minimum tensile strength of 70,000 pounds per square inch.

The *S* that follows the ER70 indicates that this electrode is a solid wire electrode. The letter *T* means that it is tubular, and those wires are used for flux cored arc welding (FCAW).

The last position in the example ER70S-2 consists of a number or letter ranging from 1 to 7, or the letter *G*. This number and letter system indicates the electrode's filler metal composition and the manufacturer's recommendations for the current setting and shielding gas.

GMAW electrodes come in a variety of packaging sizes—reels, spools, or drums. The weight of each of these packages varies depending on the type of filler metal. For example, a 7-in. (178-mm) reel of steel filler wire weighs a lot more than a 7-in. (178-mm) reel of aluminum filler wire.

The following are examples of the various types of carbon steel and low-alloy steel filler metals that are available.

ER70S-2

With ER70S-2, the 2 indicates that the electrode is a deoxidizing mild steel filler wire and can be used on metal that has a light cover of rust or oxide. This electrode is a general-purpose electrode that can be used in all weld positions. The weld pool is not very fluid, making this electrode filler a good choice for out-of-position welding or when short arc metal transfer is selected. Shielding gases such as CO_2, Ar/CO_2, and Ar/O_2 gas blends may be used with this wire electrode.

ER70S-3

With ER70S-3, the 3 indicates that this electrode does not have the deoxidizers needed for welding over a metal surface that has rust or mill scale on it. The weld area must be ground sanded or sandblasted clean. It is recommended for use on thinner steels of low to medium carbon content where the tensile strength will exceed the strength of the base metal. It maintains a degree of ductility in the weld. The weld pool is more fluid than ER70S-2, but the electrode can still be used in all positions. ER70S-3 can be used with either CO_2, Ar/CO_2, or Ar/O_2 shielding gases.

ER70S-4

With ER70S-4, the 4 indicates that the wire electrode contains a higher level of a deoxidizer like silicone than ER70S-3. This electrode is commonly used when welding on structural steel, pipe systems, ships, and in many metal fabricating job shops. Metal transfer can be accomplished by either the short-circuit axial spray, or with pulsed-arc transfer methods. The ER70S-4 electrode performs well in all welding positions, but it is difficult to weld out of position when using the axial spray transfer method. Under the same conditions, the weld bead will be flatter and slightly wider than beads made with other fillers. This electrode also works well with CO_2, Ar/CO_2, or Ar/O_2 shielding gases.

ER70S-5

With ER70S-5, the 5 indicates that this electrode is used in the flat position only. The weld pool will be extremely fluid. You can weld over mill scale or rusty surfaces and maintain weld quality with this electrode. Short-circuit weld metal transfer should not be used. This filler electrode is excellent for axial spray arc metal transfer with a larger diameter wire electrode used in the flat position on heavy or thick materials. ER70S-5 has very good deoxidizing characteristics and works best with Ar/O_2 or Ar/CO_2 shielding gases.

ER70S-6

With the ER70S-6, the 6 indicates that this is an electrode with the highest levels of manganese and silicone for strength and deoxidation. This electrode is used on thick or thin sections. It works well over rusty surfaces, mill scale or areas that cannot be easily cleaned, and light-gauge metals used in the automobile industry. It can be used in the short-circuit metal transfer method in all positions. ER70S-6 welds a smooth bead with a uniform appearance. It can be used with CO_2, Ar/CO_2, or Ar/O_2 shielding gases.

ER70S-7

With the ER70S-7, the 7 indicates that this is a high manganese carbon steel electrode with a balanced level of silicone. This electrode can be used over a wide range of welding parameters in all positions. It can be used over moderate amounts of rust and mill scale. ER70S-7 welds have excellent mechanical properties, and the electrode is widely used in the automotive industry, for welding heavy equipment and with robotic welding. It can be used with CO_2 or Ar/CO_2 shielding gases.

ER70S-G

With the ER70S-G, the *G* indicates that this is a nonstandard electrode. That means that different manufacturers can produce this electrode with special properties to meet an industry's need. You must refer to that manufacturer's literature to determine the properties, applications, and shielding gas requirements for its product. Manufacturers of electrodes in this classification usually have a trade term for their product.

ER80S-Ni1

The addition of nickel (Ni) to this electrode increases its strength and atmospheric corrosion-resistance for welding on weathering steels.

ER80S-D2

The addition of molybdenum provides strength after the weldment has been postweld heat-treated. This is an ideal filler wire for use on high-temperature piping, flanges, and pressure vessels, all of which may be postweld stress-relieved by some type of heat treatment.

Stainless Steel Wire Electrodes

The AWS specification for stainless steel bare wire electrodes and welding rods is A5.9. ER308L is an example of a GMAW filler metal electrode that uses the A5.9 system for stainless steel. As with the mild steel electrodes, the AWS uses the *ER* prefix. Following the prefix, the American Iron and Steel Institute's (AISI) three-digit stainless steel number for the specific alloy is used. This number indicates the type of stainless steel in the filler metal. The letter *L* may be added to the right of the AISI number to indicate a low-carbon stainless welding electrode. The letters *Si* may be added to indicate the addition of silicone deoxidizer.

The following are examples of the various types of stainless steel filler metals that are available.

ER308L and ER308LSi

The ER308L and ER308LSi filler metals can be used to weld on all 18-8-type stainless steels such as 301, 302, 302B, 303, 304, 305, 308, 201, and 202. The 308 stainless steels are used for food or chemical equipment, tanks, pumps, hoods, and evaporators.

ER309L and ER309LSi

The ER309L and ER309LSi filler metals can be used to weld on 309 stainless steel or to join mild steel to any 18-8-type stainless steel. The 309 stainless steels are used for high-temperature service, such as furnace parts and mufflers.

ER316L and ER316L-Si

The ER316L and ER316L-Si filler metals are used for welding tubing, chemical pumps, filters, tanks, and furnace parts. Molybdenum is added to improve these properties and to resist corrosive pitting. All E316 filler metals can be used on 316 stainless steels or when weld-resistance to pitting is required. The 316 stainless steels are used for high-temperature service when high strength with low creep is desired.

Aluminum and Aluminum Alloy Wire Electrodes

The AWS specification for aluminum and aluminum alloy filler metals is A5.10 for bare welding rods and electrodes. Filler metals classified within A5.10 use the prefix *ER* with the Aluminum Association number for the alloy.

ER1100

The ER1100 filler metal can be used to weld 1100- and 3003-grade aluminum. The filler wire is relatively pure. ER1100 produces welds that have good corrosion-resistance and high ductility with tensile strengths ranging from 11,000 to 17,000 psi. The weld deposit has a high-resistance to cracking during welding. The 1100 aluminum has the lowest percentage of alloying agents of all of the aluminum alloys, and it melts at 1215°F (657°C). It is commonly used for items such as food containers, food processing equipment, storage tanks, and heat exchangers.

ER4043

The ER4043 filler metal is a general purpose welding filler metal. It has 4.5 to 6.0% silicone added, which lowers its melting temperature to 1155°F (624°C). The lower melting temperature helps promote a free-flowing molten weld pool. The welds have high ductility and a high-resistance to cracking during welding. ER4043 can be used to weld on the following rolled or drawn alloys—2014, 3003, 3004, 4043, 5052, 6061, 6062, 6063—and these cast alloys: 43, 355, 356, and 214.

ER5356

The ER5356 filler metal has 4.5 to 5.5% magnesium added to improve the tensile strength. The weld has high ductility but only an average resistance to cracking during welding. ER5356 can be used to weld on the following rolled or drawn alloys: 5050, 5052, 5056, 5083, 5086, 5154, 5356, 5454, and 5456.

ER5556

The ER5556 filler metal has 4.7 to 5.5% magnesium and 0.5 to 1.0% manganese added to produce a weld with high strength. The weld has high ductility and average resistance to cracking during welding. ER5556 can be used to weld on the following rolled or drawn alloys: 5052, 5083, 5356, 5454, and 5456.

EQUIPMENT SETUP

The key to making quality GMA welds is the equipment setup. There are many setup variables such as voltage, current, wire size, travel speed, gas flow, and others that can affect weld quality. It is deceptively easy to make a weld that looks good but in fact has significant defects, Figure 11-31. Slight changes in setup procedures can produce major defects such as excessive

FIGURE 11-31 Weld separated from the plate; there is no fusion between the weld and plate.

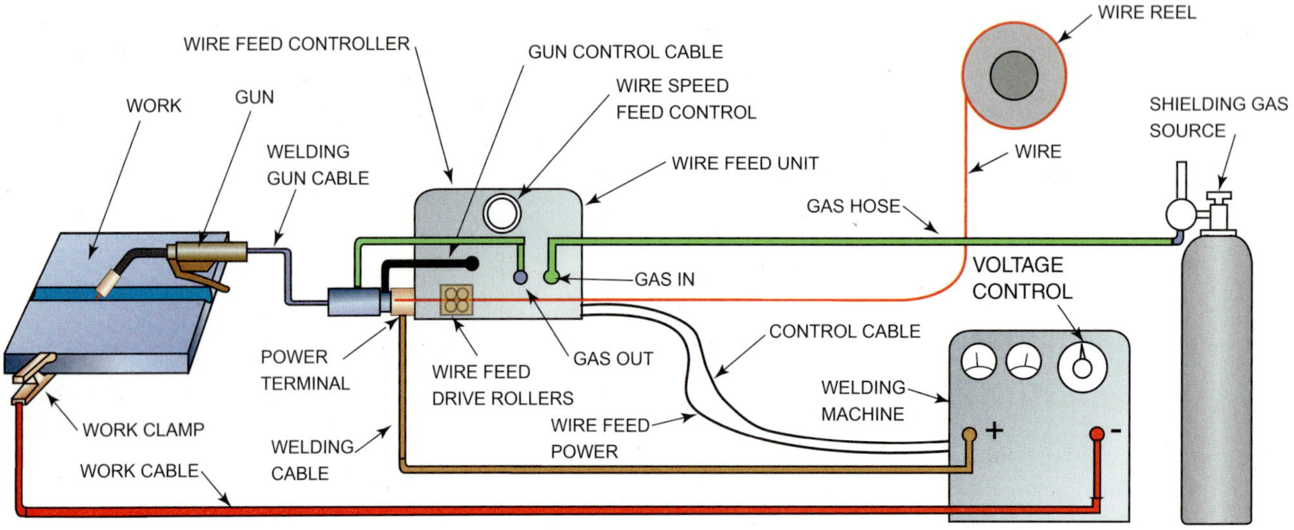

FIGURE 11-32 **Schematic layout and part identification of a GMA welding setup.** Larry Jeffus

spatter, undercut, overlap, porosity, and poor weld bead contours. Setup becomes even more important for out-of-position welds. Making quality vertical and overhead welds can be difficult for a student welder with a properly setup system, but it becomes impossible with a system that is out of adjustment.

Learning to set up and properly adjust the GMA welding system allows you to produce high-quality welds at a high level of productivity.

GMAW is set up and manipulated in a similar manner to FCAW. The results of changes in electrode extension, voltage, amperage, and torch angle are essentially the same with both processes.

Although every manufacturer designs its GMA welding equipment differently, the equipment is all set up in a similar manner. It is always best to follow the specific manufacturer's recommendations regarding setup as provided in the equipment literature. You will find, however, that in the field, the manufacturer's literature is not always available for the equipment you are asked to use. It is, therefore, important to have a good general knowledge and understanding of the setup procedure for GMA welding equipment. Figure 11-32 shows all of the various components that make up a GMA welding station.

PRACTICE 11-1

GMAW Equipment Setup

For this practice, you need a GMAW power source, welding gun, electrode feed unit, electrode supply, shielding gas supply, shielding gas flowmeter regulator, electrode conduit, power and work leads, shielding gas hoses, assorted hand tools, spare parts, and any other required materials. In this practice you will properly set up a GWA welding installation.

If the shielding gas supply is a cylinder, it must be chained securely in place before the valve protection cap is removed, Figure 11-33. Standing to one side of the cylinder, quickly crack the valve to blow out any dirt in the valve before the flowmeter regulator is attached, Figure 11-34. Attach the correct hose from the regulator to the "gas-in" connection on the electrode feed unit or machine.

FIGURE 11-33 **Make sure the gas cylinder is chained securely in place before removing the safety cap.** © Cengage Learning 2012

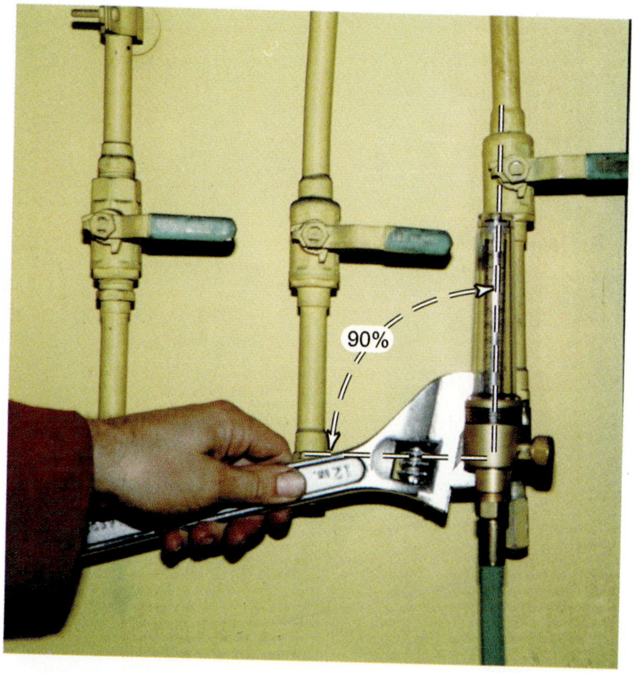

FIGURE 11-34 Attach the flowmeter regulator. Be sure the tube is vertical. Larry Jeffus

Install the reel of electrode (welding wire) on the holder and secure it. Check the feed roller size to ensure that it matches the wire size. The conduit liner size should be checked to be sure that it is compatible with the wire size. Connect the conduit to the feed unit. The conduit or an extension should be aligned with the groove in the roller and set as close to the roller as possible without touching. Misalignment at this point can contribute to a bird's nest. **Bird nesting** of the electrode wire results when the feed roller pushes the wire into a tangled ball because the wire would not go through the out feed side conduit, with the wire resembling a bird's nest.

Be sure the power is off before attaching the welding cables. The electrode and work leads should be attached to the proper terminals. The electrode lead should be attached to the electrode or positive (+). If necessary, it is also attached to the power cable part of the gun lead. The work lead should be attached to work or negative (–).

The shielding "gas-out" side of the solenoid is then also attached to the gun lead. If a separate splice is required from the gun switch circuit to the feed unit, it should be connected at this time. Check to see that the welding contractor circuit is connected from the feed unit to the power source.

The welding gun should be securely attached to the main lead cable and conduit. There should be a gas diffuser attached to the end of the conduit liner to ensure proper alignment. A contact tube (tip) of the correct size to match the electrode wire size being used should be installed. A shielding gas nozzle is attached to complete the assembly.

Recheck all the fittings and connections for tightness. Loose fittings can leak; loose connections can cause added resistance, reducing the welding efficiency. Some manufacturers include detailed setup instructions with their equipment, Figure 11-35.

Complete a copy of the performance qualification test record in Appendix IV or use form provided by your instructor.◆

PRACTICE 11-2

Threading GMAW Wire

Using the GMAW machine that was properly assembled in Practice 11-1, you will turn the machine on and thread the electrode wire through the system.

Check to see that the unit is assembled correctly according to the manufacturer's specifications. Switch on the power and check the gun switch circuit by depressing the switch. The power source relays, feed relays, gas solenoid, and feed motor should all activate.

Cut the end of the electrode wire free. Hold it tightly so that it does not unwind. The wire has a natural curve that is known as its cast. The cast is measured by the diameter of the circle that the wire would make if it were loosely laid on a flat surface. The cast helps the wire make a good electrical contact as it passes through the contact tube. However, the cast can be a problem when you thread the system. To make threading easier, straighten about 12 in. (305 mm) of the end of the wire and cut off any kinks.

Separate the wire feed rollers, and push the wire first through the guides, then between the rollers, and finally into the conduit liner. Reset the rollers so there is a slight amount of compression on the wire. Set the wire feed speed control to a slow speed. Hold the welding gun so that the electrode conduit and cable are as straight as possible.

Press the gun switch. Pressing the gun switch to start the wire feeder is called **triggering** the gun. The wire should start feeding into the liner. Watch to make certain that the wire feeds smoothly, and release the gun switch as soon as the end comes through the contact tube.

FIGURE 11-35 Example of manufacturer's setup instructions. The Lincoln Electric Company

> **CAUTION**
>
> If the wire stops feeding before it reaches the end of the contact tube, stop and check the system. If no obvious problem can be found, mark the wire with tape and remove it from the gun. It then can be held next to the system to determine the location of the problem.

With the wire feed running, adjust the feed roller compression so that the wire reel can be stopped easily by a slight pressure. If the roller pressure is to tight, it causes the wire to feed erratically. Too high a pressure can turn a minor problem into a major disaster. If the wire jams at a high roller pressure, the feed rollers keep feeding the wire, causing it to bird nest and possibly short out. With a light pressure, the wire can stop, preventing bird nesting. This is very important with soft wires. The other advantage of light pressure is that the feed will stop if something like clothing or gas hoses get caught in the reel.

With the feed running, adjust the spool drag so that the reel stops when the feed stops. The reel should not coast to a stop because the wire can easily snag. Also, when the feed restarts, a jolt occurs when the slack in the wire is taken up. This jolt can be enough to momentarily stop the wire, possibly causing a discontinuity in the weld.

When the test runs are completed, the wire can either be rewound or cut off. Some wire feed units have a retract button. This allows the feed driver to reverse and retract the wire automatically. To rewind the wire on units without this retract feature, release the rollers and turn them backward by hand. If the machine does not allow the feed rollers to be released without upsetting the tension, you must cut the wire.

Complete a copy of the performance qualification test record in Appendix IV or use form provided by your instructor.◆

CAUTION

Do not discard pieces of wire on the floor. They present a hazard to safe movement around the machine. In addition, a small piece of wire can work its way into a filter screen on the welding power source. If the piece of wire shorts out inside the machine, it could become charged with high voltage, which could cause injury or death. Always wind the wire tightly into a ball or cut it into short lengths before discarding it in the proper waste container.

SUMMARY

The speed, efficiency, and quality of welds produced with the GMA welding process make it the first choice for most welding fabrication. The ability to make long, uninterrupted, high-quality welds has significantly reduced the time and cost to manufacture weldments. In addition, you can make changes in the machine settings and shielding gases to permit you to weld on a wide range of metal thicknesses.

Properly setting up and adjusting the GMA welding equipment is the key to producing quality welds. Once you have mastered these skills, the only remaining obstacle to your producing consistent, uniform, high-quality welds is your ability to follow or track the joint consistently. Some welders find that lightly dragging their glove along the metal surface or edge of the fabrication can aid them in controlling the weld consistency.

Welder fatigue can become a problem when making long welds. To help avoid fatigue you should find a comfortable welding position that you can maintain for several minutes.

REVIEW QUESTIONS

1. Why is GMA welding called a semiautomatic process?
2. What is the purpose of the shielding gas during the GMA welding process?
3. What are some other terms that gas metal arc welding is referred to as?
4. What are the advantages of GMA welding over conventional electrode type arc (stick) welding?
5. What is the power source commonly referred to as?
6. What are three terms that refer to a measurement of electrical pressure?
7. What three terms refer to the total number of electrons flowing?
8. List the components for the welding current transferred through from welding machine to the gun assembly and back to the welder.
9. What is the purpose of the wire feeder?
10. What is burn back?
11. What are the four major types of wire feed systems?
12. What is the purpose of the groove in a wire feed roller?
13. What are knurled wire feed rollers and when should they be used?
14. What types of wire feed rollers should be used with soft wires?
15. How do you know what size of filler wire to use with a wire feed roller?
16. What type of wire feed system would be best to use when moving soft wire over a great distance?
17. Describe the push-pull-type wire feed system.
18. What is the purpose of the feed rollers?

19. What can happen if the feed roller tension is too light?
20. What can happen if the feed roller tension is too tight?
21. What are the three main parts of the GMAW gun?
22. What happens when the trigger is pressed on a GMAW gun?
23. What is the purpose of the gas diffuser on a GMAW gun?
24. Why is it important for the contact tube to be sized properly for the filler wire being used?
25. How does the size of the nozzle diameter affect the welding process?
26. How is GMA welding affected by changes in the arc voltage?
27. What are some advantages of having the wire feed assembly separate from the welding machine?
28. GMAW shielding gas can be supplied to the welding station by what two methods?
29. What are the most commonly used shielding gases?
30. What factors should you consider when selecting a shielding gas or blend?
31. What are inert gases?
32. How does argon gas shield the weld?
33. What other gases can be blended with argon?
34. Why does helium require a higher flow rate than argon?
35. What are advantages of using carbon dioxide (CO_2) as a shielding gas for GMA welding of steels?
36. What effect would adding nitrogen to a gas blend have?
37. How do you choose the correct flow rate?
38. What cost significance can weld spatter have?
39. What are four modes of metal transfer?
40. What factors should you consider when selecting the mode of metal transfer?
41. Describe the short-circuit metal transfer process.
42. What is the most common mode of transfer used with GMAW?
43. Describe the globular transfer mode of transfer.
44. Which mode of transfer uses the highest voltage and amperage settings?
45. How do higher voltage and amperage settings affect the transfer process?
46. Describe the pulsed-arc transfer mode.
47. Why is it important to wear safety glasses when looking at the surface of a cooling weld bead?
48. What kind of coating is on some wire electrodes, and what is its purpose?
49. What are the purposes of the cast and the helix that are manufactured into the wire?
50. What problems can occur if the electrode wire has been mishandled?
51. What information is provided by the series of letters and numbers that AWS uses to identify welding electrodes?
52. Identify what each of the letters and numbers represents in the following example of a carbon steel and low-alloy wire electrode: ER70S-2.
53. Identify what each of the letters and numbers represents in the following example of a stainless steel wire electrode: ER308L.
54. What equipment setup variables can affect the quality of a weld?

Chapter 12

Gas Metal Arc Welding

OBJECTIVES

After completing this chapter, the student should be able to:

- Explain the relationship between the wire feed speed and the amperage when GMA welding.
- Demonstrate how to do a wire feed speed test.
- Describe the effect that electrode manipulation has on the weld.
- Describe how changes in electrode extension affect the weld bead.
- Demonstrate the safe way to prepare a GMA welding workstation for welding.
- Describe how changing the welding gun angle changes a weld bead.
- Demonstrate how to make a tack weld.

KEY TERMS

backhand technique
depth of fusion
electrode extension
forehand technique
gun angle
joint following
joint visibility
limited visibility
perpendicular technique
postweld finishing
travel angle
travel speed
uniform bead shape
weld bead visibility
weld buildup
weld spatter
weld thickness
welding speed
wire feed speed
work angle

INTRODUCTION

Making a good gas metal arc (GMA) weld requires that you have both the welding skills and knowledge to properly set up the welding machine. The skills required to watch the molten weld pool and control its shape and penetration are similar to those needed for other welding processes. The knowledge to set up the voltage, amperage, electrode extension, and welding angle, as well as other factors, dramatically affect the weld quality that you produce. The very best welding conditions are those that allow you to produce the largest quantity of successful welds in the shortest period of time with the highest productivity.

The more cost efficient welders can be, the more competitive they and their companies become. This can make the difference between being awarded a bid or a job and having or losing work.

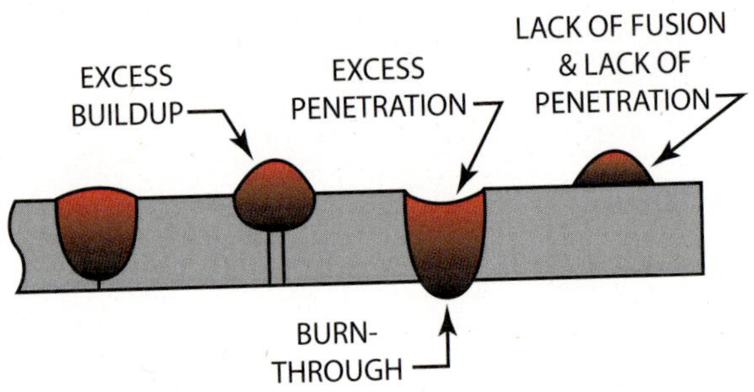

FIGURE 12-2 Weld bead defects. © Cengage Learning 2012

ARC VOLTAGE AND AMPERAGE CHARACTERISTICS

Setting up the arc voltage and amperage for GMA welding is different from most other welding processes. The voltage is set on the welder, and the amperage is set by changing the wire feed speed. At any one voltage setting the amperage required to melt the wire as it is fed into the weld will change. At any specific voltage it requires more amperage to melt the wire the faster it is fed because there is more metal that has to be melted. In other words, because the voltage remains constant in GMA welding, the amperage will rise automatically to match the faster electrode feed rate. Therefore, when the wire feed speed is slowed, the amperage required to melt the electrode at that slower rate will decrease proportionately.

The higher the wire feed speed, the higher the welding current. For the most part, as the current increases, so does the weld's temperature, and therefore, the weld bead penetration increases too, Figure 12-1 (new weld penetration). However, that is not always true for GMA welding. The wire feed speed can be set too high with so much weld metal being melted by the arc that weld penetration can actually decrease. The same thing can happen if the wire feed speed is set too low—weld penetration will decrease. The welder must set both the welding voltage and wire feed speed correctly to produce a satisfactory weld. Charts are available to provide you with the range of voltages and wire feed speeds (amperage) that work best for each size of filler metal and shielding gas combination.

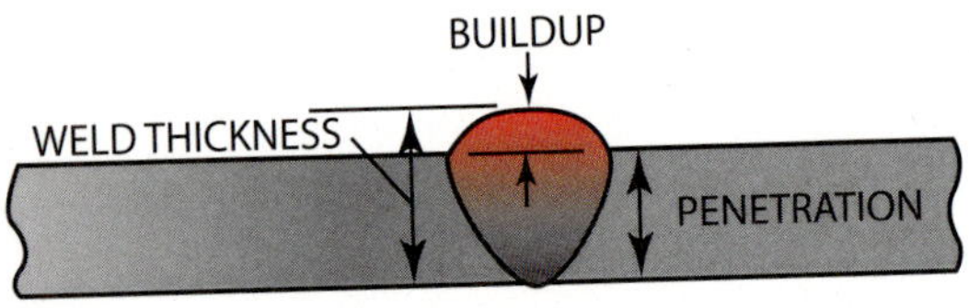

FIGURE 12-1 Weld bead identification. Larry Jeffus

The effect of having too high or too low an arc voltage is the opposite of having too high or too low a wire feed speed. That is because if the wire feed speed is too high, the voltage is too low; and if the wire feed speed is too low, then the voltage is too high. So, you could possibly correct too high a wire feed speed problem by increasing the voltage, but only if the hotter weld did not cause too much penetration and possibly a weld bead burnthrough. Too low a wire feed speed might be corrected by decreasing the arc voltage if that did not result in a loss of penetration or lack of fusion, Figure 12-2 (weld defects).

Wire Feed Speed

Because changes in the wire feed speed automatically change the amperage, it is possible to set the amperage by using a chart and measuring the length of wire fed per minute, Table 12-1. The **wire feed speed** is generally recommended by the electrode manufacturer and is selected in inches per minute (ipm), which can be measured by how fast the wire exits the contact tube. The welder uses a wire feed speed control dial on the wire feed unit to control the ipm.

Because at high wire feed speeds many feet of wire can be fed out during a full minute's wire feed speed test, a shorter time test is desirable. For example, in Table 12-1 the slowest wire feed speed for 0.030 spray arc would be 500 ipm, which is 41 ft (12.5 m) of wire per minute, and at the highest speed of 650 ipm, 54 ft (16.5 m) of wire would be fed per minute. There are two commonly used shorter-timed tests to reduce the wasting of electrode. One uses 15 secs, and the resulting

ELECTRODE DIAMETER INCH	SHORT-CIRCUIT ARC AMPS	VOLTS	WIRE FEED SPEED (IPM)	SPRAY METAL ARC AMPS	VOLTS	WIRE FEED SPEED (IPM)
0.023	45-**70**-90	14-**15**-16	150-**300**-380	100-**110**-125	23-**23**-25	400-**450**-620
0.030	60-**100**-140	14-**15**-16	150-**220**-350	160-**180**-200	24-**25**-26	500-**520**-650
0.035	90-**130**-160	15-**17**-19	180-**250**-300	180-**200**-230	25-**26**-27	400-**480**-550
0.045	130-**160**-200	17-**18**-19	125-**150**-200	260-**300**-340	25-**27**-30	300-**350**-500
0.052	150-**160**-200	17-**18**-20	135-**140**-190	275-**325**-400	26-**28**-33	265-**310**-390

RECOMMENDED GMA WELDING SETUP

BOLD values represent the optimum setting using DCEP and a shielding gas flow rate from 35 to 45 CFM.

Table 12-1 Recommended GMA Welding Setup

length is multiplied by 4 to determine the inches per minute. The second test uses 6 secs, and the resulting length is multiplied by 10 to determine the inches per minute. The 15-sec test is more accurate, but for most applications the 6-sec test is adequate.

To accurately measure wire feed ipm, snip off the wire at the contact tube. Squeeze the trigger for 6 secs, release it, and measure the length of wire. Multiply the length of the wire by 10. The result is how many inches of wire would be fed in a minute. Release the drive roller spring tensioner so that the electrode spool can be hand turned backward to draw the electrode back onto the spool.

NOTE: Rewinding the electrode rather than cutting it off both reduces electrode waste and eliminates the potential hazards that long lengths of loose electrode wire can cause. Loose electrode wire can be a trip hazard as well as an electrical hazard if it comes in contact with the welder electrical terminals.

The wire feed speed is given in a range. The range allows you to adjust the feed speed according to the welding conditions. The wire speed control dial can be advanced or slowed to control the burn weld size and deposition rate.

GUN ANGLE

The **gun angle**, **work angle,** and **travel angle** are names used to refer to the relation of the gun to the work surface, Figure 12-3. The gun angle can be used to control the weld pool. The electric arc produces an electrical force known as the arc force. The arc force can be used to counteract the gravitational pull that tends to make the liquid weld pool sag or run ahead of the arc. By manipulating the electrode travel angle for the flat and horizontal position of welding to a 20° to 90° angle from the vertical, the weld pool can be controlled. A 40° to 50° angle from the vertical plate is recommended for fillet welds.

Changes in this angle will affect the weld bead shape and penetration. Shallower angles are needed when welding thinner materials to prevent burn-through. Steeper, perpendicular angles are used for thicker materials.

Vertical up welds require a forehand gun angle. The forehand angle is needed to direct the arc deep into the groove or joint for better control of the weld pool and deeper penetration, Figure 12-4.

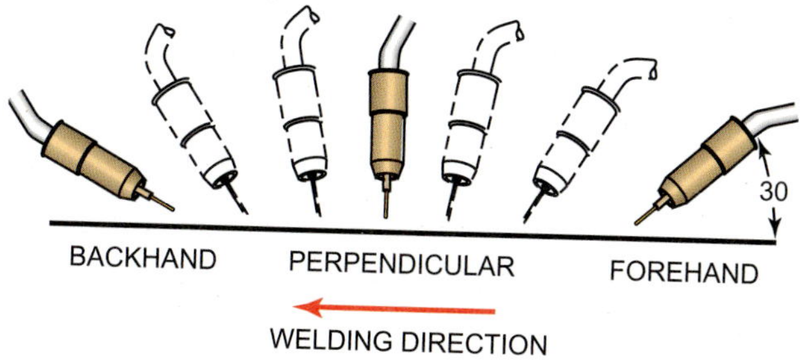

FIGURE 12-3 Welding gun angles. © Cengage Learning 2012

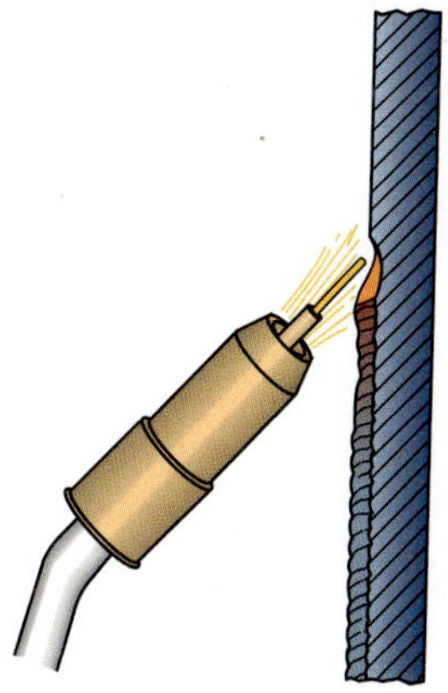

FIGURE 12-4 Vertical up gun angle. © Cengage Learning 2012

FIGURE 12-5 Overhead gun angle. © Cengage Learning 2012

A gun angle around 90° either slightly forehand or backhand works best for overhead welds, Figure 12-5. The slight angle aids with visibility of the weld, and it helps control spatter buildup in the gas nozzle.

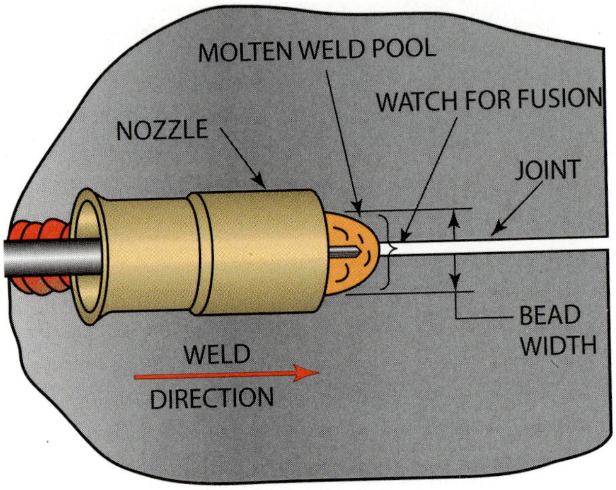

FIGURE 12-7 The shielding gas nozzle restricts the welder's view. © Cengage Learning 2012

Forehand/Perpendicular/Backhand Techniques

Forehand, *perpendicular*, and *backhand* are the terms most often used to describe the gun angle as it relates to the work and the direction of travel. The forehand technique is sometimes referred to as pushing the weld bead, Figure 12-6A; and backhand may be referred to as pulling or dragging the weld bead, Figure 12-6C. The term *perpendicular* is used when the gun angle is at approximately 90° to the work surface, Figure 12-6B.

The advantages of using the **forehand welding technique** are:

- **Joint visibility**—You can easily see the joint where the bead will be deposited, Figure 12-7.
- Electrode extension—The contact tube tip is easier to see, making it easier to maintain a constant extension length.
- Less weld penetration—It is easier to weld on thin sheet metal without melting through.
- Out-of-position welds—This technique works well on vertical up and overhead joints for better control of the weld pool.

The disadvantages of using the forehand welding technique are:

- **Weld thickness**—Thinner welds are formed because less weld reinforcement is applied to the weld joint.
- **Welding speed**—Because less weld metal is being applied, the rate of travel along the joint can be faster, which may make it harder to make a uniform weld.
- Spatter—Depending on the electrode, the amount of spatter may be slightly increased with the forehand technique.

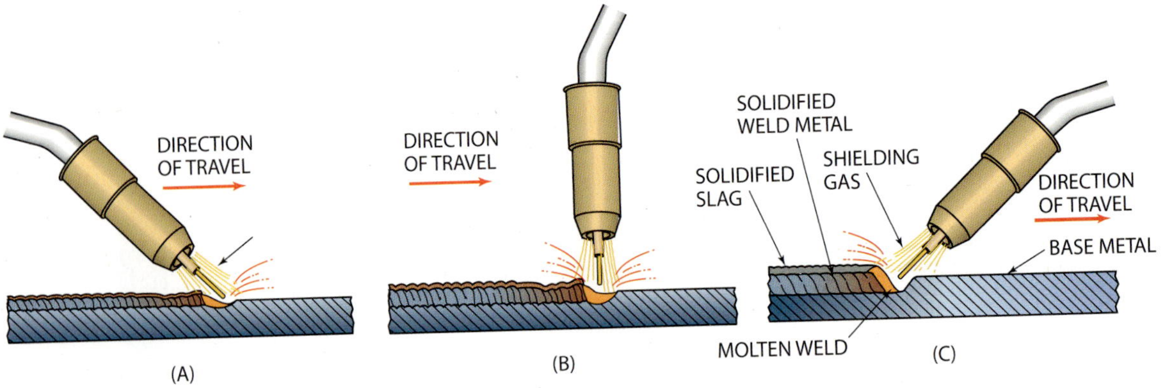

FIGURE 12-6 Changing the welding gun angle among forehand, perpendicular, or backhand angles changes the shape of the weld bead produced. © Cengage Learning 2012

The advantages of using the **perpendicular welding technique** are:

- Machine and robotic welding—The perpendicular gun angle is used on automated welding because there is no need to change the gun angle when the weld changes direction.
- **Uniform bead shape**—The weld's penetration and reinforcement are balanced between those of forehand and backhand techniques.

The disadvantages of using the perpendicular welding technique are:

- **Limited visibility**—Because the welding gun is directly over the weld, there is limited visibility of the weld unless you lean your head way over to the side.
- **Weld spatter**—Because the weld nozzle is directly under the weld in the overhead position, more weld spatter can collect in the nozzle, causing gas flow problems or even shorting the tip to the nozzle.

The advantages of the **backhand welding technique** are:

- **Weld bead visibility**—It is easy to see the back of the molten weld pool as you are welding, which makes it easier to control the bead shape.
- **Travel speed**—Because of the larger amount of weld metal being applied, the rate of travel may be slower, making it easier to make a uniform weld.
- **Depth of fusion**—The arc force and the greater heat from the slower travel rate both increase the depth of weld joint penetration.

The disadvantages of the backhand welding technique are:

- **Weld buildup**—When you use the backhand technique, the weld bead may have a convex (raised or rounded) shaped weld face.
- **Postweld finishing**—Because of the weld bead shape, more work may be required if the product has to be finished by grinding it smooth.
- **Joint following**—It is harder to follow the joint because of your hand position with the GMAW gun positioned over the joint, and you may wander from the seam, Figure 12-8.
- Loss of penetration—An inexperienced welder sometimes directs the wire too far back into the weld pool, causing the wire to build up in the face of the weld pool and reducing joint penetration.

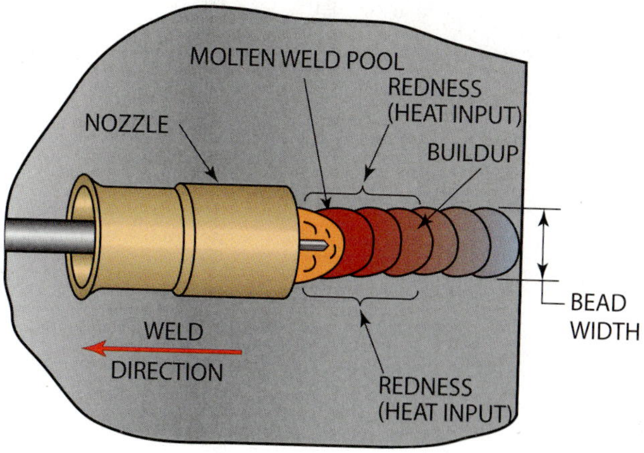

FIGURE 12-8 Watch the trailing edge of the molten weld pool. © Cengage Learning 2012

Electrode Manipulation

The movement or weaving of the welding electrode can control the following characteristics of the weld bead: buildup, width, undercut, and overlap. However, because moving the tip of the electrode off of the molten weld pool can cause a great deal of spatter, most GMA welding is done with a straight line movement. When a weave is needed for better bead control, the weaves are usually fairly small.

Practice Welds

The two most important points that a GMA welder needs to master with the practices are setting up the welding current and electrode control. The setting of the voltage and **wire feed speed** (amperage) according to the setup tables will give you a good starting point, but some adjustments may be needed. The electrode control consists primarily of keeping its movement consistent along the entire length of the joint. You must maintain a constant speed and weave along the joint as well as a constant welding gun angle and electrode extension.

Starting with the flat position allows you to build your skills slowly so that out-of-position welds become easier to do. The horizontal tee and lap welds are almost as easy to make as the fillet welds. Overhead welds are as simple to make as vertical welds, but they are harder to position. Horizontal butt welds are more difficult to perform than most other welds.

Electrode Extension

The **electrode extension** (stickout) is the length from the contact tube to the arc measured along the

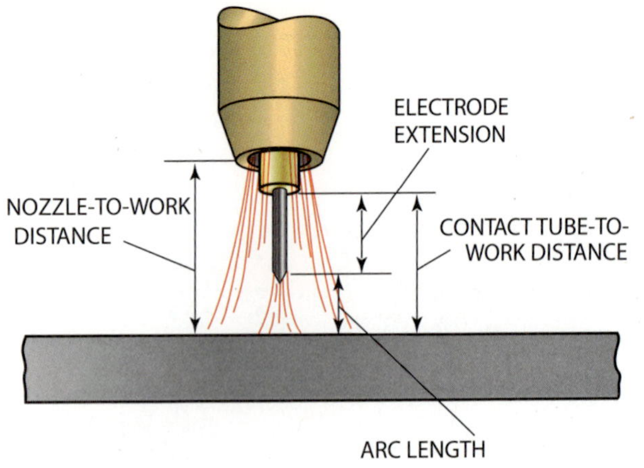

FIGURE 12-9 Electrode-to-work distances. © Cengage Learning 2012

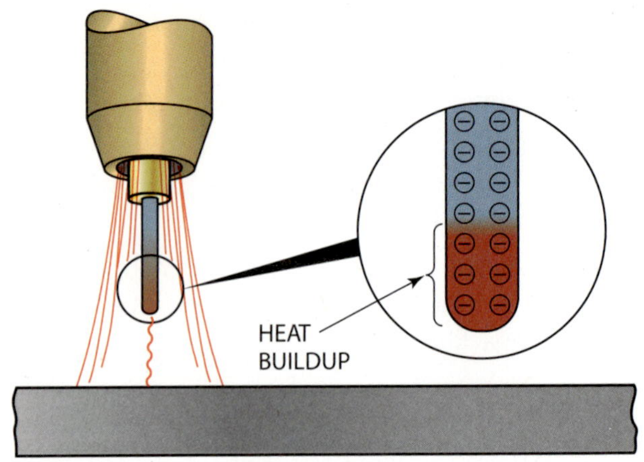

FIGURE 12-10 Heat buildup due to the extremely high current for the small electrode. © Cengage Learning 2012

wire, Figure 12-9. A change in this length causes a change in the welding arc's current and the resulting weld bead. This current change occurs because GMA welding currents are very high for the electrode wire sizes. The resistance of the very small welding wire from the contact tube to the arc results in some of the welding current being lost to heat in this short piece of wire, Figure 12-10. As the electrode extension is increased, there is a reduction in weld heat, penetration, and fusion, and an increase in buildup. On the other hand, as the electrode extension length is decreased, the weld heats up, penetrates more, and builds up less, Figure 12-11.

NOTE: Comparing the size of the welding power cable to the size of the welding electrode wire will give you an idea of how small the electrode is related to the welding current it is carrying.

PRACTICE 12-1

Gas Metal Arc Welding Safety

Skill to be learned: The safe setup of a welding station and the use of proper personal protective equipment (PPE).

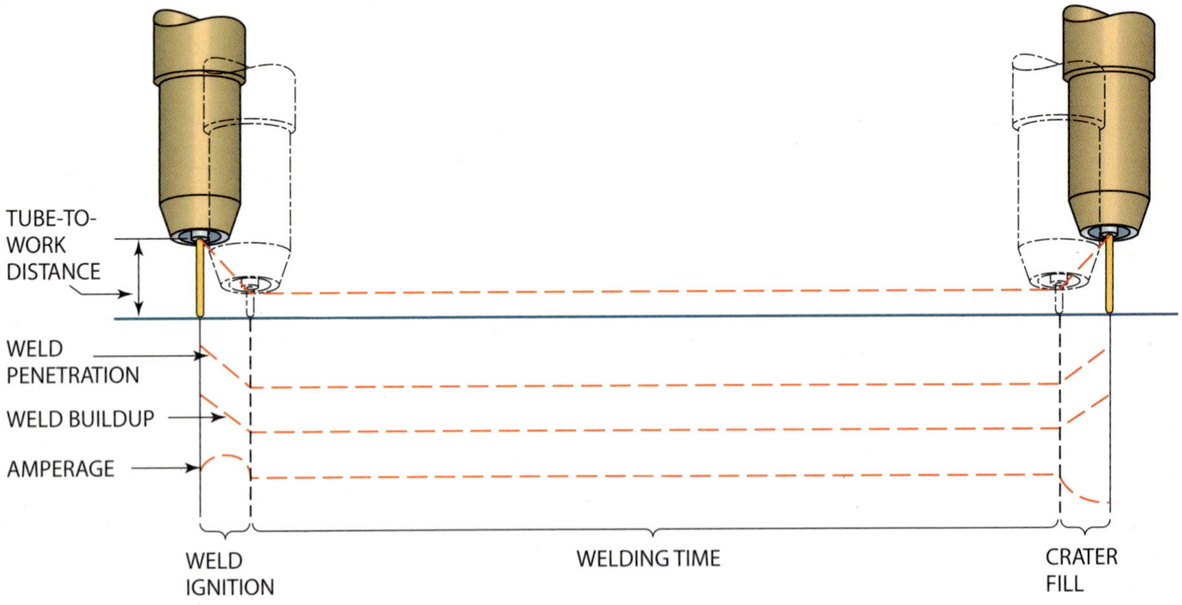

FIGURE 12-11 Using the changing tube-to-work distance to improve both the starting and stopping points of a weld. © Cengage Learning 2012

Using a welding workstation, welding machine, welding electrode wire, welding helmet, eye and ear protection, welding gloves, proper work clothing, and any special protective clothing that may be required, demonstrate to your instructor and other students the safe way to prepare yourself and the welding workstation for welding. Include in your demonstration appropriate references to burn protection, eye and ear protection, material specification data sheets, ventilation, electrical safety, general work clothing, special protective clothing, and area cleanup.

EXPERIMENT 12-1

The Effect Electrode Extension Distance Changes Have on a Weld Bead

Skill to be learned: How to change the electrode extension to control a weld's penetration and buildup and to help maintain weld bead shape during welding.

Experiment Description

You will be making GMA weld stringer beads on a steel plate in the flat position. As the weld progresses along the plate, you will be changing the electrode extension length so you can observe how this change affects the weld produced.

Project Materials and Tools

The following items are needed to complete this practice.

Materials

- 0.035 GMA welding electrode wire
- One or more 12-in. long pieces of metal ranging from 16-gauge sheet metal to 1/4-in. thick plate

Tools

- GMA welder
- PPE
- Square and/or 12-in. rule
- Soapstone

Layout

Using a square or 12-in. rule and soapstone, lay out parallel lines on the metal that are spaced 3/4 in. apart.

Welding

Start with the voltage and wire feed speed set to the lowest settings according to Table 12-1 for short-circuit arc welding. Put on all of your PPE and follow all shop and manufacturer's safety rules for welding. You will make a series of stringer bead welds along the soapstone line starting with the 16-gauge sheet metal.

Holding the welding gun at a comfortable angle and height, lower your helmet and start to weld. Make a weld approximately 2 in. long. Then reduce the distance from the gun to the work while continuing to weld. After a few inches, again shorten the electrode extension even more. Keep doing this in steps until the nozzle is as close as possible to the work. Stop when you have made a weld all the way down the entire length of the plate.

Repeat the process but now increase the electrode extension, making welds of a few inches each. Keep increasing the electrode extension until the weld will no longer fuse to the base metal or the wire becomes impossible to control.

Now change to another plate thickness and repeat the procedure.

When a series of weld beads has been completed with each plate thickness, raise the voltage and wire feed speed to a midrange setting and repeat the process. Upon completing this series of tests, adjust the voltage and wire feed speed to the highest range setting. Make a full series of tests using the same procedure as before.

Finishing

None.

Paperwork

Write a short report describing what you observed as the electrode extension length and welding voltage and wire feed speeds were changed. Include in your report the details regarding the weld buildup, width, penetration, spatter, and any other observations you made during the welding.

Complete a copy of the time sheet in Appendix I, the bill of materials in Appendix III, and the performance qualification test record in Appendix IV or use forms provided by your instructor.◆

EXPERIMENT 12-2

The Effect That Gun Angle Changes Have on a Weld Bead

Skill to be learned: How to change the welding gun angle to control a weld's penetration and buildup and to help maintain weld bead shape during welding.

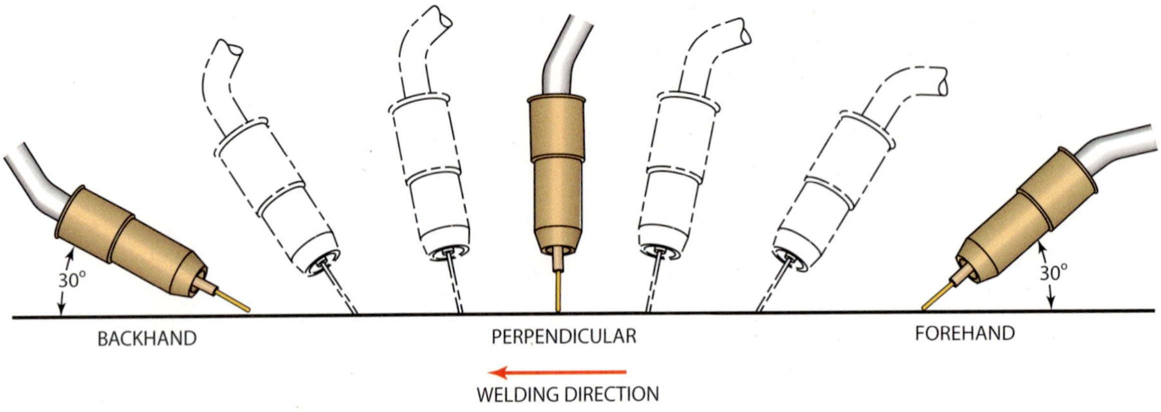

FIGURE 12-12 Welding gun angles. © Cengage Learning 2012

Experiment Description

You will be making GMA weld stringer beads on a steel plate in the flat position. As the weld progresses along the plate, you will be changing the welding gun angle so you can observe how this change affects the weld produced.

Project Materials and Tools

The following items are needed to complete this practice.

Materials

- 0.035 GMA welding electrode wire
- One or more 12-in. long pieces of 1/4-in. thick steel plate

Tools

- GMA welder
- PPE
- Square and/or 12-in. rule
- Soapstone

Layout

Using a square or 12-in. rule and soapstone, lay out parallel lines that are spaced 3/4 in. apart on the metal.

Welding

Start with the voltage and wire feed speed set to the lowest settings according to Table 12-1 for short-circuit arc welding. Put on all of your PPE and follow all shop and manufacturer's safety rules for welding. You will make a series of stringer bead welds along the soapstone line starting with the 16-gauge sheet metal.

Holding the welding gun at a comfortable height and at a 30° angle to the plate in the direction of the weld, lower your helmet and start to weld, Figure 12-12. Make a weld approximately 2 in. long. Then gradually increase the gun angle while continuing to weld. After a few inches, again increase the gun angle even more. Keep doing this in steps until the gun is at a 30° angle to the plate in the opposite direction of the weld. Stop when you have made a weld all the way down the entire length of the plate.

Raise the voltage and wire feed speed to a midrange setting and repeat the process. Upon completing this series of tests, adjust the voltage and wire feed speed to the highest range setting. Make a full series of tests using the same procedure as before.

Finishing

None.

Paperwork

Write a short report describing what you observed as the gun angle changed from a 30° pushing angle to a 30° dragging angle and welding voltage and wire feed speeds were changed. Include in your report the details regarding the weld buildup, width, penetration, spatter, and any other observations you made during the welding.

Complete a copy of the time sheet in Appendix I, the bill of materials in Appendix III, and the performance qualification test record in Appendix IV or use forms provided by your instructor.♦

INTERMITTENT BUTT WELDS

Butt welds are commonly used to join the edges of plates and structural members. Usually, butt welds are made all the way along the length of the joint;

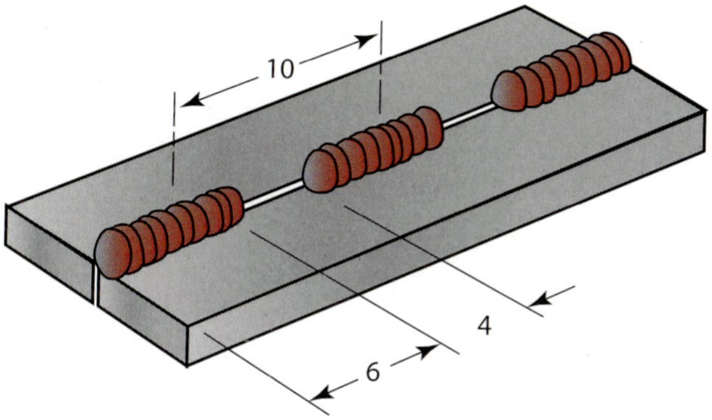

FIGURE 12-13 Intermittent butt weld. © Cengage Learning 2012

however, there are several reasons they may be made as intermittent welds. Intermittent welds are used when the strength of a full-length weld is not required, to speed up the welding process and reduce welding cost, to reduce welding heat distortion, or to help stop weld bead cracking, Figure 12-13.

PROJECT 12-1

Assembling a Birdhouse Using Intermittent Butt Welds

Skill to be learned: How to make intermittent butt welds using a GMA welder. Layout, measuring, fitting, and assembling fabrication skills and techniques will be developed.

Project Description

You will build a log birdhouse made from 3/8-in. (#3) reinforcement rod, Figure 12-14. The birdhouse will be assembled using intermittent butt welds.

Project Materials and Tools

The following items are needed to fabricate the birdhouse.

Materials

- 0.035 GMA welding electrode wire
- 50-ft 3/8-in. (#3) reinforcement rod

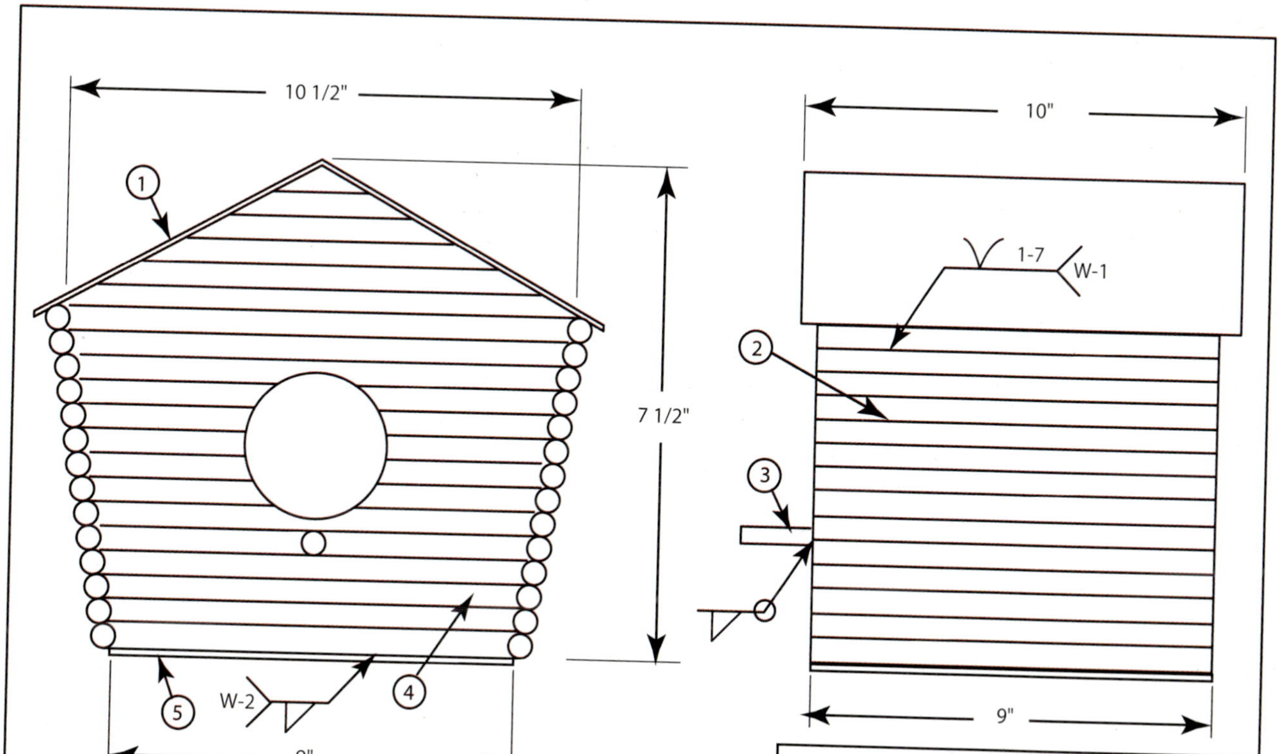

Log Bird house		
Parts List		
Item	Qty	
1	1	10" x 12" 16-Gauge Sheet Metal
2	28	3/8" (#3) Rainforest Rod
3	1	3/8" x 2" (#3) Rainforest Rod
4	40	3/8" (#3) Rainforest Rod
5	1	9" x 9" 16-Gauge Sheet Metal

FIGURE 12-14 Project 12-1. © Cengage Learning 2012

- 1 10 × 12-in. 16-gauge sheet metal
- 1 9 × 9-in. 16-gauge sheet metal
- Paint

Tools

- GMA welder
- Pencil, paper, and scissors
- Plasma cutting torch, oxyfuel cutting torch, or shear
- PPE
- Tape measure
- Hammer
- 12-in. rule
- Soapstone
- Square
- Pliers

Layout

The side walls are constructed with 14 pieces of 9-in. long 3/8-in. reinforcement rods. Using the tape measure and the soapstone, mark 28 lines every 9 in. along a piece of reinforcement rod. Make your marks by hooking the end of the tape over one end of the reinforcement rod and mark the lines at 9 in., 18 in., 27 in., and so on. Making your marks this way is much more accurate than moving a 12-in. rule down the reinforcement rod each time a mark is made.

NOTE: If you are using a tape measure that does not have inches consecutively marked but uses feet and inches, then you would mark 9 in., 1 ft 6 in., 2 ft 3 in., and so on, instead of 9 in., 18 in., 27 in.

Using a plasma cutting torch, oxyfuel cutting torch, or shear, cut the reinforcement rod to length.

Cutting Out

Using a plasma cutting torch, oxyfuel cutting torch, or shear, cut the reinforcement rod for the side walls to length.

NOTE: The ends of the front and back reinforcement rods may be cut square, and ground to the roof pitch angle after they are welded in place, Figure 12-15A; or they can be cut short enough so that the square ends will fit under the roof, Figure 12-15B.

Layout and Cutting Out

Because no two pieces of reinforcement rod will be the same length for the front wall, you must lay out and cut out each piece one at a time. To do this you need to make a paper pattern. Using a square, 12-in. rule, pencil, and paper, lay out a pattern for the front and back walls of the birdhouse as shown in Figure 12-16. Using the paper pattern and the soapstone, mark and cut the 3/8-in. rebar to length. Each piece of rebar should be cut and laid in place on the paper pattern because no two pieces of rebar will be cut to the same length.

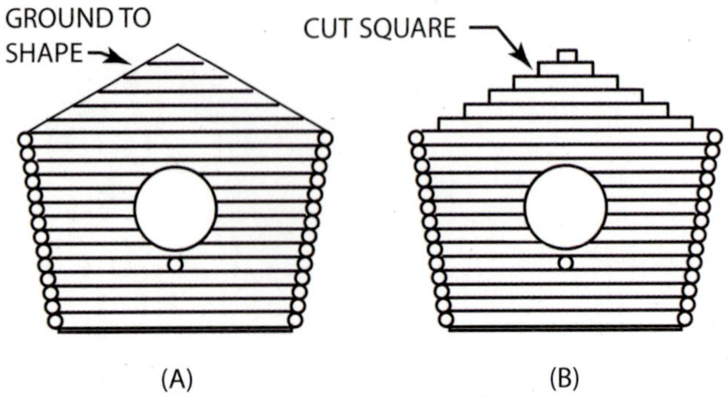

FIGURE 12-15 Roof reinforcement rods (A) cut square and ground or (B) cut square.

Fabrication

On a welding table lay out the reinforcement rod pieces for the sides of the birdhouse. Use the square to make sure the finished walls will be square. Draw a chalk line located as shown in Figure 12-17. The 1/2-in. long butt welds will be centered along these center lines.

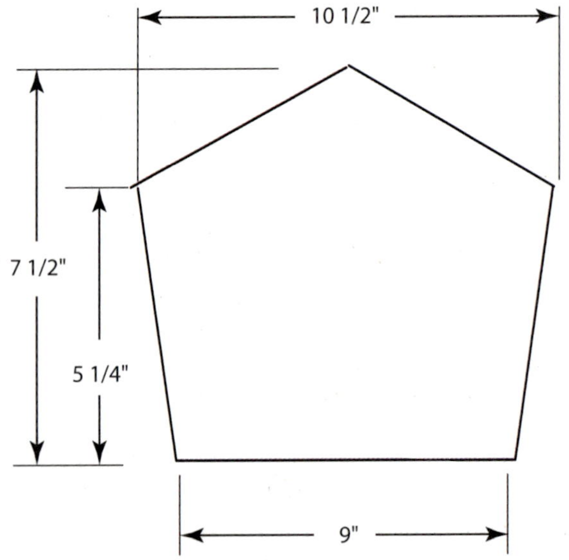

FIGURE 12-16 Birdhouse front and back wall layout.

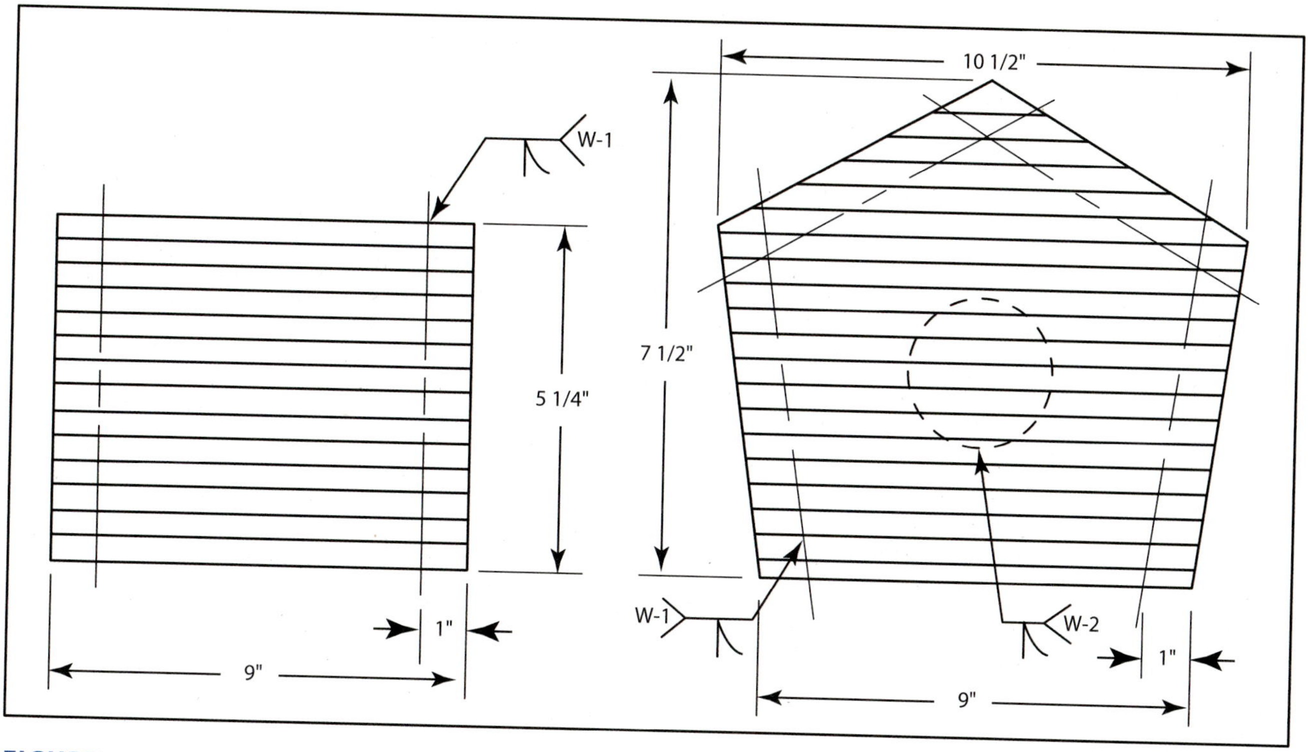

FIGURE 12-17 Location for butt welds. © Cengage Learning 2012

Welding

Set the welder voltage and wire feed speed for short-circuit arc welding according to Table 12-1. Put on all of your PPE and follow all shop and manufacturer's safety rules for welding.

Hold the tip of the electrode approximately 1/4 in. from the center line. Using your gloved hand, hold the reinforcement rod together, lower your helmet, pull the gun trigger, and make a 1/2-in. long butt weld, Figure 12-18. Raise your helmet after you have made the weld. Sometimes weld shrinkage may pull the opposite ends of the reinforcement rod apart. This is a common problem when making welds. If this happens, you can use a pair of pliers to squeeze them together.

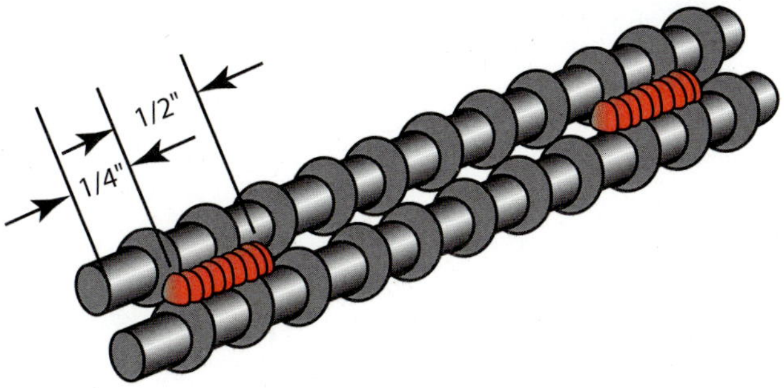

FIGURE 12-18 Butt weld locations on reinforcement rods. © Cengage Learning 2012

You must make sure that there is no gap between the reinforcement rods so that the wall will be parallel. Holding the reinforcement rod together, place the electrode tip approximately 1/4 in. from the center line on the opposite end, lower your helmet, and make a 1/2-in. long butt weld. Repeat this process for each of the additional pieces of reinforcement rods until the side panel has been completely welded together, Figure 12-19.

Assemble the other side of the birdhouse using the same assembly technique.

Fabrication

On a welding table lay out the reinforcement rod pieces for the front of the birdhouse. Use the square to make sure the finished wall will be square. Draw a chalk line located as shown in Figure 12-14. The 1/2-in. long welds will be centered along these center lines.

Welding

Use the same welding technique and PPE that you used for the side walls to weld the front and back walls of the birdhouse. Because the top reinforcement rod is short, it will have only one weld in the center, Figure 12-20.

Layout

Lay out the entrance hole in the birdhouse by first drawing the size hole on a sheet of paper. Use

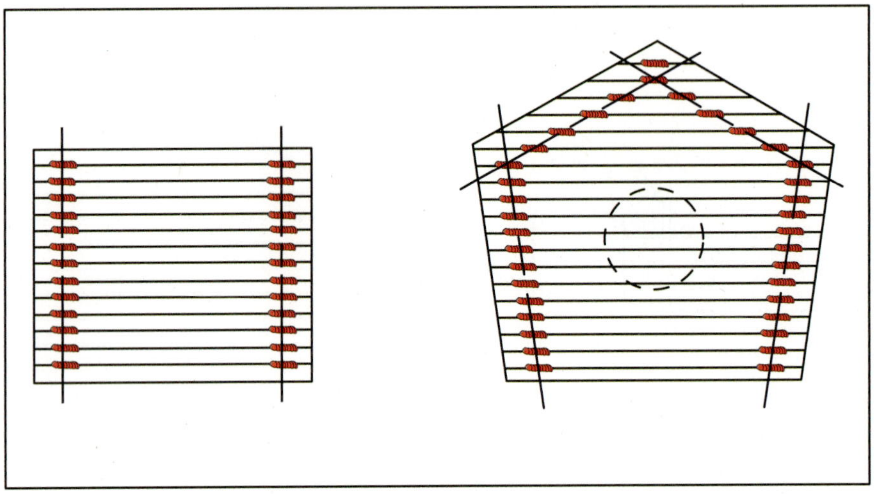

FIGURE 12-19 Intermittent butt welds on side and front panels of the birdhouse.
© Cengage Learning 2012

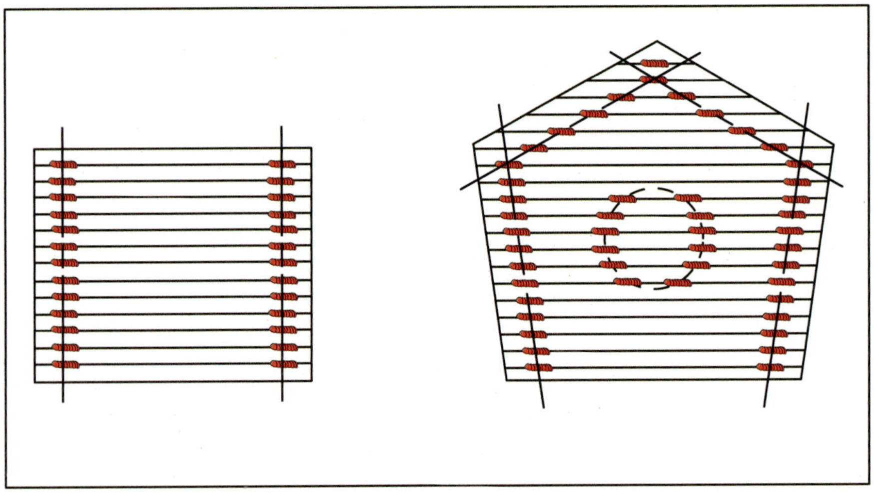

FIGURE 12-20 Make welds around the birdhouse opening to hold reinforcement rods in place after the opening is cut. © Cengage Learning 2012

Table 12-2 to determine the size hole for your birdhouse. Cut out the hole in the paper and center your template on the inside of the front of the birdhouse. Draw a chalk line around the hole.

Welding

Make 1-in. long welds across the drawn circle for the birdhouse door. These welds will keep the reinforcement rod ends aligned when the door opening is cut.

Cutting

If necessary, retrace the circle on the front wall with soapstone so that you can follow the line as you cut out the door opening. Using a plasma cutting torch or oxy-fuel cutting torch, cut out the circle for the bird door.

BIRD SPECIES	ENTRANCE SPECIFICATION	
	ENTRANCE DIAMETER IN INCHES	DISTANCE ABOVE FLOOR IN INCHES
BLUEBIRD	1 1/2	6
CAROLINE WREN	1 1/2	4–6
CHICKADEE	1 1/8	6–8
CRESTED FLYCATCHER	2	6–8
DOWNY WOODPECKER	1 1/4	6–8
HOUSE WREN	1 1/4	4–6
NUTHATCH	1 1/4	6–8
PURPLE MARTIN	2 1/2	2
TITMOUSE	1 1/4	6–8
TREE SWALLOW	1 1/2	1–5

Table 12-2 Birdhouse Hole Sizes

Assembly

Once the four walls of the birdhouse have been fabricated, you will assemble them. Start by placing one side wall and the front wall on the welding table. Use the square to align the parts as shown in Figure 12-14. Make a 1/4-in. long tack weld at the top corner of the walls and the same size tack weld at the bottom of the walls.

NOTE: The walls should be fairly flat following the tack welding; however, a hammer can be used to flatten them out if there has been some distortion from the tack welds.

Align the back wall of the birdhouse to the side; using a square to align the parts, make two 1/4-in. long tack welds at the top and bottom of the joint.

Place the remaining side wall, and adjust the alignment as necessary so that it fits in the opening between the front and back walls. Make two small tack welds at both the back corner and front corner of the side wall.

NOTE: By making the tack welds small, you will be able to square the birdhouse walls if necessary.

Using the square, check the birdhouse to see that it is square on all sides before making a longer tack weld in each of the corners at the top and bottom of the birdhouse walls.

Assemble the sheet metal bottom in the birdhouse by using intermittent butt welds to hold it in place. There should be four 1/4-in. long welds on all four sides of the sheet metal bottom.

Tack weld the 2-in. long rebar perch 1/2 in. below the birdhouse opening. Use a hammer to tap the perch into alignment so that it is perpendicular to the front wall before completing a weld around the entire joint.

Fabricate the roof of the birdhouse by bending a slight angle in the center of the 10 × 12-in. 16-gauge sheet metal roof. The roof of the birdhouse will not be attached to the birdhouse by welding so that it can be removed seasonally to clean out the debris left by the previous bird tenant.

Finishing

Cool off the metal and wire brush the surface. Paint the side surfaces with a latex paint.

Paperwork

Complete a copy of the time sheet in Appendix I, the bill of materials in Appendix III, and the performance qualification test record in Appendix IV or use forms provided by your instructor.

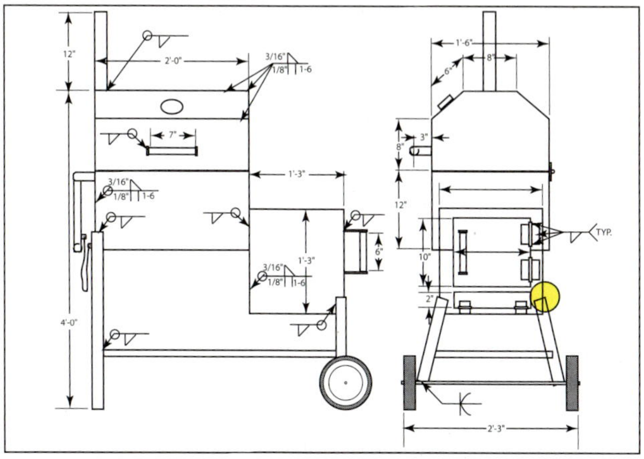

FIGURE 12-21 Butt welds made on a barbeque grill.

Butt joints are used to join plate and structural steel shapes. The advantage of using the butt joint is that the metal can be welded as it is without additional grinding or shaping, Figure 12-21.◆

SPARK YOUR IMAGINATION

Using Table 12-3, calculate the weight of the birdhouse. Use a scale to see how close your calculation of the weight was.

Lap Joint

Lap joints are made by overlapping the edges of the two plates, Figure 12-22. The joint can be welded on one side or both sides with a fillet weld. In most cases, both sides of the joint should be welded. When just one side is welded, the joint is not as

Bar Size	Weight (lb/ft)	Nominal Diameter (in)
#3	0.376	0.375 = 3/8
#4	0.668	0.500 = 1/2
#5	1.043	0.625 = 5/8
#6	1.502	0.750 = 3/4
#7	2.044	0.875 = 7/8
#8	2.670	1.000 = 1

Table 12-3 Reinforcement Bar Sizes

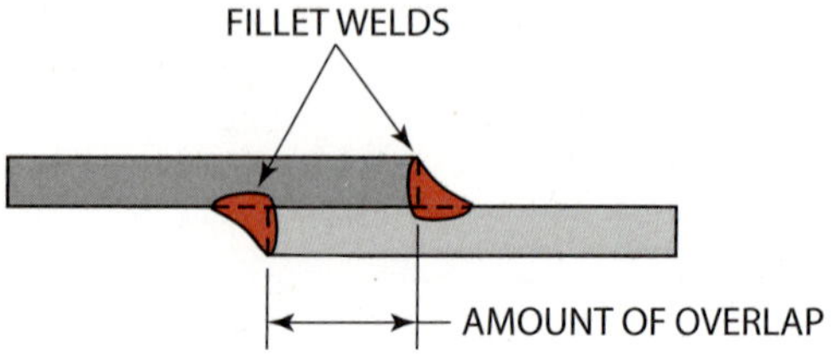

FIGURE 12-22 Lap joint. © Cengage Learning 2012

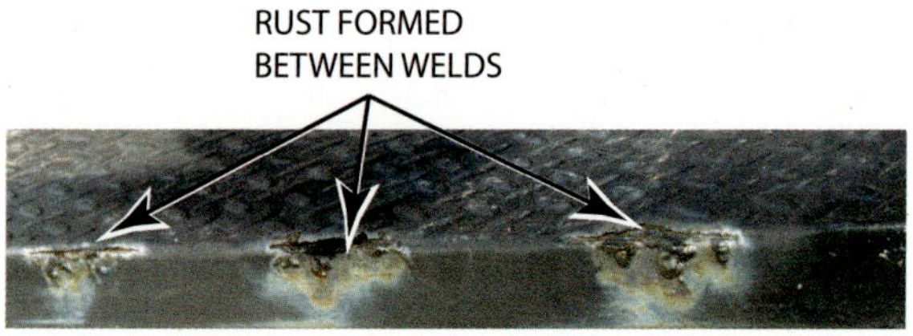

FIGURE 12-23 Rust can form when only one side of a lap joint is welded. Larry Jeffus

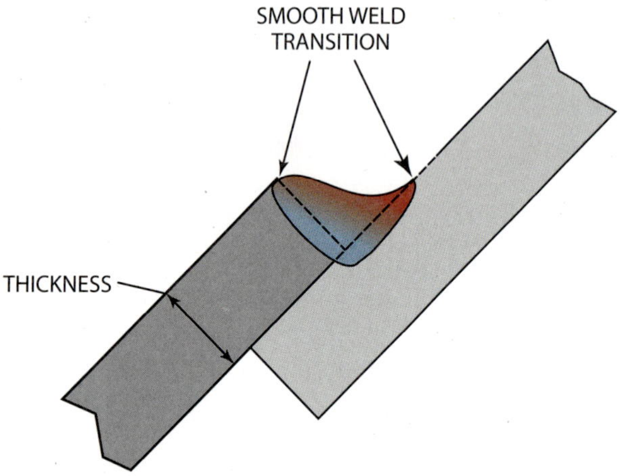

FIGURE 12-24 Buildup should be equal to the thickness of the plate on a lap joint and have a smooth transition from the plate surface to the weld. © Cengage Learning 2012

strong, and water can cause rust to form in the joint space, Figure 12-23.

As the fillet weld is made on the lap joint, the buildup should equal the thickness of the plate, Figure 12-24. A good weld will have a smooth transition from the plate surface to the weld. If this transition is abrupt, it can cause stresses that will weaken the joint.

PROJECT 12-2

Fillet Welds in Lap Joints to Build a Birdhouse Roof

Skill to be learned: The control of the weld to produce uniform fillet welds in lap joints in the flat position. Measuring, fitting, and assembling fabrication skills and techniques will be developed.

Project Description

The roof of a house is usually covered with shingles that are lapped one on top of the other. The lap joints on the roof of your birdhouse are designed to give it the look of shingles, Figure 12-25.

Project Materials and Tools

The following items are needed to fabricate the birdhouse roof.

Materials

- 0.035 GMA welding electrode wire
- 12 1 1/2-in. wide 10-in. long pieces of 3/16-in. to 1/4-in. thick mild steel plate
- Paint

Tools

- GMA welder
- Pencil, paper, and scissors
- Plasma cutting torch, oxyfuel cutting torch, or shear
- PPE
- Tape measure
- Hammer
- Square and/or 12-in. rule
- Soapstone
- Wire brush
- Pliers

Layout

Using a square, 12-in. rule, and soapstone, lay out the bar stock as shown in the project drawing, Figure 12-25.

Cutting Out

The first step is to cut the mild steel bar stock to length with a shear, or thermally.

Fabrication

The two roof panels will be constructed with five 1 1/2-in. wide and 10-in. long pieces of bar stock. The roof must be 5 1/2 in. wide, and the overlaps of each joint must be equal.

Lay out the bars so that they overlap 1/2 in. and have a total width of 5 1/2 in. Holding the bars in place, make a small tack weld on the end of each

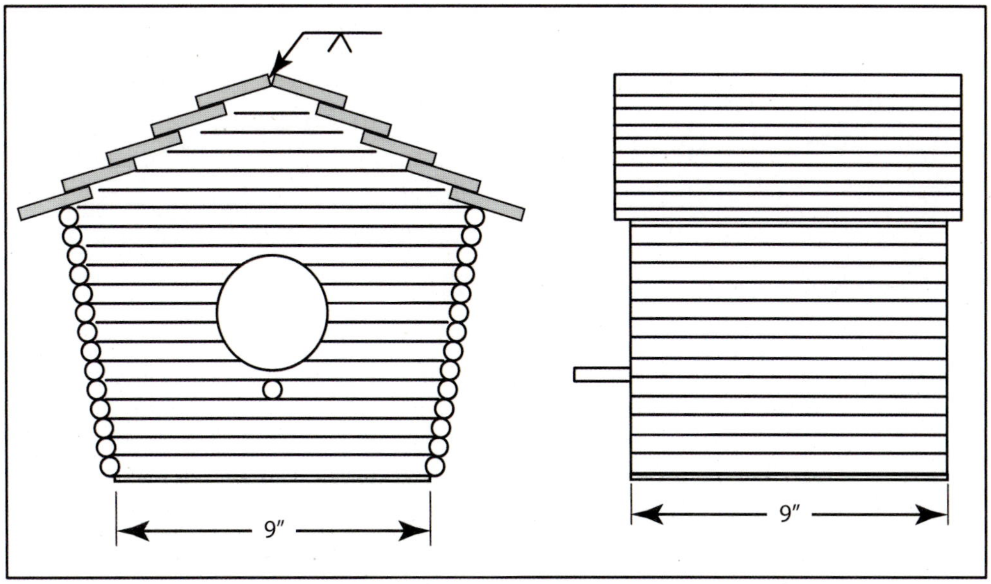

FIGURE 12-25 Birdhouse roof with overlapping joints. © Cengage Learning 2012

overlap, Figure 12-26. Placing the tack welds on the ends will keep them out of the way of the lap welds.

Once both lap weld panels are tack welded together, recheck to see that the overall panel width of 5 1/2 in. is accurate. If necessary, grind off a tack weld and adjust the overlap so that the finished panel width is correct within ±1/8 in.

Welding

Set the welder voltage and wire feed speed for short-circuit arc welding according to Table 12-1. Put on all of your PPE and follow all shop and manufacturer's safety rules for welding.

The lap joint panel will need to be braced at an approximate 45° angle so that the face of the weld will be flat. Also, the welds will have to be alternated from side to side to minimize the distortion of the roof panel.

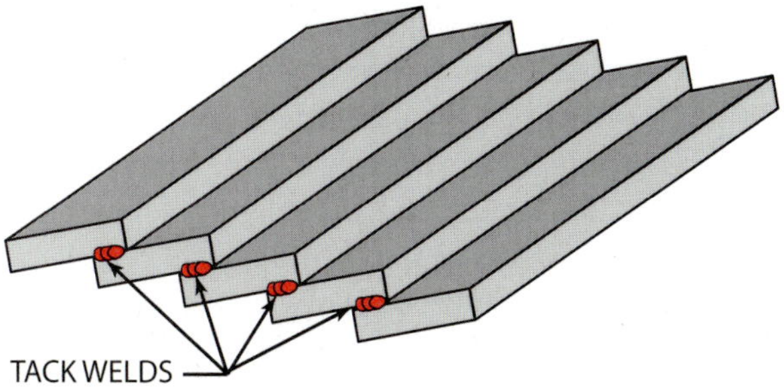

FIGURE 12-26 Tack welds on birdhouse roof. © Cengage Learning 2012

The welding heat is not evenly distributed between the two plates of a lap joint. The edge of the top plate is easily overheated as compared to the surface of the bottom plate, Figure 12-27. To keep the top plate's edge from melting back, direct most of the arc's heat on the surface of the bottom plate.

One problem that occurs with all fillet welds is lack of fusion to the root of the weld. This occurs as the result of not getting enough of the arc's heat down in the root of the joint. More heat can be directed into the joint root by keeping the arc as

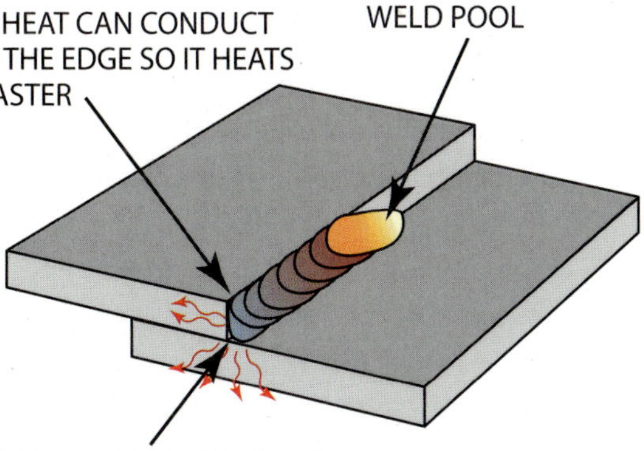

FIGURE 12-27 Direct most of the arc's heat on the surface of the bottom plate. © Cengage Learning 2012

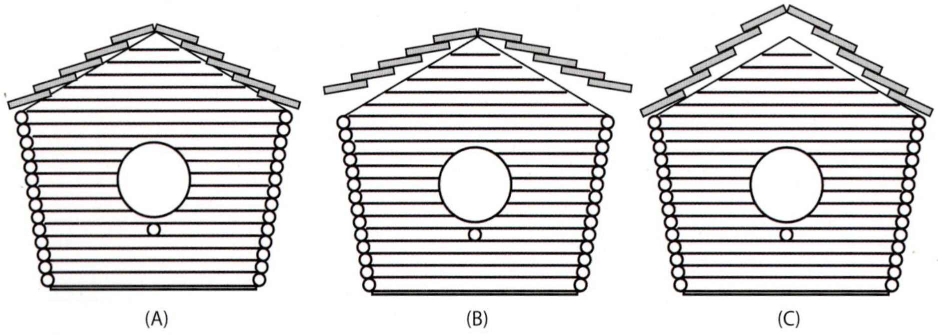

FIGURE 12-28 **Weld shrinkage can pull bars out of alignment unless the metal is restrained from warping.** © Cengage Learning 2012

close to the front leading edge of the molten welding pool as possible.

Cool the metal and look at the weld to see if it has uniform width and buildup and follows the weld joint. A lack of uniform width might indicate that you were not using a consistent weave pattern. A lack in uniformity in buildup might indicate that your rate of travel down the joint was not steady. Match your welding speed so that each weld is made with only one electrode.

When the welding is completed on the first panel, weld the second roof panel lap joints. Place the panels on the log birdhouse and tack weld them together in four places along the ridge of the roof. Turn the assembly over and make a small weld all the way along the inside of the roof panels. Now flip the panels back over and make a small weld along the ridge. The purpose of making the tack welds on the outside and then welding the inside is to reduce the effect of weld shrinkage that could pull the panels out of alignment so that they do not fit the front wall's slope, Figure 12-28.

Finishing

Cool off the metal and wire brush the surface. Paint the side surfaces with a latex paint.

Paperwork

Complete a copy of the time sheet in Appendix I, the bill of materials in Appendix III, and the performance qualification test record in Appendix IV or use forms provided by your instructor. Lap joints are commonly used to join structural shapes such as angle iron to the flat surface of plates. A lap joint can also be found when a thin plate joins a thicker plate.

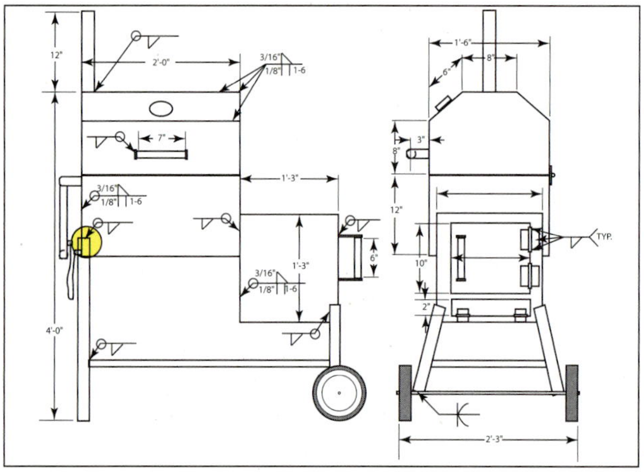

FIGURE 12-29 **Lap welds on a barbeque grill.** © Cengage Learning 2012

Because a lap joint may not be as strong in all directions, it is important to weld all sides of the joint when possible, Figure 12-29.◆

Outside Corner Joint

An outside corner joint is made by placing the plates at an angle to each other, with the edges forming a V-groove, Figure 12-30. The angle between the plates may range from a very slight 15° angle to almost flat at a 165° angle. There may or may not be a slight root opening left between the plate edges. Small tack welds should be made approximately 1/2 in. from each end of the joint. The weld bead should completely fill the V-groove formed by the plates and may have a slightly convex surface buildup.

Outside corner joints are used at the corners of tanks, boxes, and ships. Four plates can be made into a square tube; three can be made into a triangular tube, Figure 12-31.

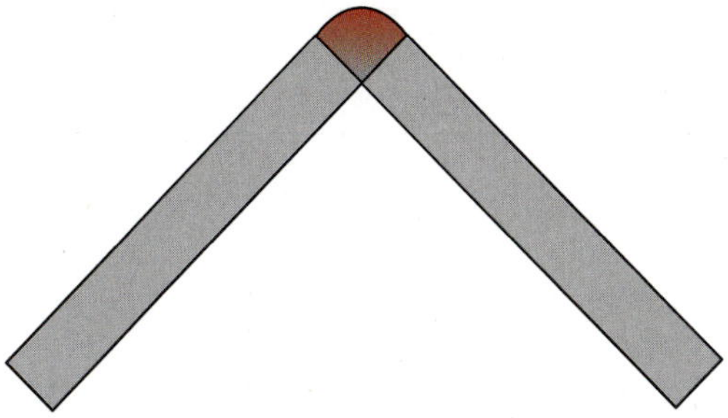

FIGURE 12-30 Outside corner joint. © Cengage Learning 2012

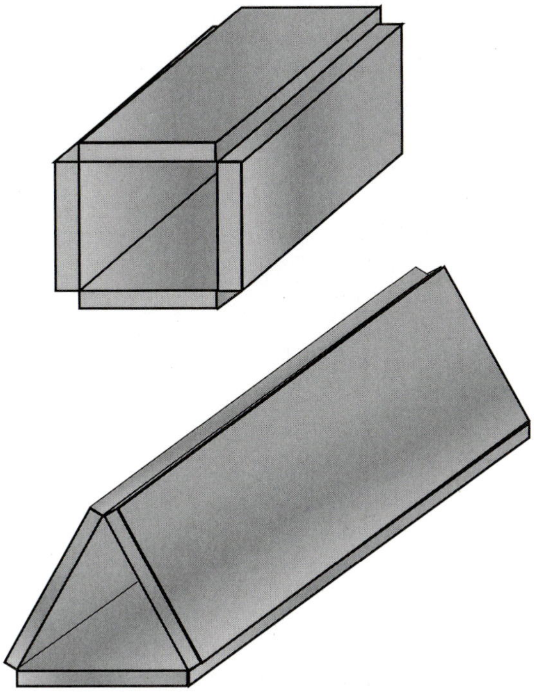

FIGURE 12-31 Four plates can be made into a square tube; three can be made into a triangular tube. © Cengage Learning 2012

PROJECT 12-3

Fillet Welds on Outside Corner Joints to Make a C-Clamp

Skill to be learned: The control of the weld to produce uniform fillet welds in outside corner joints in the flat position. Additional welding skills learned in the previous projects will be reinforced. Detailed fabrication skills and techniques required to align and square parts will be developed.

Project Description

C-clamps are one of the most commonly used fabrication hand tools. They are used to both hold parts in place until tack welds can be made and to force parts together. Sometimes enough force can be applied with one or more C-clamps to bend one part so it fits properly to another part.

As the bolt is turned, forcing parts together, it is important that the upper jaw turn on the bolt so that no twisting force is applied to the parts. If the top jaw twists, it can cause both the top and bottom C-clamp jaws to slip. This C-clamp has an upper jaw assembly of a nut and three different washers that allow the upper jaw to turn freely as the clamping pressure is applied with the C-clamp. A little light oil can be used so the jaw will turn easier.

Project Materials and Tools

The following items are needed to fabricate the C-clamp.

Materials

- 0.035 GMA welding electrode wire
- 6 2-in. wide 7 1/2-in. long pieces of 3/16-in thick mild steel strip for the back top beams
- 3 2-in. wide 9 1/2-in. long pieces of 3/16-in. thick mild steel strip for the bottom beam
- 3/8-in. diameter 4 1/2-in. long rod handle
- 3/4-10NC 10-in. bolt
- 3/4-10NC connector
- 3/4-10NC nut upper jaw part
- 3/4-in. flat washer upper jaw part
- 1-in. flat washer upper jaw part
- 3/8-in. washer upper jaw part
- 1-in. diameter 3/16-in. to 1/4-in. thick lower jaw

Tools

- GMA welder
- PPE
- Angle grinder
- Wire brush
- Square
- Soapstone
- Plasma cutting torch, oxyfuel cutting torch, or shear
- 12-in. steel rule
- Tape measure
- C-clamp

Layout

Using a square, 12-in. rule, measuring tape, and soapstone, lay out the plate as shown in the project drawing, Figure 12-32.

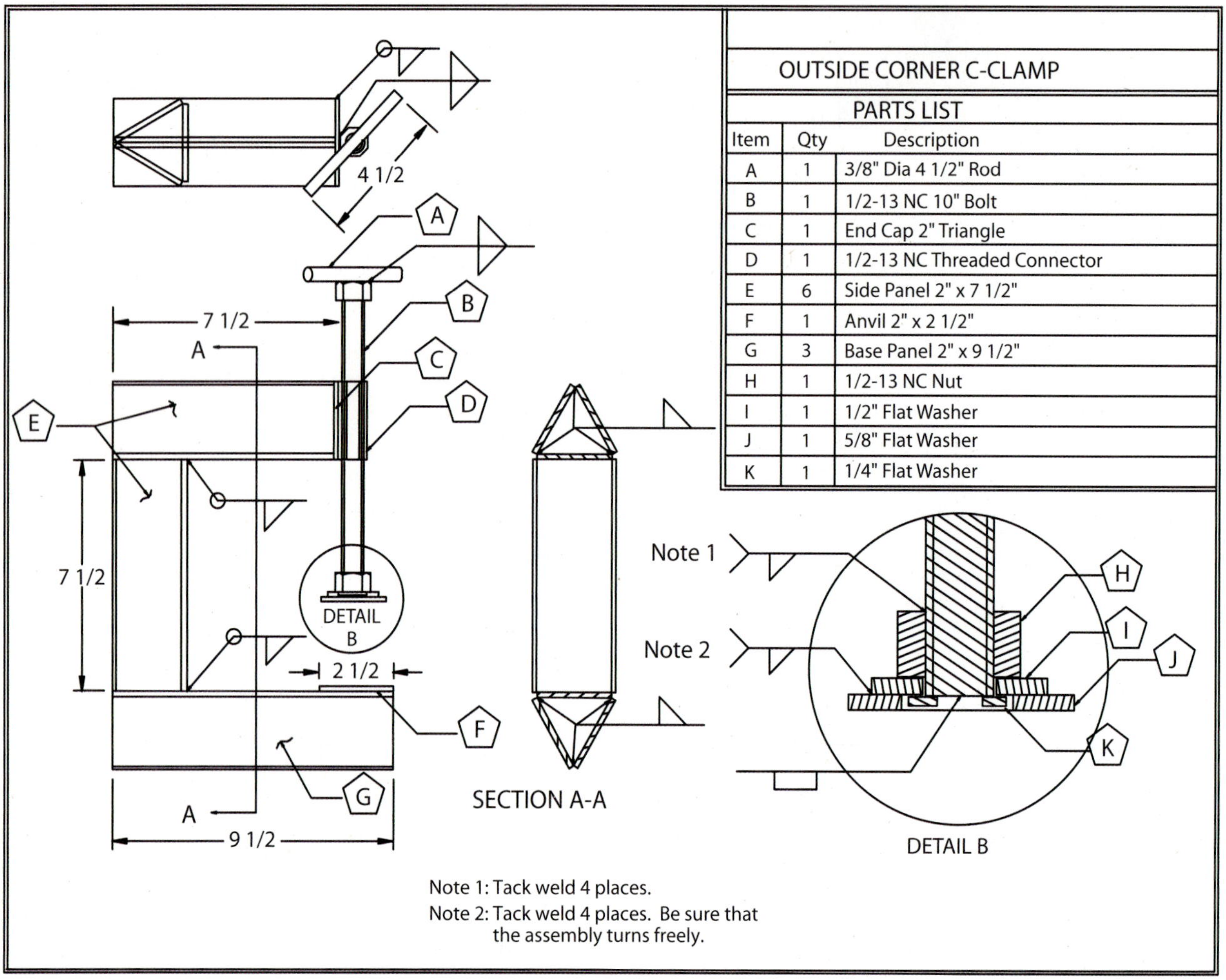

Item	Qty	Description
A	1	3/8" Dia 4 1/2" Rod
B	1	1/2-13 NC 10" Bolt
C	1	End Cap 2" Triangle
D	1	1/2-13 NC Threaded Connector
E	6	Side Panel 2" x 7 1/2"
F	1	Anvil 2" x 2 1/2"
G	3	Base Panel 2" x 9 1/2"
H	1	1/2-13 NC Nut
I	1	1/2" Flat Washer
J	1	5/8" Flat Washer
K	1	1/4" Flat Washer

FIGURE 12-32 Project 12-3. © Cengage Learning 2012

Cutting Out

The first step is to cut the mild steel bar stock to length with a shear, or thermally. Put on all of your PPE and follow all shop and manufacturer's safety rules for cutting.

Fabrication

This project will be fabricated in steps.

1st Step—Assemble the top, back, and bottom triangle beams.

- Place one 2-in. strip flat on the welding table.
- Hold the next two strips on top of the first strip so that they form a triangular shape.
- Make tack welds just inside of the ends of the triangular-shaped assembly.
- The tack welds should be approximately 1/4 in. long and about 1/2 in. from the ends of the joint.
- At the center of the joint, make an additional set of tack welds on the outside.
- Repeat this process until all three of the triangular beams have been assembled.

Welding

Set the welder voltage and wire feed speed for short-circuit arc welding according to Table 12-1. Put on all of your PPE and follow all shop and manufacturer's safety rules for welding and grinding.

Lay one of the triangular beams flat on the welding table. Holding the welding gun at a 30° pushing angle, make a fillet weld all the way down the joint. At the end of the weld, the gun trigger can be pulsed several times to fill the weld crater. Repeat this process, alternating between pushing, perpendicular, and pulling gun angles until all three beams have been welded.

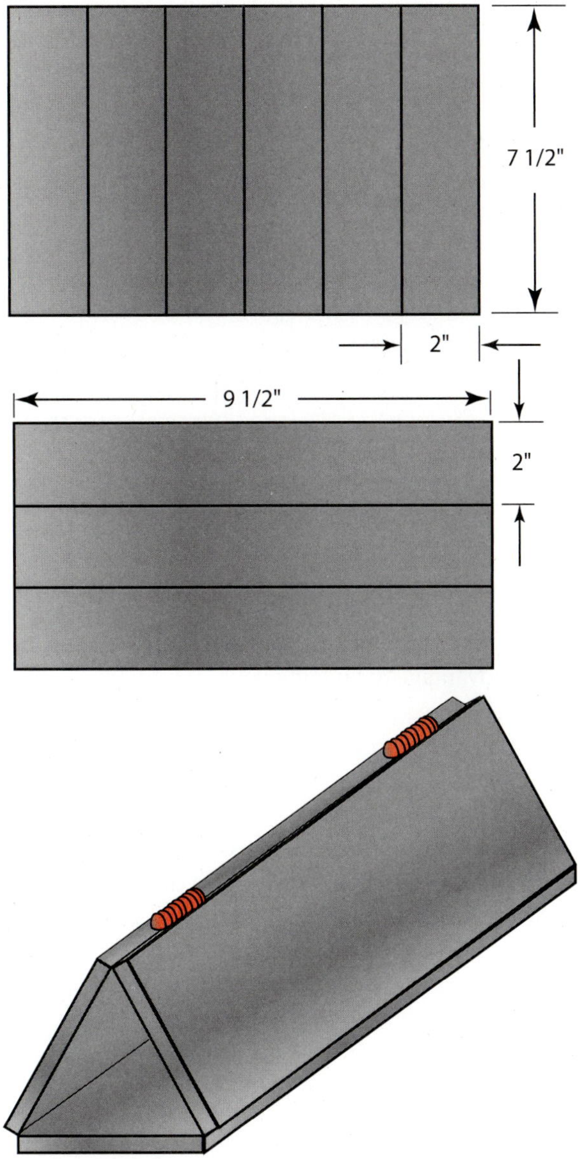

FIGURE 12-33 C-clamp assembly.

2nd Step—Assembling C-clamp beams.

- Once the three triangular beams are welded, they can be assembled according to the layout shown in Figure 12-33.
- Use the square or the magnetic clamp to hold the beams squarely together.
- Make tack welds on the beam joints to hold them in alignment for welding.

Welding

Once all of the beams have been tack welded in place, you are going to use the same welding settings and PPE as before to complete the welding on the C-clamp frame. It is important that the welds that you make to hold the beams together are strong. If you feel that any of these welds is not as strong as it should be, you can use the angle grinder to remove the weld. Grinding out a weld to reweld a joint is common practice in high-strength welding applications. When you are grinding out the weld, try to avoid grinding away any of the base metal.

Once the beams are welded, check to see that weld distortion did not cause them to twist out of shape by placing the C-clamp frame on a flat surface. The frame should lie flat and not "rock." If it is not flat, you can use a vise and twist it back into shape before attaching the screw.

3rd Step—Assemble the bolt and top jaw assembly.

- Screw the 3/4-10NC 14-in. connector onto the 3/4-10NC 14-in. bolt.
- Place the 3/4-in. flat washer on top of and in the center of the 1-in. flat washer and use the C-clamp to hold them tightly together on the welding table. Use the steel rule to check the center.
- Make four small fillet welds as shown on the number 1 welding symbol in Detail A, Figure 12-32. Cool and clean the welds.
- Screw the 3/4-10NC nut just far enough onto the bolt so that the assembled washers will fit as shown in Detail A, Figure 12-32.

Welding

Place the 3/8-in. washer on the end of the bolt and hold it in place with your gloved finger. Rest the welding gun nozzle on your gloved hand, lower your helmet, and fill the hole in the 3/8-in. washer by making a small plug weld. Screw the nut down on the bolt far enough so that the top of the plug weld can be ground flat. Screw the nut back down to the end so the washer assembly is close but not tight to the 3/8-in. washer. The washer assembly needs to be loose enough to turn but not so loose that the 3/8-in. washer extends past the washer assembly as the C-clamp is tightened, Figure 12-34.

NOTE: If the center washer touches when the C-clamp is tightened on a surface, it may not stay in place. If this happens, you can add another 1-in. washer to the one already on the top jaw assembly by tack welding the edges of the washers together.

Protect the bolt threads from weld spatter and make four small fillet welds to hold the nut in place as

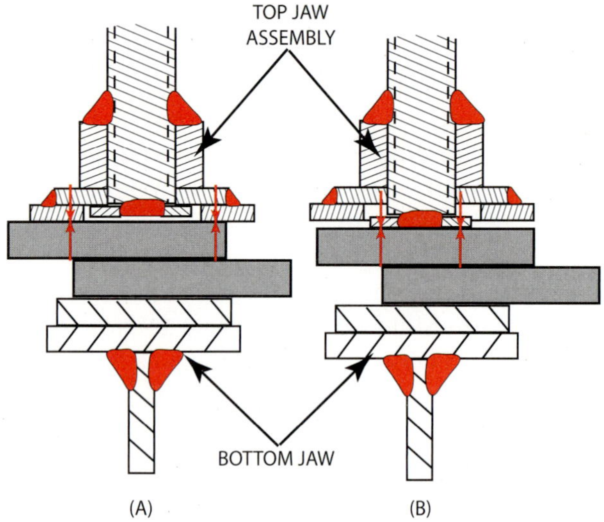

FIGURE 12-34 Washer assembly on C-clamp.

shown on the number 2 welding symbol in Detail A, Figure 12-32.

The screw threads can be protected by using one of the commercially available anti-spatter products. You can also wrap the threads with a damp shop towel. Do not use a wet towel because it could quench the welds, making them brittle and weak.

4th Step—Mounting the screw assembly to the C-clamp beams.

- Run the connector up the threads of the bolt so that the upper C-clamp jaw is touching the bottom jaw and the connector is even with the top beam.
- Align a flat side of the connector so it is centered on the top beam. Grind off any excessive weld bead material that might interfere with these parts fitting tightly together.
- Protect the bolt threads from weld spatter, and tack weld the connector to the top beam.
- Run the bolt in and out to be sure it is lined up with the lower jaw and moves freely in the top jaw assembly.

Welding

Weld the screw assembly to the top jaw. Be sure to fill the weld end crater.

Finishing

Cool off the metal, chip off any silicon slag, and wire brush the surface. As long as the C-clamp is kept dry,

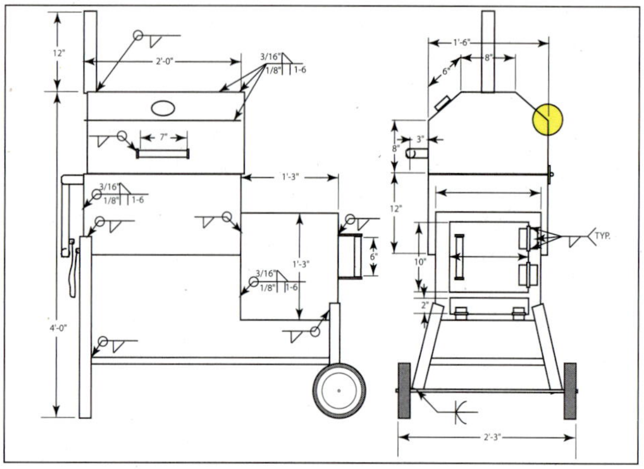

FIGURE 12-35 Outside corner joint on a barbeque grill.

it will not rust. If you are using it in a damp area, you may want to spray it with a light oil to prevent rusting.

One advantage to building your own C-clamp is that if the screw threads get damaged by weld spatter, you can cut off the end and remove and replace the bolt.

Paperwork

Complete a copy of the time sheet in Appendix I, the bill of materials in Appendix III, and the performance qualification test record in Appendix IV or use forms provided by your instructor.

Outside corner joints can be made with higher travel speeds than many other weld joints. This reduces the amount of welding heat and that reduces postweld distortion, Figure 12-35. ◆

Tee Joint

The tee joint is made by tack welding one piece of metal on another piece of metal at a right angle, Figure 12-36.

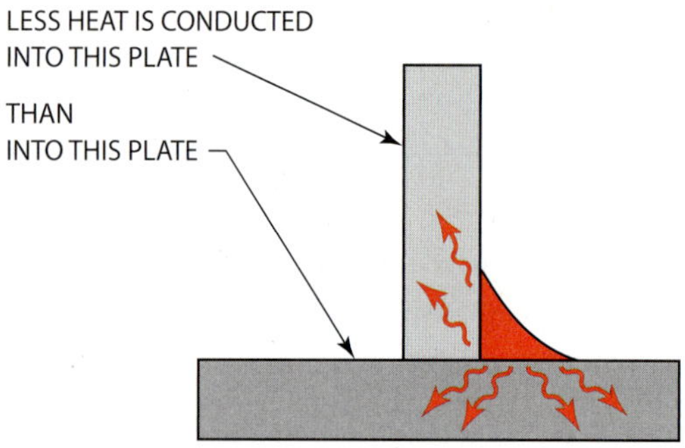

FIGURE 12-36 Heat conduction from a tee joint weld.

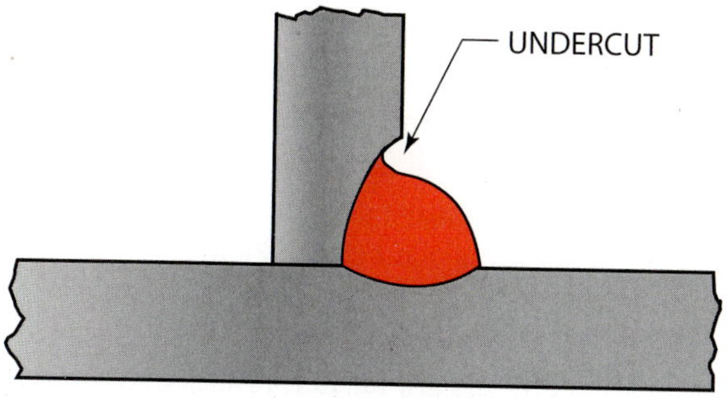

FIGURE 12-37 Undercutting along the top of a horizontal tee joint weld can be a problem. © Cengage Learning 2012

The heat is not distributed uniformly between both plates during a tee weld. The edge of the plate that forms the stem of the tee conducts heat away in only one direction while the baseplate conducts the heat in two directions. So, you need to direct more of the heat from the arc onto the baseplate to keep from melting away the stem. If the stem plate is overheated, the weld size will not be uniform, and it is difficult to impossible to control undercutting along that edge of the plate, Figure 12-37 (find the undercut on tee joint).

A welded tee joint can be strong if it is welded on both sides, even without having deep penetration, Figure 12-38. The weld will be as strong as the base plate if the size of the two welds equals the total thickness of the baseplate. The weld bead should have a flat or slightly concave appearance to ensure the greatest strength and efficiency, Figure 12-39.

Tee joints are used to join two pieces of metal at right angles. The joining of an I-beam and H-column in a building's construction is a good example of a tee joint, Figure 12-40.

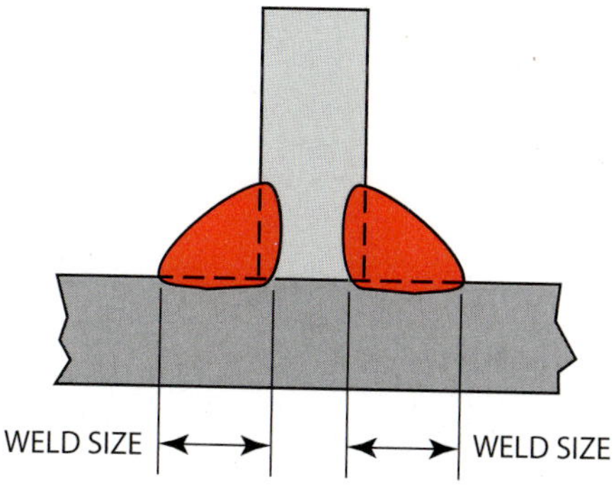

FIGURE 12-38 The tee joint is strong if the size of the two welds equals or is thicker than the thickness of the baseplate. © Cengage Learning 2012

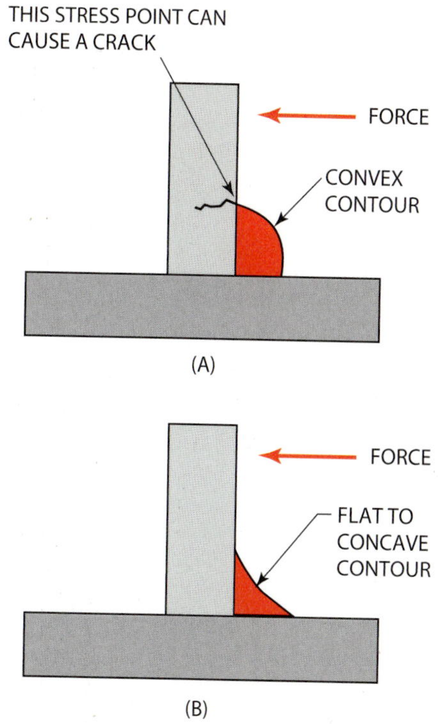

FIGURE 12-39 A fillet weld on a tee joint is stronger when the surface of a fillet weld has a flat or slightly concave surface. © Cengage Learning 2012

FIGURE 12-40 Hundreds of welded tee joints were made on this roller coaster at Silver Dollar City in Branson, Missouri. Larry Jeffus

PROJECT 12-4

Tee Joint to Make a Plant Stand

Skill to be learned: The ability to construct a tee joint that has little or no distortion due to postweld bead shrinkage. In addition, you will learn how to fit coped and mitered 90° corners, make fillet welds, and fabricate matching angled joints, Figure 12-41.

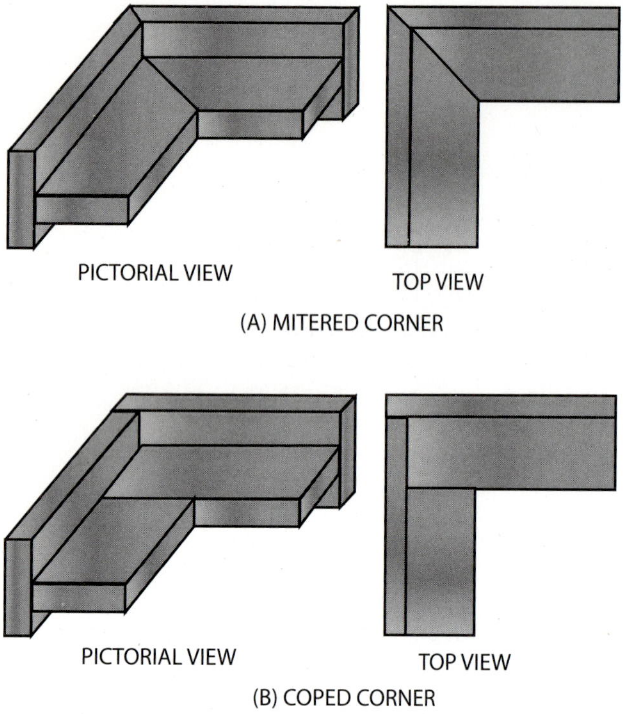

FIGURE 12-41 Two methods of fitting corners: (A) mitered and (B) coped. © Cengage Learning 2012

Project Description

The plant stand will be fabricated out of flat bar stock that is being welded into a tee joint spaced so that the ceramic tile top will fit perfectly. It will have angle iron legs. The decorative ceramic tile top adds to the designer look of this plant stand, Figure 12-42.

Project Materials and Tools

The following items are needed to fabricate the plant stand.

Materials

- 9 ft of 1 × 1/4-in. thick mild steel bar
- 32 in. of 1 × 1 × 1/8-in. angle iron
- 12-sq in. decorative ceramic tile
- Paint

Tools

- GMA welder
- Pencil and paper
- Plasma cutting torch, oxyfuel cutting torch, or shear
- PPE

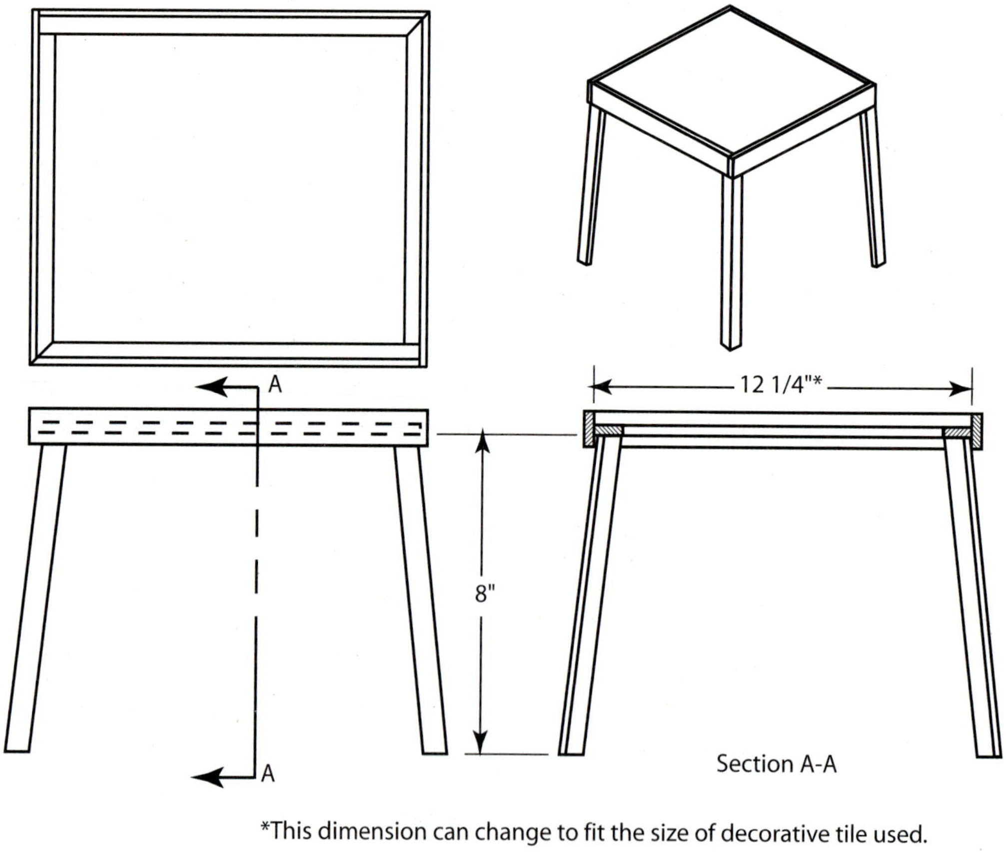

FIGURE 12-42 Project 12-4. © Cengage Learning 2012

- Awl or punch
- Tape measure
- Angle grinder
- Hammer
- Combination square
- Soapstone
- C-clamps
- Wire brush
- Square
- Pliers

Layout

Measure the side dimensions of the ceramic tile, place it flat on a table, and use the combination square to accurately measure the thickness. Add 2 in. to the measurement of the tile, and using the tape measure and soapstone, mark that distance on the 1-in. bar stock to make 12 pieces of uniform length.

Measure and mark four 8-in. long pieces of the angle iron for the legs.

Cutting the Parts

Use the plasma cutting torch, oxyfuel cutting torch, or shear, cut the bar stock and angle the iron to length.

Assemble the T Bars

On four of the 1-in. bar stocks, you are going to scribe a line to locate the stem portion of the tee joint. You are going to measure the distance from the top of the 1-in. bar stock to locate the stem. By placing the stem piece of the tee below this line, the ceramic tile surface should be even with the side rail of the plant stand. Use the combination square to accurately measure and mark the thickness of the tile on the 1-in. stock. Using the combination square and a sharp pointed tool such as an awl or punch, scribe a line along the length of the bar stock, Figure 12-43. You may need someone to hold the bar stock or put it in a vise to make the scribed line accurately. Mark a soapstone *X* on the side of the line where the stem of the tee will be set, Figure 12-44.

NOTE: Marking the side of a line where a part fits is a good practice because sometimes the line may be near the center, or someone else is placing the part to a line you drew. In that case, it may have to be remeasured to make sure everything is in the right place.

Two of the corners of the top T rail will be coped. Because all of the pieces of coped corners are cut at a 90° angle, they are often easier to make than mitered corners. The coped corners can be laid out before the T rail is welded. The easiest way to accurately lay out a coped corner is to place the parts in position and

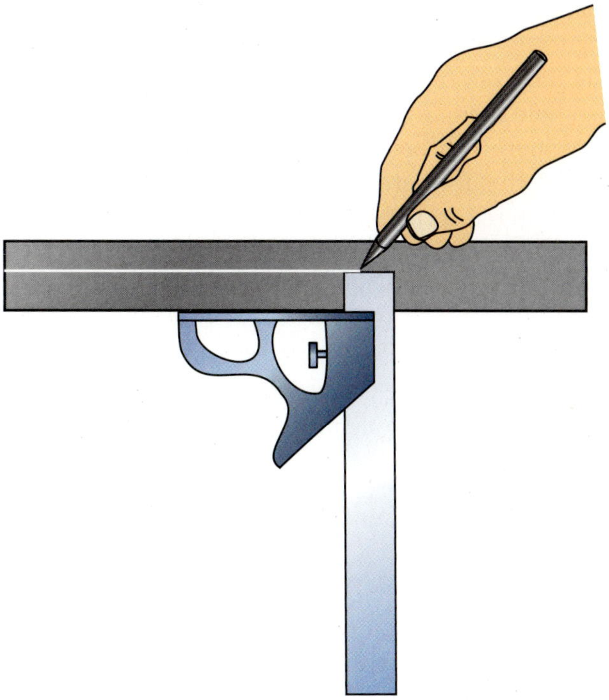

FIGURE 12-43 **Scribing a straight line using a combination square.** © Cengage Learning 2012

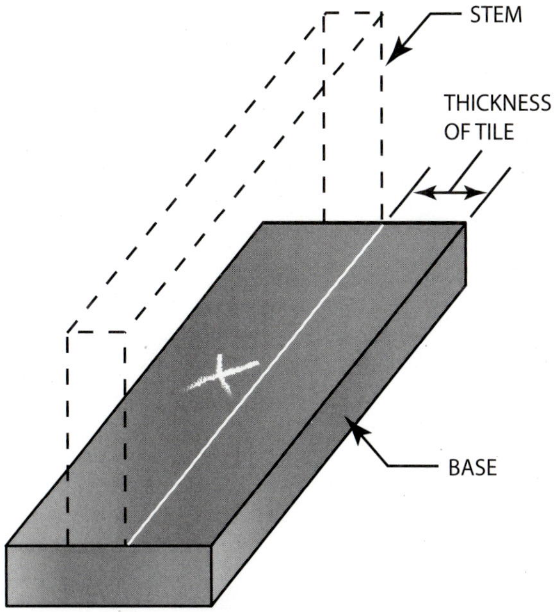

FIGURE 12-44 **Mark an *X* on the side of the line where the stem of the tee will sit.** © Cengage Learning 2012

mark them. Follow the layout for the coped corner in Figure 12-44 by offsetting the base and stem parts of the T so they fit. Mark the position of the parts so you can hold them in the correct place for welding.

There will be excessive stock left at the other end of the T where you will make the mitered corner. Mitered corners are usually made after the part is assembled, and this extra stock length will not cause a problem for you when cutting the miter joint, so it can be left in place.

Clamp the two bars together using two C-clamps. If you only snug the C-clamps, you can use a hammer to tap the parts into alignment. Once they are aligned, tighten the C-clamps.

Welding the Tee Joints

Weld shrinkage can pull the stem portion of the tee joint out of alignment after the weld has cooled. To prevent this you can offset the stem to an angle equal to the angle of the finished weld, Figure 12-45. Another way to reduce distortion is to make small intermittent welds. Since the ceramic tile needs to fit into the finished frame, small welds are needed anyway.

Set the welder voltage and wire feed speed for short-circuit arc welding according to Table 12-1. Put on all of your PPE and follow all shop and manufacturer's safety rules for welding.

Make the intermittent welds on the top side of the stem according to the specifications shown in the drawing, Figure 12-42.

When the welds have cooled, check the T for squareness. If it is within 1/16 in. of square, then make the same size welds on the back side of the stem; however, if the part is distorted, making a larger size weld on the opposed side may correct the distortion.

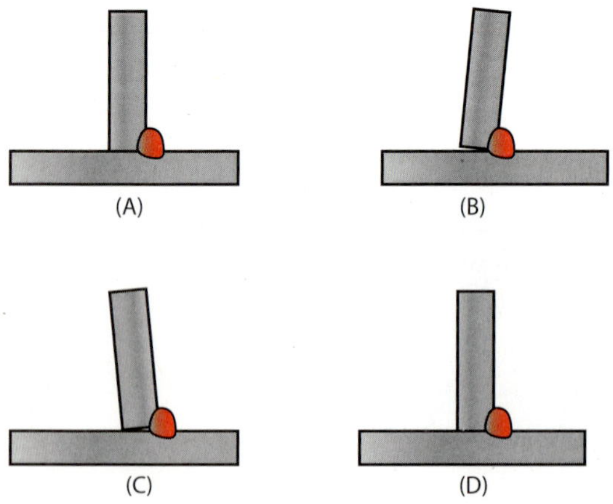

FIGURE 12-45 Weld shrinkage can pull the stem of tee joints out of alignment. © Cengage Learning 2012

Repeat this process until all four of the T side rails have been welded. Make adjustments in the weld sizes as necessary to try to make the T as close to square as you can.

Use a square to align two of the T side rails to a 90° angle. Use the angle grinder if it is necessary to remove any weld metal from the corners so that they fit squarely together.

Make a 1/4-in. tack weld on the bottom side of the T side rails. Recheck squareness and make two more small tack welds on the bottom and outside of the corner. Do not make the tack welds too large so that, if necessary, you can make a slight adjustment in the frame to fit the ceramic tile.

Fitting the Top

Once the two coped corner pieces have been tack welded, place them on the welding table and fit the ceramic tile in place. Make a mark on the base of the T side rails that is 1/4 in. longer than the tile, Figure 12-46. When the plant stand is finished, this will give you a gap of approximately 1/8 in. on all sides of the tile. Use the combination square and a sharp pointed tool such as an awl or punch to scribe a line at a 45° angle on the four corners of the T side rails, Figure 12-47.

Cutting the Mitered Corners

Cut the 45° degree marks using a plasma cutting torch or oxyfuel cutting torch. Make sure you cut on the scrap side of the line. Use the angle grinder if needed to grind any roughness off of the cut edge so that the two pieces fit evenly at the corners. Test your fitting by placing the ceramic tile on the top rails.

Welding the Corners

Tack weld the mitered corners on the outside of the miter joint. Retest the fit-up by placing the ceramic

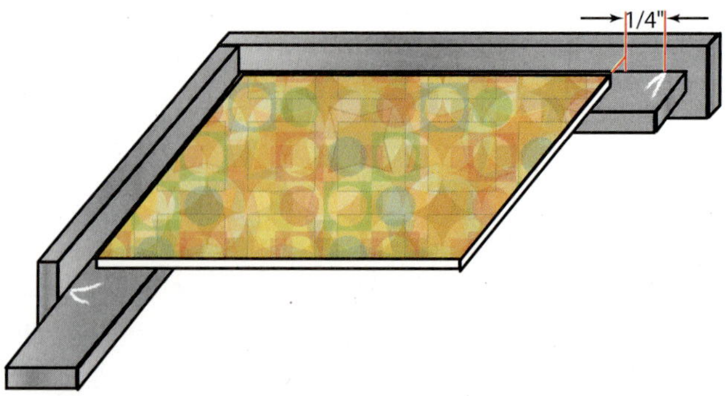

FIGURE 12-46 Marking the plant stand rails to allow for a small gap or space around the tile. © Cengage Learning 2012

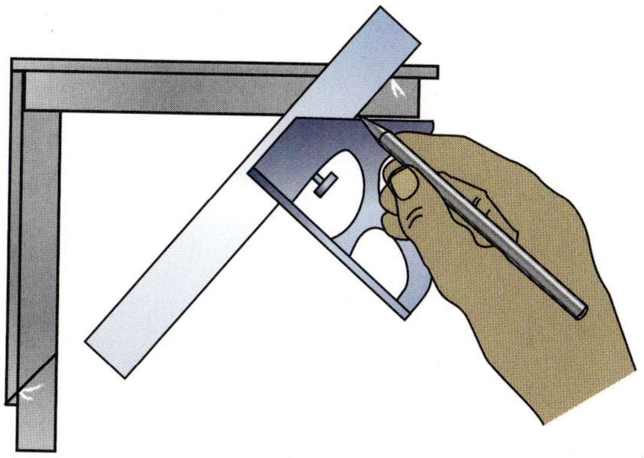

FIGURE 12-47 Marking a 45° angle. © Cengage Learning 2012

tile on the top rails. If the tile fits, place two more tack welds on the frame at each of the mitered corners. If the tile does not fit, place the frame on the long corner and tap the opposite corner with a hammer, Figure 12-48. Retest the fit with the tile and repeat the adjusting with the hammer until the tile fits into the frame.

Weld the outside and back side of the corner joints. Welds on the top of the joints will interfere with the tile fitting squarely.

Fitting and Tack Welding the Legs

The legs of the table spread out 1 in. to make the plant stand a little more stable. Place the welded top

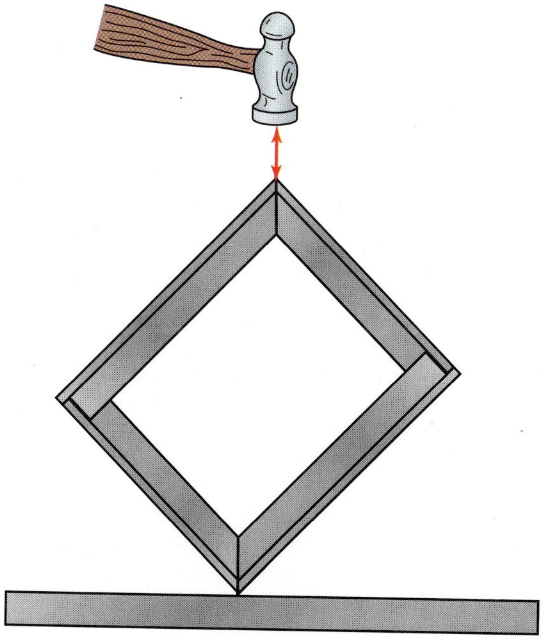

FIGURE 12-48 Adjust the frame to make it square, if necessary, by tapping with a hammer. © Cengage Learning 2012

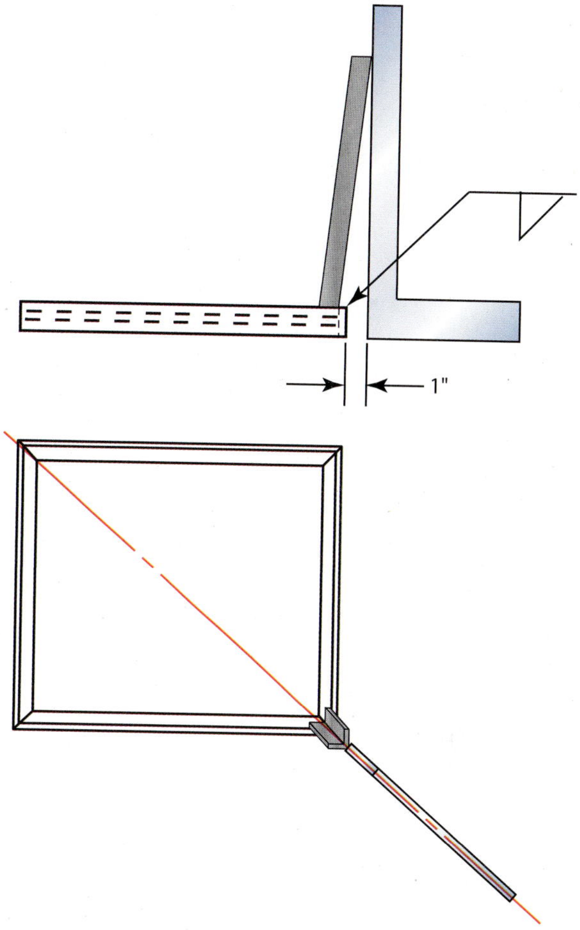

FIGURE 12-49 Fitting and tack welding the plant stand legs. © Cengage Learning 2012

T rail upside down on the welding table. The end of the leg fits inside the frame and rests on the stem of the tee. Use a square that is in a line with the opposite side of the frame as shown by the red center line in Figure 12-49. Using the try square scale, measure 1 in. from the corner of the frame to the base of the square as shown in Figure 12-49. Make a small tack weld at the corner of the angle iron and the corner of the T rail. A tack weld in this location will allow you to bend the leg slightly to adjust its angle if needed before welding.

Repeat this process until all of the legs are tack welded to the plant stand. Turn the stand over and place it on the welding table. Check to see that the table is stable and does not wobble. If one leg seems longer, it may not be at the correct angle; slightly bend it in or out to correct this problem.

Make a tack weld inside of the angle iron on the stem of the T rail. A tack weld at this location will keep the legs from moving out of alignment.

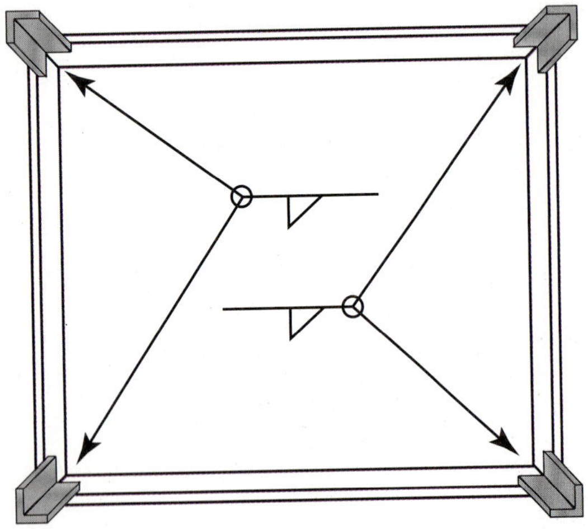

FIGURE 12-50 Make a fillet weld all the way around each of the plant stand legs. © Cengage Learning 2012

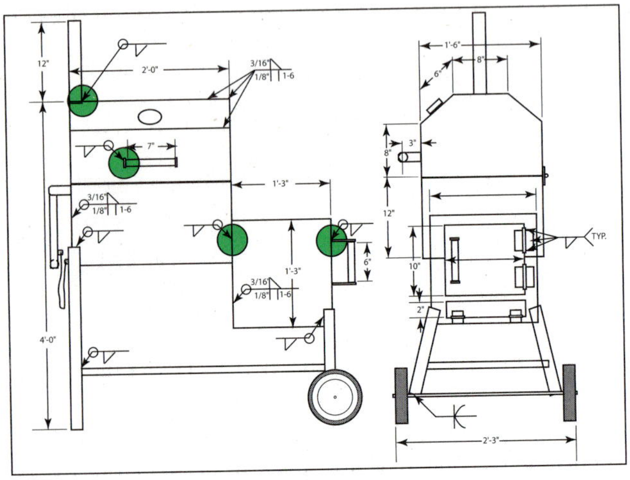

FIGURE 12-51 Tee joint on a barbeque grill. © Cengage Learning 2012

Welding

Make a fillet weld all the way around each of the legs as shown in Figure 12-50.

Finishing

Cool off the metal and wire brush the surface. Paint the side surfaces with a latex paint.

Paperwork

Complete a copy of the time sheet in Appendix I, the bill of materials in Appendix III, and the performance qualification test record in Appendix IV or use forms provided by your instructor.

The tee joint is often used to join both structural shapes and flat plate. The tee joint adds a lot of strength to the structure because the forces have two directions to work against. For that reason tee joints are often used to make a stiff corner when plates are joined, Figure 12-51.

PROJECT 12-5

Spray Metal Transfer to Fabricate a Bench Anvil

Skill to be learned: The ability to make multiple-pass GMA welds using axial spray arc metal transfer on thick sections. In addition, you will learn how to flame cut thick sections and cut through welds, Figure 12-52.

Project Description

This bench anvil is designed to be used for light to moderate service. It is capable of withstanding significant hammer blows that might damage the anvil area found on many bench vises. Even if the surface of the anvil is damaged, it can be easily welded to repair face damage. Holes can be drilled in the ears on the base of the anvil so it could be bolted to a workbench.

Project Materials and Tools

The following items are needed to fabricate the bench anvil.

Materials

- 1 3 × 8 1/2-in. piece of 1 1/4-in. thick mild steel bar
- 1 3 1/2 × 8 1/2-in. piece of 3/4-in. thick mild steel bar
- 1 5 × 8 1/2-in. piece of 3/4-in. thick mild steel bar

Tools

- GMA welder
- PPE
- Angle grinder

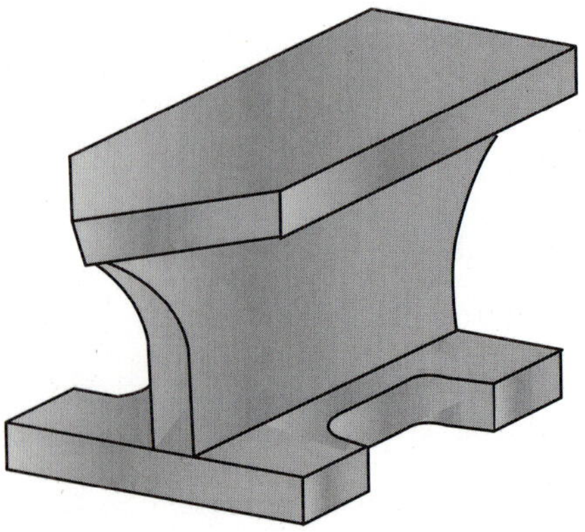

FIGURE 12-52 Bench anvil. © Cengage Learning 2012

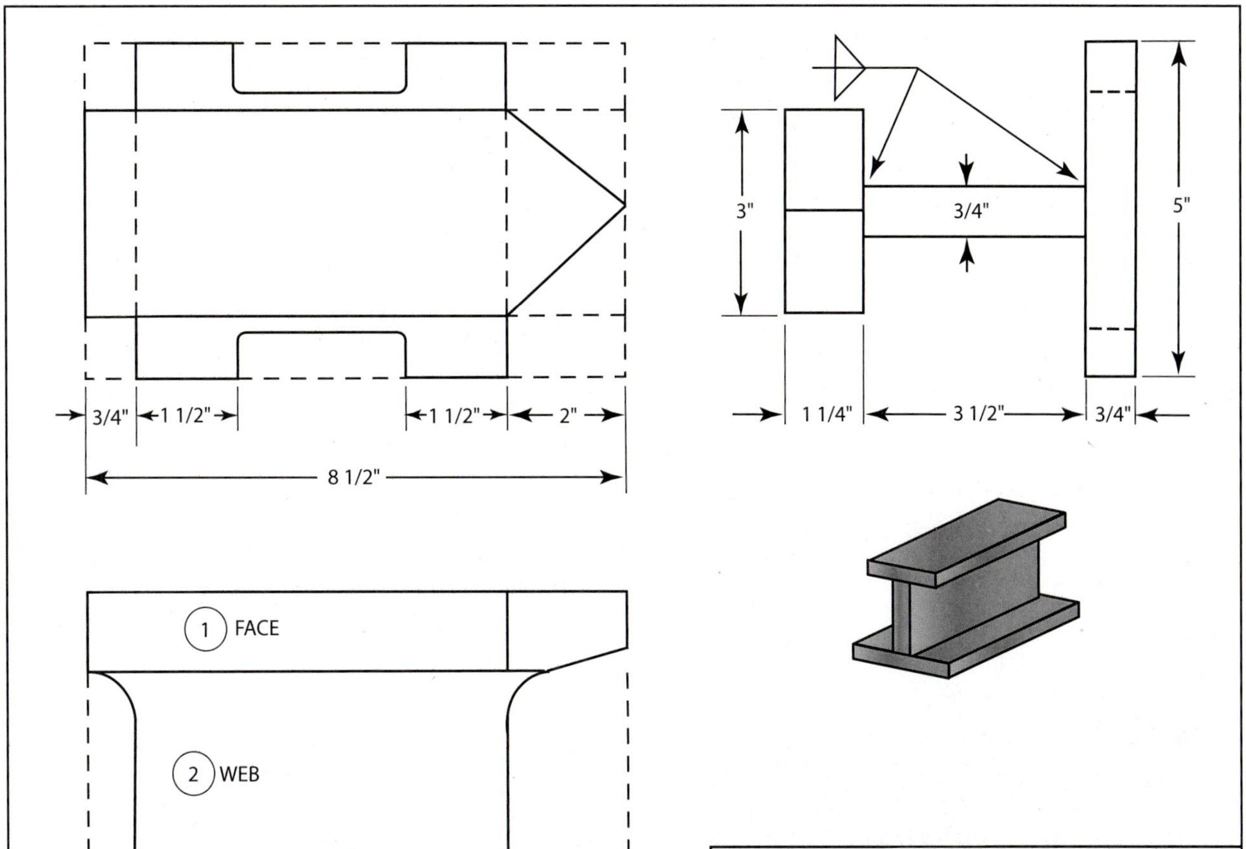

Bench Anvil		
Parts List		
Item	Qty	
1	1	3-in x 8 1/2-in x 1 1/4-in mild steel bar
2	1	3 1/2-in x 8 1/2-in x 3/4-in mild steel bar
3	1	5-in x 8 1/2-in x 3/4-in mild steel bar

FIGURE 12-53 Project 12-5.

- Wire brush
- Oxyfuel cutting torch
- Tape measure
- Square and/or 12-in. rule
- Soapstone
- Wire brush
- Hammer

Layout

Using a square, 12-in. rule, measuring tape, and soapstone, lay out the bar as shown in the project drawing, Figure 12-53.

Cutting

Cutting thicker sections of metal with an oxyfuel torch can be more difficult to accomplish than thin sections. The primary differences are starting the cut and the cutting speed. It takes longer to preheat the metal before the torch cutting lever can be depressed; and if the torch is not pointed in the correct direction, a lot of sparks can be thrown from the cut, Figure 12-54. Watch the metal's surface to see when it starts to glow red. Sometimes it may take a minute or more to preheat the metal.

Moving too fast will result in the oxygen stream not fully penetrating the complete thickness, and moving too slow will melt the top edge of the plate. Watching both the torch track and the stream of sparks from the bottom of the cut will help you adjust your cutting speed if necessary, Figure 12-55.

Assembly and Tack Welding

Set the welder voltage and wire feed speed for short-circuit arc welding according to Table 12-1. Put on all of your PPE and follow all shop and manufacturer's safety rules for welding and grinding.

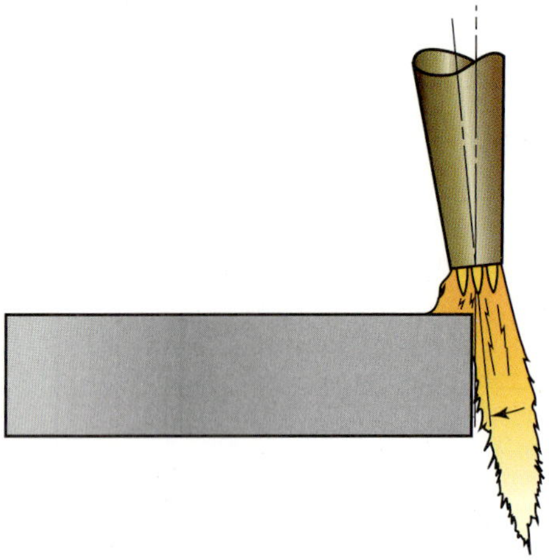

FIGURE 12-54 Thicker metal takes longer to preheat and the correct torch angle to minimize sparks. © Cengage Learning 2012

FIGURE 12-55 To make a good cut, you must watch both the torch track and the stream of sparks from the bottom. Larry Jeffus

Use the angle grinder to remove any slag that may be left from the cutting.

Place the baseplate flat on the welding table. Put the anvil web in the center of the baseplate. With

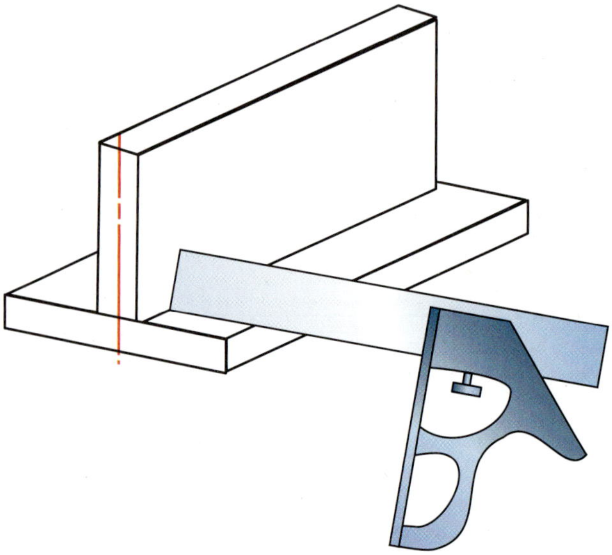

FIGURE 12-56 Center the anvil web on the baseplate. © Cengage Learning 2012

the 12-in. rule, measure the distance on both sides of the web. Adjust the web's position on the base until the measurements on both sides are the same so that the web is centered, Figure 12-56.

Use the square to check to see that the web is square to the baseplate. If it is square, make a tack weld on both ends of the web and baseplate, Figure 12-57. However, if it is not square, make a tack weld on the side with the greater angle. This will allow you to tap the web square with a hammer, Figure 12-58.

With the web square, place the faceplate on top of it. Measure it as you did before to get it to the

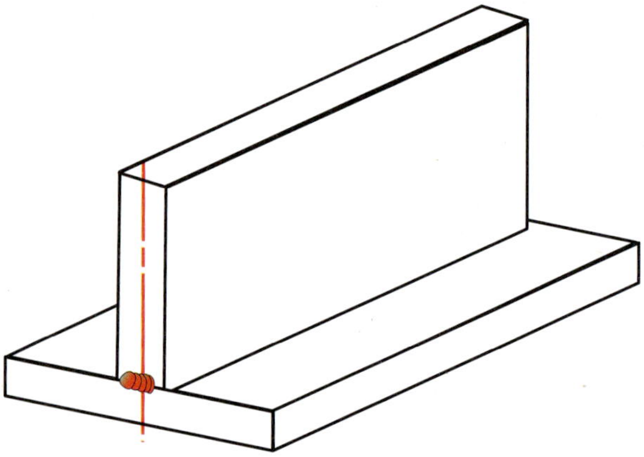

FIGURE 12-57 Place the web squarely onto the baseplate and make a tack weld on each end. © Cengage Learning 2012

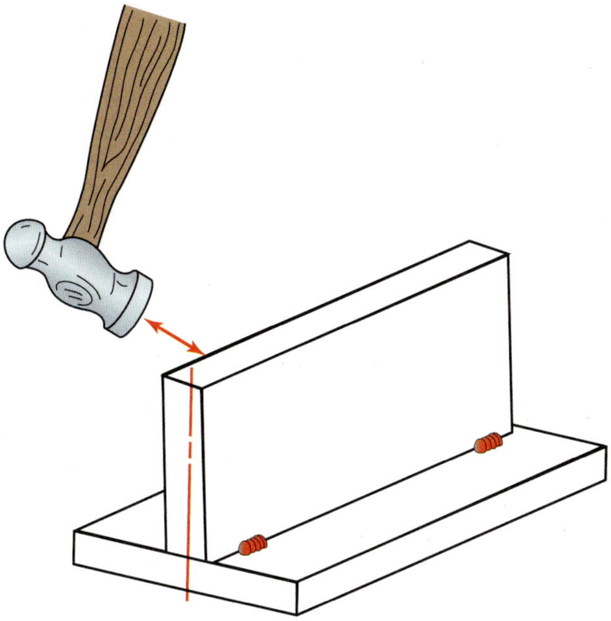

FIGURE 12-58 Making the tack welds on one side will let you tap the web to make it square if necessary. © Cengage Learning 2012

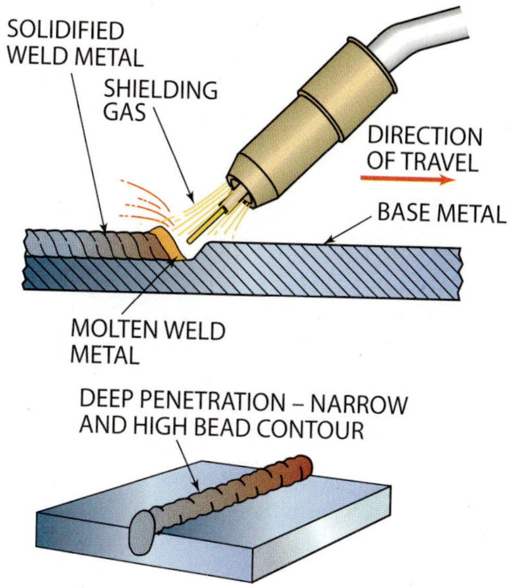

FIGURE 12-60 Use the backhand or dragging gun angle to maximize root penetration. © Cengage Learning 2012

center of the web. Measure the distance from both sides of the faceplate to the baseplate to check to see if the faceplate is square, Figure 12-59. Tack weld the faceplate to the web in the same way that the web was tack welded to the baseplate.

Welding

Set the welder voltage and wire feed speed for the spray metal arc method according to Table 12-1.

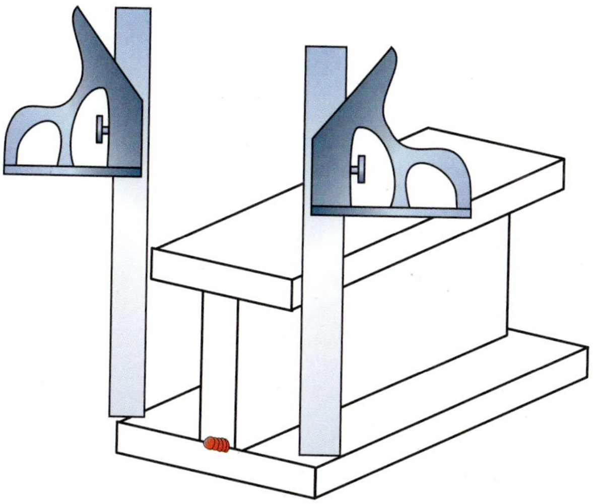

FIGURE 12-59 Measure on both sides of the faceplate to make it square and level. © Cengage Learning 2012

Starting with the joint between the baseplate and the web, brace the assembled anvil so that the tack weld will be made in the flat position. Make a 1-in. long tack weld on both ends of the joint starting about 1 in. from the end. Reposition the assembly and make the same size tack welds on the other side. Turn the anvil over and repeat the tack welding process on the joint between the faceplate and web.

NOTE: The larger tack welds are needed to hold the parts in alignment as the finished welds are made.

The first weld to be made in the multipass fillet weld is the root pass. It should be made with a backhand or dragging gun angle to maximize the root penetration, Figure 12-60. The arc of the spray metal arc transfer method can be directed at the leading edge of the molten weld pool without causing spatter like that caused by short-circuit arc welding. At the end of the weld you will need to pause and possibly pulse the gun trigger to fill the weld crater. Filling the weld crater at the end of the weld is important any time but becomes more important with multipass welds. If the crater is not filled at the end of each pass, then the end of the last weld can become farther and farther from the end of the joint, Figure 12-61.

Once the first root pass is completed, you will make the root pass on the opposite side of the same joint. Repeat this process on the other joint.

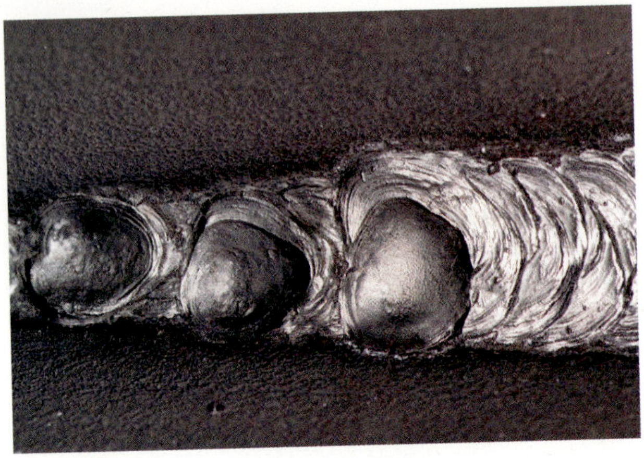

FIGURE 12-61 Each of the welds on the large multi-pass fillet weld on the anvil needs to be made all the way to the end of the joint and not stopped short of the end. Larry Jeffus

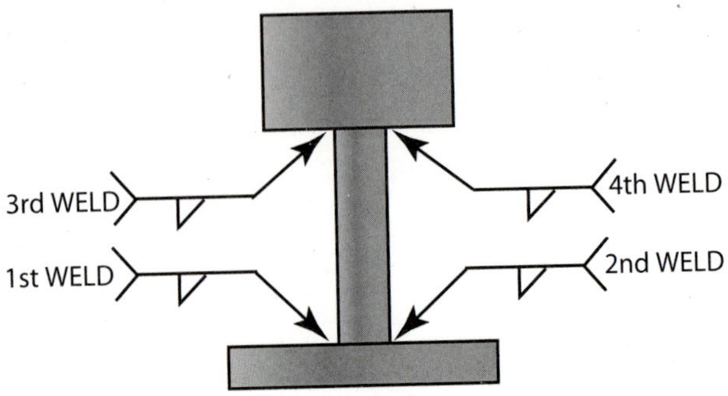

FIGURE 12-63 Alternate the welds from side to side when making the filler weld passes. © Cengage Learning 2012

The next weld passes are called the filler passes, and you will make two of these welds on both sides of each joint. The filler pass need only have complete fusion and not deep penetration, so it should be made using a forehand or pushing gun angle, Figure 12-62. The forehand gun angle makes welds with less penetration and more buildup. Alternate each filler weld pass so that only one is made at a time on any one joint, Figure 12-63. The first filler pass should be made so that most of the weld metal is deposited on the web, and the second filler weld should be directed more toward the other side of the joint, Figure 12-64.

Once all eight filler pass welds have been completed, you will make the cover pass. The cover pass, sometimes called the cap, finishes the weld. This weld will be made with some weaving so that the weld just covers both filler passes, Figure 12-64. Avoid excessive weld buildup by watching the weld surface so that the finished weld face is flat or nearly flat.

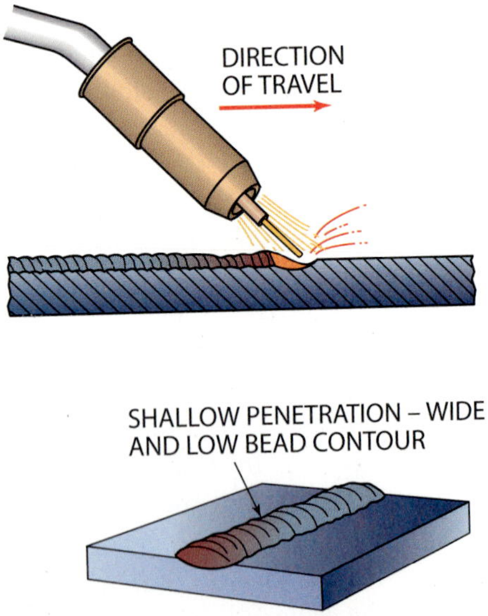

FIGURE 12-62 The forehand gun angle makes welds with less penetration and more buildup. © Cengage Learning 2012

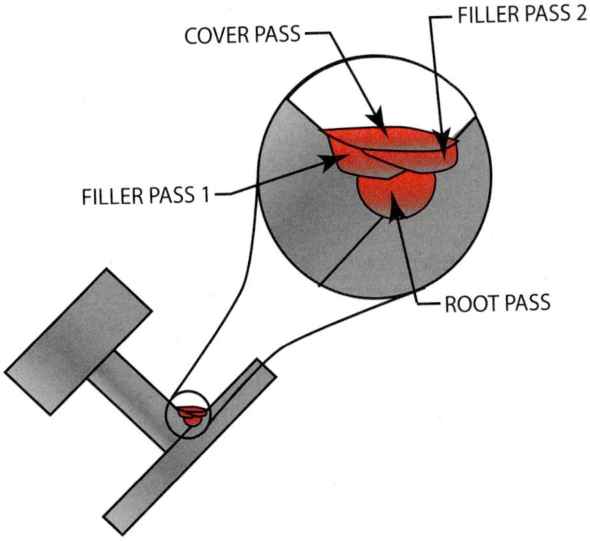

FIGURE 12-64 Filler passes sequence. © Cengage Learning 2012

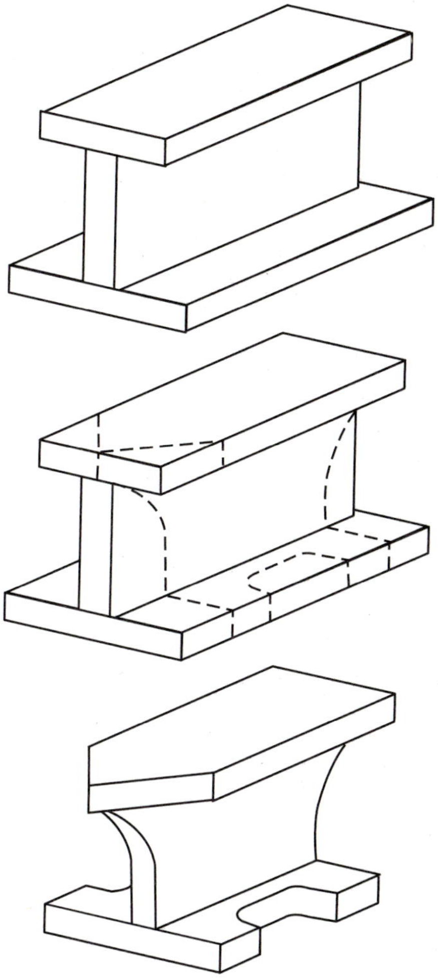

FIGURE 12-65 Anvil layout to be flame cut.

Layout

Follow the finished design layout of the anvil shown in Figure 12-65. Using the combination square, 12-in. rule, and soapstone, draw the cut lines on the finished weldment.

Cutting

Use the same oxyfuel cutting settings as you used to cut the bar stock to length to cut the weldment to its final shape.

Finishing

Grind any slag and irregularities in the cut surfaces. You need a good sharp 90° angle on the edges of the face of the anvil, so be careful as you grind the cut surfaces.

Paperwork

Complete a copy of the time sheet in Appendix I, the bill of materials in Appendix III, and the performance qualification test record in Appendix IV or use forms provided by your instructor. ◆

SUMMARY

In this chapter you have learned enough layout, cutting, fabrication, and welding skills to build the barbecue grill and smoker shown in Figure 12-66A and B. As you have built projects in this chapter, you have been learning the welding skills needed to build this grill. Similar weld joints have been highlighted in chapter figures.

Some barbecues or smokers are made from large diameter pipe. This barbecue grill smoker is made from plate. Making it from plate makes it possible to raise and lower the charcoal bed when the barbecue grill smoker is being used as a grill.

Sometimes you may want to grill a steak, chicken, or pork chop, while at other times you may want to smoke a rack of ribs, brisket, or chicken. This barbecue grill smoker is designed to let you either grill with charcoal or smoke with wood on the same cooker. This dual function makes this barbecue grill smoker very unique.

Because GMA welding has become one of the most commonly used welding processes, it is important that you learn to make high-quality welds with this process.

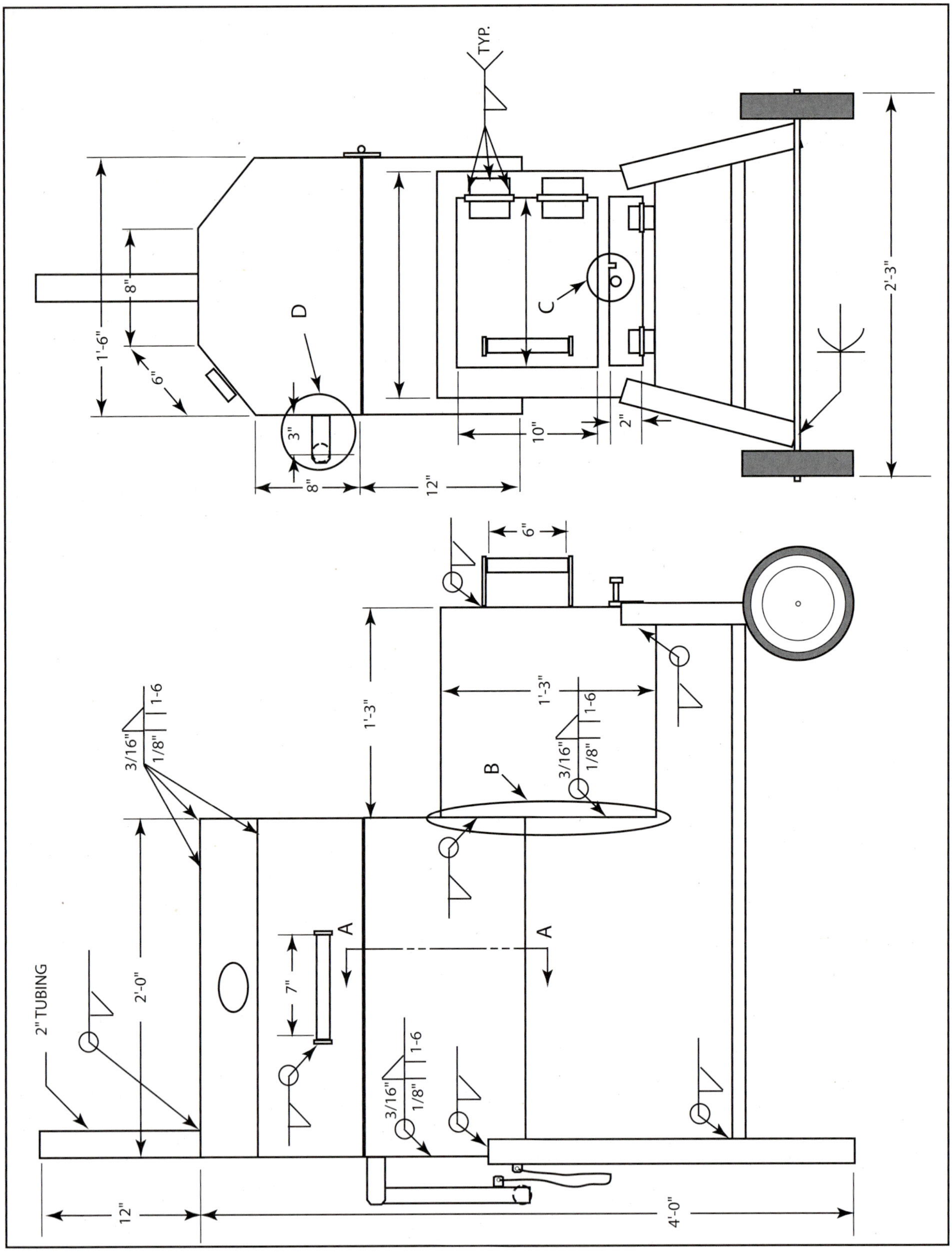

FIGURE 12-66(A) Barbeque grill and smoker plans.

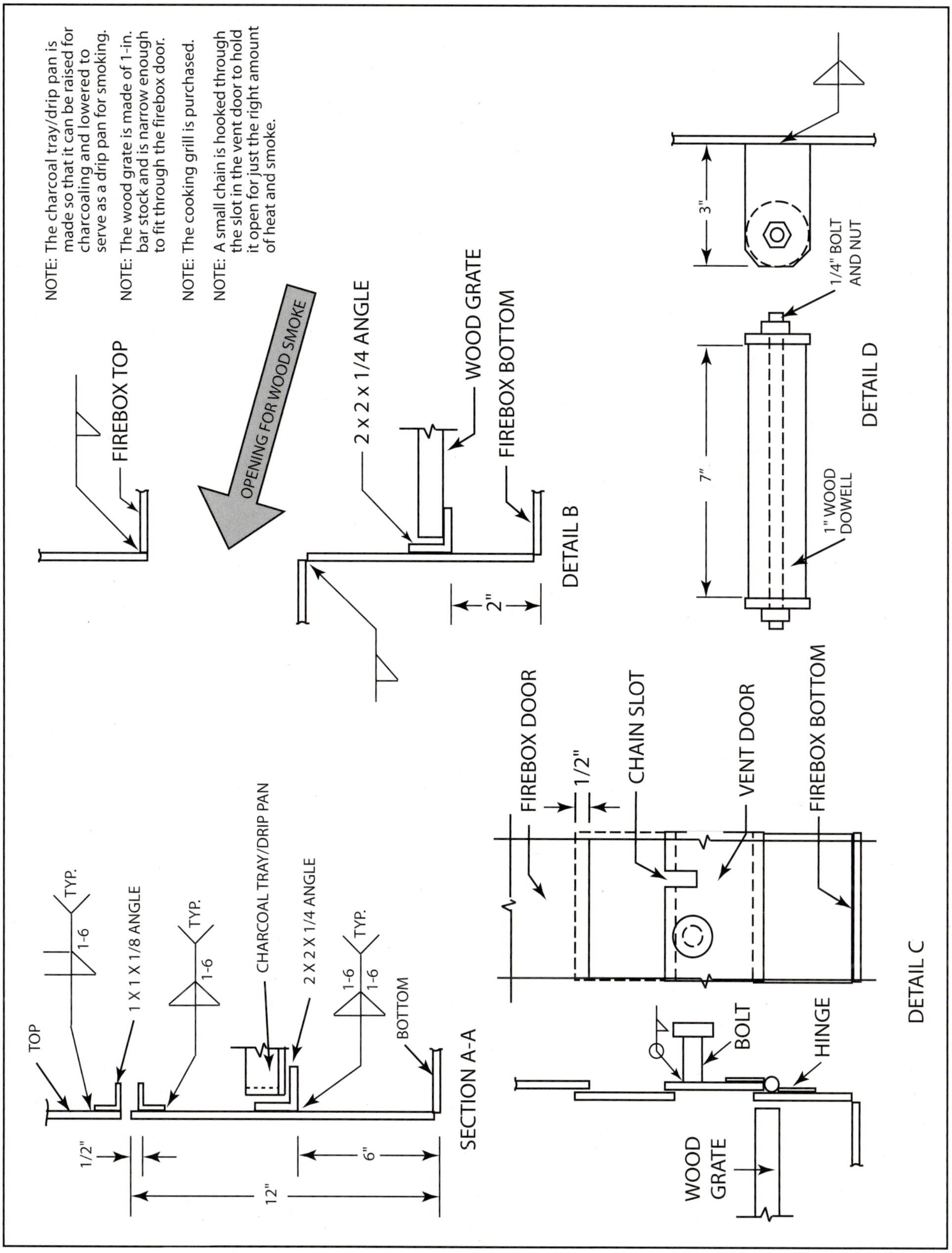

FIGURE 12-66(B) Barbeque grill and smoker plans.

REVIEW QUESTIONS

1. How does a welder set up the arc voltage and amperage for GMA welding?
2. In what unit is the wire feed speed measured?
3. What is the advantage of using a 6-second wire feed speed test rather than a 1-minute test?
4. What would be the ipm (inches per minutes) if 24 in. of wire exit the contact tube in a 6-second test?
5. Why should the electrode wire be rewound back on the spool after a wire feed speed test?
6. List three terms that all refer to the relation of the gun to the work surface.
7. What effect does arc force have on the weld?
8. Describe the gun angle needed when welding thinner materials and thicker materials.
9. What is the difference between the forehand, perpendicular, and backhand techniques?
10. Which welding gun angle would be easier to weld on thin sheet metal without melting through?
11. Which gun angle is used for automated welding?
12. What are some advantages of using the backhand welding technique?
13. Why would a welder move or weave the electrode during a weld?
14. What is electrode extension?
15. Describe how changes in electrode extension affect the weld bead.
16. What are tack welds?
17. Describe the characteristics of a good tack weld.
18. What is a lap joint?
19. How much buildup should be on a fillet weld made on a lap joint?
20. What might cause a lack of uniform width on a weld bead?
21. What might cause a lack of uniform buildup on a weld bead?
22. Describe an outside corner joint.
23. Sketch a tee joint.

Chapter 13

Flux Cored Arc Welding Equipment and Materials

OBJECTIVES

After completing this chapter, the student should be able to:

- Describe the two methods of flux cored arc (FCA) welding.
- Explain voltage and amperage characteristics of FCA welding machines.
- State the advantages and limitations of FCA welding.
- Explain the effects of travel speed, gun angle, and electrode extension on FCA welding.

KEY TERMS

carbon
cast
deoxidizers
deposition rate
dual shield
electrode feed systems
flux cored arc welding (FCAW)
forehand
helix
lime-based flux
rutile-based flux
self-shielding
slag
smoke extraction nozzles

INTRODUCTION

The **flux cored arc welding (FCAW)** process has become extremely popular for welding fabrication. Its popularity is due in part because of its portability, versatility, ease of use, and lower equipment cost. In 2005, it passed the 50% mark for all welding production, and it is still rising. The introduction of small welders, some weighing less than 30 lb (13.6 kg), has made FCA welding a very portable process. Some of these small welders can be plugged into a standard wall outlet or portable engine generator.

FCA welding is versatile. It can be used to make welds in any position in metals ranging from 24-gauge sheet metal up to an inch or more in thickness with the same diameter electrode wire. The wide range of metal thicknesses means that you may need to have only a single wire size to do most welding fabrication jobs.

Once the FCA welding equipment is set up correctly, it is easy to use. In most cases, the welder only has to follow the joint being welded. Proper setup of the equipment is the key to making good FCA welds. The popularity of the equipment has resulted in lower prices for the equipment and supplies. It has become so popular that retail stores are selling FCA welding equipment. Some larger retail stores and supply houses even carry more than one brand.

FCA welding makes a weld by having an arc between a continuously fed tubular wire electrode and the work. Heat from the arc melts the end of the wire electrode and the surface of the base metal. The molten droplets formed as the wire electrode is melted, travel across the arc and mix with the molten base metal to form the weld. The molten weld metal is protected from contamination by the gases formed as the flux core of the wire electrode vaporizes. Sometimes a shielding gas is added to help protect the molten weld. One method of FCA welding done without a gas shield is called **self-shielding,** Figure 13-1A. The other method of FCA welding that is done with a shielding gas is called dual-shielded, Figure 13-1B.

The equipment and setup for FCA welding is very similar to that of gas metal arc (GMA) welding. The major difference in the equipment is that with GMA welding, a shielding gas is always required, and FCA welding can be done with or without a shielding gas. FCA welding equipment is more portable than GMA welding equipment because no heavy shielding gas cylinder is required with FCAW. The welder can be lifted and placed near the welding job, Figure 13-2.

FCAW PROCESS

The FCA welding equipment, setup, and operation are similar to that of the GMA welding process. Both processes use the same type of constant potential (CP) or constant voltage (CV) welding machine. The terms *potential* and *voltage* when applied to electricity have the same meaning. Sometimes they are used together as *potential voltage*. So, a welder that is called a constant potential and one that is called constant voltage are the same type of welder. In this text, the term *constant potential* and *CP* will be used.

CP FCA welding machines are different from stick welding machines. Stick welding machines are constant current (CC) machines. The welder sets the amperage, and it remains constant as the weld is made. On a CP welder that is used for FCA welding, the welding voltage is set by the welder, and the voltage remains constant as the weld is made.

On CC stick welders, there are no adjustments for the arc voltage, and on a CP welder used for FCA welding there is no amperage adjustment. On FCA welders the amperage (current) is set by adjusting the wire feed speed. The higher the electrode feed speed, the higher the amperage is; and the slower the wire feed speed, the lower the amperage is.

The effects on the weld of electrode extension, gun angle, welding direction, travel speed, and other welder manipulations are similar to those experienced in GMA welding. As in GMA welding, having a correctly set welder does not ensure a good weld. The skill of the welder is an important factor in producing high-quality welds.

Some of the effects that the flux inside the electrode has on the molten weld pool include providing protection from the atmosphere, improving strength through chemical reactions and alloys, and improving the weld shape.

Without some type of protection, atmospheric contamination of molten weld metal can occur as it travels across the arc gap and within the pool before it solidifies. The major atmospheric contaminations come from oxygen and nitrogen, the major elements in air. The addition of fluxing and gas-forming elements to the core electrode reduces or eliminates their effects.

Flux additives improve strength and other physical or corrosive properties of the finished weld. Small additions of alloying elements, **deoxidizers,** and slag agents can all improve the desired weld properties. **Carbon,** chromium, and vanadium are some of the alloying elements that can be added to improve hardness, strength, creep resistance, and corrosion resistance. Aluminum, silicon, and titanium all help remove oxides and/or nitrides in the weld. Potassium, sodium, and zirconium are added to the flux and form a slag.

The flux core additives that serve as deoxidizers, shielding gas formers, and slag-forming agents either protect the molten weld pool or help to remove impurities from the base metal. Deoxidizers may convert

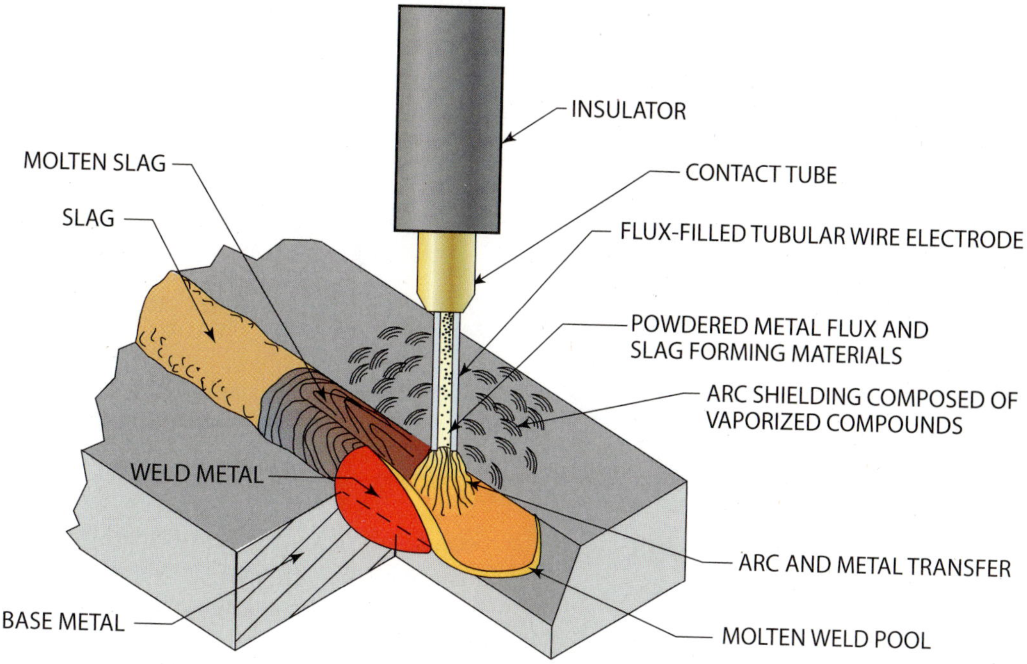

(A) SELF-SHIELDED FLUX CORED ARC WELDING (FCAW-S)

(A)

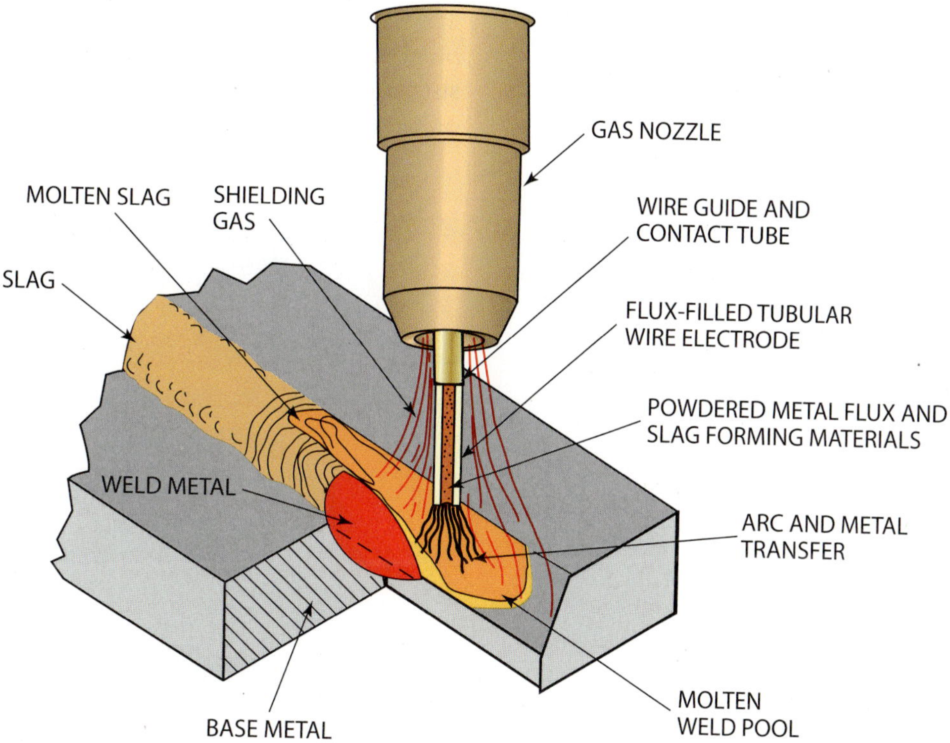

(B) DUAL-SHIELDED FLUX CORED ARC WELDING (FCAW-G)

(B)

FIGURE 13-1 (A) Self-shielded flux cored arc welding. (B) Gas-shielded flux cored arc welding.
The American Welding Society

FIGURE 13-2 FCA welder sitting on a wood frame so that it can be lifted closer to the welding. Larry Jeffus

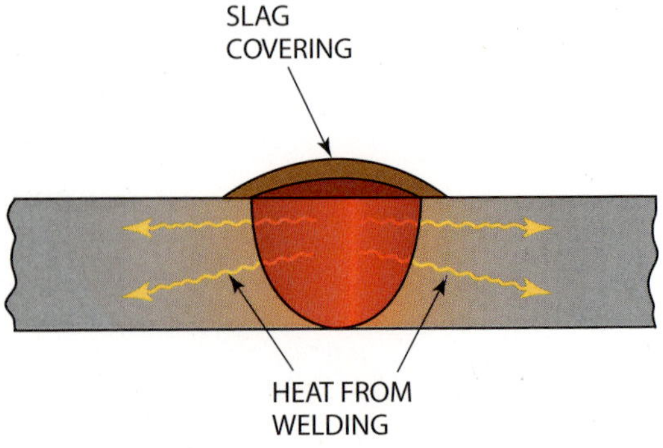

FIGURE 13-3 The slag covering keeps the welding heat from escaping quickly, thus slowing the cooling rate.

small amounts of surface oxides like mill scale back into pure metal. They work much like the elements used to refine iron ore into steel.

Gas formers rapidly expand and push the surrounding air away from the molten weld pool. If oxygen in the air were to come in contact with the molten weld metal, the weld metal would quickly oxidize. Sometimes this can be seen at the end of a weld when the molten weld metal erupts in to a shower of tiny sparks.

The slag covering of the weld is useful for several reasons. Slag helps the weld by protecting the hot metal from the effects of the atmosphere; controlling the bead shape by serving as a dam or mold; and serving as a blanket to slow the weld's cooling rate, which improves its physical properties, Figure 13-3.

Equipment

Power Supply

The FCA welding power supply must be constant potential machines. Some welding machines are capable of providing both CP and CC welding currents. These machines can be used for both FCA and stick welding. FCAW machines are available with up to 1500 amperes of welding power.

FCA Welding Guns

FCA welding guns are available as two types: water-cooled or air-cooled, Figure 13-4. Although most of the FCA welding guns that you will find in schools are air-cooled, our industry often needs water-cooled guns because of the higher heat caused by the longer welds made at higher currents. The water-cooled FCA welding gun is more efficient at removing waste heat than an air-cooled gun. The air-cooled gun is more portable because it has fewer hoses, and it may be made lighter so than it is easier to manipulate than the water-cooled gun.

Also, the water-cooled gun requires a water reservoir or another system to give the needed cooling. There are two major ways that water can be supplied to the gun for cooling. Cooling water can be supplied directly from the building's water system, or it can be supplied from a recirculation system.

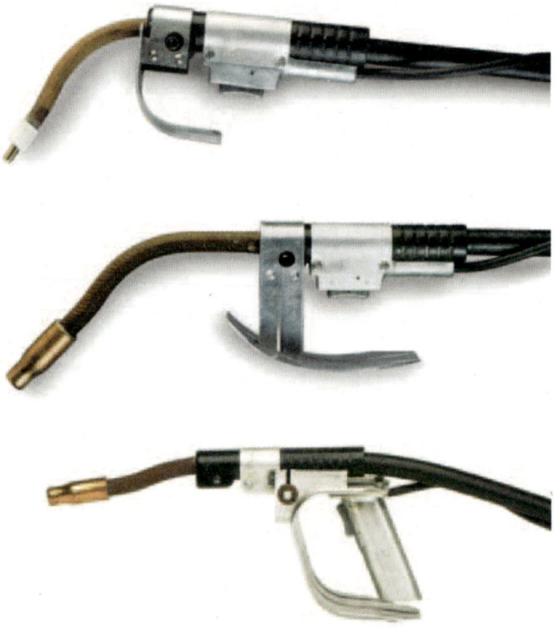

FIGURE 13-4 Typical FCA welding guns: (A) 350-ampere rating self-shielding, (B) 450-ampere rating gas-shielding, (C) 600-ampere rating gas-shielding.
Lincoln Electric Company

Cooling water supplied directly from the building's water system is usually dumped into a wastewater drain once it has passed through the gun. When this type of system is used, a pressure regulator must be installed to prevent pressures that are too high from damaging the hoses. Water pressures higher than 35 psi may cause the water hoses to burst. Check valves must also be installed in the supply line to prevent contaminated water from being drawn back into the water supply. Some cities and states have laws that restrict the use of open systems because of the need for water conservation. Check with your city or state for any restrictions before installing an open water-cooling system.

Recirculating cooling-water systems eliminate any of the problems associated with open systems. Chemicals may be added to the water in recirculating systems to prevent freezing, to aid in pump lubrication, and to prevent algae growth. Only manufacturer-approved additives should be used in a recirculation system. Read all of the manufacturer's safety and data sheets before using these chemicals.

Smoke Extraction Nozzles. Because of the large quantity of smoke that can be generated during FCA welding, systems for smoke extraction that fit on the gun have been designed, Figure 13-5. These systems

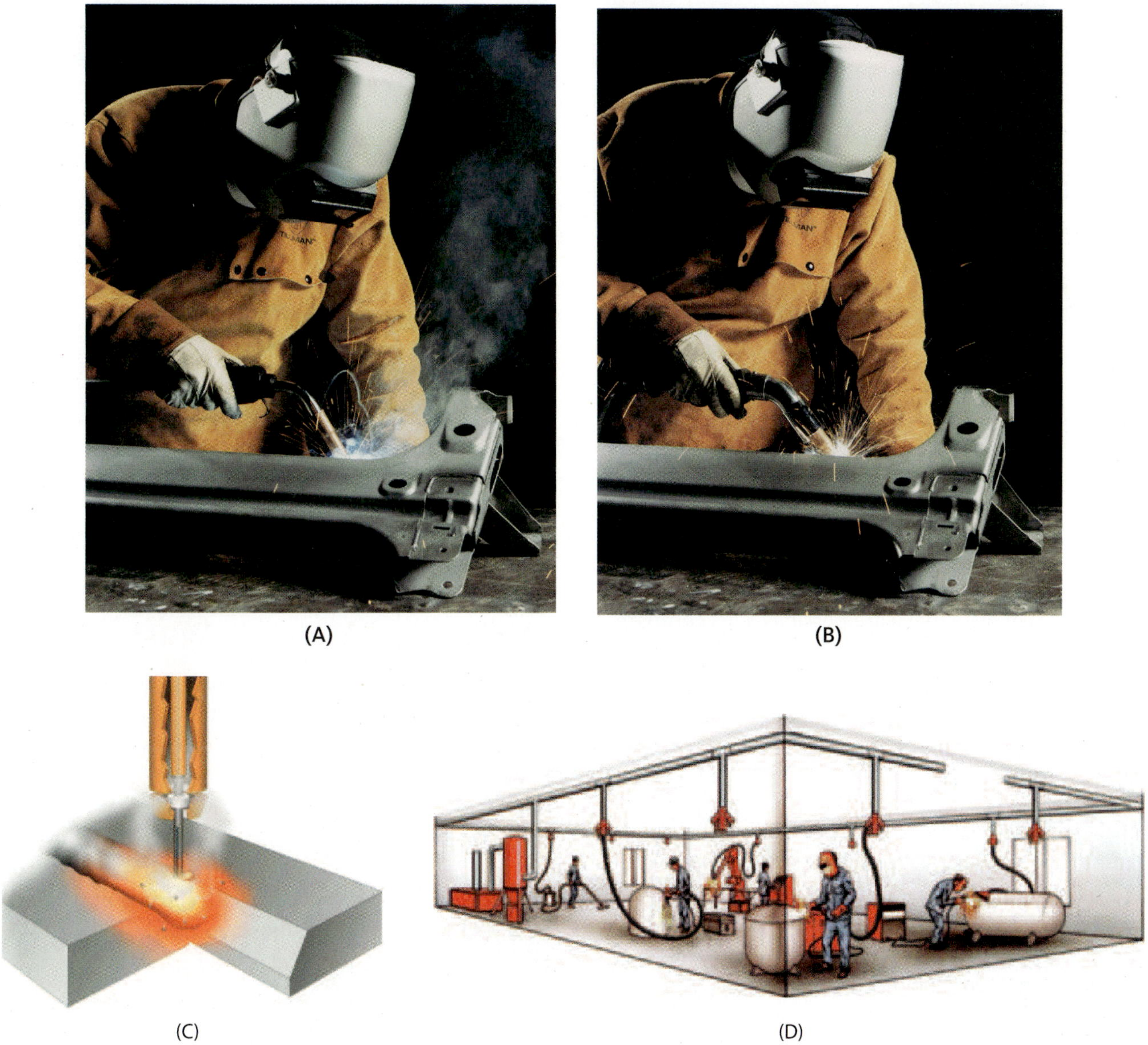

FIGURE 13-5 (A) FCA welding without smoke extraction and (B) with smoke extraction. (C) Typical FCAW smoke extraction gun. (D) Typical smoke exhaust system. The Lincoln Electric Company

use a vacuum to pull the smoke back into a specially designed **smoke extraction nozzle** on the welding gun. The disadvantage of having a slightly heavier gun is offset by the system's advantages. The advantages of the system are as follows:

- Cleaner air for the welder to breathe because the smoke is removed before it rises to the welder's face
- Reduced heating and cooling costs because the smoke is concentrated, so less shop air must be removed with the smoke

Electrode Feed. **Electrode feed systems** are sometimes called wire feeders. They use rollers to push or pull the wire through the welding cable assembly to the gun. Small feed systems may have only one driven roller while larger systems have two or more drive rollers, Figure 13-6. FCAW electrode feed systems are similar to those used for GMAW. The major difference is that larger FCAW machines that can use large diameter wire most often have two sets of feed rollers.

(A)

(B)

FIGURE 13-6 The drive roller (A) must have the wire tension roller (B) properly set with the tension spring to push the wire to the welding gun. Larry Jeffus

ADVANTAGES OF FCA WELDING

FCA welding offers the welding industry a number of important advantages.

High Deposition Rate

High rates of depositing weld metal are possible. FCA welding **deposition rates** of more than 25 lb/hr (11.3 kg/hr) of weld metal are possible. This compares to about 10 lb/hr (4.5 kg/hr) for shielded metal arc (SMA) welding using a very large diameter electrode of 1/4 in. (6 mm).

Portability

Small portable FCA welding machines can be carried almost anywhere and plugged into a standard wall outlet. Their low input power requirement means that they can be easily used with extension cords. Some even have a two-level input power switch, so that they can be used on either 15-degrees or 20-amp 120-volt circuits. Although small portable FCA welding machines have low input power requirements, they can make single-pass welds on metal as thick as 3/8 in. (10 mm) and multiple-pass welds on metal up to 1 in. (25 mm) and thicker.

Minimum Electrode Waste

The FCA method makes efficient use of filler metal; from 75 to 90% of the weight of the FCAW electrode is metal, the remainder being flux. SMAW electrodes have a maximum of 75% filler metal; some SMAW electrodes have much less. Also, a stub must be left at the end of each SMA welding electrode. The stub will average 2 in. (51 mm) in length, resulting in a loss of 11% or more of the SMAW filler electrode purchased. FCA welding has no stub loss, so nearly 100% of the FCAW electrode purchased is used.

Narrow Groove Angle

Because of the deep penetration characteristic, no edge-beveling preparation is required on some joints in metal

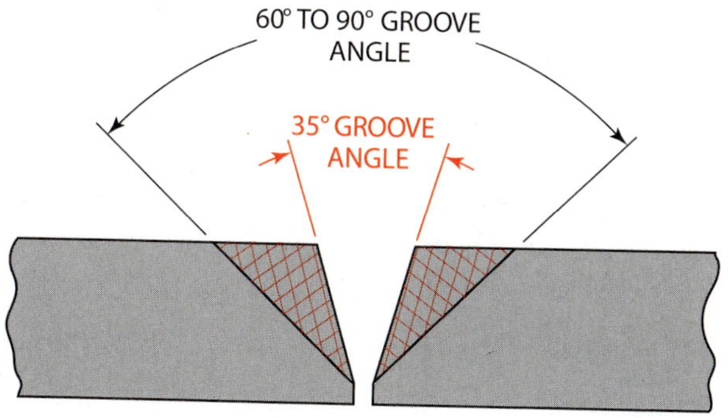

FIGURE 13-7 The narrower groove angle for FCAW saves on filler metal, welding time, and heat input into the part. © Cengage Learning 2012

up to 1/2 in. (13 mm) in thickness. When bevels are cut, the joint-included angle can be reduced to as small as 35°, Figure 13-7. The reduced groove angle results in a smaller-size weld. This can save 50% of filler metal, with about the same savings in time and weld power used.

Minimum Precleaning

The addition of deoxidizers and other fluxing agents permits high-quality welds to be made on plates with light surface oxides and mill scale. This eliminates most of the precleaning required before GMA welding can be performed. It is often possible to make excellent welds on plates in the "as cut" condition; no cleanup is needed.

All-Position Welding

Small diameter electrode sizes in combination with special fluxes allow excellent welds in all positions. The slags that are produced assist in supporting the weld metal. This process is easy to use, and when properly adjusted, it is much easier to use than other all-position arc welding processes.

Flexibility

Changes in power settings can permit welding to be made on thin-gauge sheet metals or thicker plates using the same electrode size. Multipass welds allow of joining the metals of unlimited thickness. This too, is attainable with one size of electrode.

High Quality

Many codes permit welds to be made using FCAW. The addition of the flux gives the process the high level of reliability needed for welding on boilers, pressure vessels, and structural steel.

Excellent Control

The molten weld pool is more easily controlled with FCAW than with GMAW. The surface appearance is smooth and uniform even with less operator skill. Visibility is improved by removing the nozzle when using self-shielded electrodes.

LIMITATIONS OF FCA WELDING

Types of Metals

The main limitation of flux cored arc welding is that it is confined to ferrous metals and nickel-based alloys. Generally, all low and medium carbon steels and some low-alloy steels, cast irons, and a limited number of stainless steels are weldable using FCAW.

Cost of Equipment

The equipment and electrodes used for the FCAW process are more expensive. However, the cost is quickly recoverable through higher productivity.

Postweld Cleanup

The removal of postweld slag requires another production step. The flux must be removed before the weldment is finished (painted) to prevent crevice corrosion.

Indoor Air Quality

With the increased welding output comes an increase in smoke and fume generation. The existing ventilation system in a shop might need to be increased to handle the added volume.

FCAW ELECTRODES

FCA welding electrodes are available as seamed and seamless type electrodes. Both types have tightly packed flux inside a metal outer covering. They differ in the way that they are manufactured but have little if any difference in the way they weld. Seamless electrode flux cores are less likely to absorb moisture than the seamed electrode.

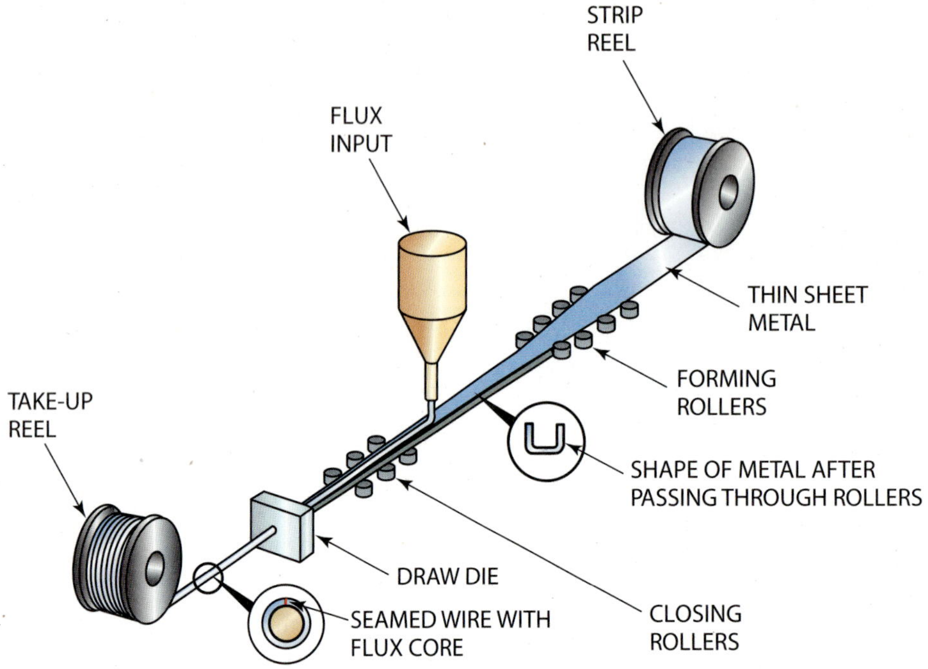

FIGURE 13-8 **Putting the flux in the flux cored wire.** © Cengage Learning 2012

Methods of Manufacturing

The electrodes have flux tightly packed inside. One method used to make them is to first form a thin sheet of metal into a U-shape, Figure 13-8. Then a measured quantity of flux is poured into the U-shape before it is squeezed shut. It is then passed through a series of dies to size it and further compact the flux.

A second method of manufacturing the electrode is to start with a seamless tube. The tube is usually about 1 in. (25 mm) in diameter. One end of the tube is sealed, and the flux powder is poured into the open end. The tube is vibrated during the filling process to ensure that it fills completely. Once the tube is full, the open end is sealed. The tube is now sized using a series of dies, Figure 13-9. In both these methods of manufacturing, the electrode, the sheet, and the tube are made up of the desired alloy. Also, in both cases the flux is compacted inside the metal skin. This compacting helps make the electrode operate smoother and more consistently.

FCAW Electrode Sizes

Electrodes are available in sizes from 0.030 in. (0.762 mm) to 5/32 in. (3.969 mm) in diameter. Smaller diameter electrodes are much more expensive per pound than the same type in a larger diameter. Larger diameter electrodes produce such large welds that they cannot be controlled in all positions. The most popular diameters range from 0.035 in. (0.889 mm) to 3/32 in. (2.381 mm).

FCAW Electrode Packaging

The finished FCA filler metal is packaged in a number of forms for purchase by the end user, Figure 13-10. The American Welding Society (AWS) has a standard for the size of each of the package units. Although the dimensions of the packages are standard, the weight of filler wire is not standard. More of a smaller diameter wire can fit into the same space than a larger diameter wire, so a package of 0.030 in. (0.762 mm) wire weighs more than the same size package of 3/32 in. (2.381 mm) wire. The standard packing units for FCAW wires are spools, coils, reels, and drums, Table 13-1.

Spools are made of plastic or fiberboard and are disposable. They are completely self-contained and are available in approximate weights from 1 lb (0.5 kg) up to around 50 lb (22.7 kg). The smaller spools, 4 in. (102 mm) and 8 in. (203 mm) weiging from 1 lb (0.5 kg) to 7 lb (3.2 kg), are most often used for smaller production runs or for home/hobby use; 12-in. (305-mm) and 14-in. (356-mm) spools are often used in welding schools and fabrication shops.

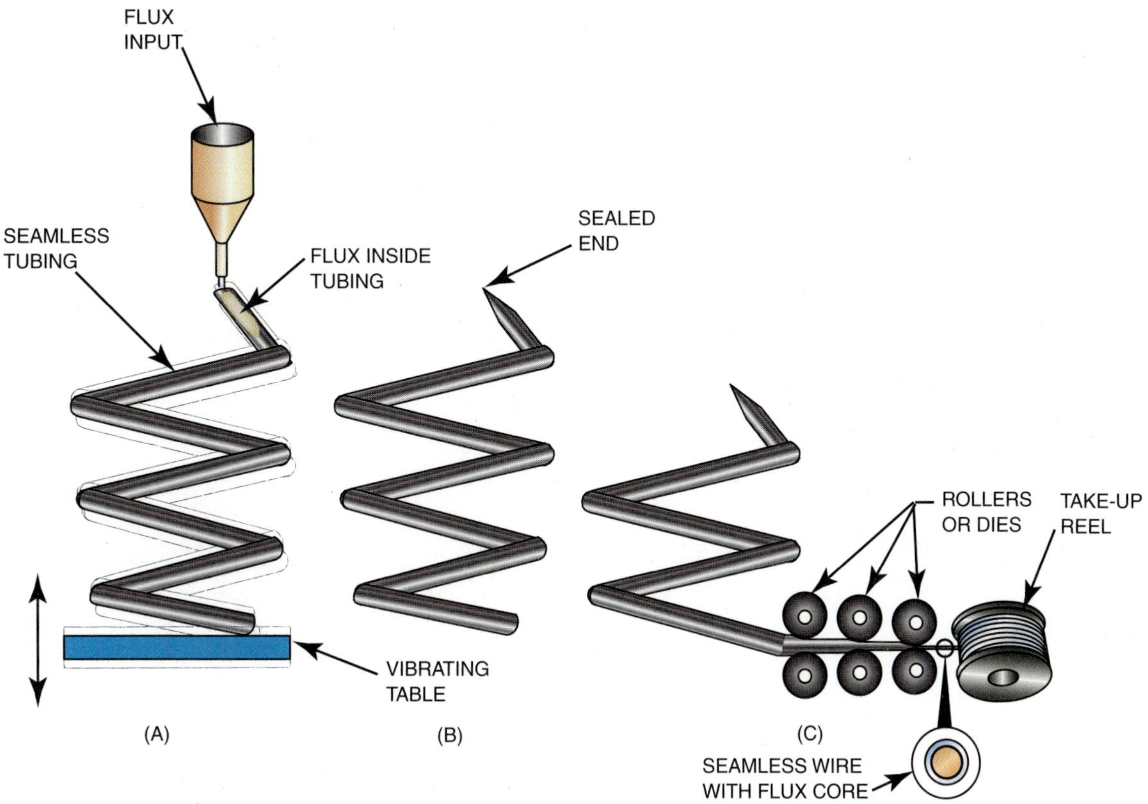

FIGURE 13-9 This shows one method of filling seamless FCA welding filler metal with flux. The vibration helps compact the granular flux inside the tube. © Cengage Learning 2012

Coils come wrapped and/or wire-tied together. They are unmounted, so they must be supported on a frame on the wire feeder to be used. Coils are available in weights around 60 lb (27.2 kg). Because FCAW wires on coils do not have the expense of a disposable core, these wires cost a little less per pound, so they are more desirable for higher-production shops.

FIGURE 13-10 FCAW filler metal weights are approximate. They will vary by alloy and manufacturer.
The Lincoln Electric Company

Reels are large wooden spools and drums are shaped like barrels. Both reels and drums are used for high-production jobs. Both can contain approximately 300 lb (136.1 kg) to 1000 lb (453.5 kg) of FCAW wire. Because of their size, they are used primarily at fixed welding stations. Such stations are often associated with some form of automation, such as turntables or robotics.

Electrode Cast and Helix

To see the **cast** and **helix** of a wire, feed out 10 ft (3.0 m) of wire electrode and cut it off. Lay it on the floor and observe that it forms a circle. The diameter of the circle is known as the cast of the wire, Figure 13-11.

Note that the wire electrode does not lay flat. One end is slightly higher than the other. This height is the helix of the wire.

The AWS has specifications for both cast and helix for all FCA welding wires.

Packaging	Outside Diameter	Width	Arbor (Hole) Diameter
Spools	4 in. (102 mm)	1 3/4 in. (44.5 mm)	5/8 in. (16 mm)
	8 in. (203 mm)	2 1/4 in. (57 mm)	2 1/16 in. (52.3 mm)
	12 in. (305 mm)	4 in. (102 mm)	2 1/16 in. (52.3 mm)
	14 in. (356 mm)	4 in. (102 mm)	2 1/16 in. (52.3 mm)
Reels	22 in. (559 mm)	12 1/2 in. (318 mm)	1 5/16 in. (33.3 mm)
	30 in. (762 mm)	16 in. (406 mm)	1 5/16 in. (33.3 mm)
Coils	16 1/4 in. (413 mm)	4 in. (102 mm)	12 in. (305 mm)
	Outside Diameter	**Inside Diameter**	**Height**
Drums	23 in. (584 mm)	16 in. (406 mm)	34 in. (864 mm)

Table 13-1 Packaging Size Specification for Commonly Used FCA Filler Wire

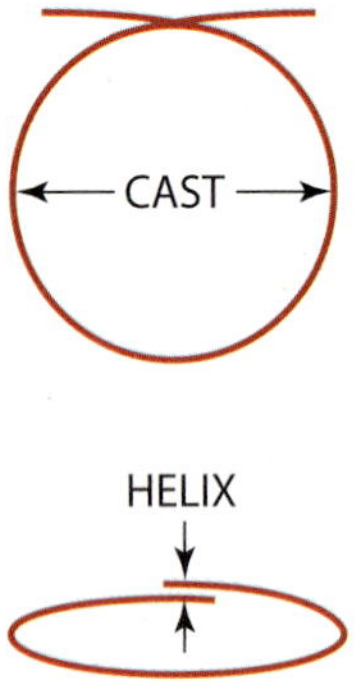

FIGURE 13-11 Method of measuring cast and helix of FCAW filler wire. © Cengage Learning 2012

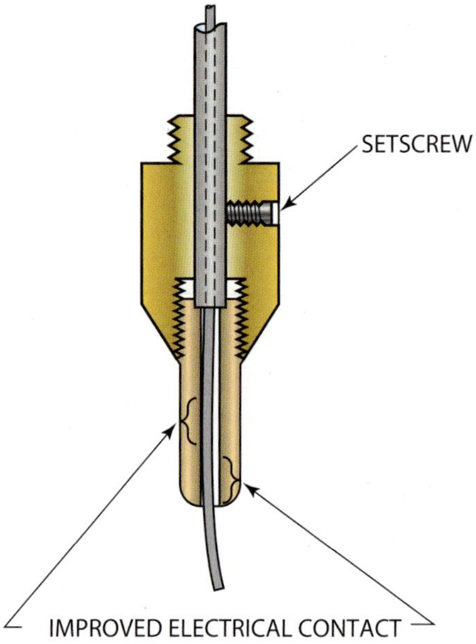

FIGURE 13-12 Cast forces the wire to make better electrical contact with the tube. © Cengage Learning 2012

The cast causes the wire to rub on the inside of the contact tube, Figure 13-12. The slight cast or bend in the electrode wire ensures a positive electrical contact between the contact tube and filler wire. The helix or twist causes the electrode to twist as it feeds through the contact tip. This rotating movement prevents the electrode from wearing a groove in one place inside the contact tip.

FCA WELDING ELECTRODE FLUX

FCA Welding Flux Actions

FCA welding fluxes provide a number of functions to the weld. Some of these functions are:

- *Deoxidizers:* Oxygen that is present in the welding zone has two forms. It can exist as free oxygen from the atmosphere surrounding the weld. Oxygen can also exist as part of a compound such as iron oxide or carbon dioxide (CO_2). In either case it can cause porosity in the weld if it is not removed or controlled. Chemicals are added that react to the presence of oxygen in either form and combine to form a harmless compound, Table 13-2. The new compound can become part of the slag that solidifies on top of the weld, or some of it may stay in the weld as very small inclusions. Both methods result in a weld with better mechanical properties with less porosity.
- *Slag formers:* Slag serves several vital functions for the weld. It can react with the molten weld metal chemically, and it can affect the weld bead physically. In the molten state, it moves through the molten weld pool and acts as a magnet or sponge to chemically combine with impurities in the metal and

Deoxidizing Element	Strength
Aluminum (Al)	Very strong
Manganese (Mn)	Weak
Silicon (Si)	Weak
Titanium (Ti)	Very strong
Zirconium (Zr)	Very strong

Table 13-2 Deoxidizing Elements Added to Filler Wire (to Minimize Porosity in the Molten Weld Pool)

remove them, Figure 13-13. Slags can be refractory, become solid at a high temperature, and solidify over the weld, helping it hold its shape and slowing its cooling rate.

- *Fluxing agents:* Molten weld metal tends to have a high surface tension, which prevents it from flowing outward toward the edges of the weld. This causes undercutting along the junction of the weld and the base metal. Fluxing agents make the weld more fluid and allow it to flow outward, filling the undercut.
- *Arc stabilizers:* Chemicals in the flux affect the arc resistance. As the resistance is lowered, the arc voltage drops and weld penetration is reduced. When the arc resistance is increased, the arc voltage increases and weld penetration is increased. Although the resistance within the ionized arc stream may change, the arc is more stable and easier to control. It also improves the metal transfer by reducing spatter caused by an erratic arc.
- *Alloying elements:* Because of the difference in the mechanical properties of metal that is formed by rolling or forging and metal that is melted to form a weld bead, the metallurgical requirements of the two also differ. Some elements change the weld's strength, ductility, hardness, brittleness, toughness, and corrosion resistance. Other alloying elements in the form of powder metal can be added to both aid in alloying the weld metal and/or to increase the **deposition rate.**
- *Shielding gas:* As elements in the flux are heated by the arc, some of them vaporize and form voluminous gaseous clouds hundreds of times larger than their original volume. This rapidly expanding cloud forces the air around the weld zone away from the molten weld metal, Figure 13-14. Without the protection this process affords the molten metal, it would rapidly oxidize. Such oxidization would severely affect the weld's mechanical properties, rendering it unfit for service.

FIGURE 13-13 Impurities being floated to the surface by slag. © Cengage Learning 2012

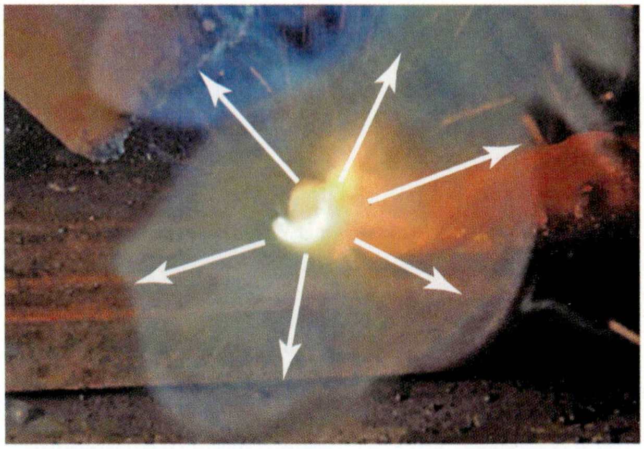

FIGURE 13-14 Rapidly expanding gas cloud.
Larry Jeffus

Types of FCAW Fluxes

All FCAW fluxes are divided into two groups based on their chemical reaction during welding. FCAW fluxes are mainly composed of two types of chemicals—rutile or lime based. The AWS classifies FCAW fluxes into two groups: T-1 and T-5.

T-1 fluxes are **rutile-based fluxes,** and chemically they are acidic. They produce a smooth, stable arc and a refractory high-temperature slag for out-of-position welding. These electrode fluxes produce a fine drop transfer, a relatively low fume, and an easily removed slag. The main limitation of the rutile fluxes is that their fluxing elements do not produce as high a quality deposit as do the T-5 systems.

T-5 fluxes are **lime-based fluxes,** and chemically they are bases. Base chemicals are the opposite of the acids. They are very good at removing certain impurities from the weld metal, but their low melting temperature slag is fluid, which makes them generally unsuitable for out-of-position welding. These electrodes produce a more globular transfer, more spatter, more fume, and a more adherent **slag** than do the T-1 systems. These characteristics are tolerated when it is necessary to deposit very tough weld metal and for welding materials having a low tolerance for hydrogen.

Some T-1 type electrodes can be used with a shielding gas. The shielding gas helps protect the weld from the atmosphere so that more alloying elements can be added to the flux. This dual shielding of the molten weld pool produces welds with the best of both flux systems, high-quality welds in all positions.

Some types of FCA welding electrodes have fluxes that cannot be used on multiple-pass welds. Some of their alloying elements can build up in each weld pass until they can cause weld strength problems. With these electrodes, it is a case where a little alloying element is good but too much can cause problems. These electrodes are listed as single-pass electrodes. Using a single-pass welding electrode for multipass welds may produce a defect-free weld, but it will not be as strong. For example, some single-pass electrodes have high levels of manganese. Manganese is added as an alloy to add strength to large single-pass welds. But the level of manganese can build up so high in a multipass weld that it makes the weld brittle.

Table 13-3 lists the shielding and polarity for the flux classifications of mild steel FCAW electrodes. The letter *G* is used to indicate an unspecified classification. The *G* means that the electrode has not been classified by the American Welding Society. Often the exact composition of fluxes is kept as a manufacturer's trade secret. Therefore, only limited information about the electrode's composition will be given. The only information often supplied is current, type of shielding required, and some strength characteristics.

As a result of the relatively rapid cooling of the weld metal, the weld may tend to become hard and brittle. This factor can be controlled by adding elements to both the flux and the slag formed by the flux, Table 13-4. Ferrite is the softer, more ductile form of iron. The addition of ferrite-forming elements can control the hardness and brittleness of a weld. Refractory fluxes are sometimes called "fast-freeze" because they solidify at a higher temperature than the weld metal. By becoming solid first, this slag can cradle the molten weld pool and control its shape. This property is very important for out-of-position welds.

The impurities in the weld pool can be metallic or nonmetallic compounds. Metallic elements that are added to the metal during the manufacturing process in small quantities may be concentrated in the weld.

Classifications		Comments		Shielding Gas
T-1		Requires clean surfaces and produces little spatter. It can be used for single- and multiple-pass welds in the flat (1G and 1F) and horizontal (2F) positions.		Carbon dioxide (CO_2)
T-2		Requires clean surfaces and produces little spatter. It can be used for single-pass welds in the flat (1G and 1F) and horizontal (2F) positions only.		Carbon dioxide (CO_2)
T-3		Used on thin-gauge steel for single-pass welds in the flat (1G and 1F) and horizontal (2F) positions only.		None
T-4		Low penetration and moderate tendency to crack for single- and multiple-pass welds in the flat (1G and 1F) and horizontal (2F) positions.		None
T-5		Low penetration and a thin, easily removed slag, used for single- and multiple-pass welds in the flat (1G and 1F) position only.		With or without carbon dioxide (CO_2)
T-6		Similar to T-5 without externally applied shielding gas.		None
T-G		The composition and classification of this electrode are not given in the preceding classes. It may be used for single- or multiple-pass welds.		With or without shielding

Table 13-3 Welding Characteristics of Seven Flux Classifications

Element	Reaction in Weld
Silicon (Si)	Ferrite former and deoxidizer
Chromium (Cr)	Ferrite and carbide former
Molybdenum (Mo)	Ferrite and carbide former
Columbium (Cb)	Strong ferrite former
Aluminum (Al)	Ferrite former and deoxidizer

Table 13-4 Ferrite-Forming Elements Used in FCA Welding Fluxes

These elements improve the grain structure, strength, hardness, resistance to corrosion, or other mechanical properties in the metal's as-rolled or formed state. But weld nugget is a small casting, and some alloys adversely affect the properties of this casting (weld metal). Nonmetallic compounds are primarily slag inclusions left in the metal from the fluxes used during manufacturing. The welding fluxes form slags that are less dense than the weld metal so that they will float to the surface before the weld solidifies.

FLUX CORED STEEL ELECTRODE IDENTIFICATION

The American Welding Society revised its A5.20 *Specification for Carbon Steel Electrodes for Flux Cored Arc Welding* in 1995 to reflect changes in the composition of the FCA filler metals. Table 13-5 lists the AWS specifications for flux cored filler metals.

Metal	AWS Filler Metal Classification
Mild steel	A5.20
Stainless steel	A5.22
Chromium–molybdenum	A5.29

Table 13-5 Filler Metal Classification Numbers

Mild Steel

The electrode number *E70T-10* is used as an example to explain the meaning of the electrode classification system as follows (Figure 13-15):

E - Electrode.

7 - Tensile strength in units of 10,000 psi for a good weld. This value is usually either *6* for 60,000 or *7* for 70,000 psi minimum weld

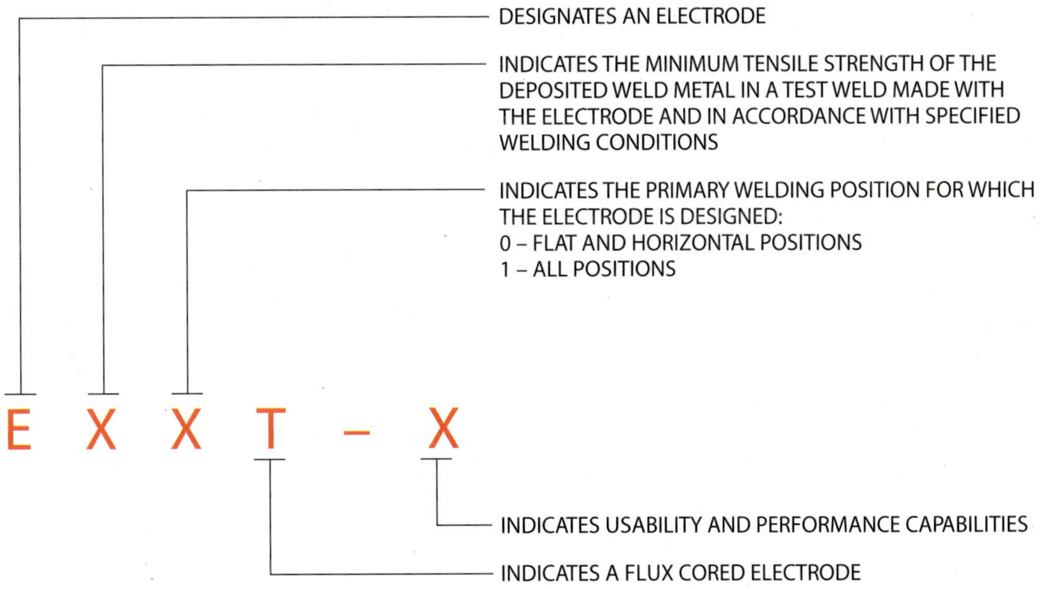

FIGURE 13-15 **Identification system for mild steel FCAW electrodes.** American Welding Society

strength. An exception is for the number *12*, which is used to denote filler metals having a range from 70,000 to 90,000 psi.

0 - 0 is used for flat and horizontal fillets only.

1 - 1 is used for all-position electrodes.

T - Tubular (flux-cored) electrode.

10 - The number in this position can range from *1* to *14* and is used to indicate the electrode's shielding gas, if any; number of passes; and other welding characteristics of the electrode. The letter *G* is used to indicate that the shielding gas, polarity, and impact properties are not specified. The letter *G* may or may not be followed by the letter *S*. *S* indicates an electrode that is only suitable for single-pass welding.

The electrode classification *E70T-10* can have some optional identifiers added to the end of the number such as *E70T-10MJH8*. These additions are used to add qualifiers to the general classification so that specific codes or standards can be met. These additions have the following meanings:

M - Mixed gas, 75 to 80% argon (Ar), balance CO_2. If there is no *M*, the shielding gas must be either CO_2 or the electrode is self-shielded.

J - Describes the Charpy V-notch impact test value of 20 ft-lb at −40°F.

H8 - Describes the residual hydrogen levels in the weld: *H4* equals less than 4 ml/100 g; *H8*, less than 8 ml/100 g; *H16*, less than 16 ml/100 g.

Stainless Steel Electrodes

The AWS classification for stainless steel for FCAW electrodes starts with the letter *E* as its prefix. Following the *E* prefix, the American Iron and Steel Institute's (AISI) three-digit stainless steel number is used. This number indicates the type of stainless steel in the filler metal.

To the right of the AISI number, the AWS adds a dash followed by a suffix number. The number 1 is used to indicate an all-position filler metal, and the number 3 is used to indicate an electrode to be used in the flat and horizontal positions only.

Metal Cored Steel Electrode Identification

The addition of metal powders to the flux core of FCA welding electrodes has produced a new classification of filler metals. These types of filler metals evolved over time, and a new identification system was established by the AWS to identify these filler metals. Some of the earlier flux cored filler metals that already had powder metals in their core had their numbers changed to reflect the new designation. The designation was changed from the letter *T* for *tubular* to the letter *C* for *core*. For example, E70T-1 became E70C-3C. The complete explanation of the cored electrode *E70C-3C* follows:

E - Electrode.

7 - Tensile strength in units of 10,000 psi for a good weld. This value is usually either *6* for 60,000 or *7* for 70,000 psi minimum weld strength. An exception is for the number *12*, which is used to denote filler metals having a range from 70,000 to 90,000 psi.

0 - *0* is used for flat and horizontal fillets only, and *1* is used for all-position electrodes.

C - Metal-cored (tubular) electrode.

3 - *3* is used for a Charpy impact of 20 ft-lb at 0°F.

6 - *6* represents a Charpy impact of 20 ft-lb at −20°F.

C - The second letter *C* indicates CO_2, and the letter *M* indicates a mixed gas, 75 to 80% Ar, with the balance being CO_2. If there is no *M* or *C*, then the shielding gas is CO_2. The letter *G* is used to indicate that the shielding gas, polarity, and impact properties are not specified. The letter *G* may or may not be followed by the letter *S*. *S* indicates an electrode that is only suitable for single-pass welding.

Care of Flux Core Electrodes

Wire electrodes may be wrapped in sealed plastic bags for protection from the elements. Others may be wrapped in a special paper, and some are shipped in cans or cardboard boxes.

A small paper bag of a moisture-absorbing material, crystal desiccant, is sometimes placed in the shipping containers. It is enclosed to protect wire electrodes from moisture. Some wire electrodes require storage in an electric rod oven to prevent contamination from excessive moisture. Read the manufacturer's recommendations located in or on the electrode shipping container for information on use and storage.

Weather conditions affect your ability to make high-quality welds. Humidity increases the chance of moisture entering the weld zone. Water (H_2O), which consists of two parts hydrogen and one part oxygen, separates in the weld pool. When only one part of hydrogen is expelled, hydrogen entrapment occurs. Hydrogen entrapment can cause weld beads to crack or become brittle. The evaporating moisture will also cause porosity.

To prevent hydrogen entrapment, porosity, and atmospheric contamination, it may be necessary to preheat the base metal to drive out moisture. Storing the wire electrode in a dry location is recommended. The electrode may develop restrictions due to the tangling of the wire or become oxidized with excessive rusting if the wire electrode package is mishandled, thrown, dropped, or stored in a damp location.

NOTE: The powdered metal added to the core flux can provide additional filler metal and/or alloys. This is one way that micro-alloys can be added in very small and controlled amounts, as low as 0.0005 to 0.005%. These are very powerful alloys that dramatically improve the metal's mechanical properties.

CAUTION

Always keep the wire electrode dry, and handle it as you would any important tool or piece of equipment.

SHIELDING GAS

FCA welding wire can be manufactured so that all of the required shielding of the molten weld pool is provided by the vaporization of some of the flux within the tubular electrode. When the electrode provides all of the shielding, it is called self-shielding. Other FCA welding wire must use an externally supplied shielding gas to provide the needed protection of the molten weld pool. When a shielding gas is added, it is called **dual shield.**

Care must be taken to use the cored electrodes with the recommended gases or not to use gas at all with the self-shielded electrodes. Using a self-shielding flux cored electrode with a shielding gas may produce a defective weld. The shielding gas will prevent the proper disintegration of much of the deoxidizers. This results in the transfer of these materials across the arc to the weld. In high concentrations, the deoxidizers can produce slags that become trapped in the welds, causing undesirable defects. Lower concentrations may cause brittleness only. In either case, the chance of weld failure is increased. If these electrodes are used correctly, there is no problem.

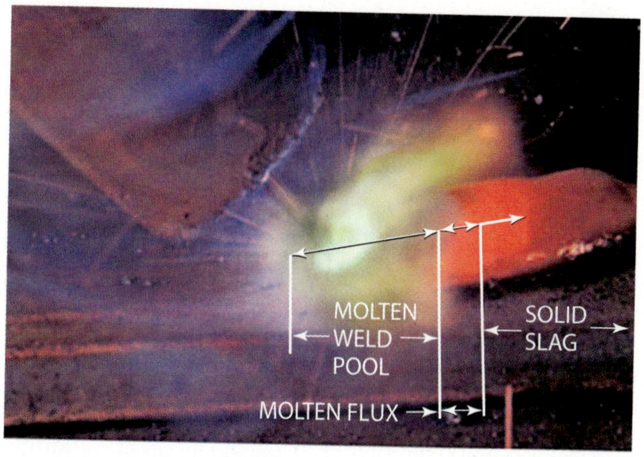

FIGURE 13-16 Axial spray transfer mode. Larry Jeffus

The selection of a shielding gas will affect the arc and weld properties. The weld bead width, buildup, penetration, spatter, chemical composition, and mechanical properties are all affected as a result of the shielding gas selection. Shielding gas comes in high-pressure cylinders. These cylinders are supplied with 2000 psi of pressure. Because of this high pressure, it is important that the cylinders be handled and stored safely. See Chapter 2 for specific cylinder safety instructions.

Gases used for FCA welding include CO_2 and mixtures of argon and CO_2. Argon gas is easily ionized by the arc. Ionization results in a highly concentrated path from the electrode to the weld. This concentration results in a smaller droplet size that is associated with the axial spray mode of metal transfer, Figure 13-16. A smooth, stable arc results, and there is a minimum of spatter. This transfer mode continues as CO_2 is added to the argon until the mixture contains more than 25% of CO_2.

NOTE: Sometimes the shielding gas is referred to as the shielding *medium*.

As the percentage of CO_2 increases in the argon mixture, weld penetration increases. This increase in penetration continues until a 100% CO_2 shielding gas is reached. But as the percentage of CO_2 is increased, the arc stability decreases. The less stable arc causes an increase in spatter. A mixture of 75% argon and 25% CO_2 works best for jobs requiring a mixed gas. This mixture is sometimes called C-25.

Straight CO_2 is used for some welding. But the CO_2 gas molecule is easily broken down in the welding arc. It forms carbon monoxide (CO) and free oxygen (O). Both gases are reactive to some alloys in the electrode. As these alloys travel from the electrode to the molten weld pool, some of them form oxides. Silicon and manganese are the primary alloys that become oxidized and lost from the weld metal.

Most FCA welding electrodes are specifically designed to be used with or without shielding gas and for a specific shielding gas or percentage mixture. For example, an electrode designed specifically for use with 100% CO_2 will have higher levels of silicon and manganese to compensate for the losses to oxidization. But if 100% argon or a mixture of argon and CO_2 is used, the weld will have an excessive amount of silicon and manganese. The weld will not have the desired mechanical or metallurgical properties. Although the weld may look satisfactory, it will probably fail prematurely.

CAUTION

Never use an FCA welding electrode with a shielding gas it is not designated to be used with. The weld it produces may be unsafe.

WELDING TECHNIQUES

A welder can control weld beads made by FCA welding by making changes in the techniques used. The following explains how changing specific welding techniques affects the weld produced.

Gun Angle

The *gun angle, work angle,* and *travel angle* are terms used to refer to the relation of the gun to the work surface, Figure 13-17. The gun angle can be used to control the weld pool. The electric arc produces an electrical force known as the arc force. The arc force can be used to counteract the gravitational pull that tends to make the liquid weld pool sag or run ahead of the arc. By manipulating the electrode travel angle for the flat and horizontal position of welding to a 20 to 45° angle from the vertical position, the weld pool can be controlled. A 40 to 50° angle from the vertical plate is recommended for fillet welds.

Changes in this angle affect the weld bead shape and penetration. Shallower angles are needed when welding thinner materials to prevent burnthrough. Steeper, perpendicular angles are used for thicker materials.

FCAW electrodes have a flux that is mineral based, often called low-hydrogen. These fluxes are refractory and become solid at a high temperature. If too steep a forehand or pushing angle is used, slag from the electrode can be pushed ahead of the weld bead and solidify quickly on the cooler plate, Figure 13-18. Because the slag remains solid at higher temperatures than the temperature of the molten weld pool, it can be trapped under the edges of the weld by the molten weld metal. To avoid this problem, most flat and horizontal welds should be performed with a backhand angle.

Vertical up welds require a forehand gun angle. The forehand angle is needed to direct the arc deep into the groove or joint for better control of the weld pool and deeper penetration, Figure 13-19. Slag

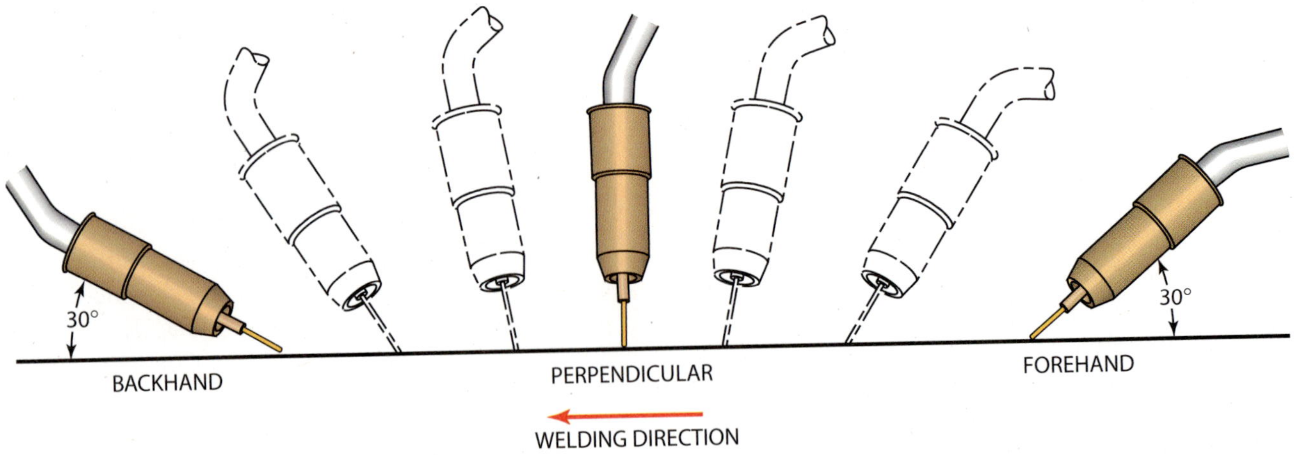

FIGURE 13-17 Welding gun angles. © Cengage Learning 2012

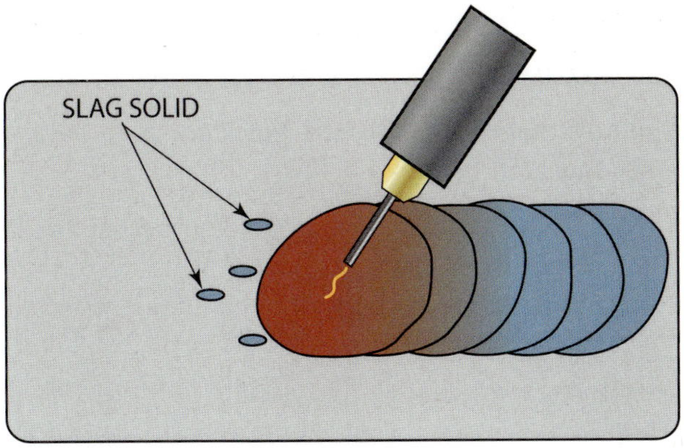

FIGURE 13-18 Large quantities of solid slag in front of a weld can cause slag to be trapped under the weld bead. © Cengage Learning 2012

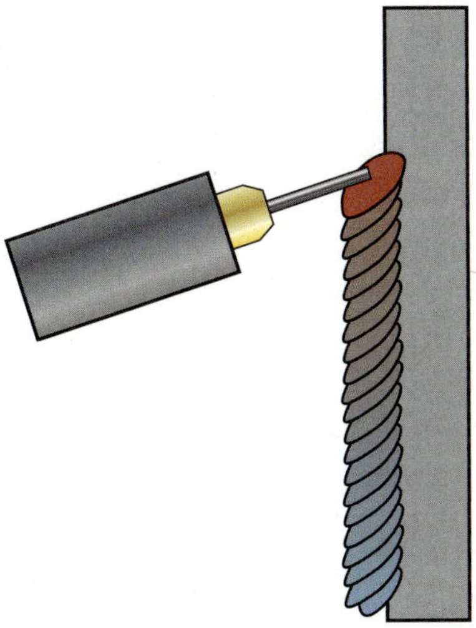

FIGURE 13-19 Vertical up gun angle. © Cengage Learning 2012

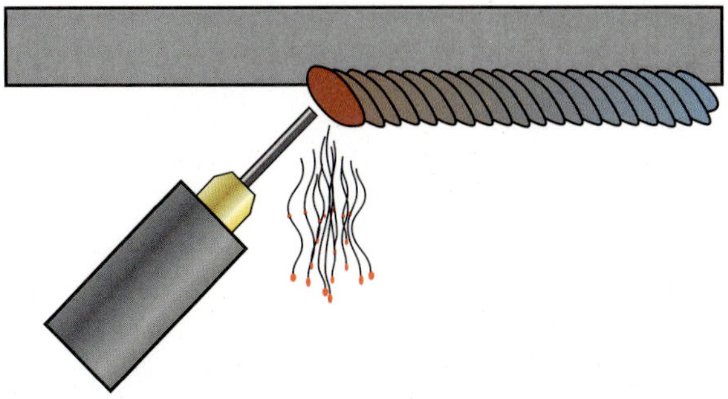

FIGURE 13-20 Weld gun position to control spatter buildup on an overhead weld. © Cengage Learning 2012

entrapment associated with most forehand welding is not a problem for vertical welds.

A gun angle around 90° to the metal surface either slightly forehand or backhand works best for overhead welds, Figure 13-20. The slight angle aids with visibility of the weld, and it helps control spatter buildup in the gas nozzle.

Forehand/Perpendicular/ Backhand Techniques

Forehand, perpendicular, and *backhand* are the terms most often used to describe the gun angle as it relates to the work and the direction of travel. The **forehand** technique is sometimes referred to as *pushing* the weld bead, and backhand may be referred to as *pulling* or *dragging* the weld bead. The term *perpendicular* is used when the gun angle is at approximately 90° to the work surface, Figure 13-21.

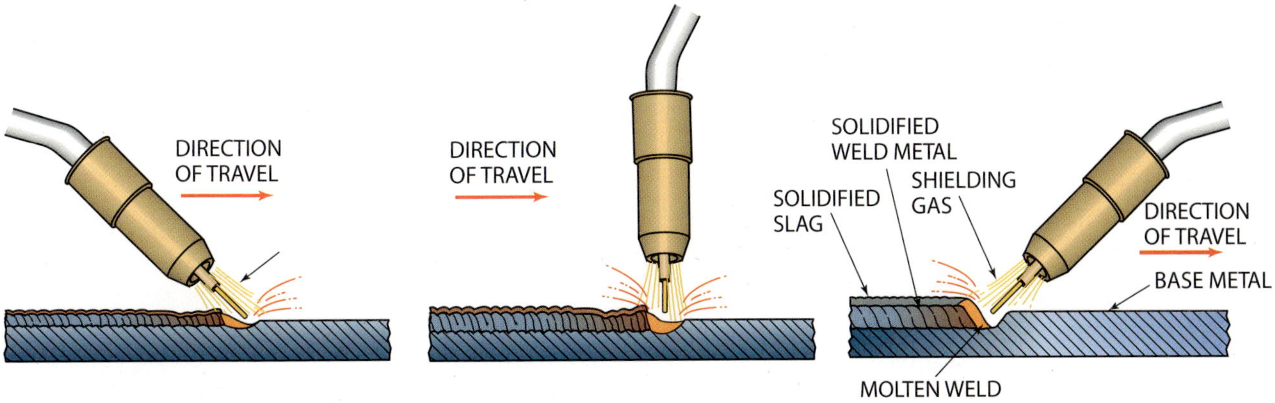

FIGURE 13-21 Changing the welding gun angle among forehand, perpendicular, or backhand angles will change the shape of the weld bead produced. © Cengage Learning 2012

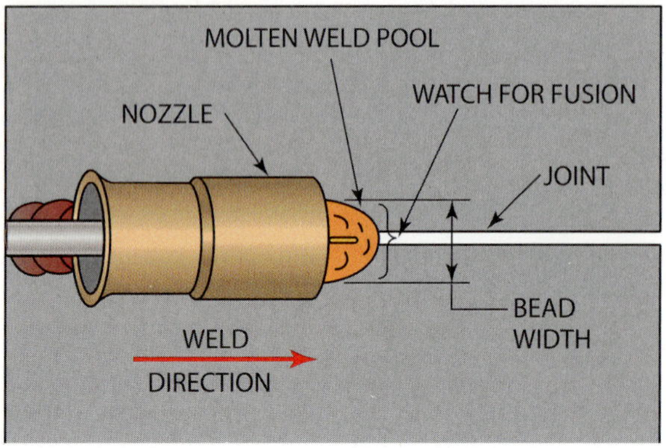

FIGURE 13-22 The shielding gas nozzle restricts the welder's view. © Cengage Learning 2012

Advantages of the Forehand Technique

The advantages of using the forehand welding technique are:

- Joint visibility—You can easily see the joint where the bead will be deposited, Figure 13-22.
- Electrode extension—The contact tube tip is easier to see, making it easier to maintain a constant extension length.
- Less weld penetration—It is easier to weld on thin sheet metal without melting through.
- Out-of-position welds—This technique works well on vertical up and overhead joints for better control of the weld pool.

Disadvantages of the Forehand Technique

The disadvantages of using the forehand welding technique are:

- Weld thickness—Thinner welds are formed because less weld reinforcement is applied to the weld joint.
- Welding speed—Because less weld metal is being applied, the rate of travel along the joint can be faster, which may make it harder to create a uniform weld.
- Slag inclusions—Some spattered slag can be thrown in front of the weld bead and be trapped or included in the weld, resulting in a weld defect.
- Spatter—Depending on the electrode, the amount of spatter may be slightly increased with the forehand technique.

Advantages of the Perpendicular Technique

The advantages of using the perpendicular welding technique are:

- Machine and robotic welding—The perpendicular gun angle is used on automated welding because there is no need to change the gun angle when the weld changes direction.
- Uniform bead shape—The weld's penetration and reinforcement are balanced between those of forehand and backhand techniques.

Disadvantages of the Perpendicular Technique

The disadvantages of using the perpendicular welding technique are:

- Limited visibility—Because the welding gun is directly over the weld, there is limited visibility of the weld unless you lean your head way over to the side.
- Weld spatter—Because the weld nozzle is directly under the weld in the overhead position, more weld spatter can collect in the nozzle, causing gas flow problems or even shorting the tip to the nozzle.

Advantages of the Backhand Technique

The advantages of the backhand welding technique are:

- Weld bead visibility—It is easy to see the back of the molten weld pool as you are welding, which makes it easier to control the bead shape.
- Travel speed—Because of the larger amount of weld metal being applied, the rate of travel may be slower, making it easier to create a uniform weld.
- Depth of fusion—The arc force and the greater heat from the slower travel rate both increase the depth of weld joint penetration.

Disadvantages of the Backhand Technique

The disadvantages of the backhand welding technique are:

- Weld buildup—The weld bead may have a convex-shaped (raised or rounded) weld face when you use the backhand technique.

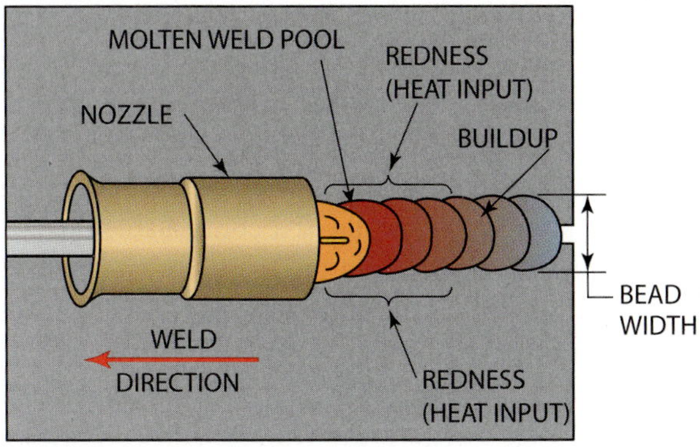

FIGURE 13-23 Watch the trailing edge of the molten weld pool. © Cengage Learning 2012

- Postweld finishing—Because of the weld bead shape, more work may be required if the product has to be finished by grinding smooth.
- Joint following—It is harder to follow the joint because your hand holding the FCAW gun is positioned over the joint, and you may wander from the seam, Figure 13-23.
- Loss of penetration—An inexperienced welder sometimes directs the wire too far back into the weld pool, causing the wire to build up in the face of the weld pool and reducing joint penetration.

Travel Speed

The American Welding Society defines *travel speed* as the linear rate at which the arc is moved along the weld joint. Fast travel speeds deposit less filler metal. If the rate of travel increases, the filler metal cannot be deposited fast enough to adequately fill the path melted by the arc. This causes the weld bead to have a groove melted into the base metal next to the weld and left unfilled by the weld. This condition is known as undercut.

Undercut occurs along the edges or toes of the weld bead. Slower travel speeds will, at first, increase penetration and increase the filler weld metal deposited. As the filler metal increases, the weld bead will build up in the weld pool. Because of the deep penetration of flux cored wire, the angle at which you hold the gun is very important for a successful weld.

If all welding conditions are correct and remain constant, the preferred rate of travel for maximum weld penetration is a travel speed that allows you to stay within the selected welding variables and still control the fluidity of the weld pool. This is an intermediate travel speed, or progression, that is not too fast or too slow.

Another way to figure out correct travel speed is to consult the manufacturer's recommendations chart for the ipm burn-off rate for the selected electrode.

Mode of Metal Transfer

The mode of metal transfer is used to describe how the molten weld metal is transferred across the arc to the base metal. The mode of metal transfer that is selected, the shape of the completed weld bead, and the depth of weld penetration depend on the welding power source, wire electrode size, type and thickness of material, type of shielding gas used, and best welding position for the task.

Spray Transfer—FCAW-G

The spray transfer mode is the most common process used with gas-shielded FCAW, Figure 13-24.

As the gun trigger is depressed, the shielding gas automatically flows and the electrode bridges the distance from the contact tube to the base metal, making contact with the base metal to complete a circuit. The electrode shorts and becomes so hot that the base metal melts and forms a weld pool. The electrode melts into the weld pool and burns back toward the contact tube. A combination of high amperage and the shielding gas along with the electrode size produces a pinching effect on the molten electrode wire, causing the end of the electrode wire to spray across the arc.

The characteristic of spray-type metal transfer is a smooth arc, through which hundreds of small droplets per second are transferred through the arc from the electrode to the weld pool. At that moment, a transfer of metal is taking place. Spray transfer can

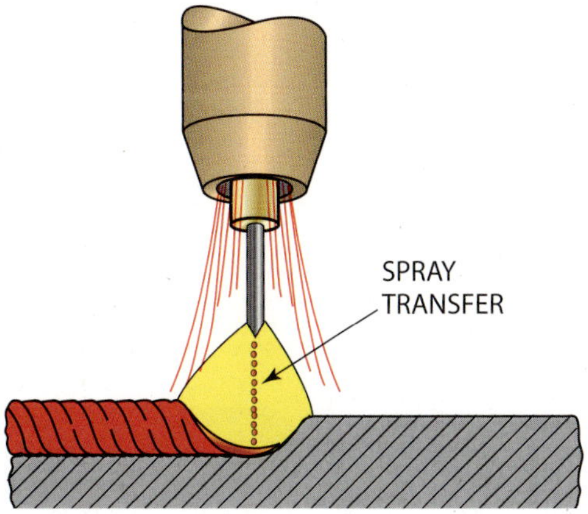

FIGURE 13-24 Axial spray transfer mode. © Cengage Learning 2012

produce a high quantity of metal droplets, up to approximately 250 per second above the transition current or critical current. This means the current is dependent on the electrode size, composition of the electrode, and shielding gas so that a spray transfer can take place. Below the transition current (critical current), globular transfer takes place.

To achieve a spray transfer, high current a and larger diameter electrode wire is needed. A shielding gas of carbon dioxide (CO_2), a mixture of carbon dioxide (CO_2) and argon (Ar), or an argon (Ar) oxygen (O_2) mixture is needed. FCAW-G is a welding process that, with the correct variables, can be used

- on thin or properly prepared thick sections of material,
- on a combination of thick to thin materials,
- with small or large electrode diameters, and
- with a combination of shielding gases.

Globular Transfer—FCAW-G

Globular transfer occurs when the welding current is below the transition current, Figure 13-25. The electrode forms a molten ball at its end that grows in size to approximately two to three times the original electrode diameter. These large molten balls are then transferred across the arc at the rate of several drops per second.

The arc becomes unstable because of the gravitational pull from the weight of these large drops. A spinning effect caused by a natural phenomenon takes place when argon gas is introduced to a large ball of molten metal on the electrode. This causes a spinning motion as the molten ball transfers across the arc to the base metal. This unstable globular transfer can produce excessive spatter.

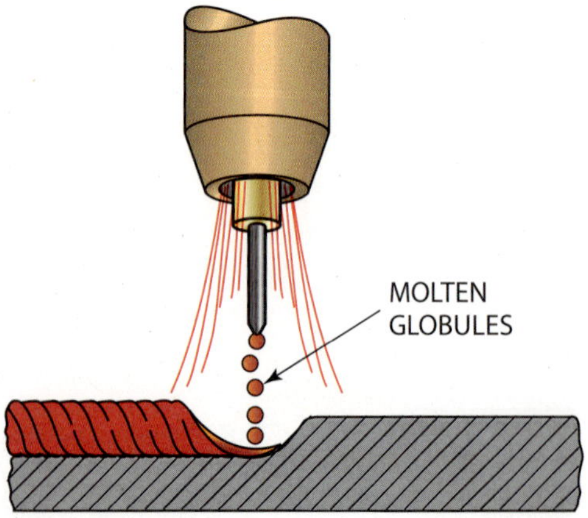

FIGURE 13-25 Globular transfer method. © Cengage Learning 2012

Both FCAW-S and FCAW-G use direct current electrode negative (DCEN) when welding on thin-gauge materials to keep the heat in the base metal and the small diameter electrode at a controllable burn-off rate. The electrode can then be stabilized, and it is easier to manipulate and control the weld pool in all weld positions. Larger diameter electrodes are welded with direct current electrode positive (DCEP) because the larger diameters can keep up with the burn-off rates.

The recommended weld position means the position in which the workpiece is placed for welding. All welding positions use either spray or globular transfer, but for now we will concentrate on the flat and horizontal welding positions.

In the flat welding position, the workpiece is placed flat on the work surface. In the horizontal welding position, the workpiece is positioned perpendicular to the workbench surface.

The amperage range may be from 30 to 400 amperes or more for welding materials from gauge thicknesses up to 1 1/2 in. (38 mm). On square groove weld joints, thicker base metals can be welded with little or no edge preparation. This is one of the great advantages of FCAW. If edges are prepared and cut at an angle (beveled) to accept a complete joint weld penetration, the depth of penetration will be greatly increased. FCAW is commonly used for general repairs to mild steel in the horizontal, vertical, and overhead welding positions, sometimes referred to as out-of-position welding.

Electrode Extension

The electrode extension is measured from the end of the electrode contact tube to the point that the arc begins at the end of the electrode, Figure 13-26. Compared to GMA welding, the electrode extension required for FCAW is much greater. The longer extension is required for several reasons. The electrical resistance of the wire causes the wire to heat up, which can drive out moisture from the flux. This preheating of the wire also results in a smoother arc with less spatter.

Porosity

FCA welding can produce high-quality welds in all positions, although porosity in the weld can be a persistent problem. Porosity can be caused by moisture

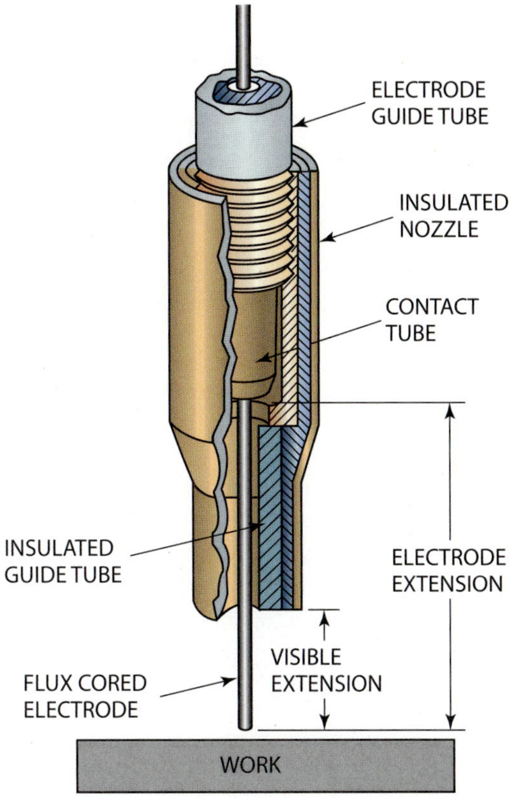

FIGURE 13-26 **Self-shielded electrode nozzle.** American Welding Society

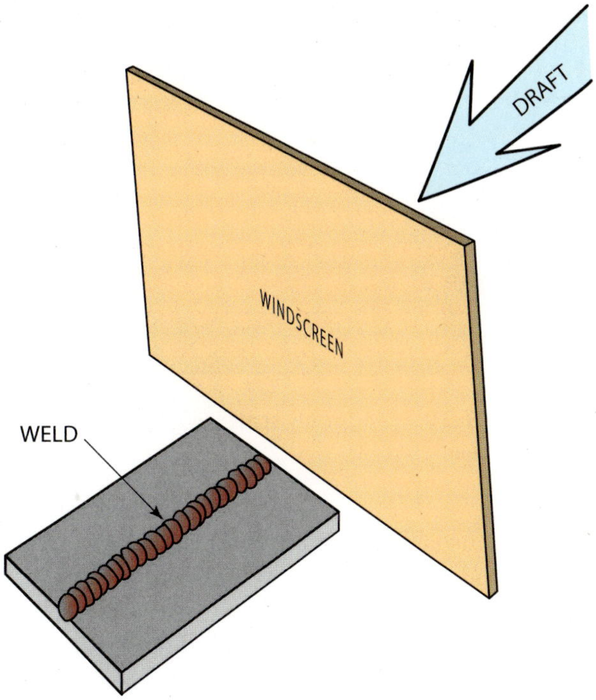

FIGURE 13-28 **A windscreen can help prevent the shielding gas from being blown away.** © Cengage Learning 2012

in the flux, improper gun manipulation, or surface contamination.

The flux used in the FCA welding electrode is subject to picking up moisture from the surrounding atmosphere, so the electrodes must be stored in a dry area. Once the flux becomes contaminated with moisture, it is very difficult to remove. Water (H_2O) breaks down into free hydrogen and oxygen in the presence of an arc, Figure 13-27. The hydrogen can be absorbed into the molten weld metal, where it can cause postweld cracking. The oxygen is absorbed into the weld metal also, but it forms oxides in the metal.

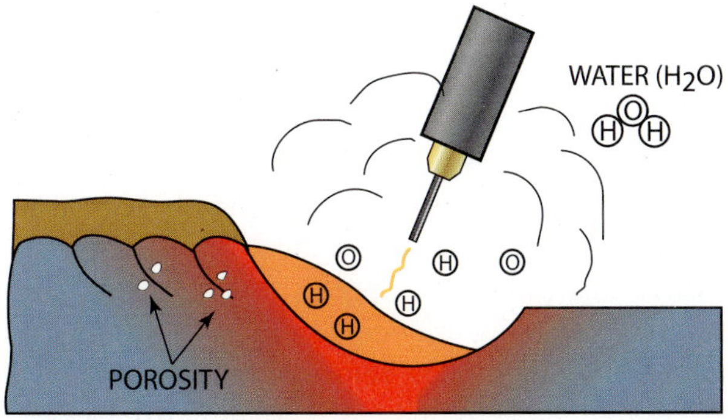

FIGURE 13-27 **Water (H_2O) breaks down in the presence of the arc and the hydrogen (H) is dissolved in the molten weld metal.** © Cengage Learning 2012

If a shielding gas is being used, the FCA welding gun gas nozzle must be close enough to the weld to provide adequate shielding gas coverage. If there is a wind or if the nozzle-to-work distance is excessive, the shielding will be inadequate and cause weld porosity. If welding is to be done outside or in an area subject to drafts, the gas flow rate must be increased or a windshield must be placed to protect the weld, Figure 13-28.

A common misconception is that the flux within the electrode will either remove or control weld quality problems caused by surface contaminations. That is not true. The addition of flux makes FCA welding more tolerant to surface conditions than GMA welding, although it is still adversely affected by such contaminations.

New hot-rolled steel has a layer of dark gray or black iron oxide called mill scale. Although this layer is very thin, it may provide a source of enough oxygen to cause porosity in the weld. If mill scale causes porosity, it is usually uniformly scattered through the weld, Figure 13-29. Unless severe, uniformly scattered

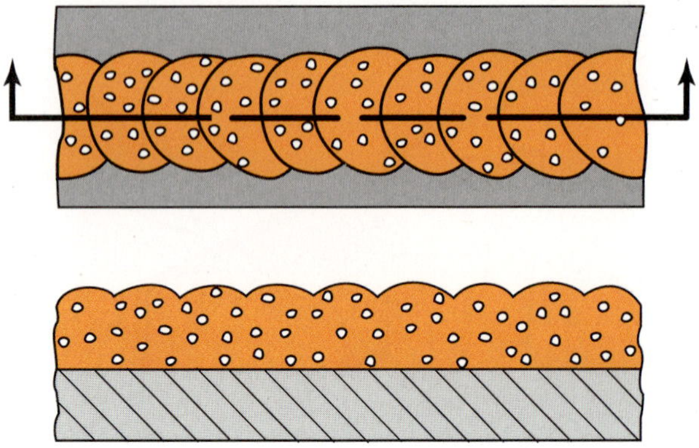

FIGURE 13-29 Uniformly scattered porosity.
© Cengage Learning 2012

porosity is usually not visible in the finished weld. It is trapped under the surface as the weld cools.

Because porosity is under the weld surface, nondestructive testing methods, including X-ray, magnetic particle, and ultrasound, must be used to locate it in a weld. It can be detected by mechanical testing such as guided-bend, free-bend, and nick-break testing for establishing weld parameters. Often, it is better to remove the mill scale before welding rather than risking the production of porosity.

FLUX CORED

All welding surfaces within the weld groove and the surrounding surfaces within 1 in. (25 mm) must be cleaned to bright metal, Figure 13-30. Cleaning may be either grinding, filing, sanding, or blasting.

Any time FCA welds are to be made on metals that are dirty, oily, rusty, wet, or that have been painted, the surface must be precleaned. Cleaning can be done chemically or mechanically.

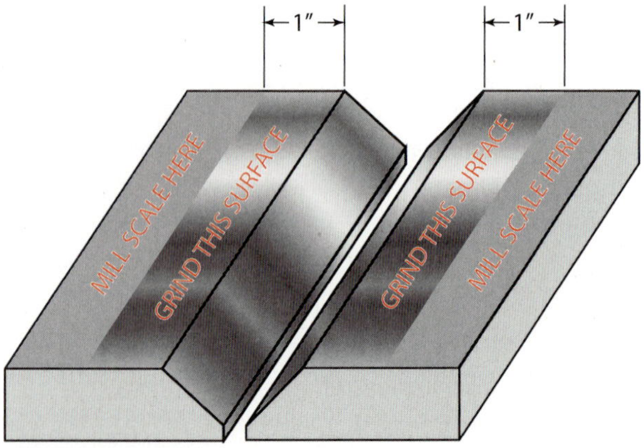

FIGURE 13-30 Grind mill scales off plates within 1 in. (25 mm) of the groove. © Cengage Learning 2012

CAUTION

Chemically cleaning oil and paint off metal must be done according to the cleaner manufacturer's directions. The work must be done in an appropriate, approved area. The metal must be dry, and all residues of the cleaner must be removed before welding begins.

One advantage of chemically cleaning oil and paint is that it is easier to clean larger areas. Both oil and paint smoke easily when heated, and such smoke can cause weld defects. They must be removed far enough from the weld so that weld heat does not cause them to smoke. In the case of small parts, the entire part may need to be cleaned.

TROUBLESHOOTING FCA WELDING

Troubleshooting FCA welding problems is often a trial-and-error process. Trial and error is where you make one adjustment or change at a time and make a trial weld to see if the problem improved. If what you tried did not improve the problem or made it worse, reset the machine and try another adjustment. Keep doing this until the problem is resolved. Make only one adjustment or change at a time. Making two or more adjustments or changes at the same time can result in one improving the weld and the others causing new problems.

The most common causes of FCA welding problems are equipment setup. However, in the field, worn and dirty parts will from time to time develop, causing problems. These worn or dirty parts can cause FCA welding problems similar to those caused by improper setup. Misdiagnosing the cause can result in the possible replacement of good parts. To reduce this time and expense, use the list of common FCA welding problems in Table 13-6 to try and solve the weld problem before replacing parts.

Additional FCA and GMA welding troubleshooting information can be found in the appendix. Often the equipment manufacturer will include a list of troubleshooting tips in the welder's instruction booklet.

Welding Problem	Cause
Gun nozzle arcs to work	1. Weld spatter buildup in nozzle 2. Contact tube bent and touching nozzle
Wire feeds but no arc	1. Bad or missing work clamp (ground) 2. Loose jumper lead in welder 3. Electrode not contacting bare metal
Arc burns off wire at contact tube end	1. Feed rollers' tension too loose 2. Wrong-sized feed rollers 3. Wire welded to contact tube 4. Wire liner worn or damaged 5. Out of wire 6. Worn or damaged contact tube
Wire feeds erratically	1. Feed rollers' tension loose 2. Dirty liner 3. Worn or dirty contact tube
Arc pops and gun jerks during welding	1. Too high a wire feed speed 2. Too low a voltage
Wire burns back and large globules of metal cross the arc	1. Too high a voltage 2. Too low a wire feed speed 3. Wire slipping in feed rollers, not feeding smoothly
Weld does not burn into base metal	1. Too long an electrode extinction 2. Too low voltage and amperage settings
Weld burns through the base metal	1. Too short an electrode extinction 2. Too high voltage and amperage settings
Porosity in weld	1. Poor shielding gas coverage on dual-shield wire 2. Wrong shielding gas type being used 3. Shielding gas used on self-shielding wire 4. Single-pass electrode used for multipass weld
Poor shielding gas coverage	1. Plugged or dirty gas diffuser 2. Too high or too low shielding gas flow rate 3. Shielding gas cylinder near empty

Table 13-6 FCA Welding Troubleshooting Chart

SUMMARY

The flexibility, portability, and range of metal thicknesses has resulted in flux cored arc welding quickly becoming the most popular arc welding process. Many FCA welders have their wire feed drive wheels and wire spools located inside the machine under a cover where they are protected from most dust and dirt, Figure 13-31. FCA welders can be low-maintenance welder equipment as long as they are stored in a dry place.

Setting up the welder is often the most difficult part of making a good FCA weld. Manufacturers provide a table of machine settings that are very helpful. These come with the machines, and replacements can often be obtained from the manufacturer's Web site.

FIGURE 13-31 FCA welders can be small and durable enough to be used where SMA welders once dominated.
Larry Jeffus

Although FCA welding can be done with or without a shielding gas, most FCA welding is done without a shielding gas. There are few differences in the welding techniques required to make welds with or without shielding gas. So, whether you learned FCA welding in school with or without a shielding gas, you should have no problem making welds in the field either way.

REVIEW QUESTIONS

1. What is the range of thickness of the metal that the FCA welding process can be used on?
2. What are the two methods of FCA welding called?
3. What other welding process's equipment is similar to FCAW equipment?
4. What is another name for a constant potential welding machine?
5. Why are FCA welding machines called CP type welding machines?
6. How is the amperage changed on a CP welder?
7. List some of the effects that the flux inside the FCA electrode has on the molten weld pool.
8. What are some of the major atmospheric contaminations to the molten weld metal?
9. List some of the alloying elements that can be added to the flux, and tell what effect they have on the finished weld.
10. What are the two types of FCA welding guns that are available?
11. Why might chemicals be added to recirculating FCA welding cooling water systems?
12. How does a smoke extraction nozzle work on an FCA welding gun?
13. What are the rollers used for in an FCA welding electrode feed system?
14. List five important advantages that FCA welding offers the welding industry.
15. List four limitations of FCA welding.
16. How are seamless FCA welding electrodes made?
17. Describe what effect cast and helix have on the electrode as it is fed through the contact tip.
18. What type of weld does a T-1 flux typically produce?
19. Why can some FCA welding electrodes not be used to make multiple-pass welds?
20. Explain each part of the FCA welding electrode classification number E70T.
21. What is different about an FCAW metal cored electrode?
22. Why is it important to store FCA welding electrodes in a dry place?
23. What problem might occur if a shielding gas is used with a self-shielding FCA welding electrode?
24. What gas and gas mixtures are used for FCA welding?
25. What are the three terms used to describe the relation of the gun to the work surface?
26. What effect does changing the gun angle have on an FCA weld?
27. What are three terms used to describe the gun angle as it relates to the weld direction of travel?
28. What is the definition of *FCAW travel speed?*
29. Describe the characteristics of the spray-type metal transfer.
30. Describe the globular method of FCA welding metal transfer.
31. Why does FCA welding require a long electrode extension?

Chapter 14

Flux Cored Arc Welding

OBJECTIVES

After completing this chapter, the student should be able to:

- Describe the effect that changing electrode extension, voltage, and wire feed speed has on flux corec arc (FCA) welding.
- Demonstrate how to safely set up and adjust an FCA welder.
- Demonstrate how to make FCA welds.

KEY TERMS

contact tube	**lap joint**	**tee joint**
crevice corrosion	**stringer bead**	**wire feed speed**

INTRODUCTION

Two very important factors play a role in the making of a good FCA weld. They are having the knowledge to properly set up the welder and the skills to control the weld pool. The technical knowledge comes from reading, studying, and attending class lectures. Over time, welding equipment and supplies change, and having a good working knowledge of FCA welding will let you master these new innovations.

The skills will come from making various types of welds. The more you practice and watch what happens during a weld, the better your welding skills will become. One of the major factors that affect FCA welding is that the welding wire is being fed out at a constant speed. This means that you must keep up your welding speed or the weld can become too large. In other words, once the welding begins, you do not have as much control over the joint travel speed as you might have with other types of welding, so make sure, you have freedom of movement along the full length of the joint.

The up side to the higher welding speeds is that FCA welding is highly productive and very cost-effective. FCA welding is one of two high-volume semiautomatic welding processes that account for most of the manual welding done in small and large shops.

ARC VOLTAGE AND AMPERAGE CHARACTERISTICS

Setting up the arc voltage and amperage for FCA welding is just like setting it up for gas metal arc (GMA) welding. The voltage is set on the welder, and the amperage is set by changing the **wire feed speed.** At any one voltage setting, the amperage required to melt the wire as it is fed into the weld will change. At any specific voltage, it requires more amperage to melt the wire the faster it is fed because there is more metal that has to be melted. In other words, because the voltage remains constant in FCA welding, the amperage rises automatically to match the faster electrode feed rate. Therefore, when the wire feed speed is slowed, the amperage required to melt the electrode at that slower rate decreases proportionately.

The higher the wire feed speed, the higher the welding current. For the most part, as the current increases, so does the weld's temperature, and, therefore, the weld bead penetration increases too, Figure 14-1. However, that is not always true for FCA welding. The wire feed speed can be set too

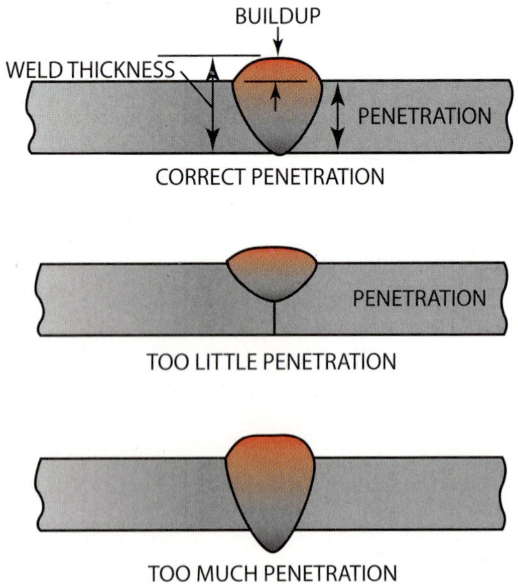

FIGURE 14-1 Weld bead terminology. © Cengage Learning 2012

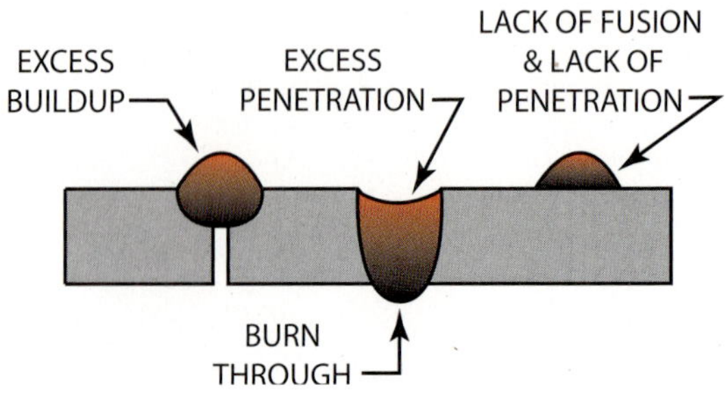

FIGURE 14-2 Weld defects. © Cengage Learning 2012

high, with so much weld metal being melted by the arc that the weld penetration can actually decrease. The same thing can happen if the wire feed speed is set too low—weld penetration decreases. The welder must set both the welding voltage and wire feed speed correctly to produce a satisfactory weld. Charts are available to provide you with the range of voltages and wire feed speeds (amperage).

The effect of having too high or too low an arc voltage is the opposite of having too high or too low a wire feed speed. That is because if the wire feed speed is too high, the voltage is too low; and if the wire feed speed is too low, then the voltage is too high. So, you could possibly correct too high a wire feed speed problem by increasing the voltage, but only if the hotter weld did not cause too much penetration and possibly a weld bead burnthrough. Too low a wire feed speed might be corrected by decreasing the arc voltage if that did not result in a loss of penetration or lack of fusion, Figure 14-2.

Wire Feed Speed

Because changes in the wire feed speed automatically change the amperage, it is possible to set the amperage by using a chart and measuring the length of wire fed per minute, Table 14-1. The wire feed speed is generally recommended by the electrode manufacturer and is selected in inches per minute (ipm), which can be measured by how fast the wire exits the **contact tube.** The welder uses a wire feed speed control dial on the wire feed unit to control the ipm.

Because at high wire feed speeds many feet of wire can be fed out during a full minute's wire feed speed test, a shorter time test is desirable. For example in Table 14-1, the slowest wire feed speed for dual shield 0.035 would be 288 ipm, which is 24 ft (7.3 m) of wire per minute, and at the highest speed of 784 ipm,

ELECTRODE TYPE	DIAMETER	VOLTS	AMPS	WIRE FEED SPEED	ELECTRODE STICKOUT (INCH)
DUAL SHIELD E70T-1* or E70T-2*	0.035	22	130	288	3/8 to 3/4
		25	150	384	3/8 to 3/4
		27	200	576	3/8 to 3/4
		30	250	784	3/8 to 3/4
	0.045	28	150	200	3/8 to 3/4
		29	210	300	3/8 to 3/4
		30	250	400	3/8 to 3/4
		33	290	500	3/8 to 3/4
		34	330	600	3/8 to 3/4
SELF SHIELD E70T-11 or 371T-GS	0.030	15	40	69	3/8
		16	100	175	3/8
		16	160	440	3/8
	0.035	15	80	81	3/8
		17	120	155	3/8
		17	200	392	3/8
	0.045	15	95	54	1/2
		17	150	118	1/2
		18	225	140	1/2

*75% ARGON/25% CO_2

Table 14-1 FCA Welding Parameters

65 ft of wire (19.8 m) would be fed per minute. There are two commonly used shorter timed tests to reduce the wasting of electrode. One uses 15 seconds, and the resulting length is multiplied by 4 to determine the inches per minute. The second test uses 6 seconds, and the resulting length is multiplied by 10 to determine the inches per minute. The 15-second test is more accurate, but for most applications the 6-second test is adequate.

To accurately measure wire feed ipm, snip off the wire at the contact tube. Squeeze the trigger for 6 seconds, release it, and measure the length of wire. Multiply the length of the wire by 10. The result is how many inches of wire would be fed in a minute. Release the drive roller spring tensioner so that the electrode spool can be hand-turned backward to draw the electrode back onto the spool.

NOTE: Rewinding the electrode rather than cutting it off reduces both electrode waste and eliminates the potential hazards that long lengths of loose electrode wire can cause. Loose electrode wire can be a trip hazard as well as an electrical hazard if it comes in contact with the welder electrical terminals.

The wire feed speed is given in a range. The range allows you to adjust the feed speed according to the welding conditions. The wire speed control dial can be advanced or slowed to control the burn weld size and deposition rate.

Electrode Manipulation

The movement or weaving of the welding electrode can control the following characteristics of the weld bead: buildup, width, undercut, and overlap. When a weave is needed for better bead control, the weaves are usually fairly small.

Practice Welds

The setting of the voltage and wire feed speed (amperage) according to the setup tables will give you a good starting point. Some adjustments may be needed in the voltage or wire feed speed. You may want to adjust the settings for the following reasons:

- Metal surface conditions—Electrodes such as E70T-1 will allow you to weld over rust, but some adjustments in setup are needed.
- Joint fit-up—Not all welding joints are fitted to the ideal, so when a joint is wider, you may want to lower the setting; and when a joint is narrower, a higher setting may be needed.
- Voltage drop—Even with good welding power and work cables, some voltage can be lost, especially when the work cables are long, so a higher voltage setting at the welder may be needed to compensate for this loss.
- Welding position—As the welding position changes, so must the welding setup.

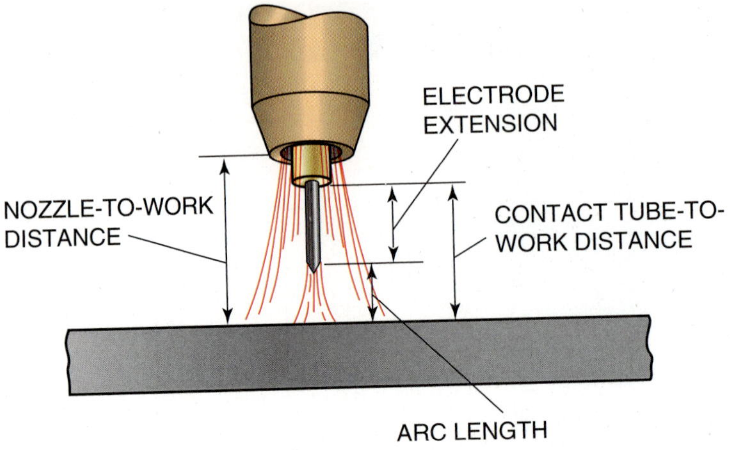

FIGURE 14-3 FCA welding terminology. American Welding Society

In addition to having the welder set correctly, you must control the electrode by keeping its movement consistent along the entire length of the joint. You must maintain a constant speed and weave along the joint as well as maintain a constant welding gun angle and electrode extension.

Starting with the flat position allows you to build your skills slowly so that out-of-position welds become easier to do. The horizontal tee and lap welds are almost as easy to make as the fillet welds. Overhead welds are as simple to make as vertical welds, but they are harder to position. Horizontal butt welds are more difficult to perform than most other welds.

Electrode Extension

The electrode extension (stickout) is the length from the contact tube to the arc measured along the wire, Figure 14-3. A change in this length causes a change in the weld. Because the length of electrode extension affects the preheating of the electrode, charts for FCA welding setup usually include a specific electrode extension or a range for that extension. Some types of FCA welding electrodes need more electrode preheating, so they have a required electrode extension of an inch or more while others need less preheating and require a fraction of an inch of extension. Using the wrong electrode extension for a specific electrode can result in very poor welding with slag inclusions, porosity, and a lack of fusion.

NOTE: Comparing the size of the welding power cable to the size of the welding electrode wire will give you an idea of how small the electrode is related to the welding current it is carrying.

PRACTICE 14-1

Flux Cored Arc Welding Safety

Skill to be learned: The safe setup of a welding station and the use of proper personal protective equipment (PPE).

Using a welding workstation, welding machine, welding electrode wire, welding helmet, eye and ear protection, welding gloves, proper work clothing, and any special protective clothing that may be required; demonstrate to your instructor and other students the safe way to prepare yourself and the welding workstation for welding. Include in your demonstration appropriate references to burn protection, eye and ear protection, material specification data sheets, ventilation, electrical safety, general work clothing, special protective clothing, and area cleanup.◆

EXPERIMENT 14-1

The Effect Electrode That Extension Distance Changes Have on a Weld Bead

Skill to be learned: How to change the electrode extension to control a weld's penetration and buildup and to help maintain weld bead shape during welding.

Experiment Description

You will be making FCA weld **stringer beads** on a steel plate in the flat position. As the weld progresses along the plate, you will be changing the electrode extension length so that you can observe how this change affects the weld produced.

Project Materials and Tools

The following items are needed to complete this practice.

Materials

- 0.035 FCA welding electrode wire
- One or more 12-in. long pieces of metal ranging from 16-gauge sheet metal to 1/4-in. thick plate

Tools

- FCA welder
- PPE
- Square and/or 12-in. rule
- Soapstone

Layout

Using a square or 12-in. rule and soapstone, lay out parallel lines on the metal that are spaced 3/4 in. apart.

Welding

Start with the voltage and wire feed speed set to the lowest settings according to the wire manufacturer or Table 14-1. Put on all of your PPE and follow all shop and manufacturer's safety rules for welding. You will make a series of stringer bead welds along the soapstone line starting with the 16-gauge sheet metal.

Holding the welding gun at a comfortable angle and height, lower your helmet, and start to weld. Make a weld approximately 2 in. long. Then reduce the distance from the gun to the work while continuing to weld. After a few inches, again shorten the electrode extension even more. Continue doing this in steps until the nozzle is as close as possible to the work. Stop when you have made a weld all the way down the entire length of the plate.

Repeat the process, but now increase the electrode extension, making welds of a few inches each. Keep increasing the electrode extension until the weld will no longer fuse to the base metal or the wire becomes impossible to control.

Now change to another plate thickness and repeat the procedure.

When a series of weld beads has been completed with each plate thickness, raise the voltage and wire feed speed to a midrange setting and repeat the process. Upon completing this series of tests, adjust the voltage and wire feed speed to the highest range setting. Make a full series of tests using the same procedure as before.

Finishing

None.

Paperwork

Write a short report describing what you observed as the electrode extension length, welding voltage, and wire feed speeds were changed. Include in your report the details regarding the weld buildup, width, penetration, spatter, and any other observations you made during the welding.

Complete a copy of the time sheet in Appendix I, the bill of materials in Appendix III, or use forms provided by your instructor.◆

EXPERIMENT 14-2

The Effect That Gun Angle Changes Have on a Weld Bead

Skill to be learned: **How to change the welding gun angle to control a weld's penetration and buildup and to help maintain weld bead shape during welding.**

Experiment Description

You will be making FCA weld stringer beads on a steel plate in the flat position. As the weld progresses along the plate, you will be changing the welding gun angle so that you can observe how this change affects the weld produced.

Project Materials and Tools

The following items are needed to complete this practice.

Materials

- 0.035 FCA welding electrode wire
- One or more 12-in. long pieces of 1/4-in. thick steel plate

Tools

- FCA welder
- PPE
- Square and/or 12-in. rule
- Soapstone

Layout

Using a square or 12-in. rule and soapstone, lay out parallel lines that are spaced 3/4 in. apart on the metal.

Welding

Start with the voltage and wire feed speed set to the lowest settings according to the wire manufacturer

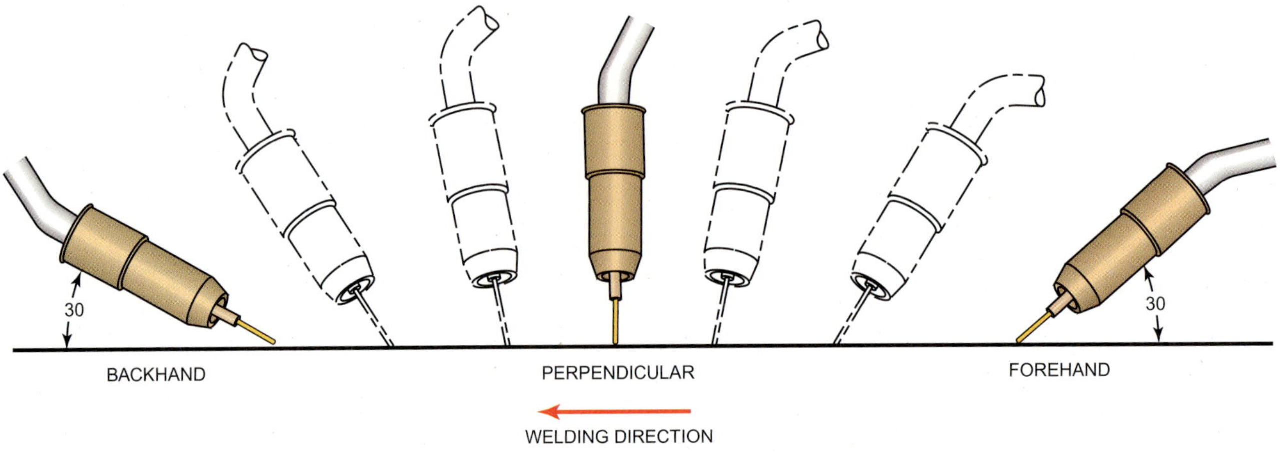

FIGURE 14-4 Welding gun angles. © Cengage Learning 2012

or Table 14-1. Put on all of your PPE and follow all shop and manufacturer's safety rules for welding. You will make a series of stringer bead welds along the soapstone line starting with the 16-gauge sheet metal.

Holding the welding gun at a comfortable height and at a 30° angle to the plate in the direction of the weld, lower your helmet and start to weld, Figure 14-4. Make a weld approximately 2 in. long. Then gradually increase the gun angle while continuing to weld. After a few inches, again increase the gun angle even more. Keep doing this in steps until the gun is at a 30° angle to the plate in the opposite direction of the weld. Stop when you have made a weld all the way down the entire length of the plate.

Raise the voltage and wire feed speed to a mid-range setting and repeat the process. Upon completing this series of tests, adjust the voltage and wire feed speed to the highest range setting. Make a full series of tests using the same procedure as before.

Finishing

None.

Paperwork

Write a short report describing what you observed as the gun angle changed from a 30° pushing angle to a 30° dragging angle and welding voltage and wire feed speeds were changed. Include in your report the details regarding the weld buildup, width, penetration, spatter, and any other observations you made during the welding.

Complete a copy of the time sheet in Appendix I, the bill of materials in Appendix III, or use forms provided by your instructor.◆

EDGE WELDS AND PLUG WELDS

The edge welds can be used to join thinner plate sections to make a thicker section. They are also used to join the flanges of structural shapes like I-beams, H-beams, channel iron, and so on. The edge joints used to make this project are very similar to V-grooved butt joints.

Plug welds are used to make a weld through the surface of a plate to a plate located directly behind the top plate. Plug welds are made by first cutting or drilling a hole through the top plate and making a weld through that hole onto the plate that is directly behind the top plate, Figure 14-5. They are also used to join a thin section to another thin section or a thin section to a thicker section, Figure 14-6. Welding on a thin section's edge often causes that edge to melt away due to overheating of the thin edge, Figure 14-7. Sometimes plug welds are used to make blind welds that would not show after the weldment is finished.

PROJECT 14-1

Assembling a Wedge Using Grooved Edge Joints, Plug Welds, and Surface Buildup

Skill to be learned: How to make grooved-edge welds, plug welds, and surfacing using the FCA welding process. Layout, measuring, pattern making, fitting, and assembling fabrication skills and techniques will be developed.

Project Description

You will build a 1 1/2-in.-wide 7-in.-long wedge out of plate sections that will be plug and edge welded

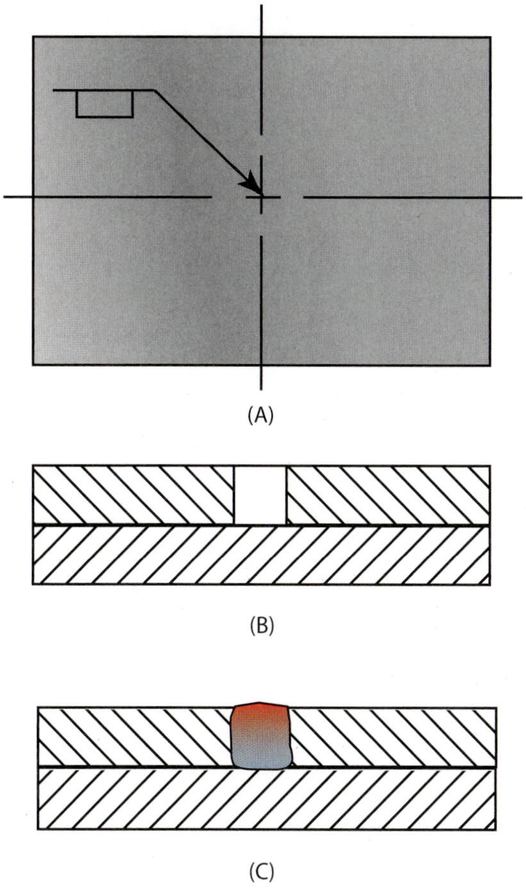

FIGURE 14-5 Plug weld. © Cengage Learning 2012

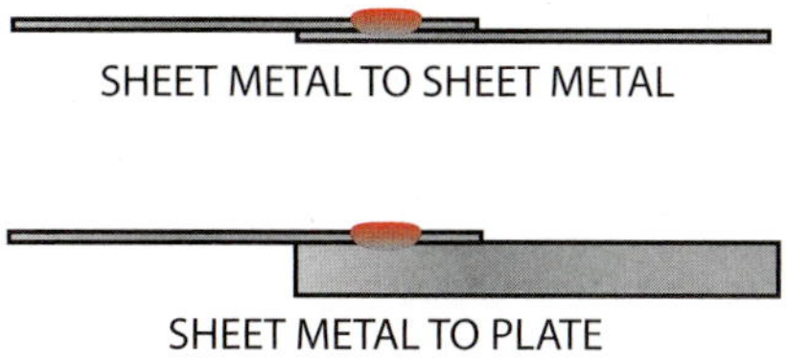

FIGURE 14-6 Plug weld application of thin to thin and thin to thick. © Cengage Learning 2012

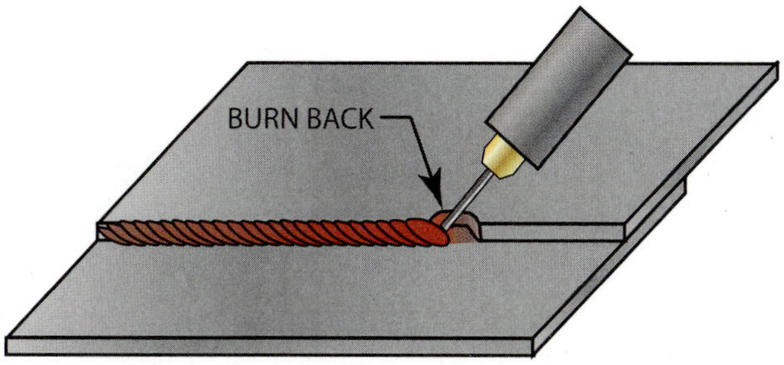

FIGURE 14-7 Burn back. © Cengage Learning 2012

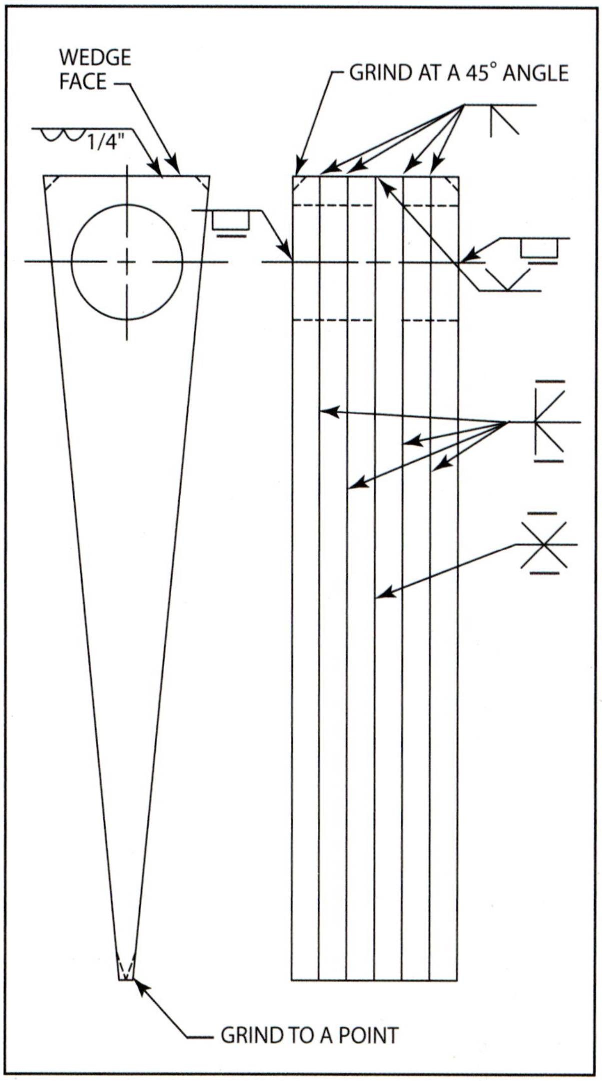

FIGURE 14-8 Project 14-1: Wedge. © Cengage Learning 2012

together, Figure 14-8. The wedge can be used to split firewood or in the shop to help with fitting edges of plates, Figure 14-9.

Project Materials and Tools

The following items are needed to fabricate the wedge.

Materials

- 0.035 FCA welding electrode wire
- 5 1/2 × 7-in. piece of 1/4-in. thick plate or 4 1/2 × 7-in. piece of 3/8-in. thick plate

Tools

- FCA welder
- Plasma cutting torch or oxyfuel cutting torch
- PPE

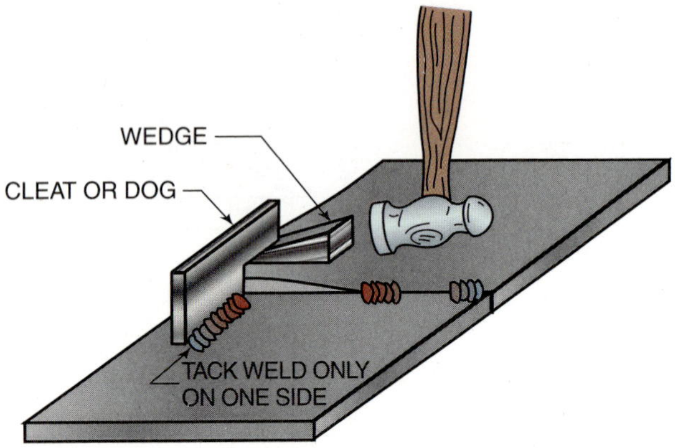

FIGURE 14-9 A use of the wedge. © Cengage Learning 2012

- Tape measure
- Chipping hammer
- 12-in. rule
- Soapstone
- Square
- Wire brush
- Angle grinder
- C-clamp
- Pliers

Layout

The total thickness of the wedge will be 1 1/2 in. If it is being made out of a 1/4-in. thick plate, you will need six blanks, but it if is being made out of 3/8-in. thick plate, only five blanks will be needed. In this project you are going to first lay out one of the blanks, and after cutting it out, you will use it as your pattern to cut out the remaining blanks. By making a pattern first, it will be easier for you to make each of the other blanks exactly the same.

Using the square, a 12-in. rule, and soapstone, lay out the pattern for the first blank on the plate according to Figure 14-8.

The wedge layout does not come to a sharp point because during welding the very thin edge would easily melt away. That is why there is a 1/8-in.-wide flat surface at the pointed end. This will be ground to a tapered point once the fabrication has been welded.

Cutting Out

Using a properly set-up plasma cutting torch or oxyfuel cutting torch, putting on all of your PPE, and following all shop and manufacturer's safety rules for welding, you will cut out the first blank

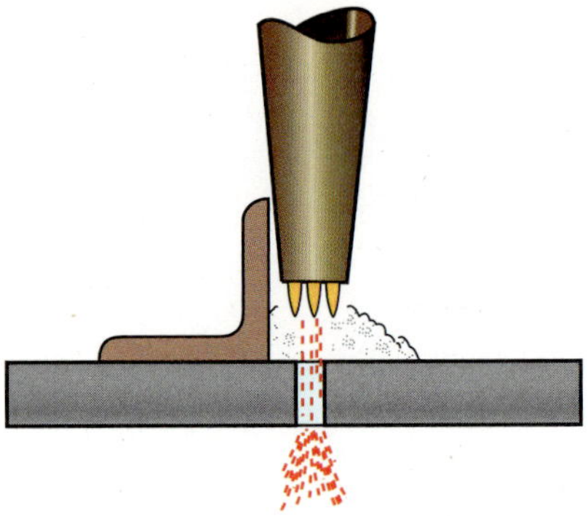

FIGURE 14-10 Using an angle iron guide to make a straighter, smoother cut. © Cengage Learning 2012

that will be used as your pattern for the remaining wedge blanks.

To make these cuts straighter and smoother, you can use a piece of angle iron as shown in Figure 14-10. Use the angle grinder and chipping hammer to remove any slag from the back side of the cutout pattern. Also use the angle grinder to remove any major roughness along the cut surface if necessary.

Make a soapstone *X* on the pattern to mark it as your master pattern. Make sure to only use the master pattern each time you mark out the next piece. Some of the common problems that can occur if you do not use the master pattern are that each piece gets a little larger in size or the straightness of the cuts decreases.

Laying Out and Cutting the Wedge Using a Pattern

Place the master pattern that you just cut out on the steel plate. Trace the pattern using a properly sharpened soapstone. Make sure you keep the point of the soapstone close to the bottom edge of the pattern, Figure 14-11.

Cut out the part using the same PPE and setup and following all of the same rules.

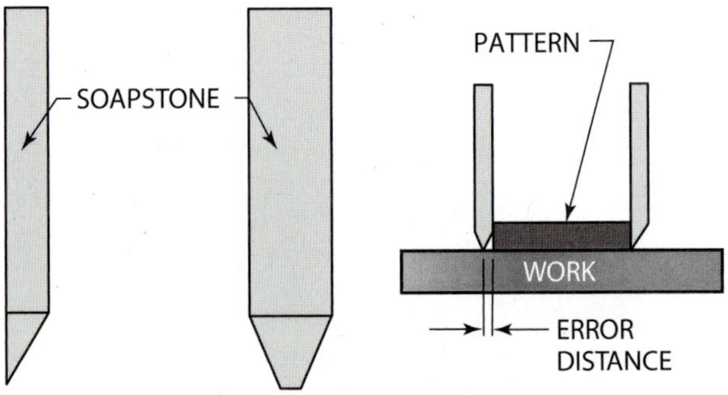

FIGURE 14-11 Proper sharpening and use of a soapstone marker. © Cengage Learning 2012

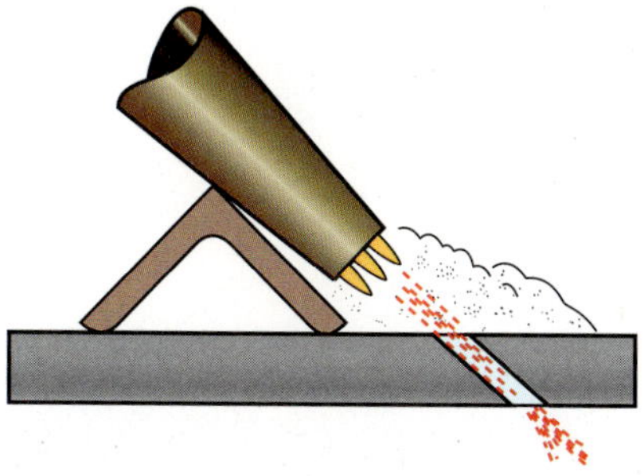

FIGURE 14-12 Using an angle iron guide to make a beveled cut. © Cengage Learning 2012

Repeat the process of tracing and cutting out the wedge blanks until you have all of them cut.

Beveling the Edges

Using the same PPE, setup, and a piece of angle iron setup as illustrated in Figure 14-12 and following all of the same rules, you are going to bevel the sides and top of the wedge blanks. A beveled edge allows the weld to penetrate into the wedge surface so that it does not interfere with the wedge being used.

NOTE: An oxyfuel gouging torch and torch tip an be used to cut a J-groove along the edge of the blanks.

Preparing the Plug Weld

Use a quarter as a template to draw a circle approximately 1 in. in diameter in the center of all but one of the blanks. The one blank not having the plug weld cut out is the center blank.

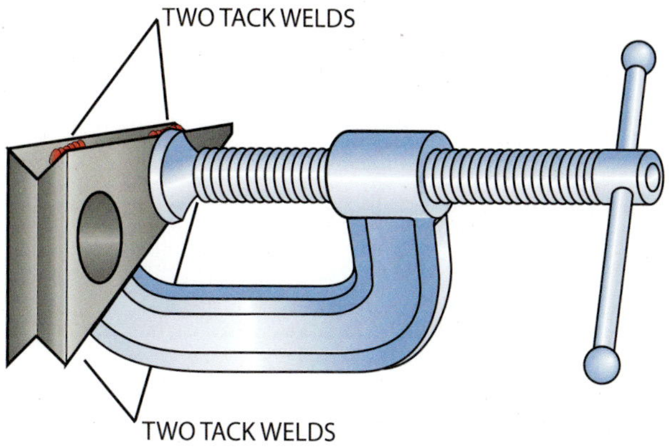

FIGURE 14-14 Using a C-clamp to hold parts for tack welding. © Cengage Learning 2012

Use the same PPE and setup, and follow all of the same rules. Follow the illustration shown in Figure 14-13, and cut out the hole for the plug weld.

Once the blanks have been beveled and the plug weld holes cut, clean off any slag with a chipping hammer or angle grinder.

Fabrication and Welding

Set the welder voltage and wire feed speed for mid-range for the type of wire being used according to the wire manufacturer or Table 14-1. Put on all of your PPE and follow all shop and manufacturer's safety rules for welding.

Use a C-clamp to hold together the first two blanks so that the weld face is in the flat position, Figure 14-14. Make a small tack weld at the face end of the wedge on

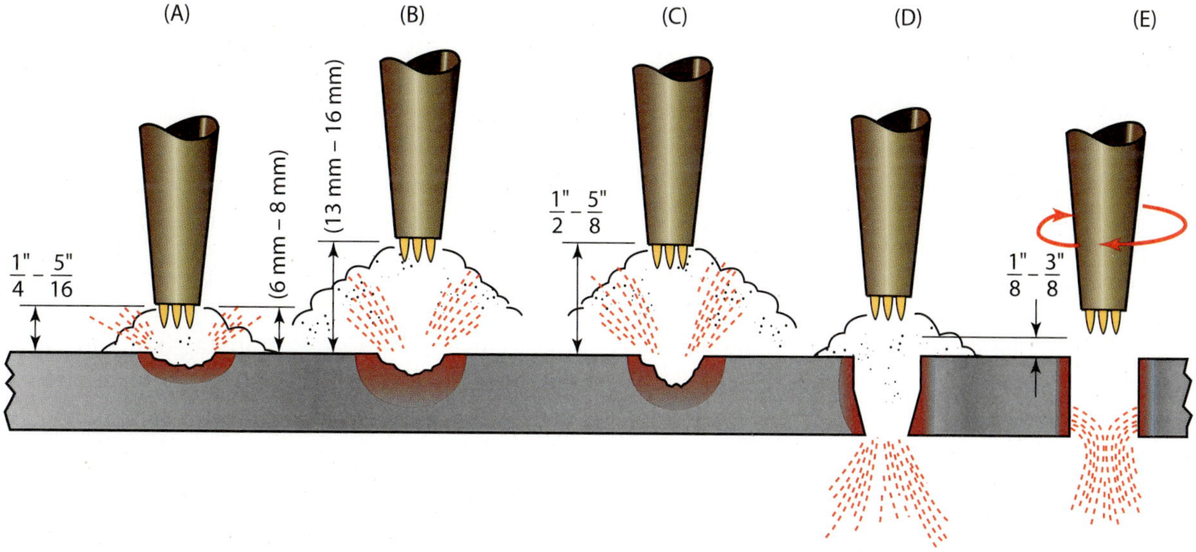

FIGURE 14-13 Sequence for cutting a hole in a plate. © Cengage Learning 2012

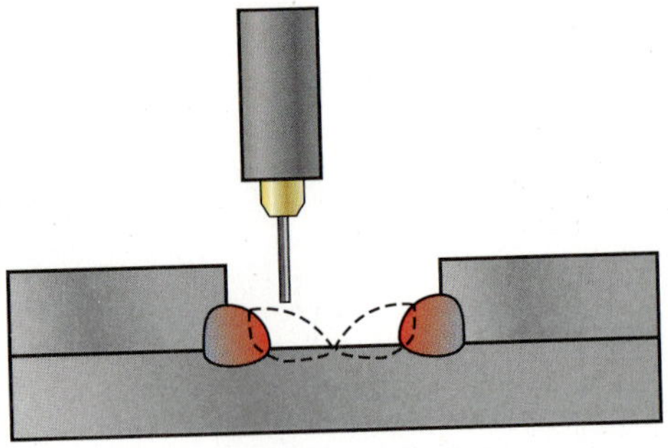

FIGURE 14-15 **Starting a plug weld.** © Cengage Learning 2012

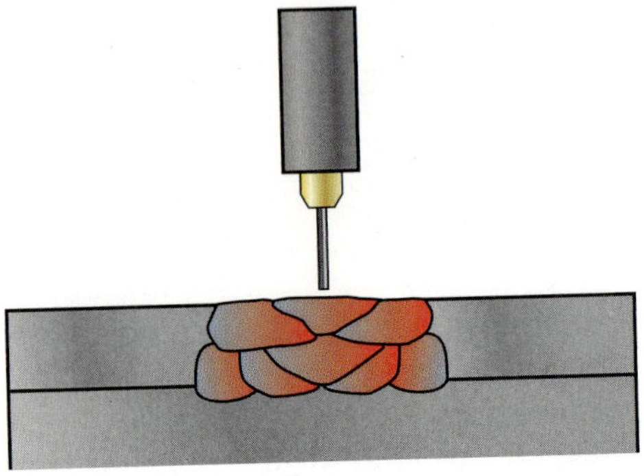

FIGURE 14-16 **Finishing a plug weld.** © Cengage Learning 2012

both sides and a 1/4-in. tack weld 1/2 in. from the point on one side.

Start the weld at the point of the wedge and weld toward the face end of the wedge. By starting at the point, less heat will have built up in the thinner section of the wedge. This will reduce the possibility of having burnthrough at this thinner section. The welding symbol calls for a flat weld bead, so you must adjust your travel speed as necessary. If you weld too fast, and you will have underfill; weld too slow, and you will have excessive buildup.

Cool the wedge, chip the slag, and evaluate your weld. The weld should be uniform in width and consistent in appearance from the beginning to the end.

Turn the wedge blank over and make the weld on the opposite side in the same manner starting at the tip, welding to the head of the wedge. Cool and chip this weld, and inspect it for uniformity.

The next weld you will make is the plug weld. Start the weld at the bottom inside edge of the hole, Figure 14-15. On thin sections of metal, the weld can sometimes be started in the center and worked outward. However, on thicker sections it is important to have the full circumference of the plug weld joined. On thicker metal the plug weld sides are typically beveled to make it easier to join the circumference of the plug plate to the baseplate.

Set the welder voltage and wire feed speed at the high range for the type of wire being used according to the wire manufacturer or Table 14-1. Put on all of your PPE and follow all shop and manufacturer's safety rules for welding.

Be sure that you have full range of movement so that you can keep the welding gun pointed at the root of the weld at an approximately 45° angle. Point the electrode tip into the groove of the joint and lower your helmet. Pull the trigger and make the weld. Try to make the weld all the way around the bottom edge of the hole, but if you cannot, stop and chip the weld before starting the next weld. Once the weld has been completed all the way around the base, stop and chip the slag. If you don't chip the slag out of a deep plug weld, then the slag can build up excessively, making it impossible for you to maintain visibility of the weld and prevent slag inclusions.

The next weld pass will be mainly on the base plate lapping halfway up on the previous weld, Figure 14-16. Hold the welding gun perpendicular to the plate, and make the weld all the way around. Continue the circular pattern, expanding it all the way out to the outside, and make two complete passes. Stop, and chip the slag. Finish the weld by starting in the center and building it up until the weld bead has come flush with the surface.

Cool the wedge blank and grind any excessive weld from the plug before the second wedge blank is clamped in place. Using the same procedure, tack weld this blank to the previous, and produce the two side welds and plug weld. Repeat this process, and alternate the side that the blank is placed on until the wedge has been completed.

NOTE: Make sure the wedge blanks are square. If the blanks are not squared to the previously welded section, it can become skewed, Figure 14-17.

Stand the wedge up so that the grooves in the face of the wedge can be welded in the flat position. Make the welds one at a time, chipping and wire brushing each before the next weld is made. Be sure to fill the end of the weld bead so that the edge of the wedge face is flat.

You are now going to make a surface weld buildup on the head of the wedge by making a series of welds starting with a perimeter weld around the face of the wedge.

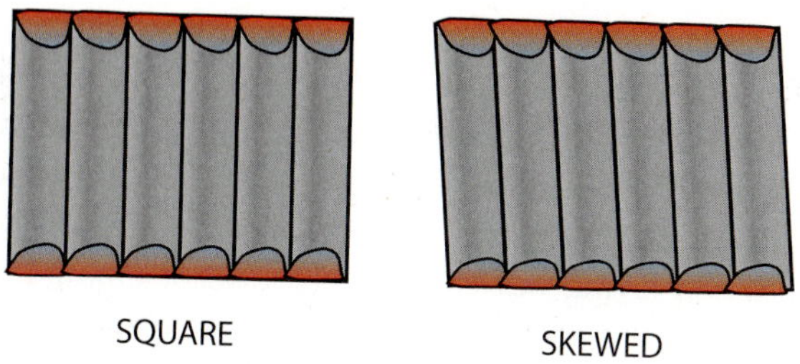

FIGURE 14-17 Wedge blanks should be square after welding. © Cengage Learning 2012

With the welding gun held perpendicular to the wedge face, make a weld continuously around the top edge of the wedge. Cool and chip the weld. The next series of welds will be made perpendicular to the wedge plate. Starting at one side, make a weld from the perimeter weld across to the other side. Repeat this process, alternating the direction of the weld until the entire wedge face has been surfaced.

Finishing

The welded surface should be ground so that there are no large protrusions of either base metal or weld metal. This surface does not have to be perfectly smooth but should be even and uniform. Using the angle grinder, grind a 45° angle around the edge of the faces of the wedge. This beveled edge will reduce the mushrooming of the head as it is hit with a hammer. Grind the point of the wedge. Make sure when you grind the point that the center of the point is as close as possible to the center of the wedge.

Paperwork

Complete a copy of the time sheet in Appendix I, the bill of materials in Appendix III, or use forms provided by your instructor.

Butt joints are used to join the edge of structural steel shapes. The advantage of using the square butt joint is that the metal can be welded as it is without additional grinding or shaping, Figure 14-18.◆

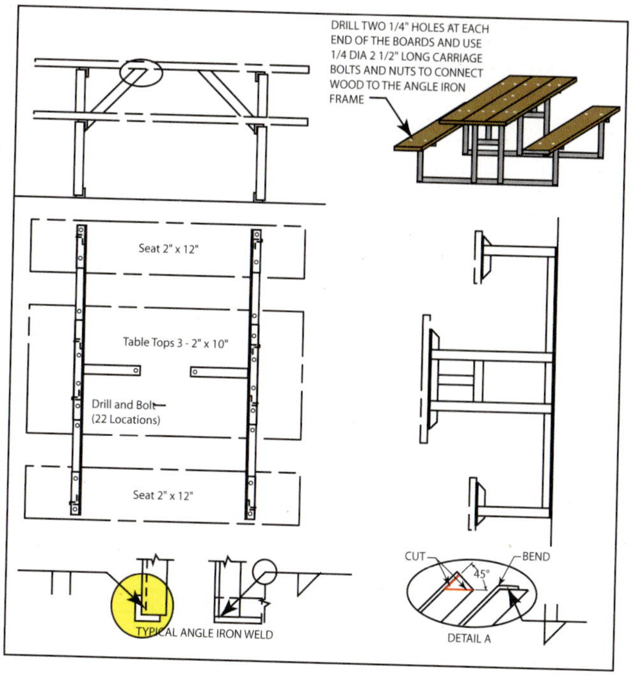

FIGURE 14-18 Application of a butt joint. © Cengage Learning 2012

PROJECT 14-2

2F Fillet Welds on Lap Joints on a Sundial

Skill to be learned: How to make various sizes of fillet welds in the horizontal positions on both tightly fitting joints and joints with gaps. Layout, linear and angular measuring, fitting, and assembling fabrication skills and techniques will be developed.

Project Description

You will build an 8-in. square sundial. The raised panels of the sundial give it the appearance of radiating sunbeams.

Project Materials and Tools

The following items are needed to fabricate the sundial.

Materials

- 0.035 FCA welding electrode wire
- 10 × 10-in. piece of 1/4-in. thick plate
- 8 × 8-in. piece of 1/4-in. thick plate
- 26 × 1-in. piece of 1/4-in. bar stock

Tools

- FCA welder
- Plasma cutting torch or oxyfuel cutting torch
- PPE
- Tape measure
- Chipping hammer
- 12-in. rule
- Soapstone
- Square
- Wire brush
- Angle grinder
- Pliers

- Protractor
- Straightedge

Lay Out the Sundial Face Segments

The sundial face will be laid out according to Figure 14-19. The layout provides for approximately a 1/2-in. additional area on one side of each of the sundial panels for a tab. This tab will provide the overlap area for the assembled sundial face.

Using a protractor and a straightedge, lay out the radial lines for each of the sundial's panels. The angle of each of the lines coincides with the hour it represents. Because of the sun's path, these lines are not uniformly spaced.

When laying out the sundial, do not change the protractor, but mark off each of the radials as shown in Figure 14-19. Rotating the protractor between measurements can result in an overall error.

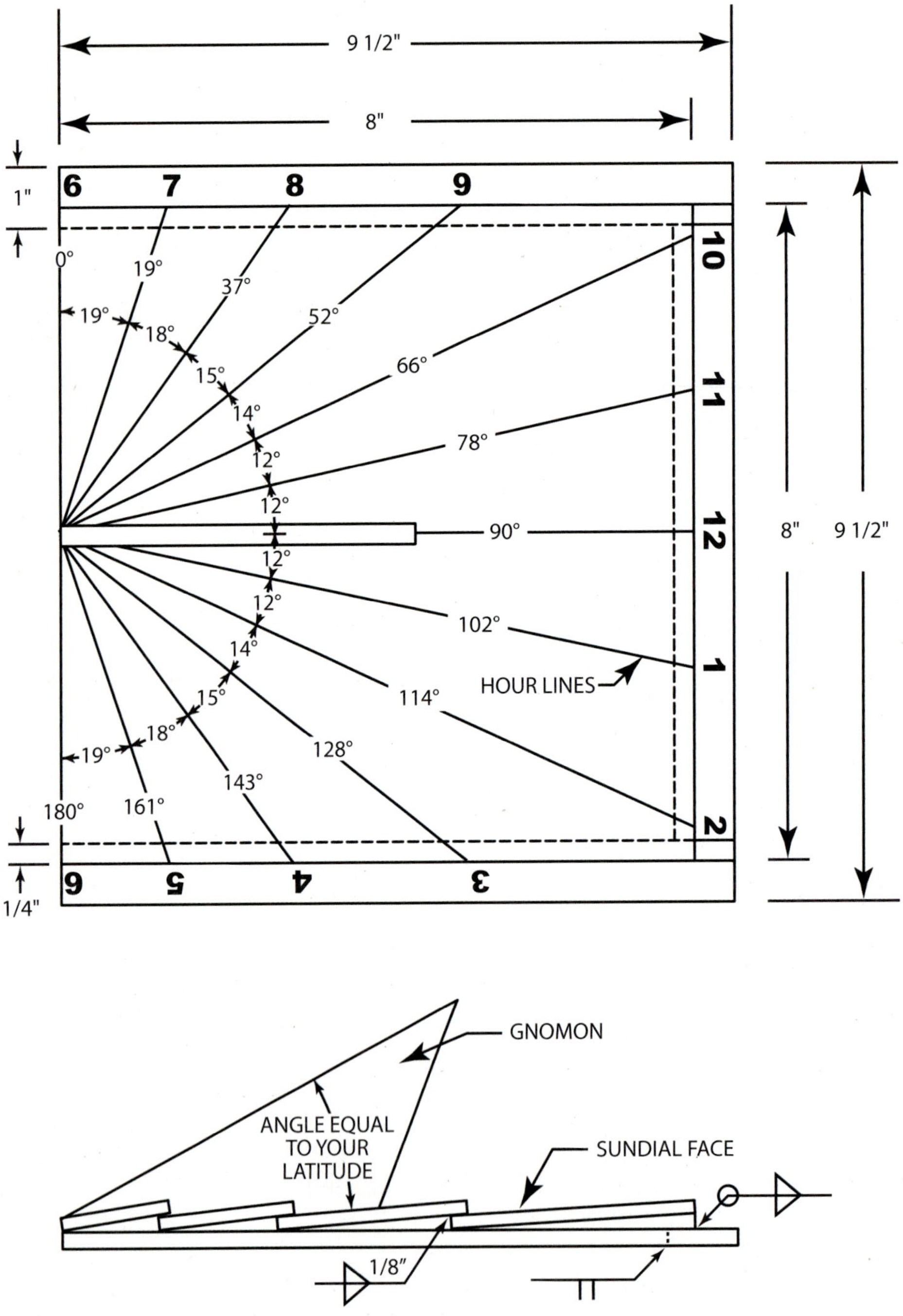

FIGURE 14-19 Project 14-2: Sundial. © Cengage Learning 2012

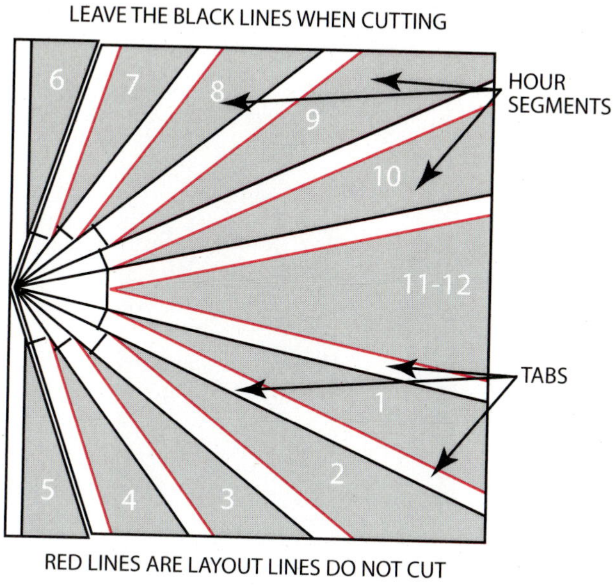

FIGURE 14-20 Laying out sundial segments.
© Cengage Learning 2012

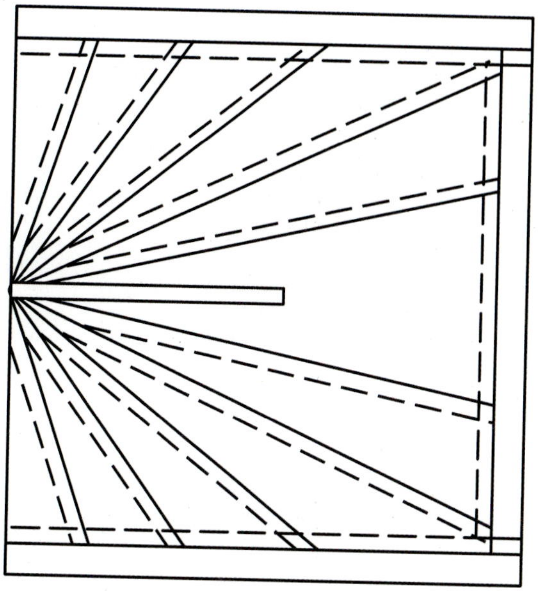

FIGURE 14-21 Assembling the segments of the sundial.
© Cengage Learning 2012

Draw the sundial face on a 10-sq in. plate as shown in Figure 14-20. Once the parts are cut out and fitted together, the finished sundial face will be 8 in. square. Check the accuracy of your angles by comparing the measurement between ends of the lines for the 6 and 7 o'clock with the measurement between the 6 and 5 o'clock. These distances should be the same, as should the distances between the 7 and 8 o'clock and the 4 and 5 o'clock, and so forth.

Cutting Out the Sundial Face Segments

Using a properly set-up plasma cutting torch or oxy-fuel cutting torch, putting on all of your PPE, and following all shop and manufacturer's safety rules for welding, you will cut out the sundial face segments.

Use an angle iron as illustrated in Figure 14-10. Make the cut on the tab side of the line as shown in Figure 14-20.

When all of the sundial face hour segments have been cut, use the chipping hammer and angle grinder to remove any slag from the cuts.

Assembling the Sundial Face

Place the 11 to 1 o'clock segment flat on the welding table. Place the 10 to 11 o'clock segment overlapping the 11 to 1 o'clock panel. Line up the edge of the 10 to 11 o'clock segment with the hour line that separates the face area from the tab on the 11 to 1 o'clock segment. Place each successive segment overlapping as shown in Figure 14-21. The outside edge of the assembly should be straight and even.

Set the welder voltage and wire feed speed for medium range for the type of wire being used according to the wire manufacturer or Table 14-1. Put on all of your PPE and follow all shop and manufacturer's safety rules for welding. Using your gloved hand, hold the segments in place, and make two small 1/4-in. long tack welds on each of the segments. The tack welds should be approximately 1 in. from the end of the joint. Repeat this process until all of the segments are tack welded together. Repeat this process to assemble the 2 to 6 o'clock segments.

NOTE: The tab on each of the hour segments may hang over the back edge of the assembled sundial. These tabs can be trimmed with a grinder after they have been welded in place, or each segment's tab can be ground to fit before they are assembled.

Check the accuracy of your fit-up by checking to see that the outside edges of the sundial face are lined up evenly, that the face is square, and that the spaces between the hours on opposite sides are equal within ±1/8 in.

SPARK YOUR IMAGINATION

Why are the radial lines for each hour on the sundial face closer together around 12 o'clock and farther apart around 6 o'clock? Can you tell anything about the season as it relates to the length of the shadow of the gnomon?

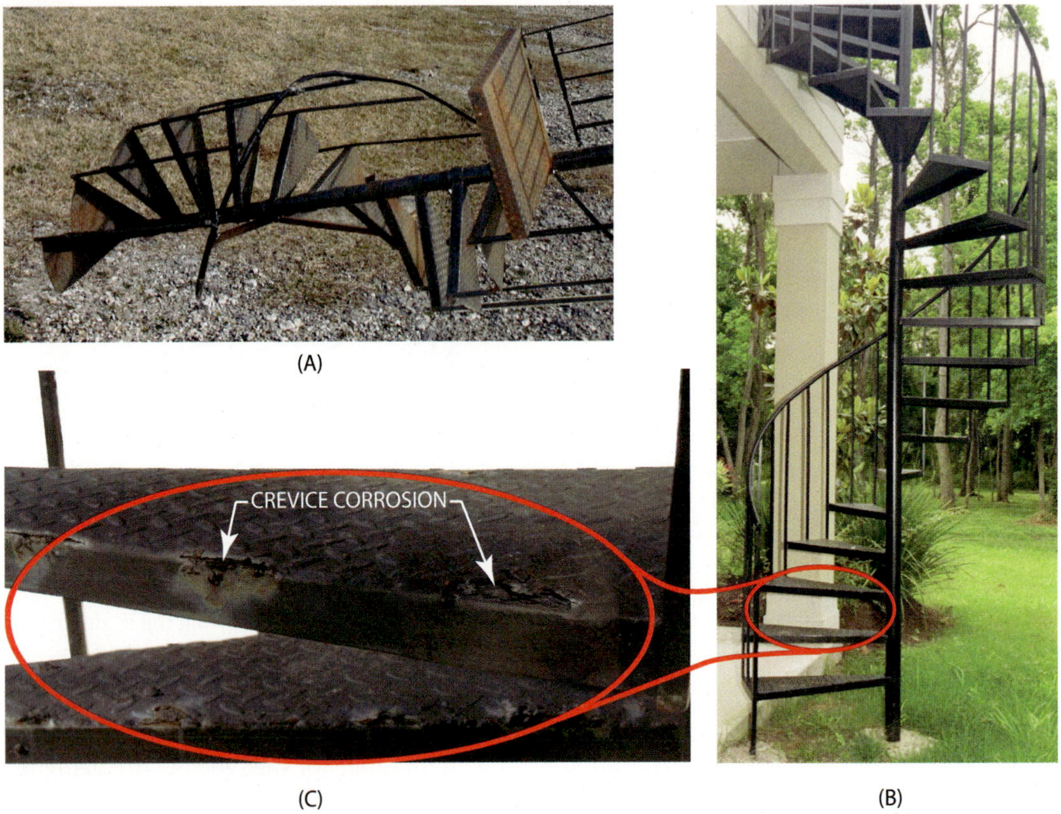

FIGURE 14-22 This spiral stair case went from the welding shop (A) to the customer's home (B) to unsafe to use (C) in five years because of crevice corrosion. Unseen in the photo is the massive amount of crevice corrosion underneath each step which made it very hazardous. © Cengage Learning 2012

Welding the Sundial Face

Set the welder voltage and wire feed speed for medium range for the type of wire being used according to the wire manufacturer or Table 14-1. Put on all of your PPE and follow all shop and manufacturer's safety rules for welding.

Clamp the sundial facedown on the welding table. Make sure that the tip ends of all of the segments are touching the weld table. Clamping the sundial down will help reduce weld distortion.

The welds will be made in the horizontal position starting at the center of the sundial and going to the outside edge. The welds are started on the narrowest portion on the segments so that excessive heat does not result in weld burn back.

Holding the gun at a 30° backhand angle with the electrode pointed at the root of the weld, lower your helmet, pull the trigger, and begin the weld. Maintain a uniform welding speed and electrode manipulation. Keep the weld size smaller than the thickness of the plate. These welds both hold the sundial in place and seal the joints to prevent **crevice corrosion,** Figure 14-22. Therefore, these welds do not have to be as large as a structural fillet weld might be. Keeping the welds smaller will also reduce the possibility of weld distortion. When the weld is completed, cool the part, chip, and wire brush the weld, and look for uniformity in width and appearance. Make any required adjustments in the welding machine or in your welding technique as you complete each of the additional welds. When all of the welds have been completed, cool, chip, and wire brush the welds.

NOTE: Crevice corrosion is the type of rusting that occurs when water seeps between two tightly fitting surfaces. Because the surfaces are so tight, the water does not evaporate so it begins to cause rust to form. The rust forms small layers that let more water seep in which accelerates the rusting process. The best way to prevent crevice corrosion is to seal lap joints with a weld.

Turn the face over and again clamp it to the welding table. Check the tack welds for size. Any tack

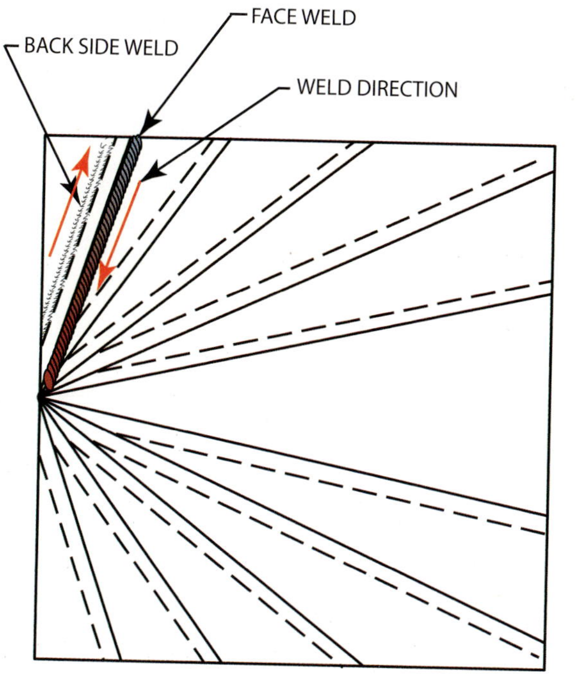

FIGURE 14-23 Welding direction for sundial segments.

welds that are too large will need to be ground down. Use the angle grinder to trim the tack welds that are so large that they will interfere with the finished weld. You want the fillet welds on the sundial face to be as uniform in appearance as possible.

Start the fillet welds on the outside edge of the sundial and weld to the center. The weld on the back side of the sundial face will conduct the welding heat away from the joint so it will not overheat, Figure 14-23. These welds will be small, filling only the bottom of the joint. To accomplish this, your travel speed will be faster than before with little or no electrode movement. You will, however, need to watch and make sure that complete fusion is occurring along the entire weld's length. Starting at the outside edge with the welding gun in the backhand angle pointed at a 45° angle to the weld joint, lower your helmet, pull the trigger, and begin the weld. When the weld is completed, cool and chip the weld, and inspect it for any imperfections. Make any necessary adjustments in the welding machine or your technique as you complete the additional welds. When all the welds are complete, wire brush and chip any slag or spatter off of the weld face.

Fit-Up and Assembly of the Frame

The outside edge of the sundial is made with 1-in. wide 1/4-in. thick bar stock. Lay out the length of the bar stock according to the drawing and adjust its length as necessary to fit the finished sundial face. The bar stock should extend 3/4 in. outside of the edge of the sundial face. Place the bar stock flat on the welding table, and then place the sundial face on the bar stock. Clamp the sundial facedown on top of the frame. Make a series of 1/4-in. long tack welds at the edge of each hour segment. Remove the clamps, turn the sundial over, and make a small tack weld at each of the sundial segments where they touch the frame.

Welding the Sundial Face to the Frame

Set the welder voltage and wire feed speed at the high range for the type of wire being used according to the wire manufacturer or Table 14-1. Put on all of your PPE and follow all shop and manufacturer's safety rules for welding.

Place the sundial face in the vertical position so that you can make the weld across one side in the horizontal position. The sundial may lean back slightly from vertical, which will make the welding a little easier. Holding the welding gun at a 30° forehand angle, start on the outside edge. You will have to adjust your welding technique as the gap between the hour segment and the frame changes. Some techniques you can try to help control the weld include using a large C pattern, extending the stick out slightly, or triggering the gun. Try each method to see which one works best for you. Cool and chip the weld, and check it for any slag inclusions or spots of incomplete fusion. Make a small weld over any spots that need to be repaired. Rotate the sundial, and complete the weld on the next two sides, Figure 14-24.

Clamp the sundial faceup on the welding table. Using a backhand gun angle, make a small sealing weld around the outside of the sundial face.

Layout and Assembly of the Gnomon

The gnomon is the part of the sundial that casts the shadow to tell the time. The angle of the top of the gnomon must be equal to the latitude of your location for the sundial to accurately tell time. To find the latitude of your area, you can go to the Internet.

Using a protractor, lay out a line that is 8 in. long at the angle equal to your area's latitude. Measure

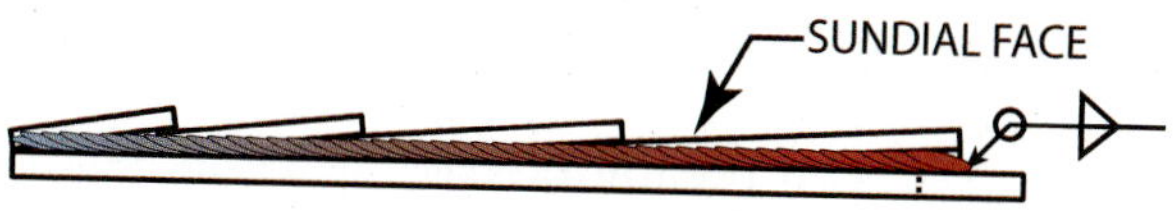

FIGURE 14-24 Small sealing weld on the outside edge of the sundial face.

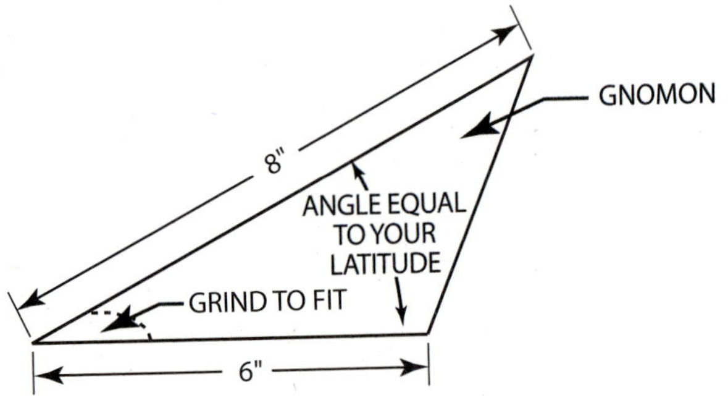

FIGURE 14-25 Gnomon layout. © Cengage Learning 2012

6 in. across the base and connect the two points, Figure 14-25. Using a properly set-up plasma cutting torch or oxyfuel cutting torch, putting on all of your PPE, and following all shop and manufacturer's safety rules for welding, you will cut out the gnomon. Use the chipping hammer and/or grinder to remove any unwanted slag.

The gnomon base must be fitted so that the angle of the top intersects the back of the sundial at the same height as the 11 to 1 o'clock segment, Figure 14-24. You can use the angle grinder to remove the necessary metal to fit this into place. Use a square to hold the gnomon vertical, and make two small tack welds on both sides. Complete the weld on both sides all the way to the back of the sundial face.

Finishing

The sundial can be wire brushed and painted using a latex paint, and the sundial numbers can be painted on, or you can weld the numbers on the outside face of the sundial. If you are not happy with the appearance of one of your welded numbers, use the angle grinder to grind it off and try again. Mount the sundial so that the 12 o'clock line is pointed to the true north and not to the magnetic north.

Paperwork

Complete a copy of the time sheet in Appendix I, the bill of materials in Appendix III, or use forms provided by your instructor.

Lap joints are commonly used to join structural shapes such as an angle iron, Figure 14-26.◆

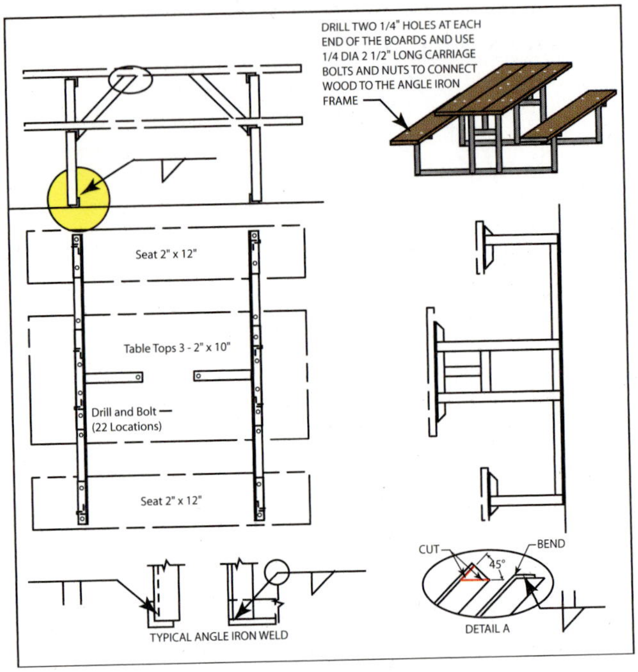

FIGURE 14-26 Application of a lap joint. © Cengage Learning 2012

PROJECT 14-3

1F Fillet Welds on Tee Joints to Make a Candlestick

Skill to be learned: How to make fillet welds in the flat position, how to make blind fillet welds, and how to make welds around corners. Basic math, layout, measuring, fitting, and assembling fabrication skills and techniques will be developed.

Project Description

The massive candlestick is designed to hold large diameter candles without looking undersized. The large top will help catch wax that might run down the candle, Figure 14-27.

Project Materials and Tools

The following items are needed to fabricate the candlestick.

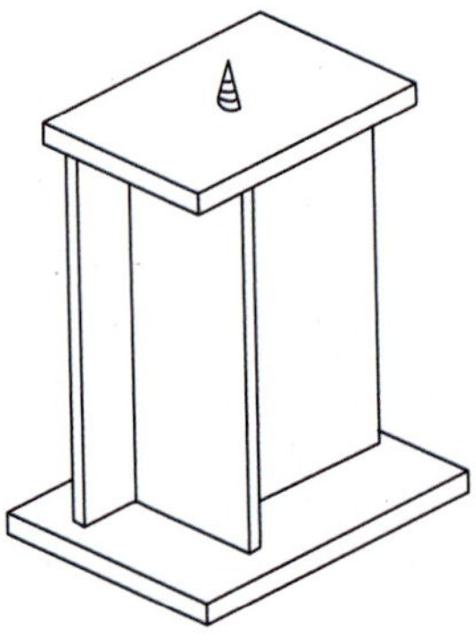

FIGURE 14-27 Candlestick holder. © Cengage Learning 2012

Materials

- 0.035 FCA welding electrode wire
- 32 × 3-in. piece of 1/4-in. bar stock
- 10 × 5-in. piece of 1/4-in. thick plate

Tools

- FCA welder
- Plasma cutting torch or oxyfuel cutting torch
- PPE
- Tape measure
- Chipping hammer
- 12-in. rule
- Soapstone
- Square
- Wire brush
- Angle grinder
- Pliers

Layout

The vertical members of the candlestick holder can be made out of 3-in. wide 1/4-in. thick bar stock, or they can be cut from 1/4-in. plate, Figure 14-28. The overall finished height of the candlestick holder is 8 in., so you must deduct from the length of the vertical members the thickness of the top and bottom plates. Using a tape measure, soapstone, and a square, lay out the four side plates. Be sure to leave space to compensate for the metal removed by the torch kerf. Lay out the top and bottom plates using a rule and square.

NOTE: **You can find out what the kerf width is for a torch by making a sample cut in scrap metal and measuring its width.**

Cutting Out

Using a properly set-up plasma cutting torch or oxyfuel cutting torch, putting on all of your PPE, and following all shop and manufacturer's safety rules for welding, you will cut out the plates.

Assembly

Draw a horizontal line 3/4 in. from one edge of each of the plates, Figure 14-29. Using a square, line up one plate on the other, and make four small tack welds to hold the plate in place. Repeat this process on the other two plates. Align the two assembled plates, and adjust them as necessary so that they

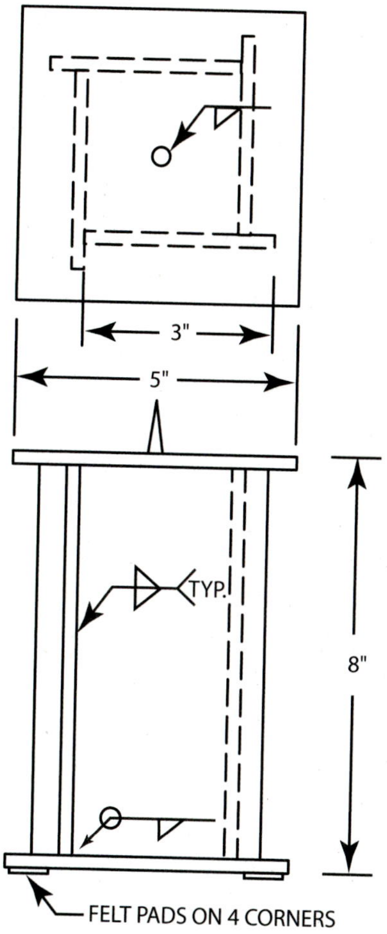

FIGURE 14-28 Project 14-3: Candlestick. © Cengage Learning 2012

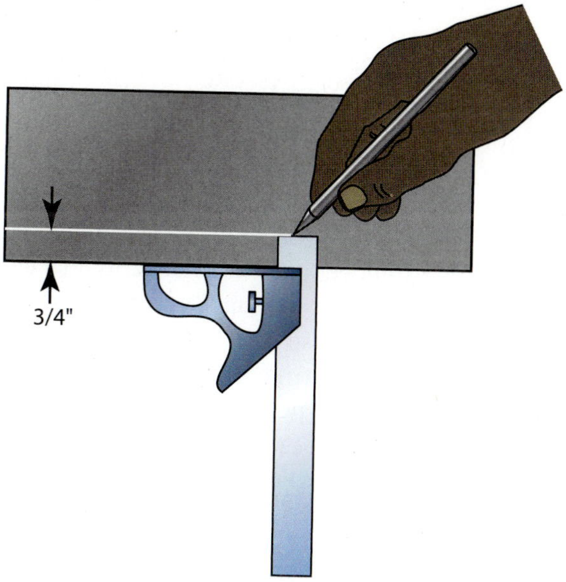

FIGURE 14-29 Using a combination square to draw a horizontal line. © Cengage Learning 2012

can be tack welded into place. After the parts have been tack welded in place, chip, and wire brush the tack welds.

Welding

Set the welder voltage and wire feed speed at the high range for the type of wire being used according to the wire manufacturer or Table 14-1. Put on all of your PPE, and follow all shop and manufacturer's safety rules for welding.

With the welding gun pointed at a 30° backhand angle, lower your helmet, pull the trigger, and start the weld at the end of the piece. Travel at a rate fast enough so that the molten weld pool is approximately 1/2 in. to 3/4 in. wide. You may not have to weave to get the desired weld width with the welding current set on the high range. At the end of the weld, trigger the gun to fill the weld crater.

Cool the weld, chip, and wire brush it, and look for any weld discontinuities. Make whatever adjustments are required, and complete the remaining three welds.

Blind Welds

Sometimes a welder is expected to make a weld in a very restricted space. These are sometimes referred to as blind welds where the welder may not actually be able to see the weld as it is being produced. You will learn by listening to the sound of the weld when it is being made and when there is a problem.

NOTE: Some welders have to make welds in very restricted spaces using a mirror.

With the welder set at the lowest range and using a self-shielding wire so that the welding nozzle can be removed, place the candlestick so that the weld inside the candlestick can be made in the flat position. Place the welding gun as far in as possible with the electrode extended approximately 1 in. Place the electrode in the base of the joint, and brace the gun with your gloved hand so that when the weld starts it does not drop and strike the inside of the tube. Lower your helmet, and pull the trigger. Maintain the same angle and travel speed to follow the joint to the end. Cool the weld, chip, and wire brush it, and look for weld shape, contour, and consistency. Rotate the box to the next weld, and repeat the process, making any necessary adjustments in technique or welding machine setup. Repeat this process on both ends of the box.

Assembly and Fit-Up

Using the angle grinder, if necessary, remove any weld metal that may have flowed over the end of the square tube so that it will fit flush on the base and top plates. Measure the width of the finished candlestick vertical tube. Subtract that distance from the baseplate's width of 5 in. Now divide that number by 2 to determine the setback distance from each side that is required to place the vertical tube in the center of the base. Using a rule, square, and soapstone, mark the setback lines on the base and top plates. Place the tube on the baseplate so it is located squarely in the center. Make a tack weld on each side to hold it in place.

Turn the assembly over, and place it squarely in the center of the top plate. Measure the overall candlestick height. It should be 8 in. ± 1/8 in. If it is too short, put a spacer in the joint to give the weld a little root opening. If it is too tall, grind a little off. Once the height is correct, make one small tack weld on all four sides, Figure 14-28.

NOTE: A gas welding rod tip can be flattened into a wedge shape as a spacer to adjust the root opening. If the wedged end of the welding rod tip gets stuck in the joint after the joint is tack welded, bend it back and forth to break it off, or strike an arc right at the edge to instantly melt off the welding rod.

Welding

Welds should not start or stop in a corner. Welds that are made through a corner are less likely to leak. Start your weld approximately two-thirds of the distance from the corner and weld through the corner, Figure 14-30. To do this you should practice moving the gun around the joint, ensuring that you have complete and free movement so that you can make the transition from one side to the other. Place the electrode in the groove of the joint, pull the trigger, and proceed to weld around through the corner and stop. Cool, chip, and wire brush the weld and look for weld consistency. Repeat this process until you have welded completely around the base and top plates of the candlestick holder.

Finishing

A small metal screw can be tack welded to the center of the top of the candlestick holder to attach a candle, or the top may be left flat if large diameter candles are going to be used. Wire brush the candlestick holder, and paint it with a latex paint. Four small adhesive

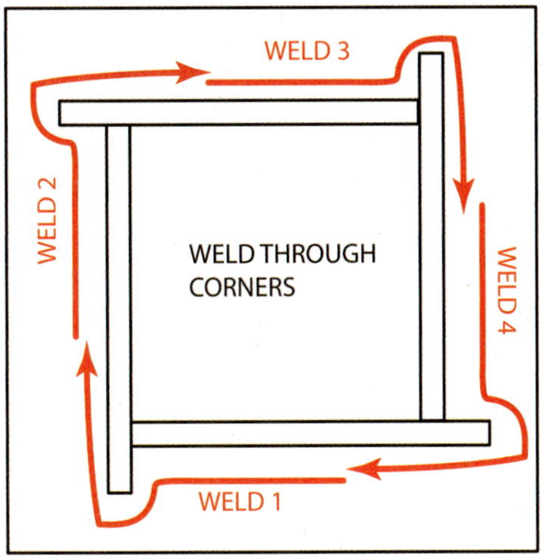

FIGURE 14-30 Welds should not stop in a corner. © Cengage Learning 2012

felt pads should be placed on the bottom, one at each corner, to prevent the candlestick holder from scratching furniture.

Paperwork

Complete a copy of the time sheet in Appendix I, the bill of materials in Appendix III, or use forms provided by your instructor.

The **tee joint** is often used to join structural shapes such as an angle iron to itself at a 90° angle. The tee joints add a lot of strength to structures because the forces have two directions to work against. For that reason, tee joints are often used to make a stiff corner with an angle iron, Figure 14-31.◆

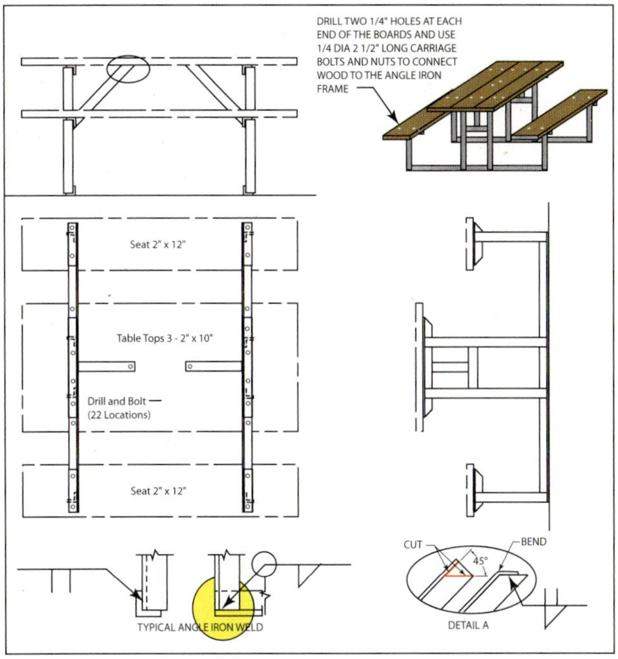

FIGURE 14-31 Application of a tee joint. © Cengage Learning 2012

PROJECT 14-4

3G Fillet Welds on Tee Joints to Make a Candlestick

Skill to be learned: How to make fillet welds in the vertical down position. Layout, measuring, fitting, and assembling fabrication skills and techniques will be developed.

Project Description

These two candlesticks are designed to be a matched pair. The weld spatter will be left on the sides of the candlesticks to give the appearance of dripped wax. Additional wax can be added to enhance that effect.

Project Materials and Tools

The following items are needed to fabricate the candlestick.

Materials

- 0.035 FCA welding electrode wire
- 36 × 2-in. piece of 16-gauge sheet metal
- 3 × 3-in. piece of 16-gauge sheet metal
- 2 1/2 × 2 1/2-in. piece of 16-gauge sheet metal
- 3/4-in. # 8 sheet metal screw

Tools

- FCA welder
- Shear
- PPE
- Tape measure
- Chipping hammer
- 12-in. rule
- Scribe or awl
- Square
- Wire brush
- Angle grinder
- Pliers

Layout

Lay out the six vertical panels for the candlesticks using a rule, square, and scribe, Figure 14-32. No additional space is needed for the kerf since these parts will be sheared. Lay out the two tops and two bottoms using

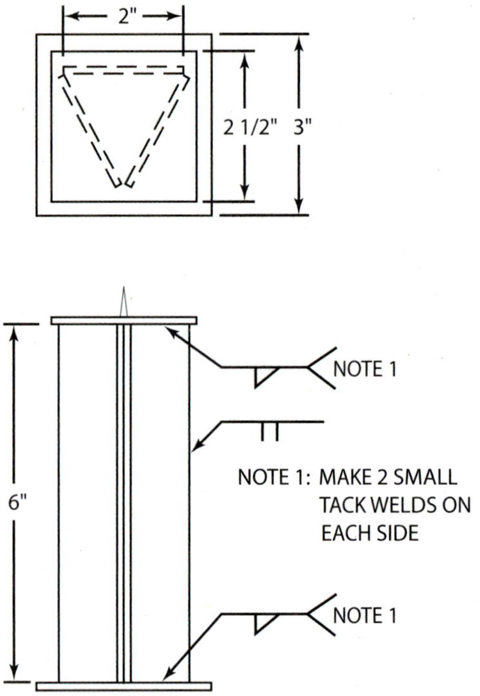

FIGURE 14-32 Project 14-4: Pair of candlesticks. © Cengage Learning 2012

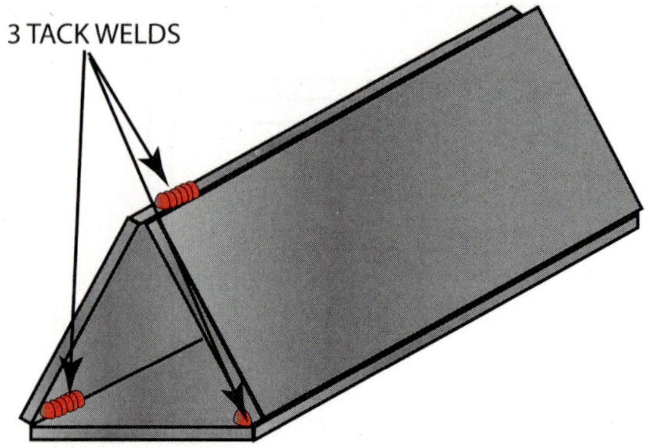

FIGURE 14-33 Tack welding candlestick. © Cengage Learning 2012

the same layout technique. A scribe or awl can be used on some layouts when accuracies closer than those afforded by soapstone are required.

Cutting Out

A shear will be used to cut out the parts.

Assembly

Place one of the sides flat on the welding table. Hold the other two in your gloved hand so that they form a triangle. Make a small tack weld to join the top two sides together. Make two tack welds inside the triangle to hold the three pieces together, Figure 14-33. Hold the other end together the same way and make three more small tack welds. Now hold the center together and make three additional tack welds. You will need to rotate the triangle between each weld to make these welds. You should now have three tack welds on each of the seams—one at each end and one in the center. Repeat this process with the other candlestick.

It is often necessary to have more tack welds on thin sheet metal so that it does not warp out of shape as the weld is produced. In this case, however, because you are making a fast vertical down weld, little distortion should occur.

Place one of the triangular columns on the base. Measure and center it, and make one small tack weld in the center of each panel. Repeat this process with the second triangle candlestick holder.

Welding

Set the welder voltage and wire feed speed at the lowest range for the type of wire being used according to the wire manufacturer or Table 14-1. Put on all of your PPE and follow all shop and manufacturer's safety rules for welding.

Vertical down welding is often used to make welds on thin-gauge sheet metal. One of the advantages of vertical down welding is that relatively high weld travel speeds are possible, and this helps reduce burnthrough and distortion. Holding the welding gun at an upward angle of approximately 30°, start the weld at the top of one of the joints and weld vertical down to the bottom. Cool, chip, and wire brush the weld to look for any discontinuity. Make adjustments as necessary and repeat the process on the remaining vertical down welds. When the welds are completed on the first candlestick holder, do the same on the second candlestick holder.

The top of the candlestick holder should be placed on the welding table and the triangular base measured square on top of it. Make one small tack weld on each of the sides of the triangular candlestick holder.

Finishing

Leave the weld spatter that has fallen on the sides and base of the candlestick holder. It is part of the rustic, dripped-wax appearance of the candlestick holder. Paint the candlestick holder with latex paint, or melt candle wax and drip it around and down the sides.

Paperwork

Complete a copy of the time sheet in Appendix I, the bill of materials in Appendix III, or use forms provided by your instructor.◆

SUMMARY

In this chapter you have learned enough layout, measuring, cutting, fabrication, and welding skills to build the picnic table shown in Figure 14-34. As you have built projects in this chapter, you have been learning the welding skills needed to build this table's frame. Similar weld joints have been highlighted in chapter figures.

This picnic table frame is designed to be stable because of the width of the base and it is easy to get into because there is not a bar to step over as you are being

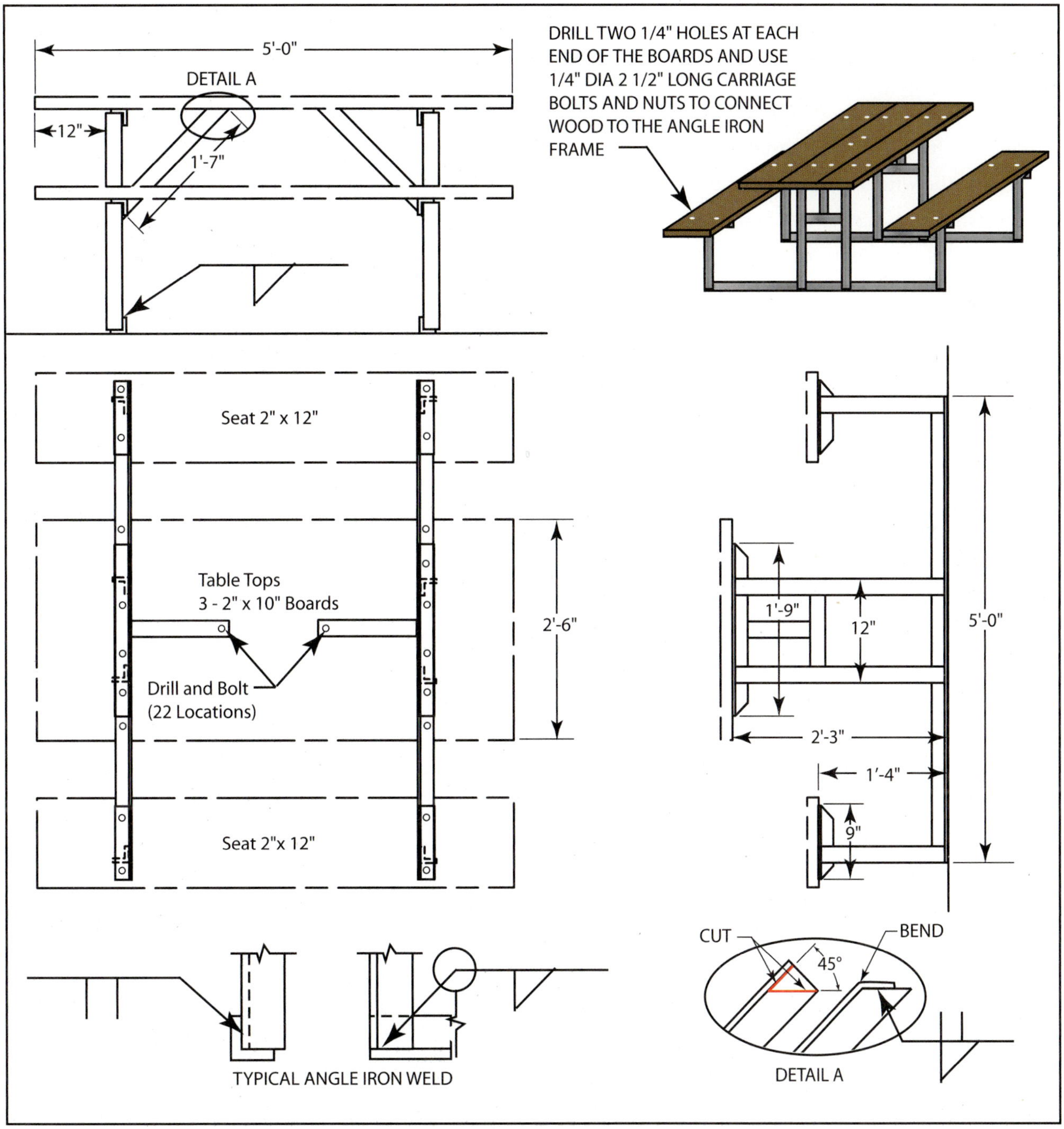

FIGURE 14-34 Welded picnic table design. © Cengage Learning 2012

seated. In addition, the length of the wood top and seats can be increased to accommodate more picnickers. The vertical angle irons can be lengthened so the picnic table could be permanently set in a concrete slab.

Because FCA welding has become one of the most commonly used welding processes, it is important that you learn to make high-quality welds with this process.

REVIEW QUESTIONS

1. What two factors are required to make a good FCA weld?
2. How is the arc voltage and amperage set up for FCA welding?
3. In FCA welding, how will the amperage change if the wire feed rate is increased?
4. How is the wire feed speed measured?
5. Explain the term *electrode manipulation.*
6. How does the length of the electrode extension affect the weld?
7. What are edge welds used for?
8. Describe a plug weld.
9. What tools can be used to remove slag?
10. Why should only the master pattern be used when tracing and cutting out duplicate pieces?
11. What problem can occur if there is too much heat buildup at a thinner edge of a section of metal?
12. If you donot chip the slag out of a deep FCAW plug weld, what problem can occur?
13. What is a blind weld?

Chapter 15

Gas Tungsten Arc Welding Equipment and Materials

OBJECTIVES

After completing this chapter, the student should be able to:

- Demonstrate how to set up a gas tungsten arc (GTA) welding station.
- Identify different types of tungsten electrodes and explain their uses.
- List the different GTA welding currents and explain their effects on welding.
- List the different GTA welding shielding gases and explain how they are used.

KEY TERMS

cleaning action
collet
flowmeter
frequency
inert gas
noble inert gases
postflow
preflow
tungsten

INTRODUCTION

The gas tungsten arc welding (GTAW) process is sometimes referred to as TIG, or heliarc. The term *TIG* is short for tungsten **inert gas** welding. Under the correct welding conditions, the tungsten electrode does not melt and is considered to be nonconsumable. The surface of the metal being welded does melt at the spot where the arc impacts its surface. This produces a molten weld pool.

To make a weld, either the edges of the metal must melt and flow together by themselves or filler metal must be added directly into the molten pool. Filler metal is added by dipping the end of a filler rod into the leading edge of the molten weld pool. Most metals oxidize rapidly in their molten state.

To prevent oxidation from occurring, an inert gas flows out of the welding torch, surrounding the hot tungsten and molten weld metal shielding it from atmospheric oxygen.

GTA welding is efficient for welding metals ranging from sheet metal up to 1/4 in. The eye–hand coordination required to make GTA welds is very similar to the coordination required for oxyfuel gas welding.

Two of the advantages of GTA welding for welding fabrication are that it can be used to produce very high-quality welds and it can be used to weld on almost any metal. Two of the limitations of GTA welding are the slow welding rate and tedious nature, both of which limit its use to small projects or high-integrity critical welds. Although most other welding processes are faster and less expensive, the clean, neat, slag-free welds GTAW produces are used because of their appearance and ease of finishing.

GTA WELDING EQUIPMENT

Four major components make up a GTA welding station. They are the welding power supply, often called the welder; the welding torch, often called a TIG torch; the work clamp, sometimes called the ground clamp; and the shielding gas cylinder, Figure 15-1. There are a variety of hoses and cables that connect all three of these components together.

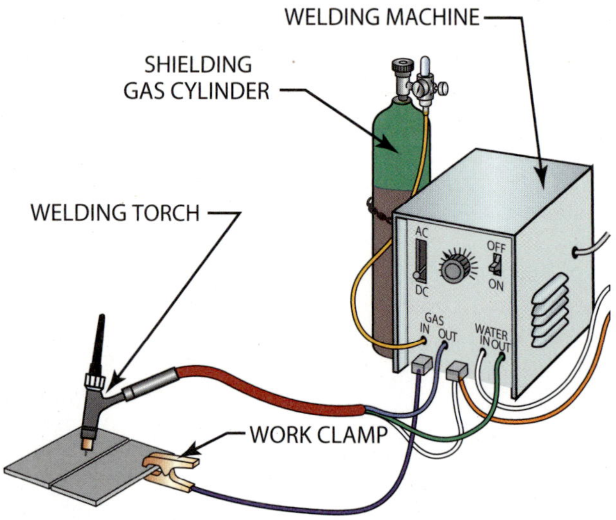

FIGURE 15-1 Gas tungsten gas welding station setup. © Cengage Learning 2012

GTA WELDING TORCHES

GTA welding torches are available water-cooled or air-cooled, Figure 15-2. The heat transfer efficiency for GTA welding may be as low as 20%. This means that 80% of the heat generated does not enter the weld. Much of this heat stays in the torch. To avoid damage to the torch, the heat must be removed by some type of cooling method. Following are some of the advantages of air-cooled GTAW torches:

- Lighter weight for the same amperage range
- Easier to manipulate without the water hoses
- More portable
- Easier to maintain
- No water supply required
- No water leakage danger

The above advantages of the air-cooled GTAW torches are the disadvantages of water-cooled GTAW torches. This is a case in which the advantages of one are the disadvantages of the other.

Some of the advantages of the water-cooled GTAW torches include the following:

- Continuous operation without overheating
- Lower torch temperatures means less tungsten erosion
- Less torch handle temperature in the welder's hands

The above advantages of the water-cooled GTAW torches are the disadvantages of air-cooled GTAW torches.

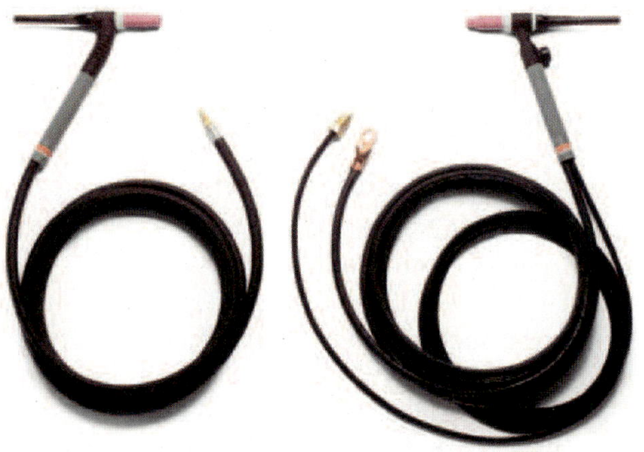

FIGURE 15-2 Power cable safety fuse. ESAB Welding and Cutting Products

The advantages of air-cooled torches make them the preferred GTA welding torch type for most small shops. The lower operating temperature and continuous operating ability of the water-cooled GTA welding torches make them the preferred torch type for most production welding.

GTA welding torch heads are available in a variety of amperage ranges and designs, Figure 15-3. The amperage listed on a torch is the maximum rating and cannot be exceeded without possible damage to the torch. The various head angles allow better access in tight places. Some of the heads can be swiveled easily to new angles. The back cap that both protects and tightens the tungsten can be short or long, Figure 15-4 and Figure 15-5.

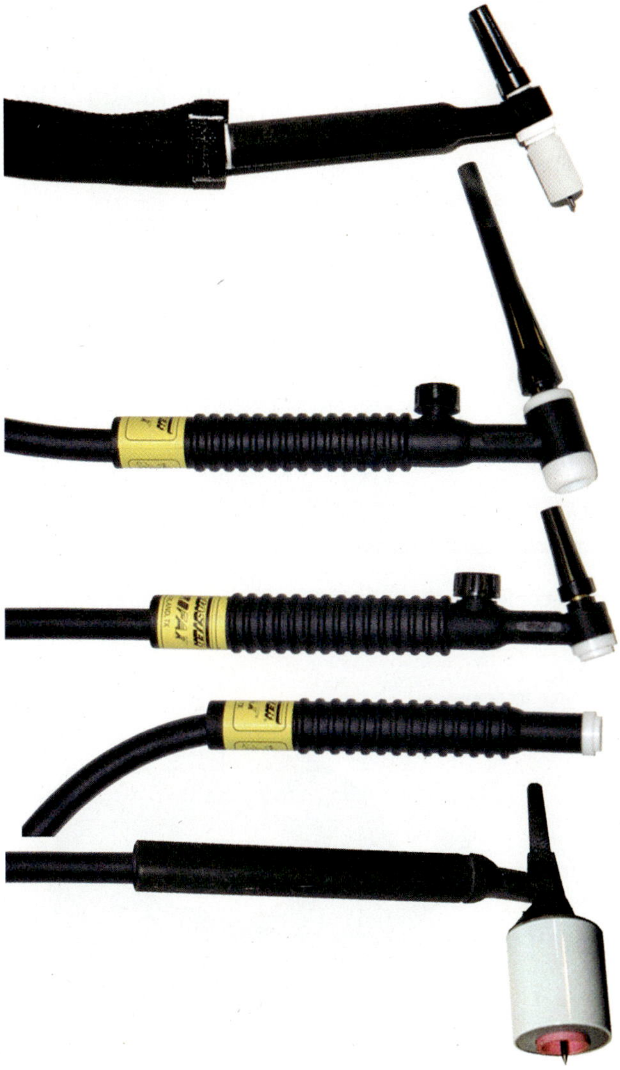

FIGURE 15-3 **GTA welding torches.** Larry Jeffus

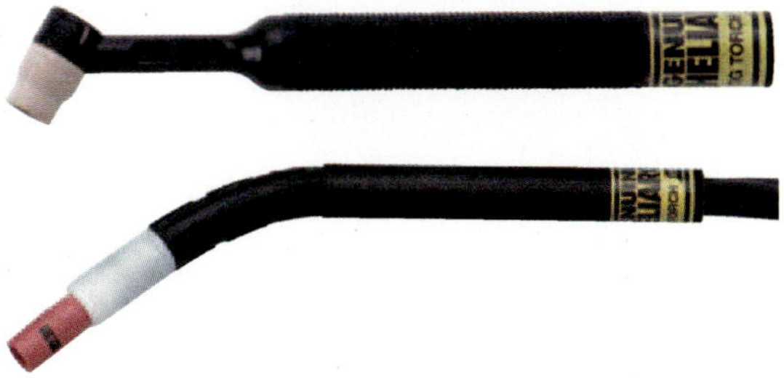

FIGURE 15-4 **Short back caps are available for torches when space is a problem.** ESAB Welding and Cutting Products

FIGURE 15-5 **Long back caps allow tungstens that are a full 7 in. (177 mm) long to be used.** ESAB Welding and Cutting Products

Shielding Gas Hose

Air-cooled torches have just the shielding gas hose that needs to be connected to the welding machine or flowmeter, Figure 15-6. The shielding gas hose must be plastic to prevent the gas from being contaminated. Rubber hoses contain oils that can be picked up by the gas, resulting in weld contamination.

Water Hoses

Water-cooled torches have three hoses connecting to the welding machine. In addition to the shielding gas hose, they have two cooling water hoses, Figure 15-7. One hose is for transporting the cooling water to the torch. This allows the head to receive the maximum cooling from the water. The power cable is usually inside the second hose, which is for the return cooling water. By running the power cable through the return water line, it is kept cool. This permits a much smaller-size cable to be used because the water keeps it cool. The smaller diameter cable is more flexible.

The water-in hose may be made of any sturdy material. Water hose fittings have left-hand threads, and gas hose fittings have right-hand threads. This prevents the water and gas hoses from accidentally being

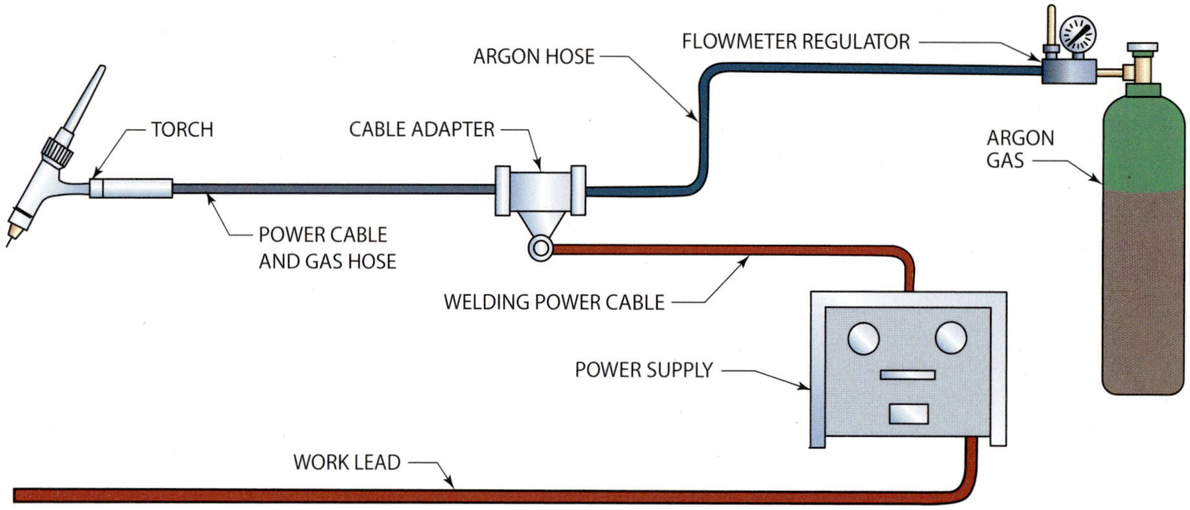

FIGURE 15-6 Schematic of GTA welding setup with air-cooled torch. © Cengage Learning 2012

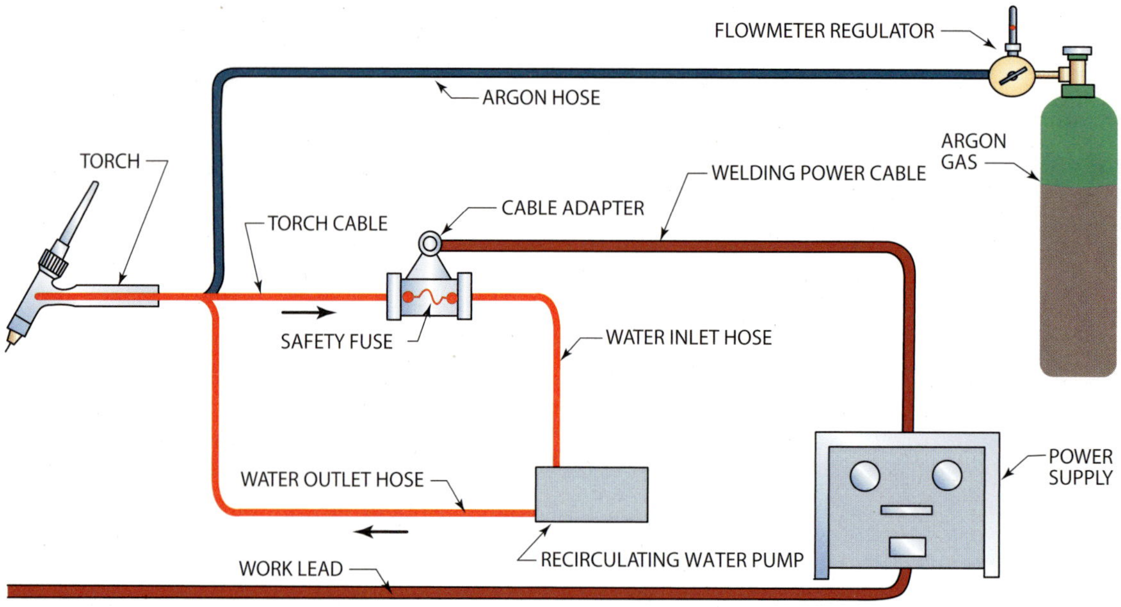

FIGURE 15-7 Schematic of GTA welding setup with water-cooled torch. © Cengage Learning 2012

reversed when attaching them to the welder. The return water hose also contains the welding power cable.

A protective covering can be used to prevent the hoses from becoming damaged by hot metal, Figure 15-8. Even with this protection, the hoses should be supported, Figure 15-9, so that they are not underfoot on the floor. Supporting the hoses reduces the chance of their being damaged by hot sparks.

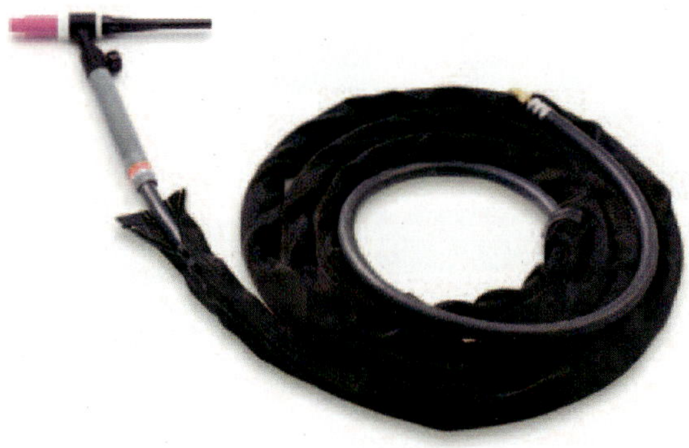

FIGURE 15-8 Zip-on protective covering also helps keep the hoses neat. ESAB Welding and Cutting Products

Cooling Water

There are two GTA welding torch water-cooling systems. One system is an open system and the other is

FIGURE 15-9 A bracket holds the leads off the floor. Larry Jeffus

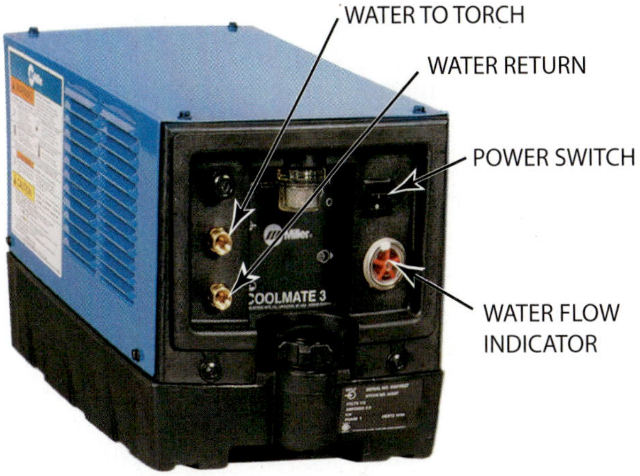

FIGURE 15-10 Typical GTA welding machine connections. © Cengage Learning 2012

a recirculating system. The open system uses potable water from the building or shop's fresh drinking water supply. The water passes through a pressure regulator, then once through the torch, and then down the drain. Water pressures higher than 35 psi may cause the water hoses to burst. These systems are not water conservative, and many local communities, cities, or states have ordinances or laws restricting their use.

Recirculating systems use water pumps to circulate water through the torch. An air-to-water coil and fan in the unit cool the water, Figure 15-10. A low conductive antifreeze solution may be added to the water to prevent freezing and corrosion. Only manufacturer-approved antifreeze solutions may be used.

Shielding Gas Nozzles

The nozzle or cup is used to direct the shielding gas directly on the welding zone. The nozzle size is determined by the diameter of the opening and its length, Table 15-1. Nozzles may be made from a ceramic such as alumina or silicon nitride (opaque) or from fused quartz (clear). The nozzle may also have a gas lens to improve the gas flow pattern.

The nozzle size, both length and diameter, is often the welder's personal preference. Occasionally, a specific choice must be made based upon joint design or location. Small nozzle diameters allow the welder to better see the molten weld pool and can be operated with lower gas flow rates. Larger nozzle diameters can give better gas coverage, even in drafty places.

Ceramic nozzles are heat-resistant and offer a relatively long life. The useful life of a ceramic nozzle is affected by the current level and proximity to the work. Silicon nitride nozzles withstand much more heat, resulting in a longer useful life.

The fused quartz (glass) used in a nozzle is a special type that can withstand the welding heat. These nozzles are no more easily broken than ceramic ones but are more expensive. The added visibility

Tungsten Electrode Diameter		Nozzle Orifice Diameter	
in.	(mm)	in.	(mm)
1/16	(2)	1/4 to 3/8	(6 to 10)
3/32	(2.4)	3/8 to 7/16	(10 to 11)
1/8	(3)	7/16 to 1/2	(11 to 13)
3/16	(4.8)	1/2 to 3/4	(13 to 19)

Table 15-1 Recommended Cup Sizes

with glass nozzles in tight, hard-to-reach places is often worth the added expense. The longer a nozzle, the longer the tungsten must be extended from the **collet**. This can cause higher tungsten temperatures, resulting in greater tungsten erosion. When using long nozzles, it is better to use low amperages or a larger-size tungsten.

TUNGSTEN ELECTRODES

The high melting temperature and good electrical conductivity make tungsten the best choice for a nonconsumable electrode. As the tungsten electrode becomes hot, the arc between the electrode and the work will stabilize. Because electrons are more freely emitted from hot tungsten, the very highest temperature possible at the tungsten electrode tip is desired. A balance must be maintained between the temperature required to have a stable arc and one too high that would melt the tungsten.

The thermal conductivity of tungsten is what allows the tungsten electrode to withstand the arc temperature well above its melting temperature. The heat of the arc is conducted away from the electrode's end so fast that it does not reach its melting temperature.

Because of the intense heat of the arc, some erosion of the electrode will occur. This eroded metal is transferred across the arc, Figure 15-11. Slow erosion of the electrode results in limited tungsten droplets entering the weld, which are acceptable. Standard codes give the size and amount of tungsten inclusions that are allowable in various types of welds. The tungsten inclusions are hard spots that cause stresses to concentrate, possibly resulting in weld failure. Although tungsten erosion cannot be completely eliminated, it can be controlled. Following are a few ways of limiting erosion:

- Have a good mechanical and electrical contact between the electrode and the collet.
- Use as low a welding current as possible.
- Use a water-cooled torch.
- Use as large a size of tungsten electrode as possible.
- Use direct-current electrode negative (DCEN) current.
- Use as short an electrode extension from the collet as possible.
- Use the proper electrode end shape.
- Use an alloyed tungsten electrode.

Types of Tungsten Electrodes

Pure tungsten has a number of properties that make it an excellent nonconsumable electrode for the GTA welding process. These properties can be improved by adding cerium, lanthanum, thorium, or zirconium to the tungsten.

The American Welding Society (AWS) classifies GTA welding as the following:

- Pure tungsten, EWP
- 1% thorium tungsten, EWTh-1
- 2% thorium tungsten, EWTh-2
- 1/4% to 1/2% zirconium tungsten, EWZr
- 2% cerium tungsten, EWCe-2
- 1% lanthanum tungsten, EWLa-1
- Alloy not specified, EWG

See Table 15-2.

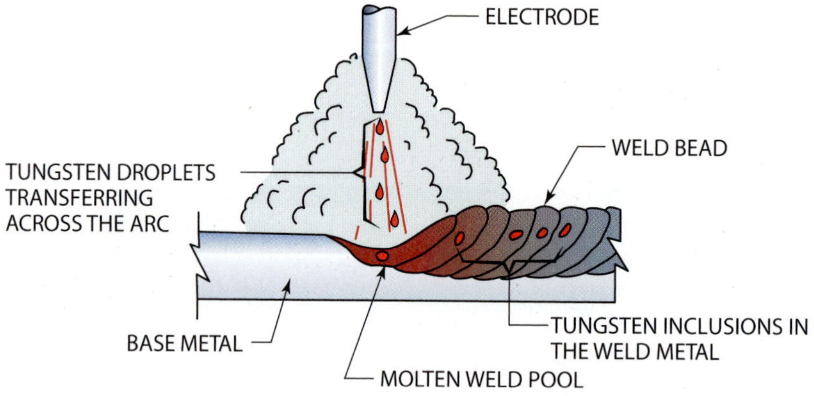

FIGURE 15-11 **Some tungsten will erode from the electrode, be transferred across the arc, and become trapped in the weld deposit.** © Cengage Learning 2012

AWS Classification	Tungsten Composition	Tip Color
EWP	Pure tungsten	Green
EWTh-1	1% thorium added	Yellow
EWTh-2	2% thorium added	Red
EWZr	1/4 to 1/2% zirconium added	Brown
EWCe-2	2% cerium added	Orange
EWLa-1	1% lanthanum added	Black
EWG	Alloy not specified	Not specified

Table 15-2 Tungsten Electrode Types and Identification

Pure Tungsten, EWP

Pure tungsten has the poorest heat resistance and electron emission characteristic of all the tungsten electrodes. It has a limited use with alternating current (AC) welding of metals, such as aluminum and magnesium.

Thoriated Tungsten, EWTh-1 and EWTh-2

Thorium oxide (ThO_2), when added in percentages of up to 0.6% to tungsten, improves its current-carrying capacity. The addition of 1 to 2% of thorium oxide does not further improve current-carrying capacities. It does, however, help with electron emission. This can be observed by a reduction in the electron force (voltage) required to maintain an arc of a specific length. Thorium also increases the serviceable life of the tungsten. The improved electron emission of the thoriated tungsten allows it to carry approximately 20% more current. This also results in a corresponding reduction in electrode tip temperature, resulting in less tungsten erosion and subsequent weld contamination.

Thoriated tungstens also provide a much easier arc-starting characteristic than pure or zirconiated tungsten. Thoriated tungstens work well with DCEN. They can maintain a sharpened point well. They are very well-suited for making welds on steel, steel alloys (including stainless), nickel alloys, and most other metals other than aluminum or magnesium.

Thoriated tungsten does not work well with AC. It is difficult to maintain a balled end, which is required for AC welding. A thorium spike, Figure 15-12, may also develop on the balled end, disrupting a smooth arc.

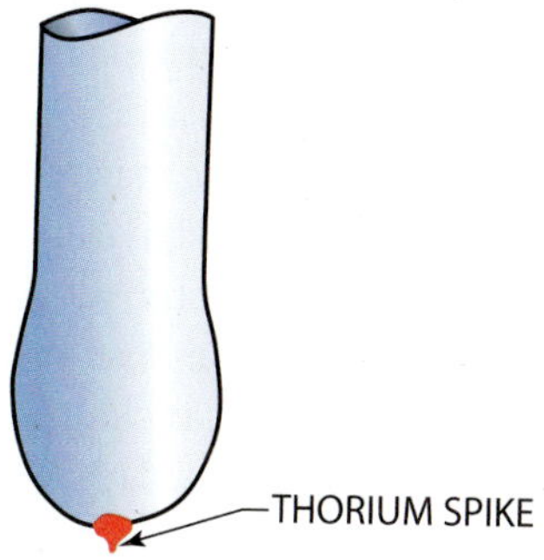

FIGURE 15-12 Thorium spike on a balled end tungsten electrode. © Cengage Learning 2012

CAUTION

Thorium is a very low-level radioactive oxide, but the level of radioactive contamination from a thorium electrode has not been found to be a health hazard during welding. It is, however, recommended that grinding dust be contained. Because of concern in other countries regarding radioactive contamination to the welder and welding environment, thoriated tungstens have been replaced with other alloys.

Zirconium Tungsten, EWZr

Zirconium oxide (ZrO_2) also helps tungsten emit electrons freely. The addition of zirconium to the tungsten has the same effect on the electrode characteristic as thorium, but to a lesser degree. Because zirconium tungstens are more easily melted than thorium tungsten, ZrO_2 electrodes can be used with both AC and direct current (DC currents). Because of the ease in forming the desired balled end on thorium versus zirconium tungstens, they are normally the electrode chosen for AC welding of aluminum and magnesium alloys. Zirconiated tungstens are more resistant to weld pool contamination than pure tungsten, thus providing excellent weld qualities with minimal contamination.

Zirconiated tungstens also have the advantage over thoriated tungsten in that they are not radioactive.

Cerium Tungsten, EWCe-2

Cerium oxide (CeO_2) is added to tungsten to improve the current-carrying capacity in the same manner as does thorium. These electrodes were developed as replacements for thoriated tungstens because they are not made of a radioactive material. Cerium oxide electrodes have a current-carrying capacity similar to that of pure tungsten; however, they have an improved arc-starting and arc-stability characteristic, similar to that of thoriated tungstens. They can also provide a longer life than most other electrodes, including thorium.

Cerium tungsten electrodes have a slightly higher arc voltage for a given length than does thoriated tungsten. This very slight increase in voltage does not cause problems for manual welding. The higher voltage, however, may require that a new weld test be performed to requalify welding procedures. Cerium tungsten may be used for both AC and DC welding. Cerium electrodes contain approximately 2% of cerium oxide.

Lanthanum Tungsten, EWLa-1

Lanthanum oxide (La_2O_3) in about 1% concentration is added to tungsten. Lanthanum oxide tungstens are not radioactive. They have current-carrying characteristics similar to those of the thorium tungstens, except that they have a slightly higher arc voltage than thorium and cerium tungstens. This does not normally pose a problem for manual arc welding; however, it will usually require that new test plates be produced to recertify weld procedures.

Alloy Not Specified, EWG

The EWG classification is for tungstens whose alloys have been modified by manufacturers. Such alloys have been developed and tested by manufacturers to meet specific welding criteria. Specific alloy compositions are not normally available from manufacturers; however, they do provide welding characteristics for these electrodes.

Tungsten Electrode Surface Finish

The type of finish on the tungsten must be specified as cleaned or ground. More information on composition and other requirements for tungsten welding electrodes is available in the AWS publication A5.12, *Specifications for Tungsten and Tungsten Alloy Electrodes for Arc Welding and Cutting.*

FLOWMETER

The **flowmeter** may be merely a flow regulator used on a manifold system, or it may be a combination flow and pressure regulator used on an individual cylinder, Figure 15-13 and Figure 15-14.

The flow is metered or controlled by opening a small valve at the base of the flowmeter. The rate of flow is then read in units of cfh (cubic feet per hour).

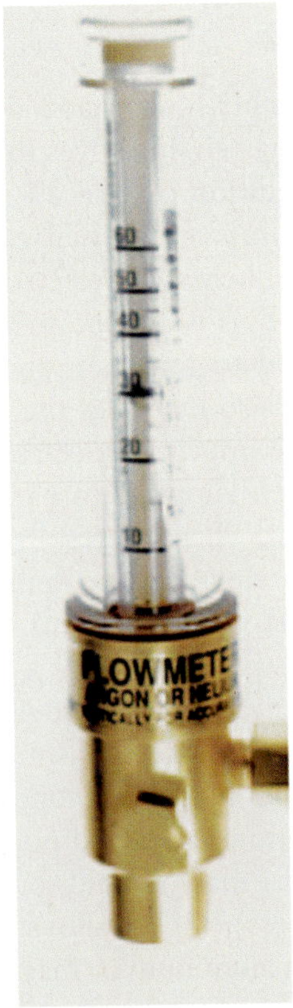

FIGURE 15-13 Flowmeter. Controls Corporation of America

FIGURE 15-14 Flowmeter regulator.
Controls Corporation of America

The reading is taken from a fixed scale that is compared to a small ball floating on the stream of gas. Meters from various manufacturers may be read differently. For example, they may read from the top, center, or bottom of the ball, Figure 15-15. The ball floats on top of the stream of gas inside a tube that gradually increases in diameter in the upward direction. The increased size allows more room for the gas flow to pass by the ball. If the tube is not vertical, the reading is not accurate, but the flow is unchanged.

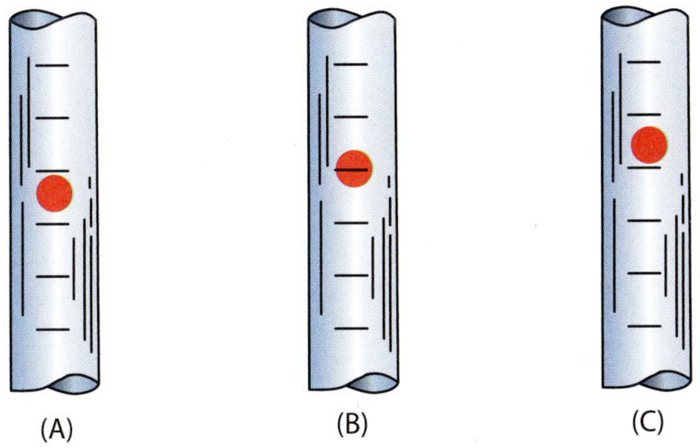

FIGURE 15-15 Three methods of reading the ball and line on a flowmeter: (A) top of the ball, (B) center of the ball, (C) bottom of the ball. © Cengage Learning 2012

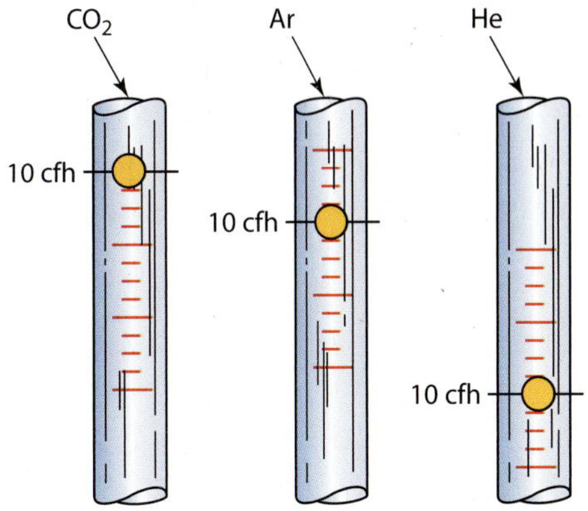

FIGURE 15-16 Each of these gases is flowing at the same cfh (L/min) rate. Because helium (He) is less dense, its indicator ball is the lowest. Be sure that you are reading the correct scale for the gas being used.
© Cengage Learning 2012

Also, when using a line flowmeter, it is important to have the correct pressure. Changes in pressure will affect the accuracy of the flowmeter reading. To get accurate readings, be sure that the gas being used is read on the proper flow scale. Less dense gases, such as helium and hydrogen, will not support the ball on as high a column with the same flow rate as a denser gas, such as argon, Figure 15-16.

Shielding Gas Flow Rate

The rate of flow should be as low as possible and still give adequate coverage. High gas flow rates waste shielding gases and may lead to contamination. The contamination comes from turbulence in the gas at high flow rates. Air is drawn into the gas envelope by a venturi effect around the edge of the nozzle. Also, the air can be drawn in under the nozzle if the torch is held at too sharp an angle to the metal, Figure 15-17.

The larger the nozzle size, the higher the permissible flow rate without causing turbulence. Table 15-3

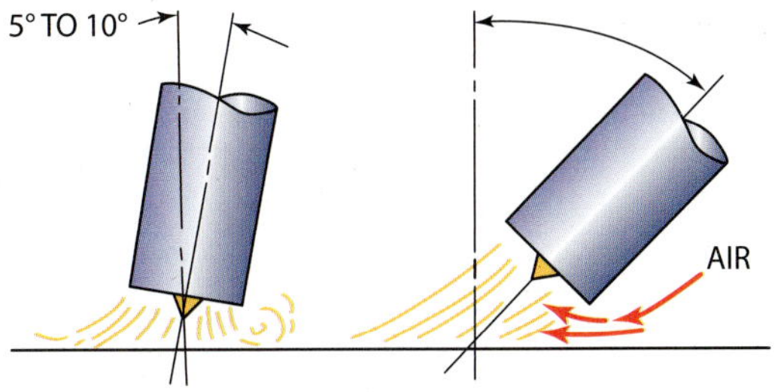

FIGURE 15-17 Too steep an angle between the torch and the work may draw in air. © Cengage Learning 2012

Nozzle Inside Diameter		Gas Flow*	
in.	(mm)	cfh	(L/min)
1/4	(6)	10–14	(4.7–6.6)
5/16	(8)	11–15	(5.2–7.0)
3/8	(10)	12–16	(5.6–7.5)
7/16	(11)	13–17	(6.1–8.0)
1/2	(13)	17–20	(8.0–9.4)
5/8	(16)	17–20	(8.0–9.4)

Table 15-3 Suggested Argon Gas Flow Rate for Given Cup Sizes

shows the average and maximum flow rates for most nozzle sizes. A gas lens can be used in combination with the nozzle to stabilize the gas flow, thus eliminating some turbulence. A gas lens will add to the turbulence problem if there is any spatter or contamination on its surface.

Preflow and Postflow

Preflow is the time during which gas flows to clear out any air in the nozzle or surrounding the weld zone. The operator sets the length of time that the gas flows before the welding current is started, Figure 15-18. Because some machines do not have preflow, many welders find it hard to hold a position while waiting for the current to start. One solution to this problem is to use the postflow for preflow. Switch on the current to engage the postflow. Now, with the current off, the gas is flowing, and the GTA torch can be lowered to the welding position. The welder's helmet should be lowered, and the current restarted before the postflow stops. This allows welders to have postflow and to start the arc when they are ready.

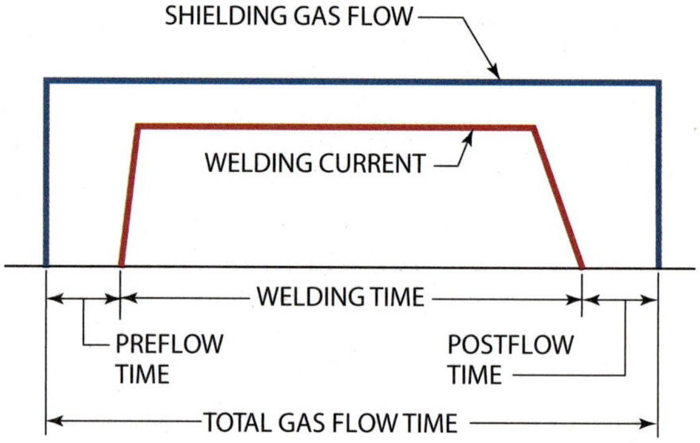

FIGURE 15-18 Welding time compared to shielding gas flow time. © Cengage Learning 2012

The **postflow** is the time during which the gas continues flowing after the welding current has stopped. This period serves to protect the molten weld pool, the filler rod, and the **tungsten** electrode as they cool to a temperature at which they will not oxidize rapidly. The time of the flow is determined by the welding current and the tungsten size, Table 15-4.

Electrode Diameter		Postwelding Gas Flow Time*
in.	(mm)	
0.01	(0.25)	5 sec
0.02	(0.5)	5 sec
0.04	(1.0)	5 sec
1/16	(2)	8 sec
3/32	(2.4)	10 sec
1/8	(3)	15 sec
5/32	(4)	20 sec
3/16	(4.8)	25 sec
1/4	(6)	30 sec

Table 15-4 Postwelding Gas Flow Times

SHIELDING GASES

The shielding gases used for the GTA welding process are argon (Ar), helium (He), hydrogen (H), nitrogen (N), or a mixture of two or more of these gases. The purpose of the shielding gas is to protect the molten weld pool and the tungsten electrode from the harmful effects of air. The shielding gas also affects the amount of heat produced by the arc and the resulting weld bead appearance.

Argon and helium are **noble inert gases**. This means that they do not combine chemically with any other material. Argon and helium may be found in mixtures but never as compounds. Because they are inert, they do not affect the molten weld pool in any way.

CAUTION

Never allow gases such as O_2, CO_2, or N to come in contact with your inert gas system. Very small amounts can contaminate the inert gas, which may result in the weld failing.

Argon

Argon is a by-product in air separation plants. Air is cooled to temperatures that cause it to liquefy; then its constituents are fractionally distilled. The primary products are oxygen and nitrogen. Before these gases were produced on a tonnage scale, argon was a rare gas. Now it is distributed in cylinders as gas or in bulk in a liquid form. Because argon is denser than air, it effectively shields welds in deep grooves in the flat position. However, this higher density can be a hindrance when welding overhead because higher flow rates are necessary. The argon is relatively easy to ionize and thus suitable for alternating-current applications and easier starts. This property also permits fairly long arcs at lower voltages, making it virtually insensitive to changes in arc length. Argon is also the only commercial gas that produces the cleaning discussed earlier. These characteristics are most useful for manual welding, especially with filler metals added, as shown in Figure 15-19.

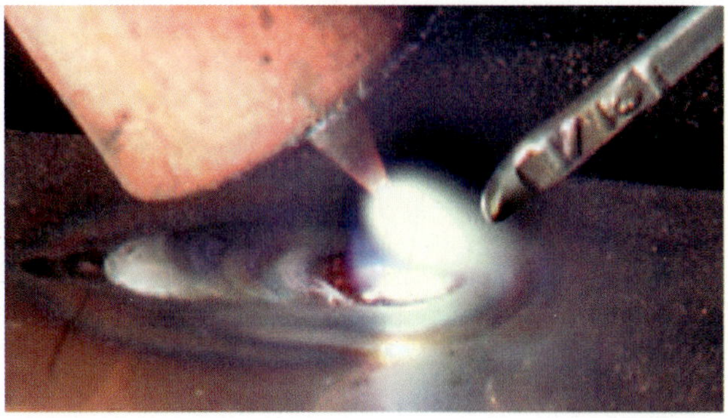

FIGURE 15-19 Highly concentrated ionized argon gas column. Larry Jeffus

Helium

Helium is a by-product of the natural gas industry. It is removed from natural gas as the gas undergoes separation (fractionation) for purification or refinement. Helium offers the advantage of deeper penetration. The arc force with helium is sufficient to displace the molten weld pool with very short arcs. In some mechanized applications, the tip of the tungsten electrode is positioned below the workpiece surface to obtain very deep and narrow penetration. This technique is especially effective for welding aged aluminum alloys prone to overaging. It is also very effective at high welding speeds, such for tube mills. However, helium is less forgiving for manual welding. With helium, penetration and bead profile are sensitive to the arc length, and the long arcs needed for feeding filler wires are more difficult to control.

Helium has been mixed with argon to gain the combined benefits of cathode cleaning and deeper penetration, particularly for manual welding. The most common of these mixtures is 75% helium and 25% argon. Although the GTA process was developed with helium as the shielding gas, argon is now used whenever possible because it is much cheaper. Helium also has some disadvantages because it is lighter than air, thus preventing good shielding. Its flow rates must be about twice as high as argon's for acceptable stiffness in the gas stream, and proper protection is difficult in drafts unless high flow rates are used. It is difficult to ionize, necessitating higher voltages to support the arc and making the arc more difficult to ignite. Alternating-current arcs are very unstable. However, helium is not used with alternating current because the cleaning action does not occur.

Hydrogen

Hydrogen is not an inert gas and is not used as a primary shielding gas. However, it can be added to argon when deep penetration and high welding speeds are

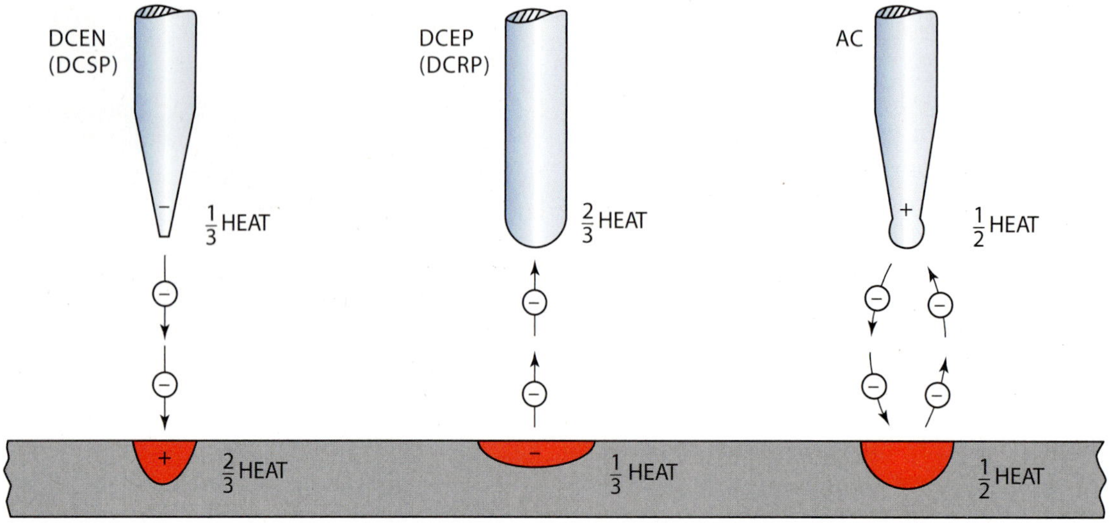

FIGURE 15-20 Heat distribution between the tungsten electrode and the work with each type of welding current. © Cengage Learning 2012

needed. It also improves the weld surface cleanliness and bead profile on some grades of stainless steel that are very sensitive to oxygen. Hydrogen additions are restricted to stainless steels because hydrogen is the primary cause of porosity in aluminum welds. It can cause porosity in carbon steels and, in highly restrained welds, underbead cracking in carbon and low-alloy steels.

Nitrogen

Nitrogen is not an inert gas. Like hydrogen, nitrogen has been used as an additive to argon. But it cannot be used with some materials, such as ferritic steels, because it produces porosity. In other cases, such as with austenitic stainless steels, nitrogen is useful as an austenite stabilizer in the alloy. It is used to increase penetration when welding copper. Unfortunately, because of the general success with inert gas mixtures and because of potential metallurgical problems, nitrogen has not received much attention as an additive for GTA welding.

TYPES OF WELDING CURRENT

All three types of welding current can be used for GTA welding. Each current has individual features that make it more desirable for specific conditions or with certain types of metals. The current used affects the heat distribution between the tungsten electrode and the weld and the degree of surface oxide cleaning that occurs. Figure 15-20 shows the heat distribution for each of the three types of currents.

Direct-Current Electrode Negative (DCEN)

DCEN, which used to be called direct-current straight polarity (DCSP), concentrates about two-thirds of its welding heat on the work and the remaining one-third on the tungsten. The higher heat input to the weld results in deep penetration. The low heat input into the tungsten means that a smaller-size tungsten can be used without erosion problems. The smaller-size electrode may not require pointing, resulting in a savings of time, money, and tungsten.

Direct-Current Electrode Positive (DCEP)

DCEP, which used to be called direct-current reverse polarity (DCRP), concentrates only one-third of the arc heat on the plate and two-thirds of the heat on the electrode. This type of current produces wide welds with shallow penetration, but it has a strong cleaning action upon the base metal. The high heat input to the tungsten indicates that a large-size tungsten is required, and the end shape with a ball must be used. The low heat input to the metal and the

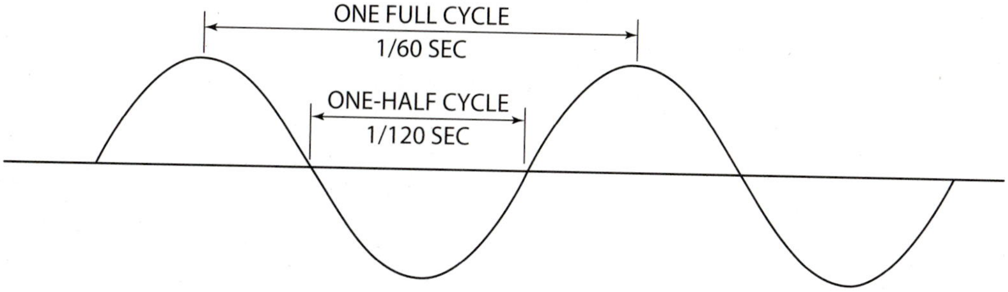

FIGURE 15-21 Sine wave of alternating current at 60 cycle. © Cengage Learning 2012

strong cleaning action on the metal make this a good current for thin, heavily oxidized metals. The metal being welded will not emit electrons as freely as tungsten, so the arc may wander or be more erratic than DCEN.

Alternating Current High Frequency (ACHF)

Alternating current (AC) concentrates about half of its heat on the work and the other half on the tungsten. Alternating current is continuously switching back and forth between DCEN and DCEP. This switching takes place once every 1/120 of a second. The complete cycle takes 1/60 of a second, so it makes 60 complete cycles per second, Figure 15-21. Alternating current would look like waves on water if you could stretch it out with the peaks of the waves coming along every 1/60 of a second. The speed at which current changes back and forth is referred to by three different names that all mean the same thing—cycles, **frequency**, or hertz.

The tungsten electrode is much better at releasing electrons than is the metal being welded. This results in an imbalance in the amount of current that flows from the electrode to the work and back. Some welding machines have internal controls that correct this problem; they are called balanced wave machines. There are several advantages of using a balanced wave GTA welder. Two of the advantages are the better arc cleaning of surface oxides and less internal heat in the welder.

Electrons flow from the tip of the tungsten electrode for 1/120 of a second. They stop flowing for a very short moment in time before reversing direction and flowing from the work to the electrode. It is this stopping and starting every 1/120 of a second that causes problems with using AC for GTA welding. To help the electrons get the arc restarted every 1/120 of a second, a very high-frequency (50,000 to 3,000,000 cycles), high-voltage (3000 volts), very low-amperage (100 milliamps) current is added to the lower-frequency (60 cycle), lower-voltage (20 volts), high-amperage (100 amps) welding current, Figure 15-22.

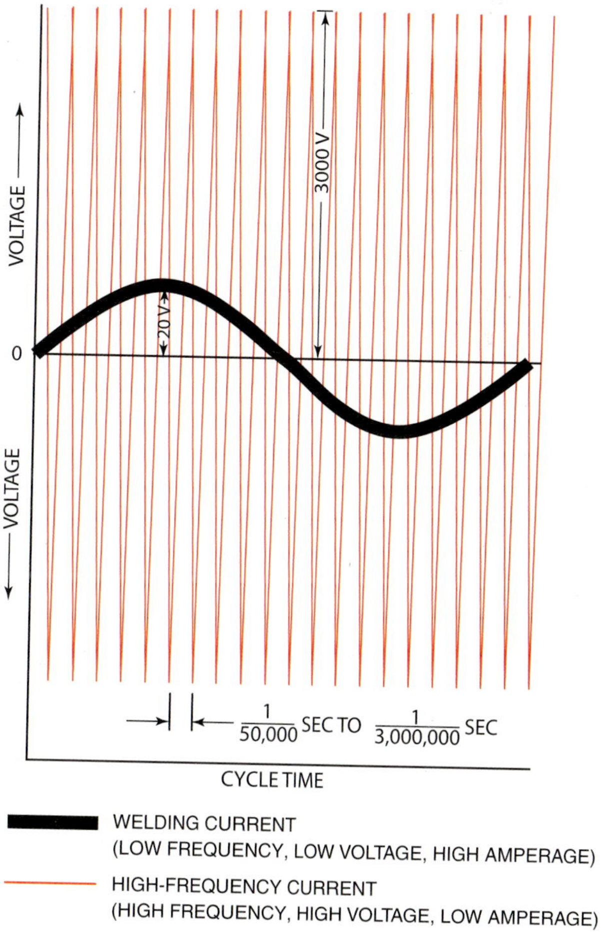

FIGURE 15-22 High-frequency arc-starting current shown over the low-frequency welding current. © Cengage Learning 2012

CAUTION

Very high-frequency, high-voltage, very low-amperage electricity is not dangerous, but it can shock you. High-frequency welding current can easily pass through lots of normal insulating materials. Be careful not to touch the electrode when the high-frequency current is turned on.

The high-frequency (HF) first current ionizes the shielding gas around the tungsten. This appears as a light blue glow, Figure 15-23. Once the shielding gas is ionized, it can conduct the lower-frequency welding current. The term *alternating current, high-frequency stabilized* (ACHF) is used to describe this GTA welding current.

The high frequency may be set so that it automatically cuts off after the arc is established when welding with DC. It is kept on continuously with AC. When used in this manner, it is referred to as alternating current, high-frequency stabilized, or ACHF.

Arc Cleaning Action

As the electrons leave the surface of the metal, they provide some surface cleaning or removal of these oxides. This **cleaning action** is most important when welding on aluminum.

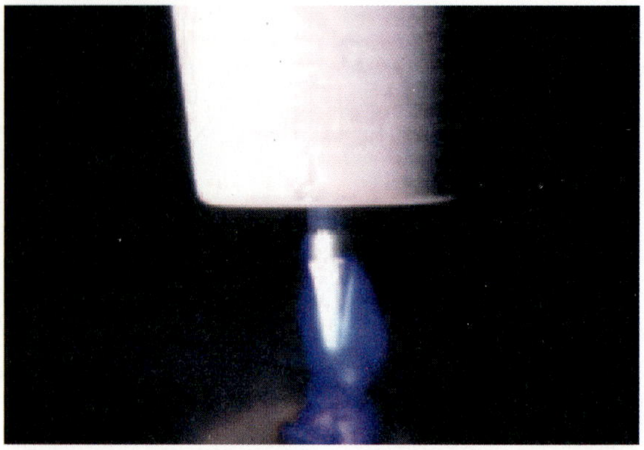

FIGURE 15-23 **The high frequency first appears as a blue glow around the tungsten before the welding current starts its arc.** Larry Jeffus

There are many theories as to why DCEP and the DCEP portion of the AC cycle have a cleaning action. The most probable explanation is that the electrons accelerated from the cathode surface lift the oxides that interfere with their movement. The positive ions accelerated to the metal's surface provide additional energy. In combination, the electrons and ions cause the surface erosion needed to produce the cleaning. Although this theory is disputed, it is important to note that cleaning does occur, that it requires argon-rich shield gases and DCEP polarity, and that it can be used to the welder's advantage, Figure 15-24.

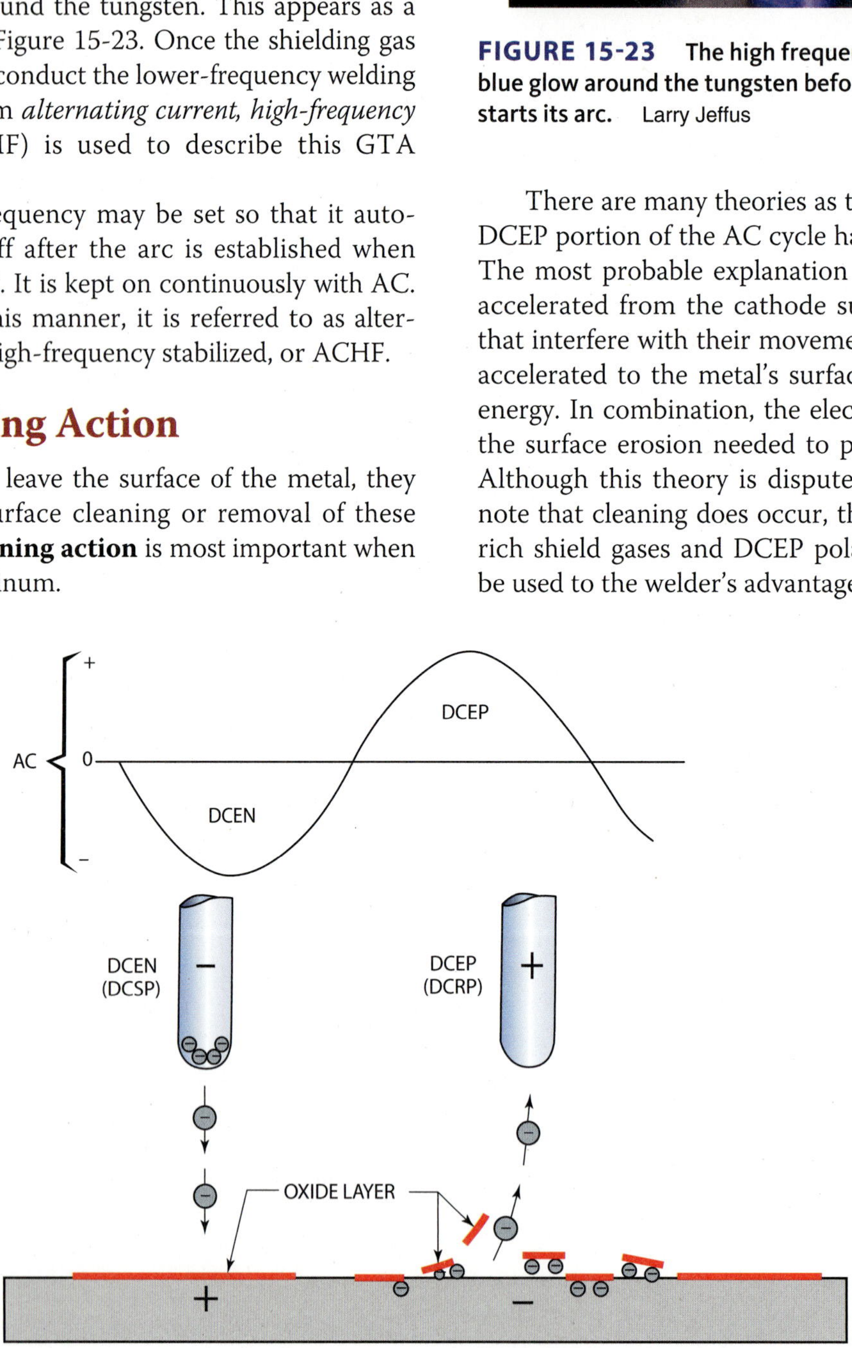

FIGURE 15-24 **Electrons collect under the oxide layer during the DCEP portion of the cycle and lift the oxides from the surface.** © Cengage Learning 2012

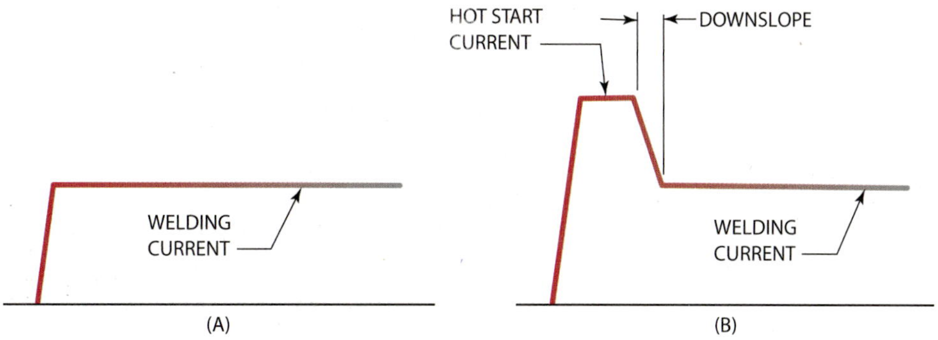

FIGURE 15-25 (A) Standard method of starting welding current. (B) Hot start method of starting welding current. © Cengage Learning 2012

Hot Start

The hot start allows a controlled surge of welding current as the arc is started to establish a molten weld pool quickly. Establishing a molten weld pool rapidly on metals with a high thermal conductivity is often hard without this higher-than-normal current. Adjustments can be made in the length of time and the percentage above the normal current, Figure 15-25.

REMOTE CONTROLS

A remote control can be used to start the weld, increase or decrease the current, and stop the weld. The remote control can be either a foot-operated or hand-operated device. The foot control works adequately if the welder can be seated. Welds that must be performed away from a welding station may use a hand or thumb control or may not have any remote welding controls.

Most remote controls have an on–off switch that is activated at the first or last part of the control movement. A variable resistor increases the current as the control is pressed more. A variable resistor works in a manner similar to the accelerator pedal on a car to increase the power (current), Figure 15-26. The operating amperage range is determined by the value that has been set on the main controls of the machine.

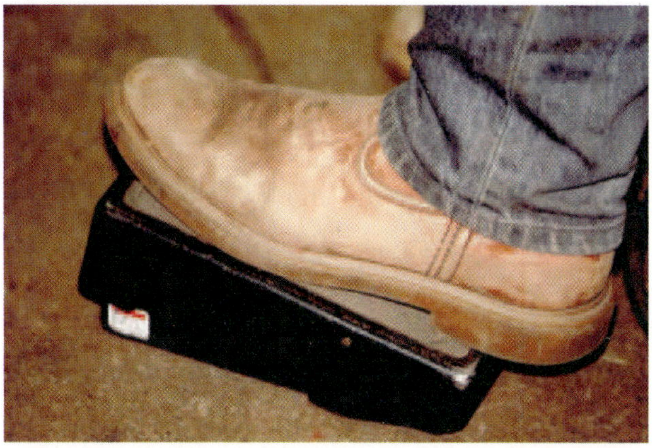

FIGURE 15-26 A foot-operated device can be used to increase the current. Larry Jeffus

SETTING UP A GTA WELDER

Each manufacturer's GTA welding machine and welding torch are assembled and set up differently. Read and follow the manufacturer's instructions and safety guidelines any time you are setting up any equipment. The following topics are intended to help explain the significance of the various steps required to assemble and set up a typical GTA welder.

Shaping the Tungsten Electrode

The desired end shape of a tungsten electrode can be obtained by grinding, breaking, remelting the end, or using chemical compounds. Tungsten is brittle and easily broken. Welders must be sure to make a smooth, square break where they want it to be located.

Grinding a Tungsten Electrode Point

A grinder is often used to clean a contaminated tungsten or to point the end of a tungsten. The grinder used to sharpen tungsten should have a fine, hard stone. It should be used for grinding tungsten only. Because of the hardness of the tungsten and its brittleness, the grinding stone chips off small particles of the electrode. A coarse grinding stone will result in more tungsten breakage and a poorer finish. If the grinder is used for metals other than tungsten, particles of these metals may become trapped on the tungsten as it is ground. The metal particles will quickly break free when the arc is started, resulting in contamination.

CAUTION

Any time you use a grinder, always wear safety glasses and follow all grinder safety instructions.

Because of the hardness of the tungsten, it will become hot. Its high thermal conductivity means that the heat will be transmitted quickly to your fingers. To prevent overheating, only light pressure should be applied against the grinding wheel. This will also reduce the possibility of accidentally breaking the tungsten.

Grind the tungsten so that the grinding marks run lengthwise, Figure 15-27 and Figure 15-28. Lengthwise grinding reduces the amount of small particles of tungsten contaminating the weld. Move the tungsten up and down as it is twisted during grinding. This will prevent the tungsten from becoming hollow-ground.

CAUTION

When holding one end of the tungsten against the grinding wheel, the other end of the tungsten must not be directed toward the palm of your hand, Figure 15-29. This will prevent the tungsten from being stuck into your hand if the grinding wheel catches it and suddenly pushes it downward.

FIGURE 15-27 Correct method of grinding a tungsten electrode. Larry Jeffus

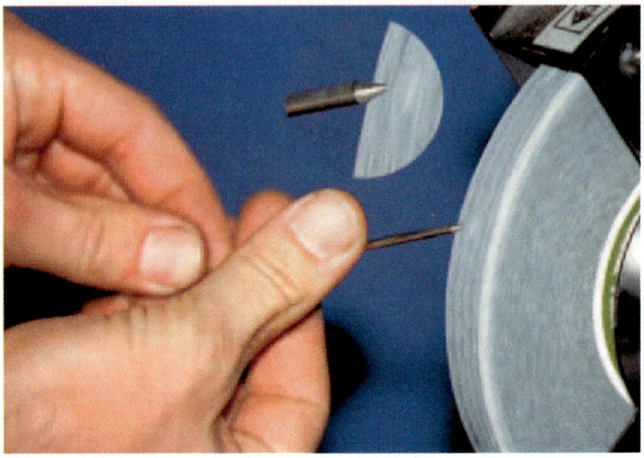

FIGURE 15-28 Incorrect method of grinding a tungsten electrode. Larry Jeffus

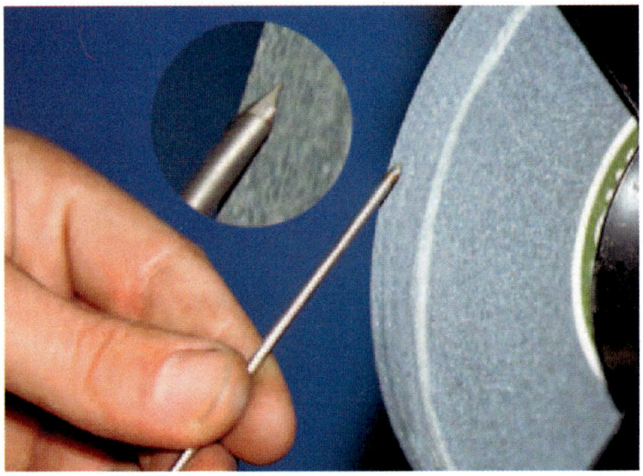

FIGURE 15-29 Correct way of holding a tungsten when grinding. Larry Jeffus

Breaking and Remelting Tungsten

Tungsten is hard but brittle, resulting in a low-impact strength. If tungsten is struck sharply, it will break without bending. When it is held against a sharp corner and hit, a fairly square break will result. Figure 15-30, Figure 15-31, and Figure 15-32 show ways to break the tungsten correctly on a sharp corner using a hammer, two pliers, and wire cutters, respectively. Observe the break; it should be square and relatively smooth, Figure 15-33. Once the tungsten has been broken squarely, the end must be melted back so that it becomes somewhat rounded. This is accomplished by switching the welding current to DCEP and striking an arc under argon shielding on a piece of copper. If copper is not available, another piece of clean metal can be used. Do not use carbon, as it will contaminate the tungsten.

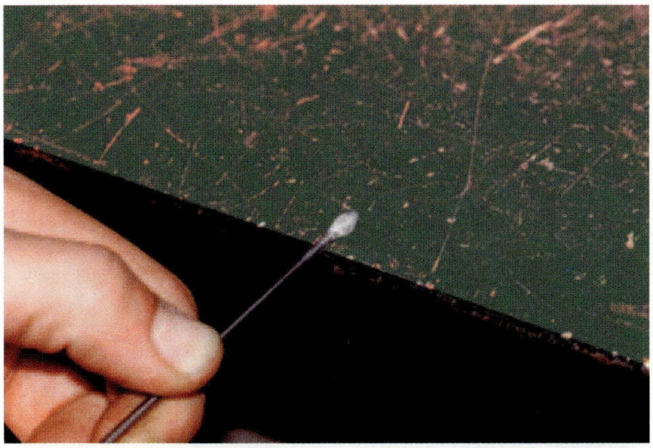

FIGURE 15-30 Breaking the contaminated end from a tungsten by striking it with a hammer. Larry Jeffus

FIGURE 15-31 Correctly breaking the tungsten using two pairs of pliers. Larry Jeffus

Chemical Cleaning and Pointing Tungsten

The tungsten can be cleaned and pointed using one of several compounds. The tungsten is heated by shorting it against the work. The tungsten is then dipped

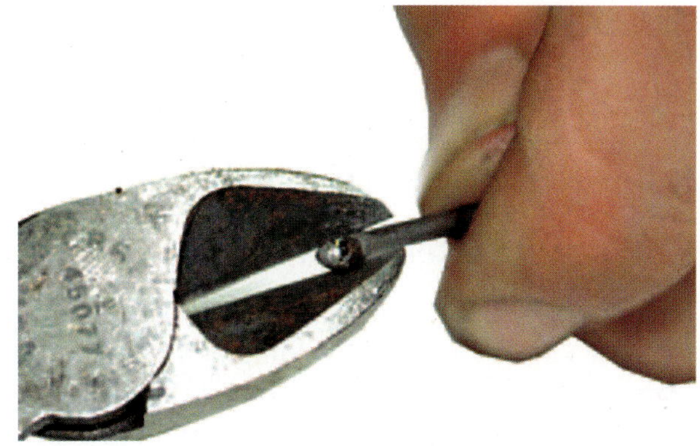

FIGURE 15-32 Using wire cutters to correctly break the tungsten. Larry Jeffus

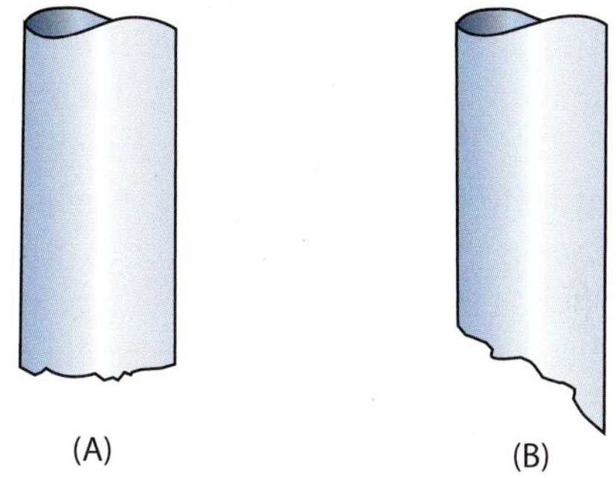

FIGURE 15-33 (A) Correctly broken tungsten electrode. (B) Incorrectly broken tungsten electrode. © Cengage Learning 2012

in the compound, a strong alkaline, which rapidly dissolves the hot tungsten. The chemical reaction is so fast that enough additional heat is produced to keep the tungsten hot, Figure 15-34. When the tungsten is removed

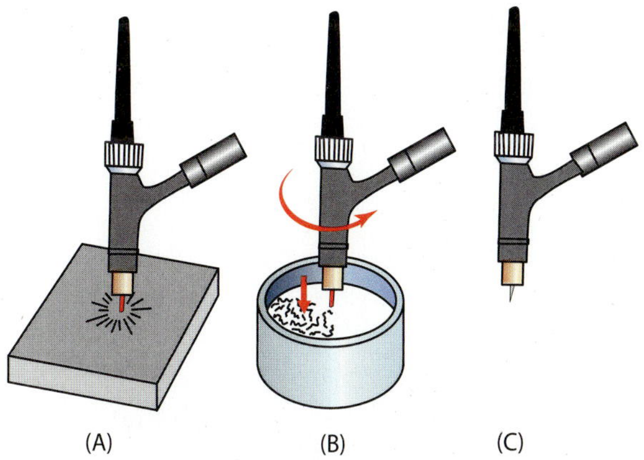

FIGURE 15-34 Chemically cleaning and pointing tungsten: (A) Shorting the tungsten against the work to heat it to red hot, (B) inserting the tungsten into the compound and moving it around, (C) cleaned and pointed tungsten ready for use. © Cengage Learning 2012

from the chemical cleaner, it must be cooled and cleaned. The chemical will have both cleaned the tungsten and produced a fine point. If the electrode was contaminated, the chemical compound dissolves the tungsten under the contamination, allowing it to fall free.

Pointing and Remelting Tungsten

The tapered tungsten with a balled end, a shape sometimes used for DCEP welding, is made by first grinding or chemically pointing the electrode. Using DCEP, as in the procedure for the remelted broken end, strike an arc on some copper under argon shielding and slowly increase the current until a ball starts to form on the tungsten. The ball should be made large enough so that the color of the end stays between dull red and bright red. If the color turns white, the ball is too small and should be made larger. To increase the size of the ball, simply apply more current until the end begins to melt. Surface tension will pull the molten tungsten up onto the tapered end. Lower the current, and continue welding. DCEP is seldom used for welding. If the tip is still too hot, it may be necessary to increase the size of the tungsten.

Assembling the GTA Welding Station

1. Start with the power switch off, Figure 15-35. Use a wrench to attach the torch hose to the

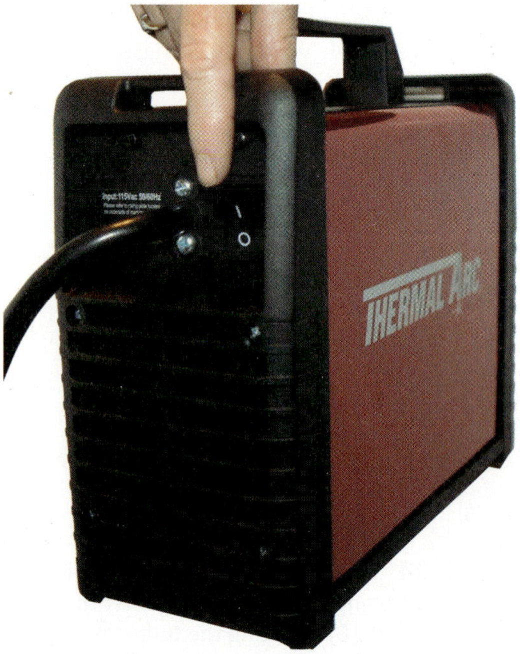

FIGURE 15-35 **Always be sure thet the power is off when making machine connections.** Larry Jeffus

FIGURE 15-36 **Tighten each fitting as it is connected to avoid missing a connection.** Larry Jeffus

machine. The water hoses should have left-hand threads to prevent incorrectly connecting them. Tighten the fittings only as tightly as needed to prevent leaks, Figure 15-36. Attach the cooling water "in" to the machine solenoid and the water "out" to the power block.

CAUTION

Never work on a welding machine when the power is on because an electrical shock or arc could occur.

2. The flowmeter or flowmeter regulator should be attached next. If a gas cylinder is used, secure it in place with a safety chain. Then remove the valve protection cap, and crack the valve to blow out any dirt, Figure 15-37. Attach the flowmeter so that the tube is vertical.
3. Connect the gas hose from the meter to the gas "in" connection on the machine.
4. With both the machine and main power switched off, turn on the water and gas so that the connection

FIGURE 15-37 **During transportation or storage, dirt may collect in the valve. Cracking the valve is the best way to remove any dirt.** © Cengage Learning 2012

to the machine can be checked for leaks. Tighten any leaking fittings to stop the leak.

5. Turn on both the machine and main power switches, and watch for leaks in the torch hoses and fittings.
6. With the power off, switch the machine to the GTA welding mode.

CAUTION

Turn off all power before attempting to stop any leaks in the water system.

7. Select the desired type of current and amperage range, Figure 15-38 and Figure 15-39. Set the fine current adjustment to the proper range, depending upon the size of tungsten used, Table 15-5.
8. Place the high-frequency switch in the appropriate position, auto (HF start) for DC or continuous for AC, Figure 15-40.
9. The remote control can be plugged in and the selector switch set, Figure 15-41.

FIGURE 15-38 Setting the current. Larry Jeffus

FIGURE 15-39 Setting the amperage range. Larry Jeffus

FIGURE 15-40 The high-frequency switch should be placed in the appropriate position. Larry Jeffus

FIGURE 15-41 Setting the remote-control switch. Larry Jeffus

Electrode Diameter		DCEN	DCEP	AC
in.	(mm)			
0.04	(1)	15–60	Not recommended	10–50
1/16	(2)	70–100	10–20	50–90
3/32	(2.4)	90–200	15–30	80–130
1/8	(3)	150–350	25–40	100–200
5/32	(4)	300–450	40–55	160–300

Table 15-5 Amperage Range of Tungsten Electrodes

FIGURE 15-42 Inserting collet and collet body.
Larry Jeffus

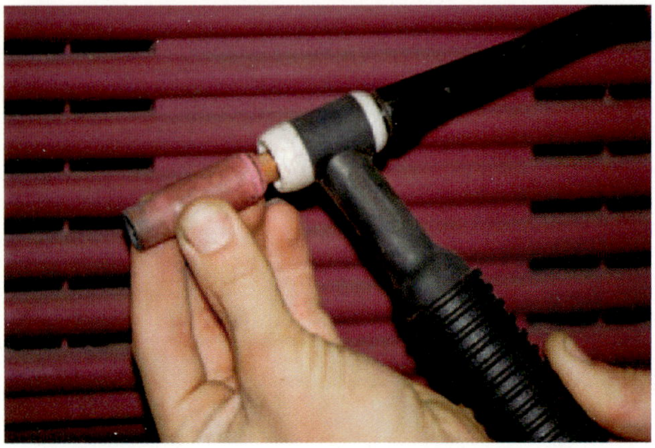

FIGURE 15-43 Install the nozzle (cup) to the torch body. Larry Jeffus

10. The collet and collet body should be installed on the torch first, Figure 15-42.
11. On the Linde or copies of Linde torches, installing the back cap first will stop the collet body from being screwed into the torch fully. A poor connection will result in excessive electrical and thermal resistance, causing a heat buildup in the head.
12. The tungsten can be installed and the end cap tightened to hold the tungsten in place.
13. Select and install the desired nozzle size, Figure 15-43. Adjust the tungsten length so that it does not stick out more than the diameter of the nozzle, Figure 15-44.
14. Check the manufacturer's operating manual for the machine to ensure that all connections and settings are correct.
15. Turn on the power, depress the remote control, and again check for leaks.
16. While the postflow is still engaged, set the gas flow by adjusting the valve on the flowmeter.

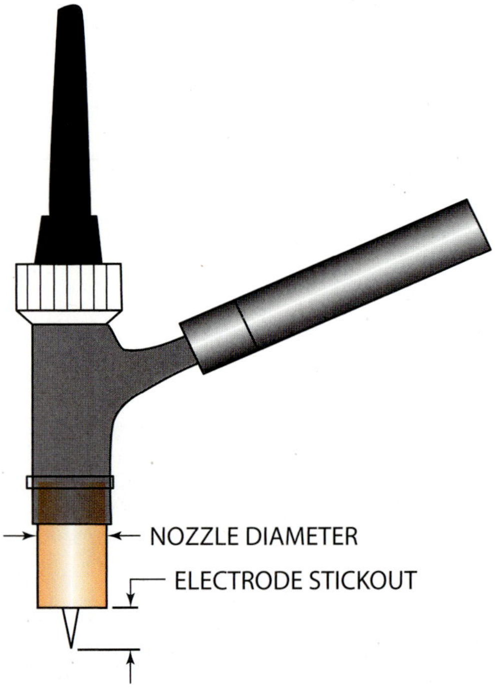

FIGURE 15-44 Electrode stickout. © Cengage Learning 2012

SUMMARY

One of the prime considerations for gas tungsten arc welding is the cleanliness of the equipment, supplies, base metal, filler metal, the welder's gloves, and so forth. When everything is clean, you will find that the welding process proceeds more easily and more successfully.

Another major factor affecting your ability to produce quality welds is the tungsten end or tip shape. As you practice making the various welds, you will find that keeping the tungsten electrode tip shaped appropriately assists you in producing uniform welds.

Often, new welders feel that there is some sort of attraction between the tungsten electrode, filler metal, and base metal during the welding process because it seems to continually become contaminated.

This almost continuous contamination can be very frustrating. At times it may seem overwhelming; however, with continued practice and diligence you will be able to control this problem. Even experienced welders in the field can be plagued from time to time with tungsten contamination. At other times, they can weld for an entire day without contaminating the tungsten. It is often beneficial for students to realize that tungsten contamination is just part of the process, and they must, therefore, try to ignore the possibility of it happening and concentrate on producing the welds.

REVIEW QUESTIONS

1. What are the two other names used for GTA welding?
2. What are two advantages of GTA welding?
3. List four major components that make up a GTA welding station.
4. List three advantages that using an air-cooled GTAW torch might have over using a water-cooled GTAW torch.
5. List three advantages that using a water-cooled GTAW torch might have over using an air-cooled GTAW torch.
6. List the hoses and cables that might be attached to a water-cooled torch.
7. What type of antifreeze solution may be used in a water-cooling system?
8. What is the purpose of the shielding gas nozzle?
9. Why is tungsten used as an electrode?
10. List five ways to limit tungsten erosion.
11. What are the AWS classifications for the following types of tungsten electrodes?
    a. Pure tungsten
    b. 1% thorium tungsten
    c. 2% thorium tungsten
    d. 1/4 to 1/2% zirconium tungsten
    e. 2% cerium tungsten
    f. 1% lanthanum tungsten
12. Which type of tungsten electrode has the poorest heat resistance and emission characteristics?
13. Which type of tungsten electrode has a very low-level radioactive oxide additive?
14. According to Table 15-2, what color tip does the EWCe-2 tungsten electrode have?
15. What is La_2O_3?
16. What are the units of shielding gas flow?
17. How high should the shielding gas flow rate be?
18. What is preflow time used for?
19. How is argon produced?
20. What is the advantage of using helium as a shielding gas?
21. Nitrogen cannot be used as a shielding gas for GTA welding for what types of materials?
22. What do the following abbreviations mean? DCEN, DCEP, and ACHF?
23. What part of the AC cycle provides surface oxide cleaning?
24. What does a hot start provide to the GTA weld?
25. What are the functions of a remote control?
26. List the ways of shaping the end of a tungsten electrode.
27. When the end of a tungsten electrode is broken off to remove contamination, how should the broken end look?
28. How does a chemical cleaner remove tungsten contamination?
29. Why must the power be off when you attach the GTA welding torch to a welding machine?
30. What is the maximum distance that the tungsten electrode should stick out of the nozzle?

Chapter 16
Gas Tungsten Arc Welding

OBJECTIVES

After completing this chapter, the student should be able to:

- Explain how the tungsten tip shape and varying the welding technique affect the weld bead's width and penetration.
- Describe carbide precipitation and explain how it can be controlled.
- Describe how the characteristics of aluminum can affect how it is welded.
- Demonstrate how to set up a gas tungsten arc (GTA) welder.
- Demonstrate how to make GTA welds on mild steel, stainless steel, and aluminum.

KEY TERMS

cleaning
collet
contamination
flowmeter
frequency
oxide layer
penetration
reinforcement
root
surface tension
tungsten

INTRODUCTION

Gas tungsten arc welding is often considered the premier welding process. It has always had a certain mystique to it. Early TIG or heliarc welders as they were known at the time often did their work under a great deal of secrecy not wanting others to know how relatively simple the process was once you have learned some basics. In this chapter you will learn the basics of making GTA welds on mild steel, stainless steel, and aluminum to produce various projects. The skills you learn in fabricating and welding will enable you to produce GTA welds in almost any welding application.

Low Carbon and Mild Steels

Low carbon and mild steel are two basic steel classifications. These steels are the most common type of steels. Carbon is the primary alloy in these classifications of steel, and it ranges from 0.15% or less for low carbon to 0.15 to 0.30% for mild steel. The GTA welding techniques required for welding steels in both classifications are the same. Start with any one of these tungsten types—EWP, EWTh-1, EWTh-2, or EWCe-2—that has a pointed end shape, Figure 16-1, with the welding machine set for direct current electrode negative [DCEN (DCSP)] welding current. Table 16-1 lists the types of filler metal used for both low carbon and mild steels.

NOTE: The point on a tungsten is usually ground so that it is twice as long as the diameter of the tungsten. However, changing the angle that a tungsten is pointed affects the weld width and depth of fusion. Long, tapered points produce narrow, deeply penetrating welds; and shallow, tapered points produce wider, shallower, penetrating welds, Figure 16-2.

Sometimes during the manufacturing process small pockets of primarily carbon dioxide gas become trapped inside low carbon and mild steels. There are only a few molecules of gas trapped inside the microscopic pockets within the steel, so they do not affect the steel strength. When these tiny gas pockets get hot and expand, they can cause weld porosity. Some GTA welding filler metals have alloys in them called deoxidizers that can help prevent porosity. Severe GTA welding porosity is usually only found when welds are being made by fusing the base metal together without adding filler metal.

Stainless Steel

Other than the need for more precleaning of the base metal and filler metal, the setup and manipulation techniques for stainless steel welds are the same as those for low carbon and mild steels. The surface of a stainless steel weld will show the **contamination** effects of dirt, surface oxides, and inadequate shielding gas more easily than steel.

The most common sign that there is a problem with a stainless steel weld is the bead color after the weld. The greater the contamination, the darker the color. The exposure of the weld bead to the atmosphere before it has cooled will also change the bead color. It is impossible, however, to determine the extent of contamination of a weld with only visual inspection. Both light-colored and dark-colored welds may not be free from oxides. Thus, it is desirable to take the time and necessary precautions to make welds that are no darker than dark blue, Table 16-2. Welds with only slight oxide layers are better for multiple passes.

SAE No.	Carbon %	AWS Filler Metal No.
Low Carbon		
1006	0.08 max	RG60 or ER70S-3
1008	0.10 max	RG60 or ER70S-3
1010	0.08 to 0.15	RG60 or ER70S-3
Mild Steel		
1015	0.11 to 0.16	RG60 or ER70S-3
1016	0.13 to 0.18	RG60 or ER70S-3
1018	0.15 to 0.20	RG60 or ER70S-3
1020	0.18 to 0.23	RG60 or ER70S-3
1025	0.22 to 0.29	RG60 or ER70S-3

Table 16-1 Filler Metals for Low-Carbon and Mild Steels

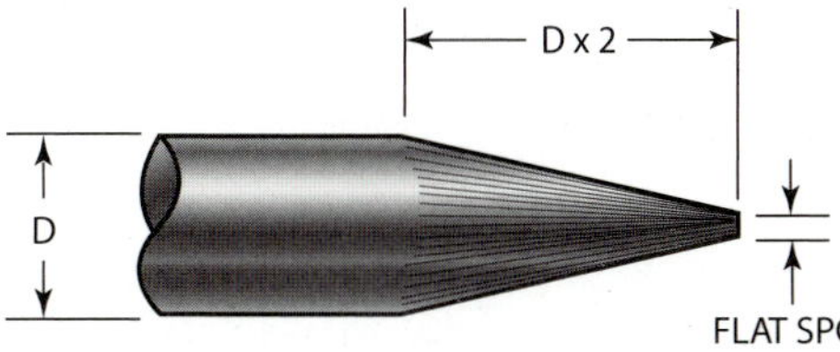

FIGURE 16-1 Tungsten tip shape for mild steel or stainless steel. © Cengage Learning 2012

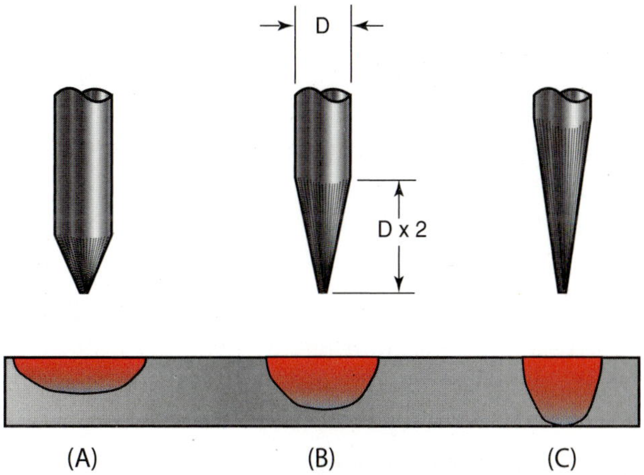

FIGURE 16-2 The length of taper on the tungsten electrode affects the weld bead shape. The bluntest point (A) has the shallowest and widest penetration, (B) is the standard shape for a pointed tungsten electrode, and (C) has the sharpest taper and produces the deepest and narrowest weld bead shape. © Cengage Learning 2012

Surface Color	Approximate Temperature at Which Color is Formed	
	°F	(°C)
Light straw	400	(200)
Tan	450	(230)
Brown	525	(275)
Purple	575	(300)
Dark blue	600	(315)
Black	800	(425)

Table 16-2 Temperature at Which Various Oxide Layers Form on Steel

Carbide precipitation occurs in some stainless steels during welding when carbon combines with chromium. If carbide precipitation is going to form, it will occur at a temperature between 800°F (427°C) and 1500°F (816°C). To reduce the possibility of carbide precipitation, use as low a welding current setting as possible and/or travel as fast as possible along the joint to help reduce carbide precipitation.

Some stainless steels are less likely to form carbide precipitation because during manufacturing of the stainless steel, alloys are added or the percentage of carbon in the stainless steel is lowered.

Table 16-3 lists some common types of stainless steels and the recommended filler metals.

Aluminum

Aluminum is GTA welded using one of these types of tungsten—EWP, EWZr, EWCe-2, or EWLa-1—that has a rounded tip shape, Figure 16-3, with the welding machine set for ACHF (alternating current high-frequency) welding current. The alternating current provides good arc **cleaning,** and the continuous high **frequency** restarts the arc as the current changes direction.

Aluminum has three major characteristics that affect how it is welded.

- High thermal conductivity—The high thermal conductivity of aluminum rapidly pulls the heat away from the welding area, making it more difficult to form a molten weld pool. The high thermal conductivity usually means that the entire part will get hot during welding. On thicker sections, so much heat can be absorbed by the part that without first preheating the base metal, a good weld cannot be produced. In these cases, the preheat temperature of the part should be around 300°F (150°C) but will vary depending on metal thickness and alloy type. Specific preheat temperatures are available from the metal supplier.
- High surface tension—The high **surface tension** of molten aluminum makes it easier to control large molten weld pools. But it can make it more difficult to transfer the filler metal into the weld pool because the molten end of the filler metal tends to ball up, Figure 16-4. Table 16-4 lists some basic types of filler metal used for aluminum welding.
- Surface oxides—Aluminum has a thin protective **oxide layer** at room temperature, but it forms a

AISI No.	AWS Filler No.	AISI No.	AWS Filler No.
303	ER308	310	ER310
304	ER308	316	ER316L
304L	ER308L	316L	ER316L
309	ER309	410	ER410

Table 16-3 Filler Metals for Stainless Steels

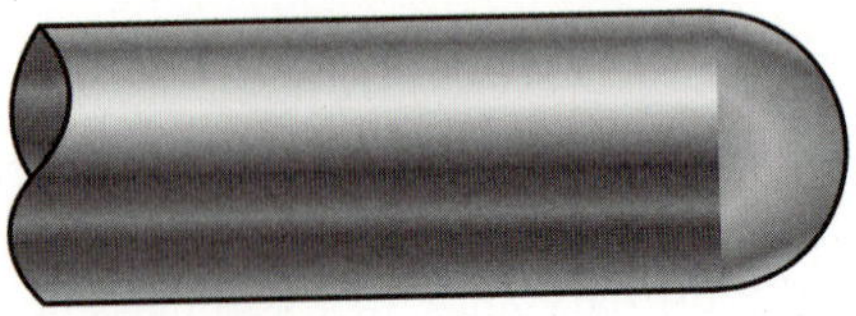

FIGURE 16-3 Tungsten tip shape for aluminum. © Cengage Learning 2012

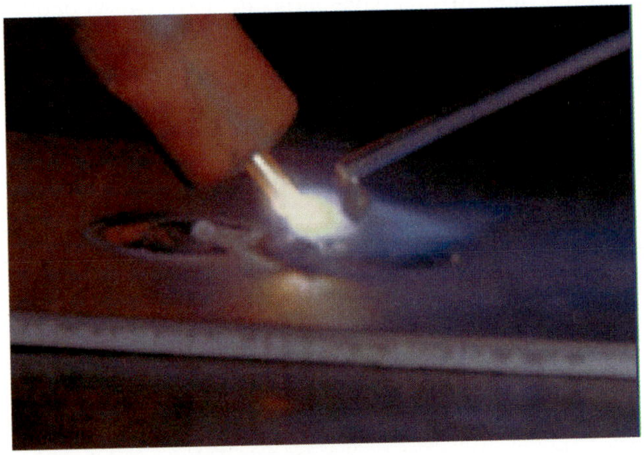

FIGURE 16-4 Filler rod being melted before it is added to the molten pool. Larry Jeffus

AISI No.	AWS Filler No.	AISI No.	AWS Filler No.
1100	ER1100	3004	ER4043
3003	ER4043	6061	ER4043

Table 16-4 Filler Metals for Aluminum Alloys

much thicker layer at welding temperatures if it is not protected from atmospheric oxygen. Any surface oxidation or contamination can cause aluminum welding problems. You must remove any oils or dirt from the metal before welding. Make sure your hands and gloves are clean and dry so you do not contaminate the surface before welding.

Metal Preparation

Both the base metal and the filler metal used in the gas tungsten arc welding (GTAW) process must be thoroughly cleaned before welding. Contamination left on the metal will be deposited in the weld because there is no flux to remove it. Oxides, oils, and dirt are the most common types of contaminants. They can be removed mechanically or chemically. Mechanical metal cleaning may be done by grinding, wire brushing, scraping, machining, or filing. Chemical cleaning may be done by using acids, alkalies, solvents, or detergents.

Setup

Typical water-cooled GTA torches are used for heavy-duty production welding and air-cooled torches are used for lighter general jobs. The equipment setup in this chapter is similar to equipment built by other manufacturers, which means that any skills developed can be transferred easily to other equipment. These are general instructions, but if the equipment manufacturer's setup instructions are available, you should follow them.

EXPERIMENT 16-1

Setting Up a GTA Welder

Using a GTA welding machine, remote-control welding torch, gas **flowmeter,** gas source (cylinder or manifold), **tungsten,** nozzle, **collet,** collet body, cap, and any other hoses, special tools, and equipment required, you will set up the machine for GTA welding, Figure 16-5.

1. Start with the power switch "off," Figure 16-6. Use a wrench to attach the torch hose to the machine. The water hoses should have left-hand threads to prevent incorrectly connecting them. Tighten the fittings only as tightly as needed to prevent leaks, Figure 16-7. Attach the cooling water "in" to the machine solenoid and the water "out" to the power block.
2. The flowmeter or flowmeter regulator should be attached next. If a gas cylinder is used, secure it in place with a safety chain. Then remove the valve protection cap and crack the valve to blow out any dirt, Figure 16-8. Attach the flowmeter so that the tube is vertical.
3. Connect the gas hose from the meter to the gas "in" connection on the machine.
4. With both the machine and main power switched "off," turn on the water and gas so that the connection to the machine can be checked for leaks. Tighten any leaking fittings to stop the leak.
5. Turn on both the machine and main power switches and watch for leaks in the torch hoses and fittings.

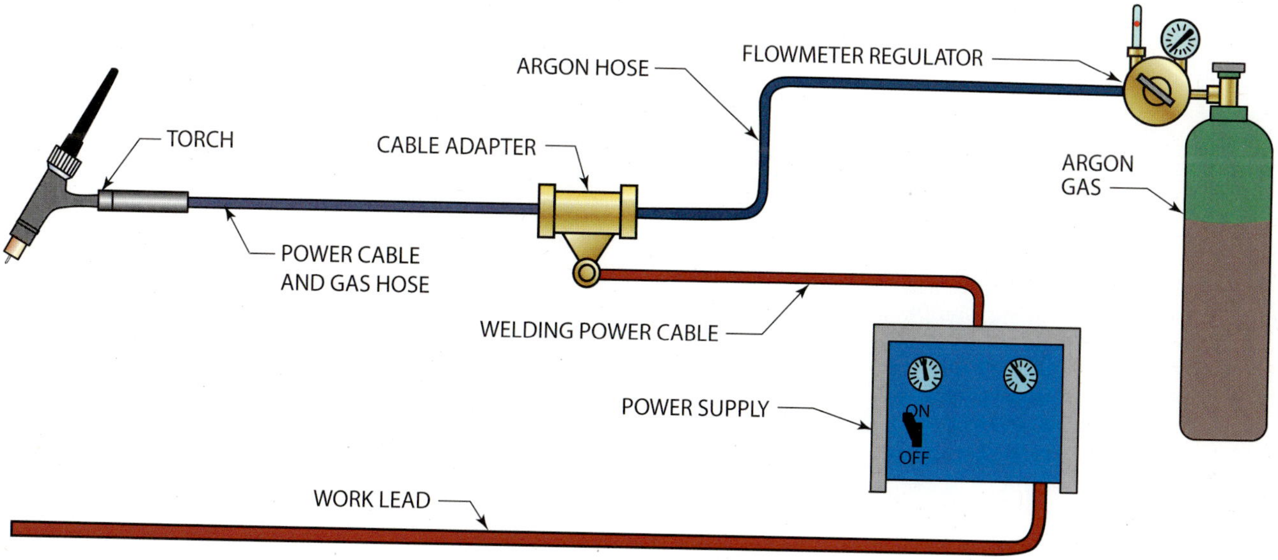

FIGURE 16-5 Schematic of a GTA welding setup with an air-cooled torch. Larry Jeffus

FIGURE 16-6 Always be sure the power is off when making machine connections. Larry Jeffus

6. With the power "off," switch the machine to the GTA welding mode.
7. Select the desired type of current and amperage range, Figure 16-9 and Figure 16-10.
8. Set the fine current adjustment to the proper range, depending upon the size of tungsten used, Table 16-5.
9. Place the high-frequency switch in the appropriate position, auto (HF start) for direct current (DC) or continuous for alternate current (AC), Figure 16-11.

FIGURE 16-7 Tighten each fitting as it is connected to avoid missing a connection. Larry Jeffus

FIGURE 16-8 During transportation or storage, dirt may collect in the valve. Cracking the valve is the best way to remove any dirt. Larry Jeffus

10. The remote control can be plugged in and the selector switch set, Figure 16-12.
11. The collet and collet body should be installed on the torch first, Figure 16-13.
12. On the Linde or copies of Linde torches, installing the back cap first will stop the collet body from

FIGURE 16-9 Setting the current. Larry Jeffus

FIGURE 16-10 Setting the amperage range. Larry Jeffus

Electrode Diameter		DCEN	DCEP	AC
in.	(mm)			
.04	(1)	15–60	Not recommended	10–50
1/16	(2)	70–100	10–20	50–90
3/32	(2.4)	90–200	15–30	80–130
1/8	(3)	150–350	25–40	100–200
5/32	(4)	300–450	40–55	160–300

Table 16-5 Amperage Range of Tungsten Electrodes

FIGURE 16-11 The high-frequency switch should be placed in the appropriate position. Larry Jeffus

FIGURE 16-13 Inserting collet and collet body. Larry Jeffus

being screwed into the torch fully. A poor connection will result in excessive electrical and thermal resistance, causing a heat buildup in the head.

13. The tungsten can be installed and the end cap tightened to hold the tungsten in place.
14. Select and install the desired nozzle size. Adjust the tungsten length so that it does not stick out more than the diameter of the nozzle, Figure 16-14.
15. Check the manufacturer's operating manual for the machine to ensure that all connections and settings are correct.
16. Turn on the power, depress the remote control, and again check for leaks.
17. While the postflow is still engaged, set the gas flow by adjusting the valve on the flowmeter.

FIGURE 16-12 Setting the remote-control switch. Larry Jeffus

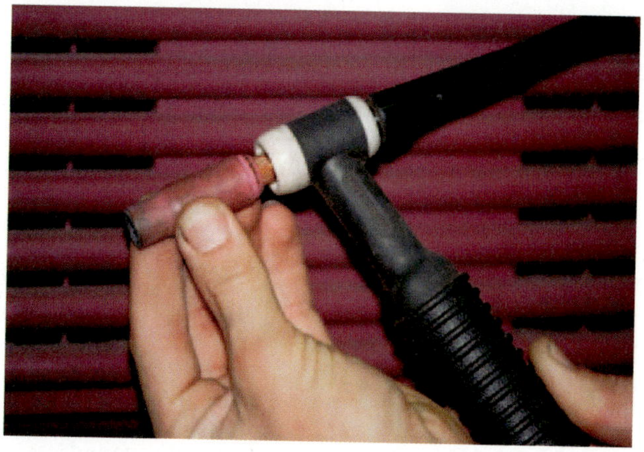

FIGURE 16-14 Install the nozzle (cup) to the torch body. Larry Jeffus

CAUTION

Turn off all power before attempting to stop any leaks in the water system.

The GTA welding system is now ready to be used.◆

Striking an Arc and Pushing a Puddle

The process of starting an arc involves melting a small spot of metal and is referred to as *striking an arc*. You need to learn how to strike an arc in the exact place you want because striking an arc outside of the area to be welded is considered a defect under most welding codes. The term *pushing a puddle* refers to the process of forming a molten weld pool and manipulating the GTA welding torch so that the molten weld pool is worked along the plate surface, Figure 16-15.

Keeping the width of the molten weld bead uniform is important, and it takes practice to be able to do that. There are several things you can do to help control the weld's width:

- Travel speed—Refers to the speed that the torch is moved across the metal's surface. Faster travel speeds result in narrower weld beads, and slower travel speeds result in wider weld beads.
- Welding power—Refers to the welding current. The higher the welding current, the wider the weld bead, and the lower the welding current, the narrower the weld bead. Unlike most other welding processes, many GTA welding machines let you change the welding current as a weld is being made. The changing current levels are accomplished by either increasing or decreasing the pressure on the foot or thumb control, Figure 16-16.
- Weave pattern—Refers to a side-to-side or circular pattern that the torch is moved in as the weld progresses along the metal surface, Figure 16-17. The wider the weave pattern, the wider the weld bead produced, and the narrower the weave pattern, the narrower the weld bead produced.

FIGURE 16-15 Notice the small, evenly shaped molten weld pool and short weld bead just to the left of the arc. Larry Jeffus

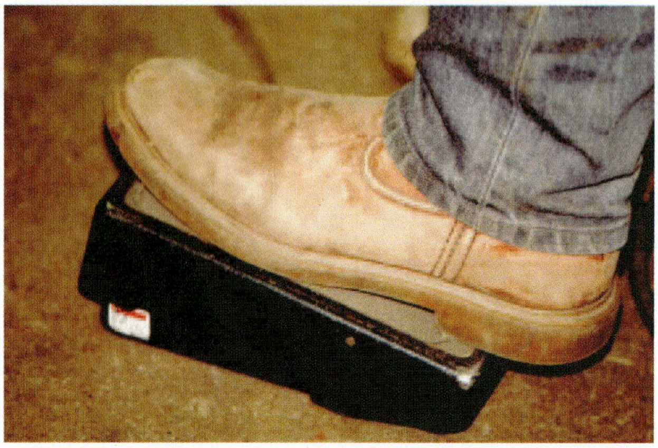

FIGURE 16-16 A foot-operated device can be used to increase the current. Larry Jeffus

PROJECT 16-1

Striking an Arc and Pushing a Puddle in the Flat Position on a Clock Face

Skills to be learned: The ability to strike an arc at a specified location and melt a small spot and the ability to strike an arc at a specific location and melt a short bead that is uniform in width.

Project Description

You will be making the face of an 8-sq in. clock. You will use the weld beads as time/mark indicators. Dots made by starting the arc and melting a small

FIGURE 16-17 Weave patterns. Larry Jeffus

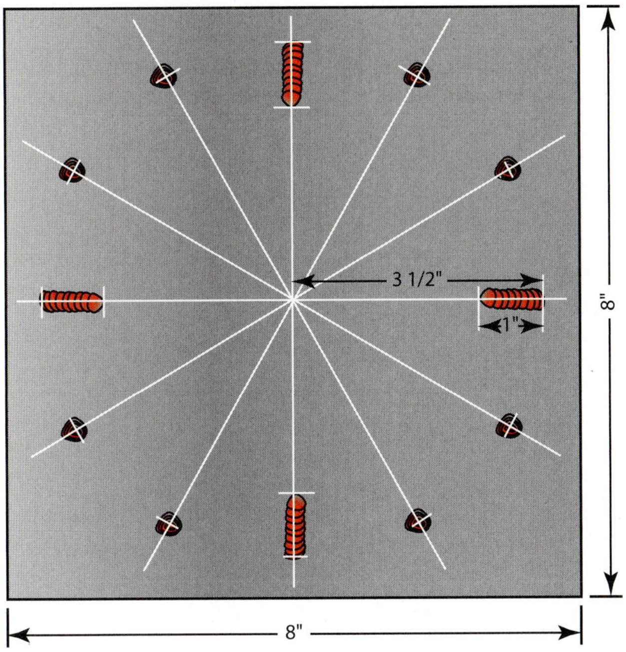

FIGURE 16-18 Project 16-1: Clock face. Larry Jeffus

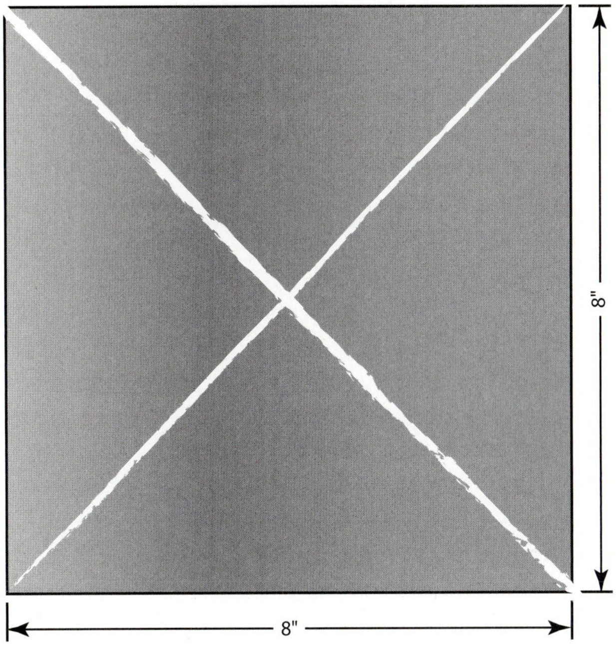

FIGURE 16-19 Project 16-1: Locating the center of the plate with a soapstone diagonal line. Larry Jeffus

spot will be used for the hours between 12 and 3, between 3 and 6, between 6 and 9, and between 9 and 12. Dashes made by starting an arc and melting a 1-in. long puddle will be made for the 12, 3, 6, and 9 o'clock marks, Figure 16-18. After the clock face has been finished, it will be framed with a fusion weld, which is part of Project 16-2. A battery-operated clock movement will be installed after all the welding is completed.

Project Materials and Tools

The following items are needed to fabricate the clock face.

Materials

- 8-in. sq., 1/8-in. to 3/16-in. thick mild steel or stainless steel
- Source of argon shielding gas

Tools

- Pencil and paper
- GTA welder
- Personal protective equipment (PPE)
- 12-in. rule
- Soapstone
- Square
- Wire brush
- Pliers
- Protractor

Layout

Locate the center of the plate by drawing a soapstone line diagonally from each corner, Figure 16-19. Use the square, and draw a line vertically and horizontally through the center point. The vertical and horizontal lines will be your 12, 3, 6, and 9 o'clock hour marks. Line up the protractor on the vertical line and mark off lines for 1, 2, 4, and 5 o'clock. The angle between each of the hour marks on a clock face is 30°. When laying out a number of angles, do not move the protractor between each mark. Sometimes students will line up the protractor and make a mark at the 30°-line then rotate the protractor so that the 0° is at that mark and then make the next 30° mark. This way they do not have to do the math to find the next angle to mark. But if this is done, you can be compounding any error from the first mark to the second or third mark and so on. Make these marks at 30°, 60°, 120°, and 150° points.

Repeat this process to mark the 11, 10, 8, and 7 o'clock positions. Draw a line through the center and out each of these angle marks. Using the 12-in. rule, measure 3 1/2 in. from the center point and make a mark on each of the lines. This will be the outer point or center point for each of the welds used to denote the hour markings, Figure 16-20.

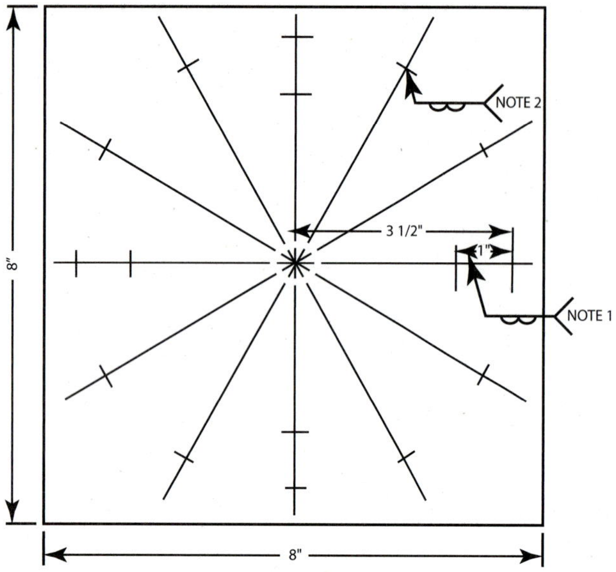
NOTE 2
3 1/2"
1"
NOTE 1
8"
8"
NOTE 1: 4 welds 1" long at the 12:00, 3:00, 6:00, and 9:00 positions.
NOTE 2: 8 round welds approximately 3/8" in diameter at the 1:00, 3:00, 4:00, 5:00, 7:00, 8:00, 10:00, and 11:00 positions.

FIGURE 16-20 Locating the outer points for the weld bead hour marks. Larry Jeffus

Welding

Set the welder for the metal type and thickness being welded according to the manufacturer's recommendation or Table 16-6, Table 16-7, and Table 16-8. Put on your PPE, and follow all shop and manufacturer's safety rules for welding.

Use a scrap piece of metal to make sure that the welding machine is set correctly. Start by holding the TIG torch vertically with the tip approximately 1/4 in. above the surface of the metal, lower your welding hood, and depress the remote control to start the current. Watch the arc to see when it stabilizes and begins to melt metal. When the diameter of the molten weld pool is approximately 3/8 in. to 1/2 in., stop the welding current, keeping the torch over the weld until it cools so postpurge gases can prevent the oxidation of the molten metal. Repeat this process several times on the scrap metal to make sure that the welding machine is set correctly.

Tungsten Electrode			Welding Power			Shielding Gas		Nozzle	Filler Metal	
Type	Size	Tip	Amp	Current	HF	Type	Flow	Size	Type	Size
EWTh-1 or EWTh-2	1/16" (2 mm)	Point	50 to 100	DCEN DCSP	Start or auto	Argon	16 cfh 7 L/min	3/8" (10 mm)	RG60 or ER70S-3	1/16"–3/32" (2-2.4 mm)
EWTh-1 or EWTh-2	3/32" (2.4 mm)	Point	70 to 150	DCEN DCSP	Start or auto	Argon	16 cfh 7 L/min	3/8" (10 mm)	RG60 or ER70S-3	1/16"–3/32" (2-2.4 mm)
EWTh-1 or EWTh-2	1/8" (3 mm)	Point	90 to 250	DCEN DCSP	Start or auto	Argon	20 cfh 9 L/min	1/2" (13 mm)	RG60 or ER70S-3	3/32"–1/8" (2.4-3 mm)

Table 16-6 Suggested Setting for GTA Welding of Mild Steel

Tungsten Electrode			Welding Power			Shielding Gas		Nozzle	Filler Metal	
Type	Size	Tip	Amp	Current	HF	Type	Flow	Size	Type	Size
EWTh-1 or EWTh-2	1/16" (2 mm)	Point	70 to 100	DCEN DCSP	Start or auto	Argon	16 cfh 7 L/min	3/8" (10 mm)	ER308 or ER316	1/16"–3/32" (2-2.4 mm)
EWTh-1 or EWTh-2	3/32" (2.4 mm)	Point	70 to 150	DCEN DCSP	Start or auto	Argon	16 cfh 7 L/min	3/8" (10 mm)	ER308 or ER316	1/16"–3/32" (2-2.4 mm)
EWTh-1 or EWTh-2	1/8" (3 mm)	Point	90 to 250	DCEN DCSP	Start or auto	Argon	20 cfh 9 L/min	1/2" (13 mm)	ER308 or ER316	3/32"–1/8" (2.4-3 mm)

Table 16-7 Suggested Setting for GTA Welding of Stainless Steel

Tungsten Electrode			Welding Power			Shielding Gas		Nozzle	Filler Metal	
Type	Size	Tip	Amp	Current	HF	Type	Flow	Size	Type	Size
EWP or EWZr	1/16"	Round 2 mm	50 to 90	AC	Continues or on	Argon	17 cfh 8 L/min	7/16" (11 mm)	ER1100 or ER4043	1/16"–3/32" (2-2.4 mm)
EWP or EWZr	3/32"	Round 2.4 mm	80 to 130	AC	Continues or on	Argon	20 cfh 9 L/min	1/2" (13 mm)	ER1100 or ER4043	1/16"–3/32" (2-2.4 mm)
EWP or EWZr	1/8"	Round 3 mm	100 to 200	AC	Continues or on	Argon	20 cfh 9 L/min	5/8" (16 mm)	ER1100 or ER4043	3/32"–1/8" (2.4-3 mm)

Table 16-8 Suggested Setting for GTA Welding of Aluminum

Next, draw a 1-in. long line on the scrap metal, and using the same procedure, establish a molten weld pool, and move the torch slowly along the line. Speeding up will make the weld bead become smaller, and slowing down will make it larger. Try to make your speed consistent.

When you feel comfortable starting and stopping the arc and making a stringer bead, it is time to start on the clock. Use the same starting technique, with the torch held at a 90° angle approximately 1/4 in. above the metal so that it is on the line at the mark. You are going to initially mark each of the following hours with a small molten weld bead: 1, 2, 4, 5, 7, 8, 10, and 11. When you are in position and your personal protective equipment is in place, depress the remote control to start the welding current and establish a molten weld pool approximately 3/8 in. to 1/2 in. in diameter, Figure 16-21. Release the remote control, keep the torch over the weld until it cools, then move to the next location. Try to make each of the weld spots the same size. Repeat this process until all of the hours have been marked.

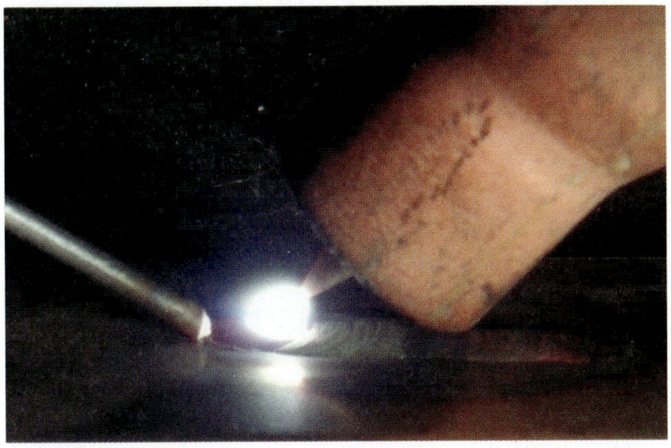

FIGURE 16-21 Surfacing weld. Larry Jeffus

Next, you are going to mark the 12, 3, 6, and 9 o'clock positions with a short-fused weld bead. Starting at 12 o'clock, use the 12-in. rule and a pencil or soapstone to mark a 1-in. line. Start your arc at the outer point of the line, and maintain a uniform speed so that the weld will be uniform and approximately 3/8 in. wide. When you have reached the end of the 1-in. mark, release the pedal, allow the weld to cool, and examine the bead to see that it is uniform in width and straight. Repeat this process on each of the other hours: 3, 6, and 9.

Allow the metal to cool, wipe off the soapstone marks, and examine the clock face. Any variation in weld bead size or width simply adds character and uniqueness to the clock. Trying to be as consistent as possible is important, but a lack of consistency should not be considered a failure in the production of this clock face.

Paperwork

Compete a copy of the time sheet in Appendix I and bill of materials in Appendix III, or as provided by your instructor.

Fusion Welds without Filler Metal

The term *autogenous weld* refers to a fusion weld made without the addition of filler metal. These welds are made by melting the edges of the metal and allowing them to flow together. Autogenous welds work best on flare, outside corner, and lap joints because as the metal is melted, it can more easily flow together than it would on butt and tee joints.

Higher welding speeds are an advantage of autogenous welds. The higher speed means less heat and less distortion. A disadvantage of autogenous welds is that severe porosity can occur when fusing the base metal together without adding filler metal if the base metal has a lot of dissolved gases.

PROJECT 16-2

Fusion Weld without Filler Metal Flat Position

Skills to be learned: The ability to control the weld bead in the flat position along an outside corner joint without the addition of filler metal.

Project Description

You will be placing a 1-in. wide band around the clock face you made in Project 16-2. This band is used to space the clock off of the wall so that the motor for the clock movement can be mounted, Figure 16-22.

Materials

- 8-in. sq. clock face from Project 16-1
- 4 8-in. long by 1-in. wide strips of the same type of metal used for the clock face
- Source of argon shielding gas
- Two different colors of spray paint and/or clear finish
- Masking tape

Tools

- GTA welder
- PPE
- 12-in. rule
- Soapstone
- Square
- Scissors
- Short piece of angle iron
- Pliers
- Locking pliers
- C-clamps
- Drill and 3/8-in. drill bit

Layout

Using the square, soapstone, and 12-in. rule, lay out four 1-in. wide 8-in. long strips of the same type and thickness of metal as the clock face.

Cutting Out

Using a properly set-up plasma cutting torch or metal shear, putting on all appropriate personal protective equipment, and following all shop and manufacturer's rules, you are going to cut out the 1-in. wide 8-in. long pieces of metal.

Fabrication

To make the fusion weld along the corner joint without the addition of filler metal, you must carefully fit the corners together so that there is no root opening, Figure 16-23. Any spacing left in the joint

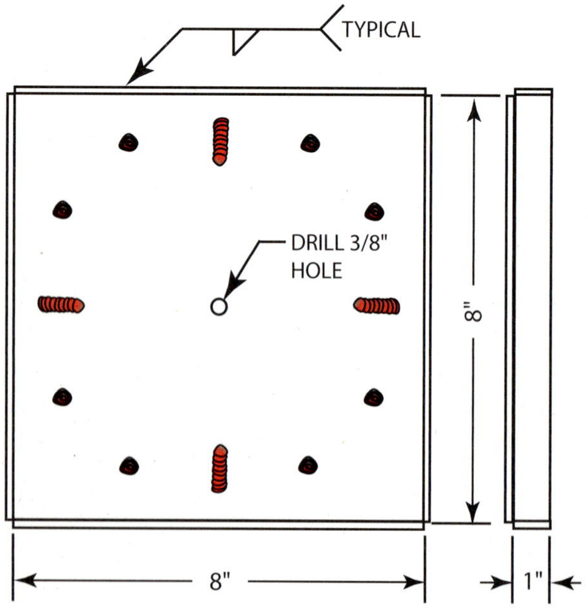

FIGURE 16-22 Project 16-2: Clock frame. Larry Jeffus

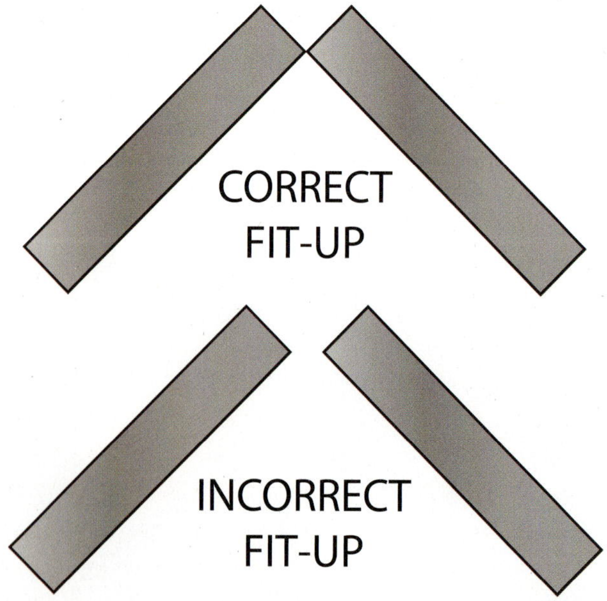

FIGURE 16-23 Properly align the plates so that the inside corners touch like the ones shown as correct fit-up above. Larry Jeffus

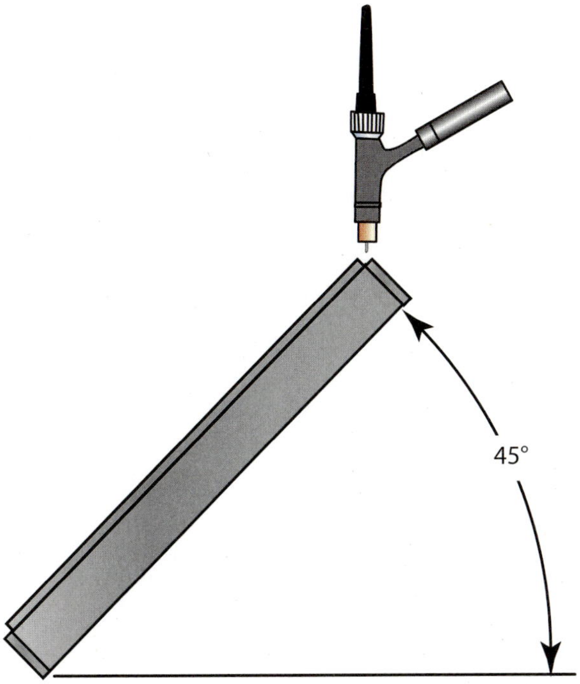

FIGURE 16-24 Bracing the weldment at a 45° angle will let you weld around the top in the flat position. Larry Jeffus

opening may melt back and not fuse together as the weld is being made. Use a short piece of angle iron, and clamp the face and strip of metal together using the self-locking pliers or C-clamps so that the corners are tight together.

Using a properly set-up and adjusted gas tungsten arc welding machine, putting on all appropriate PPE, and following all shop and manufacturer's safety rules, you are going to make a series of small tack welds along the joint.

Brace the clock face on the welding bench so that the outside corner joint is in the flat position, Figure 16-24. You may need to use a firebrick or other similar block to support the clock face in this position.

Approximately 1/4 in. from one end of the joint, hold the electrode 1/4 in. above the metal surface. Lower your welding hood and strike an arc. Move the tungsten around in a small circle as the metal begins to melt, Figure 16-25. This will help the metal flow together. Once the metal edge has melted and flowed together, stop the welding current, and keep the torch over the weld until it has cooled. Move to the opposite end, and in approximately the same location make a second tack weld. Repeat this process of making tack welds along the entire joint at approximately 1-in. intervals, Figure 16-26.

FIGURE 16-25 Tack weld on a butt joint. Larry Jeffus

NOTE: If the weld metal does not flow together because surface tension pulls the two melting edges apart, do not try to create a much larger weld pool. This will not help the metal flow together. Stop the tack weld, allow it to cool, and move to another location. During the fusion welding process, this spot should flow together. If not, it will be left open. Minor inconsistencies like this in the joint will not adversely affect the fabricated clock; therefore, this would not be considered a defect in this application.

Repeat this assembly process with the remaining three sides of the clock. When the clock band has been assembled, you must tack weld the 1-in. edges together using the same tack welding procedures as you did with the strip and face. But only one tack weld is required approximately 1/4 in. from the outside edge.

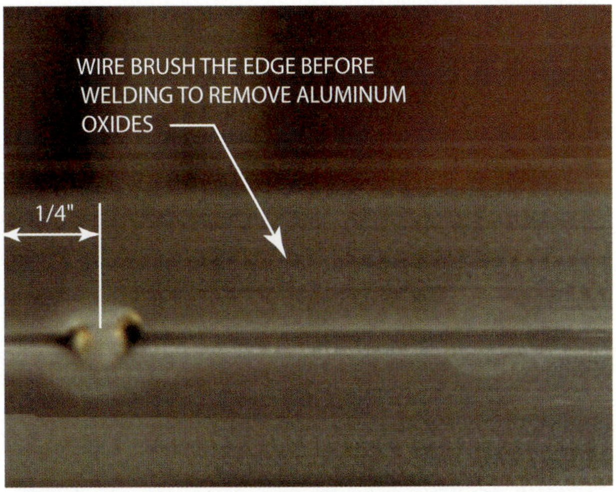

FIGURE 16-26 Outside corner tack welded together. Larry Jeffus

Welding

Set the clock face so that the outside corner joint is in the flat position. Brace yourself so that you can make a smooth uniform pass all the way along the joint without interruption. You may need to practice moving back along the joint several times before the welding power is turned on to make sure you have complete freedom of movement. Sometimes the gas tungsten arc welding leads may become caught as they rub across the table or other material in the area. Practicing moving across the joint will help ensure that you can make the weld uniformly without interruption.

You are going to start the weld at one end approximately 1/4 in. from the corner. Starting on the corner with an autogenous weld on the outside corner joint often results in the metal melting away from the corner; starting just a short distance away from the corner will eliminate that problem. Holding your torch vertically with the tip approximately 1/4 in. from the metal surface, lower your welding helmet, depress the foot control, and establish a molten weld pool. Move the torch along the joint at a smooth, even pace until you get to the other end. You may find that a small circular movement of the electrode will help melt the corners together uniformly, Figure 16-27.

Repeat this welding process on the remaining three corners of the clock face.

Once all four sides have been welded, you are going to make an autogenous weld along the edges of the band. Starting approximately 1/4 in. from the outside edge, weld to the corner of the clock face and band. At this point the corner will not be fused, but if you were to fuse the corner, it may melt back. Not fusing the corner will leave it as a sharp square corner. Either way, the clock has more than enough weld joint length to hold it without it ever coming apart; therefore, this section of unfused metal is insignificant in this application.

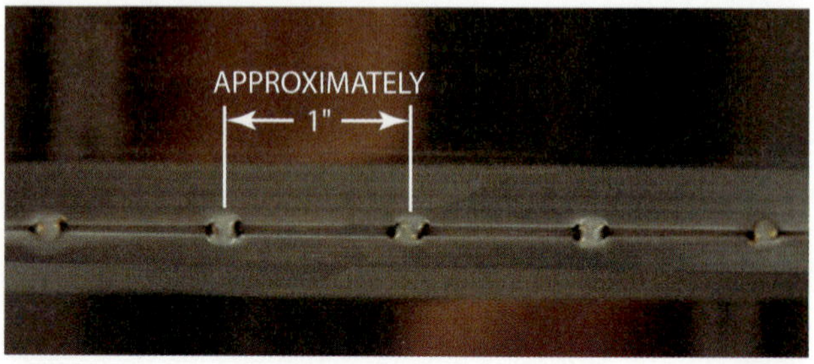

FIGURE 16-27 Outside corner joint. Note precleaning along weld. Larry Jeffus

Finishing

After the metal has been cooled, you can finish the face by spray painting the welds one color and the face another color. This will make the welds stand out. The easiest way to do this is to first spray the welds with one color. When that paint dries, cut out small pieces of masking tape to cover the welds, and spray the clock face with the second color. In some cases your welds may be clearly visible without having to paint them. If that is the case with your clock face, you may want to spray it with a clear finish.

Drill a 3/8-in. hole for mounting the clock movement. Mount the clock movement in the back of the clock face and assemble the hands.

Paperwork

Complete a copy of the time sheet in Appendix I and bill of materials in Appendix III, or as provided by your instructor.◆

Surfacing Welds

Surfacing welds, sometimes called stringer beads, should be uniform in width and **reinforcement** while having minimum **penetration.** The more penetration that these welds have, the greater the weld metal is mixed with the base metal. In some applications too much mixing can reduce the benefits of the filler metal. Following are some examples of the uses for surface welding:

- Build up the metal surface—The surface of equipment may wear down as a result of use or by accident as a result of the loss of lubrication. These surfaces can be rebuilt using GTA welding so they can be repaired.
- Apply a hard surface—A harder surface material is sometimes applied to extend the useful life of equipment that is used in heavy wear applications such as sand, gravel or rock digging, grinding, or moving.
- Apply a corrosion-resistant surface—Corrosion-resistant metals can be expensive and may not provide the mechanical strength required, so a thin layer of a corrosion-resistant material may be used on a substructure of mild steel.

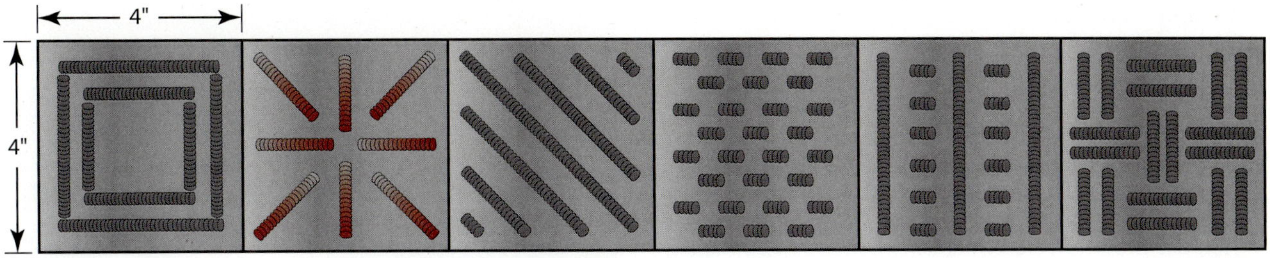

FIGURE 16-28 Project 16-3: Weld bead pattern casters. Larry Jeffus

PROJECT 16-3

1G Stringer Beads on Drink Coasters

Skill to be learned: The ability to control the weld bead on a horizontal rolled pipe.

Project Description

You will make six 4-in. square drink coasters that have weld beads that are both decorative and functional. Figure 16-28 has illustrations of some of the various decorative layouts for weld beads on the coasters. The welds support the drink slightly above the plate surface so that any condensation from a cold beverage can drain away without dripping onto the table, and the limited contact area for hot beverages helps keep them hot longer.

Project Materials and Tools

The following items are needed to fabricate the drink coasters.

Materials

- 4-in. wide 1/8-in. to 1/4-in. thick and 24-in. long strips of mild steel, stainless steel, and aluminum
- 3/32-in. mild steel, stainless steel, and aluminum filler rods
- Source of argon shielding gas
- Clear finish or paint

Tools

- Pencil and paper
- GTA welder
- Plasma cutting torch or oxyfuel cutting torch
- PPE
- Tape measure
- 12-in. rule
- Soapstone, punch or awl
- Square
- Wire brush
- Angle grinder
- File
- Pliers

Layout

You may either use the suggested weld bead patterns shown in Figure 16-28 or make a drawing on a piece of paper showing the design you want to use. Some suggestions for designs can be to draw your initials, your state outline, a map, and so on. Just make sure if you are using one of your own designs, that the welds extend close to the edges and enough of the surface of the coaster is being covered with welds so that any drink will sit securely.

Using the soapstone on the steel and the punch or awl on the stainless steel and aluminum, the 12-in. rule, and a square, lay out the center line for each of the welds that make up your coaster design on the 4-in. wide 24-in. long piece of metal.

Welding

Set the welder for the metal type and thickness being welded according to the manufacturer's recommendations or Table 16-6, Table 16-7, and Table 16-8. Put on all of your personal protective equipment, and follow all shop and manufacturer's safety rules for welding.

Starting the welds near the corner and welding toward the center reduces the possibility of overheating the corner of the metal. Overheating the corner is more of a problem when welding aluminum.

- Place the plate flat on the welding table and turn it to an angle so you are in a comfortable position to easily follow the line to be welded.
- Practice moving along the line, and adjust the plate angle if needed.
- Start by holding the torch as close as possible to a 90° angle with the tungsten tip about 1/4 in. above

the metal's surface at one end of one of the lines to be welded.

- Lower your welding hood, and depress the remote control to start the welding current.
- Watch the arc to see when the arc stabilizes and melts the metal.
- When the molten weld pool has expanded to approximately 3/8 in. in width, start adding filler metal to the leading edge of the weld pool, Figure 16-29.
- If you accidentally touch the tungsten with the filler metal or touch the molten weld pool, stop and clean the tungsten before continuing the weld.
- To keep the weld height, width, and appearance uniform, add the filler at a consistent rate by establishing a rhythm.
- Move the torch in a stepping or circular oscillation pattern at the same rate or rhythm as you add filler metal all the way down the plate to the other end, Figure 16-30.
- If the size of the weld pool changes, speed up or slow down the travel rate to keep the weld pool the same size for the entire length of the line.
- Keep the torch tip over the weld until it is cool to protect it from oxidation.
- Turn the plate and repeat this process until all of the lines on the drink coasters have been welded.

FIGURE 16-29A Establish a molten weld pool and dip the filler rod into it. Larry Jeffus

FIGURE 16-29B Keep the end of the filler rod close to the arc so that the shielding gas prevents it from oxidizing. Larry Jeffus

FIGURE 16-29C Dipping the filler metal at a constant rate will keep the weld bead produced uniform. Larry Jeffus

Cutting Out

Using a properly set-up plasma cutting torch or oxyfuel cutting torch, putting on all of your personal protective equipment, and following all shop and manufacturer's safety rules for welding, you will cut out the drink coasters.

FIGURE 16-30 Moving the torch back slightly as the filler metal is added will prevent the arc from melting the end of the filler metal causing it to form a large molten glob. Larry Jeffus

Finishing

Use the file to level any weld buildup that is so high that it would interfere with a drink sitting securely on the drink coaster. If you accidentally touched the tungsten to the metal or to the filler metal as you were welding, you can wire brush off the oxides before painting. Paint the drink coasters with a clear coat or colored coat of spray paint.

NOTE: If the height of the welds is too high to be easily filed down, you can use an angle grinder to level them off.

Paperwork

Complete a copy of the time sheet in Appendix I and bill of materials in Appendix III, or as provided by your instructor.◆

PROJECT 16-4

1G Outside Corner Joint on Bookends

Skills to be learned: The ability to produce a uniform weld bead on outside corner joints in the flat position by adding filler metal and to make it with sharp fused end corners.

Project Description

You will build three pairs of bookends: a pair made from mild steel, a pair made from stainless steel, and a pair made from aluminum, Figure 16-31.

Project Materials and Tools

The following items are needed to fabricate the bookends.

Materials

- 3-in. wide 25 3/4-in. long, 1/8 to 3/16-in. thick strips of mild steel, stainless steel, and aluminum
- 3/32-in. mild steel, stainless steel, and aluminum filler rods
- Source of argon shielding gas
- Clear finish and/or paint

Tools

- GTA welder
- Plasma cutting torch or metal shear
- PPE
- Tape measure
- Soapstone, punch or awl
- Square
- Pliers
- Locking pliers
- C-clamps
- Short piece of angle iron
- Pencil and paper

Layout

You are going to lay out the parts for Project 16-4 on the 3-in. wide, 25 3/4-in. long strips of metal following the layout illustrated in Figure 16-32.

Using a pencil and paper, calculate the distances for A, B, C, and D. These dimensions must be calculated so that the lines representing their dimensions are accurate. Using the soapstone, punch or awl, and the tape measure, measure each of the line's locations from one end starting with the 6-in. long

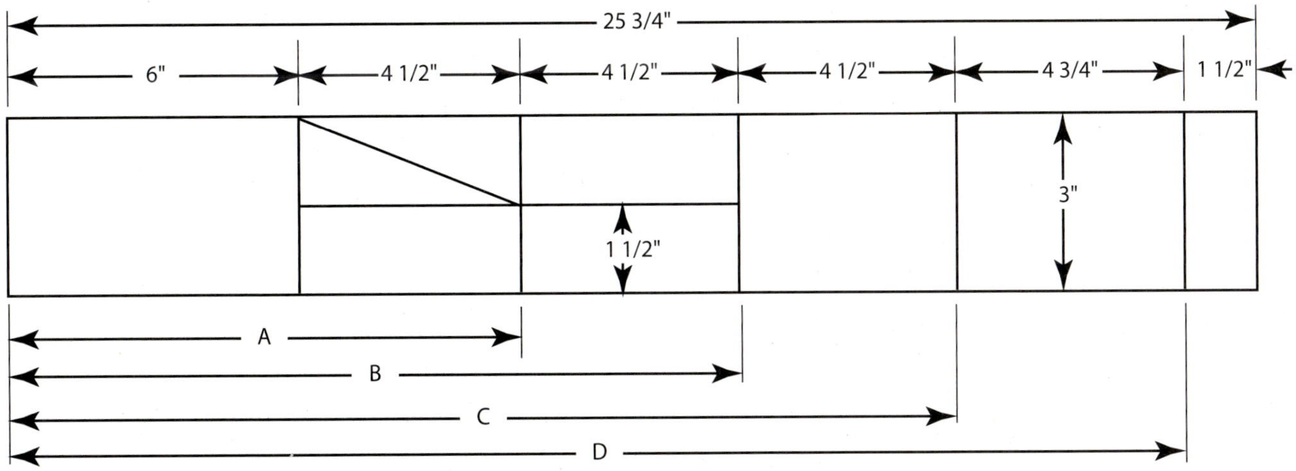

FIGURE 16-31 Project 16-4: Bookend. © Cengage Learning 2012

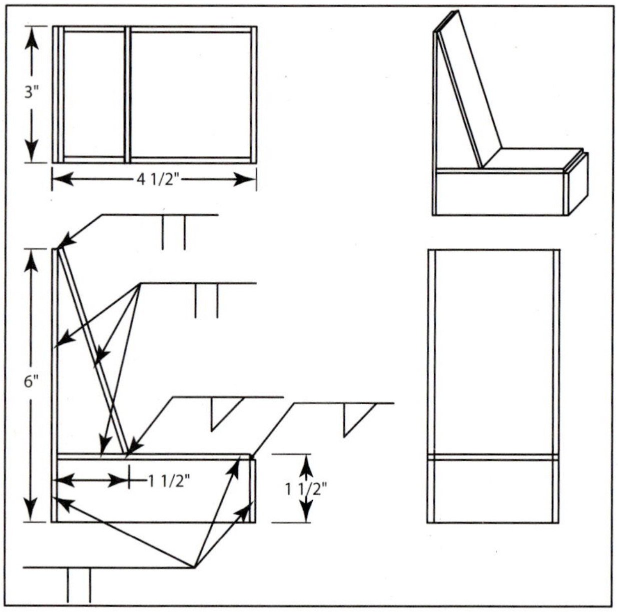

FIGURE 16-32 Bookend assembly drawing. © Cengage Learning 2012

piece. Use the square to scribe a line straight across the part.

NOTE: If you are going to use the plasma torch to cut the parts, you must add in the width of the plasma cutting torch kerf to the length of each part so that the parts' finished sizes are correct. Failure to add the kerf width to the length will result in each of the parts being slightly shorter than the overall length needed. Measure the width of the plasma kerf that the torch you are using makes by making a practice cut on some scrap metal. This will mean that the overall length of 25 3/4 in. must be increased to accommodate the material lost through the thermal cutting process.

Using the tape measure, locate the center of the width of two of the 4 1/2-in. long pieces. Use the combination square and marker as shown in Figure 16-33 to mark a line through the marks on the plate so it can be cut in half.

Cutting Out

Using a properly set-up plasma cutting torch or shear, putting on all of your personal protective equipment, and following all shop and manufacturer's safety rules for the equipment you are using, you will cut out the pieces.

Fabrication

Start by tack welding the sides of the base around the 4 1/2-in. long, 3-in. wide flat top. Using a short piece

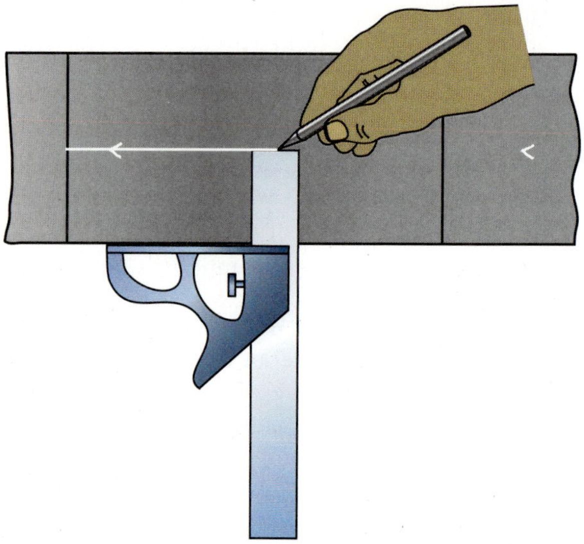

FIGURE 16-33 Drawing a parallel line using a combination square and marker. © Cengage Learning 2012

of angle iron and the locking pliers and/or C-clamps, secure one side of the base to the top. Place the assembled part on the welding table so that the outside corner joint is in the flat position. Using the same techniques to tack weld the outside corner together as described in Project 16-2, make tack welds on the edge to hold it in place. Repeat this process with the other two sides.

Place the base assembly on its side flat on the welding table, and line up the 3-by-6 in. faceplate of the bookend. Use the square to fit the faceplate to the base, Figure 16-34. Check to see that the bottom edge of the faceplate is even with the bottom of the base sides. Make two tack welds along the edge of the joint between the base side and the faceplate. Turn the assembly over, and make two tack welds along the joint on this side.

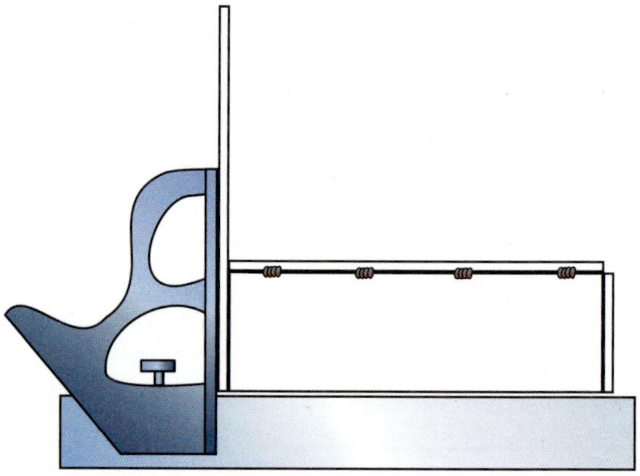

FIGURE 16-34 Squaring the faceplate and base assembly. © Cengage Learning 2012

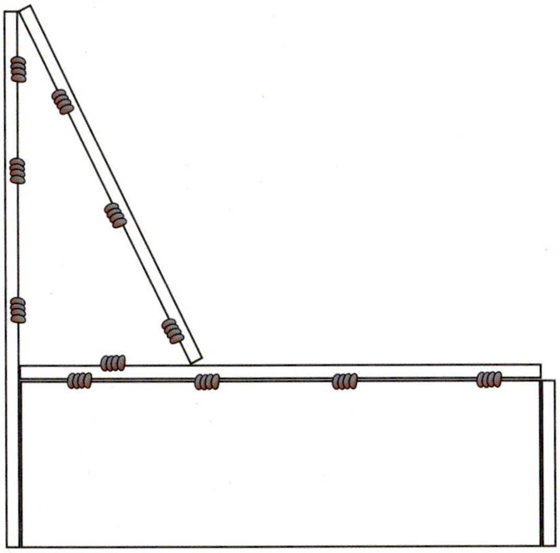

FIGURE 16-35 **Making the tack welds slightly away from the corners will make it easier to have a uniform weld at the corners that would be less likely to leak.** © Cengage Learning 2012

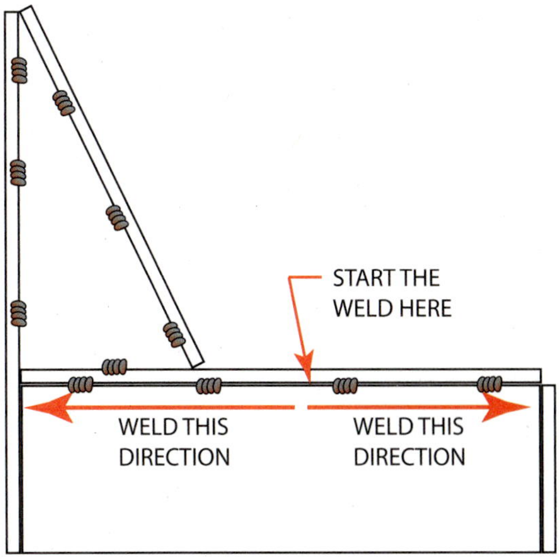

FIGURE 16-36 **Start the weld near the middle of the joint and weld to the corner.** © Cengage Learning 2012

The two triangle-shaped side pieces are assembled next. The angle iron and clamps can be used to hold these parts in place as they are tack welded. Make three tack welds along the joint between the faceplate and the triangle and one along the base joint, Figure 16-35. The base joint is a butt joint and may need to have filler metal added to the tack weld to get it to flow together.

The back plate can now be tack welded to the assembly. After clamping the plate securely in place, make tack welds on all of the joints except for the tee joint between the back plate and the top of the base plate. This joint will not need to be tack welded.

Repeat the assembly process to make a matching second bookend using the same type and thickness of metal.

Welding

Using a properly set-up and adjusted GTA welding machine, putting on all of your personal protective equipment, and following all shop and manufacturer's safety rules for the equipment you are using, you will weld the assembly, maintaining uniform weld pool size.

The easiest way to avoid rounding off or melting back the corners as the edges are welded is to start the weld in the middle of the joint and weld to the corner, Figure 16-36. Place the assembled bookend on the welding table so that the outside corner joint between the back and the top plates is in the flat position, Figure 16-37. Hold the tungsten electrode approximately 1/4 in. above the center of the joint, and practice moving along the joint before striking the arc. When you are sure you can move freely and evenly along the joint, you are ready to start welding.

Lower your helmet, and depress the foot control to establish an arc. When the molten weld pool has been established, start moving the electrode in a small circle, and add filler metal to the leading edge of the weld pool. Add filler metal to the leading edge of the weld pool as needed to keep the weld size uniform. The corners of parts may overheat because there is less metal to absorb the welding heat. For that reason you may need to reduce the welding current slightly and add a little more filler

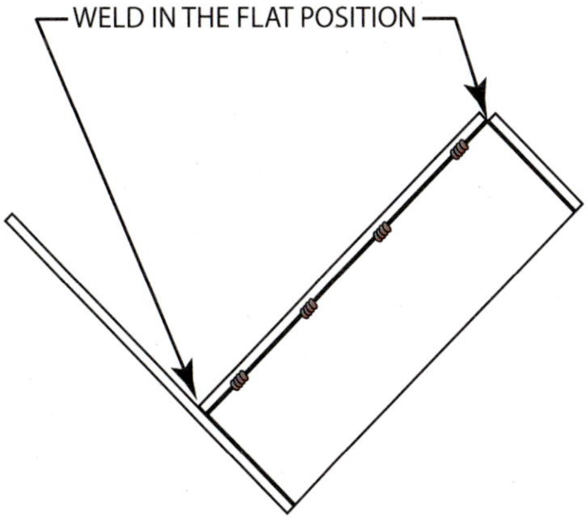

FIGURE 16-37 **Make the welds in the flat position.** © Cengage Learning 2012

metal to keep from rounding off or melting away the corner. Stop the welding current when the weld reaches the corner.

To make the other half of the weld, start the molten weld pool on top of the first weld. This will ensure that you have good fusion and penetration of the joint because starting the weld next to the end can possibly leave a small gap in the joint's fusion. This lack of fusion at the tie-in of welds can be a defect in some applications.

Repeat this process on all of the edge joints on both bookends.

The tee joint between the top plate and the angled back plate may be left unwelded.

Finishing

Place the completed bookends on a flat surface to see that they sit firmly without wobbling. Wobbling can be caused by a small spot of weld metal or a slight twist or misalignment of one of the sides. If needed, you can identify the problem and use the file or grinder to make any needed adjustments so the bookends sit firmly.

The welds are to be left as they are, and the bookends can be sprayed with a clear finish or any color.

Paperwork

Complete a copy of the time sheet in Appendix I and bill of materials in Appendix III, or as provided by your instructor.◆

Edge Joints

Edge joints are often made by simply melting the metal edges together. However, the edge joints on the try square are more like square butt joints than they are edge joints. Edge welds are made all the way across the joint, but these welds may not go all the way across the edge, Figure 16-38.

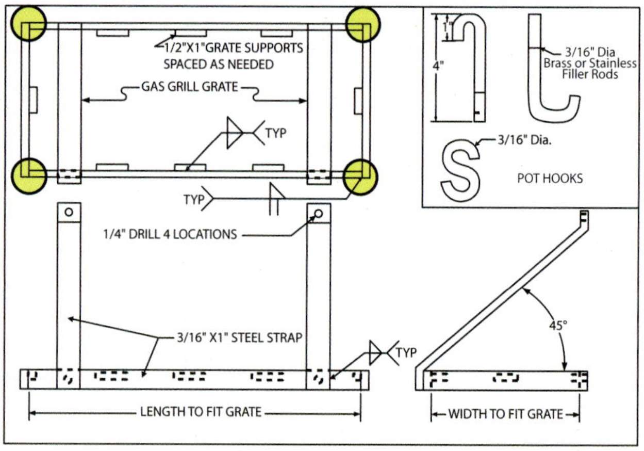

FIGURE 16-38 Edge joints on the pot rack.

PROJECT 16-5

1G Edge Joint on Try Square

Skills to be learned: The ability to produce a uniform weld bead on edge joints in the flat position by adding filler metal.

Project Description

You will build a 6-in. try square that can be used for laying out 90° angles or to check for squareness of parts or surfaces, Figure 16-39. The 45° angle on the handle allows you to lay out the parts for a 90° corner.

Project Materials and Tools

The following items are needed to fabricate the try square.

Materials

- 1 1/2-in. wide, 6-in. long, 1/8-in. thick strips of mild steel, stainless steel, and aluminum
- 1 1/4-in. wide, 2 1/4-in. long, 1/8-in. thick strips of mild steel, stainless steel, and aluminum
- 1 1/2-in. wide, 4-in. long, 1/4-in. thick strips of mild steel, stainless steel, and aluminum
- 3/32-in. mild steel, stainless steel, and aluminum filler rods
- Source of argon shielding gas
- WD-40 or other light oil

Tools

- GTA welder
- Plasma cutting torch or metal shear
- PPE
- 12-in. steel rule
- Punch or awl
- Square
- Pliers
- Bench vise
- Metal file
- Hammer
- 1/8-in. filler rod
- C-clamps
- Pencil and paper

Layout

You are going to lay out the parts for Project 16-5 on the 1/8-in. and 1/4-in. thick metal using the punch or awl, Figure 16-39. The fine scribed line made by the punch or awl will help to lay out these parts very precisely. It is important that they be precise for the square to be accurate enough to be useful. Carefully make the measurements with the 12-in. steel rule.

NOTE: **It is easier to make accurate measurements with a steel rule than it is to make them with a tape measure.**

Cutting Out

Using a properly set-up plasma cutting torch or shear, putting on all of your PPE, and following all shop and manufacturer's safety rules for the equipment you are using, you will cut out the pieces.

It is very important to follow the layout lines exactly because in order for the assembled try square to be accurate enough to be usable, the sides of the handle and blade of the square must be parallel.

Fabrication of the Handle

The try square handle will be assembled and welded first before the blade is assembled so that any distortion that may occur during the welding of the handle can be corrected before the blade is assembled.

Use a C-clamp to hold the two sides firmly but not too tightly against the center spacer. The spacer can be tapped into place by placing a piece of 1/8-in. diameter

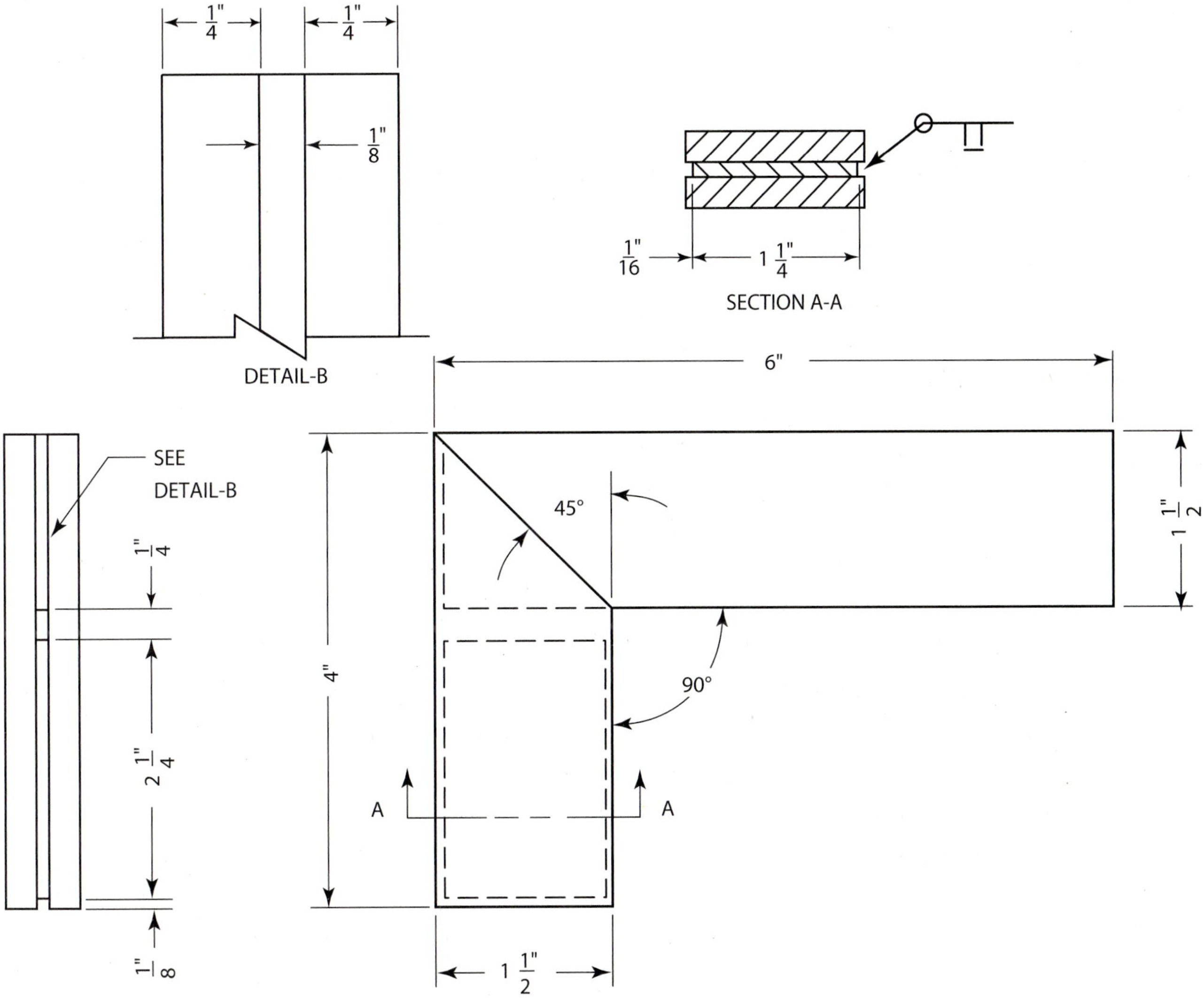

FIGURE 16-39 Project 16-5: Try square. © Cengage Learning 2012

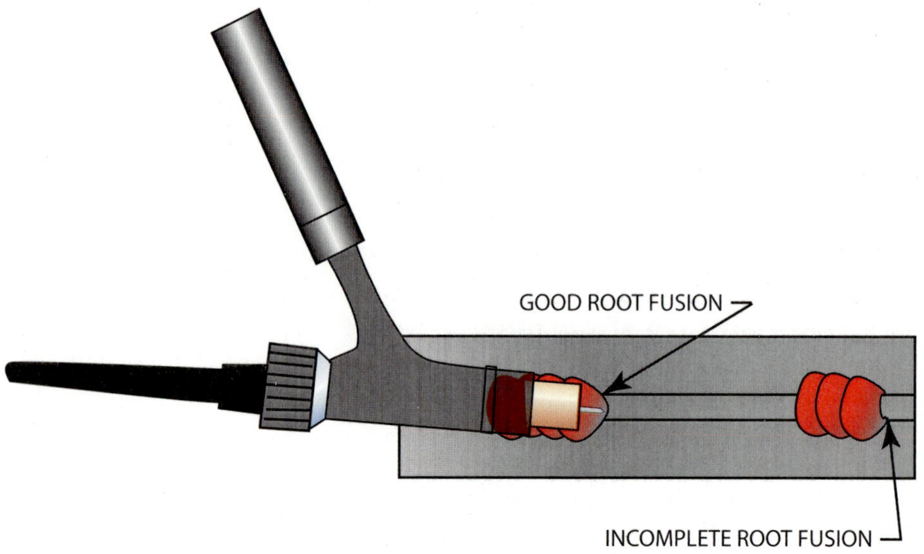

FIGURE 16-40 Hold the GTA welding torch so that you can see the root of the joint to make sure it fuses to the side plates. © Cengage Learning 2012

welding rod in the gap along one of the edges, then tapping it with a hammer to slide the center spacer back. If the center spacer does not move easily, loosen the C-clamp slightly. Use the 1/8-in. rod to set the spacing for the bottom. When all three sides are spaced evenly, tighten the C-clamp and add another C-clamp to make sure the parts do not move during welding.

Welding of the Handle

Place the try square handle on the welding table so that it can be welded in the flat position. Make two small tack welds in the groove on each side of the handle. Watch to make sure that the tack weld melts and fuses the spacer plate as well as the sides, Figure 16-40.

Leave the C-clamps on the handle as the square groove welds are made along the side of the handle. During the welding process, watch the groove **root** to see that it is fusing the spacer to the sides. Just fill the groove with filler metal. Too large a weld will interfere with the square's edge, so it would have to be filed away later.

Fabrication of the Blade

The edges of the blade need to be straight, parallel, and square, Figure 16-41. The sides should already be cut straight and parallel. The edges may need to be squared up a little. If the try square blade needs to be squared, it can be clamped securely in a bench vise and draw filed. *Draw filing* is the process of holding the file across the edge of the blade at a 90° angle and pulling or drawing the file toward you, Figure 16-42. Too much filing can cause the edges to become wavy and unparallel.

The welds made on the handle should draw the top beveled ends of the handle together slightly so the blade fits firmly. If you have to tap the blade to get it into the gap, use a piece of wood to protect the blade from damage from the hammer taps. Use a square

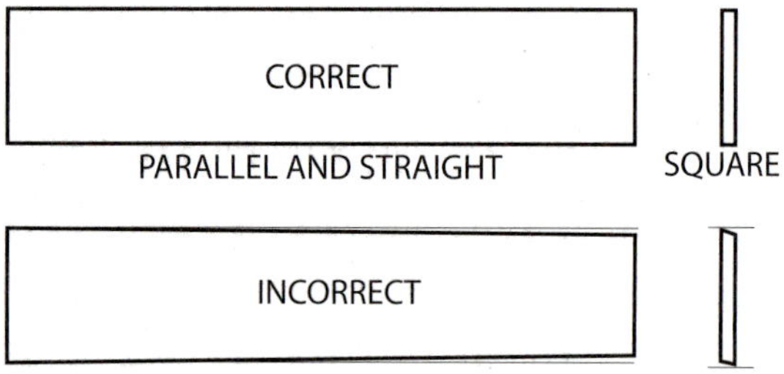

FIGURE 16-41 Check the blade for parallel edges and squareness. © Cengage Learning 2012

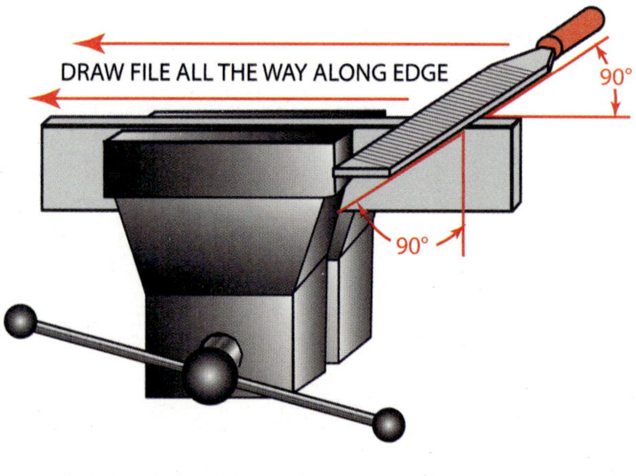

FIGURE 16-42 Draw filing. © Cengage Learning 2012

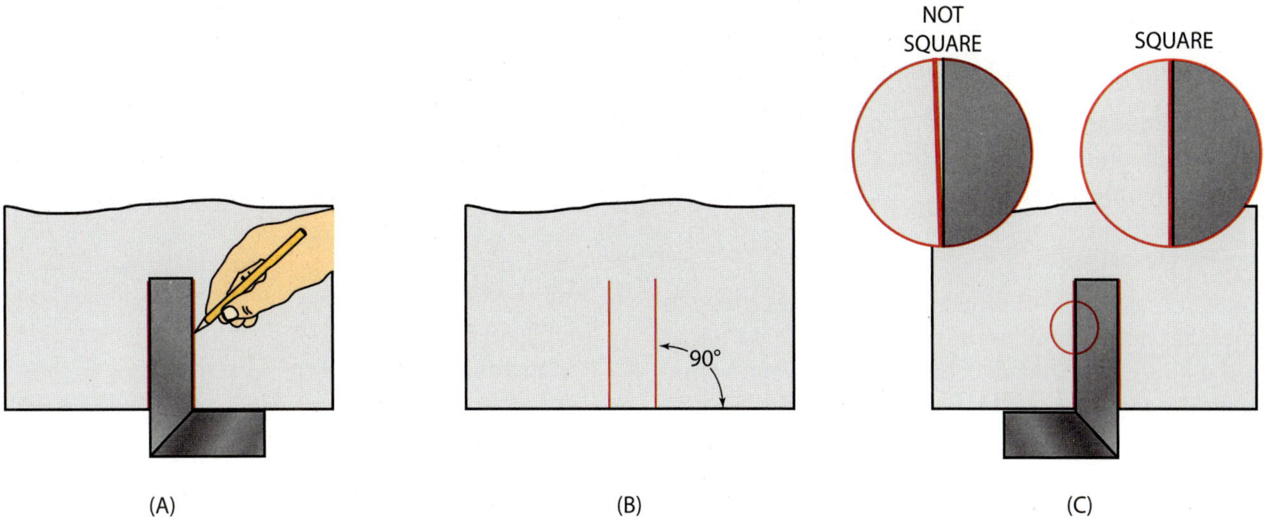

FIGURE 16-43 Checking the square for accuracy.

to align the blade at a 90° angle with the handle. To check the squareness of the blade and handle, hold the square against the straight edge of a piece of paper and draw lines along both sides of the blade, Figure 16-43A. Flip the square over to check to see that the blade lines up with the lines you drew, Figure 16-43C. After it is aligned correctly, use a C-clamp to keep the blade from moving as it is being welded.

Welding of the Blade to Handle

Place the try square handle on the welding table so that it can be welded in the flat position. Make two small tack welds in the groove on each side of the handle. As before, watch to make sure that the tack weld melts and fuses the spacer plate as well as the sides. Make a weld across the back side of the handle to fuse it and the blade together.

Finishing

Draw file any weld metal off of the handle that would keep the handle from fitting evenly against the edge of the material being squared. Use WD-40 or other light oil to cover the steel try square to keep it from rusting.

Paperwork

Complete a copy of the time sheet in Appendix I and bill of materials in Appendix III, or as provided by your instructor.◆

Tee Joint

The tee joint can be one of the more difficult joints to do with gas tungsten arc welding. The greatest difficulty is that the filler metal can stick to the base metal's surface if it accidentally comes in contact with it as it is being fed into the molten weld pool. Adding to this problem is the fact that the torch nozzle restricts the view of the molten weld pool.

Another problem with the gas tungsten arc welded tee joint is trying to ensure that 100% of the root is fused. If the tungsten is not extended directly into the corner, there is a tendency for the side of the joint to be melted and not the root. Getting the tungsten down closer to the joint eliminates this problem, but it is easier to accidentally touch the tungsten to the molten weld pool or the filler metal to the tungsten when it is extended farther from the nozzle. Practice is required to develop the skill to easily step back with the torch as filler metal is added to minimize the problems associated with GTA welding of the tee joint, Figure 16-44.

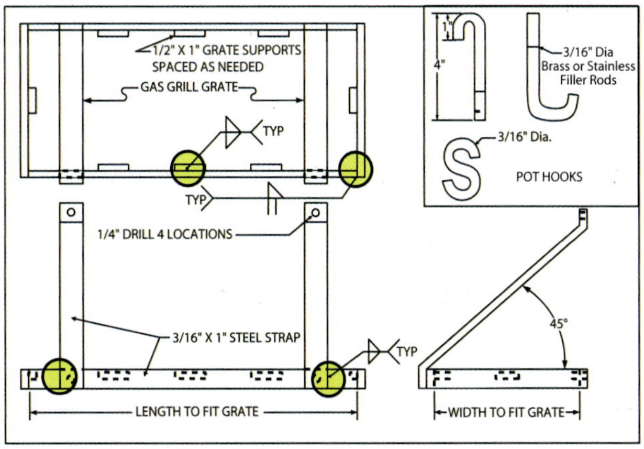

FIGURE 16-44 Tee joints on the pot rack.

PROJECT 16-6

1F Tee Joint on Letters and Numbers

Skills to be learned: The ability to produce uniform weld beads on a tee joint in the flat position while adding filler metal.

Project Description

You will make a series of letters or numbers that can be used as house numbers or to display your last name on the porch, post, or mailbox in front of your home, Figure 16-45. Some of the numbers or letters will be made out of mild steel, stainless steel, and aluminum.

Project Materials and Tools

The following items are needed to fabricate the letters and/or numbers.

Materials

- 6-in. wide, 1/8-in. thick strips of mild steel, stainless steel, and aluminum
- 1/2-in. wide, 1/8-in. thick strips of mild steel, stainless steel, and aluminum
- 3/32-in. mild steel, stainless steel, and aluminum filler rods

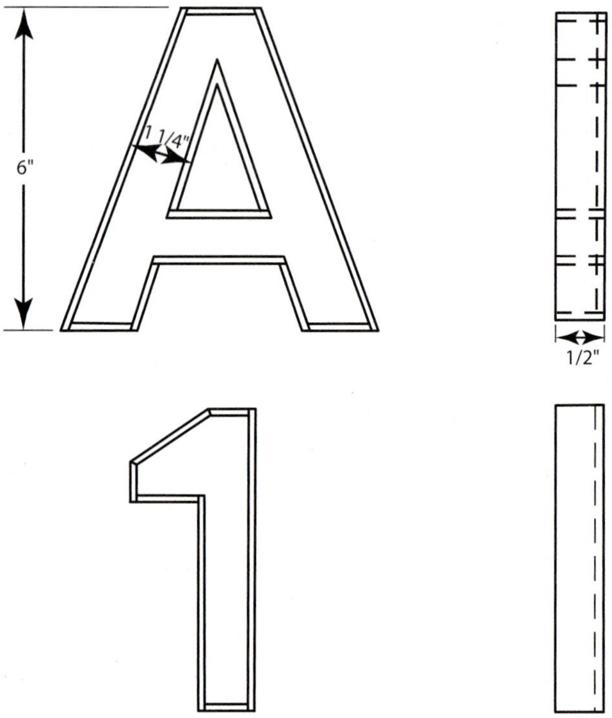

FIGURE 16-45 Project 16-6: Letters and numbers.

- Source of argon shielding gas
- Clear finish and/or paint

Tools

- GTA welder
- Plasma cutting torch or metal shear
- PPE
- 12-in. rule
- Tape measure
- Soapstone, punch or awl
- Square
- Pliers
- Locking pliers
- C-clamps
- Short piece of angle iron
- Pencil and paper

Layout

The letters and/or numbers will be laid out on paper and cut out to make a template. An advantage of the paper template is that it can be used over and over again to make additional letters. Another advantage of creating a paper template is that some letters such as *A, H, M, T, W, X,* and *V* are mirror images, one side of the other. For these letters the paper template can be folded so that these letters are uniform on both sides.

The letters will be 6 in. tall and 1 1/4 in. wide with a 1/2-in. band around the letters for accent. Use the 1-in. grid layout shown in Figure 16-46 as a guide.

NOTE: A 1 1/4-in. wide strip of construction paper or card stock can be used to lay out many of the letters that are primarily made up of straight parallel lines such as *E, F, K, L, M, W, V,* and so on.

Lay the paper templates down on the 6-in. wide 1/8-in. thick strip of metal. Position the letters and/or numbers in as tight a positioning as possible. In some cases letters will fit closer together if some are turned upside down next to those that are straight up, Figure 16-47.

Using the soapstone and punch or awl, outline the letters. Be sure to leave enough space between each letter for the cutting torch kerf.

Cutting Out

Using a properly set-up plasma cutting torch or shear, putting on all of your personal protective equipment,

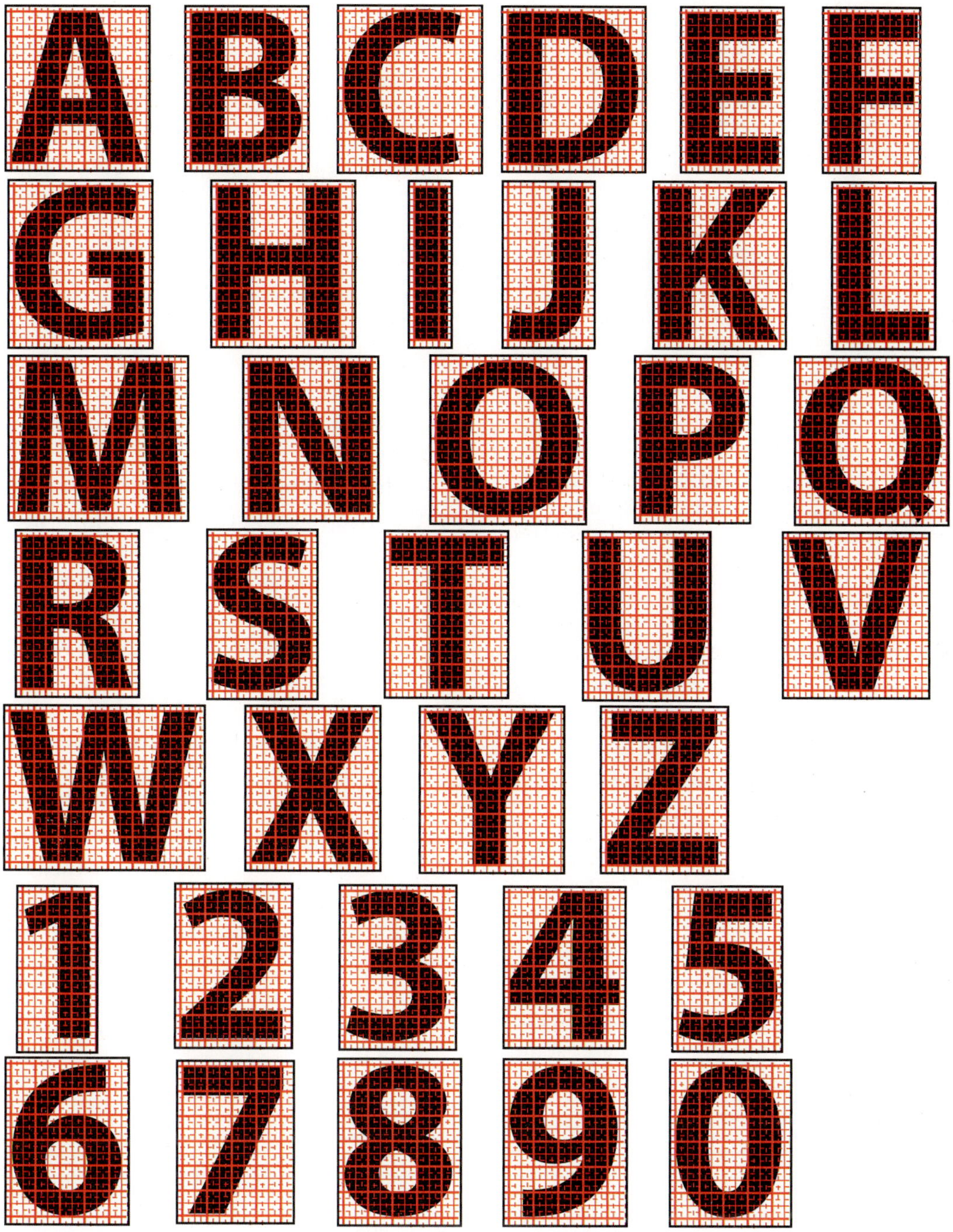

FIGURE 16-46 Grid layout for letters and numbers.

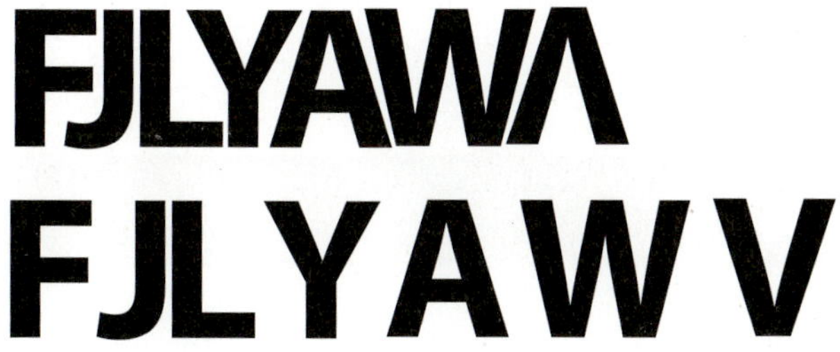

FIGURE 16-47 Nesting the parts can save material.

and following all shop and manufacturer's safety rules for the equipment you are using, you will cut out the pieces.

Fabrication

The numbers or letters can be either framed with the edge band so the letter faces are up, giving them a shadow box appearance, or they may be banded so they have a raised, more massive look. In either case the fabrication sequence is the same.

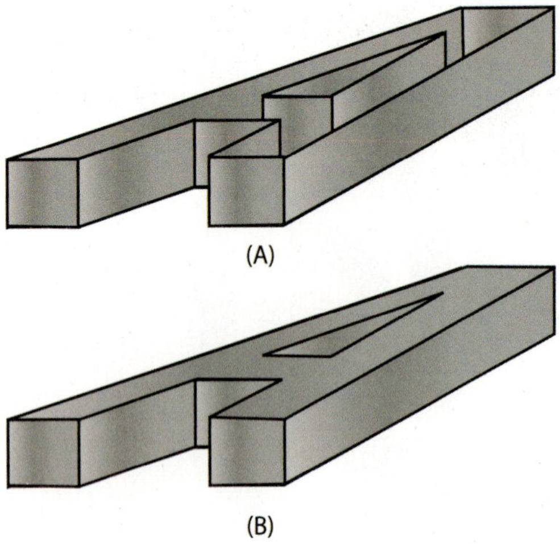

FIGURE 16-48 (A) Letters banded along the top edge of the letter can be used as small containers. (B) Letters banded along the bottom edge of the letter can be used on a front pillar, post, or mailbox. © Cengage Learning 2012

NOTE: If the letters or numbers are banded so that they form a shallow container, they can be used as a candy dish or for knickknacks on an office desk, Figure 16-48A. If they are banded so they have a massive appearance, they can be used on the front pillar, post, or mailbox of a residence, Figure 16-48B.

Using a properly set-up and adjusted GTA welding machine, putting on all of your PPE, and following all shop and manufacturer's safety rules for the equipment you are using, you will make small tack welds to hold the edge band in place.

Placing the letter outline on the welding table either faceup or facedown and using a square, stand the band up against the letter starting at a corner.

Lower your welding helmet, establish a small molten spot at the joint base, add a small amount of filler metal to fuse the two pieces together, stop the welding current, and hold the torch over the tack weld until it cools. Repeat the process along the edge of the joint at approximately every 3/4 in. When you reach a corner, bend the edge band around the corner and hold it in place with a clamp, if necessary, and repeat the tack welding process. Continue the process of tack welding and bending the band around the corners until the entire outside of the letter has been banded. If the letter contains an internal area such as the letters *A*, *B*, *D*, and so on, you will band these internal areas using the same procedure as before.

Welding

Using a properly set-up and adjusted GTA welding machine, putting on all of your personal protective equipment, and following all shop and manufacturer's safety rules for the equipment you are using, you will make small fillet welds around the joints between the letter or number and the edge band.

This welding project requires a very close control of the weld size and the torch or filler metals because of the limited space on the inside of the banding. This type of close-quarter welding is not unusual for welders to encounter in the field. You will make welds on all of the internal joint edges around the letter or number. In some cases you can rest the TIG welding torch nozzle on the metal and holding the torch at a slight angle so that the electrode does not touch the corner, drag it along that edge as filler metal is added.

This technique will help keep the tungsten electrode from touching the molten weld pool. It does make it difficult, however, to move the torch back and forth as filler metal is added, so care must be taken to be sure not to accidentally touch the filler metal to the tungsten electrode or to the hot metal next to the molten weld pool.

Make a series of small fillet welds all the way around the letter or number.

Finishing

The welds are to be left as they are, and the letters or numbers can be sprayed with a clear finish or any color.

Paperwork

Complete a copy of the time sheet in Appendix I and bill of materials in Appendix III, or as provided by your instructor.◆

SUMMARY

In this chapter you have learned enough layout, cutting, fabrication, and welding skills to build the pot rack shown in Figure 16-49. As you have built projects in this chapter, you have learned all of the necessary welding skills to fabricate this pot rack. Similar welded joints have been highlighted in figures in this chapter.

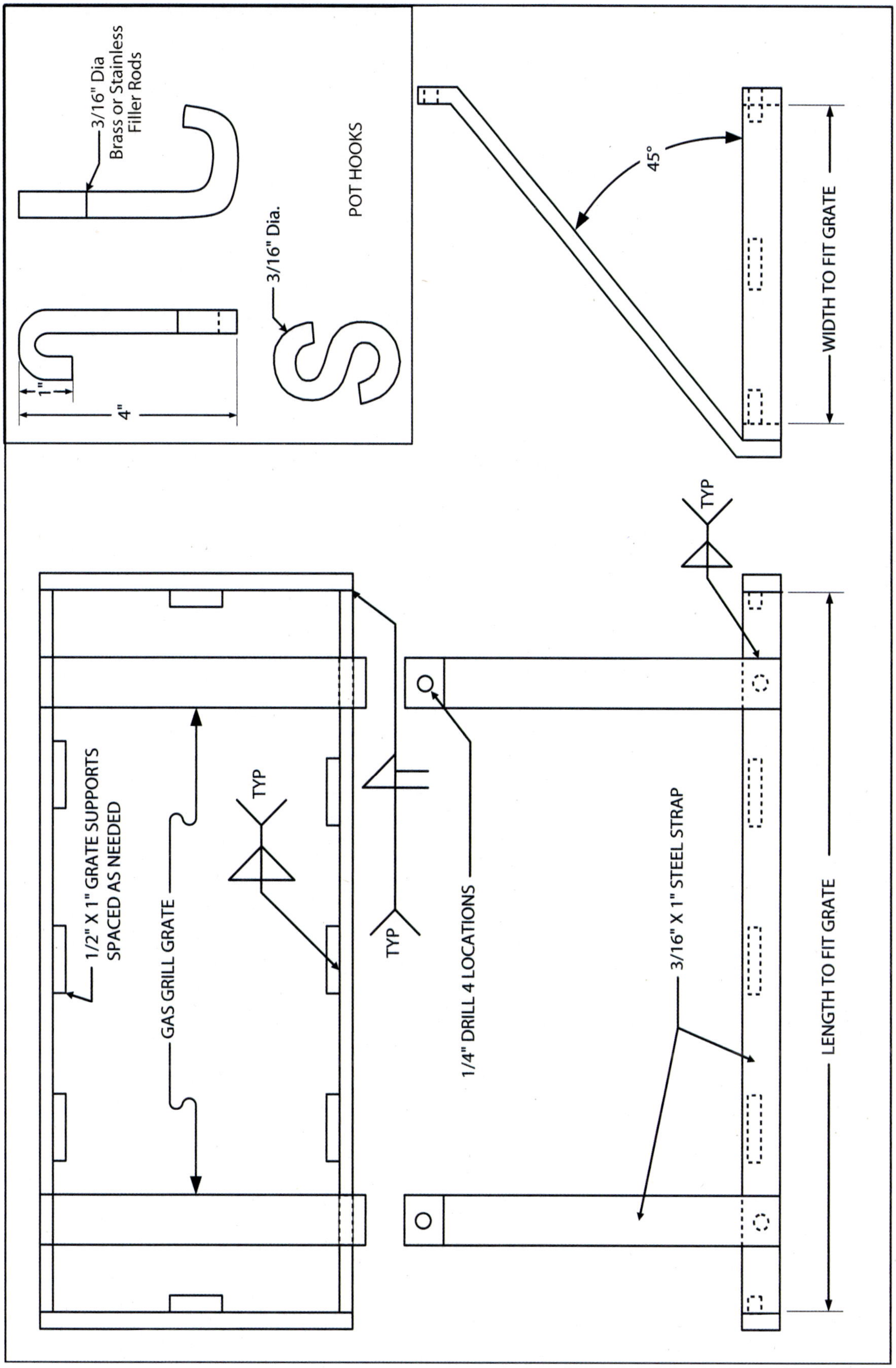

FIGURE 16-49 Pot rack drawing.

The length and width of this pot rack will be determined by the size of the gas grill grate you choose to use. Gas grill grates are available from many local home, garden, and building supply stores. Select one whose size will fit where you intend to mount it in your home. The gas grill grate allows this pot rack a great deal of flexibility in that the "S" hooks and "J" hooks may be attached to the outside band or to different spots in the grate itself for hanging utensils—pots, pans, and so on. The top of the grate is an area where cookbooks, measuring cups, and other utensils that might not hang well can be placed.

The "S" hooks and "J" hooks can be fabricated from brass or stainless steel filler rods and can vary in length depending on the utensils that you hang on them.

REVIEW QUESTIONS

1. What are the two most common types of steel?
2. How will changing the angle (length) that a tungsten is pointed affect the weld width and depth of fusion?
3. Why does stainless steel base metal and filler metal need more precleaning than steel does before welding?
4. What does the bead color of a stainless steel weld reveal?
5. What can be done when welding stainless steel to reduce the possibility of carbide precipitation?
6. What three characteristics of aluminum affect how it is welded?
7. How does the high thermal conductivity of aluminum affect the welding process?
8. How can the high surface tension of molten aluminum make it more difficult to transfer the filler metal into the weld pool?
9. How can contaminants be removed off of base metal and filler metal before welding?
10. List the common parts needed to set up a GTA welder.
11. How can the width of the molten weld bead be controlled?
12. How does the travel speed of the torch affect the weld bead?
13. How do changes in the welding current affect the width of the weld bead?
14. How does the width of the weave pattern affect the weld bead?
15. Define *autogenous weld.*
16. Autogenous welds work best on what kinds of joints?
17. What should be done when making an autogenous tack weld when the surface tension pulls the two melting edges apart?
18. What can be done before welding to ensure that a smooth uniform pass can be made all the way along the joint without interruption?
19. Give three examples of uses for surface welding.
20. When welding aluminum, what can be done to reduce the possibility of overheating the corner of the metal?
21. How can the width of the plasma kerf that the torch makes be determined before laying out the parts so that the part's finished size is correct?
22. What is draw filing?
23. What are the advantages of making and using a paper template when laying out and cutting out a part?

Chapter 17

Oxyfuel Welding and Cutting Equipment, Setup, and Operation

OBJECTIVES

After completing this chapter, the student should be able to:

- Identify all of the components and equipment found in a typical oxyfuel welding station.
- Demonstrate the proper assembly, testing, lighting, adjusting, and disassembling of an oxyfuel system.
- List the proper safety procedures for setting up and operating an oxyfuel system.

KEY TERMS

acetylene (C_2H_2)
backfire
Bourdon tube
carburizing flame
combination welding and cutting torch
creep
cutting torch
cylinder pressure
diaphragm
flashback
flashback arrestor
gauge
leak-detecting solution
line drop
methylacetylene-propadiene (MPS)
neutral flame
oxidizing flame
oxyacetylene
oxyfuel gas torch
purged
regulators
safety disk
safety release valve
seat
spark lighter
two-stage regulators
valve packing
working pressure

INTRODUCTION

Oxyfuel welding is one of the oldest welding processes. It was referred to as gas welding, although the process today is referred to as **oxyacetylene** or oxyfuel welding. Oxyfuel refers to the fact that many fuel gases other than **acetylene (C_2H_2)** can be used, although acetylene is still the most popular fuel gas.

Oxyfuel welding, cutting, brazing, hard surfacing, heating, and other similar processes use the same basic equipment. The same safety procedures must be followed for each process when storing, handling, assembling, testing, adjusting, lighting, shutting off, and disassembling this equipment. Oxyfuel processes are safe only when all of the proper setup and operating procedures have been followed. Improper or careless work habits can cause serious safety hazards.

Although numerous manufacturers produce a large variety of gas equipment, it all works on the same principle. Much of the basic equipment is common to all oxyfuel processes, and some parts, such as cylinders, regulators, hoses, hose fittings, safety valves, and torches, may be interchangeable. When welders are not sure how new equipment is operated, they should seek professional help. A welder should never experiment with any equipment.

All oxyfuel processes use a high-heat, high-temperature flame produced by burning a fuel gas mixed with pure oxygen. The gases are supplied in pressurized cylinders. The cylinder gas pressure must be reduced by using a regulator to lower working gas pressure. The gas flows from the regulator through flexible hoses to the torch. Valves on the torch body are opened and adjusted to control the gas flow through the torch to the tip. At the torch tip the gases have been mixed to produce a properly combusting flame.

Acetylene is the most widely used fuel gas because of its high temperature and concentrated flame, but about 25 other gases are available. The regulator and torch tip are usually the only equipment change required to use another fuel gas. The flame adjustment and operating skills required are often different, but the storage, handling, assembling, and testing are the same. When changing gases, make sure the tip can be used safely with that gas.

OXYFUEL EQUIPMENT

Because of the similarities in the way equipment is operated and assembled, the following information can be easily applied to all oxyfuel welding systems.

Pressure Regulators

All pressure **regulators** reduce the high cylinder or system pressure to the proper lower **working pressure**. It is important that the regulator keep the lower pressure constant over a range of flow rates. Some of the various types of pressure regulators are low-pressure regulators, high-pressure regulators, single-stage regulators, dual-stage regulators, cylinder regulators, manifold regulators, line regulators, and station regulators. Although they all work the same, they are not interchangeable.

CAUTION

Although all regulators work the same way, they cannot be safely used interchangeably on different types of gas or for different pressure ranges without the possibility of a fire or an explosion.

Regulator Operation

A regulator works by holding the forces on both sides of a **diaphragm** in balance, Figure 17-1. As the

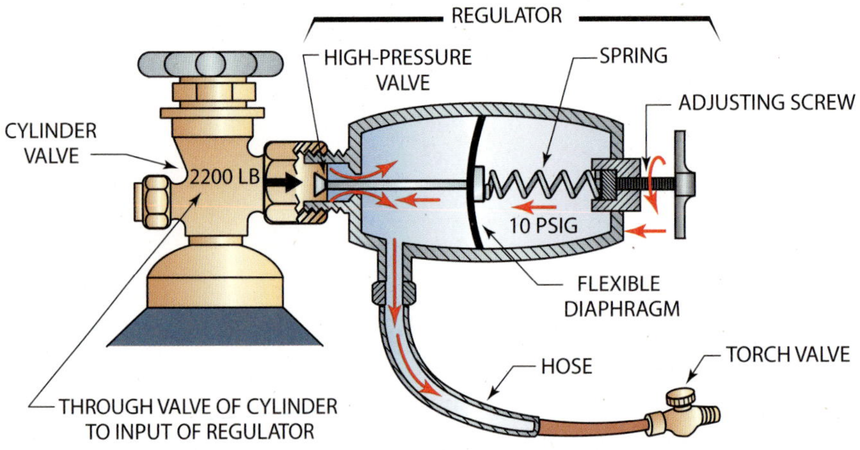

FIGURE 17-1 Force applied to the flexible diaphragm by the adjusting screw through the spring opens the high-pressure valve. © Cengage Learning 2012

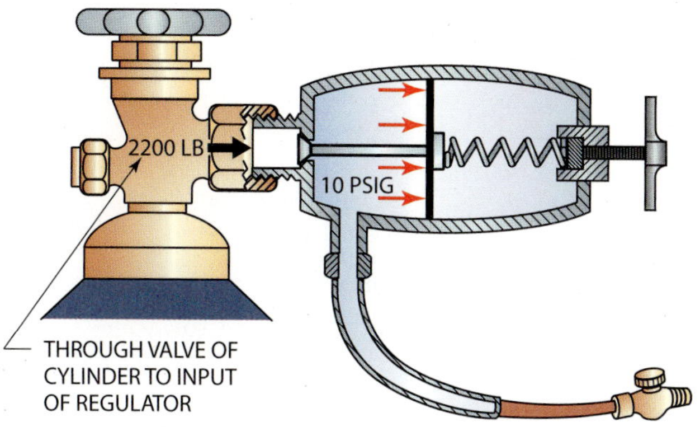

FIGURE 17-2 When the gas pressure against the flexible diaphragm equals the spring pressure, the high-pressure valve closes. © Cengage Learning 2012

pressure-adjusting screw is turned inward, it increases the force of a spring on the flexible diaphragm and bends the diaphragm away. As the diaphragm is moved, the small high-pressure valve is opened, allowing more gas to flow into the regulator. The gas pressure cancels the spring pressure, and the diaphragm returns back to its original position, closing the high-pressure valve, Figure 17-2. When the regulator is used, the gas pressure on the back side of the diaphragm is reduced, the spring again forces the valve open, and gas flows. The drop in the internal pressure can be seen on the working pressure gauge, Figure 17-3.

Because some small single-stage pressure regulators have difficulty maintaining the proper working pressure under high flow rates, **two-stage regulators** were developed. Two-stage regulators have two sets of springs, diaphragms, and valves, Figure 17-4. The first spring is preset at the factory to reduce the **cylinder pressure** to 225 psig (0.35 kg/cm^{2}g). The second spring is adjusted like other regulators. Because the second high-pressure valve has to control a maximum pressure of only around 225 psig (0.35 kg/cm^{2}g), it can be larger, thus allowing a greater flow.

Pressure Gauges

There may be one or two pressure gauges on a regulator. The working pressure **gauge** shows the regulated pressure being controlled for the torch, and the cylinder pressure gauge shows the remaining cylinder pressure, Figure 17-5. The working pressure gauge shows the pressure at the regulator, which is not exactly the same as the actual pressure at the torch tip. The torch tip pressure is always a little less than the working pressure due to some **line drop**. Line pressure drop is the greater on small diameter lines and on longer line lengths than it is on larger diameter and shorter line lengths, Table 17-1. The line drop is caused by the resistance of a gas as it flows through a line.

NOTE: The high-pressure gauge on a regulator shows cylinder pressure only. This gauge may be used to indicate the amount of gas that remains in a cylinder. However, the cylinder pressure of liquefied gases, such as CO_2, propane, and methylacetylene-propadiene (MPS), will remain fairly constant as the gas is used as long as some liquid remains in the cylinder. Therefore, the cylinder pressure on these cylinders cannot be used to determine the quantity of remaining gases. The only way to accurately determine the amount of remaining gases in these cylinders is to weigh them.

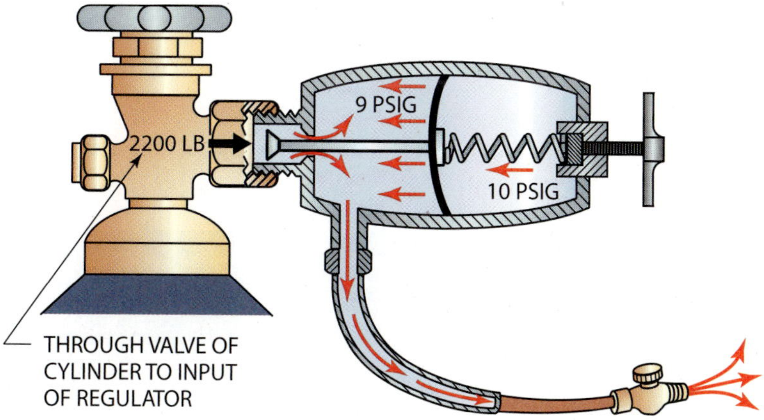

FIGURE 17-3 A drop in the working pressure occurs when the torch valve is opened and gas flows through the regulator at a constant pressure. © Cengage Learning 2012

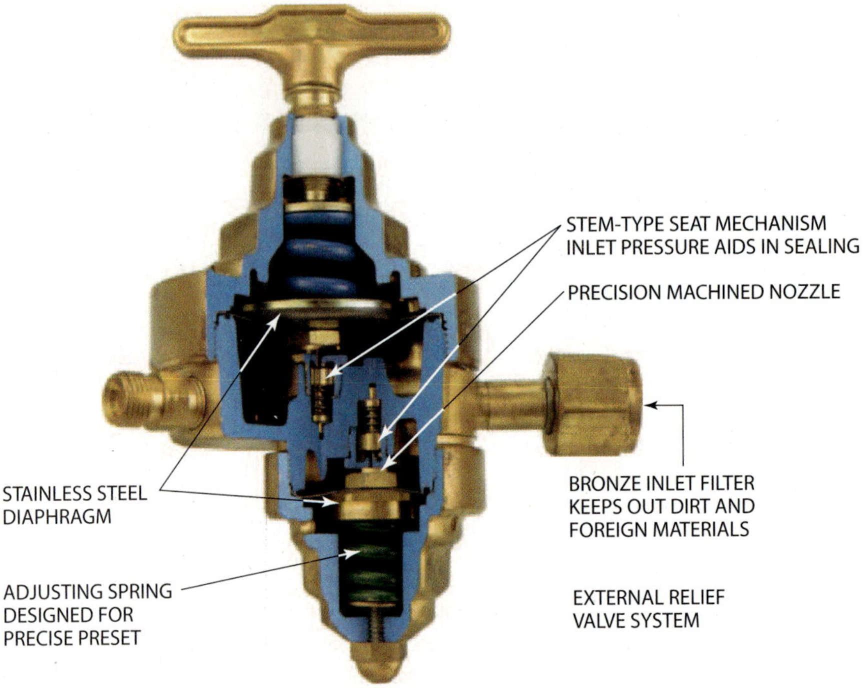

FIGURE 17-4 Two-stage oxygen regulator. Thermadyne Industries, Inc.

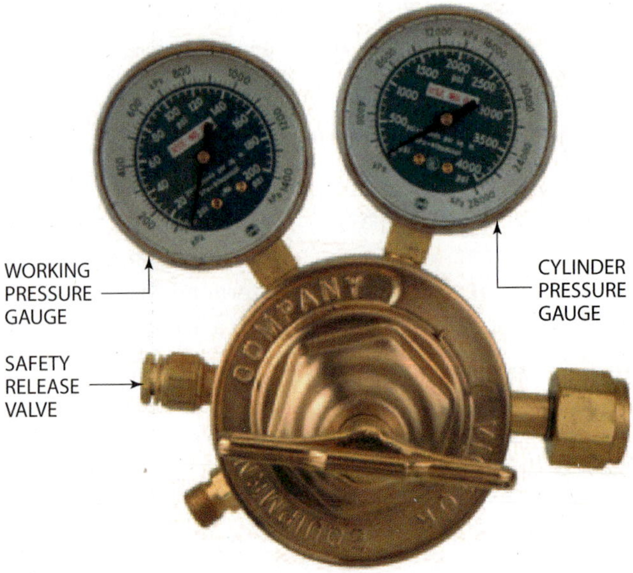

FIGURE 17-5 Safety release valve on an oxygen regulator. Thermadyne Industries, Inc.

Tip Pressure psig (kg/cm^2 G)	Regulator Pressure* for Hose Lengths ft (m)				
	10 ft (3 m)	25 ft (7.6 m)	50 ft (15.2 m)	75 ft (22.9 m)	100 ft (30.5 m)
1 (0.1)	1 (0.1)	2.25 (0.15)	3.5 (0.27)	4.75 (0.35)	6 (0.4)
5 (0.35)	5 (0.35)	6.25 (0.4)	7.5 (0.52)	8.75 (0.6)	10 (0.7)
10 (0.7)	10 (0.7)	11.25 (0.75)	12.5 (0.85)	13.75 (0.95)	15 (1.0)

*These values are for hoses with a diameter of 1/4 in. (6 mm); larger or smaller hose diameters or high flow rates will change these pressures.

Table 17-1 Regulator Pressure for Various Lengths of Hose

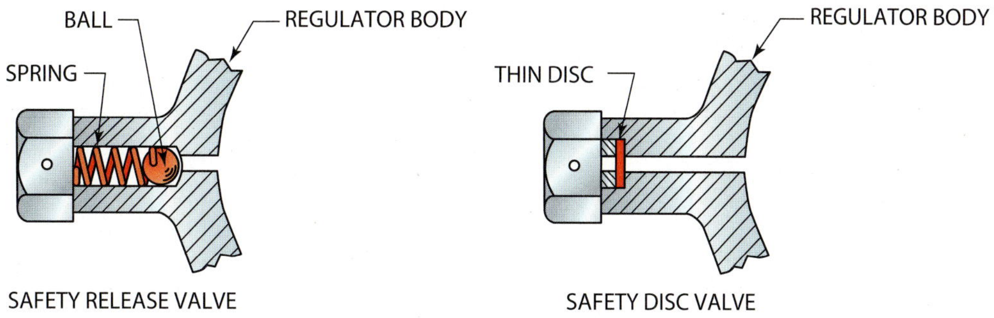

FIGURE 17-6 Pressure release valves. © Cengage Learning 2012

Regulator Safety Pressure Releases

Regulators may be equipped with either a safety release valve or a safety disk to prevent excessively high pressures from damaging the regulator. A **safety release valve** is made up of a small ball held tightly against a **seat** by a spring. Excessively high pressures will push the ball back away from the seat, safely releasing potentially explosive pressure. The release valve will reseat itself after the excessive pressure has been reduced.

A **safety disc** is a thin piece of metal held between two seals, Figure 17-6. Excessively high pressures will cause this disk to rupture or burst, releasing the pressure. Rupture disk safety devices are one-time-use devices. They release all of the cylinder pressure once ruptured; they do not reseat after the pressure is lowered to a safe level as the safety release valves do. If a rupture disk opens, it must then be replaced before the regulator can be used again.

Regulator Safety Practices

The regulator pressure-adjusting screw should be backed off each time the oxyfuel system is being shut down. This is done to release the spring and diaphragm pressures, which, over time, may cause damage. Keeping a spring compressed and the diaphragm stretched can cause the spring to weaken and the diaphragm to be permanently distorted.

In addition, when the cylinder valve is reopened, some high-pressure gas can pass by the open high-pressure valve before the diaphragm can close it. This condition may cause the diaphragm to rupture, the low-pressure gauge to explode, or both.

High-pressure valve seats that leak result in a **creep** or rising pressure on the working side of the regulator. This usually occurs when the gas pressure is set but no gas is flowing. If the leakage at the seat is severe, the maximum safe pressure can be exceeded on the working side, resulting in damage to the diaphragm, gauge, hoses, or other equipment.

CAUTION

Regulators that creep excessively or beyond the safe working pressure must not be used.

A gauge that gives a faulty reading or that is damaged can result in dangerous pressure settings. Gauges that do not read "0" (zero) pressure when the pressure is released, or those that have a damaged glass or case, must be repaired or replaced.

CAUTION

All work on regulators must be done by properly trained repair technicians.

CAUTION

Regulators should be located far enough from the actual work that flames or sparks cannot reach them.

Leak Detection

A **leak-detecting solution** should be purchased premixed. Leak-detecting solution must be free flowing so that it can seep into small joints, cracks, and other areas that may have a leak. The solution must produce a good quantity of bubbles without leaving a film. The solution can be dipped, sprayed, or brushed on the joints.

CAUTION

Some detergents are not suitable for O_2 because of an oil base. Use only O_2-approved leak-detection solutions on oxygen fittings.

Regulator Fittings

Both regulators and gas cylinder fittings have a variety of different designs for various types of gases to ensure that regulators cannot be connected to the wrong gas or pressure, Figure 17-7. A few adapters are available that will allow some regulators to be attached to a different type of fitting, provided that the gas type and pressure are similar. For example, adapters are available for an external left-hand acetylene regulator fitting to an internal right-hand cylinder fitting, or vice versa. Both of these cylinders contain a low-pressure fuel gas. Other adapters are available that will allow argon or mixed gas regulators to fit a flat washer-type CO_2 cylinder fitting. Both of these cylinders contain a high-pressure inert or semi-inert gas, Figure 17-8.

The connections to the cylinder and to the hose must be kept free of dirt and oil. Fittings should screw together freely by hand and require only light wrench pressure to be leak tight. If the fitting does not tighten freely on the connection, both parts should be cleaned. If the joint leaks after it has been tightened with a wrench, the seat should be checked. Examine the seat and threads for damage. If the seat is damaged, it can be repaired by a manufacturer-authorized regulator repair shop. Severely damaged connections must be replaced.

The outlet connection on a regulator is either a right-hand fitting for oxygen or a left-hand fitting for fuel gases. A left-hand threaded fitting has a notched nut, Figure 17-9.

Regulator Use and Servicing

There are no internal or external moving parts on a regulator or a gauge that require oiling, Figure 17-10.

(A) (B) (C) (D)

FIGURE 17-7 (A) Acetylene cylinder valve (left-hand thread). (B) Oxygen cylinder valve. (C) Argon cylinder valve. (D) Carbon dioxide (CO_2) cylinder valve. Larry Jeffus

(A)

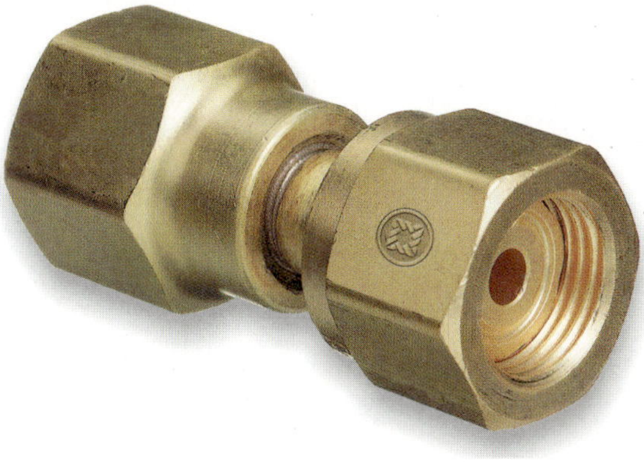

(B)

FIGURE 17-8 Cylinder to regulator adapter. Larry Jeffus

FIGURE 17-9 Left-hand threaded fittings are identified with a notch. Larry Jeffus

CAUTION

Oiling a regulator is unsafe and may cause a fire or an explosion.

FIGURE 17-10 Never oil a regulator. Thermadyne Industries, Inc.

If the adjusting screw becomes tight and difficult to turn, it can be removed and cleaned with a dry, oil-free rag. When replacing the adjusting screw, be sure it does not become cross-threaded. Many regulators use a nylon nut in the regulator body, and the nylon is easily cross-threaded.

When welding is finished and the cylinders are turned off, the gas pressure must be released and the adjusting screw backed out. This is required both by federal regulation and to prevent damage to the diaphragm, gauges, and adjusting spring if they are left under a load. A regulator that is left pressurized causes the diaphragm to stretch, the **Bourdon tube** to straighten, and the adjusting spring to compress. These changes result in a less accurate regulator with a shorter life expectancy.

DESIGN AND SERVICE OF WELDING AND CUTTING TORCHES

The oxyacetylene hand torch is the most common type of **oxyfuel gas torch** used for welding and cutting. The hand torch can be purchased as either a **combination welding and cutting torch** or a **cutting torch** only, Figure 17-11 and Figure 17-12. The

FIGURE 17-11 A torch body or handle used for welding or cutting. Thermadyne Industries, Inc.

FIGURE 17-12 A torch used for cutting only. © Cengage Learning 2012

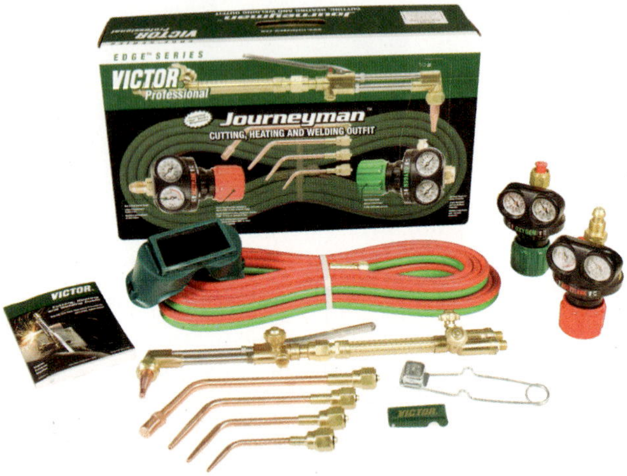

FIGURE 17-13 A combination welding and cutting torch kit. Thermadyne Industries, Inc.

combination welding and cutting torch offers more flexibility because a cutting head, welding tip, brazing tip, or heating tips can be attached to the same torch body, Figure 17-13. Combination torch sets are often used in schools, automotive repair shops, auto body shops, small welding shops, or any other situation where flexibility is needed. The combination torch sets are usually more practical for portable welding since the one unit can be used for both cutting and welding.

Dedicated cutting torches are usually longer than combination torches. The longer length helps keep the operator farther away from heat and sparks. In addition, thicker material can be cut with greater comfort.

Most manufacturers make torches in a variety of sizes for different types of work. There are small torches for jewelry work and large torches for heavy plates. Specialty torches for heating, brazing, or soldering are also available. Some of these use a fuel-air mixture. Fuel-air torches are often used by plumbers and air-conditioning technicians for brazing and soldering copper pipe and tubing.

There are no industrial standards for tip size identification, tip threads, or seats. Therefore, each style, size, and type of torch can be used only with the tips made by the same manufacturer to fit the specific torch.

Torch Care and Use

The torch body contains threaded connections for the hoses and tips. These connections must be protected from any damage. Most torch connections are external and made of soft brass that is easily damaged. Some connections, however, are more protected because they have either internal threads or stainless steel threads for the tips. The best protection against damage and dirt is to leave the tip and hoses connected when the torch is not in use. Because the hose connections are close to each other, a wrench should never be used on one nut unless the other connection is protected with a hose-fitting nut, Figure 17-14.

The hose connections should not leak after they are tightened with a wrench. If leaks are present, the seat should be repaired or replaced, Figure 17-15. The valves should be easily turned on and off and should stop all gas flowing with minimum finger pressure.

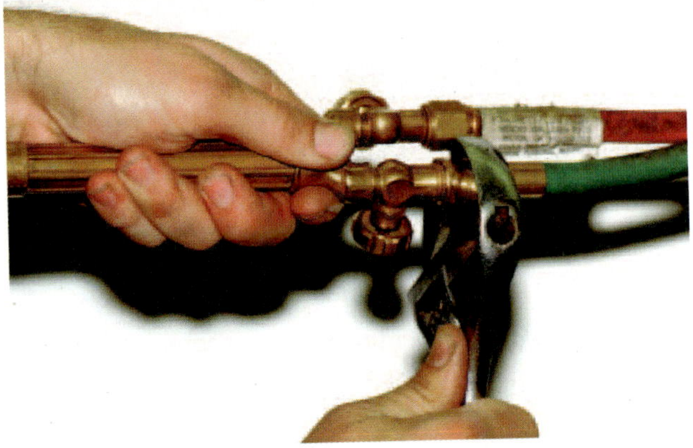

FIGURE 17-14 One hose fitting will protect the threads when the other nut is loosened or tightened. Larry Jeffus

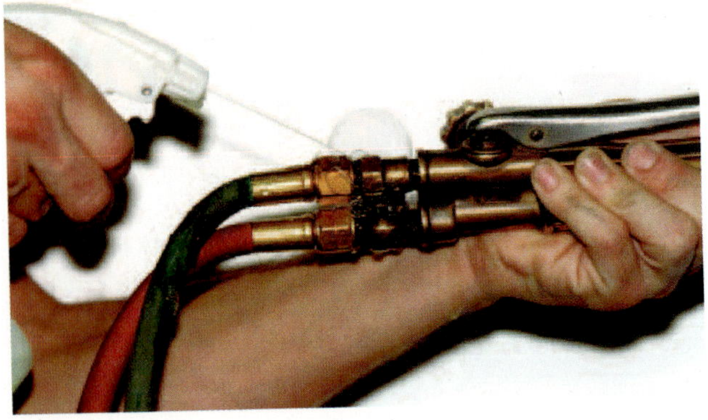

FIGURE 17-15 The torch valves should be checked for leaks, and the valve packing nut should be tightened if necessary. Larry Jeffus

To find leaking valve seats, set the regulators to a working pressure. With the torch valves off, spray the tip with a leak-detecting solution. The presence of bubbles indicates a leaking valve seat.

No gas should leak past the valve stem packing. To test for leaks around the valve stem, set the regulator to a working pressure. With the valves off, spray the valve stem with a leak-detecting solution and watch for bubbles, indicating a leaking **valve packing**. The valve stem packing can now be tested with the valve open. Open the valve and spray the stem and watch for bubbles, which would indicate a leaking valve packing. If either test indicates a leak, the valve stem packing nut can be tightened until the leak stops. After the leak stops, turn the valve knob. It should still turn freely. If it does not or if the leak cannot be stopped, replace the valve packing.

The valve packing and valve seat can be easily repaired on most torches by following the instructions given in the repair kit. On some torches, the entire valve assembly can be replaced if necessary.

WELDING AND HEATING TIPS

Because no industrial standard tip size identification system exists, the student must become familiar with the size of the orifice (hole) in the tip. The larger the diameter of the hole in a welding tip, the higher the heating capacity. The heating capacity of a tip determines the thickness range it can be used on. Comparing the overall size of tips and, therefore, their heating capacity can be done only for tips made by the same manufacturer for the same type and style of torch, Figure 17-16. Learning a specific manufacturer's system is not always the answer because on older, worn tips the orifice may have been enlarged by repeated cleaning.

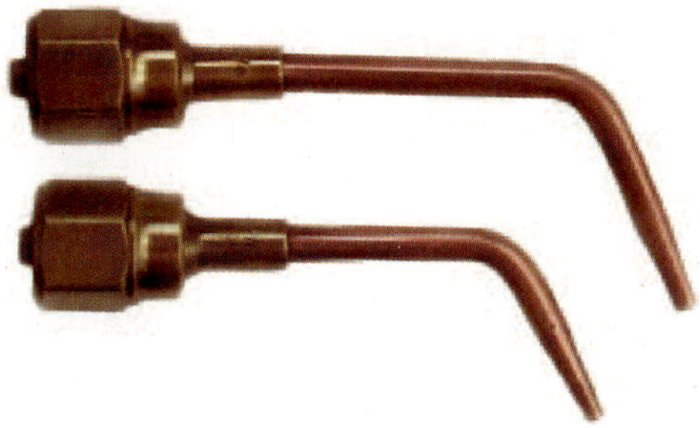

FIGURE 17-16 A variety of tip styles and sizes for one torch body. Larry Jeffus

Tip sizes can be compared to the numbered drill size used to make the hole, Table 17-2. The sizes of tip cleaners are given according to the drill size of the hole they fit. By knowing the tip cleaner size commonly used to clean a tip, the welder can find the same-size tip made by a different manufacturer. The tip size can also be determined by trial and error.

Tip Cleaner Standard Set			
	Use Cleaner	For Drill	
Smallest	1	77-76	77 = .0160" (0.4064 mm)
↓	2	75-74	
	3	73-72-71	
	4	70-69-68	
	5	67-66-65	
	6	64-63-62	
	7	61-60	
	8	59-58	
	9	57	
	10	56	
	11	55-54	
	12	53-52	
Largest	13	51-50-49	49 = .0730" (1.8542 mm)

Table 17-2 Tip Cleaner Size Compared to Drill Size as Found on Most Standard Tip-Cleaning Sets

Tip Care and Use

Torch tips may have metal-to-metal seals, or they may have O rings or gaskets between the tip and the torch seat. Metal-to-metal seal tips must be tightened with a wrench. Tips with an O ring or a gasket are tightened by hand. Follow the torch manufacturer's recommendations. Using the wrong method of tightening the tip fitting may result in damage to the torch body or the tip.

Dirty welding and cutting tips can be cleaned using a set of tip cleaners or tip drills, Figure 17-17. Using the file provided in the tip-cleaning set, file the end of the tip smooth and square, Figure 17-18. Next, select the size of tip cleaner that fits easily into the orifice. The tip cleaner is a small, round file and should be moved in and out of the orifice only a few times, Figure 17-19. Be sure the tip cleaner is straight and that it is held in a steady position to prevent it from bending or breaking off in the tip. Excessive use of the tip cleaner tends to ream the orifice, making it too large. Therefore, use the tip cleaner only as required. Once the tip is cleaned, turn on the oxygen for a moment to blow out any material loosened during the cleaning.

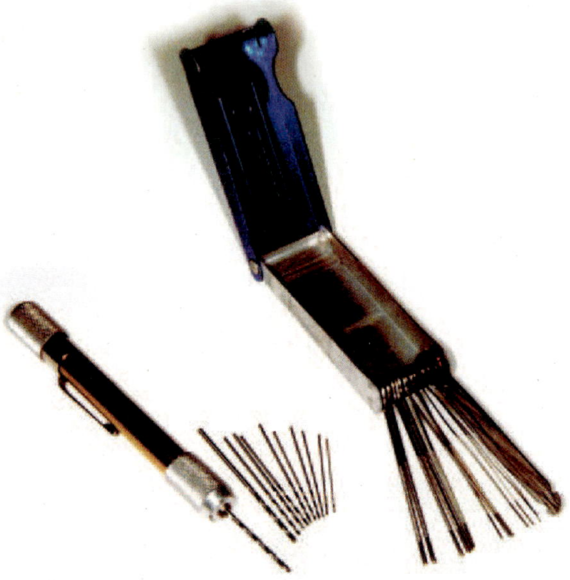

FIGURE 17-17 Tools used to repair tips. Larry Jeffus

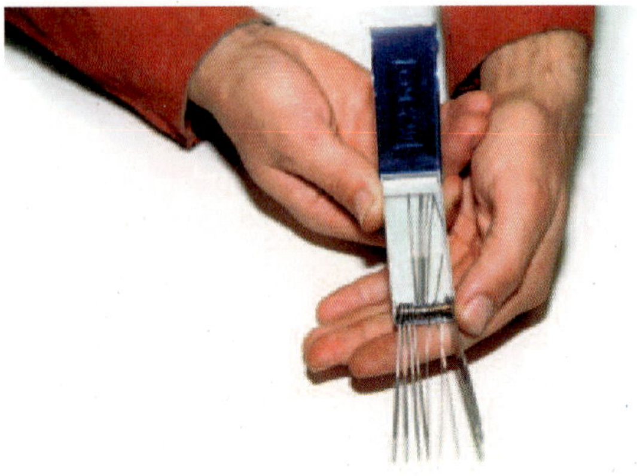

FIGURE 17-18 Standard set of tip cleaners. Larry Jeffus

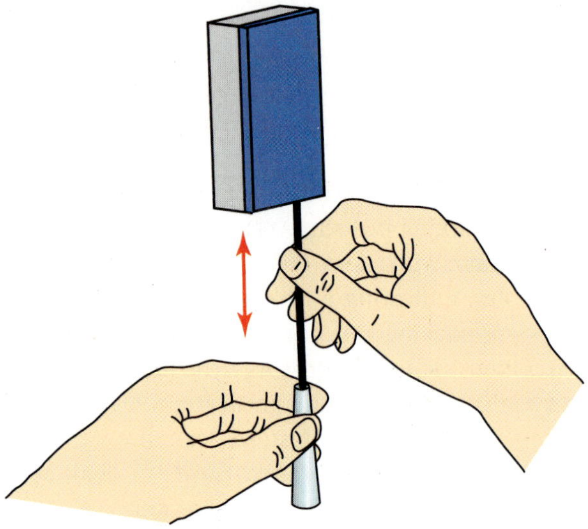

FIGURE 17-19 Cleaning a tip with a standard tip cleaner. © Cengage Learning 2012

Backfires

A **backfire** occurs when a flame goes out with a loud snap or pop. A backfire may be caused by the following:

- Touching the tip against the workpiece
- Overheating the tip
- Operating the torch when the flame settings are too low
- Loose tip
- Damaged seats
- Dirt in the tip

The problem that caused the backfire must be corrected before relighting the torch. A backfire may cause a flashback.

Flashbacks

A **flashback** occurs when the flame burns back inside the tip, torch, hose, or regulator. A flashback produces a high-pitched whistle. If the torch does flashback, close the welding torch oxygen valve at once, and then close the torch fuel valve. The order in which the valves are closed is not as important as the speed at which they are closed. A flashback that reaches the cylinder may cause a fire or an explosion.

Closing the torch oxygen valve stops the flame inside at once. Then the fuel-gas valve should be closed and the torch must be allowed to cool off before repairing the problem. When a flashback occurs, there is usually a serious problem with the equipment, and a qualified technician should be called. After locating and repairing the problem, blow gas through the tip for a few seconds to clear out any soot that may have accumulated in the passages. A flashback that burns in the hose leaves carbon deposits inside that may explode and burn in a pressurized oxygen system. Replace any hose that has been damaged by a flashback.

Reverse Flow Valves

The purpose of the reverse flow valve is to prevent gases from accidentally flowing out of one hose through the torch body and then into the other hose. If the oxygen and fuel gases are allowed to mix in the hose or regulator, they might explode. The reverse flow valve is a spring-loaded check valve that closes when gas pressure from a backflow tries to occur through the torch valves, Figure 17-20. Some torches have reverse flow valves built into the torch body, but most torches must have these safety devices added. If the torch does not come with a reverse flow valve, it must be added to either the torch end or regulator end of the hose.

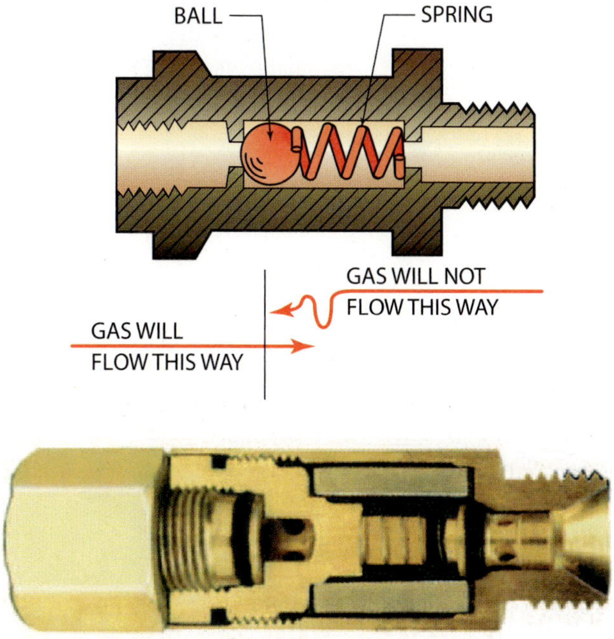

FIGURE 17-20 Reverse flow valve only.
© Cengage Learning 2012

A reverse flow of gas can occur through the torch if it is not turned off or the pressure is not properly bled off. When bleeding off the gas pressure, open one valve at a time so that the gas pressure in that hose will be vented into the atmosphere and not through the torch into the other hose, Figure 17-21. A reverse flow valve will not stop the flame of a flashback from continuing through the hoses.

CAUTION

If both valves are opened at the same time, one gas may be pushed back up the hose of the other gas.

Flashback Arrestors

A **flashback arrestor** will stop both reverse gas flow and the flame of a flashback, Figure 17-22. The flashback arrestor is designed to quickly stop the flow of gas during a flashback. These valves work on a similar principle as the gas valve at a service station. They are very sensitive to any back pressure in the hose and stop the flow if any back pressure is detected.

Servicing the Reverse Flow Valve and Flashback Arrestor

Both devices must be checked on a regular basis to see that they are working correctly. The internal valves may become plugged with dirt or they may become sticky and not operate correctly. To test the reverse flow valve, you can try to blow air backward through the valve. To test the flashback arrestor, follow the manufacturer's recommended procedure. If the safety device does not function correctly, it should be replaced.

Hoses and Fittings

Welding hoses are molded together as one piece. Fuel gas hoses are red and have left-hand threaded fittings. Oxygen hoses are green and have right-hand threaded fittings. Hoses are available in four sizes: 3/16 in. (4.7 mm), 1/4 in. (6 mm), 5/16 in. (8 mm), and 3/8 in. (10 mm). The size given is the inside diameter of the

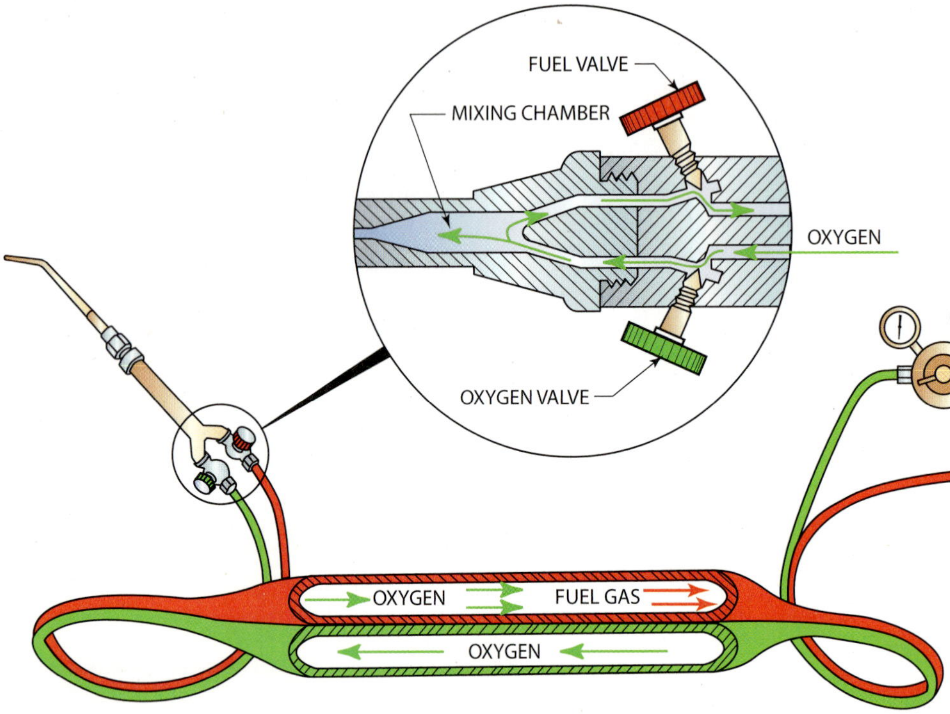

FIGURE 17-21 Gas may flow back up the hose if both valves are opened at the same time when the system is being bled down after use. Installing reverse flow valves on the torch can prevent this from occurring. ESAB Welding & Cutting Products

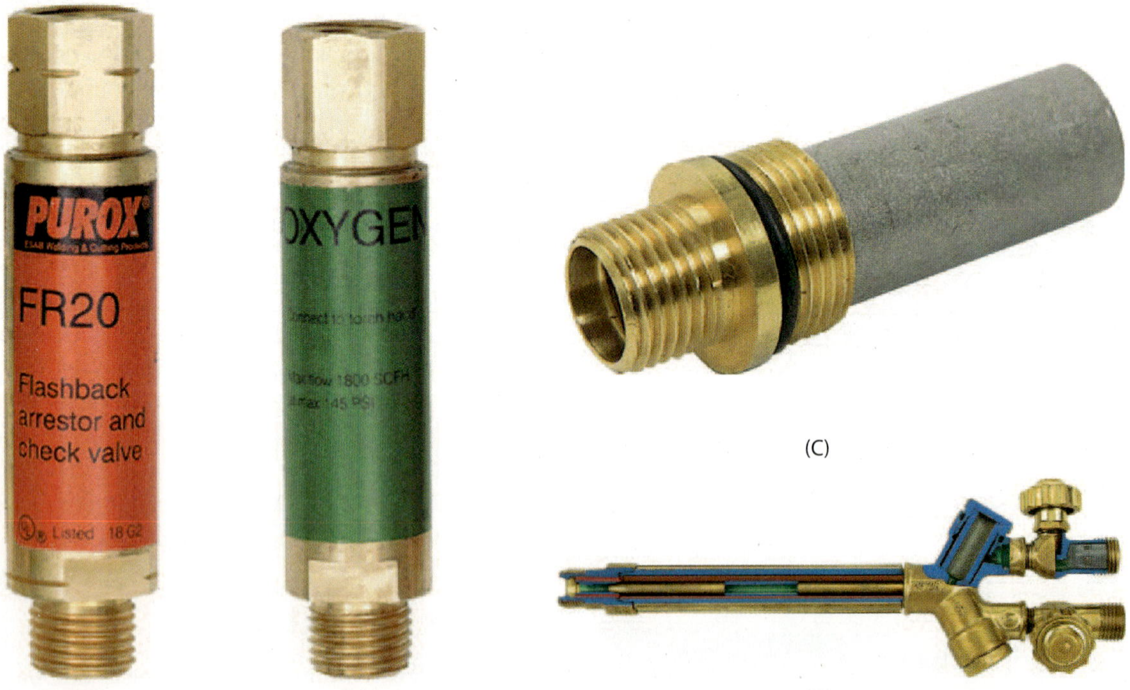

FIGURE 17-22 **(A) Acetylene.** ESAB Welding & Cutting Products **(B) Oxygen combination flashback arrestors and check valves.** ESAB Welding & Cutting Products **(C) Replacement cartridge for flashback arrestor.** ESAB Welding & Cutting Products **(D) Torch designed with flashback arrestors and check valves built into the torch body.** Thermadyne Industries, Inc.

hose. Larger diameter hoses offer less resistance to gas flow and should be used when long hose lengths are required. The smaller sizes are more flexible and easier to handle for detailed work.

The three sizes of hose end fittings available are A (small), B (standard), and C (large). The three sizes are made to fit all hose sizes.

Hose Use and Servicing

When hoses are not in use, the gas must be turned off and the pressure bled off. Turning off the equipment and releasing the pressure prevents any undetected leaks from causing a fire or an explosion. Turning the system off and bleeding off the pressure eliminates the danger of a fire or explosion if the hoses are accidentally damaged or cut while not in use and you are not there to quickly shut off the gas.

Hoses are resistant to burns, but they are not burn proof. They should be kept out of direct flame and sparks and away from hot metal. You must be especially cautious when using a cutting torch. If the hose becomes damaged, the damaged section should be removed and the hose repaired with a splice. Damaged hoses should never be taped to stop leaks. Hoses should be checked periodically for leaks. To test a hose for leaks, adjust the regulator to a working pressure with the torch valves closed. Wet the hose with a leak-detecting solution by rubbing it with a wet rag, spraying it, or dipping it in a bucket. Then watch for bubbles, which indicate that the hose leaks.

The hose fittings can be replaced if they become damaged. Several kits are available that have new nuts, nipples, ferrules, a ferrule crimping tool, and any other supplies required to replace the hose ends, Figure 17-23. To replace the hose end, the hose is first cut square. The correct-size ferrule is inserted. Then both the hose end and nipple are sprayed with a leak-detecting solution. This will help the nipple slide in more easily. Screw the nipple and nut on a torch body. This will hold the nipple deep inside the nut, and the body will act as a handle for leverage as the nipple is pushed inside the hose, Figure 17-24. After the hose is slid up to the nut, crimp the ferrule until it is tight. The crimping tool should be squeezed twice, the second time at right angles to the first, Figure 17-25. When the crimping is complete, install the hose on a torch and regulator. Then adjust the regulator to a working pressure and spray the fitting with a leak-detecting solution. Watch for any bubbles, which indicate a leaking fitting.

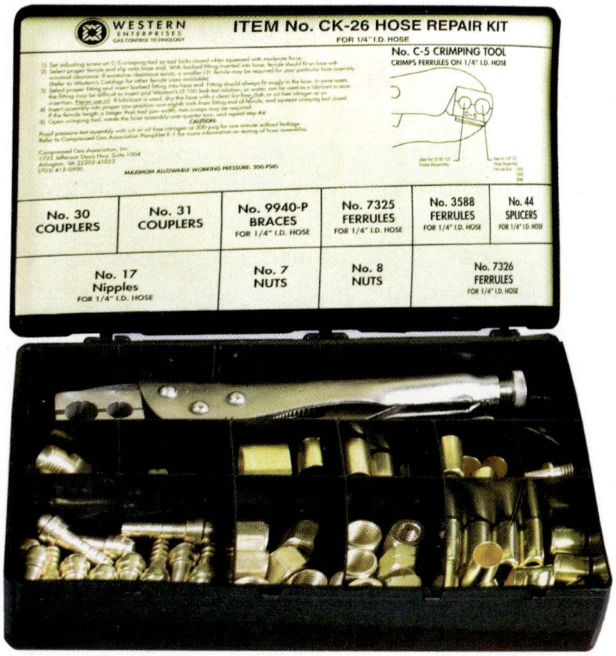

FIGURE 17-23 Hose repair kit. Western Enterprises, a Scott Fetzer Company

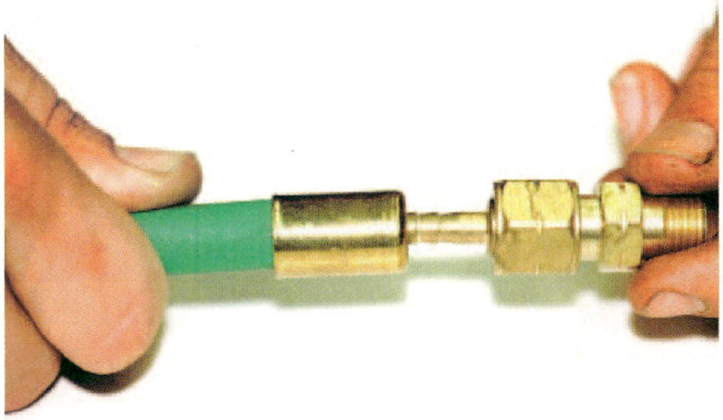

FIGURE 17-24 Screwing the hose nut onto a fitting will help when pushing the nipple into the hose. Larry Jeffus

FIGURE 17-25 Crimping hose ferrule. Larry Jeffus

OXYFUEL EQUIPMENT SETUP AND OPERATION

It is always recommended that you read and follow the equipment manufacturer's safety and operation manual before using oxyfuel equipment for the first time. Copies of these manuals are available from the manufacturer, supply houses, and/or on the Internet. The following method of assembling, testing, lighting, and adjusting the flame and disassembling an oxyfuel welding set is designed to guide you safely through the process.

Setting Up an Oxyfuel Torch Set

The following steps can be used to assemble almost any manufacturer's oxyfuel equipment if the original manual is not available.

1. Safety chain the cylinders separately to the cart or to a wall, Figure 17-26. Then remove the valve protection caps, Figure 17-27.
2. Crack the cylinder valve on each cylinder for a second to blow away dirt that may be in the valve, Figure 17-28.

CAUTION

If a fuel-gas cylinder does not have a valve hand wheel permanently attached, you must use a non-adjustable wrench to open the cylinder valve. The wrench must stay with the cylinder as long as the cylinder is on, Figure 17-29.

FIGURE 17-26 Safety chain cylinder. Larry Jeffus

CAUTION

Always stand to one side. Point the valve away from anyone in the area and be sure there are no sources of ignition when cracking the valve.

3. Attach the regulators to the cylinder valves, Figure 17-30A. The nuts can be started by hand and then tightened with a wrench, Figure 17-30B.
4. Attach a reverse flow valve or flashback arrestor, if the torch does not have them built in, Figure 17-31. Occasionally test each reverse

FIGURE 17-27 Unscrew the valve protector caps. Put the caps in a safe place as they must be replaced on empty cylinders before they are returned. Larry Jeffus

FIGURE 17-28 Cracking the oxygen and fuel cylinder valves to blow out any dirt lodged in the valves. Larry Jeffus

(A)

(B)

(C)

FIGURE 17-29 **(A) Small combination wrench. (B) Large combination wrench. (C) T-wrench.** ESAB Welding & Cutting Products

flow valve by blowing through it to make sure it works properly and follow the manufacturer's recommendation on proper flash arrester maintenance.

5. Connect the hoses. The red hose has a left-hand grooved nut and attaches to the fuel-gas regulator.

(A)

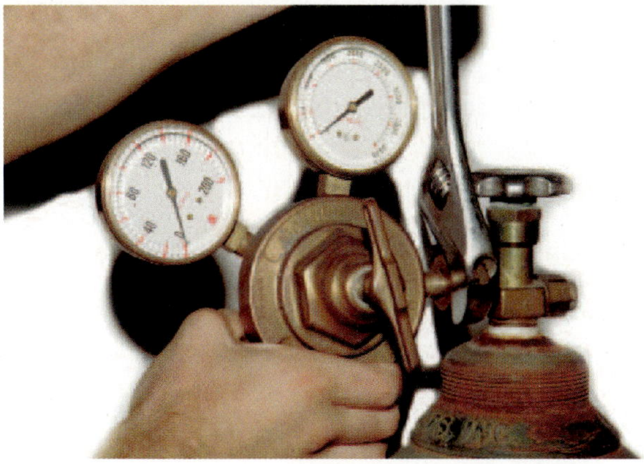

(B)

FIGURE 17-30 **Attach the oxygen regulator (A) to the oxygen cylinder valve. Using a wrench (B), tighten the nut.** Larry Jeffus

FIGURE 17-31 **Attach reverse flow valves.** Larry Jeffus

The green hose has a right-hand nut without grooves and attaches to the oxygen regulator.

6. Attach the torch to the hoses, Figure 17-32. Connect both hose nuts finger tight before using a wrench to tighten either one.

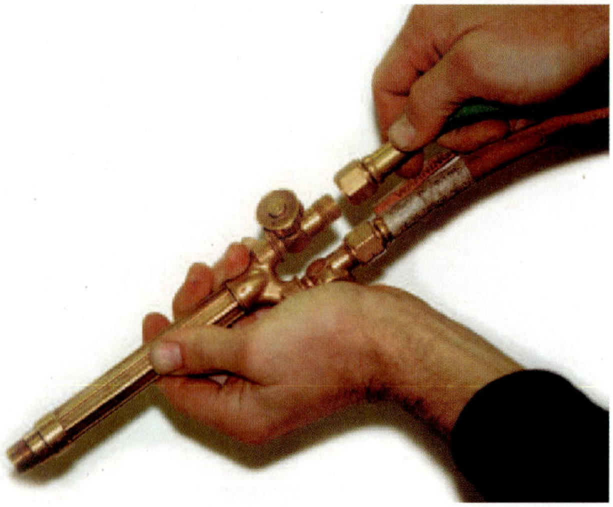

FIGURE 17-32 Connect the free ends of the oxygen (green) and the acetylene (red) hoses to the welding torch. Larry Jeffus

FIGURE 17-33 Select the proper tip or nozzle and install it on the torch body. Larry Jeffus

7. Check the tip seals for nicks or O rings, if used, for damage. In most cases tips that have O ring–type seals are hand tightened, and tips that have metal-to-metal seals are wrench tightened, but it is best to check the owner's manual, or a supplier, to determine if the torch tip should be tightened, Figure 17-33.

CAUTION

Tightening a tip the incorrect way may be dangerous and might damage the equipment.

Turning On and Testing Oxyfuel Welding Equipment

The following steps can be used to turn on and test any manufacturer's oxyfuel equipment.

1. Back out the regulator pressure adjusting the screws until they are loose, Figure 17-34.
2. Standing to one side of the regulator, open the cylinder valve slowly so that the pressure rises on the gauge slowly, Figure 17-35.

FIGURE 17-34 Back out both regulator adjusting screws before opening the cylinder valve. Larry Jeffus

FIGURE 17-35 Stand to one side when opening the cylinder valve. Larry Jeffus

CAUTION

If the valve is opened quickly, the regulator or gauge may be damaged, or the gauge may explode.

Open the oxygen valve all the way until it is sealed at the top, Figure 17-36.

3. Open the acetylene or other fuel-gas valve one quarter turn, or just enough to get gas pressure, Figure 17-37. If the cylinder valve does not have a hand wheel, use a nonadjustable wrench and leave it in place on the valve stem while the gas is on.

CAUTION

The acetylene valve should never be opened more than one and a half turns so that in an emergency it can be turned off quickly.

Open one torch valve and point the tip away from any source of ignition, including the cylinders, regulators, and hoses. Slowly turn in the pressure adjusting screw until gas can be heard escaping from the torch. The gas should flow long enough to allow the hose to be completely **purged** (emptied) of air and replaced by the gas before the torch valve is closed. Repeat this process with the other gas.

4. After purging is completed, and with both torch valves off, adjust both regulators to read 5 psig, Figure 17-38.
5. Spray a leak-detecting solution on each hose and regulator connection and on each valve stem on the torch and cylinders. Watch for bubbles, which indicate a leak. Turn off the cylinder valve before tightening any leaking connections, Figure 17-39.

CAUTION

Connections should not be overtightened. If they do not seal properly, repair or replace them.

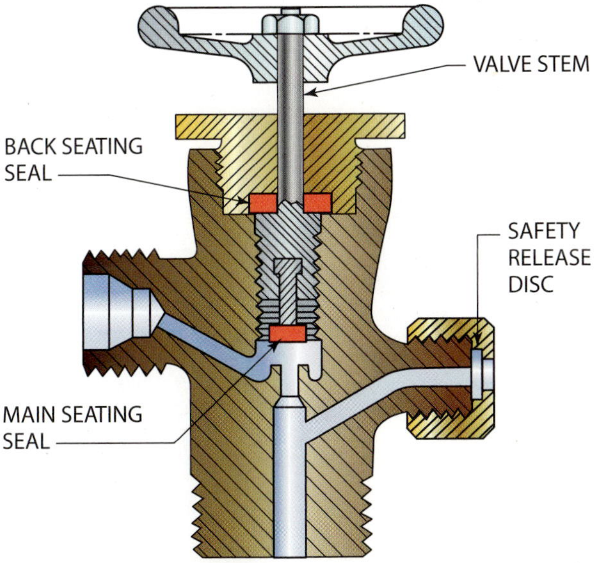

FIGURE 17-36 Cutaway of an oxygen cylinder valve showing the two separate seals. The back seating seal prevents leakage around the valve stem when the valve is open. Larry Jeffus

FIGURE 17-37 Open the cylinder valve slowly. Larry Jeffus

FIGURE 17-38 Adjust the regulator to read 5-psig working pressure. Larry Jeffus

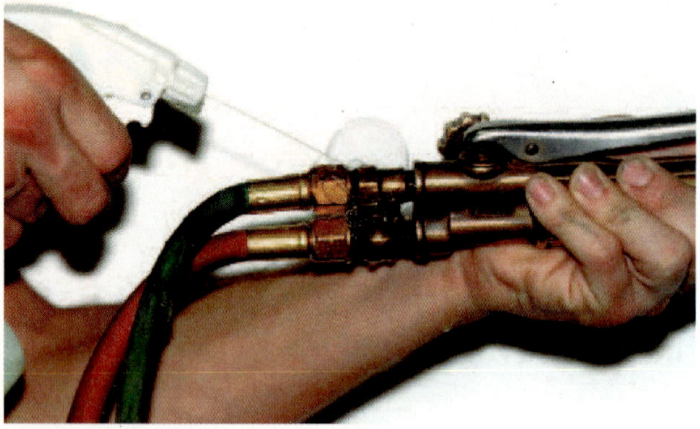

FIGURE 17-39 Spray fittings with a leak-detecting solution. Larry Jeffus

CAUTION

Leaking cylinder valve stems should not be repaired. Turn off the valve, disconnect the cylinder, mark the cylinder, move it outdoors or to a very well-ventilated area, and notify the supplier to come and pick up the bad cylinder, Figure 17-40.

Types of Flames

There are three distinctly different oxyacetylene flame settings. A **carburizing flame** has an excess of fuel gas. This flame has the lowest temperature and may put extra carbon in the weld metal. A **neutral flame** has a balance of fuel gas and oxygen. It is the most commonly used flame because it adds nothing to the weld metal. An **oxidizing flame** has an excess of oxygen. This flame has the highest temperature and may put oxides in the weld metal.

FIGURE 17-40 Identify any cylinder that has a problem by marking it. Larry Jeffus

Lighting and Adjusting an Oxyacetylene Flame

The following steps can be used to light and adjust manufacturers' oxyacetylene equipment.

1. Wearing proper clothing, gloves, and gas welding goggles, turn both regulator adjusting screws in until the working pressure gauges read 5 psig. If you mistakenly turn on more than 5 psig, open the torch valve to allow the pressure to drop as the adjusting screw is turned outward.
2. Turn on the torch fuel-gas valve just enough so that some gas escapes.

CAUTION

Be sure the torch is pointed away from any sources of ignition or any object or person that might be damaged or harmed by the flame when it is lit.

3. Using a spark lighter, light the torch. Hold the lighter near the end, Figure 17-41, of the tip but not covering the end, Figure 17-42.

CAUTION

A **spark lighter** is the only safe device to use when lighting any torch.

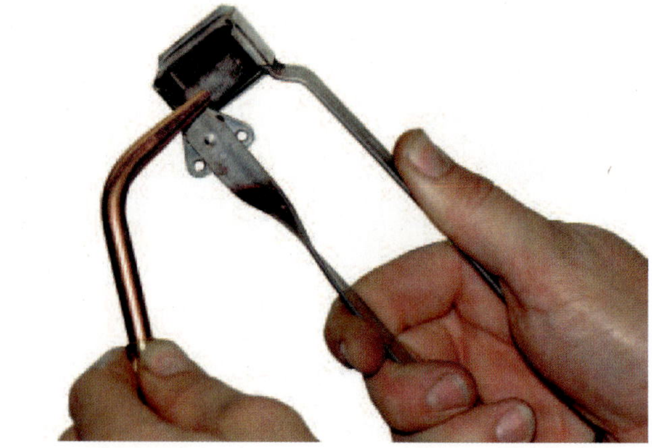

FIGURE 17-41 Correct position to hold a spark lighter. Larry Jeffus

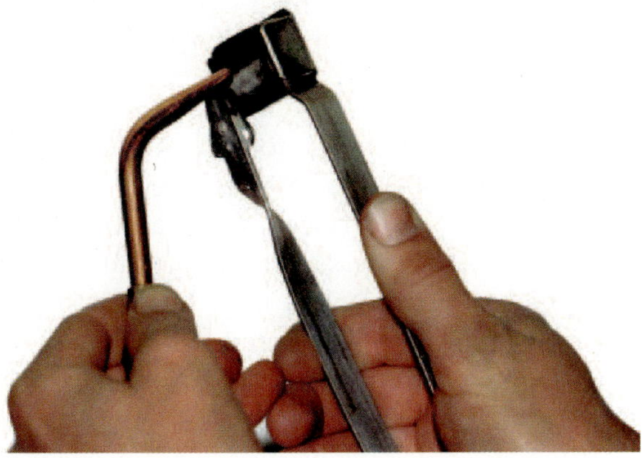

FIGURE 17-42 Spark lighter held too close over the end of the tip. Larry Jeffus

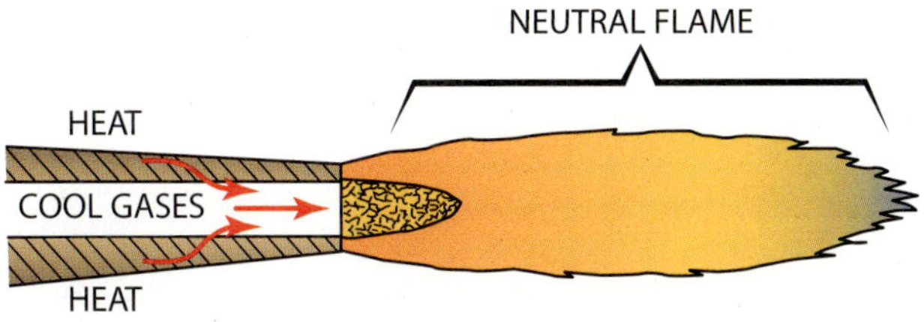

FIGURE 17-43 Enough cool gas flowing through the tip will help prevent popping. Larry Jeffus

4. With the torch lit, increase the flow of acetylene until the flame stops smoking.
5. Slowly turn on the oxygen and adjust the torch to a neutral flame.

This flame setting uses the minimum gas flow rate for this specific tip. The fuel flow should never be adjusted to a rate below the point where the smoke stops. This is the minimum flow rate at which the cool gases will pull the flame heat out of the tip, Figure 17-43. If excessive heat is allowed to build up in a tip, it can cause a backfire or flashback. The maximum gas flow rate gives a flame enough flow so that, when adjusted to the neutral setting, it does not settle back on the tip. This will help keep the tip cooler so it is less likely to backfire, Figure 17-44.

Shutting Off and Disassembling Oxyfuel Welding Equipment

The following steps can be used to shut down and disassemble any manufacturer's oxyfuel equipment.

1. First, quickly turn off the torch fuel-gas valve. This action blows the flame out and away from the tip, ensuring that the fire is out. In addition, it prevents the flame from burning back inside the torch. On large tips or hot tips, turning the fuel off first may cause the tip to pop. The pop is caused by a lean fuel mixture in the tip. If you find that the tip pops each time you turn the fuel off first, then do not turn off the fuel first, but turn the oxygen off first. Turning the oxygen off first will prevent the pop. Be sure that the flame is out before putting the torch down.

FIGURE 17-44 Explosion-proof light fixtures suitable for a manifold room. Cooper Crouse-Hinds®

2. After the flame is out, turn off the oxygen valve.
3. Turn off the cylinder valves.
4. Open one torch valve at a time to bleed off the pressure.
5. When all of the pressure is released from the system, close both torch valves and back both regulator adjusting screws out until they are loose.
6. Loosen both ends of both hoses and unscrew them.
7. Loosen both regulators and unscrew them from the cylinder valves.
8. Replace the valve protection caps.

SUMMARY

The oxygen and acetylene welding equipment and process have been around for hundreds of years. It was one of the first practical welding processes. Oxyacetylene welding was the only practical welding process for many years. During this time it was occasionally used to manufacture and repair metal sections several feet thick. The heat that welders had to endure making such large welds was almost unbearable. Fortunately, many other welding processes are available, so welders do not have to be nearly roasted alive to make oxyacetylene welds. Today, oxyfuel welding is used only on thin sheet metal.

Although changes in equipment and procedures have relegated this once popular welding process to the brink of extinction for commercial applications, it still flourishes in small shops, art studios, automotive repair services, home and farm shops, and so on. There are several reasons for it remaining popular. For example, none of the other welding systems can cut, weld, and heat using the same equipment. Another reason it is taught in schools is because learning it can serve as a great basis for learning other welding processes. Many welders and welding educators believe that learning oxyacetylene welding first makes learning other welding processes much easier.

REVIEW QUESTIONS

1. Why is acetylene the most popular fuel gas used for oxyfuel welding?
2. What do pressure regulators do?
3. List five various types of pressure regulators.
4. Why are two-stage regulators used sometimes and not a single-stage regulator?
5. Why is the pressure at the torch tip always lower than the pressure shown on the working pressure gauge?
6. When does the high-pressure regulator gauge not show an approximate quantity of remaining gas in the cylinder?
7. Which type of safety pressure release reseals once the excessive pressure has been released?
8. Why should the regulator pressure adjusting screw be backed off each time the oxyfuel system is being shut down?
9. What qualities must a leak-detecting solution have?
10. Why do gas cylinders have different design fittings?
11. Why is it unsafe to use oil on a regulator?
12. What is the advantage of using a combination welding and cutting torch?
13. What is the advantage of using a long dedicated cutting torch?
14. What should be done if the hose connections leak after they are tightened?
15. Can the size of a welding tip be used to determine the tip heating capacity?
16. What may occur if the wrong method of tightening the tip fitting is used?
17. How can dirty welding tips be cleaned?
18. List five possible causes of a welding tip backfiring?
19. What is a flashback?
20. What does a flashback arrestor do?
21. Why would you use a larger diameter welding hose?
22. Why must the welding hose be kept out of the direct flame and sparks and away from hot metal?
23. What must you read before using oxyfuel equipment for the first time?
24. When setting up an oxyfuel torch set, why should you crack the cylinder valve on each cylinder before you attach the regulators to the cylinder valves?
25. List the safety precautions for lighting and adjusting an oxyacetylene flame.

Chapter 18

Oxyacetylene Welding

OBJECTIVES

After completing this chapter, the student should be able to:

- Discuss the uses of oxyacetylene welding and its advantages and disadvantages.
- List factors that affect the weld.
- Discuss some commonly occurring problems associated with oxyacetylene welding.
- Explain what factors are affected by adjusting the flame on mild steel.
- Tell how changes in the torch angle and torch height affect the molten weld pool.
- Demonstrate tack welds and weld beads.
- Make welds on outside corner joints, butt joints, lap joints in the horizontal position, tee joints in the flat position, and vertical outside corner joints.

KEY TERMS

burnthrough
butt joint
flashing
horizontal weld
kindling temperature
lap joint
molten weld pool
out-of-position welding
outside corner joint
overhead weld
penetration
shelf
tee joint
torch angle
torch manipulation
trailing edge
vertical weld
weld crater

INTRODUCTION

The welds produced with an oxyacetylene torch can be as strong and durable as those made with any other process. For many years it was the only approved process for welding aircraft frames. Because other welding processes are seen as more modern, some people have discounted the need for gas welding and its effectiveness.

For small welding jobs and for hobbyists, gas welding is often the best welding method for thin metal and small parts because it is so easy to control the flame heat. When using a larger-size welding tip, you can easily join the ends of 1/4-in. bolts, and with a smaller tip, you can join the ends of 1/16-in.-diameter welding rods. An advantage of gas welding is that by changing the size of the tip you can change the welding speed. This will let you make the weld at a rate that is easy for you to control. Although the travel speed on other welding processes can be controlled, none has the range of speed control that gas welding has.

Often in agriculture, welding jobs need to be done away from the shop. Because a gas welding rig is so portable, it is frequently used to make these repair welds. Gas welding is used even when another welding process might have been the first choice but no electrical power is available. This means that you will be making some welds on material that is thicker than what you would normally weld with a torch. The same skills you will learn from this chapter for welding on thin sheet metal and small diameter tubing can be used to make welds on thicker material. The weld just takes longer to heat up, and it takes more time.

Gas welding can even be used on much thicker metal. For example, in the early part of the last century, very thick sections of metal—in some cases nearly a foot thick—were welded using gas welding.

Mild Steel Welds

Mild steel is the easiest metal to gas weld. With this metal, it is possible to make welds with 100% integrity (lack of porosity, oxides, or other defects), which have excellent strength, good ductility, and other positive characteristics. The secondary flame shields the molten weld pool from the air, which would cause oxidation. The atmospheric oxygen combines with the carbon monoxide (CO) from the outer flame envelope to produce carbon dioxide (CO_2). The carbon dioxide will not react with the molten weld pool. In addition, the carbon dioxide forces the surrounding atmosphere away from the weld.

Factors Affecting the Weld

Torch Tip Size

The torch tip size should be used to control the weld bead width, penetration, and speed. **Penetration** is the depth into the base metal that the weld fusion or melting extends from the surface, excluding any reinforcement. Because each tip size has a limited operating range in which it can be used, tip sizes must be changed to suit the thickness and size of the metal being welded. Never lower the size of the torch flame when the correct tip size is unavailable. An oxyacetylene welding tip may overheat if the flame is turned down below the minimum for that particular tip size. If the flame is lowered significantly, the tip can overheat even without using it to weld. The correct gas flow rate cools the tip and helps keep the flame heat off of the end of the tip. Overheated tips will backfire. Backfiring can cause dangerous flashback. Other factors that can be changed to control the weld size are the torch angle, the flame-to-metal distance, the welding rod size, and the way the torch is manipulated.

CAUTION

It is never safe to lower the flame size if the tip's flame produces too much heat for your welding job. Get a smaller tip or change your welding technique.

Torch Angle

The **torch angle** and the angle between the inner cone and the metal have a great effect on the speed of melting and size of the **molten weld pool**. The ideal angle for the welding torch is 45°. As this angle increases toward 90°, the rate of heating increases. As the angle decreases toward 0° to the plate's surface, the rate of heating decreases, as illustrated in Figure 18-1. The distance between the inner cone and the metal ideally should be 1/8 in. to 1/4 in. (3 mm to 6 mm). As this distance increases, the rate of heating decreases; as the distance decreases, the heating rate increases, Figure 18-2.

Welding Rod Size

Welding rod size and **torch manipulation** can be used to control the weld bead characteristics. A larger

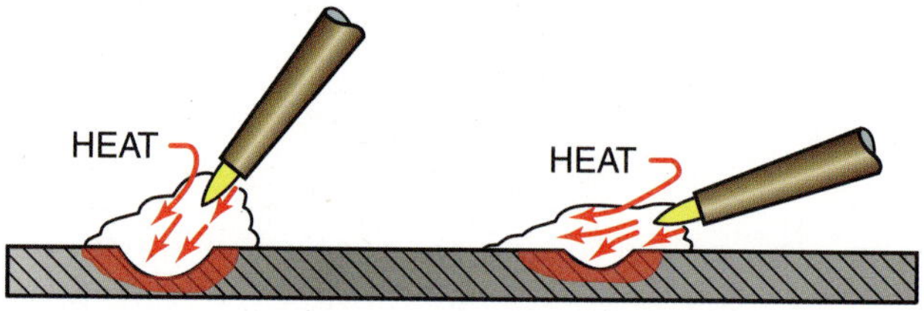

FIGURE 18-1 Changing the torch angle changes the percentage of heat that is transferred into the metal.
© Cengage Learning 2012

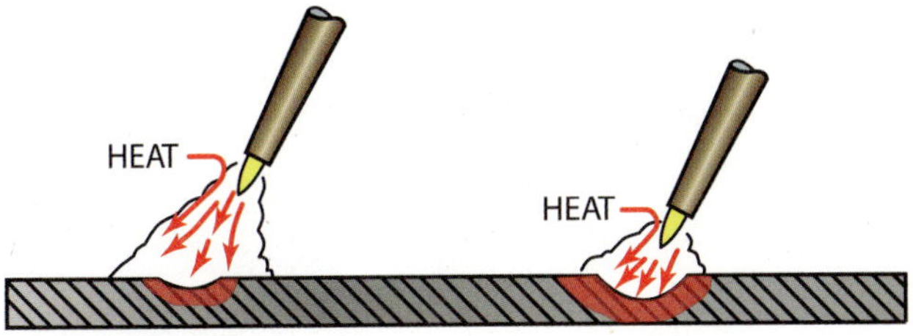

FIGURE 18-2 Changing the distance between the torch tip and the metal changes the percentage of heat input into the metal. © Cengage Learning 2012

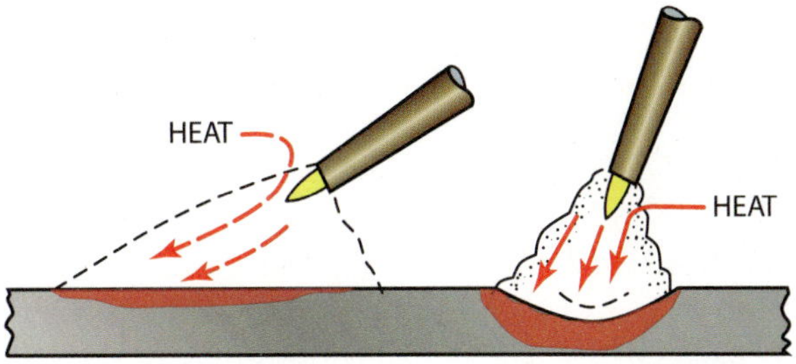

FIGURE 18-4 Flashing the flame off the metal will allow the molten weld pool to cool and reduce in size. © Cengage Learning 2012

welding rod can be used to cool the molten weld pool, increase buildup, and reduce penetration, Figure 18-3. The torch can be manipulated so that the direct heat from the flame is flashed off the molten weld pool for a moment to allow it to cool, Figure 18-4.

Characteristics of the Weld

The molten weld pool must be protected by the secondary flame to prevent the atmosphere from contaminating the metal. If the secondary flame is suddenly moved away from a molten weld pool, the pool will throw off a large number of sparks. These sparks are caused by the rapid burning of the metal and its alloys as they come into contact with oxygen in the air. This is particularly a problem when a weld is stopped. The **weld crater** is especially susceptible to cracking. This tendency is greatly increased if the molten weld pool

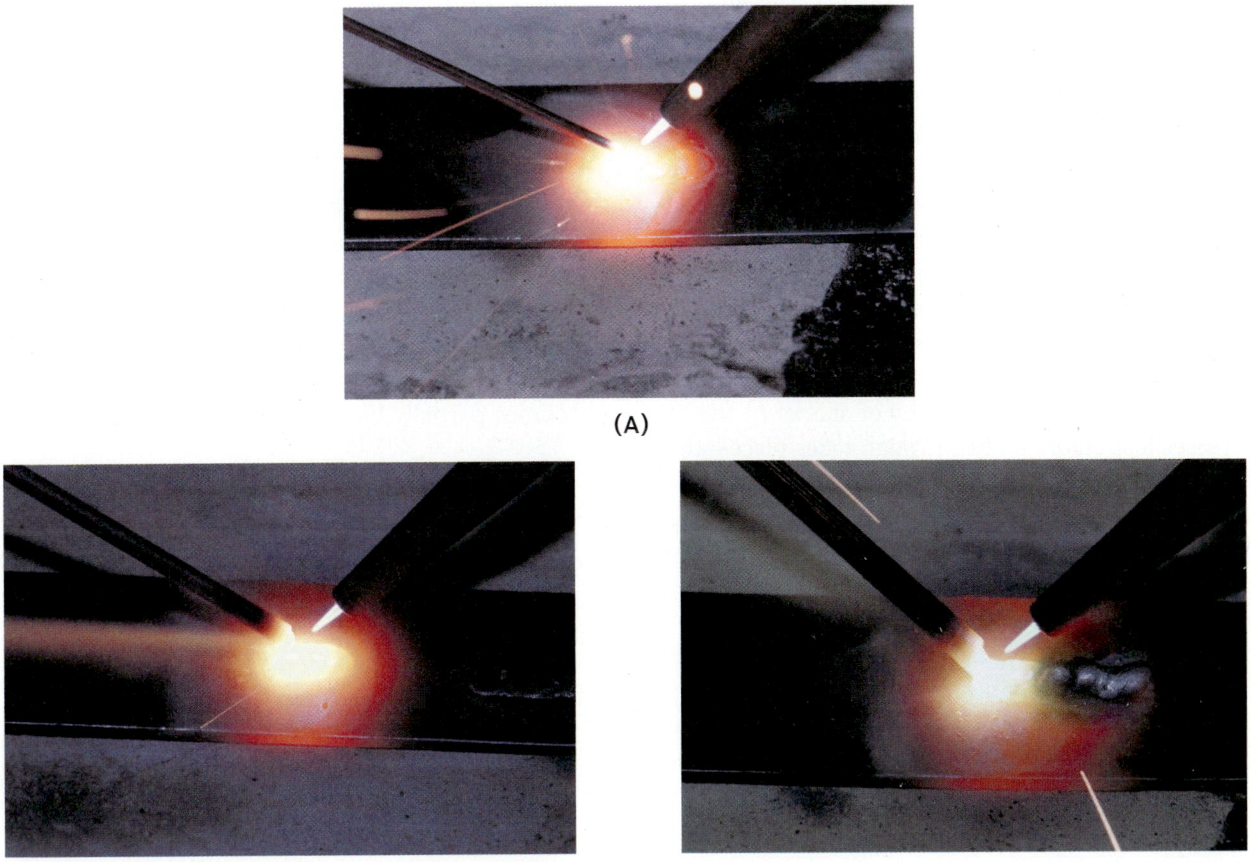

(A)

(B)

(C)

FIGURE 18-3 If all other conditions remain the same, changing the size of the filler rod will affect the weld as shown in (A), (B), and (C). Larry Jeffus

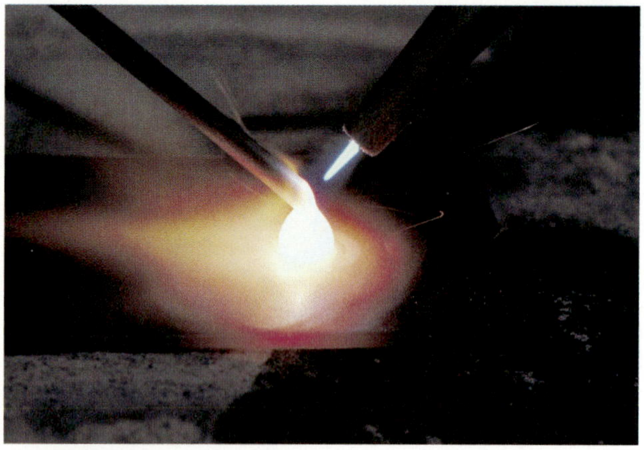

FIGURE 18-5 Building up the molten weld pool before it is ended will help prevent crater cracking. Larry Jeffus

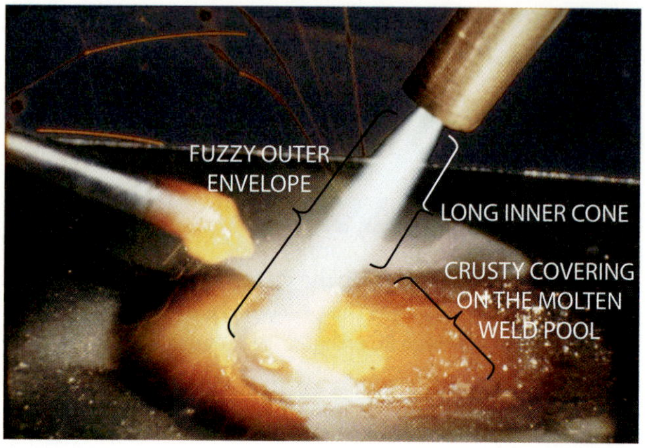

FIGURE 18-6 Carbonizing flame (excessive acetylene). Larry Jeffus

is allowed to burn out, Figure 18-5. To prevent **burnthrough**, the torch should be raised or tilted, keeping the outer flame envelope over the molten weld pool until it solidifies. As the molten weld pool is being cooled, the molten weld pool should also be filled with the welding rod so that it is a uniform height compared to the surrounding weld bead.

The sparks that occur as the weld progresses are due to metal components that are being burned out of the weldment. Silicon oxides make up most of the sparks, and extra silicon can be added by the filler metal so that the weldment retains its desired soundness. A change in the number of sparks given off by the weld as it progresses can be used as an indication of changes in weld temperature. An increase in sparks on clean metal means an increase in weld temperature. A decrease in sparks indicates a decrease in weld temperature. Often the number of sparks in the air increases just before a burnthrough takes place—that is, burning out the molten metal that appears on the back side of the plate. This burnthrough does not happen to molten metal until it reaches the **kindling temperature** (the temperature that must be attained before something begins to burn). Small amounts of total penetration usually will not cause a burnthrough. When the sparks increase quickly, the torch should be pulled back to allow the metal to cool, preventing a burnthrough.

EXPERIMENT 18-1

Flame Effect on Metal

The first experiment examines how the flame affects mild steel. After putting on all your required personal protective equipment (PPE) and safely setting up the gas welding equipment as outlined in Chapter 17 and following all manufacturer and shop safety rules, you will use a piece of 16-gauge mild steel and the proper-size torch tip. Light and adjust the flame by turning down the oxygen so that the flame has excessive acetylene. Hold the flame on the metal until it melts and observe what happens, Figure 18-6. Now adjust the flame by turning up the oxygen so that the flame is neutral. Hold this flame on the metal until it melts and observe what takes place, Figure 18-7. Next, adjust the flame by turning up the oxygen so that the flame has excessive oxygen. Hold this flame on the metal until it melts and observe what happens, Figure 18-8. During gas welding, sometimes the torch flame might come out of adjustment as the result of slight changes in the regulator pressure, the tip becomes partially blocked by sparks, or other similar

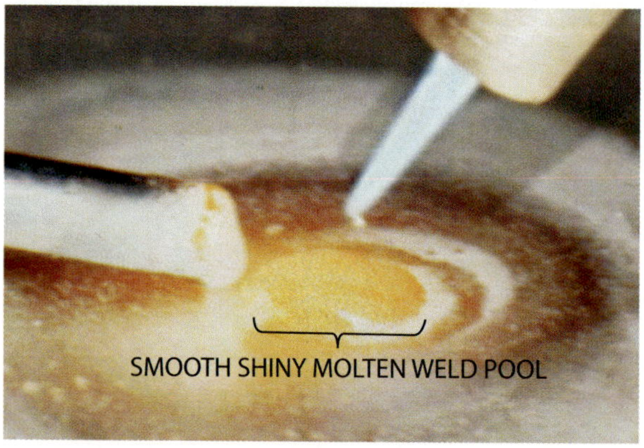

FIGURE 18-7 Neutral flame (balanced oxygen and acetylene). Larry Jeffus

FIGURE 18-8 Oxidizing flame (excessive oxygen). Larry Jeffus

problems. Knowing how the different flames affect the molten weld pool's appearance will help you spot this problem more quickly.

Repeat this experiment until you can easily identify each of the three flame settings by the flame, molten weld pool, sound, and sparks.

Paperwork

Complete a copy of the time sheet in Appendix I or as provided by your instructor.◆

Getting Ready to Weld

After putting on all your required PPE and safely setting up the gas welding equipment and following all manufacturer and shop safety rules, you will need to find a comfortable position to weld. The more comfortable or relaxed you are, the easier it will be for you to make uniform welds. The angle of the plate to your body and the direction of travel are important.

Place a plate on the table in front of you and, with the torch off, practice moving the torch in one of the suggested patterns along a straight line, as illustrated in Figure 18-9. Turn the plate and repeat this step until you determine the most comfortable direction of travel. Later, when you have mastered several joints, you should change this angle and try to weld in a less comfortable position. Welding in the field or shop must often be done in positions that are less than comfortable, so the welder needs to be somewhat versatile.

It is important to feed the welding wire into the molten weld pool at a uniform rate. Figure 18-10 shows some suggested methods of feeding the wire by hand. It is also suggested that you not cut the welding wire into two pieces for welding. Short lengths are easy to use, but this practice often results in wasted filler metal. The end of the welding wire may be rested on your shoulder so that it is easy to handle.

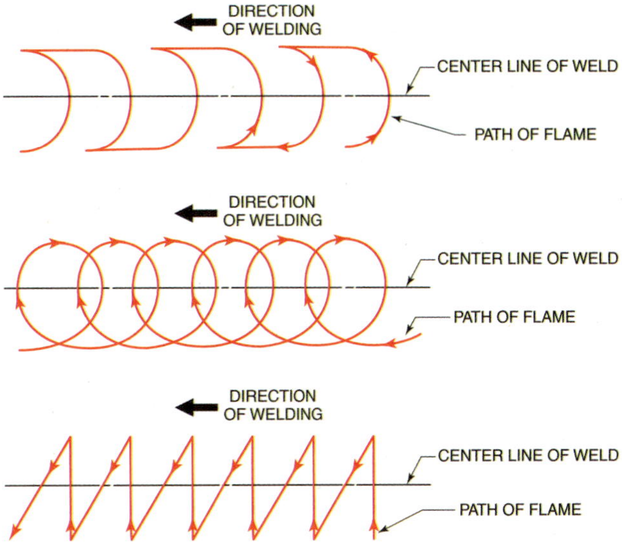

FIGURE 18-9 A few torch patterns. © Cengage Learning 2012

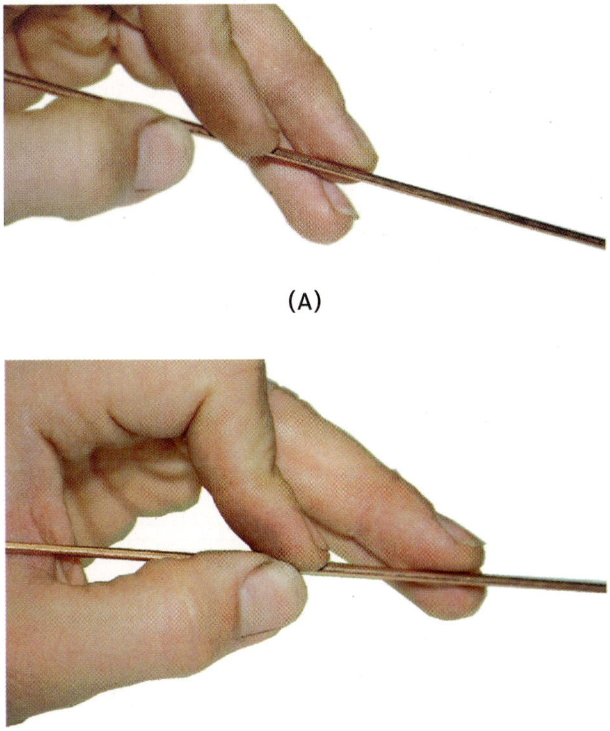

(A)

(B)

FIGURE 18-10 Feed the filler rod by using your index finger. Larry Jeffus

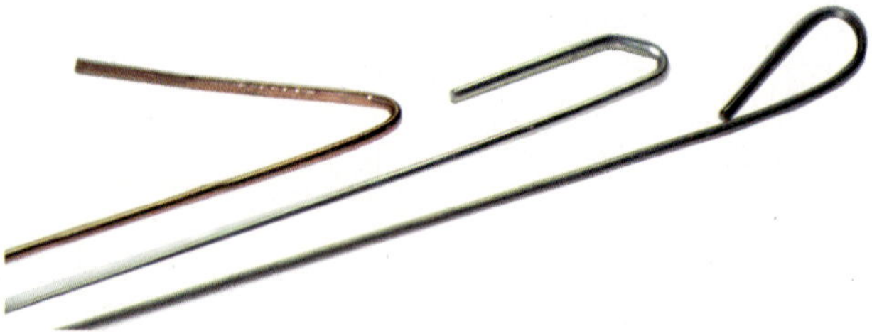

FIGURE 18-11 The end of the filler rod should be bent for safety and easy identification. Larry Jeffus

CAUTION

The end of the filler rod should have a hook bent in it so that you can readily tell which end may be hot and so that the sharp end will not be a hazard to a welder who is working next to you, Figure 18-11.

The torch hoses may be stiff and may, therefore, cause the torch to twist. Before you start welding, move the hoses so that there is no twisting of the torch. This will make the torch easer to manipulate and will be more relaxing for you.

EXPERIMENT 18-2

Effect of Torch Angle and Torch Height Changes

After putting on all your required PPE and safely setting up the gas welding equipment and following all manufacturer and shop safety rules, you will use a clean piece of 16-gauge mild steel and a torch adjusted to a neutral flame. You will be experimenting with different torch angles and torch heights to change the size of the molten weld pool, Figure 18-12.

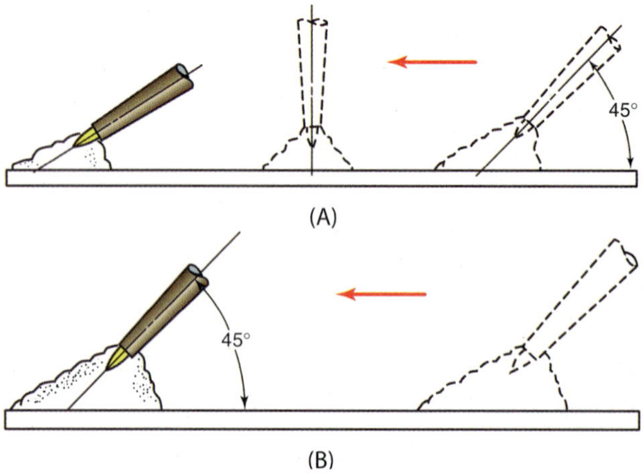

FIGURE 18-12 (A) Changing torch angle and (B) changing torch height both affect the weld pool size. © Cengage Learning 2012

To start this experiment, hold the torch at a 45° angle to the metal with the inner cone about 1/8 in. to 1/4 in. above the metal surface, Figure 18-13A, Figure 18-13B, and Figure 18-13C. As the metal starts to melt, move the torch slowly down the sheet.

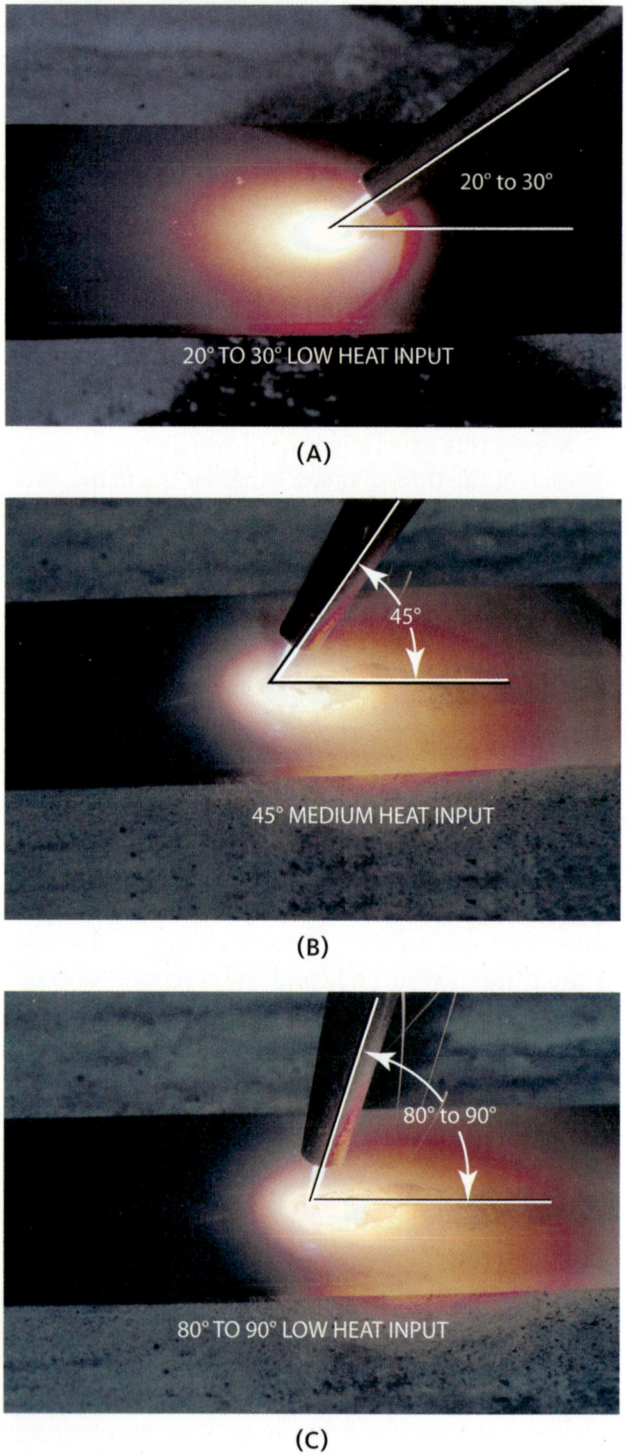

FIGURE 18-13 Changes in the angle between the torch and the work will change the molten weld pool produced. Larry Jeffus

FIGURE 18-14 Outside corner joint. Larry Jeffus

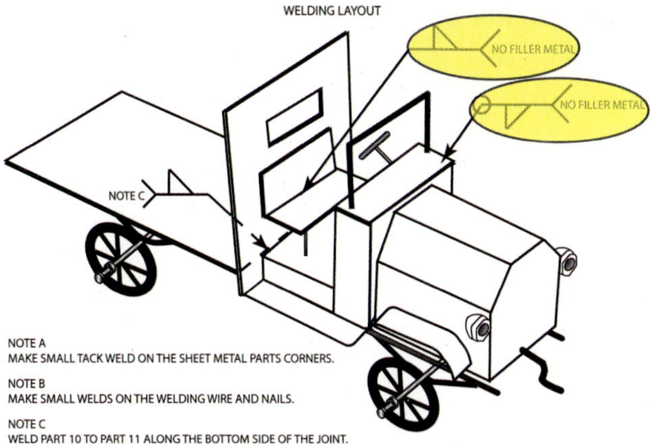

FIGURE 18-15 Outside corner joint on antique truck project. © Cengage Learning 2012

As you move, increase and decrease the angle of the torch to the sheet and observe the change in the size of the molten weld pool. Repeat the experiment, but this time as you move down the sheet, raise and lower the torch and observe the effect on the size of the molten weld pool.

Paperwork

Complete a copy of the time sheet in Appendix I or as provided by your instructor.◆

Outside Corner Joint

The flat **outside corner joint** can be made with or without the addition of filler metal. This joint is one of the easiest welded joints to make. If the sheets are tacked together properly, the addition of filler metal is not needed, Figure 18-14. However, if filler metal is added, it should be added uniformly. The antique truck model in Figure 18-15 includes outside corner joints.

PROJECT 18-1

1G Outside Corner Joint on a Candlestick

Skills to be learned: The ability to tack weld on a weldment, control the weld distortion, and develop the skills to make uniform outside corner weld beads in the flat position.

Project Description

This 6-in. tall candlestick is artistic in its simplicity, yet functional. The wide top and tall sides are designed to minimize the problem of candle wax drips reaching the table, Figure 18-16.

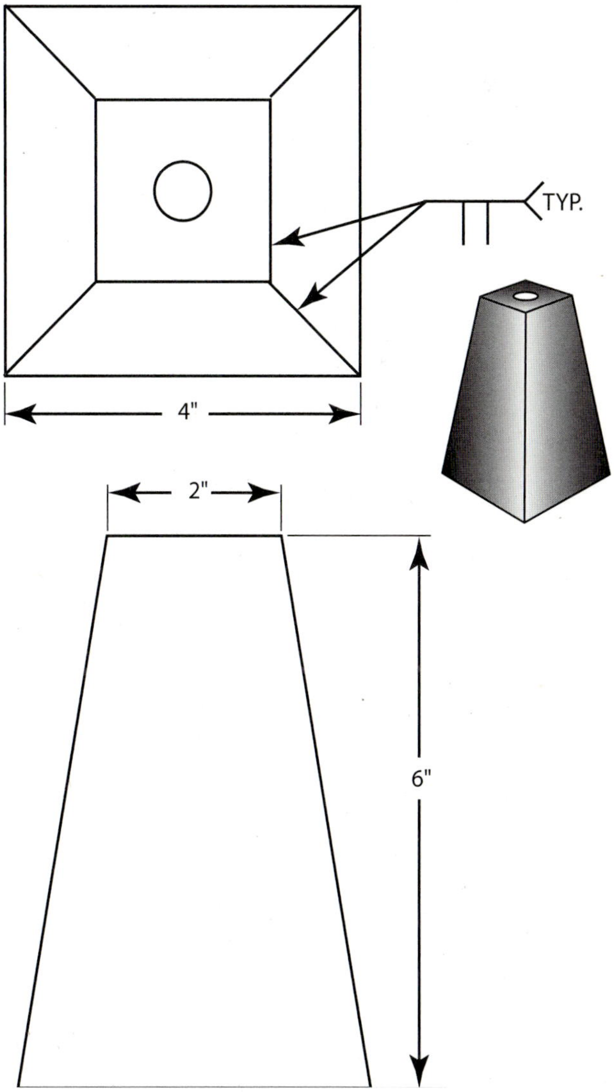

FIGURE 18-16 Project 18-1. © Cengage Learning 2012

Project Materials and Tools

The following items are needed to fabricate the candlestick.

Materials

- 14-in. long 6-in. wide 16-gauge thick mild steel sheet metal
- 2 × 2-in. square 16-gauge thick mild steel sheet metal
- 4 × 4-in. square 16-gauge thick mild steel sheet metal
- Source of oxygen and acetylene gases
- Reciprocating saw blades
- Clear or colored spray paint

Tools

- Oxyacetylene welding torch and tips
- 1/16-in. and 3/32-in. RG45 welding rods
- Saber saw and/or reciprocating saw
- Two blocks of wood
- PPE
- 12-in. rule
- Soapstone, pencil and/or marker
- Square or try square
- Wire brush
- Pliers
- 2 or more C-clamps
- 2 or more firebricks
- Drill and 3/4-in. drill bit

Layout

Start the layout by measuring 4 in. across the base and make a mark at the 4-in. and 2-in. measurements. Use the square and extend the 2-in. center point mark across the 6-in. width of the sheet metal to locate the center of the panel. On the opposite side use the 12-in. rule and measure 1 in. on each side of the center line you just drew. Using the 12-in. rule, draw a line from one side of the 4-in. line to the corner of the 2-in. mark. Repeat the process on the other side.

If the parts are going to be sheared, the next layout will be made right next to the first. However, if the parts are going to be sawed, you must leave a 1/16-in. wide space between the parts as shown in Figure 18-17. Measure the 1/16 in. perpendicular to the angled side of the first panel you drew, then repeat the layout process described previously until all four panels have been laid out.

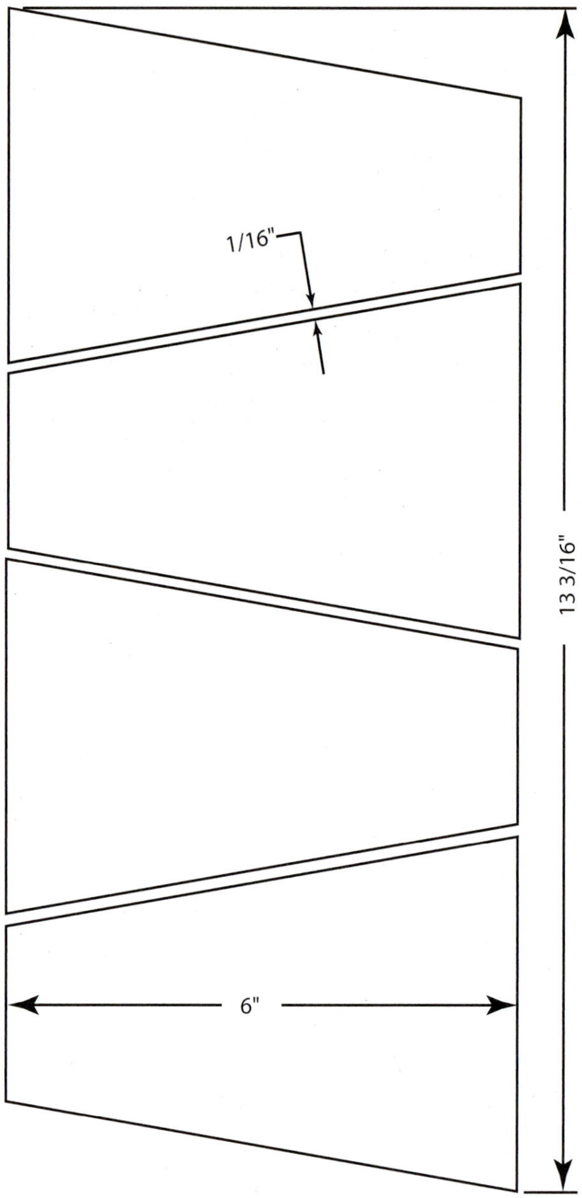

FIGURE 18-17 Candlestick layout.

Next, use the 12-in. rule and square to lay out the 2-in. square top. You can use the square to draw the two parallel sides and mark the 2-in. width so that the straightedge can be used to draw this line between the marks. Or you can use a try square as illustrated in Figure 18-18 to draw the top line parallel to the side of the sheet metal.

Cutting Out

Using a shear, saber saw, or reciprocating saw, putting on all of your PPE, and following all shop and

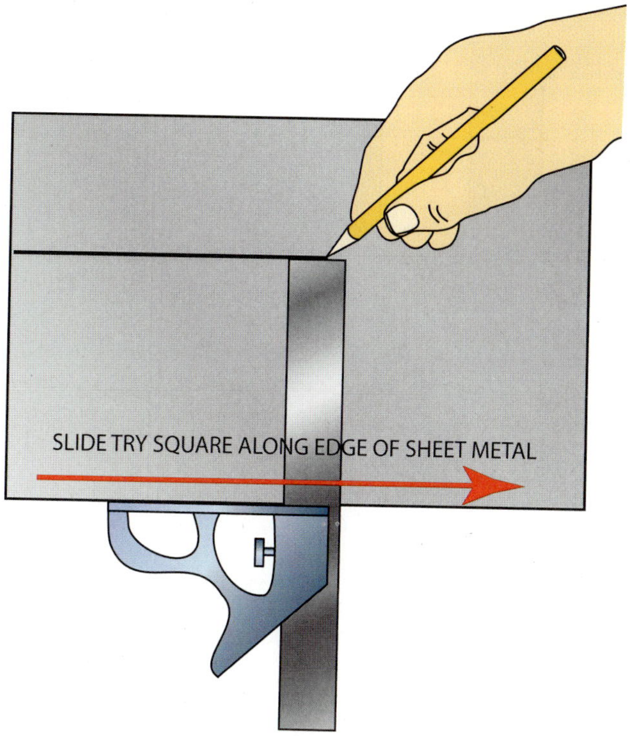

FIGURE 18-18 Drawing a straight line using a combination square and marker. © Cengage Learning 2012

manufacturer's safety rules for the power tool and/or equipment you are using, cut out the candlestick side panels and top.

NOTE: Often a band saw is used to cut out sheet metal in the shop. Band saw cutting is easier than using a reciprocating-type saw because the metal can be more easily cut without excessive vibration. However, both saber saws and reciprocating saws (Sawzall) have a reciprocating blade that can cause the part to vibrate or bounce. There are several ways of controlling this vibrating problem.

- A saw blade with the correct number of teeth per inch for the metal being cut should be selected, refer to Table 18-1.
- With a reciprocating saw, the saw can be held at a steep angle to control excessive vibration or bounce.
- The saber saw's foot can be held down with sufficient pressure against the metal to keep the vibration to a minimum.
- Blocks of wood can be clamped with a bench vise or C-clamps on both sides of the sheet metal; then the cut is made right along the edge of the wood.

Hacksaw Blade Selection

Teeth Pitch (Teeth per Inch)	Minimum Material Thickness Fractions of an Inch (Decimal Fractions)	
14 per inch	9/64 (0.140)	Less than two teeth cutting at a time. Pitch too large.
18 per inch	7/64 (0.109)	
24 per inch	3/32 (0.093)	Two or more teeth cutting at a time. Pitch OK.
32 per inch	1/16 (0.062)	Too many teeth cutting at a time has no chip space. Pitch too small.

Table 18-1 Recommended Hacksaw Blade Pitch for Material Thickness

After the cut is made, any sharp burrs should be filed off the edges so that they do not pose a safety hazard or interfere with the fit-up.

Fabrication

Set up the oxyacetylene welding equipment according to the manufacturer's recommendation, put on all of your PPE, and follow all shop and manufacturer's safety rules for welding.

You will construct three jigs with different-sized square holes in them to aid in the fabrication of the candlestick, Figure 18-19. The side panels will be slid inside the jigs, which will hold them in alignment for tack welding, Figure 18-20. Making the base and center tack welds below the jig will make it easier to remove the jig without the welds interfering with the jig. Remove the jig, and make tack welds between the bottom and center and top and center tack welds.

Stand the tack welded assembly up on the welding table and place the 2-in. square top in place. In the center of one side make a small tack weld. As this weld cools, it will shrink and raise the opposite side up. You can either push it back down by hand or use a C-clamp to pull it back before making a tack weld on that side. With these two tack welds complete, tack welds can be made on the two remaining sides. The first two tack welds will keep the top plate from moving as the last two tack welds are made.

Repeat this assembly and tack welding process to assemble the bottom 4 × 4-in. base.

Welding

Set up the oxyacetylene welding equipment according to the manufacturer's recommendation, put on

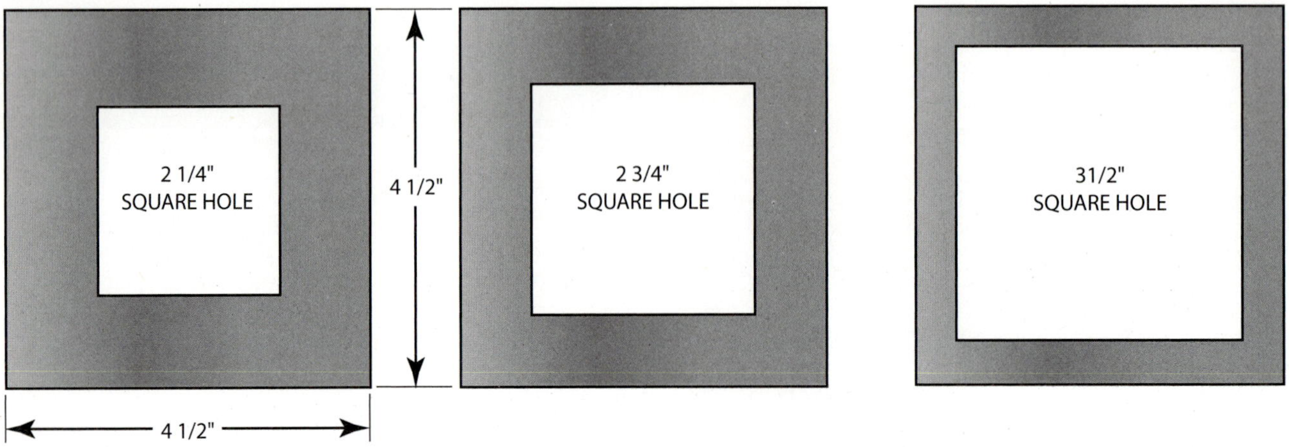

FIGURE 18-19 Layout for the jigs for candlestick. © Cengage Learning 2012

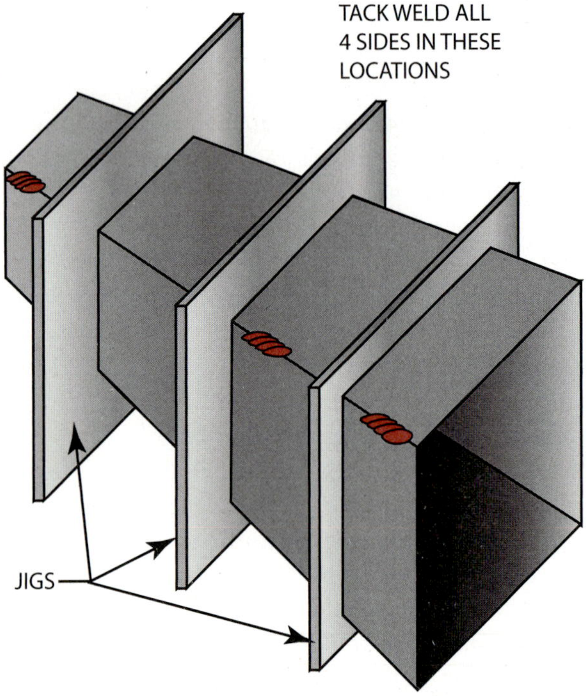

FIGURE 18-20 Using the jigs to align the side panels for the candlestick. © Cengage Learning 2012

all of your PPE, and follow all shop and manufacturer's safety rules for welding.

Place the candlestick on the welding table between two firebricks so that the corner is in a horizontal, slightly sloping position. Starting at the top corner, bring the torch tip inner flame (cone) to within 1/8 to 1/4 in. while holding the torch at a 45° angle, Figure 18-21. When the metal first melts, it will have a shiny surface appearance. Add a small amount of filler metal to the leading edge of the molten weld pool, Figure 18-22. As the weld moves along the joint, add the filler metal to the leading edge in a rhythmic manner. The faster the rhythm, the larger the weld bead produced; and the slower the rhythm, the smaller the weld bead. This rhythm will also affect the finished weld bead ripple pattern's spacing. Continue welding until you have reached the bottom. Raise the torch or angle it to reduce the heat on the corner. Add filler metal so the weld is complete all the way to the bottom edge.

Repeat this welding process on the remaining three corners. When all three corners have been welded, stand the candlestick up and make the welds around the top and side panels and the bottom and side panels using the same technique described earlier.

Finishing

Cool off the metal. Use the straightedge and marker to locate the center of the top. Make a punch mark in the center so that the drill bit will not wander, Figure 18-23. Drill the 3/4-in. diameter hole for the candle base. The candlestick can be painted with a clear coat or colored coat of latex paint.

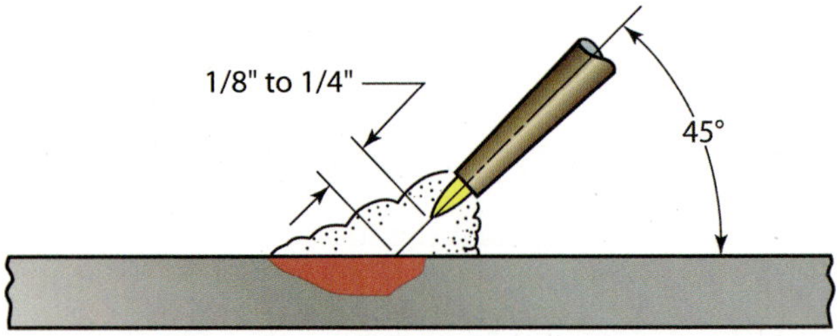

FIGURE 18-21 Hold the torch at a 45° angle in the direction of the weld. © Cengage Learning 2012

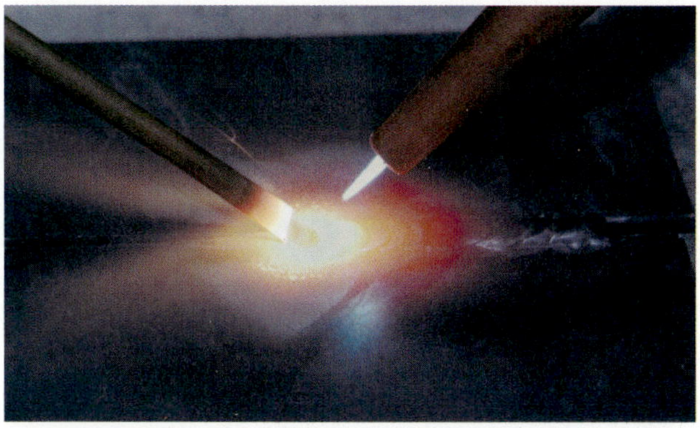

FIGURE 18-22A The hot end of the filler rod is protected from the atmosphere by the outer flame envelope.
Larry Jeffus

FIGURE 18-22B Filler metal being correctly added by dipping the rod into the leading edge of the molten weld pool.
Larry Jeffus

FIGURE 18-22C Filler metal being incorrectly added by allowing the rod to melt and drip into the molten weld pool.
Larry Jeffus

Paperwork

Complete a copy of the time sheet in Appendix I and the bill of materials in Appendix III, or as provided by your instructor.◆

FIGURE 18-23 Center punching the plate for drilling.
Larry Jeffus

Butt Joint

A **butt joint** is made by aligning the edges of two pieces of flat metal together so they are next to each other. To make the butt joint, place two clean pieces of metal flat on the table and tack weld both ends together, as illustrated in Figure 18-24. Tack welds may also be placed along the joint before welding begins. Point the torch so that the flame is distributed equally on both sheets. The flame is to be in the direction that the weld is to progress. If the sheets to be welded are of different sizes or thicknesses, the torch should be pointed so that both pieces melt at the same time, Figure 18-25.

PROJECT 18-2

1G Butt Joint on a Magnetic Bulletin Board

Skills to be learned: The ability to tack weld a weldment to control distortion and develop the skills to make uniform butt welds in the flat position.

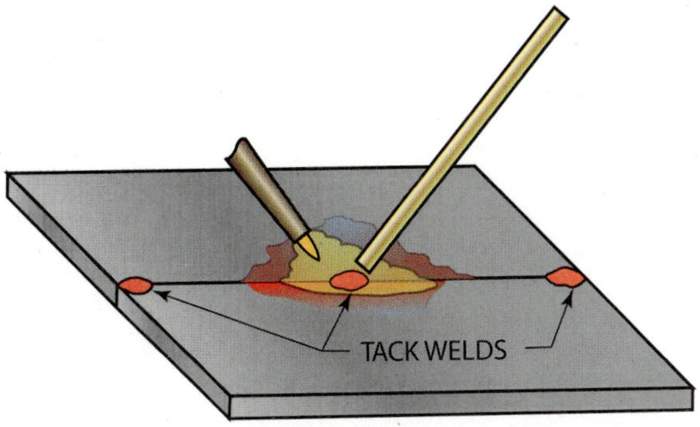

FIGURE 18-24 Making a tack weld. © Cengage Learning 2012

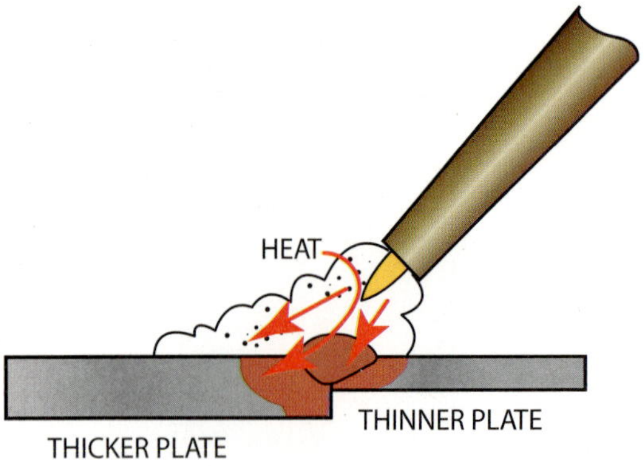

FIGURE 18-25 Direct the flame on the thicker plate.

Project Description

This 18 × 12-in. welded bulletin board is designed to hold refrigerator-type magnets and Post-it®–type notes. It can be constructed with a variety of shaped pieces of sheet metal, Figure 18-26.

Project Materials and Tools

The following items are needed to fabricate the magnetic bulletin board.

Materials

- 12-in. long 18-in. wide 16-gauge thick mild steel sheet metal
- Source of oxygen and acetylene gases
- Clear or colored spray paint

Tools

- Oxyacetylene welding torch and tips
- 1/16-in. and 3/32-in. RG45 welding rods
- Sheet metal shear
- PPE
- 12-in. rule
- Soapstone, pencil and/or marker
- Square or try square
- Wire brush
- Pliers
- Hammer and anvil
- Drill and 1/4-in. drill bit
- 4 screws or 24 in. of wire for hanging the finished project

SPARK YOUR IMAGINATION

The finished size of the magnetic bulletin board is to be 18 in. wide and 12 in. tall. To make the bulletin board the correct size, the square pieces would have to be 6 in. by 6 in., Figure 18-27. What size would the rectangle, triangle, pentagon, octagon, or strips have to be to make the 18 × 12-in. bulletin board?

Layout

Using the 12-in. rule, square, soapstone, pencil or marker, lay out the pieces for the magnetic bulletin board on the 16-gauge sheet of metal, Figure 18-26. If you have chosen the octagon or pentagon shapes,

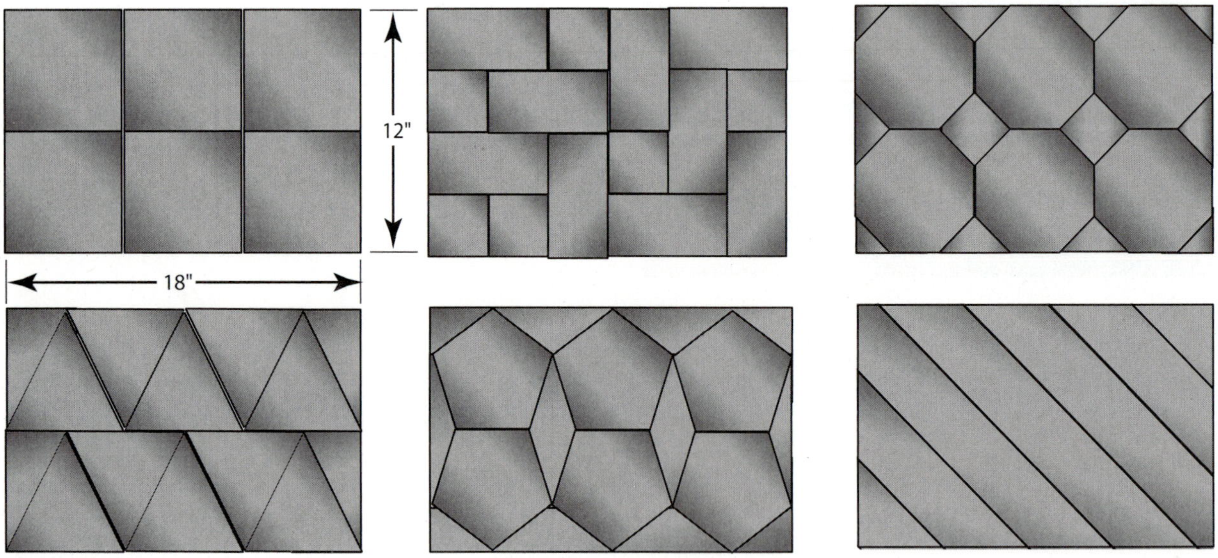

FIGURE 18-26 Project 18-2.

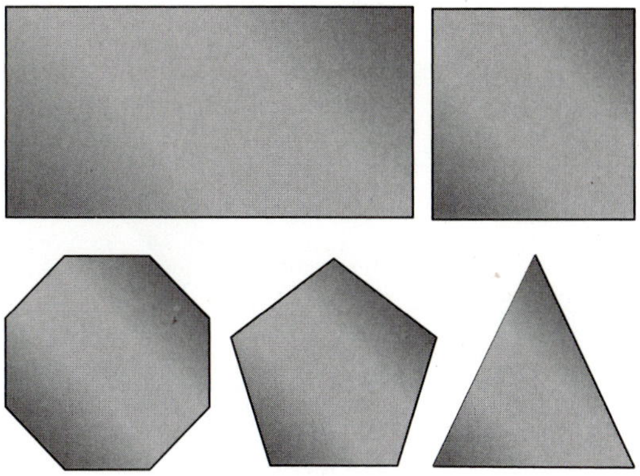

FIGURE 18-27 Spark your imagination.
© Cengage Learning 2012

they can be laid out as squares and then the corners marked and cut off. Some of the corners can be used to fill in the spaces to make the finished magnetic bulletin board square.

Cutting Out

Using a sheet metal shear, putting on all of the required PPE, and following all shop and manufacturer's rules for using the equipment, you are going to cut out the paneled pieces. Make sure that you make the cuts on the lines accurately because miscut pieces may not fit together correctly.

Fabrication

Start by laying out all of the pieces on the firebrick welding tabletop. Set up the oxyacetylene welding equipment according to the manufacturer's recommendation. Put on all your PPE and follow all shop and manufacturer's safety rules for welding. Start by making a small tack weld approximately 1/4 in. from the corner of one of the panels. Bring the torch to within 1/8 to 1/4 in., holding it at a 45° angle. Wait for the metal to begin to melt and the surface to appear shiny. Add a small amount of filler metal as soon as the metal begins to melt, Figure 18-28. As the tack weld begins to flow, add only enough filler metal to fuse the two plates together. Stop and move to the opposite end of the joint. If weld shrinkage has caused the joint to separate, use your pliers to pull the joint together. Repeat the tack welding process at this end of the joint. You will follow this procedure across the entire magnetic bulletin board surface. From time to time you will need to flatten out the paneled parts because of the weld shrinkage, which may cause them to bend and become distorted. Welding heat distortion and weld metal shrinkage are common problems caused by using an oxyacetylene torch to make large flat weldments.

FIGURE 18-28 Adding filler metal to the leading edge of the molten weld pool. Larry Jeffus

Welding

Using the same oxyacetylene equipment setup, and following the same safety procedures, you are going to make butt welds on each of the joints of the magnetic panel.

Place the panel so that you are at a comfortable angle to make a weld along one of the joints. Practice moving across the joint until you are comfortable, and make sure that the gas welding hoses do not snag as you move across the joint.

One of the ways of controlling the filler metal and welding torch in a rhythmic manner is to hold your elbows close to your sides so that you slightly rotate your body back and forth. This will move the welding torch and filler metal in a uniform, coordinated manner. Adding the filler metal this way is more consistent and can be much easier than holding your arms away from your body where slight movements can result in the filler metal accidentally touching outside of the weld zone. If this happens, the rod will stick, and it will have to be melted off by bringing the torch over to the end of the rod to melt it free.

During the welding process, the more frequently you add filler metal, the larger the weld will be. If you hold the torch closer to the metal, the weld will become wider. As you raise it up, the weld will become narrower in width and will have less joint penetration, Figure 18-29. Use these two techniques to control the weld bead size, and make the weld uniform across the joint. Lack of uniformity in the weld, however, is not a cause for concern or rejection of

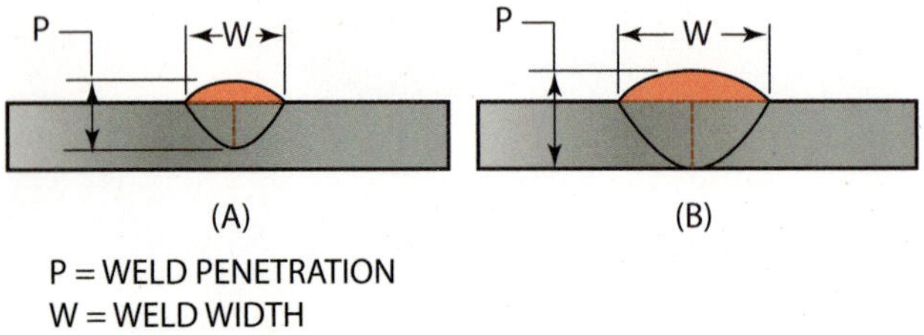

FIGURE 18-29 Weld bead shape. © Cengage Learning 2012

the finished product. This is an artistic piece, and discontinuities or differences in the weld bead width or appearance add to its aesthetics.

Weld each of the joints starting at one end, welding to the other end of the joint. With practice, you may be able to weld down one of the joints and turn the corner and continue the weld. This is a skill that, when developed, will aid in your making welds in the field on irregular-shaped sheet metal joints such as those you might find when making a repair.

Finishing

Cool off the metal, and use a hammer and anvil if necessary to flatten the panel. The panel does not have to be perfectly flat but flat enough so that magnets can adhere easily. Use a wire brush to remove all loose oxides. Drill two small holes near the top corners so that the magnetic board may be hung with a wire, or drill holes at all four corners so it may be screwed to the wall. Spray the magnetic bulletin board with a clear coat or colored paint.

Paperwork

Complete a copy of the time sheet in Appendix I and the bill of materials in Appendix III, or as provided by your instructor.◆

Lap Joint

The **lap joint** is made when one piece of metal is laid over the edge of another. When heating the two sheets, exercise caution to ensure that both sheets start melting at the same time. Heat is not distributed uniformly in the lap joint, Figure 18-30. Because of this difference in heating rate, the flame must be directed on the bottom sheet and away from the metal top sheet, Figure 18-31. The filler rod should be added to the top sheet. Gravity will pull the molten weld pool down to the bottom sheet, so it is, therefore, not necessary to put metal on

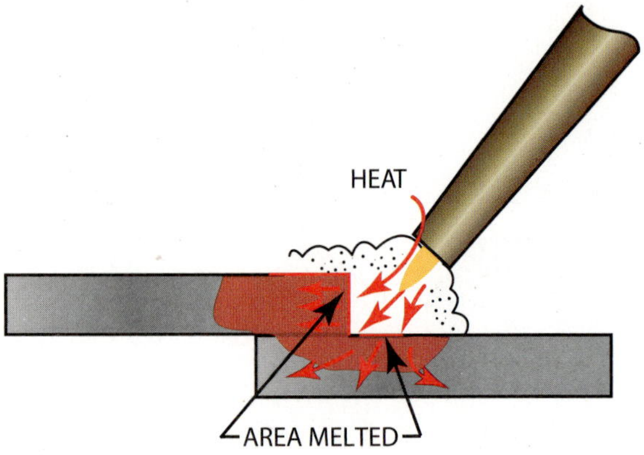

FIGURE 18-30 Heat is conducted away faster in the bottom plate, resulting in the top plate melting more quickly. © Cengage Learning 2012

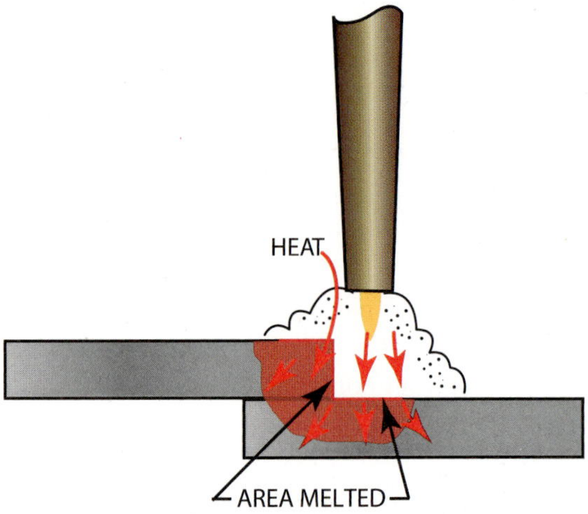

FIGURE 18-31 Flame heat should be directed at the bottom plate to compensate for thermal conductivity. © Cengage Learning 2012

the bottom sheet. If the filler metal is not added to the top sheet or if it is not added fast enough, surface tension will pull the molten weld pool back from the joint, Figure 18-32. When this happens, the rod should be added directly into this notch, and it will close. The weld appearance and strength should not be affected.

PROJECT 18-3

1G Lap Joint Welds in the Horizontal Position on a Stair-Stepped Candle Stand

Skills to be learned: The ability to bend sheet metal in a consistent manner, assemble a number of parts so that the overall shape of the finished weldment is uniform and square, and develop the skills needed to make uniform fillet welds on lap joints.

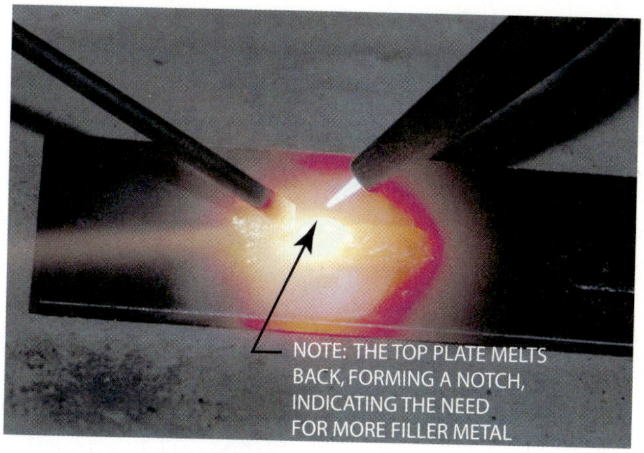

FIGURE 18-32 Add filler metal. Larry Jeffus

Project Description

This 4-in. high candle stand holds seven individual candles on a stair-stepped 1 1/2-in. wide arching frame, Figure 18-33. Two or more candle arches can be fabricated and attached at the bases to form an even more interesting effect, Figure 18-34.

Project Materials and Tools

The following items are needed to fabricate the stair-stepped candle stand.

Materials

- 40-in. long 1 1/2-in. wide 16-gauge thick mild steel sheet metal
- Source of oxygen and acetylene gases

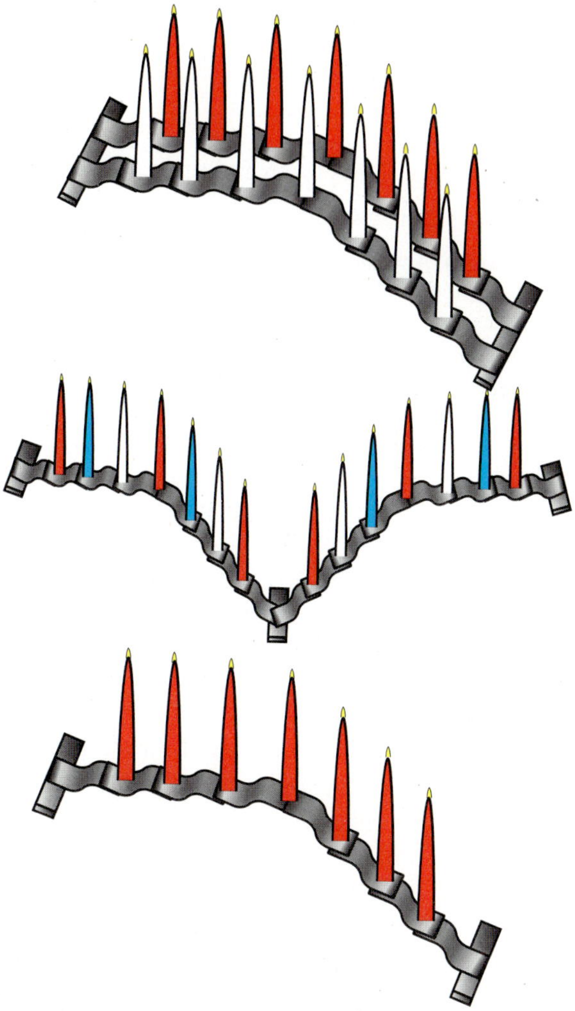

FIGURE 18-34 Alternative candlestick designs.

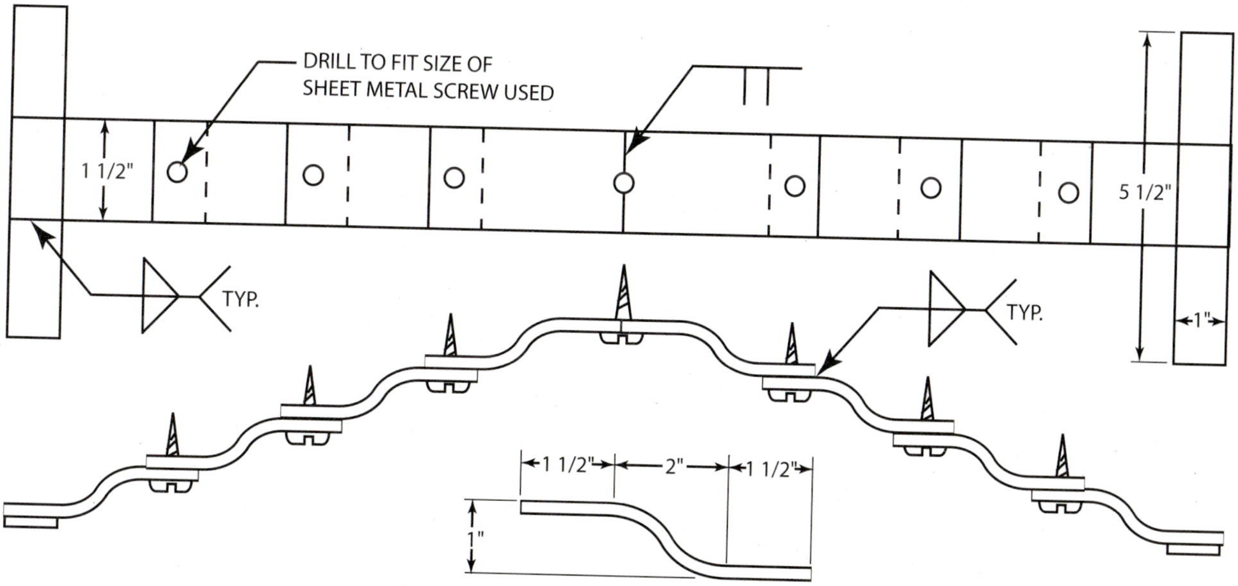

FIGURE 18-33 Project 18-3.

- 1/16-in. and 3/32-in. RG45 welding rods
- Clear or colored spray paint

Tools

- Oxyacetylene welding torch and tips
- Band saw, saber saw, and/or reciprocating saw
- Two blocks of wood
- PPE
- 12-in. rule
- Soapstone, pencil and/or marker
- Square or try square
- Wire brush
- Pliers
- 2 or more C-clamps
- 2 or more firebricks
- Drill and 3/4-in. drill bit

Layout

Mark a 1 1/2-in. wide strip along the 16-gauge sheet metal using the try square as shown in Figure 18-35. Using the tape measure, mark off the 5-in. long pieces that will make up the arch. Use the square to make a square line at each of the 5-in. marks. Once the parts are cut out, you will make a mark using the soapstone 1 1/2 in. from each end of the eight pieces.

Cutting Out

Using a sheet metal shear, putting on all of the required PPE, and following all shop and manufacturer's rules for using the equipment, you are going to cut out nine strips for the candles and two bases. The one extra strip will be used as the bending pattern.

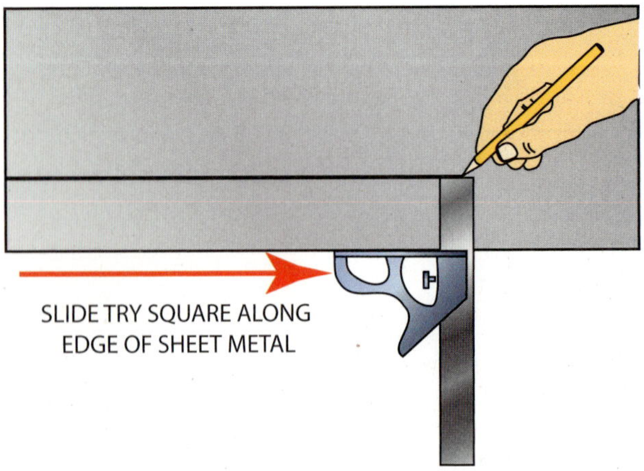

FIGURE 18-35 Drawing a straight line using a combination square and marker. © Cengage Learning 2012

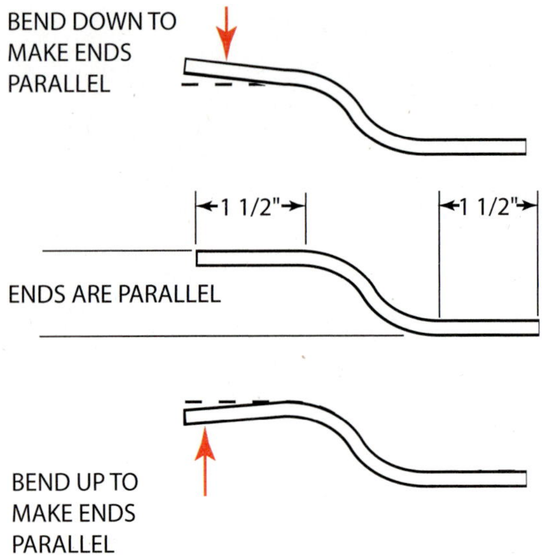

FIGURE 18-36 Freehand bending and checking the stepped candlestick base. © Cengage Learning 2012

Fabrication

One of the candlestick holder strips will be marked and bent to serve as the pattern for bending all of the other strips. Start by marking 1 in. from the end of the pattern strip and then 2 in. beyond that. Place the strip in a vise, and using the oxyacetylene welding torch flame, heat the metal between the 2-in. and 1-in. marks. When the metal is red hot, use pliers to bend it into the desired shape. Cool the part, and check to see that the 1 1/2-in. long ends are parallel, Figure 18-36. If they are not, they can be bent with the pliers once they are lined up. Place the bending pattern and one of the strips in the vise as shown in Figure 18-37. Heat the strip and using a pair of pliers, bend it to the bending pattern's shape. Be careful when heating the strip not to heat the bending pattern. Once the strip has cooled, remove it from the vise and repeat the process with the remaining seven strips.

If the 1 1/2-in. soapstone guide marks on the strips have been smudged or wiped off, re-mark them. Using a pair of locking pliers, clamp two of the strips together, lining up the marks and making sure they are parallel. Place the clamped piece on the welding table and make a small autogenous fusion weld on one side of the strip. Reverse the pliers so that they are out of the way and make a tack weld on the opposite side. Repeat this process with two more of the strips.

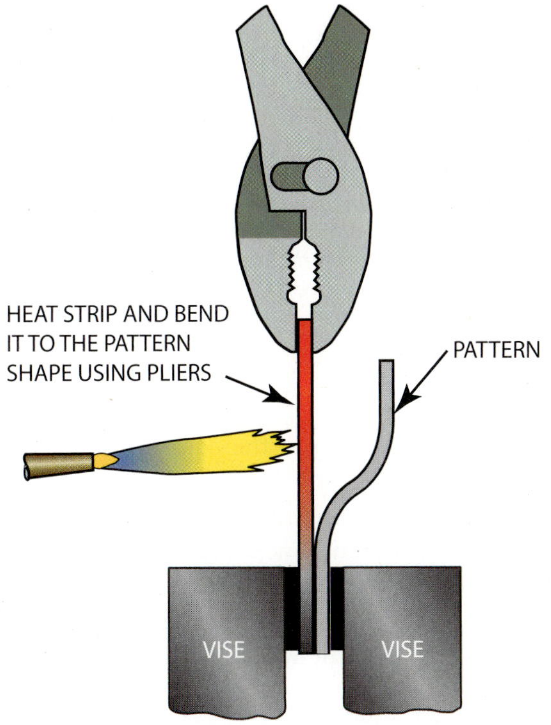

FIGURE 18-37 **Using the first strip as a guide, bend the remaining strips.** © Cengage Learning 2012

Now assemble the other half of the candlestick holder in the same manner as before, starting with two strips clamped together with locking pliers. When both halves of the arch have been fabricated, use the locking pliers to hold both halves of the arch together and make a small tack weld on one side, Figure 18-38. Repeat the process on the opposite side. Lay the arch down on a flat surface to make sure that it is lined up properly. It should not be more than ±1/16 in. out of level from one end to the other. If any one section needs to be realigned, the tack weld can be ground off using an angle grinder or pedestal grinder. Grind off the weld on only one side so that it can be twisted back into alignment before being tack welded again.

The two legs should be bent slightly at the end much like the bend in the arch strips so that the bend will raise the candlestick holder approximately 1/4 in. above the table. This clearance is needed so that the weld holding the leg on does not rest on the table, which would make the candlestick holder unstable. Also, by extending the leg out with a slight bend, a cork or felt pad can be placed at the ends to keep them from scratching the table.

Once the legs have been bent slightly using a vise and pliers, clamp them onto the arch's end with locking pliers. Make one small fillet tack weld on one side

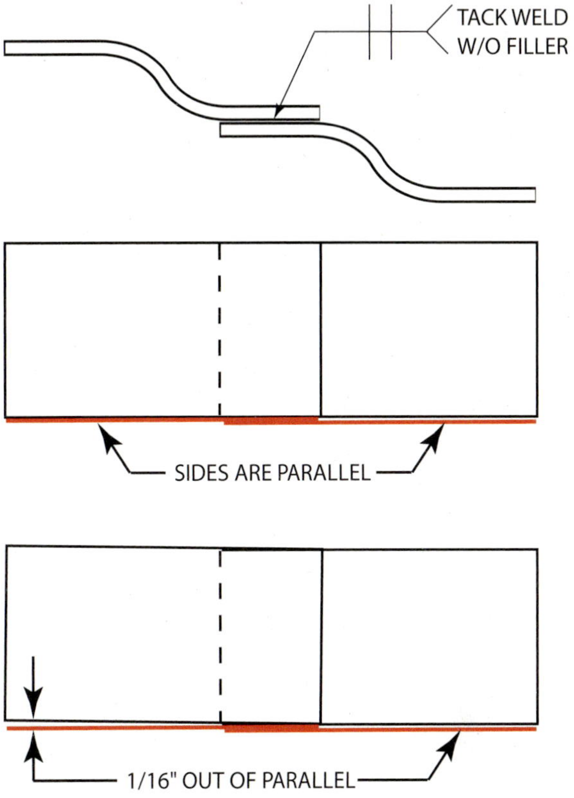

FIGURE 18-38 **Align and tack weld the candlestick steps.** © Cengage Learning 2012

using the oxyacetylene torch and filler metal. Repeat the process on the opposite side of this leg. Repeat this process on the opposite end.

Welding

Using the same oxyacetylene welding setup and following the same manufacturer's procedures, you are going to make the fillet weld on each of the joints of the candlestick holder.

Place the assembly flat on the welding table and rotate it so that the welds will be made at a comfortable angle. If you start at the top weld and work back toward yourself, you are less likely to accidentally burn yourself by touching one of your welds.

Hold the oxyacetylene torch flame approximately 1/8 to 1/4 in. above the joint. Direct most of the heat to the base side of the lap joint because the top edge will heat up and can overheat easily. Watch the metal to see that both sides of the joint melt at the same time, Figure 18-39. As soon as you see a bright, shiny molten pool form, add a small amount of filler metal, Figure 18-40. Add the filler metal at a uniform rate so that the weld will have a uniform shape. If the filler metal is not added at a sufficient rate, part of the top

FIGURE 18-39 Heating the joint before tacking.
Larry Jeffus

FIGURE 18-40 Filler rod is added after both pieces are heated to a melt. Larry Jeffus

edge plate may melt away. If this happens, add a little more filler metal to the top edge; gravity will pull it down and this notch can be filled. When you reach the end of the joint, add a little more filler metal and either raise the torch height from the plate or angle it slightly more to reduce the heat so the end of the weld can be filled. Repeat this process on each of the welds and both sides of the candlestick holder and on both sides of the legs.

The center joint is a butt joint, and it will be welded in a similar manner as the lap joints; however, care must be taken not to melt the tack welds before some of the weld has been produced. Prematurely melting the tack weld can result in the arch either coming apart or sagging. One way to avoid this problem would be to make a third tack weld in the center of this joint before welding begins.

Screw Size	Self-Tapping Drill Size*	Sheet Metal Drill Size*
#6	7/64 to 1/8	7/64 to 1/8
#8	9/64 to 5/32	1/8 to 5/32
#10	5/32 to 11/64	5/32 to 11/64
#12	3/16 to 13/64	3/16 to 7/32

* The drill sizes are recommended for the two layers of 16-gauge sheet metal. But if necessary use a slightly larger drill size so the screw will not break off.

Table 18-2 Recommended Drill Size for Sheet Metal Screws

Bring the torch up to the edge of the butt joint at an approximately 45° angle with the inner cone 1/8 to 1/4 in. above the metal. Watch and as soon as the metal begins to melt, add a small quantity of filler metal. This will keep the weld from melting back the edge of the candlestick holder. Add filler metal at a constant rate across the weld. At the end of the weld, angle the torch or raise it so that the edge is not overheated and so it can be filled.

Finishing

Cool the metal and align the candlestick holder so that the legs fit squarely onto the table, and twist them if necessary so that the candles will stand vertically.

Use the 12-in. rule and marker to locate the center of each of the steps of the arch. With a center punch and hammer, make a small punched spot at the center location. Using Table 18-2, select the drill size for the sheet metal screw size you are going to use for the candle support. Using an electric drill and bit, drill each of these holes on all seven steps.

NOTE: Sometimes with small diameter drill bits you are better off using your hand to tighten the chuck rather than using a chuck key because if the bit grabs and the chuck was tightened with a chuck key, the bit may break. However, if the bit were hand tightened, it will most likely slip in the chuck. Remove the drill and retighten with your hand to continue making a cut if it does stick.

Screw in the sheet metal screws from the bottom sides of the candlestick holder using the appropriate screw driver tip.

Wire brush any loose oxides, and spray the candlestick holder with either a clear coat or colored paint.

Paperwork

Complete a copy of the time sheet in Appendix I and the bill of materials in Appendix III, or as provided by your instructor. ◆

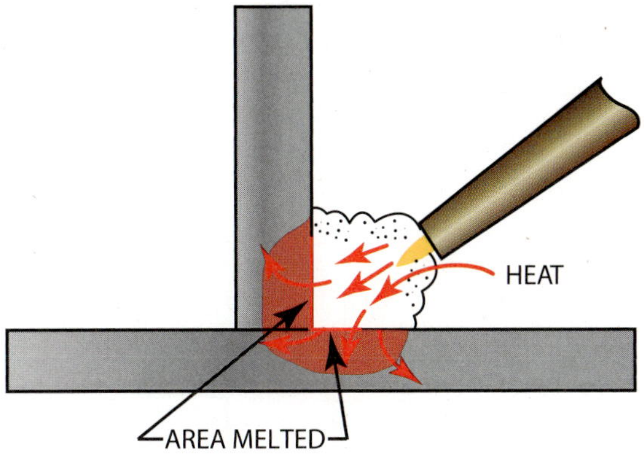

FIGURE 18-41 Direct the heat on the bottom plate to equalize the heating rates. © Cengage Learning 2012

Tee Joint

The flat **tee joint** is more difficult to make than the butt and lap joints. The tee joint has the same problem with uneven heating that the lap joint does. It is important to hold the flame so that both sheets melt at the same time, Figure 18-41. Another problem that is unique to the tee joint is that a large percentage of the welding heat is reflected back on the torch. This reflected heat can cause even a properly cleaned and adjusted torch to backfire or pop. To help prevent this from happening, angle the torch more in the direction of the weld travel. Because of the slightly restricted atmosphere of the tee joint, it may be necessary to adjust the flame so that it is somewhat oxidizing. The beginning student should not be overly concerned with this. A tee weld is used to hold the bed of the truck to the back of the cab on this antique model truck, Figure 18-42.

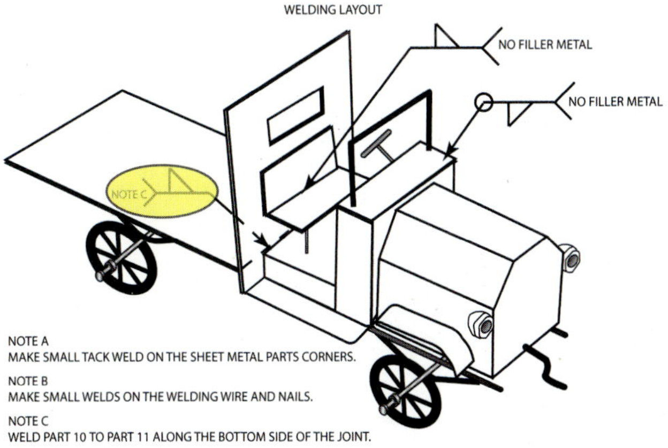

FIGURE 18-42 Tee joint on antique truck project. © Cengage Learning 2012

PROJECT 18-4

1G Tee Joint on a Sculptured Candlestick

Skills to be learned: The ability to tack weld a weldment to control weld distortion and develop the skills to make a uniform tee joint weld in the flat position.

Project Description

This 6-in. tall candlestick is designed to be both functional and decorative. The center relief gives the candlestick a shadow box appearance, Figure 18-43.

Project Materials and Tools

The following items are needed to fabricate the sculptured candlestick.

Materials

- 14-in. long 6-in. wide 16-gauge thick mild steel sheet metal
- 2 × 2-in. square 16-gauge thick mild steel sheet metal

FIGURE 18-43 Project 18-4. © Cengage Learning 2012

- Source of oxygen and acetylene gases
- Reciprocating saw blades
- 1/16-in. and 3/32-in. RG45 welding rods
- Clear or colored spray paint

Tools

- Oxyacetylene welding torch and tips
- Band saw, saber saw, and/or reciprocating saw
- Two blocks of wood
- PPE
- 12-in. rule
- Soapstone, pencil and/or marker
- Square or try square
- Wire brush
- Pliers
- 2 or more C-clamps
- 2 or more firebricks
- Drill and 3/4-in. drill bit

Layout

Use a pencil and graph paper to draw the outline of the curved side of the candlestick, Figure 18-44. The curved line is drawn by locating a series of points along the graph line and then sketching the line to connect the dots, Figure 18-45. Fold the paper in half, and using a pair of scissors, cut out the outline so that both sides of the side panel are the same. Place your template on the 6-in. wide sheet metal and use the pencil or marker to trace the outline. Flip the pattern over. Leave a 1/4-in. space between the first outline you just drew and this pattern. Follow the tracing procedures listed before, and repeat this process until you have drawn three sides. Lay out the 4-in. square base and 2-in. square top using your 12-in. rule and square.

Cutting Out

Using a band saw, saber saw, reciprocating saw, or a nibbler, putting on all your PPE, and following all shop and manufacturer's safety rules for the power tools or equipment you are using, cut out the candlestick side panels and top.

The power portable sheet metal nibbler is a tool that can be very helpful when cutting irregular shapes such as the side of this candlestick. The nibbler uses a reciprocating cutter to remove small crescent-shaped pieces of metal. The nibbler can easily change directions and make irregular turns easily. The three

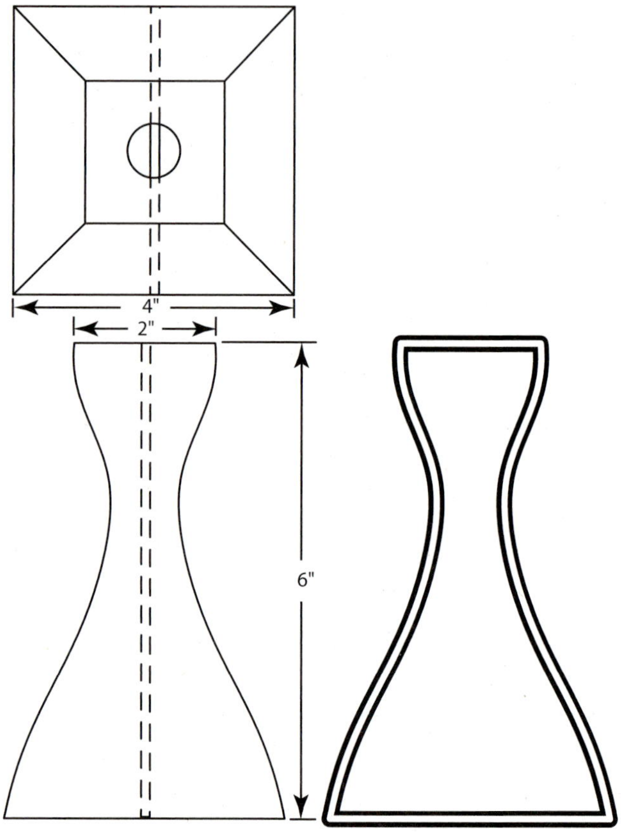

FIGURE 18-44 Laying out the candlestick. © Cengage Learning 2012

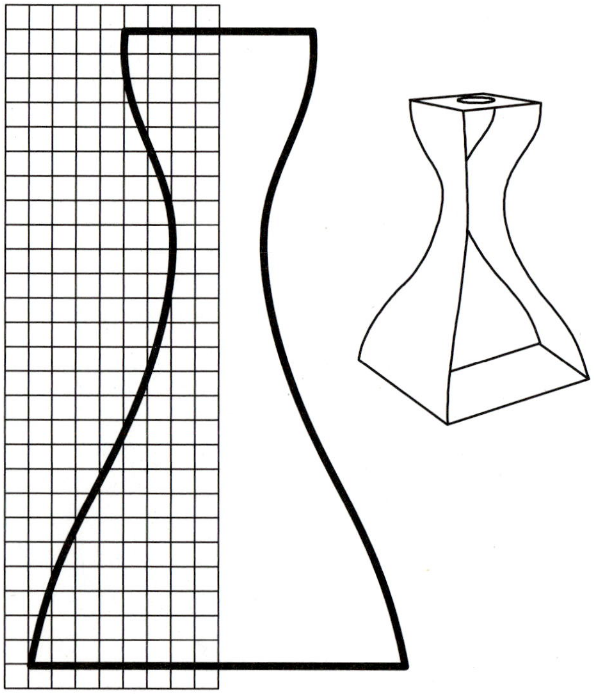

FIGURE 18-45 Sketching the curved outline of the candlestick. © Cengage Learning 2012

disadvantages of using a nibbler are that only thin sections of sheet metal can be cut and the small crescent-shaped pieces can become stuck in your shoes if you walk over them. The third problem is that it is sometimes difficult to cut a straight line with a nibbler because it so easily moves in any direction. Some nibblers can be locked so they cut straight lines easier. If this is not a feature on your nibbler, you can place a straightedge a short distance from the line you want to cut and rest the nibbler tool against the edge to aid in making a straight cut.

Use a C-clamp and block of wood to secure the sheet metal to a workbench, and using the nibbler or reciprocating saw, cut out one side of the panel. Turn the metal around, reclamp it, and repeat the process until all three panels have been cut out.

It is important that all three panels be exactly the same, so clamp them together with a C-clamp and file or grind the edges so that all three have the same shape. Do not remove too much of the metal during this straightening process or the pieces might not fit together correctly.

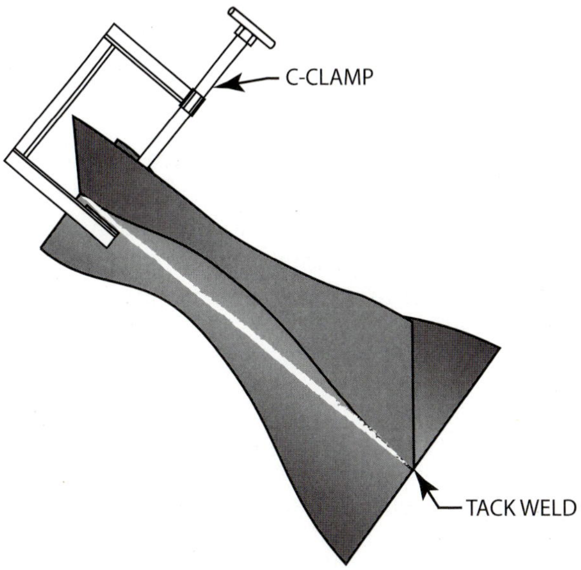

FIGURE 18-46 Using a C-clamp to hold the candlestick parts in alignment. © Cengage Learning 2012

Fabrication

The curved sides of the candlestick will be formed by squeezing them with a C-clamp tightly up against the center web. To do this you must first tack weld each of the sides to the web starting at the bottom. Locate the center of the 4-in. wide base on both side panels. Measure the center of the 2-in. wide top, and using a straightedge, draw a chalk line from the bottom to the top to create a center line for the web to be welded to.

Using a small C-clamp, hold the web in the center of the line so it is vertical, and make a small tack weld at the bottom edge, Figure 18-46. Turn this assembly over and repeat the process so that both sides are tack welded to the web. Lay the candlestick holder on its side and make one more tack weld approximately an inch from the bottom, joining the web to the side. Repeat this on both sides.

Using a C-clamp, squeeze the side panels to the center web at its narrowest point. This will bring the web in contact with the sides from the base to the center. Make several tack welds along the joint where the side and the web meet securely. Do not make a tack weld anywhere that they are not directly touching, and do not skip over any area that does not touch. Move the C-clamp up or down as needed to pull the two parts together. When they are together, make a small tack weld at that spot. Repeat this process until both side panels have been formed to the center web.

The distance from the bottom to the top of the side panels is 6 in.; however, the length of the curved line of the center web is approximately 6 1/4 in. So the center web will now extend slightly above the top. Use the nibbler or reciprocating saw to remove this top extension.

Mark a 3/4-in. wide 1-in. deep box in the center of the top of the web. Use the nibbler or reciprocating saw to cut the two vertical sides of the box. Using a pair of pliers, bend the center of this box over. This is the area for the candlestick to stick down into once the hole is drilled on the top.

Place the assembled sides and web on the 4-in. square base plate. Use a C-clamp to hold the base securely to the welded assembly. Tack weld the four corners of the candlestick to the base. Repeat this process for the 2-in. square top.

With the C-clamp in place, make one or two tack welds between the web and the base and the web and the top. This will minimize distortion.

Welding

Using the same oxyacetylene welding equipment, setup, and procedures as outlined in the tack welding of this section, you are going to make the fillet welds on the tee joint inside the candlestick. These welds will be in a somewhat restricted area where you do not have as open and free access as you might have on other welds. Working in a restricted area or an area with limited access is not uncommon when making welds in the field. As weldments begin to become assembled, parts often interfere with accessibility of the welded joint as well as making repair welds in the field.

When the filler metal is touched to the plate outside of the weld joint, it will stick to the metal. Because of the limited space for this weld on the candlestick, accidentally touching the plate with the filler metal is likely. If this happens, move the torch over to the end of the filler metal and melt it free. Bracing yourself with your elbows held tightly to your side will minimize accidental touching while making your arms and hands more steady. You may also brace yourself with your elbows on the weld table; however, this may limit some of your movement across the joint. Place the candlestick holder flat on the welding table. Holding the torch at approximately a 45° angle and about 1/8 to 1/4 in. above the surface, start in the center of one of the side panels and weld toward a corner. One of the characteristics of a tee joint is that one side of the metal will melt faster, so you will need to direct more of the heat on the side panel as opposed to the center web. Uneven heating is the result of the conductivity that can go in both directions on the vertical member and not on the horizontal web. To counter this effect, point the torch slightly more toward the side panel than toward the web. Watch the heating, and align the torch as necessary so both surfaces melt at approximately the same time.

When the metal has begun to melt, slowly add filler metal to the leading edge of the molten weld pool. Continue the weld all the way to the corner and stop. As you get to the corner, the weld flame will be reflected up and heat will tend to concentrate more because of the confined space. You may need to raise the torch slightly, also, so that you do not overheat or melt through the sheet metal corner. Repeat this welding process until all of the joints on this side have been welded. Turn the part over, and the opposite side can be welded. This weld will be slightly more difficult to heat because of the weld metal that has been deposited on the back side. Also, there may be some metal oxide flakes along the surface that will need to be wire brushed out so that they do not interfere with the welding. As before, hold the torch pointed at the corner approximately 1/8 to 1/4 in. above the surface at a 45° angle, and begin heating in the center of the side panel joint. Add filler metal as before, and make a uniform weld around the inside of the joint. You may want to try to weld through some of the corners. It is not a good practice in the field to stop a weld at the corner of a joint. This often may lead to leaks or stress points that can cause weld failure, not a problem with the candlestick, but if you can practice this skill, it will be beneficial once you begin welding in the field. Complete all of the welds on the second side.

Finishing

Use a wire brush to remove any loose oxides that have formed on the inside and/or outside of the candlestick. If the candlestick has excessive distortion, use a hammer and anvil to make slight adjustments. Drill the 3/4-in. diameter hole for the candle base in the top. The candlestick can now be painted with a clear coat or colored coat of latex paint. Use four cork or felt pads on the base, one at each corner, to prevent the candlestick from scratching the surface of the table.

Paperwork

Complete a copy of the time sheet in Appendix I and the bill of materials in Appendix III, or as provided by your instructor.◆

Out-of-Position Welding

A part to be welded cannot always be positioned so that it can be welded in the flat position. Whenever a weld is performed in a position other than flat, it is said to be **out-of-position welding**. Welds made in the **vertical**, **horizontal**, or **overhead** positions are out of position and somewhat more difficult than flat welds.

Vertical Welds

A vertical weld is the most common out-of-position weld that a welder is required to perform. When making a vertical weld, it is important to control the size of the molten weld pool. If the molten weld pool size increases beyond that which the **shelf** will support, Figure 18-47, the molten weld pool will overflow and drip down the weld. These drops, when cooled, look like the drips of wax on a candle. To prevent the molten weld pool from dripping, the trailing edge of the molten weld pool must be watched. The **trailing edge** will constantly be solidifying, forming a new shelf to support the molten weld pool as the weld progresses upward, Figure 18-48. Small molten weld pools are less likely than large ones to drip.

The less vertical the sheet, the easier the weld is to make, but the type of manipulation required is the same. Welding on a sheet at a 45° angle requires the same manipulation and skill as welding on a vertical

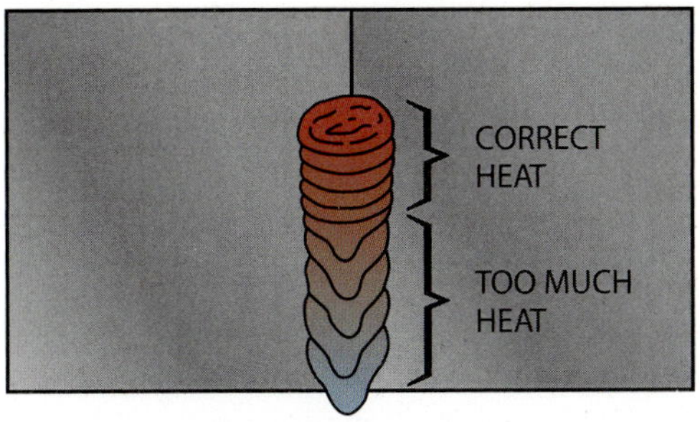

FIGURE 18-47 Vertical weld showing effect of too much heat. © Cengage Learning 2012

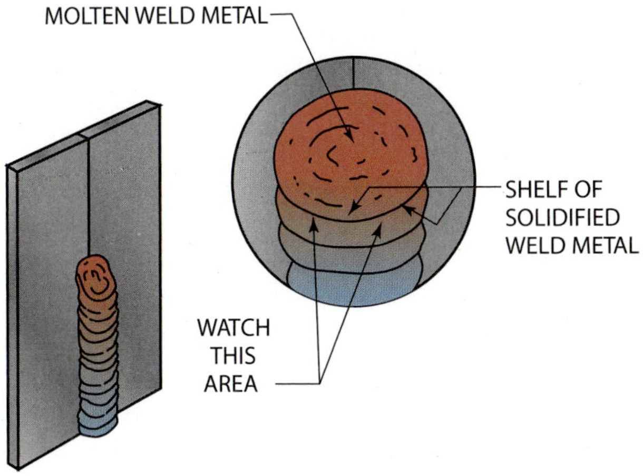

FIGURE 18-48 Watch the trailing edge to see that the molten pool stays properly supported on the shelf. © Cengage Learning 2012

sheet. However, the speed of manipulation is slower, and the skill is less critical than at 90° vertical. You may want to practice welding at a 45° angle before you start welding on this project.

Vertical Outside Corner Joint

PROJECT 18-5

3G Outside Corner Joint on a Funnel

Skill to be learned: The ability to make uniform vertical welds on an outside corner joint.

Project Description

This small metal funnel can be used in your garage or shop to help you pour liquids such as oil or paint or to help you pour powders such as fertilizer or insecticide. The nice thing about a metal funnel is that nothing will stain it like oils or solvents stain some plastic funnels. This easy cleanup means it will last for many years.

Project Materials and Tools

The following items are needed to fabricate the funnel.

Materials

- 9-in. long 6-in. wide 16-gauge thick mild steel sheet metal
- 2 × 2-in. square 16-gauge thick mild steel sheet metal
- Source of oxygen and acetylene gases
- Reciprocating saw blades
- Clear or colored spray paint

Tools

- Oxyacetylene welding torch and tips
- 1/16-in. and 3/32-in. RG45 welding rods
- Band saw, saber saw, and/or reciprocating saw
- Two blocks of wood
- PPE
- 12-in. rule
- Soapstone, pencil and/or marker
- Square or try square
- Wire brush
- Pliers
- 2 or more C-clamps
- 2 or more firebricks
- Drill and 3/4-in. drill bit

Layout

Start the layout by measuring 4 in. across the base and make a mark at the 4-in. and 2-in. measurements. Use the square and extend the 2-in. center point mark across the 6-in. width of the sheet metal to locate the center of the panel. On the opposite side, use the 12-in. rule and measure 3/8 in. on each side of the center line you just drew. Using the 12-in. rule, draw a line from one side of the 4-in. line to the corner of the 3/4-in. mark. Repeat the process on the other side.

If the parts are going to be sheared, the next layout will be made right next to the first. However, if the parts are going to be sawed, you must leave a 1/16-in. wide space between the parts as shown back in Figure 18-17. Measure the 1/16 in. perpendicular to the angled side of the first panel you drew, then repeat the layout process described previously until all three panels have been laid out.

Use the try square to lay out the 5-in. long 1/2-in. wide strip of sheet metal that will be used for the handle.

Cutting Out

Using a sheet metal shear, putting on all of the required PPE, and following all shop and manufacturer's rules for using the equipment, you are going to cut out the side pieces and the handle. This project lends itself to mass production of the parts with minimum materials waste. By placing a guide at the correct angle on the shear table, a 6-in. wide strip of stock can be fed in and cut. Flipping the stock over after each cut will let you cut out both angles of the funnel sides at one time.

Fabrication

An efficient way to tack weld the funnel sides together is to place two panels side by side vertically on the welding table. The panels can be leaned up against a firebrick to hold them in place, Figure 18-49. If the top corners touch, then you can simply melt them slightly and gravity will pull the molten metal down and it should flow together. This will result in a very small tack weld. If the metal does not flow together, bring the tip of the filler metal into the flame, and allow it to melt and drip into the molten base metal. This will both complete the tack weld while not accidentally jarring the panels out of position.

FIGURE 18-49 Outside corner joint. Larry Jeffus

Place the third side of the funnel in place and bend the tack weld with your hand or pliers to fit it into place. With funnel vertical on the table, make a tack weld as described earlier. Turn the funnel over, and align the panel as necessary by bending the tack weld. Brace the funnel as necessary with firebricks, and make a small tack weld on the three corners.

Using a bench vise, bend the funnel handle as shown in Figure 18-50. Place the handle on the side, and use a C-clamp to hold it in place so that a small tack weld can be made in the center of both of the ends.

Welding

You will make a vertical up outside corner joint weld. Place the funnel near the edge of the welding table or on a firebrick so that the welding torch can point upward at approximately a 45° angle starting at the base of the funnel. Bring the torch tip to within 1/8 to 1/4 in. and watch for the surface to melt. Once the surface has melted, begin adding filler metal. The molten weld pool will solidify and form a shelf to support the molten weld metal. If the molten weld pool begins to increase in size or if molten metal runs over the supporting ledge of the cooled weld metal, flash the torch off momentarily to allow the metal to cool slightly. **Flashing** the torch off is a process in which the torch is quickly moved away from the weld for a fraction of a second and then brought back to the welding position. Flashing the torch is a process that can help you control the molten weld pool size when welding out of position. The weld metal is not exposed to the air for a sufficient amount of time to cause oxide formation. Continue the weld up the joint until you have reached the top. Move the torch away slightly and add filler metal to the end of the weld so that it is filled completely before stopping the weld. Rotate the funnel and make the next weld in the same manner. Repeat this process on the last remaining weld.

The handle will be welded on using a fillet weld. Place the funnel so that the handle can be welded in approximately the flat position. Hold the torch so that the handle and sides of the funnel melt at the same time. The handle is much smaller and has an edge. Both of these factors will cause the handle to heat faster, so

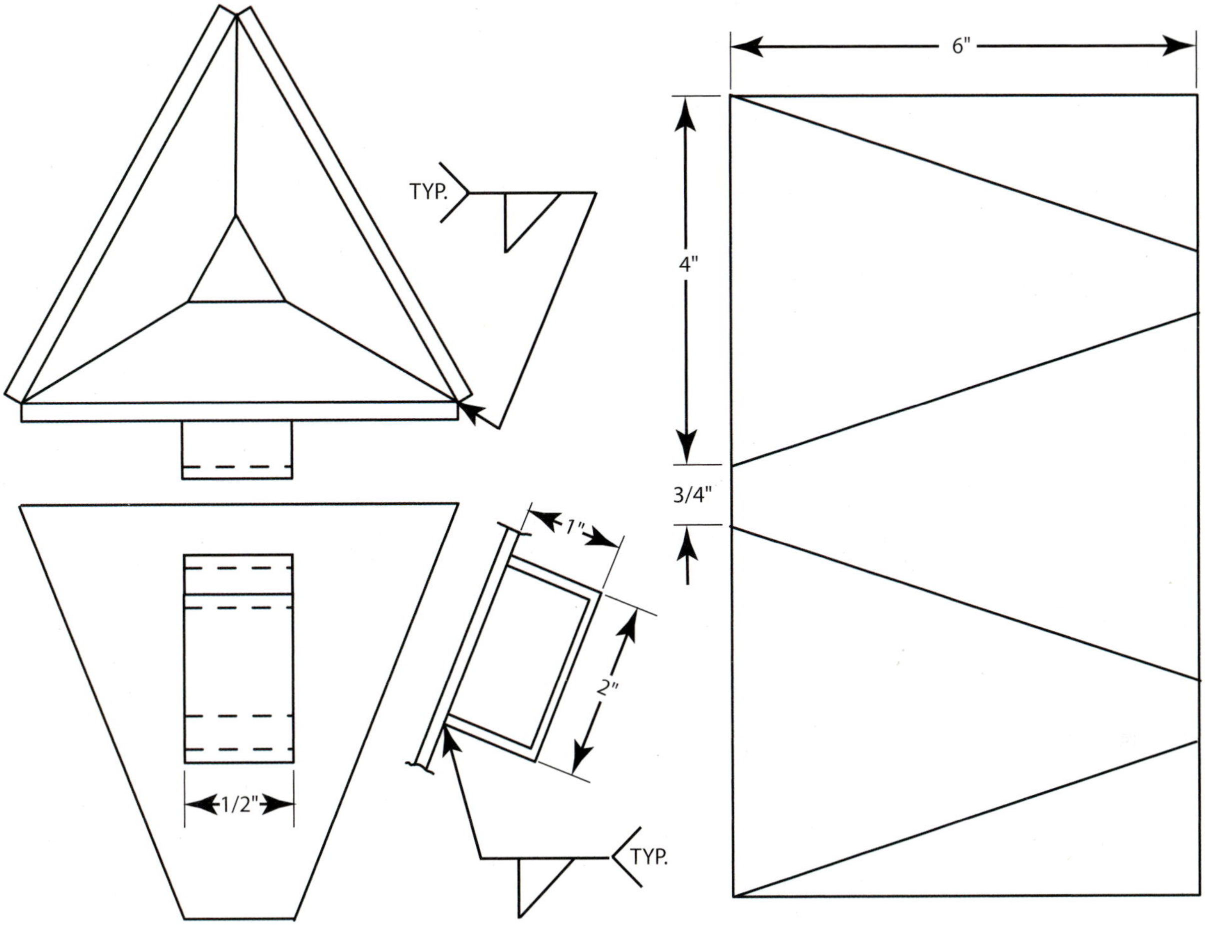

FIGURE 18-50 Project 18-5.

most of the heat will have to be directed on the side of the funnel so that melting occurs simultaneously. Once the molten weld pool has been established, add filler metal and make the weld all the way across the handle. Repeat this process for the other end of the handle.

Finishing

Use a wire brush to remove any loose oxide that has formed during the welding process. Spray the outside of the funnel with a clear finish or colored paint. Do not paint the inside of the funnel because it may be subject to oil and solvents that could remove the paint.

Paperwork

Complete a copy of the time sheet in Appendix I and the bill of materials in Appendix III, or as provided by your instructor. ◆

SUMMARY

In this chapter you have learned enough layout, cutting, fabrication, and welding skills to build the antique truck in Figure 18-51A and Figure 18-51B. As you have built projects in this chapter, you have learned all of the necessary welding skills to fabricate this truck. Similar welded joints have been highlighted in figures in this chapter. The truck shown in Figure 18-52 won the author a ribbon in the Southern Kern County Fair in 1969.

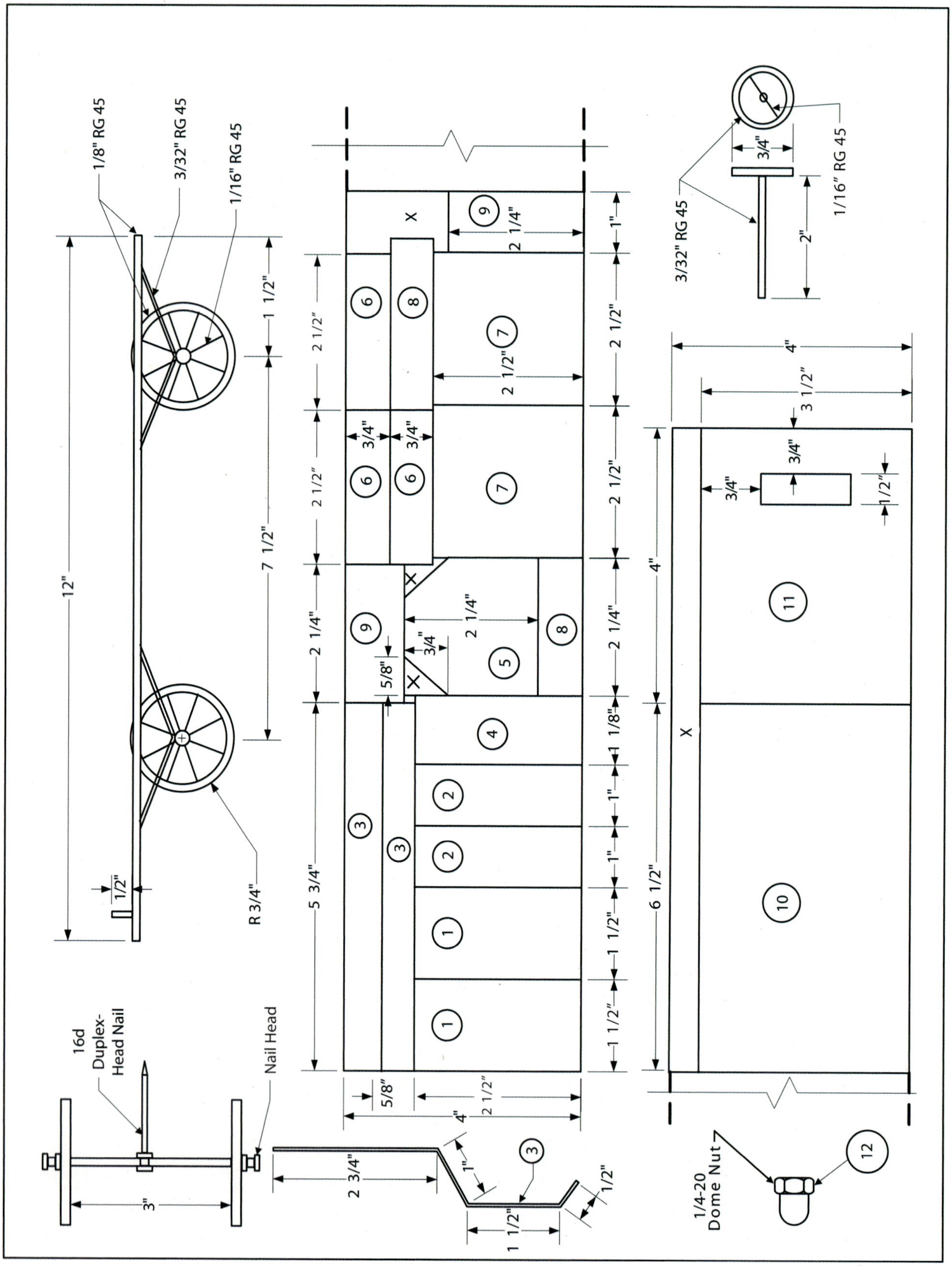

FIGURE 18-51A Antique truck layout drawing. © Cengage Learning 2012

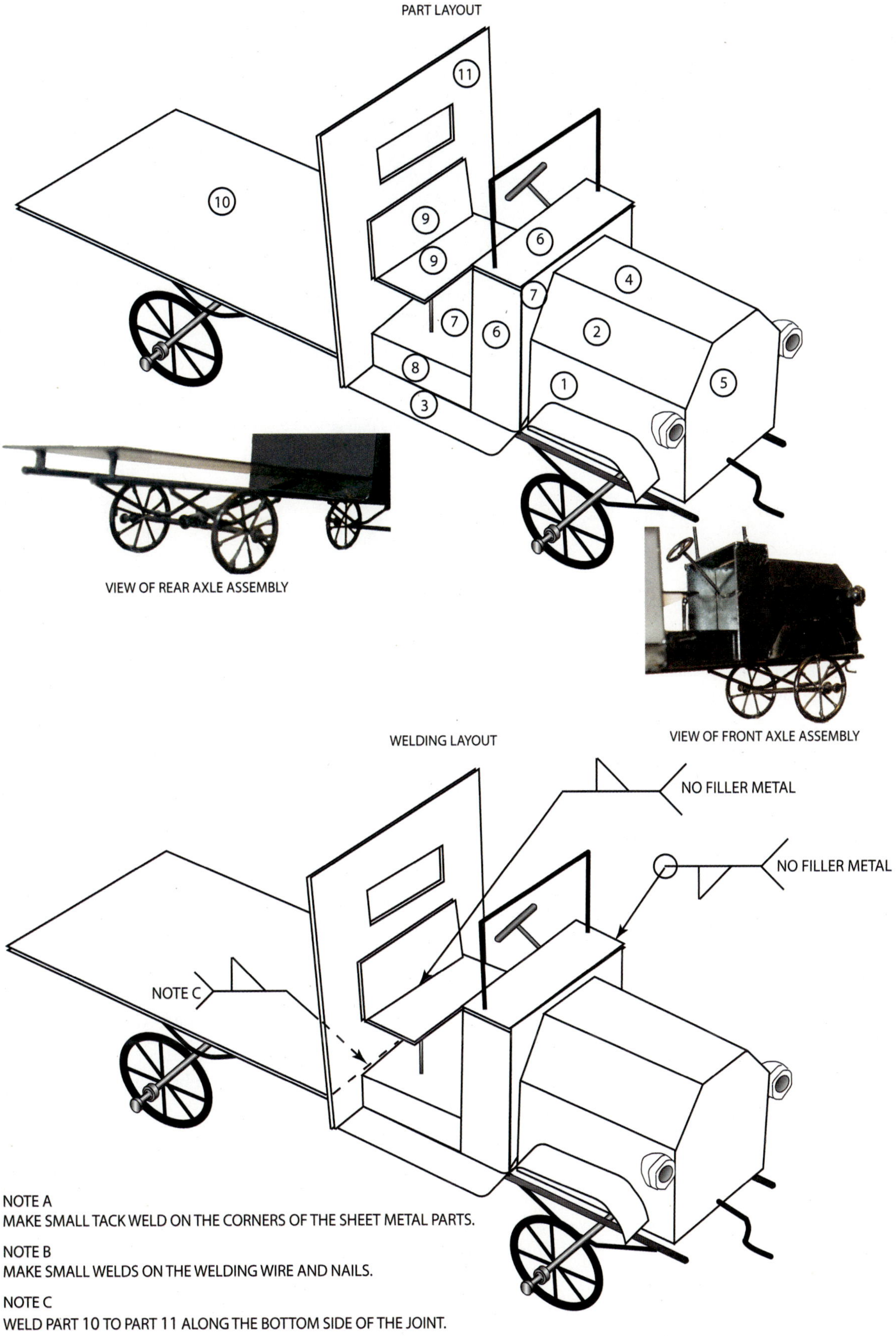

FIGURE 18-51B Antique truck assembly drawing. © Cengage Learning 2012

FIGURE 18-52 Antique truck. Larry Jeffus

REVIEW QUESTIONS

1. What are some advantages of gas welding?
2. What is the easiest metal to gas weld?
3. What characteristics of the weld bead can be controlled by the torch tip size?
4. What can cause the torch tip to overheat?
5. What can happen when a torch tip overheats?
6. How does the angle of the torch affect the rate of heating?
7. What should the distance between the inner cone and the metal ideally be?
8. How should the torch be manipulated to prevent burnthrough?
9. Describe a butt joint.
10. How does the frequency that filler metal is added affect the weld size?
11. How does the distance of the torch to the metal affect the weld bead width?
12. Where should the flame be directed when heating a lap joint?
13. What are some problems associated with welding a tee joint?
14. Why is it important to become proficient at using band saws, reciprocating saws, and saber saws to cut sheet metal?
15. When there is limited space in which to weld, how can you position your body so that you avoid accidentally touching the end of the filler metal outside of the molten weld pool?
16. Why is it not a good practice in the field to stop a weld at the corner of a joint?
17. What are out-of-position welds?
18. What is the most common out-of-position weld?
19. What can be done to prevent the molten weld pool from dripping on a vertical weld?
20. What does it mean to flash the torch, and when should this technique be used?

Chapter 19

Soldering, Brazing, and Braze Welding Processes

OBJECTIVES

After completing this chapter, the student should be able to:

- Compare the difference between soldering and brazing.
- List the advantages of soldering and/or brazing.
- Explain tensile strength, shear strength, ductility, fatigue resistance, and corrosion resistance as they relate to the strength of a joint.
- Explain why flux is used in soldering and brazing.
- Discuss the advantages and disadvantages of the five methods of heating material for soldering or brazing.
- Describe what factors must be considered when selecting a filler metal.
- Discuss the applications for common soldering and brazing alloys.
- Describe the preparation needed for a part before it is soldered or brazed.

KEY TERMS

brazing alloys
corrosion resistance
dip soldering
ductility
elastic limit
eutectic composition
fatigue resistance
furnace
induction
liquid-solid phase bonding
low-fuming alloys
paste range
phase
shear strength
soldering alloys
tensile strength

INTRODUCTION

Soldering and brazing are similar processes. Both processes can be used to produce strong, durable joints. Both processes can use similar alloying metals. Both processes can be completed with or without flux. The only difference is that soldering is done at a lower temperature, below 840°F (449°C), and brazing is done at a higher temperature, above 840°F (449°C). In both processes, the filler metal is melted, and the base metal is heated. At some point in the heating process a change occurs between the way the molten solder or braze metal and the heated metal surface react to each other. Before this temperature point is reached, the liquid metal often beads up, forming little balls sitting on the hot surface. At a high enough temperature the solder or braze metal flows out over the base metal surface and a strong bond is formed between the molten metal and base metal. The term *wetting* is used to describe the flowing out across the surface of the molten metal. Sometimes a flux may be required to remove a light surface oxide from the base metal surface and to promote wetting, but with or without a flux, the process the same.

Because soldering and brazing are similar processes, both are classified by the American Welding Society (AWS) as **liquid-solid phase bonding**. The term *liquid* is used in reference to the fact that the filler metal is melted. The term *solid* is used because the base material or materials are not melted. The **phase** refers to that temperature at which bonding takes place between the solid base material and the liquid filler metal. The term *bonding* refers to the attraction that exists between the cooled solidified filler metal and the base metal. The terms *material* or *materials* are used because both soldering and brazing can be used to join both metals and nonmetals. The most common nonmetals that are joined are ceramics. If done correctly, the bond can result in a joint that has four or five times the tensile strength of that of the filler metal that was used to join the parts.

ADVANTAGES OF SOLDERING AND BRAZING

Some advantages of soldering and brazing as compared to other methods of joining include the following:

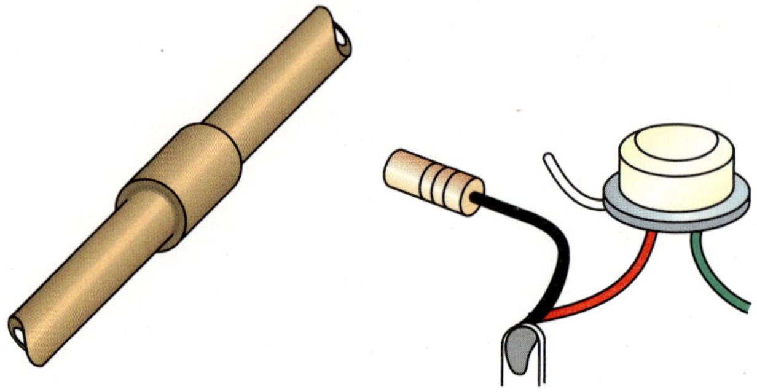

FIGURE 19-1 Examples of permanent joints that can easily be disassembled and the parts reused. © Cengage Learning 2012

- Low temperature—Since the base metal does not have to melt, a lower-temperature heat source can be used.
- May be permanently or temporarily joined—Since the base metal is not damaged, parts may be disassembled at a later time by simply reapplying heat. The parts then can be reused. However, the joint is solid enough to be permanent, Figure 19-1.
- Dissimilar materials can be joined—It is easy to join dissimilar metals, such as copper to steel, aluminum to brass, and cast iron to stainless steel, Figure 19-2. It is also possible to join nonmetals to each other or nonmetals to metals. Ceramics are easily brazed to each other or to metals.
- Speed of joining
  a. Parts can be preassembled and dipped or furnace soldered or brazed in large quantities, Figure 19-3.
  b. A lower temperature means less time in heating.
- Less chance of damaging parts—A heat source can be used that has a maximum temperature below the temperature that may cause damage to the parts, Figure 19-4.
- Slow rate of heating and cooling—Because it is not necessary to heat a small area to its melting temperature and then allow it to cool quickly to a solid,

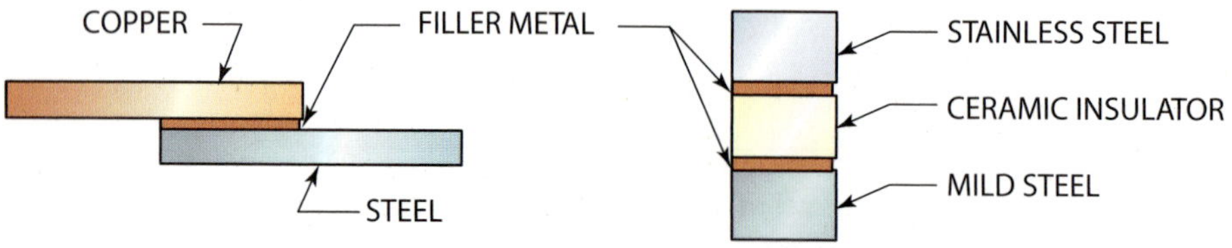

FIGURE 19-2 Dissimilar materials joined. © Cengage Learning 2012

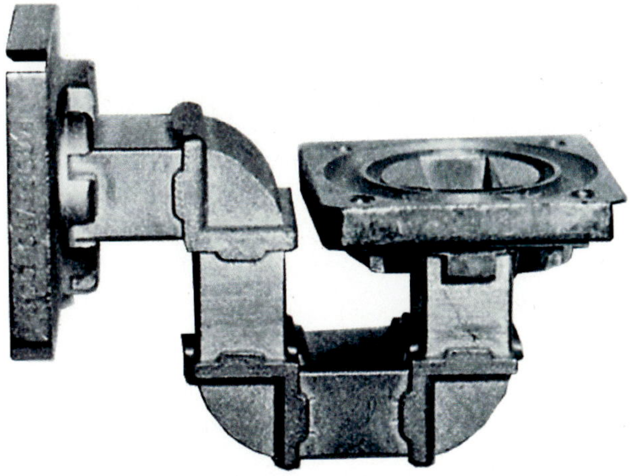

FIGURE 19-3 **Furnace-brazed part.** Larry Jeffus

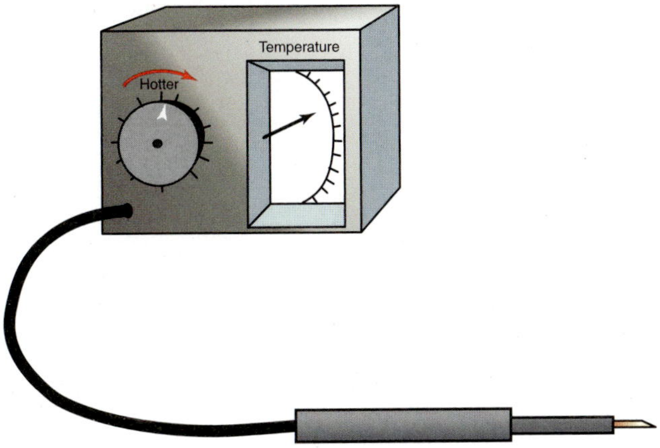

FIGURE 19-4 **Control console for resistance soldering.** © Cengage Learning 2012

the internal stresses caused by rapid temperature changes can be reduced.

- Parts of varying thicknesses can be joined—Very thin parts or a thin part and a thick part can be joined without burning or overheating them.
- Easy realignment—Parts can easily be realigned by reheating the joint and then repositioning the part.

Brazing and Braze Welding

Brazing is divided into two major categories named brazing and braze welding. In the brazing process the parts that are being joined must be fitted together very snugly. The joint spacing is very small, approximately 0.025 in. (0.6 mm) to 0.002 in. (0.06 mm), Figure 19-5A. This small spacing allows capillary action to draw the filler metal into the joint when the parts

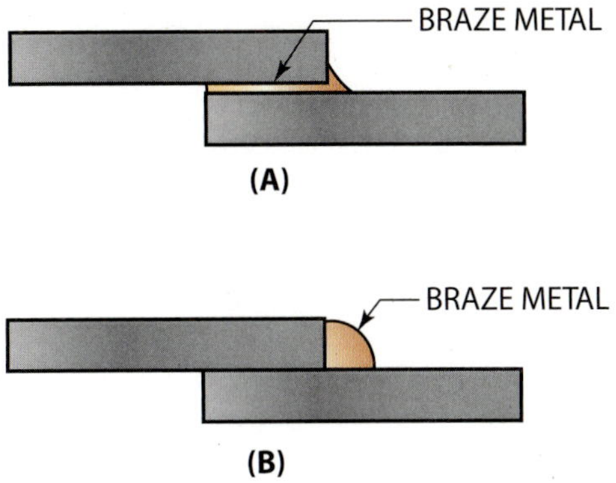

FIGURE 19-5 **A brazed lap joint (A) and a braze welded lap joint (B).** © Cengage Learning 2012

reach the proper phase temperature. Capillary action is the force that pulls water up into a paper towel or pulls a liquid into a very fine straw, Figure 19-6. This results in a very strong joint that uses very little filler metal.

Braze welding does not use capillary action to pull filler metal into a tightly fitted joint, Figure 19-5B. The same brazing alloy can be used for both brazing and braze welding. It is only how the filler metal reacts with the joint that makes the difference between brazing and braze welding. The parts joined by braze welding may have a very open or loose fitting. The filler metal may be used to fill a large joint gap or space. Braze welding uses more filler metal than brazing. Parts joined by braze welding are not as strong as those joined by brazing if they were both made with the same filler metal.

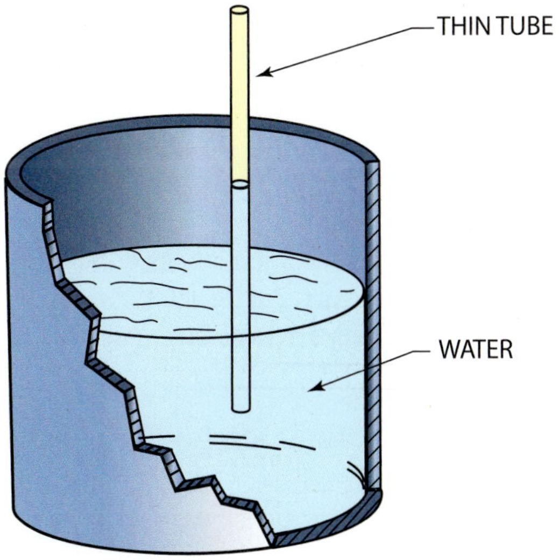

FIGURE 19-6 **Capillary action pulls water into a thin tube.** © Cengage Learning 2012

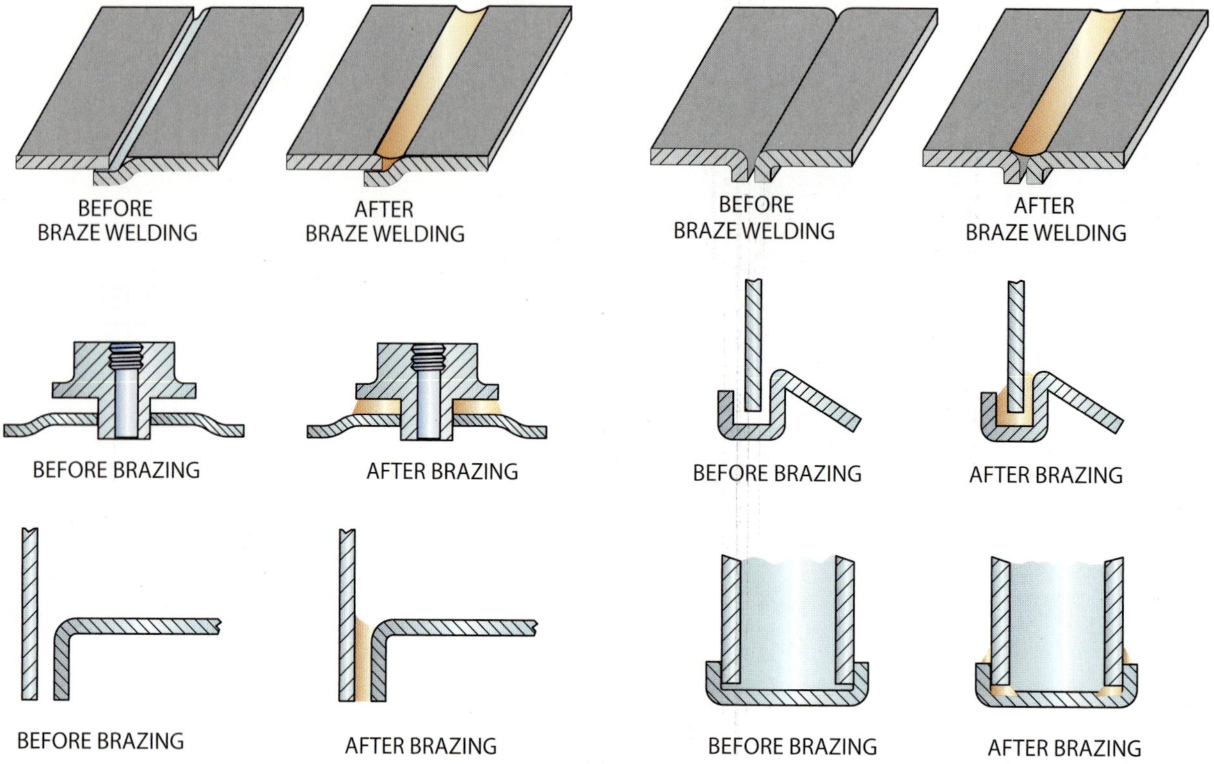

FIGURE 19-7 Examples of brazing and braze welded joints. American Welding Society

Examples of brazing and braze welding joint designs are shown in Figure 19-7. Some joints may contain areas of both brazing and braze welding. For example, the small fillet added to a copper sleeved joint could be classified as braze welding, while the filler metal that was pulled into the joint gap would be classified as brazing.

PHYSICAL PROPERTIES OF THE JOINT

Tensile Strength

The **tensile strength** of a joint is its ability to withstand being pulled apart, Figure 19-8. A brazed joint can be made that has a tensile strength four to five times higher than the filler metal itself. If a few drops of water are placed between two smooth and flat panes of glass and the panes are pressed together, a tensile force is required to pull the panes of glass apart. The water, which has no tensile strength itself, has added tensile strength to the glass joint.

The glass is being held together by the surface tension of the water. As the space between the pieces of glass decreases, the tensile strength increases. The same action takes place with a soldered or brazed joint. As the joint spacing decreases, the surface tension increases the tensile strength of the joint, Table 19-1.

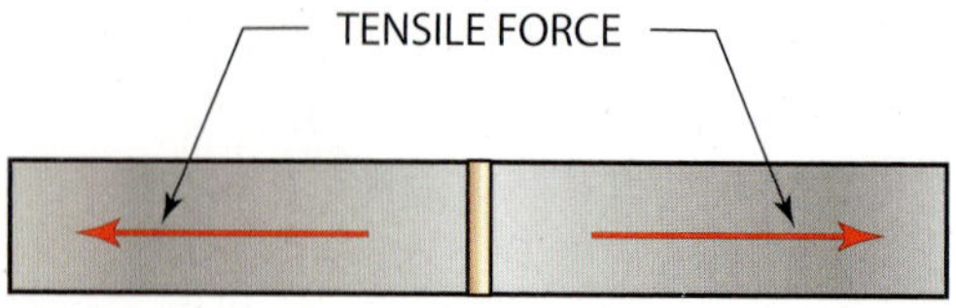

FIGURE 19-8 Joint in tension. © Cengage Learning 2012

Shear Strength

The **shear strength** of a brazed joint is its ability to withstand a force parallel to the joint, Figure 19-9. For a solder or braze joint, the shear strength depends upon the amount of overlapping area of the base parts. The greater the area that is overlapped, the greater is the strength.

Ductility

Ductility of a joint is its ability to bend without failing. Most soldering and brazing alloys are ductile metals, so the joint made with these alloys is also ductile.

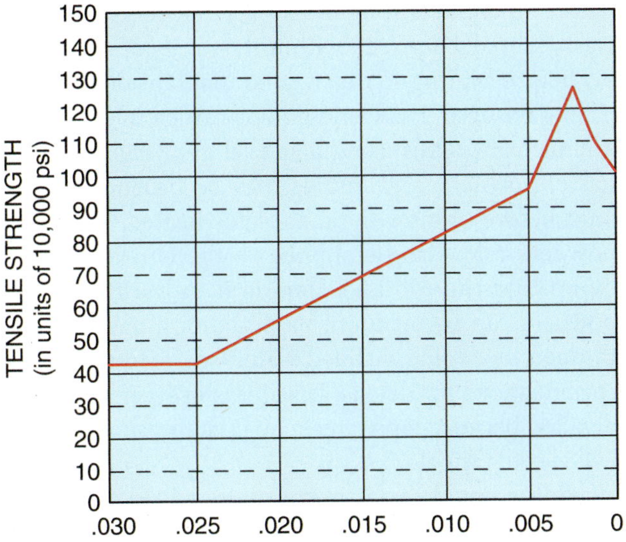

Table 19-1 Tensile Strength of Brazed Joints Increases as Joint Space Decreases

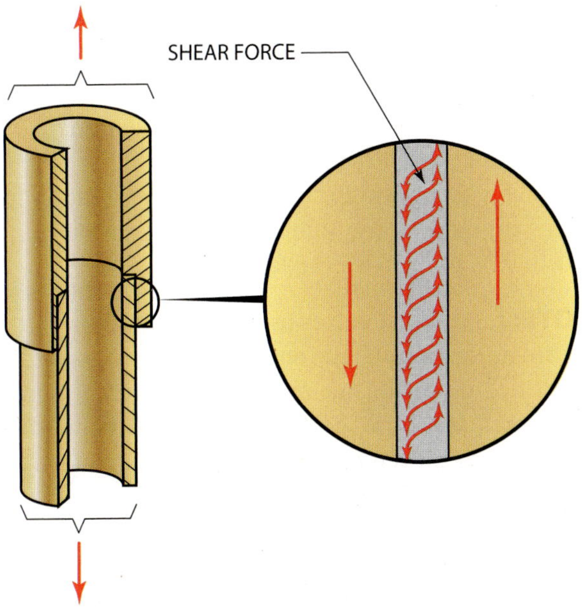

FIGURE 19-9 Effect of shear force on a joint. © Cengage Learning 2012

Fatigue Resistance

The **fatigue resistance** of a metal is its ability to be bent repeatedly without exceeding its **elastic limit** and without failure, Figure 19-10.

Corrosion Resistance

Corrosion resistance of a joint is its ability to resist chemical attack. The compatibility of the base

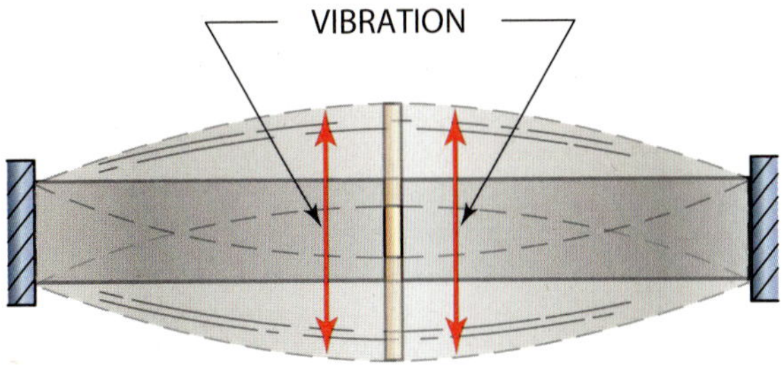

FIGURE 19-10 Vibration is a common cause of joint failure. © Cengage Learning 2012

materials to the filler metal will determine the corrosion resistance. Using the proper filler metal with the base materials that are listed in this chapter will result in corrosion-free joints. However, using filler metals on base materials that are not recommended in this chapter may result in a joint that looks good when completed but will eventually corrode. For example, a brass brazing rod that contains copper (Cu) and zinc (Zn) (BCuZn) will make a nice looking joint on stainless steel. But the zinc in the brass will combine with the nickel in the stainless steel if the part is kept hot for too long. As a result, a corrosive, brittle structure is formed in the joint, reducing strength.

FLUXES

General

Fluxes used in soldering and brazing have three major functions:

- They must remove any oxides that form as a result of heating the parts.
- They must promote wetting.
- They should aid in capillary action.

The flux, when heated to its reacting temperature, must be thin and flow through the gap provided at the joint. As it flows through the joint, the flux absorbs and dissolves oxides, allowing the molten filler metal to be pulled in behind it, Figure 19-11. After the joint is complete, the flux residue should be easily removable.

Fluxes are available in many forms, such as solids, powders, pastes, liquids, sheets, rings, and washers, Figure 19-12. They are also available mixed with the filler metal, inside the filler metal, or on the outside

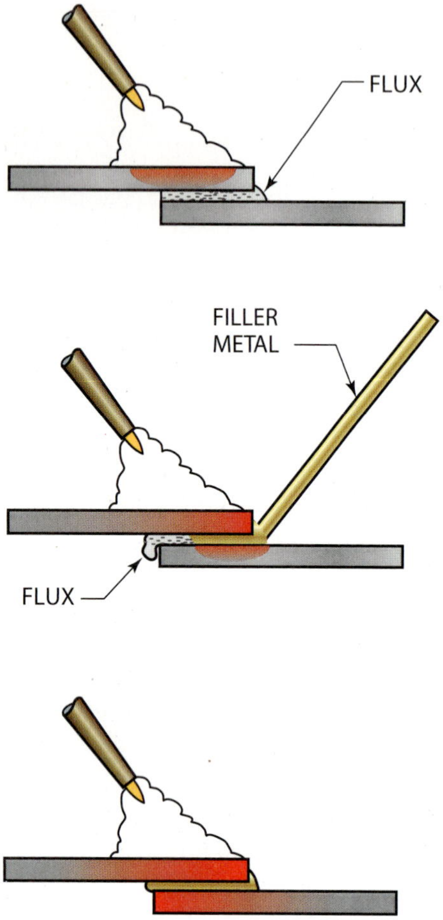

FIGURE 19-11 Flux flowing into a joint reduces oxides to clean the surfaces and gives rise to a capillary action that causes the filler metal to flow behind it. © Cengage Learning 2012

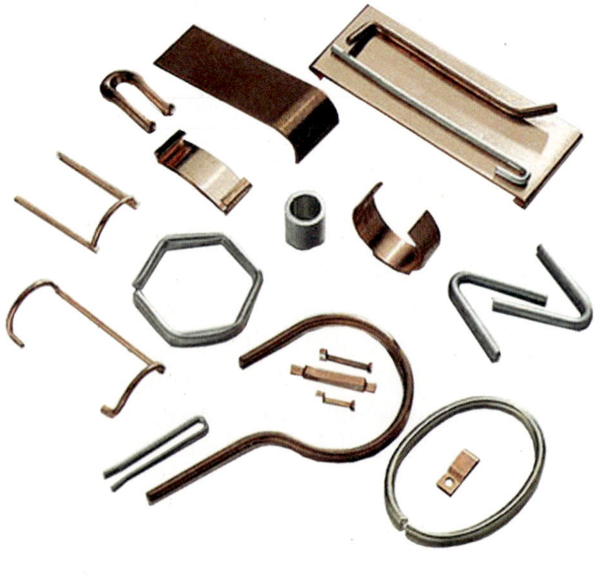

FIGURE 19-12 Braze/solder forms that can be preplaced in a braze/solder joint. Prince & Izant Co.

of the filler metal, Figure 19-13. Sheets, rings, and washers may be placed within the joints of an assembly before heating so that a good bond inside the joint can be ensured. Paste and liquids can be injected into a joint from tubes using a special gun, Figure 19-14. Paste, powders, and liquids may be brushed on the joint before or after the material is heated. Paste and powders may also be applied to the end of the rod by heating the rod and dipping it in the flux. Most powders can be made into a paste, or a paste can be thinned by adding distilled water; see manufacturers' specifications for details. If water is used, it should be distilled because tap water may contain minerals that will weaken the flux.

Some liquid fluxes may also be added to the gas when using an oxyfuel gas torch for soldering or brazing. The flux is picked up by the fuel gas as it is bubbled through the flux container and is then carried to the torch where it becomes part of the flame.

Flux and filler metal combinations are most convenient and easy to use, Figure 19-15. It may be necessary to stock more than one type of flux-filler metal combination for different jobs. In cases where the flux covers the outside of the filler metal, it may be damaged by humidity or chipped off during storage.

Using excessive flux in a joint may result in flux being trapped in the joint, weakening the joint, or causing the joint to leak or fail.

NOTE: Most of the fluxes used for brazing and soldering are not harmful to the environment. However, large quantities of even the most benign materials introduced accidentally or intentionally into the environment can cause damage. Even lemon juice, a common electronic soldering flux, in large enough quantities can cause harm to the environment. Before disposing of any soldering or brazing fluxes, read the material safety data sheet (MSDS) carefully and follow its recommended procedures. Keeping our environment clean and safe is everyone's responsibility.

Fluxing Action

The use of fluxes does not eliminate the need for good joint cleaning. Fluxes will not remove oil, dirt, paint, glues, heavy oxide layers, or other surface contaminants.

Soldering fluxes are chemical compounds such as muriatic acid (dilute hydrochloric acid), sal ammoniac (ammonium chloride), or rosin. Brazing fluxes are chemical compounds such as fluorides, chlorides, boric acids, and alkalies. These compounds

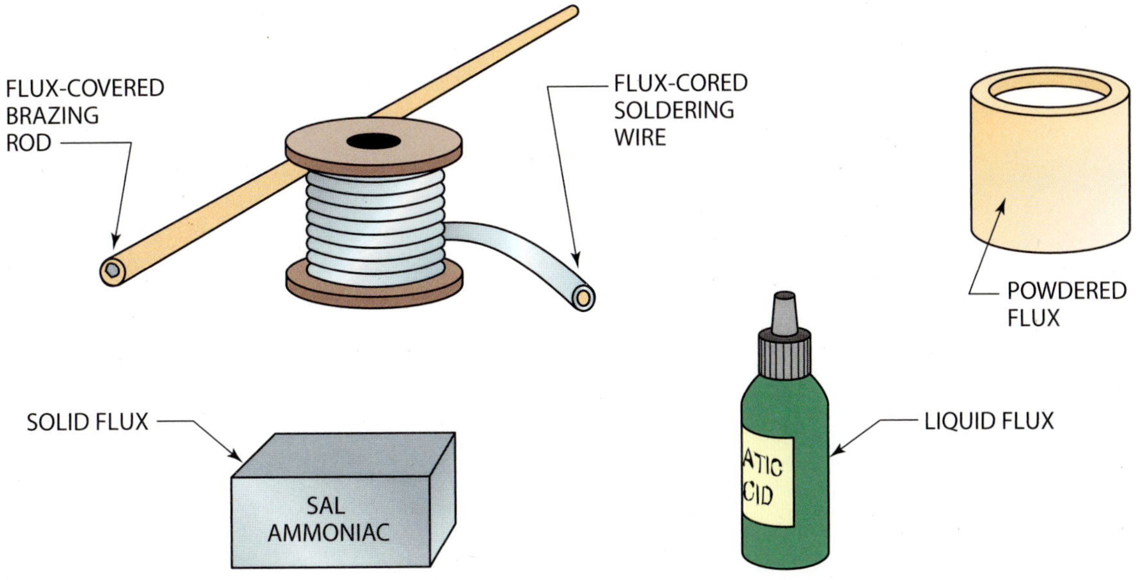

FIGURE 19-13 Flux can be purchased with the filler metal or separately. © Cengage Learning 2012

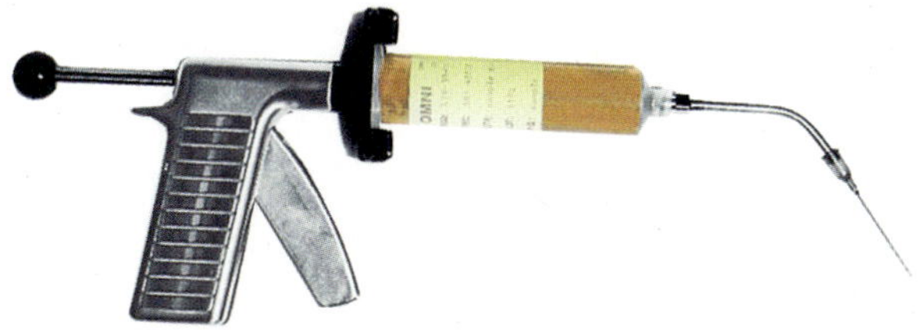

FIGURE 19-14 Gun for injecting flux into a joint
Larry Jeffus

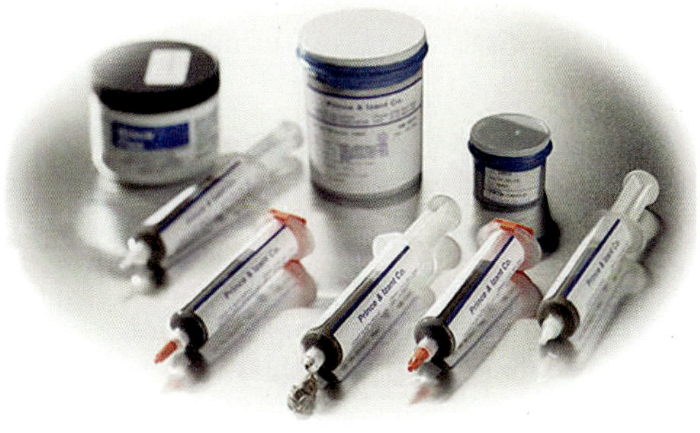

FIGURE 19-15 Tubes that contain flux filler metal mixtures. Prince & Izant Co.

react to dissolve, absorb, or mechanically break up thin surface oxides that are formed as the parts are being heated. Fluxes must be stable and remain active through the entire temperature range of the solder or braze filler metal. The chemicals in the flux react with the oxides as either acids or alkalies (bases). Some dip fluxes are salts.

The reactivity of a flux is greatly affected by temperature. As the parts are heated to the soldering or brazing temperature, the flux becomes more active. Some fluxes are completely inactive at room temperature. Most fluxes have a temperature range within which they are most effective. Care should be taken to avoid overheating fluxes. If they become overheated or burned, they will stop working as fluxes, and they become a contamination in the joint. If overheating has occurred, the welder must stop and clean off the damaged flux before continuing.

Fluxes that are active at room temperature must be neutralized (made inactive) or washed off after the job is complete. If these fluxes are left on the joint, premature failure may result due to flux-induced corrosion. Fluxes that are inactive at room temperature may not have to be cleaned off the part. However, if the part is to be painted or auto body filler is to be applied, fluxes must be removed.

SOLDERING AND BRAZING METHODS

General

Soldering and brazing methods are grouped according to the method with which heat is applied: torch, furnace, induction, dip, or resistance.

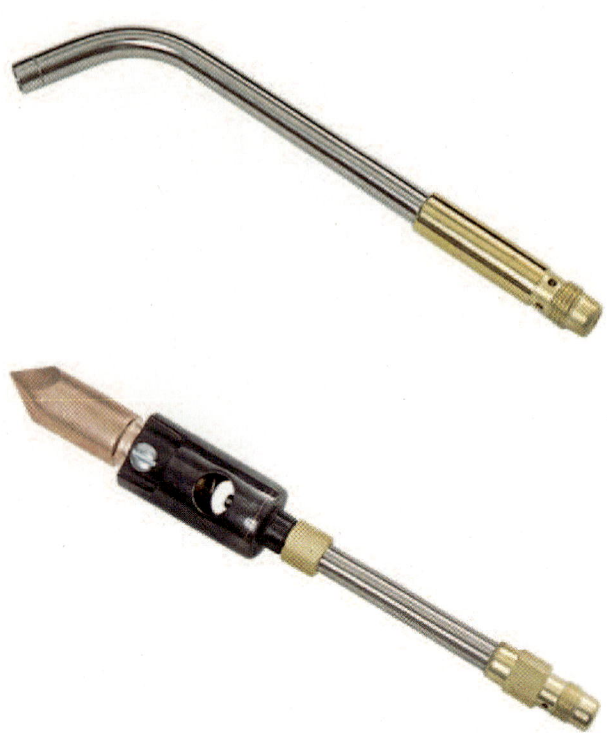

FIGURE 19-16 **An air propane torch can be used in soldering joints.** ESAB Welding & Cutting Products

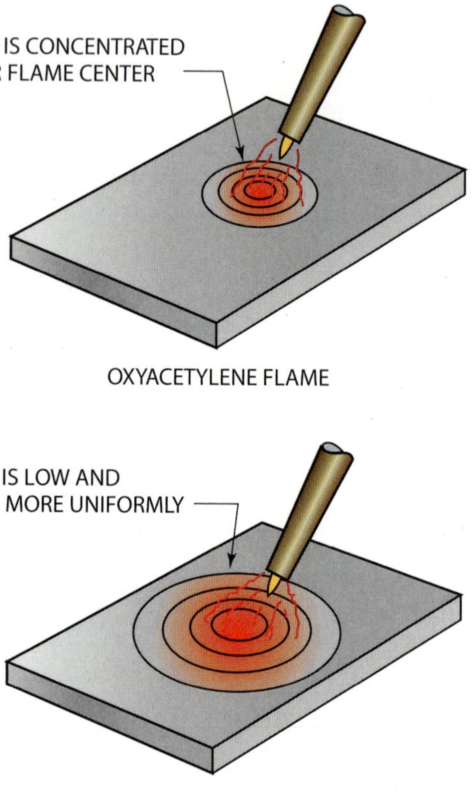

FIGURE 19-17 **The high temperature of an oxyacetylene flame may cause localized overheating.** © Cengage Learning 2012

Torch Soldering and Brazing

Oxyfuel or air-fuel torches can be used either manually or automatically, Figure 19-16. Acetylene is often used as the fuel gas, but it is preferable to use one of the other fuel gases having a higher heat level in the secondary flame, Figure 19-17. The oxyacetylene flame has a very high temperature near the inner cone, but it has little heat in the outer flame. This often results in the parts being overheated in a localized area. Such fuel gases as MAPP®, propane, butane, and natural gas have a flame that will heat parts more uniformly. Often torches are used that mix air with the fuel gas in a swirling or turbulent manner to increase the flame's temperature, Figure 19-18. The flame may even completely surround a small diameter pipe, heating it from all sides at once, Figure 19-19.

Some advantages of using a torch include the following:

- Versatility—Using a torch is the most versatile method. Both small and large parts in a wide variety of materials can be joined with the same torch.
- Portability—A torch is very portable. Any place a set of cylinders can be taken or anywhere the hoses can be pulled into, can be soldered or brazed with a torch.
- Speed—The flame of the torch is one of the quickest ways of heating the material to be joined, especially on thicker sections.

Some of the disadvantages of using a torch include the following:

- Overheating—When using a torch, it is easy to overheat or burn the parts, flux, or filler metal.
- Skill—A high level of skill with a torch is required to produce consistently good joints.
- Fires—It is easy to start a fire if a torch is used around combustible (flammable) materials.

Furnace Soldering and Brazing

In this method the parts are heated to their soldering or brazing temperature by passing them through or putting them into a **furnace**. The furnace may be heated by electricity, oil, natural gas, or any other locally available fuel. The parts may be passed through the furnace on a conveyor belt in trays or placed on the belt itself, Figure 19-20. The parts may also be

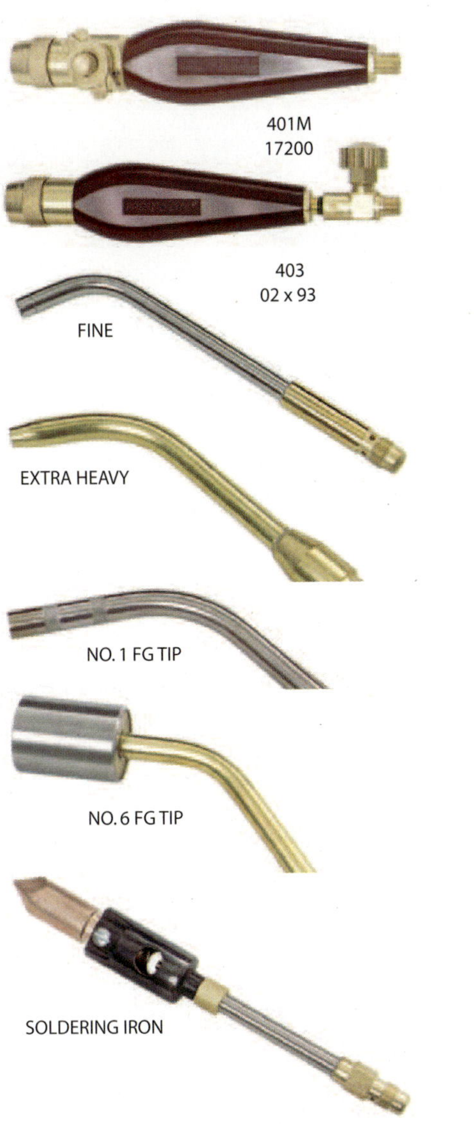

FIGURE 19-18 Examples of torch tips and handles that use air-fuel mixtures for brazing. ESAB Welding & Cutting Products

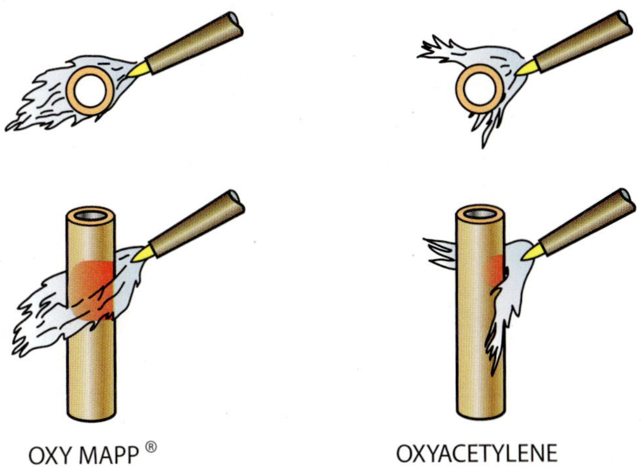

FIGURE 19-19 Heating characteristics of oxy MAPP® compared with oxyacetylene on round materials.
© Cengage Learning 2012

loaded in trays to be placed in a furnace that does not use a conveyor belt, Figure 19-21.

Some of the advantages of using a furnace include the following:

- Temperature control—The furnace temperature can be accurately controlled to ensure that the parts will not overheat.
- Controlled atmosphere—The furnace can be filled with an inert gas to prevent oxides from forming on the parts.
- Uniform heating—The uniform heating of the parts reduces stresses and distortion.
- Mass production—By using a furnace, it is easy to produce a high quantity of parts.

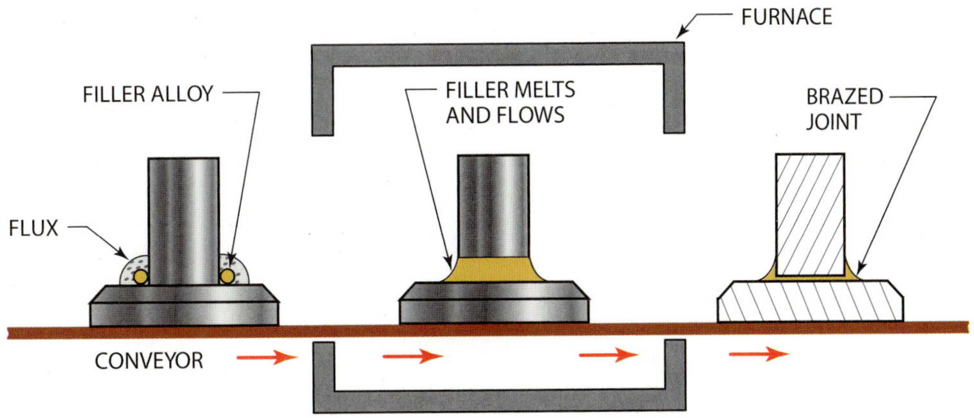

FIGURE 19-20 Furnace brazing permits the rapid joining of parts on a production basis.
American Welding Society

FIGURE 19-21 Small furnace-brazed part.
Larry Jeffus

Some of the disadvantages of using a furnace include the following:

- Size—Unless parts are small, the length of time required to heat them is extremely long.
- Heat damage—The entire part must be able to withstand heating without burning.

Induction Soldering and Brazing

The **induction** method of heating uses a high-frequency electrical current to establish a corresponding current on the surface of the part, Figure 19-22. The current on the part causes rapid and very localized heating of the surface only. There is little, if any, internal heating of the part except by conductivity of heat from the surface.

The advantage of the induction method is:

- Speed—Very little time is required for the part to reach the desired temperature.

Some of the disadvantages of the induction method include the following:

- Distortion—The very localized heating may result in some distortion.

FIGURE 19-22 Induction brazing and soldering machine. Prince & Izant Co.

- Lack of temperature control—The electrical resistance of the part increases as the part heats up. This, in turn, increases the temperature produced.
- Incomplete penetration—Because the inside of the part is not directly heated, it may be too cool to permit the filler metal to flow fully through the joint.

Dip Soldering and Brazing

Two types of **dip soldering** or brazing are used: molten flux bath and molten metal bath. With the molten flux method, the soldering or brazing filler metal in a suitable form is preplaced in the joint, and the assembly is immersed in a bath of molten flux, as shown in Figure 19-23. The bath supplies the heat needed to preheat the joint and fuse the solder or braze metal, and it provides protection from oxidation.

With the molten metal method, the prefluxed parts are immersed in a bath of molten solder or braze metal, which is protected by a cover of molten flux. This method is confined to wires and other small parts. Once they are removed from the bath, the ends of the wires and parts must not be allowed to move until the solder or braze metal has solidified. As with all soldering or brazing operations, any movement of the parts as they cool from a liquid through the paste range to become a solid will result in microfractures in the filler metal. In electronic parts these microfractures cause

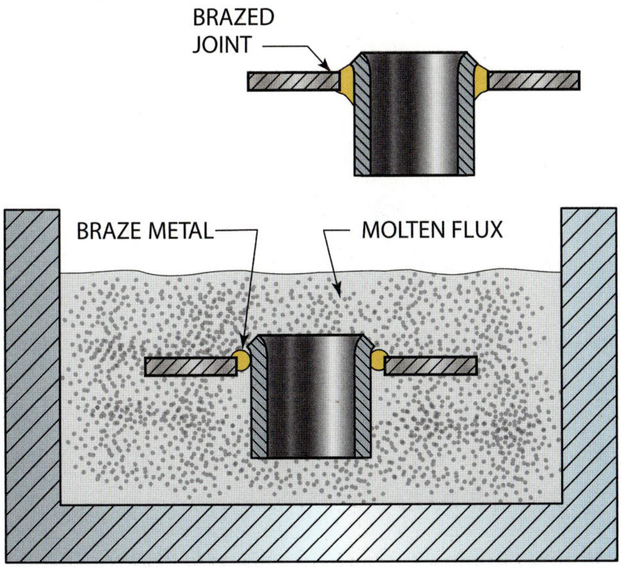

FIGURE 19-23 Dip brazing eliminates the need for a separate fluxing operation. American Welding Society

resistance to the electron flow and may render the part unfit for service. Reheating can be used to refuse the joint only if reheating will not damage the part beyond use.

Some of the advantages of dip processing include the following:

- Mass production—It is possible to dip many small parts at one time.
- Corrosion protection—The entire surface of the part can be covered with the filler metal at the same time that it is being joined. If a corrosion-resistant filler metal is used, the thin layer provided will help protect the part from corrosion.
- Distortion is minimized—The entire part is heated uniformly, which reduces distortion.

Some of the disadvantages of dip processing include the following:

- Steam explosions—Moisture trapped in the joint may cause a steam explosion that can scatter molten metal.
- Corrosion—If any of the salt is trapped in the joint or is left on the surface, corrosion may cause the part to fail at some time in the future.
- Size—Parts must be small to be effectively joined.
- Quantity—Only a large quantity of parts can justify heating the large amount of molten filler metal or salt required for dipping.

Resistance Soldering and Brazing

The resistance method of heating uses an electric current that is passed through the part. The resistance of the part surfaces at the joint face to the current flow results in the heat needed to produce the bond. The flux and filler metal are usually preplaced.

The machine used in this method resembles a spot welder.

Some of the advantages of the resistance-heating method include the following:

- Localized heating—The heat can be localized so that the entire part may not get hot.
- Speed—A wide variety of spots can be made on the same machine without having to make major adjustments on the machine.
- Multiple spots—Many spots can be joined in a small area without disturbing joints that are already made.

Some of the disadvantages of the resistance-heating method include the following:

- Distortion—Localized heating may result in distortion.
- Conductors—Parts must be able to conduct electricity.
- Joint design—Lap joints in plate are the only joint designs that can be made.

Special Methods

A few other methods of producing soldered or brazed parts are used that do not entirely depend upon heat to produce the joint. The ultrasonic method uses high-frequency sound waves to produce the bond or to aid with heat in the bonding, Figure 19-24. Another process uses infrared light to heat the part for soldering or brazing.

FILLER METALS

General

The type of filler metal used for any specific joint should be selected by considering as many of the criteria listed in Figure 19-25 as possible. Welders must decide which factors they feel are most important and then base their selection on that decision.

Soldering and brazing metals are alloys—that is, a mixture of two or more metals. Each alloy is available

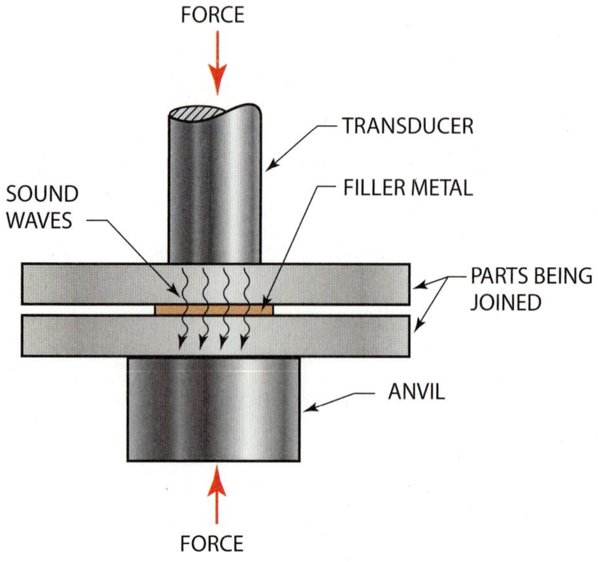

FIGURE 19-24 Ultrasonic bonding.
American Welding Society

- Material being joined
- Strength required
- Joint design
- Availability
- Appearance
- Service (corrosion)
- Heating process to be used
- Cost

FIGURE 19-25 Criteria for selecting filler metal.
© Cengage Learning 2012

in a variety of percentage mixtures. Some mixtures are stronger, and some melt at lower temperatures than other mixtures. Each one has specific properties. Almost all of the alloys used for soldering or brazing have a paste range. A **paste range** is the temperature range in which a metal is partly solid and partly liquid as it is heated or cooled. As the joined part cools through the paste range, it is important that the part not be moved. If the part is moved, the solder or braze metal may crumble like dry clay, destroying the bond.

EXPERIMENT 19-1

Paste Range

This experiment shows the effect on bonding of moving a part as the filler metal cools through its paste range. The experiment also shows how metal can be

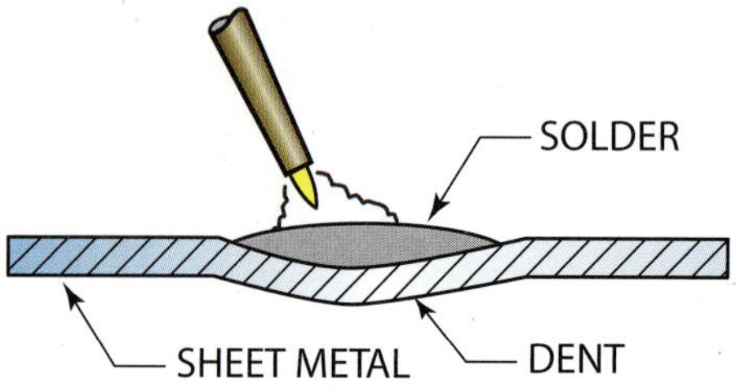

FIGURE 19-26 Partially fill the dent with molten solder.
© Cengage Learning 2012

"worked" using its paste range. You will need tin-lead solder composed of 20 to 50% tin, with the remaining percentage being lead. You also will need a properly lit and adjusted torch, a short piece of brazing rod, and a piece of sheet metal. Using a hammer, make a dent in the sheet metal about the diameter of a quarter (25¢), Figure 19-26.

In a small group, watch the effects of heating and cooling solder as it passes through the paste range. Using the torch, melt a small amount of the solder into the dent and allow it to harden. Remelt the solder slowly, frequently flashing the torch off and touching the solder with the brazing rod until it is evident that the solder has all melted. Once the solder has melted, stick the brazing rod in the solder and remove the torch. As the solder cools, move the brazing rod in the metal and observe what happens, Figure 19-27.

As the solder cools to the uppermost temperature of its paste range, it will have a rough surface appearance as the rod is moved. When the solder cools more, it will start to break up around the rod. Finally, as it becomes a solid, it will be completely broken away from the rod. Now slowly reheat the solder and work the surface with the rod until it can be shaped like clay. If the surface is slightly rough, a quick touch

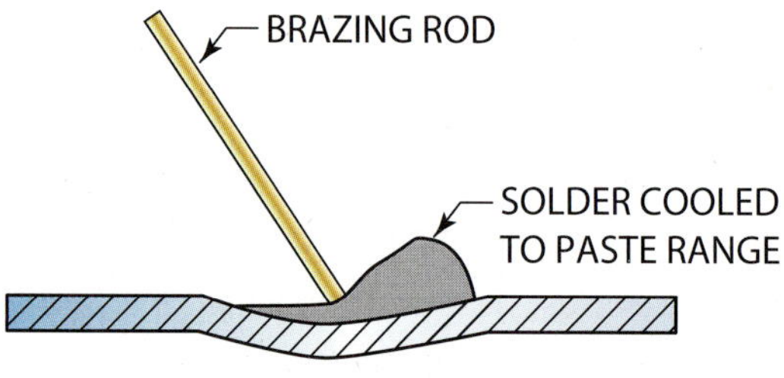

FIGURE 19-27 Solder being shaped as it cools to its paste range. © Cengage Learning 2012

Tin-lead	Copper and copper alloys Mild steel Galvanized metal
Tin-antimony	Copper and copper alloys Mild steel
Cadmium-silver	High strength for copper and copper alloys Mild steel Stainless steel
Cadmium-zinc	Aluminum and aluminum alloys

Table 19-2 Soldering Alloys

of the flame will smooth it. This is the same way in which "lead" is applied to some body panel joints on a new car so that the joints are not seen on the car when it is finished. The lead used is actually a tin-lead alloy or solder. A large area can be made as smooth as glass without sanding by simply flashing the area with the flame.◆

Soldering Alloys

Soldering alloys are usually identified by their major alloying elements. Table 19-2 lists the major types of solder and the materials they will join. In many cases, a base material can be joined by more than one solder alloy. In addition to the considerations for selecting filler metal listed in Figure 19-25, specific factors are listed in the following sections for the major soldering alloys.

Tin-Lead

This solder is available in a wide range of alloy combinations. Each alloy combination changes a characteristic, such as melting temperature or past range of the solder. An alloy of 61.9% tin and 38.1% lead melts at 362°F (183°C) and has no paste range. This is the **eutectic composition** (lowest possible melting point of an alloy) of the tin-lead solder. An alloy of 60% tin and 40% lead is commercially available and is close enough to the eutectic alloy to have the same low melting point with only a 12°F (7.8°C) paste range. The widest paste range is 173°F (78°C) for a mixture of 19.5% tin and 80.5% lead. This mixture begins to solidify at 535°F (289°C) and is totally solid at 362°F (193°C). The closest mixture that is commercially available is a 20% tin and 80% lead alloy. Table 19-3 lists the percentages, temperatures, and paste ranges for tin-lead solders.

Tin-lead solders are most commonly used on electrical connections, air-conditioning and refrigeration drain piping, and for architectural accents where good corrosion resistance is needed. Tin-lead solders must never be used for water piping. Most health and construction codes will not allow tin-lead solders for use on water or food handling equipment.

CAUTION

Tin-lead solders must not be used where lead could become a health hazard in things such as food and water.

Tin-Antimony

This family of solder alloys has a higher tensile strength and lower creep than the tin-lead solders. The most common alloy is 95/5, 95% tin and 5% antimony. This is often referred to as "hard solder." The use of "C" flux, which is a mixture of flux and small flakes of solder, makes it easier to fabricate quality joints. This mixture of flux and solder draws additional solder into the joint as it is added.

Tin-antimony solders are used for plumbing because they are lead-free and for refrigeration work.

Tin-Antimony-Lead

Antimony is added to tin-lead solders in amounts up to 6% to increase the strength and mechanical properties of the alloy. Tin-antimony-lead solder alloys can be used when higher joint strength is required. However, the higher antimony content may form a very brittle joint if used on aluminum, zinc, or zinc-coated metals.

Cadmium-Silver

These solder alloys have excellent wetting, flow, and strength characteristics, but they are expensive. The silver in this solder helps improve wetting and strength. Cadmium-silver alloys melt at a temperature of around 740°F (393°C); they are called high-temperature solders because they retain their strength at temperatures above most other solders. These solder alloys can be used to join aluminum to itself or other metals—for example, to piping that is used in air-conditioning equipment.

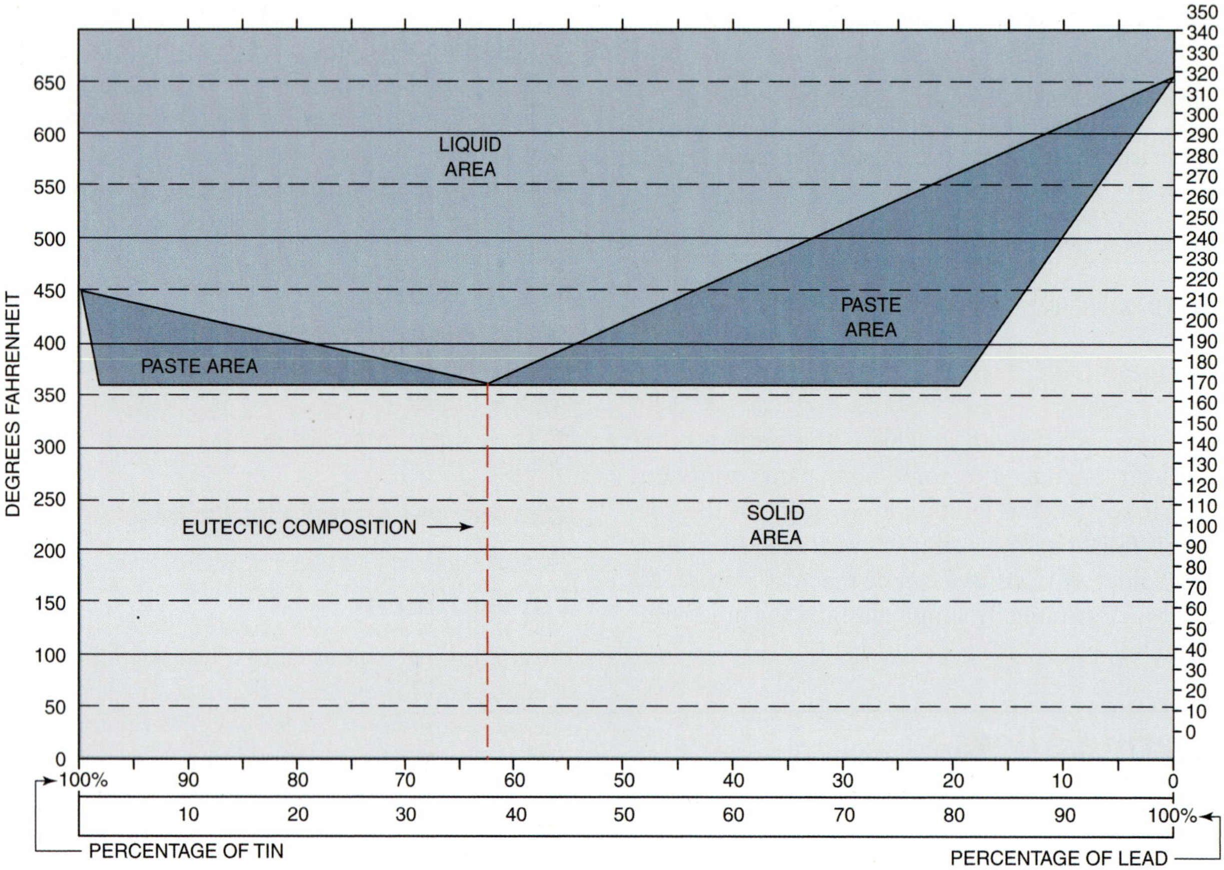

Table 19-3 Melting, Solidification, and Paste Range Temperatures for Tin-Lead Solders

CAUTION

When silver soldering on food-handling equipment, use a cadmium-free silver solder.

CAUTION

If the cadmium is overheated, the fumes can be hazardous unless the area is properly ventilated.

Cadmium-Zinc

Cadmium-zinc alloys have good wetting action and corrosion resistance on aluminum and aluminum alloys. The melting temperature is high, and some alloys have a wide paste range, Table 19-4.

Brazing Alloys

The American Welding Society's classification system for **brazing alloys** uses the letter *B* to indicate that the alloy is to be used for brazing. The next series of letters in the classification indicates the atomic symbol

Cadmium	Zinc	Completely Liquid	Completely Solid	Paste Range
82.5%	17.5%	509°F (265°C)	509°F (265°C)	No paste range
40.0%	60.0%	635°F (335°C)	509°F (265°C)	126°F (52°C)
10.0%	90.0%	750°F (399°C)	509°F (265°C)	241°F (116°C)

Table 19-4 Cadmium-Zinc Alloys

of metals used to make up the alloy, such as CuZn (copper and zinc). There may be a dash followed by a letter or number to indicate a specific alloyed percentage. The letter *R* may be added to indicate that the braze metal is in rod form. An example of a filler metal designation is BRCuZn-A, which indicates a copper-zinc brazing rod with 59.25% copper, 40% zinc, and 0.75% tin. Table 19-5 provides a list of the base metals and the most common alloys used to join the base metals. Not all of the available brazing alloys have an AWS classification. Some special alloys are known by registered trade names.

Copper-Zinc

Copper-zinc alloys are the most popular brazing alloys. They are available as regular and **low-fuming alloys**. The zinc in this braze metal has a tendency to burn out if it is overheated. Overheating is indicated by a red glow on the molten pool, which gives off a white smoke. The white smoke is zinc oxide. If zinc oxide is breathed in, it can cause zinc poisoning. Using a low-fuming alloy helps eliminate this problem. Examples of low-fuming alloys are RCuZn-B and RCuZn-C.

Base Metal	Brazing Filler Metal
Aluminum	BAlSi, aluminum-silicon
Carbon steel	BCuZn, brass (copper-zinc) BCu, copper alloy BAg, silver alloy
Alloy steel	BAg, silver alloy BNi, nickel alloy
Stainless steel	BAg, silver alloy BAu, gold base alloy BNi, nickel alloy
Cast iron	BCuZn, brass (copper-zinc)
Galvanized iron	BCuZn, brass (copper-zinc)
Nickel	BAu, gold base alloy BAg, silver alloy BNi, nickel alloy
Nickel-copper alloy	BNi, nickel alloy BAg, silver alloy BCuZn, brass (copper-zinc)
Copper	BCuZn, brass (copper-zinc) BAg, silver alloy BCuP, copper-phosphorus
Silicon-bronze	BCuZn, brass (copper-zinc) BAg, silver alloy BCuP, copper-phosphorus
Tungsten	BCuP, copper-phosphorus

Table 19-5 Base Metals and Common Brazing Filler Metals Used to Join the Base Metals

CAUTION

Breathing zinc oxide can cause zinc poisoning. If you think you have zinc poisoning, get medical treatment immediately.

Copper-Zinc and Copper-Phosphorus A5.8

The copper-zinc filler rods are often grouped together and known as brazing rods. The copper-phosphorus rods are referred to as phos-copper. Both terms do not adequately describe the metals in this group. There are vast differences among the five major classifications of the copper-zinc filler metals, as well as among the five major classifications of the copper-phosphorus filler metals. The following material describes the five major classifications of copper-zinc filler rods.

Class BRCuZn is used for the same application as BCu fillers. The addition of 40% zinc (Zn) and 60% copper (Cu) improves the corrosion resistance and aids in this rod's use with silicon-bronze, copper-nickel, and stainless steel.

CAUTION

Care must be exercised to prevent overheating this alloy, as the zinc will vaporize, causing porosity and poisonous zinc fumes.

Class BRCuZn-A is commonly referred to as naval brass and can be used to fuse weld naval brass. The addition of 17% tin (Sn) to the alloy adds strength and corrosion resistance. The same types of metal can be joined with this rod as could be joined with BRCuZn.

Class BRCuZn-B is a manganese-bronze filler metal. It has a relatively low melting point and is free flowing. This rod can be used to braze weld steel, cast iron, brass, and bronze. The deposited metal is higher than BRCuZn or BRCuZn-A in strength, hardness, and corrosion resistance.

Class BRCuZn-C is a low-fuming, high silicon (Si) bronze rod. It is especially good for general-purpose work due to the low-fuming characteristic of the silicon on the zinc.

Class BRCuZn-D is a nickel-bronze rod with enough silicon to be low fuming. The nickel gives the deposit a silver-white appearance and is referred to as white brass. This rod is used to braze and braze weld steel, malleable iron, and cast iron and for building up wear surfaces on bearings.

Copper-Phosphorus

This alloy is sometimes referred to as phos-copper. It is a good alloy to consider for joints where silver braze alloys may have been used in the past. Phos-copper has good fluidity and wettability on copper and copper alloys. The joint spacing should be from 0.001 in. (0.03 mm) to 0.005 in. (0.12 mm) for the strongest joints. Heavy buildup of this alloy may cause brittleness in the joint. Phosphorus forms brittle iron phosphide at brazing temperatures on steel. Copper-phos or copper-phos-silver should not be used on copper-clad fittings with ferrous substrates because the copper can easily be burned off, exposing the underlying metal to phosphorus embrittlement.

The copper-phosphorus (BCuP group) rods are used in air-conditioning applications and in plumbing to join copper piping. The phosphorus makes the rod self-fluxing on copper. This feature is one of the major advantages of copper-phosphorus rods. The addition of a small amount of silver, approximately 2%, helps with wetting and flow into joints.

Class BCuP-1 has a low-wetting characteristic and a lower flow rate than the other phos-copper alloys. This type of filler metal should be preplaced in the joint. The major advantage of this type of filler metal is its increased ductility.

Classes BCuP-2 and BCuP-4 both have good flow into the joint. The high phosphorus content of the rods makes them self-fluxing on copper. Both of these classes are used often for plumbing installations.

Classes BCuP-3 and BCuP-5 both have high surface tension and low flow so that they are used when close fit-ups are not available.

Copper-Phosphorus-Silver

This alloy is sometimes referred to as sil-phos. Its characteristics are similar to copper-phosphorus except the silver gives this alloy a little better wetting and flow characteristics. Often it is not necessary to use flux with alloys containing 5% or more of silver when joining copper pipe. This is the most common brazing alloy used in air-conditioning and refrigeration work. When sil-phos is used on air-conditioning compressor fittings that are copper-clad steel, care must be taken to make the braze quickly. If the fitting is heated too much or for too long, the copper cladding can be burned off. With this burn-off, the phosphorus can make the steel fitting very brittle, and embrittlement can cause the fitting to crack and leak sometime later.

Silver-Copper

Silver-copper alloys can be used to join almost any metal, ferrous or nonferrous, except aluminum, magnesium, zinc, and a few other low-melting metals. This alloy is often referred to as silver braze and is the most versatile. It is among the most expensive alloys, except for the gold alloys.

Nickel

Nickel alloys are used for joining materials that need high strength and corrosion resistance at an elevated temperature. Some applications of these alloys include joining turbine blades in jet engines, torch parts, furnace parts, and nuclear reactor tubing. Nickel will wet and flow acceptably on most metals. When used on copper-based alloys, nickel may diffuse into the copper, stopping its capillary flow.

Nickel and Nickel Alloys A5.14

Nickel and nickel alloys are increasingly used as a substitute for silver-based alloys. Nickel is generally more difficult to use than silver because it has lower wetting and flow characteristics. However, nickel has a much higher strength than silver.

Class BNi-1 is a high-strength, heat-resistant alloy that is ideal for brazing jet engine parts and for other similar applications.

Class BNi-2 is similar to BNi-1 but has a lower melting point and a better flow characteristic.

Class BNi-3 has a high flow rate that is excellent for large areas and close-fitted joints.

Class BNi-4 has a higher surface tension than the other nickel filler rods, which allows larger fillets and poor-fitted joints to be filled.

Class BNi-5 has a high oxidation resistance and high strength at elevated temperatures and can be used for nuclear applications.

Class BNi-6 is extremely free flowing and has good wetting characteristics. The high corrosion resistance gives this class an advantage when joining low chromium steels in corrosive applications.

Class BNi-7 has a high resistance to erosion and can be used for thin or honeycomb structures.

Aluminum-Silicon

BAlSi brazing filler metals can be used to join most aluminum sheet and cast alloys. The AWS type number 1 flux must be used when brazing aluminum. It is very easy to overheat the joint. If the flux is burned by overheating, it will obstruct wetting. Use standard torch brazing practices but guard against overheating.

Copper and Copper Alloys A5.7

Although pure copper (Cu) can be gas fusion welded successfully using a neutral oxyfuel flame without a flux, most copper filler metals are used to join other metals in a brazing process.

Class BCu-1 can be used to join ferrous, nickel, and copper-nickel metals with or without a flux. BCu-1 is also available as a powder that is classified as BCu-1a. This material has the same applications as BCu-1. The AWS type number 3B flux must be used with metals that are prone to rapid oxidation or with heavy oxides such as chromium, titanium, manganese, and others.

Class BCu-2 has applications similar to those for BCu-1. However, BCu-2 contains copper oxide suspended in an organic compound. Since copper oxides can cause porosity, tying up the oxides with the organic compounds reduces the porosity.

Silver and Gold

Silver and gold are both used in small quantities when joining metals that will be used under corrosive conditions, when high joint ductility is needed, or when low electrical resistance is important. Because of the ever-increasing price and reduced availability of these precious metals, other filler metals should first be considered. In many cases, other alloys can be used with great success. When substituting a different filler metal for one that has been used successfully, the new metal and joint should first be extensively tested.

JOINT DESIGN

General

The spacing between the parts being joined greatly affects the tensile strength of the finished part. Table 19-6 lists the spacing requirements at the joining temperature for the most common alloys. As the parts are heated, the initial space may increase or decrease, depending upon the joint design and fixturing. The changes due to expansion can be calculated, but trial and error also works. The strongest joints are obtained when the parts use lap or scarf joints where the joining area is equal to three times the thickness of the thinnest joint member, Figure 19-28. The strength of a butt joint can be increased if the area being joined can be increased. Parts that are 1/4-in. (6-mm) thick should not be considered for brazing or soldering if another process will work successfully.

Some joints can be designed so that the flux and filler metal may be preplaced. When this is possible, visual checking for filler metal around the outside of the joint is easy. Evidence of filler metal around the outside is a good indication of an acceptable joint.

Joint cleaning is very important to a successful soldered or brazed part. The surface must be cleaned of all oil, dirt, paint, oxides, or any other contaminants. The surface can be mechanically cleaned by using a wire brush, sanding, sandblasting, grounding,

Filler Metal	Joint Spacing in.	Joint Spacing mm
BAlSi	.006–.025	(0.15–0.61)
BAg	.002–.005	(0.05–0.12)
BAu	.002–.005	(0.05–0.12)
BCuP	.001–.005	(0.03–0.12)
BCuZn	.002–.005	(0.05–0.12)
BNi	.002–.005	(0.05–0.12)

Table 19-6 Brazing Alloy Joint Tolerance

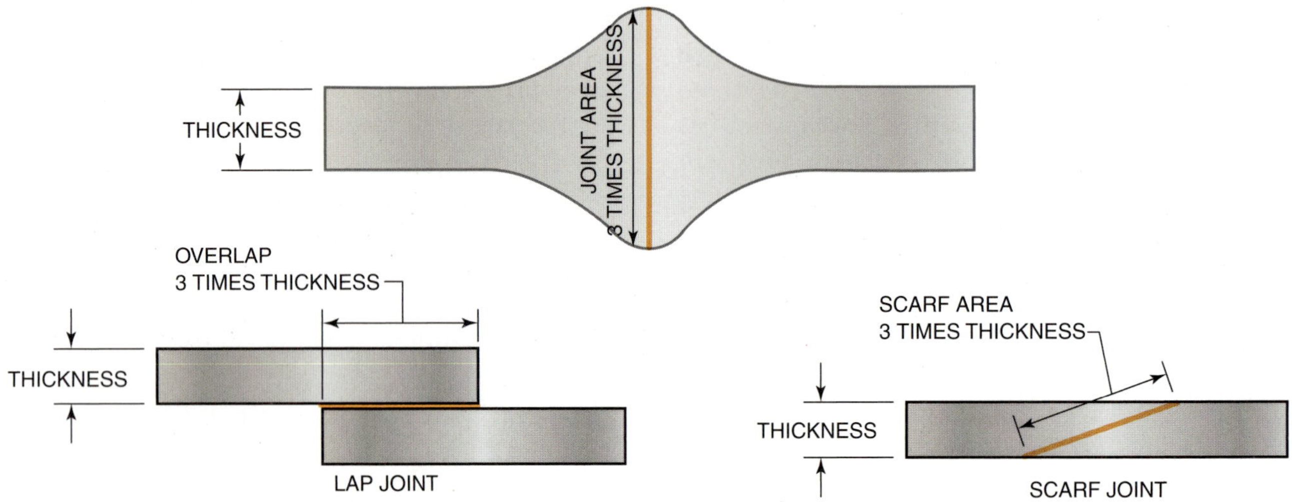

FIGURE 19-28 The joining area should be three times the thickness of the thinnest joint member. © Cengage Learning 2012

scraping, or filing. It can be cleaned chemically with an acid, alkaline, or salt bath.

Soldering or brazing should start as soon as possible after the parts are cleaned to prevent any additional contamination of the joint.

EXPERIMENT 19-2

Fluxing Action

In this experiment, as part of a small group, you will observe oxide removal by a flux as the flux reaches its effective temperature. For this experiment, you need a piece of copper, either tubing or sheet, rosin or C-flux, and a properly lit and adjusted torch.

Any paint, oil, or dirt must first be removed from the copper. Do not remove the oxide layer unless it is blue-black in color. Put some flux on the copper and start heating it with the torch. When the flux becomes active, the copper that is covered by the flux will suddenly change to a bright coppery color. The copper that is not covered by the flux will become darker and possibly turn blue-black, Figure 19-29. Continue heating the copper until the flux is burned off and the once clean spot quickly builds an oxide layer.

Repeat this experiment, but this time hold the torch further from the metal's surface. When the flux begins to clean the copper, flash the torch off the metal (quickly move the flame off and back onto the same spot). Try to control the heat so that the flux does not burn off.◆

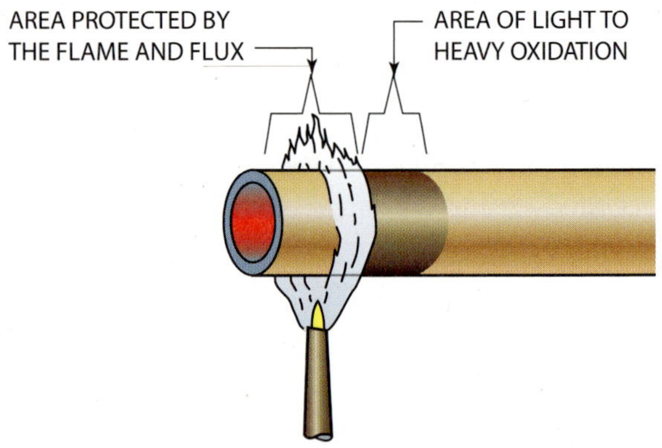

FIGURE 19-29 Copper pipe fluxed and exposed to heat. © Cengage Learning 2012

EXPERIMENT 19-3

Tinning or Phase Temperature

In this experiment, as part of a small group, you will observe the wetting of a piece of metal by a filler metal. For this experiment, you will need one piece of 16-gauge mild steel, BRCuZn filler metal rod, powdered flux, and a properly lit and adjusted torch.

Place the sheet flat on a firebrick. Heat the end of the rod and dip it in the flux so that some flux sticks on the rod, Figure 19-30A, Figure 19-30B,

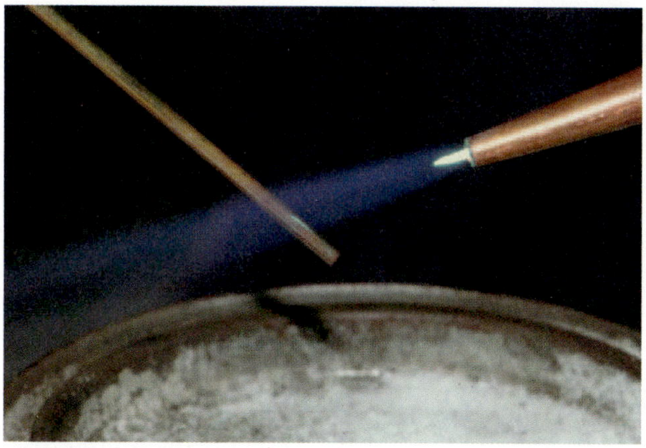

FIGURE 19-30A Heating a brazing rod.
Larry Jeffus

FIGURE 19-30B Dipping the heated rod into the flux.
Larry Jeffus

FIGURE 19-30C Flux stuck to rod ready for brazing.
Larry Jeffus

FIGURE 19-30D Prefluxed brazing rods.
Larry Jeffus

FIGURE 19-31 Deposit a spot of braze on the plate and continue heating the plate until the braze flows onto the surface. Larry Jeffus

and Figure 19-30C. BRCuZn brazing rods are available as both bare rods and prefluxed, Figure 19-30D. Direct the flame onto the plate. When the sheet gets hot, hold the brazing rod in contact with the sheet, directing the flame so that a large area of the sheet is dull red and the rod starts to melt, Figure 19-31. After a molten pool of braze metal is deposited on the sheet, remove the rod and continue heating the sheet and molten pool until the braze metal flows out. Repeat this experiment until you can get the braze metal to flow out in all directions equally at the same time. ◆

SUMMARY

Brazing and soldering are processes that have many great advantages but are often overlooked when a joining process is being selected. For example, brazing and soldering are excellent processes for portable applications. In addition, their versatility makes them great choices for many jobs in which good joint design will result in joint strength equal to or higher than that of welding. The ability to join many different materials with a limited variety of fluxes and filler metals reduces the need for a large inventory of materials, which can result in great cost savings for a small business, home shop, or farm.

For example, solder can be used on the threads of a bolt on an off-road vehicle to act as a locknut to keep it from vibrating off. To remove the nut, one need only heat the threads, and it unscrews easily. Soldering can then be either a permanent or a temporary attaching process.

Be creative in the way you apply these processes. They can be very beneficial to you and your employer.

REVIEW QUESTIONS

1. What is the temperature difference between soldering and brazing?
2. Discuss three advantages of soldering and brazing over other methods of joining.
3. What is the difference between brazing and braze welding?
4. Define *tensile strength.*
5. What does the amount of shear strength of a brazed joint depend on?
6. Explain how ductility and fatigue resistance would be important qualities to have on the brazed joints of a mountain bike.
7. What determines the corrosion resistance of a joint?
8. List three functions flux has when used in soldering and brazing.
9. What happens if fluxes are overheated?
10. Name five methods by which heat is applied when soldering or brazing.
11. Which heating method would be best for the following situations:
    - **a.** The temperature must be closely controlled so that the parts do not overheat.
    - **b.** A thick section must be heated the quickest way.
    - **c.** A large quantity of small parts must be joined.
    - **d.** It is important that the interior of the part not be heated.
12. What factors should be considered when choosing the type of filler metal for a specific joint?
13. Where must tin-lead solder never be used?
14. What type of solder is used for plumbing and why is it used?
15. Why is silver added to cadmium-silver brazing alloys?
16. Explain what the letters represent in the American Welding Society's classification system for brazing alloys.
17. What would a common use be for the following brazing alloys?
    - **a.** BRCuZn-A
    - **b.** BRCuZn-D
    - **c.** BCuP
    - **d.** BNi-1
    - **e.** BAlSi
18. How much overlap should a joint have to maximize its strength?
19. What joint preparation is needed for a successful soldered or brazed part?

Chapter 20

Soldering and Brazing

OBJECTIVES

After completing this chapter, the student should be able to:

- Compare soldering to brazing processes.
- State the advantages of soldering and brazing.
- Demonstrate how to solder.
- Demonstrate how to braze.
- Describe techniques for controlling heat from the torch.

KEY TERMS

braze buildup
braze welding
brazing
capillary action
flux
silver braze
soldering

INTRODUCTION

Brazing and **soldering** both fall under the same American Welding Society (AWS) classification, and it is only the temperature required to melt the filler metals that separates soldering from brazing. Soldering takes place at temperatures below 840°F (450°C), and brazing takes place at temperatures above 840°F (450°C). You can silver solder at a low temperature with one alloy and **silver braze** with another that melts at a higher temperature.

One of the advantages of soldering and brazing is that they can make either a permanent or a temporary joint. For example, a soldered copper water pipe fitting can last for years, or it can be easily heated and separated within a few moments. The reusability of parts can cut time and expense. For instance, suppose a bolt or nut keeps vibrating loose on a yard tractor, four-wheeler, or boat motor. The bolt can be tinned with a thin layer of solder before assembly, and a thin layer of flux can be applied inside the nut. Once the two are tight, warming them with a torch will solder them in place. Warming them again

(A)

(B)

FIGURE 20-1 The threads on a worn or damaged bolt can be tinned with solder to keep the bolt from vibrating loose. Larry Jeffus

later allows them to be removed, Figure 20-1. This works best on small diameter screws and nuts; it is an easy way to make them self-locking.

Soldering and brazing are not a weaker way of joining or repairing parts. When done correctly, both soldering and brazing can produce joints that are several times stronger than the filler metal itself. The higher strength comes from using a lap joint so that the area of the joint is greater than the thickness of the parts being joined, Figure 20-2.

To obtain this higher strength, the parts being joined must be fitted so that the joint spacing is very small. This small spacing allows capillary action to draw the filler metal into the joint when the parts reach the proper phase temperature.

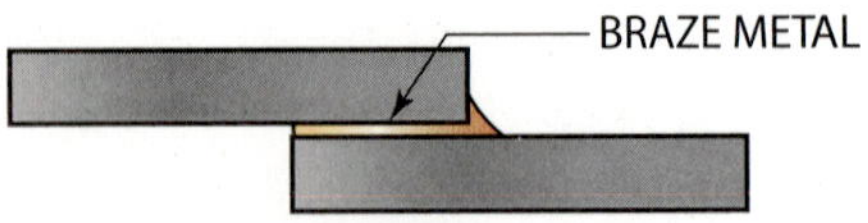

(A)

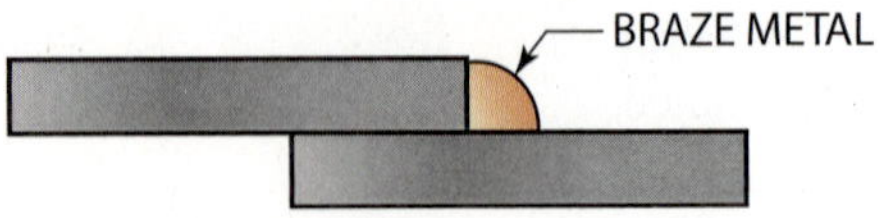

(B)

FIGURE 20-2 (A) Brazed joint. (B) Braze welded joint. © Cengage Learning 2012

Capillary action is the force that pulls water up into a paper towel or pulls a liquid into a very fine straw. **Braze welding** does not need capillary action to pull filler metal into the joint.

EXPERIMENT 20-1

Uniform Heating

In this experiment, as part of a small group, you will learn how to control the flame direction so that two pieces of metal of unequal size are heated at the same rate to the same temperature. For this experiment you will need two pieces of mild steel, one 16 gauge and the other 1/8 in. (3 mm) thick, and a properly lit and adjusted torch.

Place the two pieces of metal on a firebrick to form a butt joint. Then take the torch and point the flame toward the thicker piece of metal, moving it as needed so that both plates turn red at the same time. Now move the torch so that the red area is equal in size on both plates. Keep the spot red but do not allow it to melt. You can control the heat for both brazing and soldering by

- raising and lowering the torch flame.
- increasing or decreasing the torch angle.
- flashing the torch off or away from the metal.

Repeat this experiment until you can control the area and rate of heating of both plates at the same time.

Paperwork

Complete a copy of the time sheet listed in Appendix I or as provided by your instructor.◆

BRAZING

The brazing practices that follow use copper-zinc alloys (BRCuZn) for brazing joints on mild steel and silver alloy (BCuP) for brazing on nonferrous metals. Using prefluxed rods is easier for students than using powdered flux, which has to have the hot tip of the brazing rod dipped to apply. Because you may be asked to braze with bare brazing rods, the practices explain how to use them. If you are using prefluxed rods, just ignore this step.

Surfacing, Surface Buildup, and Filling Holes

The application of a surface layer of braze material can be used to add a decorative and durable surface

FIGURE 20-3 Brazing used to cover a mild steel house number to both protect it and give it a rustic brass look. Larry Jeffus

to mild steel. The brazed metal can be left with the bead surface, Figure 20-3, or sanded smooth as in Figure 20-4.

Worn or damaged surfaces can be built up with braze metal, Figure 20-5. Braze metal is ideal for parts that receive limited abrasive wear because the buildup is easily machinable. Unlike welding or hard surfacing, **braze buildup** has no hard spots that make remachining difficult. Braze buildups are good for both flat

FIGURE 20-4 Sanded and polished braze metal covering a steel base plate. Larry Jeffus

FIGURE 20-5 Braze metal used to repair a worn wheel bearing race. Larry Jeffus

and round stock. The lower temperature used in brazing does not tend to create hard spots where the base metal becomes hardened as it often can in welding.

Screw holes or other small holes in light-gauge metal can be filled using braze metal. The filled hole can be ground flush if it is required for clearance, leaving a strong patch with minimum distortion.

PROJECT 20-1

Brazed Stringer Beads for Appearance

Skill to be learned: The ability to control the brazing bead on a flat surface.

Project Description

Brazing beads will be made on the sides of a small welded project such as this candlestick, Figure 20-6. The brazing is being done strictly for appearance. After the brazing beads have been made on the sides, there are two ways they can be finished.

The first way is to leave them as individual beads, right side, or the second way is to use the torch and remelt the braze beads so they all flow together, left side, Figure 20-7.

Project Materials and Tools

The following items are needed to complete this project.

Materials

- A small welded project such as a candlestick, bookend, birdhouse, sundial, clock face, house numbers, and so forth.

FIGURE 20-6 Candlestick from Project 21-2 can be decorated with a braze covering. Larry Jeffus

FIGURE 20-7 Two ways of finishing the brazed surface. Larry Jeffus

- Source of oxygen and acetylene gases
- 3/32-in. BCuZn brazing rods
- Flux if brazing rods are not flux covered

Tools

- Oxyacetylene welding torch and tips
- Torch wrench
- Tip cleaners
- PPE
- Wire brush
- Pliers
- Firebricks

Brazing

Set up the oxyacetylene welding equipment according to the manufacturer's recommendation, put on all of your personal protective equipment (PPE), and follow all shop and manufacturer's safety rules for brazing.

Wire brush any loose rust or other surface contamination off of the metal. You do not have to remove the mill scale, the light oxide layer found on most hot-rolled steel, because the brazing flux will become active enough at the brazing temperature to dissolve it.

Place the candlestick on the firebricks so that the surface to be brazed is in the flat position. Apply flux to the brazing rod, if needed. Hold the torch at approximately a 45° angle to the metal and pointed in the direction you will be making the braze bead. This will let the flame preheat the metal ahead of the braze bead. You are going to start the braze bead at the bottom of the candlestick and make the first bead along the edge of the side.

Bring the torch and brazing rod together at the bottom edge of the candlestick with the inner flame (cone) of the flame to within 1/8 to 1/4 in. of the metal. Transfer a small amount of flux from the brazing rod to the surface of the metal. The flux will become clear as the metal begins to glow red, Figure 20-8. Test the temperature of the metal by touching the end of the brazing rod to the surface. When the metal is hot enough, the end of the brazing rod will melt onto the base metal, Figure 20-9.

Dip the end of the brazing rod into the leading edge of the molten pool of braze metal. As the braze moves along the surface, add the braze metal to the leading edge in a rhythmic manner. The faster the rhythm, the larger the braze bead produced, and the slower the rhythm, the smaller the braze bead. This rhythm will also affect the finished braze bead's ripple pattern and spacing. Continue the braze bead until you have

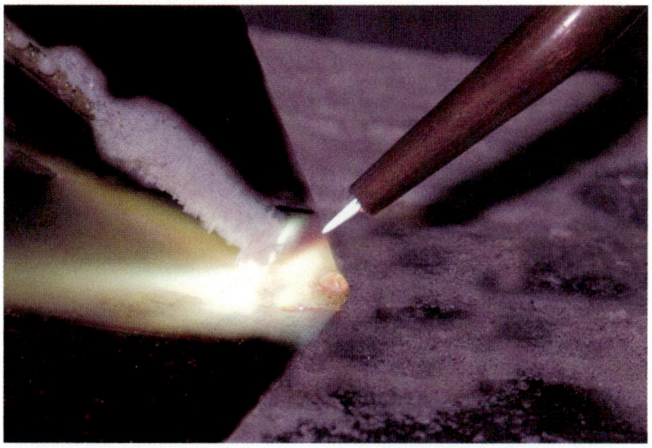

FIGURE 20-8 Heating the metal to make a tack braze.
Larry Jeffus

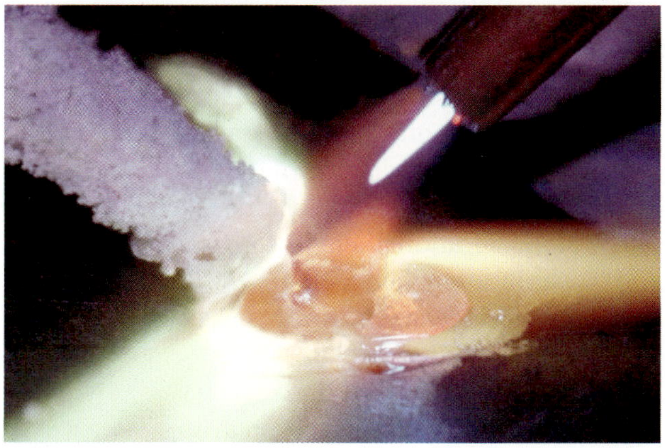

FIGURE 20-9 Starting brazing. Larry Jeffus

reached the other end. Raise the torch or angle it to reduce the heat. Add filler metal so the braze bead is completed all the way to the end. Repeat this process until all four sides of the candlestick are covered with braze metal.

NOTE: **It is easy to burn the braze metal with an oxyacetylene torch. When the braze metal begins to overheat, raise the torch tip higher above the surface, travel faster, or lower the torch angle. Some of the signs that the braze metal is being overheated include the bead spreading out over the metal surface, the flux turning black, and/or a white powdery zinc oxide forming alongside of the braze bead, Figure 20-10. In severe cases of overheating, the zinc in the brazing alloy may start to burn with an orangish-white flame.**

Once the side's braze beads are completed, stand the candlestick up so the top is in the flat position. Use the same brazing technique as before, and cover the top with brazing beads. The top braze bead can be

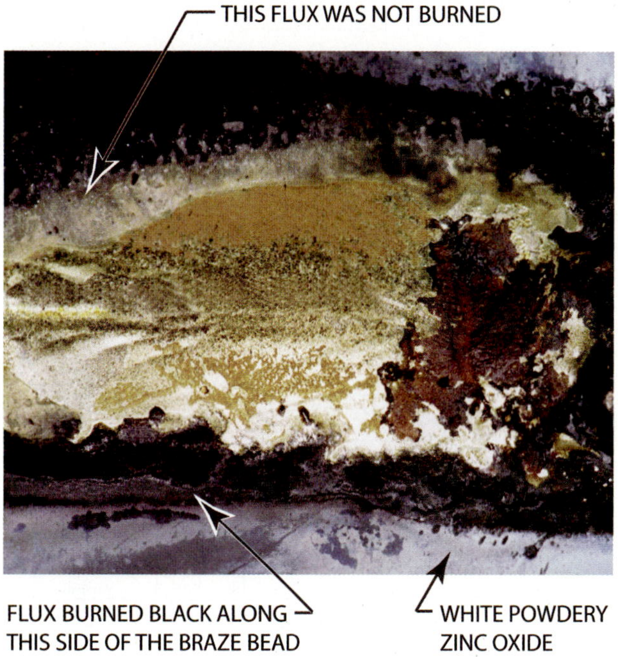

FIGURE 20-10 Overheated braze metal. Larry Jeffus

made by a series of straight beads or as a spiral starting at the outside edge and ending at the center hole.

Finishing

If you are going to leave the braze beads showing, cool off the candlestick and wash off any flux residue. However, if you are going to melt the beads together, place the candlestick on the firebrick, and use a larger welding tip to melt the braze beads together. Moving the torch flame around will help the beads to flow together. Adding a little powdered flux to the surface may also help the beads flow. Cool and wash the flux off.

Paperwork

Complete a copy of the time sheet in Appendix I and bill of materials in Appendix III, or as provided by your instructor.◆

PROJECT 20-2

Brazed Fillet on a Craft Clamp

Skill to be learned: The ability to make fillet brazed joints.

Project Description

This foot-operated craft clamp is ideal when a third hand is needed when you are working on a project. The foot-operated craft clamp easily secures to a table or workbench with a hand screw. If you are concerned about damaging the surface of your project,

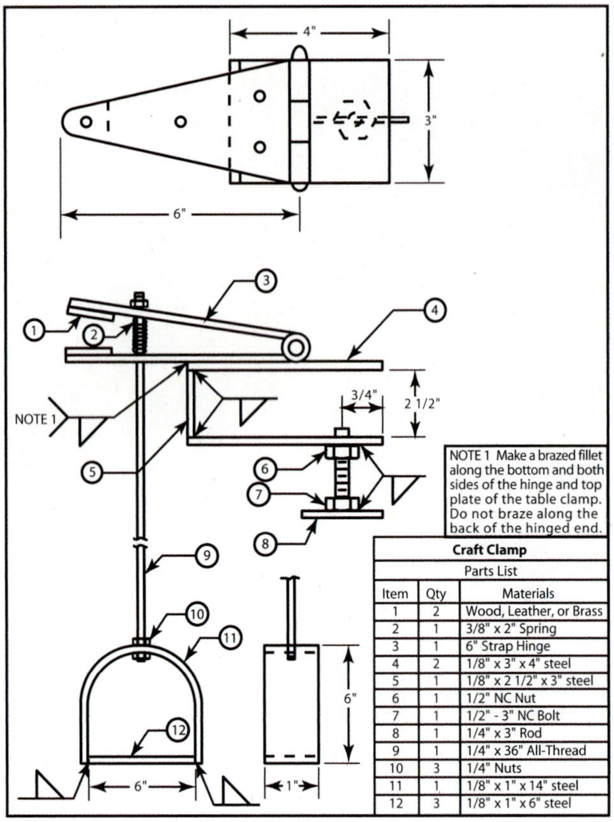

Craft Clamp

Parts List

Item	Qty	Materials
1	2	Wood, Leather, or Brass
2	1	3/8" x 2" Spring
3	1	6" Strap Hinge
4	2	1/8" x 3" x 4" steel
5	1	1/8" x 2 1/2" x 3" steel
6	1	1/2" NC Nut
7	1	1/2" - 3" NC Bolt
8	1	1/4" x 3" Rod
9	1	1/4" x 36" All-Thread
10	3	1/4" Nuts
11	1	1/8" x 1" x 14" steel
12	3	1/8" x 1" x 6" steel

FIGURE 20-11 Project 20-2. © Cengage Learning 2012

the jaws can be covered with wood, leather, brass, or other similar material, Figure 20-11.

Project Materials and Tools

The following items are needed to complete this project.

Materials

- 6-in. strap hinge
- 10 1/2-in. long 3-in. wide strip of 1/8-in. thick steel
- 20-in. long 1-in. wide strip of 1/8-in. thick steel
- 3/8-in. diameter 2-in. long compression spring
- 1/2-in. diameter 3-in. long NC bolt and nut
- 3-in. long 1/4-in. diameter rod
- 1/4-in. diameter 36-in. long all tread rod and 3 nuts
- Source of oxygen and acetylene gases
- 3/32-in. BCuZn brazing rods
- Flux if brazing rods are not flux covered

Tools

- Oxyacetylene welding torch and tips
- Torch wrench
- Tip cleaners
- PPE
- Wire brush
- Pliers
- Ball-peen hammer
- Center punch
- Bench vise
- Shear, saber saw, and/or reciprocating saw
- 12-in. rule
- Soapstone, pencil and/or marker
- Square or try square
- 2 or more C-clamps
- 2 or more firebricks
- Drill with 3/8-in. and 5/8-in. drill bits

Layout

Start by making a mark at the 4-in., 8-in., and 10 1/2-in. measurements on the long piece of the 3-in. steel strip. Next, mark off the 14-in. and 6-in. lengths on the 1-in. wide steel strip. Use the square and mark the cut line across the steel strips. Normally you would add a small additional length to each measurement to compensate for the metal removed by the cutting process kerf. However, for this project that level of accuracy is not required.

Cutting Out

Using a shear, saber saw, or reciprocating saw, putting on all of your PPE, and following all shop and manufacturer's safety rules for the power tool and/or equipment you are using, cut the strips to length.

Drilling

Locate and mark the point where the 5/8-in. hole will be drilled in one of the 4 × 3-in. metal strips. It is important to use a sharpened soapstone and make a "V"-shaped mark that points at the spot to be center-punched. Use the punch and hammer to center-punch the metal so that the drill bit will be easier to start, Figure 20-12.

NOTE: It is often easier to drill a small pilot hole first before trying to drill the larger diameter hole in metal. All metal drill bits have a small flat spot on the tip called the dead center, Figure 20-13. The larger the drill bit, the wider the dead center is. Almost all of the drill bit cutting is done by the cutting edge while the dead center basically just forces its way through. Using a smaller drill bit first to drill a pilot hole makes it easier to drill the larger hole. This is especially true if the larger drill bit is slightly dull.

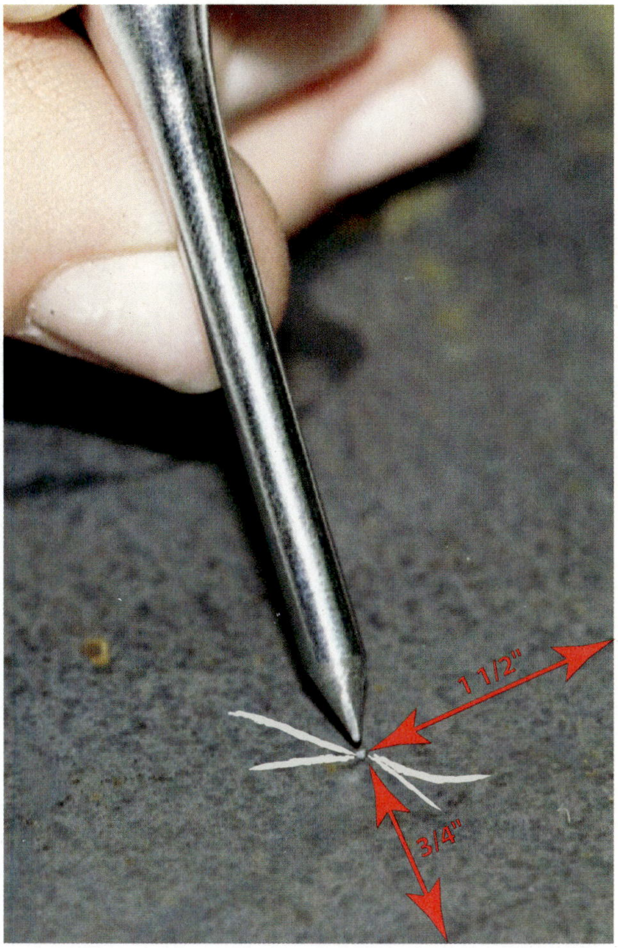

FIGURE 20-12 Center-punching to start a drilled hole.
Larry Jeffus

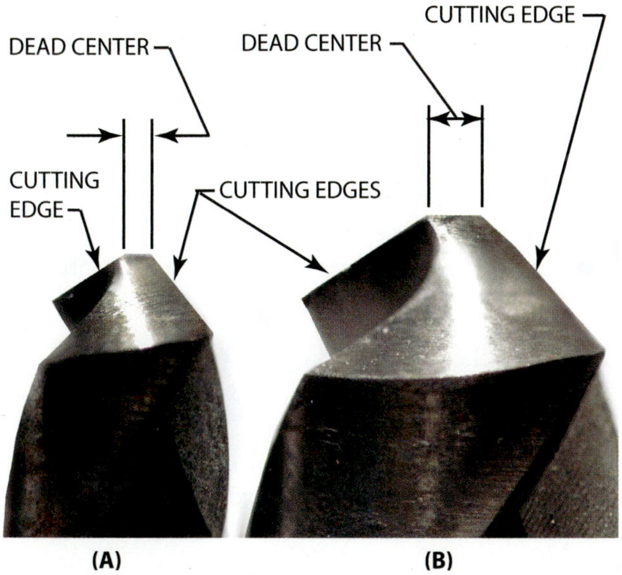

FIGURE 20-13 Drill bit tip identification. Larry Jeffus

Clamp the metal strip to a workbench with a C-clamp, or hold it in a vise so that it does not spin around if the drill bit grabs as the hole is being drilled. The metal should be clamped so that the drill can bore all the way through without hitting anything on the back side.

SAFETY: Always securely clamp metal being drilled because if the drill bit grabs the metal, it will spin around the bit. The sharp corners of metal can cause serious injuries.

There is a mounting screw hole at the end of a strap hinge and often there is a second screw hole a short distance away. Use the 3/8-in. drill bit to enlarge that second screw hole in the strap hinge or to drill a new hole at least 1 in. from the end. Fold the hinge shut so you can drill the hole through both sides of the hinge at the same time.

The 1/4-in. hole will be drilled in the foot stirrup after the part is bent. Drilling the hole first will make it more difficult to make a smooth, even bend.

Fabrication

The stirrup will be hand bent as it is heated while one end is held in a vise. The process of using heat to soften metal so it can be bent or formed has been used as long as mankind has worked with metals. Although the metal at the top of the stirrup is lightweight enough to be bent cold by hand, we will bend it with heat.

As a pattern for the bend, use a soapstone to draw a 6-in. diameter circle directly on a clean, flat area on the welding table. Refer to Chapter 4 for the technique of how to sketch a circle. Measure and mark the center of the 1-in. wide steel strap. From the center point measure 10 1/4 in. in both directions and make a mark. The radius of the bend will extend from one mark to the other. Clamp one end of the metal in the vise no closer than 1 in. from the first mark. If you clamped it right at the first mark, the vise would absorb too much of the heat, which would make it difficult to start the bend with a smooth radius. Using a properly adjusted large-bore welding torch tip or heating tip, begin heating the metal by moving the flame up and down across the strip. As the metal begins to glow dull red, put a small amount of force on the opposite end with your hand. This will begin the bend. As the metal begins to move, move the torch along the metal, heating the strip as it bends. If you heat uniformly and apply even pressure, the radius of the bend will be consistent. Bending with heat does not normally produce as accurate a bend as if it were bent in a jig. However, for

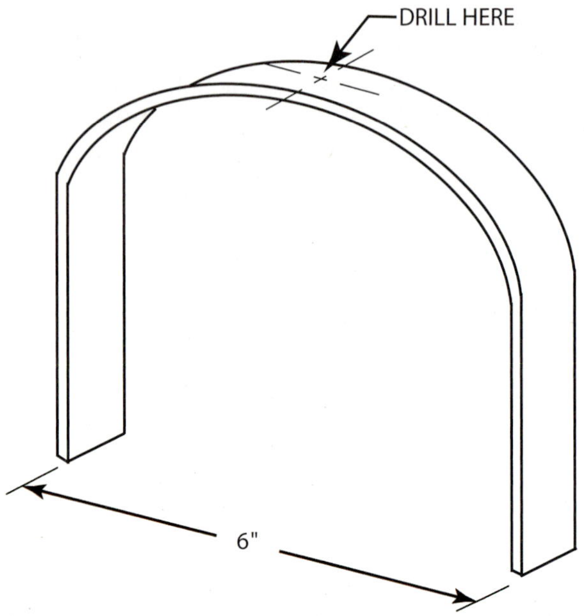

FIGURE 20-14 Bending the U-shaped strap.
© Cengage Learning 2012

this project any slight irregularities in the curve are insignificant. Learning to use the heat-bending process is more important than a smooth, uniform bend on the stirrup. Continue the process until the metal strip is bent in a complete U shape. It should measure approximately 6 in. between the two U-shaped ends, see Figure 20-14. Cool and dry the part and then measure it against the soapstone circle. If it is more than 1/4 in. off or if one side is longer than the other, you can decrease or increase the bending radius to make the part fit by reheating and reapplying the bending force. Look at the bend for areas that are not consistent. The finished bend should be within tolerance.

Stand the U-shaped stirrup on a flat surface and visually locate the top of the radius. Depending on how accurate you were, this should be the center point you previously marked. If the mark is within 3/4 in. of the center, go ahead and mark this new location to be drilled. However, if the radius is not within 3/4 in. of being centered, you need to go back and rebend and reshape the stirrup.

Clamp the stirrup in the vise. Use the hammer and punch to center-punch the stirrup before drilling the 3/8-in.-diameter hole for the all-thread.

Assembling and Brazing

Set up the oxyacetylene welding equipment according to the manufacturer's recommendation, put on all of your PPE, and follow all shop and manufacturer's safety rules for brazing.

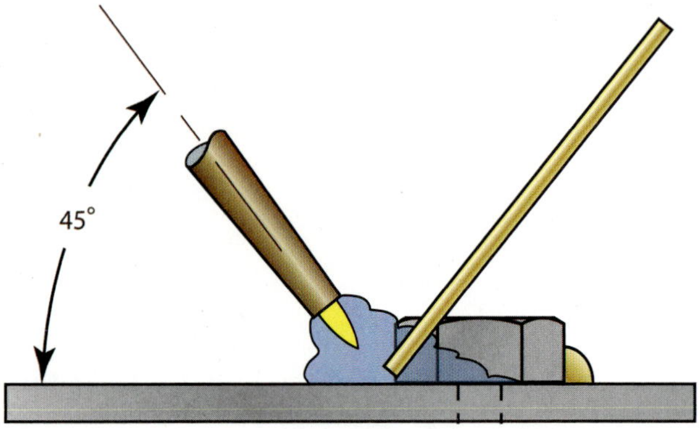

FIGURE 20-15 Tack brazing the nut to a plate.
© Cengage Learning 2012

Place the bottom of the drilled bench clamp flat on two firebricks so that the hole that was drilled is between the firebricks. Place the nut so that the drill hole lines up. Using a properly lit and adjusted oxyacetylene torch and a flux-coated brazing rod, you are going to make a small brazed fillet to join the nut to the plate. Point the torch at the joint between the nut and the plate, Figure 20-15. Watch the nut and plate as they heat up. Because the nut is smaller and is more directly into the torch flame, it will heat up faster. To prevent overheating, move the torch onto the plate or angle it more toward the plate so that both parts reach the brazing temperature at the same time. Very carefully, melt a small piece of brazed material to one side of the nut. You want to make sure that the brazing rod does not push the nut out of alignment with the hole. Once the brazed metal has flowed out onto the surfaces, move the torch to the opposite side and repeat the process. The first small braze will hold the nut in place as a fillet braze joint is made around the nut.

NOTE: Be careful not to overheat the nut. Overheating can allow the braze material to flow into the threads, which will cause a problem later. Also, overheating can burn off the cadmium coating found on many nuts and bolts. Cadmium fumes can be hazardous, so you must avoid breathing these fumes during the brazing process, see Chapter 2, "Welding Safety."

Allow the part to cool. Next, you are going to braze the bar to the top of the 3-in. bolt. Stand two firebricks on their sides and rest the head of the bolt on the gap between the firebricks, Figure 20-16A. Place the rod across the bolt head, centering it. Using the properly adjusted oxyfuel flame, heat up the nut and bar. The bar will tend to heat faster, so direct a little more heat on the bolt head so that both parts

FIGURE 20-16A Position the T handle on the bolt head.
Larry Jeffus

FIGURE 20-16B Braze the T handle to bolt head.
Larry Jeffus

come to temperature at the same time. When they reach the brazing temperature, put a small amount of braze filler metal on one side, being careful not to knock the rod out of position. Move to the opposite side and make a braze across the entire top of the nut, Figure 20-16B. Repeat this process on the other side. Cool this part.

To check the alignment of the nut and drilled hole and to make sure the threads are clear, screw the T handle bolt into the nut. The bolt should screw through the nut with little or no resistance; however, if the nut is misaligned with the hole or if brazing metal got on the threads, the bolt will get stuck and not go all the way through the nut. If that happens, back out the t-bolt.

Clean the threads with a tap, Figure 20-17. Use a little lubricant and run the tap into the nut, turn it forward and back it up approximately 1/2 to 3/4 turn each time. Backing it up 1/4 turn allows the chips that are being cut by the tap to break free and drop out of the bottom. Continue this process until the tap will easily turn all the way through the nut. Back the tap out and rethread the bolt and handle. The next step will be to braze the end plate on the bench clamp.

Lay the top piece of the bench clamp flat on a firebrick. Stand up the 2 1/2-in. wide end piece, and support it with a firebrick. Using a properly lit and adjusted oxyacetylene torch, heat up the corner of both pieces until a small spot of brazing metal can be used to join the two pieces. Now move to the opposite end, and repeat this process. Once the part has been tack brazed together, check the part for squareness. Any slight misalignment can be fixed by simply bending the brazed tacks. Now, you can position it at a comfortable angle so that the tee fillet brazed joint can be made that will join the two pieces together. Starting at one end right next to the tack braze, but being careful not to melt it, bring the metal up to brazing temperature and add the brazing metal into the molten braze pool. Add the braze material by dipping it into the pool as the fillet braze continues across the joint. When you reach the opposite end of the joint, add a small amount of additional brazing material to fill the crater. Cool the part and wire brush off any flux to see that you have a uniform brazed bead.

Next, thread in the 3-in. bolt so that 2 1/2 in. of the bolt extends through the plate. Place the bottom plate with the T handle of the bolt either hanging off the edge of a firebrick or in a gap between two firebricks. Next, place the top and end pieces that you just assembled on this piece. The bolt will support one end. Measure the gap from the back to the bolt end to be sure that it is parallel, Figure 20-18. If necessary, adjust the bolt in or out to level the plate. If the bolt is not quite long enough, then a small piece of scrap metal can be used as a shim on the tip of the bolt to give it enough additional length to level the part. Once the parts are parallel, use the same procedure to make tack brazes at both ends of this joint.

Once the parts are tack brazed together, brace them against a firebrick so that the fillet braze will be made in the flat position. This brazed joint will be made in an area with limited visibility and access. Position yourself so that the torch can be directed into the joint from one side and the filler metal from the other. Hold the flame so that the metal is heated. Be careful not to melt the tack braze, and make a

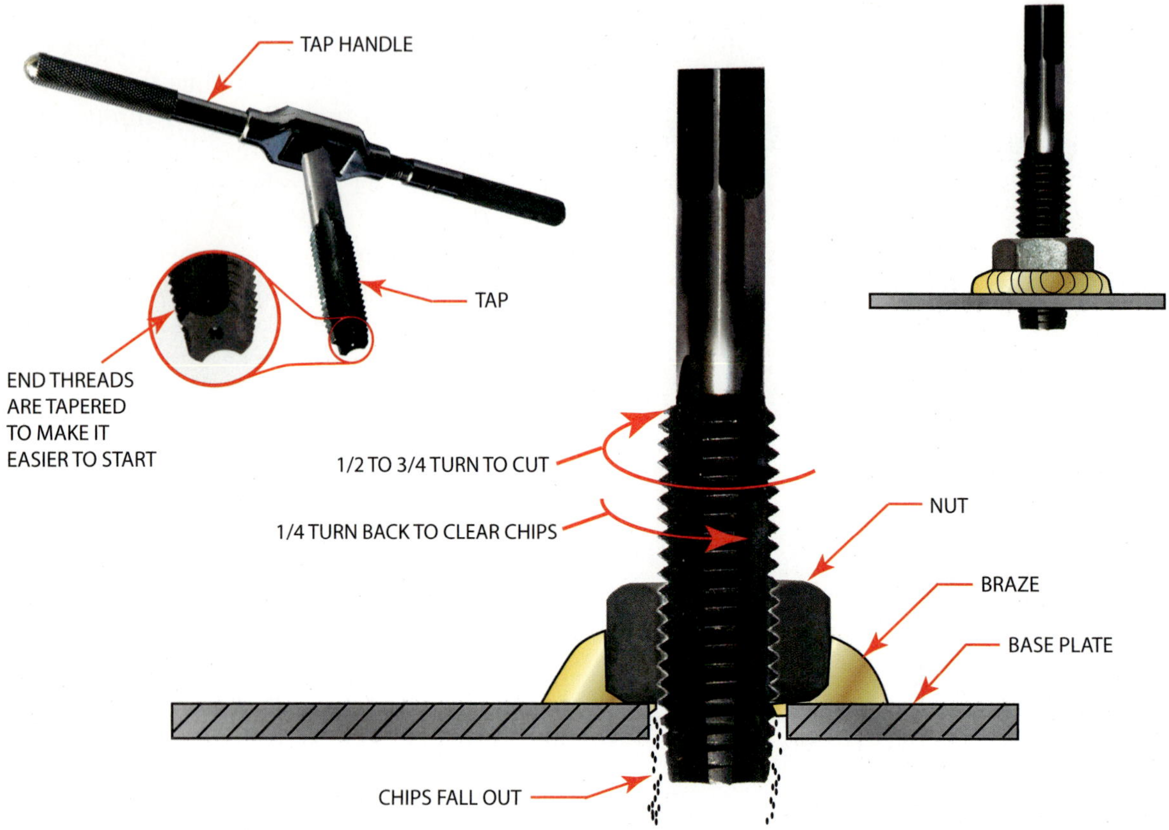

FIGURE 20-17 Using a tap to clean out the nut's threads. Larry Jeffus

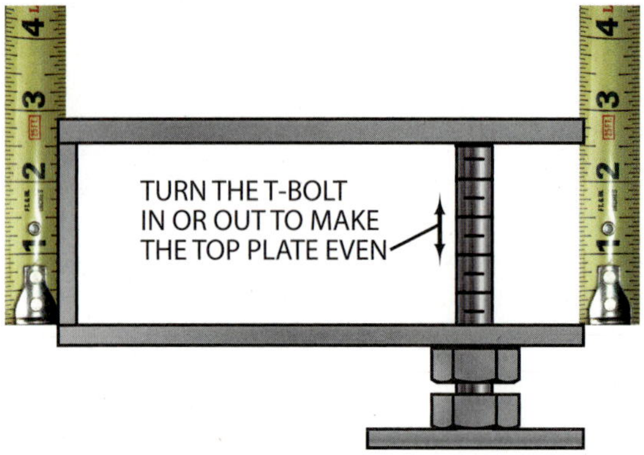

FIGURE 20-18 Checking the bracket for parallel. Larry Jeffus

brazed fillet approximately halfway across the joint. Stop and reposition the part so that the opposite end can be brazed in the same manner. It is not necessary that the brazed metal meet completely in the center. If there is a small gap, the part will still be functional.

Place the top of the stirrup over the base plate flat on a firebrick. Using a properly adjusted oxyacetylene torch, make the two fillet brazes to join the base to the top of the stirrup. Cool and wire brush any remaining flux.

Assembly

Assemble the parts as illustrated in Figure 20-11. If the protective end coverings on the hinge are glued to the hinge using a type of rubberized cement, they can be easily removed and replaced. The end coverings can be more securely attached by using a two-part epoxy; however, removing and replacing them is more difficult.

Finishing

The craft clamp can be sprayed with a thin coating of a light lubricant to prevent rusting, or it can be painted.

Paperwork

Complete a copy of the time sheet in Appendix I and bill of materials in Appendix III or as provided by your instructor.◆

Silver Brazing

The silver brazing practices that follow will use BCuP-2 to BCuP-5 alloys to make brazed joints. The melting temperature for the alloys is around 1400°F

(760°C). At these temperatures the metal being joined will glow a dull red. The best types of flame to use for this type of brazing are air acetylene, air MAPP®, air propane, or any air fuel-gas mixture. The most popular types are air acetylene and air MAPP®. If an oxyacetylene torch is used, it is easy to overheat the alloy. To prevent overheating with an oxyacetylene flame, keep the torch moving and hold the flame so that the inner cone is about 1 in. (25 mm) from the surface.

When using BCuP silver brazing alloys on clean copper, it is not necessary to use a flux. The phosphorus in these alloys makes them self-fluxing on copper, brass, and silver. It is the phosphorus that promotes the wetting and enhances the flow of the alloy into the joint space. A flux must be used when joining steel, stainless steel, or on some heavily oxidized metals.

PROJECT 20-3

Silver Brazing to Make a Coin Belt Buckle

Skill to be learned: The ability to carefully control the torch flame heat while making brazed joints.

Project Description

Handcrafted ornate belt buckles have long been a source of pride for welders. In this brazing project you are going to make a copper belt buckle covered with coins that is both attractive and functional, Figure 20-19.

Project Materials and Tools

The following items are needed to fabricate the belt buckle.

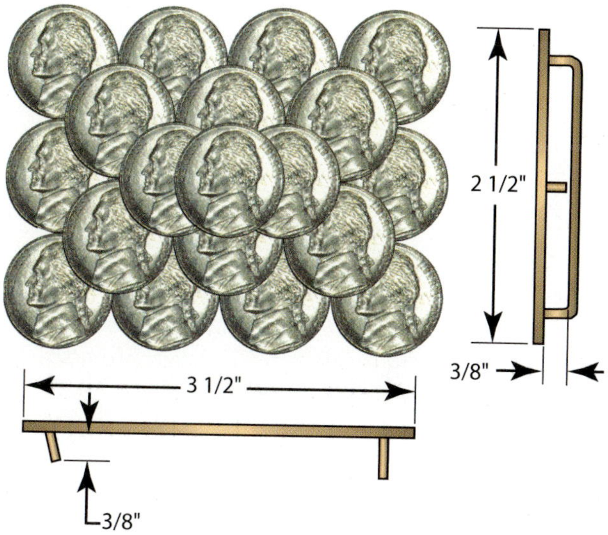

FIGURE 20-19 Project 20-3. Larry Jeffus

Materials

- 3 1/4-in. long piece of 3/4-in. copper pipe
- 20 state quarters
- 3 in. of 1/8-in. diameter BRCuZn rod
- 1 silver brazing rod
- Metal cleaner or polish
- Soft cloth and/or sponge
- Source of acetylene and/or oxygen gases

Tools

- Oxyfuel or air-fuel torch
- PPE
- Tubing cutter
- Hacksaw
- Pliers
- Aviation-type sheet metal shears
- Hammer
- File
- Tape measure
- 12-in. steel rule or straightedge
- Square
- Pencil or marker
- Table vise
- Cloth buffing wheel and compound

Layout

Using the tape measure, pencil, and/or marker, make a mark at a point 3 1/4 in. from the end of the 3/4-in. copper pipe, Figure 20-20.

FIGURE 20-20 Measuring and marking the length of copper pipe. Larry Jeffus

FIGURE 20-21 Cutting the copper pipe to length.
Larry Jeffus

Cutting Out

Clamp the copper pipe securely in a vise, being careful not to overtighten the vise. Overtightening can squeeze the pipe out of round, Figure 20-21. Use the hacksaw to cut the pipe at the 3 1/4-in. mark. The best hacksaw blade to use would be one with 32 to 24 teeth per inch. A hacksaw blade with too large a tooth will tend to catch or grab as the cut is made. Open the vise jaws wide enough to clamp the 3 1/4-in. copper pipe crosswise as shown in Figure 20-22. Cut the copper pipe lengthwise.

FIGURE 20-22 Lengthwise cutting of the copper pipe.
Larry Jeffus

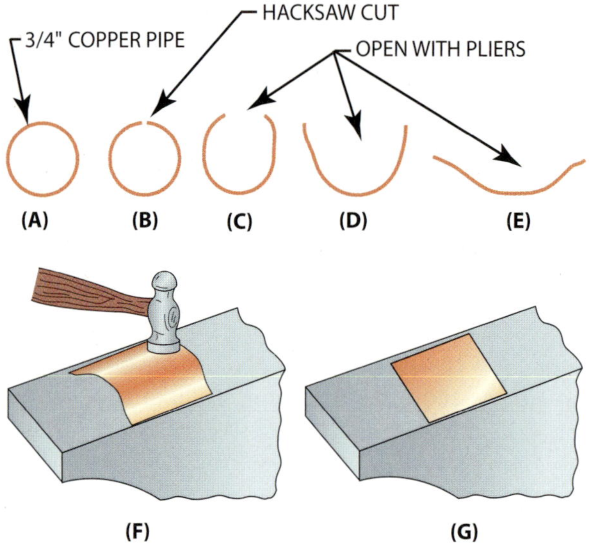

FIGURE 20-23 Open and flatten the copper pipe.
Larry Jeffus

NOTE: If the copper pipe is clamped lengthwise in the vise jaws, the hacksaw blade will bind as the lengthwise cut is made.

Fabrication

Use two pairs of pliers and open up the copper pipe so that it can be flattened with a hammer on an anvil, Figure 20-23A through Figure 20-23E. Place the opened piece of copper pipe upside down on the anvil, Figure 20-23F. Use only light hammer taps to flatten the copper, Figure 20-23G.

NOTE: Overly heavy hammer strikes will deform the copper.

Lay the copper plate flat on a firebrick and use a straightedge to line up the coins, Figure 20-24. The

FIGURE 20-24 Lining up the coins to size the plate.
Larry Jeffus

FIGURE 20-25 Mark the plate width. Larry Jeffus

copper width will need to be trimmed for the coin denomination you have chosen to use. Use a pencil and make a mark on the copper plate, Figure 20-25. Remove the coins and draw a line to be cut with sheet metal shears, Figure 20-26. The shears may slightly deform the copper plate, so it may need to be reflattened, Figure 20-27.

Sometimes shears will leave a small sharp burr. That burr and any other sharp edges must be filed off before the belt buckle is assembled, Figure 20-28. Sharp edges that are left on the plate are more difficult to file smooth once the belt buckle has been made.

Line up the quarters along a straightedge and check to see that they are straight and even, Figure 20-29. Sometimes the center two quarters are covered and not visible in the finished buckle, so it might not be necessary to have a special coin at these two locations.

Bring the torch flame over the coins and point it at the opening between four coins. Melt a small amount of silver alloy forming a mound on the copper plate,

FIGURE 20-26 Draw a line to be cut. Larry Jeffus

FIGURE 20-27A Cut the line with the sheet metal shear. Larry Jeffus

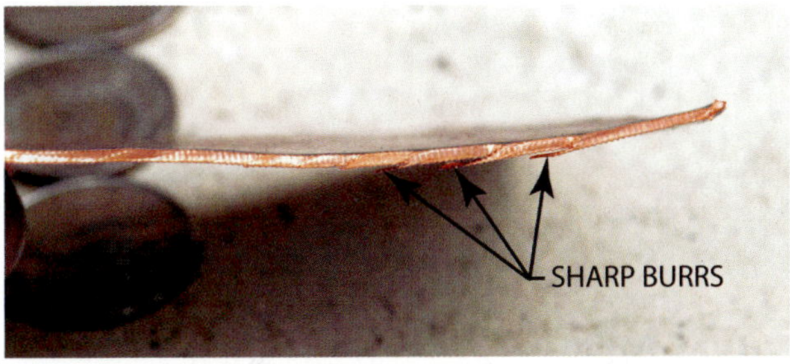

FIGURE 20-27B Shears will cause the plate to bend slightly and may leave a burr. Larry Jeffus

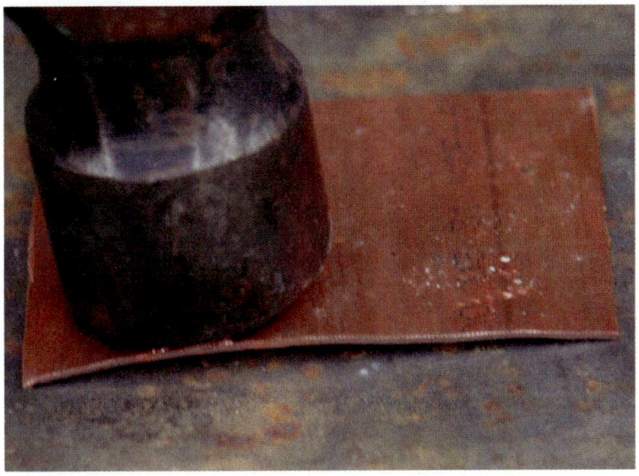

FIGURE 20-27C Use a ball-peen hammer. Larry Jeffus

FIGURE 20-27D Flatten the copper. Larry Jeffus

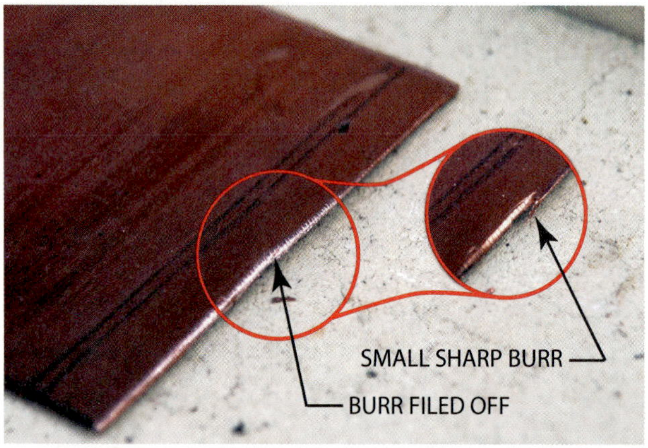

FIGURE 20-28 File the edge to remove any sharp edges or burrs. Larry Jeffus

FIGURE 20-29 Lay out the first layer of coins. Larry Jeffus

FIGURE 20-30 Heat the coins and base plate uniformly. Larry Jeffus

Figure 20-30. The coins are much smaller, and the flame can literally surround them on all sides, so they will heat up very quickly, Figure 20-31. So move the torch around to heat the copper plate and coins uniformly until the brazing temperature is reached and the silver alloy flows freely. The filler metal will begin to flow out onto the surface of the copper plate and under the coins if they are hot enough, Figure 20-32. By putting the filler metal in the spaces between the coins they will wick the braze material between the surface and not have filler metal on top of the coin.

NOTE: Avoid getting brazing material on the face of the coins because it will discolor the coins and take

FIGURE 20-31 Watch to see that you do not overheat the coins. Larry Jeffus

FIGURE 20-32 Filler metal flowing under the coins. Larry Jeffus

FIGURE 20-34 Lay out the next layer of coins. Larry Jeffus

away from the finished appearance of the belt buckle. If silver braze metal does get on the coin surface, the coin can be reheated and removed so it can be replaced with a clean coin.

The coins can float on top of the molten filler metal and can easily be knocked out of place if you accidentally touch them with the end of the filler rod. If any of the coins are accidentally knocked out of place, use a steel welding rod to gently move them back in place. You may need to let the plate cool and reheat any coins that need to be relocated.

When the first layer of coins has been brazed in place, use a needle-nose or long-nose pair of pliers to carefully pick up the hot belt buckle, grabbing the backing plate between the coins. The coins are hot and soft and easily damaged. Grabbing it between the coins will prevent damage, Figure 20-33. Cool the belt buckle so that the next layer of coins can be placed, Figure 20-34. Braze this layer of coins in place in the same manner by putting the brazing material between the row of coins.

NOTE: The longer it takes to complete the braze, the thicker the oxide layer will become. If the brazing takes an excessively long period of time, then the oxide layer on the coins may build up to a point where a flux is needed to promote tinning between the filler metal and the first layer of coins. An oxyacetylene flame has some cleaning effect, and the silver braze alloy often will compensate for a light layer of oxides on some metals.

Slightly build up the brazing material between the top layer of coins so that it is a little higher than the coins. This will allow the top two coins to be placed on the hot belt buckle and heated so that the braze alloy will bond them without showing around the edge. Hold the flame so that it heats the belt buckle and the top two coins evenly. A slight angle will allow the flame to flow between the top coins and the buckle. This can help in the heating, Figure 20-35.

FIGURE 20-33 Be careful not to damage the coin faces while they are hot. Larry Jeffus

FIGURE 20-35 Angle the torch so that the new coins and the previously brazed coins heat uniformly. Larry Jeffus

FIGURE 20-36 Use steel wool or a sand cloth to clean off the oxides. Larry Jeffus

When the belt buckle has been completely assembled, cool it, and use steel wool to remove the heavy oxide layer on the back of the buckle, Figure 20-36. The buckle loop will be bent out of 1/8-in. brazing rod. Using a pair of pliers or vise, bend one end of the rod at a right angle so that it is approximately 3/8 in. in length. Measure approximately 2 to 2 1/2 in. to locate the second bend. Make this bend the same way as before. Measure and cut both bent ends so that they are 3/8 in. long, Figure 20-37A. It is difficult to make freehand bends of the exact radius and length, so bend them first, and then cut them to length.

NOTE: There must be enough space between the buckle loop and the back of the belt buckle so that both the belt and the belt end will fit through this space, Figure 20-37B. If your belt is slightly thicker, you may need to make the buckle loop ends slightly longer than 3/8 in.

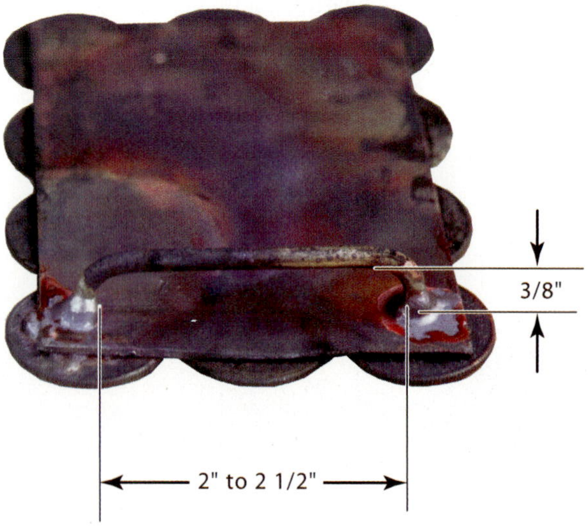

FIGURE 20-37A Locate the belt loop. Larry Jeffus

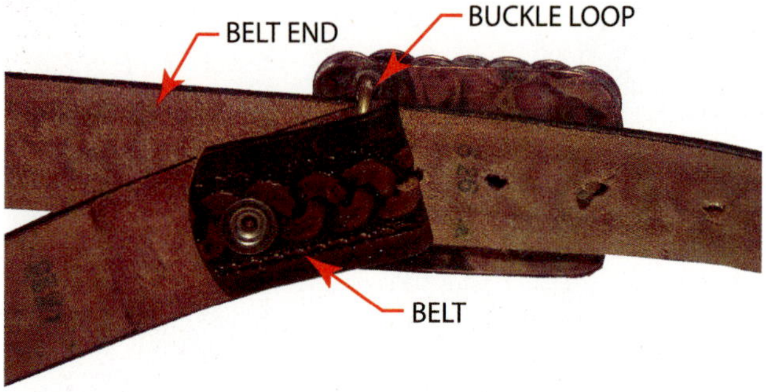

FIGURE 20-37B The loop should be just big enough to allow for two thicknesses of leather. Larry Jeffus

NOTE: Some brazing rods are too brittle to bend in a short 90° radius without cracking. If cracks occur during the bending process, you can either anneal the brazing rod by heating it and letting it cool, or make a longer radius bend.

Using a pair of locking pliers, hold the buckle loop squarely in position on the back of the belt buckle, Figure 20-38. Be sure that you are working on the correct end of the buckle so that it will be right side up when worn, Figure 20-39. The belt catch is also made out of 1/8-in. brazing rod and is brazed on the back before it is cut to length, Figure 20-40 and Figure 20-41. Brazing a longer length on, is much easier to hold and control than trying to assemble a small piece.

Finishing

Cool and clean the belt buckle using soap, hot water, and a soft cloth. Use a soft cloth or sponge and a copper or silver cleaning compound to remove the oxide on the coins. A polishing compound can be used with a cloth buffing wheel if a bright polished appearance

FIGURE 20-38 Use self-locking pliers to hold the belt loop in place. Larry Jeffus

FIGURE 20-39 Finished belt buckle. Larry Jeffus

FIGURE 20-40 Brazing the catch to the buckle. Larry Jeffus

FIGURE 20-41 Cutting the catch to length. Larry Jeffus

FIGURE 20-42 Natural and polished finishes. Larry Jeffus

is desired, Figure 20-42. After polishing, a clear coat can be sprayed on the buckle, or it can be left as is.

Paperwork

Complete a copy of the time sheet in Appendix I and bill of materials in Appendix III or as provided by your instructor.◆

SPARK YOUR IMAGINATION

What are some other things that could be brazed onto the copper backing to make an attractive belt buckle?

PROJECT 20-4

Copper Wire Bicycle

Skills to learned: The ability to sketch a scale drawing and fabricate the part to that scaled drawing.

Project Description

This scale model of a bicycle is built out of copper electrical wire and is designed as a desk or wall cabinet ornament, Figure 20-43.

Project Materials and Tools

The following items are needed to complete this project.

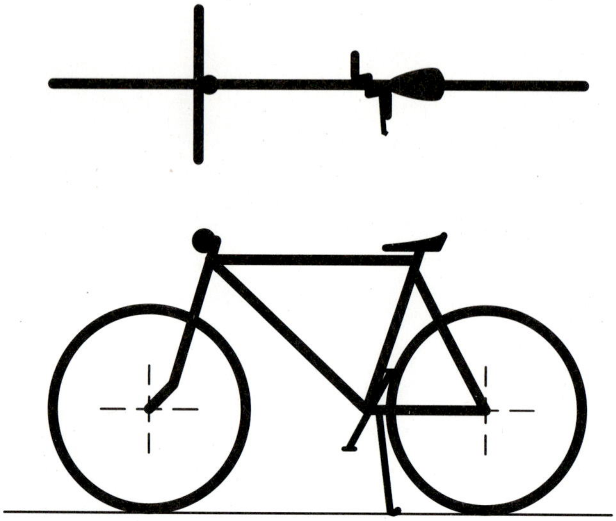

FIGURE 20-43 Project 20-3. Larry Jeffus

FIGURE 20-44 Measuring a standard bicycle. Larry Jeffus

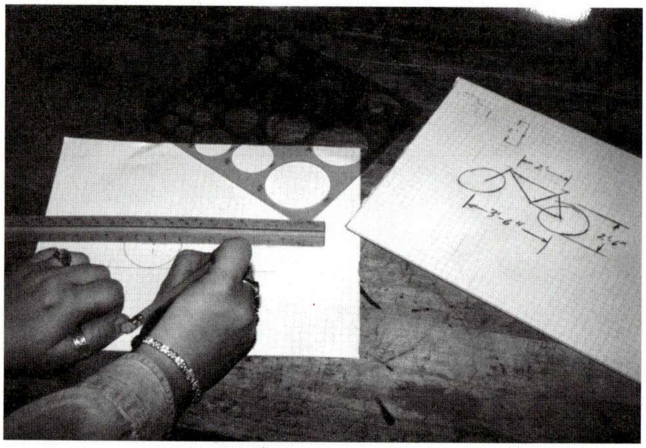

FIGURE 20-45 Use graph paper templates and a scale to make a scale drawing of the standard bicycle dimensions. Larry Jeffus

Materials

- 24-in. 10-gauge copper electrical wire
- 6-in. 14-gauge copper electrical wire
- 1 silver brazing rod
- Source of acetylene and oxygen gases
- Solid copper penny

Tools

- Oxyfuel torch
- PPE
- Pliers
- Utility knife
- Hammer
- Tape measure
- 12-in. steel rule or straightedge
- Pencil or marker and graph paper
- Circle template, drafting scale, and a compass

Layout

Start by measuring a standard bicycle, Figure 20-44. Make careful notes on a pencil and paper sketch of the bicycle's dimensions. Now you are going to convert your rough pencil sketch and dimensions into a shop drawing, Figure 20-45. Using a scale and a compass or circle template, graph paper, and pencil, measure and mark a 1/8-in.-to-the-foot scaled drawing of the bicycle. Graph paper will help you lay out the drawing more accurately than plain paper. Darken the lines as necessary to make the drawing more visible.

Cutting Out

Use a utility knife and strip the insulation off of a 24-in. long piece of solid copper #10 electrical wire. Next, transfer the wire section lengths directly from the drawing by holding the wire next to the sketch and using a pair of wire cutters to cut the wires to length, Figure 20-46. After they are cut, lay the straight wires on the drawing to check for accuracy.

Locate a round object that is approximately the same size as the two bicycle wheels, Figure 20-47. In this case a container of flux was used as the template to bend the copper wire for the bicycle's wheels. Wrap the copper wire around the cylinder at least 2 1/2 times. Use a pair of wire cutters and snip two complete circles. The front wheel has one cut, and the back wheel is cut twice so that it will fit in the frame. Lay the circles on your drawing to locate where the back wheel is to be

FIGURE 20-46 Cut the bicycle parts to length.
Larry Jeffus

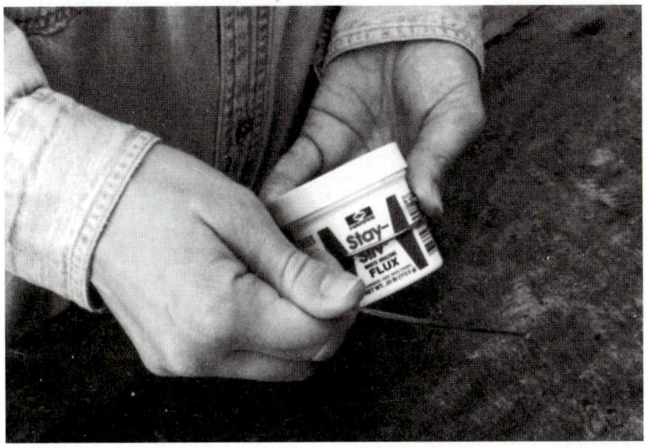

FIGURE 20-47 Bend the wheels around a cylindrical object. Larry Jeffus

cut. The bicycle seat can be made out of a copper penny or small, flattened piece of copper pipe.

NOTE: Pennies manufactured before 1982 are solid copper. All newer pennies are a composite of a soft metal core and copper covering. These pennies are not easily brazed because of the lower melting temperature of the soft metal core and are not suitable for using as the bicycle seat.

The bicycle kickstand and pedals are made out of a short section of 14-gauge copper wire. Refer to the drawing to see how to bend the pedals into shape. Flatten both pedal ends on an anvil using a hammer. The ends will flare out and must be cut off with a pair of wire cutters to square the ends up, Figure 20-48.

Fabrication and Brazing

Set up the oxyacetylene welding equipment according to the manufacturer's recommendations, put on all your PPE, and follow all shop and manufacturer's safety rules for brazing.

FIGURE 20-48 Cut the ends of the pedal square.
Larry Jeffus

The concentrated flame of an oxyacetylene torch is ideal for detailed work like this bicycle.

Lay out the bicycle frame and wheels on a firebrick. Using a properly lit small-bore oxyacetylene torch and silver braze alloy, carefully heat the joints and apply a very small quantity of silver brazing alloy to each of the joints. The torch flame will quickly heat up the joints and can easily overheat them, so pay close attention to the heating.

Once the frame has been brazed, the seat, handlebar, pedals, and kickstand can all be added. These parts may be held in place by another student, or you can pre-tin both parts and then use the torch to melt the braze metal. Pretinning parts is an excellent way of assembling small parts when you are working by yourself. Pretinning is the process of putting a small amount of braze metal on the parts before assembly and then remelting the braze metal to fuse the parts together, Figure 20-49.

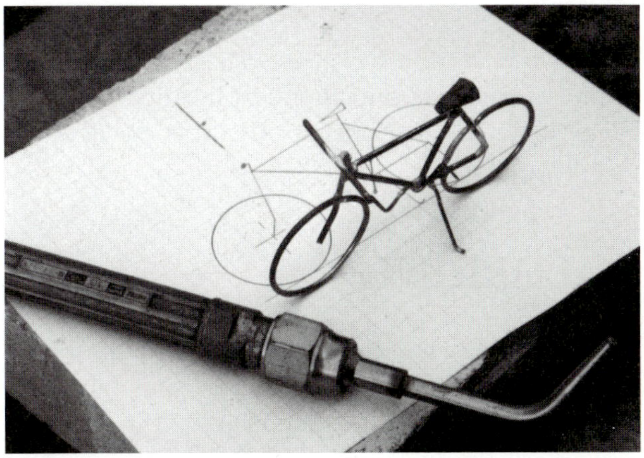

FIGURE 20-49 Be careful not to overheat the joints so that the braze metal does not run. Larry Jeffus

FIGURE 20-50A Polished finish. Larry Jeffus

FIGURE 20-50B Natural finish. Larry Jeffus

Finishing

The bicycle can be cleaned using a metal cleaning compound and then sprayed with a clear finish to keep it bright and shiny, or it can be allowed to develop a copper oxide layer, called a patina, which initially will be dark and may later turn green when heavier oxide is formed, Figure 20-50.

Paperwork

Complete a copy of the time sheet in Appendix I and bill of materials in Appendix III or as provided by your instructor.◆

SOLDERING

The soldering practices that follow use tin-antimony solders. Both solders have low-melting temperatures. If an oxyacetylene torch is used, it is very easy to overheat the solder. Caution is necessary because most of the **fluxes** used with this type of solder are easily overheated. Overheated fluxes will not only stop working as a flux but may become a barrier themselves to tinning. The best type of flame to use for this type of soldering is air acetylene, air MAPP, air propane, or any air fuel-gas mixture. The most popular types are air acetylene and air propane. If galvanized metal is being soldered or if tin-lead solder is being used, additional ventilation should be provided to prevent zinc oxide poisoning.

PROJECT 20-5

Copper Pipe Plant Stand

Skills to be learned: The ability to lay out and cut copper pipe to length and assemble it so that it is both accurate and square, and perform solder joints in all positions.

Project Description

This ceramic tile-topped copper pipe plant stand is designed to hold a houseplant up off of the floor. This plant stand is both decorative and functional, Figure 20-51.

Project Materials and Tools

The following items are needed to complete this project.

Materials

- 8 ft of 3/4-in. copper pipe
- 2 ft of 1/2-in. copper pipe
- 8 ea. 3/4-in. 90° copper elbows
- 4 ea. 3/4 × 1/2 × 3/4-in. copper tees
- 4 ea. 3/4-in. pipe caps
- Tin-antimony solder and C-flux
- Source of acetylene or propane gas
- Sand cloth

Tools

- Air-fuel torch
- PPE
- Pliers
- Hammer
- Tape measure
- 12-in. steel rule or straightedge
- Square
- Pencil or marker
- Pipe brush

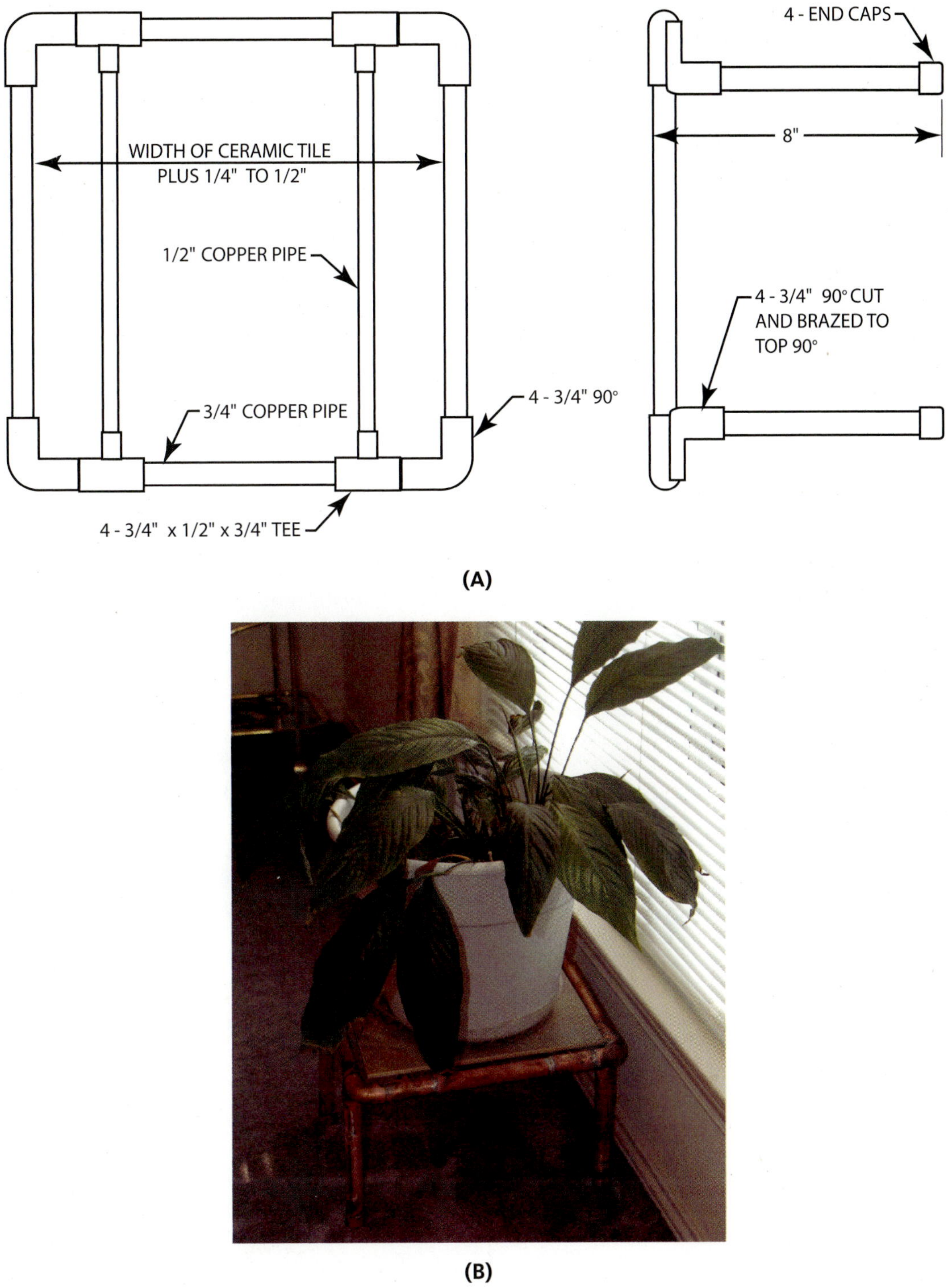

FIGURE 20-51 Project 20-4. Larry Jeffus

- Aviation tin snips and/or hacksaw
- Tubing cutter

Layout

Remove any labels from the fittings before you begin work. Heat from the torch causes labels to burn and makes them more difficult to remove later. In addition, the labels may interfere with the soldering process. Some stickers are easily removed, Figure 20-52, while others must be removed with a solvent. Use a solvent and a rag to remove stubborn stickers and adhesive, Figure 20-53.

FIGURE 20-52 Peel off any stickers before soldering. Larry Jeffus

FIGURE 20-53 Use a solvent to remove any sticker adhesive that might remain. Larry Jeffus

CAUTION

Most solvents are flammable, so they should only be used in a well-ventilated area away from any source of ignition.

The ceramic tile is 12 in. square, and depending on the type of 90° copper elbow you have selected, the length of the offset must be calculated, Figure 20-54. The offset is the difference between the edge of the pipe and the depth the side pipe will slide into the fitting. In this illustration, the offset is approximately 1/2 in. Using a measuring tape and marker, mark the 11 1/2-in. distance for the first tubing cut, Figure 20-55. Part of the distance on the front of the plant stand is taken up by the two T fittings.

Place the 90s and T fittings on the ceramic tile so that you can accurately measure the length of straight pipe that will be required to join these two pieces,

FIGURE 20-54 Use the tile and fitting to determine the length that the copper pipe should be. Larry Jeffus

FIGURE 20-55 Mark the copper to length. Larry Jeffus

Figure 20-56. Using the measuring tape and a marker, mark the 7 1/2-in. long length of this pipe.

The corner 90 and T fittings must be connected with a small section of pipe that is 1 1/4 in. in

FIGURE 20-56 Lay the pipe on the tile to check the length of the copper pipe. Larry Jeffus

length. Measure these pieces, and mark them on the 3/4-in. diameter copper pipe. Lay out the pipe on the tile, and measure the distance between the two Ts for the length of the 1/2-in. copper pipe. Measure and mark this distance on a piece of 1/2-in. copper pipe. The four legs will be 8 in. long, and they can be laid out and marked on the copper pipe.

Cutting Out

Copper pipe can be cut with a hacksaw or tubing cutter. For this exercise, a pipe tubing cutter will be used. Line up the cutter on the mark on the tube, and turn the handle on the tubing cutter to bring a light pressure on the cutter wheel. Rotate the tubing cutter around the tube, and gradually increase the pressure on the roller. Continue rotating the cutter around the copper pipe and increasing the roller pressure until the pipe is completely cut. Repeat this process until all of the sections of pipe have been cut.

To have the plant stand legs as close as possible to the corners for maximum stability, a special fitting will be fabricated using two 90° elbows. One elbow will be cut and fitted onto the other elbow making the corner piece. Using a marker, draw a line across the fitting approximately in the center, Figure 20-57. The top half of one side of the 90 will be removed. This half can be cut off using a hacksaw or aviation-type tin snips, Figure 20-58. Once the top half has been removed, mark a radius on the end of the fitting. This radius will be cut or filed into the corner of the fitting, Figure 20-59. The radius should be cut deep enough so that it will not interfere with the fitting that makes up the corner of the top frame, Figure 20-60. Trim the fitting, if necessary, to make it fit properly.

Fit the part just fabricated to the 90° elbow. Check to make sure that the fitting is square by placing two pieces of copper pipe in the fitting and checking it

FIGURE 20-57 Marking a 90° to be cut for the corner piece. Larry Jeffus

FIGURE 20-58 Use aviation sheet metal shears to cut the copper 90. Larry Jeffus

FIGURE 20-59 Trim the 90. Larry Jeffus

FIGURE 20-60 Check to see that the cut 90 will fit squarely in place. Larry Jeffus

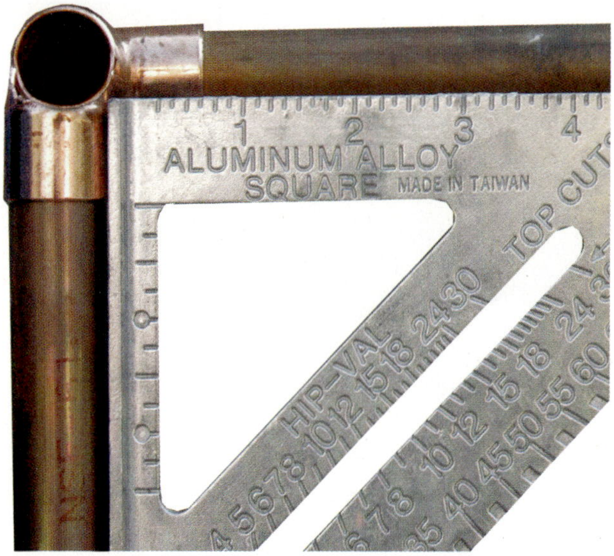

FIGURE 20-61 Use a square to align the parts. Larry Jeffus

with a square, Figure 20-61. Use a pair of pliers to hold the fitting in place as the two pieces of copper pipe are removed, Figure 20-62. Silver braze will be used to fabricate this fitting so that it can withstand the heat as the other joints are soldered, Figure 20-63. Using the same techniques as used in Project 20-3, braze the two fittings together.

Before assembling the new parts, use a tube brush to clean the oxide out, Figure 20-64. The oxides must be removed so that the solder will flow into the joint properly. Fit the two ends together, and align them next to each other to make sure that the top sides are equal in length, Figure 20-65.

Before any of the parts are assembled for the final time prior to soldering, use a solder paste brush to

FIGURE 20-62 Hold the parts in place. Larry Jeffus

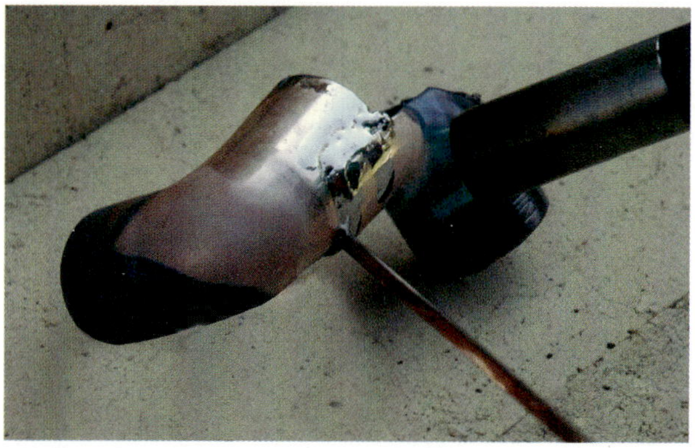

FIGURE 20-63 Silver braze the fitting. Larry Jeffus

FIGURE 20-64A The inside of the fitting will have a heavy oxide. Larry Jeffus

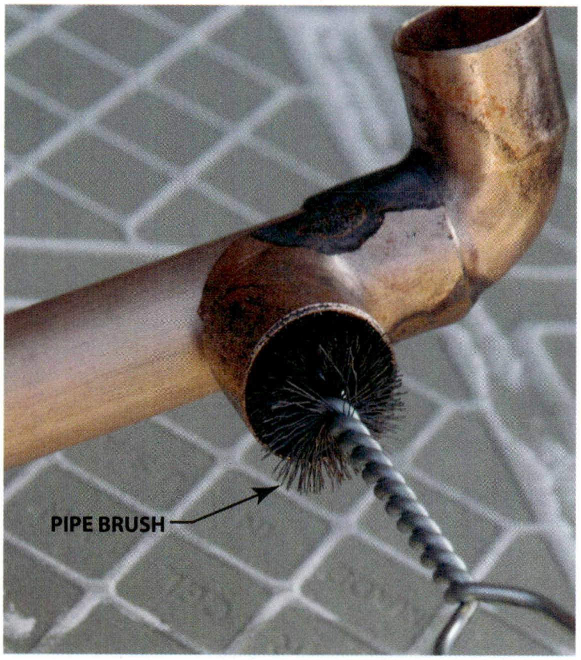

FIGURE 20-64B Use a wire tubing brush to clean the inside. Larry Jeffus

FIGURE 20-64C Check to see that the fitting is clean.
Larry Jeffus

apply a thin layer of flux on the end of the fittings, Figure 20-66. To help prevent excessive solder from flowing into the joint, do not put the solder flux all the way to the very end of the pipe.

Stand the assembly up on a flat surface such as the welding table, and use a square and measuring tape to check for accuracy of squareness and levelness before the assembly is soldered, Figure 20-67. The assembly must be within ±1/16 in. of level and ±1/8 in. of overall size. The end caps on the legs can be slid up or down to adjust the height and level as necessary, but if the overall width needs adjustment, you may have to remove and recut the pipes. Leave the assembly in the same place as it was squared and leveled to avoid accidentally shifting its alignment.

Soldering

Use a properly set up air MAPP® torch according to the manufacturer's recommendations, put on all of your PPE, and follow all shop and manufacturer's safety rules for soldering.

Heating the pipe first causes it to expand and tighten up inside the fitting. This tighter fit makes it easier to heat the entire pipe, and the smaller joint space makes it easier for the solder to wick into the joint. When the pipe is hot, start moving the flame between the pipe and fitting and occasionally touch the surface of the pipe with the end of the solder. When the pipe and joint reach the soldering temperature, the end of the

FIGURE 20-65 Check the fitting alignment.
Larry Jeffus

FIGURE 20-66 Use a flux brush to apply a thin layer of flux. Larry Jeffus

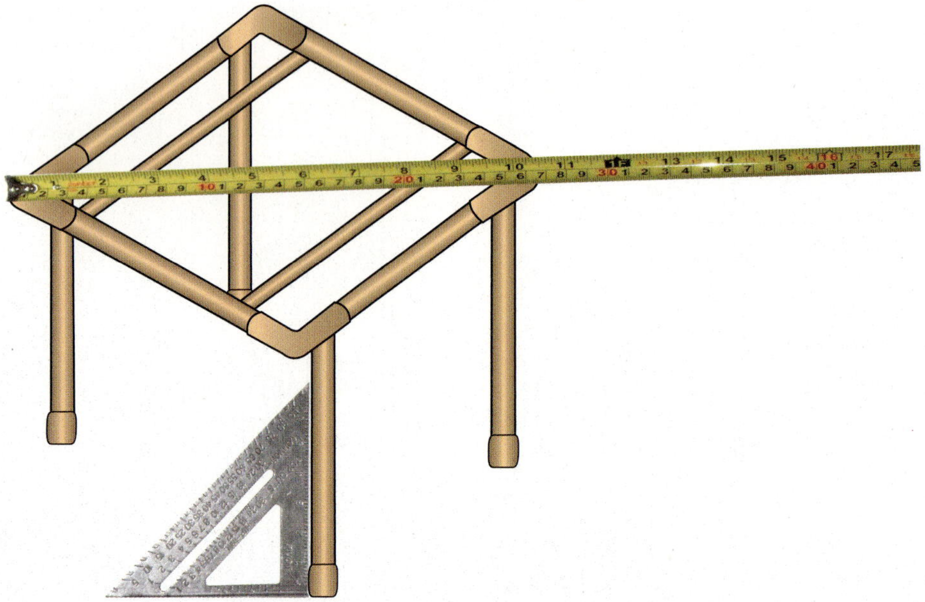

FIGURE 20-67 Check for squareness and levelness before soldering the fittings.
Larry Jeffus

solder will melt, Figure 20-68. To prevent overheating the joint, move the torch back from the joint or occasionally flash it off the joint. It should take no more than 3/4 in. of solder to complete the joint. If you overfill the joint with too much solder, it will probably just run down inside the pipe. Solder that runs down inside of copper pipe used for water can cause problems.

Building this plant stand will give you experience in soldering in the 1G, 2G, 3G, and 4G positions. Repeat the soldering process until all the joints have been completed.

FIGURE 20-68 Solder the assembly in place.
Larry Jeffus

Finishing

The solder joints may be left as they are, or excessive solder can be sanded off using sand cloth, Figure 20-69. The pipe may also be painted if so desired. Place the ceramic tile on top of the finished plant stand to complete this project.

Paperwork

Complete a copy of the time sheet in Appendix I and bill of materials in Appendix III or as provided by your instructor.◆

PROJECT 20-6

Key Hook

Skills to be learned: The ability to carefully control the soldering flame heat on a small part so that completed joints that are in close proximity are not remelted and the ability to fill a hole with solder.

Project Description

This is a four-hooked wall-mounted key holder that is made out of old brass keys. It is designed to be functional. Additional keys can be added to the base design if more than four hooks are needed.

Project Materials and Tools

The following items are needed to complete this project.

FIGURE 20-69 A sand cloth can be used to clean off any excessive solder. Larry Jeffus

FIGURE 20-70 Clean the keys with steel wool.
Larry Jeffus

Materials

- 4 brass keys
- 4 cup hooks
- Tin-antimony solder and C-flux
- Source of acetylene or propane gas
- Steel wool

Tools

- Air-fuel torch
- PPE
- Pliers

Layout

Start by cleaning the keys with steel wool. Many soldering fluxes are not as active as brazing fluxes, and, therefore, the parts must be precleaned to ensure a good bond, Figure 20-70. Use a straightedge to align the two center keys, and measure the space between the center holes of the side keys and that center line to make sure the finished key hook will be symmetrical, Figure 20-71.

FIGURE 20-71 Lay the keys out on a firebrick.
Larry Jeffus

Soldering

Set up an air propane soldering torch according to the manufacturer's recommendations, put on all your PPE, and follow all shop and manufacturer's safety rules for the torch and soldering.

Using a flux brush, apply a small quantity of flux on all of the key ring holes, except for the top key, and at the tips of the keys where they join, Figure 20-72. Use the torch to bring the keys up to soldering temperature. Place a small quantity of solder in the key ring hole of the keys. Fill the hole with solder but do

FIGURE 20-72 The small amount of flux. Larry Jeffus

FIGURE 20-73 Fill the key ring hole with solder.
Larry Jeffus

not excessively overfill it, Figure 20-73. You must carefully control the heat or the solder will run through the hole.

Using the torch, add a small quantity of solder at the ends of the keys where they join, Figure 20-74. Cool off the assembly to make it easier to control the heat as the hooks are installed.

Cut off the threaded end on three of the cup hooks using side cutters, Figure 20-75A and Figure 20-75B. Sometimes small parts can fly off when cut, so hold the cup hook with your fingers or with a pair of pliers.

Dip the end of the cup hook in solder flux, Figure 20-76. Using the long-nosed pliers, carefully heat the top of the key. Watch to see when the solder in the hole first melts. It will change from a dull to glassy appearance as it melts. Lower the cup hook end into the solder and remove the heat. Hold it very steady until the solder has cooled and solidified,

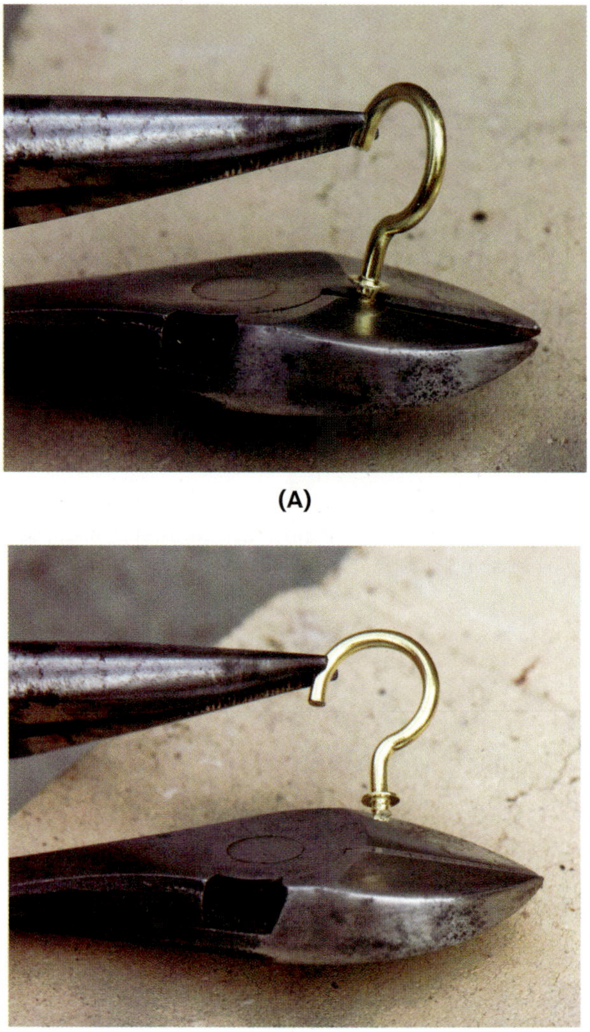

(A)

(B)

FIGURE 20-75 Cut off the end of the key hook.
Larry Jeffus

FIGURE 20-74 Solder the tips of the keys together.
Larry Jeffus

FIGURE 20-76 Dip the end of the hook in the solder flux. Larry Jeffus

FIGURE 20-77 Reheat the key ring hole solder. Larry Jeffus

FIGURE 20-78 Attach the key hook assembly to the wall using a cup hook. Larry Jeffus

Figure 20-77. Repeat this process until the remaining cup hooks have been installed.

Finishing

Cool the key hook, and use soap and water to remove any remaining soldering flux. The top cup hook screw is what will secure this to the wall, Figure 20-78. In some cases you may want to use a plastic or nylon insert in Sheetrock to allow the screw to hold more weight; however, with only four cup hooks on this small key holder, additional strength may not be required, Figure 20-79.

Paperwork

Complete a copy of the time sheet in Appendix I and bill of materials in Appendix III or as provided by your instructor.◆

FIGURE 20-79 Project 20-5. Larry Jeffus

SUMMARY

Although brazing and soldering are very similar processes, there is enough of a difference so that this chapter has two separate chapter end projects that can be built. One project requires the brazing skills you have learned, and the second project draws on your soldering skills.

The brazing project is the fabrication of a set of house numbers out of steel that will be brazed together before it is completely covered in brass. The brass covering provides both a durable long-lasting house identification and one that is very attractive, giving it a cast bronze appearance, Figure 20-80.

The frame and house numbers will be joined to the back with a brazed fillet. The difference in size and thickness of these parts will require that you control the torch flame so that each part is properly heated without overheating the smaller piece as you learned in Project 20-3. Once the steel parts are assembled, you will use the skills you learned in Project 20-1 to cover all of the surfaces with brazing beads. You can leave the beads or melt them together.

In the soldering project you will make a copper pipe garden arch, Figure 20-81. This project draws on the skills learned in soldering and the heat-forming

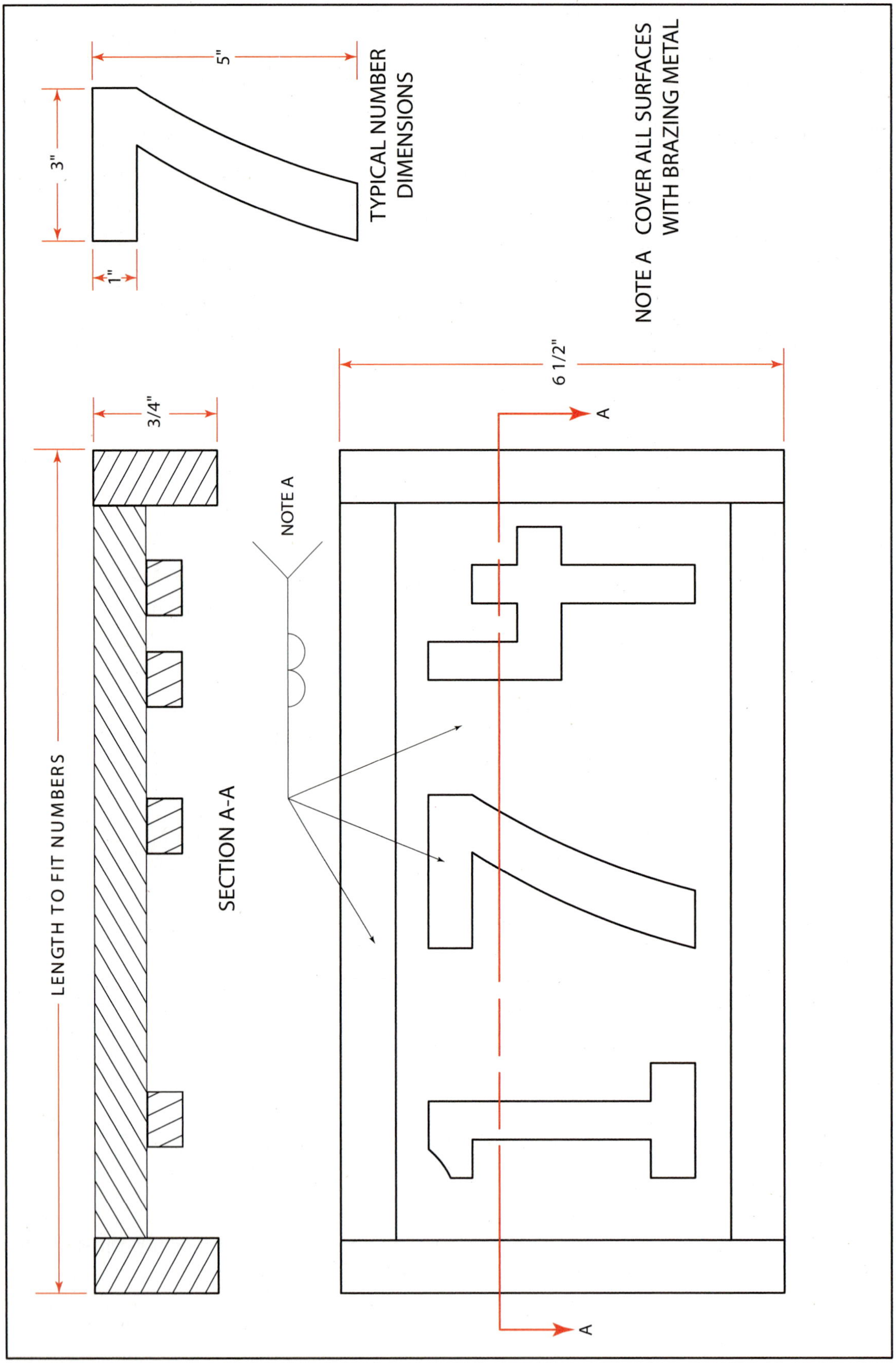

FIGURE 20-80 Brazing project using the skills you developed in this chapter.

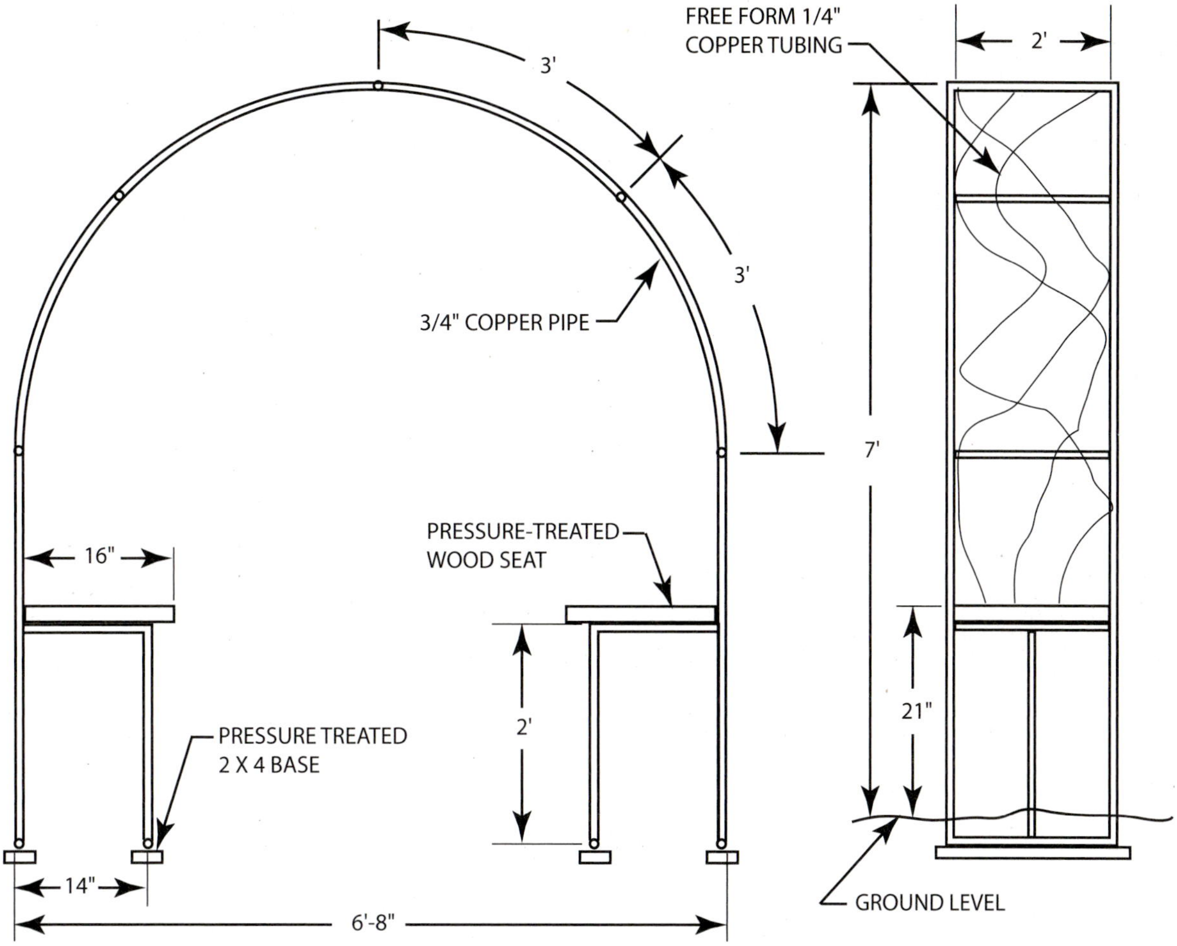

LEVEL 2 X 4 BASE

CLAMP ARCH PIPE TO BASE

PLANT CLIMBING VINE

NOW IT'S A SHADY PLACE TO SIT

FIGURE 20-81 Soldering project using the skills you developed in this chapter.

skills that you learned in Project 20-2. Using chalk or soapstone, sketch the arch on the shop floor or driveway as a template for the arch.

Copper pipe comes in 8-ft long sections. Four sections will be used to fabricate the arches. Start by heat bending one of the copper pipes to form one-half of the arch. Repeat this process until all four pipes have been formed. Lay two of the pipes down on the 6-ft 8-in. circle template. Measure and mark the location for the crossbars, and cut the pipe.

Measure and cut four 2-ft long pieces of 3/4-in. copper pipe. Lay these sections of pipe on the template to determine where to cut the bottom straight section of the arch. This will locate both the height of the bench and overall 7-ft height of the arch. Repeat this process until both arches have been cut.

Assemble the arch according to the drawing in Figure 20-81 and solder all of the joints. To give the climbing vine plant a better hold, solder three strands of free-formed 1/4-in. copper tubing or heavy-gauge copper wire across the arch framework.

Brazing and soldering are processes that can be used to fabricate a number of functional and decorative items rapidly with minimal investment in equipment. That is one of the reasons that many home hobbyists have begun using these processes for items that they produce for gifts or sales.

REVIEW QUESTIONS

1. Compare soldering to brazing.
2. What are the advantages of soldering and brazing?
3. How strong can joints that have been soldered and brazed be?
4. Describe what makes a brazed or soldered joint so strong.
5. Why might a surface layer of braze material be applied to steel?
6. How should the torch be held so that the metal ahead of the braze bead gets preheated?
7. When preheating metal for brazing, how can you tell when the temperature of the metal is hot enough to begin brazing?
8. What can you do if braze metal begins to overheat while brazing?
9. How can you tell that the braze metal is being overheated?
10. What can be done to make it easier to drill a large hole?
11. Why should metal always be clamped before being drilled?
12. How can you prevent overheating with an oxyacetylene flame?
13. What can cause the oxide layer to become thicker when brazing?
14. What is pretinning?
15. Why should labels be removed from copper pipe and fittings before you begin work?
16. Name two methods of cutting copper pipe.

Chapter 21

Oxyacetylene Cutting

OBJECTIVES

After completing this chapter, the student should be able to:

- Demonstrate the proper and safe method of setting up cylinders, regulators, hoses, and the cutting torch.
- Demonstrate how to maintain a cutting tip and torch.
- Demonstrate how to light a torch, adjust it, and make a cut.
- Describe a good oxyacetylene cut.
- Discuss safety procedures to be followed when oxyfuel cutting.

KEY TERMS

coupling distance
cutting tips
drag
drag lines
orifice
oxyacetylene hand torch
oxyfuel gas cutting (OFC)
preheat flame
preheat holes
slag

INTRODUCTION

Oxyacetylene cutting (OFC-A) is the primary cutting process in a larger group called **oxyfuel gas cutting (OFC).** OFC is a group of oxygen cutting processes that uses heat from an oxygen fuel gas flame to raise the temperature of the metal to its kindling temperature. When the metal is hot enough, a high-pressure stream of oxygen is directed onto the metal, causing it to be cut. The kindling temperature of a material is the temperature at which combustion (rapid oxidation) can begin. The kindling temperature of steel in pure oxygen is a dull red temperature of about 1600°F to 1800°F (870°C to 900°C). The processes in this group are identified by the type of fuel gas mixed with oxygen to produce the preheat flame. Oxyfuel gas cutting is most commonly

Fuel Gas	Flame (Fahrenheit)	Temperature* (Celsius)
Acetylene	5589°	3087°
MAPP	5301°	2927°
Natural gas	4600°	2538°
Propane	4579°	2526°
Propylene	5193°	2867°
Hydrogen	4820°	2660°

*Approximate neutral oxyfuel flame temperature.

Table 21-1 Fuel Gases Used for Flame Cutting

performed with OFC-A. Table 21-1 lists a number of other fuel gases used for OFC. Although other fuel gases are being used, only acetylene and oxygen can be used to weld. Because it can be used for both cutting and welding, acetylene will remain the primary fuel gas.

Most welding shops use the oxyacetylene cutting torch and cutting tips more than any other welding equipment, Figure 21-1. Unfortunately, it is one of the most commonly misused processes. Many welders know how to light the torch and make a cut, but their cuts are very poor quality. Often, in addition to making bad cuts, they use unsafe torch techniques. A good oxyacetylene cut should not only be straight and square, but it also should require little or no postcut cleanup.

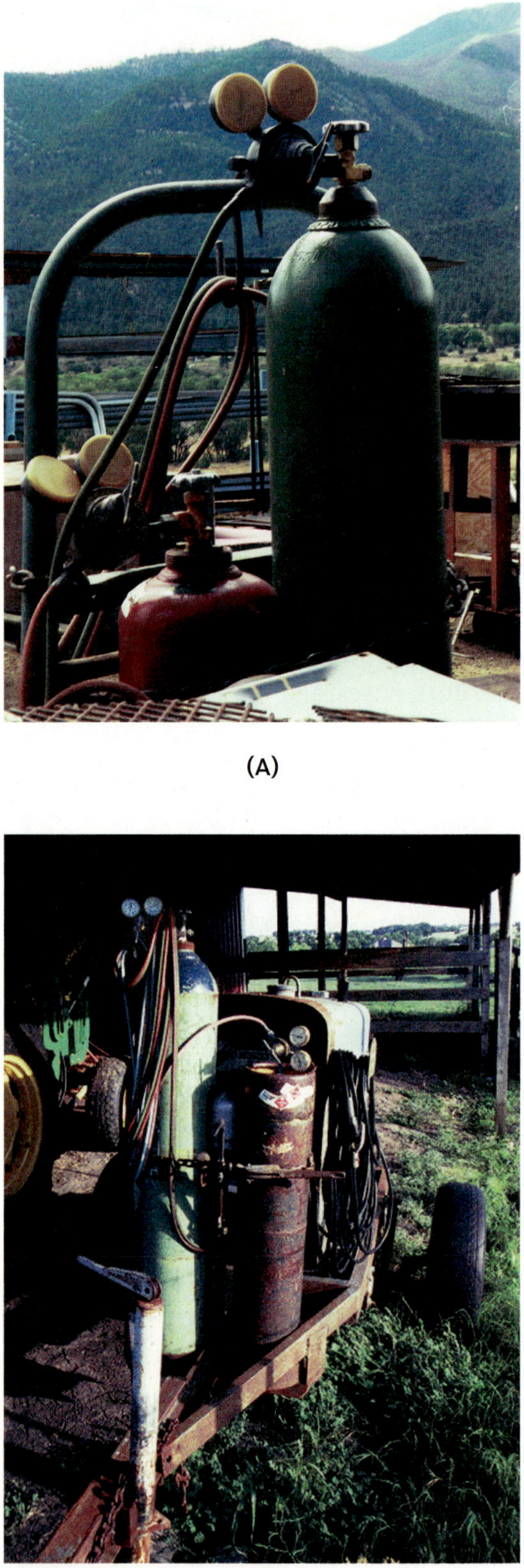

(A)

(B)

FIGURE 21-1 (A) Oxyacetylene portable cutting rig mounted on a handcart for use in a shop area. (B) Oxyacetylene setup mounted on a trailer for use outside of a shop. Larry Jeffus

Metals Cut by the Oxyfuel Process

Oxyfuel gas cutting is used to cut iron base alloys. Low carbon steels (up to 0.3% carbon) are easy to cut. High nickel steels, cast iron, and stainless steel are considered uncuttable with OCF-A. Most nonferrous metals—such as brass, copper, and aluminum—cannot be cut by oxyacetylene cutting.

Eye Protection for Flame Cutting

Cutting goggles or cutting glasses are required any time that you are using a cutting torch. You must protect your eyes from both the bright light and flying sparks. The National Bureau of Standards has identified proper filter plates and uses. The recommended filter plates are identified by shade number and are related to the type of cutting operation being performed.

Goggles, glasses, or other suitable eye protection must be used for flame cutting. Goggles should have vents near the lenses to prevent fogging. Cover lenses

Type of Cutting Operation	Hazard	Suggested Shade Number
Light cutting, up to 1 in.	Sparks, harmful rays, molten metal, flying particles	3 or 4
Medium cutting, 1–6 in.		4 or 5
Heavy cutting, over 6 in.		5 or 6

Table 21-2 A General Guide for the Selection of Eye and Face Protection Equipment

or plates should be provided to protect the filter lens. Filter lenses must be marked so that the shade number can be readily identified, Table 21-2.

CAUTION

Sunglasses or other dark shaded glasses are not safe to use for oxyacetylene cutting.

Cutting Torches

The **oxyacetylene hand torch** is the most common type of oxyfuel gas cutting torch used. The hand torch, as it is often called, may be either a part of a combination welding and cutting torch set or a cutting torch only, Figure 21-2. The combination welding and cutting torch offers more flexibility because a cutting head, welding tip, or heating tip can be attached quickly to the same torch body, Figure 21-3. Combination torch sets are often used. A cut made with either type of torch has the same quality; however, the dedicated cutting torches are usually longer and have larger gas flow passages than the combination torches. The added length of the dedicated cutting torch helps keep the operator further away from the heat and sparks and allows thicker material to be cut.

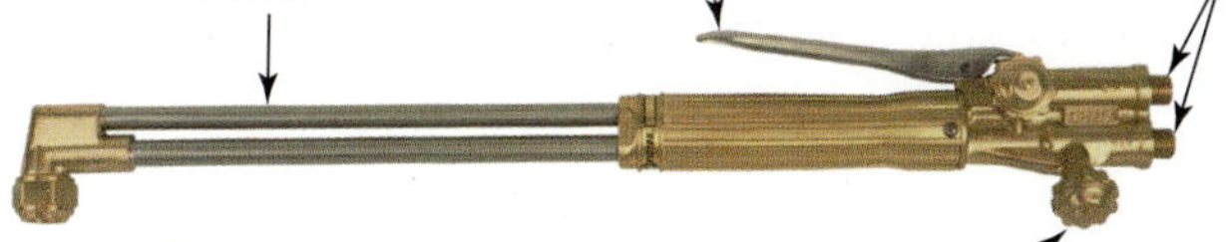

FIGURE 21-2 Oxyfuel cutting torch. Thermadyne Industries, Inc.

Oxygen is mixed with the fuel gas to form a high-temperature preheating flame. The two gases must be completely mixed before they leave the tip and create the flame. Two methods are used to mix the gases. One method uses a mixing chamber, and the other method uses an injector chamber.

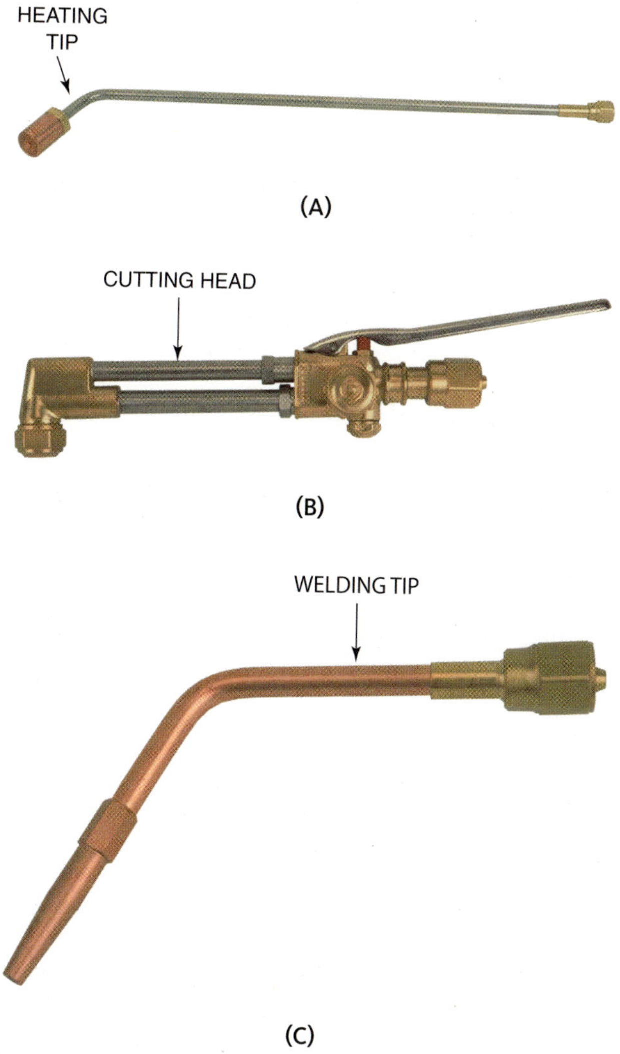

FIGURE 21-3 The attachments that are used for heating, cutting, welding, or brazing make the combination torch set flexible. Thermadyne Industries, Inc.

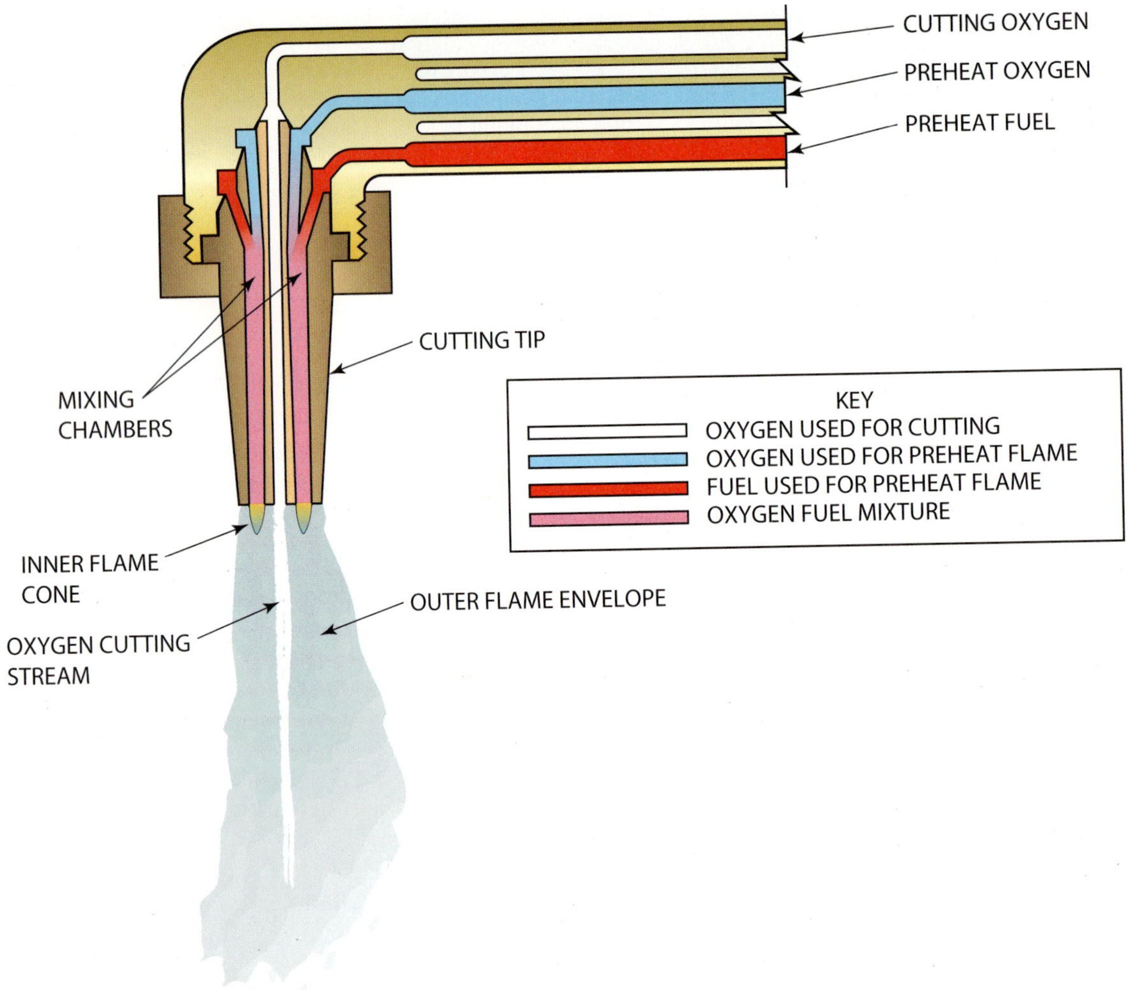

FIGURE 21-4 A mixing chamber located in the tip. American Welding Society

The mixing chamber may be located in the torch body or in the tip, Figure 21-4. Torches that use a mixing chamber are known as equal-pressure torches because the gases must enter the mixing chamber under the same pressure. The mixing chamber is larger than both the gas inlet and the gas outlet. This larger size causes turbulence in the gases, resulting in the gases mixing thoroughly.

Injector torches will work with both equal gas pressures and low fuel-gas pressures, Figure 21-5. The injector allows the oxygen to draw the fuel gas into the chamber even if the fuel gas pressure is as low as 6 oz/in.^{2} (26 g/cm^2). The injector works by passing the oxygen through a venturi, which creates a low-pressure area that pulls the fuel gases in and mixes them together. An injector-type torch must be used if a low-pressure acetylene generator or low-pressure residential natural gas is used as the fuel gas supply.

The cutting head may hold the cutting tip at a right angle to the torch body or it may be held at a slight angle. Torches with the tip slightly angled are easier for you to use when cutting a flat plate. Torches with a right-angle tip are easier to use when cutting pipe, angle iron, I beams, or other uneven material shapes. Both types of torches can be used for any type of material being cut, but practice is needed to keep the cut square and accurate.

The location of the cutting lever may vary from one torch to another, Figure 21-6. Most cutting levers pivot from the front or back end of the torch body. Personal preference will determine which one you use.

Cutting Tips

Most **cutting tips** are made of copper alloy, but some tips are chrome. Chrome plating prevents spatter

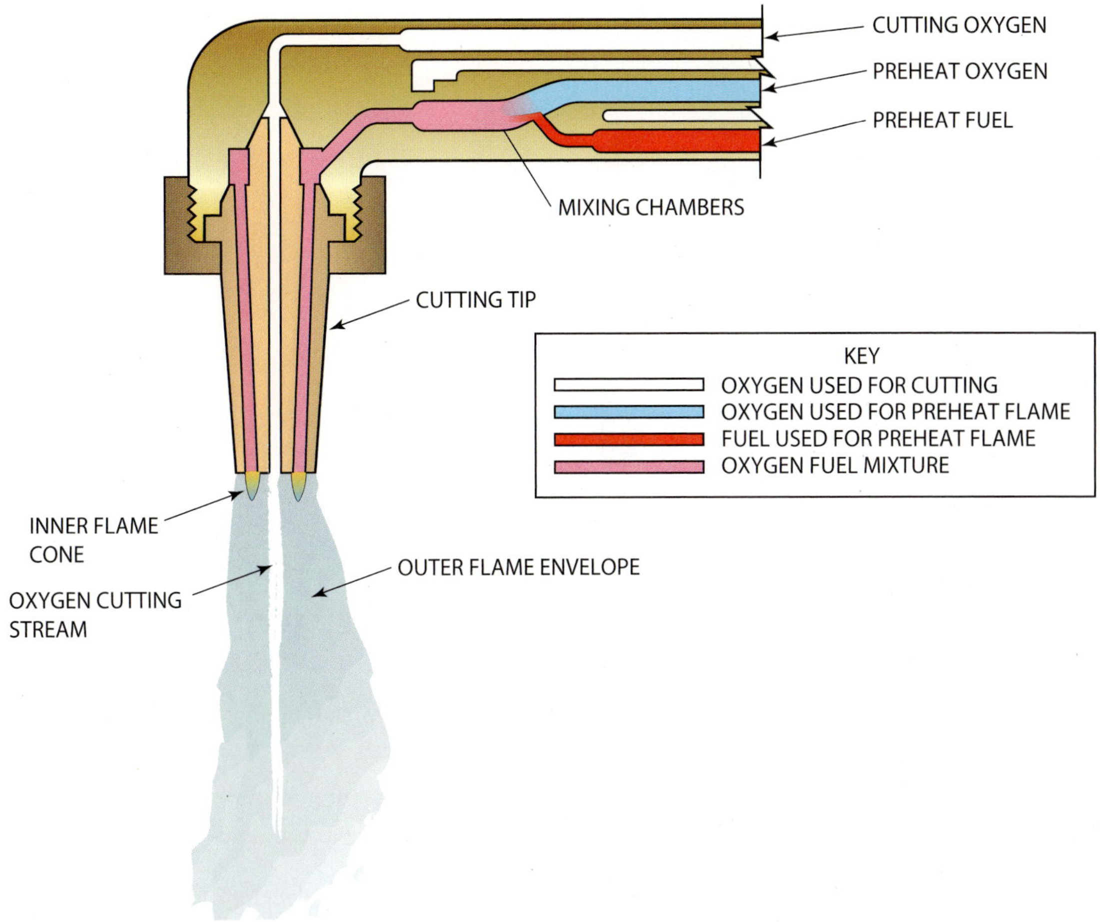

FIGURE 21-5 **Injector mixing torch.** American Welding Society

from sticking to the tip, thus prolonging its usefulness. Tip designs change for the different types of uses and gases, and from one torch manufacturer to another, Figure 21-7.

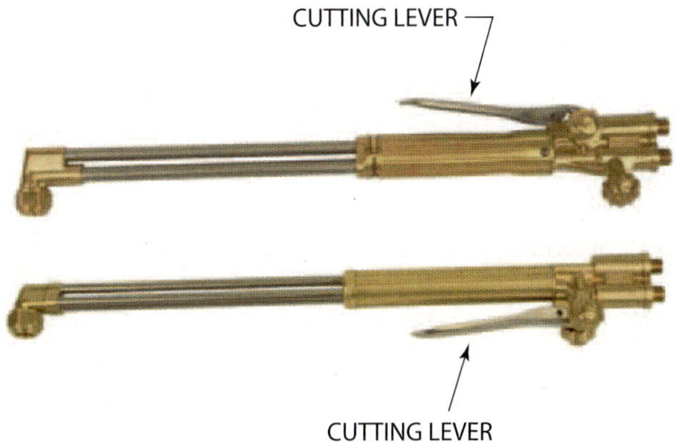

FIGURE 21-6 **The cutting lever may be located on the front or back of the torch body.** Thermadyne Industries, Inc.

The diameter, or size of the center cutting **orifice**, determines the thickness of the metal that can be cut. A larger diameter oxygen orifice is required for cutting thick metal. There is no standard numbering system for sizing cutting tips. Each manufacturer uses its own system. Some systems are similar; some are not. Table 21-3 lists several manufacturers' tip numbering systems. As a way of comparing the size of one manufacturer's tip size to another, the center hole diameter in inches is given below the tip number. For example, in Table 21-3 you can see that Airco's tip number 00 has a center orifice size equal to a number 70 drill size. You can also see that this cutting tip is designed for cutting metal approximately 1/8 in. (3 mm) thick. Other manufacturer tip numbers designed for this thickness have the following numbers: 000, 00, 1/4, 2, and 3.

Finding the correctly sized tip for a job can be confusing, especially if you are using the cutting unit for the first time. To make it easier to select a tip, you can use a standard set of tip cleaners to find the size

FIGURE 21-7 Different manufacturers have differently designed cutting torch tip seats. Larry Jeffus

Manufacturer	Metal Thickness, Inches (mm)										
	1/8 (3)	1/4 (6)	1/2 (13)	3/4 (19)	1 (24)	1 1/2 (37)	2 (49)	2 1/2 (61)	3 (74)	4 (98)	5 (123)
Cutting orifice drill number	70	68	60	56	54	53	50	47	45	39	31
Airco	00	0	1	1	2	2	3	4	4	5	6
ESAB	1/4	1/4	1/2	1 1/2	1 1/2	1 1/2	4	4	4	4	8
Harris	000	00	0	1	1	2	2	3	3	3	4
Oxweld	2	3	4	6	6	6	8	8	8	8	8
Purox	3	3	4	4	5	5	7	7	7	7	9
Smith	00	0	1	2	2	3	3	4	4	4	5
Victor	000	00	0	1	2	2	3	4	4	5	6

Table 21-3 Comparison of Some Manufacturers' Oxyacetylene Cutting Tip Identifications

of the center cutting orifice. Table 21-4 lists the material thickness being cut with the tip cleaner size.

If the manufacturers' recommendations for gas pressure are not available, you can use Table 21-4 to find the approximate pressures to be used with the tip. Actual gas pressures vary, depending on a number of factors, such as the equipment manufacturer, the condition of the equipment, hose length, hose diameter, regulator size, and operator skill. In all cases start out with the pressure recommended by the particular manufacturer of the equipment being used. Adjust the pressure to fit the job being cut.

A wide variety of tip shapes are also available for specialized cutting jobs. Each tip, of course, also comes in several sizes. Some tips are specialized for the kind of fuel gas being used. Different means are used to attach the cutting tip to the torch head. Some tips screw in; others have a push fitting.

Always choose the correct type and size of tip for the specific cutting job. Check the manufacturer's literature for tip size and type recommendations. Make sure the tip is designed for the type of fuel gas being used. Inspect the tip before using it. If the tip is clogged or dirty, clean the tip and clean out the orifices with the proper size drill. Check to make sure there is no damage to the threads. If the threads or the tapered seat is damaged, do not use the tip.

Tip Size	Metal Thickness, Inches (mm)										
	1/8 (3)	1/4 (6)	1/2 (13)	3/4 (19)	1 (24)	1 1/2 (37)	2 (49)	2 1/2 (61)	3 (74)	4 (98)	5 (123)
Cutting orifice drill number	70	68	60	56	54	53	50	47	45	39	31
WYPO tip cleaner number*	10	10	15	18	22	24	26				
Campbell Hausfeld tip cleaner number*	3	3	6	9	10	11	12				
Oxygen pressure, psi**	20 25	20 25	25 30	30 35	35 40	35 40	40 45	40 45	40 45	45 55	45 55
Oxygen pressure, kPa**	140 170	140 170	170 200	200 240	240 275	240 275	275 310	275 310	275 310	310 380	310 380
Acetylene pressure, psi**	3 5	3 5	3 5	3 5	3 5	3 5	4 8	4 8	5 11	6 13	8 14
Acetylene pressure, kPa**	20 35	20 35	20 35	20 35	20 35	20 35	30 55	30 55	35 75	40 90	55 95

*There is no standard numbering system for tip cleaners, so numbers can differ from one manufacturer to another.
**Tip size and pressures are approximate. Use the manufacturer's specification for equipment being used when available.

Table 21-4 Center Cutting Orifice Size, Metal Thickness, and Gas Pressures for Oxyacetylene Cutting

The amount of **preheat flame** required to make a perfect cut is determined by the type of fuel gas used and by the material thickness, shape, and surface condition. Materials that are thick, are round, or have surfaces covered with rust, paint, oil, and so on, require more preheat flame, Figure 21-8.

Different cutting tips are available for each of the major types of fuel gases. The differences in the type or number of **preheat holes** determine the type of fuel gas to be used in the tip. Table 21-5 lists the fuel gas and range of preheat holes or tip designs used with each gas. Acetylene is used in tips having from

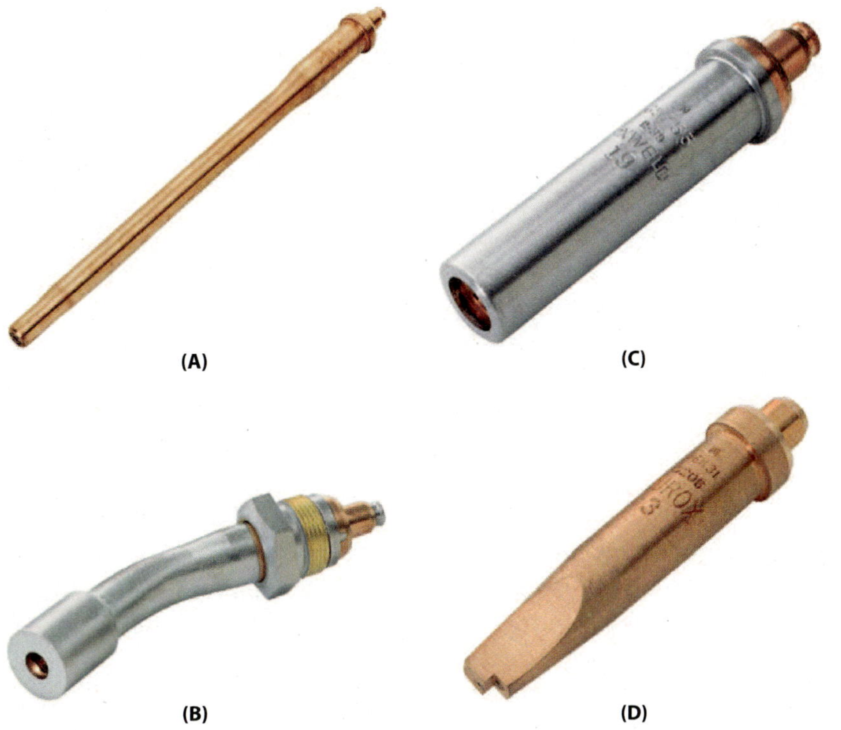

FIGURE 21-8 Special cutting tips: (A) 10-in. long cutting tip, (B) water-cooled cutting tip, (C) two-piece cutting tip, (D) sheet metal cutting tip.
ESAB Welding & Cutting Products

Fuel Gas	Number of Preheat Holes
Acetylene	One to six
MPS (MAPP)	Eight- to two-piece tip
Propane and natural gas	Two-piece tip

Table 21-5 Fuel Gas and Number of Preheat Holes Needed in the Cutting Tip

one to six preheat holes. Some large acetylene cutting tips may have eight or more preheat holes.

MPS, propane, and natural gas may be used in two-piece tips, Figure 21-9A and B. The MPS gas two-piece tip's inner part is flush with the end of the outer tip part, Figure 21-9C. The inner part on many propane and natural gas two-piece tips is deeply recessed, Figure 21-9D. Because propane and natural gases burn at a relatively slow rate, the flame may lift off of tips not specifically designed for these gases. The only problem with trying to use propane or natural gas in the wrong tip is that the flame may not stay lit.

CAUTION

If acetylene is used in a tip that was designed to be used with one of the other fuel gases, the tip may overheat, causing a backfire or the tip to explode.

MPS gases are used in tips having eight preheat holes or in a two-piece tip that is not recessed. These gases have a slower flame combustion rate than acetylene. For tips with less than eight preheat holes, there may not be enough heat to start a cut, or the flame may pop out when the cutting lever is pressed.

CAUTION

If MPS gases are used in a deeply recessed, two-piece tip, the tip will overheat, causing a backfire or the tip to explode.

Some cutting tips have metal-to-metal seals. When they are installed in the torch head, a wrench must be used to tighten the nut. Other cutting tips have fiber packing seats to seal the tip to the torch. If a wrench is used to tighten the nut for this type of tip, the tip seat may be damaged, Figure 21-10. A torch owner's manual should be checked or a welding supplier should be asked about the best way to tighten various torch tips.

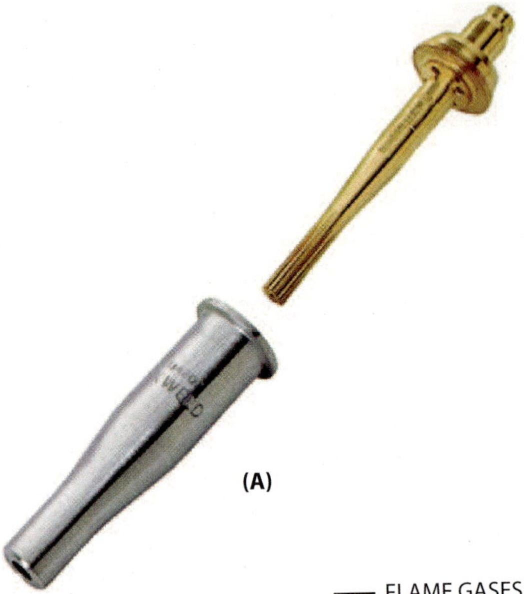

(A)

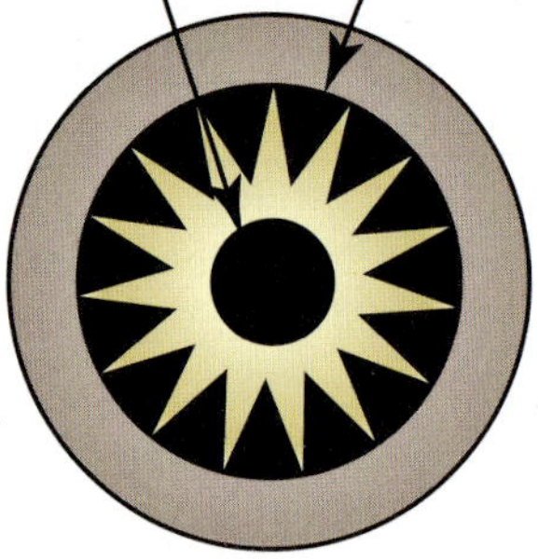

(B)

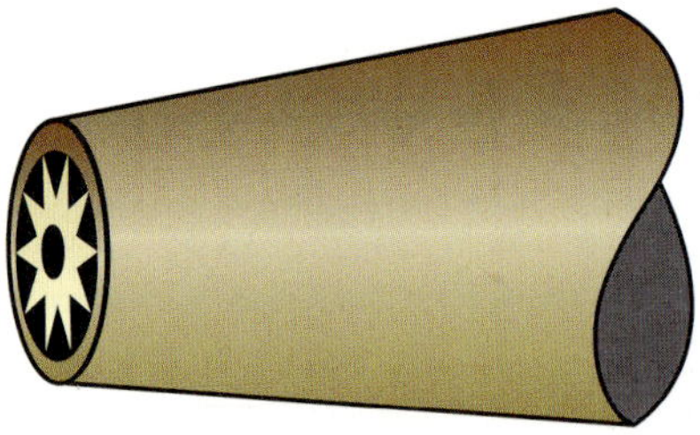

(C)

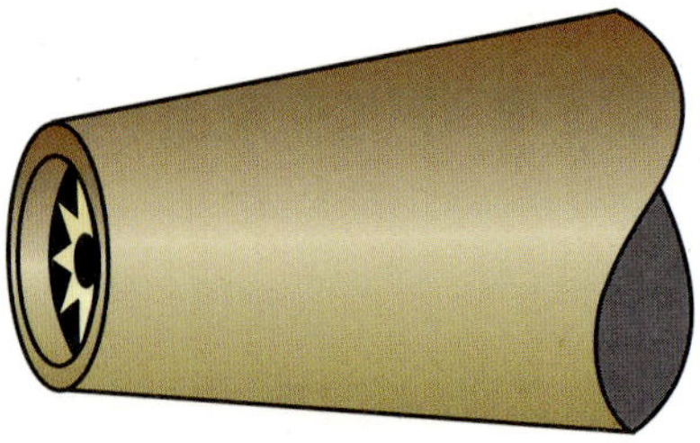

(D)

FIGURE 21-9 (A, B) Parts of a two-piece cutting tip. (ESAB Welding & Cutting Products) (C) MPS two-piece cutting tip. (D) Propane natural gas two-piece cutting tip. © Cengage Learning 2012

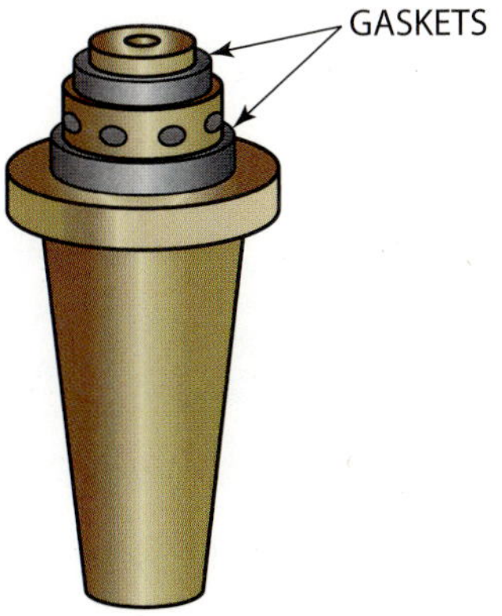

FIGURE 21-10 Some cutting tips use gaskets to make a tight seal. © Cengage Learning 2012

When removing a cutting tip, if the tip is stuck in the torch head, tap the back of the head with a plastic hammer, Figure 21-11. Any tapping on the side of the tip may damage the seat.

To check the assembled torch tip for a good seal, turn on the oxygen valve and spray the tip with a leak-detecting solution, Figure 21-12.

CAUTION

Carefully handle and store the tips to prevent damage to the tip seats and to keep dirt from becoming stuck in the small holes.

If the cutting tip seat or the torch head seat is damaged, it can be repaired by using a reamer designed for the specific torch tip and head, Figure 21-13, or it can be sent out for repair. New fiber packings are available

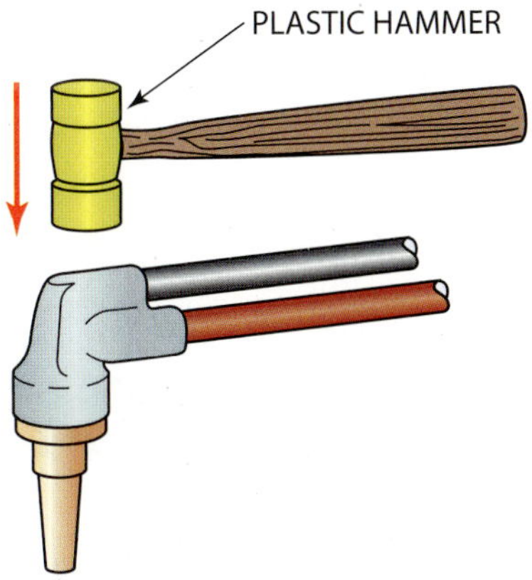

FIGURE 21-11 Tap the back of the torch head to remove a tip that is stuck. The tip itself should never be tapped. © Cengage Learning 2012

FIGURE 21-12 Checking a cutting tip for leaks. Larry Jeffus

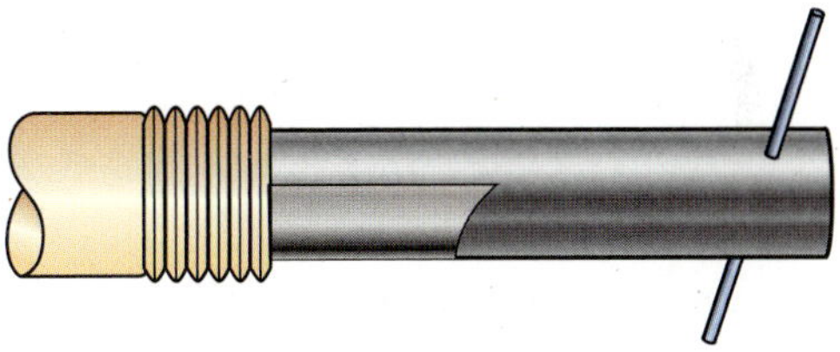

FIGURE 21-13 Damaged torch seats can be repaired by using a reamer. © Cengage Learning 2012

for tips with packings. The original leak-checking test should be repeated to be sure the new seal is good.

Oxyfuel Cutting, Setup, and Operation

The setting up of a cutting torch system is exactly like setting up oxyfuel welding equipment except for the adjustment of gas pressures. This chapter covers gas pressure adjustments and cutting equipment operations. Chapter 17, "Oxyfuel Welding and Cutting Equipment, Setup, and Operation," gives detailed technical information and instructions for oxyfuel systems. Chapter 17 covers the following topics:

- Safety
- Pressure regulator setup and operation
- Welding and cutting torch design and service
- Reverse flow and flashback valves
- Hoses and fittings
- Types of flames
- Leak detection

PRACTICE 21-1

Setting Up a Cutting Torch

Putting on your personal protective equipment (PPE) and following all shop and manufacturer's safety rules for OFC, demonstrate to other students and your instructor the proper method of setting up cylinders, regulators, hoses, and the cutting torch.

1. The oxygen and acetylene cylinders must be securely chained to a cart or wall before the safety caps are removed.
2. After removing the safety caps, stand to one side and crack (open and quickly close) the cylinder valves, being sure there are no sources of possible ignition that may start a fire. Cracking the cylinder valves is done to blow out any dirt that may be in the valves.
3. Visually inspect all of the parts for any damage, needed repair, or cleaning.
4. Attach the regulators to the cylinder valves and tighten them securely with a wrench.
5. Attach a reverse flow valve or flashback arrestor, if the torch does not have them built in, to the hose connection on the regulator or to the hose connection on the torch body, depending on the type of reverse flow valve in the set. Occasionally, test each reverse flow valve by blowing through it to make sure it works properly.
6. If the torch you will be using is a combination-type torch, attach the cutting head at this time.
7. Last, install a cutting tip on the torch.
8. Before the cylinder valves are opened, back out the pressure regulating screws so that when the valves are opened the gauges will show zero pounds working pressure.
9. Stand to one side of the regulator's face as the cylinder valves are opened slowly.
10. The oxygen valve is opened all the way until it becomes tight, but not overtight, and the acetylene valve is opened no more than one-half turn.
11. Open one torch valve and then turn the regulating screw in slowly until 2 psig to 4 psig (14 kPag to 30 kPag) shows on the working pressure gauge. Allow the gas to escape so that the line is completely purged.
12. If you are using a combination welding and cutting torch, the oxygen valve nearest the hose connection must be opened before the flame adjusting valve or cutting lever will work.

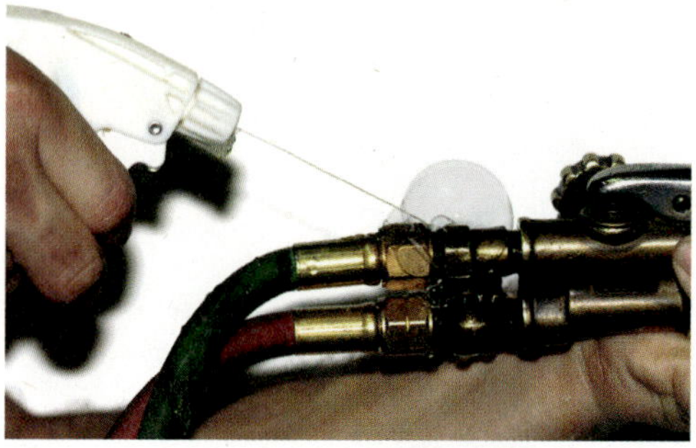

FIGURE 21-14 Leak-check all gas fittings. Larry Jeffus

13. Close the torch valve and repeat the purging process with the other gas.
14. Be sure there are no sources of possible ignition that may result in a fire.
15. With both torch valves closed, spray a leak-detecting solution on all connections, including the cylinder valves. Tighten any connection that shows bubbles, Figure 21-14.

Paperwork

Complete a copy of the time sheet in Appendix I and bill of materials in Appendix III, or as provided by your instructor. ◆

PRACTICE 21-2

Cleaning a Cutting Tip

Putting on your PPE, following all shop and manufacturer's safety rules for OFC, and using a cutting torch set that is assembled and adjusted as described in Practice 21-1, and a set of tip cleaners, you will clean the cutting tip.

1. Turn on a small amount of oxygen, Figure 21-15. This procedure is done to blow out any dirt loosened during the cleaning.
2. The end of the tip is first filed flat, using the file provided in the tip cleaning set, Figure 21-16.

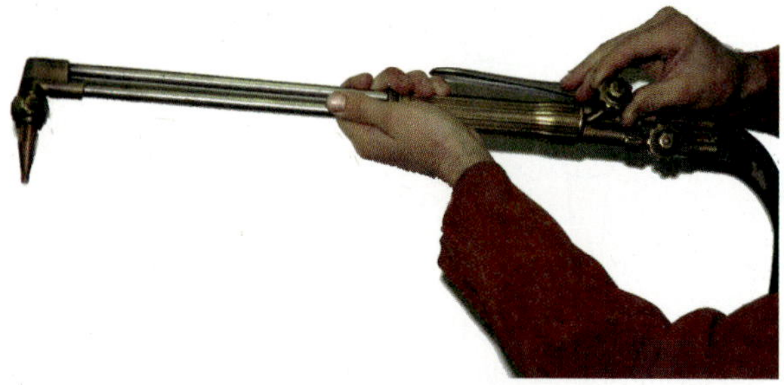

FIGURE 21-15 Turn on the oxygen valve. Larry Jeffus

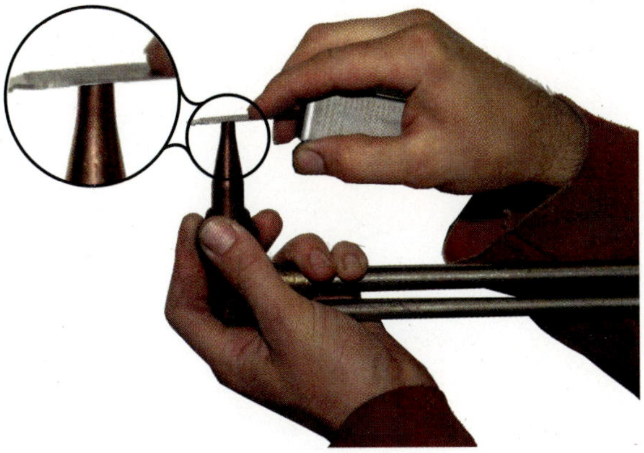

FIGURE 21-16 File the end of the tip flat. Larry Jeffus

3. Try several sizes of tip cleaners in a preheat hole until the correct size cleaner is determined. It should easily go all the way into the tip, Figure 21-17.
4. Push the cleaner in and out of each preheat hole several times. Tip cleaners are small, round files. Excessive use of them will greatly increase the orifice (hole) size.
5. Next, depress the cutting lever and, by trial and error, select the correct size tip cleaner for the center cutting orifice. Push the cleaner in and out several times.

A tip cleaner should never be forced. If the tip needs additional care, a tip cleaning drill set can be used.

Paperwork

Complete a copy of the time sheet in Appendix I and bill of materials in Appendix III, or as provided by your instructor.◆

FIGURE 21-17 A tip cleaner should be used to clean the flame and center cutting holes. Larry Jeffus

PRACTICE 21-3

Lighting the Torch

Putting on your PPE, following all shop and manufacturer's safety rules for cutting, and using a cutting torch set that is safely assembled, you will light the torch.

1. Set the regulator working pressure for the tip size. If you do not know the correct pressure for the tip, start with the fuel set at 5 psig (35 kPag) and the oxygen set at 25 psig (170 kPag).
2. Point the torch tip upward and away from any equipment or other students.
3. Turn on just the acetylene valve and use only a spark lighter to ignite the acetylene. The torch may not stay lit. If this happens, close the valve slightly and try to relight the torch.
4. If the flame is small, it will produce heavy black soot and smoke. In this case, turn up the flame to stop the soot and smoke. You need not be concerned if the flame jumps slightly away from the torch tip.
5. With the acetylene flame burning smoke-free, slowly open the oxygen valve and by using only the oxygen valve adjust the flame to a neutral setting, Figure 21-18.
6. When the cutting oxygen lever is depressed, the flame may become slightly carbonizing. This may occur because of a drop in line pressure due to the high flow of oxygen through the cutting orifice.
7. With the cutting lever depressed, readjust the preheat flame to a neutral setting.

The flame will become slightly oxidizing when the cutting lever is released. Since an oxidizing flame is hotter than a neutral flame, the metal being cut will be preheated faster. When the cut is started by depressing the lever, the flame automatically returns to the neutral setting and does not oxidize the top of the plate. Extinguish the flame by first turning off the oxygen and then the acetylene.

Paperwork

Complete a copy of the time sheet in Appendix I and bill of materials in Appendix III, or as provided by your instructor.◆

CAUTION

Sometimes with large cutting tips, the tip will pop when the acetylene is turned off first. If that happens, turn the oxygen off first.

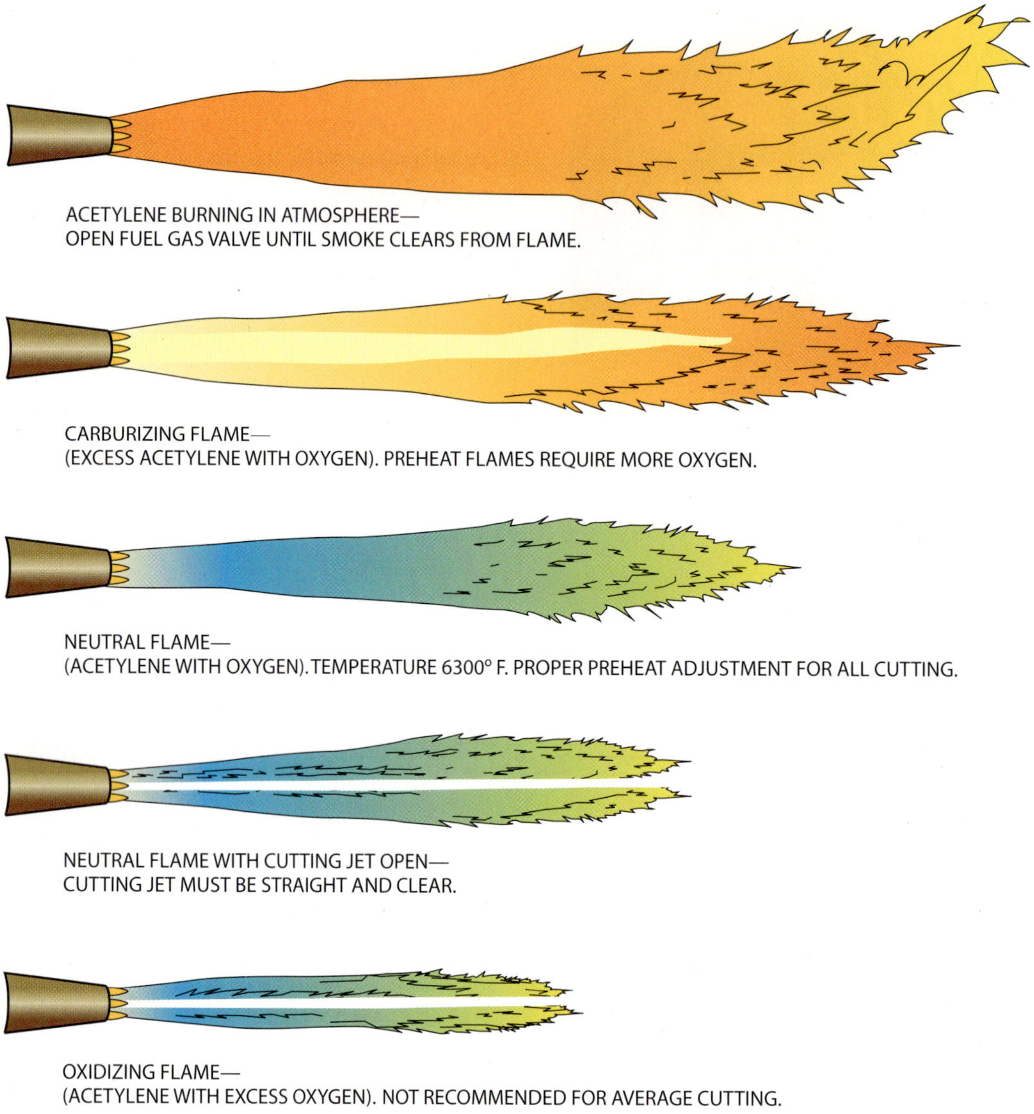

FIGURE 21-18 Oxyacetylene flame adjustments for the cutting torch.
© Cengage Learning 2012

Torch Tip Alignment

The proper alignment of the preheat holes will speed up and improve the cut. The holes should be aligned so that one is directly on the line ahead of the cut and another is aimed down into the cut when making a straight line square cut, Figure 21-19. The flame is directed toward the smaller piece and the sharpest edge when cutting a bevel. For this reason, the tip should be changed so that at least two of the flames are on the larger plate and none of the flames is directed on the sharp edge, Figure 21-20. If the preheat flame is directed at the edge, it will be rounded off as it is melted off.

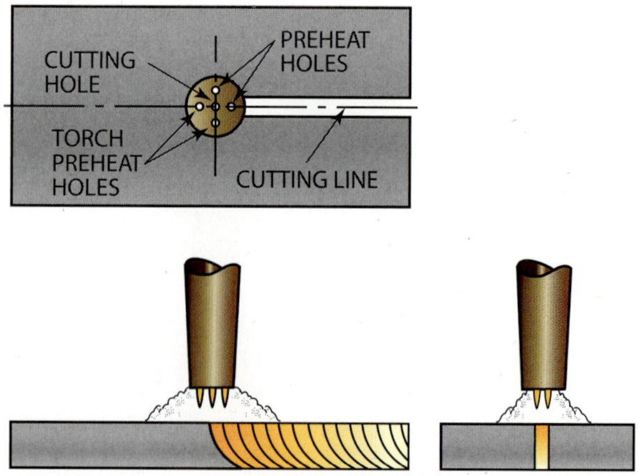

FIGURE 21-19 Tip alignment for a square cut.
American Welding Society

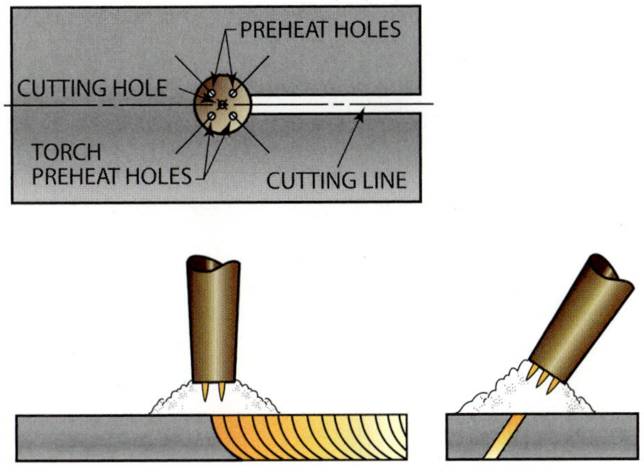

FIGURE 21-20 Tip alignment for a bevel cut. American Welding Society

FIGURE 21-22 Holding the punch slightly above the surface allows the punch to be struck rapidly and moved along a line to mark it for cutting. Larry Jeffus

Layout

Laying out a line to be cut can be done with a piece of soapstone or a chalk line. To obtain an accurate line, a scribe or a punch can be used. If a piece of soapstone is used, it should be sharpened properly to increase accuracy, Figure 21-21. A chalk line will make a long, straight line on metal and is best used on large jobs. The scribe and punch can both be used to lay out an accurate line, but the punched line is easier to see when cutting. A punch can be held as shown in Figure 21-22, with the tip just above the surface of the metal. When the punch is struck with a lightweight hammer, it will make a mark. If you move your hand along the line and rapidly strike the punch, it will leave a series of punch marks for the cut to follow.

Selecting the Correct Tip and Setting the Pressure

Each welding equipment manufacturer uses its own numbering system to designate the tip size. It would be impossible to remember each of the systems. Each manufacturer, however, does relate the tip number to the numbered drill size used to make the holes. On the back of most tip cleaning sets, the manufacturer lists the equivalent drill size of each tip cleaner. By remembering approximately which tip cleaner was used on a particular tip for a metal thickness range, a welder can easily select the correct tip when using a new torch set. Using the tip cleaner that you are familiar with, try it in the various torch tips until you find the correct tip that the tip cleaner fits. Table 21-6 lists the tip drill size, pressure range, and metal thickness range for which the tip can be used.

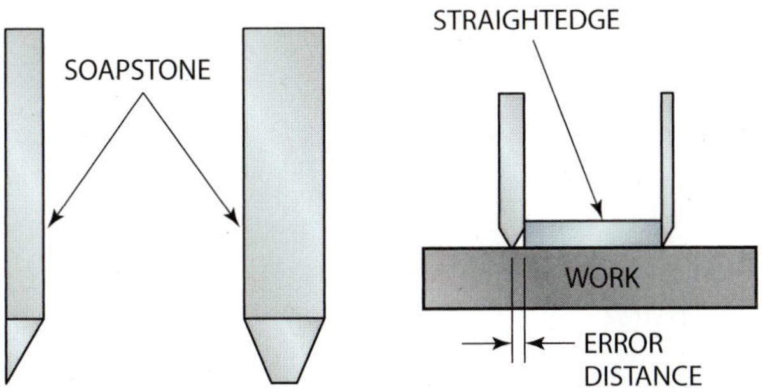

FIGURE 21-21 Proper method of sharpening a soapstone. © Cengage Learning 2012

PRACTICE 21-4

Setting the Gas Pressures

Put on your PPE, and follow all shop and manufacturer's safety rules for OFC. You are going to set the working pressure of the regulators which can be done by either following a table or by watching the flame.

1. To set the regulator by watching the flame, first set the acetylene pressure at 2 psig to 4 psig (14 kPag to 30 kPag) and then light the acetylene flame.
2. Open the acetylene torch valve one to two turns and reduce the regulator pressure by backing out the setscrew until the flame starts to smoke.
3. Increase the pressure until the smoke stops and then increase it just a little more.

This is the maximum fuel gas pressure the tip needs. With a larger tip and a longer hose, the pressure must be set higher. This is the best setting, and it is the

Metal Thickness in. (mm)	Center Orifice Size No. Drill Size	Tip Cleaner No.*	Oxygen Pressure lb/in^2 (kPa)	Acetylene lb/in^2 (kPa)
1/8 (3)	60	7	10 (70)	3 (20)
1/4 (6)	60	7	15 (100)	3 (20)
3/8 (10)	55	11	20 (140)	3 (20)
1/2 (13)	55	11	25 (170)	4 (30)
3/4 (19)	55	11	30 (200)	4 (30)
1 (25)	53	12	35 (240)	4 (30)
2 (51)	49	13	45 (310)	5 (35)
3 (76)	49	13	50 (340)	5 (35)
4 (102)	49	13	55 (380)	5 (35)
5 (127)	45	**	60 (410)	5 (35)

*The tip cleaner number when counted from the small end toward the large end in a standard tip cleaner set.

**Larger than normally included in a standard tip cleaner set.

Table 21-6 Cutting Pressure and Tip Size

safest one to use. With this lowest possible setting, there is less chance of a leak. If the hoses are damaged, the resulting fire will be much smaller than a fire burning from a hose with a higher pressure. There is also less chance of a leak with the lower pressure.

4. With the acetylene adjusted so that the flame just stops smoking, slowly open the torch oxygen valve.
5. Adjust the torch to a neutral flame. When the cutting lever is depressed, the flame will become carbonizing, not having enough oxygen pressure.
6. While holding the cutting lever down, increase the oxygen regulator pressure slightly. Readjust the flame, as needed, to a neutral setting by using the oxygen valve on the torch.
7. Increase the pressure slowly and readjust the flame as you watch the length of the clear cutting stream in the center of the flame, Figure 21-23. The center stream will stay fairly long until a pressure is reached that causes turbulence, disrupting the cutting stream. This turbulence will cause the flame to shorten in length considerably, Figure 21-23.
8. With the cutting lever still depressed, reduce the oxygen pressure until the flame lengthens once again. This is the maximum oxygen pressure that this tip can use without disrupting turbulence in the cutting stream. This turbulence will cause a

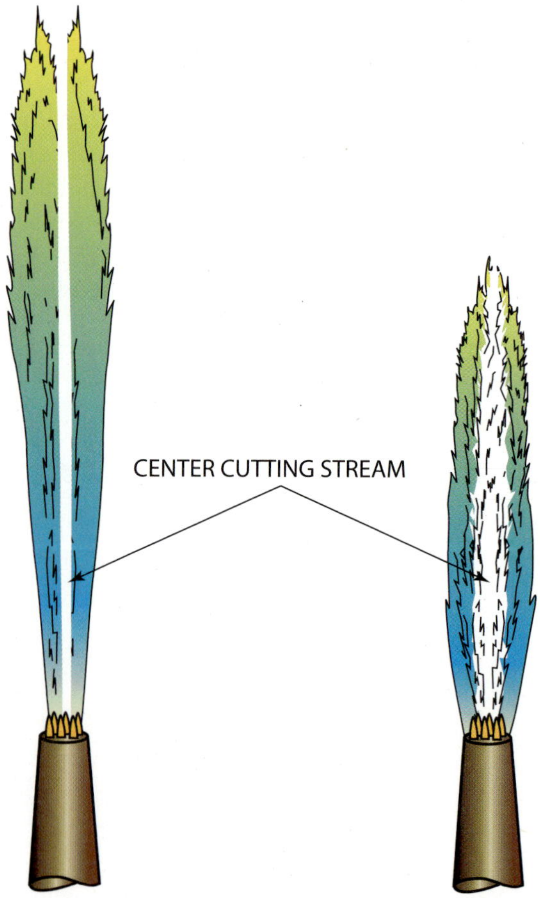

FIGURE 21-23 A clean cutting tip will have a long, well-defined oxygen stream.

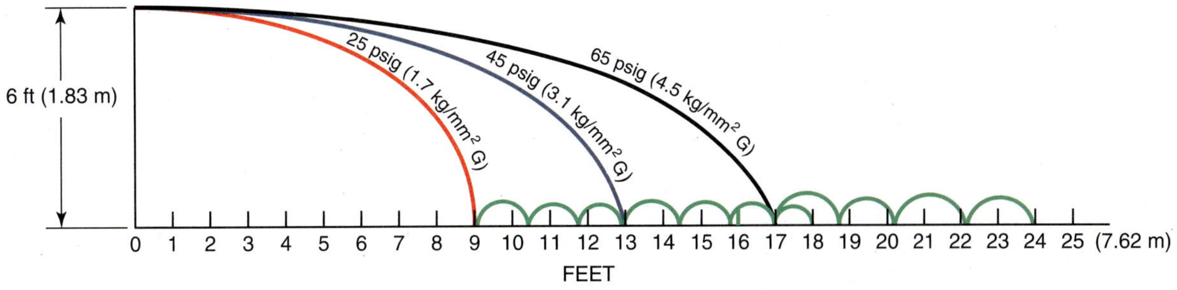

FIGURE 21-24 The sparks from cutting a mild steel plate, 3/8 in. (10 mm) thick, 6 ft (1.8 m) from the floor, will be thrown much further if the cutting pressure is too high for the plate thickness. These cuts were made with a Victor cutting tip no. 0-1-101 using 25 psig (1.7 kg/mm²) as recommended by the manufacturer and by excessive pressures of 45 psig (3.1 kg/mm²) and 65 psig (4.5 kg/mm²). © Cengage Learning 2012

very poor cut. The lower pressure will also keep the sparks from being blown a longer distance from the work, Figure 21-24.

Paperwork

Complete a copy of the time sheet listed in Appendix I or as provided by your instructor.◆

The Chemistry of a Cut

The oxyfuel gas cutting torch works when the metal being cut rapidly oxidizes or burns. This rapid oxidization or burning occurs when a high-pressure stream of pure oxygen is directed on the metal after it has been preheated to a temperature above its kindling point. Kindling point is the lowest temperature at which a material will burn. The kindling temperature of iron is 1600°F (870°C), which is a dull red color. Note that iron is the pure element and cast iron is an alloy consisting primarily of iron and carbon. The process will work easily on any metal that will rapidly oxidize, such as iron, low carbon steel, magnesium, titanium, and zinc.

CAUTION

Some metals release harmful oxides when they are cut. Extreme caution must be taken when cutting used, oily, dirty, or painted metals. They often produce very dangerous fumes when they are cut. You may need extra ventilation and a respirator to be safe.

The process is most often used to cut iron and low carbon steels. Few or no oxides are left on the metal, and it can easily be welded after cutting.

The burning away of the metal is a chemical reaction with iron (Fe) and oxygen (O). The oxygen forms an iron oxide, primarily Fe_3O_4, that is light gray in color. Heat is produced by the metal as it burns. This heat helps carry the cut along. On thick pieces of metal, once a small spot starts burning (being cut), the heat generated helps the cut continue quickly through the metal. With some cuts the heat produced may overheat small strips of metal being cut from a larger piece. As an example, the center piece of a hole being cut will quickly become red hot and will start to oxidize with the surrounding air. This heat produced by the cut makes it difficult to cut out small or internal parts.

Cutting Applications

It is important to know that not all the damage that can be done with a torch is to your machinery. Newspapers often have stories stating "the sparks from welders' torches are thought to have started the fire." The torch they are referring to is the cutting torch. The torch throws out so many sparks that almost anything will catch fire. In the welding industry, the cutting torch is used in a shop or yard that contains almost nothing that would catch fire. However, when working on construction or agriculture machinery, you may find dry grass, dry crop stubble, brush, and many other combustible items lying around your work areas. When possible, move anything that might catch fire well away from your work because cutting sparks can easily travel 30 ft (10 m) or more. In some cases, such as working in a field or in the timber, it is impossible to remove all combustibles from around your work area. If you must work around combustibles, wet the area first and keep a bucket of water and a fire extinguisher handy, Figure 21-25. Keeping someone as a fire watch will also help.

FIGURE 21-25 Have plenty of water and a fire extinguisher available before starting to cut or weld. Larry Jeffus

CAUTION

Never weld or cut in a dry area or an area that has been posted by the county or state as a fire zone, Figure 21-26. You have too great an investment in the land to accidentally start a fire while welding. Also, you could be held legally responsible for a fire that starts if you ignore the warning.

Making practice cuts on a piece of metal that will only become scrap is a good way to learn the proper torch techniques. If a bad cut is made, there is no loss. But when you are making a repair on your equipment and make a bad cut, that can be disastrous. It is easy to do a lot of damage very quickly with a cutting torch. Once a cut has gone wrong, a lot of metal can be removed quickly that will take hours to repair. When you see a cut is going bad—and bad cuts happen to everyone—stop cutting. Do not assume the cut will get better. If you cannot make a good cut with a torch, use a grinder or saw, or get help.

A number of factors that do not exist during practice cuts can affect your ability to make a quality cut on a part. The following are some problems that can occur during cutting:

- *Brace yourself:* The tip of a cutting torch is nearly a foot from your hands as you cut. This distance amplifies the slightest movement of your hand, which translates to grooves and notches in the cut. It is easier to brace yourself so you can make good cuts in the shop rather than in the field. In the field, you may be working in awkward positions and small spaces. One way of steadying yourself is to lean against the equipment you are working on. Even just resting your hand can help, Figure 21-27. Before even lighting the torch, practice how you can move across the cut and which way would be best to brace yourself.
- *Changing positions:* Often, parts are too large to be cut from one position, so you may have to move to complete the cut. Stopping and restarting a cut can result in a small flaw in the cut surface. To avoid this problem, always try to stop at corners if the cut cannot be completed without moving.
- *Sparks:* All cuts create sparks that bounce around. You must always know where the sparks from a cut are being thrown. Make sure they are not being directed toward the fuel tanks, dead grass or straw, oil, or other combustible materials. Sparks striking glass or mirrors can cause pitting. Cutting on equipment where the sparks hit another surface behind the one being cut makes the problem of bouncing sparks much worse. Sparks often find their way into your glove, under your arm, or any other place

FIGURE 21-26 In most cases this sign also means no welding or cutting outdoors. Larry Jeffus

FIGURE 21-27 Bracing your hand on the tractor tongue will help keep you steady as you make the cut. Larry Jeffus

that will become uncomfortable. Experienced welders usually keep working if the sparks are not too large or too uncomfortable. With experience you will learn how to angle the torch, direct the cut, and position your body to minimize this problem. It is your responsibility to make sure the sparks do not start a fire or damage the equipment.

- *Hot surfaces:* As you continue making cuts to complete the part, the part will begin to heat up. Depending on the size of the part, the number of cuts per part, and the number of parts being cut, this heat can become uncomfortable. You may find it necessary to hold the torch further back from the tip, but this will affect the quality of your cuts, Figure 21-28. Sometimes you might be able to rest your hand on a block to keep it off the plate. Another problem with heat buildup is that it may become high enough to affect the equipment's hydraulic or other systems. Planning your cutting sequence and allowing cooling time will help control this potential problem.
- *Tip cleaning:* The cutting tip will catch small sparks and become dirty or clogged. You must decide how dirty or clogged you will let the tip get before you stop to clean it. Time spent cleaning the tip reduces productivity, unfortunately. On the other hand, if you do not stop occasionally to clean up, the quality of the cut will become bad. It is your responsibility to decide when and how often to clean the tip.

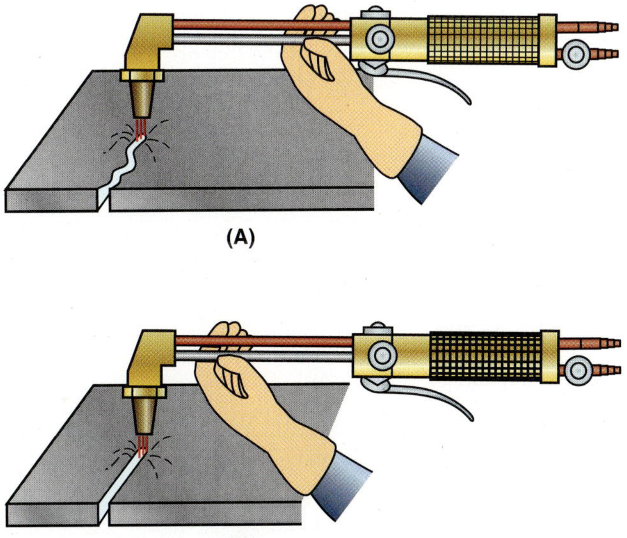

FIGURE 21-28 It is easier to make straight, smooth cuts if you can brace the torch closer to the tip, as in cut (B).

- *Blowback:* As a cut progresses across the surface, it may cross over parts underneath. During practice cuts this seldom if ever happens, but, depending on the part being cut, it will occur. If the part is small, the blowback may not cover you with sparks, plug the cutting tip, or cause a major flaw in the cut surface. If the part is large, then one or all of these events can occur. If you see that the blowback is not clearing quickly, it may be necessary to stop the cut. Stopping the cut halts the shower of sparks but leaves you with a restart problem.

EXPERIMENT 21-1

Observing Heat Produced during a Cut

This experiment may require more skill than you have developed by this time. You may wish to observe your instructor performing the experiment or try it at a later time.

Using a properly lit and adjusted cutting torch, putting on your PPE, following all shop and manufacturer's safety rules for cutting, and using one piece of clean mild steel plate 6 in. (152 mm) long by 1/4 in. (6 mm) to 1/2 in. (13 mm) thick, you will make an oxyfuel gas cut without the preheat flame.

Place the piece of metal so that the cutting sparks fall safely away from you. With the torch lit, pass the flame over the length of the plate until it is warm, but not hot. Brace yourself and start a cut near the edge of the plate. When the cut has been established, have another student turn off the acetylene regulator. The cut should continue if you remain steady and the plate is warm enough. Hint: Using a slightly larger tip size will make this easier.

Paperwork

Write a short paragraph describing what you observed about the heat produced during the cutting and complete a copy of the time sheet in Appendix I and bill of materials in Appendix III, or as provided by your instructor.◆

The Physics of a Cut

As a cut progresses along a plate, a record of what happened during the cut is preserved along both sides of the kerf. This record indicates what was correct or incorrect with the preheat flame, cutting speed, and oxygen pressure.

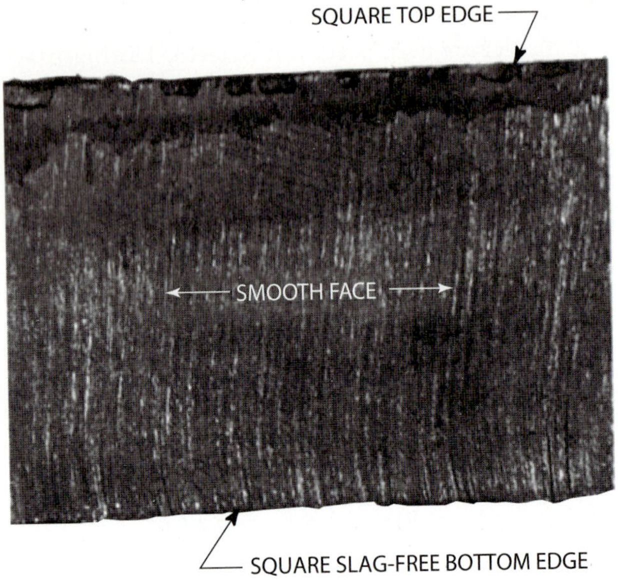

FIGURE 21-29 Correct cut. Larry Jeffus

FIGURE 21-30 Too fast a travel speed resulting in an incomplete cut; too much preheat and the tip is too close, causing the top edge to be melted and removed. Larry Jeffus

Preheat

The size and number of preheat holes in a tip has an effect on both the top and bottom edges of the metal. An excessive amount of preheat flame results in the top edge of the plate being melted or rounded off. In addition, an excessive amount of hard-to-remove slag is deposited along the bottom edge. If the flame is too small, the travel speed must be slower. A reduction in speed may result in the cutting stream wandering from side to side. The torch tip can be raised slightly to eliminate some of the damage caused by too much preheat. However, raising the torch tip causes the cutting stream of oxygen to be less forceful and less accurate.

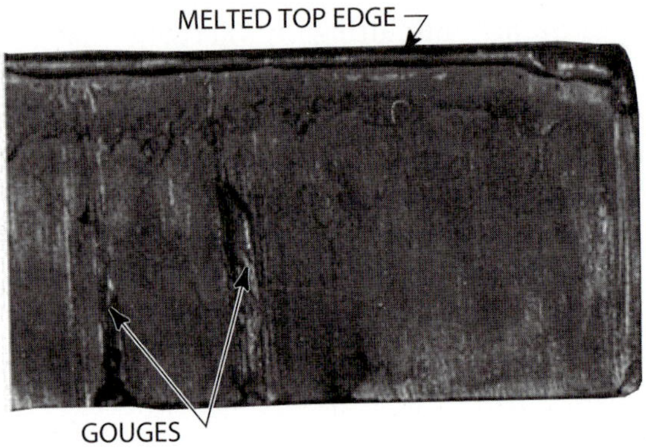

FIGURE 21-31 Too slow a travel speed results in the cutting stream wandering, thus causing gouges in the surface; preheat flame is too close, melting the top edge. Larry Jeffus

Speed

The cutting speed should be fast enough so that the **drag lines** have a slight slant backward if the tip is held at a 90° angle to the plate, Figure 21-29. If the cutting speed is too fast, the oxygen stream may not have time to go completely through the metal, resulting in an incomplete cut, Figure 21-30. Too slow a cutting speed results in the cutting stream wandering, thus causing gouges in the side of the cut, Figure 21-31 and Figure 21-32.

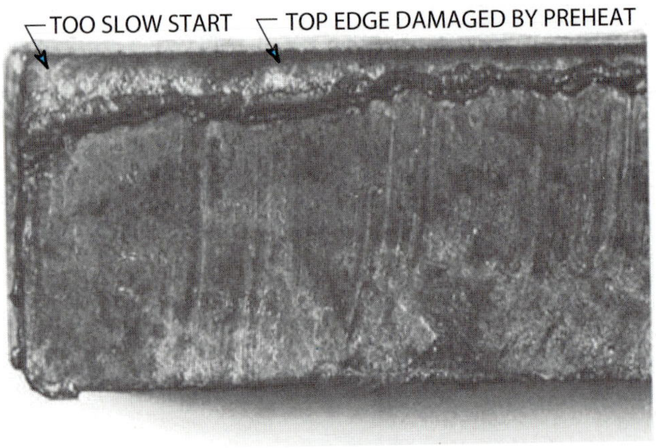

FIGURE 21-32 Too slow a travel speed at the start; too much preheat. Larry Jeffus

Pressure

A correct pressure setting results in the sides of the cut being flat and smooth. A pressure setting that

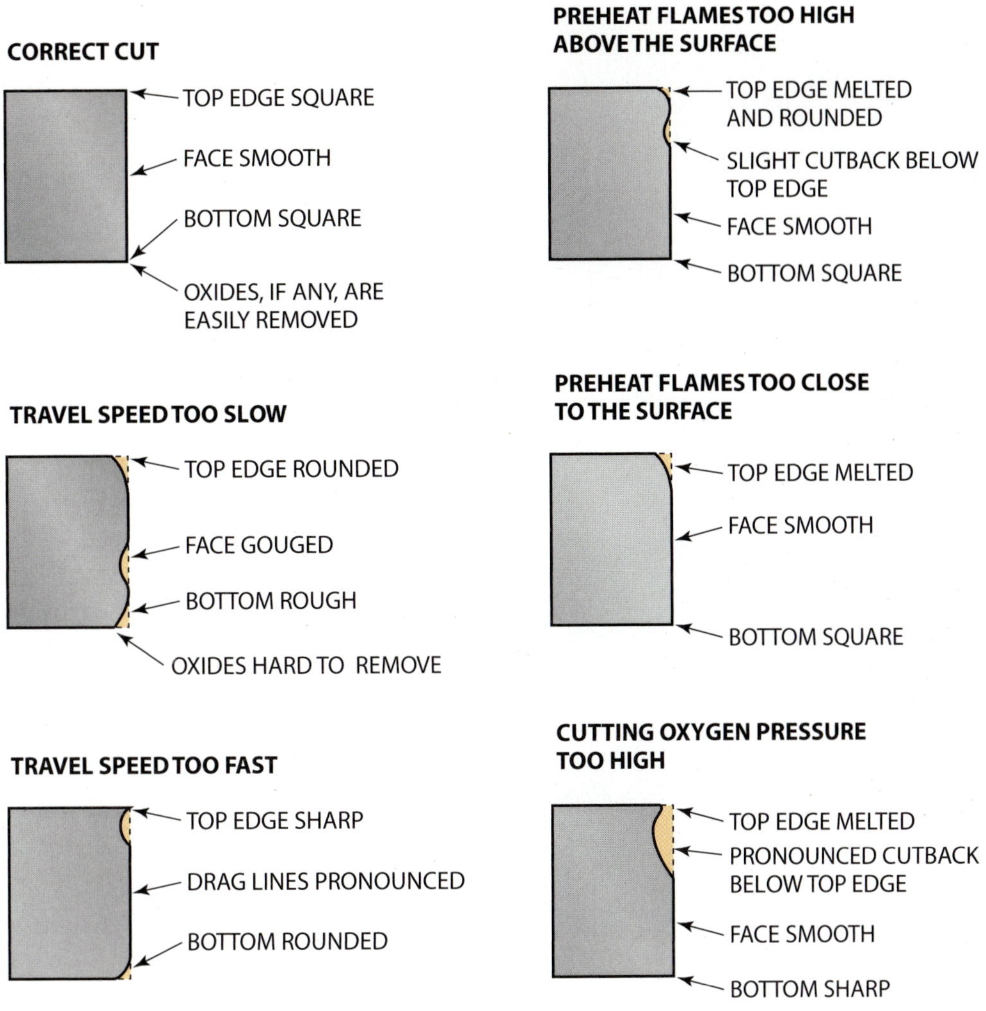

FIGURE 21-33 Profile of flame-cut plates. © Cengage Learning 2012

is too high causes the cutting stream to expand as it leaves the tip, resulting in the sides of the kerf being slightly dished, Figure 21-33. When the pressure setting is too low, the cut may not go completely through the metal.

EXPERIMENT 21-2

Effect of Flame, Speed, and Pressure on a Machine Cut

Using a properly lit and adjusted automatic cutting machine, putting on your PPE, following all shop and manufacturer's safety rules for cutting, and using a variety of tip sizes and one piece of mild steel plate 6 in. (152 mm) long by 1/2 in. (13 mm) to 1 in. (25 mm) thick, you will observe the effect of the preheat flame, travel speed, and pressure on the metal being cut.

Using the variety of tips, speeds, and oxygen pressures, make a series of cuts on the plate. As the cut is being made, listen to the sound it makes. Also look at the stream of sparks coming off the bottom. A good cut should have a smooth, even sound, and the sparks should come off the bottom of the metal more like a stream than a spray, Figure 21-34. When the cut is complete, look at the drag lines to determine what was correct or incorrect with the cut, Figure 21-35.

Repeat this experiment until you know a good cut by the sound it makes and the stream of sparks. A good cut has little or no slag left on the bottom of the plate.

Paperwork

Write a short paragraph describing what you observed about the flame effect, speed, and pressure on machine cutting and complete a copy of the time sheet in Appendix I and bill of materials in Appendix III, or as provided by your instructor.◆

FIGURE 21-34 A good cut showing a steady stream of sparks flying out from the bottom of the cut. Larry Jeffus

FIGURE 21-35 Poor cut. The slag is backing up because the cut is not going through the plate. Larry Jeffus

EXPERIMENT 21-3

Effect of Flame, Speed, and Pressure on a Hand Cut

Using a properly lit and adjusted hand torch, putting on your PPE, following all shop and manufacturer's safety rules for cutting, and using the same tip sizes and mild steel plate, repeat Experiment 21-2 to note the effects of the preheat flame, travel speed, and pressure on hand cutting.

Paperwork

Write a short paragraph describing what you observed about the flame effect, speed, and pressure on hand cutting and complete a copy of the time sheet in Appendix I and bill of materials in Appendix III, or as provided by your instructor.◆

Slag

The two types of slag produced during a cut are soft slag and hard slag. Soft slag is very porous, brittle,

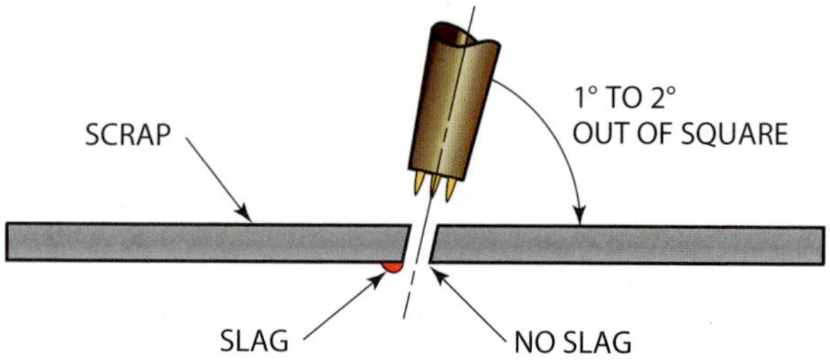

FIGURE 21-36 A slight angle on the torch will put the slag on the scrap side of the cut. © Cengage Learning 2012

and easily removed from a cut. There is little or no unoxidized iron in it. It may be found on some good cuts. Hard slag may be mixed with soft slag. Hard slag is attached solidly to the bottom edge of a cut, and it requires a lot of chipping and grinding to be removed. There is 30 to 40% or more unoxidized iron in hard slag. The higher the unoxidized iron content, the more difficult the slag is to remove. **Slag** is found on bad cuts, due to dirty tips, too much preheat, too slow a travel speed, too short a coupling distance, or incorrect oxygen pressure.

The slag from a cut may be kept off one side of the plate being cut by slightly angling the cut toward the scrap side of the cut, Figure 21-36. The angle needed to force the slag away from the good side of the plate may be as small as 2° or 3°. This technique works best on thin sections; on thicker sections the bevel may show.

Plate Cutting

Low carbon steel plate can be cut quickly and accurately, whether thin-gauge sheet metal or sections more than 4 ft (1.2 m) thick are used. It is possible to achieve cutting speeds as fast as 32 in. per minute (13.5 mm/s), in a 1/8-in. (3-mm) plate, and accuracy on machine cuts of ± 3/64 in. Some very large hand-cutting torches with an oxygen cutting volume of 600 cfh (2830 L/min) can cut metal that is 4 ft (1.2 m) thick, Figure 21-37. Most hand torches will not easily

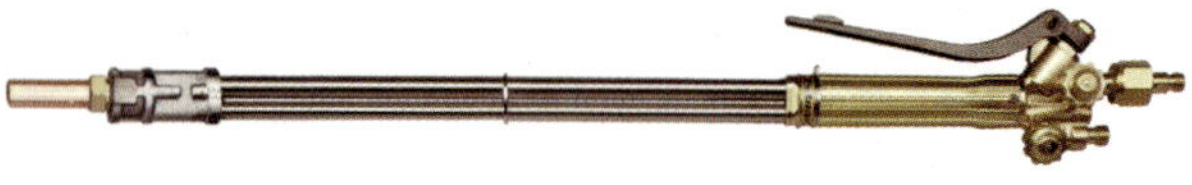

(A)

(B)

FIGURE 21-37 Hand torches for thick sections. Thermadyne Industries, Inc.

cut metal that is more than 7 in. (178 mm) to 10 in. (254 mm) thick.

The thicker the plate, the more difficult the cut is to make. Thin plate, 1/4 in. (6 mm) or less, can be cut and the pieces separated even if poor techniques and incorrect pressure settings are used. Thick plate, 1/2 in. (13 mm) or thicker, often cannot be separated if the cut is not correct. For very heavy cuts, on a plate 12 in. (305 mm) or thicker, the equipment and operator technique must be near perfection or the cut will be faulty.

Plate that is properly cut can be assembled and welded with little or no postcut cleanup. Poor-quality cuts require more time to clean up than is needed to make the required adjustments to make a good weld.

Cutting Table

Because of the nature of the torch cutting process, special consideration is given to the flame-cutting support. Any piece being cut should be supported so the torch flame will not cut through the piece and into the table. Special cutting tables are used that expose only a small metal area to the torch flame. Some tables use parallel steel bars of metal and others use cast-iron pyramids. All cutting should be set up so the flame and oxygen stream runs between the support bars or over the edge of the table.

If an ordinary welding table or another steel table is used, special care must be taken to avoid cutting through the tabletop. The piece being cut may be supported above the support table by firebrick. Another method is to cut the metal over the edge of the table.

Torch Guides

In manual torch cutting a guide or support is frequently used to allow for better control and more even cutting. It takes a very skilled welder to make a straight, clean cut even when following a marked line. It is even more difficult to make a radius cut to any accuracy. Guides and supports allow the height and angle of the torch head to remain constant. The speed of the cut, which is very important to making a clean, even kerf, must be controlled.

Since the torch must be held in an exact position while making any accurate cut, you will normally support the torch weight with your hand. Supporting the torch weight this way not only allows for more accurate work but also cuts down on fatigue. A rest, such as a firebrick, is also used to support the torch.

Various types of guides can be used to direct the torch in a straight line. Figure 21-38 shows one type of guide using angle iron. The edge of the angle is followed to make the straight cut. Bevel cuts can be made freehand with the torch, but it is very difficult to keep them uniform. More accurate bevel cuts are made by resting the torch against the angle side of an angle iron.

Special roller guides, Figure 21-39A, can also be attached to the torch head. The attachment holds the torch cutting tip at an exact height.

When cutting circles, a circle cutting attachment is used. Figure 21-39B shows how the attachment fits on the torch head. The radius can be preset to any required distance. The cutter revolves around the center point when making the cut. The roller controls the torch tip height above the plate surface.

Stopping and Starting Cuts

Starts and stops can be made better and more easily if one side of the metal being cut is scrap. When it is necessary to stop and reposition oneself before continuing the cut, the cut should be turned out a short distance into the scrap side of the metal, Figure 21-40. The extra space that this procedure provides will allow a smoother and more even start with less chance that slag will block the cut. If neither side of

FIGURE 21-38 Using angle irons to aid in making cuts.

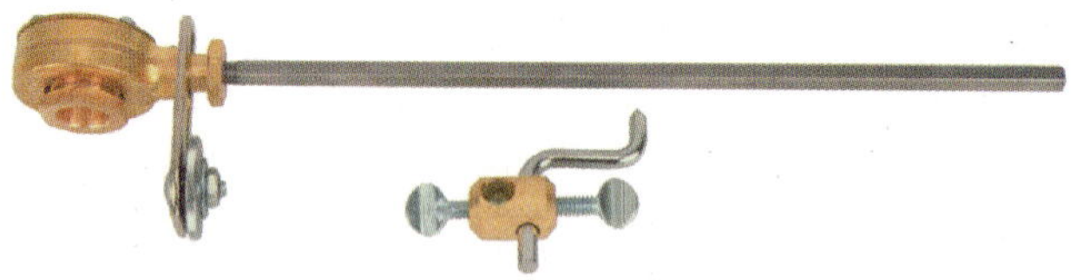

FIGURE 21-39 Devices that are used to improve hand cutting. Thermadyne Industries, Inc.

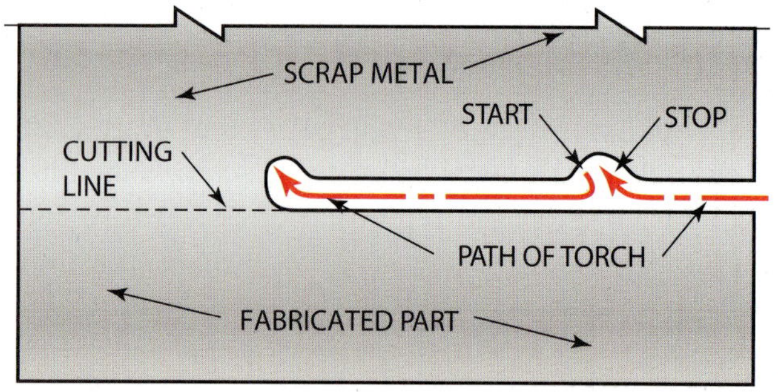

FIGURE 21-40 Turning out into the scrap to make stopping and starting points smoother. © Cengage Learning 2012

the cut is to be scrap, the forward movement should be stopped for a moment before releasing the cutting lever. This action will allow the **drag**, or the distance that the bottom of the cut is behind the top, to catch up before stopping, Figure 21-41. To restart, use the same procedure that was given for starting a cut at the edge of the plate.

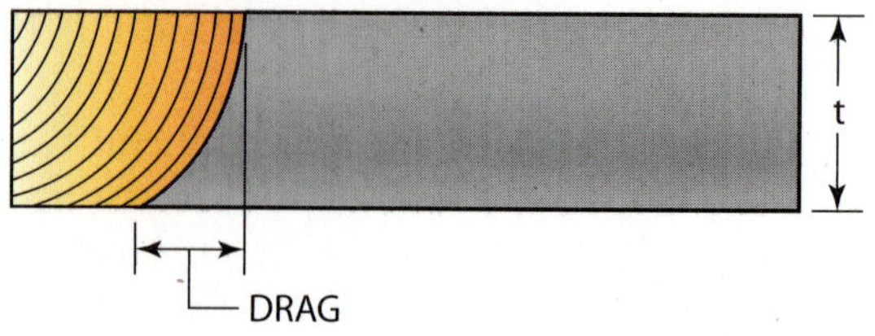

FIGURE 21-41 Drag is the distance by which the bottom of the cut lags behind the top. © Cengage Learning 2012

PROJECT 21-1

Flat, Straight Cut in a Thin Plate to Make a Candlestick

Skills to be learned: How to select the correct cutting tip and set the pressures to make straight OF cuts on a thin plate that are straight, square, smooth, and slag-free using a handheld oxyacetylene torch. In addition, you will need to do some basic math to determine the length of the vertical sidepieces so that the overall height of the candlestick meets the specifications.

Project Description

The massive candlestick is designed to hold large diameter candles without looking undersized. The large top will help catch wax that might run down the candle, Figure 21-42.

Project Materials and Tools

The following items are needed to fabricate the candlestick.

Materials

- 32-in. × 3-in. piece of 1/4-in. bar stock
- 10-in. × 5-in. piece of 1/4-in. thick plate
- Source of oxygen and acetylene gases

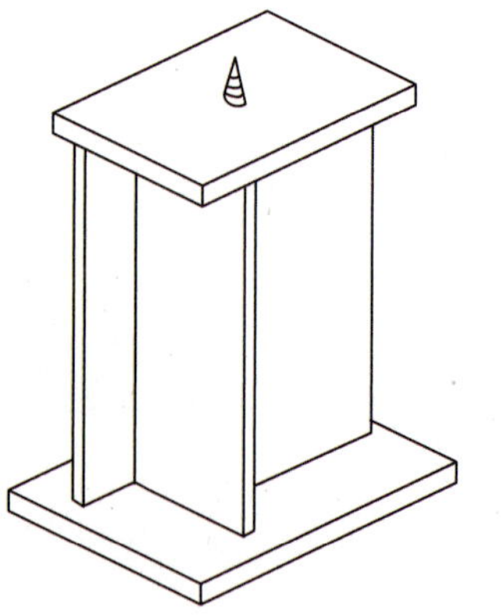

FIGURE 21-42 Project 21-1. © Cengage Learning 2012

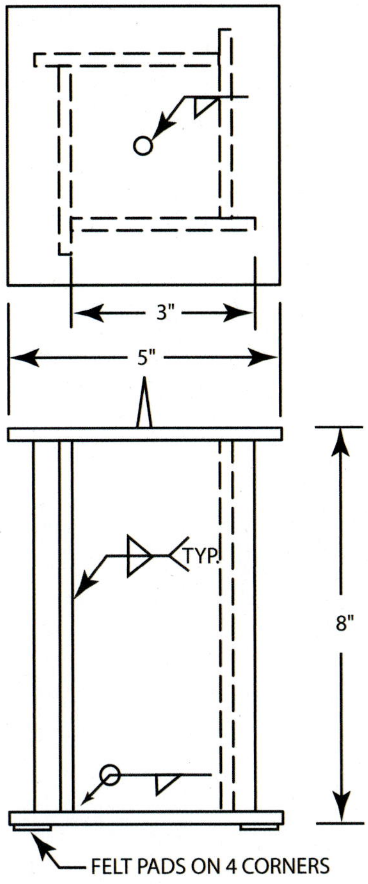

FIGURE 21-43 Candlestick holder drawing.

Tools

- Oxyacetylene cutting torch and cutting tips
- Tip cleaners
- Torch lighter
- Wrenches
- PPE
- Tape measure
- Chipping hammer
- 12-in. rule
- Soapstone, pencil and/or marker
- Square
- Pliers

Layout

The vertical members of the candlestick holder can be made out of 3-in. wide 1/4-in. thick bar stock, or they can be cut from a 1/4-in. plate, Figure 21-43. The overall finished height of the candlestick holder is 8 in., so you must deduct from the length of the vertical members the thickness of the top and bottom plates. Using a tape measure, soapstone, and a square, lay out the four side plates. Be sure to leave space to compensate for the metal removed by the torch. Lay out the top and bottom plates using a rule and square.

NOTE: You can find out what the kerf width is for a torch by making a sample cut in scrap metal and measuring its width.

Cutting Out

Using a properly set up and adjusted OF cutting system, proper safety protection, and one or more pieces of mild steel, you will cut out the candlestick parts.

Put on your PPE, and following all shop and manufacturer's safety rules for cutting, you will cut out the parts for the candlestick.

To make this cut with a hand torch, it is important for you to be steady to make the cut as smooth as possible. You must also be comfortable and free to move the torch along the line to be cut. It is a good idea for you to get into position and practice the cutting movement a few times before lighting the torch. Even when you and the torch are braced properly, a tiny movement such as a heartbeat will cause a slight ripple in the cut. Attempting a cut without leaning on the work to brace yourself is tiring and causes inaccuracies.

The torch should be braced with the left hand if you are right-handed or with the right hand if you are left-handed. The torch may be moved by sliding it toward you over your supporting hand, Figure 21-44 and Figure 21-45. The torch can also be pivoted on the supporting hand. If the pivoting method is used, care must be taken to prevent the cut from becoming a series of arcs.

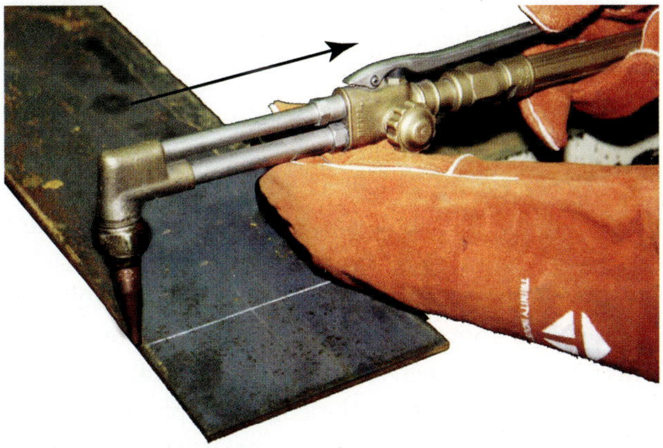

FIGURE 21-44 The torch may be moved by sliding it toward you over your supporting hand. Larry Jeffus

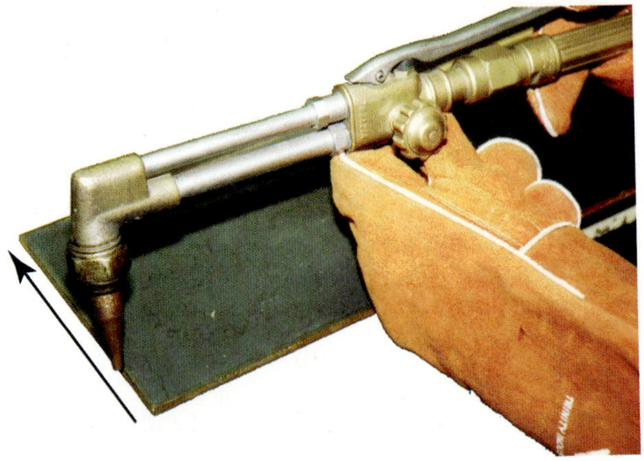

(A)

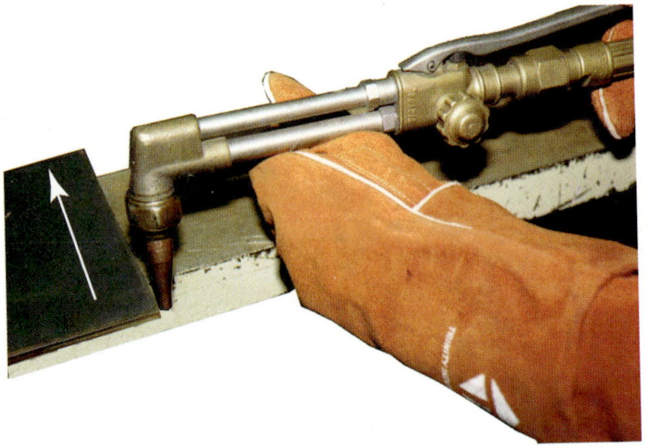

(B)

FIGURE 21-45 For longer cuts, the torch can be moved by sliding your gloved hand along the plate parallel to the cut: (A) start and (B) finish. Always check for free and easy movement before lighting the torch. Larry Jeffus

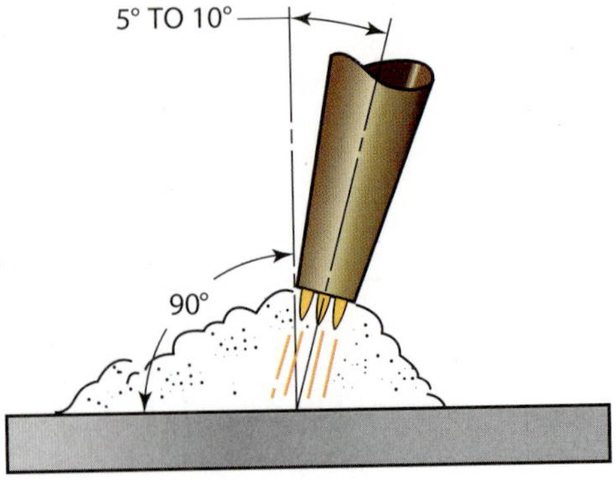

FIGURE 21-46 A slight forward angle helps when cutting thin material. © Cengage Learning 2012

FIGURE 21-47 Keep the inner cones of the flame approximately 1/8 in. to 1/4 in. from the surface of the plate. Larry Jeffus

FIGURE 21-48 Starting a cut on the edge of a plate. Notice how the torch is pointed at a slight angle away from the edge. Larry Jeffus

Bring the torch flame up to the line with a slight forward torch angle. This slight forward angle helps the flame preheat the metal, keeps some of the reflected flame heat off the tip, aids in blowing dirt and oxides away from the cut, and keeps the tip clean for a longer period of time because slag is less likely to be blown back on it, Figure 21-46. The forward angle can be used only for a straight line square cut. If shapes are cut using a slight angle, the part will have beveled sides.

Keep the inner cones of the flame approximately 1/8 in. to 1/4 in. from the surface of the plate, Figure 21-47. This distance is known as the **coupling distance**.

To start the cut on the edge of a plate, hold the torch at a right angle to the surface or pointed slightly away from the edge, Figure 21-48. The torch must also be pointed so that the cut is started at the very

FIGURE 21-49 The torch is rotated to allow the preheating of the plate ahead of the cut. This speeds the cutting and also provides better visibility of the line being cut.
Larry Jeffus

edge. The edge of the plate heats up more quickly and allows the cut to be started sooner. Also, fewer sparks will be blown around the shop. Once the cut is started, the torch should be rotated back to a right angle to the surface or to a slight leading angle.

A good cut will have a stream of sparks being sprayed out of the bottom of the cut, Figure 21-49. When the sparks are spraying out like this, there is little slag left to get stuck on the bottom edge of the cut.

CAUTION

NEVER USE A CUTTING TORCH TO CUT OPEN A USED CAN, DRUM, TANK, OR OTHER SEALED CONTAINER. The sparks and oxygen cutting stream may cause even nonflammable residue inside to burn or explode. If a used container must be cut, it must first have one end removed and all residue cleaned out. In addition to the possibility of a fire or an explosion, you might be exposing yourself to hazardous fumes. Before making a cut, check the material specification data sheet (MSDS) for the chemical that was in the drum or tank for a listing of any safety concerns.

When the cut is made correctly, the part will fall free, be slag-free, and be within ±3/32 in. (2 mm) of a straight line and ±5° of being square.

Repeat this procedure until the cut can be made straight and slag-free. If the torch tip is too large, the top edge of the cut will be melted back and the kerf may be wider than normal. If the tip size is too small, the travel speed will be too slow, and the oxygen stream may not cut all the way through the plate. Using the manufacturer's tip chart or Table 21-6, make any changes in tip size and adjustments in pressure to make a slag-free cut.

Turn off the cylinder valves, bleed the hoses, back out the pressure regulators, and clean up your work area when you are finished cutting.

Fabrication and Welding

The flux core arc welding process can be used for this project which is covered in Project 14-3 in Chapter 14. The welding procedures are also covered in that section; if another welding process is to be used, refer to the chapter that discusses that process.

Paperwork

Complete a copy of the time sheet in Appendix I and bill of materials in Appendix III, or as provided by your instructor.◆

PROJECT 21-2

Flat, Straight Cut in Sheet Metal to Make a Candlestick

Skills to be learned: How to select the correct cutting tip and set the pressures to make straight OF cuts on sheet metal that are straight, smooth, and slag-free using a handheld oxyacetylene torch.

Skills to be learned: The ability to tack weld on a weldment, control the weld distortion, and develop the skills to make uniform outside corner weld beads in the flat position.

Project Description

This 6-in. tall candlestick is artistic in its simplicity, yet functional. The wide top and tall sides are designed to minimize the problem of candle wax drips reaching the table, Figure 21-50.

Project Materials and Tools

The following items are needed to fabricate the candlestick.

Materials

- 14-in. long 6-in. wide 16-gauge to 3/32-in. thick mild steel sheet metal
- 2-in. × 2-in. square 16-gauge to 3/32-in. thick mild steel sheet metal
- 4-in. × 4-in. square 16-gauge to 3/32-in. thick mild steel sheet metal
- Source of oxygen and acetylene gases

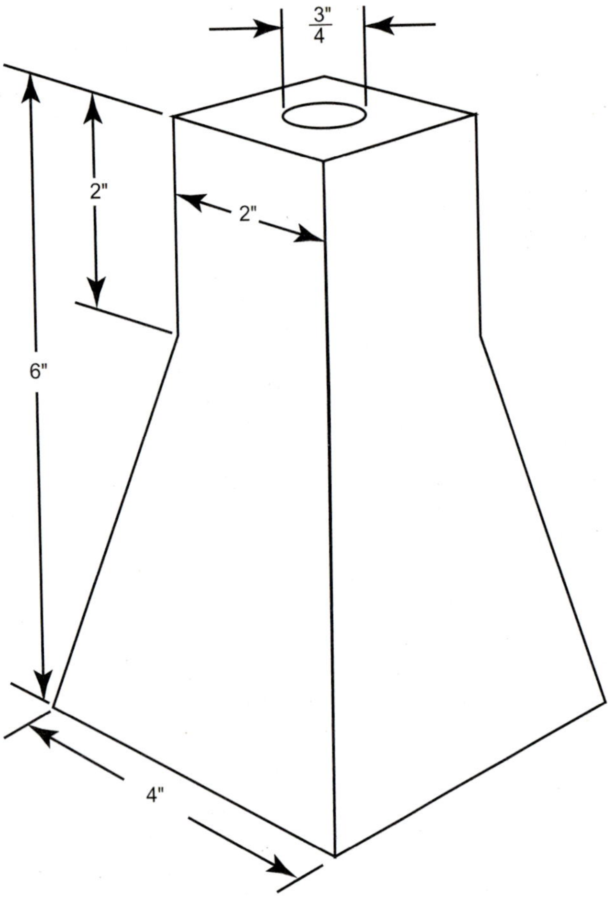

FIGURE 21-50 Project 21-2. © Cengage Learning 2012

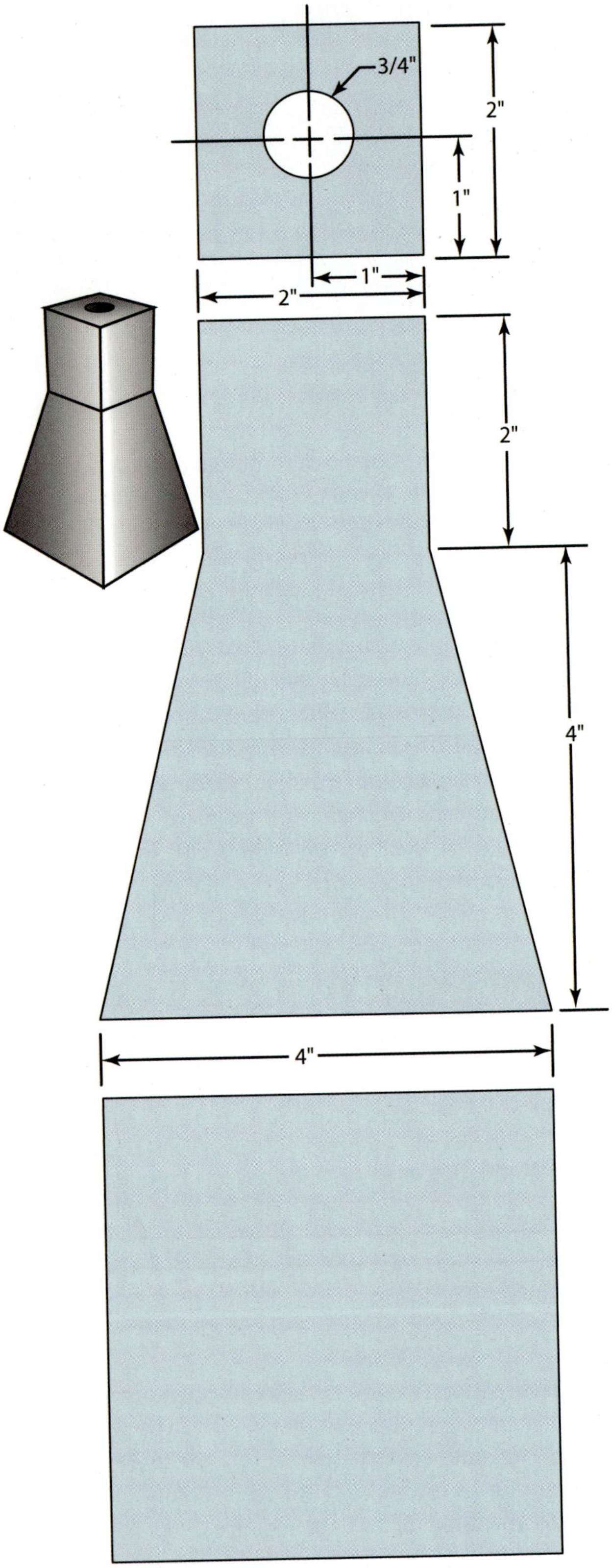

FIGURE 21-51 Candlestick holder layout. © Cengage Learning 2012

Tools

- Oxyacetylene cutting torch and cutting tips
- Tip cleaners
- Torch lighter
- Wrenches
- PPE
- Tape measure
- Chipping hammer
- 12-in. rule
- Soapstone, pencil and/or marker
- Square
- Pliers

Layout

You are going to lay out the candlestick shown in Figure 21-51. Start the layout by measuring 4 in. across the base, and make a mark at the 4-in. and 2-in. measurements. Use the square and extend the 2-in. center point mark across the 6-in. width of the sheet metal to locate the center of the panel. On the opposite side

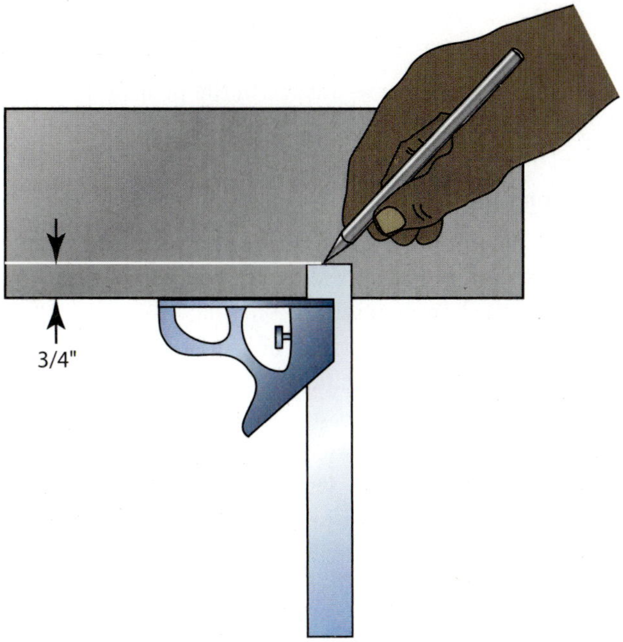

FIGURE 21-52 Use the try square to draw parallel lines. © Cengage Learning 2012

use the 12-in. rule and measure 1 in. on each side of the center line you just drew. Use the square to draw this line 2 in. down to locate the point where the angled side meets the straight side of the top. Using the 12-in. rule, draw a line from one side of the 4-in. line to the corner of the 2-in. mark. Repeat the process on the other side.

You must leave a gap between the parts for the torch kerf. Make a practice cut and measure the kerf to determine the gap width. Repeat the layout process described above until all four panels have been laid out.

Next, use the 12-in. rule and square to lay out the 2-in. square top and a 4-in. square for the bottom. You can mark the sides, then use the square to draw the two parallel sides. A try square, as illustrated in Figure 21-52, can be used to draw the opposite side parallel line.

Cutting Out

Putting on your PPE and following all shop and manufacturer's safety rules for cutting, you will cut out the parts for the candlestick.

The torch preheat flame will quickly heat up the thin sheet metal, so the cut can be started almost instantly. Holding the torch at a very sharp leading angle, Figure 21-53, cut the sheet along the line.

Thin sheet metal heats up quickly, which can cause it to warp. To reduce warpage you must travel as fast as possible along the cut.

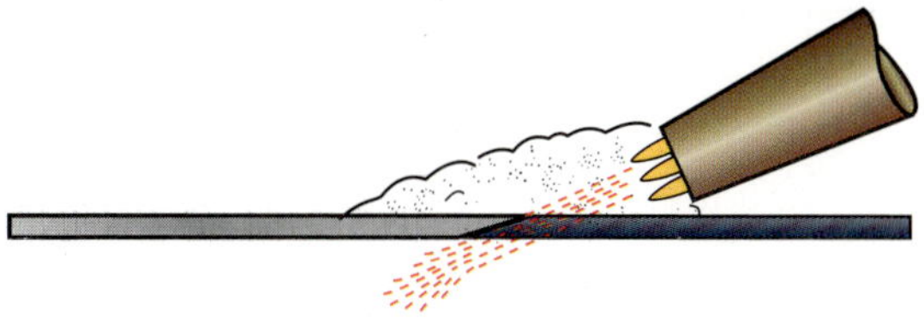

FIGURE 21-53 Cut the sheet metal at a very sharp angle. © Cengage Learning 2012

The cut must be smooth and straight with as little slag as possible. If the torch tip is too large, there will be more heat distortion; and if the tip size is too small, the travel speed will be too slow. Using the manufacturer's tip chart or Table 21-6, make any changes in tip size and adjustments in pressure to make a slag-free cut.

Repeat this procedure until the cut can be made flat, straight, and slag-free. Turn off the cylinder valves, bleed the hoses, back out the pressure regulators, and clean up your work area when you are finished cutting.

Fabrication and Welding

A project very similar to the fabrication procedures for this project is covered in Project 18-1 in Chapter 18. The welding procedures are also covered in that section; if another welding process is to be used, refer to the chapter that discusses that process.

Paperwork

Complete a copy of the time sheet in Appendix I and bill of materials in Appendix III, or as provided by your instructor.◆

PROJECT 21-3

Flat, Straight Cut in a Thick Plate to Fabricate a Bench Anvil Blank

Skills to be learned: How to select the correct cutting tip and set the pressures to make straight OF cuts on thick plate that are straight, square, smooth, and slag-free using a handheld oxyacetylene torch, Figure 21-54.

Project Description

The bench anvil, when completed, is designed for light to moderate service. It is capable of withstanding significant hammer blows that might damage the anvil area found on many bench vises. Even if the surface of the anvil is damaged, it can be easily welded to repair

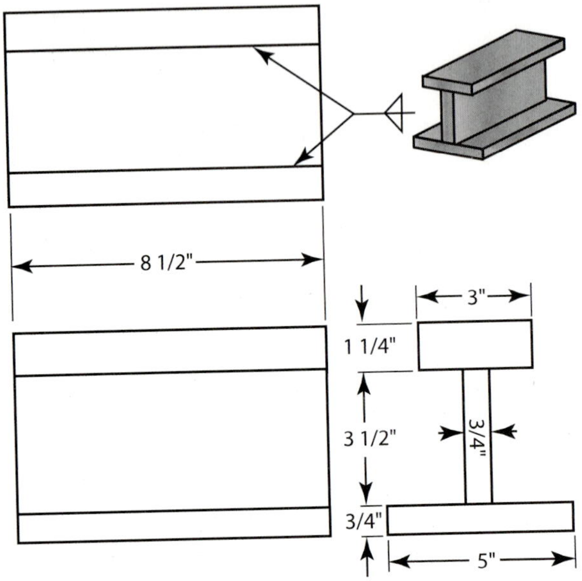

FIGURE 21-54 Project 21-3. © Cengage Learning 2012

face damage. Holes will be cut in the ears on the base of the anvil so it could be bolted to a workbench.

Project Materials and Tools

The following items are needed to fabricate the bench anvil.

Materials

- 1 3-in. wide 8 1/2-in. long piece of 1 1/4-in. thick mild steel bar
- 1 3 1/2-in. wide 8 1/2-in. long piece of 3/4-in. thick mild steel bar
- 1 5-in. wide 8 1/2-in. long piece of 3/4-in. thick mild steel bar
- Source of oxygen and acetylene gases

Tools

- Oxyacetylene cutting torch and cutting tips
- Tip cleaners
- Torch lighter
- Wrenches
- PPE
- Chipping hammer
- Wire brush
- Tape measure
- Square and/or 12-in. rule
- Soapstone, pencil and/or marker
- Wire brush
- Hammer

Layout

Using a square, 12-in. rule, measuring tape, and soapstone, lay out the bar as shown in the project drawing, Figure 21-55.

Cutting Out

Putting on your PPE and following all shop and manufacturer's safety rules for cutting, you will cut out the parts for the bench anvil.

Cutting thicker sections of metal with an oxyfuel torch can be more difficult than making thin sections. The primary differences are starting the cut and the cutting speed. It takes longer to preheat the metal before the torch cutting lever can be depressed; and if the torch is not pointed in the correct direction, a lot of sparks can be thrown from the cut, Figure 21-48. Watch the metal's surface to see when it starts to glow red. Sometimes it may take up to a minute to preheat the metal.

Moving too fast will result in the oxygen stream not fully penetrating the complete thickness, and moving too slow will melt the top edge of the plate. Watching both the torch track and the stream of sparks from the bottom of the cut will help you adjust your cutting speed if necessary, Figure 21-56. Using the manufacturer's tip chart or Table 21-6, make any changes in tip size and adjustments in pressure to make a slag-free cut.

Fabrication and Welding

The fabrication procedures for this project are covered in Project 12-5 in Chapter 12. The welding procedures are also covered in that section; if another welding process is to be used, refer to the chapter that discusses that process.

Paperwork

Complete a copy of the time sheet in Appendix I and bill of materials in Appendix III, or as provided by your instructor.◆

PROJECT 21-4

Straight Cut in Thick Sections to Make a Bench Anvil

Skills to be learned: How to select the correct cutting tip and set the pressures to make straight, square, smooth, and slag-free cuts using a handheld oxyacetylene torch, Figure 21-57.

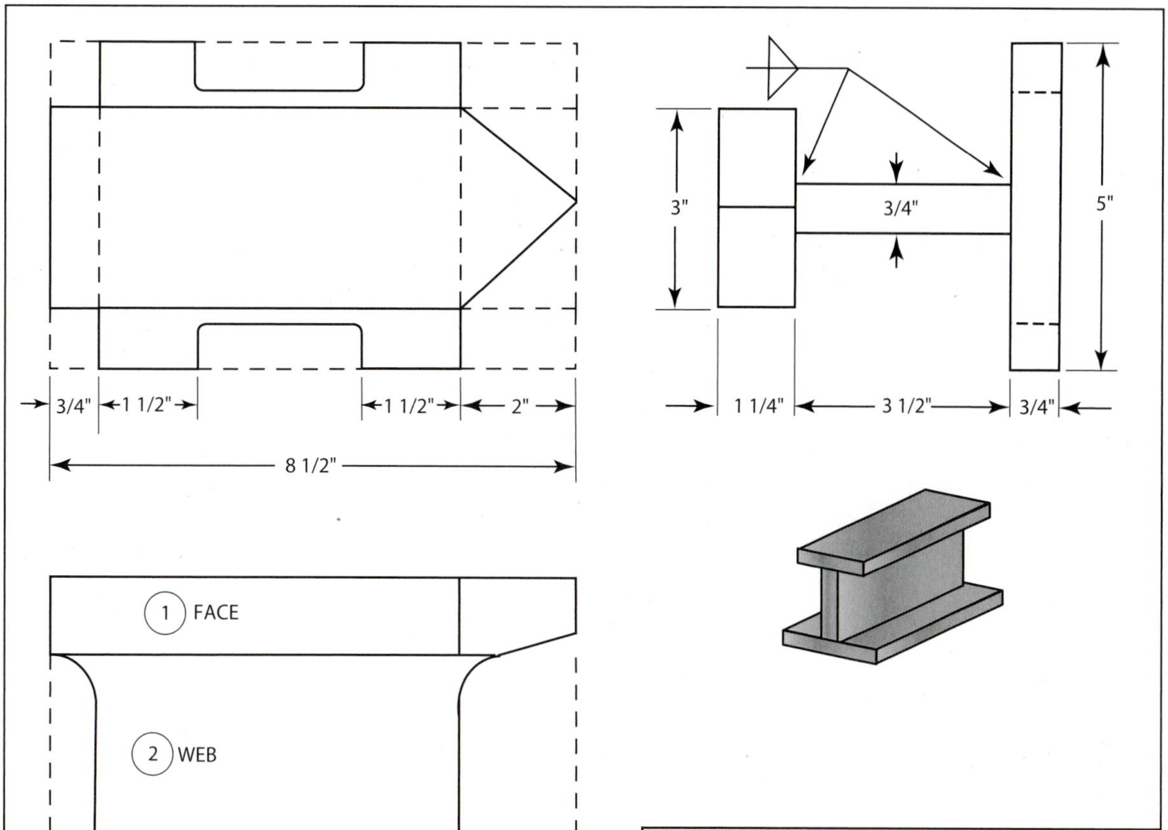

Bench Anvil		
Parts List		
Item	Qty	
1	1	3-in x 8 1/2-in x 1 1/4-in mild steel bar
2	1	3 1/2-in x 8 1/2-in x 3/4-in mild steel bar
3	1	5-in x 8 1/2-in x 3/4-in mild steel bar

FIGURE 21-55 Bench anvil layout. © Cengage Learning 2012

FIGURE 21-56 Watching both the torch track and the stream of sparks from the bottom of the cut will help you adjust your cutting speed. Larry Jeffus

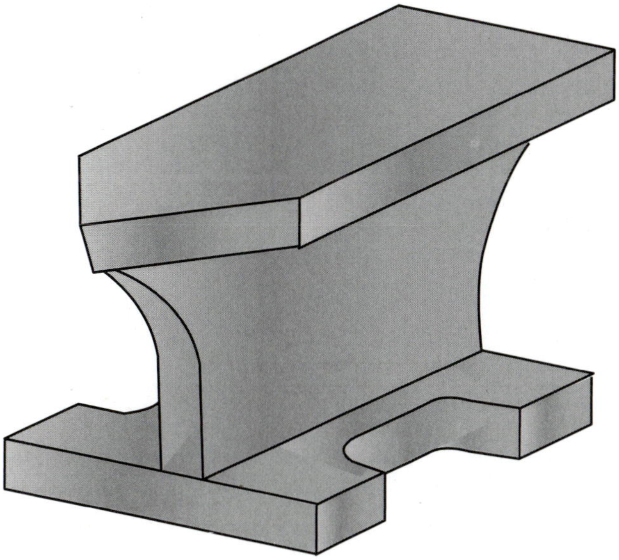

FIGURE 21-57 Bench anvil. © Cengage Learning 2012

FIGURE 21-58 Project 21-4. Larry Jeffus

Project Description

This bench anvil blank has been welded and now needs to be cut into its final shape, Figure 21-58. In Project 21-5 the bolt holes will be flame cut in the ears so the anvil can be bolted to a workbench.

Project Materials and Tools

The following items are needed to flame cut the final shape of the bench anvil.

Materials

- Anvil blank from Project 21-3
- Source of oxygen and acetylene gases

Tools

- Oxyacetylene cutting torch and cutting tips
- Tip cleaners
- Torch lighter
- Wrenches
- PPE
- Chipping hammer
- Wire brush
- Tape measure
- Square and/or 12-in. rule
- Soapstone, pencil and/or marker
- Wire brush

Layout

Following the finished design, lay out the anvil shown in Figure 21-55. Using the combination square, 12-in. rule, and soapstone, draw the cut lines on the finished weldment.

Cutting Out

Using a properly set up and adjusted OF cutting system, proper safety protection, and one or more pieces of mild steel, you will cut out the bench anvil final shape from the anvil blank started in Project 21-3.

These cuts will be irregular in shape. Also, you may have to make the cuts while working in an awkward position because of the size and shape of the anvil blank. Any time you are working in a position where bracing yourself is difficult, it is harder to make a clean, neat square cut. Because you may be working in an awkward position, you may need to stop and readjust yourself; so remember how to stop and start a cut.

Make any needed adjustments, and repeat the cut until a straight, slag-free cut can be made.

Turn off the cylinder valves, bleed the hoses, back out the pressure regulators, and clean up your work area when you are finished cutting.

Paperwork

Complete a copy of the time sheet in Appendix I and bill of materials in Appendix III, or as provided by your instructor.◆

Flame Cutting Holes

If a cut is to be started in a place other than the edge of the plate, the inner cones should be held as close as possible to the metal. Having the inner cones touch the metal will speed up the preheat time. When the metal is hot enough to allow the cut to start, the torch should be raised as the cutting lever is slowly depressed. When the metal is pierced, the torch should be lowered again, Figure 21-59. By raising the torch tip away from the metal, the amount of sparks blown into the air is reduced, and the tip is kept cleaner. If the metal being cut is thick, it may be necessary to move the torch tip in a small circle as the hole pierces the metal. If the metal is to be cut in both directions from the spot where it was pierced, the torch should be moved backward a short distance and then forward, Figure 21-60. This prevents slag from refilling the kerf at the starting point, thus making it difficult to cut in the other direction. The kerf is the space produced during any cutting process. The process of piercing a plate to flame cut a hole may be referred to in the field as burning a hole in the plate.

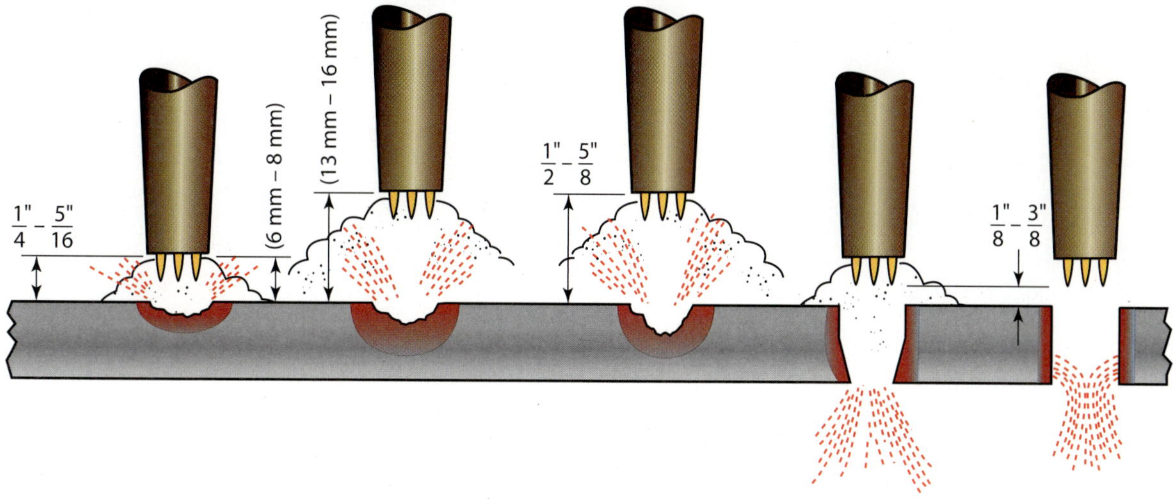

FIGURE 21-59 Sequence for piercing a plate. © Cengage Learning 2012

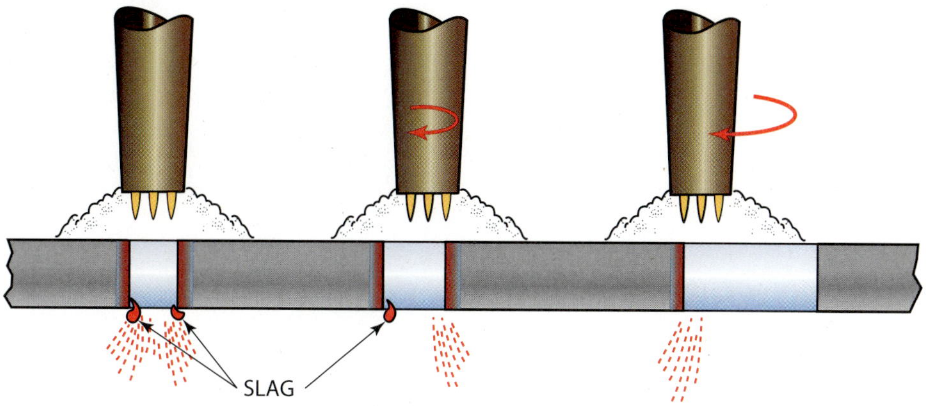

FIGURE 21-60 A short, backward movement (A) before the cut is carried forward (B) clears the slag from the kerf (C). Slag left in the kerf may cause the cutting stream to gouge into the base metal, resulting in a poor cut. © Cengage Learning 2012

PROJECT 21-5

Flame Cutting the Bolt Holes in the Anvil Ears

Skills to be learned: How to select the correct cutting tip and set the pressures to make OF pierces through a plate to make a round hole with a handheld oxyacetylene torch, Figure 21-61.

Project Description

The bolt holes being cut through the 3/4-in. thick ears on the base of the bench anvil will allow the anvil to be permanently attached to a workbench. Drilling four 5/8-in. holes would take much more time than flame cutting these holes. If the part was being machined, the time to drill the holes could be justified. But this anvil is being flame cut to shape, so a good, round, flame-cut hole can provide as long a service in this application as a drilled hole.

Project Materials and Tools

The following items are needed to flame cut the holes in the bench anvil.

Materials

- Anvil blank from Project 21-4
- 1/2" diameter 3" long bolt
- Source of oxygen and acetylene gases

Tools

- Oxyacetylene cutting torch and cutting tips
- Tip cleaners

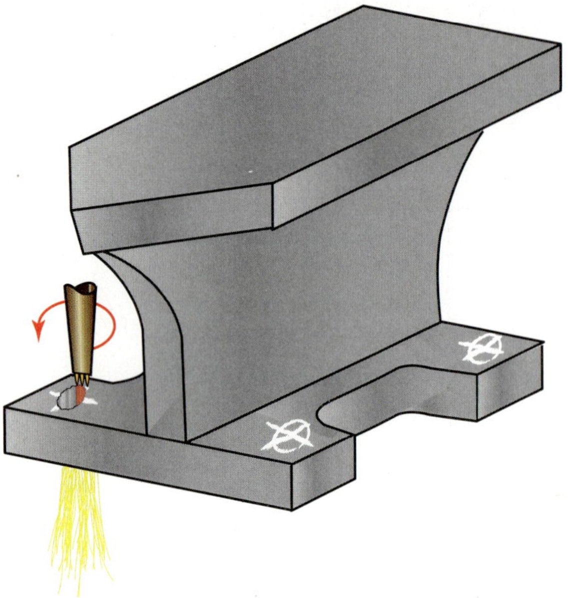

FIGURE 21-61 Oxyacetylene torch piercing the plate to make a round hole.

- Torch lighter
- Wrenches
- PPE
- Chipping hammer
- Wire brush
- Tape measure
- Square and/or 12-in. rule
- Soapstone, pencil and/or marker
- Wire brush
- Light oil

Layout

Use the tape measure, square, and soapstone to locate and mark the 5/8-in. diameter holes on the anvil ears.

Cutting Out

Putting on your PPE and following all shop and manufacturer's safety rules for cutting, you will cut the holes.

Using the technique described previously for piercing a hole, start in the center and make an outward spiral until the hole is the desired size, Figure 21-62.

Try the bolt in the hole to see that it fits. If necessary, the hole can be trimmed to make it large enough for the bolt to fit. The finished hole must be within ±3/32 in. (2 mm) of being round and ±5° of being square. The hole may have slag on the bottom. Using

(A)

(B)

(C)

FIGURE 21-62 As a hole is cut, the center may be overheated. Larry Jeffus

the manufacturer's tip chart or Table 21-6, make any changes in tip size and adjustments in pressure to make a slag-free cut.

Finishing

Chip any slag from the cuts with the chipping hammer, and wire brush off any loose oxides or mill scale. The point and square back surfaces of the top of the anvil may be ground a little to smooth out any major unevenness of the hand flame cuts. It is not necessary that these surfaces be square. Wipe the anvil with a light oil to prevent rusting.

Paperwork

Complete a copy of the time sheet in Appendix I and bill of materials in Appendix III, or as provided by your instructor.◆

PROJECT 21-6

Cutting Angle Iron to Make a Tee Joint for a Plant Stand

Skill to be learned: How to make straight OF cuts on angle iron that are square, smooth, and slag-free using a handheld oxyacetylene torch, Figure 21-63.

Project Description

The plant stand will be fabricated out of flat bar stock that is being welded into a tee joint, spaced so that the ceramic tile top will fit perfectly. It will have angle iron legs. The decorative ceramic tile top adds to the designer look of this plant stand, Figure 21-64.

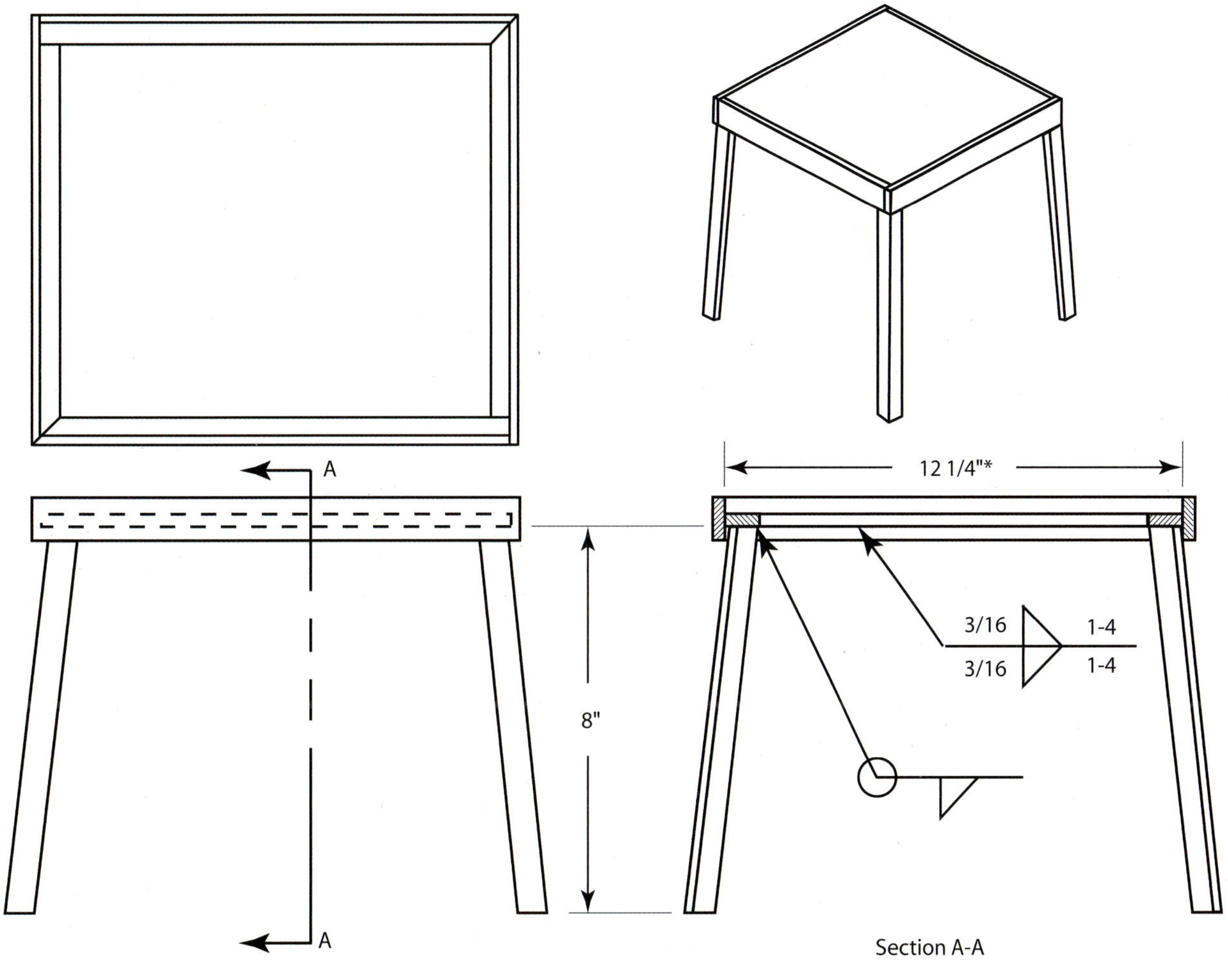

FIGURE 21-63 Project 21-6. © Cengage Learning 2012

FIGURE 21-64 Plant stand. © Cengage Learning 2012

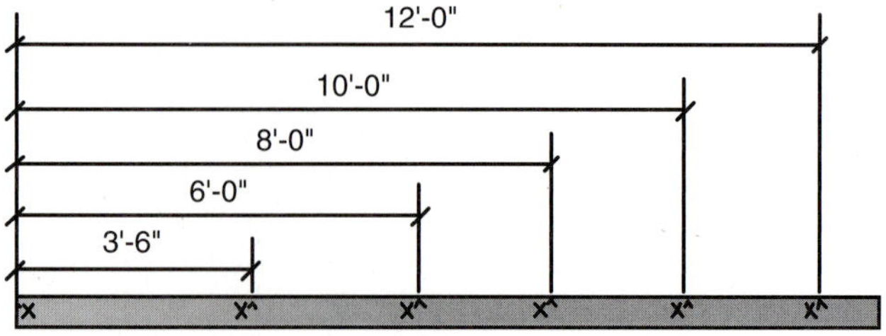

FIGURE 21-65 Measuring from one end is a more accurate way to lay out parts. © Cengage Learning 2012

Project Materials and Tools

The following items are needed to fabricate the plant stand.

Materials

- 9 ft of 1-in. × 1/4-in. thick mild steel bar
- 32 in. of 1 × 1 × 1/8-in. angle iron
- Source of oxygen and acetylene gases

Tools

- Oxyacetylene cutting torch and cutting tips
- Tip cleaners
- Torch lighter
- Wrenches
- Pencil and paper
- PPE
- Awl or punch
- Tape measure
- Chipping hammer
- Combination square
- Soapstone, pencil and/or marker
- Wire brush
- Square
- Pliers

Layout

Measure the side dimensions of the ceramic tile, place it flat on a table, and use the combination square to accurately measure the thickness. Add 2 in. to the measurement of the tile, and using the measuring tape and soapstone, mark that distance on the 1-in. bar stock to make 12 pieces of uniform length.

When laying out parts on a section of structural steel, you can measure the pieces one at a time or make all of the measurements from one end. In most cases making the measurements from the end is the more accurate way to lay out parts, Figure 21-65.

Using the tape and soapstone, mark the 8-in. lengths for the four legs. Remember to leave a space for the cut's kerf.

Use the square to draw a line all the way across all the surfaces that will be cut. On the angle, you will draw a line on two sides, Figure 21-66.

Cutting Out

Put on your PPE, and following all shop and manufacturer's safety rules for cutting, you will cut out the parts for the plant stand parts.

Place the metal to be cut on a cutting table at a comfortable height. Check to make certain there are not combustible materials in the area and that it is safe to cut in the area.

The parts can be cut so they remain on the cutting table or fall to the ground. If they are being allowed to fall, make sure they will not damage anything when they drop.

Start the cut on the edge of the metal and cut across one side. Move the torch to the other side and restart the cut. By cutting from the edges to the center, the part being cut will drop off more easily, Figure 21-67.

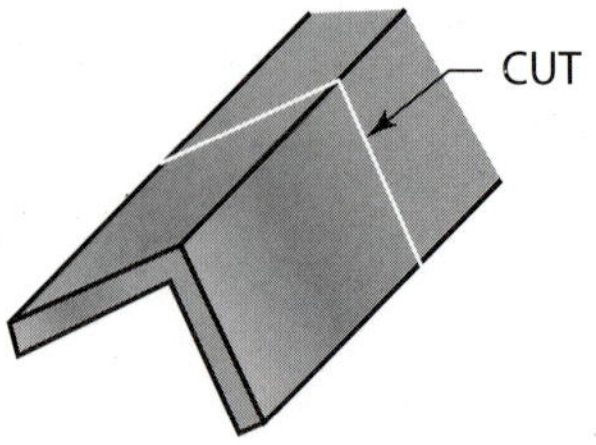

FIGURE 21-66 Use a square to draw a line on both sides of the angle. © Cengage Learning 2012

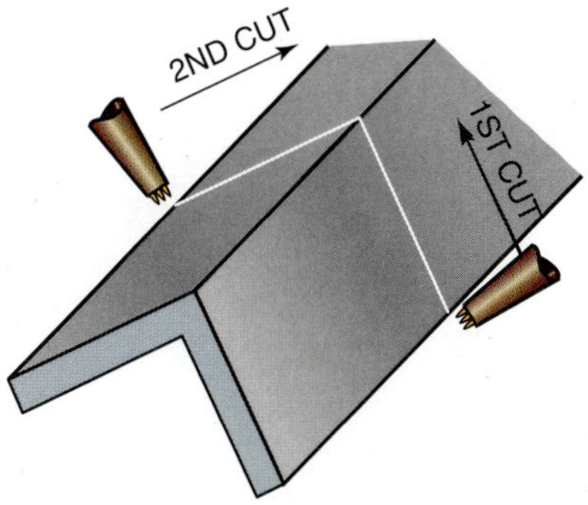

FIGURE 21-67 The cut part will drop more easily if you cut from the edges to the center.

The cut must be smooth and straight with as little slag as possible.

Fabrication and Welding

The fabrication procedures for this project are covered in Project 12-4 in Chapter 12. The welding procedures are also covered in that section; if another welding process is to be used, refer to the chapter that discusses that process.

Paperwork

Complete a copy of the time sheet in Appendix I and bill of materials in Appendix III, or as provided by your instructor.◆

Distortion

Distortion is when the metal bends or twists out of shape as a result of being heated during the cutting process. This is a major problem when cutting a plate. If the distortion is not controlled, the end product might be worthless. There are two major methods of controlling distortion. One method involves making two parallel cuts on the same plate at the same speed and time, Figure 21-68. Because the plate is heated evenly, distortion is kept to a minimum, Figure 21-69.

The second method involves starting the cut a short distance from the edge of the plate, skipping a short tab every 2 ft (0.6 m) to 3 ft (0.9 m) to keep the cut from separating. Once the plate cools, the remaining tabs are cut, Figure 21-70.◆

EXPERIMENT 21-4

Minimizing Distortion

Putting on your PPE and following all shop and manufacturer's safety rules for cutting, you will flame cut four pieces of metal to observe the effect the cuts have on distortion.

Using two pieces of steel plates approximately 10 in. long and 1/4 in. thick, you will make one cut on each plate as shown in Figure 21-71. After the plates have air cooled to a temperature where you can handle them, compare the distortion between the two cut plates.

Paperwork

Write a short paragraph describing what you observed between the two plates based on how they were cut, and complete a copy of the time sheet in Appendix I and bill of materials in Appendix III, or as provided by your instructor.◆

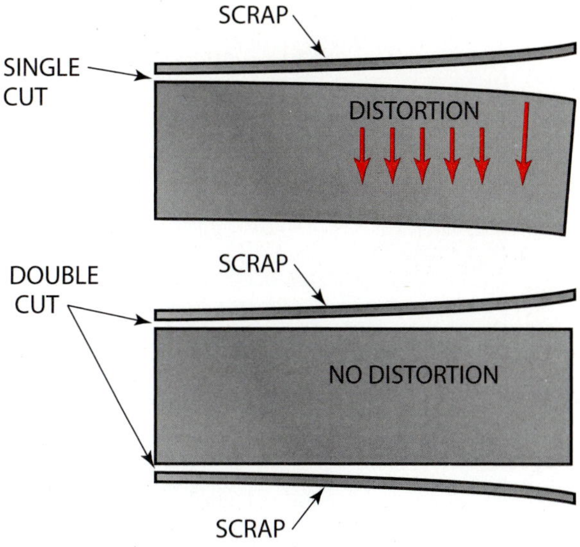

FIGURE 21-68 Making two parallel cuts at the same time will control distortion.

FIGURE 21-69 Slitting adapter for a cutting machine. It can be used for parallel cuts from 1 in. (38mm) to 12 in. (500 mm). Ideal for cutting test coupons.
Thermadyne Industries, Inc.

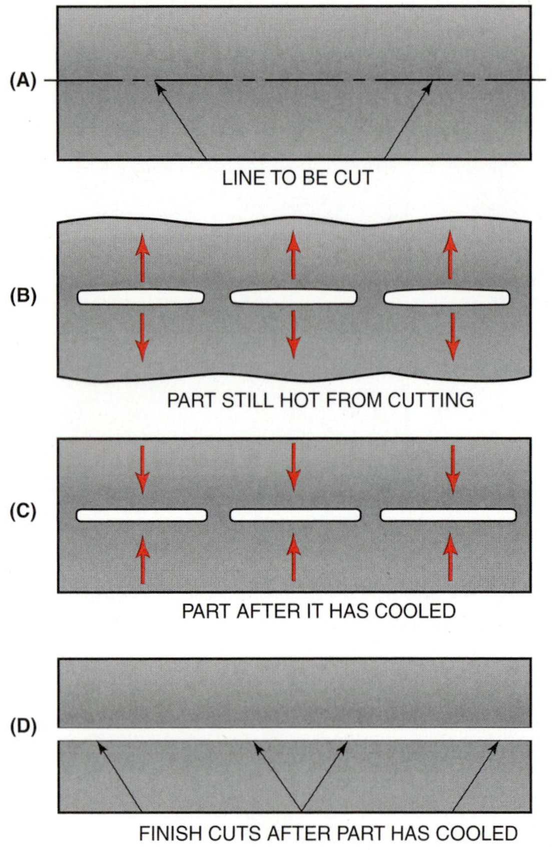

FIGURE 21-70 **Steps used during cutting to minimize distortion.** © Cengage Learning 2012

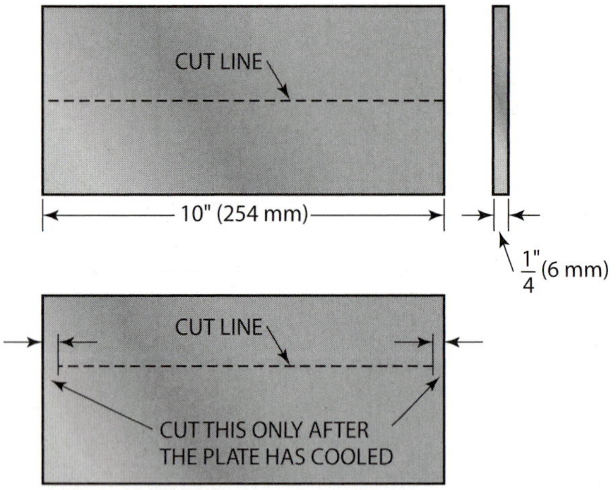

FIGURE 21-71 **Making two cuts with minimum distortion.** © Cengage Learning 2012

Machine Cutting Torch

The flame on the cutting tip of a machine cutting torch is set up in a similar manner as that on a hand cutting torch. A major difference is that a machine cutting torch may let you use two oxygen regulators, one for the preheat oxygen and the other for the cutting oxygen stream. The addition of a separate cutting oxygen supply allows the flame to be more accurately adjusted. It also allows the pressures to be adjusted during a cut without disturbing the preheat flame setting. Various machine cutting torches are shown in Figure 21-72, Figure 21-73, and Figure 21-74.

FIGURE 21-72 **Portable oxyfuel cutting machine.** Thermadyne Industries, Inc.

PROJECT 21-7

Beveling Welding Certification Test Plates

Skill to be learned: How to make a beveled cut with an OF cutting torch that is at the correct angle, smooth, and slag-free using a handheld oxyacetylene torch or machine cutting torch, Figure 21-75.

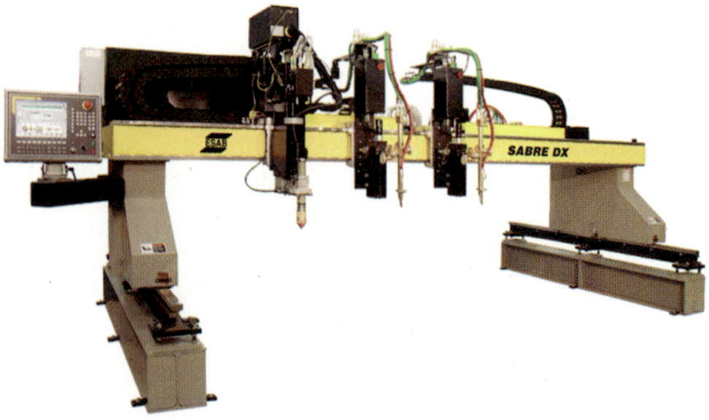

FIGURE 21-73 **Multiple head cutting machine.** ESAB Welding & Cutting Products

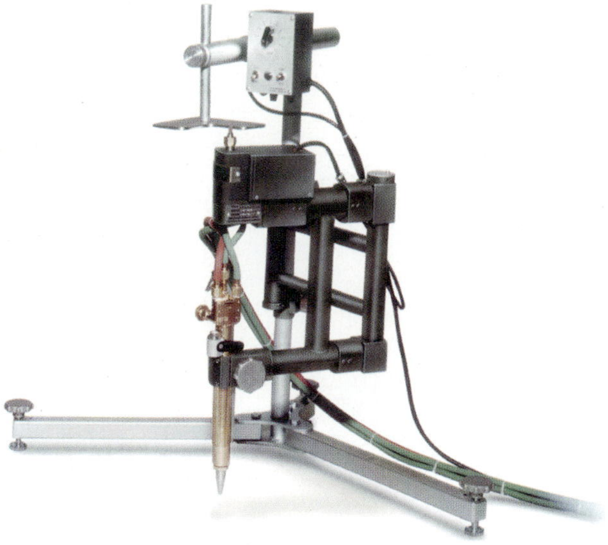

FIGURE 21-74 Portable cutting machine for highly complex shapes. ESAB Welding & Cutting Products

Project Description

The beveled plates are used for welder certification testing.

Project Materials and Tools

The following items are needed to fabricate welder certification testing plates.

Materials

- 3/8-in. thick steel, large enough to cut out a 7-in. × 3-in. test plate
- Source of oxygen and acetylene gases

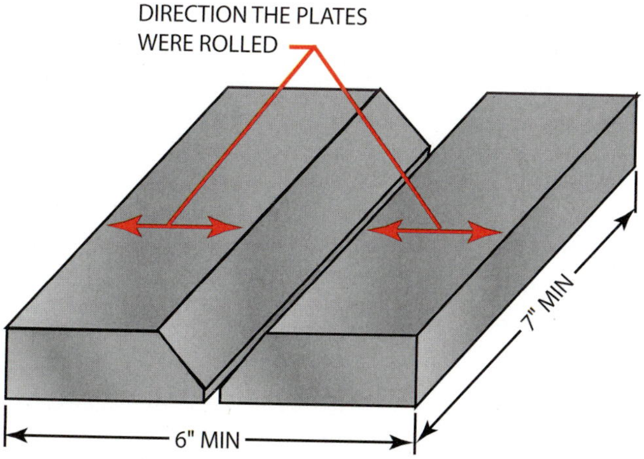

FIGURE 21-75 Carbon steel test plates. © Cengage Learning 2012

Tools

- Hand oxyacetylene cutting torch and/or machine cutting torch and cutting tips
- Tip cleaners
- Torch lighter
- Wrenches
- Angle iron
- PPE
- 12-in. rule
- Chipping hammer
- Soapstone
- Wire brush
- Square
- Pliers

Layout

All rolled metal plates have a grain structure that is strongest when pulled in the direction that the plate was rolled. If a welding test specimen is made with the grain running in the wrong direction, it could possibly crack and fail because of the weaker grain structure.

NOTE: Wood is the most recognized material that has a visible grain structure. All wood structural building materials such as two-by-fours and four-by-sixes are sawed from trees so the grain runs the length of the boards. For example, a 1-in. thick 12-in. long strip of wood that is sawed with the grain is harder to break than the same size strip that is cut across the grain, Figure 21-76. Metal is stronger than wood, but it is definitely weaker in one direction just like wood.

Using the 12-in. rule and soapstone, mark out two test specimens that are 3-in. wide 7 in. long, Figure 21-77.

Cutting Out

Put on your PPE, and following all shop and manufacturer's safety rules for cutting, you will make a 37 1/2° bevel down the 7 in. length of the plate.

Cutting Machine Specimens

Set the oxygen and acetylene pressures and torch travel speed according to the torch manufacturer's recommendations. Make a test cut on scrap metal to see if any adjustments to the flame, cutting pressure, or travel speed are needed, Figure 21-33.

FIGURE 21-76 Metal, like wood, is stronger with the grain than it is across the grain. Larry Jeffus

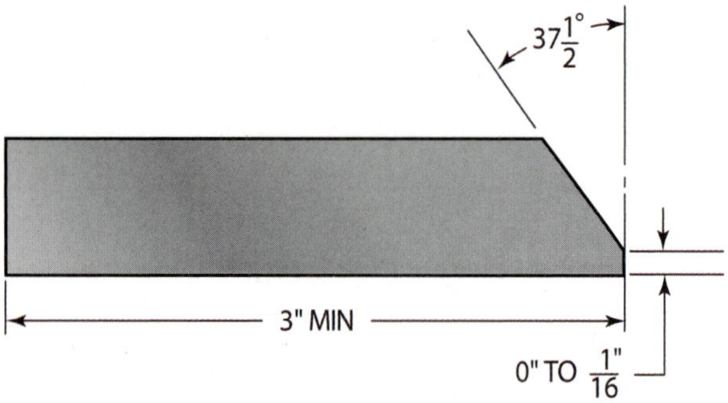

FIGURE 21-77 Test plate bevel dimensions.
© Cengage Learning 2012

Any time beveled cuts are made, there may be some soft slag left on the bottom edge of the bevel.

CAUTION

Cut only one of the 7-in. long pieces at a time because the heat from the flame and cut can cause the metal to warp as the cut is being made, resulting in an unacceptable cut.

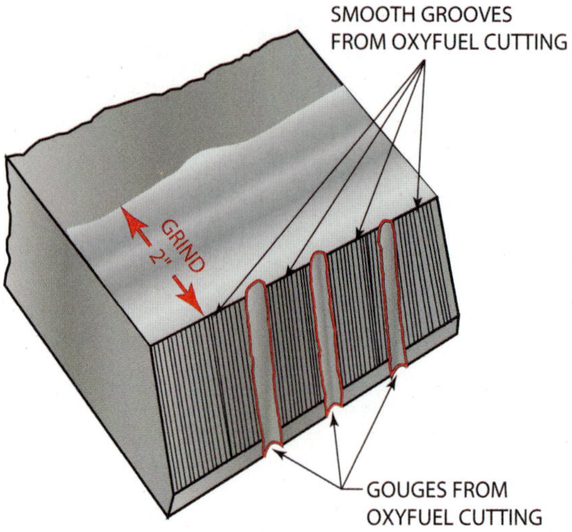

FIGURE 21-78 The flame-cut bevel must be smooth and free of any notches. © Cengage Learning 2012

Hand Cutting Specimens

The hand cut can be made freehand or with a guide. Two common ways to guide the cut are to use a piece of angle iron or a commercially available roller guide, Figure 21-38 and Figure 21-39A.

If an angle iron guide is used, the torch must be rotated to the correct 37 1/2° angle. The roller guide angle is set by lowering the height of one of the roller legs.

Turn off the cylinder valves, bleed the hoses, back out the pressure regulators, and clean up your work area when you are finished cutting.

Finishing

The flame-cut bevel must be smooth and free of any notches, Figure 21-78. Any bevel surface roughness or notches that are deeper than 1/64 in. must be ground smooth. In addition, a band of surface oxides 2 in. wide must be ground off next to the bevel.

Fabrication and Welding

Welding certification, test specification fabrication, and welding procedures are covered in Chapter 29.

Paperwork

Complete a copy of the time sheet in Appendix I and bill of materials in Appendix III, or as provided by your instructor. ◆

Irregular Shapes

Cutting out irregular-shaped parts presents several problems. At times, your cutting torch tip will be in the way, so it may be difficult or impossible to clearly see the line that you are following. Another problem is that you will have to stop cutting from time to time so you can reposition yourself. These stops and starts can leave noticeable changes in the cut if not done correctly. On larger cuts the part may become hot, so you may find it difficult to find a place that is not hot to brace your hand or arm to steady yourself.

PROJECT 21-8

Cutting Out Shapes

Skills to be learned: Develop the skills to follow an irregular line with the cutting stream and making the occasional stop and starting spots as smooth and uniform as the rest of the cut, Figure 21-79.

Project Description

The turtle plant stand, frog card holder, and boot scraper can be used as soon as they cool and are painted. The turtle and frog require some postcut heating so they can be bent into shape. The pine tree on snow, trailer hitch cover, and letters and numbers require some postcutting welding. You can cut out any design such as your school or college logo, state silhouette, or any other design for the trailer hitch cover.

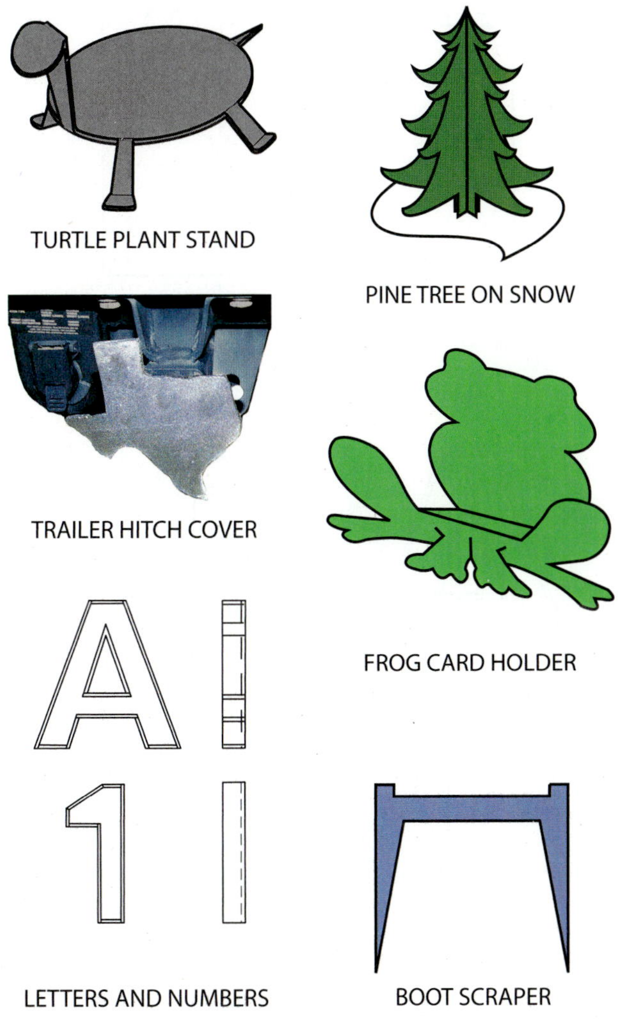

FIGURE 21-79 Project 21-8.

Project Materials and Tools

The following items are needed to cut out the irregular shape.

Materials

- 1/8-in. to 1/4-in. thick steel large enough to cut out a silhouette
- Source of oxygen and acetylene gases

Tools

- Hand oxyacetylene cutting torch tips
- Tip cleaners
- Torch lighter
- Wrenches
- Angle iron
- PPE
- 12-in. rule
- Chipping hammer
- Pencil and soapstone
- Graph paper
- Scissors
- Wire brush
- Square
- Pliers

Layout

Refer to Chapter 4's "Sketching" section and Practice 4-5 for more instructions on how to sketch an outline on graph paper. Use a pencil and graph paper to draw the outline of the irregular shape that you want to cut out. You may make up your own design or choose one of the designs shown in Figure 21-80. The curved line is drawn by locating a series of points along the graph line and then sketching the line to connect the dots. Use the scissors and cut out the outline. For the designs that are symmetrical such as the turtle, pine tree, and frog, you can fold the paper in half and cut out the outline so that both sides are the same. For this project, the exact size and shape of the layout are not as important as the accuracy of the cut.

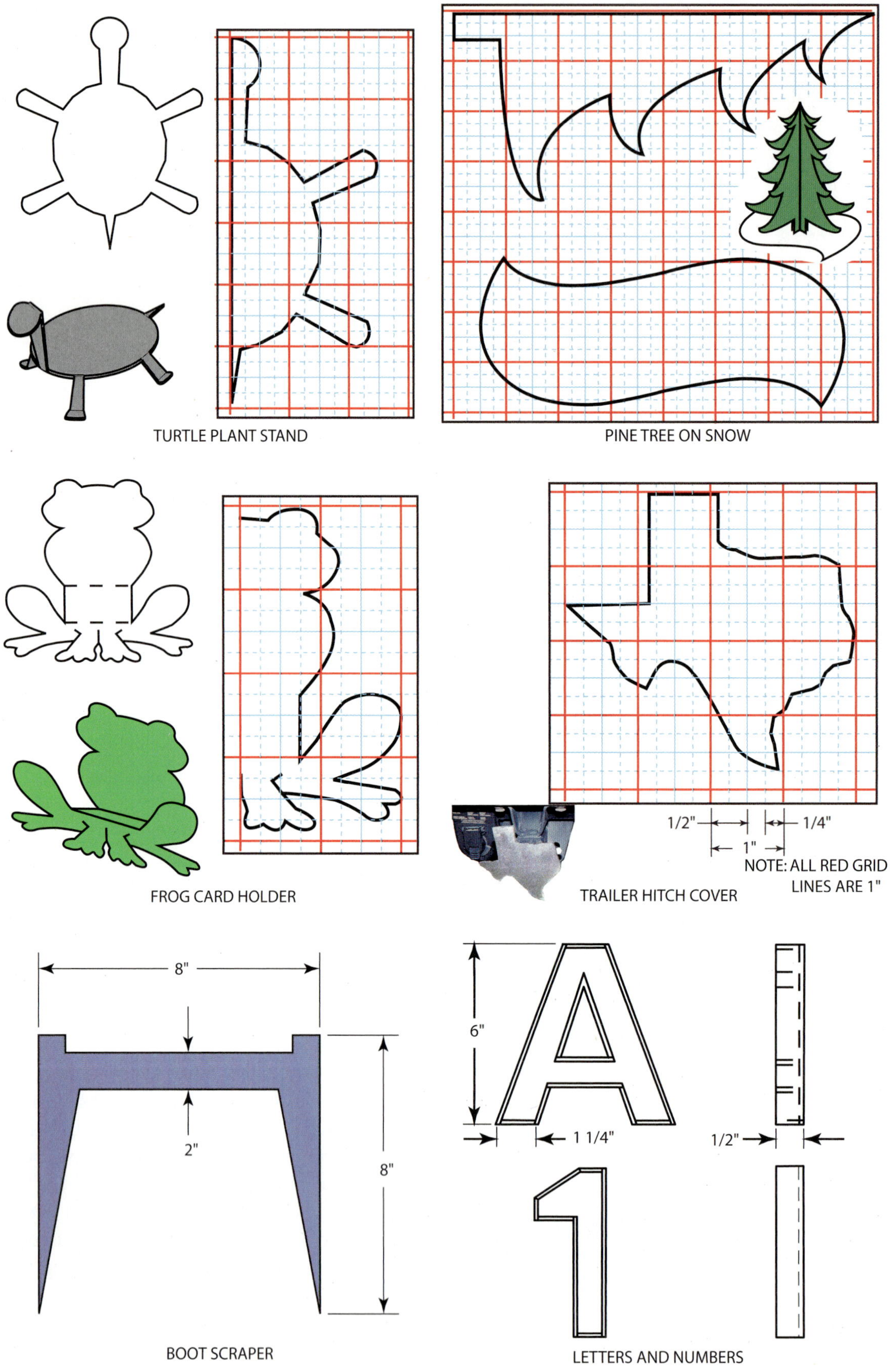

FIGURE 21-80 Irregular shapes for cutting. © Cengage Learning 2012

FIGURE 21-81 Beginning a cut with the torch concentrating the flame on the top edge to speed starting. Larry Jeffus

FIGURE 21-82 The torch is rotated to allow the preheating of the plate ahead of the cut. This speeds the cutting and provides better visibility of the line being cut. Larry Jeffus

Place your template on the metal and use the soapstone or marker to trace the outline. Sometimes soapstone lines can be smudged by your glove during cutting. To avoid this problem you may want to center-punch the line as described in the preceding "Layout" section.

Flame-Cut Layout

Putting on your PPE and following all shop and manufacturer's safety rules for cutting, you will cut out the pattern. Use the manufacturer's tip chart or Table 21-6, and make any changes in tip size and adjustments in pressure to make a slag-free cut.

Brace yourself so that you can cut as steadily as possible. The cut should be made so that the marked or center-punched line is left on the part but with no more than 1/8 in. (3 mm) between the cut edge and the line, Figure 21-81 and Figure 21-82.

You want to make as long a cut as possible, but when you can no longer move freely or you are losing sight of the line, you need to stop. The best way to stop a cut so it can more easily be restarted is to turn out slightly into the scrap area before stopping the cut, Figure 21-40. To restart the cut, point the torch preheat flame at the end of the slight turnout cut you just made. When you depress the oxygen cutting lever, there may be a slight burst of sparks thrown from the cut. This can be caused by an oxide deposit left in the kerf as the cut was stopped. If this happens, there will be a small area where the kerf was blown wider. That is why you stop in the scrap so that the blowout does not affect the part being cut. When the cutting stream and cut have stabilized, move the torch back to the line and continue the cut.

It is not always possible to make irregular cuts that are slag-free, so a little slag on the back side of the cut can be expected.

Fabrication and Welding

The welding procedures you might use to complete your fabrication of this project are covered in other chapters.

Paperwork

Complete a copy of the time sheet in Appendix I and bill of materials in Appendix III, or as provided by your instructor.◆

Pipe Cutting

Freehand pipe cutting may be done in one of two ways. On small diameter pipe, usually under 3 in. (76 mm), the torch tip is held straight up and down and moved from the center to each side, Figure 21-83.

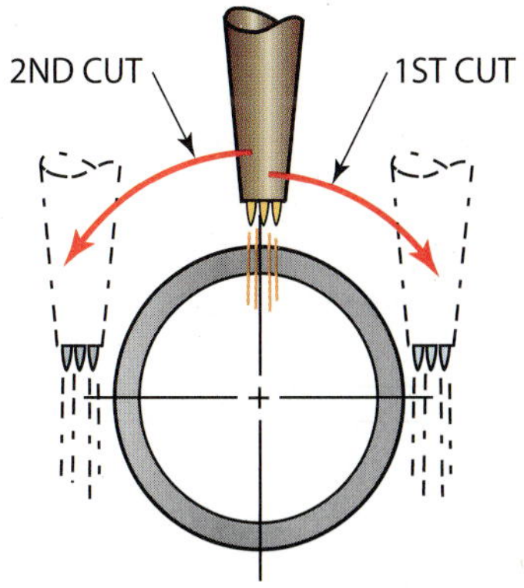

FIGURE 21-83 Small diameter pipe can be cut without changing the angle of the torch. After the top is cut, roll the pipe to cut the bottom. © Cengage Learning 2012

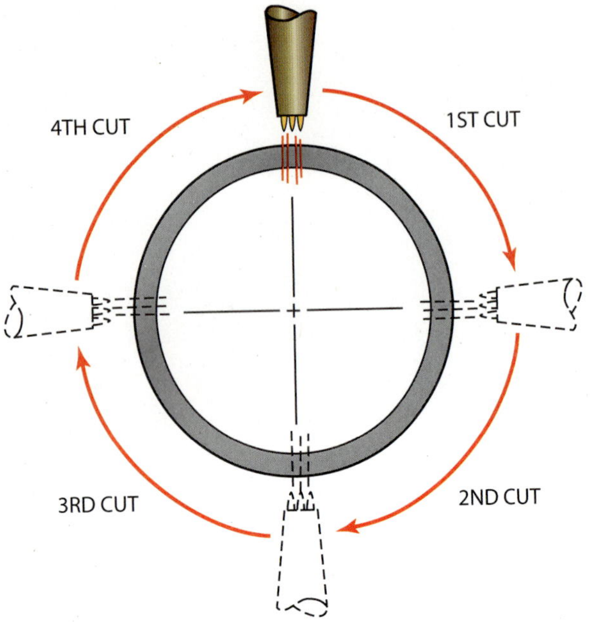

FIGURE 21-84 On large diameter pipe, the torch is turned to keep it at a right angle to the pipe. The pipe should be cut as far as possible before stopping and turning it. © Cengage Learning 2012

This technique can also be used successfully on larger pipe.

For large diameter pipe, 3 in. (76 mm) and larger, the torch tip is always pointed toward the center of the pipe, Figure 21-84. This technique is also used on all sizes of heavy-walled pipe and can be used on some smaller pipe sizes.

The torch body should be held so that it is parallel to the center line of the pipe. Holding the torch parallel helps to keep the cut square.

CAUTION

When you cut pipe, hot sparks can come out of the end of the pipe nearest you, causing severe burns. For protection from hot sparks, plug up the open end of the pipe nearest you, put up a barrier to the sparks, or stand to one side of the material being cut.

PROJECT 21-9

Square Cut on Pipe, 1G (Horizontal Rolled) Position

Skills to be learned: Develop the skills to make straight, square, slag-free OF cuts on a horizontal pipe that can be rolled so that all of the cuts are made on the top side of the pipe.

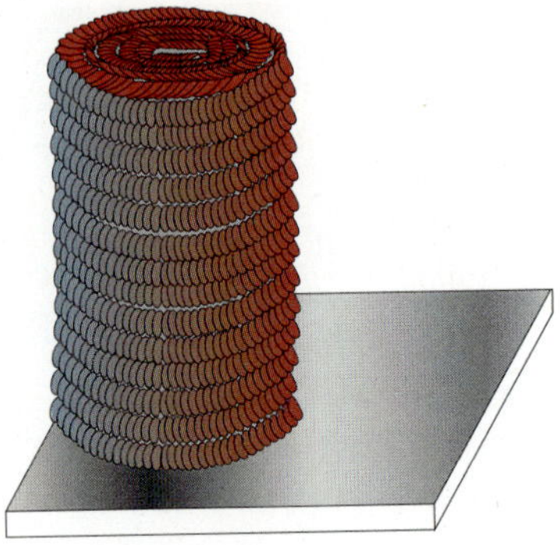

FIGURE 21-85 Project 21-9. © Cengage Learning 2012

Project Description

This decorative bookend will have the appearance of being wrapped by a rope, Figure 21-85.

Project Materials and Tools

The following items are needed to cut out the parts for the rope bookend.

Materials

- 1 4-in. long piece of 2-in. diameter steel pipe
- 1 4-in. × 3-in. piece of 1/4-in. thick mild steel plate
- Source of oxygen and acetylene gases

Tools

- Oxyacetylene cutting torch and cutting tips
- Tip cleaners
- Torch lighter
- Wrenches
- Contour marker or pipe wrap
- PPE
- Wire brush
- Chipping hammer
- Soapstone, pencil and/or marker

Layout

Using a square, 12-in. rule, and soapstone, make two marks, one at 4 in. and another 8 in. from the end of the pipe, Figure 21-86.

To use a contour marker, first set the angle to 0°, Figure 21-87. Line up the soapstone with the 4-in. mark, then hold the base of the marker securely on the pipe, Figure 21-88. The arm has three flexible joints that allow it to mark a little more than halfway around the pipe, Figure 21-89. When one side is marked, you can flex the arm so you can mark the other side of the pipe. You have to avoid putting sideward force on the

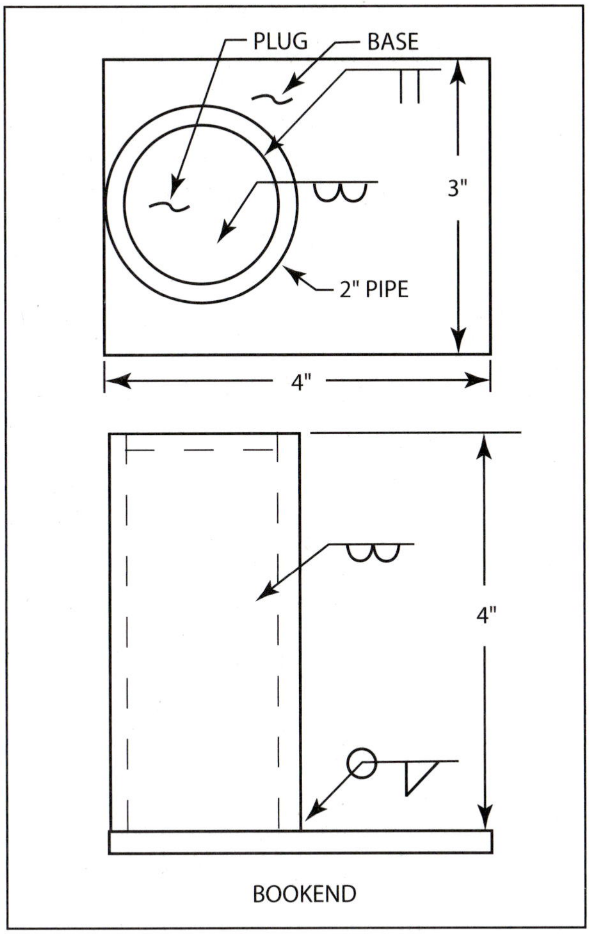

FIGURE 21-86 Bookend layout.
© Cengage Learning 2012

FIGURE 21-87 Contour marker set at o° for a square cut. Larry Jeffus

FIGURE 21-88 Hold the base securely so that the arm can be swung around the pipe. Larry Jeffus

FIGURE 21-89 The flexible arms allow a mark to be drawn completely around the pipe. Larry Jeffus

flexible arm because it can easily be moved so the line would not be straight.

NOTE: Contour markers can be used to lay out pipe for cutting elbows and tees.

To use a pipe Wrap-A-Round, Figure 21-90, you will wrap the tool around the pipe as if you were wrapping it with a strip of tape, Figure 21-91. Draw a line around the pipe next to the pipe Wrap-A-Round with a marker.

Cutting Out

Putting on your PPE and following all shop and manufacturer's safety rules for cutting, you will cut out the parts for the bookend.

Putting on your PPE and following all shop and manufacturer's safety rules for cutting, you will

FIGURE 21-90 Wrap-A-Round. Larry Jeffus

cut out the pattern. Use the manufacturer's tip chart or Table 21-6, and make any changes in tip size and adjustments in pressure to make a slag-free cut.

Hold the torch perpendicular to the top of the pipe, Figure 21-83. Preheat the pipe, and pierce the pipe. Move the torch backward to keep the slag out of the cut and then forward around the pipe, stopping when you have gone as far as you can comfortably. Restart the cut at the top and proceed with the cut in the other direction. Roll the pipe and continue the cut until the pipe falls free.

Stand the pipe on a flat surface. Check to see that the pipe stands within 5° of vertical and has no gaps higher than 1/8 in. (3 mm). Using the manufacturer's tip chart or Table 21-6, make any changes in tip size and adjustments in pressure to make a slag-free cut.

Turn off the cylinder valves, bleed the hoses, back out the pressure regulators, and clean up your work area when you are finished cutting.

Fabrication and Welding

The fabrication procedures for this project are covered in Project 10-1 in Chapter 10. The welding procedures are also covered in that section; if another welding process is to be used, refer to the chapter that discusses that process.

Paperwork

Complete a copy of the time sheet in Appendix I and bill of materials in Appendix III, or as provided by your instructor.◆

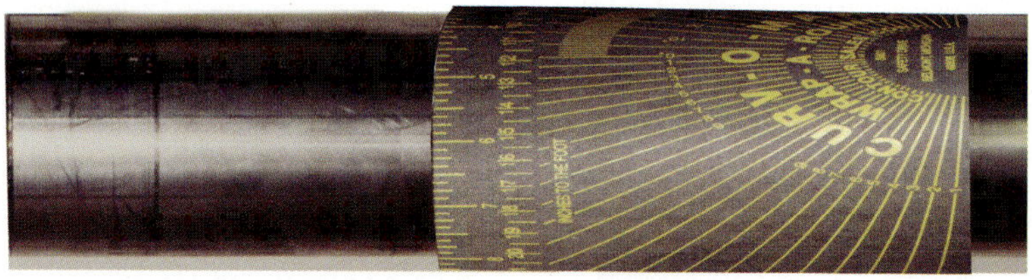

FIGURE 21-91 Using a Wrap-A-Round. Larry Jeffus

SUMMARY

The speed and accuracy with which an oxyacetylene torch can cut off damaged parts or cut out new parts makes it an indispensable tool in any welding shop. The torch's ability to heat and loosen rusty nuts or simply cut them off has earned it the nickname "smoke wrench." Frequently, nuts and bolts on equipment that have been left out in the weather could not be removed without the torch, Figure 21-92. Anyone who has worked on equipment or vehicles that have been stored outside for a long time knows how much time using the cutting torch can save when trying to remove rusted bolts, nuts, and other parts. It is even possible to

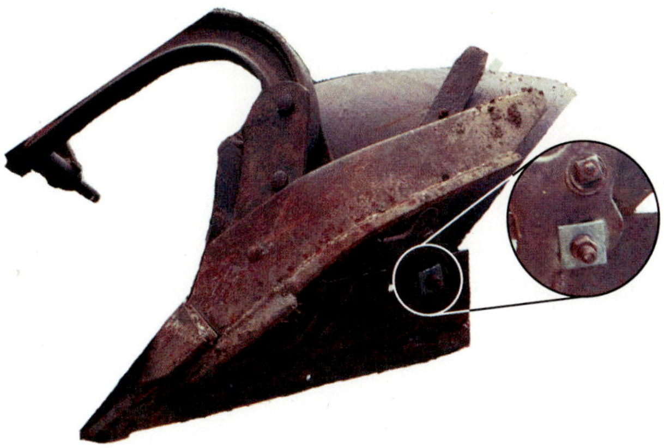

FIGURE 21-92 Nuts and bolts on this rusty plow could not be taken off without a torch. Larry Jeffus

cut a really rusty nut off without damaging the threads on the bolt. The thick layer of rust acts like an insulator, keeping the bolt from getting heated as quickly as the nut. Because the bolt does not get heated to its kindling temperature, the threads will remain intact. You can even cut out a 1/2-in. (13-mm) rusty stud that is broken off in a cast-iron housing if you first drill a 1/4-in. (6-mm) hole through the stud. The hole allows the stud to be heated quickly and cut almost cleanly out. With good torch control, this will work when an easy out would not.

In the hands of a skilled welder, a properly set up and cleaned oxyacetylene torch can make a cut that is nearly as accurate as can be made with a hacksaw but with a lot less work. Taking your time to set up, clean, and adjust the torch before you begin working will save more time than it will take to clean up a poor-quality cut made with a badly set up, dirty torch.

REVIEW QUESTIONS

1. Why is acetylene the primary fuel gas used with oxyfuel gas cutting (OFC)?
2. Describe a good oxyacetylene cut.
3. List the metals that can and cannot be cut with the oxyfuel process.
4. Why must cutting goggles or cutting glasses be used any time that you are using a cutting torch?
5. On a cutting tip what does the diameter of the center cutting orifice determine?
6. When preparing to cut, what gas pressure should you start out with?
7. What should you do to the torch tip before using it to cut?
8. What can happen if acetylene is used in a tip that was designed to be used with one of the other fuel gases?
9. Why should you check with the owner's manual or a welding supplier before tightening a cutting tip to the torch head?
10. How can you check an assembled torch tip to make sure it has a good seal?
11. Why should the oxygen and acetylene cylinder valves be cracked before using?
12. Why should tip cleaners not be forced or used excessively?
13. When lighting the torch, what regulator working pressure should be set if you do not know the correct pressure for the tip size?
14. Why is it important to brace yourself when making a cut with a hand torch?
15. Why should a slight forward torch angle be used when making a straight line square cut?
16. What is coupling distance?
17. Why should a cutting torch never be used to cut open a used can, drum, tank, or other sealed container?
18. Name different methods of laying out a line on a part to be cut.
19. How can you determine the correct tip size when using a new torch set?
20. What two methods can be used to set the gas pressure of the regulator?
21. Describe how an oxyfuel gas cutting torch cuts.
22. What can you tell about a cut by looking at both sides of the kerf?

23. How can you tell if an automatic cutting machine cut that is in the process of being made is going to be a good cut?
24. Compare soft slag to hard slag.
25. How can you avoid cutting through the top of a welding table?
26. What can be used to support and guide the cutting torch so that you have better control and more even cutting?
27. What is distortion?
28. Describe two methods used to control distortion.
29. If you must cut in the field around combustibles, what safety measures should you follow?
30. List some things that can become problems when cutting.
31. What safety precautions must be taken when cutting pipe?

Chapter 22

Plasma Arc Cutting

OBJECTIVES

After completing this chapter, the student should be able to:

- Explain how a plasma arc cutting torch works.
- Demonstrate how to assemble a plasma arc cutting torch.
- Demonstrate how to safely set up and use a plasma arc cutting system.
- Demonstrate how to safely make a variety of cuts using a plasma arc cutting torch.
- List the common plasma cutting gases and the metals they can be used to cut.

KEY TERMS

arc cutting
arc plasma
cup
dross
electrode setback
electrode tip
heat-affected zone
high-frequency alternating current
ionized gas
joules
kerf
nozzle
pilot arc
plasma
plasma arc
plasma arc gouging
stack cutting
standoff distance

INTRODUCTION

The plasma arc cutting (PAC) process has become very popular as the result of the introduction of smaller, less expensive plasma equipment to the welding field, Figure 22-1. A typical portable plasma cutting system can cut mild steel up to 1 1/2 in. (38 mm) thick. Plasma cutters have the unique ability to cut metals without making them very hot. This means that there is less distortion and heat damage than would be caused with an oxyacetylene cutting torch. Very intricate shapes can be cut out without warping.

FIGURE 22-1 Plasma arc cutting machine. This unit can have additional power modules added to the base of its control module to give it more power.
ESAB Welding & Cutting Products

Plasma can cut sheet metal so easily that it has become popular to use it to cut out the smallest of decorations. Most often the smallest of animals, people, buildings, and scenery used on gates, fences, barns, and so forth have been cut out using PAC, Figure 22-2.

Small plasma cutting machines can do many of the same cutting jobs that are done with an oxyacetylene torch, but without the expense of renting gas cylinders. Small plasma cutting machines can use 120-volt electrical power from any standard wall plug or auxiliary power plug on a portable welder, Figure 22-3.

Plasma machines can cut any type of metal, including aluminum, stainless steel, and cast iron. They can cut mild steel ranging from sheet metal up to about 3/8 in., which means that the plasma torch can do most of the cutting required in any welding shop. Larger, more powerful machines are available that can cut an inch (25 mm) or more, but their expense for most small welding shops would be hard to justify.

FIGURE 22-3 Portable welder's auxiliary power plug can be used for plasma cutting. Larry Jeffus

Plasma

Plasma is a state of matter that can be found in the region of an electrical discharge (arc). All states of matter have their own characteristics. A solid has shape and form, a liquid seeks its own level and takes the shape of its container, and a gas has no distinct shape or volume and fills its container. Plasma is a matter that is highly conductive and easily controlled and shaped by a magnetic field. Plasma is created by an arc and is an **ionized gas** that has both electrons and positive ions whose charges are nearly equal to each other. The glowing mass of a star and the bright light from a fluorescent lightbulb are both plasma. One is very hot and the other is almost cool to the touch.

FIGURE 22-2 Plasma-cut sheet metal panels decorate this gate.
Larry Jeffus

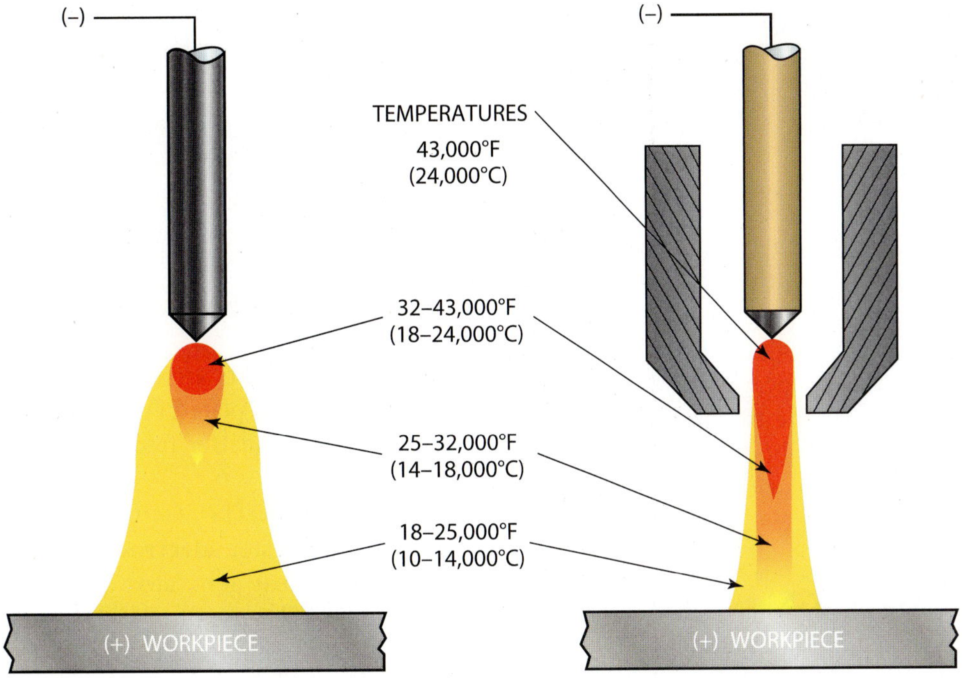

FIGURE 22-4 **Approximate temperature differences between a standard arc and a plasma arc.** American Welding Society

A welding plasma results when a gas is heated to a high enough temperature to convert into positive and negative ions, neutral atoms, and negative electrons. The temperature of an unrestricted arc is about 11,000°F (6,000°C), but the temperature created when the arc is concentrated to form a plasma stream is about 43,000°F (24,000°C), Figure 22-4. This is hot enough to rapidly melt or vaporize any metal it comes in contact with.

Arc Plasma

The term **arc plasma** is defined as gas that has been heated to at least a partially ionized condition, enabling it to conduct an electric current. The term **plasma arc** is the term most often used in the welding industry when referring to the arc plasma used in welding and cutting processes. The plasma arc produces both the high temperature and intense light associated with all forms of arc welding and **arc cutting** processes.

Plasma Torch

The plasma torch is a device that allows for the creation and control of the plasma for cutting processes. The plasma is created in the cutting torch head. A plasma torch supplies electrical energy to a gas that is in a constricted space where the gas is changed into the high-energy state of a plasma.

Torch Body

The torch body is made of a special plastic that is resistant to high temperatures, ultraviolet light, and impact. It provides a good grip area and protects the cable and hose connections to the head. The torch body is available in a variety of lengths and sizes. Generally, the longer, larger torches are used for the higher-capacity machines; however, sometimes you might want a longer or larger torch to give yourself better control or a longer reach.

Torch Head

The torch head is attached to the torch body where the cables and hoses attach to the electrode tip, nozzle tip, and nozzle. Torches are available with heads that are fixed at a 75° angle, 90° angle, or an 180° angle (straight), or they may have a flexible head that can be adjusted to any desired angle. The 75° and 90° angles are popular for manual operations, and the 180° straight torch heads are most often used for machine operations. Because of the heat in the head produced by the arc, some provisions for cooling the head and its internal parts must be made. The cooling for low-powered torches is typically done by allowing the air to continue flowing for a short period of time after the cutting power has stopped. On larger high-powered torches, cooling is typically provided by circulating water through the head. It is possible to replace just the torch head on most torches if it becomes worn or damaged.

Power Switch

Most handheld torches have a manual power switch that is used to start and stop the power source, gas, and cooling water (if used). The switch most often used is a thumb switch located on the torch body, but it may be a foot control or located on the panel for machine-type equipment. The thumb switch is typically molded into the torch body. The foot control must be rugged enough to withstand the welding shop environment.

Common Torch Parts

Some of the parts are designed to be replaced as they are worn or damaged. The following are the parts that are most commonly replaced: electrode tip, nozzle insulator, nozzle tip, and nozzle, Figure 22-5. Some manufacturers have combined some of these parts to simplify the replacement of worn or damaged parts. Refer to the torch manufacturer's literature for specific information. Special tools may be required to replace the parts on some torches.

CAUTION

Improper use of the torch or assembly of torch parts may result in damage to the torch body as well as the frequent replacement of these parts.

The metal parts are usually made out of copper, and they may be plated. The plating of copper parts helps them stay spatter-free longer.

Electrode Tip. The electrode is often made of copper with a tungsten **electrode tip** attached. By using copper, the heat generated at the tip is conducted away faster. The use of a copper/tungsten tip in the torch increases the quantity of work it can produce. Keeping the tip as cool as possible lengthens the life of the tip and allows for better-quality cuts for a longer time. Minimizing the amount of time that the pilot arc is on before the cutting plasma is initiated will also increase the operational life of the electrode tip and nozzle.

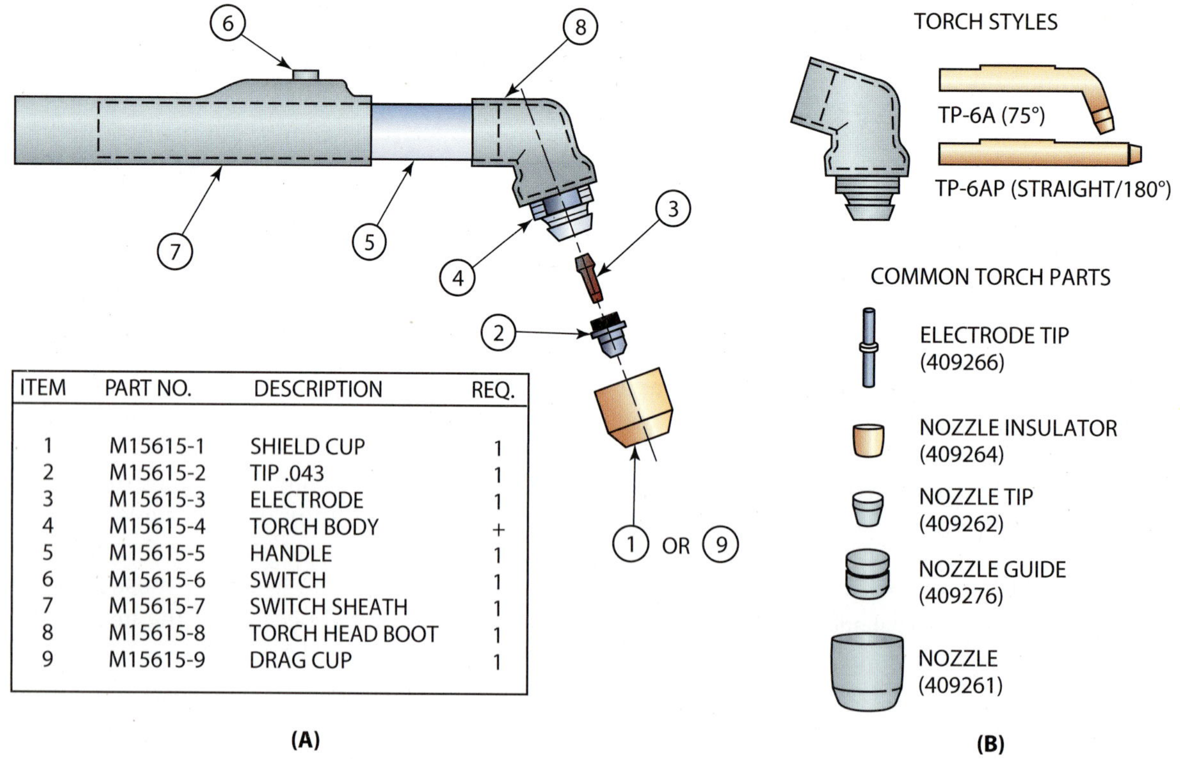

ITEM	PART NO.	DESCRIPTION	REQ.
1	M15615-1	SHIELD CUP	1
2	M15615-2	TIP .043	1
3	M15615-3	ELECTRODE	1
4	M15615-4	TORCH BODY	+
5	M15615-5	HANDLE	1
6	M15615-6	SWITCH	1
7	M15615-7	SWITCH SHEATH	1
8	M15615-8	TORCH HEAD BOOT	1
9	M15615-9	DRAG CUP	1

FIGURE 22-5 Replaceable torch parts. Larry Jeffus

FIGURE 22-6 Nozzles are available in a variety of shapes for different types of cutting jobs. Larry Jeffus

FIGURE 22-7 Different torches use different types of nozzle tips. Larry Jeffus

Nozzle Insulator. The nozzle insulator is between the electrode tip and the nozzle tip. The nozzle insulator provides the critical gap spacing and the electrical separation of the parts. The spacing between the electrode tip and the nozzle tip, called **electrode setback**, is critical to the proper operation of the system. Some plasma arc welding power supplies have an automatic safety check to prevent damage to the torch that will not allow the torch to be operated if this gap is not set correctly.

Nozzle Tip. The nozzle tip has a small, cone-shaped, constricting orifice in the center. The electrode setback space, between the electrode tip and the nozzle tip, is where the electric current forms the plasma. The preset, close-fitting parts provide the restriction of the gas in the presence of the electric current so the plasma can be generated, Figure 22-6. The diameter of the constricting orifice and the electrode setback are major factors in the operation of the torch. As the diameter of the orifice changes, the plasma jet action will be affected.

Nozzle. The **nozzle**, sometimes called the **cup**, is made out of high temperature-resistant material such as ceramic, Figure 22-7. The nozzle can serve three main purposes. The most important is that it helps to protect the internal parts from accidentally shorting out to the work. It may be used as a guide so that the torch can be dragged across the work surface. Nozzles designed to be dragged have cut outs to allow sparks to escape without damaging the nozzle tip. When a shielding gas is used, the nozzle directs the gas around the cut to protect it from oxidation.

Cables and Hoses

A number of power and control cables and gas and cooling water hoses may be used to connect the power supply with the torch, Figure 22-8. This multipart cable is usually covered to provide some protection to the cables and hoses inside and to make handling the cable easier. This covering is heat-resistant but may not prevent damage to the cables and hoses inside if it

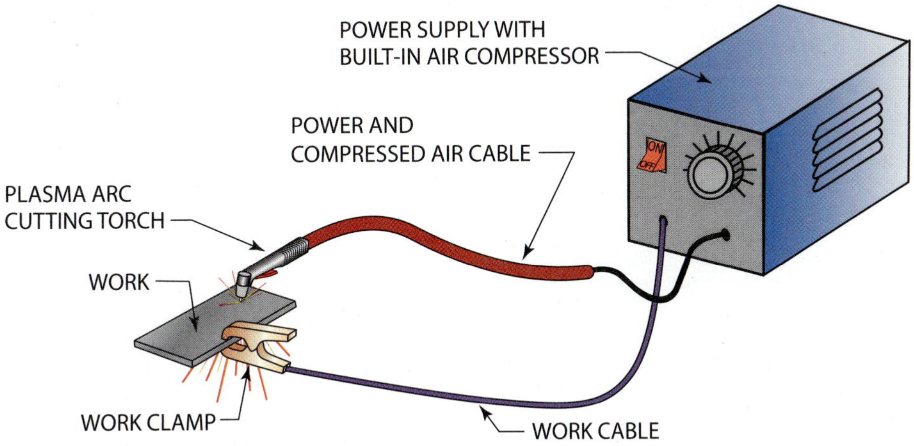

FIGURE 22-8 Typical manual plasma arc cutting setup. © Cengage Learning 2012

comes in contact with hot metal or is exposed directly to the cutting sparks.

Power Cable

The power cable must have a high-voltage rated insulation, and it is made of finely stranded copper wire to allow for maximum flexibility of the torch, Figure 22-9. For all nontransfer-type torches and those that use a high-frequency pilot arc, there are two power conductors, one positive (+) and one negative (–). The size and current-carrying capacity of this cable are controlling factors to the power range of the torch. As the capacity of the equipment increases, the cable must be made large enough to carry the increased current. The larger cables are less flexible and more difficult to manipulate. To make the cable smaller on water-cooled torches, the cable is run inside the cooling water return line. Putting the power cable inside the return water line allows a smaller cable to carry more current. The water prevents the cable from overheating.

Compressed Air

Most small shop plasma arc cutting torches use compressed air to form the plasma and to make the cut. Compressed air must be clean and dry, so a filter dryer must be used to prevent contaminants like oil, dirt, or moisture from entering the plasma torch. Any contamination entering the torch can cause internal arcing between the electrode and the nozzle. Compressed air can be supplied by either an external compressor or an internal compressor. Many of these PA cutting machines have air compressors built into the power supply. These internal compressors are designed to provide the correct airflow and pressure. Cutting machines with internal compressors are very convenient.

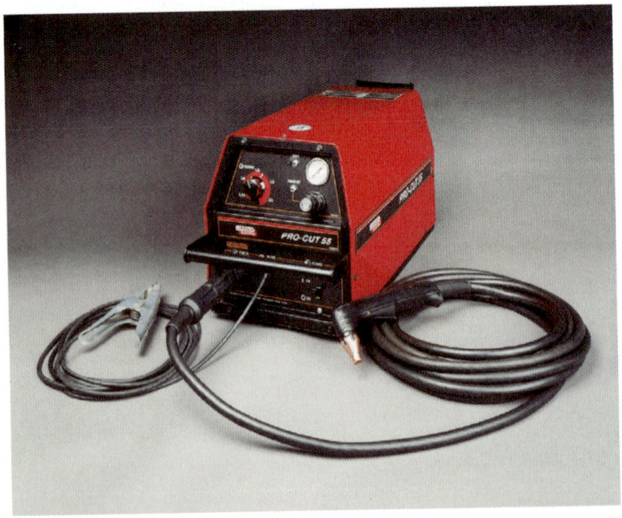

FIGURE 22-9 Thin strands of copper help to make the cables more flexible. Larry Jeffus

Gas Hose

The gas hose carries compressed air from the plasma machine to the torch and is made of a special plastic that is resistant to heat and ultraviolet light. Most plastic tubing will not withstand the heat or ultraviolet light, so if this hose gets damaged, replace it only with the same hose type and size from the equipment manufacturer.

Control Wire

The control wire is a two-conductor, low-voltage, stranded copper wire that connects the power switch to the power supply. This will allow you to start and stop the plasma power and gas as needed during the cut.

Water Tubing

High-amperage torches may be water cooled. If cooling water is required, it must be switched on and off at the same time as the plasma power. Allowing the water to circulate continuously might result in condensation in the torch.

Power Requirements

Voltage

The production of the plasma requires a direct-current (DC), high-voltage, constant-current (drooping arc voltage) power supply. A constant-current–type machine allows for a rapid start of the plasma arc at the high open circuit voltage and a more controlled plasma arc as the voltage rapidly drops to the lower closed voltage level. The voltage required for most welding operations, such as shielded metal arc, gas metal arc, gas tungsten arc, and flux cored arc, ranges from 18 to 45 volts. The voltage for a plasma arc process ranges from 50 to 200 volts closed circuit and 150 to 400 volts open circuit. This higher electrical potential is required because the resistance of the gas increases as it is forced through a small orifice. The potential voltage of the power supplied must be high enough to overcome the resistance in the circuit in order for electrons to flow, Figure 22-10.

FIGURE 22-10 Inverter-type plasma arc cutting power supply. Thermadyne Industries, Inc.

Amperage

Although the voltage is higher, the current (amperage) flow is much lower than it is with most other welding processes. Some low-powered PAC torches will operate with as low as 10 amps of current flow. High-powered plasma cutting machines can have amperages as high as 200 amps, and some very large automated cutting machines may have 1000-ampere capacities. The higher the amperage capacity, the faster and thicker the machine will cut.

Heat Input

Although the total power used by both plasma and nonplasma processes is similar, the actual energy input into the work per linear foot is less with plasma. The very high temperatures of the plasma process allow much higher traveling rates so that the same amount of heat input is spread over a much larger area. This has the effect of lowering the **joules** per inch of heat that the weld or cut will receive. Table 22-1 shows the cutting performance of a typical plasma torch. Note the relationship among amperage, cutting speed, and metal thickness. The lower the amperage, the slower the cutting speed or the thinner the metal that can be cut.

A high travel speed with plasma cutting will result in a heat input that is much lower than that of any oxyfuel cutting process. A steel plate cut using the plasma process may have only a slight increase in temperature following the cut. It is often possible to pick up a part only moments after it is cut using plasma. The same part cut with oxyfuel would be much hotter and require a longer time to cool off.

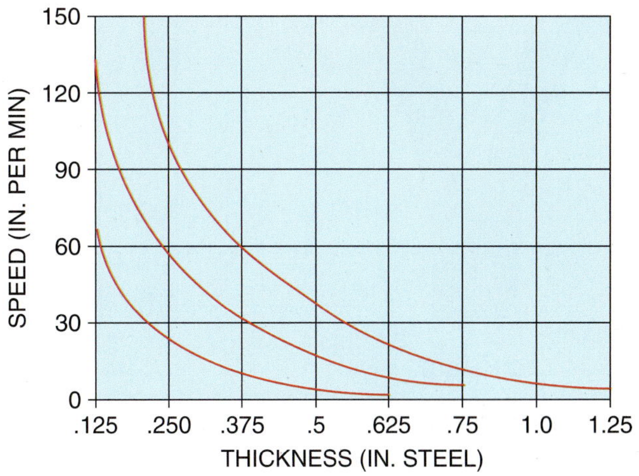

Table 22-1 Plasma Arc Cutting Perimeters
ESAB Welding & Cutting Products

Distortion

Any time metal is heated in a localized zone or spot, it expands in that area and, after the metal cools, it is no longer straight or flat, Figure 22-11. If a piece of metal is cut, there will be localized heating along the edge of the cut, and, unless special care is taken, the part will not be usable as a result of its distortion, Figure 22-12. This distortion is a much greater problem with thin metals. By using a plasma cutter, a worker can cut the thin sheet metal of damaged equipment with little problem from distortion.

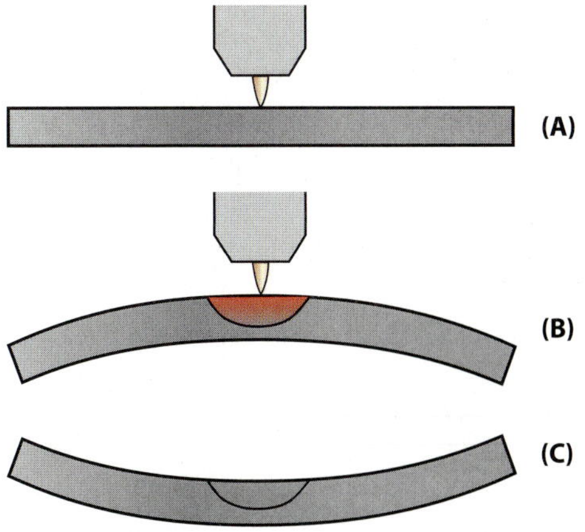

FIGURE 22-11 (A) When a flat piece of metal is heated, (B) it expands, bending the edges away from the heat. (C) When the spot cools, it shrinks, bending the edges toward the heat. © Cengage Learning 2012

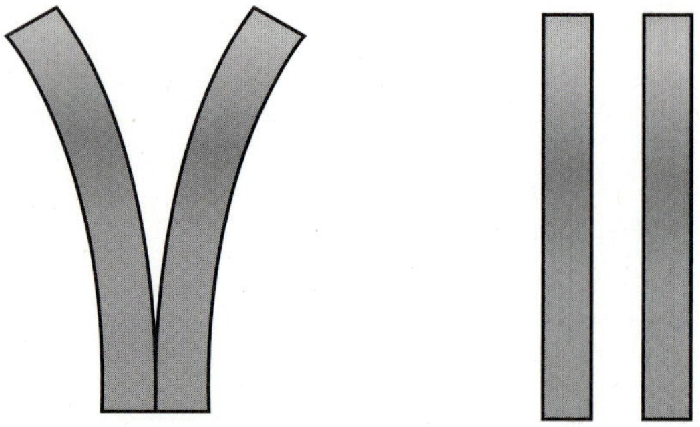

FIGURE 22-12 The heat of a torch cut causes metal to bend, but the plasma cut is so fast, little or no bending occurs. © Cengage Learning 2012

On thicker sections, the hardness zone along the edge of a cut will be reduced so much that it is not a problem. When using oxyfuel cutting of thick plate, especially higher-alloyed metals, this hardness zone can cause cracking and failure if the metal is shaped after cutting, Figure 22-13. Often the plates must be preheated before they are cut using oxyfuel to reduce the **heat-affected zone**. This preheating adds greatly to the cost of fabrication both in time and fuel costs. By being able to make most cuts without preheating, the plasma process greatly reduces the fabrication cost.

Applications

Plasma cutting equipment has rapidly replaced oxyacetylene cutting equipment in welding shops. A major problem with keeping oxyacetylene cylinders is that they are usually rented. You have to pay for the cylinders month after month even if you are not using them. Plasma, however, does not need compressed gas cylinders, so once you purchase the equipment there are no additional monthly charges. Small 120-volt plasma equipment can cut metal up to 3/8 in. thick. That is within the range for most small jobs, so plasma can do most metal cutting jobs around most welding shops.

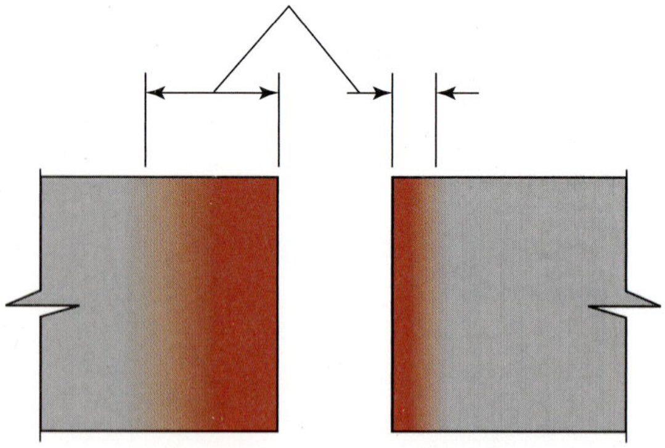

FIGURE 22-13 A smaller heat-affected zone will result in less hardness or brittleness along the cut edge. © Cengage Learning 2012

Sometimes it might be necessary to cut a thicker piece of material than your plasma torch is designed to cut. For example, you may need to cut out a bent shaft, axle, or guard plate when a tractor, harvester, or other equipment is damaged in the field in order to make a repair. You can make a thicker cut by traveling more slowly and, in some cases, weaving the torch back and forth to make a wider kerf so that the cut can carry all the way through. Thick cuts may not be neat or clean, but when you need to get something cut out so that a repair can be made, neatness may not be as important as speed. The surface can always be ground smooth if needed.

Cutting Speed

The plasma arc cutting process can produce very high cutting rates when an automated process is used. However, most manual plasma arc cutting speeds are around 10 in. per minute. Any faster and it would be difficult for you to accurately follow the line. The current setting affects the cutting speed. For example, a higher amperage allows you to make cuts faster on thicker material, while a lower amperage setting allows the same cutting speed on thin metal.

Metals

Any material that is conductive can be cut using the PAC process. In a few applications nonconductive materials can be coated with conductive material so they can also be cut. Although it is possible to make cuts in metal as thick as 7 in., it is not cost effective. The most popular materials cut are carbon steel, stainless steel, aluminum, and sheet metal.

The PAC process is also used to cut expanded metals, screens, and other items that would require frequent starts and stops, Figure 22-14.

Stack Cutting. Because the PAC process does not rely on the thermal conductivity between stacked parts, thin sheets can be stacked and cut efficiently. **Stack cutting** is helpful when you are cutting out a number of duplicate sheet metal parts. It is also helpful when you are trying to cut through several straps that

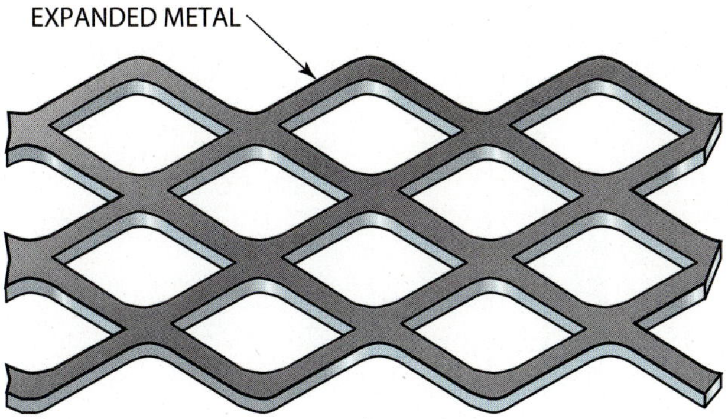

FIGURE 22-14 Expanded metal.
© Cengage Learning 2012

might be clamped or welded together, such as the top of some three-point hitch connections on a bush hog.

The PAC process does not have these limitations. It is recommended that the sheets be held together for cutting, but this can be accomplished by using standard C-clamps. The clamping needs to be tight because if the space between layers is excessive, the sheets may stick together. The only problem that will be encountered is that, because of the kerf bevel, the parts near the bottom might be slightly larger if the stack is very thick. This problem can be controlled by using the same techniques as described for making the kerf square.

Dross. **Dross** is the metal compound that resolidifies and attaches itself to the bottom of a cut. This metal compound is made up mostly of unoxidized metal, metal oxides, and nitrides. It is possible to make cuts dross-free if the PAC equipment is in good operating condition and the metal is not too thick for the size of torch being used. Because dross contains more unoxidized metal than most oxyfuel gas cutting (OFC) slag, often it is much harder to remove if it sticks to the cut. The thickness that a dross-free cut can be made is dependent on a number of factors, including the gas(es) used for the cut, travel speed, standoff distance, nozzle tip orifice diameter, wear condition of the electrode tip and nozzle tip, gas velocity, and plasma stream swirl.

Stainless steel and aluminum are easily cut dross-free. Carbon steel, copper, and nickel-copper alloys are much more difficult to cut dross-free.

Standoff Distance

The **standoff distance** is the distance from the nozzle tip to the work, Figure 22-15. This distance is very critical to producing quality plasma arc cuts. As the distance increases, the arc force is diminished and tends to spread out. This causes the kerf to be wider, the top

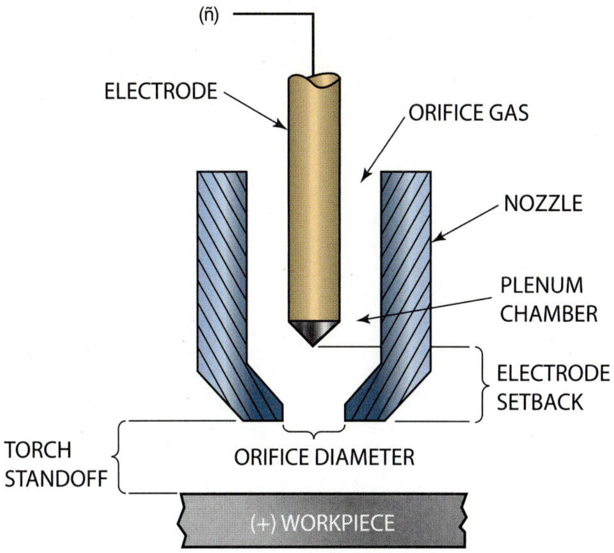

FIGURE 22-15 Conventional plasma torch terminology.
American Welding Society

edge of the plate to become rounded, and the formation of more dross on the bottom edge of the plate. However, if this distance becomes too close, the working life of the nozzle tip will be reduced. In some cases, an arc can form between the nozzle tip and the metal that instantly destroys the tip.

On some torches it is possible to drag the nozzle tip along the surface of the work without shorting it out. This is a large help when working on metal out of position or on thin sheet metal. Before you use your torch in this manner, you must check the owner's manual to see if it will operate in contact with the work, Figure 22-16. This technique allows the nozzle tip orifice to become contaminated more quickly.

Starting Methods

Because the electrode tip is located inside the nozzle tip, and a high initial resistance to current flow exists in the gas flow before the plasma is generated, it is necessary to have a specific starting method. Two methods are used to establish a current path through the gas.

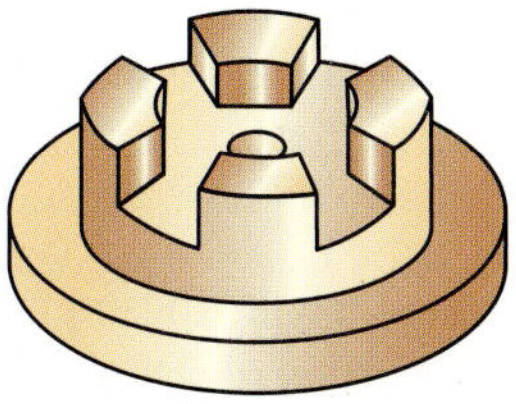

FIGURE 22-16 A castle nozzle tip can be used to allow the torch to be dragged across the surface.
© Cengage Learning 2012

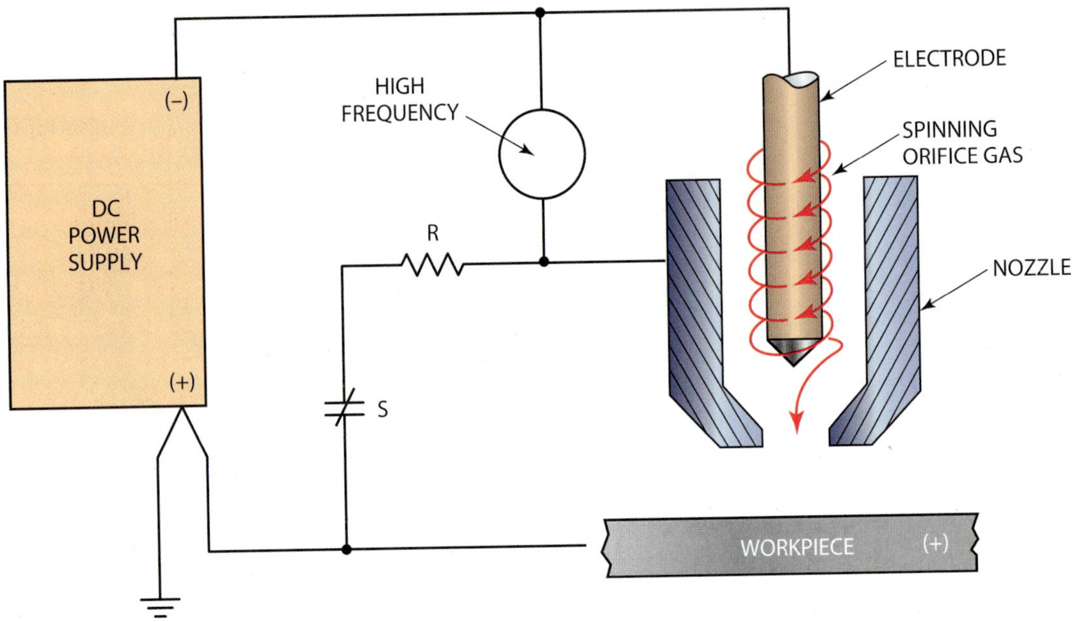

FIGURE 22-17 Plasma arc torch circuitry. American Welding Society

One method uses a high-frequency current, called high-frequency start, and the other method brings the tungsten electrode in contact with the nozzle tip, called contact start. The contact start method is primarily used on automated or robotic cutting to eliminate high-frequency disruption of the computer controls.

The most common method uses a **high-frequency alternating current** carried through the conductor, the electrode, and back from the nozzle tip. This high-frequency current ionizes the gas and allows it to carry the initial current to establish a pilot arc, Figure 22-17. After the pilot arc has been started, the high-frequency starting circuit can be stopped. A **pilot arc** is an arc between the electrode tip and the nozzle tip within the torch head. This is a nontransfer arc, so the workpiece is not part of the current path. The low current of the pilot arc, although it is inside the torch, does not create enough heat to damage the torch parts as long as it is not left on too long before the arc is started.

When the torch is brought close enough to the work, the primary arc will follow the pilot arc across the gap, and the main plasma is started. Once the main plasma is started, the pilot arc power can be shut off.

Kerf

The **kerf** is the space left in the metal as the metal is removed during a cut. The width of a PAC kerf is often wider than that of an oxyfuel cut. Several factors affect the width of the kerf. A few of the factors are as follows:

- Standoff distance—The closer the torch nozzle tip is to the work, the narrower the kerf will be, Figure 22-18.

FIGURE 22-18 When the standoff distance is correct, as with this machine cut, almost no sparks bounce back on the cutting tip. ESAB Welding & Cutting Products

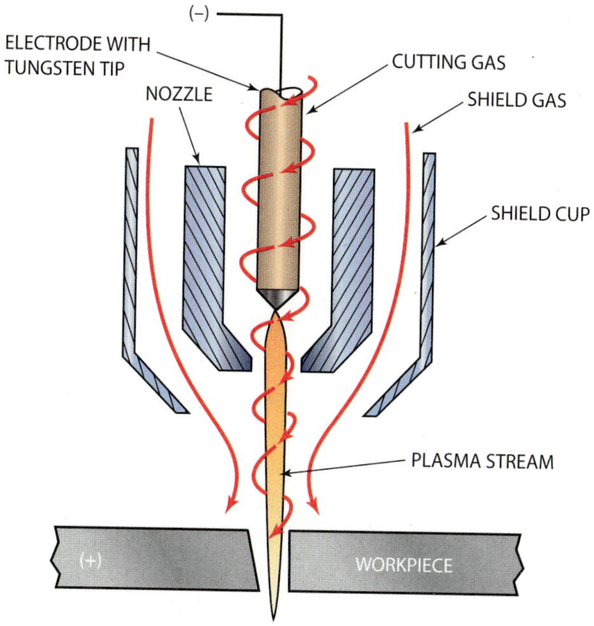

FIGURE 22-19 The cutting gas can swirl around the electrode to produce a tighter plasma column.
American Welding Society

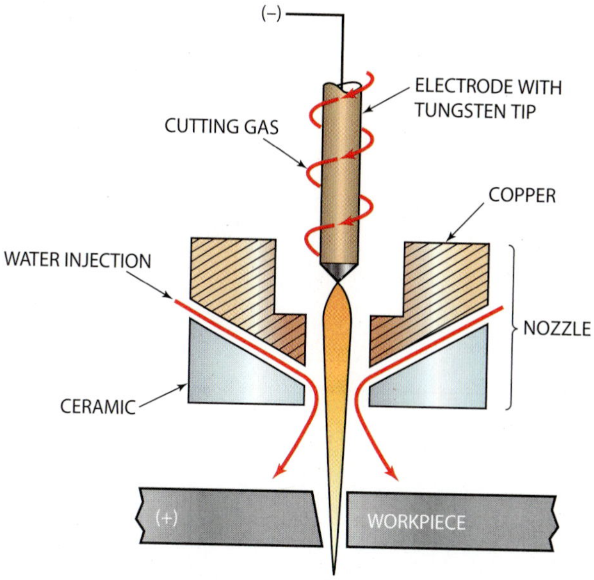

FIGURE 22-20 Water injection plasma arc cutting. Notice that the kerf is narrow, and one side is square.
American Welding Society

- Orifice diameter—Keeping the diameter of the nozzle orifice as small as possible will keep the kerf smaller.
- Power setting—Too high a power setting will cause an increase in the kerf width.
- Travel speed—As the travel speed is increased, the kerf width will decrease; however, the bevel on the sides and the dross formation will increase if the speeds are excessive.
- Gas—The type of gas or gas mixture will affect the kerf width as the gas change affects travel speed, power, concentration of the plasma stream, and other factors.
- Electrode and nozzle tip—As these parts begin to wear out from use or are damaged, the PAC quality and kerf width will be adversely affected.
- Swirling of the plasma gas—On some torches, the gas is directed in a circular motion around the electrode before it enters the nozzle tip orifice. This swirling causes the plasma stream that is produced to be more dense with straighter sides. The result is an improved cut quality, including a narrow kerf, Figure 22-19.
- Water injection—The injection of water into the plasma stream as it leaves the nozzle tip is not the same as the use of a water shroud. Water injection into the plasma stream will increase the swirl and further concentrate the plasma. This improves the cutting quality; lengthens the life of the nozzle tip; and makes a squarer, narrower kerf, Figure 22-20.

Table 22-2 lists some standard kerf widths for several metal thicknesses. These are to be used as a guide for nesting of parts on a plate to maximize the material used and minimize scrap. The kerf size may vary from this depending on a number of the variables with your PAC system as listed in Table 22-1. You should make test cuts to verify the size of the kerf before starting any large production cuts.

Because the sides of the plasma stream are not parallel as they leave the nozzle tip, there is a bevel left on the sides of all plasma cuts. This bevel angle is from 1/2° to 3° depending on metal thickness, torch speed, type of gas, standoff distance, nozzle tip condition, and other factors affecting a quality cut. On thin metals, this bevel is undetectable and offers no problem in part fabrication or finishing.

When possible, make your cut so that the piece you want is on the right side of the cut because the right side will have the squarer edge and have less dross. This technique is effective provided only that one side of the cut is to be scrap. Make a clockwise cut if you

Plate Thickness		Kerf Allowance	
in.	mm	in.	mm
1/8 to 1	3.2 to 25.4	+3/32	+2.4
1 to 2	25.4 to 51.0	+3/16	+4.8
2 to 5	51.0 to 127.0	+5/16	+8.0

Table 22-2 Standard Kerf Widths for Several Metal Thicknesses

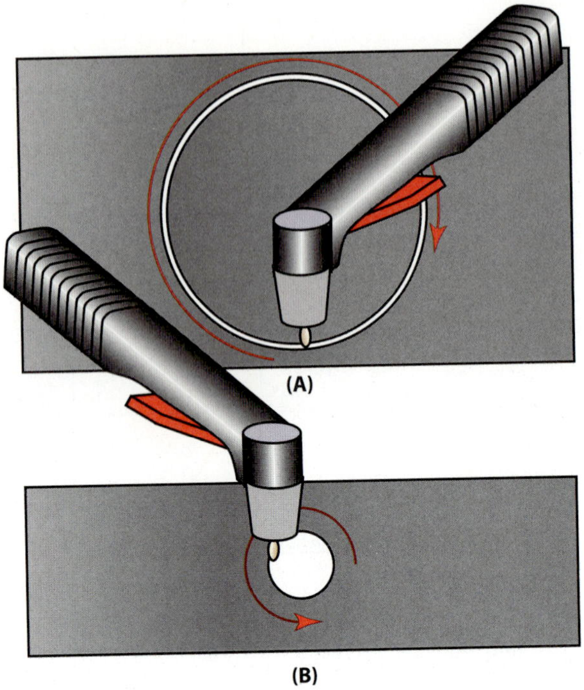

FIGURE 22-21 Different cutting directions for a round part and a hole. © Cengage Learning 2012

are cutting out a circular piece, but make a counterclockwise cut if you are cutting a hole, Figure 22-21.

Gases

The most popular gas for PA cutting is compressed air. Other gas and gas mixtures are used. The selection of the type of gas or gas mixture will significantly affect the quality of the cut and type of material that can be cut. The following are some of the effects on the cut that changing the PAC gas(es) will have:

- Force—The amount of mechanical impact on the material being cut; the density of the gas and its ability to disperse the molten metal.
- Central concentration—Some gases have a more compact plasma stream. This factor greatly affects the kerf width and cutting speed.
- Heat content—As the electrical resistance of a gas or gas mixture changes, it will affect the heat content of the plasma it produces. The higher the resistance, the higher the heat produced by the plasma.
- Kerf width—The ability of the plasma to remain in a tightly compact stream produces a deeper cut with less of a bevel on the sides.
- Dross formation—The dross that may be attached along the bottom edge of the cut can be controlled or eliminated.

Metal	Gas
Carbon and low alloy steel	Compressed air Nitrogen Argon with 0 to 35% hydrogen
Stainless steel	Nitrogen Argon with 0 to 35% hydrogen
Aluminum and aluminum alloys	Nitrogen Argon with 0 to 35% hydrogen

Table 22-3 Gases for a Plasma Arc

- Top edge rounding—The rounding of the top edge of the plate can often be eliminated by correctly selecting the gas(es) that are to be used.
- Metal type—Because of the formation of undesirable compounds on the cut surface as the metal reacts to elements in the plasma, some metals may not be cut with specific gas(es).

Table 22-3 lists some of the popular gases and gas mixtures used for various PAC metals. The selection of a gas or gas mixture for a specific operation to maximize the system performance must be tested with the equipment and setup being used. With constant developments and improvements in the PAC system, new gases and gas mixtures are continuously being added to the list. Although compressed air is recommended for cutting only mild steel, when making some repairs in the field it is used to cut any metal. It is often much easier and cheaper to clean up a PA cut on stainless steel or aluminum than it would be to get the correct gas.

In addition to the type of gas, it is important to have the correct gas flow rate for the size tip, metal type, and thickness. Too low a gas flow results in a cut having excessive dross and sharply beveled sides. Too high a gas flow produces a poor cut because of turbulence in the plasma stream and waste gas. Most machines have a table that lists the proper gas flow and/or pressure settings for various thicknesses of metal to be cut.

Machine Cutting

Almost any plasma torch can be attached to some type of semiautomatic or automatic device to allow it to make machine cuts. The simplest devices are machines that run on tracks, Figure 22-22. These portable machines are good for mostly straight or circular cuts. Complex shapes can be cut with a pattern cutter that follows a template's shape, drawing, or computer program, Figure 22-23.

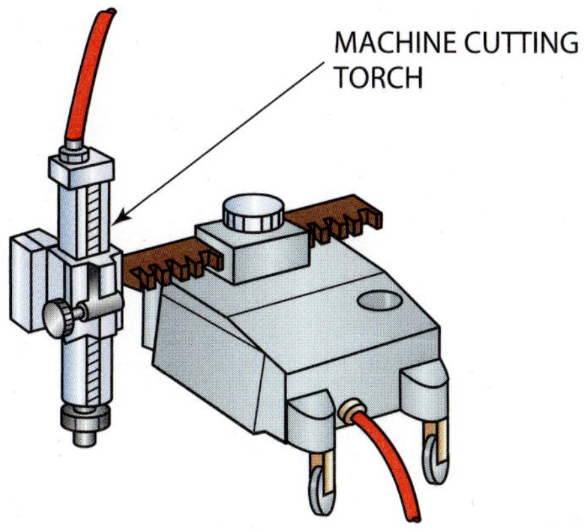

FIGURE 22-22 Machine cutting tool.
ESAB Welding & Cutting Products

FIGURE 22-23 Portable pattern cutter can cut shapes, circles, and straight lines. ESAB Welding & Cutting Products

Water Tables

Machine cutting lends itself to the use of water cutting tables, although they can be used with most hand torches. The water table is used to reduce the noise level, control the plasma light, trap the sparks, eliminate most of the fume hazard, and reduce distortion.

Water tables either support the metal just above the surface of the water or they submerge the metal about 3 in. below the water's surface. Both types of water tables must have some method of removing the cut parts, scrap, and slag that build up in the bottom. Often the surface-type tables will have the PAC torch connected to a water shroud nozzle, Figure 22-24. By using a water shroud nozzle, the surface table offers the same advantages to the PAC process as the submerged table offers. In most cases, the manufacturers of this type of equipment have made provisions for a special dye to be added to the water. This dye helps control the harmful light produced by the PAC. Check with the equipment's manufacturer for limitations and application of the use of dyes.

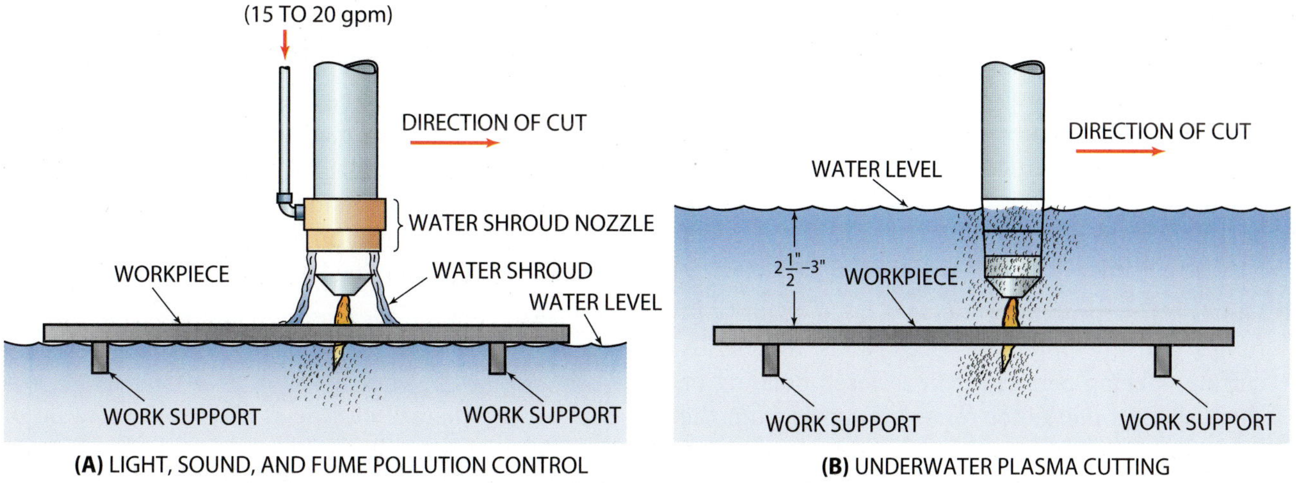

FIGURE 22-24 A water table can be used either with (A) a water shroud or (B) underwater torches.
ESAB Welding & Cutting Products

Safety

PAC has many of the same safety concerns as do most other electric welding or cutting processes. Some special concerns are specific to this process.

- Electrical shock—Because the open circuit voltage is much higher for this process than for any other, extra caution must be taken. The chance that a fatal shock could be received from this equipment is much higher than for any other welding equipment.
- Moisture—Often water is used with PAC torches to cool the torch or improve the cutting characteristic, or as part of a water table. Any time water is used, it is very important that there are no leaks or splashes. The chance of electrical shock is greatly increased if there is moisture on the floor, cables, or equipment.
- Noise—Because the plasma stream is passing through the nozzle orifice at a high speed, a loud sound is produced. The sound level increases as the power level increases. Even with low-power equipment the decibel (dB) level is above safety ranges. Some type of ear protection is required to prevent damage to the operator and other people in the area of the PAC equipment when it is in operation. High levels of sound can have a cumulative effect on one's hearing. Over time, one's ability to hear will decrease unless proper precautions are taken. See the owner's manual for the recommendations for the equipment in use.
- Light—The PAC process produces light radiation in all three spectrums. This large quantity of visible light, if the eyes are unprotected, will cause night blindness. The most dangerous of the lights is ultraviolet. As in other arc processes, this light can cause burns to the skin and eyes. The third light, infrared, can be felt as heat, and it is not as much of a hazard. Some type of eye protection must be worn when any PAC is in progress. Table 22-4 lists the minimum lens shade numbers for various power-level machines. It is always a good idea to use the darkest possible shade lens that will still let you see what you are cutting.
- Fumes—This process produces a large quantity of fumes that are potentially hazardous. When PA cutting is being performed in a closed building or shop, specific means for removing fumes from the work space should be in place. A downdraft table is ideal for manual work, but some special pickups may be required for larger applications. The use of a water table and/or a water shroud nozzle greatly helps to control fumes. Often the fumes cannot be exhausted into the open air without first being filtered or treated to remove dangerous levels of contaminants. Before installing an exhaust system, you must first check with local, state, and federal officials to see if specific safeguards are required.
- Gases—Some of the plasma gas mixtures include hydrogen; because this is a flammable gas, extra care must be taken to ensure that the system is leakproof.
- Sparks—As with any process that produces sparks, the danger of an accidental fire is always present. This is a larger concern with PAC because the sparks are often thrown some distance from the work area and the operator's vision is restricted by a welding helmet. If there is any possibility that sparks will be thrown out of the immediate work area, a fire watch must be present. A fire watch is a person whose sole job is to watch for the possible starting of a fire. This person must know how to sound the alarm and have appropriate firefighting equipment handy. Never cut in the presence of combustible materials.
- Operator checkout—Never operate any PAC equipment until you have read the manufacturer, owner, and operator's manual for the specific equipment to be used. It is a good idea to have someone who is familiar with the equipment go through the operation with you after you have read the manual.

Amperage Range	Minimum Shade # Clearly Visible Arc*	Minimum Shade # Hidden Arc**
Below 20	8	4
20 to 40	8	5
40 to 60	8	6
60 to 80	8	8
80 to 300	8	8
300 to 400	9	9
400 plus	10	10

* These values apply where the actual arc is clearly seen.
** These values apply where the arc is hidden by the workpiece or torch nozzle.

Table 22-4 Recommended Shade Densities for Filter Lenses

Manual Cutting

Manual plasma arc cutting is the most versatile of the PAC processes. It can be used in all positions, on almost any surface, and on most metals. This process is limited to low-power plasma machines; however, even these machines can cut up to 1 1/2-in. thick metals. The limitation to low power, 100 amperes or less, is primarily for safety reasons. The higher-powered machines have extremely dangerous open circuit voltages that can kill a person if accidentally touched.

The setup of most plasma equipment is similar, but do not ever attempt to set up a system without the manufacturer's manual for the specific equipment.

CAUTION

Consult the owner's manual. Be sure all of the connections are tight and that there are no gaps in the insulation on any of the cables. Check the water and gas lines for leaks. Visually inspect the complete system for possible problems.

CAUTION

Before you touch the nozzle tip, be sure that the main power supply is off. The open circuit voltage on even low-powered plasma machines is high enough to kill a person. Replace all parts to the torch before the power is restored to the machine.

Setup

Wearing all of the required personal protective equipment (PPE) and following all of the manufacturer's safety rules, most equipment can be set up using the following steps:

- Make any adjustments and changes to the electrode tip, nozzle tip, nozzle, or other torch component before the machine power is turned on because you can be shocked if the gun trigger is accidentally activated while you are servicing these parts.

CAUTION

The open circuit voltage on a plasma machine can be high enough to cause severe electrical shock or death.

- Make sure that the work clamp is connected to a clean, unpainted spot on the metal you will be cutting.
- Check to see that there is nothing behind the cut that will prevent the sparks from falling free of the cut.
- Check to see that there is nothing that will be damaged or set on fire by the sparks.
- Set the cutting amperage to maximum.
- Make a practice cut to see that the material can be cut cleanly.
- Reduce the amperage, and make another practice cut. Repeat this process until you have the amperage set as low as possible while still making a clean cut.

NOTE: Setting the amperage to the lowest possible level will extend the life of the torch parts.

Straight Cuts

Straight cuts are the most common type of cuts made with PAC torches. You can hold the torch close to the head because it does not get as hot as an oxyacetylene torch. This will help you keep the cut smoother. One common problem with making long, straight cuts is a tendency to make the cut with a slight arc when the torch is held at a 90° angle to the cut, Figure 22-25. If you slide your hand along the plate surface, you can eliminate some of this arcing. Another technique to help you keep your line straight is to use a guide such as an angle iron or straightedge, Figure 22-26. Because there are so few sparks when cutting thin metal, you may even be able to use a square as a guide without damaging it with sparks or heat. If you use a straightedge as a guide, make sure that your cut is on the line as it is hard to see the line as it is being cut because of the size of the nozzle on many plasma torches.

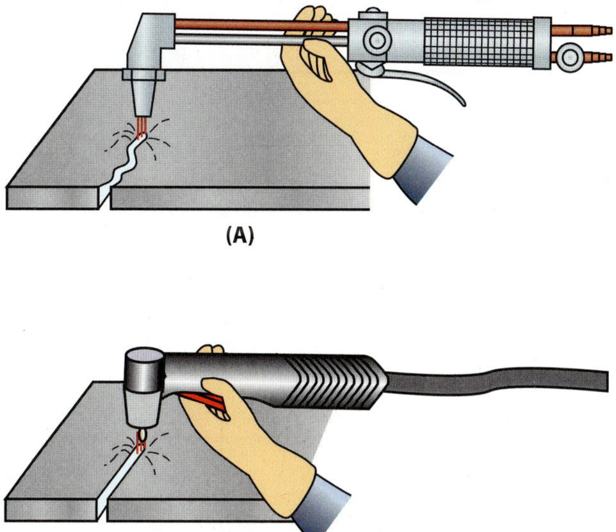

(A)

(B)

FIGURE 22-25 It is easier to make straight, smooth cuts if you can brace the torch closer to the tip, as in cut (B). American Welding Society

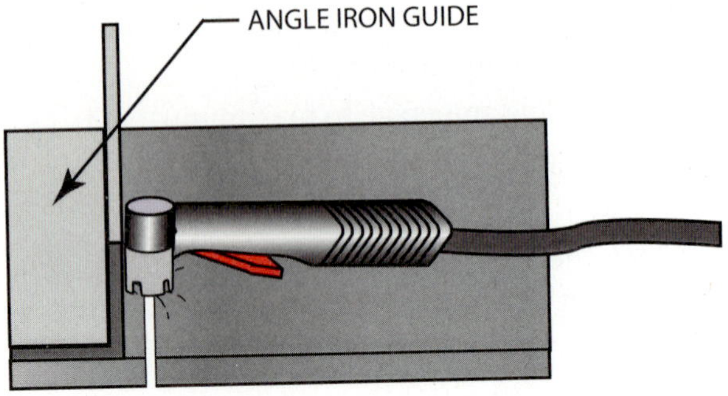

FIGURE 22-26 Use an angle iron as a guide to make a straight cut. © Cengage Learning 2012

PROJECT 22-1

Flat, Straight Cuts in Thin Sheet Metal to Make a Candlestick

Skill to be learned: The ability to make a straight, clean cut on thin-gauge metal with a handheld plasma cutting torch.

Project Description

The material cut will be used to make a 6-in. tall candlestick, Figure 22-27.

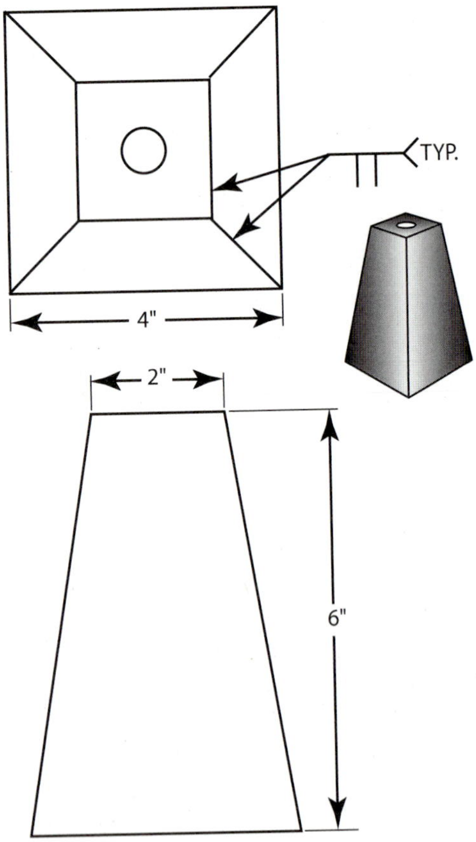

FIGURE 22-27 Project 22-1. © Cengage Learning 2012

Project Materials and Tools

The following items are needed to lay out and cut out the candlestick.

Materials

- 14-in. long 6-in. wide 16-gauge thick mild steel sheet metal
- 2-in. × 2-in. square 16-gauge thick mild steel sheet metal
- 4-in. × 4-in. square 16-gauge thick mild steel sheet metal
- Source of plasma cutting gas or compressed air

Tools

- Plasma cutting torch
- PPE
- 12-in. rule
- Soapstone, pencil and/or marker
- Square or try square
- Wire brush
- Pliers

Layout

Start the layout by measuring 4 in. across the base and make a mark at the 4-in. and 2-in. measurements. Use the square and extend the 2-in. center point mark across the 6-in. width of the sheet metal to locate the center of the panel. On the opposite side use the 12-in. rule and measure 1 in. on each side of the center line you just drew. Using the 12-in. rule, draw a line from one side of the 4-in. line to the corner of the 2-in. mark. Repeat the process on the other side.

If the parts are going to be sheared, the next layout will be made right next to the first. However, if the parts are going to be sawed, you must leave a 1/16-in. wide space between the parts as shown in Figure 22-28. Measure the 1/16 in. perpendicular to the angled side of the first panel you drew, then repeat the layout process described previously until all four panels have been laid out.

Next, use the 12-in. rule and square to lay out the 2-in. square top. You can use the square to draw the two parallel sides and mark the 2-in. width so that the straightedge can be used to draw this line between the marks. Or you can use a try square as illustrated in Figure 22-29 to draw the top line parallel to the side of the sheet metal.

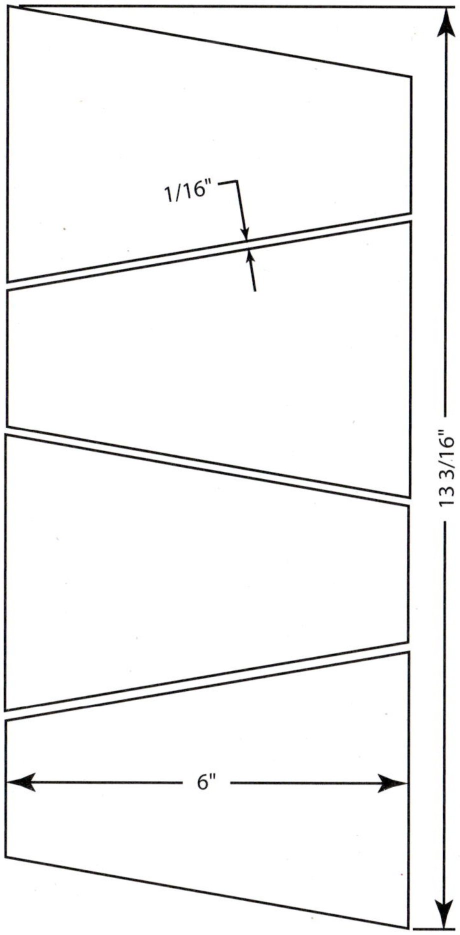

FIGURE 22-28 Start the plasma right next to the edge of the plate, and slowly bring it to the edge of the plate to start the cut. © Cengage Learning 2012

Cutting Out

Use a properly set up and adjusted PA cutting system, put on your personal protection equipment (PPE), and follow all shop and manufacturer's safety rules for cutting. Using one or more pieces of mild steel, stainless steel, and/or aluminum, you will cut out the candlestick parts.

- Before you start the cut, make several practice moves along the line to be cut to make sure that you can freely move along the line.
- Remember that the kerf on the right side will be the squarest, so when possible, make your cut so that the angled side of the cut is on the scrap side.
- Hold the torch tip close to the edge of the plate at a 90° angle to the plate unless a beveled cut is being made.
- Lower your hood.
- Pull the gun trigger, and wait for the pilot arc to start.

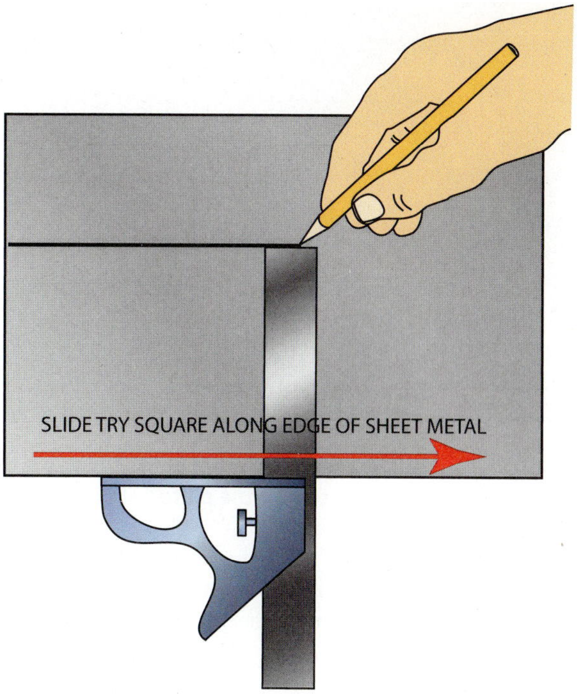

FIGURE 22-29 Use a combination square to mark a straight line. Larry Jeffus

NOTE: Do not let the torch nozzle touch the plate as the cut is started or during the cut because this can cause arcing between the nozzle and the work, resulting in damage to the nozzle unless the torch is equipped with a drag-type nozzle or cup.

- Keep the nozzle 1/8 in. to 1/4 in. above the plate surface unless a drag-type nozzle is being used.
- Bring the torch forward slowly until the cutting plasma stream is established on the edge of the plate, Figure 22-30.

FIGURE 22-30 When the plasma cut begins, move the torch at a constant speed down the plate. Larry Jeffus

FIGURE 22-31 The spark stream should be blowing the metal away from the bottom of the plate. Larry Jeffus

- Once the cut starts, move the torch along the cut line at a consistent speed, watching the spark stream to see that you are traveling at the correct speed, Figure 22-31.
- When your cutting speed is correct, the stream of sparks will be forceful and have a 15° to 20° forward angle.
- Release the gun trigger when the cut is completed; some torches have a postflow of air through the torch to cool it.
- When the cut is complete, check the bottom edge of the cut for dross. If the cutting speed was correct, the bottom edge of the cut will be free of any significant dross. If the cutting speed is too slow, the dross on the bottom edge of the cut can be chipped off. If the cut is too fast, the dross on the bottom edge of the cut will be hard to remove and may even need to be ground off.

Make any needed adjustments in the amperage setting and travel speed, and repeat the cut until a straight, dross-free cut can be made. Repeat the cuts until you can consistently make smooth dross-free cuts that are within ±3/32 in. of straight.

Fabrication and Welding

The fabrication procedures for this project are covered in Project 18-1 in Chapter 18. The welding procedures are also covered in that section; if another welding process is to be used, refer to the chapter that discusses that process.

Paperwork

Complete a copy of the time sheet in Appendix I and bill of materials in Appendix III, or as provided by your instructor.◆

PROJECT 22-2

Flat, Straight Cuts in a Thick Plate to Make a Sundial

Skill to be learned: The ability to make a beveled, clean cut on a metal plate with a handheld plasma cutting torch.

Project Description

The material cut will be used to make a sundial, Figure 22-32.

Project Materials and Tools

The following items are needed to lay out and cut out the sundial.

Materials

- 10-in. × 10-in. piece of 1/4-in. thick plate
- 8-in. × 8-in. piece of 1/4-in. thick plate
- 26-in. × 1-in. piece of 1/4-in. bar stock
- Source of plasma cutting gas or compressed air

Tools

- Plasma cutting torch
- PPE
- 12-in. rule
- Chipping hammer
- Soapstone, pencil and/or marker
- Square or try square
- Wire brush
- Pliers

Layout

The sundial face will be laid out according to Figure 22-33. The layout provides for approximately

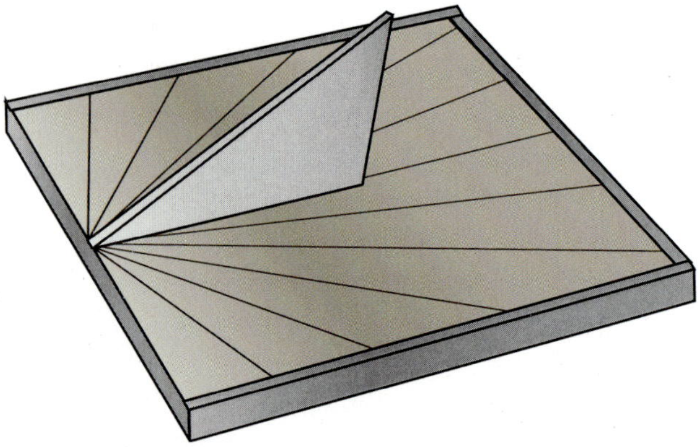

FIGURE 22-32 Project 22-2. © Cengage Learning 2012

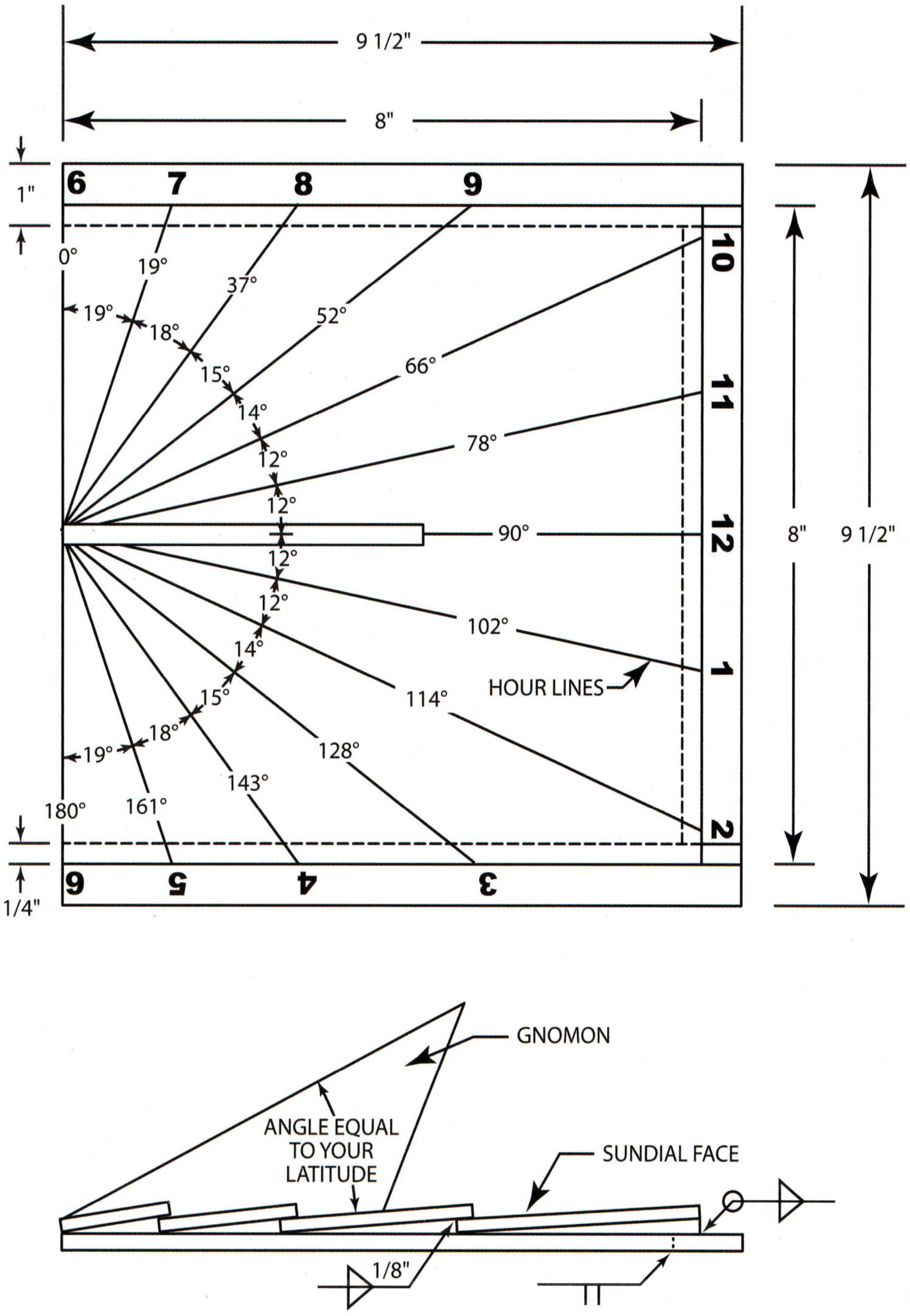

FIGURE 22-33 **Sundial layout.** © Cengage Learning 2012

a 1/2-in. additional area on one side of each of the sundial panels for a tab. This tab will provide the overlap area for the assembled sundial face.

Using a protractor and a straightedge, lay out the radial lines for each of the sundial's panels. The angle of each of the lines coincides with the hour they represent. Because of the sun's path, these lines are not uniformly spaced.

When laying out the sundial, do not change the protractor, but mark off each of the radials as shown in Figure 22-33. Rotating the protractor between measurements can result in an overall error.

Draw the sundial face on a 10-in. square plate as shown in Figure 22-34. Once the parts are cut out and fitted together, the finished sundial face will be 8 in. square. Check the accuracy of your angles by

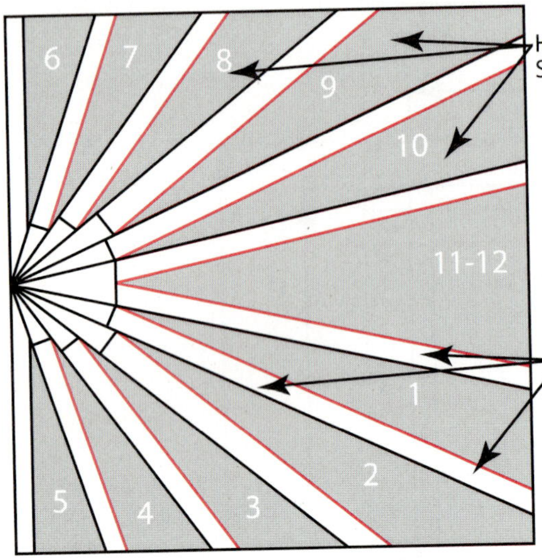

FIGURE 22-34 Sundial detailed drawing.
© Cengage Learning 2012

comparing the measurement between the ends of the lines for the 6 and 7 o'clock with the measurement between the 6 and 5 o'clock. These distances should be the same, as should the distances between the 7 and 8 o'clock and the 4 and 5 o'clock, and so forth.

Cutting Out

Use a properly set up and adjusted PA cutting system, put on your PPE, and follow all shop and manufacturer's safety rules for cutting. Using one or more pieces of mild steel, stainless steel, and/or aluminum, you will cut out the sundial parts.

- Follow all of the same procedures as outlined in Project 22-1.
- At the end of the cut, slow down your travel speed, and angle the torch slightly forward to allow the cut to be completed through the bottom edge of the plate.

Make any needed adjustments in the amperage setting and travel speed, and repeat the cut until a straight, dross-free cut can be made. Repeat the cut using both thicknesses and all three types of metal until you can make consistently smooth dross-free cuts that are within ±3/32 in. of straight and ±5° of being square. Turn off the PAC equipment, and clean up your area when you are finished cutting.

Fabrication and Welding

The fabrication procedures for this project are covered in Project 14-2 in Chapter 14. The welding procedures are also covered in that section; if another welding process is to be used, refer to the chapter that discusses that process.

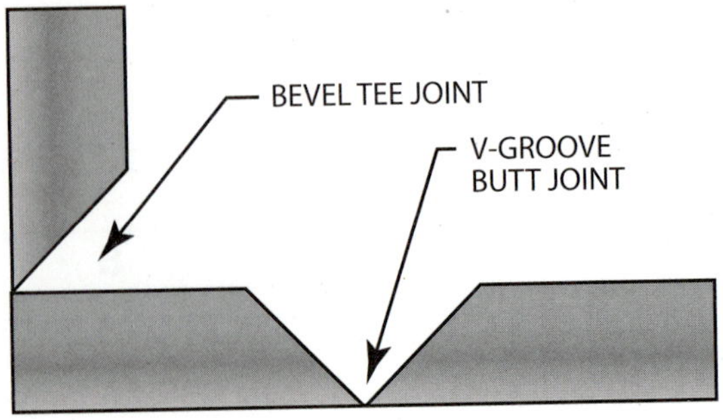

FIGURE 22-35 Beveled plate. © Cengage Learning 2012

Paperwork

Complete a copy of the time sheet in Appendix I and bill of materials in Appendix III, or as provided by your instructor.♦

Beveling of a Plate

The edge of a plate can be beveled so that a full thickness weld can be made, Figure 22-35. PAC beveling of a plate often leaves a hard dross strip along the edge. Grinding may be required to remove the dross.

PROJECT 22-3

Beveling of a Plate to Make a Wedge

Skill to be learned: The ability to make a straight, clean cut on a metal plate with a handheld plasma cutting torch.

Project Description

The material cut will be used to make a wedge, Figure 22-36.

Project Materials and Tools

The following items are needed to lay out and cut out the wedge.

Materials

- 5 1/2-in. × 7-in. piece of 1/4-in. thick plate or 4 1/2-in. × 7-in. piece of 3/8-in. thick plate
- Source of plasma cutting gas or compressed air

Tools

- Plasma cutting torch
- PPE
- 12-in. rule

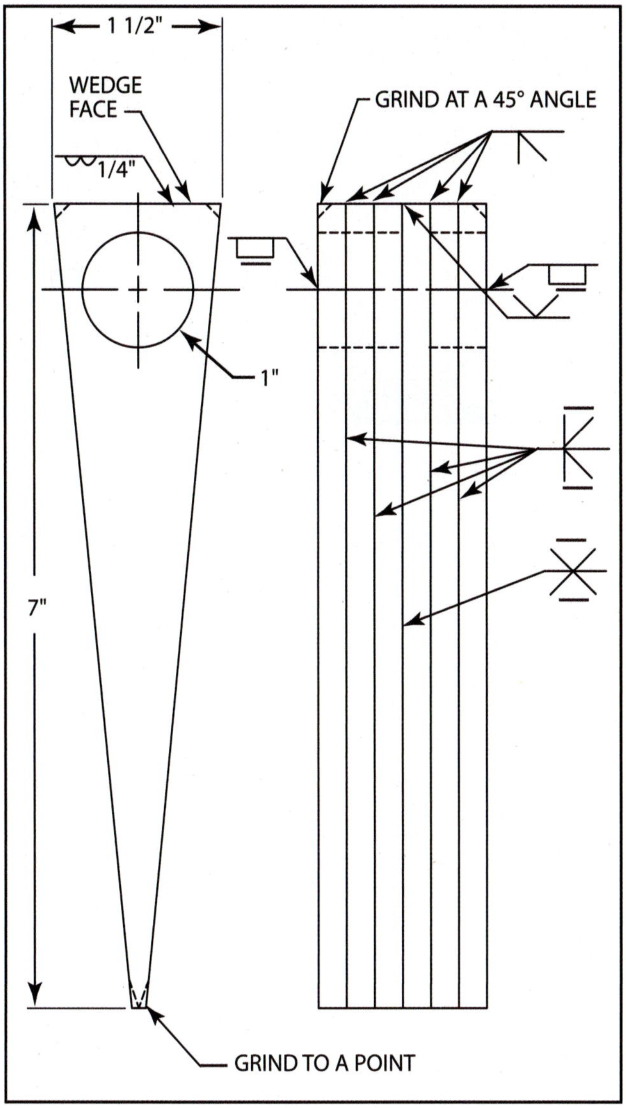

FIGURE 22-36 Project 22-3. © Cengage Learning 2012

- Chipping hammer
- Soapstone, pencil and/or marker
- Square or try square
- Wire brush
- Pliers

Layout

The total thickness of the wedge will be 1 1/2 in. If it is being made out of 1/4-in. thick plate, you will need six blanks; but it if is being made out of 3/8-in. thick plate, only five blanks will be needed. In this project you are going to first lay out one of the blanks, and after cutting it out, you will use it as your pattern to cut out the remaining blanks. By making a pattern first, it will be easier for you to make each of the other blanks exactly the same.

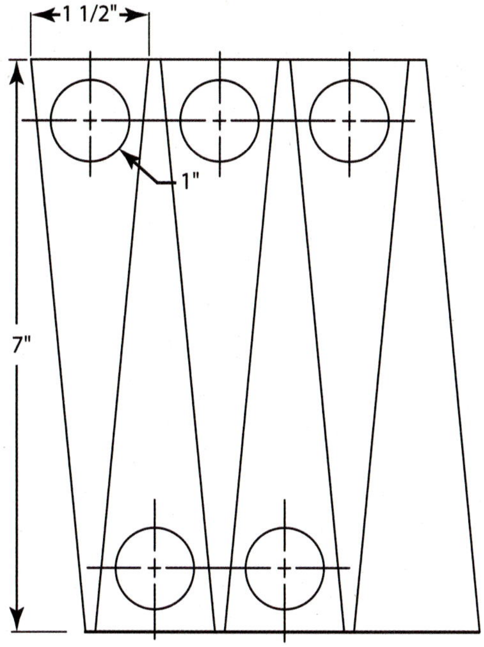

FIGURE 22-37 Wedge layout.
© Cengage Learning 2012

Using the square, a 12-in. rule, and soapstone, lay out the pattern for the first blank on the plate according to Figure 22-37.

The wedge layout does not come to a sharp point because during welding, the very thin edge would easily melt away. That is why there is a 1/8-in. wide flat surface at the pointed end. This will be ground to a tapered point once the fabrication has been welded.

Cutting Out

Using a properly set up and adjusted PA cutting system, putting on your PPE, following all shop and manufacturer's safety rules for cutting, and using one or more pieces of mild steel, you will cut a 45° beveled angle on the plates to make a wedge.

- Follow the setup and starting of the plasma procedures as outlined in Project 22-1 but hold the torch at a 45° angle.

NOTE: A piece of angle iron can be used as a guide to keep the torch at a 45° angle and to help make the cut straight, Figure 22-38.

- Lower your hood, and pull the trigger to start the pilot arc and the cutting plasma stream.
- Move the torch at a constant speed until you reach the end of the cut. Slow down so that the cut can be completed all the way to the end of the bevel, Figure 22-39.

FIGURE 22-38 Use an angle iron as a guide to make a beveled cut. © Cengage Learning 2012

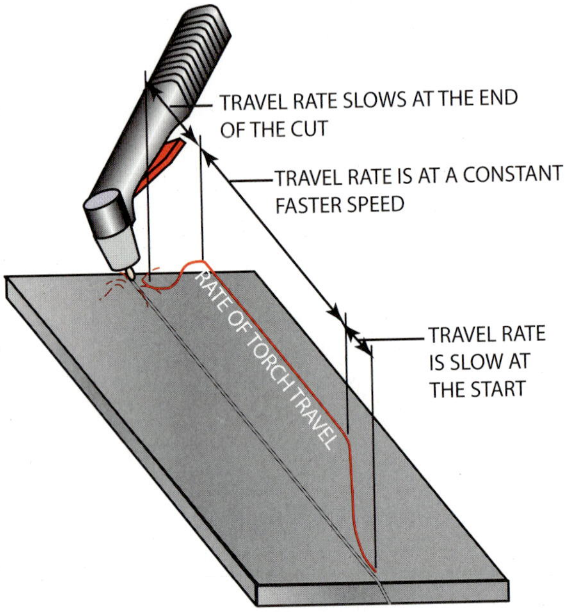

FIGURE 22-39 Travel speed changes to make a complete cut. © Cengage Learning 2012

Place the master pattern that you just cut out on the steel plate. Trace the pattern using a properly sharpened soapstone. Make sure you keep the point of the soapstone close to the bottom edge of the pattern, Figure 22-40.

Follow the preceding steps to cut out the remaining wedge parts.

Make any needed adjustments in the amperage setting and travel speed, and repeat the cut using both thicknesses and all three types of metal until you can make a consistently round cut that is within ±3/32 in. of being straight and ±5° of being at a 45° angle. Turn off the PAC equipment, and clean up your area when you are finished cutting.

FIGURE 22-40 Tracing a pattern. Larry Jeffus

Fabrication and Welding

The fabrication procedures for this project are covered in Project 14-1 in Chapter 14. The welding procedures are also covered in that section; if another welding process is to be used, refer to the chapter that discusses that process.

Paperwork

Complete a copy of the time sheet in Appendix I and bill of materials in Appendix III, or as provided by your instructor.

Cutting Holes

Sometimes it is necessary to start a cut in the center of a plate to cut a hole or to cut out a circular part. You must pierce the plate to make a hole through it to begin the cut. The major problem encountered when starting the cut is that all of the sparks and molten metal will not be able to be blown out the back side of the plate until the cut is made completely through the plate. This initial shower of hot metal can damage the torch cutting tip unless the sparks are directed away from the torch by angling it as the cut is started. On a thick plate a hole can be drilled to allow the sparks to exit the back side of the plate, which makes starting easier.♦

PROJECT 22-4

Cutting a Hole in a Plate to Make the Plug Welds on a Wedge

Skill to be learned: The ability to cut a round hole in a metal plate with a handheld plasma cutting torch.

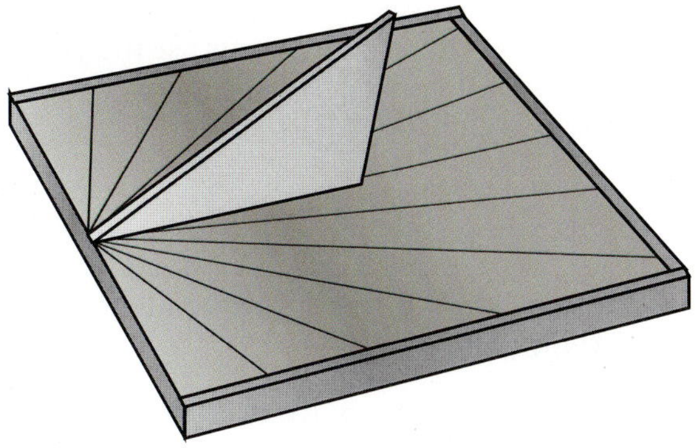

FIGURE 22-41 Project 22-4. © Cengage Learning 2012

Project Description

The material cut will be used to make a wedge, Figure 22-41.

Project Materials and Tools

The following items are needed to lay out and cut out the wedge.

Materials

- Blanks from cutting Project 22-3
- Source of plasma cutting gas or compressed air

Tools

- Plasma cutting torch
- PPE
- 12-in. rule
- Chipping hammer
- Soapstone, pencil and/or marker
- Compass or circle template
- Square or try square
- Wire brush
- Pliers

Layout

Using the 12-in. rule and soapstone, pencil and/or marker, locate the center of the 1-in. circle. Using the compass or circle template, draw the 1-in. circle on five of the six wedge blanks. Sometimes it is easier to keep the center point in place when drawing a circle if you put a small piece of masking tape at the center. The compass point can stick into the tape so it is less likely to slip as you are drawing the circle.

Cutting Holes in the Flat Position

Using a properly set up and adjusted PA cutting system, put on your PPE and follow all shop and manufacturer's safety rules for cutting. Using wedge project planks of mild steel, you will cut a 1-in. round hole in the plates to make a wedge.

- Follow the setup and starting of the plasma procedures as outlined in Project 22-1 but hold the torch at a 20° to 40° angle.
- Lower your hood, and pull the trigger to start the pilot arc and the cutting plasma stream.

NOTE: Some PA cutting machines do not have a high enough open circuit voltage to start the plasma stream when the torch is held at an angle. On these machines you may have to start the plasma arc with the torch nearly vertical to the plate surface and quickly tilt it to a 20° to 40° angle to minimize the possible damage to the nozzle tip.

- Hold the torch slightly higher than normal as you rotate the gun to a 90° angle. This will allow the sparks to be blown away and not back up onto the torch nozzle.
- Once the cut has been made all the way through the plate, lower the torch nozzle to the normal cutting height, and begin the cut.
- Move the torch in a counterclockwise direction in an outward spiral until the hole is the desired size, Figure 22-42.

Make any needed adjustments in the amperage setting and travel speed, and repeat the cut using both thicknesses and all three types of metal until you can

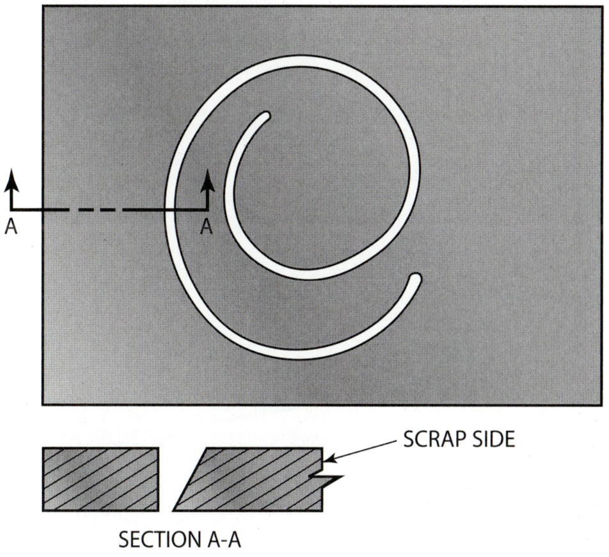

FIGURE 22-42 **When cutting a hole, spiral the torch outward in a counterclockwise direction.**
© Cengage Learning 2012

make a consistently round cut that is within ±3/32 in. of being round and ±5° of being square. Turn off the PAC equipment, and clean up your area when you are finished cutting.

Fabrication and Welding

The fabrication procedures for this project are covered in Project 14-1 in Chapter 14. The welding procedures are also covered in that section; if another welding process is to be used, refer to the chapter that discusses that process.

Paperwork

Complete a copy of the time sheet in Appendix I and bill of materials in Appendix III, or as provided by your instructor.

Irregular Shapes

Two of the advantages of using a plasma cutting torch are that it can easily cut in any direction, which makes cutting out irregular shapes easier, and the cut sheet or plate will have little distortion. These two cutting characteristics are why so many artists use the plasma torch to cut out things such as silhouettes or other detailed artistic sheet metal pieces.◆

PROJECT 22-5

Flat, Irregular Cuts in a Thick Plate to Make a Weather Vane

Skill to be learned: **The ability to make a beveled, clean cut on a metal plate with a handheld plasma cutting torch.**

Project Description

The material cut will be used to make a weather vane, Figure 22-43.

Project Materials and Tools

The following items are needed to lay out and cut out the weather vane.

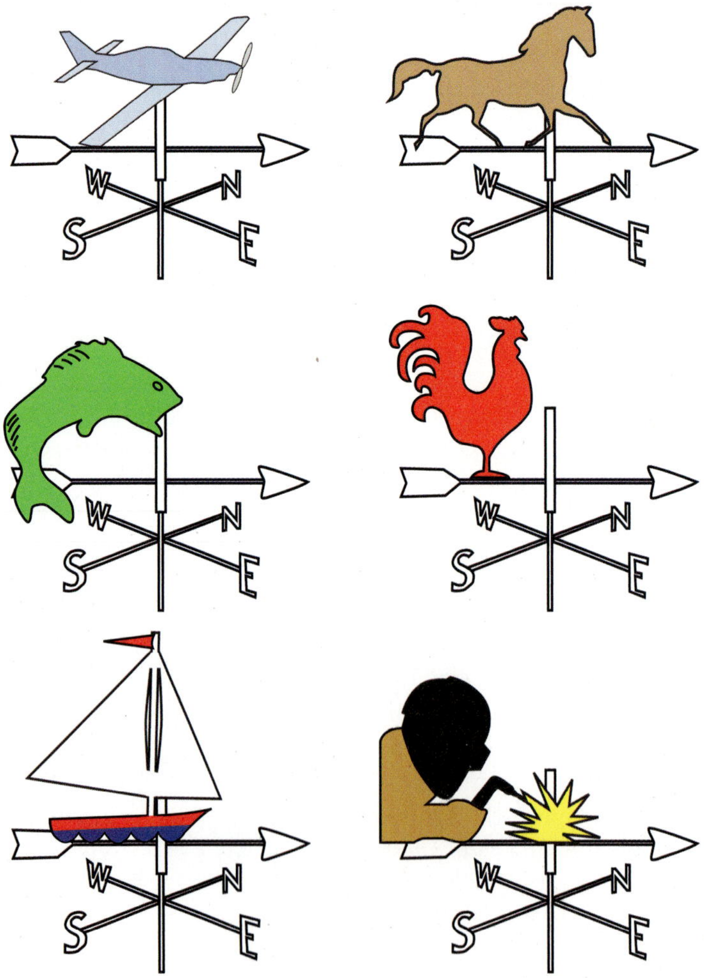

FIGURE 22-43 Project 22-5.

Materials

- 12-in. × 7-in., 16-in. × 10-in., or 12-in. × 10-in. piece of 16-gauge to 1/8-in. thick metal
- 3-in. × 12-in. piece 16-gauge to 1/8-in. thick metal
- Source of plasma cutting gas or compressed air

Tools

- Plasma cutting torch
- PPE
- 12-in. rule
- Chipping hammer
- Soapstone, pencil and/or marker
- Scissors
- Graph paper
- Masking tape
- Center punch
- Hammer
- Square or try square
- Wire brush
- Pliers

Layout

Refer to Chapter 4's "Sketching" section and Practice 4-5 for more instructions on how to sketch an outline on graph paper. Use a pencil and graph paper to draw the outline of the weather vane silhouette, Figure 22-44. The curved line is drawn by locating a series of points along the graph line and then sketching the line to connect the dots.

Tape the pattern on the metal. Use a center punch to make a series of punch marks through the paper onto the metal by placing the punch on the pencil line and hitting it with the hammer, Figure 22-45. On the straight line sections, make the punch marks about 1/4 in. apart. But on the curve line sections, you will need to make them much closer together. When the pattern is removed, you can see the punch marks on the metal.

The *N*, *E*, *S*, and *W* letters can be laid out directly on the metal using the square and marker. The entire arrow can be cut from a 20-in. long piece of the same metal stock as the silhouette, or just the point and feathers can be cut out of that stock. If only the point and feathers are cut out, then they and the directional markers will be welded to 1/4-in. to 3/8-in. diameter 20-in. long rods.

NOTE: Balance the silhouette and arrow by hanging the assembly on a rod between the arrow and silhouette. The pivoting point for the weather vane must be at this balance point for the weather vane to move freely with the changing wind direction. The pivoting point must also be above the vertical point, so a tube will be used to both raise it to the required height and to serve as a guide.

Close one end of a 6-in. to 10-in. long piece of 3/8-in. guide tubing by welding or brazing a plug. Weld this tube vertically to the silhouette and arrow assembly at the balance point. Slide the assembly over a 1/4-in. pointed rod and check for balance and freedom of movement, Figure 22-46. If necessary, the assembly can be balanced by welding a weight on the light end.

Cutting Out

Use a properly set up and adjusted PA cutting system, put on your PPE, and follow all shop and manufacturer's safety rules for cutting. Using one or more pieces of mild steel, stainless steel, and/or aluminum, you will cut out the weather vane parts.

- Follow all of the same procedures as outlined in Project 22-1.
- At the end of the cut, slow down your travel speed, and angle the torch slightly forward to allow the cut to be completed through the bottom edge of the plate.

Make any needed adjustments in the amperage setting and travel speed, and repeat the cut until a straight, dross-free cut can be made. Repeat the cut using both thicknesses and all three types of metal until you can make consistently smooth dross-free cuts that are within ±3/32 in. of straight and ±5° of being square. Turn off the PAC equipment, and clean up your area when you are finished cutting.

Fabrication and Welding

The welding procedures you might use to weld this weather vane are covered in other chapters.

Paperwork

Complete a copy of the time sheet in Appendix I and bill of materials in Appendix III, or as provided by your instructor.◆

Cutting Round Stock

Often it is necessary to PA cut a round piece of metal such as a pipe, shaft, rod, or bolt. Round pieces of metal can be a challenge to cut because the cut starts

FIGURE 22-44 Weather vane silhouettes.

out much like a gouged groove and transitions to something like piercing a hole. In addition, it is important to keep the plasma stream straight and in line with the line that is being cut. The plasma torch will cut in the direction it is pointed, so if it is not straight, the cut may have a beveled edge.

PROJECT 22-6

Cutting Round Pipe on a Birdhouse

Skill to be learned: The ability to cut round solid metal pipe and rods with a handheld plasma cutting torch.

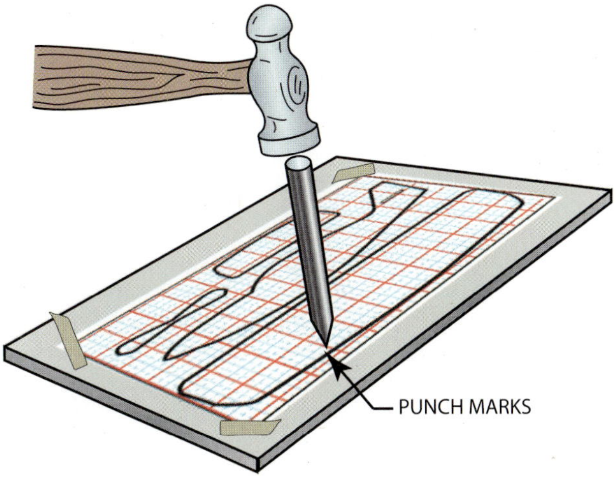

FIGURE 22-45 Use a center punch to transfer the drawing to the metal plate. © Cengage Learning 2012

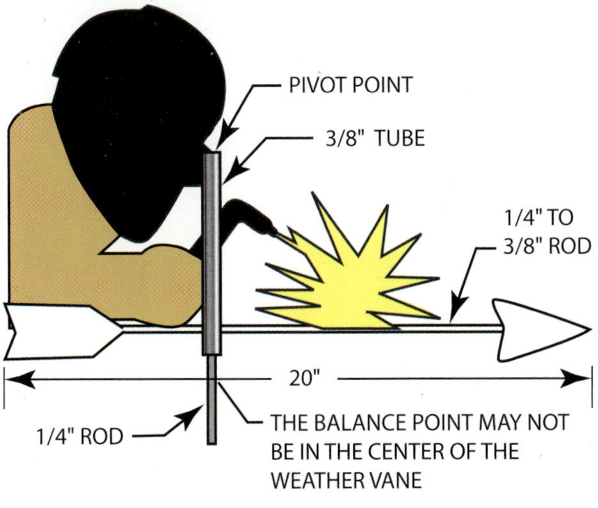

FIGURE 22-46 Assembling a weather vane. © Cengage Learning 2012

Project Description

The material cut will be used to make a birdhouse, Figure 22-47.

Project Materials and Tools

The following items are needed to lay out and cut out the birdhouse.

Materials

- 5 2-in. long pieces of 8-in. diameter steel pipe
- 1 10-in. × 10-in. piece of 3/16-in. to 1/4-in. thick mild steel plate
- 1 7 1/2-in. diameter piece of 3/16-in. to 1/4-in. thick mild steel plate

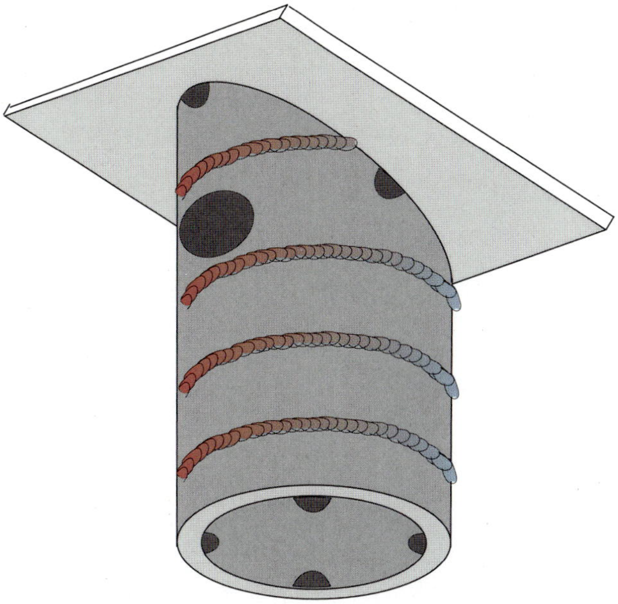

FIGURE 22-47 Project 22-6. © Cengage Learning 2012

- 1 1 1/2-in. wide 8-in. long piece of 3/16-in. to 1/4-in. thick mild steel plate
- 4 1 1/2-in. wide 1-in. long pieces of 3/16-in. to 1/4-in. thick mild steel plate
- Source of plasma cutting gas or compressed air

Tools

- Plasma cutting torch
- PPE
- 12-in. rule
- Chipping hammer
- Soapstone, pencil and/or marker
- Compass or circle template
- Square or try square
- Wire brush
- Pliers

Layout

Using a square, 12-in. rule, and soapstone, lay out the plate as shown in the project drawing, Figure 22-48. Lay out the 10-in. square piece, and use a hand or automated torch to cut out the piece. Remember to account for the cut's kerf so the part is not under- or oversized.

Mark the pipe to length; if it is going to be cut using a hand thermal torch, a line must be drawn all the way around the pipe. In this case, you do not have to allow for the kerf because the pipe will be beveled and the joint will have a 1/8-in. root gap. The root gap will be very close in size to the kerf.

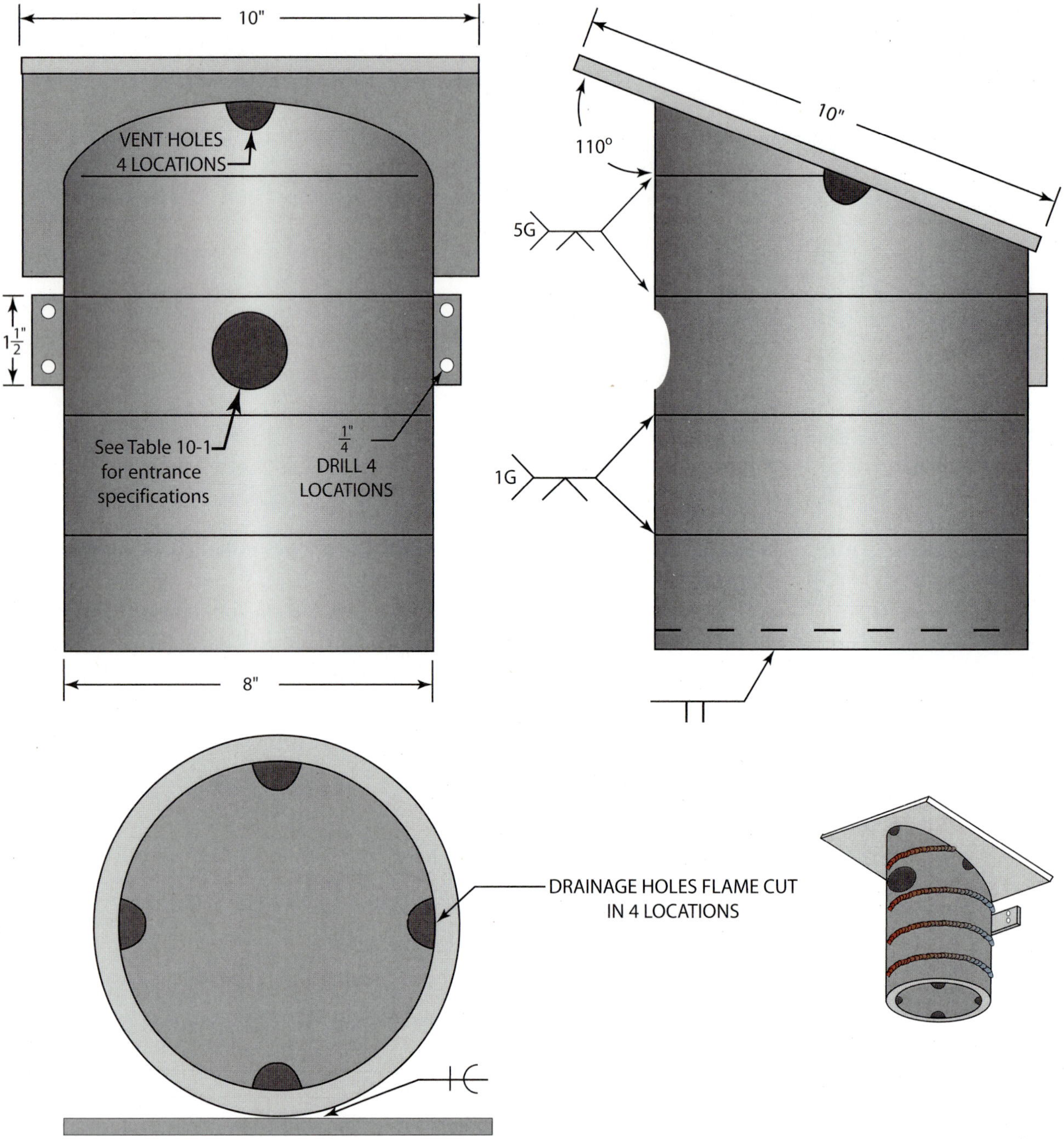

FIGURE 22-48 **Birdhouse detailed drawing.** © Cengage Learning 2012

Measure the inside diameter of the pipe and use the compass to lay out the circle for the bottom plug.

Cutting Out

Use a properly set up and adjusted PA cutting system, put on your PPE, and follow all shop and manufacturer's safety rules for cutting. Using one or more pieces of 8-in. diameter pipe, you will cut off the pipe sections as shown in Figure 22-48.

- Follow the setup and starting of the plasma procedures as outlined in Project 22-1 but hold the torch so it is pointed parallel to the round piece and in line with the line to be cut, Figure 22-49.
- Lower your hood, and pull the trigger to start the pilot arc and the cutting plasma stream.
- Move the torch onto the side of the round surface so that a groove is started.

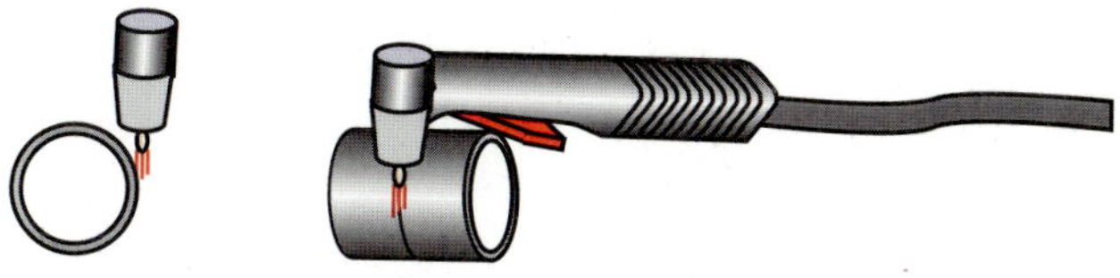

FIGURE 22-49 Cutting pipe with a plasma torch.
© Cengage Learning 2012

- If the round stock is 1/2 in. or smaller, keep the torch pointed in the same direction, and move the torch across the piece.
- If the round stock is thicker than 1/2 in. or thicker than the PAC torch can cut through easily, then you must move the torch back and forth to make the kerf wider to allow the cut to go all the way through the round stock.

Make any needed adjustments in the amperage setting and your technique, and repeat the cut using all the diameter pieces you have until you can make a cut that is within 5° of being square. Turn off the PAC equipment, and clean up your area when you are finished cutting.

Fabrication and Welding

The fabrication procedures for this project are covered in Project 10-3 in Chapter 10. The welding procedures are also covered in that section; if another welding process is to be used, refer to the chapter that discusses that process.

Paperwork

Complete a copy of the time sheet in Appendix I and bill of materials in Appendix III, or as provided by your instructor.◆

Plasma Arc Gouging

Plasma arc gouging is a recent introduction to the PAC processes. The process is similar to that of air carbon arc gouging in that a U-groove can be cut into the metal's surface. The removal of metal along a joint before the metal is welded or the removal of a defect for repairing can easily be done using this variation of PAC, Figure 22-50. An advantage of using PA cutting to remove a weld is that any slag trapped in the weld will not affect the PAC gouging process.

The torch is set up with a less-concentrated plasma stream. This will allow the washing away of the molten metal instead of thrusting it out to form a cut. The torch is held at approximately a 30° angle to the metal surface. Once the groove is started, it can be controlled by the rate of travel, torch angle, and side-to-side torch movement.

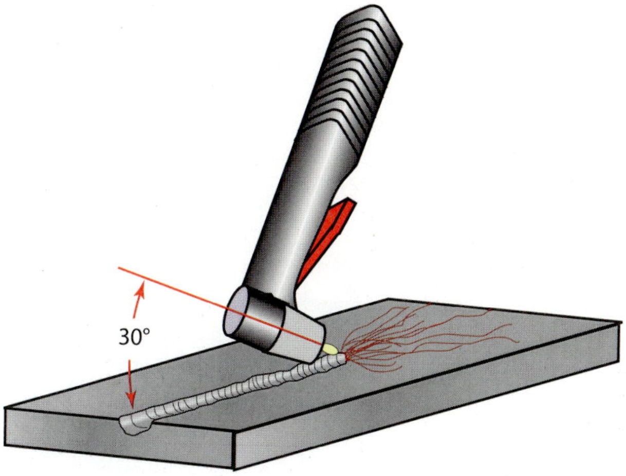

FIGURE 22-50 Plasma arc gouging a U-groove in a plate. © Cengage Learning 2012

Plasma arc gouging is effective on most metals. Stainless steel and aluminum are especially good metals to gouge because there is almost no cleanup. The groove is clean, bright, and ready to be welded. Plasma arc gouging is especially beneficial with these metals because there is no reasonable alternative available. The only other process that can leave the metal ready to weld is to have the groove machined, and machining is slow and expensive compared to plasma arc gouging.

It is important to try not to remove too much metal in one pass. The process will work better if small amounts are removed at a time. If a deeper groove is required, multiple gouging passes can be used.

PROJECT 22-7

U-Grooving of a Plate for a Cooking Press

Skill to be learned: The ability to control the plasma gouging process to make grooves that are uniform in width and depth.

Project Description

This 4-in. wide 8-in. long cooking weight is great for cooking bacon because with the weight on top, all of the bacon cooks evenly without leaving some spots undercooked. It is also handy when cooking burgers or other meats on your outdoor grill, Figure 22-51.

Project Materials and Tools

The following items are needed to cut out and groove the cooking press.

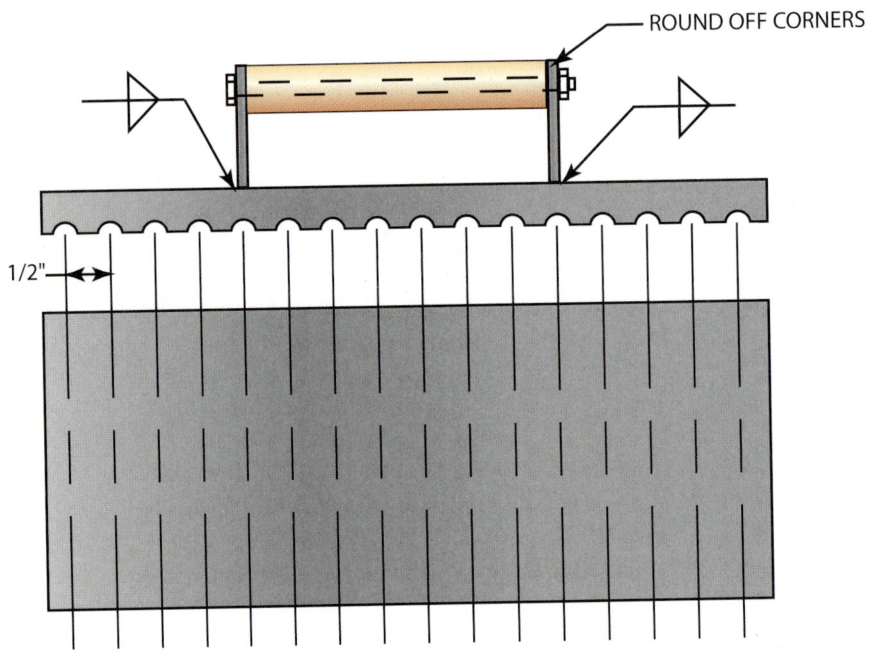

FIGURE 22-51 Project 22-7. © Cengage Learning 2012

Materials

- 4-in. wide 8-in. long piece of 1/4-in. to 3/8-in. thick mild steel plate
- 3/4-in. wide 2-in. long piece of 1/8-in. to 3/16-in. thick mild steel
- 4-in. long 3/4-in. to 1-in. diameter wooden dowel rod
- 4 1/2-in. long 1/4-in. diameter bolt nut and washer
- Source of plasma cutting gas or compressed air

Tools

- Plasma cutting torch
- PPE
- 12-in. rule
- Chipping hammer
- Soapstone, pencil and/or marker
- Square or try square
- Wire brush
- Pliers
- Drill and 1/4-in. drill bit

Layout

Using the square and the soapstone or marker, lay out the center lines for the grooves on the face of the cooking press. The grooves will be cut 1/2 in. on center all the way across the face.

NOTE: These grooves serve two main purposes. First, they give the grease and cooking juices a way to drain away from the food. Second, they reduce the area that the press contacts the food. This lets less heat transfer from the food to the press.

Gouging

Use a properly set up and adjusted PA cutting system, put on your PPE, and follow all shop and manufacturer's safety rules for cutting. Using a piece of mild steel or stainless steel, you will gouge grooves in the face of the cooking press.

- Follow the setup and starting of the plasma procedures as outlined in Project 22-1 but hold the torch at a 30° angle pointed in the direction you will be making the groove, Figure 22-50.
- Lower your hood, and pull the trigger to start the pilot arc and the cutting plasma stream.
- Move the torch along the plate in a straight line down the plate toward the other end.
- If the width of the U-groove is too wide, move faster; if it is too narrow, slow down.
- If the depth of the U-groove is too deep, decrease the torch angle; if it is too shallow, increase the torch angle.

Make any needed adjustments in the amperage setting and travel speed, and repeat gouging the U-groove using both thicknesses and all three types of metal until you can consistently make U-grooves within ±3/32 in. of being straight and uniform in depth and width. Turn off the PAC equipment, and clean up your area when you are finished cutting.

Fabrication and Welding

The welding procedures you might use to weld this weather vane are covered in other chapters. The easiest way to locate and weld the handle brackets onto the cooking press is to drill the hole through the wood and brackets at one time. That way, if the drill drifts off center slightly, the handle will still line up. The second thing to do is to use the bolt and nut to hold the assembled wood dowel rod and the brackets together. Place the assembly on the cooking press, and tack weld the brackets. This way the brackets will fit the wood dowel exactly. Remove the bolt and dowel rod before welding.

Paperwork

Complete a copy of the time sheet in Appendix I and bill of materials in Appendix III, or as provided by your instructor.◆

SUMMARY

A typical plasma cutting system power supply weighs about 30 lb (13 k). It will be powered with 120-volt and/or 220-volt AC electrical power. It needs an external air supply capable of supplying approximately 65 psi with a flow rate of 4.5 cubic feet per minute (cfm). In this small system you have a machine that can provide cutting for almost every project. Plasma cutting equipment is in a continuing state of development, which has provided significant breakthroughs in design. It is expected that this research and development will produce even smaller, more powerful, versatile, and portable systems.

It is hoped that research and development will make advancements in consumable supplies. Research and development (R & D) has already made a difference in the durability of the electrode, nozzle insulator, nozzle tip, and nozzle, although in your training program you have probably seen that the plasma cutting torch's consumable supplies are often destroyed very quickly. Before the recent R & D, the useful life of torch consumables, even if you were a skilled welder, was very short. However, as you have developed better cutting skills, these torch consumables have lasted longer.

The need to make cuts in shops and factories will always exist. Plasma cutting will probably never completely replace oxyacetylene cutting, but it certainly offers a great alternative.

REVIEW QUESTIONS

1. Define ***plasma***.
2. What is a welding plasma?
3. Name the common torch parts.
4. How does the use of a copper/tungsten tip lengthen the life of the tip and help to produce a better quality cut?
5. What is the purpose of the nozzle insulator?
6. What two aspects of the nozzle tip affect the plasma jet action and the current flow?
7. What are the three main purposes of the nozzle?
8. On water-cooled torches, why is the power cable sometimes placed inside the return water line?
9. How does the compressed air get from the plasma machine to the torch?
10. What is the purpose of the control wire?
11. Why is a part cut with plasma not nearly as hot afterward as it would be if it were cut with an oxyfuel process?
12. Describe distortion.
13. What would be a typical manual plasma arc cutting speed?
14. What are the most popular materials cut with the PAC process?
15. Define *dross*.

16. What is standoff distance?
17. What factors affect the width of the kerf?
18. List three effects on the cut that changing the PAC gas can cause.
19. What effect will too high a gas flow have on the cut?
20. What is the purpose of a water table?
21. List some PAC safety concerns.
22. Why is manual plasma arc cutting the most versatile of the PAC processes?
23. What safety precautions must be observed when setting up plasma equipment?

Chapter 23

Arc Cutting, Gouging, and Related Cutting Processes

OBJECTIVES

After completing this chapter, the student should be able to:

- Describe the different types of lasers used for cutting, drilling, and welding.
- Discuss the advantages of laser beam cutting and drilling.
- Explain how an oxygen lance is used.
- Describe how water jet cutting works.
- Demonstrate how to make an air arc cut.

KEY TERMS

abrasives
air carbon arc cutting
arc cutting electrodes
carbon electrode
exothermic gases
gas laser
gouging
graphite
laser beam cutting (LBC)
laser beam drilling (LBD)
laser beam welding (LBW)

INTRODUCTION

The number of specialized cutting processes being developed and improved increases every year. A number of specialized cutting processes have been perfected and are in use. The new cutting processes can be used to cut a wide variety of things, such as glass, plastic, printed circuit boards, cloth, and fiberglass insulation, and more things are being added almost daily. Because of the increased accuracy, speed, cost effectiveness, and reliability, material cutting and hole drilling using a welding-related process have become common in the workplace.

Only a few of the more common cutting processes are covered in this chapter. These are the ones that a welder might be required to either be able to perform or have a working knowledge of.

Lasers

Lasers have developed from the first bright red spot generated by the ruby rod into a multibillion-dollar industry. The use of lasers has become very common today. We see their use in our world every day. They help speed our checking out at stores when they are used to read the uniform product code (UPC) on our purchases, Figure 23-1.

The term *laser* is an acronym for light amplification stimulation emission of radiation. A laser is an amplified form of light that is a single-color wavelength that travels in parallel waves, Figure 23-2. This form of light is generated when certain types of materials are excited either by intense light or with an electric current. The atoms or molecules of the lasing material release their energy in the form of light. The laser light is produced as the atom or molecule vibrates between a high-energy state to a lower-energy state, Figure 23-3. The laser light is then bounced back and forth between one fully reflective and one partially reflective surface at the ends of the lasing material. As the light is reflected, it begins to form a synchronized waveform. The light that passes through the partially reflective mirror is the laser light, and it is ready to do its job, Figure 23-4. The frequency of this vibration determines the color of the laser's light beam. Most lasers used for cutting, drilling, and welding produce a laser light beam in the infrared range.

FIGURE 23-1 **Bar codes are read by lasers to input information into computers.** © Cengage Learning 2012

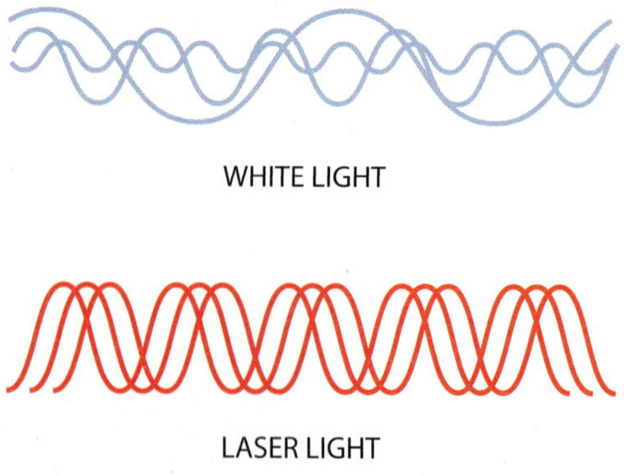

FIGURE 23-2 **White light is made up of different frequencies (colors) of light waves. Laser light is made up of single-frequency light waves traveling parallel to each other.** © Cengage Learning 2012

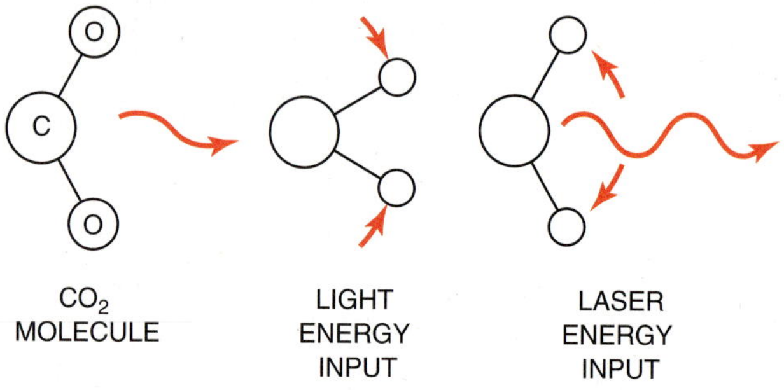

FIGURE 23-3 **Energy is stored in a carbon dioxide (CO_2) molecule until the molecule can not hold any more. The energy is released suddenly like when a balloon pops. This quick release results in a burst of light energy.** © Cengage Learning 2012

Manufacturers use the laser to do everything from burning information on products as small and hard as diamonds to guiding machines to grind, cut, punch, drill, and cut to accuracies within a few thousandths of an inch. Fabricators are using lasers to cut, trim, weld, heat treat, clad, vapor deposit coatings, anneal, and shock harden metals.

When the laser beam exits the laser generator, it can be treated as any other type of light. It is possible to focus, reflect, absorb, or defuse the light. By having the light waves traveling in such a uniform manner, they exhibit some specialized characteristics. The laser light, unlike ordinary light that spreads out, tends to remain in a very tight beam without spreading out, Figure 23-5. This characteristic allows the laser beam to stay focused as it travels without significantly being affected by the distance it travels.

Laser Types

The early lasers all used a synthetic ruby rod as the material that produced the laser light. Today, a large number of materials can be used to produce the laser. These materials include such common items as glass

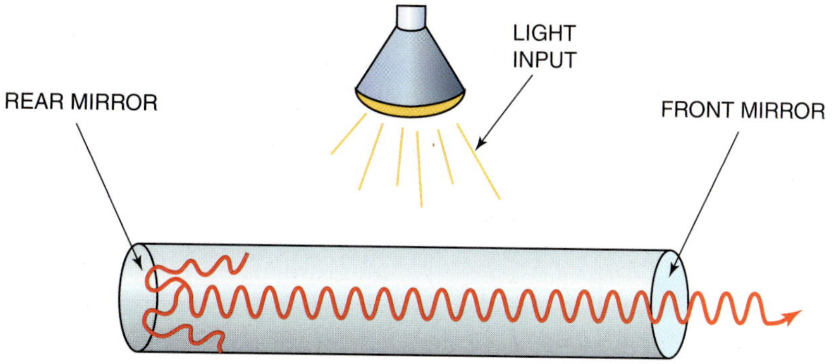

FIGURE 23-4 Light energy reflects back and forth between the end mirrors until it forms a parallel laser light beam. © Cengage Learning 2012

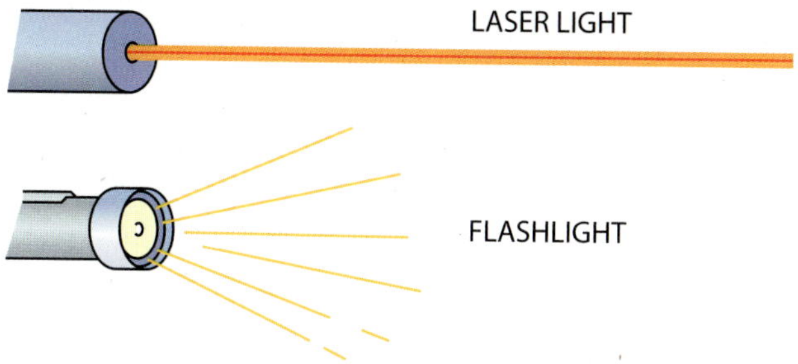

FIGURE 23-5 Laser light stays in a very tight column unlike most other lights. © Cengage Learning 2012

and such exotic items, such as neodymium-doped yttrium aluminum garnet (Nd:YAG), often referred to as a YAG laser, Figure 23-6.

Lasers can be divided into two major types: lasers using a solid material for the laser and lasers using a gas. Each of these two types is divided into two groups based on their method of operation: lasers that operate continuously and lasers that are pulsed.

Solid State Lasers

The first lasers were all solid state lasers. They used a synthetic ruby rod to produce the laser. Today, the most popular industrial solid laser is YAG. This is a synthetic crystal that when exposed to the intense light from the flash tubes can produce high quantities of laser energy. The high-powered solid lasers have a problem because the internal temperatures of the laser rod increase with operation time. These lasers are most often used with a low-power continuous or high-powered pulse. The solid state laser is capable of generating the highest-powered laser pulses.

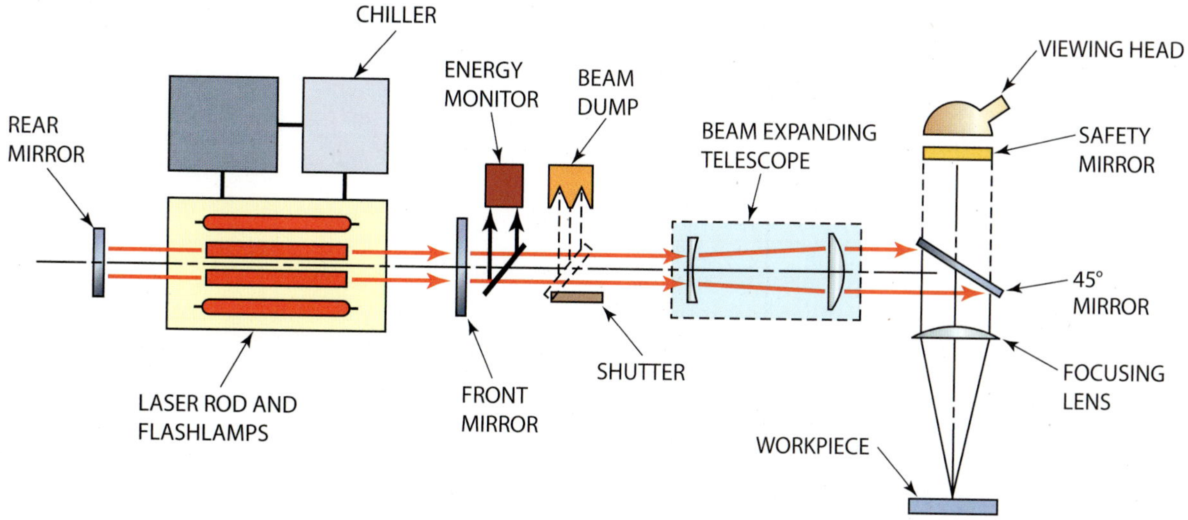

FIGURE 23-6 Schematic representation of the elements of a Nd:YAG laser. American Welding Society

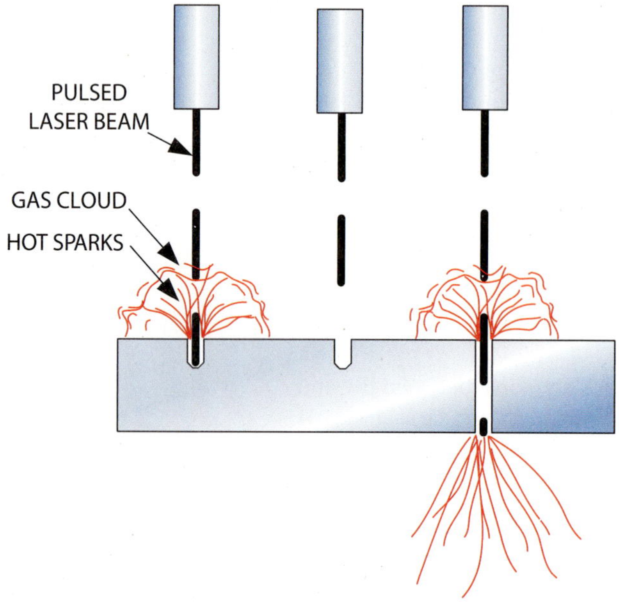

FIGURE 23-7 The gas cloud and hot sparks caused by the cutting laser beam dissipate between the pulses of the laser beam, which results in a better cut.
© Cengage Learning 2012

Gas Laser

The **gas laser** uses one or more gases for the laser. Popular gas-type lasers use nitrogen, helium, and carbon dioxide (CO_2) or mixtures of helium and nitrogen with CO_2. Gas lasers use either a gas-charged cylinder where the gas is static or a chamber that has a means of circulating the laser gas. The highest continuous power output-type lasers are the gas type with a blower to circulate the gas through a heat exchanger; they can be rapidly pulsed to improve their cutting capacity, Figure 23-7.

Applications

Lasers are versatile tools used to make products that could not be produced without the laser's light. A laser can be used to drill holes (LBD) through the hardest materials, such as synthetic diamonds, tungsten carbide cutting tools, quartz, glass, or ceramics. Lasers are used for welding (LBW) materials that are too thin or too hard to be welded with other heat sources. They can be used to cut (LBC) materials that must be cut without overheating delicate parts that might be located just a few thousandths of an inch away.

The benefits and capacities of the laser have made it one of the most rapidly growing areas in manufacturing. This is a useful tool that will continue to be developed and improved for years to come.

Because the laser light beams can be focused into a very compact area, the light energy can be converted to heat energy as it strikes the surface of a material. The highly concentrated energy from a high-powered laser can cause the instantaneous melting or vaporization of the material being struck by the laser beam. Most laser welding and cutting operations use a gas laser, and most drilling operations use a solid state laser.

The ability of a material's surface to absorb or reflect the laser light affects the laser's operating efficiency. All materials will reflect some of the laser's light, some more than others. The absorption rate increases for any material once the laser beam begins to heat the surface. Once the threshold of the surface temperature is reached, the process will continue at a much higher level of efficiency.

Laser Beam Cutting

Laser beam cutting (LBC) uses a high-pressure jet of gas to blow the molten or vaporized material out the bottom of the cut. Lasers are primarily used to cut very thin materials, but some powerful machines can cut an inch or more of steel.

The high-pressure jet can either be a nonreactive gas or an exothermic gas. Table 23-1 lists the various gases and the materials that they are used to cut. Nonreactive gases do not add any heat to the cutting process; they simply remove the material by blowing it out of the kerf. **Exothermic gases** react with the material being cut like an oxyfuel cutting torch. The additional heat produced as the exothermic cutting

Assist Gases	Material
Air	Aluminum Plastic Wood Composites Alumina Glass Quartz
Oxygen	Carbon steel Stainless steel Copper
Nitrogen	Stainless steel Aluminum Nickel alloys
Argon	Titanium

Table 23-1 Various Cutting Assist Gases and the Materials They Cut

FIGURE 23-8 Detailed small part not much larger than a dime made with a laser. Larry Jeffus

gas reacts with the metal being cut helps blow the molten material out of the kerf.

Some advantages of laser cutting include the following:

- High speed—Depending on the power rating and material thickness, some laser cutting machines can cut 800 in. per minute in 1/8-in. steel.
- Accuracy—Some laser cutting equipment can make cuts that are within ± a hundred thousandths of an inch, which is as accurate as some rough machining operations. This accuracy allows many parts to be used as cut without the need for postcutting machining.
- Narrow heat-affected zone—Little or no heating of the surrounding material is observed. It is possible to make very close parallel cuts without damaging the strip that is cut out, Figure 23-8.
- No electrical conductivity is required—The part being cut does not have to be electrically conductive, so materials like glass, quartz, wood, and plastic can be cut. There is also no chance that a stray electrical charge might damage delicate computer chips while they are being cut using a laser.
- No contact—Nothing comes in contact with the part being cut except the laser beam. Small parts that may have finished surfaces or small surface details can be cut without the danger of disrupting or damaging the surface. It is also not necessary to hold the parts securely as it is when a cutting tool is used.
- Narrow, smooth, accurate kerf—The width of the kerf is very small, which allows the nesting of parts in close proximity to each other, which will reduce waste of expensive materials, Figure 23-9.
- Clean cuts—Little or no postcutting cleanup is required because there is little dross or slag left on the back edge of the cut.
- Computer numerically controlled (CNC) and robotics—The laser beam can easily be directed through an articulated guide to the working end of a CNC machine or a robot to produce very accurate complex cuts.
- Top edge—The top edge will be smooth and square without being rounded.

FIGURE 23-9 Because of the narrow kerf, small, detailed parts can be nested one inside the other. Larry Jeffus

Some of the limitations of laser cutting include the following:

- Stacked materials that have significantly different melting temperatures
- Composite or sandwiched materials
- Cuts on fabrications that limit the beam's access due to component placement
- Materials thicker than 0.40 in. can often be cut more efficiently with another process.

Laser Beam Drilling

Most **laser beam drilling (LBD)** operations use a pulsed laser because the cloud of vaporized material caused by the laser will diffuse a continuous laser beam.

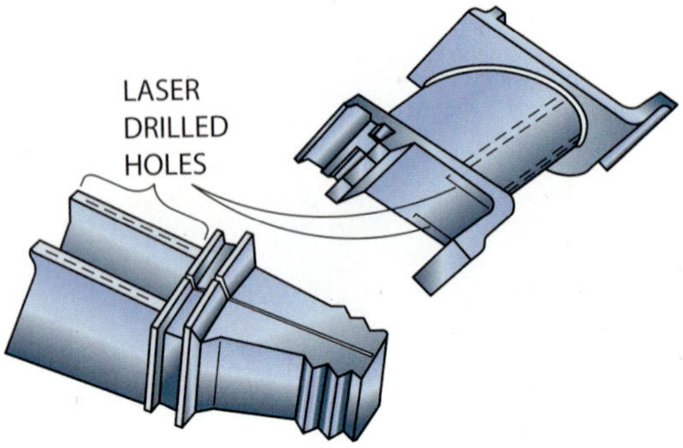

FIGURE 23-10 Jet engine blades and a rotor component showing laser drilled holes. American Welding Society

A very short burst of high laser energy concentrated on a small spot vaporizes the small spot with enough force to thrust the material out, leaving a small crater. The vaporized cloud of material dissipates in a fraction of a second so that the repeated pulses of laser are not diffused and result in the crater becoming increasingly deeper. This process continues until the hole is drilled through the part or to a desired depth, Figure 23-10.

A laser can be used to drill holes through the hardest materials, such as tungsten carbide cutting tools, quartz, glass, ceramics, or even synthetic diamonds. No other drilling process can match the precision, speed, or economy of the laser drill.

Lasers can drill up to 600 holes a minute through a 1/8-in. thick steel plate, and holes as small as 0.0001 in. in diameter can be drilled. The limitation on the hole's depth is the laser's focal length. Most holes are less than 1 in. in depth.

Laser Beam Welding

The laser beam can be used to melt the surface of the materials so that they can flow together and cool to form the weld metal. Most **laser beam welding (LBW)** is performed on thin materials. Following are some of the advantages of laser beam welding:

- The highly focused laser beam's heat allows it to make welds on very small, thin parts.
- Welds can be made that are narrow and have deep penetration, Figure 23-11.
- Laser's high welding speeds produce very narrow heat-affected zones.
- The very fast localized heating results in little or no postweld distortion.
- There is no electric current that might damage sensitive electronic parts that are being welded on or are nearby.

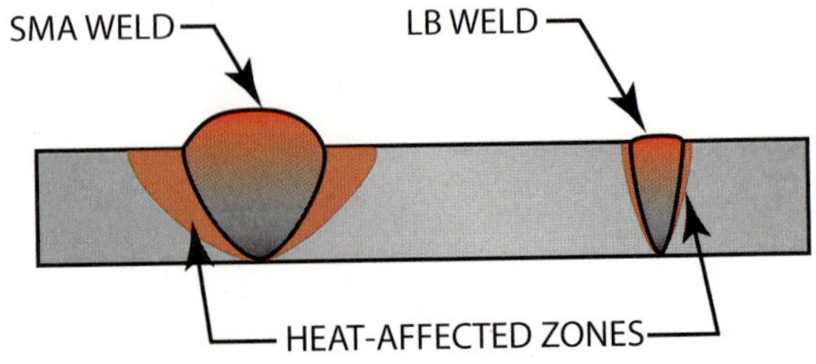

FIGURE 23-11 Comparison of weld size and heat-affected zone between SMAW and LBW processes. © Cengage Learning 2012

Laser Equipment

Most **lasers** range from 400 to 1500 watts in power. Some large machines have as much as 25 kW of power. Although the power of most lasers is relatively small, when compared to other welding processes, it is the laser's ability to concentrate the power into a small area that makes it work so well. The power density of a cutting laser can be equal to 65 million watts per square inch.

Laser equipment is physically much larger than most of the other welding or cutting power supplies. A typical unit will require about as much floor space as a large desk or about 10 sq ft.

Oxygen Lance Cutting

The oxygen lance cutting (OLC) process uses a consumable alloy tube, Figure 23-12. The tip of the tube is heated to its kindling temperature. A high-pressure oxygen flow is started through the lance. The oxygen reacts with the hot lance tip, releasing sufficient heat to sustain the reaction, Figure 23-13.

The rod tip is heated up to a red-hot temperature using an oxyfuel torch or electric resistance. Once the oxygen stream is started, it reacts with the lance material, which results in the creation of both a high temperature and heat-releasing reaction.

The intense reaction of the lance allows it to be used to cut through a variety of materials. The hot metal leaving the lance tip has not completed its exothermic reaction. As this reactive mass impacts the surface of the material being cut, it releases a large quantity of energy into that surface. Thermal conductivity between the molten metal and the base material is a very efficient method of heat transfer. This, along with the continued burning of the lance material on the surface, causes the base material to become molten.

FIGURE 23-12 Oxygen lance rods may be rolled out of flat strips of special alloys to form a tube.
Larry Jeffus

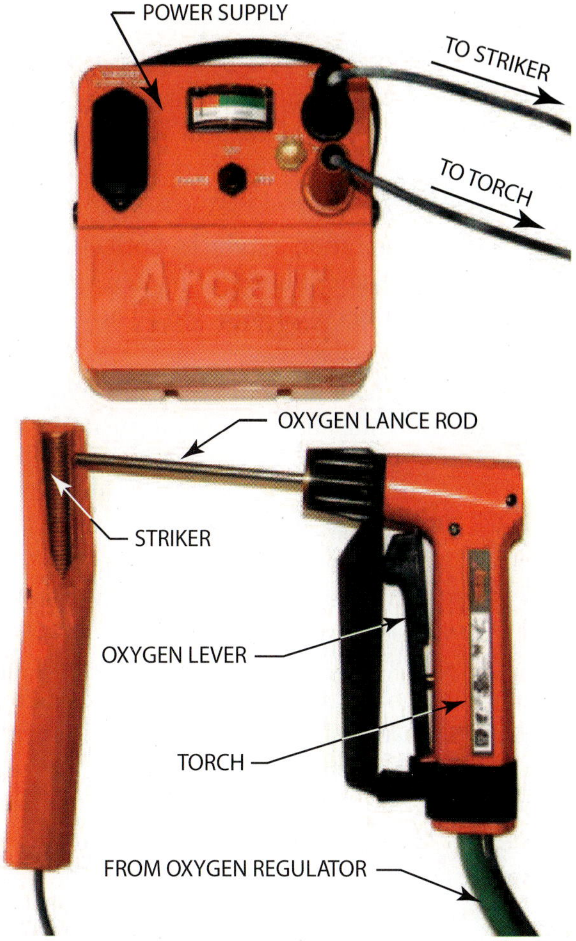

FIGURE 23-13 Arcair SLICE portable cutting system that uses sparks created between the striker and tube to ignite the oxygen lance rod for cutting.
Larry Jeffus

Once the base material is molten, it may react with the burning lance material, forming fumes or slag, which is then blown from the cut. Any molten material that does not become reactive is carried out of the cut with the slag or blown out with the oxygen stream.

The addition of steel rods or other metals to the center of the oxygen lance tube has increased the lance's productivity. The improved lances last longer and cut faster.

Application

The oxygen lance's unique method of cutting allows it to be used to cut material not normally cut using a thermal process. Films have portrayed the oxygen lance as a tool used by thieves to cut into safes. In reality, this would result in the valuables in the safe being destroyed. The oxygen lances can be used to cut reinforced concrete.

Cutting concrete has been used in the demolition of buildings. It allows the quick removal of thick sections of the building without the dangerous vibration caused by most conventional methods. This has been a lifesaving factor in the use of the oxygen lance for rescue work following earthquakes. The oxygen lance saved thousands of hours and countless lives in Mexico City following the devastation from the city's worst earthquake. Large sections of concrete that fell from buildings were cut into manageably sized pieces using oxygen lances. Local and national news agencies showed building rubble being cut away by rescue workers using the oxygen lance.

The oxygen lance is also used to cut thick sections of cast iron, aluminum, and steel. Often in the production of these metals, thick sections must be cut. Occasionally, equipment failure will stop metal production. If the metal in production is allowed to cool, it may need to be cut in sections so that it can be removed from the machine. The oxygen lance's process is very effective in this type of work.

Safety

It is important to follow all safety procedures when using this process. Manufacturers list specific safety precautions for the oxygen lances they produce. Read and follow those instructions carefully. Following are the major safety concerns:

- Fumes—The large quantity of fumes generated is often a health hazard. An approved ventilation system must be provided if this work is to be done in a building or any other enclosed area.
- Heat—This operation produces both high levels of radiant heat and plumes of molten sparks and slag. The operators must wear special heat-resistant clothing.
- Noise—Sound is produced well above safety levels. Ear protection must be worn by anyone in the area.

Water Jet Cutting

Water jet cutting is not a thermal cutting process. The cut is accomplished by the rapid erosion of the material by a high-pressure jet of water, Figure 23-14. The water jet may have a pressure as high as 60,000 psi. An abrasive powder such as garnet may be added to the stream of water. **Abrasives** are added when hard materials such as metals are being cut. Water jets can be used to cut many metals, including carbon steel, tool steel, stainless steel, Inconel, titanium, aluminum, clad materials, hard-surfaced metal, and armored plated metal. They can also cut nonmetals such as ceramics, granite, plastic, glass, and laminated materials such as printed circuit boards.

Water jet cutting does not put any heat into the material being cut, and it is this lack of heat input that makes this process unique. Heat can make materials distort, become harder, or cause delamination. The lack of heat distortion allows thin material to be cut with the edge quality of a laser cut and as distortion-free as a shear cut. Delamination is not a problem when cutting composite or laminated materials such as carbon fibers, resins, or computer circuit boards, Figure 23-15.

Applications

Kerf widths as small as 0.02 in. are possible, and the kerf width does not tend to change unless too high a travel speed is being used. The top edge of the cut is square, and the cut surface is flat and smooth with a sandblasted appearance. Because of the slight lagging behind of the cut along the lower edge when thicker

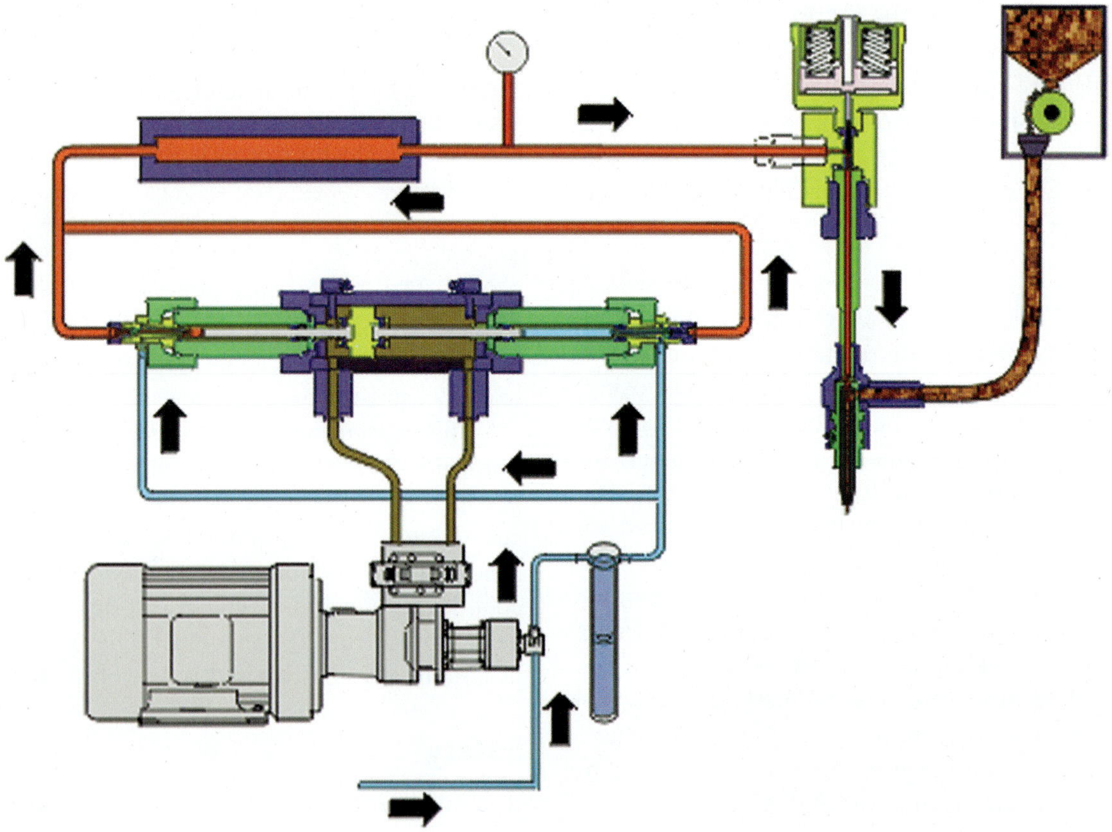

FIGURE 23-14 Basic diagram illustrating the elements of the water jet cutting system. Larry Jeffus

FIGURE 23-15 **Water jet cutting is very versatile. A machine with multiple cutting jets is cutting out printed circuit boards.** Ingersoll-Rand

sections are being cut, there may be some slight offset when curves are being cut, Figure 23-16. Postcutting cleanup of the parts is totally eliminated for most materials, and only slight work is needed on a few others, Figure 23-17. The quality of the cut surface can be controlled so that even parts for the aerospace industry can often be assembled as cut.

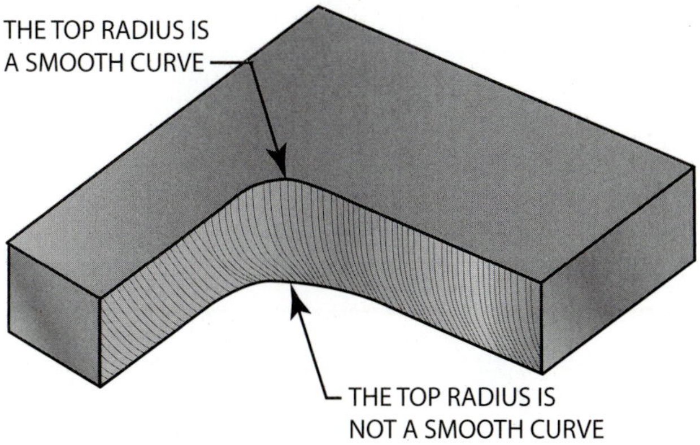

FIGURE 23-16 **Offset bottom surface from a water jet cut.** © Cengage Learning 2012

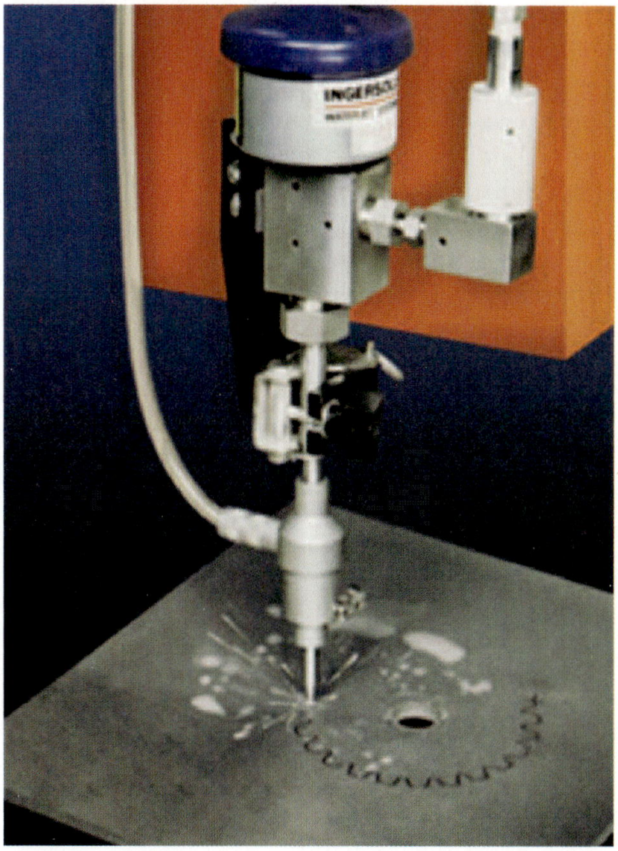

FIGURE 23-17 **Notice how narrow and clean this cut is.** Ingersoll-Rand

The addition of an abrasive powder can speed up the cutting, allow harder materials to be cut, and improve the surface finish of a cut. The powder most often used is garnet. It is also commonly used as an abrasive on sandpaper. If an abrasive is used, the small water jet orifice will wear out faster.

Materials that often gum up a cutting blade, such as plexiglass, ABS plastic, and rubber, can be cut easily. There is nothing for the material to adhere to that would disrupt the cut. The lack of heat also reduces the tendency of the material's cut surface to become galled.

Most of the water jet cutting is performed by some automated or robotic system. There are a few band-saw type, hand-fed cutting machines that are used for single cuts or when limited production is required.

Arc Cutting Electrodes

Arc cutting electrodes do not require any special equipment other than a standard shielded metal arc (SMA) welding machine, and they fit into the standard electrode holder. Because the arc cutting electrodes work with any standard arc welder, this makes them the ideal choice for many cutting jobs. For example, if only an occasional cut is needed, the expense of having either an oxyfuel cutting (OFC) or a plasma arc cutting (PAC) piece of equipment cannot be justified. Many types of arc cutting electrodes can make clean cuts, although they may not make as square, smooth, and as clean a cut as many other cutting processes. But they are much easier and faster to use than a metal saw for cutting or to drill a hole in thick metal.

Arc cutting electrodes are available for cutting steel, stainless steel, cast iron, aluminum, copper alloys, and many other metals. Steel up to 1 1/2-in. thick can be cut using this process. Some arc cutting electrodes are available for underwater cutting where using an acetylene torch might be difficult or impossible to use.

Arc cutting electrodes differ from standard arc welding electrodes in that they burn back inside the outer flux covering, creating a small cavity, Figure 23-18. The heat of the arc causes the metal core and inner layer of flux to vaporize and rapidly expand. The small cavity acts like a small combustion chamber, and the hot vaporized material is blasted out like a small jet engine. Combined with the heat of the arc and the jetting action, the metal is cut.

Application

This cutting process is ideal any time that it is not practical or cost effective to bring along an oxygen acetylene cutting set and cylinders or a plasma cutting machine. Arc cutting electrodes can be used to cut metal, pierce holes, or to gouge a groove for welding or gouge out a defective weld.

Air Carbon Arc Cutting

The **air carbon arc cutting** (CAC-A) process was developed in the early 1940s, and was originally named air arc cutting (AAC). Air carbon arc cutting was an improvement of the carbon arc process. The carbon arc process was used in the vertical and overhead positions and removed metal by melting a large enough spot so that gravity would cause it to drip off the base plate, Figure 23-19. This process was slow and could not be accurately controlled. It was found that by using a stream of air the molten metal could be blown away. This greatly improved the speed, quality, and controllability of the process.

The torch both holds the **carbon electrode** and it has air nozzles that direct the cutting stream into the arc. This basic design is still in use today, Figure 23-20.

Unlike the oxyfuel process, the air carbon arc cutting process does not require that the base metal react with the cutting stream. Oxyfuel cutting can be used only on metals that will rapidly oxidize when a stream of oxygen is directed onto the hot surface. The ACA air stream blows the molten metal away. This greatly increases the list of metals that can be cut, Table 23-2.

Manual Torch Design

The air carbon arc cutting torch is designed differently than the shielded metal arc electrode holder. The major differences between an electrode holder and an air carbon arc torch include the following:

- The lower electrode jaw has a series of air holes.
- The jaw has only one electrode-locating groove.

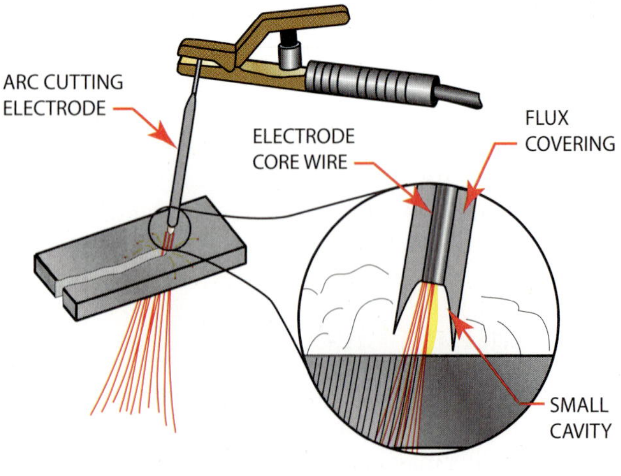

FIGURE 23-18 Arc cutting electrode showing jetting action from a small tip cavity.

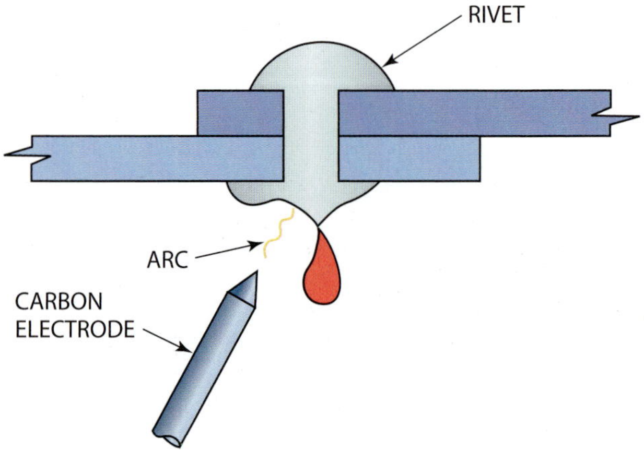

FIGURE 23-19 A carbon electrode was used to melt the head off rivets.

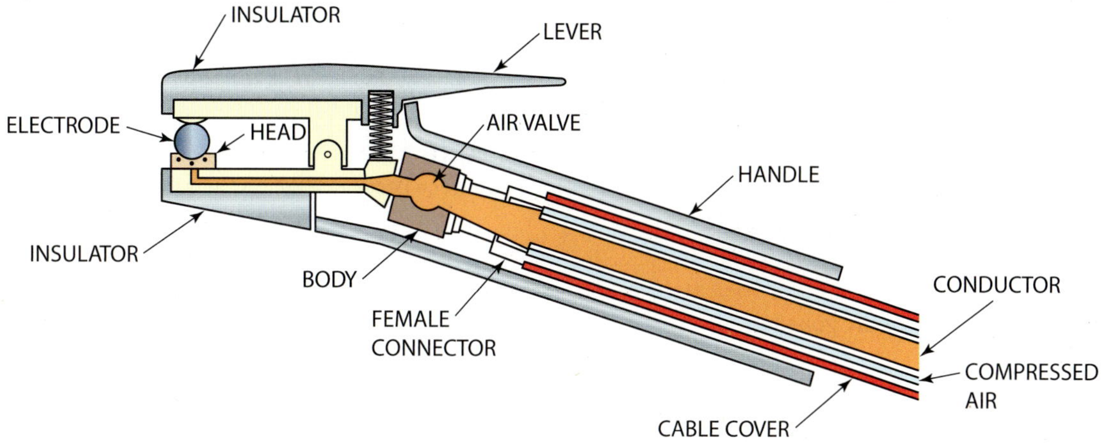

FIGURE 23-20 Typical cross section of an air carbon arc gouging torch. American Welding Society

- The electrode jaw can pivot.
- There is an air valve on the torch lead.

By having only one electrode-locating groove in the jaw and pivoting the jaw, the air stream will always be aimed correctly. The air must be aimed just under and behind the electrode and always in the same direction, Figure 23-21. This ensures that the air stream will be directed at the spot where the electrode arcs to the base metal.

Torches are available in a number of amperage sizes. The larger torches have a greater capacity but are less flexible to use on small parts.

The torch can be permanently attached to a welding cable and air hose, or it can be attached to welding power by gripping a tab at the end of the cable with

Base Metals	Recommendations
Carbon steel and low alloy steel	Use DC electrodes with DCEP current. AC can be used but with a 50% loss in efficiency.
Stainless steel	Same as for carbon steel.
Cast iron, including malleable and ductile iron	Use of 1/2 in. or larger electrodes at the highest rated amperage is necessary. There are also special techniques that need to be used when gouging these metals. The push angle should be at least 70° and depth of cut should not exceed 1/2 in. per pass.
Copper alloys (copper content 60% and under)	Use DC electrodes with DCEN (electrode negative) at maximum amperage rating of the electrode.
Copper alloys (copper content over 60%, or size of workpiece is large)	Use DC electrodes with DCEN at maximum amperage rating of the electrode or use AC electrodes with AC.
Aluminum bronze and aluminum nickel bronze (special naval propeller alloy)	Use DC electrodes with DCEN.
Nickel alloys (nickel content is over 80%)	Use AC electrodes with AC.
Nickel alloys (nickel content less than 80%)	Use DC electrodes with DCEP.
Magnesium alloys	Use DC electrodes with DCEP. Before welding, surface of groove should be wire brushed.
Aluminum	Use DC electrodes with DCEP. Wire brushing with stainless wire brushes is mandatory prior to welding. Electrode extension (length of electrode between electrode torch and workpiece) should not exceed 3 in. for good-quality work. DC electrodes with DCEN can also be used.
Titanium, zirconium, hafnium, and their alloys	Should not be cut or gouged in preparation for welding or remelting without subsequent mechanical removal of surface layer from cut surface.

Note: Where preheat is required for welding, similar preheat should be used for gouging.

Table 23-2 Recommended Procedures for Air Carbon Arc Cutting of Different Metals

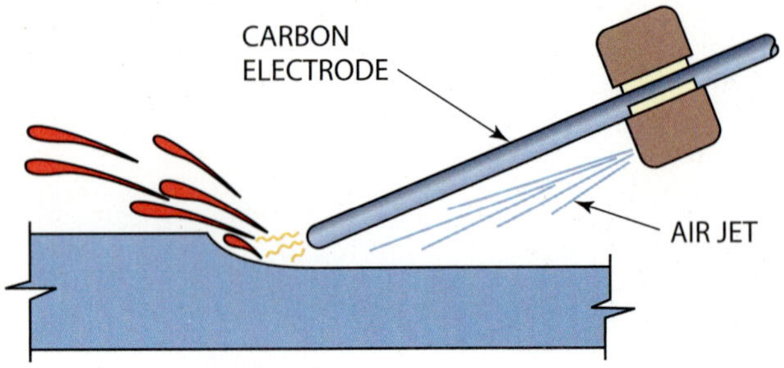

FIGURE 23-21 **Air carbon arc gouging.** © Cengage Learning 2012

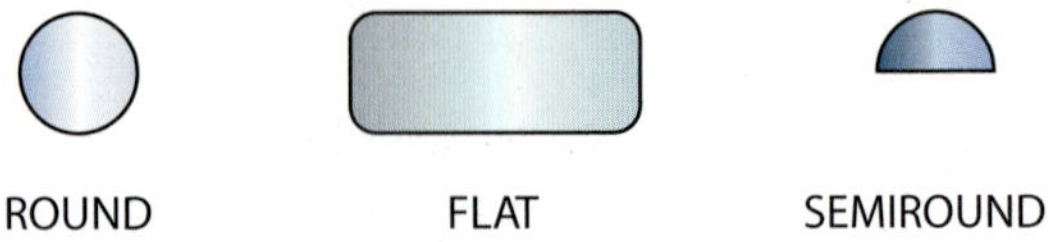

FIGURE 23-23 **Cross sections of carbon electrodes.** © Cengage Learning 2012

the shielded metal arc electrode holder, Figure 23-22. The temporary attachment can be made easier if the air hose is equipped with a quick disconnect. A quick disconnect on the air hose will allow it to be used for other air tools such as grinders or chippers. Greater flexibility for a workstation can be achieved with this arrangement.

Electrodes

Air carbon arc cutting electrodes are available as copper-coated or plain (without a coating). The copper coating helps decrease the carbon electrode overheating by increasing its ability to carry higher currents, and it improves the heat dissipation. The copper coating provides increased strength to reduce accidental breakage.

Electrodes come in round, flat, and semiround, Figure 23-23. The round electrodes are used for most **gouging** operations, and the flat electrodes are most often used to scarf off a surface. Round electrodes are also available in sizes ranging from 1/8 in. to 1 in. in diameter. Flat electrodes are available in 3/8-in. and 5/8-in. sizes.

Electrodes are available to be used on both direct-current electrode positive (DCEP) and alternating current (AC). The DCEP electrodes are the most commonly used, and they are made of carbon in the form of **graphite**. The AC electrodes are less common; they have some elements added to the carbon to stabilize the arc, which is needed for the AC power.

To reduce waste, electrodes are made so that they can be joined together. The joint consists of a female tapered socket at the top end and a matching tang on the bottom end, Figure 23-24. The connection of

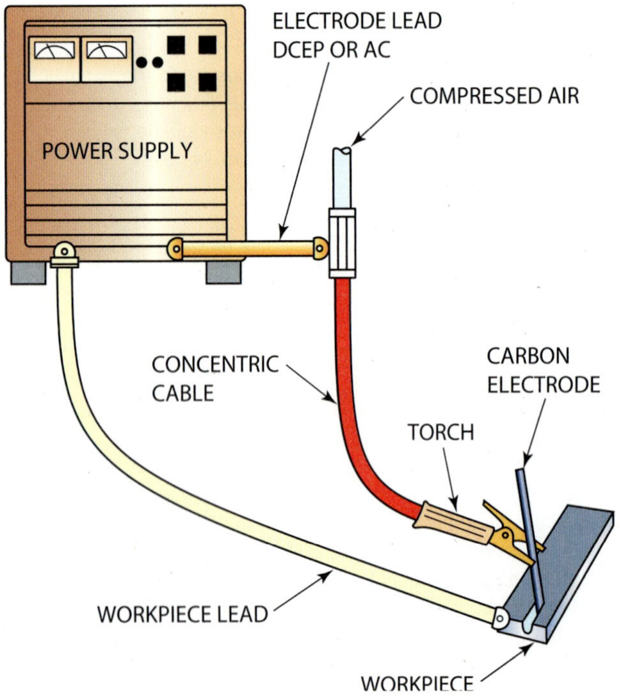

FIGURE 23-22 **Air carbon arc gouging equipment setup.** American Welding Society

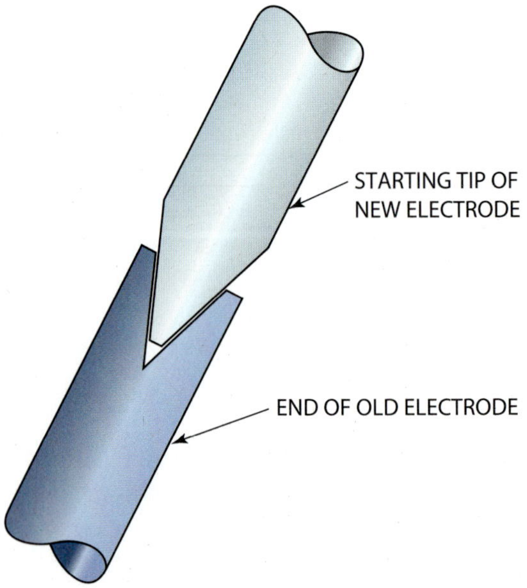

FIGURE 23-24 **Air carbon arc electrode joint.** © Cengage Learning 2012

the new electrode to the remaining setup will allow the stub to be consumed with little loss of electrode stock. This connecting of electrodes is required for most track-type air carbon arc cutting operations to allow for longer cuts.

Power Sources

Most shielded metal arc welding power supplies can be used for air carbon arc cutting. The operating voltage required for air carbon arc cutting needs to be 28 volts or higher. This voltage is slightly higher than that required for most SMA welding, but most welders will meet this requirement. Check the manufacturer's owner manual to see if your welder is approved for air carbon arc cutting. If the voltage is lower than the minimum, the arc will tend to sputter out, and it will be hard to make clean cuts.

Because most carbon arc cutting requires a high amperage setting, it may be necessary to stop some cuts so that the duty cycle of the welder is not exceeded. On large industrial welders this is not normally a problem.

Air Supply

Air pressure supplied to the torch should be between 80 and 100 psi. The minimum pressure is around 40 psi. The correct air pressure will result in cuts that are clean, smooth, and uniform. The airflow rate is also important. If the air line is too small or the compressor does not have the required capacity, there will be a loss in air pressure at the torch tip. This line loss will result in a lower-than-required flow at the tip. The resulting cut will be less desirable in quality.

Application

Air carbon arc cutting can cut a variety of materials. It is a relatively low-cost way of cutting most metals, including steel, cast iron, stainless steel, aluminum, nickel alloys, and copper. Air carbon arc cutting is most often used for repair work. Few cutting processes can match the speed, quality, and cost savings of this process for repair or rework. In repair or rework the most difficult part is the removal of the old weld or cutting a groove so that a new weld can be made. The air carbon arc can easily remove the worst welds even if they contain slag inclusions or other defects. For repairs, the arc can cut through thin layers of paint, oil, or rust and make a groove that needs little, if any, cleanup.

CAUTION

Never cut on any material that might produce fumes that would be hazardous to your health without proper safety precautions, including adequate ventilation and/or the wearing of a respirator.

The highly localized heat results in only slight heating of the surrounding metal. As a result, usually there is no need to preheat hardenable metals to prevent hardness zones. Cast iron is a metal that can be carbon arc gouged to prepare a crack for welding without causing further damage to the part by inputting excessive heat.

Air carbon arc cutting can be used to remove a weld from a part. The removal of welds can be accomplished with such success that often the part needs no postcut cleanup. The root of a weld can be back gouged so that a backing weld can be made, ensuring 100% weld penetration, Figure 23-25.

The electrode should extend approximately 6 in. from the torch when starting a cut, and as the cut progresses, the electrode is consumed. Stop the cut, and readjust the electrode when its end is approximately 3 in. from the electrode holder. This will reduce the damage to the torch caused by the intense heat of the operation.

Gouging. is the most common application of the air carbon arc cutting processes. Arc gouging is the removal of a quantity of metal to form a groove or

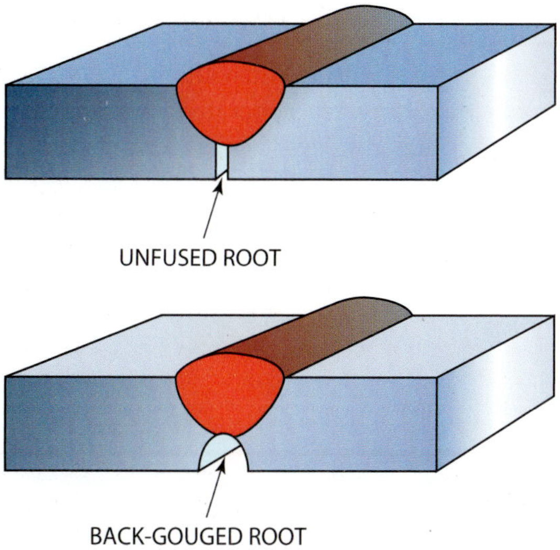

FIGURE 23-25 **Back gouging the root of a weld made in thick metal can ensure that a weld with 100% joint fusion can be made.** © Cengage Learning 2012

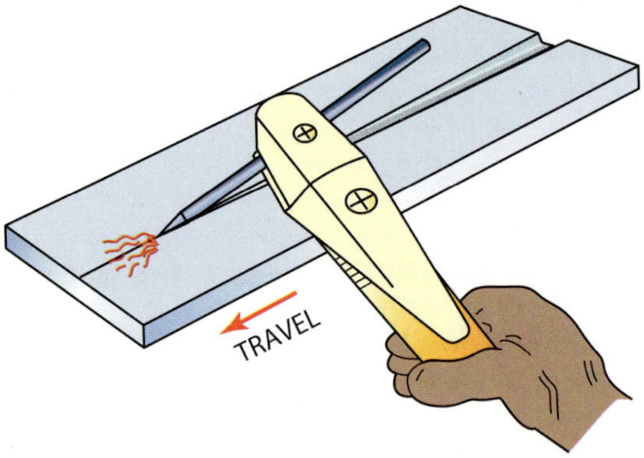

FIGURE 23-26 Manual air carbon arc gouging operation in the flat position. American Welding Society

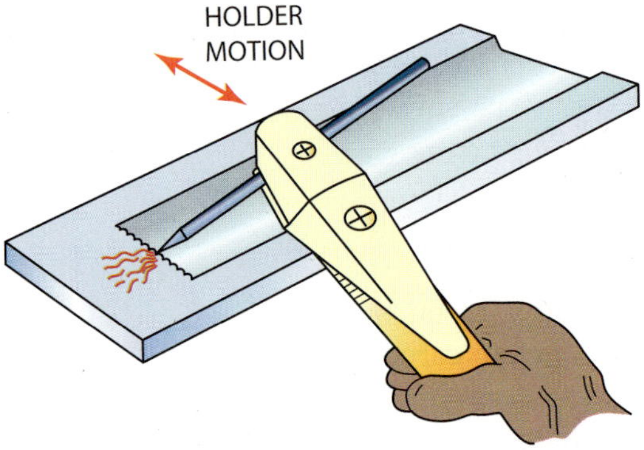

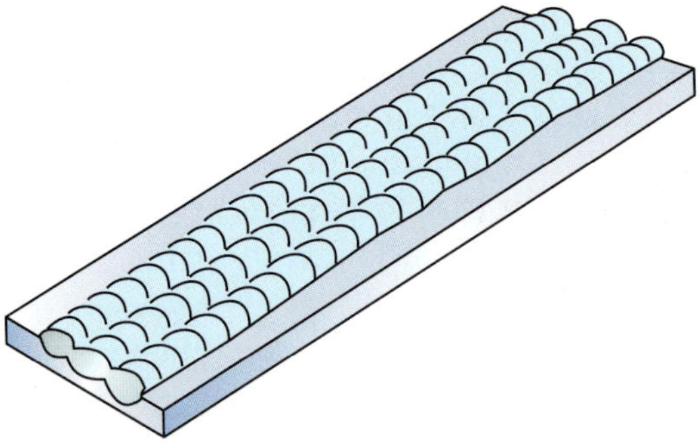

FIGURE 23-28 Hard surfacing weld applied in the carbon arc cut groove. American Welding Society

bevel, Figure 23-26. The groove produced along an edge of a plate is usually a J-groove. The groove produced along a joint between plates is usually a U-groove, Figure 23-27. Both grooves are used as a means to ensure that the weld applied to the joint will have the required penetration into the metal.

Washing. Washing is the process sometimes used to remove large areas of metal so that hard surfacing can be applied, Figure 23-28. Washing can be used to remove large areas that contain defects, to reduce the transitional stresses of unequal-thickness plates, or to allow space for the capping of a surface with a wear-resistant material.

Safety

In addition to the safety requirements of shielded metal arc, air carbon arc requires several special precautions, such as the following:

- Sparks—The quantity and volume of sparks and molten metal spatter generated during this process are a major safety hazard. Extra precautions must be taken to ensure that other workers, equipment, materials, or property in the area will not be affected by the spark stream.
- Noise—This process produces a high level of sound. The sound level is high enough to cause hearing damage if proper ear protection is not used.
- Light—The arc light produced is the same as that produced by the shielded metal arc welding process. But because the arc has no smoke to defuse the light and the amperages are usually much higher, the chances of receiving arc burns are much higher. Additional protection should be worn, such as thicker clothing, a leather jacket, and leather aprons.
- Eyes—Because of the intense arc light, a darker welding filter lens for the helmet should be used.
- Fumes—The combination of the air and the metal being removed results in a high volume of fumes. Special consideration must be made for the removal of these fumes from the work area. Before installing a ventilation system, check with local, state, and federal laws. Some of the fumes may have to be filtered before they can be released into the air.
- Surface contamination—Often this process is used to prepare damaged parts so that they can be repaired.

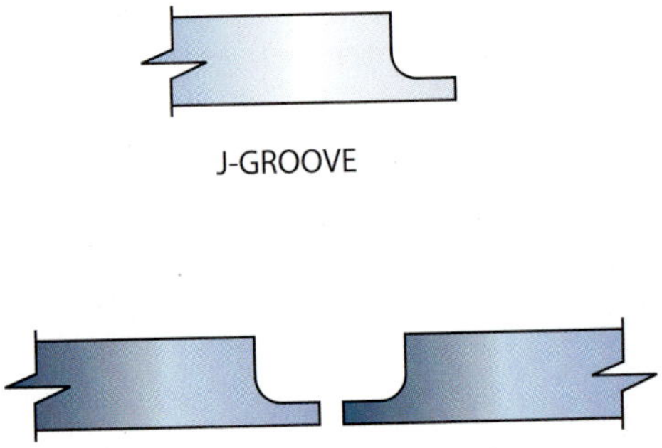

FIGURE 23-27 Air carbon arc gouging groove shapes. © Cengage Learning 2012

If the used parts have paint, oils, or other contamination that might generate hazardous fumes, they must be removed in an acceptable manner before any cutting begins.

- Equipment—Check the manufacturer's owner manual for specific safety information concerning the power supply and the torch before you start any work with each piece of equipment for the first time.

U-Grooves

U-grooving is used to remove defective welds and to prepare a thick metal joint so that a full penetration weld can be made. Often the easiest way to prepare a joint for a full penetration weld is to use one of the thermal processes and cut a V-groove or bevel along the edge. However, sometimes parts must be fitted and assembled to ensure everything fits as designed before a groove can be cut. In these cases, a U-groove can easily be cut along the assembled edges of the joint.

PROJECT 23-1

Air Carbon Arc Straight Cut in the Flat Position to Make a Checkerboard

Skills to be learned: The ability to make straight air carbon arc gouges and control the width and depth of the U-grooves.

Project Description

This 15-in. square checkerboard will have the squares marked off with air carbon arc grooves. The slight variations and texture of the U-grooves make this an attractive, functional game board that everyone can enjoy, Figure 23-29.

Project Materials and Tools

The following items are needed to fabricate the checkerboard.

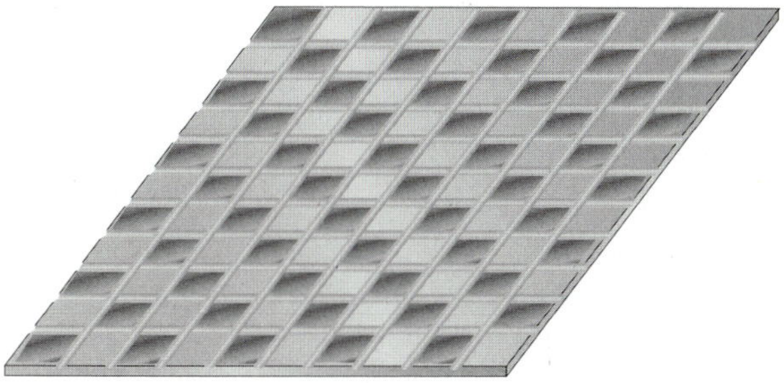

FIGURE 23-29 Project 23-1. © Cengage Learning 2012

Materials

- 15-in. × 15-in. piece of 1/4-in. to 3/8-in. thick plate
- Air carbon arc electrodes
- Air carbon arc torch
- Source of 80-psi air

Tools

- Arc welding helmet
- SMA welder
- PPE
- 24-in. rule or tape measure
- Soapstone or metal marker
- Center punch and hammer
- Square
- Pliers
- Right angle grinder with wire brush and grinding disk
- Chipping hammer
- Hand wire brush

Layout

Using the square, soapstone, and the rule or tape, lay out the nine horizontal and nine vertical lines on 1.5-in. centers that are used to divide the metal plate into the standard checkerboard layout, Figure 23-30. Because the lines can be smudged by your gloved hand or covered by slag from nearby groove cuts, you should make a large center-punched mark at the end of each of the lines so they can be easily redrawn if needed.

Cutting Out

Using a properly set up shielded metal arc welder and an air carbon arc torch, putting on all of your personal protective equipment (PPE), and following all shop and manufacturer's safety rules for welding, you will cut the grooves in the checkerboard.

1. Adjust the air pressure to approximately 80 psi.
2. Set the amperage within the range for the diameter electrode you are using by referring to the box the electrodes came in.
3. Check to see that the stream of sparks will not start a fire or cause any damage to anyone or anything in the area.
4. Make sure the area is safe, and turn on the welder.
5. Using a good dry leather glove to avoid electrical shock, insert the electrode in the torch jaws so

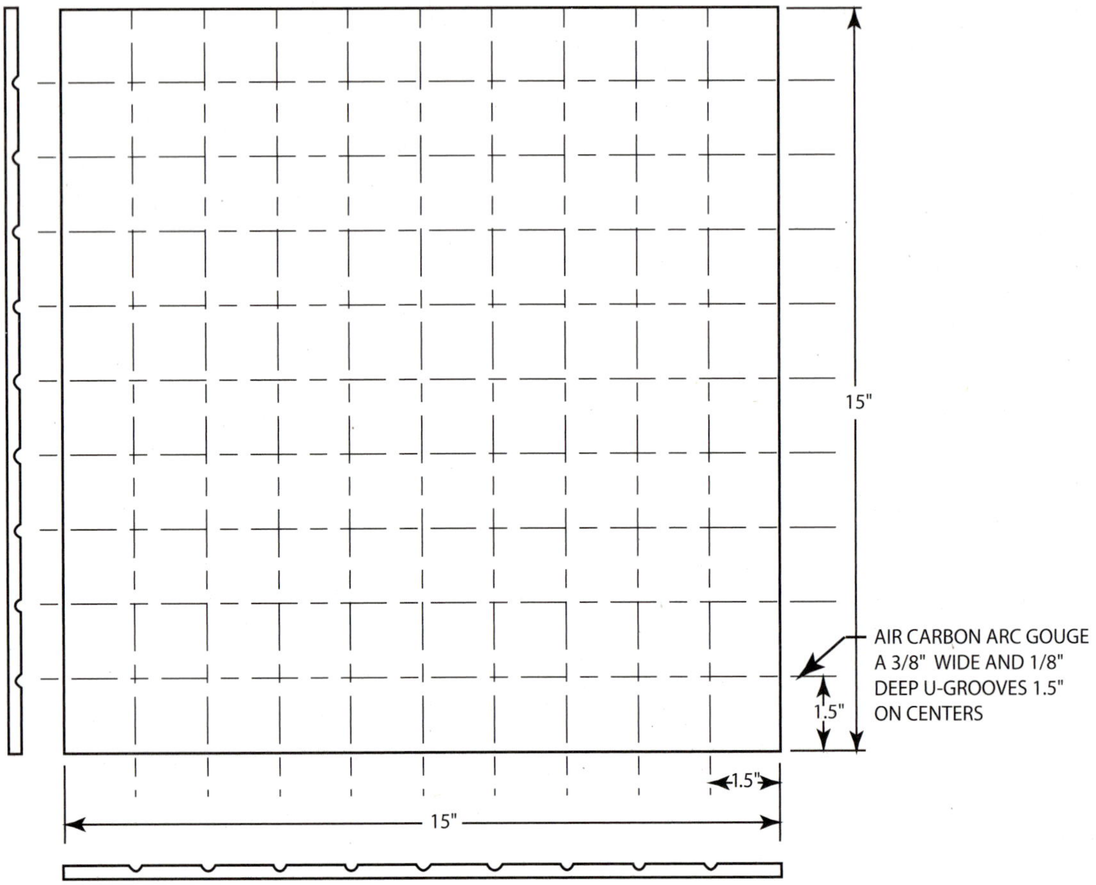

FIGURE 23-30 Project 23-1 layout. © Cengage Learning 2012

that about 6 in. is extending outward. Be sure not to touch the electrode to any metal parts because it may short out.

6. Turn on the air at the torch head.
7. Lower your arc welding helmet.
8. Slowly bring the electrode down at about a 30° angle so it will make contact with the plate near the starting edge, Figure 23-31. Be prepared for a loud, sharp sound when the arc starts.
9. Once the arc is struck, move the electrode in a straight line down the plate toward the other end.
10. Keep the speed and angle of the torch constant. You can rest your gloved hand on the metal as a way of keeping the cut more even and straighter.
11. As the carbon electrode becomes shorter because the end is being burned off, you have to keep lowering the torch so that the approximate 30° electrode angle will remain constant.
12. When you reach the other end, lift the torch so the arc will stop.
13. Raise your helmet, and stop the air.
14. Remove the remaining electrode from the torch so that it will not accidentally touch anything.
15. When the metal is cool, chip or brush any slag or dross off of the plate. This material should remove easily.
16. Examine the groove. It should be within ±1/8 in. of being straight and within ±3/32 in. of uniformity in width and depth.
17. If the air pressure was too low, you will be able to see where the groove metal was melted. If the air pressure was too high, the groove width will be too wide.
18. Make any adjustments to the setup or your technique, and repeat the grooving process until all of the U-grooves have been cut.

Finishing

Cool off the metal, chip the slag, and wire brush the surface. A power wire brush can be used to give the welds a polished, shiny appearance. The project can be painted with a clear coat or colored coat of latex paint.

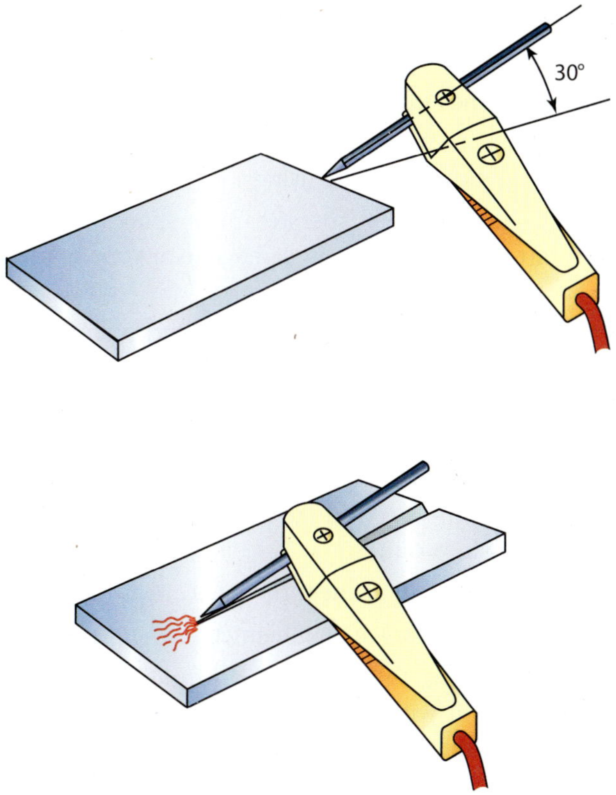

FIGURE 23-31 Air carbon arc U-groove gouging. © Cengage Learning 2012

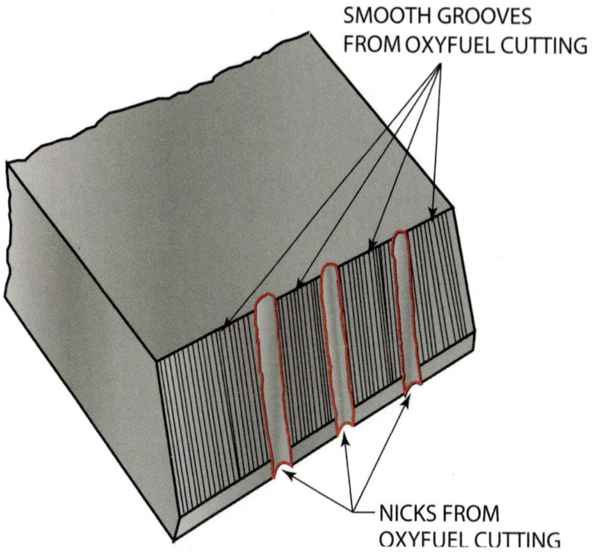

FIGURE 23-32 Oxyfuel beveled cut with nicks in the surface. © Cengage Learning 2012

Paperwork

Complete a copy of the time sheet in Appendix I and bill of materials in Appendix III, or as provided by your instructor.◆

J-Groove

The J-groove is easier to hand produce for a grooved weld as compared to using an oxyacetylene torch. A hand oxyacetylene beveled cut will often have small nicks in the beveled surface, which are made as the torch is being moved along the edge, Figure 23-32. These nicks may have to be ground out before the joint can be welded. A gouged J-groove is much smoother. Less postcutting cleanup or nick repair is needed.

PROJECT 23-2

Air Carbon Arc Straight Cut in the Flat Position to Make a J-Groove around the Edge of the Checkerboard Made in Project 23-1

Skill to be learned: The ability to make a uniform J-groove along the edge of a plate.

Project Description

The 15-in. square checkerboard made in Project 23-1 will have a J-grooved edge made using air carbon arc gouging. The J-groove is used to give the checkerboard a finished look, Figure 23-33.

Project Materials and Tools

The following items are needed to fabricate the checkerboard.

Materials

- 15-in. × 15-in. piece checkerboard from Project 23-1
- Air carbon arc electrodes
- Air carbon arc torch

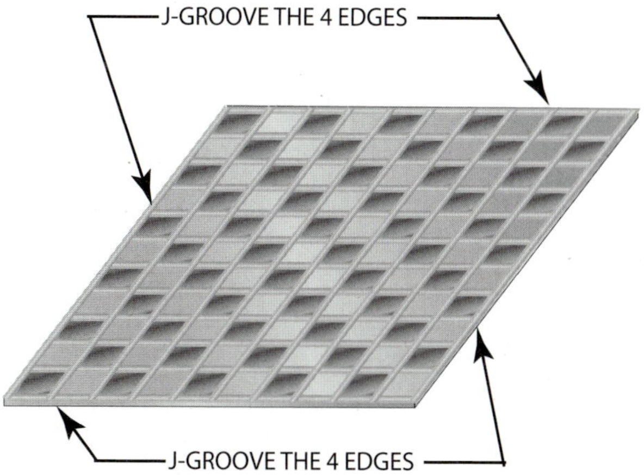

FIGURE 23-33 Project 23-2. © Cengage Learning 2012

- Source of 80-psi air
- Clear finish or paint

Tools

- Arc welding helmet
- SMA welder
- PPE
- Pliers
- Right angle grinder with wire brush and grinding disk
- Chipping hammer
- Hand wire brush

Layout

No layout is required because the J-groove will follow the edge of the checkerboard, Figure 23-34.

Cutting Out

Using a properly set up shielded metal arc welder and an air carbon arc torch, putting on all of your PPE, and following all shop and manufacturer's safety rules for welding, you will cut the grooves in the checkerboard.

1. Adjust the air pressure to approximately 80 psi.
2. Set the amperage within the range for the diameter electrode you are using.
3. Check to see that the stream of sparks will not start a fire or cause any damage to anyone or anything in the area.
4. Make sure that the area is safe, and turn on the welder.
5. Using a good, dry, leather glove to avoid electrical shock, insert the electrode in the torch jaws so that about 6 in. is extending outward. Be sure not to touch the electrode to any metal parts because it may short out.
6. Turn on the air at the torch head.
7. Lower your arc welding helmet.
8. Slowly bring the electrode down at about a 30° angle so it will make contact with the plate near the starting edge, Figure 23-35.

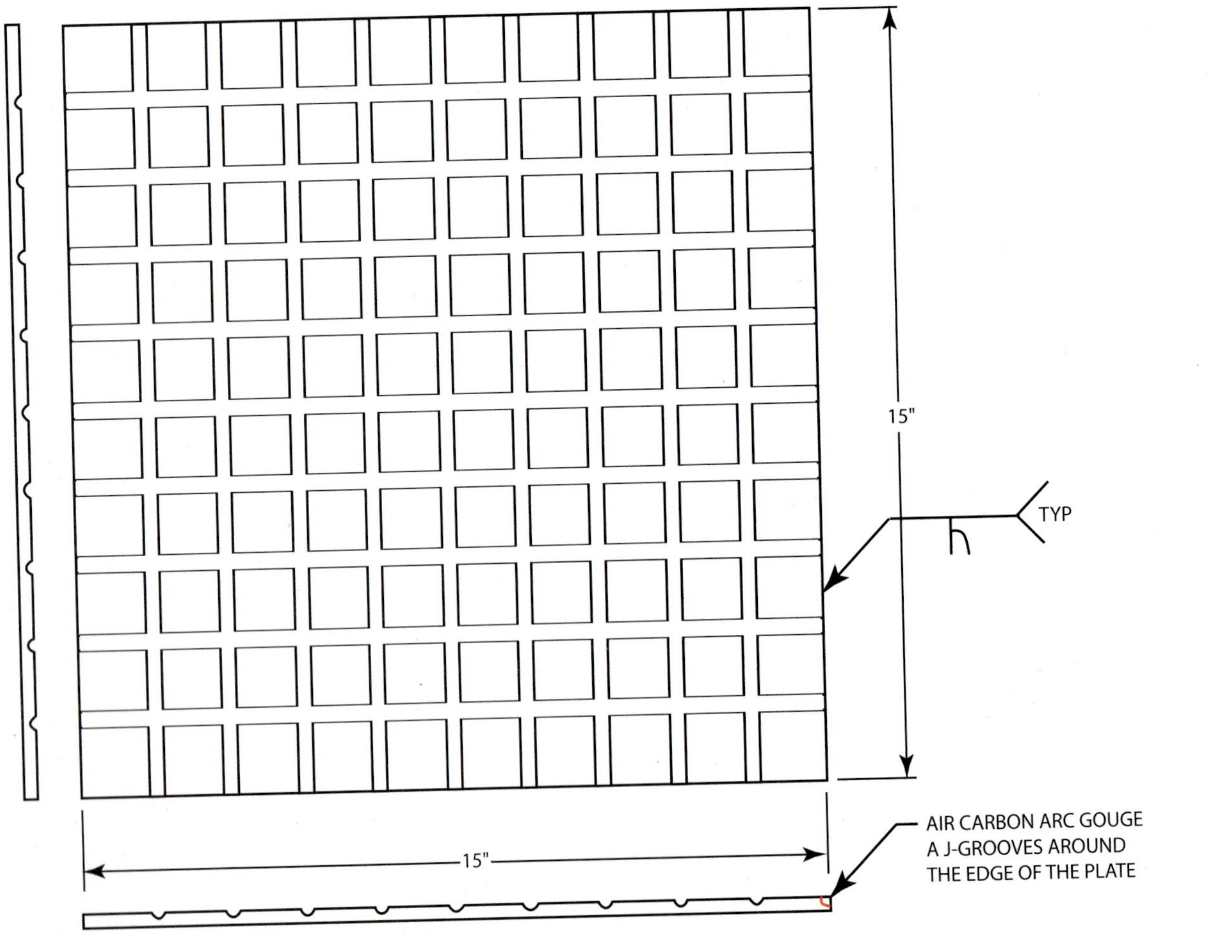

FIGURE 23-34 Project 23-2 layout. © Cengage Learning 2012

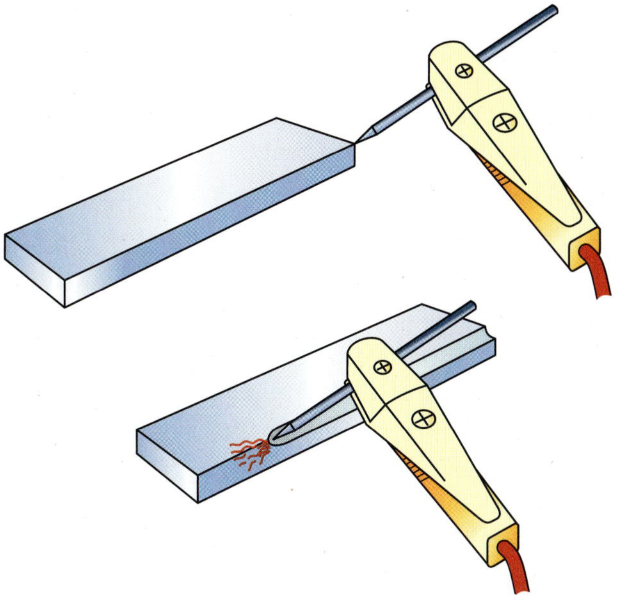

FIGURE 23-35 Air carbon arc J-grooving.

9. Once the arc is struck, move the electrode in a straight line down the edge of the plate toward the other end. Keep the speed and angle of the torch constant.
10. When you reach the other end, lift the torch so the arc will stop.
11. Raise your helmet, and stop the air.
12. Remove the remaining electrode from the torch so it will not accidentally touch anything.
13. When the metal is cool, chip or brush any slag or dross off of the plate. This material should remove easily. The groove must be within ±1/8 in. of being straight and within ±3/32 in. of uniformity in width and depth.
14. Make any adjustments to the setup or your technique, and repeat the grooving process until all of the J-grooves have been cut.

Finishing

Cool off the metal, chip the slag, and wire brush the surface. Use the angle grinder and grinding disk to grind the mill scale or oxides off alternating squares on the checkerboard to make the light and dark-colored squares. The project can be painted with a clear coat.

Paperwork

Complete a copy of the time sheet in Appendix I and bill of materials in Appendix III, or as provided by your instructor.

Weld Metal Removal

Sometimes it is necessary to remove a weld. It might be because the weld did not pass inspection, it was made in the wrong place, the weld was used to temporarily hold parts together, or the weld was used to fill a defect in the part. ACA gouging is ideal for this because it is not affected by slag inclusions in the weld; also, there is less heat input to the part as compared to oxyfuel gouging.◆

PROJECT 23-3

Welding to Fill a Poorly Made U-groove in the Checkerboard and Then Air Carbon Arc Gouging Can Be Used to Remove the Weld to Fix the Groove

Skills to be learned: The ability to make a repair to a part by welding and then re cutting the grooves using the ACA process.

Project Description

Repairing of any of the U-grooves made on the checkerboard that are too wide, deep, or were not made in a straight line. Any arc welding processes can be used to fill a portion of the groove. Once the weld is made, a new U-groove can be made and the remaining weld metal then ground off.

Project Materials and Tools

The following items are needed to repair the checkerboard U-grooves.

Materials

- 15-in. × 15-in. piece checkerboard from Project 23-1
- Air carbon arc electrodes
- Air carbon arc torch
- Source of 80-psi air
- Clear finish or paint

Tools

- Arc welding helmet
- SMA welder
- PPE
- Pliers
- Right angle grinder with wire brush and grinding disk
- Chipping hammer
- Hand wire brush

Layout

Mark any U-grooves that you would like to repair with a piece of soapstone.

Welding

Set the welder for the metal type and thickness being welded according to the manufacturer's recommendations. Put on your PPE, and follow all shop and manufacturer's safety rules for welding.

Follow the recommended procedures for welding found in the chapter covering the process you selected to make the repair weld. Shielded metal arc welding (SMAW) is covered in Chapters 8 and 9, gas metal arc welding (GMAW) is covered in Chapters 11 and 12, flux core arc welding (FCAW) is covered in Chapters 13 and 14, and gas tungsten arc welding (GTAW) is covered in Chapters 15 and 16.

Cutting Out

Using a properly set up shielded metal arc welder and an air carbon arc torch, putting on all of your PPE, and following all shop and manufacturer's safety rules for welding, you will recut the repaired places to make new grooves in the checkerboard.

Follow the same starting procedure as you did in Project 23-1.

1. Start the arc on the weld.
2. Once the arc is struck, move the electrode in a straight line down the weld toward the other end. Watch the bottom of the cut to see that it is deep enough. If there is not a line along the bottom of the groove, it needs to be deeper. Once the groove depth is determined, keep the speed and angle of the torch constant.
3. When you reach the other end, break the arc off.
4. Raise your helmet, and stop the air.
5. Remove the remaining electrode from the torch so that it will not accidentally touch anything.
6. When the metal is cool, chip or brush any slag or dross off of the plate. This material should remove easily.
7. The groove must be within ±1/8 in. of being straight, but it may vary in depth so that all of the weld metal has been removed.
8. Make any adjustments to the setup or your technique, and repeat the grooving process until all of the J-grooves have been cut.

Finishing

Cool off the metal, chip the slag, and wire brush the surface. Follow the finishing instructions in Project 23-2.

Paperwork

Complete a copy of the time sheet in Appendix I and bill of materials in Appendix III, or as provided by your instructor.◆

SUMMARY

The welding field's ability to provide the industry with a wide variety of cutting processes has increased productivity for a variety of industries. An example of this diversity is the use of a laser beam to cut apart computer chips in the electronics field. Without these cutting processes, the industry would have to rely on the much slower and more expensive mechanical or abrasive cutting process. There are no mechanical or abrasive cutting processes that can compete with the speed of these new cutting processes and none that can compete with their versatility.

Somewhat less attractive but nonetheless efficient processes such as carbon arc gouging and oxygen lance cutting fulfill specific industrial applications, such as salvage or scrap work. Few cutting processes can rival the metal-removing capacity of the air carbon arc or oxygen lance processes. In addition, the low cost of the equipment and the flexibility and application of these processes have lent themselves very successfully to the salvage and scrap industry. Air carbon arc gouging is extensively used for weld removal and repair work. A skilled technician can produce a groove that requires little or no postcut cleanup prior to rewelding.

REVIEW QUESTIONS

1. List some of the ways that fabricators can use lasers.
2. What characteristic of a laser light beam allows it to travel a great distance?
3. What is the most popular industrial solid lasing material?
4. What are some of the most popular industrial gas lasing materials?
5. What do the abbreviations LBD, LBC, and LBW stand for?
6. What is the high pressure jet of gas used for when laser beam cutting?
7. List some of the advantages of laser cutting.
8. Why is a pulsed laser beam used for drilling?
9. List some of the materials that can be drilled with a laser.
10. What is the power density that a laser cutting machine can produce?
11. How is the oxygen lance lit so it can start cutting?
12. List some of the materials that an oxygen lance can be used to cut.
13. List some of the materials that a water jet can cut.
14. Why is an abrasive powder added to the water for some water jet cuts?
15. What equipment is needed to make a cut with an arc cutting electrode?
16. What improvements were made in the air arc cutting process?
17. What is the purpose of the copper coating on some air carbon arc cutting electrodes?
18. What is the range of the air pressure for air carbon arc cutting?
19. List some of the materials that can be cut using an air carbon arc torch.
20. What is the difference between gouging and washing?

Chapter 24

Other Welding Processes

OBJECTIVES

After completing this chapter, the student should be able to:

- Describe how resistance, ultrasonic, inertia, laser beam, plasma arc, and stud welding processes work.
- Describe the different applications of hardfacing.
- List some of the applications or uses for resistance, ultrasonic, inertia, laser beam, plasma arc, and stud welding processes.

KEY TERMS

electron beam welding (EBW)
hardfacing
inertia welding
laser beam welding (LBW)
plasma arc welding (PAW)
resistance welding (RW)
resistance spot welding (RSW)
stud welding (SW)
thermal spraying (THSP)
ultrasonic welding (USW)

INTRODUCTION

There are some 67 welding and joining processes listed by the American Welding Society (AWS). Most of the welding and joining processes are very unique, but some are slight variations of other processes. Of the seven processes briefly covered in this chapter, resistance welding and hardfacing are the ones you are most likely to see in a welding shop. You might have an opportunity to observe or use the five other processes only in a large production-type welding shop because of the high cost of the equipment.

RESISTANCE WELDING (RW)

In the **resistance welding (RW)** process the weld is made by clamping the parts together between the welding machine's electrodes. Then an electric current is passed through the parts to heat up the surfaces so they will fuse together.

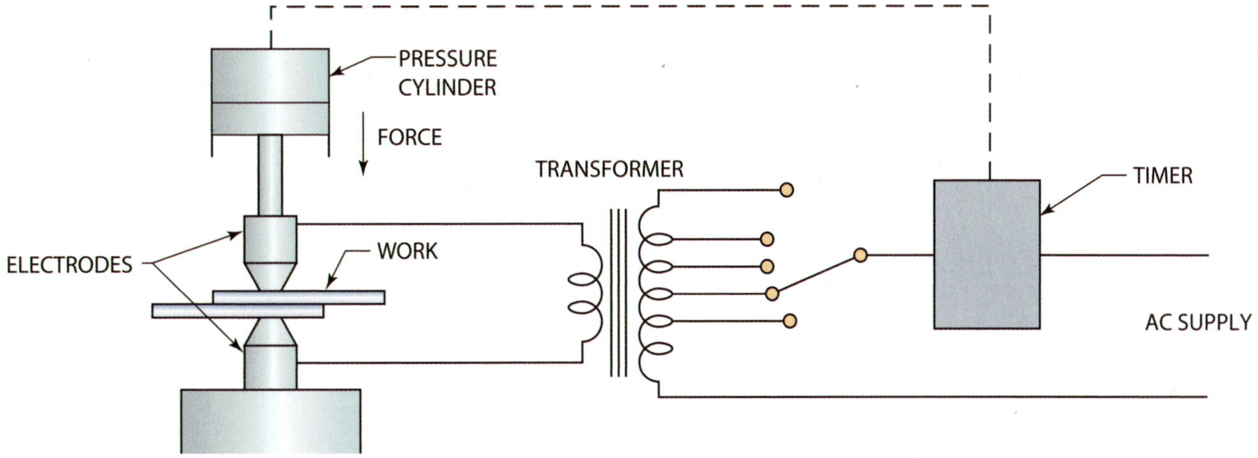

FIGURE 24-1 Fundamental resistance welding machine circuit. American Welding Society

The welding current required to make a resistance weld must be at a very low voltage but high amperage, Figure 24-1. The pressure is applied to ensure a continuous electrical circuit and to force the heated parts together. The parts are usually joined as a result of heat and pressure and are not simply melted together. Fluxes or filler metals are not needed for this welding process.

The current for resistance welding is usually supplied by either a transformer or a transformer/capacitor arrangement. The transformer, in both power supplies, is used to convert the high line voltage (low-amperage) power to the welding high-amperage current at a low voltage. A capacitor, when used, stores the welding current until it is used. This storage capacity allows such machines to use a smaller size transformer. The required pressure, or electrode force, is applied to the workpiece by pneumatic, hydraulic, or mechanical means. The pressure applied may be as little as a few ounces for very small welders to tens of thousands of pounds for large spot welders.

Most resistance welding machines consist of the following three components:

- The mechanical system to hold the workpiece and to apply the electrode force
- The electrical circuit made up of a transformer and, if needed, a capacitor, a current regulator, and a secondary circuit to conduct the welding current to the workpiece
- The control system to regulate the time of the welding cycle

There are several basic resistance welding processes. These processes include spot (RSW), seam (RSEW), high-frequency seam (RSEW-HF), projection (PW), flash (FW), upset (UW), and percussion (PEW).

Resistance welding is one of the most useful and practical methods of joining metal. This process is ideally suited to high-production methods.

Resistance Spot Welding (RSW)

Resistance spot welding (RSW) is the most common of the various resistance welding processes. In this process, the weld is produced by the heat obtained at the interface between the workpieces. This heat is due to the resistance to the flow of electric current through the workpieces, which are held together by pressure from the electrode, Figure 24-2. The size and shape of the formed welds are controlled somewhat by the size and contour of the electrodes.

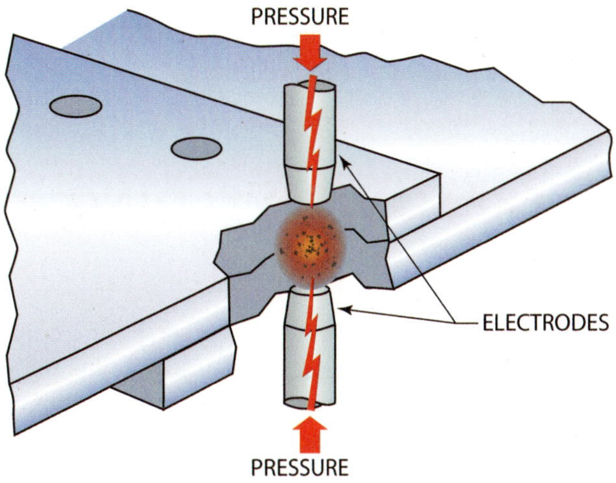

FIGURE 24-2 Heat resulting from resistance of the current through the metal held under pressure by the electrodes creates fusion of the two workpieces during spot welding. © Cengage Learning 2012

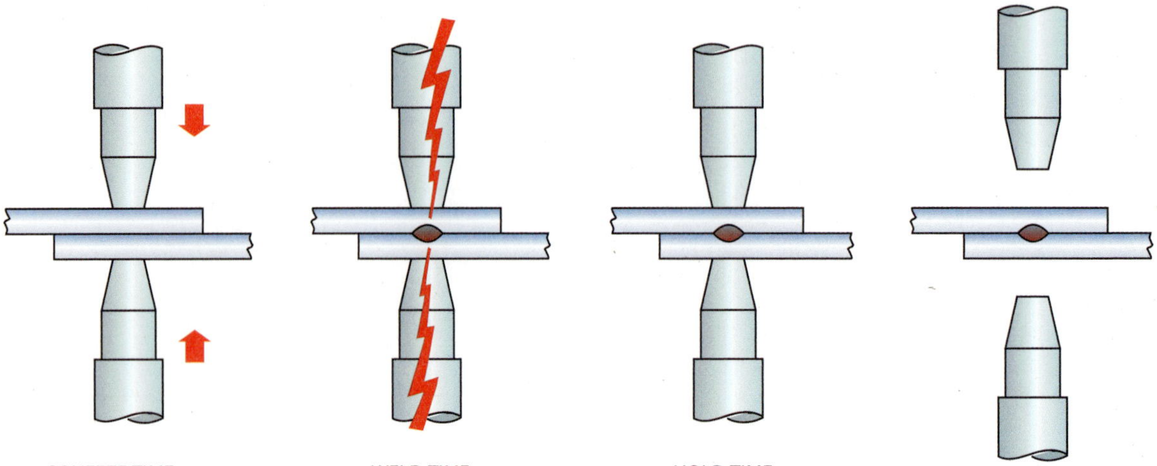

FIGURE 24-3 Basic periods of spot welding. © Cengage Learning 2012

The welding time is controlled by a timer built into the machine. The timer controls four different steps, Figure 24-3. The steps are as follows:

- Squeeze time, or the time between the first application of electrode force and the first application of welding current
- Weld time, or the actual time the current flows
- Hold time, or the period during which the electrode force is applied and the welding current is shut off
- Off period, or the time during which the electrodes are not contacting the workpieces

Tables supplied by the machine manufacturer provide information for the exact time for each stage for different types and thicknesses of metal.

Material from 0.001 in. (0.0254 mm) to 1 in. (25 mm) thick may be joined by spot welding.

ULTRASONIC WELDING (USW)

Ultrasonic welding (USW) is a process for joining similar and dissimilar metals by introducing high-frequency vibrations into the overlapping metals in the area to be joined. Fluxes and filler metals are not required, electrical current does not pass through the weld metal, and only localized heating is generated. The temperature produced is below the melting point of the materials being joined. Thus, no melting occurs during the welding cycle.

Ultrasonic welding has many applications in the assembly of electrical products. Typical applications include the following:

- Attaching oxide-resistant contact surface buttons to switches
- Attaching leads to coils of foil, sheet, or wire made of aluminum
- Attaching very fine wire leads and elements to other components

Figure 24-4 shows an ultrasonic spot welder used to perform the types of welds just listed.

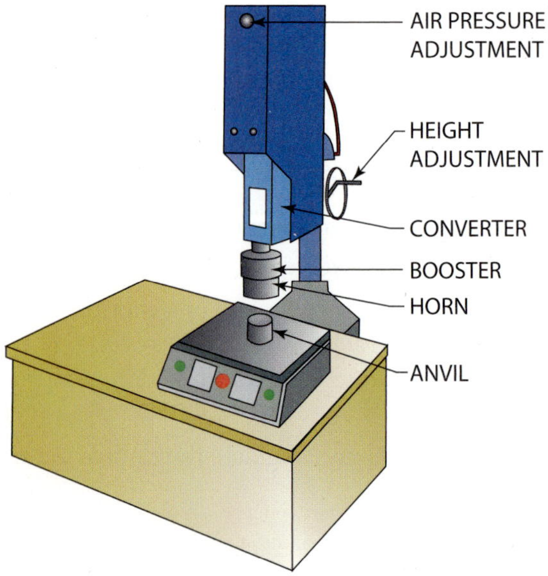

FIGURE 24-4 Ultrasonic spot welder.
© Cengage Learning 2012

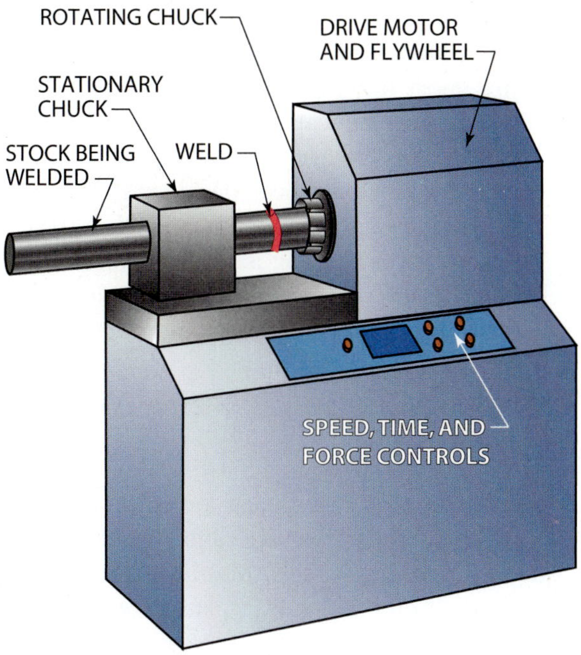

FIGURE 24-5 Inertia welder. © Cengage Learning 2012

INERTIA WELDING PROCESS

Inertia welding is a form of friction welding. In inertia welding, one workpiece is fixed in a stationary holding device, Figure 24-5. The other is clamped in a spindle chuck, which is accelerated rapidly. At a predetermined speed, power is cut, as shown in Figure 24-6A. As a result, one part is then thrust against the other piece. Friction between the parts causes the spindle to decelerate, converting stored energy to frictional heat. Enough heat is formed to soften, but not melt, the faces of the part, Figure 24-6B.

Some of the advantages of the inertia welding process are as follows:

- Superior weld
- A very narrow heat-affected zone adjacent to the weld
- Uniform production welds
- Fast production welds
- Clean operation
- Lowest cost of energy
- Minimum skill required to operate the welder
- The amount of upset of parts can be controlled to close tolerances
- A complete interface weld can be obtained
- Safe to operate

LASER BEAM WELDING (LBW)

In **laser beam welding (LBW)** fusion is obtained by directing a highly concentrated beam of coherent light to a very small spot, Figure 24-7. Laser beams combine

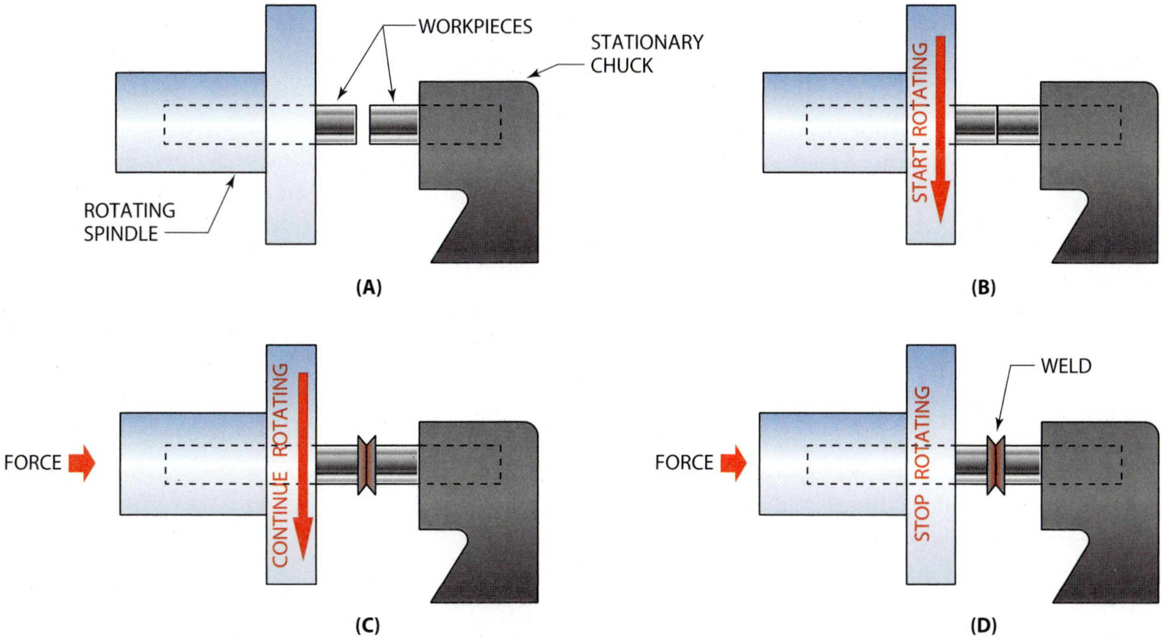

FIGURE 24-6 Inertia welding process. American Welding Society

FIGURE 24-7 Gas filter, laser welded. Preco, Inc.

low-heat input (0.1 to 10 joules) with high-power intensity of more than 10,000 watts per square centimeter (considerably more than the electron beam). Because the heat is provided by a beam of light, there is no physical contact between the workpiece and the welding equipment. It is possible to make welds through transparent materials.

The ease with which the beam can be directed to any area of the work makes laser welding very flexible. In the manufacture of complex forms, for example, it is possible to move the focused laser beam under digital control to seam weld any desired shape. The proven high-quality laser welds, along with the flexibility and the comparatively moderate cost of laser welding equipment, indicate that laser welding plays an increasing role in microelectronics and other light-gauge metal welding applications.

Laser welding of materials with high thermal conductivity, such as copper, is not difficult to do. The extremely concentrated laser heat will melt the metal locally to make a weld or will vaporize the metal to drill a hole before it can be conducted away by the copper, which happens in most other forms of welding.

Advantages and Disadvantages of Laser Welding

Laser welding has some distinct advantages and disadvantages when compared to other welding processes. **Electron beam welding (EBW)** is the only method that rivals the heat output of a laser. Generally, however, electron beam welding must operate in a vacuum. Since the laser beam is a light beam, it can operate in air or any transparent material, and the source of the beam need not even be close to the work. The material being welded need not be an electrical conductor that limits most other processes or even part of an electrical or a mechanical circuit. However, the light may be diffused by the welding vapors, so techniques to bypass the vapors have been developed.

Laser welds are small, sometimes less than 0.001 in. (0.0254 mm). Laser welding is used to connect leads to elements in integrated circuitry for electronics. Lead wires insulated with polyurethane can be welded without removing the insulation.

Using the laser welding process, it is possible to weld heat-treated alloys without undoing the heat treatment.

This method of welding can be used to join dissimilar metals. Metals that are difficult to weld, such as tungsten, stainless steels, titanium alloys, Kovar, nickel alloy, aluminum, and tin-plated steels, can also be successfully welded by this process.

An optical system is used to focus the beam on the workpiece. A switch controls the welding energy.

PLASMA ARC WELDING (PAW) PROCESS

The term *plasma* should be defined in its electrical sense. A gas, or plasma, is present in any electrical discharge if sufficient energy is present. The plasma consists of charged particles that transport the charge across the gap.

The two outstanding advantages of plasmas are higher temperature and better heat transfer to other objects. The higher the temperature differential between the heating fluid and the object to be heated, the faster the object can be heated.

In **plasma arc welding (PAW)**, a plasma jet is produced by forcing gas to flow along an arc restricted electromagnetically as it passes through a nozzle, Figure 24-8. The stiffness of the arc is increased by its decreased cross-sectional area. As a result, you have better control of the weld pool. By forcing the gas into the arc stream, it is heated to its ionization temperature, where it forms free electrons and positively charged ions. The plasma jet produced resembles a brilliant flame. The tip of the electrode is situated above the opening in the torch nozzle, which constricts the arc. A plasma welding installation is shown in Figure 24-9. The plasma jet passing through the restraining orifice has an accelerated velocity.

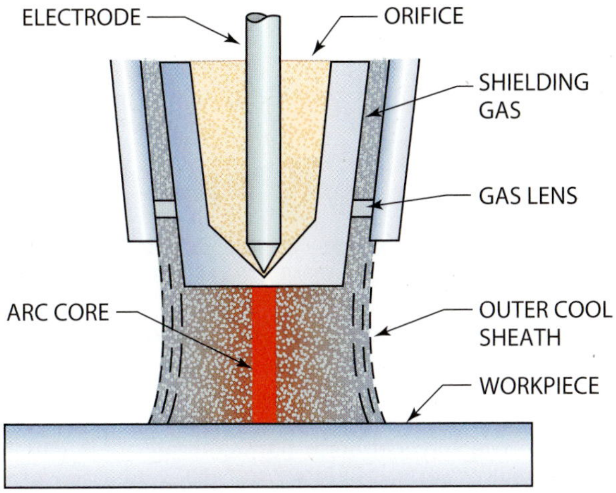

FIGURE 24-8 Schematic diagram of plasma welding process. American Welding Society

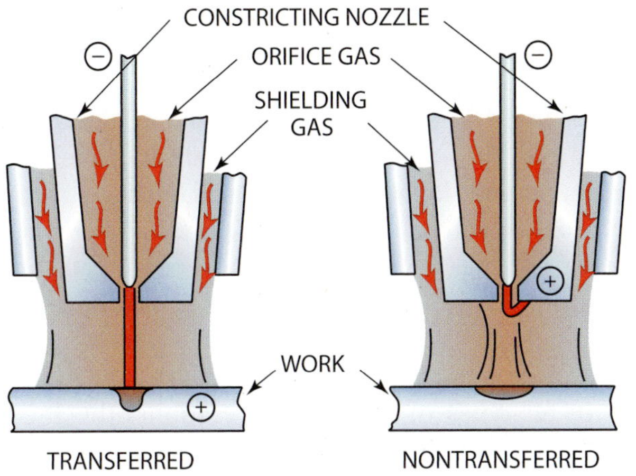

FIGURE 24-9 Transferred and nontransferred plasma arc modes. American Welding Society

When this intense arc is directed on the workpiece, it is possible to make a welded butt joint in metal having a thickness of up to 1/2 in. (13 mm) in a single pass. No edge preparation or filler metal is required.

Any known metal can be melted, even vaporized, by the plasma jet process, making it useful for many welding operations. This process can be used to weld carbon steels, stainless steels, Monel, Inconel, aluminum, copper, and brass alloys.

STUD WELDING (SW)

Stud welding (SW) is a semiautomatic or automatic arc welding process. An arc is drawn between a metal stud and the surface to which it is to be joined. When the end of the stud and the underlying spot on the surface of the work have been properly heated, they are brought together under pressure.

The process uses a pistol-shaped welding gun, which holds the stud or fastener to be welded. When the trigger of the gun is pressed, the stud is lifted to create an arc and is then forced against the molten pool by a backing spring. The operation is controlled by a timer. The arc is shielded by surrounding it with a ceramic ferrule, which confines the metal to the weld area.

In the welding operation, a stud is loaded in the chuck of the gun, and the ferrule is fastened over the stud. The gun is then placed on the workpiece. The action of the gun when the trigger is squeezed causes the stud to pull away from the workpiece, resulting in an arc. The arc melts the end of the stud and an area on the workpiece. At the correct moment, a timing device shuts off the current and causes the spring to plunge the stud into the molten pool, which freezes instantly. The gun is then released from the stud and the ferrule knocked off.

HARDFACING

Hardfacing is defined as the process of obtaining desired properties or dimensions by applying, using oxyfuel or arc welding, an integral layer of metal of one composition onto a surface, an edge, or the point of a base metal of another composition. The hardfacing operation makes the surface highly resistant to abrasion.

There are various techniques of hardfacing. Some apply a hard surface coating by fusion welding. In other techniques, no material is added but the surface metal is changed by heat treatment or by contact with other materials.

Several properties are required of surfaces that will be subjected to severe wearing conditions, including hardness, abrasion resistance, impact resistance, and corrosion resistance.

Hardfacing may involve building up surfaces that have become worn. Therefore, it is necessary to know how the part will be used and the kind of wear to expect. In this way, the proper type of wear-resistant material can be selected for the hardfacing operation.

When a part is subjected to rubbing or continuous grinding, it undergoes abrasion wear. When metal is deformed or lost by chipping, crushing, or cracking, impact wear results.

Selection of Hardfacing Metals

Many different types of metals and alloys are available for hardfacing applications. Most of these materials can be deposited by any conventional manual or automatic arc or oxyfuel welding method. Deposited layers may be as thin as 1/32 in. (0.794 mm) or as thick as necessary. The proper selection of hardfacing materials will yield a wide range of characteristics.

Steel or special hardfacing alloys should be used where the surface must resist hard or abrasive wear. Where surfacing is intended to withstand corrosion-type or friction-type wear, bronze or other suitable corrosion-resistant alloys may be used.

Most hardfacing metals have a base of iron, nickel, copper, or cobalt. Other elements that can be added include carbon, chromium, manganese, nitrogen, silicon, titanium, and vanadium. The alloying elements have a tendency to form carbides. Hardfacing metals are provided in the form of rods for oxyacetylene welding, electrodes for shielded metal arc welding, or in hard wire form for automatic welding. Tubular rods containing a powdered metal mixture, powdered alloys, and fluxing ingredients can be purchased from various manufacturers.

Many hardfacing materials are designated by manufacturers' trade names. Some of the materials have AWS designations. AWS materials are classified into the following designations:

- High-speed steel
- Austenitic manganese steel
- Austenitic high chromium iron
- Cobalt-base metals
- Copper-base alloy metals
- Nickel-chromium boron metals
- Tungsten carbides

The coding system identifies the important elements of the hardfacing metal. The prefix *R* is used to designate a welding rod, and the prefix *E* indicates an electrode. Certain materials are further identified by the addition of digits after a suffix.

Hardfacing Welding Processes

Oxyfuel Welding

In hardfacing operations, oxyfuel welding permits the surfacing layer to be deposited by flowing molten filler metal into the underlying surface. This method of surfacing is called sweating or tinning, Figure 24-10.

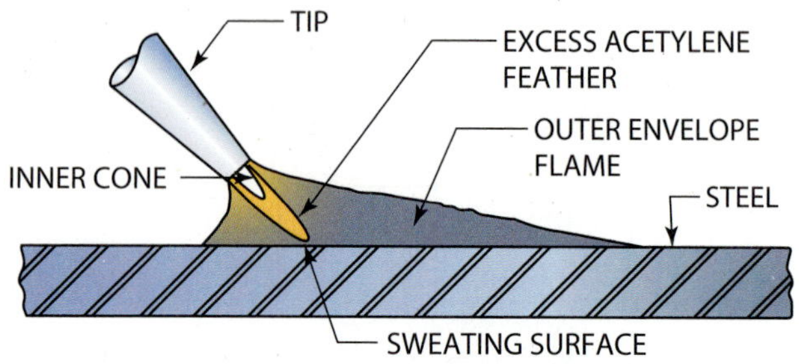

FIGURE 24-10 **An example of how to produce sweating.**
American Welding Society

With the oxyacetylene flame, small areas can be hardfaced by applying thin layers of material. In addition, the alloy can be easily flowed to the corners and edges of the workpiece without overheating or building up deposits that are too thick. Placement of the metal can be controlled accurately.

The size of the weld is affected by many factors. These factors include the rate of travel, degree of preheat, type of metal being deposited, and thickness of the work.

Figure 24-11 shows the approximate relationship of the tip, rod, molten pool, and base metal during the hardfacing operation for both backhand and forehand travel.

Iron, nickel, and cobalt-base alloys require a carburizing flame. Copper alloys and bronze call for a neutral or slightly oxidizing flame. Laps, blowholes, and poor adhesion of deposits can be prevented by a flame characteristic that is soft and quiet.

In all types of surfacing operations, the metal should be cleaned of all loose scale, rust, dirt, and other foreign substances before the alloy is applied. The best method of removing these impurities is by grinding or machining the surface. Fluxes may be used to maintain a clean surface. They also help to overcome oxidation that may develop during the operation.

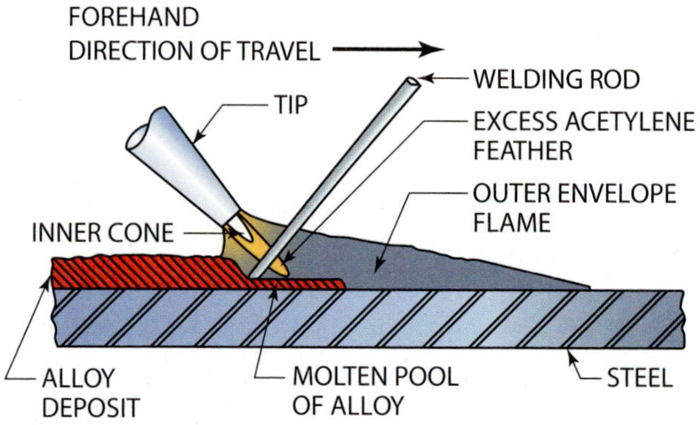

FIGURE 24-11 **Approximate relationship of the tip, rod, and molten weld pool for forehand hardfacing.**
American Welding Society

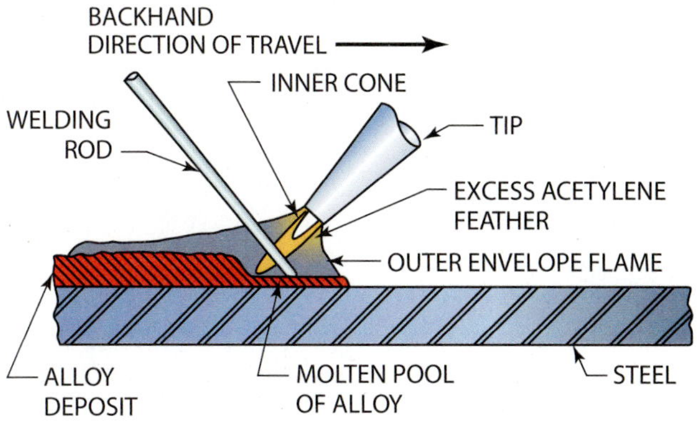

FIGURE 24-12 Backhand method of hardfacing.
American Welding Society

Conventional methods may be used in holding the torch and rod. Figure 24-12 shows the backhand method of hardfacing.

If the base metal is cast iron, it will not "sweat" like steel. Therefore, slightly less acetylene should be used. Alloys do not flow as readily on cast iron as they do on steel. Usually, it is necessary to break the surface crust on the metal with the end of the rod. A cast-iron welding flux is generally necessary. The best method is to apply a thin layer of the alloy and then build on top of it.

The oxyacetylene process is preferred for small parts. Cracking can be minimized by using adequate preheat, postheat, and slow cooling. Shielded metal arc welding is preferred for large parts.

Arc Welding

Hardfacing by arc welding may be accomplished by shielded metal arc, gas metal arc, gas tungsten arc, submerged arc, plasma arc, or other processes.

The techniques employed for any one of these processes are similar to those used in welding for joining. The factor of dilution must be carefully considered because the composition of the added metal will differ from the base metal. The least amount of dilution of filler metal with base metal is an important goal, especially where the two metals differ greatly. Little dilution means that the deposited metal maintains its desired characteristics. When using alloys with high melting points, dilution of the weld metal is usually kept well below 15%.

Hardfacing by the arc welding method has many advantages, including high rates of deposition, flexibility of operation, and ease of mechanization.

Hardfacing may be applied to many types of metals, including low and medium carbon steels, stainless steels, manganese steels, high-speed steels, nickel alloys, white cast iron, malleable cast iron, gray and alloy cast iron, brass, bronze, and copper.

Quality of Surfacing Deposit

The type of service to which a part is to be exposed governs the degree of quality required of the surfacing deposit. Some applications require that the deposited metal contain no pinholes or cracks. For other applications, these requirements are of little importance. In most cases, the quality of the deposited metals can be very high. Steel-base alloys do not tend to crack, while other materials, such as high-alloy cast steels, are subject to cracking and porosity.

Hardfacing Electrodes

The proper type of surfacing electrode must be selected, as one type of electrode will not meet all requirements. Most electrodes are sold under manufacturers' trade names.

Electrodes may be classified into the following three general groups:

- Resistance to severe abrasion
- Resistance to both impact and moderate abrasion
- Resistance to severe impact and moderately severe abrasion

Tungsten carbide and chromium electrodes are included in the first group. The material deposited is very hard and abrasive resistant. These electrodes can be one of two types, either coated tubular or regular coated cast alloy. The tubular types contain a mixture of powdered metal, powdered ferroalloys, and fluxing materials. The tubes are the coated type. These electrodes are used with the electric arc.

Electrodes contain small tungsten carbide crystals embedded in the steel alloy. After this material is applied to a surface, the steel wears away with use, leaving the very hard tungsten carbide particles exposed. This wearing away of steel results in a self-sharpening ability of the surfacing material. Cultivator sweeps and scraping tools are among parts that are surfaced with this material, Figure 24-13.

Chromium carbide electrodes are tougher than tungsten carbide–type electrodes. However, chromium carbide electrodes are not as hard and are less abrasion resistant. This material is too hard to be machined, but it has good corrosion-resistant qualities.

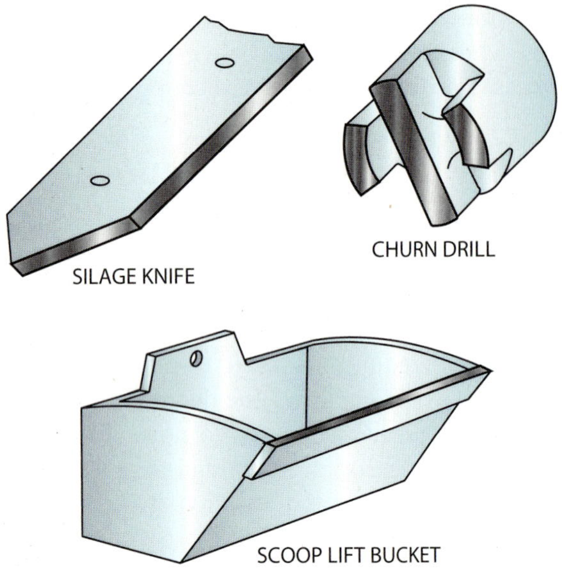

FIGURE 24-13 Farm tools that can be hardfaced with tungsten carbide electrodes to increase the life of the tools. © Cengage Learning 2012

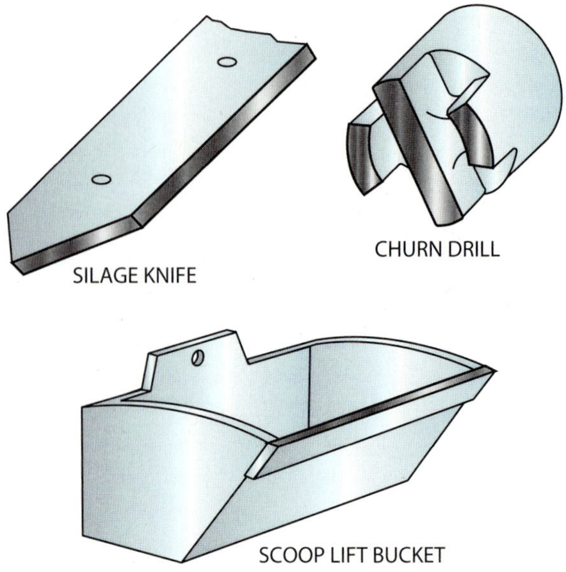

FIGURE 24-14 Products that are hardfaced to produce moderate impact resistance and severe abrasion resistance. © Cengage Learning 2012

The electrodes in the second group are the high carbon type. When used for surfacing, these electrodes leave a tough and very hard deposit. Examples of hardfaced products in this group include gears, tractor lugs, and scraper blades.

The third group of electrodes is used for surfacing rock-crusher parts, links, pins, railroad track components, and parts where severe abrasion resistance is a requirement, Figure 24-14. Deposits from these electrodes are very tough but not hard. The surface toughness protects the softer base metal below from damage, and the soft base metal seems to prevent the surfacing material from cracking.

Shielded Metal Arc Method

- Start the process by cleaning the surface.
- Since most hardfacing electrodes are too fluid for out-of-position welding, the work should be arranged in the flat position.
- Set the amperage so that just enough heat is provided to maintain the arc. Too much heat will cause excessive dilution.
- Hold a medium-long arc, using either a straight or weaving pattern. When a thin bead is required, use the weave pattern and keep the weave to a width of 3/4 in. (19 mm).
- If more than one layer is required, remove all slag before placing other layers.

Hardfacing with gas tungsten arc (GTA), gas metal arc (GMA), and flux core arc (FCA) welding processes may be used in hardfacing operations. These three processes, in many instances, are better methods of hardfacing because of the ease with which the metal can be deposited. In addition, the hardfacing materials may be deposited to form a porosity-free, smooth, and uniform surface.

Where the job calls for cobalt-base alloys, the GTA method does an effective job. Very little preheating of the base is required. The GMA and FCA welding processes are somewhat faster than surfacing by GTA due to the fact that continuous wire is used.

Care must be exercised when using the GMA, FCA, and GTA welding processes for hardfacing in order to avoid dilution of the weld. Helium or a mixture of helium-argon normally produces a higher arc voltage than pure argon. For this reason, the dilution of the weld metal increases. An argon and oxygen mixture should be used for surfacing with the gas metal arc processes and argon used with gas tungsten arc processes. When using FCAW, shielding may be provided as either shielding gas or self-shielding. The self-shielding hardsurfacing process is used when working outdoors because of its ability to better resist the effect of light winds.

THERMAL SPRAYING (THSP)

Thermal spraying (THSP) is the process of spraying molten metal onto a surface to form a coating. Pure or alloyed metal is melted in a flame or an electric arc and atomized by a blast of compressed air. The resulting fine spray builds up on a previously prepared surface to form a solid metal coating. Because the molten metal is accompanied by a blast of air, the object being sprayed does not heat up very much. Therefore, thermal spraying is known as a "cold" method of building up metal, Figure 24-15.

Thermal Spraying Equipment

A thermal spraying installation requires, at a minimum, the following equipment: air compressor, air control unit, air flowmeter, oxyfuel gas or arc equipment, and exhaust equipment, Figure 24-16.

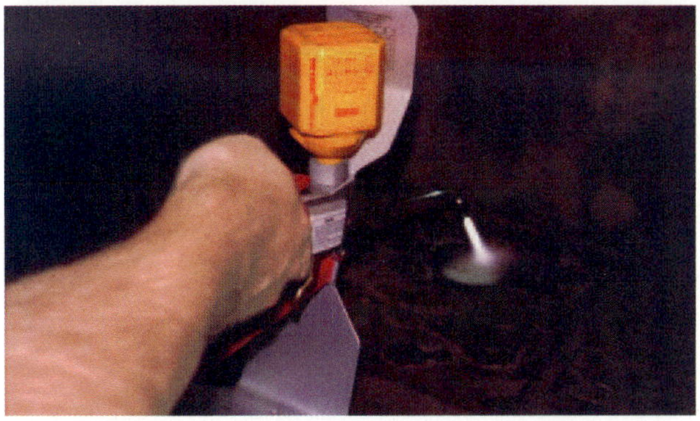

FIGURE 24-15 **Rebuilding a worn crankshaft bearing with a Mogul Turbo-jet thermal spraying gun.** Eutectic Corporation

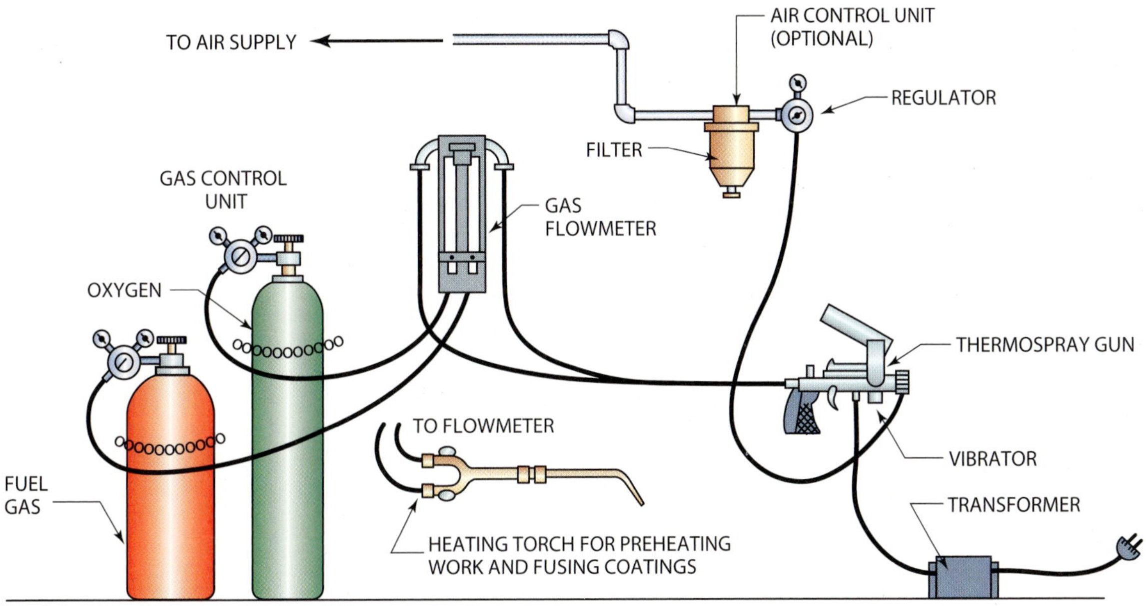

FIGURE 24-16 **Complete thermal spray installation.** American Welding Society

SUMMARY

Of the nearly 70 welding processes in use today, only a few are commonly used. Often there are processes, not commonly used, that if applied to a weldment could increase productivity and reduce cost on a specific job. Having studied this chapter, you are in a better position to select the most cost-effective process. Understanding the opportunities that are afforded by these various processes will make you far more competitive in the labor market and business world. For example, a company may be using a torch brazing process, when a furnace braze may be more effective. Sometimes we become comfortable with our knowledge and abilities with a single process and fail to look at all of the emerging technology's opportunities.

Good sources of current knowledge are welding applications, manufacturers' literature, and the Internet. You should become a member of a professional organization, such as the American Welding Society; this will provide you with up-to-date process information. Being a top-notch professional welder is a lifelong learning activity.

REVIEW QUESTIONS

1. What two forces cause parts to be joined together in resistance welding?
2. What is the function of a transformer in resistance welding?
3. What three components do most resistance welding machines have?
4. How is the weld produced in spot welding?
5. What are some applications of ultrasonic welding in the assembly of electrical products?
6. How is heat generated in the inertia welding process?
7. List advantages of the inertia welding process.
8. How is heat generated in laser beam welding?
9. How is the plasma jet produced in the plasma arc welding process?
10. What metals can be melted by the plasma jet process?
11. Describe the stud welding process.
12. Why is hardfacing used?
13. What should be considered when selecting hardfacing metals?
14. Describe the sweating or tinning method of surfacing.
15. Why must the factor of dilution be carefully considered when hardfacing by arc welding?
16. In what three general groups can electrodes be classified?
17. Why should the work be arranged in the flat position for the shielded metal arc method?
18. Why can GTA, GMA, and FCA be better methods of hardfacing?
19. What is thermal spraying?

Chapter 25

Welding Automation and Robotics

OBJECTIVES

After completing this chapter, the student should be able to:

- Describe the manual, semiautomatic, machine, automatic, and automated joining processes.
- Explain the role that the welder plays in the operation of the manual, semiautomatic, machine, automatic, and automated joining processes.
- Give examples of the types of applications the manual, semiautomatic, machine, automatic, and automated joining processes are used for.

KEY TERMS

automated joining
automatic joining
cycle time
industrial robot
machine joining
manipulator
manual joining
pick-and-place robot
semiautomatic joining
sensors

INTRODUCTION

The first industrial robots were **pick-and-place robots** that were used to move material with little repetitive accuracy required. Today, computers have improved the accuracy, reliability, and functionality so that robots now serve as intelligent controllers for automation. The decreasing cost and advancements have made this technology possible, even for small businesses.

Automation has allowed manufacturers to increase productivity and cut costs, which makes their products more competitively priced.

This chapter gives you a general overview of automatic welding processes and robotics.

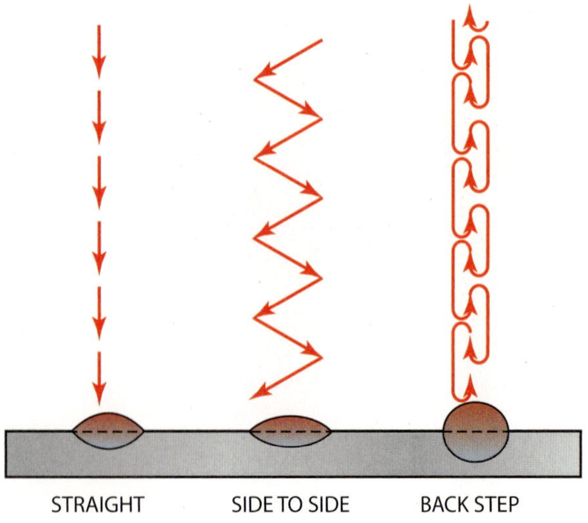

FIGURE 25-1 The electrode manipulation affects the size and shape of the weld bead. © Cengage Learning 2012

Manual Joining Processes

A **manual joining** process is one that is completely performed by hand. You control all of the manipulation, rate of travel, joint tracking, and, in some cases, the rate at which filler metal is added to the weld. The manipulation of the electrode or torch in a straight line or oscillating pattern affects the size and shape of the weld, Figure 25-1. The manipulation pattern may also be used to control the size of the weld pool during out-of-position welding. The rate of travel or speed at which the weld progresses along the joint affects the width, reinforcement, and penetration of the weld, Figure 25-2. The placement or location of the weld bead within the weld joint affects the strength, appearance, and possible acceptance of the joint. The rate at which filler metal is added to the weld affects the reinforcement, width, and appearance of the weld, Figure 25-3.

The most commonly used manual arc (MA) welding process is shielded metal arc welding (SMAW). The flexibility you have in performing the weld makes this process one of the most versatile. By changing the manipulation, rate of travel, or joint tracking, you can make an acceptable weld on a variety of material thicknesses, Figure 25-4.

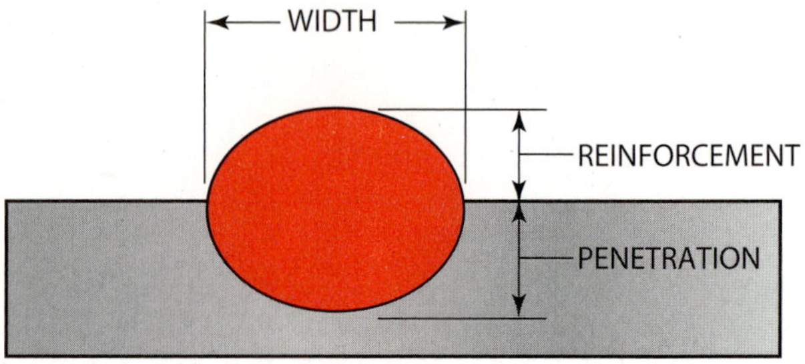

FIGURE 25-2 The travel rate of the weld affects width, reinforcement, and penetration of the weld bead. © Cengage Learning 2012

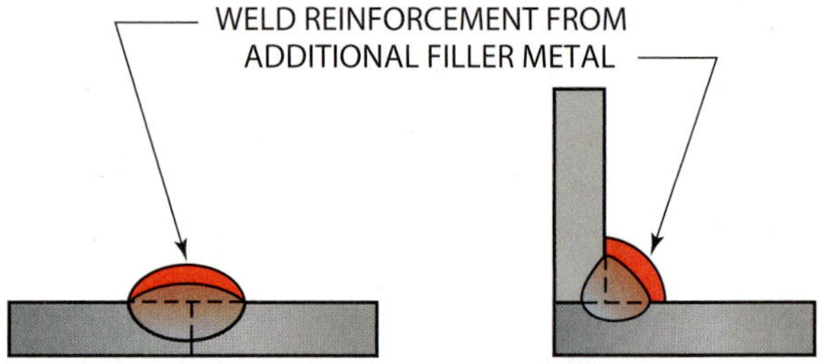

FIGURE 25-3 Addition of filler metal. © Cengage Learning 2012

Arc
Shielded metal arc welding (SMAW)
Gas tungsten arc welding (GTAW)
Gas
Oxyacetylene welding (OAW)
Brazing
Torch brazing (TB)

Table 25-1 Manual Joining Processes

The most commonly used manual arc welding, gas welding, and brazing processes are listed in Table 25-1.

Semiautomatic Joining Processes

A **semiautomatic joining** process is one in which the filler metal is fed into the weld automatically. You control most other functions. The addition of filler metal to the weld by an automatic wire-feeder system enables you to increase the uniformity of welds, productivity,

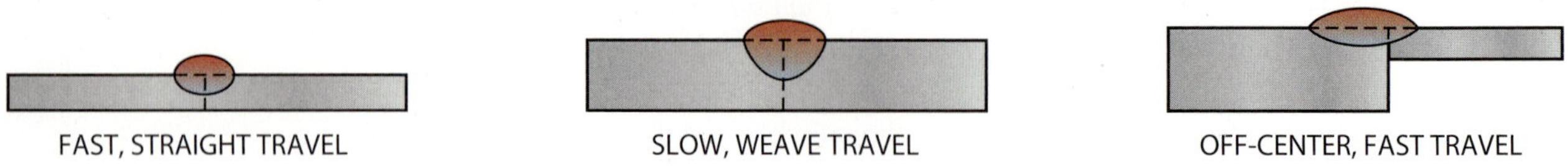

FIGURE 25-4 Controlling weld bead size by adjusting welding parameters. © Cengage Learning 2012

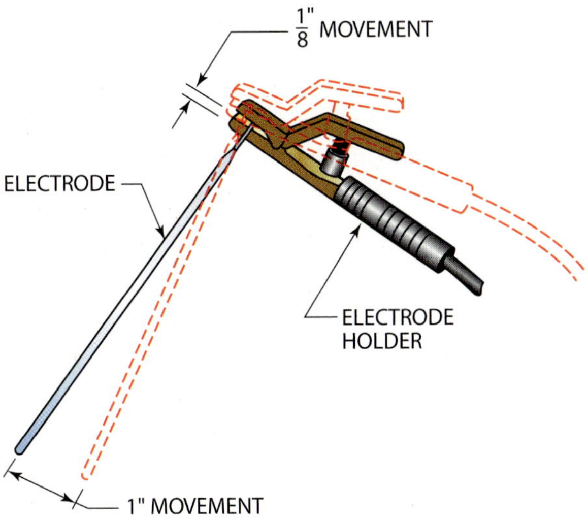

FIGURE 25-5 A 1/8-in. (3-mm) movement of the electrode holder results in a 1-in. (25-mm) movement of the electrode at the surface of the work.

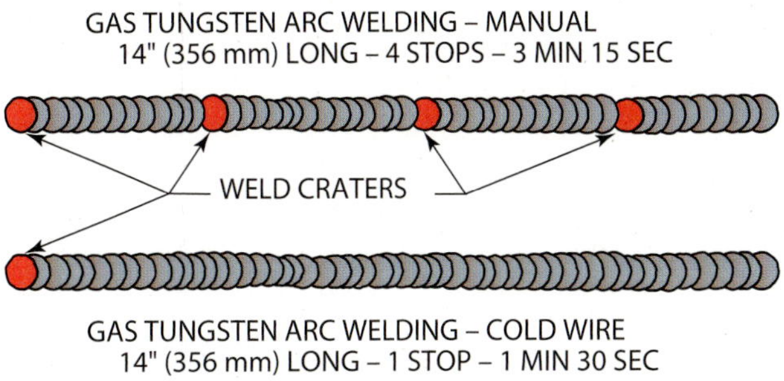

FIGURE 25-7 Fewer weld craters occur in continuous welding and it is much faster.

and weld quality. The distance of the welding gun or torch from the work remains constant. This gives you better manipulative control as compared to, for example, shielded metal arc welding, in which the electrode holder starts at a distance of 14 in. (356 mm) from the work. This distance exaggerates the slightest accidental movement made during the first part of the weld, Figure 25-5. In the SMAW process, the electrode holder must be lowered steadily as the weld progresses to feed the electrode and maintain the correct arc length, Figure 25-6. This constant changing of the distance above the work causes you to shift your body position frequently. This change, too, may affect the consistency of the weld.

Because the filler metal is being fed from a large spool, you do not have to stop welding to change filler electrodes or filler metal. SMA electrodes cannot be used completely as they have a waste stub of approximately 2 in. (51 m). This waste stub represents approximately 15% of the filler metal that must be discarded. The frequent stopping for rod and electrode changes, followed by restarting, wastes time and increases the number of weld craters. These craters are often a source of cracks and other weld discontinuities, Figure 25-7. In some welding procedures each weld crater must be chipped and ground before the weld can be restarted. These procedures can take up to 10 minutes—time that can be used for welding in a semiautomatic welding process.

The most commonly used semiautomatic (SA) arc welding processes are gas metal arc welding (GMAW)

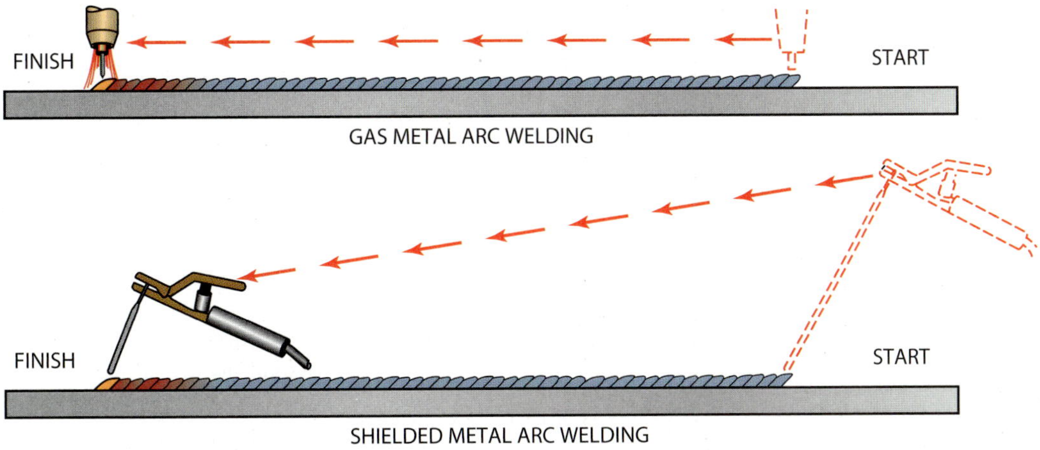

FIGURE 25-6 In GMAW, the torch height remains constant above the work surface. In SMAW, the height of the electrode holder steadily decreases from the beginning to end of the weld.

Welding Processes
Gas metal arc welding (GMAW)
Flux cored arc welding (FCAW)
Submerged arc welding (SAW)
Gas tungsten arc welding (GTAW)
Cold-hot wire feed

Table 25-2 Semiautomatic Joining Processes

and flux cored arc welding (FCAW). Table 25-2 lists several other semiautomatic processes.

Machine Joining Processes

A **machine joining** process is one in which the welding is performed by equipment and you control the welding progress by making adjustments as required. The parts being joined may or may not be loaded and unloaded automatically. You may monitor the joining progress by watching it directly, observing instruments only, or using a combination of both methods. Adjustments in travel speed, joint tracking, work-to-gun or work-to-torch distance, and current settings may be needed to ensure that the joint is made according to specifications.

The work may move past a stationary welding or joining station, Figure 25-8, or it may be held stationary and the welding machine moves on a beam or track along the joint, Figure 25-9. On some large machine welds, the operator may ride with the welding head along the path of the weld. During the assembly of the external fuel tanks used for the space shuttle, two operators were required for a few of the machine welds. One operator watched the root side of the weld while the other observed the face side of the weld. They were able to communicate with each other so that any needed changes could be made.

To minimize adjustments during machine welds, a test weld is often performed just before the actual weld is produced. This practice weld helps increase the already high reliability of machine welds.

Automatic Joining Processes

An **automatic joining** process is a dedicated process (designed to do only one type of welding on a specific part) that does not require you to make adjustments during the actual welding cycle. All operating guidelines are preset, and you may or may not have to load or unload parts. Automatic equipment is often dedicated to one type of product or part. A large investment is usually required in jigs and fixtures used to hold the parts to be joined in the proper alignment. The operational cycle is computer controlled. The cycle may be as simple as starting and stopping points, or it may be more complex. A more complex cycle may include such steps as prepurge time, hot start, initial current, pulse power, downslope, final current, and postpurge time, Figure 25-10.

Automatic welding or brazing is best suited to large-volume production runs because of the expense involved in special jigs and fixtures.

Automated Joining

Automated joining processes are similar to automatic joining except that they are flexible and more easily adjusted or changed. Unlike automatic joining, there is no dedicated machine for each product. The equipment can be easily adapted or changed to produce a wide variety of high-quality welds.

The industrial robot is rapidly becoming the main component in automated welding or joining stations. The welding or joining cycles are controlled by computers or microprocessors. The flexibility provided by automated workstations makes it possible for even small companies with limited production runs to invest in automated equipment. The equipment is controlled by programs, or a series of machine commands expressed in numerical codes, that direct the welding, cutting, assembling, or any other activities. The programs can be stored and quickly changed. Some systems can store and retrieve many different programs internally. Other systems are controlled by a host computer. Both types of systems can speed up production when frequent changes are required.

Industrial Robots

An **industrial robot** is a "reprogrammable, multifunctional **manipulator** designed to move material, parts, tools, or specialized devices through variable programmed motions for the performance of a variety of tasks."* Industrial robots are primarily powered by electric stepping motors, hydraulics, or pneumatics and are controlled by a program, Figure 25-11.

Robots can be used to perform a variety of industrial functions, including grinding, painting, assembling,

**Robot Institute of America.*

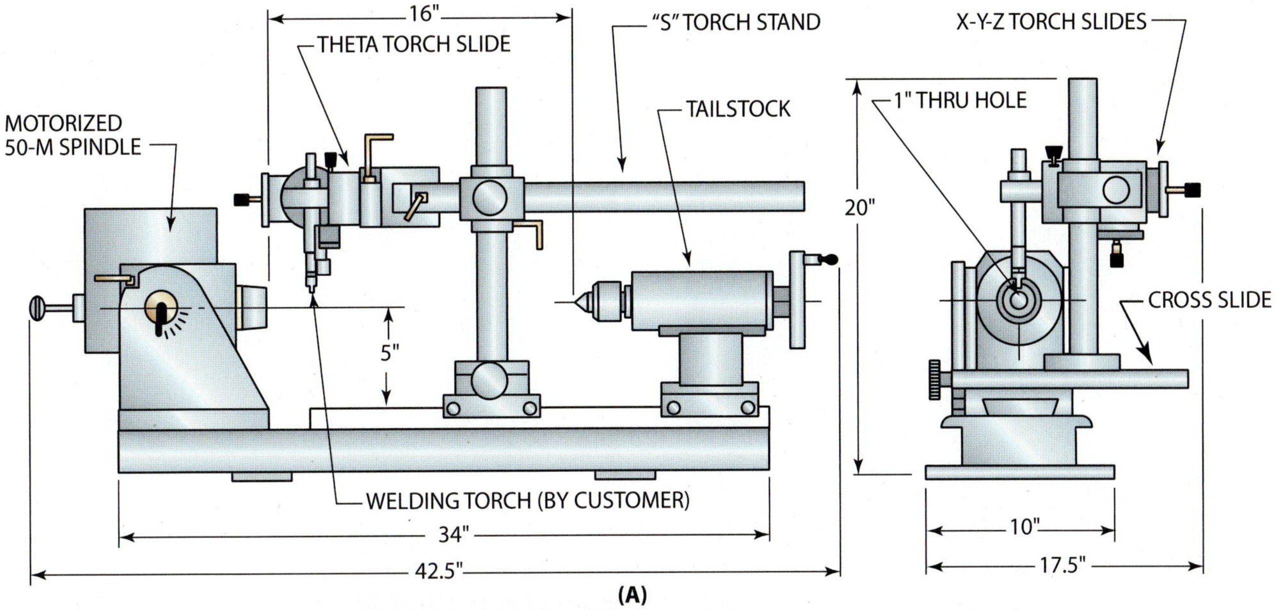

(A)

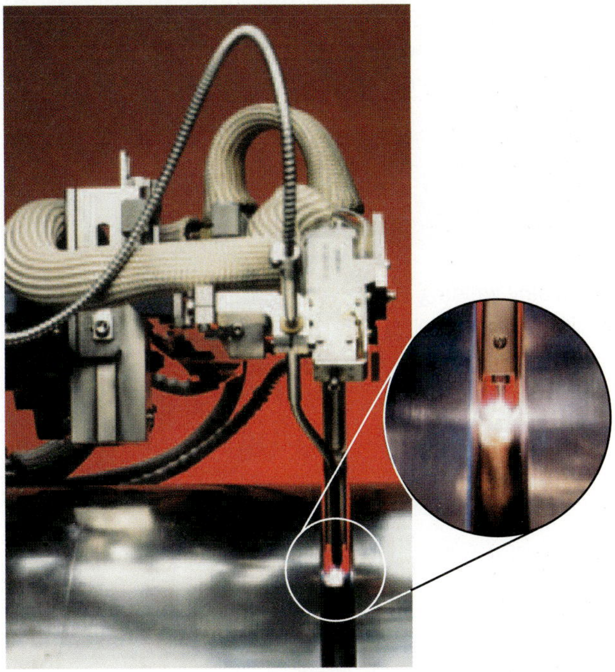
(B)

FIGURE 25-8 **Two precision bench-welding systems.** (A) Larry Jeffus; (B) ARC Machines, Inc.

machining, inspecting, flame cutting, product handling, and welding.

Robots range in size and complexity from small desktop units capable of lifting only a few ounces to large floor models capable of lifting tons. Most robots can perform movements in three basic directions: longitudinal (X), transverse (Y), and vertical (Z), Figure 25-12. The tool end of the robot arm may also be jointed so that it can tilt and rotate, Figure 25-13.

The robot may be used with other components to increase production and the flexibility of the system. A computer or microprocessor can synchronize the robot's operation to petitioners, conveyors, automatic fixtures, and other production machines. Parallel or

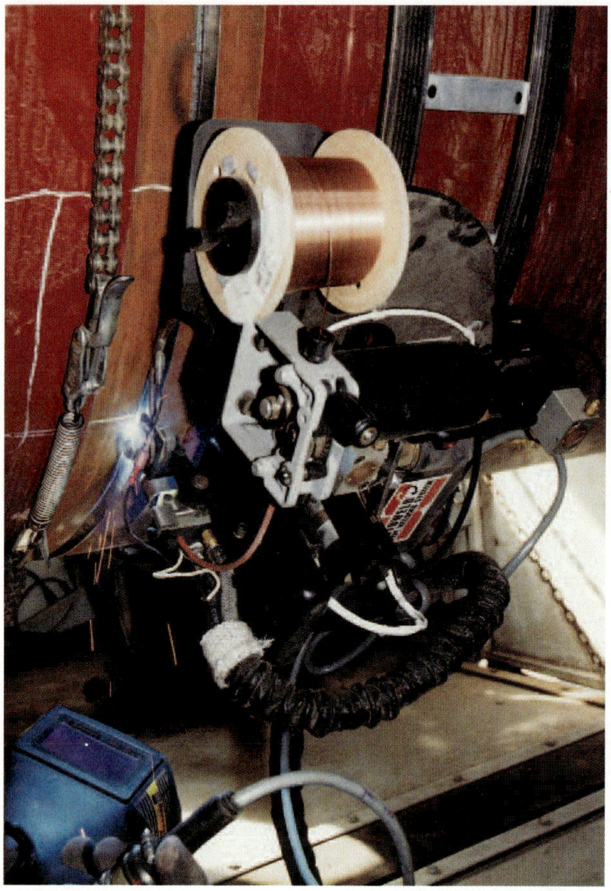

FIGURE 25-9 Automatic GTA machine welding along the seam of a stationary pipe. The Lincoln Electric Company

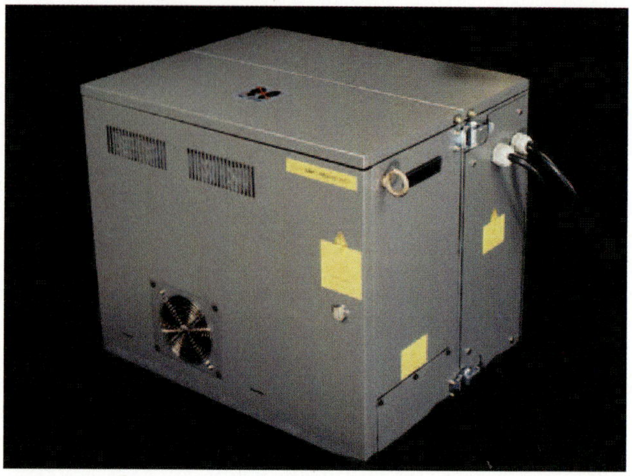

FIGURE 25-11 Robot control unit. The Lincoln Electric Company

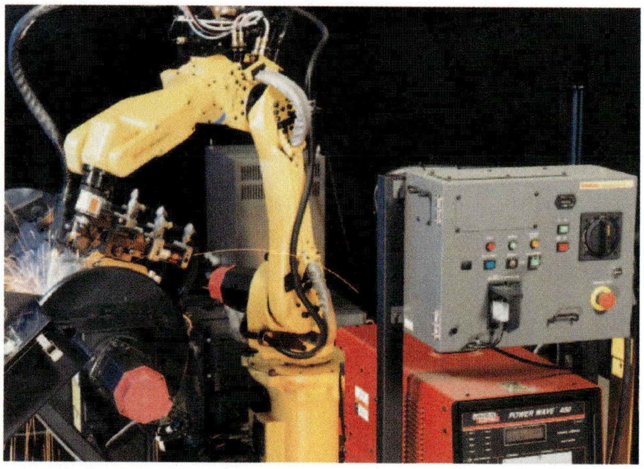

FIGURE 25-12 Machine axis. Reproduced with permission from FANUC Robotics America Corporation © FANUC Robotics America Corporation. All rights reserved.

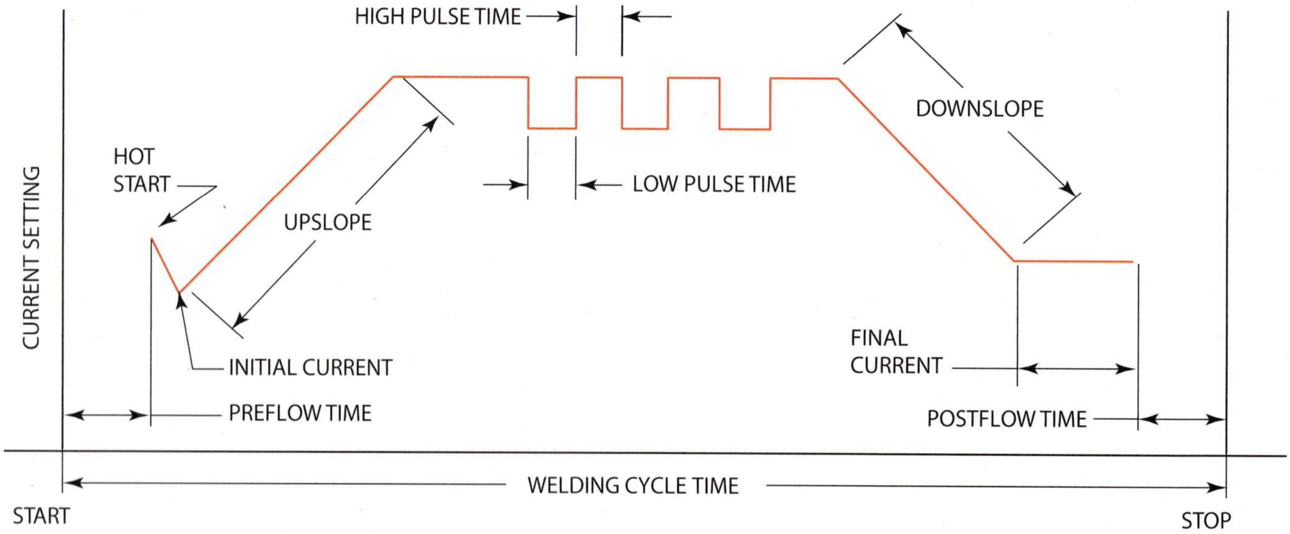

FIGURE 25-10 Typical GTAW automatic welding program. © Cengage Learning 2012

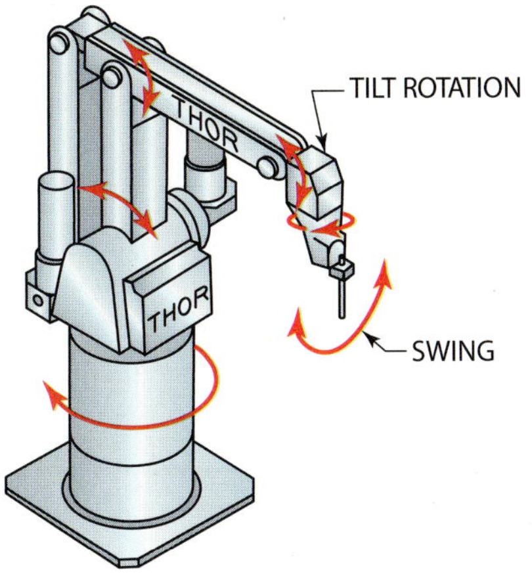

(A)

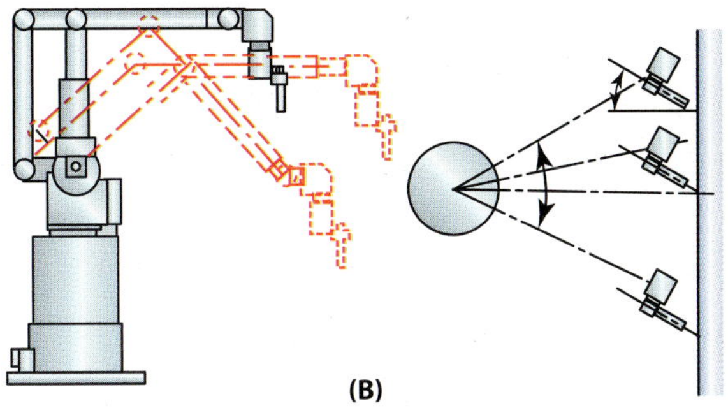

(B)

FIGURE 25-13 The tool end may be jointed so that it can tilt and rotate. Larry Jeffus

multiple workstations increase the duty cycle (the fraction of time during which welding or work is being done) and reduce **cycle time** (the period of time from starting one operation to starting another), Figure 25-14. Parts can be loaded or unloaded by the operator at one station while the robot welds at another station.

Safety

The following precautions are recommended for the use of automatic welding equipment and robots:

- All personnel should be instructed in the safe operation of the robot.
- All personnel should be instructed in the location of an emergency power shutoff.
- The work area should be restricted to authorized persons only.
- The work area should have fences, gates, or other restrictions to prevent access by unauthorized personnel.

FIGURE 25-14 Rotating worktable increases the work zone. Eutectic Corporation

- **Sensors** should be mounted around the floor and work area to stop all movement when unauthorized personnel are detected in the work area during the operation.
- The arc welding light should be screened from other work areas.
- A breakaway toolholder should be used in case of accidental collision with the part, Figure 25-15.
- A signal should sound or flash before the robot starts moving.

FIGURE 25-15 When using automatic equipment and robots, a break away collision sensor should be used. Applied Robotics, Inc.

SUMMARY

The automation of welding does not necessarily have to be in large shops. You may find that in some production applications it would be beneficial to use turntables or positioners. These can either fully or partially automate the welding process so that welder fatigue will be reduced. As you reduce welder fatigue with simple automation equipment, you can increase productivity significantly. It is important, therefore, not to look at automation as a major undertaking in every case. Many welding applications can be improved by using automated equipment, thereby reducing welder fatigue, increasing productivity, and reducing weld defects.

As the number of companies using automated and robotic equipment has increased, the cost of equipment has significantly decreased, and the versatility of the equipment has improved. Some equipment companies provide you with test runs of your product to help you determine the appropriateness of their equipment for your application. Such sample welds will help you determine which equipment is the most cost effective and appropriate for your application.

REVIEW QUESTIONS

1. Describe a manual joining process.
2. What is the most commonly used manual arc welding process?
3. What is a semiautomatic joining process?
4. Compare the distance of the welding gun or torch from the work in the semiautomatic joining process compared to the SMAW process.
5. What are the advantages of using filler metal that is fed from a large spool rather than filler electrodes?
6. What are the most commonly used semiautomatic arc welding processes?
7. Describe a machine joining process.
8. What type of adjustments to the equipment might be needed in a machine joining process?
9. Describe an automatic joining process.
10. Why is automatic welding or brazing best suited to large-volume production runs?
11. How are automated joining processes different from automatic joining?
12. What are automated joining processes controlled by?
13. What functions can industrial robots perform?
14. What three basic directions can most robots perform movements in?
15. What is a duty cycle?
16. Discuss several safety precautions that should be followed when using automatic welding equipment and robots.

Chapter 26

Filler Metal Selection

OBJECTIVES

After completing this chapter, the student should be able to:

- Explain various identification systems for filler metals.
- List what technical information is provided by electrode manufacturers and how this information can be used by a welder.
- Describe the parts of an shielded metal arc welding (SMAW) electrode and their function.
- Name some of the factors to consider when selecting an SMAW filler metal.
- Name the characteristics of a particular filler metal by interpreting the series of letters and numbers used to describe the filler metal.

KEY TERMS

alloying elements	**core wire**	**filler metals**
arc blow	**fast freezing**	**flux covering**

INTRODUCTION

Manufacturers of filler metals may use any one of a variety of identification systems. There is not a mandatory identification system for filler metals. Manufacturers may use their own numbering systems, trade names, color codes, or a combination of methods to identify filler metal. They may voluntarily choose to use any one of several standardized systems.

The most widely used numbering and lettering system is the one developed by the American Welding Society (AWS). Other numbering and lettering systems have been developed by the American Society for Testing and Materials (ASTM) and the American Iron and Steel Institute (AISI). A system of using colored dots has also been developed by the National Electrical

Manufacturers Association (NEMA). Some manufacturers have produced systems that are similar to the AWS system. Most major manufacturers include both the AWS identification and their own identification on the box, on the package, and/or directly on the filler metal.

Information that pertains directly to specific filler metals is readily available from most electrode manufacturers. The information given in charts, pamphlets, and pocket electrode guides is specific to their products, and they may or may not include standard AWS tests, terms, or classifications within their identification systems.

The AWS publishes a variety of books, pamphlets, and charts showing the minimum specifications for filler metal groups. It also publishes comparison charts that include all of the information manufacturers provide to the AWS regarding their filler metals. Both the literature on filler metal specifications and filler metal comparisons may be obtained directly from the AWS.

The AWS classification system is for minimum requirements within a grouping. Filler metals manufactured within a grouping may vary but still be classified under that grouping's classification.

A manufacturer may add elements to the metal or flux, such as more arc stabilizers. When one characteristic is improved, another characteristic may also change. The added arc stabilizer may make a smoother weld with less penetration. Other changes may affect the strength and ductility or other welding characteristics.

Because of the variables within a classification, some manufacturers make more than one type of filler metal that is included in a single classification. This and other information may be included in the data supplied by manufacturers.

Manufacturers' Electrode Information

The type of information given by different manufacturers ranges from general information to technical, chemical, and physical information. A mixture of different types of information may be given.

General information given by manufacturers may include some or all of the following: welding electrode manipulation techniques, joint design, prewelding preparation, postwelding procedures, types of equipment that can be welded, welding currents, and welding positions.

Understanding the Electrode Data

Technical procedures, physical properties, and chemical analysis information given by manufacturers include the following:

- Number of welding electrodes per pound
- Number of inches of weld per welding electrode
- Welding amperage range setting for each size of welding electrode
- Welding codes for which the electrode can be used
- Types of metal that can be welded
- Ability to weld on rust, oil, paint, or other surface contamination
- Weld joint penetration characteristics
- Preheating and postheating temperatures
- Weld deposit physical strengths: ultimate tensile strength, yield point, yield strength, elongation, and impact strength
- Percentages of such alloys as carbon, sulfur, phosphorus, manganese, silicon, chromium, nickel, molybdenum, and other alloys

The information supplied by the manufacturer can be used for a variety of purposes, including the following:

- Estimates of the pounds of electrodes needed for a job
- Welding conditions under which the electrode can be used—for example, on clean or dirty metal
- Welding procedure qualification information regarding amperage, joint preparation, penetration, and welding codes
- Physical and chemical characteristics affecting the weld's strengths and metallurgical properties

Data Resulting from Mechanical Tests

Most of the technical information supplied is self-explanatory and easily understood. The mechanical properties of the weld are given as the results of standard tests. The following are some of the standard tests and the meaning of each test:

- Minimum tensile strength, psi—The load in pounds that would be required to break a section of sound weld that has a cross-sectional area of 1 sq in.

- Yield point, psi—The point in low and medium carbon steels at which the metal begins to stretch when force (stress) is applied and after which it will not return to its original length.
- Elongation, percent in 2 in. (51 mm)—The percentage that a 2-in. (51-mm) piece of weld will stretch before it breaks.
- Charpy V notch, foot-pound—The impact load required to break a test piece of weld metal. Because some metals are used at low temperatures where they might become brittle, this test may be performed on metal below room temperature.

Data Resulting from Chemical Analysis

Chemical analysis of the weld deposit may also be included in the information given by manufacturers. It is not so important to know what the different percentages of the alloys do, but it is important to know how changes in the percentages of the alloys affect the weld. Chemical composition can easily be compared from one electrode to another. The following are the major elements and the effects of their changes on the iron in carbon steel:

- Carbon (C)—As the percentage of carbon increases, the tensile strength increases, the hardness increases, and ductility is reduced. Carbon also causes austenite to form.
- Sulfur (S)—It is usually a contaminant, and the percentage should be kept as low as possible below 0.04%. As the percentage of carbon increases, sulfur can cause hot shortness and porosity.
- Phosphorus (P)—It is usually a contaminant, and the percentage should be kept as low as possible. As the percentage of phosphorus increases, it can cause weld brittleness, reduced shock resistance, and increased cracking.
- Manganese (Mn)—As the percentage of manganese increases, the tensile strength, hardness, resistance to abrasion, and porosity all increase; hot shortness is reduced. It is also a strong austenite former.
- Silicon (Si)—As the percentage of silicon increases, tensile strength increases, and cracking may increase. It is used as a deoxidizer and ferrite former.
- Chromium (Cr)—As the percentage of chromium increases, tensile strength, hardness, and corrosion resistance increase with some decrease in ductility. It is also a good ferrite and carbide former.
- Nickel (Ni)—As the percentage of nickel increases, tensile strength, toughness, and corrosion resistance increase. It is also an austenite former.
- Molybdenum (Mo)—As the percentage of molybdenum increases, tensile strengthens at elevated temperatures; creep resistance and corrosion resistance increase. It is also a ferrite and carbide former.
- Copper (Cu)—As the percentage of copper increases, the corrosion resistance and cracking tendency increases.
- Columbium (Cb)—As the percentage of columbium (niobium) increases, the tendency to form chrome-carbides is reduced in stainless steels. It is also a strong ferrite former.
- Aluminum (Al)—As the percentage of aluminum increases, the high-temperature scaling resistance improves. It is also a good oxidizer and ferrite former.

SMAW Electrode Operating Information

Shielded metal arc welding electrodes, sometimes referred to as welding rods, arc welding rods, stick electrodes, or simply electrodes, have two parts. These two parts are the inner core wire and a flux covering, Figure 26-1.

The functions of the core wire include the following:

- To carry the welding current
- To serve as most of the filler metal in the finished weld

The functions of the **flux covering** may include the following:

- To provide some of the alloying elements
- To provide an arc stabilizer (optional)
- To serve as an insulator

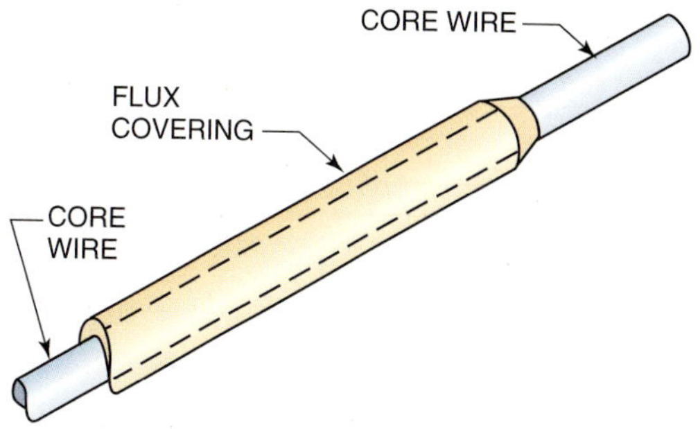

FIGURE 26-1 The two parts of a welding electrode.

- To provide a slag cover to protect the weld bead and slow cooling rate
- To provide a protective gaseous shield during welding

CORE WIRE

A **core wire** is the primary metal source for a weld. For fabricating structural and low alloy steels, the core wires of the electrode use inexpensive rimmed or low carbon steel. For more highly alloyed materials, such as stainless steel, high nickel alloys, or nonferrous alloys, the core wires are of the approximate composition of the material to be welded. The core wire also supports the coating that carries the fluxing and alloying materials to the arc and weld pool.

Functions of the Flux Covering

Provides Shielding Gases

Heat generated by the arc causes some constituents in the flux covering to decompose and others to vaporize, forming shielding gases. These gases prevent the atmosphere from contaminating the weld metal as it transfers across the arc gap. They also protect the molten weld pool as it cools to form solid metal. In addition, shielding gases and vapors greatly affect both the drops that form at the electrode tip and their transfer across the arc gap, Figure 26-2. They also cause the spatter from the arc and greatly determine arc stiffness and penetration. For example, the E6010 electrode contains cellulose. Cellulose decomposes into the hydrogen responsible for the deep electrode penetration so desirable in pipeline welding.

Alloying Elements

Elements in the flux are mixed with the filler metal. Some of these elements stay in the weld metal as alloys.

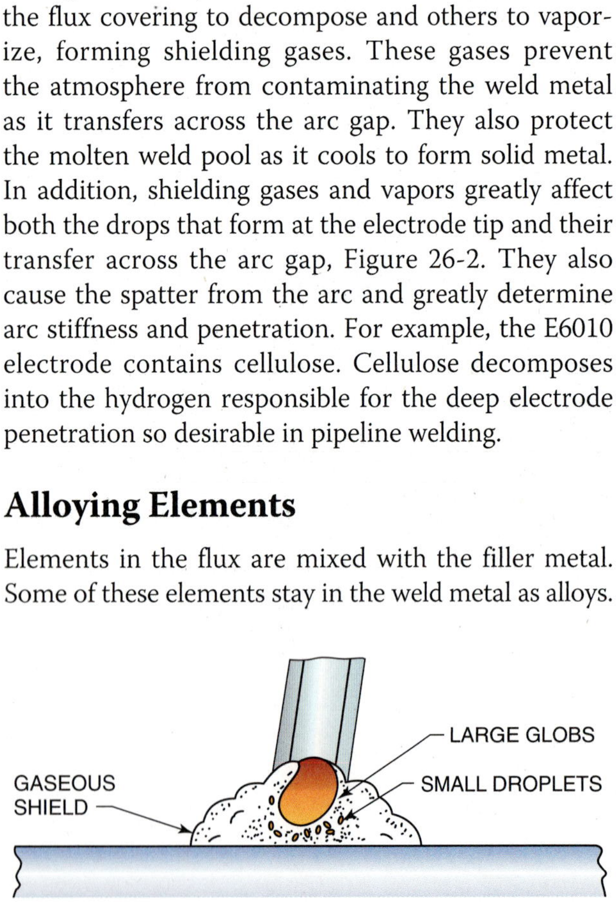

FIGURE 26-2 **Methods of metal transfer during an arc.** © Cengage Learning 2012

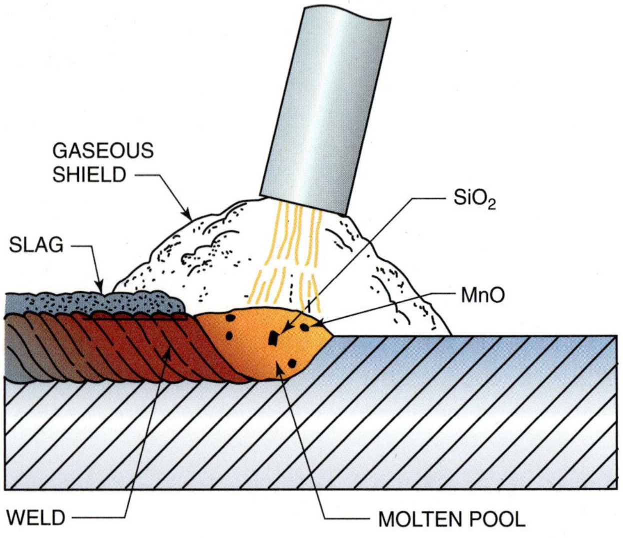

FIGURE 26-3 **Silicon and manganese act as scavengers that combine with contaminants and float to the slag on top of the weld.** American Welding Society

Other elements pick up contaminants in the molten weld pool and float them to the surface. At the surface, these contaminants form part of the slag, Figure 26-3.

Effect on Weld

Welding fluxes can affect the penetration and contour of the weld bead. Penetration may be pushed deeper if the core wire is made to melt off faster than the flux melts. This forms a small chamber or crucible at the end of the electrode that acts as the combustion chamber of a rocket. As a result, the molten metal and hot gases are forced out very rapidly. The effect of this can be seen on the surface of the molten weld pool as it is blown back away from the end of the electrode. Some electrodes do not use this jetting action, and the resulting molten weld pool is much calmer (less turbulent) and may be rounded in appearance. In addition, the resulting bead may have less penetration, Figure 26-4.

Weld bead contour can also be affected by the slag formed by the flux. Some high-temperature slags, called refractory, solidify before the weld metal solidifies, forming a mold that holds the molten metal in place. These electrodes are sometimes referred to as **fast freezing** and are excellent for vertical, horizontal, and overhead welding positions.

Filler Metal Selection

Selecting the best filler metal for a job in large shops is seldom delegated to you. The selection of the correct process and filler metal is a complex undertaking. If

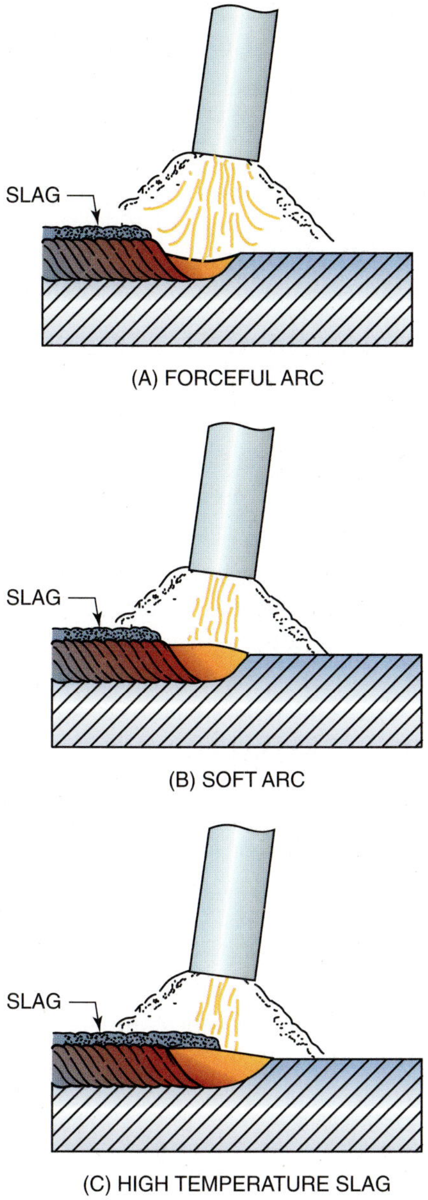

FIGURE 26-4 Arc force. American Welding Society

the choice is given to you, it is one of the most important decisions you will make.

Covering all of the variables for selecting a filler metal would be well beyond the scope of this text. A sample of the factors that must be considered for the selection of an SAW electrode follow. To further complicate things, welding electrodes have more than one application, and many welding electrodes may be used for the same type of work.

The following conditions that you should consider when choosing a welding electrode are not in order of importance. They are also not all of the factors that must be considered.

Shielded Metal Arc Welding Electrode Selection

- Type of electrode—What electrode has been specified in the blueprints or in the contract for this job?
- Type of current—Can the welding power source supply AC only, DC only, or both AC and DC?
- Power range—What is the amperage range on the welder and its duty cycle? Different types of electrodes require different amperage settings even for the same size welding electrode. For example, the amperage range for a 1/8-in. (3-mm) diameter E6010 electrode is 75 A to 125 A, and the amperage range for a 1/8-in. (3-mm) diameter E7018 electrode is 90 A to 165 A.
- Type of metal—Some welding electrodes may be used to join more than one similar type of metal. Other electrodes may be used to join together two different types of metal. For example, an E309-15 electrode can be used to join 305 stainless steel to 1020 mild steel.
- Thickness of metal—The penetration characteristics of each welding electrode may differ. Selecting one electrode that will weld on a specific thickness of material is important. For example, E6013 has very little penetration and is therefore good for welding on sheet metal.
- Weld position—Some welding electrodes can be used to make welds in all positions. Other electrodes may be restricted to making flat, horizontal, and/or vertical position welds; a few electrodes may be used to make flat position welds only.
- Joint design—The type of joint and whether it is grooved or not may affect the performance of the welding electrode. For example, the E7018 electrode does not produce a large, gaseous cloud to protect the molten metal. For this reason, the electrode movement is restricted so that the molten weld pool is not left unprotected by the gaseous cloud.
- Surface condition—It is not always possible to work on new, clean metal. Some welding electrodes will weld on metal that is rusty, oily, painted, dirty, or galvanized.
- Number of passes—The amount of reinforcement needed may require more than one welding pass. Some welding electrodes will build up faster, and others will penetrate deeper. The slag may be removed more easily from some welds than from

others. For example, E6013 will build up a weld faster than E6010, and the slag is also more easily removed between weld passes.

- Distortion—Welding electrodes that will operate on low-amperage settings will have less heat input and cause less distortion. Welding electrodes that have a high rate of deposition (fill the joint rapidly) and can travel faster will also cause less distortion. For example, the flux on an E7024 has 50% iron powder, which gives it a faster fill rate and allows it to travel faster, resulting in less distortion of the metal being welded.
- Preheat or postheat—On low carbon steel plate 1 in. (25 mm) thick or more, thick preheating is required with most welding electrodes. Postheating may be required to keep a weld zone from hardening or cracking when using some welding electrodes. However, no postheating may be required when welding low alloy steel using E310-15.
- Temperature service—Weld metals react differently to temperature extremes. Some welds become very brittle and crack easily in cryogenic (low-temperature) service. A few weld metals resist creep and oxidation at high temperatures. For example, E310Mo-15 can weld on most stainless and mild steels without any high-temperature problems.
- Mechanical properties—Mechanical properties such as tensile strength, yield strength, hardness, toughness, ductility, and impact strength can be modified by the selection of specific welding electrodes.
- Postwelding cleanup—The hardness or softness of the weld greatly affects any grinding, drilling, or machining. The ease with which the slag can be removed and the quantity of spatter will affect the time and amount of cleanup required.
- Shop or field weld—The general working conditions such as wind, dirt, cleanliness, dryness, and accessibility of the weld will affect the choice of the welding electrode. For example, the E7018 electrode must be kept very dry, but the E6010 electrode is not as affected by moisture.
- Quantity of welds—If a few welds are needed, a more expensive welding electrode requiring less skill may be selected. For a large production job requiring a higher skill level, a less expensive welding electrode may be best.

After deciding the specific conditions that may affect the welding, you have most likely identified more than one condition that needs to be satisfied. Some of the conditions will not interfere with others. For example, the type of current and whether a welder makes one or more weld passes have little or no effect on each other. However, if a welder needs to machine the finished weld, hardness is a consideration. When two or more conditions conflict, you are seldom the person who will make the decision. It may be necessary to choose more than one welding electrode. When welding pipe, E6010 and E6011 are often used for the root pass because of their penetration characteristics, and E7018 is used for the cover pass because of its greater strength and resistance to cracking.

Each AWS electrode classification has its own welding characteristics. Some manufacturers have more than one welding electrode in some classifications. In these cases, the minimum specifications for the classification have been exceeded. An example of more than one welding electrode in a single classification is Lincoln's Fleetweld 35, Fleetweld 35LS, and Fleetweld 180R. These electrodes are all in AWS classification E6011. For the manufacturer's complete description of these electrodes, consult Table 26-1.

The characteristics of each manufacturer's **filler metals** can be compared to one another by using data sheets supplied by the manufacturer. General comparisons can be made easily using an electrode comparison chart. When making an electrode selection, many variables must be kept in mind, and the performance characteristics must be compared before making a final choice.

AWS Filler Metal Classifications

The AWS classification system uses a series of letters and numbers in a code that gives the important information about the filler metal. The prefix letter is used to indicate the filler's form, a type of process the filler is to be used with, or both. The prefix letters and their meanings are as follows:

- *E*—Indicates an arc welding electrode. The filler carries the welding current in the process. We most often think of the *E* standing for an SMA "stick" welding electrode. It also is used to indicate wire electrodes used in GMAW, FCAW, SAW, ESW, EGW, etc.
- *R*—Indicates a rod that is heated by some source other than electric current flowing directly through it. Welding rods are sometimes referred to as being "cut length" or "welding wire." It is often used with OFW and GTAW.

Electrode Identification and Operating Data

Coating Color	AWS Number on Coating	(L) Lincoln	Electrode	Electrode Polarity	Sizes and Current Ranges (Amps) (Electrodes Are Manufactured in These Sizes for Which Current Ranges Are Given) 3/32" Size	1/8" Size	5/32" Size	3/16" Size	7/32" Size	1/4" Size
Brick red	6010		Fleetweld 5P	DC+[2]	40–75	75–130	90–175	140–225	200–275	220–325[1]
Gray	6011		Fleetweld 35	AC	50–85	75–120	90–160	120–200	150–260	190–300
				DC+	40–75	70–110	80–145	110–180	135–235	170–270
Red brown	6011	Green	Fleetweld 35LS	AC		80–130	120–160			
				DC±		70–120	110–150			
Brown	6011		Fleetweld 180	AC	40–90	60–120	115–150			
				DC±	40–80	56–110	105–135			
Pink	7010-A1		Shield-arc 85	DC+	50–90	75–130	90–175	140–225		
Pink	7010-A1	Green	Shield-arc 85P	DC+				140–225		
Tan	7010-G		Shield-arc Hyp	DC+		75–130	90–185	140–225	160–250	
Gray	8010-G		Shield-arc 70+	DC+		75–130	90–185	140–225		

[1]Range for 5/16″size is 240–400 amps. DC + is Electrode Positive. DC – is Electrode Negative.

All tests were performed in conformance with specifications AWS A5.5 and ASME SFA.5.5 in the aged condition for the E7010-G and E8010-G electrodes and in the stress-relieved condition for Shield-Arc 85 & 85P. Tests for the other products were performed in conformance with specifications AWS A5.1 and ASME SF A.5.1 for the as-welded condition.

Typical Mechanical Properties

Low figures in the stress-relieved tensile and yield strength ranges below for Shield-Arc 85 and 85P are AWS minimum requirements.
Low figures in the as-welded tensile and yield strength ranges below for the other products are AWS minimum requirements.

	Fleetweld 5P	Fleetweld 35	Fleetweld 35LS	Fleetweld 180	Shield-arc 85	Shield-arc 85P	Shield-arc Hyp	Shield-arc 70+
As-welded tensile strength—psi	62–69,000	62–68,000	62–67,000	62–71,000	70–78,000	70–78,000	70–84,000	80–92,000
Yield point—psi ductility—% elong. in 2"	52–62,000 22–32	50–62,000 22–30	50–60,000 22–31	50–64,000 22–31	60–71,000 22–26	57–63,000 22–27	60–77,000 22–23	67,000–83,000 19–24
Charpy V-notch toughness —ft. lb	20–60 @–20°F	20–90 @–20°F	20–57 @–20°F	20–54 @–20°F	68 @ 70°F	68 @ 70°F	30 @ 20°F	40 @ 50°F
Hardness, Rockwell B (avg)[3]	76–82	76–85	73–88	75–85			83–92	88–93
Stress-relieved @ 1150°F tensile strength—psi	60–69,000	60–66,000	60-65,000		70–83,000	70–74,000	80–82,000	80–84,000
Yield point—psi ductility—% elong. in 2"	46–56,000 28–36	46–56,000 28–36	46–51,000 28–33		57–69,000 22–28	57–65,000 22–27	72–76,000 24–27	71–76,000 22–26
Charpy V-notch toughness —ft. lb	71 @ 70°F		120 @ 70°F		64 @ 70°F	68 @ 70°F	30 @ –20°F	30 @ –50°F
Hardness, Rockwell B (avg)[3]					80–89	80–87		

Conformances and Approvals

See *Lincoln Price Book* for certificate numbers, size, and position limitations, and other data.

Conforms to test requirements of AWS—A5.1 and ASME—SFA5.1 AWS—A5.5 and ASME—SFA5.5	*FW-5P* E6010	*FW-35* E6011	*FW-35LS* E6011	*FW-180* E6011	*SA-85* E7010-A1[4]	*SA-85P* E7010-A1	*SA-HYP* E7010-G	*SA-70+* E8010-G
ASME boiler code Group	F3	F3	F3	F3	F3	F3	F3	F3
Analysis	A1		A1	A1	A2	A2	A2	
American Bureau of Shipping and U.S. Coast Guard	Approved	Approved	Approved		Approved			
Conformance certificate available[4]	Yes	Yes	Yes	Yes	Yes	Yes	Yes	Yes
Lloyds	Approved	Approved						
Military specifications	MIL-QQE-450	MIL-QQE-450			MIL-E-22200/7			

[3]Hardness values obtained from welds made in accordance with AWS A5.1.

[4]Certificate of Conformance to AWS classification test requirements is available. These are needed for Federal Highway Administration projects.

Table 26-1 Fleetweld 35®, Fleetweld 35LS®, and Fleetweld 180® Lincoln Electrodes (*Lincoln Electric Company*)

- *ER*—Indicates a filler metal that is supplied for use as either a current-carrying electrode or in rod forms. The same alloys are used to produce the electrodes and the rods. This filler metal may be supplied as a wire on a spool for GMAW or as a rod for OFW or GTAW.
- *EC*—Indicates a composite electrode. These electrodes are used for SAW. Do not confuse an ECu, copper arc welding electrode, for an ECNi2, which is a composite nickel submerged arc welding wire.
- *B*—Indicates a brazing filler metal. This filler metal is usually supplied as a rod, but it can come in a number of other forms. Some of the forms it comes in are powder, sheets, washers, and rings.
- *RB*—Indicates a filler metal that is used as a current-carrying electrode, as a rod, or both. The form the filler is supplied in for each of the applications may be different. The composition of the alloy in the filler metal will be the same for all of the forms supplied. This filler can be used for processes like arc braze welding or oxyfuel brazing.
- *RG*—Indicates a welding rod used primarily with oxyfuel welding. This filler can be used with all of the oxyfuels, and some of the fillers are used with the GTAW process.
- *IN*—Indicates a consumable insert. These are most often used for welding on pipe. They are preplaced in the root of the groove to supply both filler metal and support for the root pass. The inserts may provide for some joint alignment and spacing.

The next two classifications are not filler metal. They are classified under the same system because they are welding consumables. The GTA welding tungsten is not a filler, but it is consumed, very slowly, during the welding process.

- *EW*—Indicates a nonconsumable tungsten electrode. The GTAW electrode is obviously not a filler metal, but it falls under the same classification system.
- *F*—Indicates a flux used for SAW. The composition of the weld metal is influenced by the flux. There are alloys and agents in the flux used for SAW that are dissolved into the weld metal. For this reason, the filler metal and flux are specified together with the filler metal identification first and the flux second.

In addition to the prefix, there are some suffix identifiers. The suffix may be used to indicate a change in the alloy in a covered electrode or the type of welding current to be used with stainless steel covered electrodes.

Carbon Steel

Carbon and Low Alloy Steel Covered Electrodes

The AWS specification for carbon steel covered arc electrodes is A5.1, and for low alloy steel covered arc electrodes it is A5.5. Filler metals classified within these specifications are identified by a system that uses the letter *E* followed by a series of numbers to indicate the minimum tensile strength of a good weld, the position(s) in which the electrode can be used, the type of flux coating, and the type(s) of welding current, Figure 26-5.

The tensile strength is given in pounds per square inch (psi). The actual strength is obtained by adding three zeros to the right of the number given. For example, E60XX is 60,000 psi, E110XX is 110,000 psi, and so on.

The next number located to the right of the tensile strength—1, 2, or 4—designates the welding position capable—for example:

- *1*—in an E601X means all positions flat, horizontal, vertical, and overhead.
- *2*—in an E602X means horizontal fillets and flat.
- *3*—is an old term no longer used; it meant flat only.
- *4*—in an E704X means flat, horizontal, overhead, and vertical-down.

The last two numbers together indicate the major type of covering and the type of welding current. For example, EXX10 has an organic covering and uses DCEP polarity. The AWS classification system for A5.1 and A5.5 covered arc welding electrodes is shown in Table 26-2. The type of welding current for

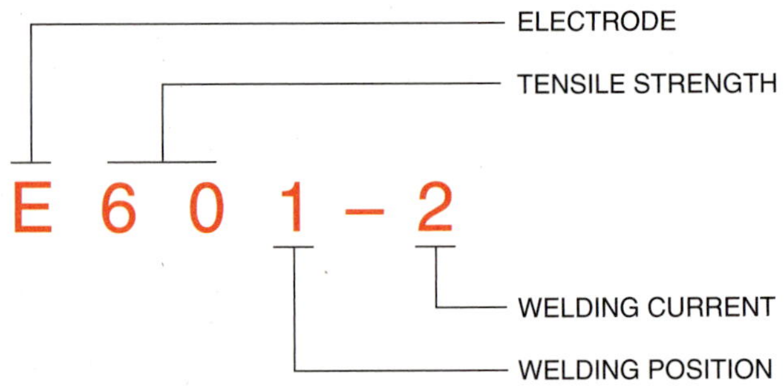

FIGURE 26-5 AWS numbering system for A5.1 and A5.5 carbon and low alloy steel covered electrodes.
American Welding Society

AWS Classification	Type of Covering	Capable of Producing Satisfactory Welds in Positions Shown[a]	Type of Current[b]
	E60 Series Electrodes		
E6010	High cellulose sodium	F, V, OH, H	DCEP
E6011	High cellulose potassium	F, V, OH, H	AC or DCEP
E6012	High titania sodium	F, V, OH, H	AC or DCEN
E6013	High titania potassium	F, V, OH, H	AC or DC, either polarity
E6020		H-fillets	AC or DCEN
E6022[c]	High iron oxide	F	AC or DC, either polarity
E6027	High iron oxide, iron powder	H-fillets, F	AC or DCEN
	E70 Series Electrodes		
E7014	Iron powder, titania	F, V, OH, H	AC or DC, either polarity
E7015	Low hydrogen sodium	F, V, OH, H	DCEP
E7016	Low hydrogen potassium	F, V, OH, H	AC or DCEP
E7018	Low hydrogen potassium, iron powder	F, V, OH, H	AC or DCEP
E7024	Iron powder, titania	H-fillets, F	AC or DC, either polarity
E7027	High iron oxide, iron powder	H-fillets, F	AC or DCEN
E7028	Low hydrogen potassium, iron powder	H-fillets, F	AC or DCEP
E7048	Low hydrogen potassium, iron powder	F, OH, H, V-down	AC or DCEP

[a]The abbreviations, F, V, V-down, OH, H, and H-fillets indicate the welding positions as follows:

F = Flat
H = Horizontal
H-fillets = Horizontal fillets
V-down = Vertical down
V = Vertical
OH = Overhead

[b]Reverse polarity means the electrode is positive; straight polarity means the electrode is negative.

[c]Electrodes of the E6022 classification are for single-pass welds.

For electrodes 3/16 in. (4.8 mm) and under, except 5/32 in. (4.0 mm) and under for classifications E7014, E7015, E7016, and E7018.

Table 26-2 Electrode Classification (*American Welding Society*)

any electrode may be expanded to include currents not listed if a manufacturer adds additional arc stabilizers to the electrode covering, Table 26-3.

On some covered arc electrodes, a suffix may be added to indicate the approximate alloy in the deposit as welded. For example, the letter *A* indicates a 1/2% molybdenum addition to the weld metal deposited. Table 26-4 is a complete list of the major **alloying elements** in electrodes.

Electrode Number	
EXXX0	DCRP only
EXXX1	AC and DCRP
EXXX2	AC and DCSP
EXXX3	AC and DC
EXXX4	AC and DC
EXXX5	DCRP only
EXXX6	AC and DCRP
EXXX8	AC and DCRP

Table 26-3 Welding Currents

Some of the more popular arc welding electrodes and their uses in these specifications are as follows.

E6010

The E6010 electrodes are designed to be used with DCEP polarity and have an organic-based flux (cellulose, C6H10O5). They have a forceful arc that results in deep penetration and good metal transfer in the vertical and overhead positions, Figure 26-6. The electrode is usually used with a whipping or stepping motion. This motion helps remove unwanted surface materials such as paint, oil, dirt, and galvanizing. Both the burning of the organic compound in the flux to form CO_2, which protects the molten metal, and the rapid expansion of the hot gases force the atmosphere away from the weld. A small amount of slag remains on the finished weld, but it is difficult to remove, especially along the weld edges. E6010 electrodes are commonly used for welding on pipe and in construction jobs, and for doing repair work.

Suffix Symbol	Molybdenum (Mo) %	Nickel (Ni) %	Chromium (Cr) %	Manganese (Mn) %	Vanadium (Va) %
A 1	0.5				
B 1	0.5		0.50		
B 2	0.5		1.25		
B 3	1.0		2.25		
B 4	0.5		2.00		
C 1		2.5			
C 2		3.5			
C 3		1.0			
D 1	0.3			1.5	
D 2	0.3			1.75	
G	0.2	0.5	0.30	1.00	0.1*
M					

*Only one of these alloys may be used.

Table 26-4 Major Alloying Elements in Electrodes

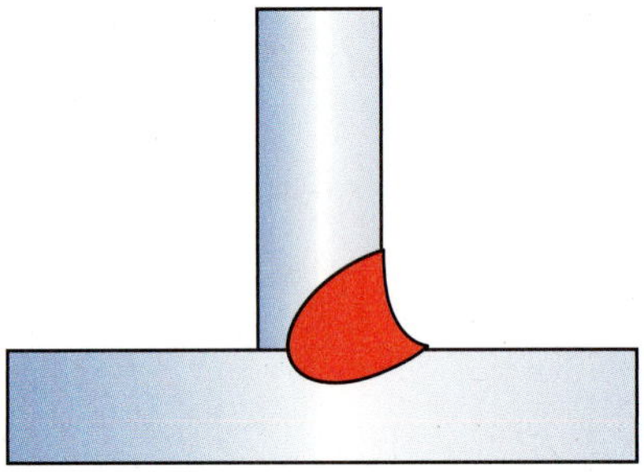

FIGURE 26-6 E6010. © Cengage Learning 2012

E6011

The E6011 electrodes are designed to be used with AC or DCEP reverse polarity and have an organic-based flux. These electrodes have many of the welding characteristics of E6010 electrodes, Figure 26-7. In most applications, the E6011 is preferred. The E6011 has added arc stabilizers, which allow it to be used with AC. Using this welding electrode on AC only slightly reduces its penetration but will help control any arc blow problem. **Arc blow**, or arc wander, is the magnetic deflection of the arc from its normal path. When welding with either E6010 or E6011, the weld pool may be slightly concave from the forceful action of the rapidly expanding gas. This forceful action also results in more spatter and sparks during welding. E6011 is the most commonly used electrode for agricultural welding.

E6012

The E6012 electrodes are designed to be used with AC or DCEN polarity and have rutile-based flux (titanium dioxide, TiO_2). This electrode has a very stable arc that is not very forceful, resulting in a shallow penetration characteristic, Figure 26-8. This limited penetration characteristic helps with poor-fitting joints or thin materials. Thick sections can be welded, but the joint must be grooved. Less smoke is generated with this welding electrode than with E6010 or E6011, but a thicker slag layer is deposited on the weld. If the weld is properly made, the slag can be removed easily and may even free itself after cooling. Spatter can be held to a minimum when using both AC and DC. E6012 electrodes are commonly used for all new work, including storage tanks, machinery fabrication, ornamental iron, and general repair work.

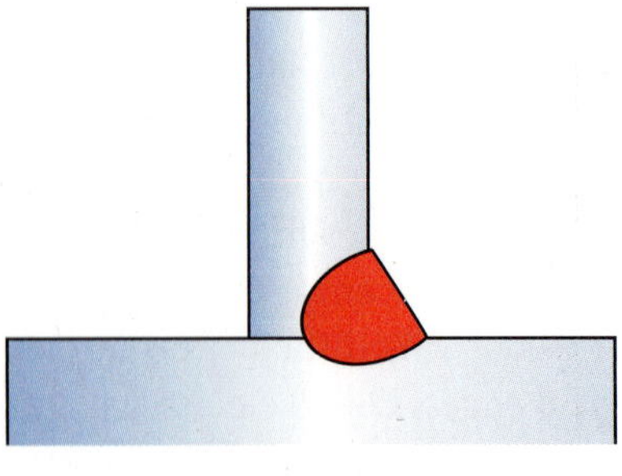

FIGURE 26-7 E6011. © Cengage Learning 2012

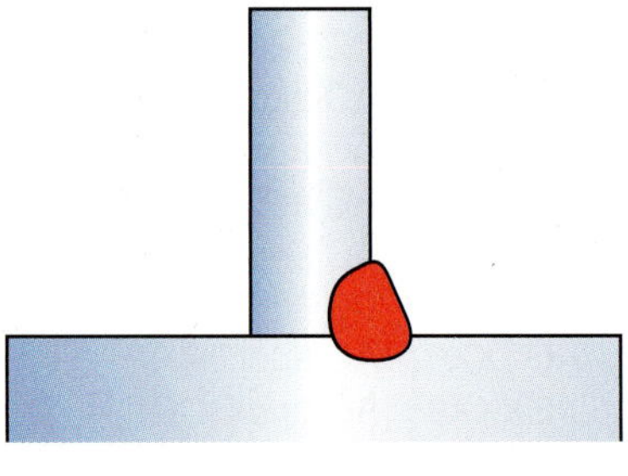

FIGURE 26-8 E6012. © Cengage Learning 2012

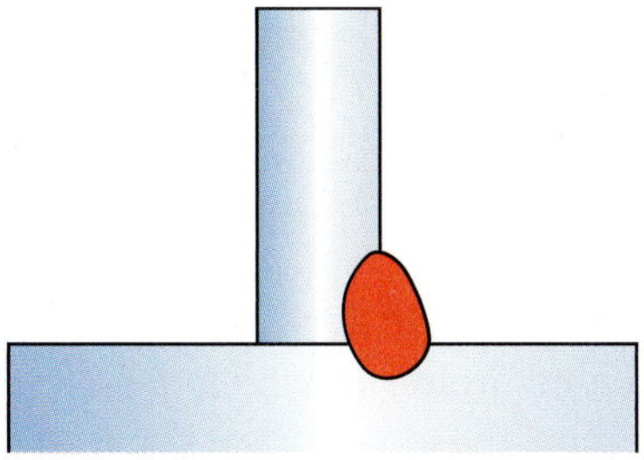

FIGURE 26-9 E6013. © Cengage Learning 2012

E6013

The E6013 electrodes are designed to be used with AC or DC, either polarity. They have a rutile-based flux. The E6013 electrode has many of the same characteristics of the E6012 electrode, Figure 26-9. The slag layer is usually thicker on the E6013 and is easily removed. The arc of the E6013 is as stable, but there is less penetration, which makes it easier to weld very thin sections. The weld bead will also be built up slightly higher than the E6012. E6013 electrodes are commonly used for sheet metal fabrication, metal buildings, surface buildup, truck and tractor work, and other farm equipment.

E7014

The E7014 electrodes are designed to be used with AC or DC, either polarity. They have a rutile-based flux with iron powder added. The E7014 electrode has many arc and weld characteristics that are similar to those of the E6013 electrode, Figure 26-10. Approximately 30% iron powder is added to the flux to allow it to build up a weld faster or have a higher travel speed. The penetration characteristic is light. This welding electrode can be used on metal with a light coating of rust, dirt, oil, or paint. The slag layer is thick and hard but can be completely removed with chipping. E7014 electrodes are commonly used for welding on heavy sheet metal, ornamental iron, machinery, frames, and general repair work.

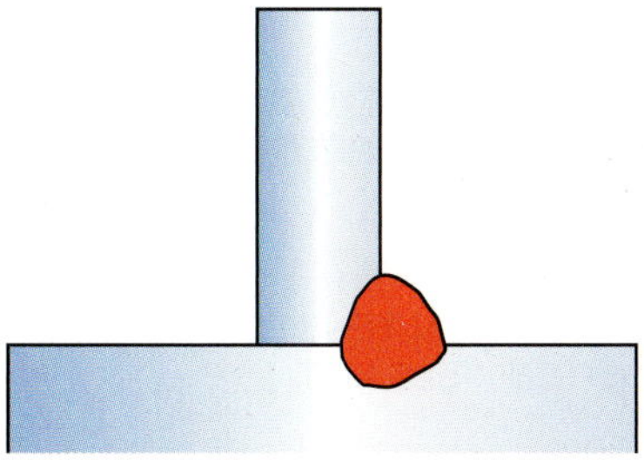

FIGURE 26-10 E7014. © Cengage Learning 2012

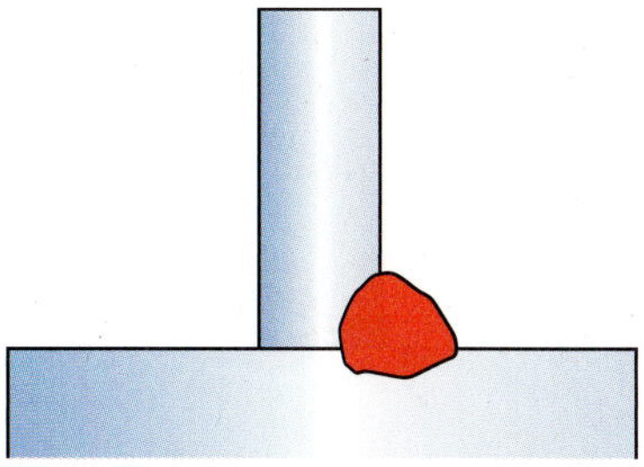

FIGURE 26-11 E7024. © Cengage Learning 2012

E7024

The E7024 electrodes are designed to be used with AC or DC, either polarity. They have a rutile-based flux with iron powder added. This welding electrode has a light penetration and fast-fill characteristic, Figure 26-11. The flux contains about 50% iron powder, which gives the flux its high rate of deposition. The heavy flux coating helps control the arc and can support the electrode so that a drag technique can be used. The drag technique allows this electrode to be used by welders with less skill. The slag layer is heavy and hard but can easily be removed. If the weld is performed correctly, the slag may remove itself. Because of the large, fluid molten weld pool, this electrode is equally used in the flat and horizontal position only, although it can be used on work that is slightly vertical. E7024 electrodes are commonly used in welding new equipment.

E7016

The E7016 electrodes are designed to be used with AC or DCEP polarity. They have a low-hydrogen–based (mineral) flux. This electrode has moderate penetration and little buildup, Figure 26-12. There is no iron powder in the flux, which helps when welding in the vertical or overhead positions. Welds on high sulfur and cold-rolled metals can be made with little porosity. Low alloy and mild steel heavy plates

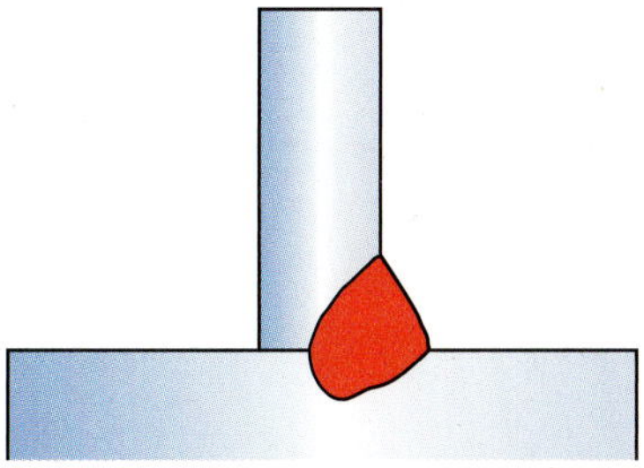

FIGURE 26-12 E7016. © Cengage Learning 2012

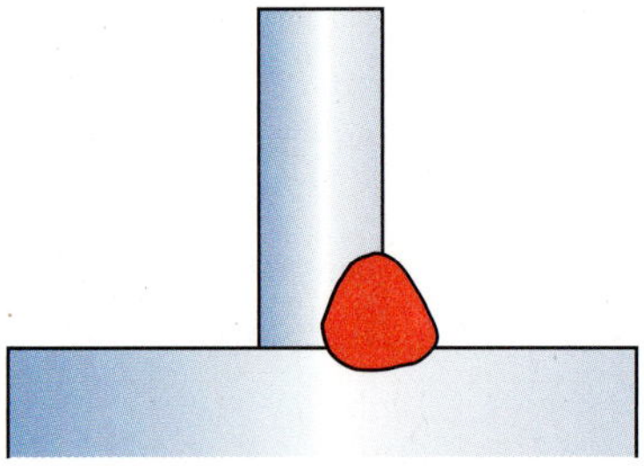

FIGURE 26-13 E7018. © Cengage Learning 2012

can be welded with minimum preheating. E7016 electrodes are commonly used for building construction and equipment fabrication.

E7018

The E7018 electrodes are designed to be used with AC or DCEP, either polarity. They have a low-hydrogen–based flux with iron powder added. The E7018 electrodes have moderate penetration and buildup, Figure 26-13. The slag layer is heavy and hard but can be removed easily by chipping. The weld metal is protected from the atmosphere primarily by the molten slag layer and not by rapidly expanding gases. For this reason, these electrodes should not be used for open root welds. The atmosphere may attack the root, causing a porosity problem. The E7018 welding electrodes are very susceptible to moisture, which may lead to weld porosity. These electrodes are commonly used for pipe, plate, trailer axles, and low-temperature equipment. E7018 electrodes are sometimes referred to as Lo-Hi rods because they allow very little hydrogen into the weld pool.

Wire-Type Carbon Steel Filler Metals

Solid Wire

The AWS specification for carbon steel filler metals for gas shielded welding wire is A5.18. Filler metal classified within these specifications can be used for GMAW, GTAW, and PAW processes. Because in GTAW and PAW the wire does not carry the welding current, the letters *ER* are used as a prefix. The *ER* is followed by two numbers to indicate the minimum tensile strength of a good weld. The actual strength is obtained by adding three zeros to the right of the number given. For example, ER70S-x is 70,000 psi.

The *S* located to the right of the tensile strength indicates that this is a solid wire. The last number—2, 3, 4, 5, 6, or 7—or the letter *G* is used to indicate the filler metal composition and the weld's mechanical properties, Figure 26-14.

ER70S-2

This is a deoxidized mild steel filler wire. The deoxidizers allow this wire to be used on metal that has light coverings of rust or oxides. There may be a slight reduction in the weld's physical properties if the weld is made on rust or oxides, but this reduction is only slight, and the weld will usually still pass the classification test standards. This is a general purpose filler that can be used on killed, semikilled, and rimmed steels. Argon-oxygen, argon-CO_2, and CO_2 can be used as shielding gases. Welds can be made in all positions.

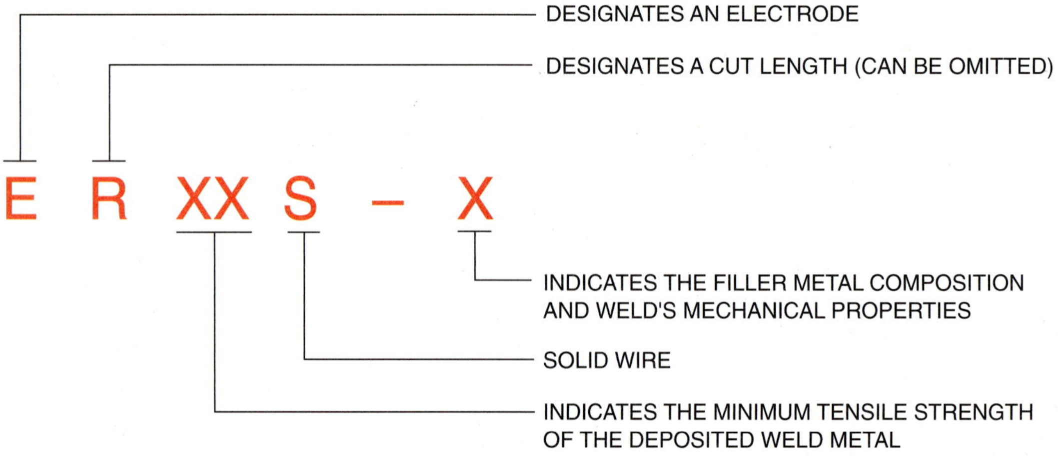

FIGURE 26-14 AWS numbering system for carbon steel filler metal for GMAW.
American Welding Society

ER70S-3

This is a popular agricultural filler wire. It can be used in single- or multiple-pass welds in all positions. ER70S-3 does not have the deoxidizers required to weld over rust, over oxides, or on rimmed steels. It produces high-quality welds on killed and semikilled steels. Argon-oxygen, argon-CO_2, and CO_2 can be used as shielding gases.

ER70S-6

This is a good general purpose filler wire. It has the highest levels of manganese and silicon. The wire can be used to make smooth welds on sheet metal or thicker sections. Welds over rust, oxides, and other surface impurities will lower the mechanical properties, but not normally below the specifications of this classification. Argon-oxygen, argon-CO_2, and CO_2 can be used as shielding gases. Welds can be made in all positions.

Tubular Wire

The AWS specification for carbon steel filler metals for flux cored arc welding wire is A5.20. Filler metal classified within this specification can be used for the FCAW process. The letter *E*, for electrode, is followed by a single number to indicate the minimum tensile strength of a good weld. The actual strength is obtained by adding four zeros to the right of the number given. For example, E6xT-x is 60,000 psi, and E7xT-x is 70,000 psi.

The next number, 0 or 1, indicates the welding position. Ex0T is to be used in a horizontal or flat position only. Ex1T is an all-position filler metal.

The *T* located to the right of the tensile strength and weld position numbers indicates that this is a tubular, flux cored wire. The last number—2, 3, 4, 5, 6, 7, 8, 10, or 11—or the letter *G* or *GS* is used to indicate if the filler metal can be used for single- or multiple-pass welds. The electrodes with the numbers ExxT-2, ExxT-3, ExxT-10, and ExxT-GS are intended for single-pass welds only, Figure 26-15.

E70T-1, E71T-1

E70T-1 and E71T-1 have a high-level deoxidizer in the flux core. They have high levels of silicon and manganese, which allow it to weld over some surface contaminations such as oxides or rust. This filler metal can be used for single- or multiple-pass welds. Argon 75% with 25% CO_2 or 100% CO_2 can be used as the shielding gas. It can be used on ASME A36, A106, A242, A252, A285, A441, and A572 or similar metals. Applications include railcars, heavy equipment, earth-moving equipment, shipbuilding, and general fabrication. The weld metal deposited has a chemical and physical composition similar to that of E7018 low-hydrogen electrodes.

E70T-2, E71T-2

E70T-2 and E71T-2 are highly deoxidized flux cored filler metal that can be used for single-pass welds only. The high levels of deoxidizers allow this electrode to

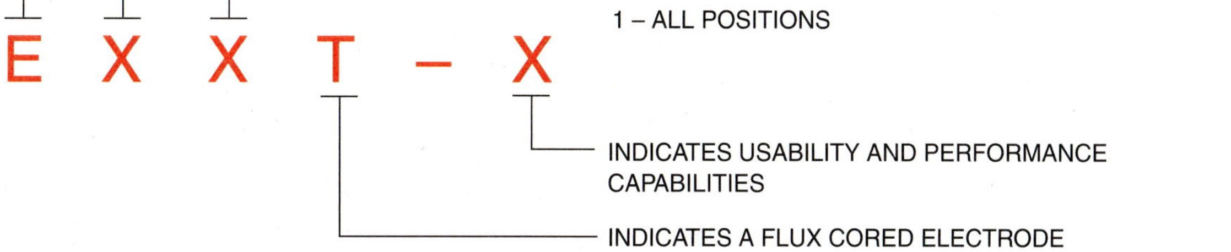

FIGURE 26-15 **Identification system for mild steel FCAW electrodes.** American Welding Society

be used over mill scale and light layers of rust and still produce sound welds. Because of the high level of manganese, if the filler is used for multiple-pass welds, there might be manganese-caused center-line cracking of the weld; 100% CO_2 can be used as the shielding gas. E70T-2 can be used on ASME A36, A106, A242, A252, A285, A441, and A572 or similar metals. Applications include repair and maintenance work and general fabrication.

E70T-4, E71T-4

E70T-4 and E71T-4 are self-shielding, flux cored filler metal. The fluxing agents produce a slag, which allows a larger-than-usual molten weld pool. The large weld pool permits high deposition rates. Weld deposits are ductile and have a high resistance to cracking. E70T-4 can be used to weld joints that have larger-than-usual root openings. Applications include large weldments and earth-moving equipment.

E71T-7

E71T-7 is a self-shielding, all-position, flux cored filler metal. The fluxing system allows the control of the molten weld pool required for out-of-position welds. The high level of deoxidizers reduces the tendency for cracking in the weld. It can be used for single- or multiple-pass welds.

Stainless Steel Electrodes

The AWS specification for stainless steel covered arc electrodes is A5.4 and for stainless steel bare, cored, and stranded electrodes and welding rods is A5.9. Filler metal classified within the A5.4 uses the letter *E* as its prefix, and the filler metal within the A5.9 uses the letters *ER* as its prefix, Table 26-5.

Following the prefix, the American Iron and Steel Institute's (AISI) three-digit stainless steel number is used. This number indicates the type of stainless steel in the filler metal.

To the right of the AISI number, the AWS adds a dash followed by a suffix number. The number 15 is used to indicate that there is a lime base coating, and the DCEP polarity welding current should be used. The number 16 is used to indicate there is a titania-type coating, and AC or DCEP polarity welding currents can be used. Examples of this classification system are E308-15 and E308-16 electrodes.

The letter *L* may be added to the right of the AISI number before the dash and suffix number to indicate a low-carbon stainless welding electrode. E308L-15 and E308L-16 arc welding electrodes and ER308L and ER316L are examples of the use of the letter *L*, Table 26-6.

Stainless steel may be stabilized by adding columbium (Cb) as a carbide former. The designation Cb is added after the AISI number for these electrodes, such as E309Cb-16. Stainless steel filler metals are stabilized to prevent chromium-carbide precipitation.

E308-15, E308-16, E308L-15, E308-16, ER308, and ER308L

All are filler metals for 308 stainless steels, which are used for food or chemical equipment, tanks, pumps, hoods, and evaporators. All E308 and ER308 filler metals can be used to weld on all 18-8-type stainless steels such as 301, 302, 302B, 303, 304, 305, 308, 201, and 202.

E309-15, E309-16, E309Cb-15, E309Cb-16, ER309, and ER309L

All are filler metals for 309 stainless steels, which are used for high-temperature service, such as furnace parts and mufflers. All E309 filler metals can be used to weld on 309 stainless or to join mild steel to any 18-8-type stainless steel.

E310-15, E310-16, E309Cb-15, E309Cb-16, E310Mo-15, E310Mo-16, and ER310

All are filler metals for 310 stainless steels, which are used for high-temperature service where low creep is desired, such as for jet engine parts, valves, and furnace parts. All E310 filler metals can be used to weld 309 stainless or to join mild steel to stainless or to weld most hard-to-weld carbon and alloy steels. E310Mo-15 and 16 electrodes have molybdenum added to improve their strength at high temperatures and to resist corrosive pitting.

E316-15, E316-16, E316L-15, E316L-16, ER316, ER316L, and ER316L-Si

All are filler metals for 316 stainless steels, which are used for high-temperature service where high strength with low creep is desired. Molybdenum is added to improve these properties and to resist corrosive pitting.

AISI TYPE NUMBER	442 446	430F 430FSE	430 431	501 502	403 405 416 416SE	321 410 420 414	348 347	317	316L	316	314	310 310S	309 309S	303 304 L	201 202 301 303S E 305 308	302 302B 304	Mild Steel
201–202–301	**310**	**310**	**310**	**310**	**309**	**309**	**308**	**308**	**308**	**308**	**308**	**308**	**308**	**308**	**308**	**308**	**312**
302–3028–304	312	312	312	312	310	310											310
305–308	309	309	309	309	312	312											309
303	**310**	**310**	**310**	**310**	**309**	**309**	**308**	**308**	**308**	**308**	**308**	**308**	**308**	**308**	**308-15**	**308**	**312**
303SE	309	309	309	309	310	310											310
	312	312	312	312	312	312											309
	310	**310**	**310**	**310**	**309**	**309**	**308**	**308**	**308-L**	**308**	**308**	**308**	**308**	**308-L**	**308**	**308**	**312**
304L	309	309	309	309	310	310											310
	312	312	312	312	312	312											309
309	**310**	**310**	**310**	**310**	**309**	**309**	**308**	**317**	**316**	**316**	**309**	**309**	**309**	**308**	**308**	**308**	**309**
309S	309	309	309	309	310	310		316									310
	312	312	312	312	312	312		309									312
310	**310**	**310**	**310**	**310**	**310**	**310**	**308**	**317**	**316**	**316**	**310**	**310**	**309**	**308**	**308**	**308**	**310**
310S	309	309	309	309	309	309		316					310				309
	312	312	312	312	312	312		309									312
	310	**310**	**310**	**310**	**310**	**310**	**309**	**309**	**309**	**309**	**310-15**	**310**	**309**	**309**	**309**	**309**	**310**
314	312	312	312	312	312	309	310	310	310	310			310	310	310	310	309
	309	309	309	309	309	312	308										312
	310	**310**	**310**	**310**	**309**	**309**	**308**	**316**	**316**	**316**	**309**	**310**	**309**	**308**	**308**	**308**	**309**
316	309	309	309	309	310	310					310	309	310	316	316	316	310
	312	312	312	312	312	312					316	316	316				312
	310	**310**	**310**	**310**	**309**	**309**	**308**	**316**	**316-L**	**316**	**309**	**310**	**316**	**308**	**308**	**308**	**309**
316L	309	309	309	309	310	310		317			310	309	309	316	316	316	310
	312	312	312	312	312	312		308			316	316					312
	310	**310**	**310**	**310**	**309**	**309**	**308**	**317**	**316**	**316**	**309**	**317**	**317**	**308**	**308**	**308**	**309**
317	309	309	309	309	310	310			308	308	310	316	316	316	316	316	310
	312	312	312	312	312	312					317	309	309	317	317	317	312
321	**310**	**310**	**310**	**310**	**309**	**309**		**308**	**347**	**347**	**309**	**347**	**347**	**347**	**347**	**347**	**309**
348	309	309	309	309	310	310	347	347	308	308	310	308	308	308-L	308	308	310
347	312	312	312	312	312	312					347						312
403–405	**310**	**310**	**310**	**310**	**309**	**410†**	**309**	**309**	**309**	**309**	**310**	**310**	**309**	**309**	**309**	**309**	**309**
410–420	309	309	309	309	310	309††	310	310	310	310	309	309	310	310	310	310	310
414	312	312	312	312													312
416	**310**	**310**	**310**	**310**	**410-15†**	**410-15†**	**309**	**309**	**309**	**309**	**309**	**310**	**309**	**309**	**309**	**309**	**309**
416SE	309	309	309			309††	310	310	310	310	310	309	310	310	310	310	310
						310††		312	312	312	312	312	312	312	312	312	312
501	**310**	**310**	**310**	**502†**	**310**	**310**	**310**	**310**	**310**	**310**	**310**	**310**	**310**	**310**	**310**	**310**	**310**
502				310††			309	309	309	309	309	309	309	309	309	309	312
																	309
430	**310**	**310**	**430-15†**	**310**	**310**	**310**	**310**	**310**	**310**	**310**	**310**	**310**	**310**	**310**	**310**	**310**	**310**
431	309	309	310††			309	309	309	309	309	309	309	309	309	309	309	309
			309††														312
430F	**310**	**410-15†**	**310**	**310**	**310**	**310**	**309**	**309**	**310**	**310**	**310**	**310**	**310**	**310**	**310**	**310**	**310**
430FSE	309		309	309	309	309	310	310	309	309	309	309	309	309	309	309	309
					312	312	312	312	312	312	312	312	312	312	312	312	312
442	**309**	**309**	**310**	**310**	**310**	**310**	**310**	**310**	**310**	**310**	**310**	**310**	**310**	**310**	**310**	**310**	**310**
443	310	310	309	309	309	309	309	309	309	309	309	309	309	309	309	309	309
		312	312	312	312	312	312	312	312	312	312	312	312	312	312	312	312

†Preheat.

††No preheat necessary.

Note: Bold numbers indicate first choice; light numbers indicate second and third choices. This choice can vary with specific applications and individual job requirements.

Table 26-5 Filler Metal Selector Guide for Joining Different Types of Stainless Steel to the Same Type or Another Type of Stainless Steel (*Thermacote Welco*)

UTP Designation	AWS/SFA5.4 Covered	AWS/SFA5.9 TIG and MIG	Description and Applications
6820	E 308-16	ER 308	For welding conventional 308 type SS
68 Kb	E 308-15		Low hydrogen coating
6820 Lc	E 308 L-16	ER 308L	Low carbon grade, prevents carbide precipitation adjacent to weld
308L Fe Hp	E 308 L-16		Fast depositing for maintenance and production coating
68 LcHL	E 308 L-16		High-performance electrode with rutile-acid coating, core wire alloyed, for stainless and acid-resisting CrNi steels
68 LcKb	E 308 L-15		Low carbon electrode for stainless, acid-resisting CrNi steels
6824	E 309-16	ER 309	For welding 309 type SS and carbon steel to SS
6824 Kb	E 309-15		Special lime-coated electrode for corrosion and heat-resistant 22/12 CrNi steels
6824 Lc	E 309 L-16	ER 309L	Same as 309 but with low carbon content
6824 Nb	E 309 Cb-16		Corrosion and heat-resistant 22/12 CrNi steels
6824 MoNb	E309MoCb-16		Corrosion and heat-resistant 22/12 CrNi steels
309L Fe Hp	E 309 L-16		High deposition rate, easy to use
6824 Mo Lc	E 309 L-16		For welding similar and dissimilar SS
68H	E 310-16	ER 310	For high-temperature service and cladding steel
6820 Mo	E 316-16	ER 316	For welding acid-resistant stainless steels
6820 Mo Lc	E 316 L-16	ER 316L	Low carbon grade, prevents intergranular corrosion
68 TI Mo	E 316 L-16		Most efficient type, for maintenance and production, high performance
68 MoLcHL	E 316 L-16		High-performance electrode with rutile-acid coating, core wire alloyed, for stainless and acid-resisting CrNiMo steels
68 MoLcKb	E 316 L-15		Low carbon electrode for stainless and acid-resisting CrNiMo steels
317 Lc Titan	E 317 L-16	ER 317L	Deposit resists sulphuric acid corrosion
317LFe Hp	E 317 L16		Fast melt-off rate, excellent for overlays, easy to use, high performance
68 Mo	E 318-16		Versatile stainless all-position electrode
320 Cb	E 320-15		For welding similar acid-resistant SS
3320 Lc	E 320		A rutile-coated electrode for welding in all positions except vertical-down
6820 Nb	E 347-16	ER 347	Stabilized grade, prevents carbide precipitation
347 FeHp	E 347-16		High-performance stainless steel electrode of class E 347-16 for welding stabilized CrNi alloys
66	E 410-15		Low-hydrogen electrode for corrosion and heat-resistant 14% Cr-steels
1915 HST			Low-hydrogen, fully austenitic electrode with 0% ferrite content
1925			Extremely corrosion-resistant to phosphoric and sulfuric acids
68 Hcb	E 310 Cb	ER 310 Cb	For high heat applications and joining steels to stainless steel
2535 NbSn			Lime-type special electrode used for surfacing and joining heat-resistant base metals, especially cast steel
E 330-16	E 330-16		Excellent for welding furnace parts
6805			Base-coated electrode age-hardenable
6808 Mo			Rutile lime type austenitic–ferritic electrode with low carbon content, suitable for joining and surfacing on corrosion-resistant steels and cast steel type with austenitic–ferritic structure (duplex steels)
6809 Mo			Rutile-basic austenitic–ferritic electrode with low carbon content suited for joining and surfacing on corrosion-resistant steels and cast steel types with an austenitic–ferritic structure (duplex steels)

Table 26-6 Stainless Steel Electrodes, Filler Metals, and Wires (*UTP Welding Materials, Inc.*)

E316 filler metals are used for welding tubing, chemical pumps, filters, tanks, and furnace parts. All E316 filler metals can be used on 316 stainless steels or when weld resistance to pitting is required.

Nonferrous Electrode

The AWS identification system for covered nonferrous electrodes is based on the atomic symbol or symbols of the major alloy(s) or the metal's identification number. The alloy having the largest percentage appears first in the identification. The atomic symbol is prefixed by the letter *E*. For example, ECu is a covered copper arc welding electrode, and ECuNiAl is a copper-nickel-aluminum alloy covered arc welding electrode. A letter, number, or letter-number combination may be added to the right of the atomic symbol to indicate some special alloys. For example, ECuAl-A2 is a copper-aluminum welding electrode that has 1.59% iron added.

Aluminum and Aluminum Alloys

The AWS specifications for aluminum and aluminum alloy filler metals are A5.3 for covered arc welding electrodes and A5.10 for bare welding rods and electrodes. Filler metal classified within the A5.3 uses the atomic symbol Al, and in the A5.10 the prefix *ER* is used with the Aluminum Association number for the alloy, Table 26-7.

Aluminum Covered Arc Welding Electrodes

Al-2 and Al-43

The aluminum electrodes do not use the letter *E* before the electrode number. Aluminum covered arc welding electrodes are designed to weld with DCEP polarity. These electrodes can be used on thin or thick sections, but thick sections must be preheated to between 300°F (150°C) and 600°F (315°C). The preheating of these thick sections allows the weld to penetrate immediately when the weld starts. Aluminum arc welding electrodes can be used on 2024, 3003, 5052, 5154, 5454, 6061, and 6063 aluminum. When welding on aluminum, a thin layer of surface oxide may not prevent welding. Thicker oxide layers must be removed mechanically or chemically. Excessive penetration can be supported by carbon plates or carbon paste. Most arc welding electrodes can also be used for oxyfuel gas welding of aluminum.

Aluminum Bare Welding Rods and Electrodes

ER1100

1100 aluminum has the lowest percentage of alloy agents of all of the aluminum alloys, and it melts at 1215°F (657°C). The filler wire is also relatively pure. ER1100 produces welds that have good corrosion resistance and high ductility, with tensile strengths ranging from 11,000 to 17,000 psi. The weld deposit has a high resistance to cracking during welding. This wire can be used with OFW, GTAW, and GMAW. Preheating to 300°F to 350°F (150°C to 175°C) is required for GTA welding on plate or pipe 3/8 in. (10 mm) and thicker to ensure good fusion. Flux is required for OFW. 1100 aluminum is commonly used for items such as food containers, food-processing equipment, storage tanks, and heat exchangers. ER1100 can be used to weld 1100 and 3003 grade aluminum.

ER4043

ER4043 is a general purpose welding filler metal. It has 4.5 to 6.0% silicon added, which lowers its melting temperature to 1155°F (624°C). The lower melting temperature helps promote a free-flowing molten weld pool. The welds have high ductility and a high resistance to cracking during welding. This wire can be used with OFW, GTAW, and GMAW. Preheating to 300°F to 350°F (150°C to 175°C) is required for GTA welding on plate or pipe 3/8 in. (10 mm) and thicker to ensure good fusion. Flux is required for OFW. ER4043 can be used to weld on 2014, 3003, 3004, 4043, 5052, 6061, 6062, and 6063 and cast alloys 43, 355, 356, and 214.

ER5356

ER5356 has 4.5 to 5.5% magnesium added to improve the tensile strength. The weld has high ductility but only an average resistance to cracking during welding. This wire can be used for GTAW and GMAW. Preheating to 300°F to 350°F (150°C to 175°C) is required for GTA welding on plate or pipe 3/8 in. (10 mm) and thicker to ensure good fusion. ER5356 can be used to weld on 5050, 5052, 5056, 5083, 5086, 5154, 5356, 5454, and 5456.

ER5556

ER5556 has 4.7 to 5.5% magnesium and 0.5 to 1.0% manganese added to produce a weld with high strength.

Base Metal	319 355	43 356	214	6061 6063 6151	5456	5454	5154 5254	5086	5083	5052 5652	5005 5050	3004	1100 3003	1060
1060	4145 4043 4047	4043 4047 4145	4043 5183 4047	4043 4047	5356 4043	4043 5183 4047	4043 5183 4047	5356 4043	5356 4043	4043 4047	1100 4043	4043	1100 4043	1260 4043 1100
1100 3003	4145 4043 4047	4043 4047 4145	4043 5183 4047	4043 4047	5356 4043	4043 5183 4047	4043 5183 4047	5356 4043	5356 4043	4043 5183 4047	4043 5183 5356	4043 5183 5356	1100 4043	
3004	4043 4047	4043 4047	5654 5183 5356	4043 5183 5356	5356 5183 5556	5654 5183 5356	5654 5183 5356	5356 5183 5556	5356 5183 5556	4043 5183 4047	4043 5183 5356	4043 5183 5356		
5005 5050	4043 4047	4043 4047	5654 5183 5356	4043 5183 5356	5356 5183 5556	5654 5183 5356	5654 5183 5356	5356 5183 5556	5356 5183 5556	4043 5183 4047	4043 5183 5356			
5052 5652	4043 4047	4043 5183 4047	5654 5183 5356	5356 5183 4043	5356 5183 5556	5654 5183 5356	5654 5183 5356	5356 5183 5556	5356 5183 5556	5654 5183 4043				
5083	NR	5356 4043 5183	5356 5183 5556	5356 5183 5556	5183 5356 5556	5356 5183 5556	5356 5183 5556	5356 5183 5556	5183 5356 5556					
5086	NR	5356 4043 5183	5356 5183 5556	5356 5183 5556	5356 5183 5556	5356 5183 5554	5356 5183 5554	5356 5183 5556						
5154 5254	NR	4043 5183 4047	5654 5183 5356	5356 5183 4043	5356 5183 5554	5654 5183 5356	5654 5183 5356							
5454	4043 4047	4043 5183 4047	5654 5183 5356	5356 5183 4043	5356 5183 5554	5554 4043 5183								
5456	NR	5356 4043 5183	5356 5183 5556	5356 5183 5556	5556 5183 5356									
6061 6063 6151	4145 4043 4047	4043 5183 4047	5356 5183 4043	4043 5183 4047										
214	NR	4043 5183 4047	5654 5183 5356											
43 356	4043 4047	4145 4043 4047												
319 355	4145 4043 4047													

Note: First filler alloy listed in each group is the all-purpose choice. "NR" means that these combinations of base metals are not recommended for welding.

Table 26-7 Recommended Filler Metals for Joining Different Types of Aluminum to the Same Type or a Different Type of Aluminum (*Thermacote Welco*)

The weld has high ductility and only average resistance to cracking during welding. This wire can be used for GTAW and GMAW. Preheating to 300°F to 350°F (150°C to 175°C) is required for GTA welding on plate or pipe 3/8 in. (10 mm) and thicker to ensure good fusion. ER5556 can be used to weld on 5052, 5083, 5356, 5454, and 5456.

Special Purpose Filler Metals

ENi

The nickel arc welding electrodes are designed to be used with AC or DCEP polarity. These arc welding electrodes are used for cast-iron repair. The carbon in cast iron will not migrate into the nickel weld metal, thus preventing cracking and embrittlement. The cast iron may or may not be preheated. A very short arc length and a fast travel rate should be used with these electrodes.

ECuAl

The aluminum bronzed welding electrodes are designed to be used with DCEP polarity. This welding electrode has copper as its major alloy. The aluminum content is at a much lower percentage. Iron is usually added but at a percentage that is very low. These electrodes are sometimes referred to as arc brazing electrodes, although this is not an accurate description. Stringer beads and a short arc length should be used with these electrodes. Aluminum bronze welding electrodes are used for overlaying bearing surfaces; welding on castings of manganese, bronze, brass, or aluminum bronze; or assembling dissimilar metals.

Surface and Buildup Electrode Classification

Hardfacing or wear-resistant electrodes are the most popular special purpose electrodes; however, there are also cutting and brazing electrodes. Specialty electrodes may be identified by manufacturers' trade names. Most manufacturers classify or group hardfacing or wear-resistant electrodes according to their resistance to impact, abrasion, or corrosion. Occasionally, electrode resistance to wear at an elevated temperature is listed. One electrode may have more than one characteristic or type of service listed.

EFeMn-A

The EFeMn-A electrodes are designed to be used with AC or DCEP polarity. This electrode is an impact-resistant welding electrode. It can be used on hammers, shovels, and spindles and in other similar applications.

ECoCr-C

The ECoCr-C electrodes are designed to be used with AC or DCEP polarity. This electrode is a corrosion- and abrasion-resistant welding electrode. It also maintains its resistance to elevated temperatures. ECoCr-C is commonly used for engine cams, seats and valves, chain saw bars, bearings, and dies.

Magnesium Alloys

The joining of magnesium alloys by torch welding or brazing is possible without a fire hazard because the melting point of magnesium is 1202°F (651°C) to 858°F (459°C) below its boiling point, where magnesium may start to burn.

ER AZ61A

The ER AZ61A filler metal can be used to join most magnesium wrought alloys. This filler has the best weldability and weld strength for magnesium alloys AZ31B, HK31A, and HM21A.

ER AZ92A

The ER AZ92A filler metal can be used on cast alloys, Mg-Al-Zn and AM 100A. This filler metal has a somewhat higher resistance to cracking.

SUMMARY

Proper filler metal selection is one of the most important factors affecting the successful welding of a joint. Many factors affect the selection of the most appropriate filler metal for a job. In some cases, cost is the greatest factor and in others it is structural strength. For example, if you were building an ultralight aircraft, you would be more concerned with strength than cost. However, if you were building an iron fence, you might be more concerned with cost. Every application is different, so it may be a help for you to list the items you feel are most important for selecting a filler metal. This will help you select the most appropriate filler metal for your needs.

Manufacturers' literature on filler metals can be divided into two general sections. One section of the literature is technical and the other is advertisement. In the technical section you are provided with specific information on each filler metal's operation, performance, and uses. In the advertisement section you are provided with marketing information and claims regarding performance. Knowing the types of information in both sections will help you evaluate new material as you select filler metal.

If you are considering a large purchase of filler metals, it is advantageous for you to request samples of the various filler metals from manufacturers so that you can test their performance in your applications. Pretesting of the products in your applications gives you an opportunity to determine which filler metal is going to give you the best value for your money. It may also be necessary to qualify the filler metal for your welding certification program before you make the purchase and begin using the product.

REVIEW QUESTIONS

1. List various identification methods by which manufacturers may identify their filler metals.
2. What is the most widely used numbering and lettering system for filler metals?
3. Where is the identification number located?
4. Where would you go to find information that pertains directly to a specific filler metal?
5. What general information about the electrode may be given by the manufacturer?
6. How can you use the information supplied by the manufacturer about the electrodes?
7. What does minimum tensile strength psi mean?
8. What kind of information about the chemical analysis of the electrodes is provided by the manufacturer?
9. What are some of the other names for shielded metal arc welding electrodes?
10. What are the two parts of a shielded metal arc welding electrode?
11. What are the functions of the core wire in an electrode?
12. What is the purpose of the shielding gas?
13. What is the purpose of alloying elements in the flux?
14. What effect can welding fluxes have on the weld bead?
15. Why is the selection of the best filler metal for a job in large shops seldom delegated to the welder?
16. If a filler metal was labeled by the AWS classification system with the letters *RG*, what information does that provide about the filler metal?
17. Using the AWS classification system, what do the series of numbers following the letter *E* indicate on a carbon and low alloy steel covered electrode?
18. What would be the minimum tensile strength of a good weld for a wire-type carbon steel filler metal labeled ER70S-x?
19. In the AWS identification system for covered nonferrous electrodes, what is the meaning of the symbols immediately following the *E*?
20. What type of polarity are aluminum covered arc welding electrodes designed to be welded with?

Chapter 27

Welding Metallurgy

OBJECTIVES

After completing this chapter, the student should be able to:

- Explain why it is important to understand the properties of the materials being welded.
- Explain the importance of preheating and postheating.
- Describe what happens when metal is cooled too quickly.
- Describe the mechanical properties of metals.
- Explain the heat-affected zone's effect on metal.

KEY TERMS

acicular structure
allotropic metals
allotropic transformation
austenite
body-centered cubic (BCC)
cementite
crystal lattices
crystalline structures
eutectic composition
face-centered cubic (FCC)
ferrite
grain refinement
heat treatments
heat-affected zone (HAZ)
martensite
metallurgy
pearlite
phase diagrams
precipitation hardening
recrystallization temperature
solid solutions
tempering
unit cell

INTRODUCTION

You need to know more than just how to establish an arc and manipulate the electrode to consistently deposit uniform, high-quality welds. It is important for a competent welder to understand the materials being welded. With this knowledge, you can select the best processes and procedure to produce a weldment that is as strong and tough as possible. You need to learn **metallurgy** to recognize that special attention might be needed when welding certain types of steel and to understand the kind of welding procedure that may be required.

Metals gain their desirable mechanical and chemical properties as a result of alloying and heat treating. Welding operations heat the metals, and that heating certainly changes not only the metal's initial structure but its properties as well. A skilled welder can minimize the effects that these changes have on the metal and its properties.

Heat, Temperature, and Energy

Heat and *temperature* are both terms used to describe the quantity and level of thermal energy that is present. To better comprehend what takes place during a weld, you must understand the differences between heat and temperature. Heat is the quantity of thermal energy, and temperature is the level of thermal activity.

Although both *heat* and *temperature* are used to describe the thermal energy in a material, they are independent values. A material can have a large quantity of heat energy in it, but the material can be at a low temperature. Conversely, a material can be at a high temperature but have very little heat.

Heat

Heat is the amount of thermal energy in matter. All matter contains heat down to absolute zero (–460°F or –273°C). The basic U.S. unit of measure for heat is the British thermal unit (BTU). One BTU is defined as the amount of heat required to raise 1 lb of water to 1° F.

There are two forms of heat. One is called sensible because as it changes, the change in temperature can be sensed or measured. The other form of heat is called latent. Latent heat is the heat required to change matter from one state to another, and it does not result in a temperature change. For example, if a pot containing water is heated on a stove, the water picks up heat from the burner and the water's temperature increases. The more heat put into the water, the higher its temperature until it begins to boil. Once the water reaches 212°F (100°C), its temperature stops rising. As long as the pot is on the burner, heat is being put into the water, but there is no increase in sensible heat. The heat is all going into the latent heat required to change the water from the liquid state to a gaseous state, Figure 27-1.

Latent heat is absorbed by a material as it changes from a solid to a liquid state and from a liquid to a gaseous state. When matter changes from a gaseous to a liquid state or from a liquid to a solid state, latent heat must be removed, Figure 27-2. A change

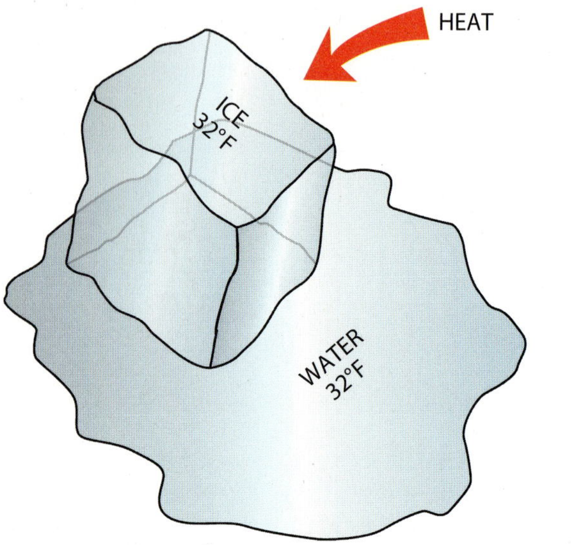

FIGURE 27-1 There is no change in temperature when there is a change in state.

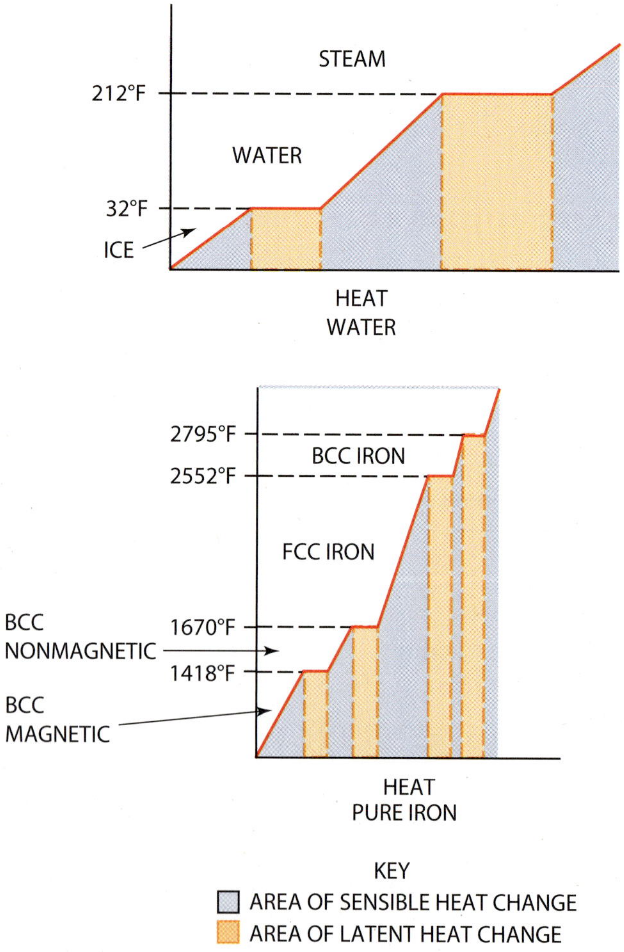

FIGURE 27-2 Sensible and latent heat.

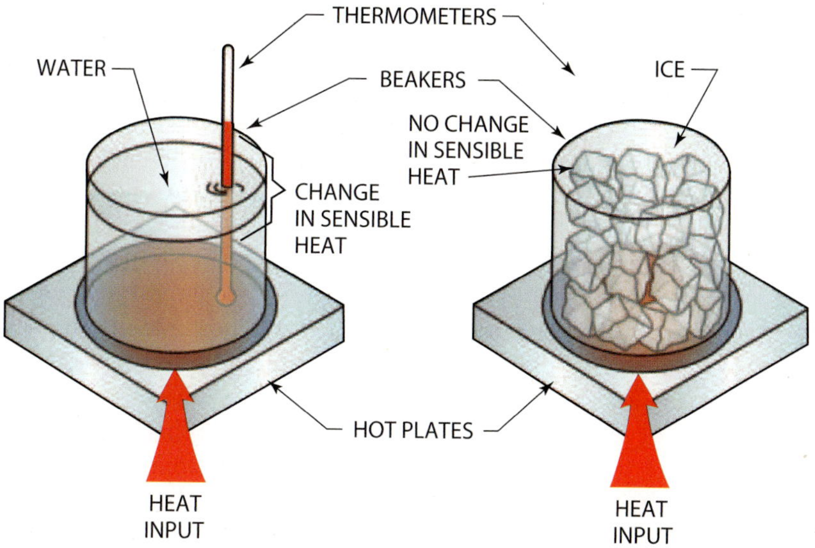

FIGURE 27-3 Both beakers have heat being added, but only one has a change in temperature. © Cengage Learning 2012

in a material's latent heat also occurs when there is a change in the structure of the material. For example, when the crystal lattice of a metal changes, a change in latent heat occurs.

EXPERIMENT 27-1

Latent and Sensible Heat

In this experiment you will be working in a small group to observe latent and sensible heat. Using two beakers, two hot plates, two thermometers, a cup of ice, a cup of ice water, gloves, safety glasses, and any other required safety protection, you are going to observe the effect of latent heat on the temperature rise of water, Figure 27-3.

Put 1 cup of ice water 32°F (0°C) in one beaker and 1 cup of ice 32°F (0°C) in the other beaker. Place both beakers on a hot plate. Slowly stir the contents of each beaker using the thermometers. Observe the change in temperatures each thermometer measures. Record the following information:

1. What was the temperature when you started?
2. What was the temperature after 1 minute?
3. What was the temperature of the water without the ice when the ice in the other cup was all gone?
4. How long did it take for all of the ice to melt?

Complete a copy of the time sheet in Appendix I or as provided by your instructor.◆

Temperature

Temperature is a measurement of the vibrating speed or frequency of the atoms in matter. The atoms in all matter vibrate down to absolute zero. The basic unit of measure is the degree. The U.S. unit is degrees Fahrenheit.

As matter becomes warmer, its atoms vibrate at a higher frequency. As matter cools, the vibrating frequency slows. This vibration of the atoms is what gives off the infrared light that comes from all objects that are above absolute zero. When the object becomes hot enough, the vibrating frequency of the atoms gives off visible light. We see that light as a dull red glow when the surface reaches a little above 1000°F (540°C). As the surface becomes even hotter, we can see the color light it gives off change until it glows "white hot," Figure 27-4.

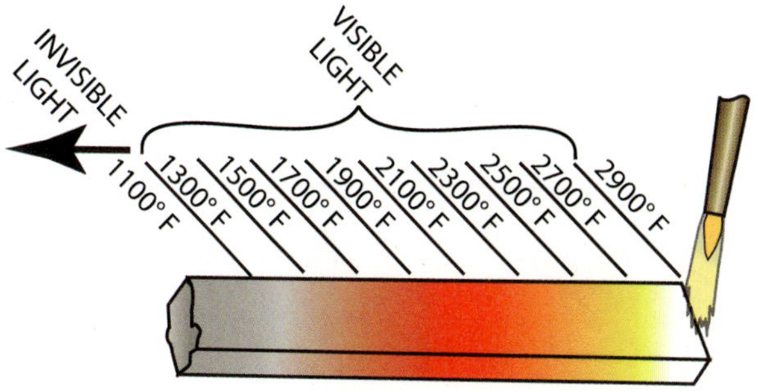

FIGURE 27-4 Visible and invisible light. © Cengage Learning 2012

We can tell the temperature of an object by the frequency of the light its vibrating atoms produce. That is how scientists tell the temperature of distant stars.

EXPERIMENT 27-2

Temper Colors

In this experiment you will be working in a small group to observe the formation of temper colors as metal is heated. Using a piece of mild steel, a safely set-up oxyfuel torch, gloves, safety glasses, and required personal protection, you are going to observe the changing temperature's effect on the color of the steel, Figure 27-5.

Light the torch and hold it near one side of the mild steel. Observe the other side of the metal to see when it begins to change color. Record the following information:

1. What was the first color that you could see?
2. What were the second, third, fourth, and so forth, colors that you could see?
3. What was the last color that appeared before the metal melted?
4. Compare the colors you saw with the colors shown in Figure 27-4.

Repeat this experiment using different metals and see what colors they produce.

Complete a copy of the time sheet in Appendix I or as provided by your instructor.◆

Mechanical Properties of Metal

In learning welding skills, an understanding of the mechanical properties of metals is most important. The mechanical properties of a metal can be described as those quantifiable properties that enable the metal to resist externally imposed forces without failing. If the mechanical properties of a metal are known, a product can be constructed that will meet specific engineering specifications. As a result, a safe and sound structure can be constructed.

All of a metal's properties interact with one another. Some properties are similar and complement each other, but others tend to be opposites. For example, a metal cannot be both hard and ductile. Some metals are hard and brittle, and others are hard and tough. Probably the most outstanding property of metals is the ability to have their properties altered by some form of heat treatment. Metals can be soft and then made hard, brittle, or strong by the correct heat treatment, yet other heat treatments can return the metal back to its original, soft form.

It is the responsibility of the metallurgist or engineer to select a metal that has the best group of properties for any specific job. Except in very unusual cases, a metallurgist would not create a new alloy for a job but merely select one from the tens of thousands of metal alloys already available. Often metallurgists must make difficult choices when designs call for properties that are usually not found together. Additionally, the more unique the alloy, the greater its cost and often the more difficult it is to weld.

This section describes some of the significant mechanical properties of metals. The next section describes how various heat treatments can be used to change a metal's properties.

The sections that follow describe some of the significant mechanical properties that you should be familiar with to do a successful job of welding fabrication.

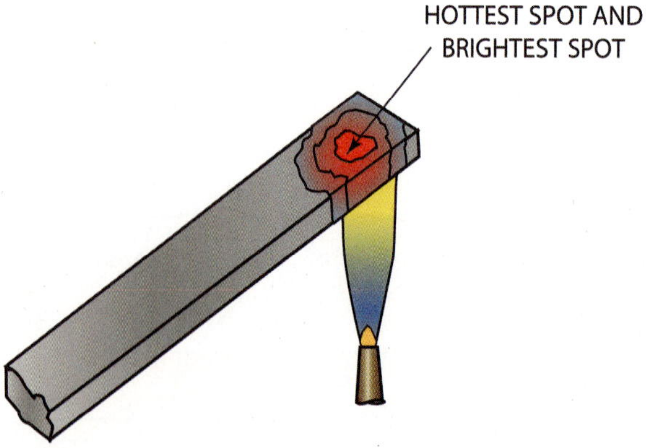

FIGURE 27-5 As the temperature of the metal increases, it begins to glow. © Cengage Learning 2012

Hardness

Hardness may be defined as resistance to penetration. Files and drills are made of metals that rank high in hardness when properly heat treated. The hardness property may in many metals be increased or decreased by heat-treating methods and increased in other metals by cold working. Since hardness is proportional to strength, it is a quick way to determine strength. It is also useful in determining whether the metal received the proper heat treatment, since heat treatment also affects strength. Hardness is measurable in a number of ways. Most methods quantify a metal's resistance to highly localized deformation.

Brittleness

Brittleness is the ease with which a metal will crack or break apart without noticeable deformation. Glass is brittle, and when broken, all of the pieces fit back together because it did not bend (deform) before breaking. Some types of cast iron are brittle and once broken will fit back together like a puzzle's parts. Brittleness is related to hardness in metals. Generally, as the hardness of a metal is increased, the brittleness is also increased. Brittleness is not measured by any testing method. It is the absence of ductility.

Ductility

Ductility is the ability of a metal to be permanently twisted, drawn out, bent, or changed in shape without cracking or breaking. Ductile metals include aluminum, copper, zinc, and soft steel. Ductility is measurable in a number of ways. Ductility in tensile tests is usually measured as a percentage of elongation and as a percentage of reduction in area. It also can be measured with bend tests.

Toughness

Toughness is the property that allows a metal to withstand forces, sudden shock, or bends without fracturing. Toughness may vary considerably with different methods of load application and is commonly recognized as resistance to shock or impact loading.

Toughness is measured most often with the Charpy test. This test yields information about the resistance of a metal to sudden loading in the presence of a severe notch. Because only a small specimen is required, the test is faulted for not providing a general picture of a component's toughness. Unfortunately, tests on a larger scale require very expensive equipment and are very time consuming.

Strength

Strength is the property of a metal to resist deforming. Common types of strength measurements are tensile, compressive, shear, and torsional, Figure 27-6.

- *Tensile strength: Tensile strength* refers to the property of a material that resists forces applied to pull metal apart. Tension has two parts: yield strength and ultimate strength. *Yield strength* is the amount of strain needed to permanently deform a test specimen. The *yield point* is the point during tensile loading when the metal stops stretching and begins to be permanently made longer by deforming. Like a rubber band that stretches and returns to its original size, metal that stretches before the yield point is reached will return to its original shape. After the yield point is reached the metal is usually longer and thinner. Some metals stretch a great deal before they yield, and others stretch a great deal before and after the yield point. These metals are considered to have high ductility. *Ultimate strength* is a measure of the load that breaks a specimen. Some metals may become work-hardened as they are stretched during a tensile test. These metals will actually become stronger and harder as a result of being tested. Other metals lose strength once they pass the yield point and fail at a much lower force.

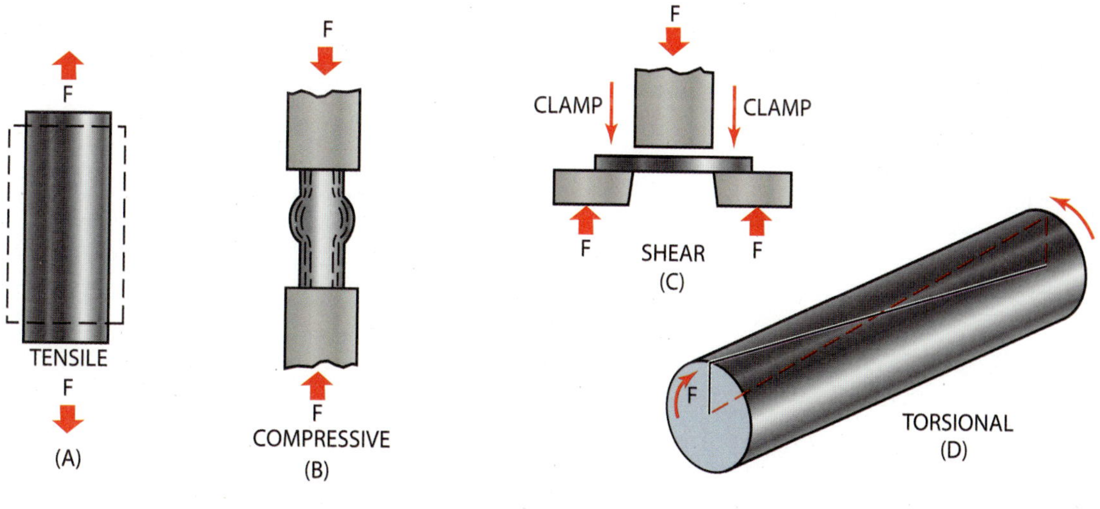

FIGURE 27-6 Types of forces (F) applied to metal.

Metals that do not stretch much before they break are brittle. The tensile strength of a metal can be determined by a tensile testing machine.

- *Compressive strength:* Compressive strength is the property of a material to resist being crushed. The compressive strength of cast iron, rather brittle material, is three to four times its tensile strength.
- *Shear strength:* Shear strength of a material is a measure of how well a part can withstand forces acting to cut or slice it apart.
- *Torsional strength:* Torsional strength is the property of a material to withstand a twisting force.

Other Mechanical Concepts

Strain is deformation caused by stress. The part shown in Figure 27-7 is under stress and was strained (deformed) by the external load. The deformation is in the form of a bend.

Elasticity is the ability of a material to return to its original form after removal of the load. The yield point of a material is the limit to which the material can be loaded and still return to its original form after the load has been removed, Figure 27-8.

Elastic limit is defined as the maximum load, per unit of area, to which a material will respond with a deformation directly proportional to the load. When the force on the material exceeds the elastic limit, the material will be deformed permanently. The amount of permanent deformation is proportional to the stress level above the elastic limit. When stressed below its elastic limit, the metal returns to its original shape.

Impact strength is the ability of a metal to resist fracture under a sudden load. An example of a material that is ductile and yet has low-impact strength is Silly Putty. If it is pulled slowly, it stretches easily, but if it is pulled quickly, it forms a brittle fracture.

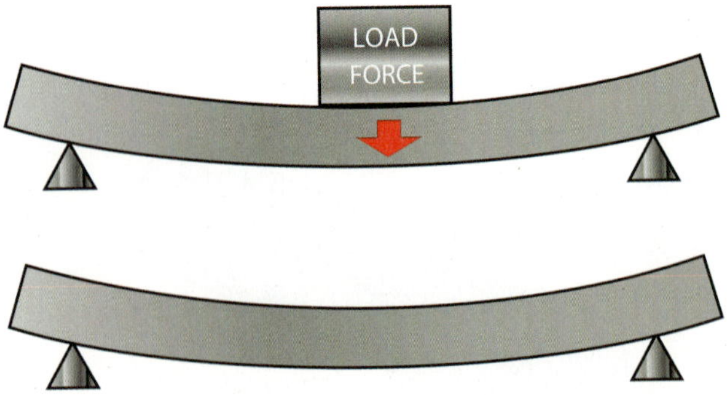

FIGURE 27-7 Effect of excessive stress causing a permanent strain in a beam. © Cengage Learning 2012

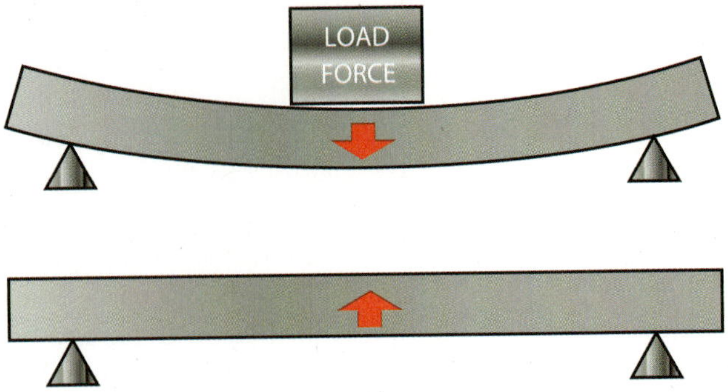

FIGURE 27-8 Reaction of an elastic beam to a force. © Cengage Learning 2012

Structure of Matter

All solid matter exists in one of two basic forms. Solid matter is either crystalline or amorphic in form. Solids that are crystalline in form have an orderly arrangement of their atoms. Each crystal making up the solid can be very small, too small to be seen without a microscope. Examples of materials that are crystalline in form include metals and most minerals, such as table salt, Table 27-1. Amorphic materials have no orderly arrangement of their atoms into crystals. Examples of amorphic materials include glass and silicon, Table 27-2. Both crystalline and amorphic materials look and feel like solids, so without sophisticated testing equipment, you cannot tell the difference between them.

Metal	Crystal Type
Aluminum	fcc
Chromium	bcc
Copper	fcc
Gold	fcc
Iron (alpha)	fcc
Iron (gamma)	fcc
Iron (delta)	bcc
Lead	fcc
Nickel	fcc
Silver	fcc
Tungsten	bcc
Zinc	hcp

KEY
fcc = Face-centered cubic
bcc = Body-centered cubic
hcp = Hexagonal close packed

Table 27-1 Crystal Structure of Common Metals

Materials		Crystal Type
Copper acetylide	Cu_2C_2	None
Gadolinium oxide	Gd_2O_3	None
Iron hydroxide	$Fe(OH)_2$	None
Lead oxide	Pb_2O	None
Nickel monosulfide	NiS	None
Silicone dioxide	SiO_2	None

Table 27-2 Amorphic Materials

Crystalline Structures of Metal

The fundamental building blocks of all metals are atoms arranged in very precise three-dimensional patterns called **crystal lattices**. Each metal has a characteristic pattern that forms these crystal lattices. The smallest identifiable group of atoms is the **unit cell**. The unit cells that characterize all commercial metals are illustrated in Figure 27-9, Figure 27-10, and Figure 27-11. It may take millions of these individual unit cells to form one crystal. Although some metals have identical atomic arrangements, the dimensions between individual atoms vary from metal to metal. Some metals change their lattice structure when heated above a specific temperature.

Crystals develop and grow by the attachment of atoms to the submicroscopic unit cells forming the metal's characteristic crystal structure. The final individual crystal size within the metal depends on the length of time that the material is at the crystal-forming temperature and any obstructions to its growth. In most cases, solid crystals continue to grow larger randomly from a liquid as it cools, until they encounter other crystals growing in a similar fashion. The result is solid metal composed of microscopic crystals with the same structure but different orientations. Their crystalline shapes depend on how they grew and how other crystals interrupted that growth.

The crystal structures are studied by polishing small pieces of the metals, etching them in dilute acids, and examining the etched structures with a microscope. These examinations enable metallurgists to determine how the metal formed and to observe changes caused by heat treating and alloying.

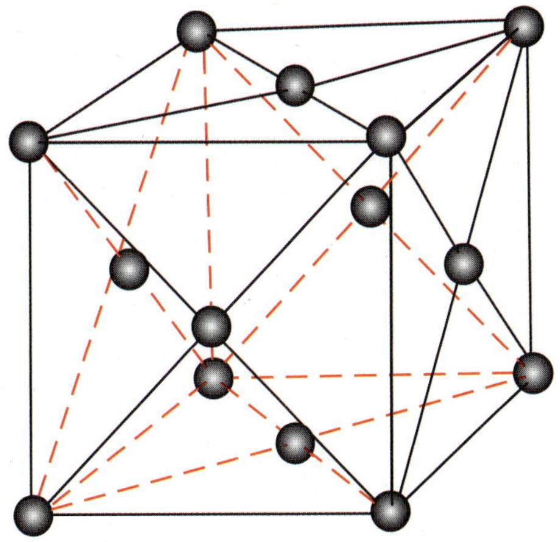

FIGURE 27-10 Face-centered cubic unit cell.

Phase Diagrams

If the metalworking industries used only pure metals, the only required information about their crystalline

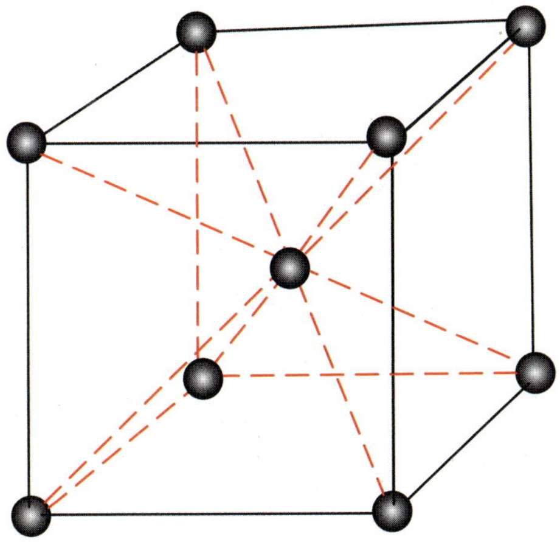

FIGURE 27-9 Body-centered cubic unit cell.

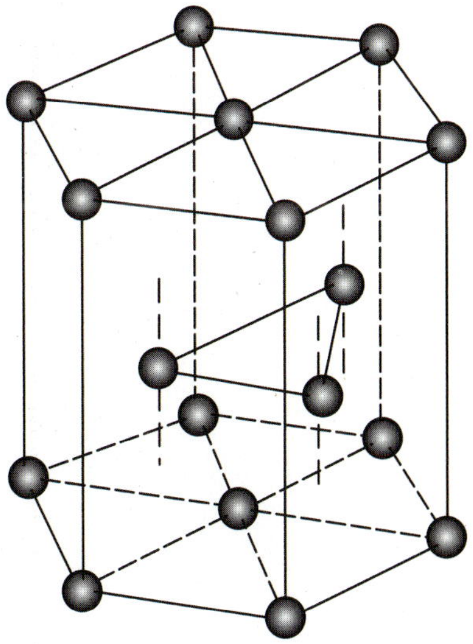

FIGURE 27-11 Hexagonal close-packed cubic unit cell.

structure could be a list of their melting temperatures and **crystalline structures**. However, most engineering materials are alloys, not pure metals. An *alloy* is a metal with one or more elements added to it, resulting in a significant change in the metal's properties. It is inconvenient, if not impossible, to list all phases and temperatures at which alloys exist. That kind of information is summarized in graphs called **phase diagrams**. Phase diagrams are also known as equilibrium or constitution diagrams, and the terms are used interchangeably. These diagrams do not necessarily describe what happens with rapid changes in temperature since metals are sluggish in response to temperature fluctuations. They do describe the constituents present at temperature equilibrium.

Lead–Tin Phase Diagram

The chart in Figure 27-12 is a phase diagram representing the changes brought about by alloying lead (Pb) with tin (Sn). Although this phase diagram is simpler than the iron–carbon phase diagram used for steel, it has many similarities. On the charts, temperature is vertical and alloy percent; in this case, lead and tin are horizontal.

Metallurgy uses Greek letters to identify different crystal structures. On this chart the Greek letter α (alpha, or *a*) is used for one crystal form and the Greek letter β (beta, or *b*) is used to represent the other crystal form. The chart has four different areas identified:

1. *Liquid phase:* The area at the top with the highest temperatures is where all the metal is a molten liquid.
2. *Solid phase:* The area in the lower center with the lowest temperatures is a solid mixture of α- and β-type crystals.
3. *Liquid–solid phase:* The two triangular-shaped areas that touch in the center contain a slurry or paste made up of liquid and a specific type of solid crystals.
4. *Solid-solution phase:* The two triangular-shaped areas that stand vertical, one on each side, next to the temperature scales are solid crystals in either α form on the left side or β form on the right side.

Notice on the chart that, although 100% lead becomes a liquid at 620°F (327°C) and 100% tin becomes a liquid at 420°F (232°C), a mixture of 38.1% lead and 61.9% tin becomes a liquid at 362°F (183°C). The mixture has a lower melting temperature than either of the two metals in the mixture. The mixture melts at a temperature 258°F (144°C) below 100% lead and 58°F (49°C) below 100% tin. This mixture is called the eutectic composition of lead and tin. A **eutectic composition** is the lowest possible melting temperature of an alloy.

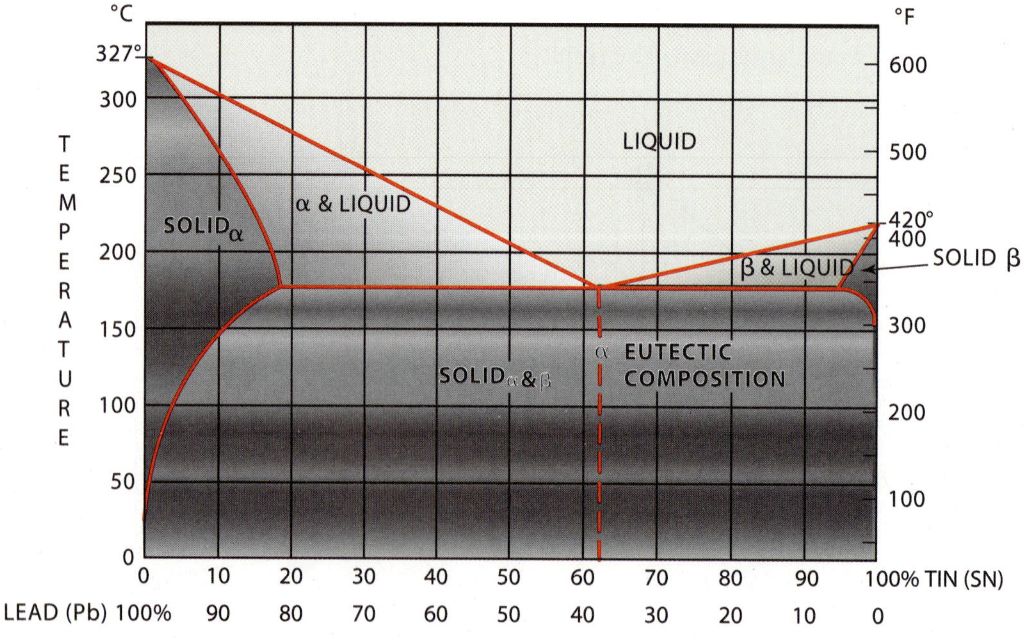

FIGURE 27-12 Lead–tin phase diagram.

On the chart, the broadest temperature for a slurry or paste is a mixture of approximately 80% (80.5%) lead and 20% (19.5%) tin. This mixture remains partially liquid and solid during a 173°F (78°C) temperature change. While a metal is in the liquid–solid phase, it is very weak and any movement will cause cracks to form. For this reason, some aluminum alloys crack when a gas tungsten arc (GTA) weld is made without adding filler metal, and many other metals form crater cracks at the end of a weld. In both cases, as the metal cools it shrinks and pulls itself apart in the center of the weld. The addition of filler metal changes the alloy so it is not as subject to hot cracking. The only metals that do not go through the liquid–solid phase are pure metals and those eutectic composition alloys.

As a 90% lead and 10% tin mixture cools from the 100% liquid phase, it forms an α solid crystal in a liquid. As cooling continues all of the liquid forms into the α crystal. But as solid α crystals cool to around 300°F (150°C), some of them change into the β crystal form. This type of solid-solution phase change occurs in many metals. Steel goes through several such changes as it is heated and cooled even though it never melted. It is these changes in steel that allow it to be hardened and softened by heating, *quenching*, **tempering**, and *annealing*.

Phase diagrams for other metal alloys are more complicated, but they are used in exactly the same way. They describe the effects of changes in temperature or alloying on different phases.

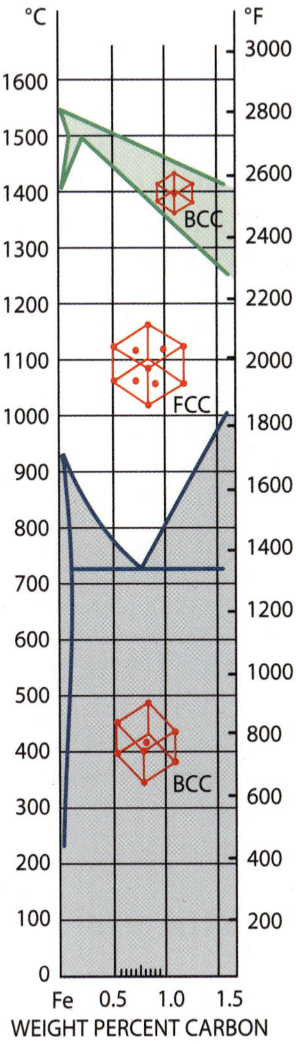

FIGURE 27-13 Iron–carbon phase diagram.

Iron–Carbon Phase Diagram

The iron–carbon phase diagram is illustrated in Figure 27-13. The iron–carbon phase diagram is more complex than that for lead–tin—it has more lines with more solid-solution phase changes—but it is read in the same way. In the iron–carbon diagram used here, the percentage of iron starts at 100% and goes down to 99.1%, while the percentage of carbon goes from 0.0 to 0.9%. Unlike the lead–tin alloys that go from 0.0 to 100% mixtures, very small changes in the percentage of carbon produce major changes in the alloy's properties.

Iron is a pure metal element containing no measurable carbon and is relatively soft. This soft metal when alloyed with as little as 0.80% carbon can become tool steel, Table 27-3. Other alloying

Alloy Name	% Carbon*	Major Properties
Iron	0.0% to 0.03%	Soft, easily formed, not hardenable
Low carbon	0.03% to 0.30%	Strong, formable
Medium carbon	0.30% to 0.50%	High strength, tough
High carbon	0.50% to 0.90%	Hard, tough
Tool steel	0.80% to 1.50%	Hard, brittle
Cast iron	2% to 4%	Hard, brittle, most types resist oxidation

Table 27-3 Iron–Carbon Alloys
**Carbon is not the only alloying element added to iron.*

elements are added to iron to enhance its properties. No other alloying element has such a dramatic effect as carbon. Iron is called an **allotropic** metal because it exists in two different crystal forms in the solid state. It changes between the different crystal forms as its temperature changes. The changes in the crystal structure occur at very precise temperatures. Pure iron forms the **body-centered cubic (BCC)** crystal below a temperature of 1675°F (913°C). Iron changes to form the **face-centered cubic (FCC)** crystal above 1675°F (913°C). The body-centered cubic form of iron is called *alpha ferrite,* abbreviated α-Fe. The face-centered cubic form of iron is called **austenite**, abbreviated γ-Fe.

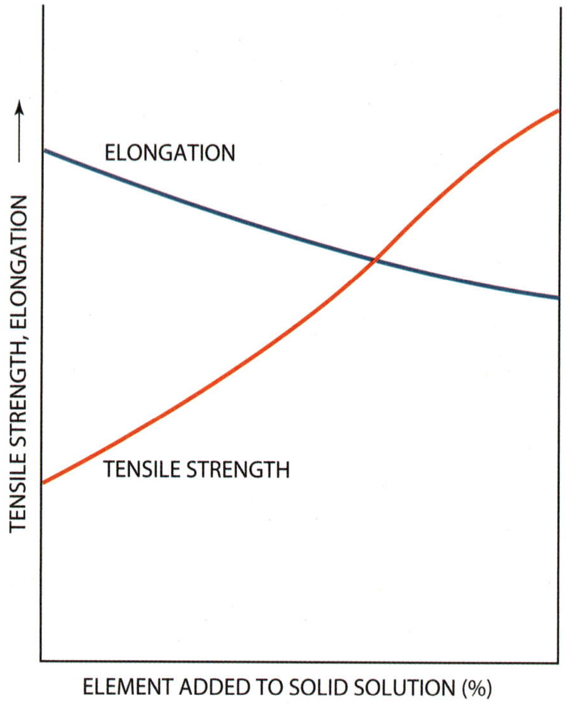

FIGURE 27-14 The general change in properties caused by the addition of other atoms in alloys, producing solid solutions. © Cengage Learning 2012

Strengthening Mechanisms

Perhaps the most important physical characteristic of a metal is strength. Most pure metals are relatively weak. Structures built of pure metals would be more massive and heavier than those built with metals strengthened by alloying and heat treating. Welders must understand numerous methods used to strengthen metals. Some of the strengthening methods used and how welding affects them are described next.

Solid-Solution Hardening

It is possible to replace some of the atoms in the crystal lattice with atoms of another metal in a process called solid-solution hardening. In such cases, the lattices of the **solid solution** and the pure metal are the same except that the lattice of the solid solution is strained as more foreign atoms dissolve. These alloyed prestressed crystals are stronger and less ductile. The amount of change in these properties depends on the number of second atoms introduced, Figure 27-14.

Although an important metallurgical tool, solid-solution hardening has its drawbacks. Not all metals have lattice dimensions that allow significant substitution of other atoms. The amount that can be introduced this way is thus limited. Solid-solution hardening does have the advantage of not changing lattice structure as the result of thermal treatments. Thus, solid-solution–strengthened alloys are generally considered very weldable, Figure 27-15. Many aluminum alloys are strengthened in this way.

Precipitation Hardening

Alloy systems that show a partial solubility in the solid phase generally have very low solubility at room temperature. Solubility increases with temperature until the alloy system reaches its solubility limit. For example, the aluminum–copper system has a solid-solution solubility of 0.25% copper at room temperature, which increases to a maximum of 5.6% copper at 1019°F (548°C). For this reason, very few, if any, aluminum alloys have more than 4.5% copper as an alloying element. These alloys reach their saturation just like a sponge saturated with water; when more water is added, it just runs off. When more alloy is added to metals, it will not be combined with the base metal.

Precipitation hardening is a heat treatment involving three steps: (1) heating the alloy enough to dissolve the second phase and form a single solid solution, (2) quenching the alloy rapidly from the solution temperature to keep the second phase in solution, thus producing a supersaturated solution, and (3) reheating the alloy with careful control of time and temperature to precipitate the second phase as very fine crystals that strengthen the lattice in which they had dissolved. This process, called precipitation hardening or age hardening, is the heat treatment used to strengthen many aluminum alloys.

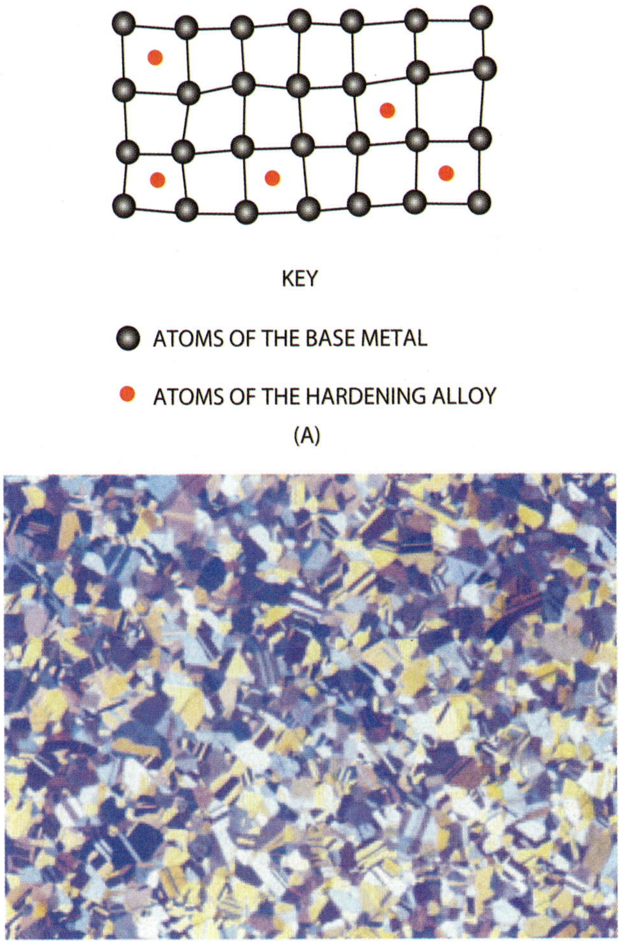

FIGURE 27-15 Solid solution. (A) © Cengage Learning 2012 (B) George F. Vander Voort, Consultant

EXPERIMENT 27-3

Crystal Formation

In this experiment you will be working in a small group to observe the formation of salt crystals. Using a beaker, water, salt, a spoon, a hot plate, gloves, safety glasses, and any other required safety protection, you are going to observe the ability of salt to be dissolved into water as the water is heated.

The dissolving of salt in water shows an increase in solubility as the temperature increases. Add a spoonful of salt to the beaker of cold water and stir vigorously until the salt is dissolved. This is an example of a liquid solution. If more salt is slowly added and stirring is continued, at some point, the salt will no longer dissolve into the solution. The water has now reached its solubility limit. No more salt will dissolve no matter how long the mixture is stirred.

If the beaker is heated, the salt crystals that did not dissolve at room temperature soon disappear. Now even more salt could be added, and it would dissolve. The increasing of the solubility limits by heating has resulted in a supersaturated solution. A supersaturated solution is one in which, because of heating, a higher percentage of another material can be dissolved.

When the beaker of water is cooled to room temperature, with time, the reverse action occurs. The supersaturated solution can no longer hold the salt in solution and rejects it from the solution. In time, salt crystals reappear in the beaker.

Complete a copy of the time sheet in Appendix I or as provided by your instructor.◆

Mechanical Mixtures of Phases

Two phases or constituents may exist in equilibrium, depending on the alloy's temperature and composition. The mixture may consist of two different crystals, which are solid solutions of the two metals of the alloy (see the α & β area in Figure 27-12), or the mixture may consist of a single solid solution and an intermetallic compound, such as the **pearlite** in Figure 27-16. The properties of alloys that are mechanical mixtures of two phases are generally related linearly to the relative amounts of the metal's two constituents. In Figure 27-17, you can see that as the tensile strength increases, the ductility decreases. As shown, the percentage of elongation decreases proportionately to the increase in tensile strength.

FIGURE 27-16 Pearlite.
George F. Vander Voort, Consultant

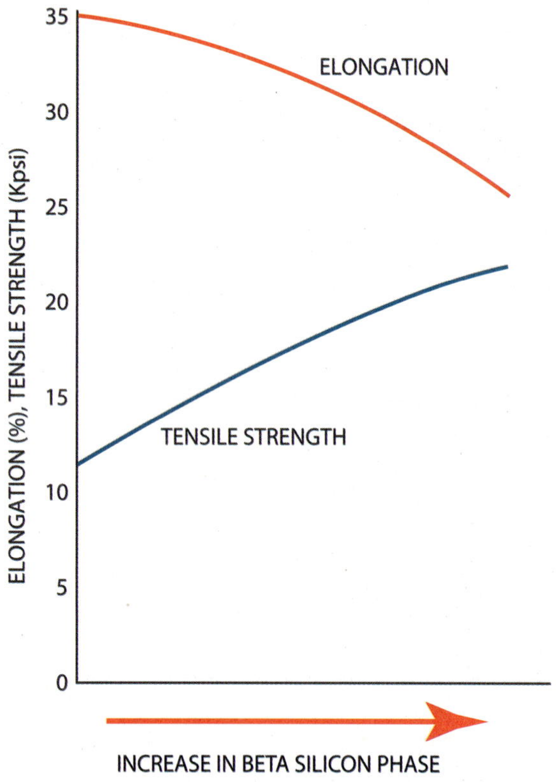

FIGURE 27-17 Change in mechanical properties caused by beta (silicon phase) in mechanical mixture with alpha (aluminum phase). © Cengage Learning 2012

At room temperature, the iron–carbon alloy has two forms: alpha iron **ferrite** and **cementite**. The ferrite is very ductile but weak; the cementite is very strong but brittle. In a careful combination, the cementite stiffens the soft ferrite crystals, increasing their strength without causing them to lose too much ductility.

Quench, Temper, and Anneal

Quenching is the process of rapidly cooling a metal by one of several methods. The quicker the metal cools, the greater the quenching effect. The most common methods of quenching are listed from the slowest to the fastest:

- Air quenching—Air is blown across the part, cooling it only slightly faster than it would cool in still air.
- Oil quenching—The part is immersed in a bath of oil; because oil is a poor conductor of heat, the part cools more slowly than it would in water.
- Water quenching—The part is immersed in a bath of water and cools rapidly.

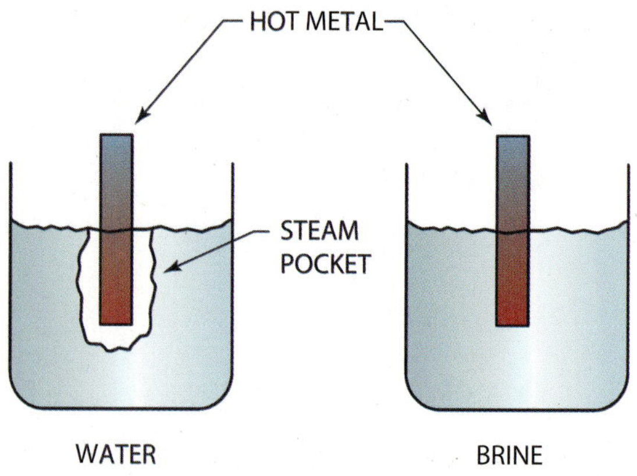

FIGURE 27-18 Effect of brine on quenching rate. © Cengage Learning 2012

- Brine quenching—The part is immersed in a salt-water solution. The salt does not allow a steam pocket to form around the part, as happens with straight water. This results in the fastest quenching time, Figure 27-18.
- Molten salt quenching—This type of quenching results in very slow cooling.

NOTE: Agitating the metal (moving it around rapidly) when it is immersed in a cooling liquid will speed up the cooling rate.

Tempering is the process of reheating a part that has been hardened through heating and quenching. The reheating reduces some of the brittle hardness caused by the quenching, replacing it with toughness and increased tensile strength.

EXPERIMENT 27-4

Effect of Quenching and Tempering on Metal Properties

In this experiment you will be working in a small group to identify the effect of quenching and tempering on metal properties. Using three or more pieces of mild steel approximately 1/4 in. thick, 1 in. wide, and 6 in. long, a vise with a hammer or tensile tester, a hacksaw, a file, water for quenching, two or more firebricks, a safely assembled and properly lit oxyfuel torch, pliers, safety glasses, gloves, and any other required safety equipment, you are going to observe the effect that quenching and tempering has on metal.

Heat the pieces of metal one at a time to a bright red color. Place one of them, while still red hot, between hot firebricks. Immerse the other two into the water while they are still glowing bright red. Moving the metal in the water will ensure a faster quench.

CAUTION

The steam given off from the quenching of the hot metal can cause severe burns. Use your gloves and be careful not to allow the steam to burn you.

Set one of the pieces aside. File a smooth, clean area on the other part. Slowly heat this piece on the opposite side from the filed area using the oxyfuel torch, Figure 27-19. Watch the clear spot, and it will begin to change colors. It will first become a very pale yellow and gradually change to a dark blue. When the surface is dark blue, stop heating it and allow it to cool as slowly as possible between hot firebricks. The surface colors formed are called temper colors, and they indicate the temperature of the metal's surface, Table 27-4. The color comes from the layer of oxide formed on the hot metal's surface. The bluing on a gun barrel is the same thing, and it is used both to make the barrel stronger and to protect it from rusting. You could also use an 800°F (425°C) colored temperature-indicating crayon such as a Tempil Temperature Indicator.

Degrees Fahrenheit	Color of Steel
430	Very pale yellow
440	Light yellow
450	Pale straw-yellow
460	Straw-yellow
470	Deep straw-yellow
480	Dark yellow
490	Yellow-brown
500	Brown-yellow
510	Spotted red-brown
520	Brown-purple
530	Light purple
540	Full purple
550	Dark purple
560	Full blue
570	Light blue
640	Dark blue

Table 27-4 Temperatures Indicated by the Colors of Mild Steel

After both specimens have cooled to room temperature, they can be tested. If you have a tensile testing machine, test each specimen and record its failure strength. If you do not have a tensile tester, you can perform a nick-break test by making a 1/4-in. (6-mm) deep saw cut on both edges of the center line of the weld on each specimen, Figure 27-20. Place the specimens in a vise and break them. Note how they break.

Look at the fractured surface and record which has the light-colored surface and which has the darker surface. Also note which one has the smallest grain sizes shown.

This experiment can be repeated using different metals and alloys.

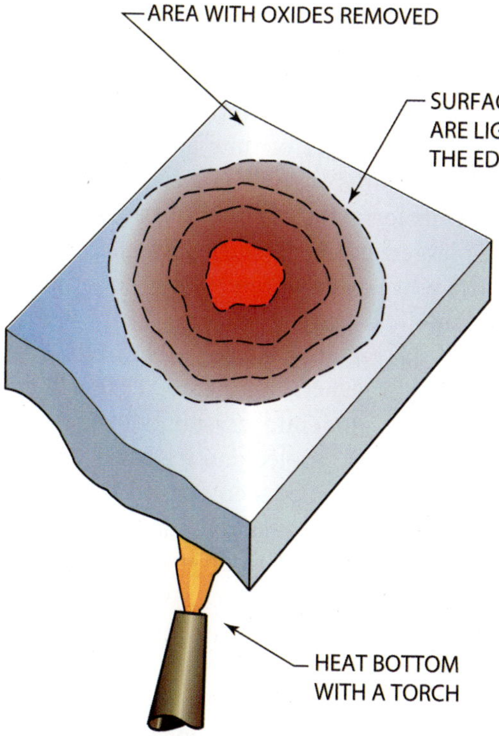

FIGURE 27-19 Surface oxide temper colors.

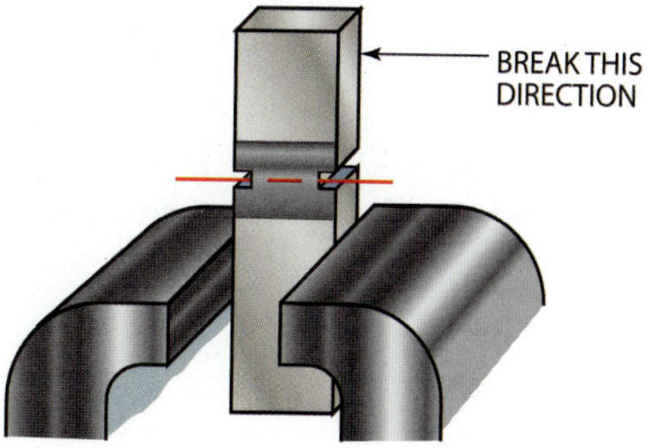

FIGURE 27-20 Nick-break test specimen.

Complete a copy of the time sheet in Appendix I or as provided by your instructor.◆

Carbon-Iron Alloy Reactions

Heating a carbon-iron alloy to a specific temperature and then quenching it rapidly in water can cause a crystalline structure called martensite. **Martensite** is the hardest carbon-iron crystal formation and it is too hard and brittle and useless for most engineering applications. When martensite is viewed under a microscope, it has a (needlelike) **acicular structure**, Figure 27-21.

Martensite can be tempered to a more useful structure at a temperature below its lower critical temperature. The exact temperature, determined by the carbon content and other alloying elements, is generally furnished by the steelmaker.

When a carbon-iron alloy is heated above a specific temperature and cooled slowly, it will form the softer austenite crystal. When medium and high carbon steels are welded, the cooling rates can be fast enough to produce the undesirable martensite. Martensite formation can be minimized and austenite formation maximized by preheating the steel to slow the cooling rates. If the surrounding metal is warmed by preheating, the weld does not lose its temperature as quickly, Figure 27-22.

If a carbon-iron alloy is heated above its austenite crystal range for a longer time, the austenitic crystals will grow larger. When this metal is cooled, the austenite can transform into a pearlite grain structure.

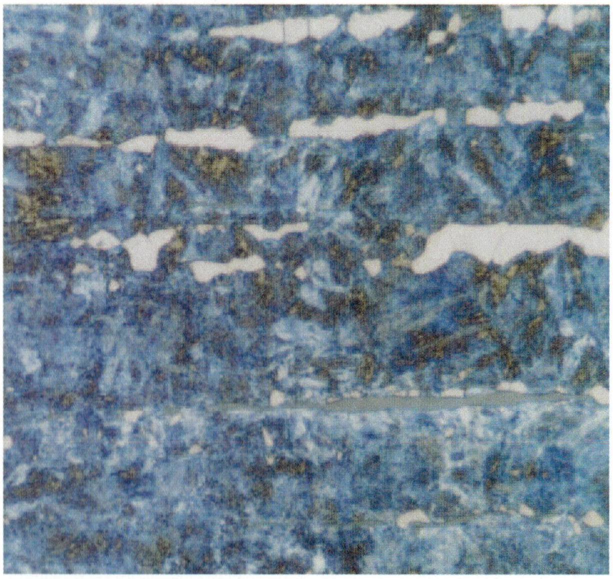

FIGURE 27-21 Microstructure of type 416 martensitic stainless steel. George F. Vander Voort, Consultant

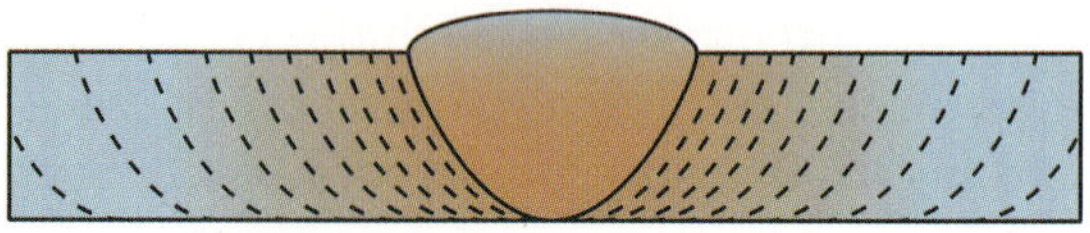

WITHOUT PREHEAT—The temperature starts close together and spreads out quicker (a large temperature change between weld and sides of plate).

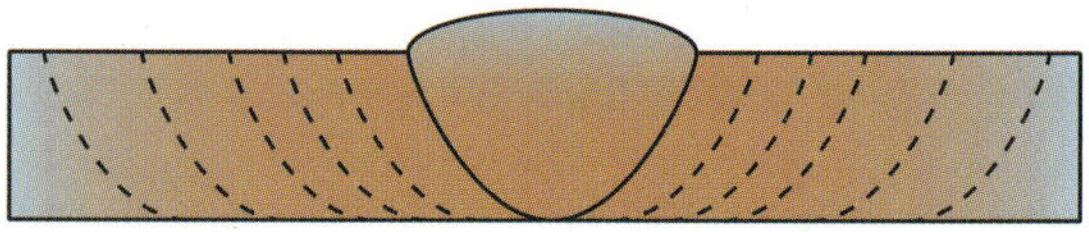

WITH PREHEAT—The temperature starts further apart and spacing changes slowly (little temperature difference between weld and sides of plate).

FIGURE 27-22 © Cengage Learning 2012

NOTE: For all practical purposes plain carbon steels with less than 0.30% carbon cannot be hardened through heat treating and quenching.

Grain Size Control

One of the few metallurgical effects common to all metals and their alloys is grain or crystal growth. When metals are heated, grain growth is expected. The rate of growth increases with temperature and the length of time at that temperature. There are charts that show the effect of time and temperature on the austenite grain growth. These are called time-temperature-transformation (TTT) charts. The process involves larger grains devouring smaller grains. Coarse grains are weaker and tend to be more ductile than fine grains.

The longer the metal is held at a high temperature above its grain growth temperature, the larger the grain size. The austenite crystals will grow so large that when they are cooled and transformed into pearlite their large size can be seen easily in a fracture. Some welders know that this large grain is detrimental to the metal's strength and often refer to the metal as having been "crystallized."

The production of fine grains is not as simple because large grains cannot be shrunk. Reducing the grain size requires the creation of fresh grains by heating the metal to the austenite range and quenching it to recrystallize it. This technique is common to all metals and alloys.

To obtain a fine-grained structure, the metal must be heated quickly above the critical temperature and cooled quickly in a process called **grain refinement**. Not all metals exhibit this transformation characteristic. Fortunately, iron does, and this is one reason that steels are such versatile materials.

Cold Work

Cold working is when metals are deformed at room temperature by cold rolling, drawing, or hammering, and the grains are flattened and elongated. Complex movements occur within and between the grains. These movements distort and disrupt the crystalline structure of the metal by markedly increasing its strength and decreasing its ductility. The presence of impurities or alloying elements in metals causes them to work harder more quickly. Sheets, bars, and tubes are intentionally cold worked to increase their strength, since cold working will strengthen almost all metals and their alloys.

The cold-worked structure can be annealed by heating the metal above the **recrystallization temperature**. Above that temperature, new crystals grow. The size of the new grains depends on the severity of the prior cold working and the time above the recrystallization temperature. The final annealed structure is weaker and more ductile than the cold-worked structure. The final properties depend primarily on the alloy and on the grain size. The coarse-grained metals are weaker.

HEAT TREATMENTS ASSOCIATED WITH WELDING

Welding specifications frequently call for heat-treating joints before welding or after fabrication. To avoid mistakes in their application, welders should understand the reasons for these **heat treatments**.

Preheat

Preheat is used to reduce the rate at which welds cool. Generally, it provides two beneficial effects—lower residual stresses and reduced cracking. The lowest possible temperature should be selected because preheat increases the size of the heat-affected zone and can damage some grades of quenched and tempered steels. The amount of preheat generally is increased when welding stronger plates or in response to higher levels of hydrogen contamination.

	Plate Thickness (inches)			
ASTM/ AIS	1/4 or less	1/2	1	2 or more
A36	70° F	70° F	70° F	150° F to 225° F
A572	70°	70°	150°	225° F to 300° F
1330	95°	200°	400°	450°
1340	300°	450°	500°	550°
2315	Room temp.	Room temp.	250°	400°
3140	550°	625°	650°	700°
4068	750°	800°	850°	900°
5120	600°	200°	400°	450°

Table 27-5 Recommended Minimum Preheat Temperatures for Carbon Steel

As the carbon percentage increases, the preheating temperature required must also increase. This is because as the percentage of carbon increases, the alloy becomes more susceptible to hardening and cracking due to rapid cooling. Preheating charts are available for recommended temperatures for various steels and nonferrous metals, Table 27-5.

The most commonly used preheat temperature range is between 250°F and 400°F (121°C and 204°C) for structural steel. The preheat temperatures can be as high as 600°F (316°C) when welding cast irons.

Care must be taken to soak heat into the region of the intended weld. Superficial heating is not enough because the purpose is to affect the rate at which a relatively large weld cools. To prevent problems and ensure uniform heating, the temperature of the preheated section should be measured at least 10 and 20 minutes after heating.

Stress Relief, Process Annealing

Residual stresses are unsuitable in welded structures, and their effects can be significant. The maximum stresses generally equal the yield strength of the weakest material associated with a specific weld. Such stresses can cause distortion, especially if the component is to be machined after welding. They can also reduce the fracture strength of welded structures under certain conditions.

Stress relieving consists of heating to a point below the lower transformation temperature where new grain growth would occur and holding it for a sufficiently long period to relieve locked-up stresses, then slowly cooling. This process is sometimes called process annealing.

The yield strength of steels decreases at higher temperatures. When heated, the residual stresses will drop to conform to the lower yield strength; thus, the higher the temperature, the better. But significant changes caused by overheating, overtempering, grain growth, or even a phase change must be avoided. Therefore, the temperatures selected must be less than the tempering temperature used to heat treat the plate. Regardless of the plate's metallurgical structure, these temperatures must be kept below the critical temperature.

The most commonly used temperature range for stress relief steel is between 1100°F and 1150°F (593°C and 620°C). This range is high enough to drop the yield residual stresses by 80% and low enough to prevent any harmful metallurgical changes in most steels. While risky, heating to just under the critical temperature does offer a stress reduction of about 90%. Caution should be exercised, however, because some steel will become brittle after thermal stress relief.

Time at temperature is also an important factor since it takes time to bring a weldment to temperature. One hour at temperature is the minimum, while six hours offers an additional, costly 10% drop in stress. Before cooling, the component must be uniformly heated. That requirement alone generally ensures the component will be at temperature long enough to relieve the stresses. The rate of cooling must also be considered. Rapid cooling results in uneven cooling (as in welding) and causes a new set of residual stresses. Ideally, the cooling rate is slow enough to cool the entire mass of the metal uniformly.

Annealing

Annealing, frequently referred to as full annealing, involves heating the structure of a metal to a high enough temperature to turn it completely austenitic. After soaking to equalize the temperature throughout the part, it is cooled in the furnace at the slowest possible rate. On cooling slowly the austenite transforms to ferrite and pearlite. The metal is now its softest, with small grain size, good ductility, excellent machinability, and other desirable properties.

Normalizing

Normalizing consists of heating steels to slightly above a specific temperature and holding for austenite to form, then followed by cooling (in still air). On cooling, austenite transforms, giving somewhat higher strength and hardness and slightly less ductility than in annealing.

Thermal Effects Caused by Arc Welding

During arc welding, liquid weld metal is deposited on the base metal, which is at or near room temperature. In the process, some of the base metal melts from contact with the liquid weld metal and arc, flame, and so on. Conduction, convection, and radiation pull the heat out of the weld metal. This causes the temperature of the deposited weld metal to cool until it becomes solid. Within a few seconds the weld and base metal have gone from being a solid at room temperature to a liquid and back to a solid near room temperature.

Metallurgical changes in the heated region are inevitable. The lowest temperature at which any such changes occur defines the outer extremity of the zone of change. That zone of change is called the **heat-affected zone (HAZ)**, Figure 27-23. The exact size and shape of the HAZ are affected by a number of factors:

- *Type of metal or alloy:* Some metals are easily affected even by small temperature changes, while others are more resistant. In work-hardened metals, the heat-affected zone is defined by the recrystallization temperature, Figure 27-24. In age-hardened alloys, this temperature is the lowest temperature at which evidence of overaging is seen, Figure 27-25. In quenched and tempered steels, it is the temperature at which overtempering is seen.
- *Method of applying the welding heat:* Some heat sources, such as plasma arc welding (PAW), are very concentrated. This high-intensity welding process can have an HAZ area that is only a few thousandths of an inch wide. Oxyacetylene welding (OFW) is a much less intense heating source, and the resulting HAZ will be very large.
- *Mass of the part:* The larger the piece of metal being welded, the greater its ability to absorb heat without a significant change in temperature. Very large weldments may not have a noticeable temperature rise, while small parts may almost completely reach the melting temperature. The more metal that gets hot, the larger the HAZ.
- *Pre- and postheating:* As the temperature of the base metal increases—whether from pre- or postheating or the welding process itself—the larger the HAZ. A cold plate may have an extremely narrow HAZ.

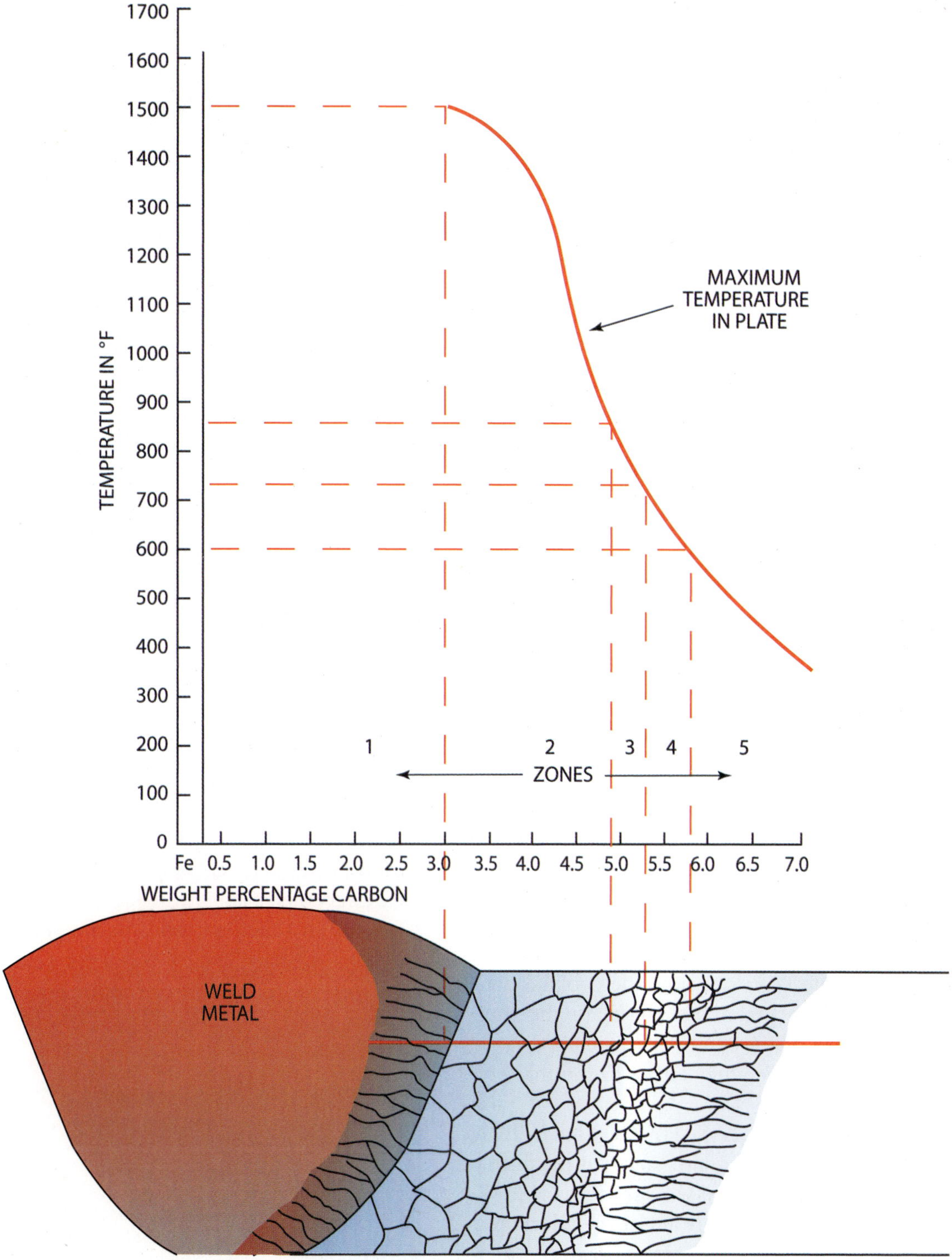

Zone 1 – Liquid metal and the beginning of grain growth
Zone 2 – Austenitic; grain growth at high temperature, fine grain at low temperature
Zone 3 – Austenite + ferrite; grain refined and grain growth
Zone 4 – Recrystallization
Zone 5 – Cold-worked steel 0.2% carbon

FIGURE 27-23 Changes in grain structure caused by heating steel plate into different zones of the iron–carbon phase diagram.

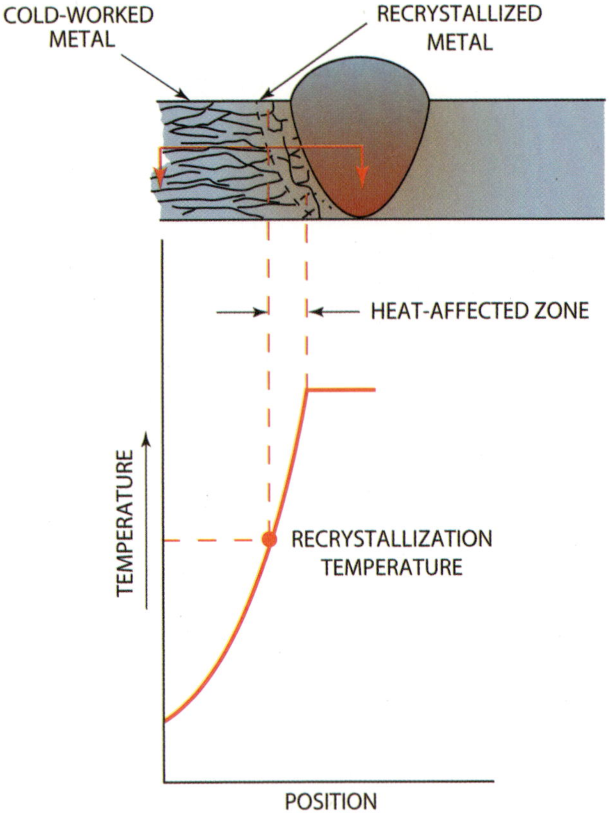

FIGURE 27-24 A heat-affected zone in cold-worked metal. The zone is defined by a change from the cold-worked grains to the recrystallized grains.

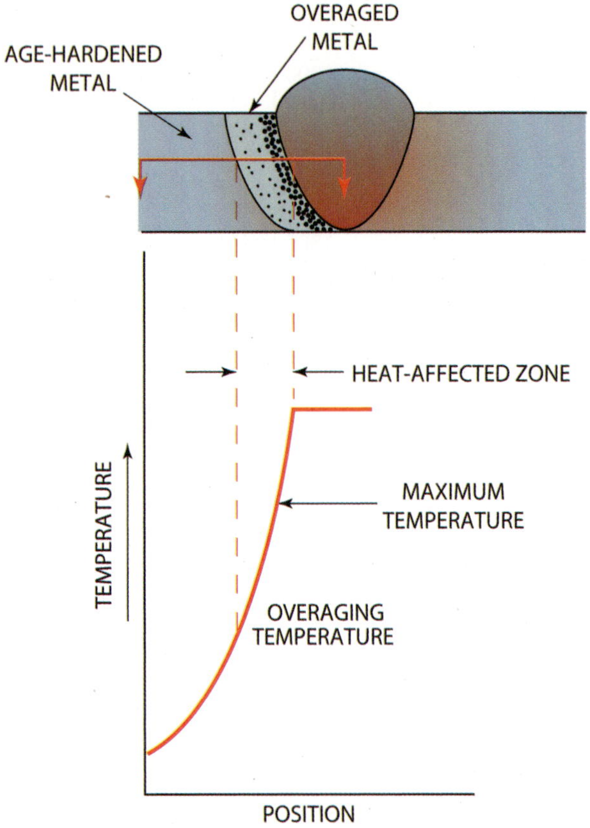

FIGURE 27-25 Heat-affected zone in age-hardened metal.

The HAZ need not always be small. A larger HAZ is desirable for welding steels that can produce martensite at the same cooling rates as those produced when welding heavy plate. You can slow the cooling rates to tolerable levels by preheating the plate and thus increasing the size of the HAZ. In this case, a large HAZ is safer than contending with a brittle martensitic structure.

An important feature of the HAZ, caused by high temperatures, is the severe grain growth at the fusion line. With steels at this critical temperature, the HAZ also produces fine grains as a result of the **allotropic transformation**. Figure 27-26 is an example of a weld that shows the HAZ in cross section. Regardless of the alloy, you must control the HAZ, whether that control is to make it large or small.

Gases in Welding

Many welding problems and defects result from undesirable gases that can dissolve in the weld metal, Figure 27-27. Except for the inert gases argon and helium, the other common gases either react with or dissolve in the molten weld pool. Those that dissolve have a very high solubility in liquid metal but a very low solubility in solidified weld metal. Thus, during the

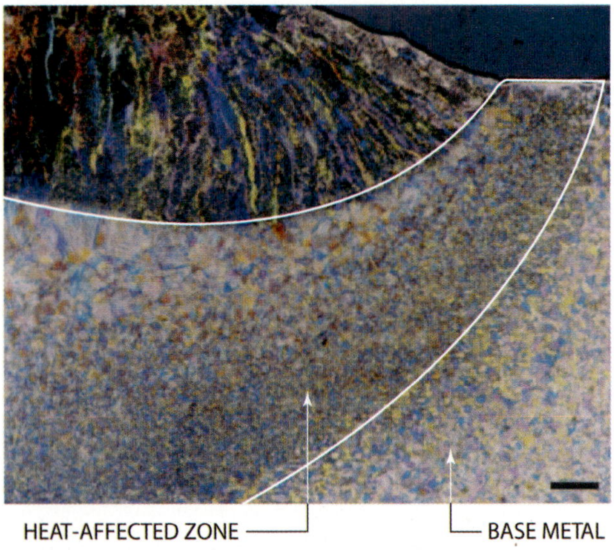

FIGURE 27-26 Heat-affected zone.
George F. Vander Voort, Consultant

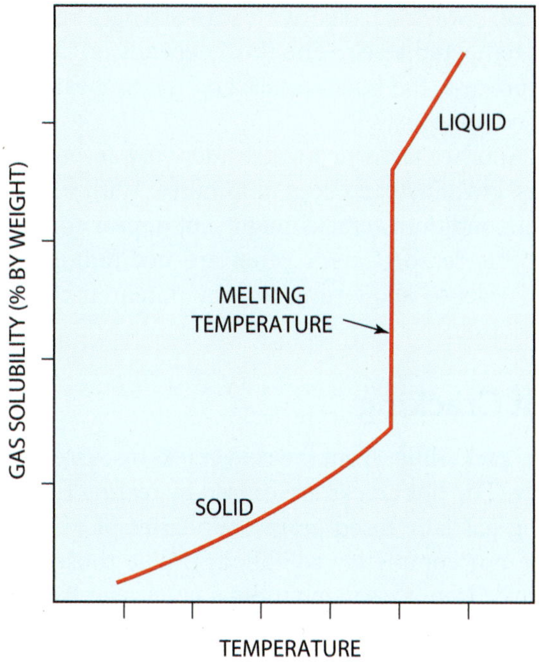

FIGURE 27-27 Typical change of solubility of "active" gases in metals. © Cengage Learning 2012

freezing process, the dissolved gases try to escape. With very slow solidification rates, the gases escape. With very high solidification rates, they become trapped in the metal as supersaturated solutes. At intermediate rates of solidification, they become trapped as gas bubbles, causing porosity. Some, such as hydrogen, produce other problems as well.

Hydrogen

Hydrogen has many sources, including moisture in electrode coatings, fluxes, very humid air, damp weld joints, organic lubricants, rust on wire or on joint surfaces or in weld joints, organic items such as paint, cloth fibers, dirt in weld joints, and others.

Hydrogen is the principal cause of porosity in aluminum welds and with GMAW welds on stainless steels. It can cause random porosity in most metals and their alloys.

Hydrogen can be troublesome in steels. Even in amounts as low as five parts per million, it can cause underbead cracking in high-strength steels. This type of cracking, called delayed cracking, cold cracking, or hydrogen-induced cracking, requires three conditions: (1) a high-stress state, (2) a martensitic microstructure, and (3) a critical level of hydrogen. The first two conditions are typical of quenched and tempered steels. With them, the critical amount of hydrogen is inversely proportional to the stress state; that is, the stronger steels can tolerate less hydrogen. You must use dry electrodes and dry submerged arc fluxes to avoid the hydrogen, in the form of moisture, when welding steels with strength levels above 80,000 psi (5624 kg/cm^2).

Hydrogen problems are avoidable by keeping organic materials away from weld joints, keeping the welding consumables dry, and preheating the components to be welded. Preheating slows the cooling rate of weldments, allowing more time for hydrogen to escape. Generally, the recommended preheat temperatures range between 250°F and 350°F (121°C and 176°C). In the case of very high-strength materials, the acceptable preheat temperatures will exceed 400°F (204°C). Low-temperature preheating of materials before welding can also remove the moisture condensed on the weld joint.

Nitrogen

Nitrogen comes from air drawn into the arc stream. In GMAW processes, it results from poor shielding or strong drafts that disrupt the shield. In SMAW processes, nitrogen can result from carrying an excessively long arc. The primary problem with nitrogen is porosity. In high-strength steels, it can cause a degree of embrittlement that results in marginal toughness.

Some alloys such as austenitic stainless steels and metals such as copper have a high solubility for nitrogen in the solid state. This high solubility does not cause porosity in those materials. Because nitrogen improves the strength of stainless steels, it sometimes is added intentionally to shield gases used when making GTA or GMAW welds in them.

Oxygen

As with nitrogen, the common source of oxygen contamination, air, reaches the weld because of poor shielding or excessively long arcs. But metallurgical changes, not porosity, cause most of the effects of oxygen. Oxygen causes the loss of oxidizable alloys such as manganese and silicon, which reduces strength; produces inclusions in weld metals, which reduce their toughness and ductility; and causes an oxide formation on aluminum welds, which affects appearance and complicates multipass welding.

About 2% of oxygen is added intentionally to stabilize the GMAW process when welding steels with argon shielding. At this concentration, oxygen does not cause the metallurgical problems listed previously. Nevertheless, the amount used must be very carefully controlled because the amount needed varies, depending on the alloys being welded. Mixtures containing only enough oxygen to do an effective stabilizing job should be used. Any more could cause problems, particularly with sensitive alloys.

Carbon Dioxide

Carbon dioxide is an oxygen substitute for stabilizing GMAW processes using argon shields, although about 5 to 8% carbon dioxide is usually added to produce the same effects achieved with 2% oxygen. However, the carbon in carbon dioxide is a potential contaminant that can cause problems with corrosion resistance in the low carbon grades of stainless steels. Even straight carbon dioxide shielding has this problem. In most cases, carbon dioxide levels below 5% do not seem to increase the carbon content of stainless steels enough to cause difficulty.

Metallurgical Defects

Cold Cracking

Cold cracking is the result of hydrogen dissolving in the weld metal and then diffusing into the heat-affected zone. The cracks develop long after the weld metal solidifies. For that reason, it is also known as hydrogen-induced cracking and delayed cracking. This cracking is most commonly found in the coarse grains of the HAZ, just under the fusion zone. For this reason, it is also called underbead cracking. Generally, these cracks do not surface and can be seen only in radiographs or by sectioning welds. Sometimes they surface in a region of the weld running parallel to the fusion zone. With high-strength steels, hydrogen-induced cracking occurs as transverse cracks in the weld metal that are seen easily with very low magnification, Figure 27-28.

Hydrogen-induced cracking requires (1) a high stress, (2) a microstructure sensitive to hydrogen, and, of course, (3) hydrogen. The first two are almost always satisfied with high-strength quenched and tempered steels. The third depends on the welding process, the filler metals, and the preheat, as discussed previously.

Another factor is time. Under very severe conditions, cracking can occur in minutes. With very marginal conditions, cracks might not appear for weeks. For this reason, welds often are not radiographed for weeks to allow time for any potential cracks to develop.

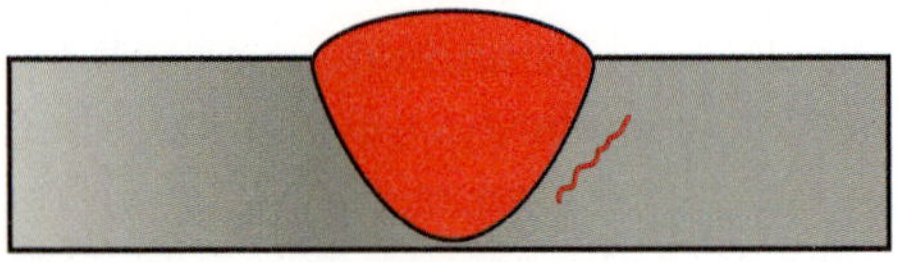

FIGURE 27-28 Underbead cracking.

Hot Cracking

Hot cracks differ from the cold cracks discussed previously. The hot cracks are caused by tearing the metal along partially fused grain boundaries of welds that have not completely solidified. Unlike those caused by hydrogen, these longitudinal cracks are located in the centers of weld beads. They develop immediately after welding, unlike the delayed nature of hydrogen-induced cracking. In the process of freezing, low-melting materials in the weld metal are rejected as the columnar grains solidify, leaving a high concentration of the low-melting materials where the grains intersect at the center of the weld. These partially melted and weak grain boundaries are stressed while the weld metal shrinks, causing them to rupture.

A high sulfur content is most often responsible for hot cracking in steel. It forms a low-melting iron sulfide on the grain boundaries. Hot cracking is more likely to occur in steels that contain higher levels of carbon and phosphorus and in steels that are high in sulfur and low in manganese, Figure 27-29.

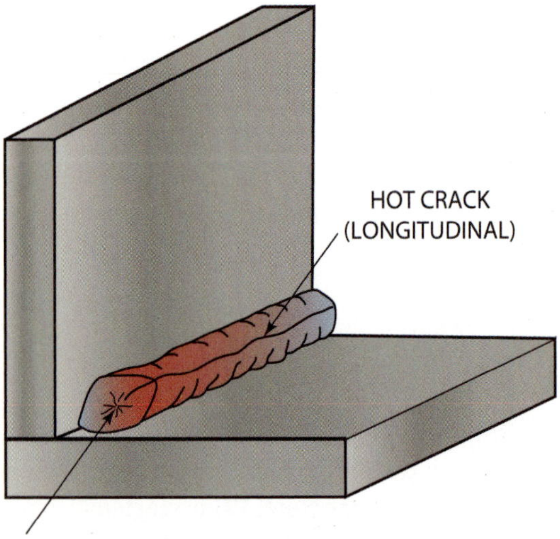

FIGURE 27-29 Hot cracking.

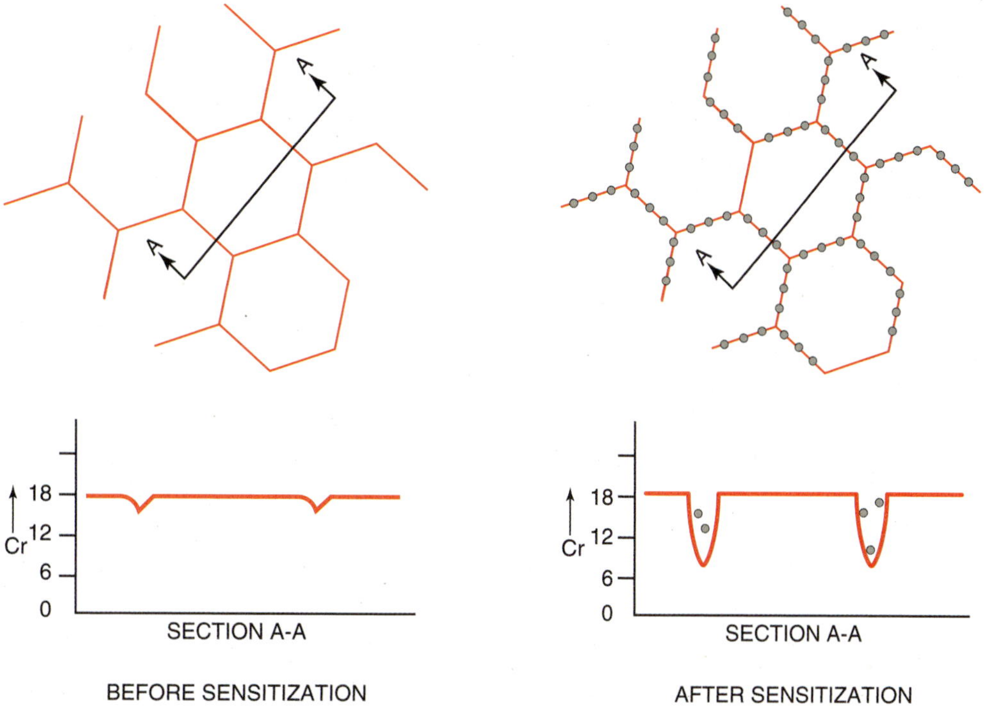

FIGURE 27-30 Depletion of chromium in grain boundaries due to the precipitation of chromium carbide. American Welding Society

Severely concave welds may also cause hot cracking because the welds are not as strong. As welds cool, they shrink. If the weld is not thick enough, when it shrinks it cannot pull the metal in, and so it cracks. Even if the weld is larger, it may crack if the weldment is rigid and cannot be pulled inward by the weld. In these situations, even large welds can have hot cracks.

Carbide Precipitation

Stainless steels rely on free chromium for their resistance to corrosion. When carbon is present and the steel is heated to temperatures between about 800°F and 1500°F (427°C and 816°C), carbon combines with the chromium to form chromium carbides in the grain boundaries. The formation of chromium carbides depletes the steel of the free chromium needed for protection. Thus, for steels, every effort is made to use low carbon stainless steel or a special stabilized grade made for welding.

Figure 27-30 illustrates how the chromium in solution can drop from the nominal 18% in the most commonly used stainless steels to levels well under the 12% needed for even minimal protection. When in contact with corrosive materials, the low chromium grain boundaries can dissolve and the weld fail, Figure 27-31.

Problems are minimized by using very low carbon steel called extra low carbon (ELC) steels. Without carbon, the chromium carbides cannot form. Another way is to tie up the carbon by forming very stable carbides of titanium or columbium.

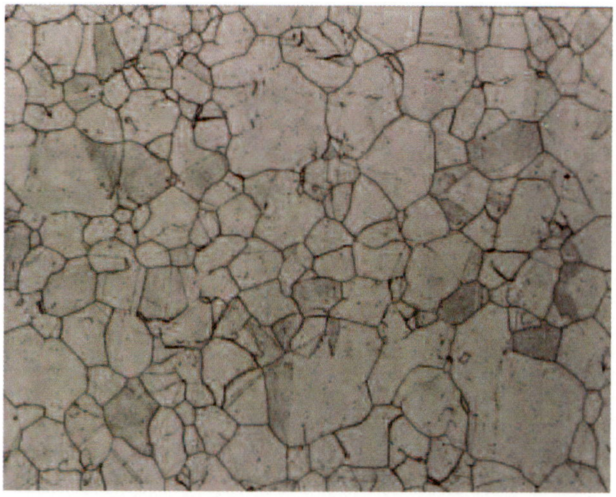

FIGURE 27-31 Microstructure of sensitized solution annealed type 304 stainless steel revealing austenite grain boundaries. George F. Vander Voort, Consultant

This also prevents chromium carbides from forming. Weld metals are alloyed similarly with low carbon levels or by stabilizing the carbides. Titanium is not generally used for that purpose in electrodes because it is lost in the transfer. Instead, most electrodes are alloyed with columbium to avoid the corrosion problems associated with carbide precipitation.

Carbon dioxide shield gases can cause a similar problem, especially with the ELC grades. These gases supply the carbon that impairs corrosion resistance by depleting the free chromium. The small amounts used in argon to stabilize the arc are often tolerable. But carbon dioxide additions should be avoided unless used with caution and an awareness that problems can develop.

SUMMARY

Welding engineers' understanding of science, physics, and metallurgy enables them to design better weldments. Welding engineers must know how the chemical elements that make up a metal alloy will react to changes in physical and thermal stresses. Such stresses are applied to metal products during welding fabrication and as part of their normal service. As metals are thermally cycled, their physical and mechanical properties change. Thermal cycling, of course, occurs every time a metal is welded. Often, as part of a postweld procedure, a welding engineer will have the part heat treated. Welding engineers use charts and graphs to determine the optimal thermal cycling that will allow for the greatest strength in the weldment design.

You do not have to have as deep an understanding of the thermal effects on metals as a welding engineer does, but you must know the importance of controlling temperature cycles during welding. Weldment failures may be a result of welder-created problems or welding metallurgical problems. A good understanding of metallurgy will aid you in avoiding welding problems.

REVIEW QUESTIONS

1. Why is it important for a welder to understand the materials being welded?
2. What is the difference between heat and temperature?
3. Define *BTU*.
4. What is the difference between sensible heat and latent heat?
5. Define *temperature*.
6. Define *mechanical properties of metal*.
7. List the most significant mechanical properties of metals that a welder should be familiar with.
8. Name four common types of strength measurements.
9. What property allows a metal to return to its original form after removal of the load?
10. Why do metallurgists study the crystalline structure of metals?
11. Define *alloy*.
12. Using the chart in Figure 27-12, what is the lowest melting temperature of the lead–tin solder at its eutectic point?
13. What is the purpose of solid-solution hardening and precipitation hardening?
14. What is the hardest carbon-iron crystal formation?
15. How can metal be cold worked?
16. Why is preheating used?
17. How is metal stress relieved?
18. Describe the difference between annealing and normalizing.
19. What do the letters *HAZ* represent?
20. List the sources of hydrogen contamination in a weld.
21. What is the difference between hot cracking and cold cracking?
22. What is the temperature range in which most carbide precipitation occurs?

Chapter 28

Weldability of Metals

OBJECTIVES

After completing this chapter, the student should be able to:

- Describe the problems that can occur when metal is not properly preheated and/or postheated.
- Explain the effect that the amount of carbon in a metal has on its properties and weldability.
- Describe the procedure that must be followed to repair cast iron with arc welding and brazing.
- Describe the properties of aluminum that make it difficult to weld.

KEY TERMS

alloy steels
aluminum
cast iron
copper
malleable cast iron
manganese
plain carbon steels
stainless steels
steel classification systems
titanium
weldability

INTRODUCTION

All metals can be welded, although some metals require far more care and skill to produce acceptably strong and ductile joints. The term **weldability** has been coined to describe the ease with which a metal can be welded properly. Good weldability means that almost any process can be used to produce acceptable welds and that little effort is needed to control the procedures. Poor weldability means that the processes used are limited and that the preparation of the joint and the procedure used to fabricate it must be controlled very carefully or the weldment will not function as intended.

Knowing why a part broke is as important to the repair process as knowing the type of metal it is made with. Parts break because they are worn out, damaged in an accident, underdesigned for the work, or defective. Welders have the greatest success fixing parts that are worn out or damaged in an accident. These parts were giving good service before they broke. Parts that were underdesigned or defective are more likely to break again if they are welded without fixing the design problem or defect first. For example, a bracket that vibrates too much and cracks will continue to vibrate and will crack again if you just weld the crack. Instead, see what can be done to stop the vibration to prevent it from cracking again. In other words, fix the problem before fixing the part—or you will be fixing the part again and again.

Welding processes produce a thermal cycle in which the metals are heated over a range of temperatures. Cooling of the metal to ambient temperatures then follows. The heating and cooling cycles can set up stresses and strains in the weld. Whatever the welding process used, certain metallurgical, physical, and chemical changes also take place in the metal. A wide range of welding conditions can exist for welding methods when joining metals with good weldability. However, if weldability is a problem, adjustments usually will be necessary in one or more of the following factors:

- Filler metal—If the wrong filler metal is selected, the weld can have major defects and not be fit for service. Common defects include porosity, cracks, and filler metal that just will not stick, Figure 28-1. The cracks can be in the filler metal or in the base metal alongside the weld. Chapter 26 lists various types of filler metals and their applications. If you are not sure which filler metal to use, a good general rule is that a little stronger or higher grade can be used successfully, but a lower strength or grade seldom works.

FIGURE 28-1 Improper welding technique for filler metal resulting in lack of fusion. Larry Jeffus

	Plate Thickness (inches)			
ASTM/ AISI	1/4 or less	1/2	1	2 or more
A36	70° F	70° F	70° F	150° F to 225° F
A572	70°	70°	150°	225° F to 300° F
1330	95°	200°	400°	450°
1340	300°	450°	500°	550°
2315	Room temp.	Room temp.	250°	400°
3140	550°	625°	650°	700°
4068	750°	800°	850°	900°
5120	600°	200°	400°	450°

Table 28-1 Recommended Minimum Preheat Temperature for Carbon Steel

- Preheat and postheat—Cracking is a common problem when welding on brittle metals such as cast iron or some high-strength alloys. Preheating the part before starting the weld reduces the stress caused by the weld and helps the filler metal flow. The most commonly used preheat temperature range is between 250°F and 400°F (120°C and 200°C) for most steel, Table 28-1. The preheat temperature can be as high as 1200°F (650°C) when welding cast iron. Preheating is required any time that the metal to be welded is below 70°F (20°C) because the cold metal quenches the weld. Postheating slows the cooldown rate following welding, which prevents postweld cracking of brittle metals. Postheating also reduces weld stresses that can result in cracks forming some time after the part is repaired.
- Welding procedure—The size of the weld bead, the number of welds, and the length of the welds all affect the weld. When large welds are needed, it is better to make more small welds than a few large welds. The small welds serve to postheat the weld. They reduce stresses and result in less distortion. Sometimes a series of short back-stepping welds can be made, Figure 28-2. For example these short welds can be used on very brittle metals like cast iron.

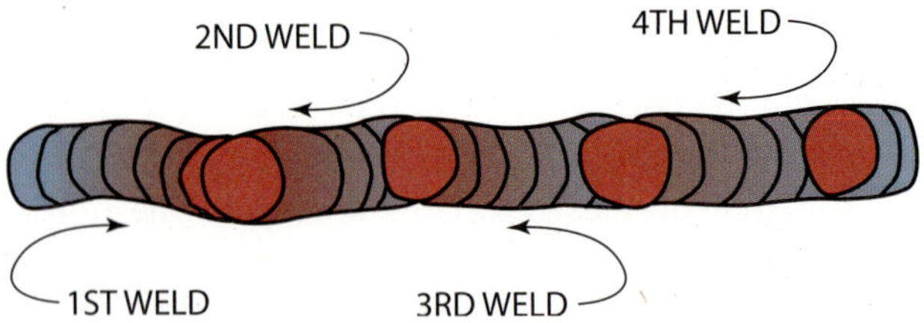

FIGURE 28-2 Welding sequence to produce the minimum weld stresses. © Cengage Learning 2012

Steel Classification and Identification

SAE and AISI Classification Systems

Two primary numbering systems have been developed to classify the standard construction grades of steel, including both carbon and alloy steels. These systems classify the types of steel according to their basic chemical composition. One classification system was developed by the Society of Automotive Engineers (SAE). The other system is sponsored by the American Iron and Steel Institute (AISI).

The numbers used in both systems are now just about the same. However, the AISI system uses a letter before the number to indicate the method used in the manufacture of the steel.

Both numbering systems usually have a four-digit series of numbers. In some cases a five-digit series is used for certain alloy steels. The entire number is a code to the approximate composition of the steel.

In both **steel classification systems**, the first number often, but not always, refers to the basic type of steel, as follows:

1XXX Carbon

2XXX Nickel

3XXX Nickel chrome

4XXX Molybdenum

5XXX Chromium

6XXX Chromium vanadium

7XXX Tungsten

8XXX Nickel chromium vanadium

9XXX Silicomanganese

The first two digits together indicate the series within the basic alloy group. There may be several series within a basic alloy group, depending upon the amount of the principal alloying elements. The last two or three digits refer to the approximate permissible range of carbon content. For example, the metal identified as 1020 would be 1XXX carbon steel with an XX20 0.20% range of carbon content, and 5130 would be 5XXX chromium steel with an XX30 0.30% range of carbon content.

The letters in the AISI system, if used, indicate the manufacturing process as follows:

- *C*—Basic open-hearth or electric furnace steel and basic oxygen furnace steel
- *E*—Electric furnace alloy steel

Table 28-2 shows the AISI and SAE numerical designations of **alloy steels**.

Unified Numbering System (UNS)

A unified numbering system is presently being promoted for all metals. This system will eventually replace the AISI and other systems.

Carbon and Alloy Steels

Steels alloyed with carbon and only a low concentration of silicon and **manganese** are known as **plain carbon steels**. These steels can be classified as low carbon, medium carbon, and high carbon steels. The division is based upon the percentage of carbon present in the material.

Plain carbon steel is basically an alloy of iron and carbon. Small amounts of silicon and manganese are added to improve its working quality. Sulfur and phosphorus are present as undesirable impurities. All steels contain some carbon, but steels that do not include alloying elements other than low levels of manganese or silicon are classified as plain carbon steels. Alloy steels contain specified larger proportions of alloying elements.

The AISI has adopted the following definition of carbon steel: "Steel is classified as carbon steel when no minimum content is specified or guaranteed for aluminum, chromium, columbium, molybdenum, nickel, titanium, tungsten, vanadium, or zirconium; and when the minimum content of copper which is specified or guaranteed does not exceed 0.40%; or when the maximum content which is specified or guaranteed for any of the following elements does not exceed the respective percentages hereinafter stated: manganese 1.65%, silicon 0.60%, copper 0.60%." Under this classification will be steels of different composition for various purposes.

Many special alloy steels have been developed and sold under various trade names. These alloy steels usually have special characteristics, such as high tensile strength, resistance to fatigue, corrosion resistance, or the ability to perform at high temperatures. Basically, the ability of carbon steel to be welded is a function of the carbon content, Table 28-3. (Other factors to be considered include thickness and the geometry of the joint.) All carbon steels can be welded by at least one method. However, the higher the carbon content of the metal, the more difficult it is to weld the steel. Special instructions must be followed in the welding process.

13XX	Manganese 1.75
23XX**	Nickel 3.50
25XX**	Nickel 5.00
31XX	Nickel 1.25; chromium 0.65
E33XX	Nickel 3.50; chromium 1.55; electric furnace
40XX	Molybdenum 0.25
41XX	Chromium 0.50 or 0.95; molybdenum 0.12 or 0.20
43XX	Nickel 1.80; chromium 0.50 or 0.80; molybdenum 0.25
E43XX	Same as above, produced in basic electric furnace
44XX	Manganese 0.80; molybdenum 0.40
45XX	Nickel 1.85; molybdenum 0.25
47XX	Nickel 1.05; chromium 0.45; molybdenum 0.20 or 0.35
50XX	Chromium 0.28 or 0.40
51XX	Chromium 0.80, 0.88, 0.93, 0.95, or 1.00
E5XXXX	High carbon; high chromium; electric furnace bearing steel
E50100	Carbon 1.00; chromium 0.50
E51100	Carbon 1.00; chromium 1.00
E52100	Carbon 1.00; chromium 1.45
61XX	Chromium 0.60, 0.80, or 0.95; vanadium 0.12, or 0.10, or 0.15 minimum
7140	Carbon 0.40; chromium 1.60; molybdenum 0.35; aluminum 1.15
81XX	Nickel 0.30; chromium 0.40; molybdenum 0.12
86XX	Nickel 0.55; chromium 0.50; molybdenum 0.20
87XX	Nickel 0.55; chromium 0.50; molybdenum 0.25
88XX	Nickel 0.55; chromium 0.50; molybdenum 0.35
92XX	Manganese 0.85; silicon 2.00; 9262-chromium 0.25 to 0.40
93XX	Nickel 3.25; chromium 1.20; molybdenum 0.12
98XX	Nickel 1.00; chromium 0.80; molybdenum 0.25
14BXX	Boron
50BXX	Chromium 0.50 or 0.28; boron
51BXX	Chromium 0.80; boron
81BXX	Nickel 0.33; chromium 0.45; molybdenum 0.12; boron
86BXX	Nickel 0.55; chromium 0.50; molybdenum 0.20; boron
94BXX	Nickel 0.45; chromium 0.40; molybdenum 0.12; boron

Note: The elements in this table are expressed in percent.

*Consult current AISI and SAE publications for the latest revisions.

**Nonstandard steel.

Table 28-2 AISI and SAE Numerical Designations of Alloy Steels

Common Name	Carbon Content	Typical Use	Weldability
Ingot iron	0.03% max.	Enameling, galvanizing, and deep drawn sheet and strip	Excellent
Low carbon (mild) steel	0.00% to 0.30%	Welding electrodes, sheet, strip, structural shapes, plate and bar	Excellent to Good
Medium carbon steel	0.30% to 0.50%	Machinery parts	Fair*
High carbon steel	0.50% to 1.00%	Springs, dies, and railroad rails	Poor**

**Preheat and frequently postheat required.*

***Difficult to weld without adequate preheat and postheat.*

Table 28-3 Iron carbon Alloys, Uses, and Weldabilities

Low Carbon and Mild Steel

Low carbon steel has a carbon content of 0.15% or less, and mild steel has a carbon content range of 0.15 to 0.30%. Both steels can be welded easily by all welding processes. The resulting welds are of extremely high quality. Oxyacetylene welding of these steels can be done by using a neutral flame. Joints welded by this process are of high quality, and the fusion zone is not hard or brittle.

Both low carbon and mild steels can be welded readily by the shielded metal arc (stick welding)

method. The selection of the correct electrode for the particular welding application helps to ensure high strength and ductility in the weld.

The gas metal arc (GMA) and flux cored arc welding processes are used for welding both low and medium carbon steels due to the ease of welding and because they prevent contamination of the weld. The high productivity and lower cost make them increasingly popular welding processes.

Medium Carbon Steel

The welding of medium carbon steels, having 0.30 to 0.50% carbon content, is best accomplished by the various fusion processes, depending upon the carbon content of the base metal. The welding technique and materials used are dictated by the metallurgical characteristics of the metal being welded. For steels containing more than 0.40% carbon, preheating and subsequent heat treatment are generally required to produce a satisfactory weld. Stick welding electrodes of the type used on low carbon steels can be used for welding this type of steel. The use of an electrode with a special low hydrogen coating may be necessary to reduce the tendency toward underbead cracking.

NOTE: SMA welding rods for welding medium and high carbon steels, stainless steel, and cast iron are available. Some of these rods are very expensive, costing more than $100 a pound. They are often prepackaged in small quantities that are available from some hardware stores and welding suppliers. When a single rod can cost $20 or more, being able to purchase just the number of rods you need for a repair weld or small job is important. However, GMAW and FCAW wires come on spools and are not available in very small quantities. Therefore, GMAW and FCAW wires are more suited for large production welding jobs on medium and high carbon steels, stainless steel, and cast iron but are not practical for use on small welding jobs.

High Carbon Steel

High carbon steels usually have a carbon content of 0.50 to 0.90%. These steels are much more difficult to weld than either the low or medium carbon steels. Because of the high carbon content, the heat-affected zone next to the weld can become very hard and brittle. You can avoid this by using preheating and by selecting procedures that produce high-energy inputs to the weld. Refer to Figure 28-3 for the preheat temperature for the specific carbon content. The martensite that does form is tempered by postweld heat treatments

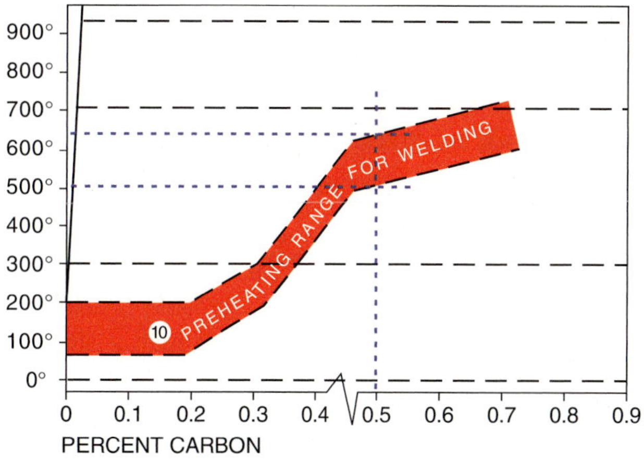

FIGURE 28-3 **Preheating range for welding. For example, 0.5% carbon steel would have a preheat range of approximately 510° F to 640° F.**
Tempil, an Illinois Tool Works Co.

such as stress relief anneal. Refer to Figure 28-4 for the temperature for stress-relief annealing between 1125°F and 1250°F (600°C and 675°C).

In arc welding high carbon steel, mild-steel shielded arc electrodes are generally used. However, the weld metal does not retain its normal ductility because it absorbs some of the carbon from the steel.

Welding on high carbon steels is often done to build up a worn surface to original dimensions or to develop a hard surface. In this type of welding, preheating or heat

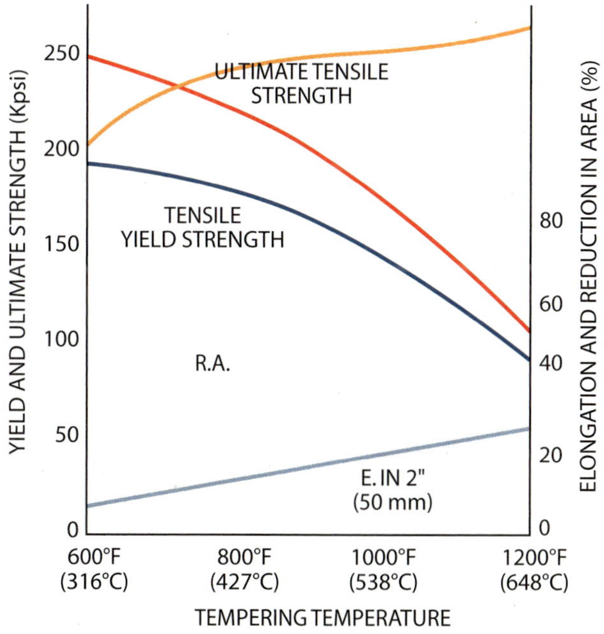

FIGURE 28-4 **Stress-relieving range.**
Tempil, an Illinois Tool Works Co.

treatment may not be needed if heat builds up in the part during continuous welding.

Stainless Steels

Stainless steels consist of four groups of alloys: austenitic, ferritic, martensitic, and precipitation hardening. The austenitic group is by far the most common. Its chromium content provides corrosion resistance, while its nickel content produces the tough austenitic microstructure. These steels are relatively easy to weld, and a large variety of electrode types are available.

The most widely used stainless steels are the chromium-nickel austenitic types. They are used for items such as dairy equipment, including milk tanks; hand sprayer tanks; and poultry medicine infusion system pumps; and are usually referred to by their chromium-nickel content as 18/8, 25/12, 25/20, and so on. For example, 18/8 contains 18% chromium and 8% nickel, with 0.08 to 0.20% carbon. To improve weldability, the carbon content should be as low as possible. Carbon should not be more than 0.03%, with the maximum being less than 0.10%.

Keeping the carbon content low in stainless steel will also help reduce carbide precipitation. Carbide precipitation occurs when alloys containing both chromium and carbon are heated. The chromium and carbon combine to form chromium carbide (Cr_3C_2).

The combining of chromium and carbon lowers the chromium that is available to provide corrosion resistance in the metal. This results in a metal surrounding the weld that will oxidize or rust. The amount of chromium carbide formed is dependent on the percentage of carbon, the time that the metal is in the critical range, and the presence of stabilizing elements.

If the carbon content of the metal is very low, little chromium carbide can form. Some stainless steel types have a special low carbon variation. These low carbon stainless steels are the same as the base type but with much lower carbon content. To identify the low carbon from the standard AISI number, the *L* is added as a suffix. See examples 304 and 304L, Table 28-4.

Chromium carbides form when the metal is between 800°F and 1500°F (625°C and 815°C). The quicker the metal is heated and cooled through this range, the less time that chromium carbides can form. Since austenitic stainless steels are not hardenable by quenching, the weld can be cooled using a chill plate. The chill plate can be water-cooled for larger welds.

Some filler metals have stabilizing elements added to prevent carbide precipitation. Columbium and **titanium** are both commonly found as chromium stabilizers. Examples of the filler metals are E310Cb and ER309Cb.

In fusion welding, stainless austenitic steels may be welded by all of the methods used for plain carbon steels.

Since *ferritic stainless* steels contain almost no nickel, they are cheaper than austenitic steels. They are used for ornamental or decorative applications such as architectural trim and at elevated temperatures for heat exchanging. However, ferritic stainless steels also tend to be brittle unless specially deoxidized. Special high-purity, high-toughness ferritic stainless steels have been developed, but careful welding procedures must be used with them to prevent embrittlement. This means very carefully controlling nitrogen, carbon, and hydrogen.

Martensitic stainless steels are also low in nickel but contain more carbon than the ferritic. They are used in applications requiring both wear resistance and corrosion resistance. Items such as surgical knives and razor blades are made of them. Quality welding requires very careful control of both preheating and tempering immediately after welding.

Precipitation hardening stainless steels can be much stronger than the austenitic, without losing toughness. Their strength is the result of a special heat treatment used to develop the precipitate. They can be solution treated prior to welding and given the precipitation treatment after welding.

Nominal Composition of Stainless Steels						
AISI Type	Nominal Composition % C	Mn Max	Si Max	Cr	Ni	Other
304	0.08 max.	2.0	1.0	18–20	8–12	
304L	0.03 max.	2.0	1.0	18–20	8–12	
316	0.08 max.	2.0	1.0	16–18	10–14	2.0–3.0 Mo
316L	0.03 max.	2.0	1.0	16–18	10–14	2.0–3.0 Mo

Table 28-4 Comparison of Standard-Grade and Low Carbon Stainless Steels

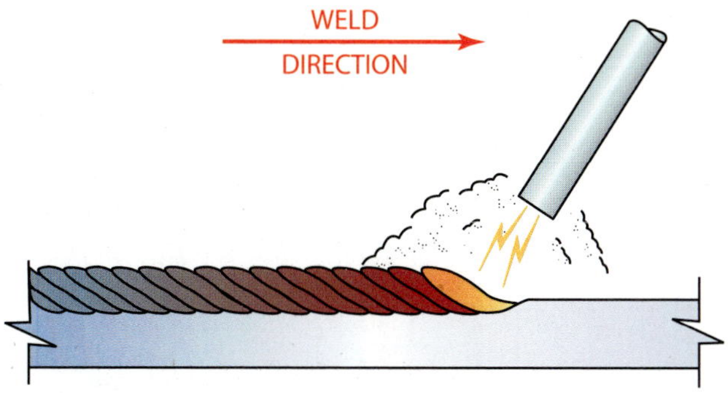

FIGURE 28-5 Backhand or drag angle.

The closer the characteristics of the deposited metal match those of the material being welded, the better is the corrosion resistance of the welded joint. The following precautions should be noted:

- Any carburization or increase in carbon must be avoided, unless a harder material with improved wear resistance is actually desired. In this case, there will be a loss in corrosion resistance.
- It is important to prevent all inclusions of foreign matter, such as oxides, slag, or dissimilar metals.

In welding with the metal arc process, direct current is more widely used than alternating current. Generally, reverse polarity is preferred where the electrode is positive and the workpiece is negative.

The diameter of the electrode used to weld steel that is thinner than 3/16 in. (4.8 mm) should be equal to, or slightly less than, the thickness of the metal to be welded.

When setting up for welding, material 0.050 in. (1.27 mm) and less in thickness should be clamped firmly to prevent distortion or warpage. The edges should be butted tightly together. All seams should be accurately aligned at the start. It is advisable to tack weld the joint at short intervals as well as to use clamping devices.

The electrode should always point into the weld in a backhand or drag angle, Figure 28-5. Avoid using a figure-eight pattern or an excessive side weaving motion such as that used in welding carbon steel. Best results are obtained with a stringer bead with a little or very slight weaving motion and steady forward travel, with the electrode holder leading the weld pool at about 60° in the direction of travel.

To weld stainless steels, the arc should be as short as possible. Table 28-5 can be used as a guide.

Cast Iron

Cast iron is widely used for engine components such as blocks, heads, and manifolds; for drive components such as transmission cases, gearboxes, transfer cases, and differential cases; and for equipment such as pumps, planters, drills, and pulleys. Cast iron is hard and ridged, which makes it ideal for any size casing or frame that must hold its shape even under heavy loads. For example, if a transmission case were to bend under a load, the gears and shafts inside would bind and stop turning.

All five types of cast iron have high carbon contents, usually ranging from 1.7 to 4%. The most common grades contain about 2.5 to 3.5% total carbon. The carbon in cast iron can be combined with iron or be in a free state. As more of the carbon atoms in the cast iron combine with iron atoms, the cast iron becomes harder and more brittle. The five common types of cast iron are as follows:

- Gray cast iron—Is the most widely used type. It contains so much free carbon that a fracture surface has a uniform dark gray color. Gray cast iron is easily welded but because it is somewhat porous it can absorb oils into the surface, which must be baked out before welding.
- White cast iron—Is the hardest and most brittle of the cast irons because almost all of the carbon atoms are combined with the iron atoms. The surface of a fractured piece of this cast iron looks silvery white and may appear shiny. White cast iron is practically unweldable.

Metal Thickness		Electrode Diameter		Current	Voltage
in.	(mm)	in.	(mm)	(Amperes)	Open Circuit
0.050	(1.27)	5/64	(1.98)	25–50	30–35
0.050–0.0625	(1.27–1.58)	3/32	(2.38)	30–90	35–40
0.0625–0.1406	(1.58–3.55)	3/32–1/8	(2.38–3.17)	50–100	40–45
0.1406–0.1875	(3.55–4.74)	1/8–5/32	(3.17–3.96)	80–125	45–50
0.250 and up	(6.35 and up)	3/16	(4.76)	100–175	55–60

Table 28-5 Shielded Metal Arc Welding Electrode Setup for Stainless Steel

- **Malleable cast iron**—Is white cast iron that has undergone a transformation as the result of a long heat-treating process to reduce the brittleness. The fractured surface of malleable cast iron has a light, almost white, thin rim around the dark gray center. If malleable cast iron is heated above its critical temperature, about 1700°F (925°C), the carbon will recombine with the iron, transforming it back into white cast iron. Malleable cast iron can easily be welded. To prevent it from reverting back to white cast iron, do not preheat it above 1200°F (650°C).
- Alloy cast iron—Has alloying elements such as chromium, **copper**, manganese, molybdenum, or nickel added to obtain special properties. Various quantities and types of alloys are added to improve alloy cast iron's tensile strength and heat and corrosion resistance. Almost all grades of alloy cast iron can be easily welded if care is taken to slowly preheat and postcool the part to prevent changes in the carbon and iron structure.
- Nodular cast iron—Sometimes called ductile cast iron, has its carbon formed into nodules or tiny round balls. These nodules are formed by adding an alloy. Nodular cast iron has greater tensile strength than gray cast iron and some of the corrosion resistance of alloy cast iron. Nodular cast iron is weldable, but proper preheating and postweld cooling temperatures and rates must be maintained or the nodular properties will be lost.

Not all cracks or breaks in cast iron present the same degree of difficulty to making welded repairs. Breaks across ears or tabs do not have nearly as much stresses as a crack in a surface, Figure 28-6. Cracks may increase in length when you try to weld them unless you drill a small hole, about 1/8 in. (3 mm), at both ends of the crack, Figure 28-7.

Preweld and Postweld Heating of Cast Iron

The major purpose of preheating and postheating of cast iron is to control the rate of temperature change. The level of temperature and the rate of change of temperature affect the hardness, brittleness, ductility, and strength of iron-carbon–based metals such as steel and cast iron, Table 28-6.

Preheating the casting before welding reduces the internal stresses resulting from the rapid or localized heating caused by welding, Table 28-7. Welding stresses occur because as metal is heated and cooled, it expands and contracts. Unless the heating and cooling cycles are slow and uniform, stresses within brittle materials will cause them to crack. In some aspects, the brittleness of cast iron is much like that of glass. Both cast iron and

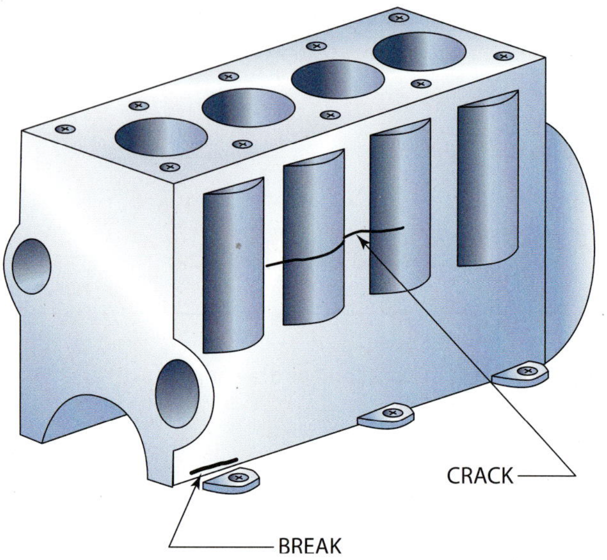

FIGURE 28-6 Cracks on engines can occur in the water jacket due to freezing, and ears can be broken off as a result of accidents or overtightening of misaligned parts.
© Cengage Learning 2012

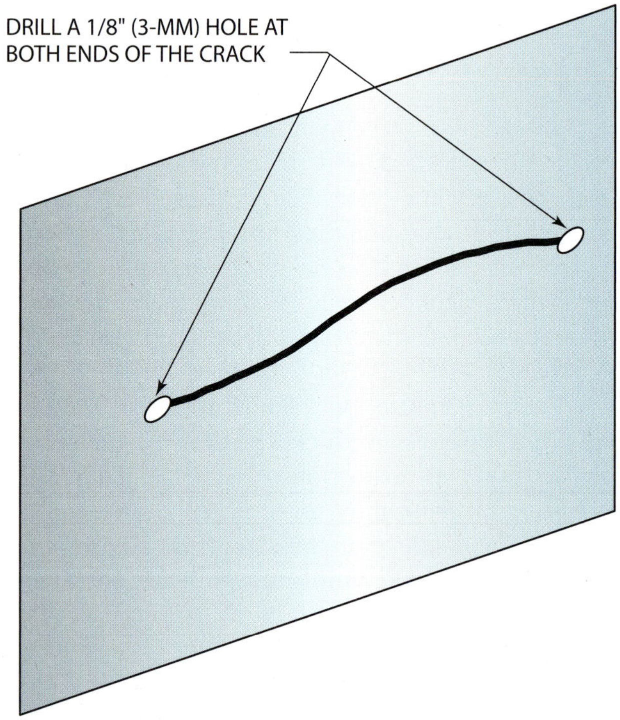

FIGURE 28-7 Stop drill the ends of a crack.
© Cengage Learning 2012

Property	Description
Hardness	The resistance to penetration or shaping by machining or drilling.
Brittleness	The ease with which a metal will crack or break apart without noticeable deformation.
Ductility	The ability of a metal to be permanently twisted, drawn out, bent, or changed in shape without cracking or breaking.
Strength	The ability of a metal to resist deformation or reshaping due to tensile, compression, shear, or torsional forces.

Table 28-6 Mechanical Properties of Metal

	Preheat Temperatures for Weldable Cast Irons		
Joining Process	Temperature Range	Preferred Temperature	Minimum Temperature
Stick Welding	600°F to 1500°F (315°C to 815°C)	900°F to 1200°F (480°C to 650°C)	400°F (200°C)
Gas Welding	400°F to 1100°F (200°C to 600°C)	500°F (260°C)	400°F (200°C)
Braze Welding	500°F to 900°F (260°C to 480°C)	900°F (480°C)	500°F (260°C)

Note: Maintain the preheat temperature for 30 minutes after it is first reached to allow the core of the casting to reach this temperature.

Table 28-7 Welding Should Be Performed at or Above the Preferred Temperature (It can be performed at the minimum temperature, but some hardening and cracking may occur.)

glass will crack if they are heated or cooled unevenly or too quickly.

Postweld heating changes the rate of cooling. Rapid cooling of a metal from a high temperature is called quenching. The faster an iron carbon metal is quenched, the harder, more brittle, less ductile, and higher in strength the metal will become. The slow cooling of an iron carbon metal from a high temperature is called annealing. The slower an iron carbon metal is cooled from a high temperature, the softer, less brittle, more ductile, and lower in strength the metal will become.

To reduce welding stresses, maintain the casting at the same temperature used for preheating or higher for 30 minutes following welding. The casting should cool slowly over the next 24 hours. Cover the casting to prevent the part from being cooled too rapidly by the surrounding air following welding. A firebrick or heavy metal box can be used to keep cool air away from the casting.

Practice Welding Cast Iron

Because there are a few differences between repairing a break and a crack, the following practices alternate between repairing breaks and cracks in cast iron. For example, other than not clamping the parts together, there would be little difference between using Practice 28-1 for welding a break and for welding a crack. In the field, you can use whichever welding procedure is most appropriate for repairing breaks or cracks.

PRACTICE 28-1

Arc Welding a Cast-Iron Break with Preheating and Postheating

Using a properly set up stick welding station, a gas torch with a heating tip, firebricks, a right angle grinder, nickel (ENi) electrodes, a C-clamp, a 900°F (482°C) temperature marking crayon, a chipping hammer, a wire brush, a broken piece of gray cast iron, a welding helmet, gas welding goggles, safety glasses, and all other required safety equipment, you are going to repair a cast-iron break.

Grind the brake into a V-groove, leaving a 1/8-in. (3-mm) root face on the broken surface, Figure 28-8.

NOTE: **Because cast iron is brittle, it does not bend before it breaks, so broken parts can usually be fitted back together like the pieces of a puzzle.**

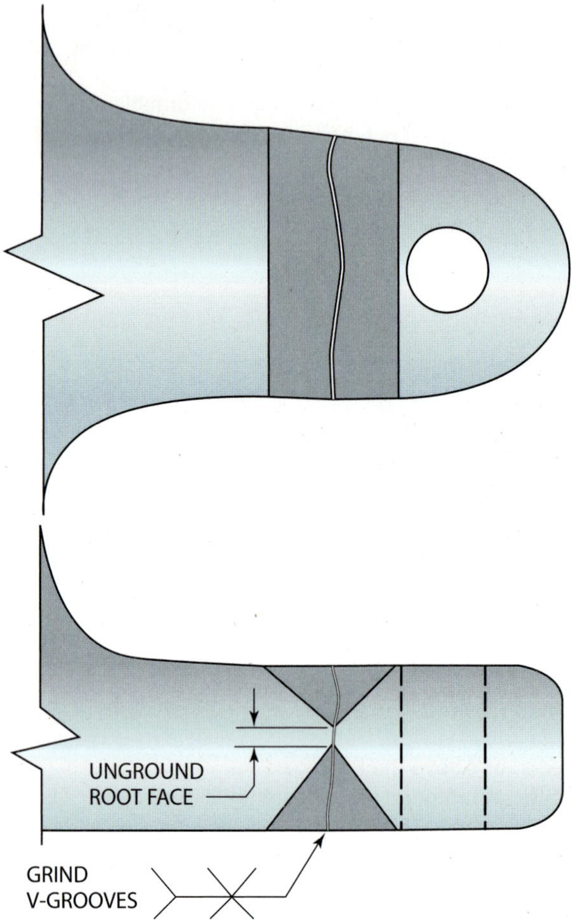

FIGURE 28-8 **Grind the V-groove, leaving a small root face so that the part can be realigned.** © Cengage Learning 2012

Use a C-clamp to hold the broken piece in place. Mark the parts with a 900°F (482°C) temperature marking crayon.

Using a properly lit and adjusted oxyacetylene heating torch, begin preheating the part. Keep the flame moving all around the cast-iron part so that it heats evenly. Do not point the flame directly onto the temperature indicator mark or it will turn color before the part is actually preheated. When the temperature mark turns black, the part is properly preheated. With the flame off the part, re-mark the part to check it for proper preheat.

On thick castings, keep the flame on the part for 30 more minutes so the inside of the part is properly preheated.

NOTE: Starting the welds on the ends of the break or crack and welding to the center concentrates the weld stresses in the center of the weld. This reduces the chances of a crack forming at the end of the weld.

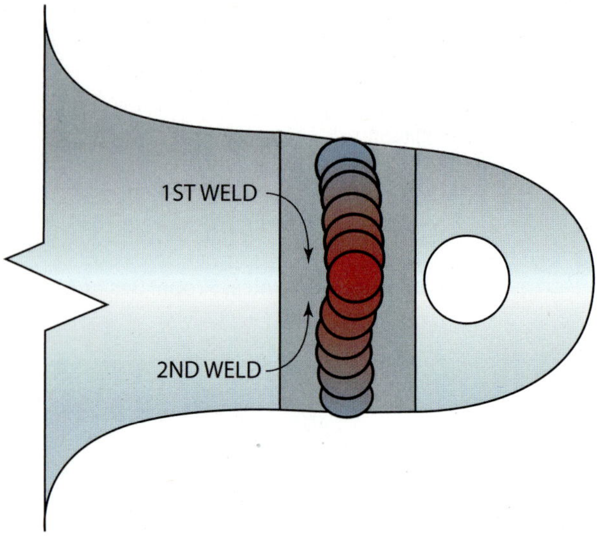

FIGURE 28-9 **Start all welding beads on the edge of the break and end them in the middle.** © Cengage Learning 2012

Strike an arc and make the first weld bead, starting at one edge. Weld to the center of the break. Break the arc, and chip and wire brush the flux off the weld. Strike another arc on the opposite end of the V-groove and make another weld back to the crack of the center, Figure 28-9. Turn the part over and make the same two welds on the back side of the break.

NOTE: A number of small welds are better than one or two large welds because the small welds do not have as much stress and are less likely to cause postweld cracking.

Repeat the process of making welds by alternating sides between weld passes until the V-groove is filled to no more than 1/8 in. (3 mm) above the surface.

Build a firebrick box around the part and place the torch so that the flame will fill the box and the part can be postweld heated. Keep heating the part for 30 minutes after completing the weld. Close any gaps in the firebrick box and allow the part to cool slowly over the next 24 hours. Turn off the welder and torch set, and clean up your work area when you are finished welding.

Complete a copy of the time sheet in Appendix I or as provided by your instructor. ◆

Welding without Preheating or Postheating

Cracks in large castings and castings that are to be repaired in place cannot be preheated and postheated to the desired temperatures but can still be repaired.

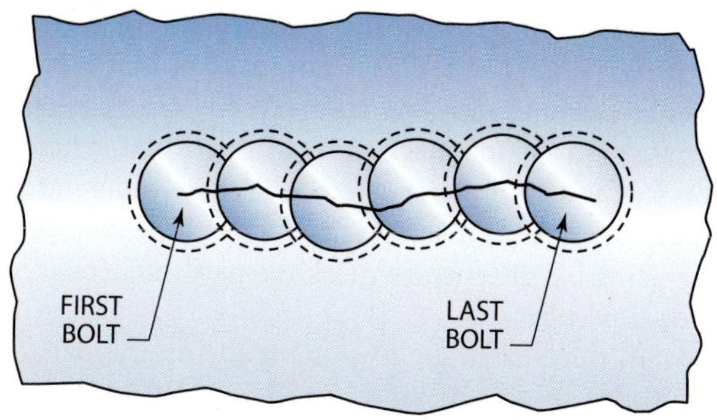

FIGURE 28-10 A crack in cast iron can be plugged by drilling and tapping overlapping holes. Each bolt overlaps the previous bolt. Only the last one must have a locking compound to prevent it from loosening. © Cengage Learning 2012

One method of repairing these cracks is to drill and tap a series of overlapping holes along the crack, Figure 28-10. This is an excellent way to repair nonload-bearing cracks like those in the water jacket of an engine. Two-part epoxy patch material can also be used on nonload-bearing cracks. Read and follow the manufacturer instructions and safety rules when using epoxy repair kits.

Cracks in parts that cannot be preheated to the desired level can still be welded, but the welds will be very hard and are more likely to recrack. However, welding cracks in engine blocks, pump housings, and other large expensive castings, even if they might crack again, is more desirable than simply buying a new part. The new cracks that might form may even be small enough to be patched with epoxy.

PRACTICE 28-2

Arc Welding a Cast-Iron Crack without Preheating or Postheating

Using a properly set up stick welding station, a gas torch with a heating tip, firebricks, a portable drill with a 1/8-in. (3-mm) drill bit, a right angle grinder, ENi electrodes, a chipping hammer, a wire brush, a cracked piece of gray cast iron, a welding helmet, gas welding goggles, safety glasses, and all other required safety equipment, you are going to repair a cast-iron crack.

Locate the ends of the crack and drill stop the crack by drilling 1/8-in. (3-mm) holes at both ends of the crack. Use the edge of the grinding disk to cut a V- or U-groove into the crack.

NOTE: Even though the part cannot be preheated to the desired level, it cannot be welded cold. It must be heated to at least 75°F (24°C) or higher before starting to weld. Engine blocks can be preheated by letting them run for a short time until they are hot.

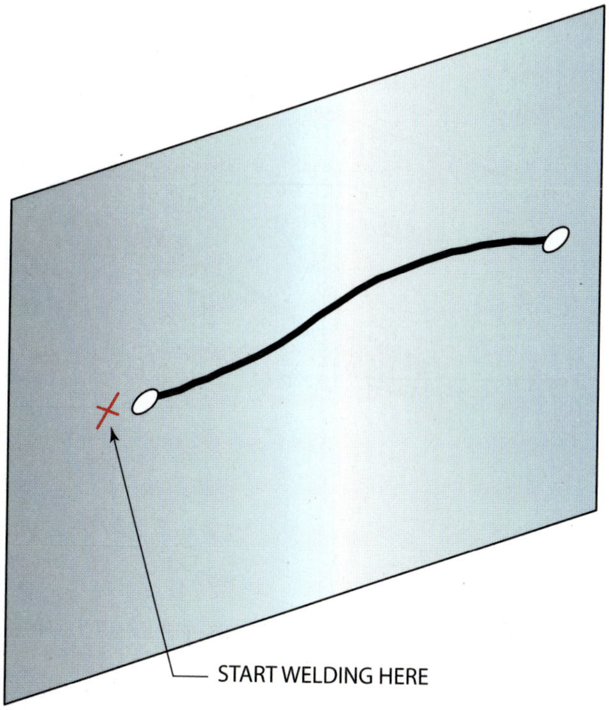

FIGURE 28-11 Start the weld on the casting's surface outside of the crack. © Cengage Learning 2012

Strike the arc on the casting just before the end of the crack and make a 1-in. (25-mm) long weld, Figure 28-11. Stop the weld and repeat the process starting at the other end of the crack.

NOTE: A series of short welds will create less stress than one large weld. The small welds are less likely to crack.

Chip and wire brush the welds.

NOTE: The next series of welds that are made to close the crack are done in a back-stepping sequence. Back-stepping welds are short welds that start ahead of the ending point of the first weld and go back to the end of the first weld. Back-stepping welds are used because they reduce weld stresses and are less likely to crack.

Start the third weld about 1 in. (25 mm) in front of the end of the first weld and move back to the end of the first weld, Figure 28-12. Vigorously chip the slag off the weld immediately after the arc stops. This both cleans the weld and mechanically works the weld surface, called peening, to reduce weld stresses. Repeat the back-step sequence of

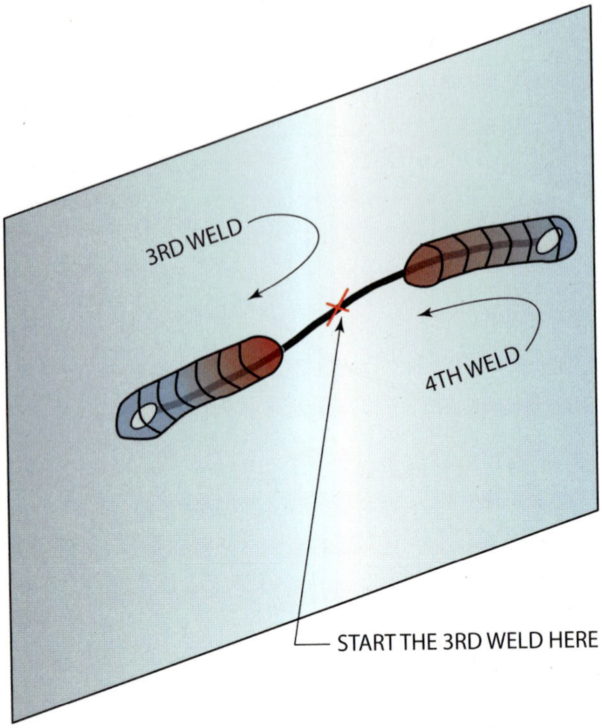

FIGURE 28-12 After both ends have been welded, use a back-step welding process to complete the weld.

welding and peening until the crack is completely covered with welds. Turn off the welder and torch set, and clean up your work area when you are finished welding.

Complete a copy of the time sheet in Appendix I or as provided by your instructor.◆

PRACTICE 28-3

Gas Welding a Cast-Iron Break with Preheating or Postheating

Using a properly set up gas torch with welding and heating tips, firebricks, a right angle grinder, cast-iron gas welding rods, high-temperature cast-iron welding flux, a C-clamp, a 500°F (260°C) temperature marking crayon, a chipping hammer, a wire brush, a broken piece of gray cast iron, gas welding goggles, safety glasses, and all other required safety equipment, you are going to repair a cast-iron break.

Grind the break into a V-groove, leaving a 1/8-in. (3-mm) root face on the broken surface. Use a C-clamp to hold the broken piece in place. Mark the parts with a 500°F (260°C) temperature marking crayon. Using a properly lit and adjusted oxyacetylene heating torch, begin preheating the part. Keep the flame moving all around the cast-iron part so that it heats evenly. Do not point the flame directly onto the temperature indicator mark or it will turn color before the part is actually preheated. When the temperature mark turns black, the part is properly preheated. With the flame off the part, re-mark the part to check it for proper preheat.

On thick castings, keep the flame on the part for 30 more minutes so that the inside of the part is properly preheated. Flux the end of the cast-iron filler rod by heating it with the torch and dipping it into the flux. As the weld progresses, occasionally re-dip the end of the filler rod back into the flux so new flux can be added to the weld. Start welding at one end of the crack. The molten weld pool formed by cast iron is not bright and shiny like that formed on mild steel. The cast-iron molten weld pool looks dull and a little lumpy. As the weld progresses, move the tip of the filler rod around in the molten weld pool to keep it stirred up and to make it flatter.

Fill the crater at the end of the weld with a little extra filler metal. Turn the part over and make the same weld on the back side of the break. Build a firebrick box around the part and place the torch so the flame will fill the box and the part can be postweld heated. Keep heating the part for 30 minutes after completing the weld.

Close any gaps in the firebrick box and allow the part to cool slowly over the next 24 hours. Turn off the torch set and clean up your work area when you are finished welding.

Complete a copy of the time sheet listed in Appendix I or as provided by your instructor.◆

PRACTICE 28-4

Braze Welding a Cast-Iron Crack with Preheating or Postheating

Using a properly set up gas torch with welding and heating tips, firebricks, a right angle grinder, a brazing rod (BRCuZn), brazing flux, a C-clamp, a 900°F (480°C) temperature marking crayon, a chipping hammer, a wire brush, a cracked piece of gray cast iron, gas welding goggles, safety glasses, and all other required safety equipment, you are going to repair a cast-iron crack.

Locate the ends of the crack and drill stop the crack by drilling 1/8-in. (3-mm) holes at both ends of the crack. Use the edge of the grinding disk to cut a

V- or U-groove into the crack. Mark the parts with a 900°F (480°C) temperature marking crayon. Using a properly lit and adjusted oxyacetylene heating torch, begin preheating the part. Keep the flame moving all around the cast-iron part so that it heats evenly. Do not point the flame directly onto the temperature indicator mark or it will turn color before the part is actually preheated. When the temperature mark turns black, the part is properly preheated. With the flame off the part, re-mark the part to check it for proper preheat. On thick castings, keep the flame on the part for 30 more minutes so that the inside of the part is properly preheated.

Flux the end of the brazing rod by heating it with the torch and dipping it into the flux. As the weld progresses, occasionally re-dip the end of the filler rod back into the flux so new flux can be added to the weld. Start welding at one end of the crack. Heat the groove until it is dull red, to about 1800°F (980°C). Touch the tip of the brazing rod into the groove occasionally to test it for the proper brazing temperature. When the braze metal begins to flow, move the tip of the rod around in the molten braze pool to help it wet the surface of the groove. When the braze reaches the end of the groove, add a little extra fill to the crater at the end of the braze weld. Turn the part over and make the same weld on the back side of the break.

Build a firebrick box around the part and place the torch so that the flame will fill the box and the part can be postweld heated. Keep heating the part for 30 minutes after completing the weld. Close any gaps in the firebrick box and allow the part to cool slowly over the next 24 hours. Turn off the torch set and clean up your work area when you are finished welding.

Complete a copy of the time sheet in Appendix I or as provided by your instructor.◆

Aluminum Weldability

One of the characteristics of **aluminum** and its alloys is that it has a great affinity for oxygen. Aluminum atoms combine with oxygen in the air to form an oxide with a high melting point that covers the surface of the metal. This feature, however, is the key to the high resistance of aluminum to corrosion. It is because of this resistance that aluminum can be used in applications where steel is rapidly corroded.

Pure aluminum melts at 1200°F (650°C). The oxide that protects the metal melts at 3700°F (2037°C). This means that the oxide must be cleaned from the metal before welding can begin.

When the GMA welding process is used, the stream of inert gas covers the weld pool, excluding all air from the weld area. This prevents reoxidation of the molten aluminum. GMA welding does not require a flux.

Aluminum can be arc welded using aluminum welding rods. These rods must be kept in a dry place because the flux picks up moisture easily. Because aluminum melts so easily, use a piece of clean steel plate as a backing to weld on thin sections. The steel backing plate can support the root of the weld without the aluminum weld sticking to the steel plate. Thick aluminum casting must be preheated to about 400°F (200°C) before welding. The preheating helps the weld flow and reduces weld spatter.

Aluminum has high thermal conductivity. Aluminum and its alloys can rapidly conduct heat away from the weld area. For this reason, it is necessary to apply the heat much faster to the weld area to bring the aluminum to the welding temperature. Therefore, the intense heat of the electric arc makes this method best suited for welding aluminum.

When aluminum welds solidify from the molten state, they will shrink about 6% in volume. The stress that results from this shrinkage may create excessive joint distortion unless allowances are made before joining the metal. Cracking can occur because the thermal contraction is about two times that of steel. The heated parent metal expands when welding occurs. This expansion of the metal next to the weld area can reduce the root opening on butt joints during the process. The contraction that results upon cooling, plus the shrinkage of the weld metal, creates tension and increases cracking.

The shape of the weld groove and the number of beads can affect the amount of distortion. Less distortion occurs with two-pass square butt welds. Other factors that have an influence on the weld are the speed of welding, the use of properly designed jigs and fixtures to support the aluminum while it is being welded, and tack welding to hold parts in alignment.

Repair Welding

Repair, or maintenance, welding is one of the most difficult types of welding. Some of the major problems include preparing the part for welding, identifying the material, and selecting the best repair method.

The part is often dirty, oily, and painted, and it must be cleaned before welding. The material on the part may include many hazardous compounds. These compounds may or may not be hazardous on the part, but when they are heated or burned during welding they can become life threatening.

CAUTION

It is never safe to start welding on any part that has not been cleaned. All surface contamination must be removed before welding to prevent the possibility of injury from exposure to materials released during welding. Some chemicals can be completely safe until they are exposed to the welding. The smoke or fumes they produce can be an irritant to the skin or eyes, and they may be absorbed through the skin or lungs. If you are exposed to an unknown contaminant, get professional help immediately.

Contamination can be removed by sandblasting, grinding, or using solvents. If a solvent is used, be sure it does not leave a dangerous residue. Clean the entire part if possible or a large enough area so that any remaining material is not affected by the welding.

Before the joint can be prepared for welding, you must try to identify the type of metal. There are several ways to determine metal type before welding. One method is to use a metal identification kit. These kits use a series of chemical analyses to identify the metal. Some kits cannot only identify a type of metal but can also tell the specific alloy.

Another way to identify metal is to look at its color, test for magnetism, and do a spark test. The spark test should be done using a fine grinding stone. With experience, it is often possible to determine specific types of alloys with great accuracy. The sparks given off by each metal and its alloy are so consistent that the U.S. Bureau of Mines uses a camera connected to a computer to identify metals. Microprocessor-controlled testing units are used to identify metals for recycling. For the beginner it is best if you use samples of a known alloy and compare the sparks to your unknown metal samples. The test specimen and the unknown metal samples should be tested using the same grinding wheel and the same force against the wheel.

EXPERIMENT 28-1

Identifying Metal Using a Spark Test

In this experiment you will be working in a small group to identify various metals using a spark test. Examples of the types of sparks you can expect to see for different metals are shown in Figure 28-13. You will need to use proper eye safety equipment, a grinder, several different known and unknown samples of metal, and a pencil and paper to identify the unknown metal samples. Starting with the known samples, make several tests and draw the spark patterns as described in the following paragraphs. Next, test the unknown samples and compare the drawings with the drawings from the known samples. See how many of the unknown metal samples you can identify. You must carefully observe several areas of the spark test pattern, Figure 28-14. The first area is the grinding stone: Are there sparks that are being carried all the way around the grinding stone as shown in area

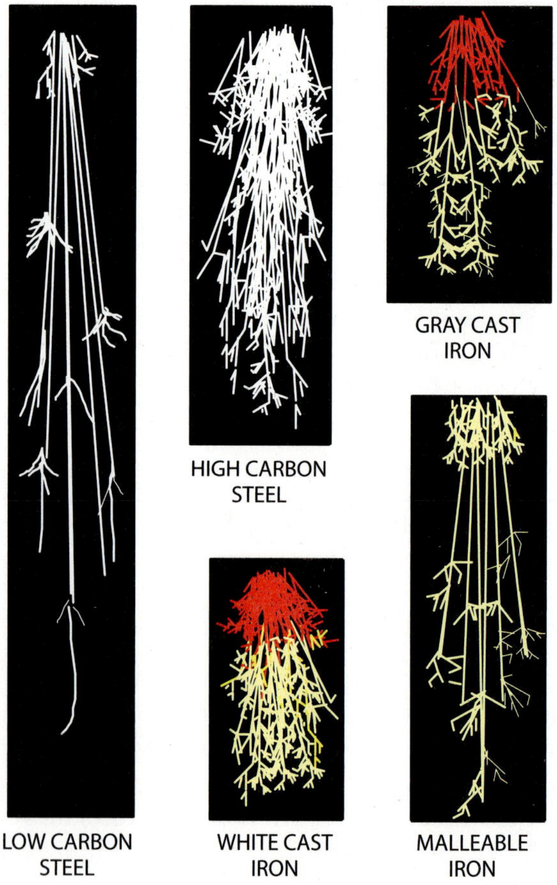

FIGURE 28-13 Spark test patterns for five common metals. © Cengage Learning 2012

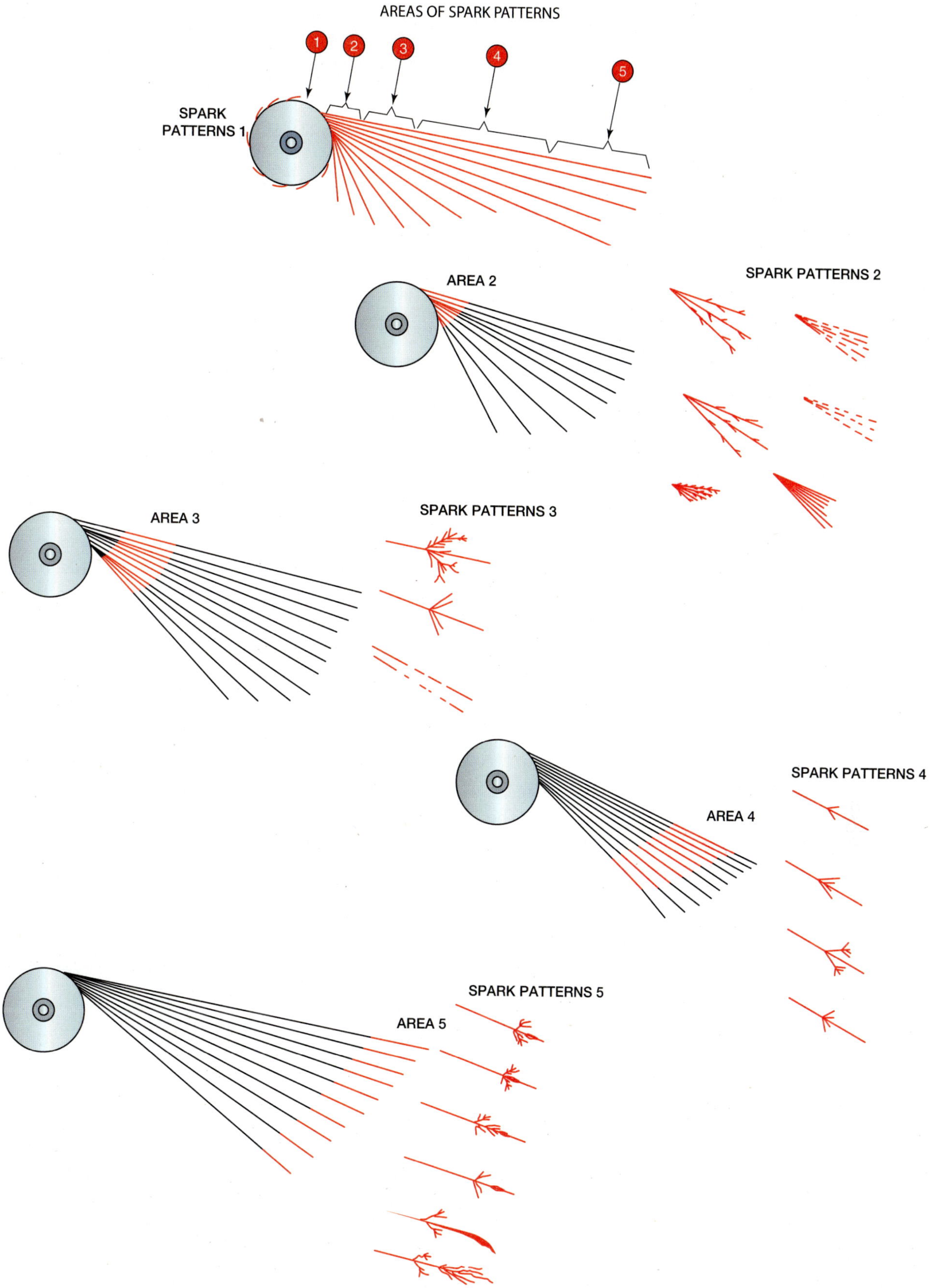

FIGURE 28-14 Spark test.

	Thickness			
Metal	1/8–1/4 in.	1/4–1/2 in.	1/2–1 in.	Over 1 in.
Mild steel	Square	V	V	V
Aluminum	Square	U	U	U
Magnesium	U	U	U	U
Stainless steel	Square	V	V	V
Cast iron	V	V	U	U

Table 28-8 Groove Shapes

1 in Figure 28-14, or are all or most of the sparks leaving the grinding stone as shown in areas 2, 3, 4, and 5?

Many high carbon steels and cast irons have very short spark patterns that do not extend past area 2 as shown in Figure 28-14. The color of the sparks in this area may vary from white to dull red. Note the stream to see if the sparks are small, medium, or large and whether the column of sparks is tightly packed or spread out. If the sparks end in this area, draw the end shape. Lower carbon steels will have longer spark patterns extending into the next area(s) than higher carbon steels.

The next area is immediately adjacent to area 2. Draw a sketch of what the spark stream looks like here, Figure 28-14 (area 3).

As the spark stream moves away from the wheel a few inches, it will begin to change. This change may be in its color, speed of the sparks, or size of the sparks; the sparks may divide into smaller separate streams, explode in a burst of tiny fragments, or just stop glowing, Figure 28-14 (area 4). Sketch these changes you see in the sparks.

Even though some sparks may fragment or divide in areas 2, 3, or 4, they can carry through area 4, where they may simply stop glowing, change color, change shape, explode, or divide into smaller parts as shown in area 5 in Figure 28-14 (area 5). Sketch these changes you see in the sparks.

Repeat this experiment with other types of metal as they become available.

NOTE: The results of the spark test will vary depending on the speed of the grinding stone and its coarseness; however, the relationship between the various areas will remain constant. Coarser, faster stones will tend to have a longer overall spark pattern than finer, slower grinding stones.

SPARK YOUR IMAGINATION

If ferrous metals create sparks when ground and nonferrous metals do not create sparks, then it must be the iron that creates most of the sparks; but what material(s) do you think cause(s) most of the difference in the sparks?

Once the type of metal to be repaired is determined, to the best of your ability decide on the type of weld groove that is needed. Some breaks may need to be ground into a V- or U-groove, and others may not need to be grooved at all, Table 28-8.

Often, thin, sections of most metals can be repaired without the need for the break to be grooved. Thicker sections of most metals will be easier to repair if the crack is ground into a groove. To help in the realignment of the part, it is a good idea to leave a small part of the crack unground so that it can be used to align the parts, Figure 28-15.

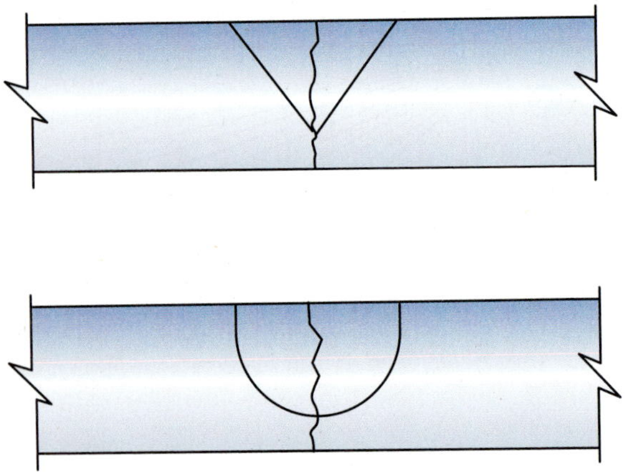

FIGURE 28-15 A small section of the bottom of the break is left so that the parts can be realigned.

After the part is ready to be welded, a test tab can be welded onto the part. This test tab should be welded in a spot that will not damage the part. Once the tab is welded on, it can be broken off to see if the weld will hold, Figure 28-16. If the test tab shows good strength, the repair should continue. If the test tab fails, a new welding procedure should be tried.

Complete a copy of the time sheet in Appendix I or as provided by your instructor.◆

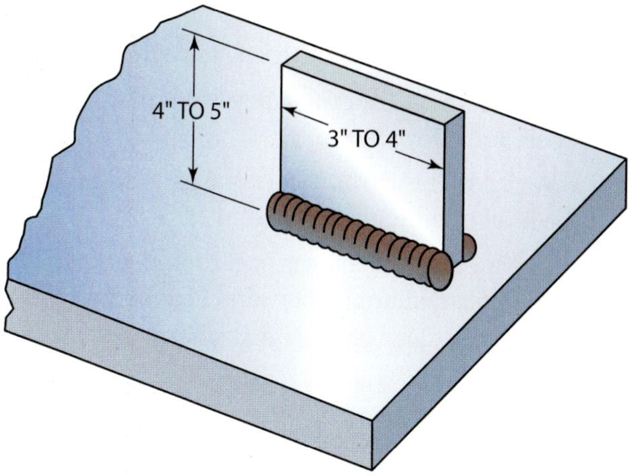

FIGURE 28-16 Tab test.

SUMMARY

All metals are weldable. The only limitation in the fabrication and repair of parts is the cost. It takes a skilled welder with an understanding of all the various characteristics of the metals and types of welding to fix worn or damaged parts. Being able to recognize the differences among the various classifications of metal will allow you to select the most appropriate welding repair procedure. Because of the complexity of this process, you may often be required to research through the original manufacturer of the equipment the types of processes and materials used in the weldment's construction.

Processes change, so sometimes after a part is placed in service, there is a need to change the welding procedures used in the part's original construction for the repair welding. In addition, the original manufacturer may no longer have the welding procedure. To make a successful weld in such cases, you must be able to establish a new welding procedure. As part of your welding procedure, you may perform tests to enhance the longevity of your repairs. Your welding experience will help you to be more efficient in producing the new welding procedure.

Repair welds do not have to look pretty; they only have to hold. Years ago, when the author was welding in a small agricultural welding shop in Madisonville, Tennessee, a local farmer said, "Go on and try and weld it. It's broken anyway, and if you can't fix it it's no use to me anyway." He was right—if you have a broken part, and it is going to cost more to have someone else weld it than to buy a new part, you might as well try welding it yourself. You have nothing to lose and everything to gain; if you fix it, you are in luck, and if you burn it up, you would have had to buy a replacement part anyway. Make sure you can get a replacement part before you start welding on the broken one. If the original part is damaged during the weld, a professional might not be able to repair it. Last, a tendency of many welders is to overweld when making a repair. Keeping the weld sized correctly will make the part stronger. A weld that is too large can cause the part to be brittle and break.

REVIEW QUESTIONS

1. What does it mean if a metal is described as having good weldability?
2. Why is it important to fix design defects before fixing the damaged part?
3. If the wrong filler metal is used, what major defects can result?
4. What is the purpose of preheating and postheating when welding?

5. When large welds are needed, why is it better to make more small welds than a few large welds?
6. What is the difference between plain carbon steels and alloy steels?
7. How does the carbon content of a metal affect the level of difficulty in welding it?
8. What are the percentages of carbon in low carbon, mild steel, medium carbon, and high carbon steel?
9. Stainless steels consist of what four groups of alloys?
10. List five common types of cast iron.
11. When welding cast iron, how does the level of temperature and the rate of temperature change affect the weld?
12. What are back-stepping welds and why are they used?
13. Why is it important to clean all parts before welding?
14. How can surface contamination be removed from a metal surface before welding?
15. How can metal type be determined before welding?

Chapter 29

Welder Certification

OBJECTIVES

After completing this chapter, the student should be able to:

- Explain welder qualification and certification.
- Outline the steps required to certify a welder.
- Make welds that meet a standard.
- Explain the information found on a typical welding procedure specification.

KEY TERMS

certified welders
entry-level welder
interpass temperature
postheating
transverse face bend
transverse root bend
weld test
welder certification
welder performance qualification

INTRODUCTION

Welding, in most cases, is one of the few professions that requires job applicants to demonstrate their skills even if they are already certified. Some other professions—doctor, lawyer, and pilot, for example—do take a written test or require a license initially. But welders are often required to demonstrate their knowledge and their skills before being hired since welding, unlike most other occupations, requires a high degree of eye–hand coordination.

A method commonly used to test a welder's ability is the qualification or certification test. Welders who have passed such a test are referred to as qualified welders; if proper written records are kept of the test results, they are referred to as **certified welders.** Not all welding jobs require that the welder be certified. Some merely require that a basic **weld test** be passed before applicants are hired.

Welder certification can be divided into two major areas. The first area covers the traditional welder certification that has been used for years. This certification is used to demonstrate welding skills for a specific process on a specific weld, to qualify for a welding assignment, or as a requirement for employment.

The American Welding Society (AWS) has developed the second, newer area of certification. This certification has three levels. The first level is primarily designed for the new welder needing to demonstrate **entry-level welder** skills. The other levels cover advanced welders and expert welders. This chapter covers the traditional certification and the AWS QC10 Specification for Qualification and Certification for Entry-Level Welder.

Qualified and Certified Welders

Welder qualification and **welder certification** are often misunderstood. Sometimes it is assumed that a qualified or certified welder can weld anything. Being certified does not mean that a welder can weld everything, nor does it mean that every weld that is made is acceptable. It means that the welder has demonstrated the skills and knowledge necessary to make good welds. To ensure that a welder is consistently making welds that meet the standard, welds are inspected and tested. The more critical the welding, the more critical the inspection and the more extensive the testing of the welds.

All welding processes can be tested for qualification and certification. The testing can range from making spot welds with an electric resistant spot welder to making electron beam welds on aircraft. Being qualified or certified in one area of welding does not mean that a welder can make quality welds in other areas. Most qualifications and certifications are restricted to a single welding process, position, metal, and thickness range.

Individual codes control test requirements. Within these codes, changes in any one of a number of essential variables, such as the ones that follow, can result in the need to recertify.

Process. Welders can be certified in each welding process such as SMAW, GMAW, FCAW, GTAW, EBW, and RSW. Therefore, a new test is required for each process.

Material. The type of metal—such as steel, aluminum, stainless steel, and titanium—being welded will require a change in the certification. Even a change in the alloy within a base metal type can require a change in certification.

Thickness. Each certification is valid on a specific range of thickness of base metal. This range is dependent on the thickness of the metal used in the test. For example, if a 3/8-in. (9.5-mm) plain carbon steel plate is used, then under some codes the welder would be qualified to make welds in plate thickness ranges from 3/16 in. to 3/4 in. (4.8 mm to 19 mm).

Filler metal. Changes in the classification and size of the filler metal can require recertification.

Shielding gas. If the process requires a shielding gas, then changes in gas type or mixture can affect the certification.

Position. In most cases, a weld test taken in the flat position would limit certification to flat and possibly horizontal welding. A test taken in the vertical position, however, would usually allow the welder to work in the flat, horizontal, and vertical positions, depending on the code requirements.

Joint design. Changes in weld type such as groove or fillet welds require a new certification. Additionally, variations in joint geometry, such as groove type, groove angle, and number of passes, can also require retesting.

Welding current. In some cases changing from AC to DC or changes such as to pulsed power and high frequency can affect the certification.

Welder Performance Qualification

Welder performance qualification is the demonstration of a welder's ability to produce welds meeting very specific, prescribed standards. The form used to document this test is called the Welding Qualification Test Record. The detailed, written instructions to be followed by the welder are called the welder qualification procedure. Welders passing this certification are often referred to as being a qualified welder or as qualified.

Welder Certification

A welder certification document is the written verification that a welder has produced welds meeting a prescribed standard of welder performance. A welder holding such a written verification is often referred to as being certified or as a certified welder.

AWS Entry-Level Welder Qualification and Welder Certification

The AWS entry-level welder qualification and certification program specifies a number of requirements not normally found in the traditional welder qualification and certification process. The additions to the AWS program have broadened the scope of the test. Areas such as practical knowledge have long been an assumed part of most certification programs but have not been a formal part of the process. Most companies have assumed that welders who could produce code-quality welds could understand enough of the technical aspects of welding.

Practical Knowledge

A written test must be passed with a minimum grade of 75% on all areas except safety. The safety questions must be answered with a minimum accuracy of 90%. The following subject areas, covered in the given chapters of this text, are included in the test:

Welding and cutting theory (Chapters 8, 11, 13, 15, 17, 19, 21, 22, 23, and 24)

Welding and cutting inspection and testing (Chapters 29 and 30)

Welding and cutting terms and definitions (Glossary)

Base and filler metal identification (Chapters 26 and 27)

Common welding process variables (Chapters 9, 10, 12, 14, 16, and 29)

Electrical fundamentals (Chapter 8)

Drawing and welding symbol interpretation (Chapters 4 and 5)

Fabrication principles and practices (Chapters 6, 7, 9, 10, 12, 14, 16, and 29)

Safe practices (Chapter 2)

Refer to the specific chapters that relate to each of the required knowledge areas for the information required to pass that area of the test.

Welder Qualification and Certification Test Instructions for Practices

After you have mastered the welding and cutting knowledge and skills, you should be ready to start the required assemblies and welding and become certified. You can now take a qualification welding and cutting skills test for the AWS entry-level certification. If you have passed the knowledge and safety test and successfully pass the two bend tests and/or any one of the several workmanship assembly weldments, you can be certified. Figure 29-1 and Figure 29-2 are pictorial representations of the workmanship assembly weldments used in the following certifying practices.

Preparing Specimens for Testing

The detailed preparation of specimens for testing in this chapter is based on the structural welding code AWS D1.1 and the ASME BPV Code, Section IX. The maximum allowable size of fissures (crack or opening)

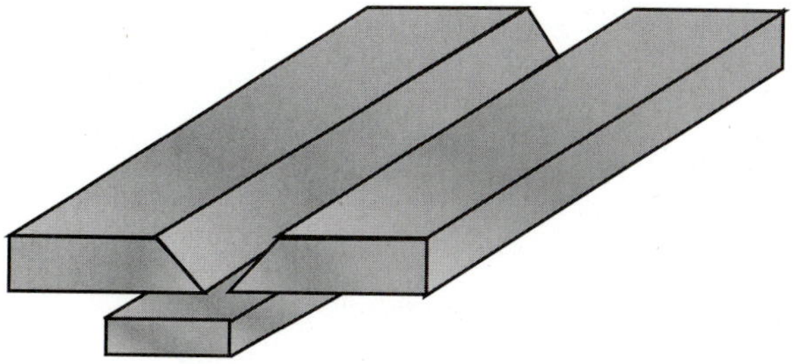

(A)

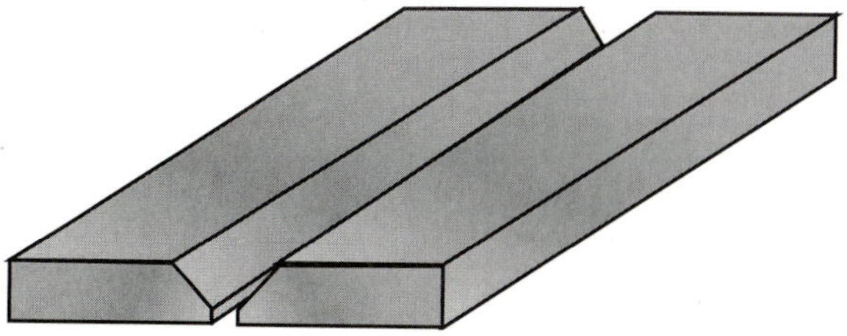

(B)

FIGURE 29-1 Carbon steel test plate (A) with backing and (B) open root (without backing).
American Welding Society

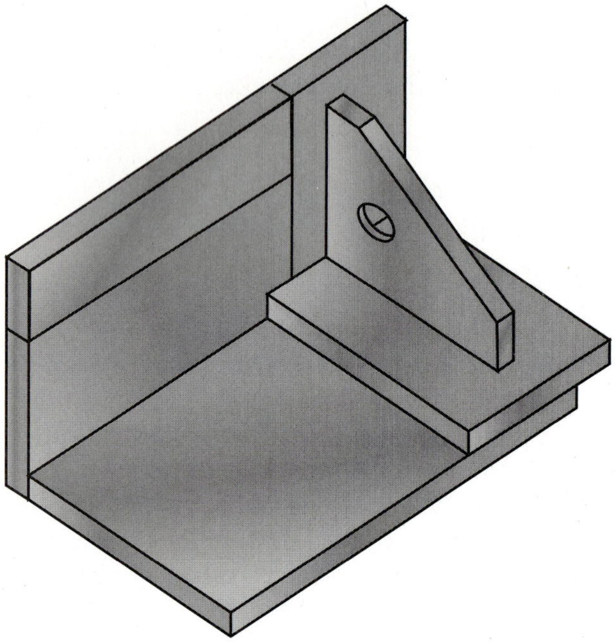

FIGURE 29-2A FCAW carbon steel workmanship sample. © Cengage Learning 2012

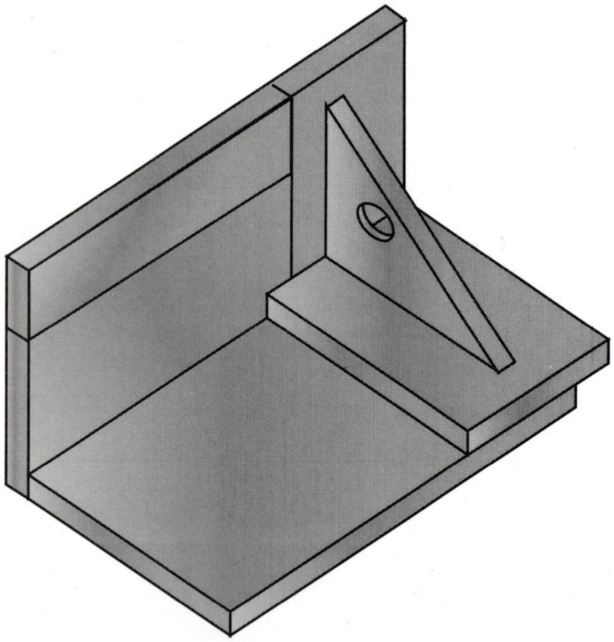

FIGURE 29-2B GMAW-S and GTAW carbon steel workmanship sample. American Welding Society

in a guided-bend test specimen is given in codes for specific applications. Some of the standards are listed in ASTM E190 or AWS B4.0, AWS QC10, AWS QC11, and others. Copies of these publications are available from the appropriate organizations. More information on tests and testing can be found in Chapter 30.

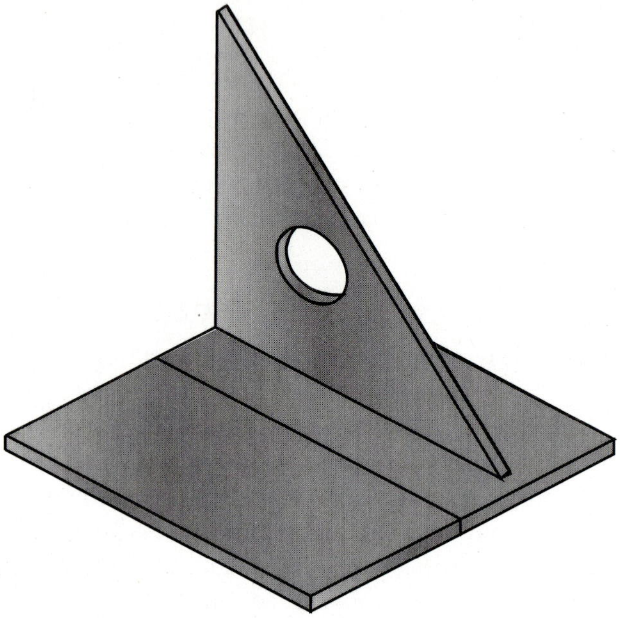

FIGURE 29-2C GMAW spray metal transfer on carbon steel and GTAW on stainless steel and aluminum workmanship samples. American Welding Society

Acceptance Criteria for Face Bends and Root Bends

The weld specimen must first pass visual inspection before it can be prepared for bend testing. Visual inspection looks to see that the weld is uniform in width and reinforcement. There should be no arc strikes on the plate other than those on the weld itself. The weld must be free of both incomplete fusion and cracks. The joint penetration must be either 100% or as stated by the specifications. The weld must be free of overlap, and undercut must not exceed either 10% of the base metal or 1/32 in. (0.8 mm), whichever is less.

Correct weld specimen preparation is essential for reliable results. The weld specimen must be uniform in width and reinforcement and have no undercut or overlap. The weld reinforcement and backing strip, if used, must be removed flush to the surface, Figure 29-3. They can be machined or ground off. The plate thickness after removal must be a minimum of 3/8 in. (9.5 mm), and the pipe thickness must be equal to the pipe's original wall thickness. The specimens may be cut out of the test weldment using an abrasive disc, by sawing, or by cutting with a torch. Flame-cut specimens must have the edges ground or machined smooth after cutting. This procedure is done to remove the heat-affected zone caused by the cut, Figure 29-4.

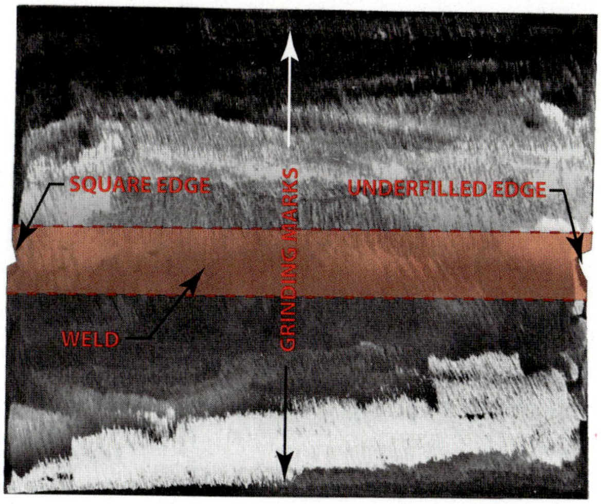

FIGURE 29-3 Plate ground in preparation for removing test specimens. Larry Jeffus

FIGURE 29-6 Guided-bend tested specimen. Larry Jeffus

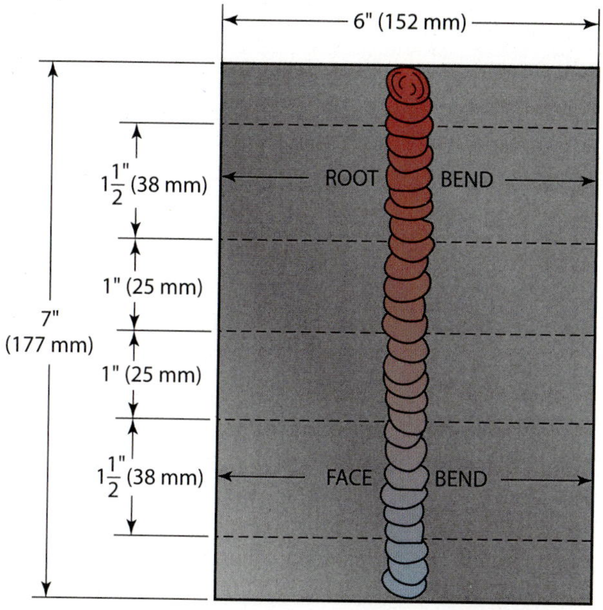

FIGURE 29-4 Sequence for removing guided-bend specimens from the plate once welding is complete. American Welding Society

All corners must be rounded to a radius of 1/8 in. (3.2 mm) maximum, and all grinding or machining marks must run lengthwise on the specimen, Figure 29-5 and Figure 29-6. Rounding the corners and keeping all marks running lengthwise reduce the chance of good weld specimen failure due to poor surface preparation.

The weld must pass both the face and root bends to be acceptable. After bending there can be no single defects larger than 1/8 in. (3.2 mm), and the sum of all defects larger than 1/32 in. (0.8 mm) but less than 1/8 in. (3.2 mm) must not exceed a total of 3/8 in. (9.5 mm) for each bend specimen. An exception is made for cracks that start at the edge of the specimen and do not start at a defect in the specimen.

Restarting a Weld Bead

On all but short welds, the welding bead will need to be restarted after a welder stops to change electrodes.

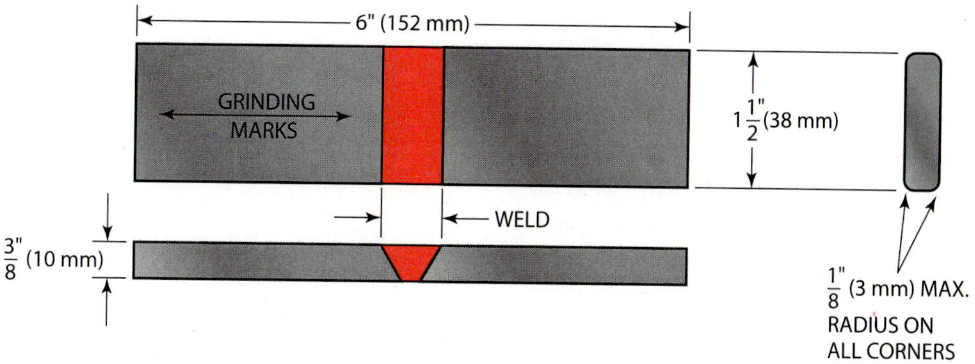

FIGURE 29-5 Guided-bend specimen. © Cengage Learning 2012

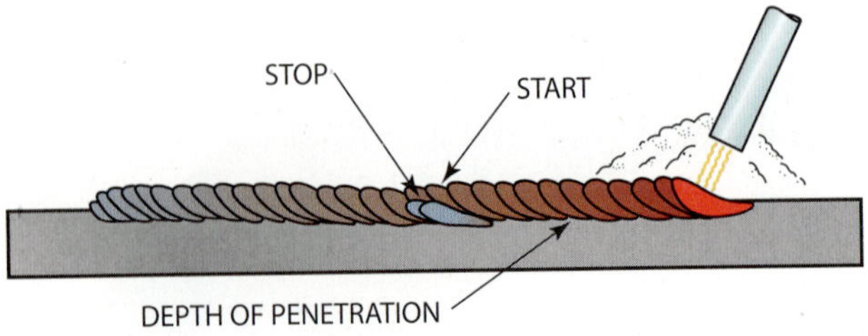

FIGURE 29-7 Tapering the size of the weld bead helps keep the depth of penetration uniform.
© Cengage Learning 2012

Because the metal cools as a welder changes electrodes and chips slag when restarting, the penetration and buildup may be adversely affected.

When a weld bead is nearing completion, it should be tapered so that when it is restarted the buildup will be more uniform. To taper a weld bead, the travel rate should be increased just before welding stops. A 1/4-in. (6.4-mm) taper is all that is required. The taper allows the new weld to be started and the depth of penetration reestablished without having excessive buildup, Figure 29-7.

The slag should always be chipped and the weld crater should be cleaned each time before restarting the weld. This is important to prevent slag inclusions at the start of the weld.

The arc should be restarted in the joint ahead of the weld. The electrodes must be allowed to heat up so that the arc is stabilized and a shielding gas cloud is reestablished to protect the weld. Hold a long arc as the electrode heats up so that metal is not deposited. Slowly bring the electrode downward and toward the weld bead until the arc is directly on the deepest part of the crater where the crater meets the plate in the joint, Figure 29-8. The electrode should be low enough to start transferring metal. Next, move the electrode in a semicircle near the back edge of the weld crater. Watch the buildup and match your speed in the semicircle to the deposit rate so that the weld is built up evenly, Figure 29-9. Move the electrode ahead and continue with the same weave pattern that was being used previously.

The movement to the root of the weld and back up on the bead serves both to build up the weld and reheat the metal so that the depth of penetration will remain the same. If the weld bead is started too quickly, penetration is reduced and buildup is high and narrow. Starting and stopping weld beads in corners should be avoided because this often results in defects, Figure 29-10.

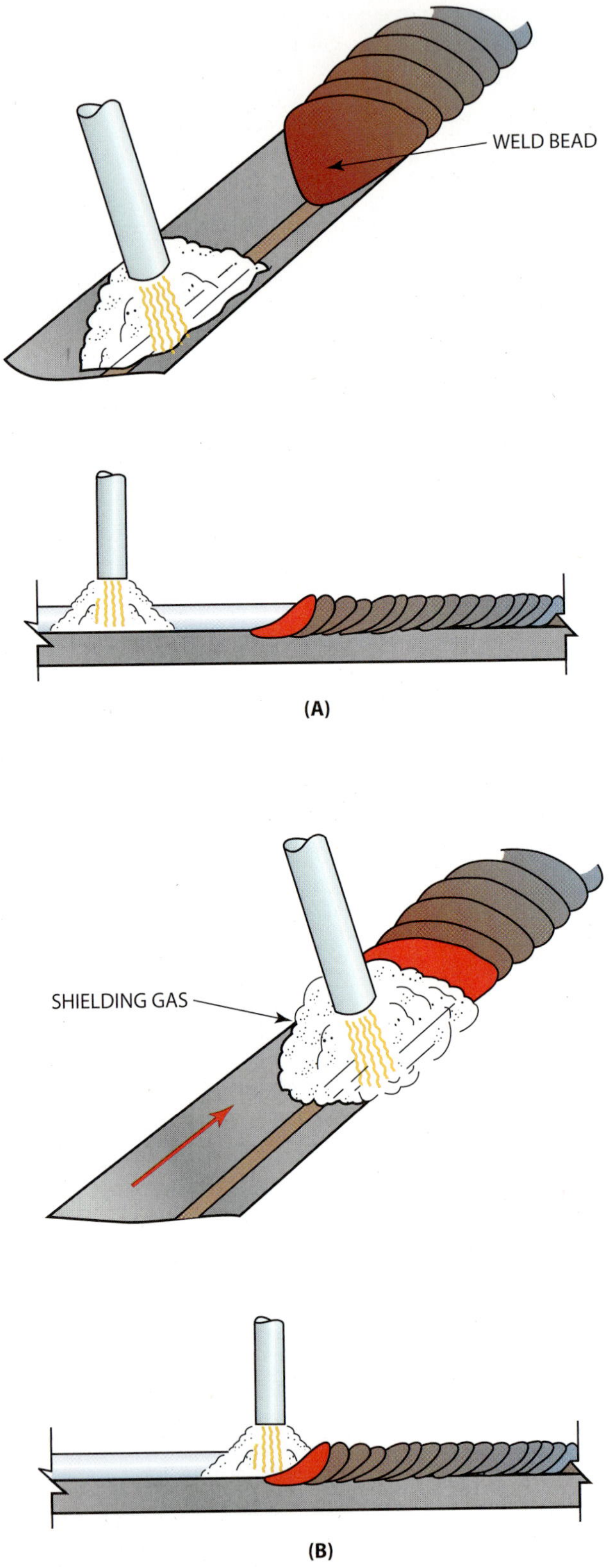

FIGURE 29-8 When restarting the arc, strike the arc ahead of the weld in the joint (A). Hold a long arc and allow time for the electrode to heat up, forming the protective gas envelope. Move the electrode so that the arc is focused directly on the leading edge (root) of the previous weld crater (B). © Cengage Learning 2012

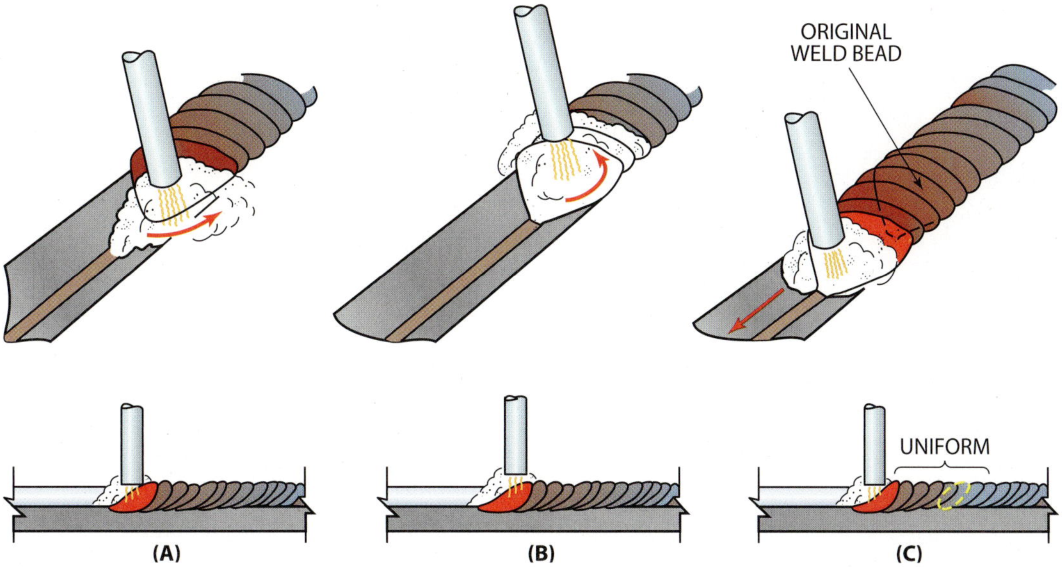

FIGURE 29-9 When restarting the weld pool after the root has been heated to the melting temperature, move the electrode upward along one side of the crater (A). Move the electrode along the top edge, depositing new weld metal (B). When the weld is built up uniformly with the previous weld, continue along the joint (C). © Cengage Learning 2012

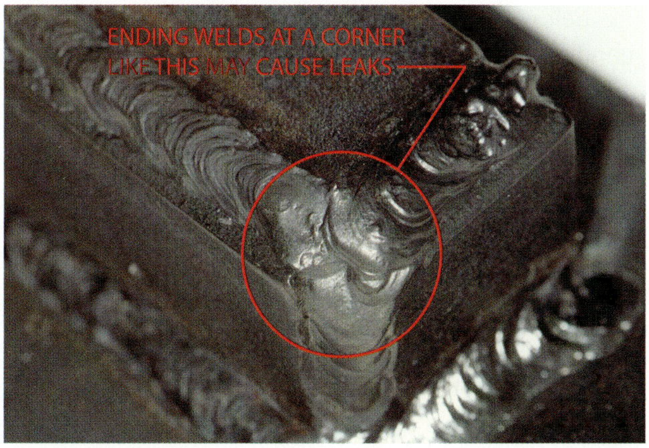

FIGURE 29-10 Incorrect method of welding a corner. Stopping on a corner like this may cause leaks. Larry Jeffus

PRACTICE 29-1

SMAW Workmanship Sample and Welder Qualification Test Plate for Limited Thickness Horizontal and Vertical Positions with E7018 Electrodes

Welding Procedure Specification (WPS) No.: Practice 29-1.

Title: Welding SMAW of plate to plate.

Scope: This procedure is applicable for single-bevel or V-groove plate with a backing strip within the range of 3/16 in. (4.8 mm) through 3/4 in. (20 mm).

Welding May Be Performed in the Following Positions: 2G and 3G (uphill).

Base Metal: The base metal shall conform to carbon steel M-1 or P-1, Group 1 or 2.

- 2G test plates: two (2); each 3/8 in. (9.5 mm) thick, 3 in. (75 mm) wide, and 7 in. (175 mm) long, one having a 45° bevel along one edge.
- 3G test plates: two (2); each 3/8 in. (9.5 mm) thick, 3 in. (76 mm) wide, and 7 in. (175 mm) long, having a 45° included bevel.

Backing Strip Specification: Carbon steel M-1 or P-1, Group 1, 2, or 3, each either 1/4 in. (6.4 mm) or 3/8 in. (9.5 mm) thick, 1 in. (25 mm) wide, and 9 in. (228 mm) long.

Filler Metal: The filler metal shall conform to AWS specification no. E7018 from AWS specification A5.1. This filler metal falls into F-number F-4 and A-number A-1.

Shielding Gas: N/A.

Joint Design and Tolerances: Refer to the drawing in Figure 29-11 for the joint layout specifications.

Preparation of Base Metal: The bevel is to be flame or plasma cut on the edge of the plate before the parts are assembled. The beveled surface must be smooth and free of notches. Any roughness or notches that are

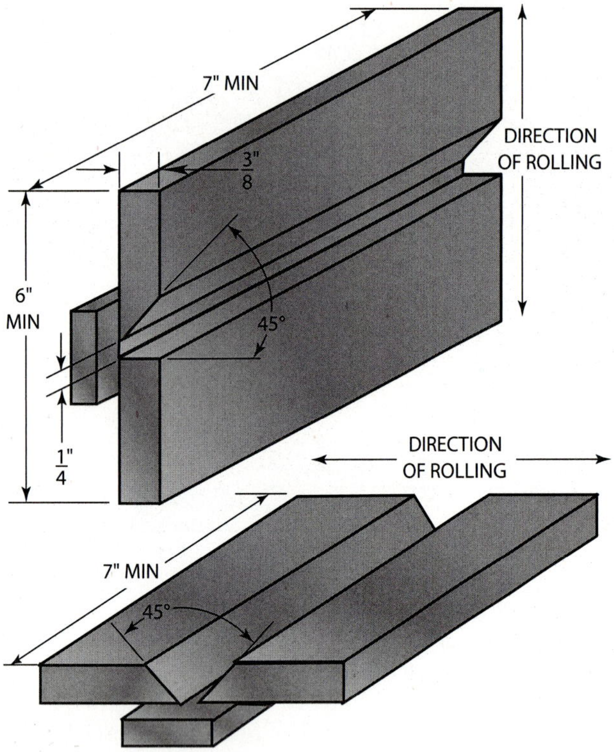

FIGURE 29-11 Practice 29-1 joint design.
American Welding Society

deeper than 1/64 in. (0.4 mm) are unacceptable and must be ground smooth.

All hydrocarbons and other contaminants, such as cutting fluids, grease, oil, and primers, must be cleaned off all parts and filler metals before welding. This cleaning can be done with any suitable solvents or detergents. The backing strip, groove face, and inside and outside plate surface within 1 in. (25 mm) of the joint must be mechanically cleaned of slag, rust, and mill scale. Cleaning must be done with a wire brush or grinder down to bright metal.

Electrical Characteristics: The current shall be direct-current electrode positive (DCEP), Table 29-1. The base metal shall be on the negative side of the line.

Preheat: The parts must be heated to a temperature higher than 50°F (10°C) before any welding is started.

Weld	Filler Metal Dia.	Current	Amperage Range
Tack	3/32 in. (2.4 mm)	DCEP	70 to 110
Root	1/8 in. (3.2 mm)	DCEP	90 to 165
Filler	5/32 in. (4 mm)	DCEP	125 to 220

Table 29-1 E7018 Current Settings

Backing Gas: N/A.

Safety: Proper protective clothing and equipment must be used. The area must be free of all hazards that may affect the welder or others in the area. The welding machine, welding leads, work clamp, electrode holder, and other equipment must be in safe working order.

Welding Technique: Tack weld the plates together with the backing strip. There should be about a 1/4-in. (6.4-mm) root gap between the plates. Use the E7018 arc welding electrodes to make a root pass to fuse the plates and backing strip together. Clean the slag from the root pass, being sure to remove any trapped slag along the sides of the weld.

Using the E7018 arc welding electrodes, make a series of stringer or weave filler welds, no thicker than 1/4 in. (6.4 mm), in the groove until the joint is filled.

Interpass Temperature: The plate should not be heated to a temperature higher than 350°F (175°C) during the welding process. After each weld pass is completed, allow it to cool but never to a temperature below 50°F (10°C). The weldment must not be quenched in water.

Cleaning: The slag must be cleaned off between passes. The weld beads may be cleaned by a hand wire brush, a chipping hammer, a punch and hammer, or a needle-scaler. All weld cleaning must be performed with the test plate in the welding position.

Visual Inspection Criteria for Entry-Level Welders*:

1. There shall be no cracks, no incomplete fusion.
2. There shall be no incomplete joint penetration in groove welds except as permitted for partial joint penetration welds.
3. The Test Supervisor shall examine the weld for acceptable appearance and shall be satisfied that the welder is skilled in using the process and procedure specified for the test.
4. Undercut shall not exceed the lesser of 10% of the base metal thickness or 1/32 in. (0.8 mm).
5. Where visual examination is the only criterion for acceptance, all weld passes are subject to visual examination at the discretion of the Test Supervisor.
6. The frequency of porosity shall not exceed one in each 4 in. (100 mm) of weld length, and the

**Courtesy of American Welding Society.*

maximum diameter shall not exceed 3/32 in. (2.4 mm).

7. Welds shall be free from overlap.

Bend Test: The weld is to be mechanically tested only after it has passed the visual inspection. Be sure that the test specimens are properly marked to identify the welder, the position, and the process.

Specimen Preparation: For 3/8-in. (9.5-mm) test plates, two specimens are to be located in accordance with the requirements in Figure 29-12. One is to be prepared for a **transverse face bend**, and the other is to be prepared for a **transverse root bend**, Figure 29-13.

- *Transverse face bend.* The weld is perpendicular to the longitudinal axis of the specimen and is bent so that the weld face becomes the tension surface of the specimen.
- *Transverse root bend.* The weld is perpendicular to the longitudinal axis of the specimen and is bent so that the weld root becomes the tension surface of the specimen.

Acceptance Criteria for Bend Test*: For acceptance, the convex surface of the face- and root-bend specimens shall meet both of the following requirements:

1. No single indication shall exceed 1/8 in. (3.2 mm) measured in any direction on the surface.

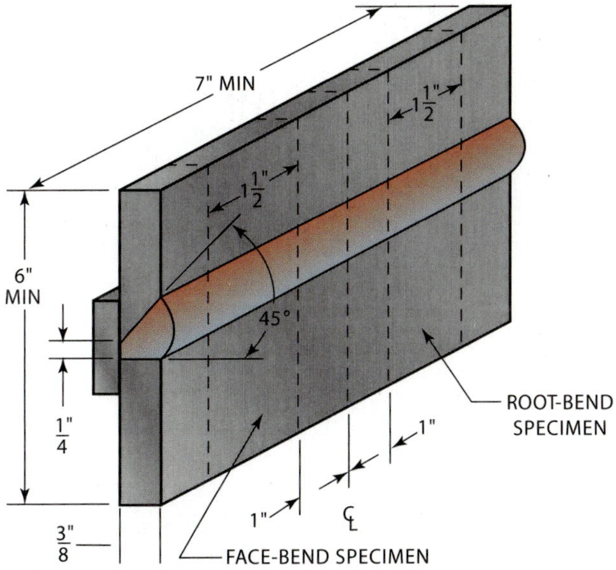

FIGURE 29-12 Test plate specimen location.
American Welding Society

*Courtesy of American Welding Society.

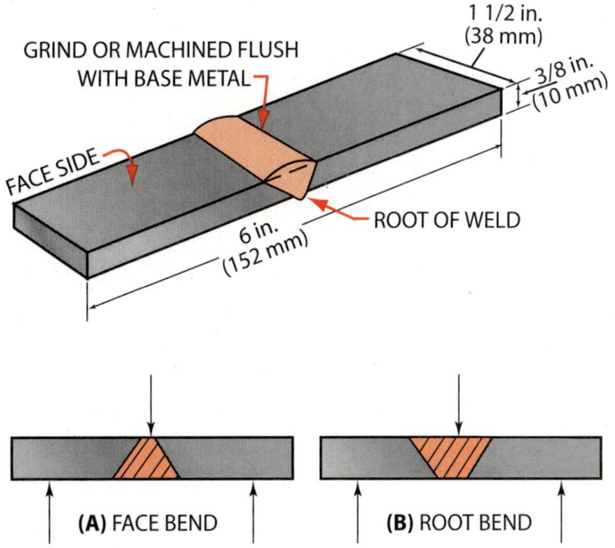

FIGURE 29-13 Test specimen specifications.
American Welding Society

2. The sum of the greatest dimensions of all indications on the surface which exceed 1/32 in. (0.8 mm) but are less than or equal to 1/8 in. (3.2 mm) shall not exceed 3/8 in. (9.5 mm).

Cracks occurring at the corner of the specimens shall not be considered unless there is definite evidence that they result from slag or inclusions or other internal discontinuities.

Sketches: 2G and 3G test plate drawings, Figure 29-14.

Paperwork: Complete a copy of the time sheet in Appendix I, the bill of materials in Appendix III, and the performance qualification test record in Appendix IV, or use forms as provided by your instructor.◆

PRACTICE 29-2

Limited Thickness Welder Performance Qualification Test Plate without Backing

Welding Procedure Specification (WPS) No.: Practice 29-2.

Title: Welding SMAW of plate to plate.

Scope: This procedure is applicable for bevel and V-groove plate with a backing strip within the range of 3/8 in. (9.5 mm) through 3/4 in. (19 mm).

Welding May Be Performed in the Following Positions: 1G, 2G, 3G, and 4G.

Base Metal: The base metal shall conform to M1020 or A36.

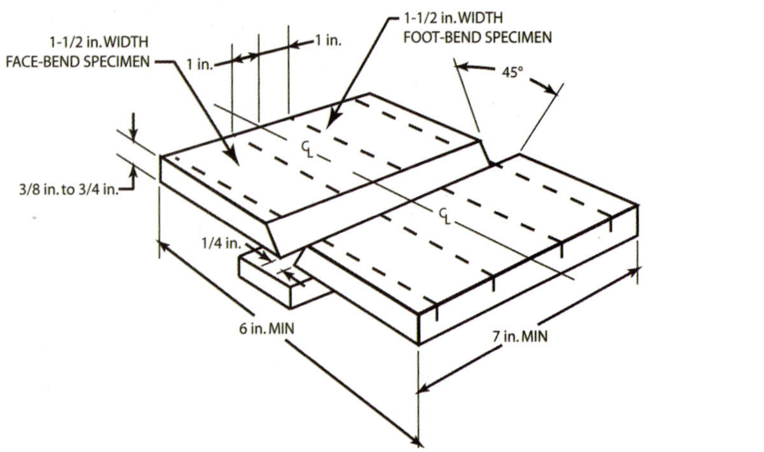

NOTES:
1. 3/8 in. thickness carbon steel material.
2. Performance Qualification #1 = 2G. Performance Qualification #2 = 3G, uphill.
3. All welding done in position, according to applicable performance qualification requirements.
4. The backing thickness shall be 1/4 in. min. to 3/8 in. max.; backing width 1 in. min.
5. All parts may be mechanically cut or machine OFC.
6. Use WPS AWS EDU SMAW-01 for PQ#1-2G, and AWS EDU SMAW-02 for PQ#2-3G uphill. (See AWS QC10, Table 2.)
7. Visual examination in accordance with requirements of AWS QC10, Table 3.
8. Bend test in accordance with the requirements of QC10, Table 4.

ID	QTY	SIZE	METRIC CONVERSION

American Welding Society

Entry Welder Performance Qualification

SMAW Carbon Steel Test Plates

DATE:	SCALE:	DWG #: AWS EDU-6
DR BY:	Tolerances: (Unless otherwise specified) DRAWING NOT TO SCALE Fractions: ± 1/16" Angles: + 10°, −5°	
APP BY:		

FIGURE 29-14 AWS EDU-6 SMAW carbon steel plate workmanship sample. American Welding Society

Backing Material Specification: M1020 or A36.

Filler Metal: The filler metal shall conform to AWS specification no. E6010 or E6011 root pass and E7018 for the cover pass from AWS specification A5.1. This filler metal falls into F-number F-3 and F-4 and A-number A-1.

Shielding Gas: N/A.

Joint Design and Tolerances: Refer to the drawing in Figure 29-15 for the joint layout specifications.

Preparation of Base Metal: The V-groove is to be flame or plasma cut on the edge of the plate before the parts are assembled. Prior to welding, all parts must be cleaned of all contaminants, such as paints, oils, grease, or primers. Both inside and outside surfaces within 1 in. (25 mm) of the joint must be mechanically cleaned using a wire brush or grinder.

Electrical Characteristics: The current shall be AC or DC with the electrode positive (DCEP), Table 29-2.

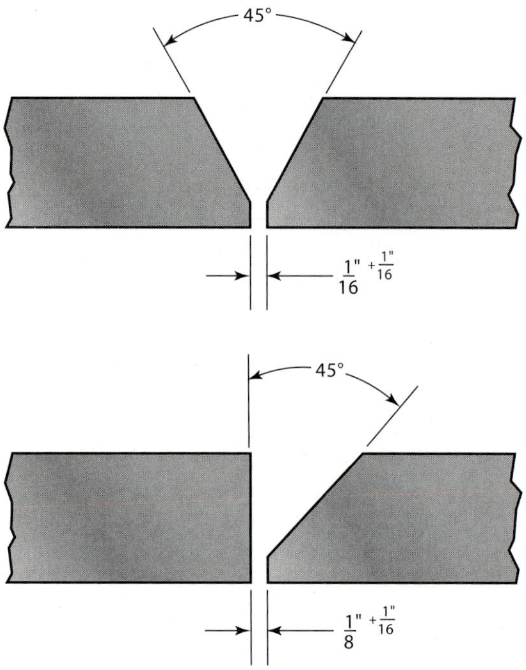

FIGURE 29-15 Practice 29-2 joint design. American Welding Society

AWS CLASSIFICATION AND POLARITY	ELECTRODE DIAMETER AND AMPERAGE RANGE		
	3"/32	1"/8	5"/32
E6010 — DCEP	40–80	70–130	110–165
E6011 — AC, DCEP	50–70	85–125	130–160
E7018 — AC, DCEP	70–110	90–165	125–220

Table 29-2 E6010, E6011, and E7018 Current Settings

Preheat: The parts must be heated to a temperature higher than 70°F (21°C) before any welding is started.

Backing Gas: N/A.

Welding Technique: You can use a small piece of metal across the ends of the plate as a spacer to make it easier to tack weld the plates together; there should be about a 1/8-in. (3.2-mm) root gap between the plates. Use the E6010 or E6011 arc welding electrodes to make a root pass to fuse the plates together. Clean the slag from the root pass, and use either a hot pass or grinder to remove any trapped slag.

Using the E7018 arc welding electrodes, make a series of filler welds in the groove until the joint is filled.

Interpass Temperature: The plate, outside of the heat-affected zone, should not be heated to a temperature higher than 400°F (205°C) during the welding process. After each weld pass is completed, allow it to cool; the weldment must not be quenched in water.

Cleaning: The slag can be chipped and/or ground off between passes but can only be chipped off of the cover pass.

Inspection: Visually inspect the weld for uniformity and discontinuities by using the criteria found in Practice 29-1. If the weld passes the visual inspection, then it is to be prepared and guided-bend tested according to the "guided-bend test" criteria found in Practice 29-1. Repeat each of the welds as needed until you can pass this test.

Postheating is the application of heat to the metal after welding. Postheating is used to slow the cooling rate and reduce hardening.

Interpass temperature is the temperature of the metal during welding. The interpass temperature is given as a minimum and maximum. The minimum temperature is usually the same as the preheat temperature. If the plate cools below this temperature during welding, it should be reheated. The maximum temperature may be specified to keep the plate below a certain phase change temperature for the mild steel used in these practices. The maximum interpass temperature occurs when the weld bead cannot be controlled because of a slow cooling rate. When this happens, the plate should be allowed to cool down but not below the minimum interpass temperature.

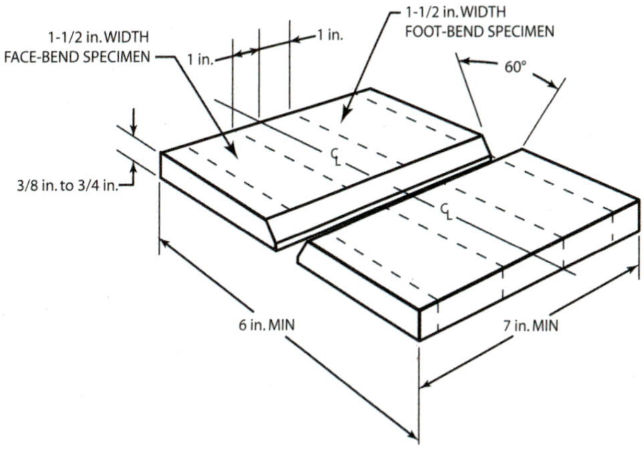

FIGURE 29-16 Open root test plate specification and specimen location for carbon steel sample.
American Welding Society

If, during the welding process, a welder must allow a practice weldment to cool so that the weld can be completed later, the weldment should be cooled slowly and then reheated before welding is started again. A weld that is to be tested or that is done on any parts other than scrap should not be quenched in water.

Sketches: 1G, 2G, 3G, and 4G test plate drawing, Figure 29-16.

Paperwork: Complete a copy of the time sheet in Appendix I, the bill of materials in Appendix III, and the performance qualification test record in Appendix IV, or use forms as provided by your instructor.◆

PRACTICE 29-3

Gas Metal Arc Welding—Short-Circuit Metal Transfer (GMAW-S) Workmanship Sample

Welding Procedure Specification (WPS) No.: Practice 29-3.

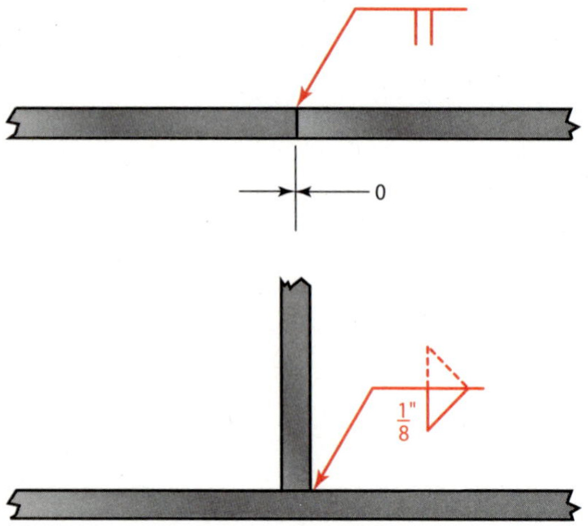

FIGURE 29-17 Practice 29-3 joint design.
American Welding Society

Title: Welding GMAW-S of plate to plate.

Scope: This procedure is applicable for square groove and fillet welds within the range of 10 gauge (3.4 mm) through 14 gauge (1.9 mm).

Welding May Be Performed in the Following Positions: All.

Base Metal: The base metal shall conform to carbon steel M-1, P-1, and S-1, Group 1 or 2.

Backing Material Specification: None.

Filler Metal: The filler metal shall conform to AWS specification no. E70S-X from AWS specification A5.18. This filler metal falls into F-number F-6 and A-number A-1.

Shielding Gas: The shielding gas, or gases, shall conform to the following compositions and purity: CO_2 at 30 to 50 cfh or 75% Ar/25% CO_2 at 30 to 50 cfh.

Joint Design and Tolerances: Refer to the drawing in Figure 29-17 for the joint layout specifications.

Preparation of Base Metal: All parts may be mechanically cut or machine PAC unless specified as manual PAC.

All hydrocarbons and other contaminants, such as cutting fluids, grease, oil, and primers, must be cleaned off all parts and filler metals before welding. This cleaning can be done with any suitable solvents or detergents. The groove face and inside and outside plate surface within 1 in. (25 mm) of the joint must be mechanically cleaned of slag, rust, and mill scale. Cleaning must be done with a wire brush or grinder down to bright metal.

Electrical Characteristics: Set the voltage, amperage, wire feed speed, and shielding gas flow according to Table 29-3.

Preheat: The parts must be heated to a temperature higher than 50°F (10°C) before any welding is started.

Backing Gas: N/A.

Safety: Proper protective clothing and equipment must be used. The area must be free of all hazards that may affect the welder or others in the area. The welding machine, welding leads, work clamp, electrode holder, and other equipment must be in safe working order.

Welding Technique: Using a 1/2-in. (13-mm) or larger gas nozzle for all welding, first tack weld the plates together according to the drawing. Use the E70S-X filler metal to fuse the plates together. Clean any silicon slag, being sure to remove any trapped silicon slag along the sides of the weld.

Using the E70S-X arc welding electrodes, make a series of stringer beads, no thicker than 3/16 in. (4.8 mm). The 1/8-in. (3.2-mm) fillet welds are to be made with one pass. All welds must be placed in the orientation shown in the drawing.

Electrode		Welding Power			Shielding Gas		Base Metal	
Type	Size	Amps	Wire Feed Speed IPM (cm/min)	Volts	Type	Flow	Type	Thickness
E70S-X	0.035 in. (0.9 mm)	90 to 120	180 to 300 (457 to 762)	15 to 19	CO_2 or 75% Ar/25% CO_2	30 to 50	Low carbon steel	1/4 in. to 1/2 in. (6 mm to 13 mm)
E70S-X	0.045 in. (1.2 mm)	130 to 200	125 to 200 (318 to 508)	17 to 20	CO_2 or 75% Ar/25% CO_2	30 to 50	Low carbon steel	1/4 in. to 1/2 in. (6 mm to 13 mm)

Table 29-3 GMAW-S Machine Settings

Interpass Temperature: The plate should not be heated to a temperature higher than 350°F (175°C) during the welding process. After each weld pass is completed, allow it to cool but never to a temperature below 50°F (10°C). The weldment must not be quenched in water.

Cleaning: Any slag must be cleaned off between passes. The weld beads may be cleaned by a hand wire brush, a chipping hammer, a punch and hammer, or a needle-scaler. All weld cleaning must be performed with the test plate in the welding position.

Visual Inspection*: Visually inspect the weld for uniformity and discontinuities.

1. There shall be no cracks, no incomplete fusion.
2. There shall be no incomplete joint penetration in groove welds except as permitted for partial joint penetration welds.
3. The Test Supervisor shall examine the weld for acceptable appearance and shall be satisfied that the welder is skilled in using the process and procedure specified for the test.
4. Undercut shall not exceed the lesser of 10% of the base metal thickness or 1/32 in. (0.8 mm).
5. Where visual examination is the only criterion for acceptance, all weld passes are subject to visual examination at the discretion of the Test Supervisor.
6. The frequency of porosity shall not exceed one in each 4 in. (100 mm) of weld length, and the maximum diameter shall not exceed 3/32 in. (2.4 mm).
7. Welds shall be free from overlap.

Sketches: GMAW Short-Circuit Metal Transfer Workmanship Sample drawing, Figure 29-18.

Paperwork: Complete a copy of the time sheet in Appendix I, the bill of materials in Appendix III,

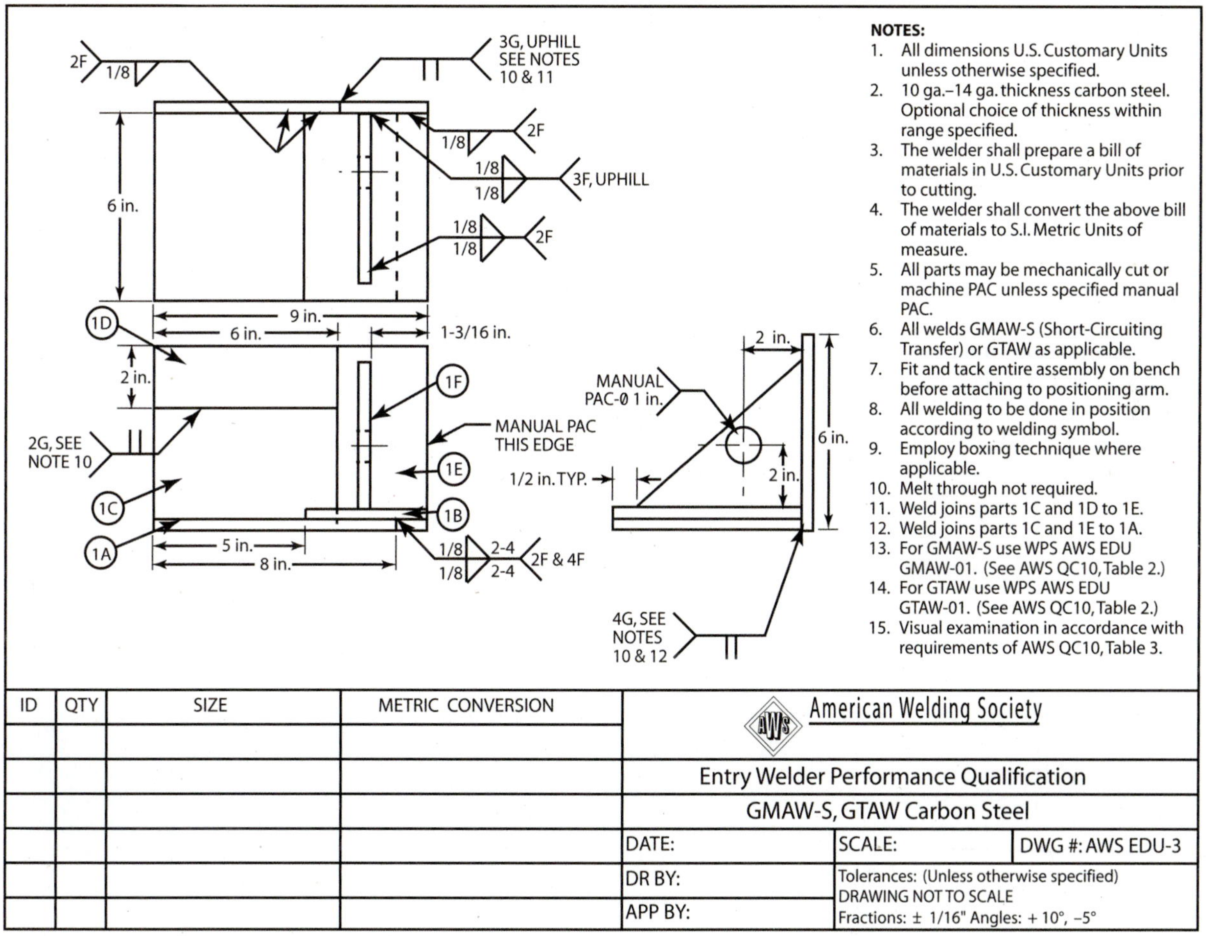

FIGURE 29-18 AWS EDU-3 GMAW Short-Circuit Metal Transfer Workmanship for Carbon Steel
American Welding Society

Courtesy of American Welding Society.

and the performance qualification test record in Appendix IV, or use forms as provided by your instructor.◆

PRACTICE 29-4

Gas Metal Arc Welding (GMAW) Spray Transfer Workmanship Sample

Welding Procedure Specification (WPS) No.: Practice 29-4.

Title: Welding GMAW of plate to plate.

Scope: This procedure is applicable for fillet welds within the range of 1/8 in. (3.2 mm) through 1 1/2 in. (38 mm).

Welding May Be Performed in the Following Positions: 1F and 2F.

Base Metal: The base metal shall conform to carbon steel M-1, P-1, and S-1, Group 1 or 2.

Backing Material Specification: None.

Filler Metal: The 0.035- to 0.045-diameter filler metal shall conform to AWS specification no. E70S-X from AWS specification A5.18. This filler metal falls into F-number F-6 and A-number A-1.

Shielding Gas: The shielding gas, or gases, shall conform to the following compositions: 98% Ar/2% O_2 or 90% Ar/10% CO_2.

Joint Design and Tolerances: Refer to the drawing in Figure 29-19 for the joint layout specifications.

Preparation of Base Metal: The bevels are to be flame or plasma cut on the edges of the plate before the parts are assembled. The beveled surface must be smooth and free of notches. Any roughness or notches deeper than 1/64 in. (0.4 mm) must be ground smooth.

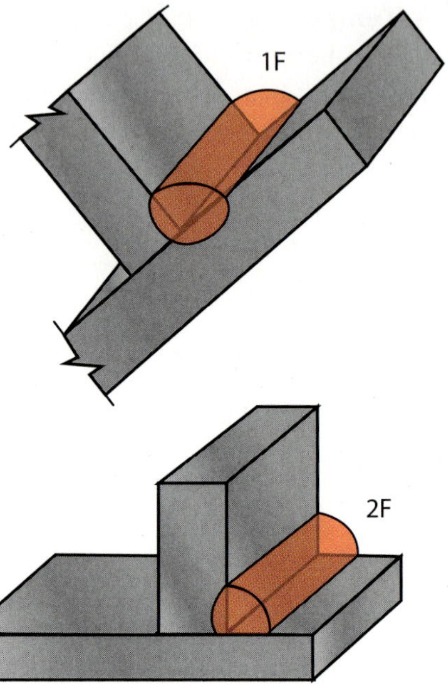

FIGURE 29-19 Practice 29-4 joint design.
American Welding Society

All hydrocarbons and other contaminants, such as cutting fluids, grease, oil, and primers, must be cleaned off all parts and filler metals before welding. This cleaning can be done with any suitable solvents or detergents. The groove face and inside and outside plate surface within 1 in. (25 mm) of the joint must be mechanically cleaned of slag, rust, and mill scale. Cleaning must be done with a wire brush or grinder down to bright metal.

Electrical Characteristics: Set the voltage, amperage, wire feed speed, and shielding gas flow according to Table 29-4.

Preheat: The parts must be heated to a temperature higher than 50°F (10°C) before any welding is started.

Electrode		Welding Power			Shielding Gas		Base Metal	
Type	Size	Amps	Wire Feed Speed IPM (cm/min)	Volts	Type	Flow	Type	Thickness
E70S-X	0.035 in. (0.9 mm)	90 to 120	180 to 300 (457 to 762)	15 to 19	CO_2 or 75% Ar/25% CO_2	30 to 50	Low carbon steel	1/4 in. to 1/2 in. (6 mm to 13 mm)
E70S-X	0.045 in. (1.2 mm)	130 to 200	125 to 200 (318 to 508)	17 to 20	CO_2 or 75% Ar/25% CO_2	30 to 50	Low carbon steel	1/4 in. to 1/2 in. (6 mm to 13 mm)

Table 29-4 GMAW-S Machine Settings

Backing Gas: N/A.

Safety: Proper protective clothing and equipment must be used. The area must be free of all hazards that may affect the welder or others in the area. The welding machine, welding leads, work clamp, electrode holder, and other equipment must be in safe working order.

Welding Technique: Using a 3/4-in. (19-mm) or larger gas nozzle for all welding, first tack weld the plates together according to the drawing. There should be about a 1/16-in. (1.6-mm) root gap between the plates with V-grooved or beveled edges. Use the E70S-X arc welding electrodes to make a root pass to fuse the plates together. Clean any silicon slag from the root pass, being sure to remove any trapped silicon slag along the sides of the weld.

Using the E70S-X arc welding electrodes, make a series of stringer or weave filler welds, no thicker than 1/4 in. (6.4 mm) in the groove until the joint is filled. The 1/4-in. (6.4-mm) fillet welds are to be made with one pass.

Interpass Temperature: The plate should not be heated to a temperature higher than 350°F (175°C) during the welding process. After each weld pass is completed, allow it to cool but never to a temperature below 50°F (10°C). The weldment must not be quenched in water.

Cleaning: Any slag must be cleaned off between passes. The weld beads may be cleaned by a hand wire brush, a chipping hammer, a punch and hammer, or a needle-scaler. All weld cleaning must be performed with the test plate in the welding position.

Inspection*: Visually inspect the weld for uniformity and discontinuities.

1. There shall be no cracks, no incomplete fusion.
2. There shall be no incomplete joint penetration in groove welds except as permitted for partial joint penetration welds.
3. The Test Supervisor shall examine the weld for acceptable appearance and shall be satisfied that the welder is skilled in using the process and procedure specified for the test.
4. Undercut shall not exceed the lesser of 10% of the base metal thickness or 1/32 in. (0.8 mm).
5. Where visual examination is the only criterion for acceptance, all weld passes are subject to visual examination at the discretion of the Test Supervisor.
6. The frequency of porosity shall not exceed one in each 4 in. (100 mm) of weld length, and the maximum diameter shall not exceed 3/32 in. (2.4 mm).
7. Welds shall be free from overlap.

Courtesy of the American Welding Society.

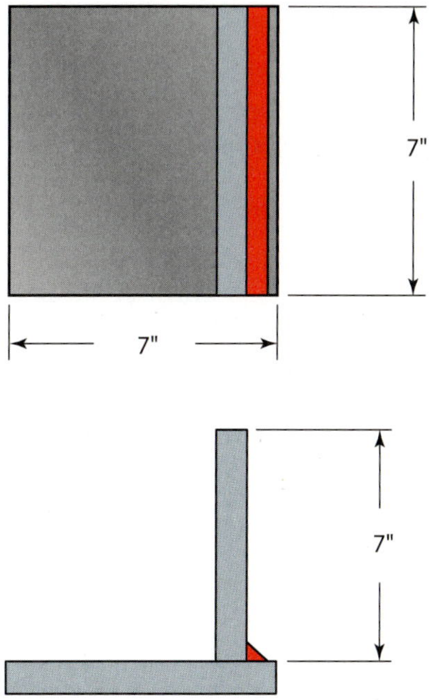

FIGURE 29-20 GMAW spray transfer fillet weld carbon steel workmanship sample. American Welding Society

Sketches: Gas Metal Arc Welding (GMAW) Spray Transfer Workmanship Sample drawing, Figure 29-20.

Paperwork: Complete a copy of the time sheet in Appendix I, the bill of materials in Appendix III, and the performance qualification test record in Appendix IV, or use forms as provided by your instructor.

PRACTICE 29-5

Gas Metal Arc Welding—Short-Circuit Metal Transfer (GMAW-S) Limited Thickness Welder Performance Qualification Test Plate All Positions without Backing

Welding Procedure Specification (WPS) No.: Practice 29-5.

Title: Welding GMAW-S of plate to plate.

Scope: This procedure is applicable for V-groove or single-bevel welds within the range of 1/8 in. (3.2 mm) through 3/4 in. (19 mm).

Welding May Be Performed in the Following Positions: All.

Base Metal: The base metal shall conform to carbon steel M-1, P-1, and S-1, Group 1 or 2.

Backing Material Specification: None.

Filler Metal: The filler metal shall conform to AWS specification no. E70S-X from AWS specification A5.18. This filler metal falls into F-number F-6 and A-number A-1.

Shielding Gas: The shielding gas, or gases, shall conform to the following compositions and purity: CO_2 at 30 to 50 cfh or 75% Ar/25% CO_2 at 30 to 50 cfh.

Joint Design and Tolerances: Refer to the drawing in Figure 29-21 for the joint layout specifications.

Preparation of Base Metal: The bevels are to be flame or plasma cut on the edges of the plate before the parts are assembled. The beveled surface must be smooth and free of notches. Any roughness or notches deeper than 1/64 in. (0.4 mm) must be ground smooth.

All hydrocarbons and other contaminants, such as cutting fluids, grease, oil, and primers, must be cleaned off all parts and filler metals before welding. This cleaning can be done with any suitable solvents or detergents. The groove face and inside and outside plate surface within 1 in. (25 mm) of the joint must be mechanically cleaned of slag, rust, and mill scale. Cleaning must be done with a wire brush or grinder down to bright metal.

Electrical Characteristics: Set the voltage, amperage, wire feed speed, and shielding gas flow according to Table 29-5.

Preheat: The parts must be heated to a temperature higher than 50°F (10°C) before any welding is started.

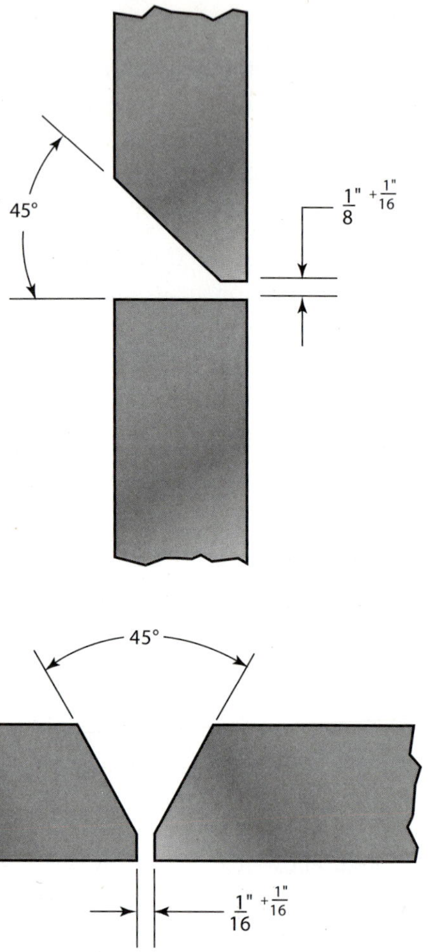

FIGURE 29-21 Practice 29-5 joint design.
American Welding Society

Backing Gas: N/A.

Safety: Proper protective clothing and equipment must be used. The area must be free of all hazards that may affect the welder or others in the area. The welding machine, welding leads, work clamp, electrode

Electrode		Welding Power			Shielding Gas		Base Metal	
Type	Size	Amps	Wire Feed Speed IPM (cm/min)	Volts	Type	Flow	Type	Thickness
E70S-X	0.035 in. (0.9 mm)	90 to 160	180 to 300 (457 to 762)	15 to 19	Ar plus 2% O_2 or 90% Ar/10% CO_2	30 to 50	Low carbon steel	1/4 in. to 1/2 in. (6 mm to 13 mm)
E70S-X	0.045 in. (1.2 mm)	130 to 200	125 to 200 (317 to 508)	17 to 19	Ar plus 2% O_2 or 90% Ar/10% CO_2	30 to 50	Low carbon steel	1/4 in. to 1/2 in. (6 mm to 13 mm)

Table 29-5 GMAW Spray Metal Transfer Machine Settings

holder, and other equipment must be in safe working order.

Welding Technique: Using a 1/2-in. (13-mm) or larger gas nozzle for all welding, tack weld the plates together according to the drawing. Using a small piece of metal across the ends of the plate as a spacer will make it easier to assemble. There should be about a 1/8-in. (3.2-mm) root gap between the plates with V-grooved or beveled edges and 1/16-in. (1.6-mm) root faces. Use the E70S-X filler wire to make a root pass to fuse the plates together. Clean any silicon slag from the root pass, being sure to remove any trapped silicon slag along the sides of the weld.

Using the E70S-X filler wire, make a series of stringer or weave filler welds, no thicker than 1/4 in. (6.4 mm) in the groove until the joint is filled. Note: The horizontal (2G) weldment should be made with stringer beads only.

Interpass Temperature: The plate should not be heated to a temperature higher than 350°F (175°C) during the welding process. After each weld pass is completed, allow it to cool but never to a temperature below 50°F (10°C). The weldment must not be quenched in water.

Cleaning: Any slag must be cleaned off between passes. The weld beads may be cleaned by a hand wire brush, a chipping hammer, a punch and hammer, or a needle-scaler. All weld cleaning must be performed with the test plate in the welding position.

Visual Inspection*: Visually inspect the weld for uniformity and discontinuities.

1. There shall be no cracks, no incomplete fusion.
2. There shall be no incomplete joint penetration in groove welds except as permitted for partial joint penetration welds.
3. The Test Supervisor shall examine the weld for acceptable appearance and shall be satisfied that the welder is skilled in using the process and procedure specified for the test.
4. Undercut shall not exceed the lesser of 10% of the base metal thickness or 1/32 in. (0.8 mm).
5. Where visual examination is the only criterion for acceptance, all weld passes are subject to visual examination at the discretion of the Test Supervisor.

Courtesy of the American Welding Society.

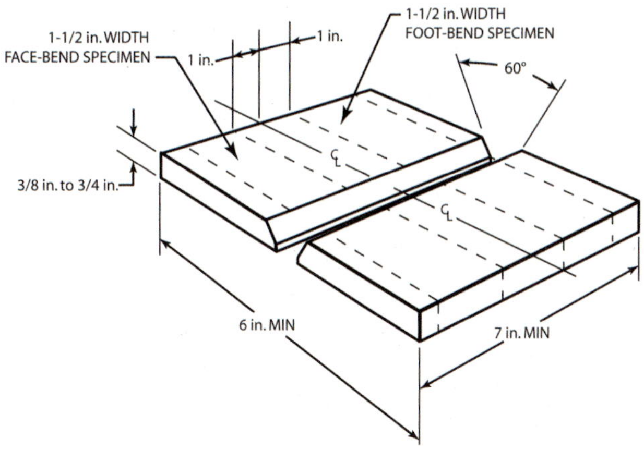

FIGURE 29-22 GMAW-S V-groove open root carbon steel workmanship sample. American Welding Society

6. The frequency of porosity shall not exceed one in each 4 in. (100 mm) of weld length, and the maximum diameter shall not exceed 3/32 in. (2.4 mm).
7. Welds shall be free from overlap.

Bend Test: The weld is to be mechanically tested only after it has passed the visual inspection. Be sure that the test specimens are properly marked to identify the welder, the position, and the process.

Specimen Preparation: For 3/8-in. (9.6-mm) test plates, two specimens are to be located in accordance with the requirements of Figure 29-22. One is to be prepared for a "transverse face bend," and the other is to be prepared for a "transverse root bend."

- *Transverse face bend.* The weld is perpendicular to the longitudinal axis of the specimen and is bent so that the weld face becomes the tension surface of the specimen.
- *Transverse root bend.* The weld is perpendicular to the longitudinal axis of the specimen and is bent so that the weld root becomes the tension surface of the specimen.

Acceptance Criteria for Face and Root Bends*: For acceptance, the convex surface of the face- and root-bend specimens shall meet both of the following requirements:

1. No single indication shall exceed 1/8 in. (3.2 mm) measured in any direction on the surface.
2. The sum of the greatest dimensions of all indications on the surface which exceed 1/32 in. (0.8 mm)

but are less than or equal to 1/8 in. (3.2 mm) shall not exceed 3/8 in. (9.6 mm).

Cracks occurring at the corner of the specimens shall not be considered unless there is definite evidence that they result from slag inclusion or other internal discontinuities.

Sketches: Gas Metal Arc Welding Short-Circuit Metal Transfer (GMAW-S) Workmanship Sample drawing, Figure 29-22.

Paperwork: Complete a copy of the time sheet in Appendix I, the bill of materials in Appendix III, and the performance qualification test record in Appendix IV, or use forms as provided by your instructor.♦

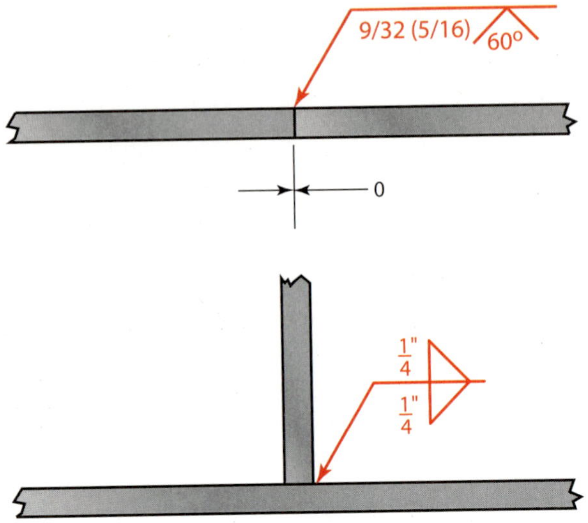

FIGURE 29-23 Practice 29-6 joint design.
American Welding Society

PRACTICE 29-6

Gas Metal Arc Welding (GMAW) Spray Transfer Workmanship Sample

Welding Procedure Specification (WPS) No.: Practice (29-6).

Title: Welding GMAW of plate to plate.

Scope: This procedure is applicable for V-groove and fillet welds.

Welding May Be Performed in the Following Positions: 1G and 2F.

Base Metal: The base metal shall conform to carbon steel M-1, P-1, and S-1, Group 1 or 2.

Backing Material Specification: None.

Filler Metal: The filler metal shall conform to AWS specification no. E70S-X for 0.035 to 0.045 diameter as listed in AWS specification A5.18. This filler metal falls into F-number F-6 and A-number A-1.

Shielding Gas: The shielding gas, or gases, shall conform to the following compositions and purity: 98% Ar/2% O_2 or 90% Ar/10%CO_2.

Joint Design and Tolerances: Refer to the drawing in Figure 29-23 for the joint layout specifications.

Preparation of Base Metal: The bevels are to be flame or plasma cut on the edges of the plate before the parts are assembled. The beveled surface must be smooth and free of notches. Any roughness or notches deeper than 1/64 in. (0.4 mm) must be ground smooth.

All hydrocarbons and other contaminants, such as cutting fluids, grease, oil, and primers, must be cleaned off all parts and filler metals before welding. This cleaning can be done with any suitable solvents or detergents. The groove face and inside and outside plate surface within 1 in. (25 mm) of the joint must be mechanically cleaned of slag, rust, and mill scale. Cleaning must be done with a wire brush or grinder down to bright metal.

Electrical Characteristics: Set the voltage, amperage, wire feed speed, and shielding gas flow according to Table 29-6.

Preheat: The parts must be heated to a temperature higher than 50°F (10°C) before any welding is started.

Backing Gas: N/A.

Safety: Proper protective clothing and equipment must be used. The area must be free of all hazards that may affect the welder or others in the area. The welding machine, welding leads, work clamp, electrode holder, and other equipment must be in safe working order.

Welding Technique: Using a 3/4-in. (19-mm) or larger gas nozzle for all welding, first tack weld the plates together, Figure 29-24. Use the E70S-X arc welding electrodes to make the welds.

Using the E70S-X arc welding electrodes, make a series of stringer filler welds, no thicker than 1/4 in. (6.4 mm), in the groove until the joint is filled.

Electrode		Welding Power			Shielding Gas		Base Metal	
Type	Size	Amps	Wire Feed Speed IPM (cm/min)	Volts	Type	Flow	Type	Thickness
E70S-X	0.035 in. (0.9 mm)	180 to 230	400 to 550 (1016 to 1397)	25 to 27	Ar plus 2% O_2 or 90% Ar/10% CO_2	30 to 50	Low carbon steel	1/4 in. to 1/2 in. (6 mm to 13 mm)
E70S-X	0.045 in. (1.2 mm)	260 to 340	300 to 500 (762 to 1270)	25 to 30	Ar plus 2% O_2 or 90% Ar/10% CO_2	30 to 50	Low carbon steel	1/4 in. to 1/2 in. (6 mm to 13 mm)

Table 29-6 GMAW Spray Metal Transfer Machine Settings

Interpass Temperature: The plate should not be heated to a temperature higher than 350°F (175°C) during the welding process. After each weld pass is completed, allow it to cool but never to a temperature below 50°F (10°C). The weldment must not be quenched in water.

Cleaning: Any slag must be cleaned off between passes. The weld beads may be cleaned by a hand wire brush, a chipping hammer, a punch and hammer, or a needle-scaler. All weld cleaning must be performed with the test plate in the welding position.

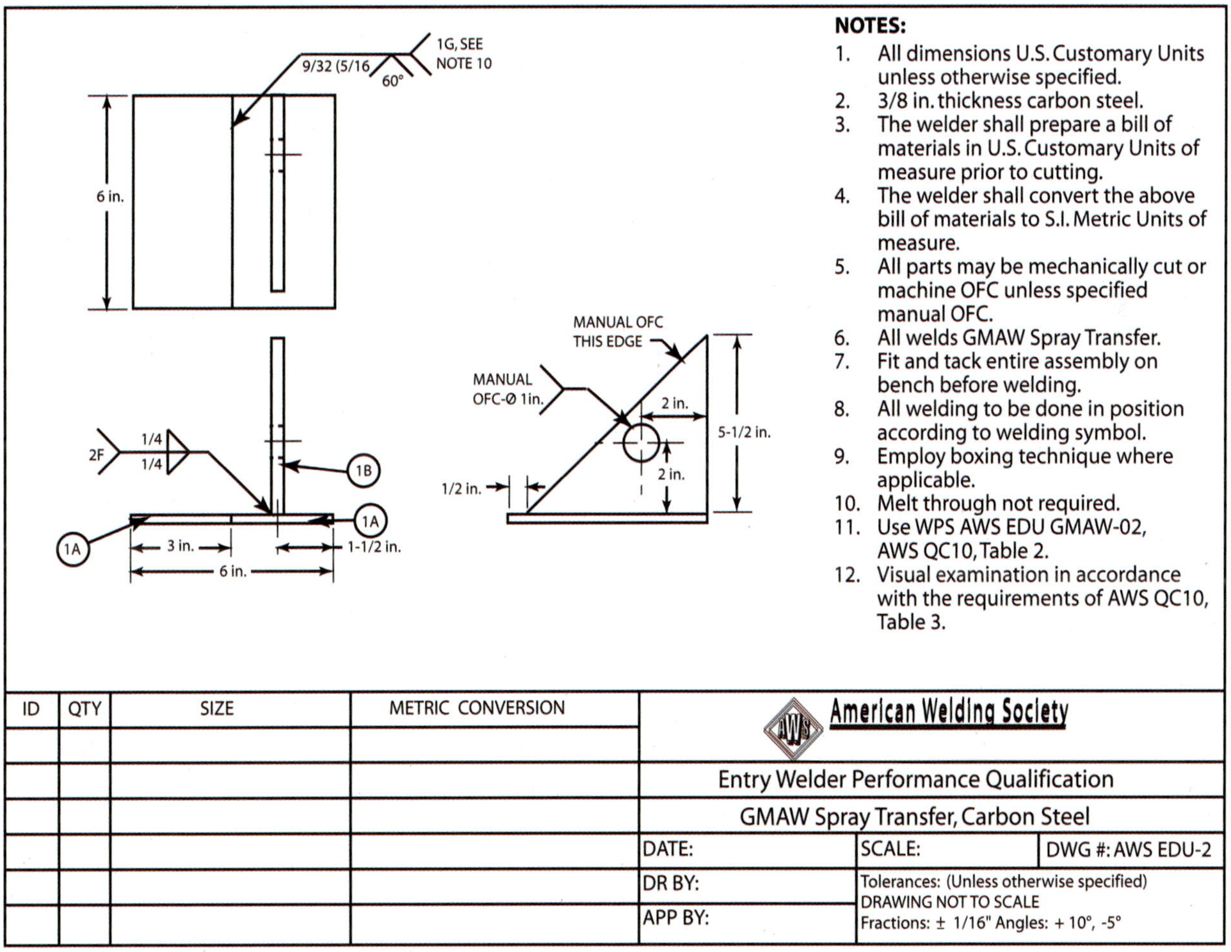

FIGURE 29-24 AWS EDU-2 GMAW spray transfer carbon steel workmanship sample. American Welding Society

Visual Inspection*: Visually inspect the weld for uniformity and discontinuities.

1. There shall be no cracks, no incomplete fusion.
2. There shall be no incomplete joint penetration in groove welds except as permitted for partial joint penetration welds.
3. The Test Supervisor shall examine the weld for acceptable appearance and shall be satisfied that the welder is skilled in using the process and procedure specified for the test.
4. Undercut shall not exceed the lesser of 10% of the base metal thickness or 1/32 in. (0.8 mm).
5. Where visual examination is the only criterion for acceptance, all weld passes are subject to visual examination at the discretion of the Test Supervisor.
6. The frequency of porosity shall not exceed one in each 4 in. (100 mm) of weld length, and the maximum diameter shall not exceed 3/32 in. (2.4 mm).
7. Welds shall be free from overlap.

Sketches: Gas Metal Arc Welding Spray Transfer (GMAW) Workmanship Sample drawing, Figure 29-25.

Paperwork: Complete a copy of the time sheet in Appendix I, the bill of materials in Appendix III, and the performance qualification test record in Appendix IV, or use forms as provided by your instructor.◆

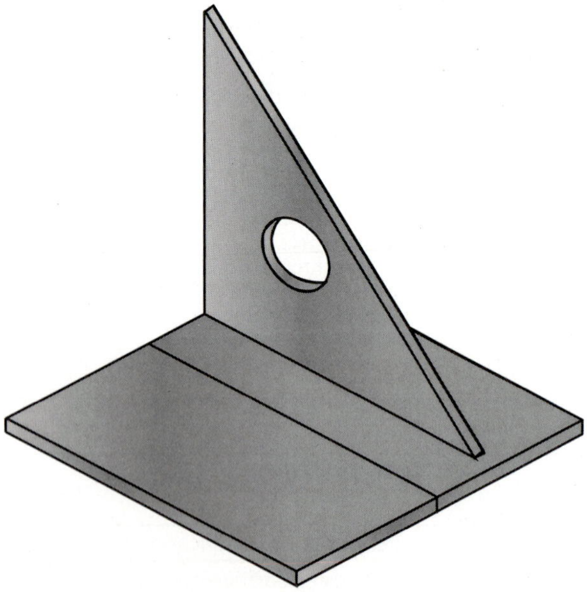

FIGURE 29-25 Practice 29-6 workmanship sample.
American Welding Society

*Courtesy of the American Welding Society.

PRACTICE 29-7

AWS SENSE Entry-Level Welder Workmanship Sample for Flux Cored Arc Welding (FCAW), Gas-Shielded

Welding Procedure Specification (WPS) No.: Practice 29-7.

Title: Welding FCAW of plate to plate.

Scope: This procedure is applicable for V-groove, bevel, and fillet welds within the range of 1/8 in. (3.2 mm) through 1 1/2 in. (38 mm).

Welding May Be Performed in the Following Positions: All.

Base Metal: The base metal shall conform to carbon steel M-1, P-1, and S-1, Group 1 or 2.

Backing Material Specification: None.

Filler Metal: The filler metal shall conform to AWS specification no. E71T-1 from AWS specification A5.20. This filler metal falls into F-number F-6 and A-number A-1.

Shielding Gas: The shielding gas, or gases, shall conform to the following compositions and purity: CO_2 at 30 to 50 cfh or 75% Ar/25% CO_2 at 30 to 50 cfh.

Joint Design and Tolerances: Refer to the drawing and specifications in Figure 29-26 for the workmanship sample layout.

Preparation of Base Metal: The bevels are to be flame or plasma cut on the edges of the plate before the parts are assembled. The beveled surface must be smooth and free of notches. Any roughness or notches deeper than 1/64 in. (0.4 mm) must be ground smooth.

All hydrocarbons and other contaminants, such as cutting fluids, grease, oil, and primers, must be cleaned off all parts and filler metals before welding. This cleaning can be done with any suitable solvents or detergents. The groove face and inside and outside plate surface within 1 in. (25 mm) of the joint must be mechanically cleaned of slag, rust, and mill scale. Cleaning must be done with a wire brush or grinder down to bright metal.

Electrical Characteristics: Set the voltage, amperage, wire feed speed, and shielding gas flow according to Table 29-7.

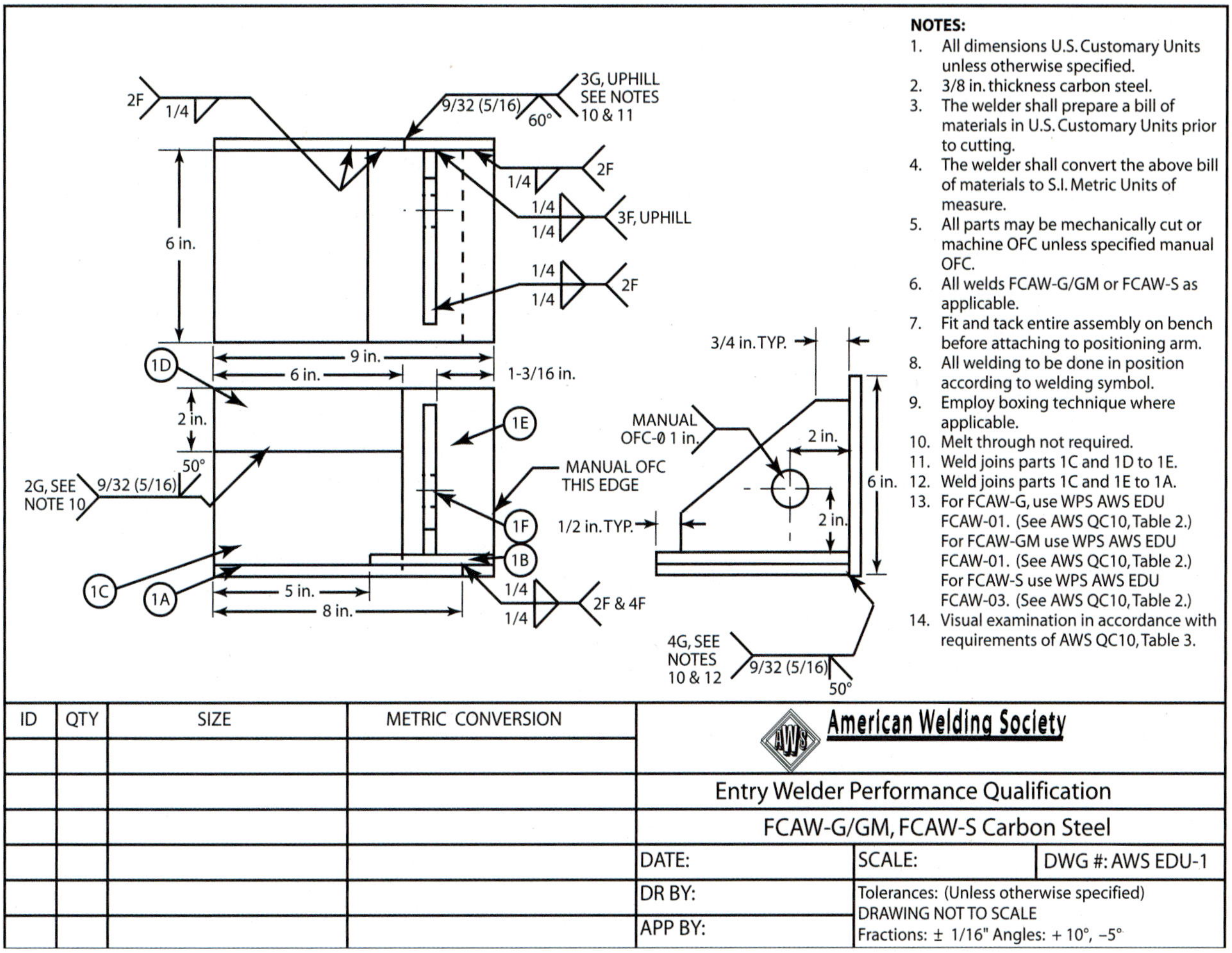

FIGURE 29-26 AWS EDU-1 FCAW carbon steel workmanship sample. American Welding Society

Preheat: The parts must be heated to a temperature higher than 50°F (10°C) before any welding is started.

Backing Gas: N/A.

Safety: Proper protective clothing and equipment must be used. The area must be free of all hazards that may affect the welder or others in the area. The welding machine, welding leads, work clamp, electrode holder, and other equipment must be in safe working order.

Welding Technique: Using a 1/2-in. (13-mm) or larger gas nozzle and a distance from contact tube to work of approximately 3/4 in. (19 mm) for all welding, first

Electrode		Welding Power			Shielding Gas		Base Metal	
Type	Size	Amps	Wire Feed Speed IPM (cm/min)	Volts	Type	Flow	Type	Thickness
E71T-1	0.035 in. (0.9 mm)	130 to 150	288 to 380 (732 to 975)	22 to 25	CO_2 or 75% Ar/25% CO_2	30 to 50	Low carbon steel	1/4 in. to 1/2 in. (6 mm to 13 mm)
E71T-1	0.045 in. (1.2 mm)	150 to 210	200 to 300 (508 to 762)	28 to 29	CO_2 or 75% Ar/25% CO_2	30 to 50	Low carbon steel	1/4 in. to 1/2 in. (6 mm to 13 mm)

Table 29-7 FCAW Gas-Shielded Machine Settings

tack weld the plates together according to Figure 29-26. There should be a root gap of about 1/8 in. (3.2 mm) between the plates with V-grooved or beveled edges. Use an E71T-1 arc welding electrode to make a weld. If multiple pass welds are going to be made, a root pass weld should be made to fuse the plates together. Clean the slag from the root pass, being sure to remove any trapped slag along the sides of the weld.

Using an E71T-1 arc welding electrode, make a series of stringer or weave filler welds, no thicker than 1/4 in. (6.4 mm) in the groove until the joint is filled. The 1/4-in. (6.4-mm) fillet welds are to be made with one pass.

Interpass Temperature: The plate should not be heated to a temperature higher than 350°F (175°C) during the welding process. After each weld pass is completed, allow it to cool but never to a temperature below 50°F (10°C). The weldment must not be quenched in water.

Cleaning: The slag must be cleaned off between passes. The weld beads may be cleaned by a hand wire brush, a chipping hammer, a punch and hammer, or a needle-scaler. All weld cleaning must be performed with the test plate in the welding position. A grinder may not be used to remove weld control problems such as undercut, overlap, or trapped slag.

Inspection: Visually inspect the weld for uniformity and discontinuities. There shall be no cracks, no incomplete fusion, and no overlap. Undercut shall not exceed the lesser of 10% of the base metal thickness or 1/32 in. (0.8 mm). The frequency of porosity shall not exceed one in each 4 in. (100 mm) of weld length, and the maximum diameter shall not exceed 3/32 in. (2.4 mm).

Sketches: Flux Cored Arc Welding (FCAW) Gas-Shielded Workmanship Sample drawing, Figure 29-27.

Paperwork: Complete a copy of the time sheet in Appendix I, the bill of materials in Appendix III, and the performance qualification test record in Appendix IV, or use forms as provided by your instructor.◆

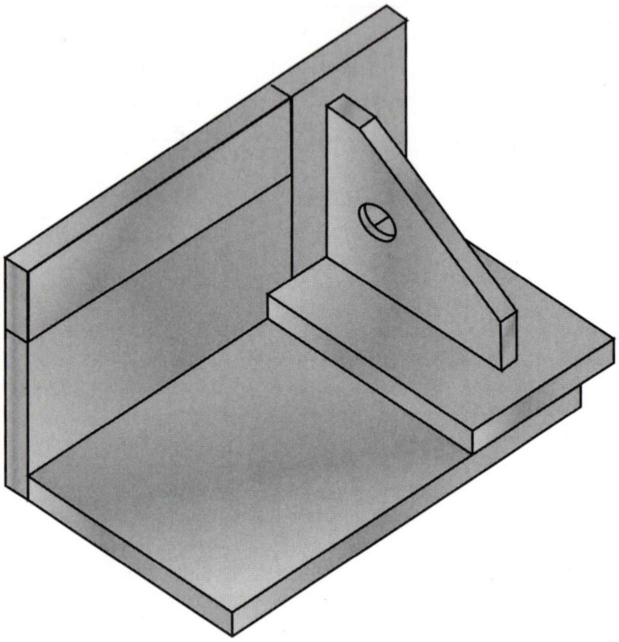

FIGURE 29-27 Practice 29-7 workmanship sample.
American Welding Society

PRACTICE 29-8

AWS SENSE Entry-Level Welder Workmanship Sample for Flux Cored Arc Welding (FCAW) Self-Shielded

Welding Procedure Specification (WPS) No.: Practice 29-8.

Title: Welding FCAW of plate to plate.

Scope: This procedure is applicable for V-groove, bevel, and fillet welds within the range of 1/8 in. (3.2 mm) through 1 1/2 in. (38 mm).

Welding May Be Performed in the Following Positions: All.

Base Metal: The base metal shall conform to carbon steel M-1, P-1, and S-1, Group 1 or 2.

Backing Material Specification: None.

Filler Metal: The filler metal shall conform to AWS specification no. E71T-11 for 0.035 to 0.0415 diameter, as listed in AWS specification A5.20. This filler metal falls into F-number F-6 and A-number A-1.

Shielding Gas: None.

Joint Design and Tolerances: Refer to the drawing and specifications in Figure 29-28 for the workmanship sample layout.

Preparation of Base Metal: The bevels are to be flame or plasma cut on the edges of the plate before the parts are assembled. The beveled surface must be smooth and free of notches. Any roughness or notches deeper than 1/64 in. (0.4 mm) must be ground smooth.

All hydrocarbons and other contaminants, such as cutting fluids, grease, oil, and primers, must be cleaned off all parts and filler metals before welding.

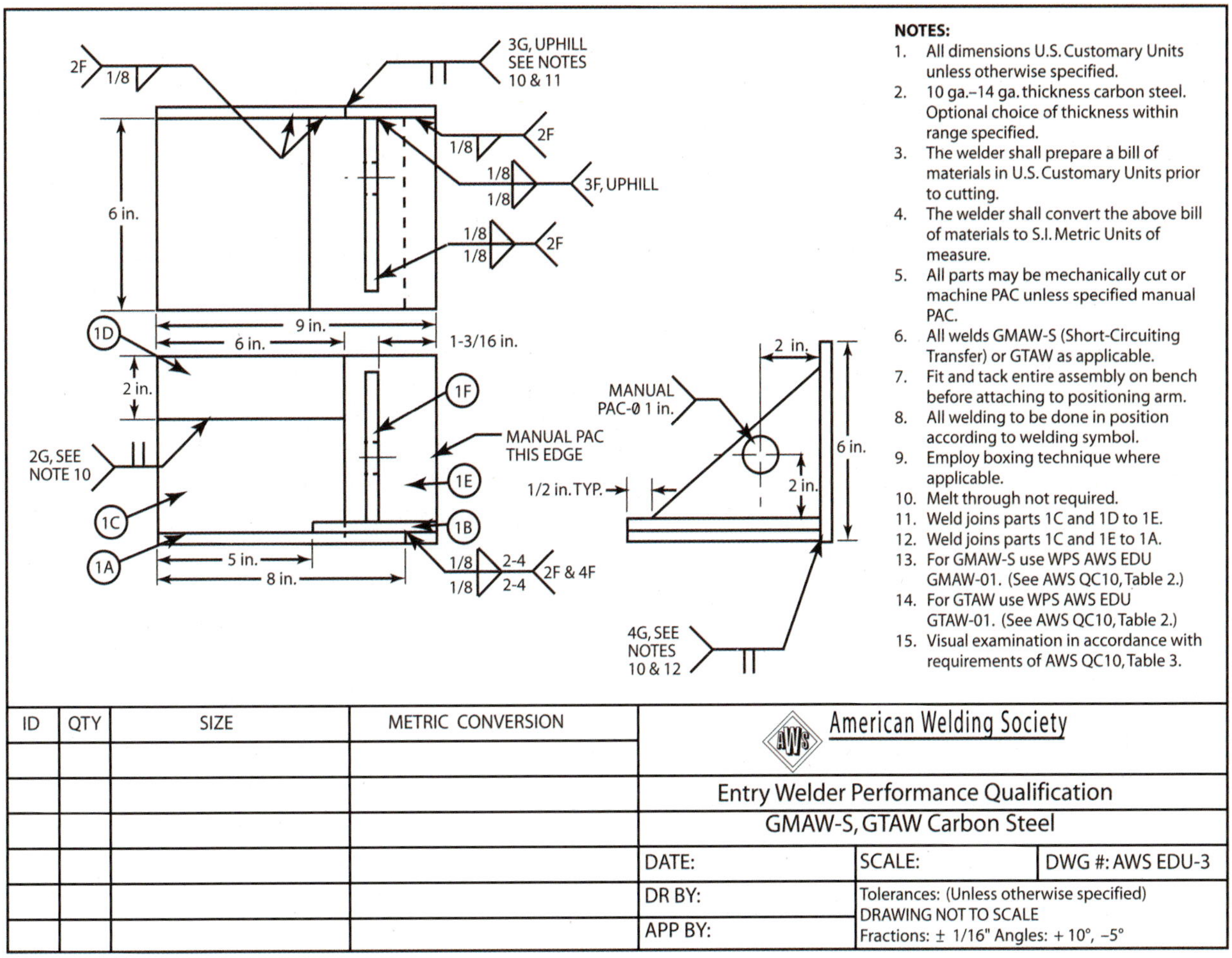

FIGURE 29-28 AWS EDU-3 FCAW carbon steel workmanship sample. American Welding Society

This cleaning can be done with any suitable solvents or detergents. The groove face and inside and outside plate surface within 1 in. (25 mm) of the joint must be mechanically cleaned of slag, rust, and mill scale. Cleaning must be done with a wire brush or grinder down to bright metal.

Electrical Characteristics: Set the voltage, amperage, and wire feed speed flow according to Table 29-8.

Preheat: The parts must be heated to a temperature higher than 50°F (10°C) before any welding is started.

Backing Gas: N/A.

Safety: Proper protective clothing and equipment must be used. The area must be free of all hazards that may affect the welder or others in the area. The welding machine, welding leads, work clamp, electrode holder, and other equipment must be in safe working order.

ELECTRODE TYPE	DIAMETER	VOLTS	AMPS	WIRE FEED SPEED IPM (cm/min)	ELECTRODE STICKOUT (INCH)
SELF-SHIELD E70T-11 or E71T-11	0.030	15	40	69 (175)	3/8
		16	100	175 (445)	3/8
		16	160	440 (1118)	3/8
	0.035	15	80	81 (206)	3/8
		17	120	155 (394)	3/8
		17	200	392 (996)	3/8
	0.045	15	95	54 (137)	1/2
		17	150	118 (300)	1/2
		18	225	140 (356)	1/2

Table 29-8 FCAW Self-Shielded Machine Settings

Welding Technique: Using a 1/2-in. (13-mm) or larger gas nozzle and a distance from the contact tube to work of approximately 3/4 in. (19 mm) for all welding, first tack weld the plates together according to Figure 29-28. There should be a root gap of about 1/8 in. (3.2 mm) between the plates with V-grooved or beveled edges. Use an E71T-11 arc welding electrode to make a weld. If multiple pass welds are going to be made, a root pass weld should be made to fuse the plates together. Clean the slag from the root pass, being sure to remove any trapped slag along the sides of the weld.

Using an E71T-11 arc welding electrode, make a series of stringer or weave filler welds, no thicker than 1/4 in. (6.4 mm) in the groove until the joint is filled. The 1/4-in. (6.4-mm) fillet welds are to be made with one pass.

Interpass Temperature: The plate should not be heated to a temperature higher than 350°F (175°C) during the welding process. After each weld pass is completed, allow it to cool but never to a temperature below 50°F (10°C). The weldment must not be quenched in water.

Cleaning: The slag must be cleaned off between passes. The weld beads may be cleaned by a hand wire brush, a chipping hammer, a punch and hammer, or a needle-scaler. All weld cleaning must be performed with the test plate in the welding position. A grinder may not be used to remove weld control problems such as undercut, overlap, or trapped slag.

Inspection: Visually inspect the weld for uniformity and discontinuities. There shall be no cracks, no incomplete fusion, and no overlap. Undercut shall not exceed the lesser of 10% of the base metal thickness or 1/32 in. (0.8 mm). The frequency of porosity shall not exceed one in each 4 in. (100 mm) of weld length, and the maximum diameter shall not exceed 3/32 in. (2.4 mm).

Sketches: Flux Cored Arc Welding (FCAW) Self-Shielded Workmanship Sample drawing, Figure 29-29.

Paperwork: Complete a copy of the time sheet in Appendix I, the bill of materials in Appendix III, and the performance qualification test record in Appendix IV, or use forms as provided by your instructor.◆

PRACTICE 29-9

Gas Tungsten Arc Welding (GTAW) on Plain Carbon Steel Workmanship Sample

Welding Procedure Specification (WPS) No.: Practice 29-9.

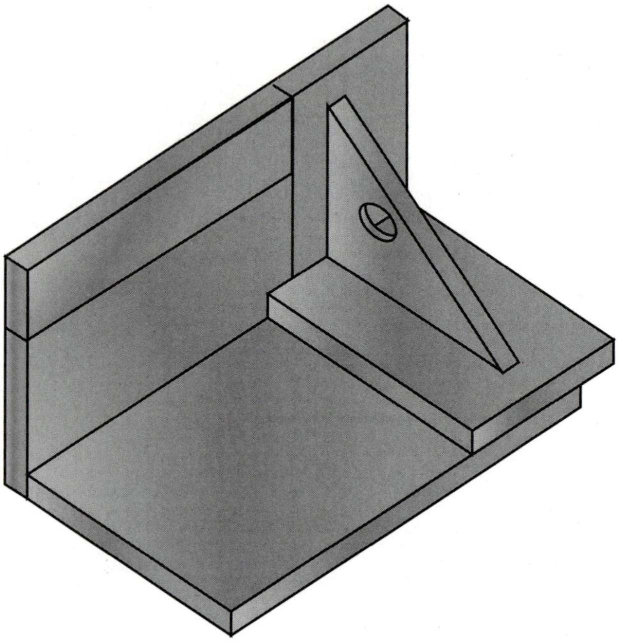

FIGURE 29-29 Practice 29-8 workmanship sample.
American Welding Society

Title: Welding GTAW of sheet to sheet.

Scope: This procedure is applicable for square groove and fillet welds within the range of 18 gauge through 10 gauge.

Welding May Be Performed in the Following Positions: 1G and 2F.

Base Metal: The base metal shall conform to carbon steel M-1, Group 1.

Backing Material Specification: None.

Filler Metal: The filler metal shall conform to AWS specification no. E70S-3 for 1/16 in. (1.6 mm) to 3/32 in. (2.4 mm) diameter, as listed in AWS specification A5.18. This filler metal falls into F-number F-6 and A-number A-1.

Electrode: The tungsten electrode shall conform to AWS specification no. EWTh-2, EWCe-2, or EWLa from AWS specification A5.12. The tungsten diameter shall be 1/8 in. (3.2 mm) maximum.

The tungsten end shape shall be tapered at two to three times its length to its diameter.

Shielding Gas: The shielding gas, or gases, shall conform to the following compositions and purity: welding grade argon.

Joint Design and Tolerances: Refer to the drawing and specifications in Figure 29-30 for the workmanship sample layout.

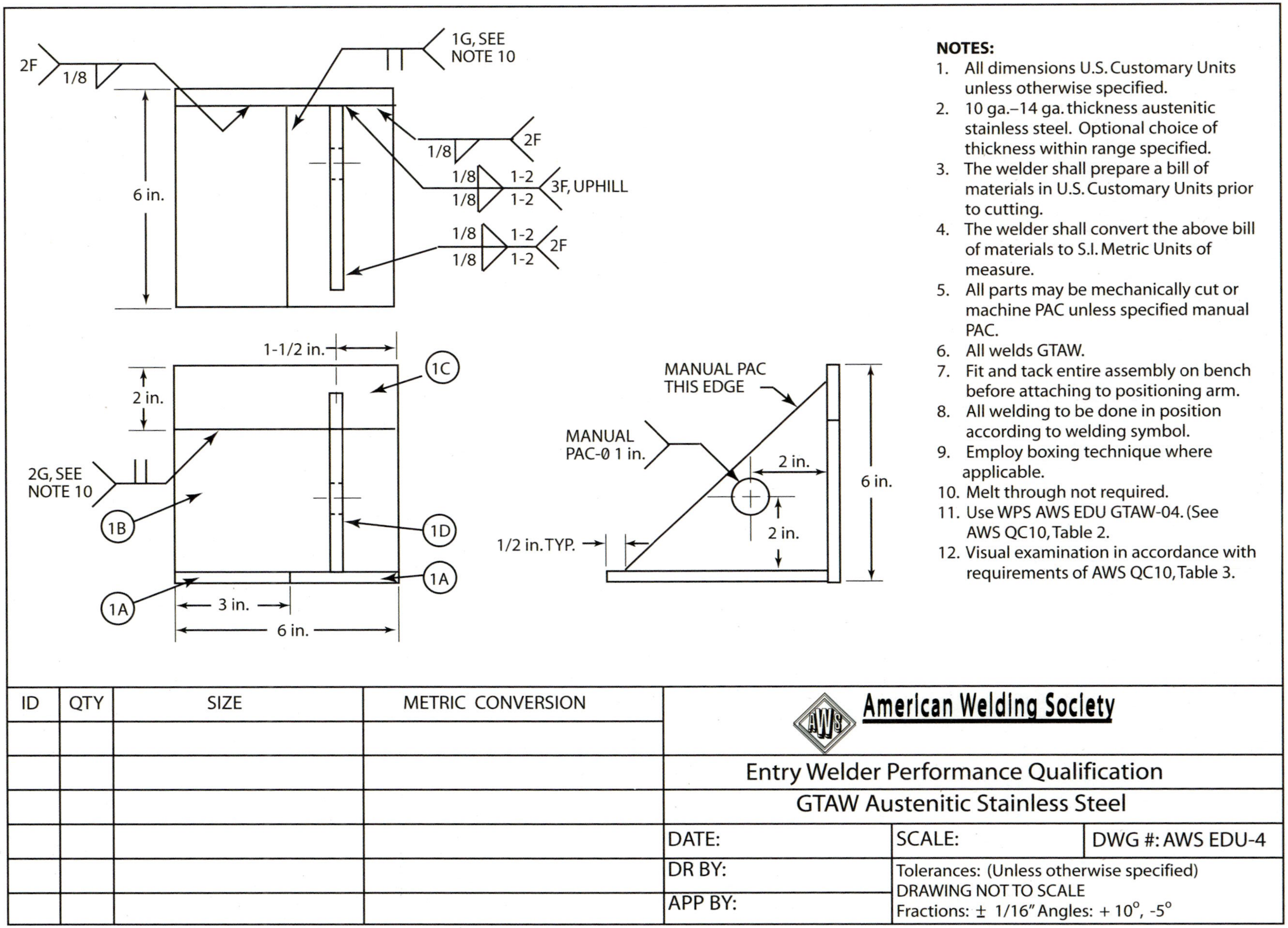

FIGURE 29-30 AWS EDU-4 GTAW carbon steel workmanship sample. American Welding Society

Metal Specifications		Gas Flow			Nozzle Size in. (mm)	Amperage Min. Max.
Thickness	Diameter of E70S-3*	Rates cfm (L/min)	Preflow Times	Postflow Times		
18 ga	1/16 in. (2 mm)	15 to 20 (7 to 9)	10 to 15 sec.	10 to 25 sec.	1/4 to 3/8 (6 to 10)	45 to 65
17 ga	1/16 in. (2 mm)	15 to 20 (7 to 9)	10 to 15 sec.	10 to 25 sec.	1/4 to 3/8 (6 to 10)	45 to 70
16 ga	1/16 in. (2 mm)	15 to 20 (7 to 9)	10 to 15 sec.	10 to 25 sec.	1/4 to 3/8 (6 to 10)	50 to 75
15 ga	1/16 in. (2 mm)	15 to 20 (7 to 9)	10 to 15 sec.	10 to 25 sec.	1/4 to 3/8 (6 to 10)	55 to 80
14 ga	3.32 in. (2.4 mm)	20 to 25 (9 to 12)	10 to 20 sec.	10 to 30 sec.	3/8 to 5/8 (10 to 16)	60 to 90
13 ga	3.32 in. (2.4 mm)	20 to 25 (9 to 12)	10 to 20 sec.	10 to 30 sec.	3/8 to 5/8 (10 to 16)	60 to 100
12 ga	3.32 in. (2.4 mm)	20 to 25 (9 to 12)	10 to 20 sec.	10 to 30 sec.	3/8 to 5/8 (10 to 16)	60 to 110
11 ga	3.32 in. (2.4 mm)	20 to 25 (9 to 12)	10 to 20 sec.	10 to 30 sec.	3/8 to 5/8 (10 to 16)	65 to 120
10 ga	3.32 in. (2.4 mm)	20 to 25 (9 to 12)	10 to 20 sec.	10 to 30 sec.	3/8 to 5/8 (10 to 16)	70 to 130

*Other E70S-X filler metal may be used.

Table 29-9 GTAW Carbon Steel Machine Settings

Preparation of Base Metal: All hydrocarbons and other contaminants, such as cutting fluids, grease, oil, and primers, must be cleaned off all parts and filler metals before welding. This cleaning can be done with any suitable solvents or detergents. The joint face and inside and outside plate surface within 1 in. (25 mm) of the joint must be mechanically cleaned of slag, rust, and mill scale. Cleaning must be done with a wire brush or grinder down to bright metal.

Electrical Characteristics: Set the welding current to DCEN and the amperage and shielding gas flow according to Table 29-9.

Preheat: The parts must be heated to a temperature higher than 50°F (10°C) before any welding is started.

Backing Gas: None.

Safety: Proper protective clothing and equipment must be used. The area must be free of all hazards that may affect the welder or others in the area. The welding machine, welding leads, work clamp, electrode holder, and other equipment must be in safe working order.

Welding Technique: TACK WELDS: With the parts securely clamped in place with the correct root gap, the tack welds are to be performed. Holding the electrode so that it is very close to the root face but not touching, slowly increase the current until the arc starts and a molten weld pool is formed. Add filler metal as required to maintain a slightly convex weld face and a flat or slightly concave root face. When it is time to end the tack weld, lower the current slowly so that the molten weld pool can be tapered down in size. When all tack welds are complete, allow the parts to cool as needed before assembling the remaining parts. Repeat the tack welding procedure until the entire part is assembled.

Square Groove and Fillet Welds: Holding the electrode so that it is very close to the metal surface but not touching, slowly increase the current until the arc starts and a molten weld pool is formed. As the weld progresses, add filler metal as required to maintain a flat or slightly convex weld face. If it is necessary to stop the weld or to reposition yourself or if the weld is completed, the current must be lowered slowly so that the molten weld pool can be tapered down in size.

Interpass Temperature: The plate should not be heated to a temperature higher than 120°F (49°C) during the welding process. After each weld pass is completed, allow it to cool but never to a temperature below 50°F (10°C). The weldment must not be quenched in water.

Cleaning: Recleaning may be required if the parts or filler metal becomes contaminated or reoxidized to a degree that the weld quality will be affected. Reclean using the same procedure used for the original metal preparation.

Visual Inspection: Visual inspection criteria for entry-level welders*:

1. There shall be no cracks, no incomplete fusion.
2. There shall be no incomplete joint penetration in groove welds except as permitted for partial joint penetration groove welds.
3. The Test Supervisor shall examine the weld for acceptable appearance and shall be satisfied that the welder is skilled in using the process and procedure specified for the test.
4. Undercut shall not exceed the lesser of 10% of the base metal thickness or 1/32 in. (0.8 mm).

**Courtesy of the American Welding Society.*

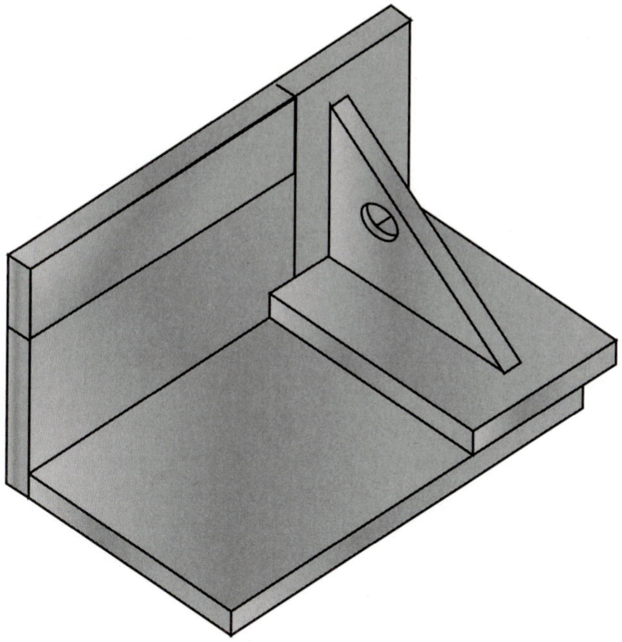

FIGURE 29-31 Practice 29-9 workmanship sample.
American Welding Society

5. Where visual examination is the only criterion for acceptance, all weld passes are subject to visual examination at the discretion of the Test Supervisor.
6. The frequency of porosity shall not exceed one in each 4 in. (100 mm) of weld length, and the maximum diameter shall not exceed 3/32 in. (2.4 mm).
7. Welds shall be free from overlap.

Sketches: Gas Tungsten Arc Welding (GTAW) Workmanship Sample drawing for Carbon Steel, Figure 29-31.

Paperwork: Complete a copy of the time sheet in Appendix I, the bill of materials in Appendix III, and the performance qualification test record in Appendix IV, or use forms as provided by your instructor.◆

PRACTICE 29-10

Gas Tungsten Arc Welding (GTAW) on Stainless Steel Workmanship Sample

Welding Procedure Specification (WPS) No.: Practice 29-10.

Title: Welding GTAW of sheet to sheet.

Scope: This procedure is applicable for square groove and fillet welds within the range of 18 gauge through 10 gauge.

Welding May Be Performed in the Following Positions: 1G and 2F.

Base Metal: The base metal shall conform to austenitic stainless steel M-8 or P-8.

Backing Material Specification: None.

Filler Metal: The filler metal shall conform to AWS specification no. ER3XX from AWS specification A5.9. This filler metal falls into F-number F-6 and A-number A-8.

Electrode: The tungsten electrode shall conform to AWS specification no. EWTh-2, EWCe-2, or EWLa from AWS specification A5.12. The tungsten diameter shall be 1/8 in. (3.2 mm) maximum. The tungsten end shape shall be tapered at two to three times its length to its diameter.

Shielding Gas: The shielding gas, or gases, shall conform to the following compositions and purity: welding grade argon.

Joint Design and Tolerances: Refer to the drawing and specifications in Figure 29-32 for the workmanship sample layout.

Preparation of Base Metal: All hydrocarbons and other contaminants, such as cutting fluids, grease, oil, and primers, must be cleaned off all parts and filler metals before welding. This cleaning can be done with any suitable solvents or detergents. The joint face and inside and outside plate surface within 1 in. (25 mm) of the joint must be cleaned of slag, oxide, and scale. Cleaning can be mechanical or chemical. Mechanical metal cleaning can be done by grinding, stainless steel wire brushing, scraping, machining, or filing. Chemical cleaning can be done by using acids, alkalies, solvents, or detergents. Cleaning must be done down to bright metal.

Electrical Characteristics: Set the welding current to DCEN and the amperage and shielding gas flow according to Table 29-10.

Preheat: The parts must be heated to a temperature higher than 50°F (10°C) before any welding is started.

Backing Gas: None.

Safety: Proper protective clothing and equipment must be used. The area must be free of all hazards that may affect the welder or others in the area. The welding machine, welding leads, work clamp, electrode holder, and other equipment must be in safe working order.

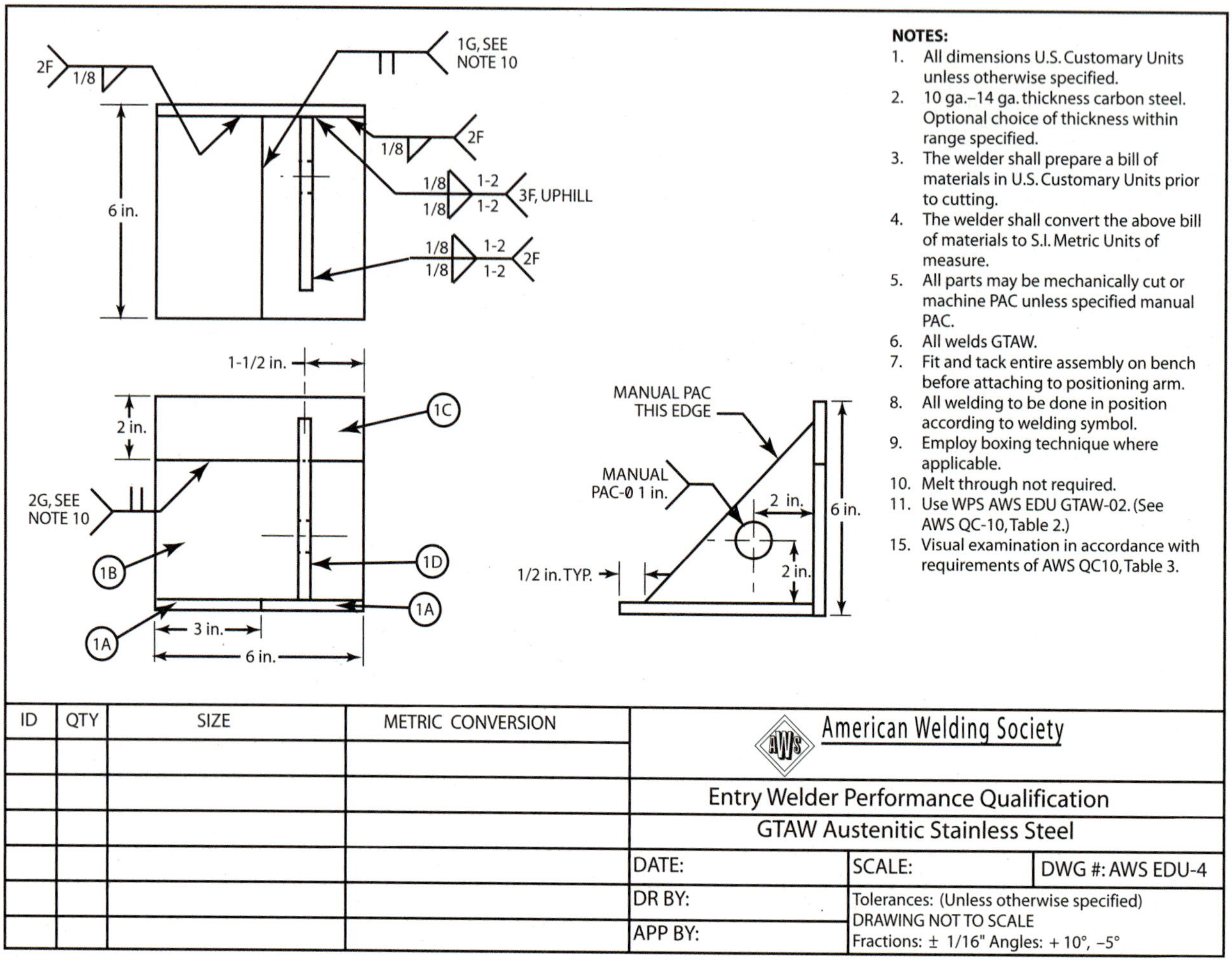

FIGURE 29-32 AWS EDU-4 GTAW austenitic stainless steel workmanship sample. American Welding Society

Welding Technique: TACK WELDS: With the parts securely clamped in place with the correct root gap, the tack welds are to be performed. Holding the electrode so that it is very close to the root face but not touching, slowly increase the current until the arc starts and a molten weld pool is formed. Add filler metal as required to maintain a slightly convex weld face and a flat or slightly concave root face. When it is time to end the tack weld, lower the current slowly so that the molten weld pool can be tapered down

Metal Specifications		Gas Flow			Nozzle Size in. (mm)	Amperage Min. Max.
Thickness	Diameter of ER3XX*	Rates cfm (L/min)	Preflow Times	Postflow Times		
18 ga	1/16 in. (2 mm)	15 to 20 (7 to 9)	10 to 15 sec.	10 to 25 sec.	1/4 to 3/8 (6 to 10)	35 to 60
17 ga	1/16 in. (2 mm)	15 to 20 (7 to 9)	10 to 15 sec.	10 to 25 sec.	1/4 to 3/8 (6 to 10)	40 to 65
16 ga	1/16 in. (2 mm)	15 to 20 (7 to 9)	10 to 15 sec.	10 to 25 sec.	1/4 to 3/8 (6 to 10)	40 to 75
15 ga	1/16 in. (2 mm)	15 to 20 (7 to 9)	10 to 15 sec.	10 to 25 sec.	1/4 to 3/8 (6 to 10)	50 to 80
14 ga	3.32 in. (2.4 mm)	20 to 25 (9 to 12)	10 to 20 sec.	10 to 30 sec.	3/8 to 5/8 (10 to 16)	50 to 90
13 ga	3.32 in. (2.4 mm)	20 to 25 (9 to 12)	10 to 20 sec.	10 to 30 sec.	3/8 to 5/8 (10 to 16)	55 to 100
12 ga	3.32 in. (2.4 mm)	20 to 25 (9 to 12)	10 to 20 sec.	10 to 30 sec.	3/8 to 5/8 (10 to 16)	60 to 110
11 ga	3.32 in. (2.4 mm)	20 to 25 (9 to 12)	10 to 20 sec.	10 to 30 sec.	3/8 to 5/8 (10 to 16)	65 to 120
10 ga	3.32 in. (2.4 mm)	20 to 25 (9 to 12)	10 to 20 sec.	10 to 30 sec.	3/8 to 5/8 (10 to 16)	70 to 130

*Other ER3XX stainless steel A5.9 filler metal may be used.

Table 29-10 GTAW Stainless Steel Machine Settings

in size. When all tack welds are complete, allow the parts to cool as needed before assembling the remaining parts. Repeat the tack welding procedure until the entire part is assembled.

Square Groove and Fillet Welds: Holding the electrode so that it is very close to the metal surface but not touching, slowly increase the current until the arc starts and a molten weld pool is formed. As the weld progresses, add filler metal as required to maintain a flat or slightly convex weld face. If it is necessary to stop the weld or to reposition yourself or if the weld is completed, the current must be lowered slowly so that the molten weld pool can be tapered down in size.

Interpass Temperature: The plate should not be heated to a temperature higher than 350°F (180°C) during the welding process. After each weld pass is completed, allow it to cool but never to a temperature below 50°F (10°C). The weldment must not be quenched in water.

Cleaning: Recleaning may be required if the parts or filler metal become contaminated or oxidized to a degree that the weld quality will be affected. Reclean using the same procedure used for the original metal preparation.

Visual Inspection: Visual inspection criteria for entry-level welders*:

1. There shall be no cracks, no incomplete fusion.
2. There shall be no incomplete joint penetration in groove welds except as permitted for partial joint penetration groove welds.
3. The Test Supervisor shall examine the weld for acceptable appearance and shall be satisfied that the welder is skilled in using the process and procedure specified for the test.
4. Undercut shall not exceed the lesser of 10% of the base metal thickness or 1/32 in. (0.8 mm).
5. Where visual examination is the only criterion for acceptance, all weld passes are subject to visual examination at the discretion of the Test Supervisor.
6. The frequency of porosity shall not exceed one in each 4 in. (100 mm) of weld length, and the maximum diameter shall not exceed 3/32 in. (2.4 mm).
7. Welds shall be free from overlap.

Courtesy of the American Welding Society.

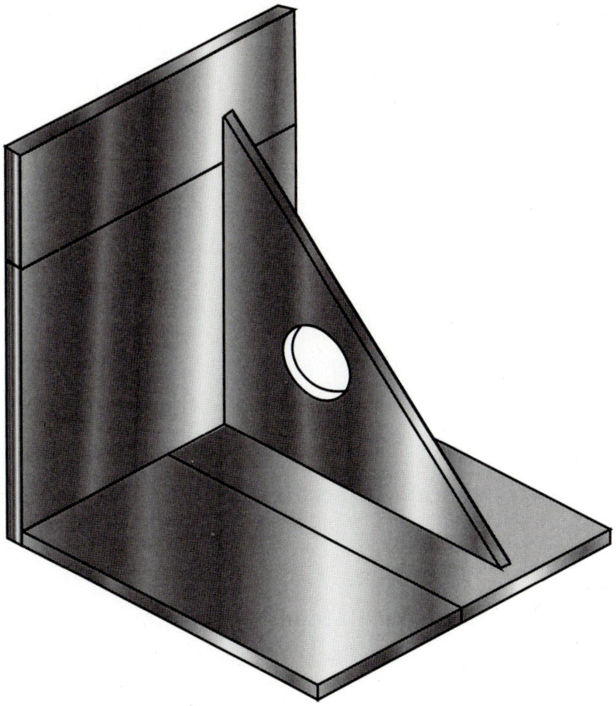

FIGURE 29-33 Practice 29-10 workmanship sample.
American Welding Society

Sketches: Gas Tungsten Arc Welding (GTAW) Workmanship Sample drawing for Stainless Steel, Figure 29-33.

Paperwork: Complete a copy of the time sheet in Appendix I, the bill of materials in Appendix III, and the performance qualification test record in Appendix IV, or use forms as provided by your instructor.◆

PRACTICE 29-11

Gas Tungsten Arc Welding (GTAW) on Aluminum Workmanship Sample

Welding Procedure Specification (WPS) No.: Practice 29-11.

Title: Welding GTAW of sheet to sheet.

Scope: This procedure is applicable for square groove and fillet welds within the range of 18 gauge through 10 gauge.

Welding May Be Performed in the Following Positions: 1G and 2F.

Base Metal: The base metal shall conform to aluminum M-22 or P-22.

Backing Material Specification: None.

Filler Metal: The filler metal shall conform to AWS specification no. ER4043 from AWS specification A5.10. This filler metal falls into F-number F-22 and A-number A-5.10.

Electrode: The tungsten electrode shall conform to AWS specification no. EWCe-2, EWZr, EWLa, or EWP from AWS specification A5.12. The tungsten diameter shall be 1/8 in. (3.2 mm) maximum. The tungsten end shape shall be rounded.

Shielding Gas: The shielding gas, or gases, shall conform to the following compositions and purity: welding grade argon.

Joint Design and Tolerances: Refer to the drawing and specifications in Figure 29-34 for the workmanship sample layout.

Preparation of Base Metal: All hydrocarbons and other contaminants, such as cutting fluids, grease, oil, and primers, must be cleaned off all parts and filler metals before welding. This cleaning can be done with any suitable solvents or detergents. The joint face and inside and outside plate surface within 1 in. (25 mm) of the joint must be mechanically or chemically cleaned of oxides. Mechanical cleaning may be done by stainless steel wire brushing, scraping, machining, or filing. Chemical cleaning may be done by using acids, alkalies, solvents, or detergents. Because the oxide layer may re-form quickly and affect the weld, welding should be started within 10 minutes of cleaning.

Electrical Characteristics: Set the welding current to AC high-frequency stabilized and the amperage and shielding gas flow according to Table 29-11.

Preheat: The parts must be heated to a temperature higher than 50°F (10°C) before any welding is started.

Backing Gas: N/A.

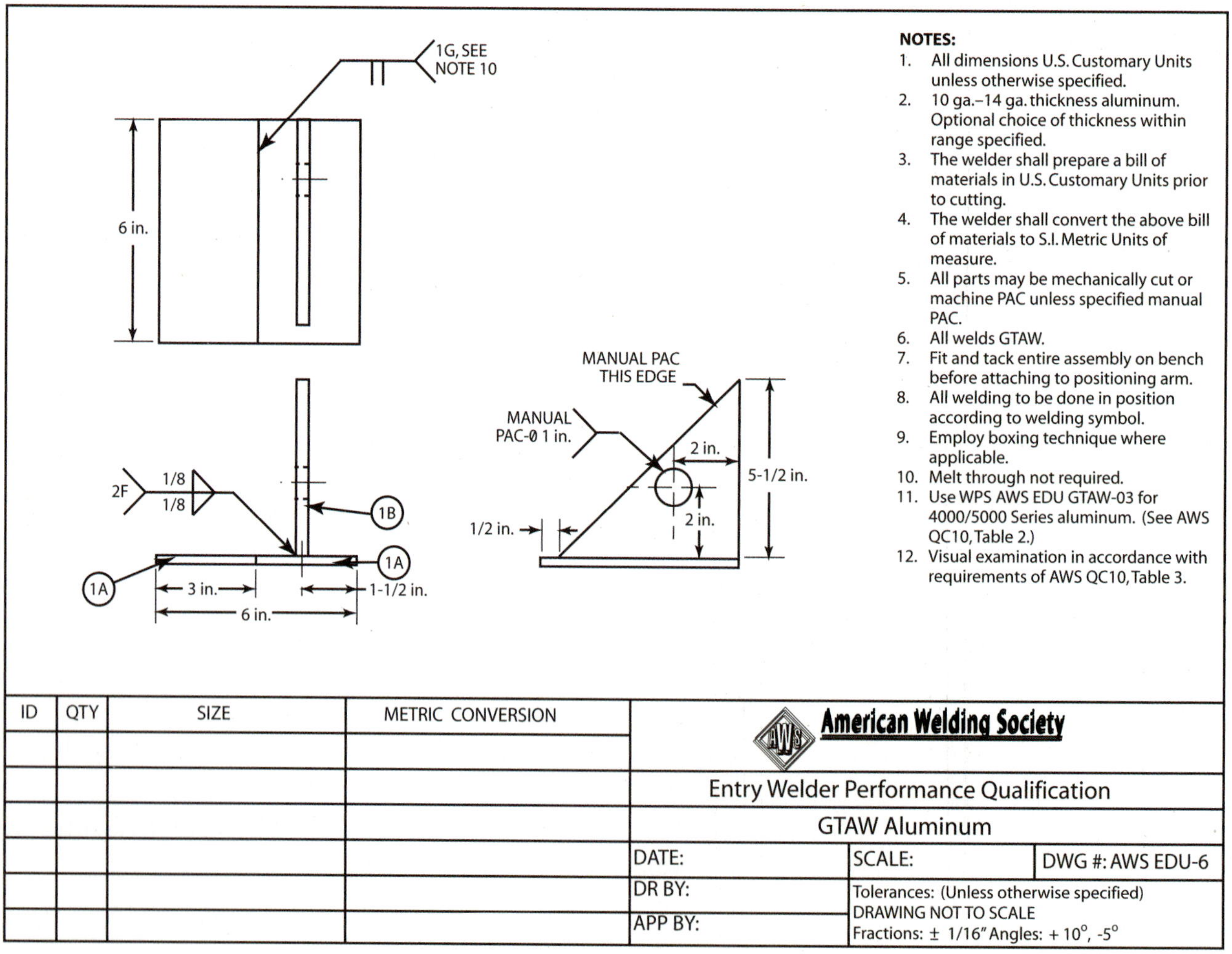

FIGURE 29-34 AWS EDU-5 GTAW aluminum workmanship sample. American Welding Society

Metal Specifications		Gas Flow			Nozzle Size in. (mm)	Amperage Min. Max.
Thickness	Diameter of ER4043*	Rates cfm (L/min)	Preflow Times	Postflow Times		
18 ga	3/32 in. (2.4 mm)	20 to 30 (9 to 14)	10 to 15 sec.	10 to 25 sec.	1/4 to 3/8 (6 to 10)	40 to 60
17 ga	3/32 in. (2.4 mm)	20 to 30 (9 to 14)	10 to 15 sec.	10 to 25 sec.	1/4 to 3/8 (6 to 10)	50 to 70
16 ga	3/32 in. (2.4 mm)	20 to 30 (9 to 14)	10 to 15 sec.	10 to 25 sec.	1/4 to 3/8 (6 to 10)	60 to 75
15 ga	3/32 in. (2.4 mm)	20 to 30 (9 to 14)	10 to 15 sec.	10 to 25 sec.	1/4 to 3/8 (6 to 10)	65 to 85
14 ga	3/32 in. (2.4 mm)	20 to 30 (9 to 14)	10 to 15 sec.	10 to 25 sec.	1/4 to 3/8 (6 to 10)	75 to 90
13 ga	1/8 in. (3 mm)	25 to 40 (12 to 19)	10 to 20 sec.	10 to 30 sec.	3/8 to 5/8 (10 to 16)	85 to 100
12 ga	1/8 in. (3 mm)	25 to 40 (12 to 19)	10 to 20 sec.	10 to 30 sec.	3/8 to 5/8 (10 to 16)	90 to 110
11 ga	1/8 in. (3 mm)	25 to 40 (12 to 19)	10 to 20 sec.	10 to 30 sec.	3/8 to 5/8 (10 to 16)	100 to 115
10 ga	1/8 in. (3 mm)	25 to 40 (12 to 19)	10 to 20 sec.	10 to 30 sec.	3/8 to 5/8 (10 to 16)	100 to 125

*Other aluminum AWS A5.10 filler metal may be used if needed.

Table 29-11 GTAW Aluminum Machine Settings

Safety: Proper protective clothing and equipment must be used. The area must be free of all hazards that may affect the welder or others in the area. The welding machine, welding leads, work clamp, electrode holder, and other equipment must be in safe working order.

Welding Technique: The welder's hands or gloves must be clean and oil free to prevent contaminating the metal or filler rods.

Tack Welds: With the parts securely clamped in place with the correct root gap, the tack welds are to be performed. Holding the electrode so that it is very close to the root face but not touching, slowly increase the current until the arc starts and a molten weld pool is formed. Add filler metal as required to maintain a slightly convex weld face and a flat or slightly concave root face. When it is time to end the tack weld, lower the current slowly so that the molten weld pool can be tapered down in size. When all tack welds are complete, allow the parts to cool as needed before assembling the remaining parts. Repeat the tack welding procedure until the entire part is assembled.

Square Groove and Fillet Welds: Holding the electrode so that it is very close to the metal surface but not touching, slowly increase the current until the arc starts and a molten weld pool is formed. As the weld progresses, add filler metal as required to maintain a flat or slightly convex weld face. If it is necessary to stop the weld or to reposition yourself or the weld is completed, the current must be lowered slowly so that the molten weld pool can be tapered down in size.

Interpass Temperature: The plate should not be heated to a temperature higher than 120°F (49°C) during the welding process. After each weld pass is completed, allow it to cool but never to a temperature below 50°F (10°C). The weldment must not be quenched in water.

Cleaning: Recleaning may be required if the parts or filler metal becomes contaminated or oxidized to a degree that the weld quality will be affected. Reclean using the same procedure used for the original metal preparation.

Visual Inspection: Visual inspection criteria for entry-level welders*:

1. There shall be no cracks, no incomplete fusion.
2. There shall be no incomplete joint penetration in groove welds except as permitted for partial joint penetration groove welds.
3. The Test Supervisor shall examine the weld for acceptable appearance and shall be satisfied that the welder is skilled in using the process and procedure specified for the test.
4. Undercut shall not exceed the lesser of 10% of the base metal thickness or 1/32 in. (0.8 mm).
5. Where visual examination is the only criterion for acceptance, all weld passes are subject to visual examination at the discretion of the Test Supervisor.
6. The frequency of porosity shall not exceed one in each 4 in. (100 mm) of weld length, and the maximum diameter shall not exceed 3/32 in. (2.4 mm).
7. Welds shall be free from overlap.

**Courtesy of the American Welding Society.*

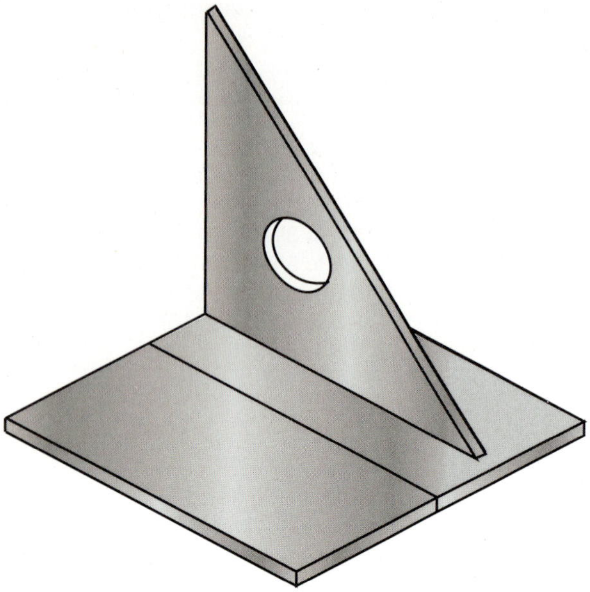

FIGURE 29-35 Practice 29-11 workmanship sample. American Welding Society

Sketches: Gas Tungsten Arc Welding (GTAW) Workmanship Sample drawing for Aluminum, Figure 29-35.

Paperwork: Complete a copy of the time sheet in Appendix I, the bill of materials in Appendix III, and the performance qualification test record in Appendix IV, or use forms as provided by your instructor.◆

PRACTICE 29-12

Welder and Welder Operator Qualification Test Record (WPS)

Using a completed weld such as the ones from Practice 29-1 or 29-2 and the following list of steps, you will complete the test record shown in Figure 29-36. This form is a composite of sample test recording forms provided by AWS, ASME, and API codes. You may want to obtain a copy of one of the codes or standards and compare a weld you made to the standard. This form is useful when you are testing one of the practice welds in this text.

NOTE: Not all of the blanks will be filled in on the forms. The forms are designed to be used with a large variety of weld procedures, so they have spaces that will not be used each time.

1. Welder's name: The person who performed the weld.
2. Identification (WPS) No.: On a welding job, every person has an identification number that is used on the time card and paycheck. In this space, you can write the class number or section number since you do not have a clock number.
3. Welding process(es): Was the weld performed with SMAW, GMAW, or GTAW?

Welder and Welding Operator Qualification Test Record (WQR)

Welder or welding operator's name ____(1)____ Identification no. ____(2)____

Welding process ____(3)____ Manual ____(4)____ Semiautomatic ____(4)____ Machine ____(4)____

Position ____(5)____

(Flat, horizonal, overhead or vertical—if vertical, state whether up or down In accordance with welding procedure specification no.) ____(6)____

Material specification ____(7)____

Diameter and wall thickness (if pipe)—otherwise, joint thickness ____(8)____

Thickness range this qualifies ____(9)____

Filler Metal

Specification No. ____(10)____ Class ification ____(11)____ F-number ____(12)____

Describe filler metal (if not covered by AWS specification) ____(13)____

Is backing strip used? ____(14)____

Filler metal diameter and trade name ____(15)____ Flux for submerged arc or gas for gas metal arc or flux cored arc welding ____(16)____

FIGURE 29-36 Welder and Welding Operator Qualification Test Record (WQR). American Welding Society

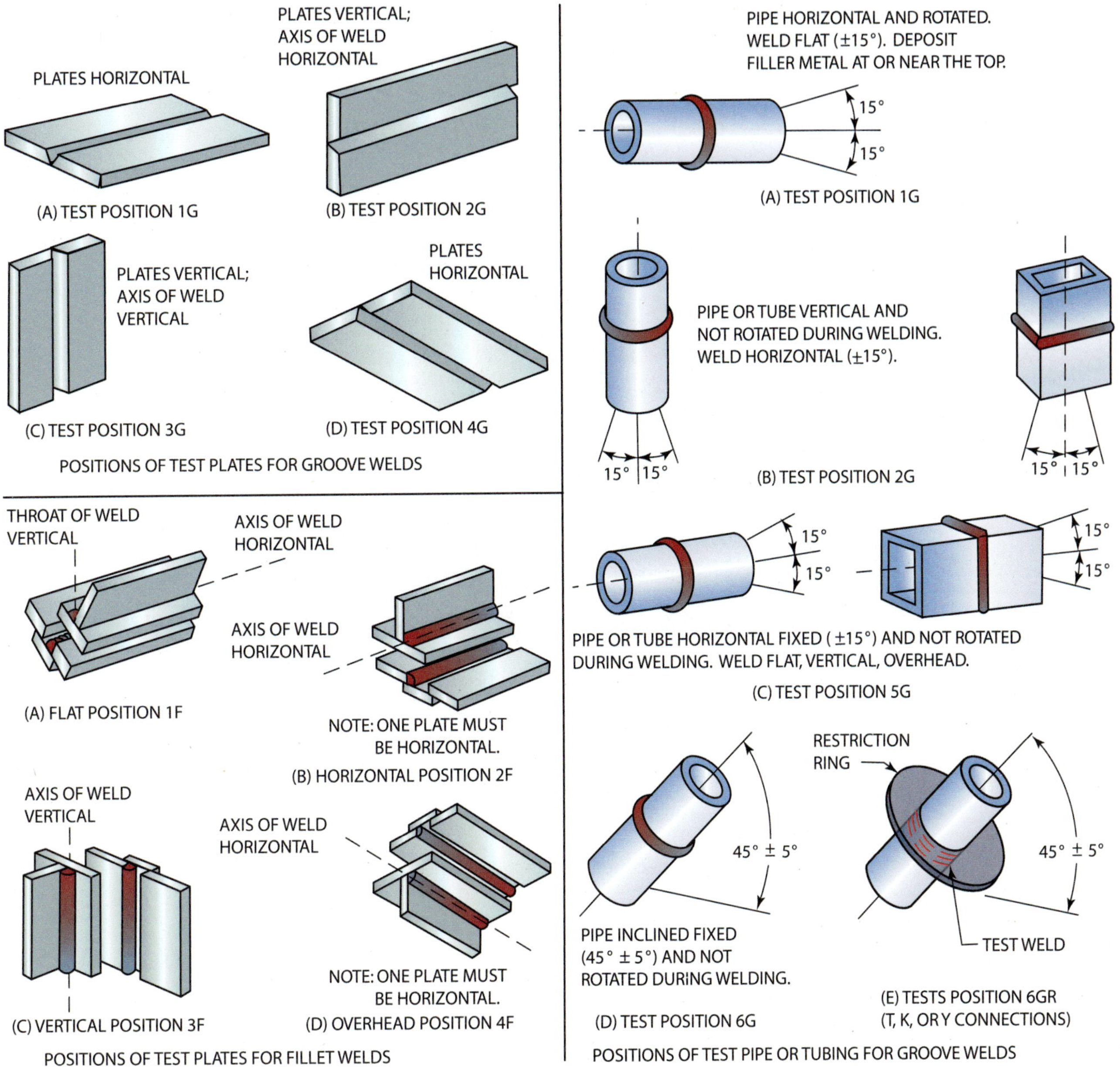

FIGURE 29-37 **Welding positions.** American Welding Society

4. How was the weld accomplished: Manually, semi-automatically, or automatically?
5. Test position: 1G, 2G, 3G, 4G, 1F, 2F, 3F, 4F, 5G, 6G, 6GR, Figure 29-37.
6. What WPS was used for this test?
7. Base metal specification: This is the ASTM specification number, Table 29-12.
8. Test material thickness (or) test pipe diameter (and) wall thickness: The actual thickness of the welded material or pipe diameter and wall thickness.
9. Thickness range qualified (or) diameter range qualified: For both plate and pipe, a weld performed

Type of Material	
P-1	Carbon Steel
P-3	Low Alloy Steel
P-4	Low Alloy Steel
P-5	Alloy Steel
P-6	High Alloy Steel — Predominantly Martensitic
P-7	High Alloy Steel — Predominantly Ferritic
P-8	High Alloy Steel — Austenitic
P-9	Nickel Alloy Steel
P-10	Specialty High Alloy Steels
P-21	Aluminum and Aluminum — Base Alloys
P-31	Copper and Copper Alloy
P-41	Nickel

Table 29-12 P Numbers

Plate Thickness (T) Tested in. (mm)	Plate Thickness (T) Qualified in. (mm)	
1/8 ≤ T < 3/8* (3.1 ≤ T < 9.5)	1/8 to 2T (3.1 to 2T)	
3/8 (9.5)	3/4 (19.0)	
3/8 < T < 1 (9.5 < T < 25.4)	2T 2T	
1 and over (25.4 and over)	Unlimited Unlimited	
Pipe Size of Sample Weld		
Diameter in. (mm)	**Wall Thickness, T**	
2 (50.8) or 3 (76.2)	Sch. 80 Sch. 40	
6 (152.4) or 8 (203.2)	Sch. 120 Sch. 80	
Pipe Size Qualified		
Diameter in. (mm)	**Wall Thickness, in. (mm)**	
	Minimum	Maximum
3/4 (19.0) through 4 (101.6)	0.063 (1.6)	0.674 (17.1)
4 (101.6) and over	0.187 (4.7)	Any

*Thickness (T) is equal to or greater than 1/8 in. (≤) and thickness (T) is less than 3/8 in. (<).

Table 29-13 Test Specimen and Range of Thickness Qualified

successfully on one thickness qualifies a welder to weld on material within that range. See Table 29-13 for a list of thickness ranges.

10. Filler metal specification number: The AWS has specifications for chemical composition and physical properties for electrodes. Some of these specifications are listed in Table 29-14.
11. Classification number: This is the standard number found on the electrode or electrode box, such as E6010, E7018, E316-15, ER1100, and so on.
12. F-number: A specific grouping number for several classifications of electrodes having similar composition and welding characteristics. See Table 29-15 for the F-number corresponding to the electrode used.
13. Give the manufacturer's chemical composition and physical properties as provided.
14. Backing strip material specification: This is the ASTM specification number.
15. Give the diameter of electrode used and the manufacturer's identification name or number.
16. Flux for SAW or shielding gas(es) and flow rate for GMAW, FCAW, or GTAW.

A-Number	Metal and Process(es)
A5.10	Aluminum—bare electrodes and rods
A5.3	Aluminum—covered electrodes
A5.8	Brazing filler metal
A5.1	Steel, carbon, covered electrodes
A5.20	Steel, carbon, flux cored electrodes
A5.17	Steel-carbon, submerged arc wires and fluxes
A5.18	Steel-carbon, gas metal arc electrodes
A5.2	Steel—oxyfuel gas welding
A5.5	Steel—low alloy covered electrodes
A5.23	Steel—low alloy electrodes and fluxes—submerged arc
A5.28	Steel—low alloy filler metals for gas shielded arc welding
A5.29	Steel—low alloy, flux cored electrodes

Table 29-14 Specification Numbers

Group Designation	Metal Types	AWS Electrode Classification
F1	Carbon steel	EXX20, EXX24, EXX27, EXX28
F2	Carbon steel	EXX12, EXX13, EXX14
F3	Carbon steel	EXX10, EXX11
F4	Carbon steel	EXX15, EXX16, EXX18
F5	Stainless steel	EXXX15, EXXX16
F6	Stainless steel	ERXXX
F22	Aluminum	ERXXXX

Table 29-15 F-Numbers

If the weld is a groove weld, follow steps 17 through 22 and then skip to step 27. If the weld test is a fillet weld, skip to step 22.

17. Visually inspect the weld and record any flaws.
18. Record the weld face, root face, and reinforcement dimensions.
19. Four (4) test specimens are used for 3/8-in. (10-mm) or thinner metal. Two (2) will be root bent and two (2) face bent. For thicker metal all four (4) will be side bent.
20. Visually inspect the specimens after testing, and record any discontinuities.
21. Who witnessed the welding for verification that the WPS was followed?
22. The identification number assigned by the testing.

If the weld is a fillet weld, follow steps 23 through 28.

23. Visually inspect the weld, and record any flaws.
24. Record the legs and reinforcement dimensions.
25. Measure and record the depth of the root penetration.
26. Polish the side of the specimens and apply an acid to show the complete outline of the weld.
27. Who witnessed the welding for verification that the WPS was followed?
28. The identification number assigned by the testing.

If a radiographic test is used, follow steps 29 through 33. If this test is not used, go to step 34.

29. The number the lab placed on the X-ray film before it was exposed on the weld.
30. Record the results of the reading of the film.
31. Whether the test passed or failed the specific code.
32. Who witnessed the welding for verification that the WPS was followed?
33. The number assigned by the testing.
34. The name of the company that requested the test.
35. The name of the person who interpreted the results. This is usually a certified welding inspector (CWI) or other qualified person.
36. Date the results of the test were completed.◆

SUMMARY

Becoming a certified welder establishes your credentials in the industry. Not every industry requires certification; however, all of the welding fields recognize the importance of being certified. Advertisements in the newspaper, the Internet, and outside of welding shops prominently display the words *certified welder.* Even people outside of the welding industry recognize the significance of someone having obtained the educational level and proficiency required to become a certified welder.

An important part of passing a certification test is your ability to follow all of the very specific details required by the certification process for which you are being tested. Read the qualification test procedures carefully. There may be specific requirements in this test you have not experienced before; do not assume that you know what is required. As you have learned in this chapter, there are many specific things that must be done in preparation for the certification test to ensure that the test results are valid and to ensure its successful completion. By diligently following the procedures, you will certainly enhance your chance of passing the certification. Often your first certification is the most difficult, but it is part of the learning process. Once you have learned the proper techniques and methods of performing certification welding, additional certification tests will become much easier. Experienced certified welders in the field have no difficulty routinely passing certification testing.

REVIEW QUESTIONS

1. Which welding processes can be tested for qualification and certification?
2. What controls certification and qualification test requirements?
3. What is implied by a certification document?
4. How do you become certified as an AWS entry-level welder?
5. What is the advantage of tapering a weld bead size down slightly when you are going to have to restart and continue the weld with a new electrode?
6. Why should the slag always be chipped and the weld crater cleaned each time before restarting the weld?
7. Why should starting and stopping weld beads in corners be avoided?
8. What is the acceptable limit for roughness or notches on the surface of a bevel?
9. Give examples of hydrocarbons and other contaminants that must be cleaned off all parts and filler metals before welding.
10. Describe how to mechanically clean a plate surface prior to welding.
11. How can SMAW slag be cleaned off between passes?
12. What is the minimum and maximum root opening allowed for the bevel and V-groove joints shown in Figure 29-15?
13. According to Table 29-2, what polarity and amperage range could be used for a 1/8-in. diameter E6011 electrode?
14. What is the maximum undercut allowed on a short-circuit metal transfer workmanship sample?
15. What is the maximum interpass temperature for a GMAW spray transfer workmanship sample?
16. What shielding gas and flow rates should be used with SMAW-S test plate welds without a backing strip?
17. According to Table 29-7, what is the amperage range and wire feed speed range for 0.035 E71T-1 electrodes?
18. What is the minimum temperature that a plate must be heated to before making a FCAW self-shielded weld?
19. What is the maximum interpass temperature for a plain carbon steel GTAW workmanship sample?
20. What is the AWS identification for the type of tungsten electrode that should be used for welding on the stainless steel workmanship sample?
21. What is the AWS identification for the type of tungsten electrode that should be used for welding on the aluminum workmanship sample?

Chapter 30

Testing and Inspecting Welds

OBJECTIVES

After completing this chapter, the student should be able to:

- Explain the importance of testing and inspecting welds.
- Compare mechanical testing to nondestructive testing.
- Compare discontinuities and defects.
- Describe various weld discontinuities and defects.
- List problems caused by the metal being fabricated.
- List the various types of tests for weld quality.

KEY TERMS

Brinell hardness tester
defect
discontinuities
eddy current inspection (ET)
mechanical testing (MT)
nondestructive testing (NDT)
quality control (QC)
radiographic inspection (RT)
Rockwell hardness tester
shearing strength
tolerance
ultrasonic inspection (UT)

INTRODUCTION

It is important to know that a weld will meet the requirements of the company and/or codes or standards. It is also necessary to ensure the quality, reliability, and strength of a weldment. To meet these demands, an active inspection program is needed. The extent to which you and the product you made are subjected to testing and inspection depends upon the intended service of the product. Items that are to be used in light, routine-type service, such as ornamental iron, fence posts, gates, and so forth, are not inspected as critically as products in critical use. Some of the items in critical use include a main nuclear reactor containment vessel, oil refinery high-pressure vessels, aircraft airframes, bridges, and so on. The type of inspection required, then, is very much dependent upon the type of service the welded part will be required to withstand. The quality of the weld that will pass or be acceptable for one welding application may not meet the needs of another.

Quality Control (QC)

Once a code or standard has been selected, a method is chosen for ensuring that the product meets the specifications. The two classifications of methods used in product **quality control (QC)** are destructive, or mechanical testing and nondestructive testing. These methods can be used individually, or a combination of the two methods can be used. **Mechanical testing (MT)** methods, except for hydrostatic testing, result in the product being destroyed. **Nondestructive testing (NDT)** does not destroy the part being tested.

Mechanical testing is commonly used to qualify you or the welding procedure you are using. It can be used in a random sample testing procedure in mass production. In many cases, a large number of identical parts are made, and a chosen number are destroyed by mechanical testing. The results of such tests are valid only for welds made under the same conditions because the only weld strengths known are the ones resulting from the tested pieces. It is then assumed that the strengths of the nontested pieces are the same.

Nondestructive testing is used for weld qualification, welding procedure qualification, and product quality control. Since the weldment is not damaged, all the welds can be tested and the part can actually be used for its intended purpose. Because the parts are not destroyed, more than one testing method can be used on the same part. Frequently, only part of the welds are tested to save time and money. The same comparison of random sampling applies to these tests as it does for mechanical testing. Welds on critical parts are usually 100% NDT tested.

Discontinuities and Defects

Discontinuities and flaws are interruptions in the typical structure of a weld. They may be a lack of uniformity in the mechanical, metallurgical, or physical characteristics of the material or weld. All welds have discontinuities and flaws, but they are not necessarily defects.

A **defect**, according to AWS, is "a discontinuity or **discontinuities**, which by nature or accumulated effect (for example, total porosity or slag inclusion length that renders a part or product unable to meet minimum applicable acceptance standards or specifications). This term designates rejectability."

American Bureau of Shipping
American Petroleum Institute
ASME International
American Society for Testing and Materials
American Welding Society
British Welding Institute
United States Government

Table 30-1 Major Code Issuing Agencies

In other words, many acceptable products may have welds that contain discontinuities. But no products may have welds that contain defects. A discontinuity becomes a defect when the discontinuity becomes so large or when there are so many small discontinuities that the weld is not acceptable under the standards for the code for that product. Some codes are more strict than others, so that the same weld might be acceptable under one code but not under another.

Ideally, a weld should not have any discontinuities, but that is practically impossible. The difference between what is acceptable, fit for service, and perfection is known as **tolerance**. In many industries, the tolerances for welds have been established and are available as codes or standards. Table 30-1 lists a few of the agencies that issue codes or standards. Each code or standard gives the tolerance that changes a discontinuity to a defect.

When evaluating a weld, it is important to note the type of discontinuity, the size of the discontinuity, and the location of the discontinuity. Any one of these factors or all three can be the deciding factors that, based on the applicable code or standard, change a discontinuity to a defect.

Table 30-2 lists the nine most common discontinuities and the welding processes that can cause them.

Porosity

Porosity results from gas that was dissolved in the molten weld pool, forming bubbles that are trapped as the metal cools to become solid. The bubbles that make up porosity form within the weld metal; for that reason they cannot normally be seen as they form. These gas pockets form in the same way that bubbles form in a carbonated drink as it warms up or as air dissolved in water forms bubbles in the center

	Shielded metal arc welding (SMAW)	Gas metal arc welding (GMAW)	Flux cored arc welding (FCAW)	Gas tungsten arc welding (GTAW)	Oxyacetylene welding (OAW)	Oxyhydrogen welding (OHW)	Submerged arc welding (SAW)	Laser beam welding (LBW)	Plasma arc welding (PAW)	Electron beam welding (EBW)	Carbon arc welding (CAW)	Pressure gas welding (PGW)	Electroslag welding (ESW)	Thermite welding (TW)
Porosity	X	X	X	X	X	X	X	X	X	X	X	X	X	X
Inclusions	X	X	X				X				X		X	X
Inadequate joint penetration	X	X	X		X		X		X	X	X		X	
Incomplete fusion	X	X	X	X	X	X	X	X	X	X	X	X	X	X
Arc strikes	X	X	X	X										
Overlap (cold lap)	X	X	X	X	X	X	X	X	X	X	X	X	X	X
Undercut	X	X	X	X	X	X	X		X		X			
Crater Cracks	X	X	X	X	X	X	X		X		X			
Underfill	X	X	X	X	X	X	X		X		X			

Table 30-2 The Nine Most Common Discontinuities and the Welding Processes That Might Cause Them

of a cube of ice. Porosity takes either a spherical (ball-shaped) or cylindrical (tube- or tunnel-shaped) form. The cylindrical porosity is called a wormhole, and it is the most likely type of porosity to reach the weld surface and be seen. The rounded edges tend to reduce the stresses around them. Therefore, unless porosity is extensive, there is little or no loss in strength.

Porosity is most often caused by improper welding techniques, contamination, or an improper chemical balance between the filler and base metals.

Improper welding techniques may result in shielding gas not properly protecting the molten weld pool. For example, the E7018 electrode should not be weaved wider than two and a half times the electrode diameter because very little shielding gas is produced. As a result, parts of the weld are unprotected. Nitrogen from the air that dissolves in the weld pool and then becomes trapped during escape can produce porosity.

The intense heat of the weld can decompose paint, dirt, or oil from machining and rust or other oxides, producing hydrogen. This gas, like nitrogen, can also become trapped in the solidifying weld pool, producing porosity. Hydrogen can also diffuse into the heat-affected zone, producing underbead cracking in some steels. The level needed to crack welds is below that necessary to produce porosity.

Porosity can be grouped into the following four major types:

- Uniformly scattered porosity is most frequently caused by poor welding techniques or faulty materials, Figure 30-1.
- Clustered porosity is most often caused by improper starting and stopping techniques, Figure 30-2.

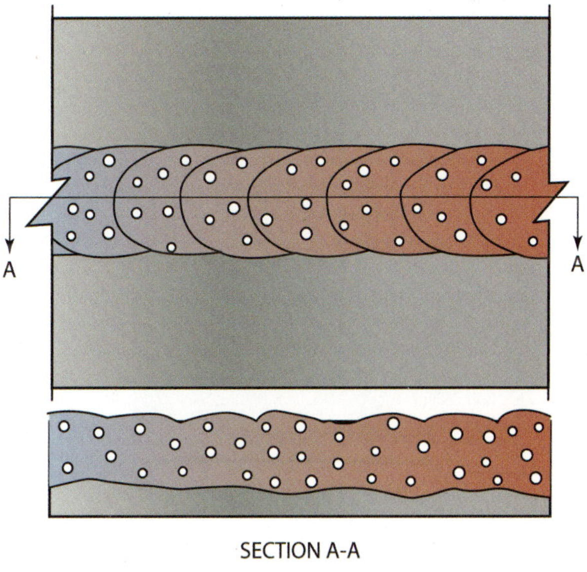

FIGURE 30-1 Uniformly scattered porosities.

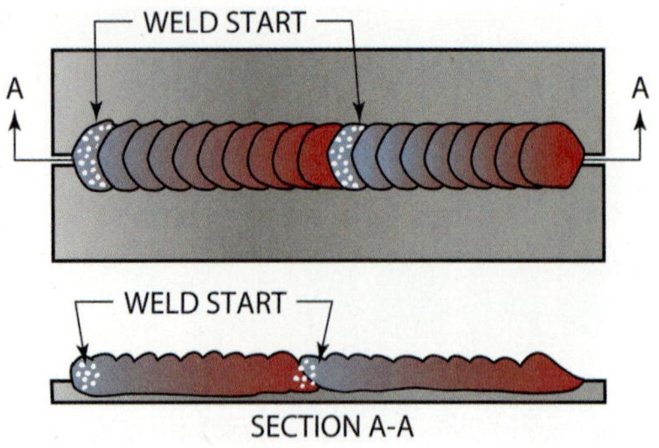

FIGURE 30-2 Clustered porosity.

- Linear porosity is most frequently caused by contamination within the joint, root, or interbead boundaries, Figure 30-3.
- Piping porosity, or wormhole, is most often caused by contamination at the root, Figure 30-4. This porosity is unique because its formation depends on the gas escaping from the weld pool at the same rate as the pool is solidifying.

Refer to Table 30-2 for a list of welding processes that might cause weld porosity.

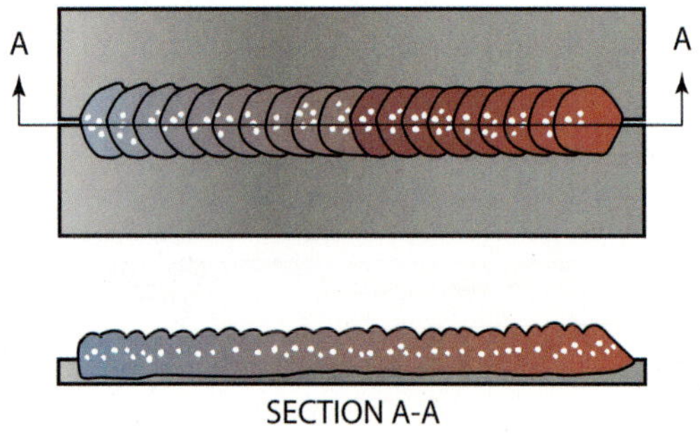

FIGURE 30-3 Linear porosity.

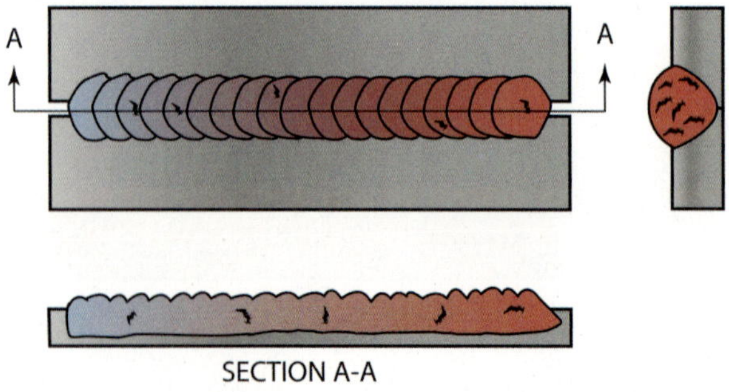

FIGURE 30-4 Piping or wormhole porosity.

Inclusions

Inclusions are nonmetallic materials, such as slag and oxides, that are trapped in the weld metal, between weld beads, or between the weld and the base metal. Inclusions sometimes are jagged and irregularly shaped. Also, they can form in a continuous line. This causes stresses to concentrate and reduces the structural integrity (loss in strength) of the weld.

Although not visible, their development can be expected if prior welds were improperly cleaned or had a poor contour. Unless care is taken in reading radiographs, the presence of slag inclusions can be interpreted as other defects.

Linear slag inclusions in radiographs generally contain shadow details; otherwise, they could be interpreted as lack-of-fusion defects. These inclusions result from a lack of slag control caused by poor manipulation that allows the slag to flow ahead of the arc, by not removing all the slag from previous welds, or by welding highly crowned, incompletely fused welds.

Scattered inclusions can resemble porosity but, unlike porosity, they are generally not spherical. These inclusions can also result from inadequate removal of earlier slag deposits and poor manipulation of the arc. Additionally, heavy mill scale or rust serves as their source, or they can result from unfused pieces of damaged electrode coatings falling into the weld. In radiographs some detail will appear, unlike linear slag inclusions.

Nonmetallic inclusions, Figure 30-5, are caused under the following conditions:

- Slag and/or oxides do not have enough time to float to the surface of the molten weld pool.
- There are sharp notches between weld beads or between the weld bead and the base metal that trap the material so that it cannot float out.
- The joint was designed with insufficient room for the correct manipulation of the molten weld pool. Refer to Table 30-2 for a listing of welds that may produce nonmetallic inclusions.

Inadequate Joint Penetration

Inadequate joint penetration occurs when the depth that the weld penetrates the joint, Figure 30-6, is less than that needed to fuse through the plate or into the preceding weld. A defect usually results that could reduce the required cross-sectional area of the joint or become a source of stress concentration that leads

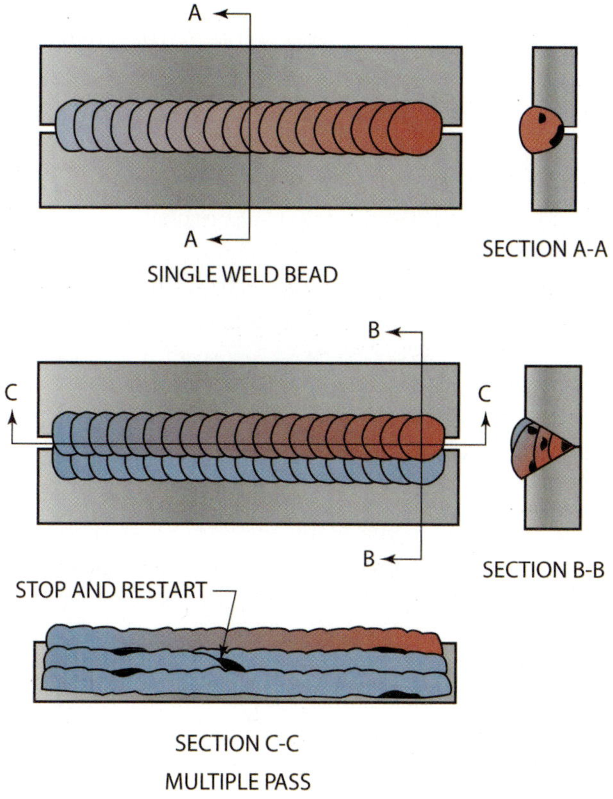

FIGURE 30-5 Nonmetallic inclusions.

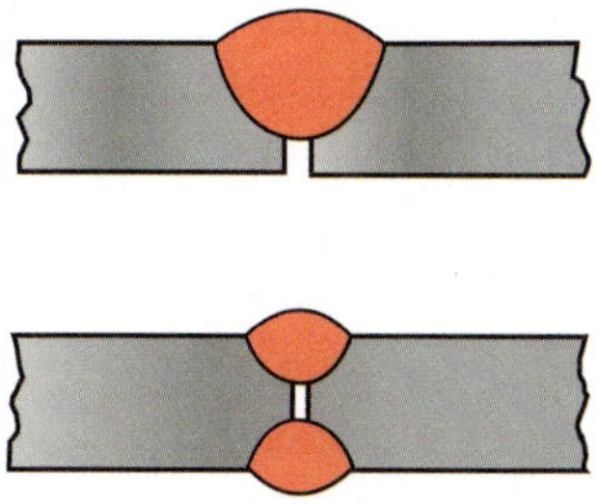

FIGURE 30-6 Inadequate joint penetration.

to fatigue failure. The importance of such defects depends on the notch sensitivity of the metal and the factor of safety to which the weldment has been designed. Generally, if proper welding procedures are developed and followed, such defects do not occur.

Following are the major causes of inadequate joint penetration:

- Improper welding technique—The most common cause is a misdirected arc. Also, the welding technique may require that both starting and run-out tabs be used so that the molten weld pool is well established before it reaches the joint. Sometimes, a failure to back gouge the root sufficiently provides a deeper root face than allowed for, Figure 30-7.

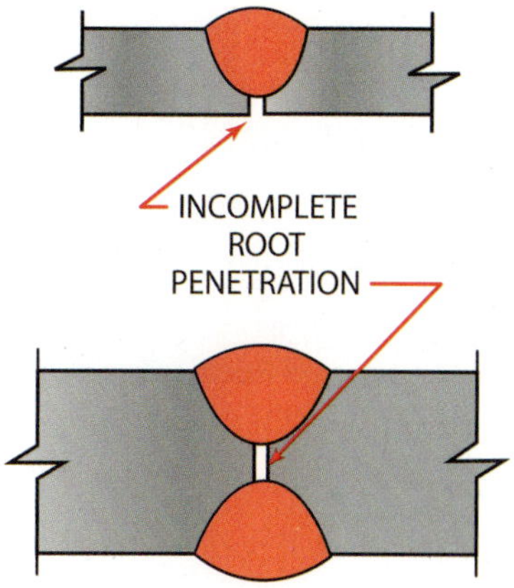

FIGURE 30-7 Incomplete root penetration.

- Not enough welding current—Metals that are thick or have a high thermal conductivity are often preheated so that the weld heat is not drawn away so quickly by the metal that it cannot penetrate the joint.
- Improper joint fit-up—This problem results when the weld joints are not prepared or fitted accurately. Too small a root gap or too large a root face will keep the weld from penetrating adequately.
- Improper joint design—When joints are accessible from both sides, back gouging is often used to ensure 100% root fusion.

Refer to Table 30-3 for a list of welding processes that may produce inadequate joint penetration.

Incomplete Fusion

Incomplete fusion is the lack of coalescence between the molten filler metal and previously deposited filler metal and/or the base metal, Figure 30-8. The lack of fusion between the filler metal and previously deposited weld metal is called *interpass cold lap*. The lack of fusion between the weld metal and the joint face is called *lack of sidewall fusion*. Both of these problems usually travel along all or most of the weld's length.

Following are some major causes of lack of fusion:

- Inadequate agitation—Lack of weld agitation to break up oxide layers. The base metal or weld filler metal may melt, but a thin layer of oxide may prevent coalescence from occurring.

TYPE OF DISCONTIUNITY	JOINT TYPE: BUTT JOINT	LAP JOING	TEE JOINT	OUTSIDE CORNER JOINT
CRACKS				
Longitudinal cracks	X	X		X
Transverse cracks	X	X	X	X
Crater cracks	X	X	X	X
Throat cracks	X	X	X	X
Toe cracks	X	X		X
Root cracks	X	X	X	X
Underbead and HAZ cracks	X	X	X	X
Fissures				
POROSITY				
Uniformly scattered porosity	X	X	X	X
Cluster porosity	X	X	X	X
Linear porosity	X	X	X	X
Piping porosity	X	X	X	X
INCLUSIONS				
Slag inclusions	X	X	X	X
Tungsten inclusions	X	X	X	X
INCOMPLETE FUSION	X	X	X	X
INADEQUATE JOINT PENETRATION	X	X	X	X
UNDERCUT	X	X	X	X
UNDERFILL	X		X	X
OVERLAP	X	X	X	X
LAMINATIONS	X	X	X	X
DELAMINATION	X	X	X	X
SEAMS AND LAPS	X	X	X	X
LAMELLAR TEARS			X	X

Table 30-3 Common Discontinuities and the Joint Types They Might Be Found On

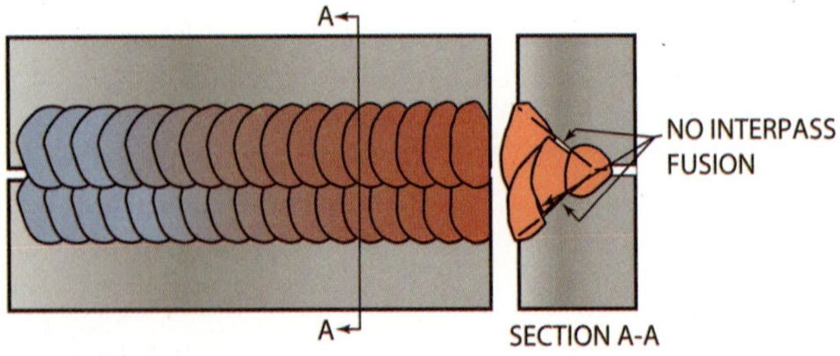

FIGURE 30-8 Incomplete fusion.

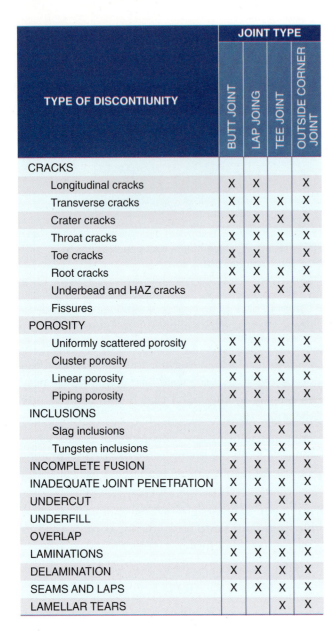

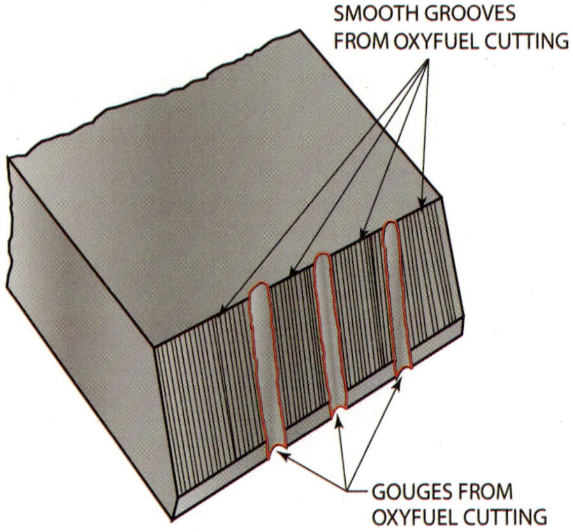

FIGURE 30-9 Remove gouges along the surface of the joint before welding.

- Improper welding techniques—Poor manipulation, such as moving too fast or using an improper electrode angle.
- Wrong welding process—For example, the use of short-circuiting transfer with GMAW to weld plate thicker than 1/4 in. (6 mm) can cause the problem because of the process's limited heat input to the weld.
- Improper edge preparation—Any notches or gouges in the edge of the weld joint must be removed. For example, if a flame-cut plate has notches along the cut, they could result in a lack of fusion in each notch, Figure 30-9.
- Improper joint design—Incomplete fusion may also result from not enough heat to melt the base metal, or too little space allowed by the joint designer for correct molten weld pool manipulation.
- Improper joint cleaning—Failure to clean oxides from the joint surfaces resulting from the use of an oxyfuel torch to cut the plate, or failure to remove slag from a previous weld.

Incomplete fusion can be found in welds produced by all major welding processes.

Arc Strikes

Figure 30-10 shows arc strikes, which are small, localized points where surface melting occurred away from the joint. These spots may be caused by accidentally striking the arc in the wrong place and/or by faulty ground connections. Even though arc strikes can be

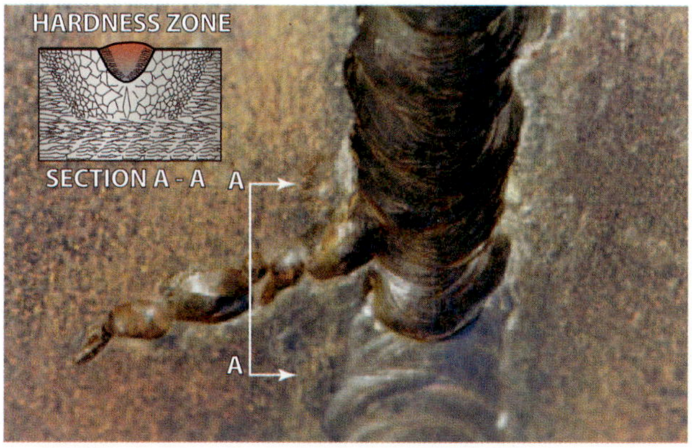

FIGURE 30-10 Arc strikes. Larry Jeffus

ground smooth, they cannot be removed. These spots will always appear if an acid etch is used. They also can be localized hardness zones or the starting point for cracking. Arc strikes, even when ground flush for a guided bend, can open up to form small cracks or holes.

Overlap

Overlap is also called cold lap, and it occurs in fusion welds when weld deposits are larger than the joint is conditioned to accept. The weld metal then flows over the surface of the base metal without fusing to it, along the toe of the weld bead, Figure 30-11. It generally occurs on the horizontal leg of a horizontal fillet weld under extreme conditions. It can also occur on both sides of flat-positioned capping passes. With GMA welding, overlap occurs when the welder uses too much electrode extension to deposit metal at low power. Misdirecting the arc into the vertical leg and keeping the electrode nearly vertical will also cause overlap. To prevent overlap, the fillet weld must be correctly sized to less than 3/8 in. (9.5 mm), and the arc must be properly manipulated.

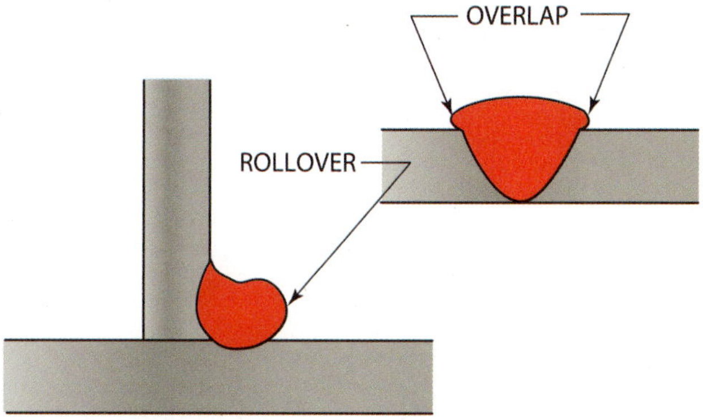

FIGURE 30-11 Rollover or overlap.
© Cengage Learning 2012

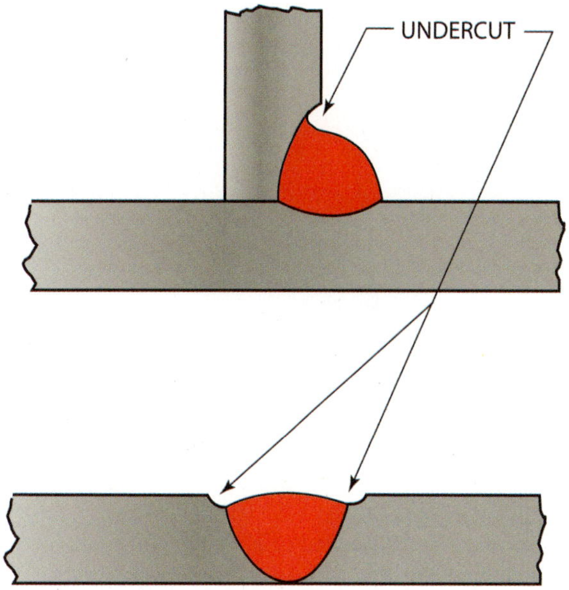

FIGURE 30-12 Undercut. © Cengage Learning 2012

Undercut

Undercut is the result of the arc force removing metal from a joint face which is not replaced by weld metal. Undercut occurs along the toe of the weld bead, Figure 30-12. It can result from excessive current. It is a common problem with GMA welding when insufficient oxygen is used to stabilize the arc. Incorrect welding technique, such as incorrect electrode angle or excessive weave, can also cause undercut. To prevent undercutting, you can weld in the flat position by using multiple instead of single passes, change the shield gas, and improve manipulative techniques to fill the removed base metal along the toe of the weld bead.

Crater Cracks

Crater cracks are the tiny cracks that develop in the weld craters as the weld pool shrinks and solidifies, Figure 30-13. Materials with a low melting temperature are rejected toward the crater center while freezing. Since these materials are the last to freeze, they are pulled apart or separated as a result of the weld metal's shrinking as it cools. The high shrinkage stresses aggravate crack formation. Crater cracks can be minimized, if not prevented, by not interrupting

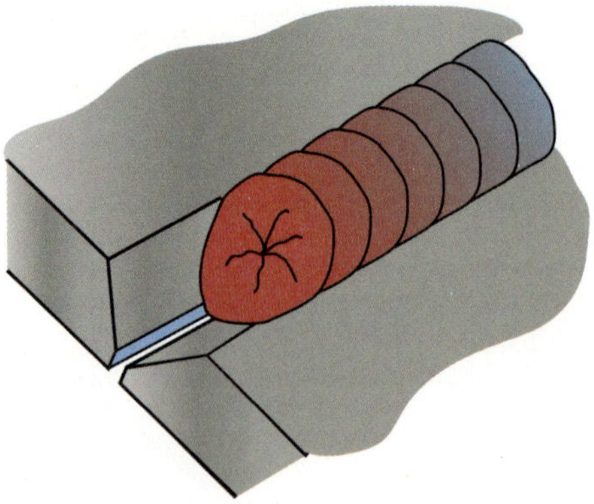

FIGURE 30-13 Crater or star cracks.

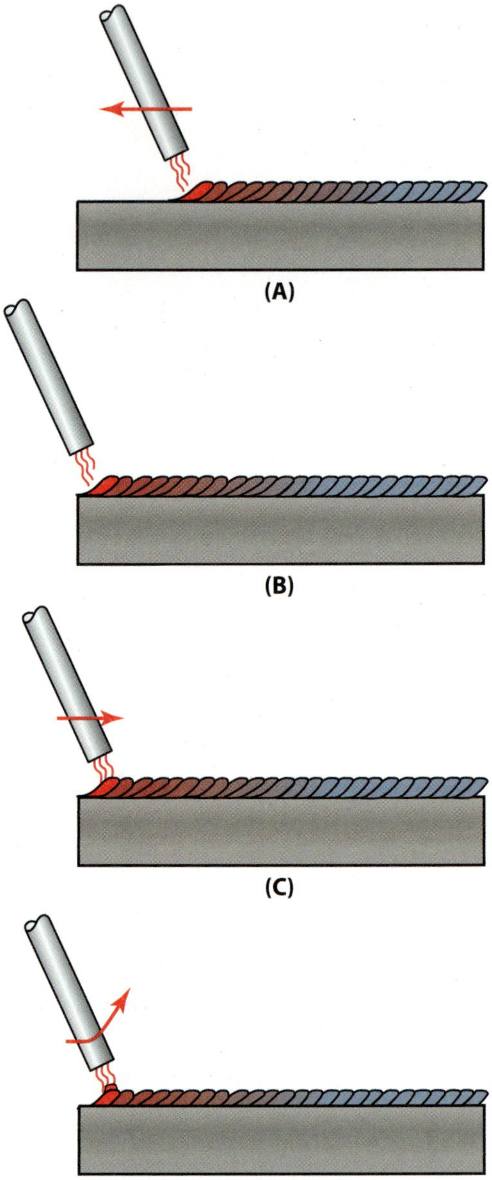

FIGURE 30-14 (A) Make a uniform weld. (B) Weld to the end of the plate. (C) Hold the arc for a second at the end of the weld. (D) Quickly, so you deposit as little weld metal as possible, move the arc back up onto the weld and break the arc.

the arc quickly at the end of a weld, which allows the arc to lengthen, the current to drop gradually, and the crater to fill and cool more slowly. Some GMAW equipment has a crater filling control that automatically and gradually reduces the wire feed speed at the end of a weld. For most welding processes an effective way of preventing crater cracking is to pull the weld slightly back over the top of the end of the weld, allowing the pool to end up on top of the weld where cracking can be minimized, Figure 30-14.

Underfill

Underfill on a groove weld appears when the weld metal deposited is inadequate to bring the weld's face or root surfaces to a level equal to that of the original plane or plate surface. For a fillet weld it occurs when the weld deposit has an insufficient effective throat, Figure 30-15. This problem can usually be corrected by slowing down the travel rate or making more weld passes.

WELD PROBLEMS CAUSED BY PLATE PROBLEMS

Not all welding problems are caused by weld metal, the process, or your lack of skill in depositing that metal. The material being fabricated can be at fault, too. Some problems result from internal plate defects that you cannot control. Others are the result of improper welding procedures that produce undesirable hard metallurgical structures in the heat-affected zone, as discussed in other chapters. The internal defects are the result of poor steelmaking practices. Steel producers try to keep their steels as sound as possible, but the

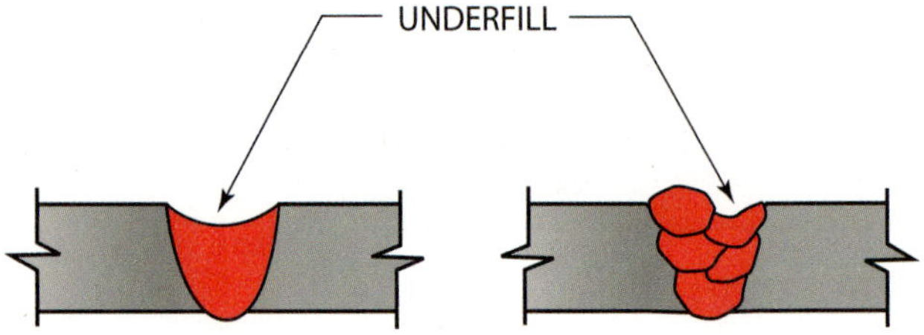

FIGURE 30-15 Underfill.

mistakes that occur in steel production are blamed, too frequently, on the welding operation.

Lamellar Tears

These tears appear as cracks parallel to and under the steel surface. In general, they are not in the heat-affected zone, and they have a steplike configuration. They result from the thin layers of nonmetallic inclusions that lie beneath the plate surface and have very poor ductility so they do not bend but pull apart under the welding stresses. Although barely noticeable, these inclusions separate when overly stressed, producing laminated cracks. These cracks are evident if the plate edges are exposed, Figure 30-16.

Lamination

Laminations differ from lamellar tearing because they are more extensive and involve thicker layers of nonmetallic contaminants. Located toward the center of the plate thickness, Figure 30-17, laminations are caused by insufficient cropping (removal of defects) of the pipe in ingots. The slag and oxidized steel in the pipe are rolled out with the steel, producing the lamination. Laminations can also be caused when the ingot is rolled at too low a temperature or pressure.

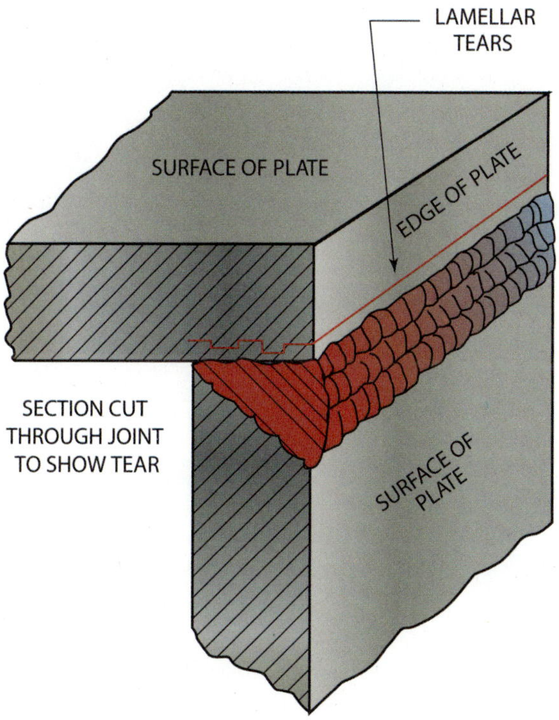

FIGURE 30-16 Example of lamellar tearing.

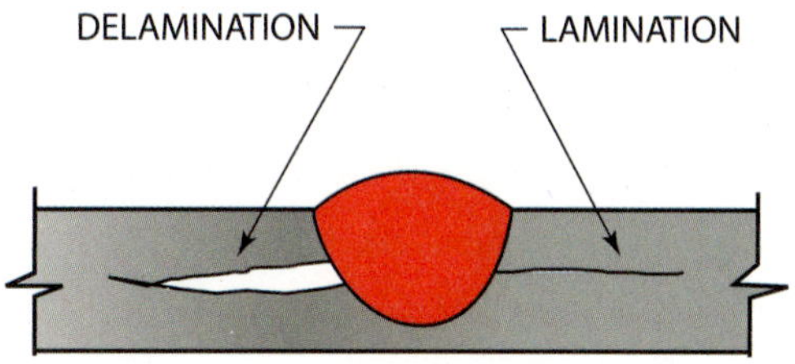

FIGURE 30-17 Lamination and delamination.

Delamination

When laminations intersect a joint being welded, the heat and stresses of the weld may cause some laminations to become delaminated. Contaminating of the weld metal may occur if the lamination contained large amounts of slag, mill scale, dirt, or other undesirable materials. Such contamination can cause wormhole porosity or lack-of-fusion defects.

The problems associated with delaminations are not easily corrected. If a thick plate is installed in a compression load, an effective solution can be to weld over the lamination to seal it. A better solution is to replace the steel.

A solution to the problem is to redesign the joints in order to impose the lowest possible strain throughout the plate thickness. This can be accomplished by making smaller welds so that each subsequent weld pass heat-treats the previous pass to reduce the total stress in the finished weld, Figure 30-18. The joint design can be changed to reduce the stress on the through thickness of the plate, Figure 30-19.

FIGURE 30-18 Using multiple welds to reduce weld stresses.

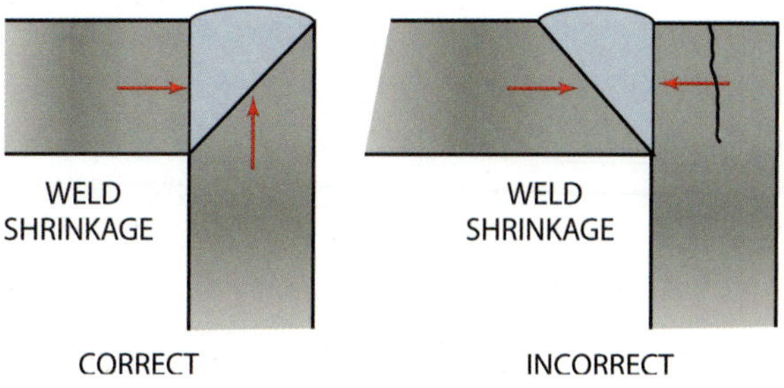

FIGURE 30-19 Correct joint design to reduce lamellar tears.

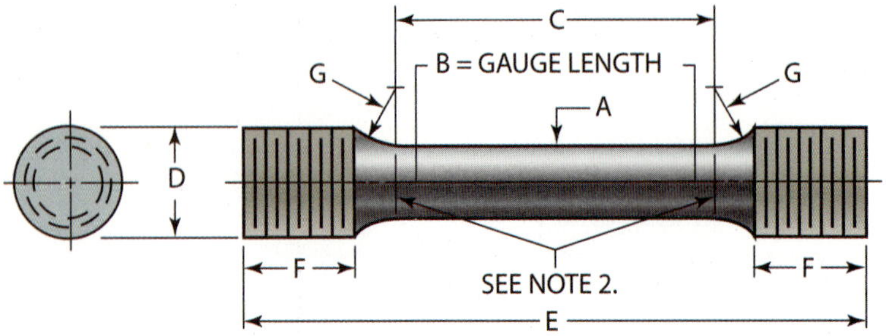

NOTE 1: DIMENSION A, B, AND C SHALL BE AS SHOWN, BUT ALTERNATE SHAPES OF ENDS MAY BE USED AS ALLOWED BY ASTM SPECIFICATION E-8.

NOTE 2: IT IS DESIRABLE TO HAVE THE DIAMETER OF THE SPECIMEN WITHIN THE GAUGE LENGTH SLIGHTLY SMALLER AT THE CENTER THAN AT THE ENDS. THE DIFFERENCE SHALL NOT EXCEED 1% OF THE DIAMETER.

FIGURE 30-20 Tensile testing specimen. American Welding Society

Destructive Testing (DT)

Tensile Testing

Tensile tests are performed with specimens prepared as round bars or flat strips. The simple round bars are often used for testing only the weld metal, sometimes called "all weld metal testing." This test can be used on thick sections where base metal dilution into all of the weld metal is not possible. Round specimens are cut from the center of the weld metal. The flat bars are often used to test both the weld and the surrounding metal. Flat bars are usually cut at a 90° angle to the weld, Figure 30-20. Table 30-4 shows how a number of standard smaller-size bars can be used, depending on the thickness of the metal to be tested. Bar size also depends on the size of the tensile testing equipment available for the testing, Figure 30-21.

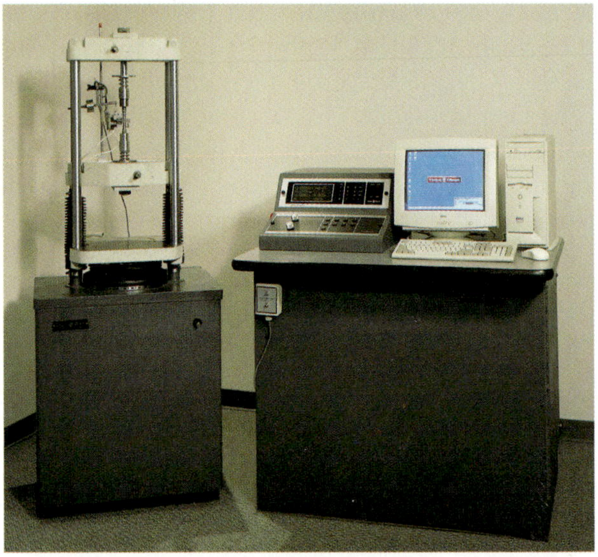

FIGURE 30-21 Typical tensile tester used for measuring the strength of welds (60,000-lb universal testing machines). American Welding Society

Specimen	Dimensions of Specimen						
	in./mm A	in./mm B	in./mm C	in./mm D	in./mm E	in./mm F	in./mm G
C-1	.500/12.7	2/50.8	2.25/57.1	.750/19.05	4.25/107.9	.750/19.05	.375/9.52
C-2	.437/11.09	1.750/44.4	2/50.8	.625/15.8	4/101.6	.750/19.05	.375/9.52
C-3	.357/9.06	1.4/35.5	1.750/44.4	.500/12.7	3.500/88.9	.625/15.8	.375/9.52
C-4	.252/6.40	1.0/25.4	1.250/31.7	.375/9.52	2.50/63.5	.500/12.7	.125/3.17
C-5	.126/3.2	.500/12.7	.750/19.05	.250/6.35	1.750/44.4	.375/9.52	.125/3.17

Table 30-4 Dimensions of Tensile Testing Specimens

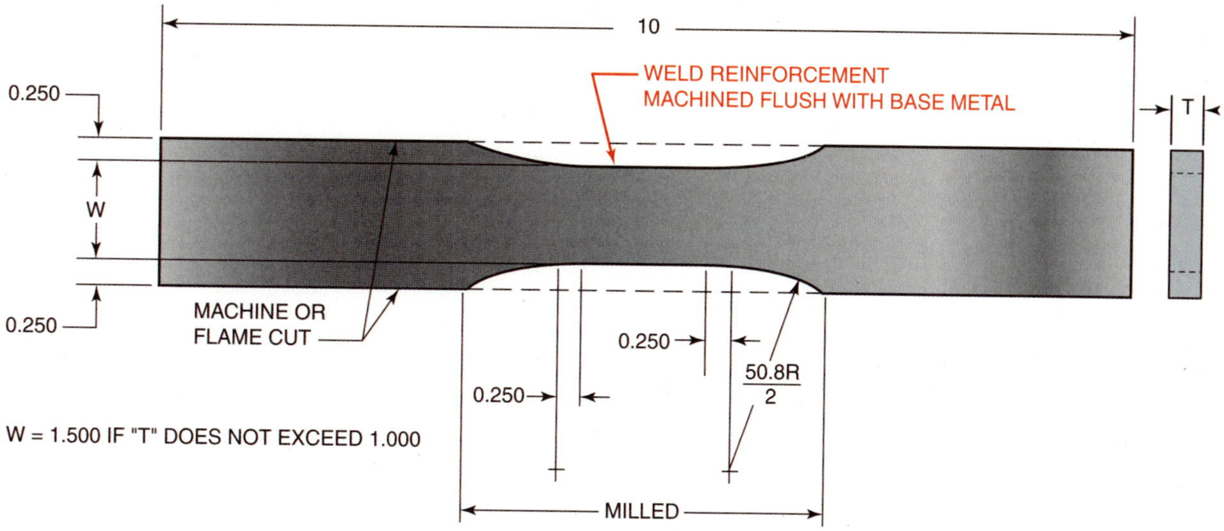

FIGURE 30-22 Tensile specimen for flat plate weld. American Welding Society

Two flat specimens are used, commonly for testing thinner sections of metal. When testing welds, the specimen should include the heat-affected zone and the base plate. If the weld metal is stronger than the plate, failure occurs in the plate; if the weld is weaker, failure occurs in the weld.

After the weld section is machined to the specified dimensions, it is placed in the tensile testing machine and pulled apart. A specimen used to determine the strength of a welded butt joint for plate is shown in Figure 30-22.

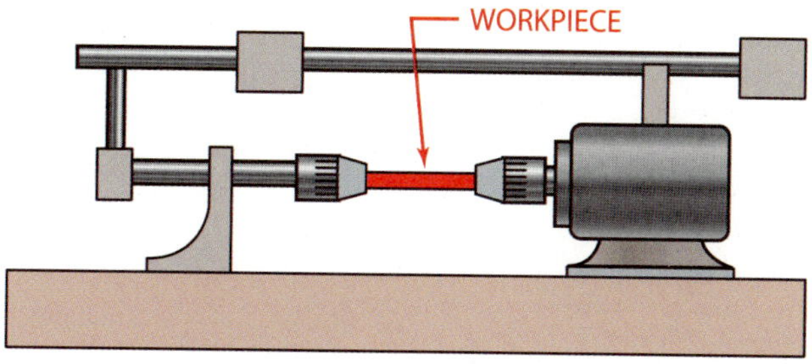

FIGURE 30-23 Fatigue testing. The specimen is placed in chucks of the machine. The machine is turned on, and as it rotates, the specimen is alternately bent twice for each revolution. American Welding Society

Fatigue Testing

Fatigue testing is used to determine how well a weld can resist repeated fluctuating stresses or cyclic loading. The maximum value of the stresses is less than the tensile strength of the material. Fatigue strength can be lowered by improperly made weld deposits, which may be caused by porosity, slag inclusions, lack of penetration, or cracks. Any one of these discontinuities can act as a point of stress, eventually resulting in the failure of the weld.

In the fatigue test, the part is subjected to repeated changes in applied stress. This test may be performed in one of several ways, depending upon the type of service the tested part must withstand. The results obtained are usually reported as the number of stress cycles that the part will resist without failure and the total stress used.

In one type of test, the specimen is bent back and forth. This test subjects the part to alternating compression and tension. A fatigue testing machine is used for this test, Figure 30-23. The machine is turned on, and, as it rotates, the specimen is alternately bent twice for each revolution. In this case, failure is usually rapid.

Shearing Strength of Welds

The two forms of **shearing strength** of welds are transverse shearing strength and longitudinal shearing strength. To test transverse shearing strength, a specimen is prepared as shown in Figure 30-24. The width of the specimen is measured in inches or millimeters. A tensile load is applied, and the specimen is ruptured. The maximum load in pounds or kilograms is then determined.

To test longitudinal shearing strength, a specimen is prepared as shown in Figure 30-25. The length

CONVERSION TABLE – MILLIMETERS TO INCHES

DIM-mm	TOL	DIM-in.
9.52		0.375
9.52	± 1.58	0.375
12.70		0.500
19.05		0.750
50.60		2.000
63.50		2.500
228.60		9.000
114.30		4.500

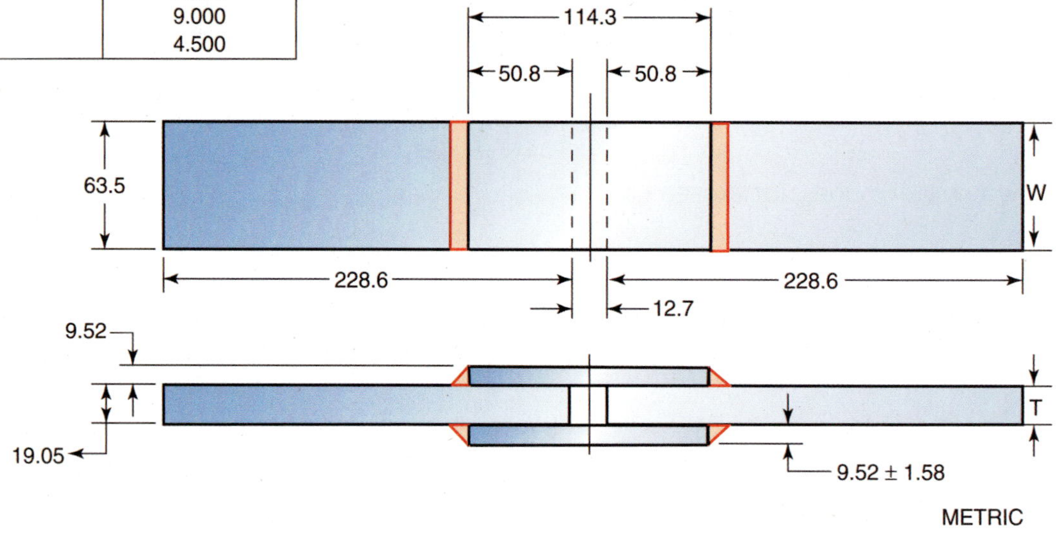

FIGURE 30-24 **Transverse fillet weld shearing specimen after welding.** American Welding Society

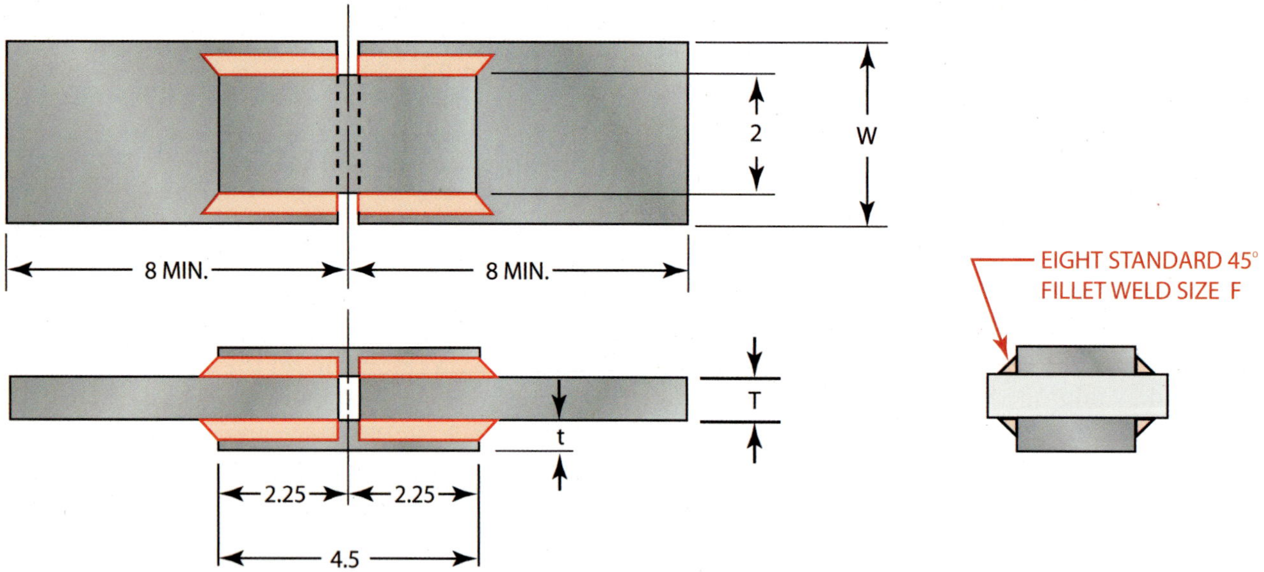

DIMENSIONS			
	in. mm.	in. mm.	in. mm.
THICKNESS - F	1/8 - 3.17	1/4 - 6.35	3/8 - 9.52
THICKNESS - t	3/8 - 9.52	1/2 - 12.7	3/4 - 19.02
THICKNESS - T	3/8 - 9.52	3/4 -19.05	1 - 25.4
WIDTH - W	3 - 76.2	3 - 76.2	3 - 76.2

FIGURE 30-25 **Longitudinal fillet weld shear specimen.** American Welding Society

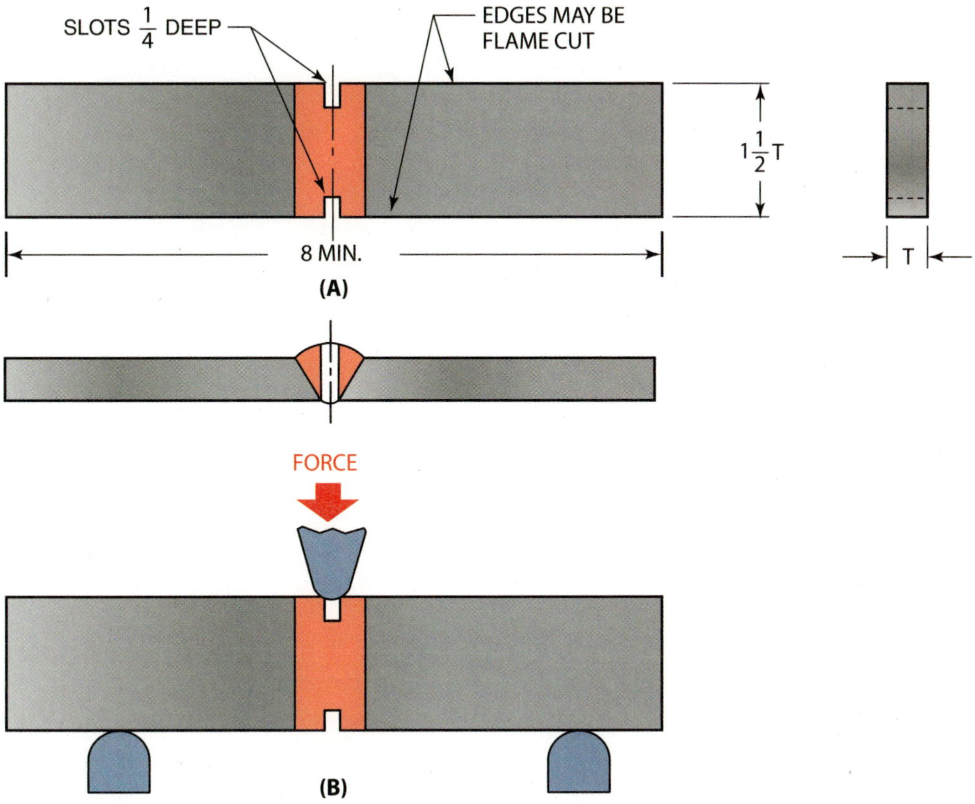

FIGURE 30-26 (A) Nick-break specimen for butt joints in plate. (B) Method of rupturing nick-break specimen. American Welding Society

of each weld is measured in inches or millimeters. The specimen is then ruptured under a tensile load, and the maximum force in pounds or kilograms is determined.

Welded Butt Joints

The three methods of testing welded butt joints are (1) the nick-break test, (2) the guided-bend test, and (3) the free-bend test. It is possible to use variations of these tests.

Nick-Break Test

A specimen for this test is prepared as shown in Figure 30-26A. The specimen is supported as shown in Figure 30-26B. A force is then applied, and the specimen is ruptured by one or more blows of a hammer. Theoretically, the rate of application could affect how the specimen breaks, especially at a critical temperature. Generally, however, there is no difference in the appearance of the fractured surface due to the method of applying the force. So for all practical purposes, striking the specimen with a small hammer swung rapidly would not affect the results as compared to a strike with a much heavier hammer swung slower. The force may be applied slowly or suddenly. The surfaces of the fracture should be checked for soundness of the weld.

Guided-Bend Test

To test welded, grooved butt joints on metal that is 3/8 in. (10 mm) thick or less, two specimens are prepared and tested—one face bend and one root bend, Figure 30-27A and B. If the welds pass this test, you are qualified to make groove welds on plate having a thickness range of from 3/8 in. to 3/4 in. (10 mm to 19 mm). These welds need to be machined as shown in Figure 30-28A. If these specimens pass, you will also be qualified to make fillet welds on materials of any (unlimited) thicknesses. For welded, grooved butt joints on metal 1/2 in. (13 mm) thick, two side-bend specimens are prepared and tested, Figure 30-28B. If the welds pass this test, you are qualified to weld on metals of unlimited thickness.

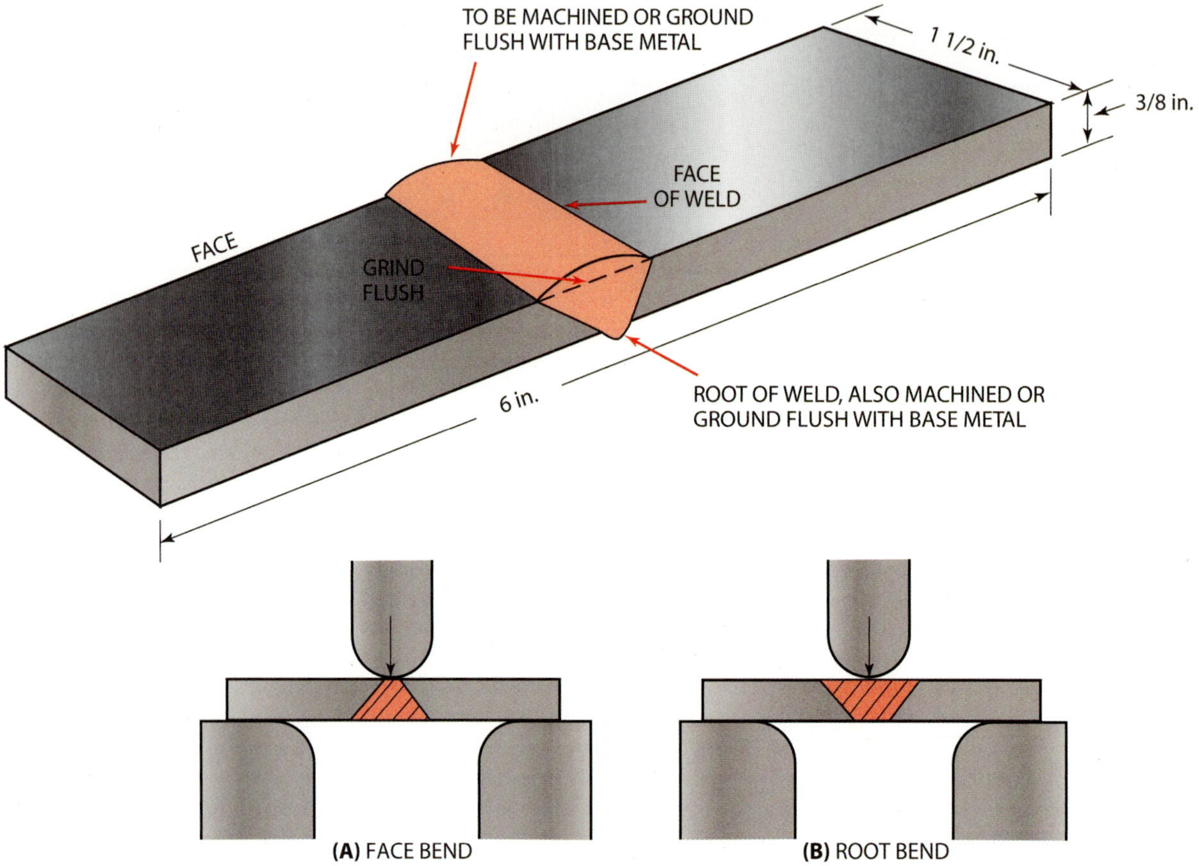

FIGURE 30-27 Root- and face-bend specimens for 3/8-in. (10-mm) plate. American Welding Society

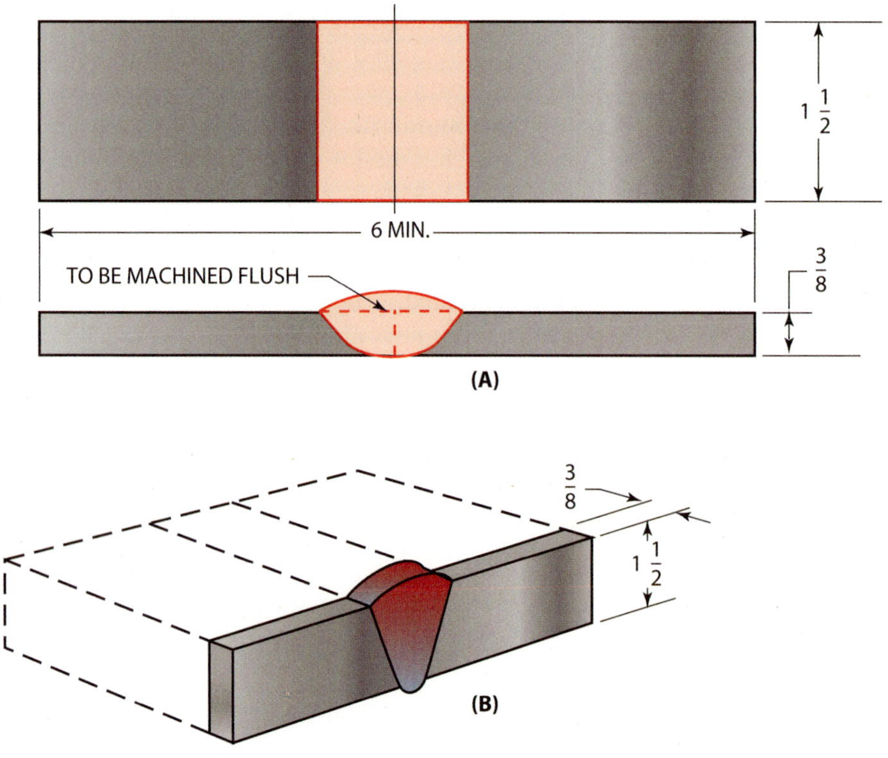

mm	CONVERSION in.
1.5	1/16
3.1	1/8
9.5	3/8
12.7	1/2
38.0	1 1/2
152.0	6

FIGURE 30-28 (A) Root- and face-bend specimens. (B) Side-bend specimen. American Welding Society

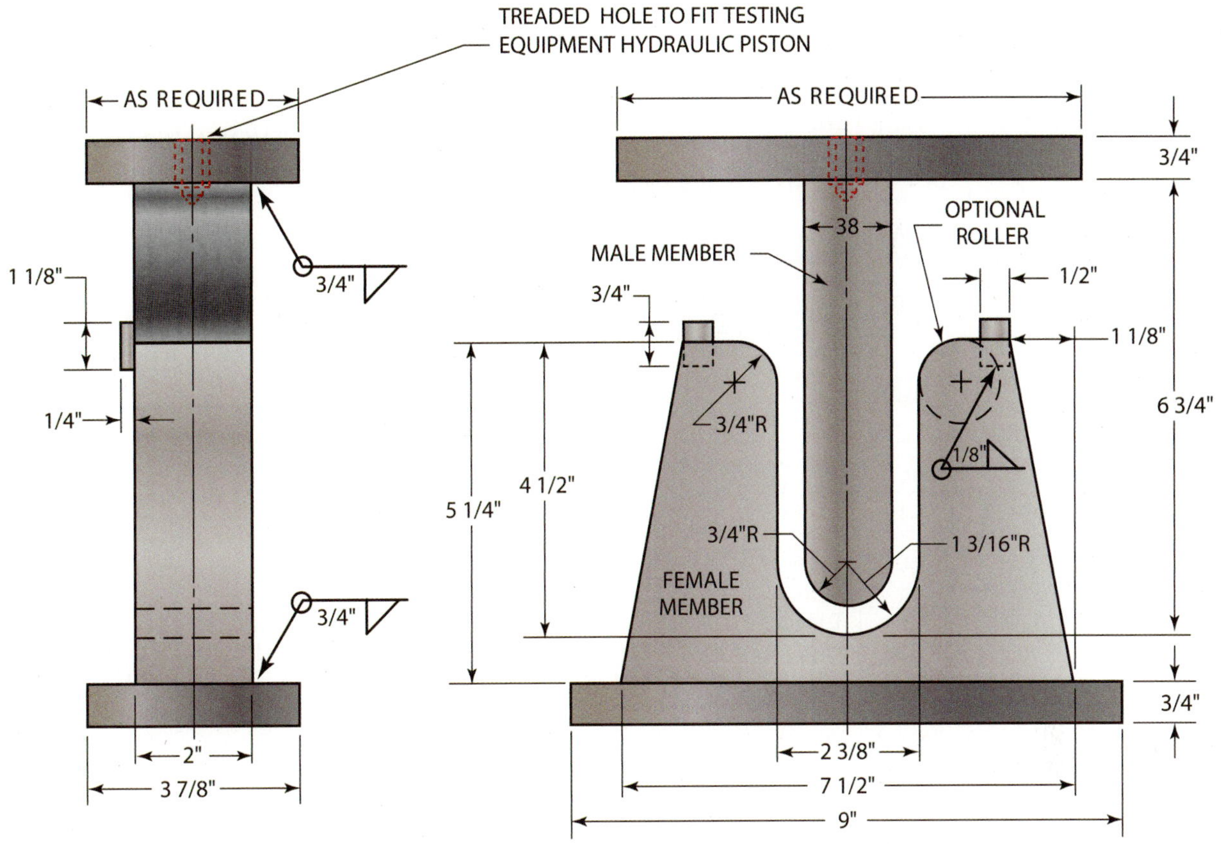

FIGURE 30-29 Fixture for guided-bend test. American Welding Society

When the specimens are prepared, caution must be taken to ensure that all grinding marks run longitudinally to the specimen so that they do not cause stress cracking. In addition, the edges must be rounded to reduce cracking that tends to radiate from sharp edges. The maximum ratio of this rounded edge is 1/8 in. (3 mm).

Procedure

The jig shown in Figure 30-29 is commonly used to bend most specimens. Not all guided-bend testers have the same bending radius. Codes specify different bending radii depending on material type and thickness. Place the specimens in the jig with the weld in the middle. Face-bend specimens should be placed with the face of the weld toward the gap. Root-bend specimens should be positioned so that the root of the weld is directed toward the gap. Side-bend specimens are placed with either side facing up. The guided-bend specimen must be pushed all the way through open (roller-type) bend testers and within 1/8 in. (3 mm) of the bottom on fixture-type bend testers.

Once the test is completed, the specimen is removed. The convex surface is then examined for cracks or other discontinuities and judged acceptable or unacceptable according to specified criteria. Some surface cracks and openings are allowable under codes.

Free-Bend Test

The free-bend test is used to test welded joints in a plate. A specimen is prepared as shown in Figure 30-30. Note that the width of the specimen is 1.5 multiplied by the thickness of the specimen. Each corner lengthwise should be rounded in a radius not exceeding one-tenth the thickness of the specimen. If the surfaces are ground, hold the grinder so that the grinder marks on the plate run lengthwise on the specimen. Grinder marks that run across the specimen can start cracks that might not have occurred with lengthwise grinder marks.

CONVERSION TABLE							
in.	mm	in.	mm	in.	mm	in.	mm
1/4	6.35	9/16	14.2	1 1/8	28.5	2	50.8
3/8	9.52	3/4	19.05	1 1/4	31.7	6	152.4
1/2	12.7	15/16	23.8	1 1/2	38.1	8	203.2
5/8	15.8	1	25.4	1 7/8	47.6	9	228.6

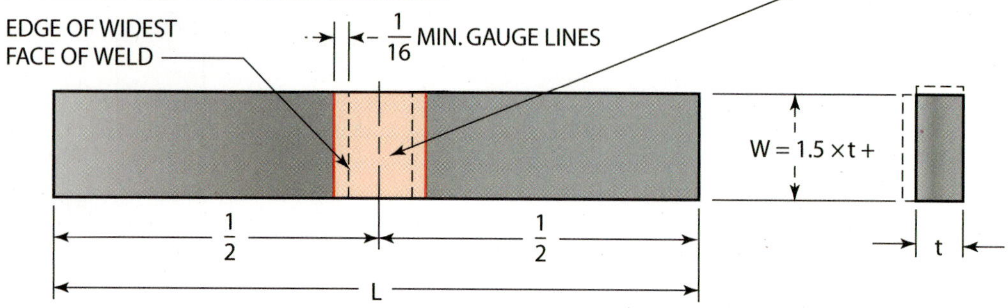

DIMENSIONS							
T, inches	1/4	3/8	1/2	5/8	3/4	1	1 1/4
W, inches	3/8	9/16	3/4	15/16	1 1/8	1 1/2	1 7/8
L, min. inches	6	8	9	10	12	13 1/2	15
B, min. inches	1 1/4	1 1/4	1 1/4	2	2	2	2

FIGURE 30-30 **Free-bend test specimen.** American Welding Society

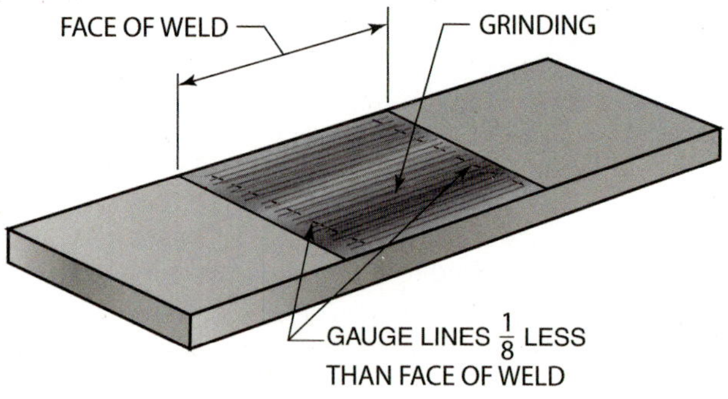

FIGURE 30-31 **Gauge lines are drawn on the weld face of a free-bend specimen.** American Welding Society

Gauge lines are drawn on the face of the weld, Figure 30-31. The distance between the gauge lines is 1/8 in. (3.17 mm) less than the face of the weld. The initial bend of the specimen is completed in the device illustrated in Figure 30-32. The gauge line surface should be directed toward the supports. The weld is located in the center of the supports and loading block.

Alternate Bend

The initial bend may be made by placing the specimen in the jaws of a vise with one-third the length projecting from the jaws. The specimen is then bent away from the gauge lines through an angle of 30° to 45° by blows of a hammer. The specimen is then inserted into the jaws of a vise and pressure is applied by tightening the vise. The pressure is continued until a crack or depression appears on the convex face of the specimen. The load is then removed.

The elongation is determined by measuring the minimum distance between the gauge lines along the convex surface of the weld to the nearest 0.01 in. (0.254 mm) and subtracting the initial gauge length. The percent of elongation is obtained by dividing the elongation by the initial gauge length and multiplying by 100.

Fillet Weld Break Test

The specimen for this test is made as shown in Figure 30-33A. In Figure 30-33B, a force is applied to the specimen until the weld ruptures or the base metal breaks. Any convenient means of applying the force may be used, such as an arbor press, a testing machine, or hammer blows. The break surface should then be examined for soundness—that is, slag inclusions, overlap, porosity, lack of fusion, or other discontinuities.

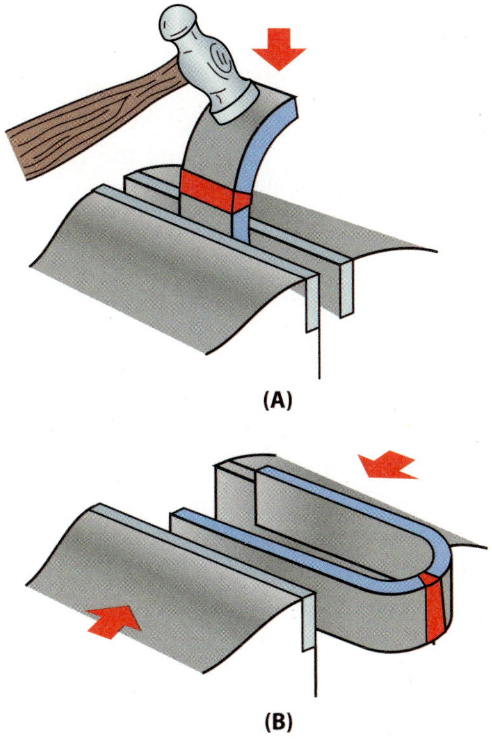

CONVERSION TABLE	
mm	in.
12.7	0.500
20.0	0.787
32.0	1.25
76.0	3.000

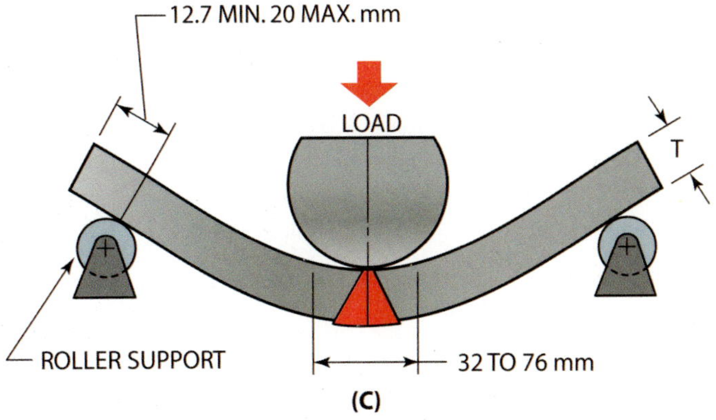

FIGURE 30-32 Free-bend test: (A) initial bend can be made in this manner; (B) a vise can be used to make the final bend; and (C) another method used to make the bend. American Welding Society

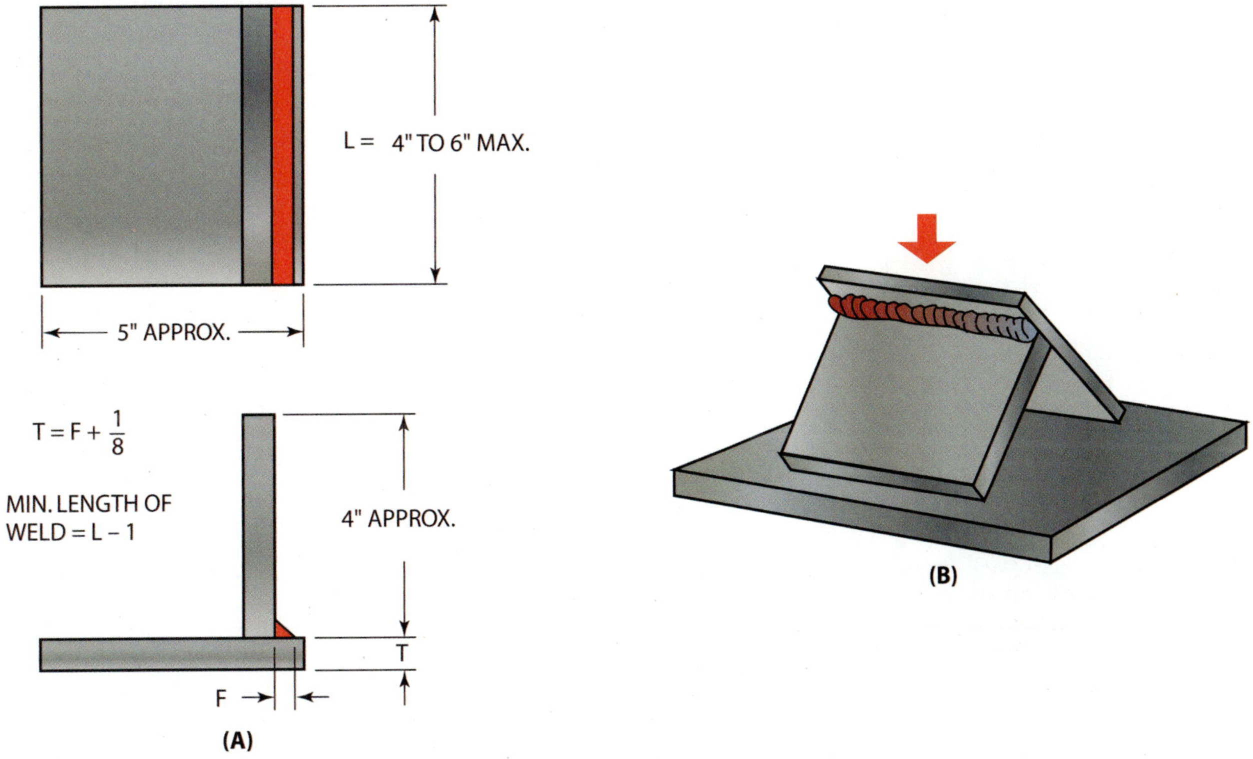

FIGURE 30-33 (A) Fillet weld break test. (B) Method of rupturing fillet weld break specimen.
American Welding Society

Impact Testing

A number of tests can be used to determine the impact capability of a weld. One common test is the Izod test, Figure 30-34A, in which a notched specimen is struck by an anvil mounted on a pendulum. The energy in foot-pounds read on the scale mounted on the machine required to break the specimen is an indication of the impact resistance of the metal. This test compares the toughness of the weld metal with the base metal.

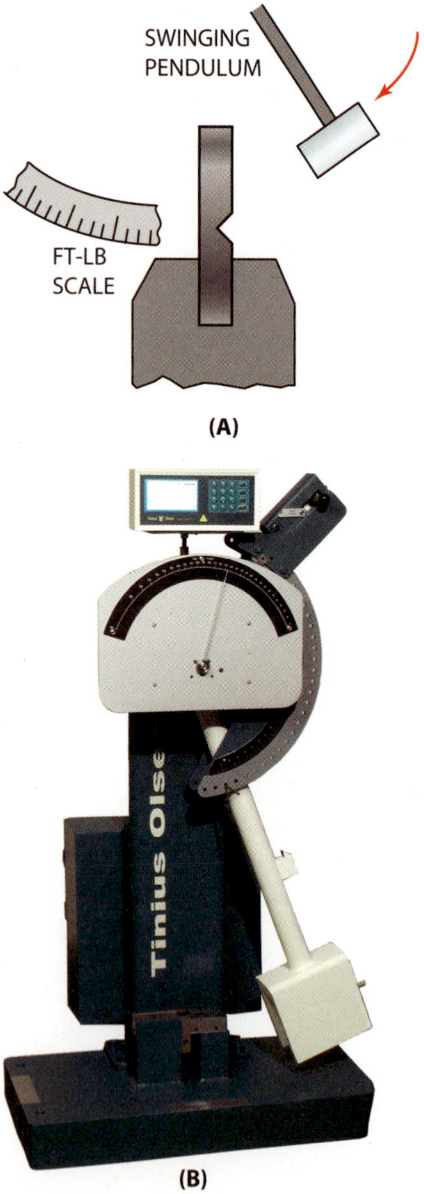

FIGURE 30-34 Impact testing: (A) specimen mounted for Izod impact toughness, and (B) a typical impact tester used for measuring the toughness of metals. (A) © Cengage Learning 2012 (B) Tinius Olsen Testing Machine Co., Inc.

Another type of impact test is the Charpy test. This test is similar to the Izod test. The major differences between the tests are that the Izod test specimen is gripped on one end and is held vertically and usually tested at room temperature, and the Charpy test specimen is held horizontally, supported on both ends and is usually tested at a specific temperature. All impact test specimens must be produced according to ASTM specifications. A typical impact tester is shown in Figure 30-34B.

Nondestructive Testing (NDT)

Nondestructive testing of welds is a method used to test materials for surface defects such as cracks, arc strikes, undercuts, and lack of penetration. Internal or subsurface defects can include slag inclusions, porosity, and unfused metal in the interior of the weld.

Visual Inspection (VT)

Visual inspection is the most frequently used nondestructive testing method and is the first step in almost every other inspection process. The majority of welds receive only visual inspection. In this method, if the weld looks good, it passes; if it looks bad it is rejected. This procedure is often overlooked when more sophisticated nondestructive testing methods are used. However, it should not be overlooked.

An active visual inspection schedule can reduce the finished weld rejection rate by more than 75%. Visual inspection can easily be used to check for fit-up, interpass acceptance, your technique, and other variables that will affect the weld quality. Minor problems can be identified and corrected before a weld is completed. This eliminates costly repairs or rejection.

Visual inspection should be used before any other nondestructive or mechanical tests are used to eliminate (reject) the obvious problem welds. Eliminating welds that have excessive surface discontinuities that will not pass the code or standards being used saves preparation time.

Penetrant Inspection (PT)

Penetrant inspection is used to locate minute surface cracks and porosity. Two types of penetrants are now in use, the color-contrast and the fluorescent versions. Color-contrast, often red, penetrants contain a colored dye that shows under ordinary white light. Fluorescent penetrants contain a more effective fluorescent dye that shows under black light.

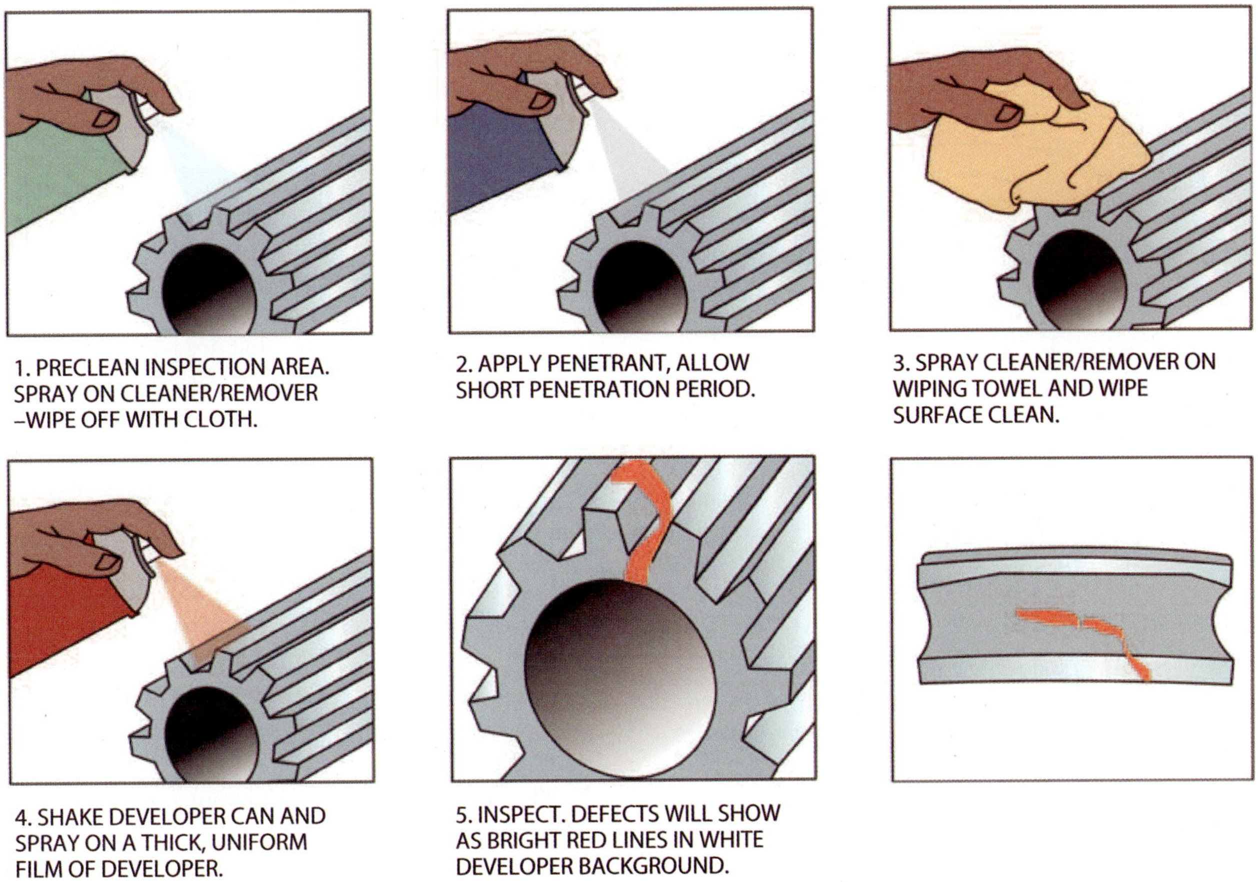
1. PRECLEAN INSPECTION AREA. SPRAY ON CLEANER/REMOVER –WIPE OFF WITH CLOTH.
2. APPLY PENETRANT, ALLOW SHORT PENETRATION PERIOD.
3. SPRAY CLEANER/REMOVER ON WIPING TOWEL AND WIPE SURFACE CLEAN.
4. SHAKE DEVELOPER CAN AND SPRAY ON A THICK, UNIFORM FILM OF DEVELOPER.
5. INSPECT. DEFECTS WILL SHOW AS BRIGHT RED LINES IN WHITE DEVELOPER BACKGROUND.

FIGURE 30-35 Penetrant testing. Magnaflux, A Division of ITW

The following steps outline the procedure to be followed when using a penetrant.

1. Precleaning. The test surface must be clean and dry. Suspected flaws must be cleaned and dried so that they are free of oil, water, or other contaminants.
2. The test surface must be covered with a film of penetrant by dipping, immersing, spraying, or brushing.
3. The test surface is then gently wiped, washed, or rinsed free of excess penetrant. It is dried with cloths or hot air.
4. A developing powder applied to the test surface acts as a blotter to speed the tendency of the penetrant to seep out of any flaws open to the test surface.
5. Depending upon the type of penetrant applied, visual inspection is made under ordinary white light or near-ultraviolet black light, Figure 30-35. When viewed under this light, the penetrant fluoresces to a yellow-green color, which clearly defines the defect.

Magnetic Particle Inspection (MT)

Magnetic particle inspection uses finely divided ferromagnetic particles (powder) to indicate defects open to the surface or just below the surface on magnetic materials.

A magnetic field is induced in a part by passing an electric current through or around it. The magnetic field is always at right angles to the direction of current flow. Ferromagnetic powder registers an abrupt change in the resistance in the path of the magnetic field, such as would be caused by a crack lying at an angle to the direction of the magnetic poles at the crack. Finely divided ferromagnetic particles applied to the area will be attracted and outline the crack.

In Figure 30-36, the flow or discontinuity interrupting the magnetic field in a test part can be either longitudinal or circumferential. A different type of magnetization is used to detect defects that run down the axis, as opposed to those occurring around the girth of a part. For some applications you may need to test in both directions.

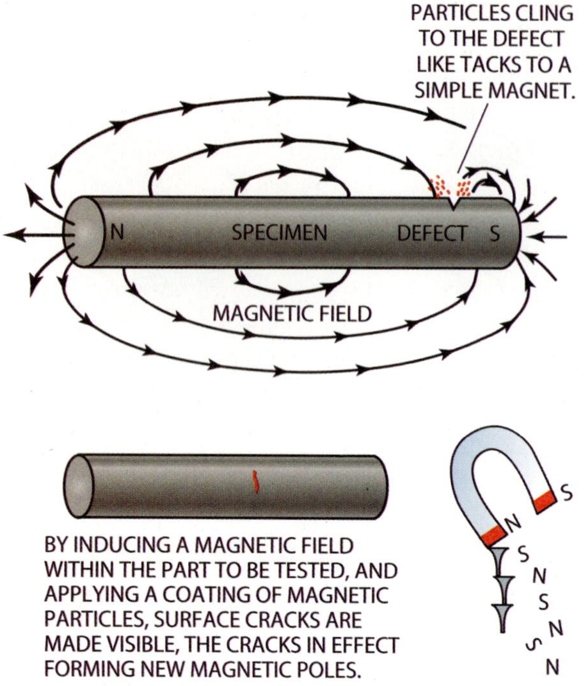

FIGURE 30-36 Flaws and discontinuities interrupt magnetic fields. Magnaflux, A Division of ITW

In Figure 30-37A, longitudinal magnetization allows detection of flaws running around the circumference of a part. The user places the test part inside an electrified coil. This induces a magnetic field down the length of the test part. In Figure 30-37B, circumferential magnetization allows detection of flaws occurring down the length of a test part. An electric current is sent down the length of the part to be inspected. The magnetic field thus induced allows defects along the length of the part to be detected.

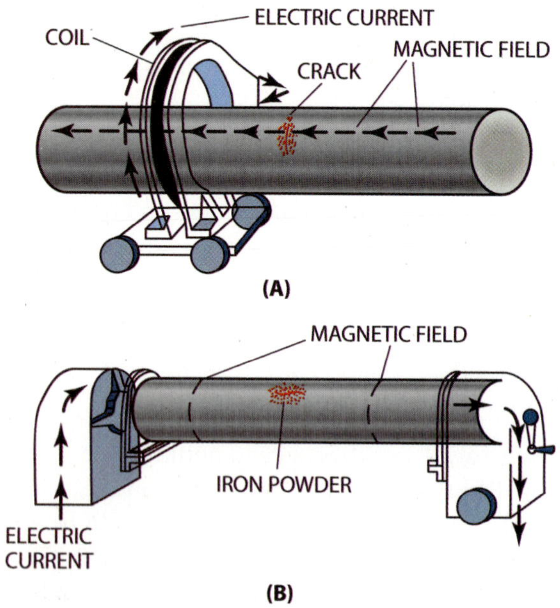

FIGURE 30-37 (A) Longitudinal magnetic field. (B) Circumferential magnetic field.
Magnaflux, A Division of ITW

Radiographic Inspection (RT)

Radiographic inspection (RT) is a method for detecting flaws inside weldments. Radiography gives a picture of all discontinuities that are parallel (vertical) or nearly parallel to the source. Discontinuities that are perpendicular (flat) or nearly perpendicular to the source may not be seen on the X-ray film. Instead of using visible light rays, the operator uses invisible, short-wavelength rays developed by X-ray machines, radioactive isotopes (gamma rays), and variations of these methods. These rays are capable of penetrating solid materials and reveal most flaws in a weldment on an X-ray film or a fluorescent screen. Flaws are revealed on films as dark or light areas against a contrasting background after exposure and processing, Figure 30-38.

The defect images in radiographs measure differences in how the X-rays are absorbed as they penetrate

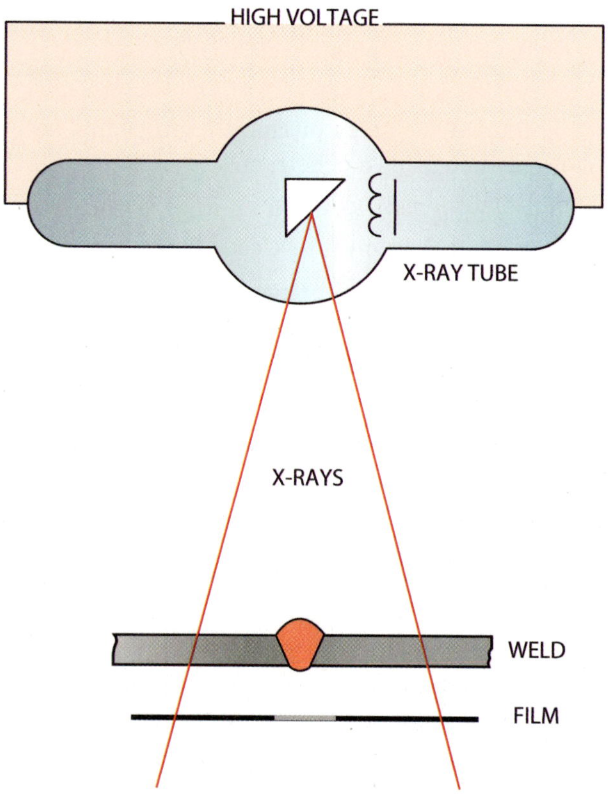

FIGURE 30-38 Schematic of an X-ray system.
© Cengage Learning 2012

the weld. The weld itself absorbs most X-rays. If something less dense than the weld is present, such as a porosity or lack-of-fusion defect, fewer X-rays are absorbed, darkening the film. If something more dense is present, such as heavy ripples on the weld surface, more X-rays will be absorbed, lightening the film.

Therefore, the foreign material's relative thickness (or lack of it) and differences in X-ray absorption determine the radiograph image's final shape and shading. Skilled readers of radiographs can interpret the significance of the light and dark regions by their shape and shading. The X-ray image is a shadow of the flaw. The farther the flaw is from the X-ray film, the fuzzier the image appears. Those skilled at interpreting weld defects in radiographs must also be very knowledgeable about welding.

Ultrasonic Inspection (UT)

Ultrasonic inspection (UT) is fast and uses few consumable supplies, which makes it inexpensive for schools to use. However, because of the time required for most UT testing, it is not as economical in the field for nondestructive testing. This inspection method employs electronically produced high-frequency sound waves that penetrate metals and many other materials at speeds of several thousand feet (meters) per second. A portable ultrasonic inspection unit is shown in Figure 30-39.

The two types of ultrasonic equipment are pulse and resonance. The pulse-echo system, most often employed in the welding field, uses sound generated in short bursts or pulses. Since the high-frequency sound used is at a relatively low power, it has little ability to travel through air, so it must be conducted from the probe into the part through a medium such as oil or water.

Sound is directed into the part with a probe held in a preselected angle or direction so that flaws will reflect some energy back to the probe. These ultrasonic devices operate very much like depth sounders, or "fish finders." The speed of sound through a material is a known quantity. These devices measure the time required for a pulse to return from a reflective surface. Internal computers calculate the distance and present the information on a display screen where an operator can interpret the results. The signals can be "monitored" electronically to operate alarms, print systems, or recording equipment. Sound not reflected by flaws continues into the part. If the angle is correct, the sound energy will be reflected back to the probe from the opposite side. Flaw size is determined by plotting the length, height, width, and shape using trigonometric rules.

Figure 30-40 shows the path of the sound beam in butt welding testing. The operator must know the exit point of the sound beam, the exact angle of the refracted beam, and the thickness of the plate when using shear wave and compression wave forms.

Leak Checking

Leak checking can be performed by filling the welded container with either a gas or liquid. There may or may not be additional pressure applied to the material in the weldment. Water is the most frequently used liquid, although sometimes a liquid with a lower viscosity is used. If gas is used, it may be either a gas that can be detected with an instrument when it escapes through a flaw in the weld or an air leak that is checked with bubbles.

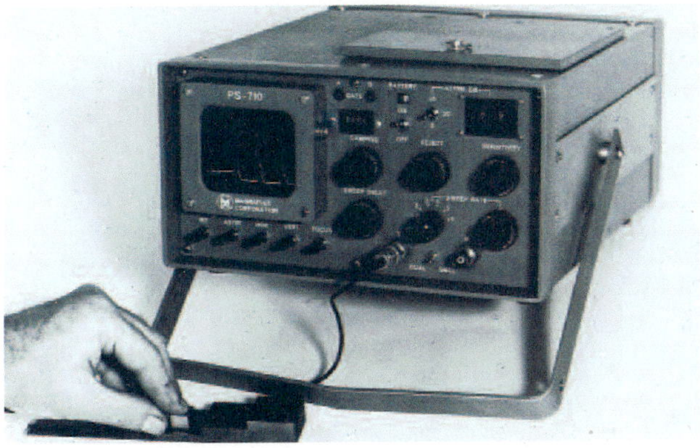

FIGURE 30-39 Portable ultrasonic inspection unit.
Magnaflux, A Division of ITW

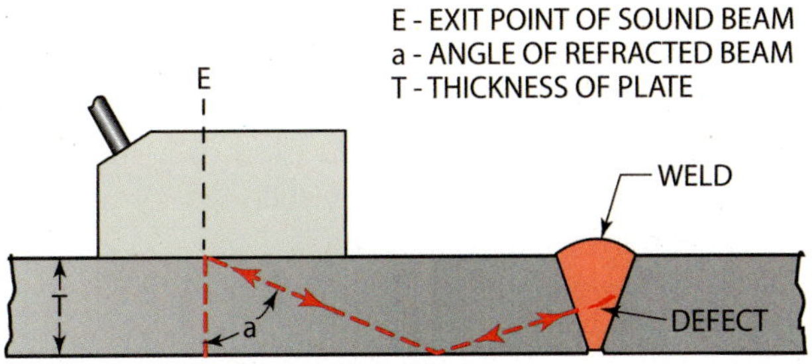

FIGURE 30-40 Ultrasonic testing.
© Cengage Learning 2012

Eddy Current Inspection (ET)

Eddy current inspection (ET) is another nondestructive test. This method is based upon magnetic induction in which a magnetic field induces eddy currents within the material being tested. An eddy current is an induced electric current circulating wholly within a mass of metal. This method is effective in testing nonferrous and ferrous materials for internal and external cracks, slag inclusions, porosity, and lack of fusion that are on or very near the surface. Eddy current cannot locate flaws that are not near the surface.

A coil-carrying high-frequency alternating current is brought close to the metal to be tested. A current is produced in the metal by induction. The part is placed in or near high-frequency, alternating-current coils. The magnitude and phase difference of these currents is, in turn, indicated in the actual impedance value of the pickup coil. Careful measurement of this impedance is the revealing factor in detecting defects in the weld.

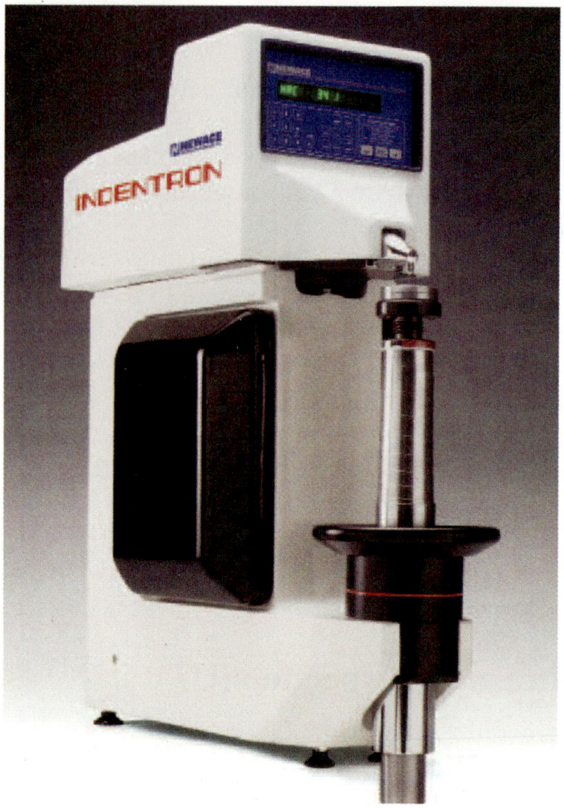

FIGURE 30-41 Rockwell hardness tester.
Newage Testing Instruments, an AMETEK Co.

Hardness Testing

Hardness is the resistance of metal to penetration and is an index of the wear resistance and strength of the metal. Hardness tests can be used to determine the relative hardness of the weld with the base metal. The two types of hardness testing machines in common use are the Rockwell and the Brinell testers.

The **Rockwell hardness tester**, Figure 30-41, uses a 120° diamond cone for hard metals and a 1/16-in. (1.58-mm) and 1/8-in. (3.175-mm) diameter hardened steel ball for softer metals. The method measures resistance to penetration. The hardness is read directly from a dial on the tester. The depth of the impression is measured instead of the diameter. The tester has two scales for reading hardness, known as the B-scale and the C-scale. The C-scale is used for harder metals, the B-scale for softer metals.

The **Brinell hardness tester** measures the resistance of material to the penetration of a steel ball under constant pressure (about 3000 kg) for a minimum of approximately 30 seconds, Figure 30-42. The diameter is measured microscopically, and the Brinell number is checked on a standard chart. Brinell hardness numbers are obtained by dividing the applied load by the area of the surface indentation.

FIGURE 30-42 Brinell hardness tester.
Newage Testing Instruments, an AMETEK Co.

SUMMARY

It is impossible to "inspect in" quality; no amount of inspection will produce a quality product. Quality is something that must be built into a product. The purpose of testing and inspection is to verify that the required level of quality is being maintained. The most important standard that a weld must meet is that it must be fit for service. A weld must be able to meet the demands placed on the weldment without failure. No weld is perfect. It is important for both you and the welding inspector to know the appropriate level of weld discontinuities. Producing or inspecting welds to an excessively high standard will result in a product that is excessively expensive.

Knowing the cause of weld defects and discontinuities can aid you in producing higher-quality welds by recognizing the cause and effect of such problems. Some people believe that high-quality welds are a matter of luck, which is not true. Producing high-quality welds is a matter of skill and knowledge, which you must develop. As you get better at welding skills, you will often know whether the welds you are producing will or will not pass inspection. Inspecting your welds as part of your training program will help you develop this skill.

REVIEW QUESTIONS

1. What are the two methods used in weld quality control?
2. What is the difference between mechanical and nondestructive testing?
3. What percent of critical welds or parts are usually tested?
4. What term is used to describe a weld discontinuity or flaw that is bad enough to cause the part to be rejected?
5. Define *tolerance.*
6. What three factors should be considered when evaluating a weld discontinuity?
7. Which type of discontinuity forms bubbles that are trapped in the weld as the metal cools?
8. What is porosity most often caused by?
9. What is an inclusion?
10. What is the most common cause of inadequate joint penetration?
11. Describe incomplete fusion.
12. What problem can arc strikes cause?
13. When does overlap occur?
14. Draw a sketch showing an undercut weld on a T joint.
15. At what point during the welding process can crater cracks form?
16. Sketch a drawing of underfill on a groove weld.
17. What happens when lamellar tears are severely stressed?
18. What is the purpose of tensile testing?
19. What is the purpose of fatigue testing?
20. After rupturing a specimen in a nick-break test, what should then be done?
21. Describe the guided-bend test.
22. How is the force applied for a fillet weld break test?
23. What surface and subsurface defects does nondestructive testing reveal?
24. Why should welds be visually inspected before any other tests are used?
25. Penetrant inspection is used to locate what kinds of weld defects?
26. How is a defect revealed with magnetic particle inspection?
27. What types of flaws are revealed by radiographic inspection?
28. What are some advantages of using the ultrasonic inspection method?
29. When leak checking, what is the welded container filled with?
30. List two instruments that can be used to test metal hardness.

Glossary

The terms and definitions in this glossary are extracted from the American Welding Society publication AWS A3.0-80 Welding Terms and Definitions. The terms with an asterisk are from a source other than the American Welding Society. Note: The English term and definition are given first, followed by the same term and definition in Spanish.

A

***abrasives.** Materials that are usually sharp and are used to clean or grind a surface. They may be used as a powder such as sand to blast the surface or they may be formed into disks or stones to be used by a grinder.

***abrasivos.** Materiales que por lo general son afilados o cortantes y que se utilizan para limpiar o esmerilar una superficie. Se pueden utilizar en forma de polvo, como la arena, para desgastar la superficie o se pueden fabricar con forma de discos o piedras para usar con una esmeriladora.

***absolute pressure.** The sum of the gauge pressure and the atmospheric pressure.

***presión absoluta.** La suma de la presión manómetro y la presión atmosférica.

absorptive lens. A filter lens designed to attenuate the effects of glare and reflected and stray light.

lente absorbente. Un lente de filtro diseñado para disminuir los efectos de la luz y la reflexión de la luz extraviada.

acceptable criteria. Agreed upon standards that must be satisfactorily met.

criterios aceptables. Las normas sobre las que se ha llegado a un acuerdo y que deben cumplirse en forma satisfactoria.

acceptable weld. A weld that meets all the requirements and the acceptance criteria prescribed by welding specifications.

soldadura aceptable. Una soldadura que satisface los requisitos y el criterio aceptable prescribida por las especificaciónes de la soldadura.

***acetone.** A fragrant liquid chemical used in acetylene cylinders. The cylinder is filled with a porous material and acetone is then added to fill. Acetylene is then added and absorbed by the acetone, which can absorb up to 28 times its own volume of the gas.

***acetona.** Un liquido fragante químico que se usa en los cilindros del acetileno. El cilindro se llena de un material poroso y luego se le agrega la acetona hasta que se llene. El acetileno es absorbido por la acetona, la cual puede absorber 28 veces el propio volumen del gas.

***acetylene.** A fuel gas used for welding and cutting. It is produced as a result of the chemical reaction between calcium carbide and water. The chemical formula for acetylene is C_2H_2. It is colorless, is lighter than air, and has a strong garlic-like smell. Acetylene is unstable above pressures of 15 psig (1.05 kg/cm^2 g). When burned in the presence of oxygen, acetylene produces one of the highest flame temperatures available.

***acetileno.** Un gas combustible que se usa para soldar y cortar. Es producido a consecuencia de una reacción química de agua y calcio y carburo. La fórmula química para el acetileno es C_2H_2. No tiene color, es más ligero que el aire, y tiene un olor fuerte como a ajo. El acetileno es inestable en presiones más altas de 15 psig (1.05 kg/cm^2 g). Cuando se quema en presencia del oxígeno, el acetileno produce una de las llamás con una temperatura más alta que la que se utiliza.

***acicular structure.** A fine micrograin structure found in rapidly cooled steel.

***estructura acicular.** Una estructura micro granulada fina que se encuentra en el acero que se ha enfriado con rapidez.

actual throat. See **throat of a fillet weld.**

garganta actual. Vea **garganta de soldadura filete.**

***adaptable.** Capable of making self-directed corrections; in a robot, this is often accomplished with visual, force, or tactile sensors.

***adaptable.** Capaz de hacer correcciones por instrucción propia de un robot, esto se lleva a cabo con sensores tangibles visuales, o de fuerza.

air acetylene welding (AAW). An oxyfuel gas welding process that uses an air-acetylene flame. The process is used without the application of pressure. This is an obsolete or seldom used process.

soldadura de aire acetileno. Un proceso de soldar con gas (oxi/combustible) que usa aire-acetileno sin aplicarse presión. Un proceso anticuado que es una rareza.

air carbon arc cutting (CAC-A). A carbon arc cutting process variation that removes molten metal with a jet of air.

arco de carbón con aire. Un proceso de cortar con arco de carbón variante que quita el metal derretido con un chorro de aire.

allotropic metals. Metals that have specific lattice or crystal structures that form when the metal is cool and that change within the solid metal as it is heated and before it melts.

metales alotrópicos. Metales que tienen un determinado enrejado o estructuras cristalinas que se forman cuando el metal está frío y cambian dentro del metal sólido mientras se lo calienta y antes de que éste se derrita.

allotropic transformation. A change in the crystalline lattice pattern of a metal due to a change in temperature.

transformación alotropico. Un cambio en el modelo cristalino enrejado del metal debido a un cambio en la temperatura.

***alloy.** A metal with one or more elements added to it, resulting in a significant change in the metal's properties.

***aleación.** Un metal en que se le agrega uno o más elementos resultando en un cambio significante en las propiedades del metal.

***alloy steels.** Steels that may contain any number of a variety of elements that are used to change the mechanical properties of steel. The amount of the added elements can range from 1 to 50%.

***aceros de aleación.** Aceros que pueden contener cualquier cantidad de una variedad de elementos que se utilizan para cambiar las propiedades mecánicas del acero. La cantidad de los elementos agregados puede variar del 1 al 50%.

***alloying elements.** Elements in the flux that mix with the filler metal and become part of the weld metal. Major alloying elements are molydenum, nickel, chromium, manganese, and vanadium.

***elementos de mezcla.** Elementos en el flujo que se mezclan con el metal para rellenar y formar parte del metal soldado. Los elementos principales de mezcla son molibdeno, niquel, cromo, manganeso y vanadio.

all-thread. A rod that is threaded from one end to the other. They are available in a wide range of materials, diameters, and lengths.

all-thread. Una barra que se coloca de un extremo a otro. Están disponibles en una amplia gama de materiales, diámetros y longitudes.

all-weld-metal test specimen. A test specimen with the reduced section composed wholly of weld metal.

prueba de metal soldado. Una prueba con una sección reducida compuesta totalmente del metal de la soldadura.

***alphabet of lines.** Lines are the language of drawing; they are used to represent various parts of the object being illustrated. The various line types are collectively known as the Alphabet of Lines.

***alfabeto de líneas.** Las líneas son el idioma del dibujo: se utilizan para representar diversas partes del objeto que se está ilustrando. Los diversos tipos de línea se conocen, de manera colectiva, como el Alfabeto de Líneas.

***aluminum.** A soft silvery-gray highly thermally and electrically conductive metal that naturally resists corrosion.

***aluminio.** Un metal blando de color gris metalizado con gran conducción térmica y eléctrica y que resiste la corrosión de manera natural.

***American Welding Society (AWS).** Organization that promotes the art and science of welding and that publishes international codes and standards.

***Asociación de Soldadura Estadounidense (American Welding Society, AWS).** Organización que promueve el arte y la ciencia de la soldadura y que publica códigos y normas internacionales.

***amperage.** A measurement of the rate of flow of electrons; amperage controls the size of the arc.

***amperaje.** Una medida de la proporción de la corriente de electrones; el amperaje controla el tamaño del arco.

amperage range. The lower and upper limits of welding power, in amperage, that can be produced by a welding machine or used with an electrode or by a process.

rango de amperaje. Los límites máximos y mínimos de poder de soldadura (en amperaje) que puede tener una máquina para soldar o que pueden usarse con un electrodo o a través de un proceso.

angle of bevel. See preferred term **bevel angle.**

ángulo del bisel. Es preferible que vea el término **ángulo del bisel.**

***anode.** Material with a lack of electrons; thus, it has a positive charge.

***ánodo.** Un material que carece electrones; por eso tiene una carga positiva.

arc blow. The deflection of arc from its normal path because of magnetic forces.

soplo del arco. Desviación de un arco eléctrico de su senda normal a causa de fuerzas magnéticas.

arc brazing (AB). A brazing process in which the heat required is obtained from an electric arc.

soldadura fuerte aplicada por arco. Un proceso de soldadura fuerte donde el calor requerido es obtenido de un arco eléctrico.

arc cutting (AC). A group of thermal cutting processes that severs or removes metal by melting with the heat of an arc between an electrode and the workpiece.

corte con arco. Un grupo de procesos termales para cortar que desúne o quita el metal derretido con el calor del arco en medio del electrodo y la pieza de trabajo.

***arc cutting electrode.** An electrode that has a flux covering that allows the core wire to burn back inside so it creates a strong jetting action which blows out the molten metal created by the arc. See **arc cutting.**

***electrodo para corte por arco.** Un electrodo que cuenta con una cobertura de fundente que permite que el núcleo de alambre se vuelva a quemar en el interior, por lo que crea una fuerte acción de chorro que sopla el metal derretido creado por el arco. Consulte **corte por arco.**

arc force. The axial force developed by a plasma.

fuerza del arco. La fuerza axial desarrollada por la plasma.

arc gouging. Thermal gouging that uses an arc cutting process variation to form a bevel or groove.

gubiadura con arco. Gubiadura termal que usa un proceso variante de corte con arco para formar un bisel o ranura.

arc length. The length from the tip of the welding electrode to the adjacent surface of the weld pool.

largura del arco. La distancia de la punta del electrodo a la superficie que colinda con el charco de la soldadura.

arc plasma. A state of matter found in the region of an electrical discharge (arc). See also **plasma.**

arco de plasma. Un estado de la materia encontrado en la región de una descarga eléctrica (arco). Vea también **plasma.**

arc spot weld. A spot weld made by an arc welding process.

soldadura de puntos por arco. Una soldadura de punto hecha por un proceso de soldadura de arco.

arc strike. A discontinuity consisting of any localized remelted metal, heat-affected metal, or change in the surface profile of any part of a weld or base metal resulting from an arc.

golpe del arco. Una discontinuidad que consiste de cualquier rederretimiento del metal localizado, metal afectado por el calor, o cambio en el perfil de la superficie de cualquier parte de la soldadura o metal base resultante de un arco.

arc time. The time during which an arc is maintained in making an arc weld.

tiempo del arco. El tiempo durante el que el arco se mantiene al hacer una soldadura de arco.

arc voltage. The voltage across the welding arc.

voltaje del arco. El voltaje a través del arco de soldar.

arc welding (AW). A group of welding processes that produces coalescence of workpieces by heating them with an arc. The processes are used with or without the application of pressure and with or without filler metal.

soldadura de arco. Un grupo de procesos de soldadura que producen una unión de piezas de trabajo calentándolas con un arco. Los procesos se usan con o sin la aplicación de presión y con o sin metal para rellenar.

arc welding deposition efficiency. The ratio of the weight of filler metal deposited in the weld metal to the weight of filler metal melted, expressed in percent.

eficiencia de deposición de soldadura de arco. La relación del peso del metal depositado en la soldadura al peso del metal para rellenar, y el metal derretido, expresado en por ciento.

arc welding electrode. A component of the welding circuit through which current is conducted between the electrode holder and the arc. See also **arc welding (AW).**

electrodo para soldadura de arco. Un componente del circuito de soldadura que conduce la corriente a través del porta-electrodo y el arco. Vea también **soldadura de arco.**

arc welding gun. A device used to transfer current to a continuously fed consumable electrode, guide the electrode, and direct the shielding gas.

pistola de soldadura de arco. Aparato que se usa para transferir corriente eléctrica continuamente a un alimentador de electrodo consumible. También se usa para guiar al electrodo y dirigir el gas de protección.

***argon (Ar).** Shielding gas commonly used for gas tungsten arc welding. See **shielding gas.**

***argón (Ar).** Gas de protección que se utiliza comúnmente para la soldadura por arco con gas de tungsteno. Consulte **gas de protección.**

***argon-CO_2.** A shielding gas mixture of argon and carbon dioxide commonly used for GMAW and FCAW welding. See **shielding gas.**

***argón-CO_2.** Una mezcla de gas de protección de argón y dióxido de carbono que se utiliza comúnmente para la soldadura GMAW y FCAW. Consulte **gas de protección.**

***argon-oxygen.** A shielding gas mixture of argon and oxygen commonly used for GMAW and FCAW welding. See **shielding gas.**

***argón-oxígeno.** Una mezcla de gas de protección de argón y oxígeno que se utiliza comúnmente para la soldadura GMAW y FCAW. Consulte **gas de protección.**

***arm.** An interconnected set of links and powered joints comprising a manipulator, which supports or moves a wrist and hand or end effector.

***brazo.** Una entreconexión que une un juego de eslabones y coyunturas de potencia conteniendo un manipulador, que apolla o mueve una muñeca o mano o el que efectúa al final.

as-welded. Pertaining to the weld metal, welded joints, and weldments after welding but prior to any subsequent thermal, mechanical, or chemical treatments.

como-soldado. Pertenece a metal soldado, juntas soldadas, o soldaduras ya soldadas pero antes de que se les hagan tratamientos termales, mecánicos, o químicos.

***atmospheric pressure.** The pressure at sea level resulting from the weight of a column of air on a specified area; expressed for an area of 1 square inch or square centimeter; normally given as 14.7 psi (1.05 kg/cm^2).

***presion atmosferica.** La presión al nivel del mar que resulta del peso de una columna de aire en una area especificada; expresada para una área de una pulgada cuadrada o un centímetro cuadrado. Normalmente dado como 14.7 psi (1.05 kg/cm^2).

atomic hydrogen. A single free, unbounded hydrogen atom (H) usually formed when molecular hydrogen is exposed to an arc.

hidrógeno atómico. Un solo átomo de hidrógeno (H) libre, que normalmente se forma cuando se expone hidrógeno molecular a un arco.

atomic hydrogen welding (AHW). An arc welding process that uses an arc between two metal electrodes in a shielding atmosphere of hydrogen and without the application of pressure. This is an obsolete or seldom used process.

soldadura atómica hidrógena. Un proceso de soldadura de arco que usa un arco entre dos electrodos de metal en una atmósfera de protección de hidrógeno y sin la aplicación de presión. Esto es un proceso anticuado y es rara la vez que se use.

***austenite.** A coarse micro grain structure found in some steels as the result of the rate of cooling.

***austenita.** Una estructura micro granulada gruesa que se encuentra en algunos aceros como consecuencia de la tasa de enfriamiento.

austenitic manganese steel. A steel alloy with a high carbon content containing 10% or more manganese that is very tough and that will harden when cold worked.

acero al manganeso austenítico. Aleación de acero con un alto contenido de carbono, que contiene un 10% o más de manganeso, y que es muy tenaz y endurece cuando se lo trabaja en frío.

autogenous weld. A fusion weld made without filler metal.

soldadura autógena. Una soldadura fundida sin metal de rellenar.

***automated operation.** Welding operations are performed repeatedly by a robot or another machine that is programmed to perform a variety of processes.

***operación automatizada.** Operaciones de soldaduras que se ejecutan repetidamente por un robot u otra maquina que está programada para hacer una variedad de procesos.

automatic arc welding downslope time. The time during which the current is changed continuously from final taper current or welding current to final current.

tiempo del pendiente con descenso en soldadura de arco automático. El tiempo durante en que la corriente cambia continuamente disminuyendo la corriente final o la corriente de la soldadura a la corriente final.

automatic arc welding upslope time. The time during which the current changes continuously from the initial current to the welding current.

tiempo de pendiente con ascenso en soldadura de arco automático. El tiempo durante en que la corriente cambia continuamente de donde se inició la corriente a la corriente de la soldadura.

***automatic operation.** Welding operations are performed repeatedly by a machine that has been programmed to do an entire operation without the interaction of the operator.

operación automática. Operaciones de soldadura que se ejecutan repetidamente por una maquina que ha sido programada para hacer una operación entera sin influencia del operador.

automatic oxygen cutting. Oxygen cutting with equipment that performs the cutting operation without constant observation and the adjustment of the controls by an operator. The equipment may or may not perform loading and unloading of the work.

corte del oxígeno automático. Cortadura del oxígeno con equipo que hace la operación de cortar sin observación y ajuste de los controles por el operador. El equipo puede que ejecute o no el trabajo de cargar o descargar las piezas.

automatic welding. Welding with equipment that requires only occasional or no observation of the welding and no manual adjustment of the equipment controls. Variations of this term are *automatic brazing, automatic soldering, automatic thermal cutting,* and *automatic thermal spraying.*

soldadura automática. Soldadura con equipos que requieren ocasional o ninguna observación, y ningun ajuste manual de los controles de los equipos. Variaciones de éste término son *soldadura automática, corte termal automático, rociadura termal automático.*

***axial spray metal transfer.** Method of metal transfer used by GMAW and FCAW processes.

***transferencia de metal por pulverización axial.** Método de transferencia de metal que utilizan los procesos GMAW y FCAW.

axis of a weld. A line through the length of a weld, perpendicular to and at the geometric center of its cross section.

eje de la soldadura. Una linea a lo largo de la soldadura perpendicula a y al centro geométrico de su corte transversal.

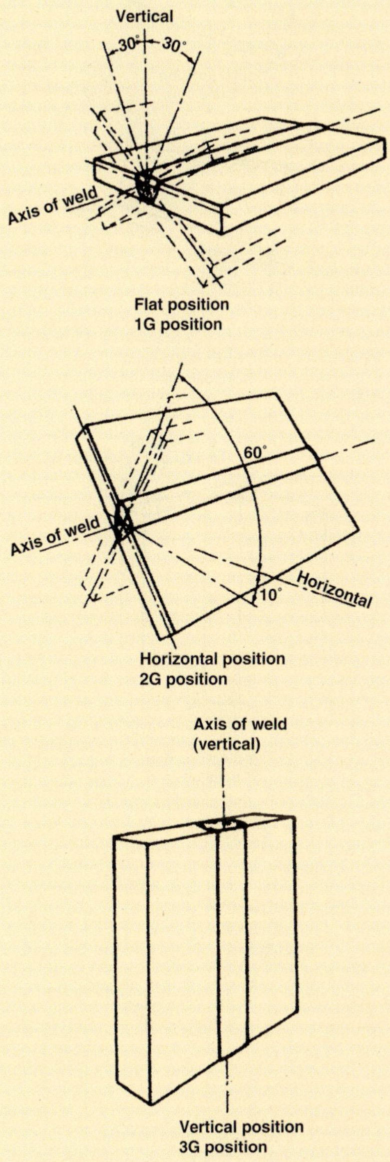

Positions of welding — groove welds

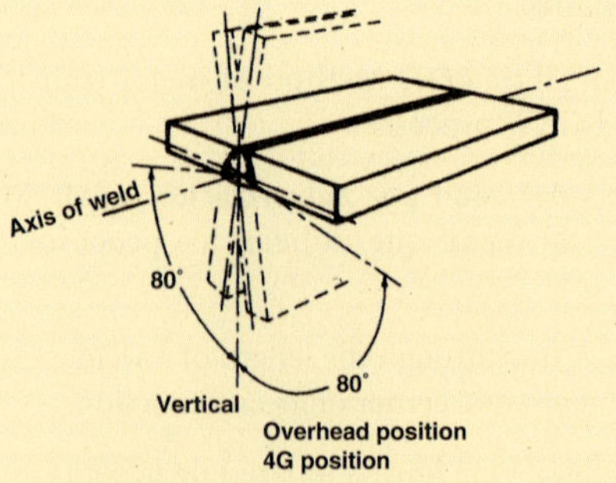

Positions of welding — groove welds

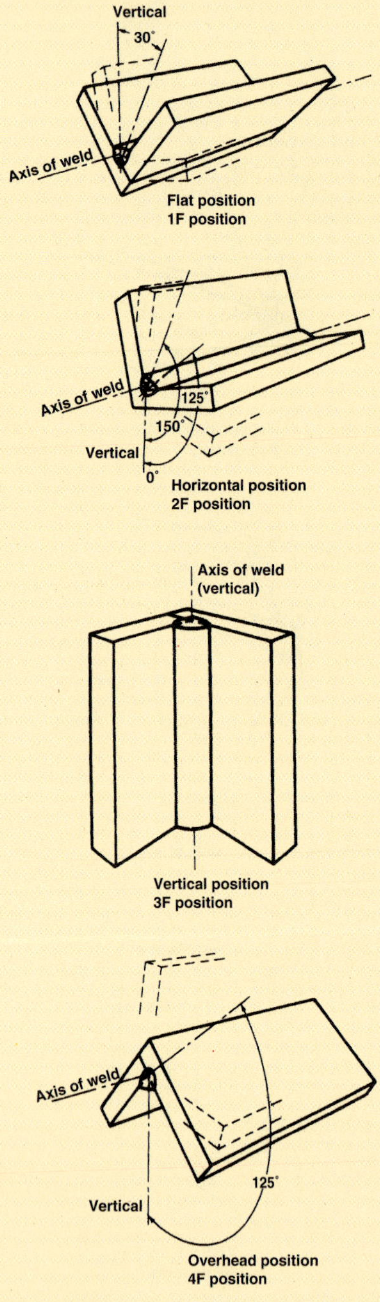

Positions of welding — fillet welds

B

back gouging. The removal of weld metal and base metal from the weld root side of a welded joint to facilitate complete fusion and complete joint penetration upon subsequent welding from that side.

gubia trasera. Quitar el metal soldado y el metal base del lado de la raíz de una junta soldada para facilitar una fusión completa y penetración completa de la junta soldada subsecuente a soldar de ese lado.

back weld. A weld deposited at the back of a single groove weld.

soldadura de atrás. Una soldadura que se deposita en la parte de atrás de una soldadura de ranura sencilla.

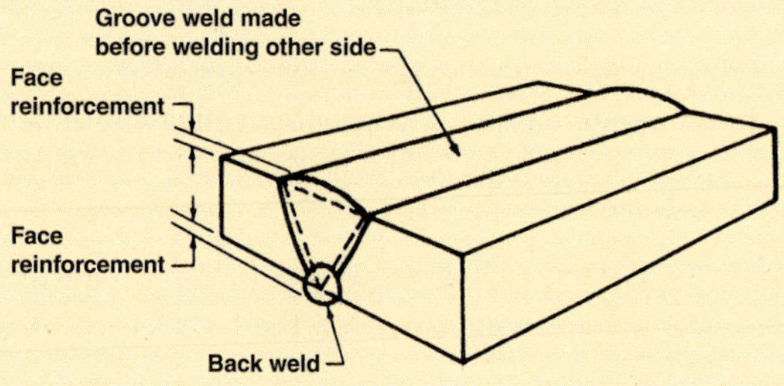

backfire. The momentary recession of the flame into the welding tip, cutting tip, or flame spraying gun, followed by immediate reappearance or complete extinction of the flame.

llama de retroceso. El retroceso momentaneo de la llama dentro de la punta para soldar, punta para cortar, o pistola para rociar con llama. La llama puede reaparecer inmediatamente o apagarse completamente.

***backhand technique.** A welding technique in which the welding torch or gun is directed opposite to the progress of welding.

***técnica de revés.** Una técnica de soldadura en la cual el soplete de soldar se coloca en dirección opuesta al progreso de la soldadura.

backhand welding. A welding technique in which the welding torch or gun is directed opposite to the progress of welding. Sometimes referred to as the "pull gun technique" in GMAW and FCAW. See also **travel angle, work angle,** and **drag angle.**

soldadura en revés. Una técnica de soldar la cual el soplete o pistola es guiada en la dirección contraria al adelantamiento de la soldadura. A veces se refiere como una tecnica de "estirar la pistola" en GMAW y FCAW. Vea también **ángulo de avance, ángulo de trabajo,** y **ángulo del tiro.**

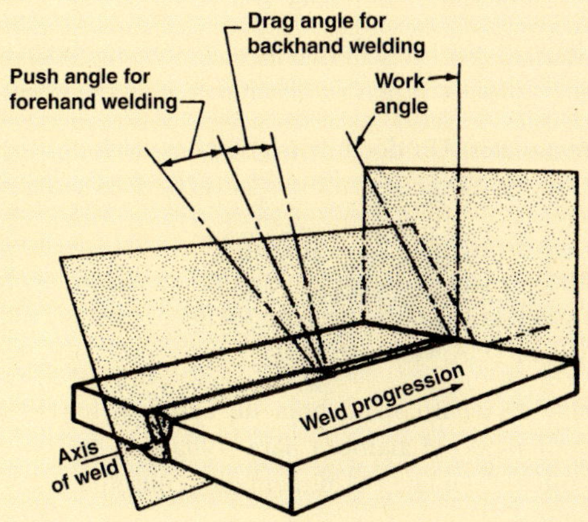

Groove weld

backing. A material (base metal, weld metal, carbon, or granular material) placed at the root of a weld joint for the purpose of supporting molten weld metal.

respaldo. Un material (metal base, metal de soldadura, carbón o material granulado) puesto en la raíz de la junta soldada con el proposito de sostener el metal de la soldadura que está derretido.

backing pass. A pass made to deposit a backing weld.

pasada de respaldo. Una pasada hecha para depositar la pasada del respaldo.

backing ring. Backing in the form of a ring, generally used in the welding of piping.

anillo o argolla de respaldo. Respaldo en forma de argolla, generalmente se usa en soldaduras de tubos.

backing strip. Backing in the form of a strip.

tira de respaldo. Un respaldo en la forma de una tira.

backing weld. Backing in the form of a weld.

soldadura de respaldo. Respaldo en la forma de soldadura.

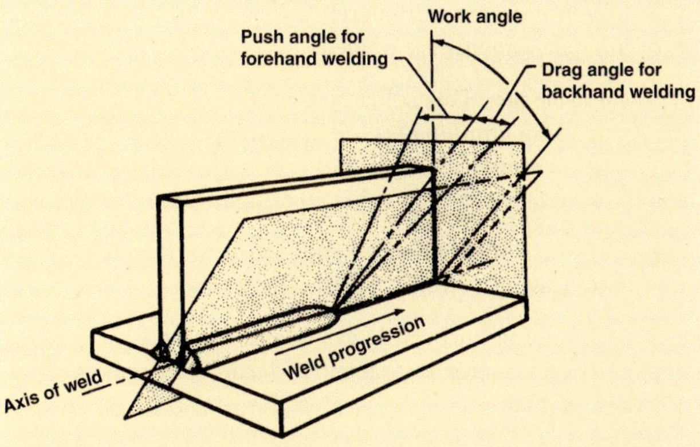

Fillet weld

Position of electrode or torch

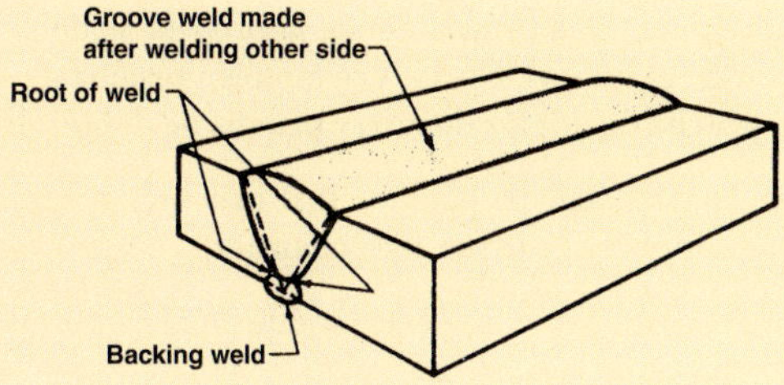

backstep sequence. A longitudinal sequence in which the weld bead increments are deposited in the direction opposite to the progress of welding the joint.

secuencia a la inversa. Una serie de soldaduras en secuencia longitudinal hechas en la dirección opuesta del progreso de la soldadura.

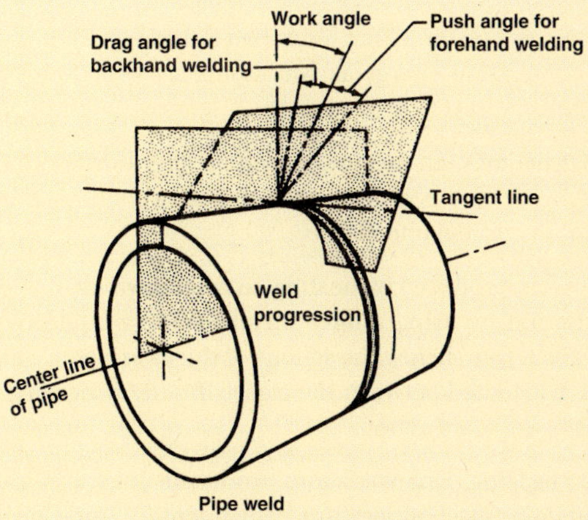

Pipe weld

Position of electrode or torch

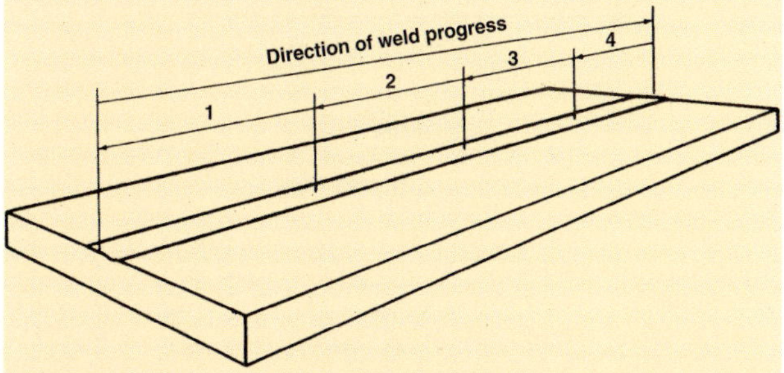

Backstep sequence

bare electrode. A filler metal electrode that has been produced as a wire, strip, or bar with no coating or covering other than that which is incidental to its manufacture or preservation.

electrodo descubierto. Un electrodo de metal para rellenar que se ha producido como alambre, tira, o barra sin revestimiento o cubierto con solo lo necesario para su fabricación y conservación.

base material. The material that is welded, brazed, soldered, or cut. See also **base metal** and **substrate.**

material base. El material que está soldado, soldado con soldadura fuerte, soldado con soldadura blanda, o cortado. Vea también **metal base** y **substrato.**

base metal. The metal or alloy that is welded, brazed, soldered, or cut. See also **base material** and **substrate.**

metal base. El metal que está soldado con soldadura fuerte, soldado con soldadura blanda, o cortado. Vea también **material base** y **substrato.**

base metal test specimen. A test specimen composed wholly of base metal.

probeta para metal base. Una probeta totalmente compuesta de metal base.

bend test. A test in which a specimen is bent to a specified bend radius. See also **face-bend test, root-bend test,** and **side-bend test.**

prueba de dobléz. Una prueba donde la probeta se dobla a una vuelta con un radio specificado. Vea también **prueba de dobléz de cara, prueba de dobléz de raíz,** y **prueba de dobléz de lado.**

bevel. An angular edge preparation that may or may not extend across the entire edge surface. (See **chamfer**)

bisel. Una preparación borde angular que puede o no se extienden por el borde de la superficie entera. (Ver **chaflán**)

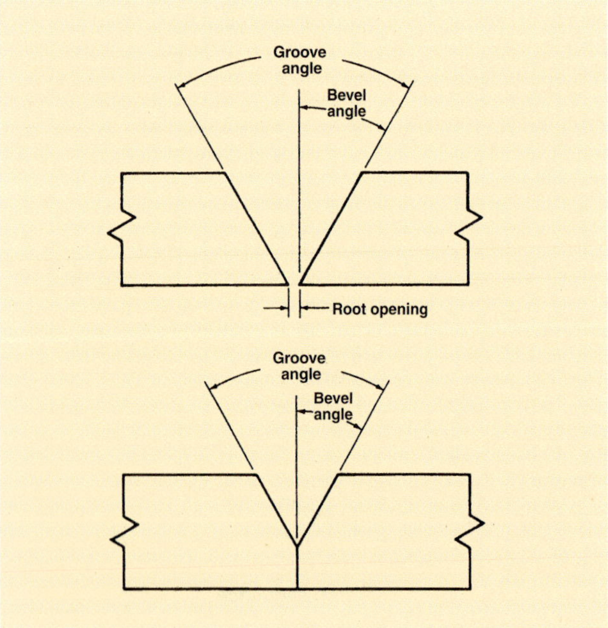

bevel angle. The angle formed between the prepared edge of a member and a plane perpendicular to the surface of the member. Refer to drawings for **bevel.**

ángulo del bisel. El ángulo formado entre el corte preparado de un miembro y la plana perpendicular a la superficie del miembro. Refiera a los dibujos del **bisel.**

***bill of materials.** A document that lists all of the material required to construct a project. It may or may not contain the prices.

***lista de materiales.** Un documento que detalla todos los materiales necesarios para construir un proyecto. Puede o no incluir los precios.

***bird-nesting.** A ball of tangled wire that results when welding wire gets stuck in the gun liner and the feeder keeps feeding on either GMA or FCA welding.

***efecto nido de pájaro.** Una bola de alambre enredado que se produce cuando el alambre para soldadura se atasca en el revestimiento del soplete y el alimentador continúa alimentando soldadura GMA o FCA.

blind joint. A joint, no portion of which is visible.

junta ciega. Una junta en que no hay porción visible.

block sequence. A combined longitudinal and cross-sectional sequence for a continuous multiple-pass weld in which separated increments are completely or partially welded before intervening increments are welded.

secuencia de bloques. Una soldadura continua de pasadas multiples en sucesión combinadas longitudinal y sección transversa donde los incrementos separados son completamente o parcialmente soldados antes que los incrementos sean soldados.

blow hole. A gas bubble that escapes to the surface as the weld metal cools leaving a circular hole in the surface.

golpe agujero. Una burbuja de gas que se escapa a la superficie como el metal de soldadura se enfría dejando un agujero circular en la superficie.

***body-centered cubic (BCC).** A crystalline structure called ferrite iron that forms as it changes from a liquid to a solid. See also **crystalline structures.**

***cúbico de cuerpo centrado (BCC, por sus siglas en inglés).** Una estructura cristalina llamada ferrita que forma el hierro a medida que cambia de estado líquido a sólido. Consulte también **estructuras cristalinas.**

bonding force. The force that holds two atoms together; it results from a decrease in energy as two atoms are brought closer to one another.

fuerza ligamentosa. La fuerza que detiene dos átomos juntos; es el resultado del decrecimiento en la energía cuando dos átomos son traídos cerca del uno al otro.

braze. A weld produced by heating an assembly to suitable temperatures and by using a filler metal having a liquidus above 840°F (450°C) and below the solidus of the base metal. The filler metal is distributed between the closely fitted faying surfaces of the joint by capillary action.

soldadura de latón. Una soldadura producida cuando se calienta un montaje a una temperatura conveniente usando un metal de relleno que se liquida arriba de 840°F (450°C) y abajo del estado sólido del metal base. El metal de relleno es distribuido por acción capilar en una junta entre las superficies empalmadas montadas muy cerca.

***braze buildup.** Braze metal added to the surface of a part to repair wear.

***formación con bronce.** Reparación de partes gastadas donde se agrega bronce.

braze metal. That portion of a braze that has been melted during brazing.

latón. La porción del bronce que se derrite cuando se solda.

braze welding. A welding process variation that uses a filler metal with a liquidus above 840°F (450°C) and below the solidus of the base metal. Unlike brazing, in braze welding the filler metal is not distributed in the joint by capillary action.

soldadura con bronce. Es un proceso de soldar variante que usa un metal de relleno con un liquido arriba de 840°F (450°C) y abajo del estado del metal base. Distinto a la soldadura fuerte, el metal de relleno no es distribuido por acción capilar.

brazeability. The capacity of a metal to be braced under the fabrication conditions imposed into a specific suitably designed structure and to perform satisfactorily in the intended service.

soldabilidad fuerte. La capacidad de un metal de refuerzo bajo las condiciones impuestas en la fabricación de una estructura diseñada especificamente para funcionar satisfactoriamente en los servicios intentados.

brazement. An assembly whose component parts are joined by brazing.

montaje de soldadura fuerte. Un montaje donde las partes son unidas por soldadura fuerte.

brazing (B). A group of welding processes that produces coalescence of materials by heating them to the brazing temperature in the presence of a filler metal with a liquidus above 840°F (450°C) and below the solidus of the base metal. The filler metal is distributed between the closely fitted faying surfaces of the joint by capillary action.

soldadura fuerte (B). Un grupo de procesos de soldadura que produce coalescencia de materiales calentándolos a una temperatura de soldar en la presencia de un material de relleno el cual se derrite a una temperatura de 840°F (450°C) y bajo del estado sólido del metal base. El metal de relleno se distribuye por acción capilar de una junta entre las superficies empalmadas montadas muy cerca.

***brazing alloys.** A nonstandard term for brazing filler metal.

***aleaciones de soldadura fuerte.** Un término no estándar para el metal de relleno de la soldadura fuerte.

brazing filler metal. The metal that fills the capillary joint clearance and has a liquidus above 840°F (450°C) but below the solidus of the base metal.

metal de relleno para soldadura fuerte. El metal que rellena el espacio libre en la junta capilar y se derrite a una temperatura de 840°F (450°C) y bajo del estado sólido del metal base.

brazing procedure qualification record (BPQR). A record of brazing variables used to produce an acceptable test brazement and the results of tests conducted on the brazement to qualify a brazing procedure specification.

registro del procedimiento de calificación de la soldadura fuerte (BPQR). Un registro de variables de la soldadura fuerte que se usan para producir una probeta bronceada aceptable y los resultados de la prueba conducidas en el bronceamiento para calificar la especificación del procedimiento de la soldadura fuerte.

brazing procedure specification (BPS). A document specifying the required brazing variables for a specific application.

especificación del procedimiento de la soldadura fuerte (BPS). Un documento especificando los variables requeridos de la soldadura fuerte para una aplicación especificada.

brazing rod. Filler metal used in the brazing process is supplied in the form of rods; the filler metal is usually an alloy of two or more metals; the percentages and types of metals used in the alloy impart different characteristics to the braze being made. Several classification systems are in use by manufacturers of filler metals. See also **brazing filler metal.**

varilla de latón. Metal de relleno usado en procesos de soldadura fuerte es surtida en forma de varillas; el metal para rellenar es usualmente un aleación de dos or más metales; los porcentajes y tipos de metales usados en la aleación imparten diferentes características a la soldadura que se está haciendo. Varios sistemás de clasificación se están usando por los fabricantes de metales de relleno. Vea también **metal de relleno para soldadura fuerte.**

brazing temperature. The temperature to which the base metal is heated to enable the filler metal to wet the base metal and form a brazed joint.

temperatura de soldadura fuerte. La temperatura a la cual se calienta el metal base para permitir que el metal de relleno moje al metal base y forme la soldadura fuerte.

***break line.** A line on a drawing that shows that part of an object has been removed.

***línea incompleta.** Una línea en un dibujo que muestra una parte de un objeto se ha eliminado.

buckling. The bending or warping of a metal surface caused by heating from welding, cutting, brazing, etc.

pandeo. La flexión o deformación de una superficie metálica causadas por el calor de la soldadura, corte, soldadura, etc.

buildup. The material deposited by the welding filler metal to a weld. Also, surfacing variation in which surfacing material is deposited to achieve the required dimensions. See also **buttering, cladding,** and **hardfacing.**

recubrimiento. La materia depositada por el metal de masilla de soldadura a una soldadura. También, una variación en la superficie donde el metal es depositado para que pueda obtener las dimensiones requeridas. Vea también **recubrimiento antes de terminar una soldadura, capa de revestimiento,** y **endurecimiento de caras.**

***buried arc transfer.** In gas metal arc welding, a method of transfer in which the wire tip is driven below the surface of the weld pool due to the force of the carbon dioxide shielding gas. The shorter arc reduces the size of the drop, and any spatter is trapped in the cavity produced by the arc.

***traslado de arco enterrado.** En soldaduras de arco metalico con gas, un método de transferir en la cual la punta del alambre es enterrado debajo de la superficie del metal derretido debido a la fuerza del carbón bióxido del gas protector. Lo corto del arco reduce el tamaño de la gota y la salpicadura es atrapada en el hueco producido por el arco.

***burnthrough.** Burning out of molten metal on the back side of the plate.

***metal quemado que pasa al otro lado.** Metal derretido que se quema en el lado de atrás del plato.

buttering. A surfacing variation in which one or more layers of weld metal are deposited on the groove face of one member (for example, a high alloy weld deposit on steel base metal that is to be welded to a dissimilar base metal). The buttering provides a suitable transition weld deposit for subsequent completion of the butt joint.

recubrimiento antes de terminar una soldadura. Una variación de la superficie donde se deposita una o más capas de metal soldado en la ranura de un miembro (por ejemplo, un deposito de soldadura de aleación alta en un metal base de acero la cual será soldada a un metal base diferente). El recubrimiento proporciona una transición conveniente a la soldadura depositada para el acabamiento subsiguiente de una junta tope.

butt joint. A joint between two members aligned approximately in the same plane.

junta a tope. Una junta entre dos miembros alineados aproximadamente en el mismo plano.

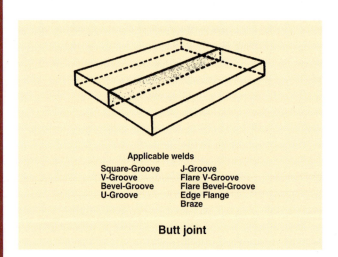

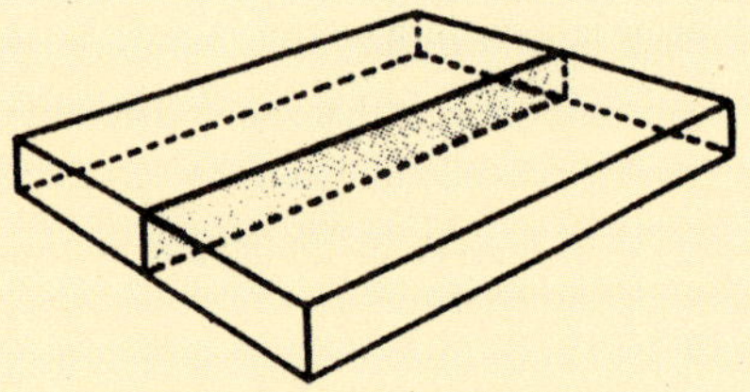

Applicable welds

Square-Groove	J-Groove
V-Groove	Flare V-Groove
Bevel-Groove	Flare Bevel-Groove
U-Groove	Edge Flange
	Braze

Butt joint

C

capillary action. The force by which liquid, in contact with a solid, is distributed between closely fitted faying surfaces of the joint to be brazed or soldered.

acción capilar. La fuerza por la que el liquido, en contacto con un sólido, es distribuido entre el empalme de las juntas del superficie para soldadura fuerte o blanda.

***carbon.** A nonmetallic element that is often added to iron to change its mechanical properties.

***carbono.** Un elemento no metálico que con frecuencia se agrega al hierro para cambiar sus propiedades mecánicas.

carbon arc cutting (CAC). An arc cutting process that uses a carbon electrode. See also **air carbon arc cutting.**

corte con arco y carbón. Un proceso de corte con arco que usa un electrodo de carbón. Vea también **arco de carbón con aire.**

carbon arc welding (CAW). An arc welding process that uses an arc between a carbon electrode and the weld pool. The process is used with or without shielding and without the application of pressure.

soldadura con arco de carbón. Un proceso de soldar de arco en que se usa un arco entre el electrodo de carbón y el metal derretido. El proceso se usa con o sin protección y sin aplicación de presión.

***carbon dioxide.** Shielding gas commonly used for GMA and FCA welding. See **shielding gas.**

***dióxido de carbono.** Gas de protección que se utiliza comúnmente en la soldadura GMA y FCA. Consulte **gas de protección.**

carbon electrode. A nonfiller metal electrode used in arc welding and cutting, consisting of a carbon or graphite rod, which may be coated with copper or other materials.

electrodo de carbón. Un electrodo de metal que no se rellena usado en soldaduras de arco y para cortes consistiendo de varillas de carbón o grafito que pueden ser cubiertas de cobre u otros materiales.

***carbon steel.** Steel whose physical properties are primarily the result of the percentage of carbon contained within the alloy. Carbon content ranges from 0.04 to 1.4%, often referred to as plain carbon steel, low carbon steel, or straight carbon steel.

***acero al carbono.** Acero cuyas propiedades físicas son primariamente el resultado del porcentaje de carbón que es contenido dentro de la aleación. El contenido del carbón es clasificado entre 0.04 a 1.4%, frecuentemente es referido como el carbono de acero liso, acero de bajo carbón, o carbon de acero recto.

carbonizing (carburizing). A reducing oxyfuel gas flame in which there is an excess of fuel gas, resulting in a carbon-rich zone extending around and beyond the cone.

llama carburante. Una llama minorada de gas combustible a oxígeno donde hay un exceso de gas combustible, resultando en una zona rica de carboón extendiendose alrededor y al otro lado del cono.

***cast.** The natural curve in the electrode wire for gas metal arc welding as it is removed from the spool; cast is measured by the diameter of the circle that the wire makes when it is placed on a flat surface without any restraint.

***distancia.** La curva natural en el alambre electrodo para soldadura de arco metálico para gas cuando se aparta del carrete; la distancia es medida en el círculo que hace el alambre cuando es puesto en una superficie plana sin restricción.

***cast iron.** Gray cast iron is the most common form of cast iron. It is used to make castings such as engine blocks and water pumps. Other forms of cast iron are: nodular, white, and alloy.

***hierro fundido.** La fundición gris es la forma más común de hierro fundido. Se utiliza para hacer objetos fundidos tales como bloques de motor y bombas de agua. Entre las otras formas del hierro fundido son: nodular, blanco y aleación.

cathode. A natural curve material with an excess of electrons, thus having a negative charge.

cátodo. Un material de curva natural con un exceso de electrones, por eso tiene una carga negativa.

***cell.** A manufacturing unit consisting of two or more workstations and the material transport mechanisms and storage buffers that interconnect them.

***celda.** Una unidad manufacturera la cual consiste de dos o más estaciones de trabajo y mecanismos para trasladar el material y los amortiguadores del almacén que los entreconecta.

cellulose-based electrode fluxes. Fluxes that use an organic-based cellulose ($C_6H_{10}O_5$) (a material commonly used to make paper) held together with a lime binder. When this flux is exposed to the heat of the arc, it burns and forms a rapidly expanding gaseous cloud of CO_2 that protects the molten weld pool from oxidation. Most of the fluxing material is burned, and little slag is deposited on the weld. E6010 is an example of an electrode that uses this type of flux.

fundentes para electrodos celulósicos. Fundentes que usan celulosa de base orgánica ($C_6H_{10}O_5$) (un material normalmente utilizado para fabricar papel), y que se mantienen unidos con un aglomerante de cal. Cuando a este fundente se lo expone al calor del arco, se consume y forma una nube gaseosa de CO_2 que se expande rápidamente y protege de la oxidación al charco de soldadura derretido. La mayor parte del material del fundente se consume, y se deposita poca escoria en la soldadura. El E6010 es un ejemplo de un electrodo que utiliza este tipo de fundente.

***cementite.** A crystalline form of iron and carbon that is hard and brittle.

***cementita.** Una forma cristalina de hierro y carbono que es dura y quebradiza.

***center.** A manufacturing unit consisting of two or more cells and the materials transport and storage buffers that interconnect them.

***centro.** Una unidad manufacturera la cual consiste de dos o más celdas y el traslado de materiales y los amortiguadores del almacén que los entreconecta.

***center line.** Lines on a drawing that show the center point of circles and arcs and round or symmetrical objects. They also locate the center point for holes, irregular curves, and bolts.

***eje longitudinal.** Líneas en un dibujo que muestran el punto central de círculos y arcos y de objetos redondos o simétricos. También ubican el punto central de orificios, curvas irregulares y pernos.

***certification.** See **certified welder.**

***certificación.** Consulte **soldador certificado.**

certified welders. Individuals who have demonstrated their welding skills for a process by passing a specific welding test.

soldador certificado. Personas que han demostrado, mediante una prueba específica de soldadura, su habilidad para soldar en un proceso.

chain intermittent welds. Intermittent welds on both sides of a joint in which the weld increments on one side are approximately opposite those on the other side.

soldadura intermitente de cadena. Soldadura intermitente en los dos lados de una junta en cual los incrementos de soldadura están aproximadamente opuestos a los del otro lado.

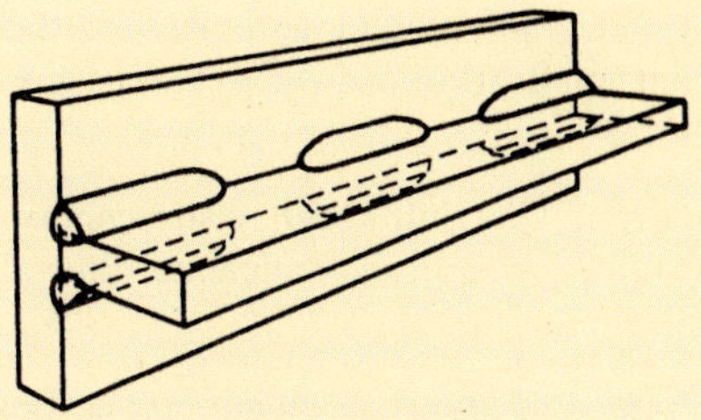

Chain intermittent fillet weld

chalk line mark. A mark made on the surface as powdered chalk is transferred from the string to the surface.

marca de línea de tiza. Una marca hecha en la superficie de tiza en polvo es transferido de la cadena a la superficie.

chalk line tool. A tool containing a cotton string and powdered chalk used to mark a straight line on a surface when the string is drawn tight and snapped.

herramienta de línea de tiza. Una herramienta que contiene una cadena de algodón y tiza en polvo utilizada para marcar una línea recta en una superficie cuando la cadena es de apretar y puesto en libertad.

chamfer. An angled edge preparation that extends from one surface to another. (See **bevel**)

chaflán. Una preparación borde en ángulo que se extiende desde una superficie a otra. (Ver **bisel**)

***chill plate.** A large piece of metal used in welding to correct overheating.

***plato desalentador.** Una pieza de metal grande que se usa para corregir el sobrecalentamiento.

cladding. A relatively thick layer (0.04 in. [>1 mm]) of material applied by surfacing for the purpose of improved corrosion resistance or other properties. See also **coating, surfacing,** and **hardfacing.**

capa de revestimiento. Una capa de material relativamente grueso (0.04 pulgadas [>1 mm]) aplicada por la superficie con el objeto de mejorar la resistencia a la corrosión u otras propiedades. Vea también **revestimiento, recubrimiento superficial,** y **endurecimiento de caras.**

coalescence. The growing together or growth into one body of the materials being welded.

coelescencia. El crecimiento o desarrollo de un cuerpo de los materiales los cuales se están soldando.

coating. A relatively thin layer (0.04 in. [>1 mm]) of material applied by surfacing for the purpose of corrosion prevention, resistance to high-temperature scaling, wear resistance, lubrication, or other purposes. See also **cladding, surfacing,** and **hardfacing.**

revestimiento. Una capa de material relativamente delgado (0.04 pulgadas [>1 mm]) aplicada por la superficie con el propósito de prevenir corrosión, resistencia a las altas temperaturas, resistencia a la deterioración, lubricación, o para otros propósitos. Vea también **capa de revestimiento, recubrimiento superficial,** y **endurecimiento de caras.**

cold soldered joint. A joint with incomplete coalescence caused by insufficient application of heat to the base metal during soldering.

junta soldada fría. Una junta con coalescencia incompleta causada por no haber aplicado suficiente calor al metal base durante la soldadura.

***combustion rate.** Also known as rate of propagation of a flame, this is the speed at which the fuel gas burns, in ft/sec (m/sec). The ratio of fuel gas to oxygen affects the rate of burning: a higher percentage of oxygen increases the burn rate.

***velocidad de combustión.** También es conocida como velocidad de propagación de una llama, ésta es la velocidad en la cual se quema el gas combustible, en pies/sec (m/sec). La proporción del gas combustible al oxígeno afecta la proporción de quemadura: un porcentaje más alto del oxígeno aumenta la proporción de quemarse.

complete fusion. Fusion over the entire fusion faces and between all adjoining weld beads.

fusión completa. Fusión sobre todas las caras de fusión y en medio de todos los cordónes de soldadura inmediatos.

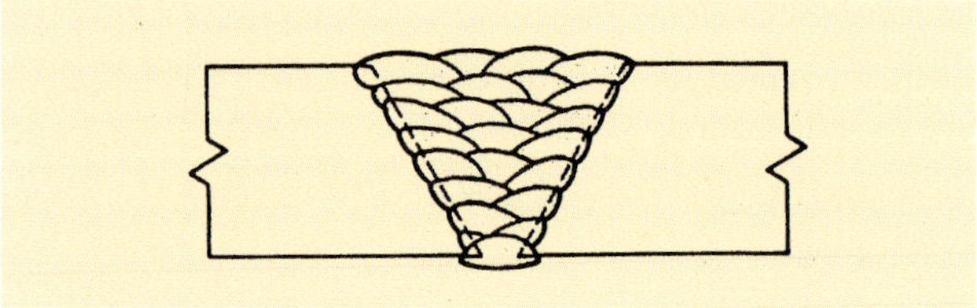

composite electrode. A generic term for multicomponent filler metal electrodes in various physical forms, such as stranded wires, tubes, and covered wire. See also **covered electrode, flux cored electrode,** and **stranded electrode.**

electrodo compuesto. Un término genérico para componentes múltiples para electrodos de metal de aporte en varias formas físicas, como cable de alambre, tubos, y alambre cubierto. Vea también **electrodo cubierto, electrodo de núcleo de fundente,** y **electrodo cable.**

***computer control.** Control involving one or more electronic digital computers.

***control de computadora.** Un control que incluye una o más computadoras electrónicas dactilares.

***computer drafting programs.** Programs that typically use vector lines to produce a mechanical type drawing.

***programas de trazado por computadora.** Programa que por lo general utilizan líneas de vectores para producir un dibujo de tipo mecánico.

***computer drawing programs.** Programs that typically use raster lines to produce picture-like drawings.

***programas de dibujo por computadora.** Programas que por lo general utilizan líneas de barrido para producir dibujos que parecen fotos.

concave fillet weld. A fillet weld with a concave face.

soldadura de filete cóncava. Soldadura de filete con cara cóncava.

concave root surface. A root surface that is concave.

superficie raíz cóncava. La superficie del cordón raíz con cara cóncava.

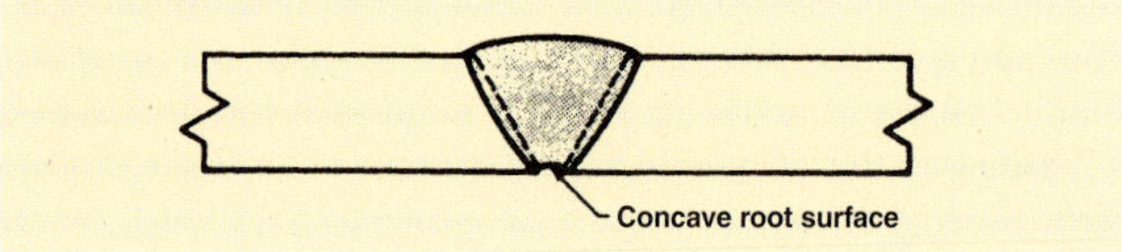

concavity. The maximum distance from the face of a concave fillet weld perpendicular to a line joining the toes.

concavidad. La distancia máxima de la cara de una soldadura de filete cóncava perpendicular a una linea que une con los pies.

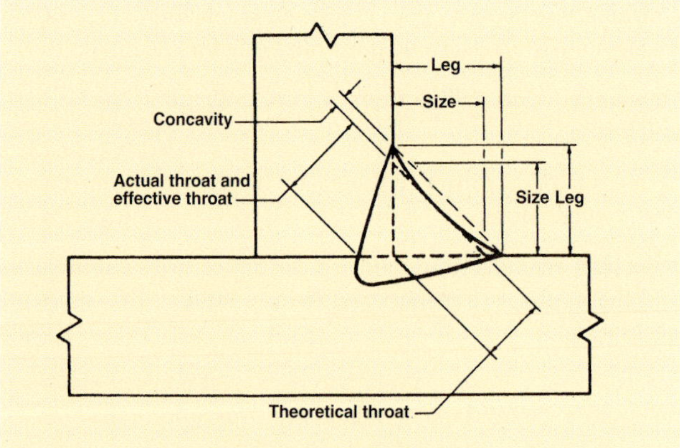

Concave fillet weld

conduit liner. A flexible steel tube that guides the welding wire from the feed rollers through the welding lead to the gun used for GMA and FCA welding. The steel conduit liner may have a nylon or Teflon® inner surface for use with soft metals such as aluminum.

revestimiento de conducto. Un tubo flexible de acero que guía el alambre para soldar desde los rodillos de alimentación, a través de los cables para soldar, hasta la pistola, usado en soldaduras de tipo GMA y FCA. El revestimiento del conducto de acero puede tener una superficie interior de Teflon® o nylon para su uso con metales blandos como el aluminio.

cone. The conical part of an oxyfuel gas flame adjacent to the orifice of the tip.

cono. La parte cónica de la llama del gas de oxígeno combustible que colinda con la abertura de la punta.

***constant current (CC).** Welding current commonly used for SMA and GTA welding.

***corriente constante (CC).** Corriente de soldadura que se utiliza comúnmente para la soldadura SMA y GTA.

***constant voltage (CV).** Welding current commonly used for GMA and FCA welding.

***voltaje constante (CV, por sus siglas en inglés).** Corriente de soldadura que se utiliza comúnmente para la soldadura GMA y FCA.

constricted arc. A plasma arc column that is shaped by the constricting orifice in the nozzle of the plasma arc torch or plasma spraying gun.

arco constreñido. Una columna de arco plasma que está formada por el constreñimiento del orificio en la lanza de la antorcha del arco plasma o pistola de rociado plasma.

constricting nozzle. A device at the exit end of a plasma arc torch or plasma spraying gun containing the constricting orifice.

boquilla de constreñimiento. Un aparato a la salida de la antorcha de un arco plasma o la pistola de rociado plasma que contiene la boquilla de constreñimiento.

constricting orifice. The hole in the constricting nozzle of the plasma arc torch or plasma spraying gun through which the arc plasma passes.

orificio de constreñimiento. El agujero en la boquilla del constreñimiento en la antorcha de arco plasma o de la pistola de rociado plasma por donde pasa el arco de plasma.

consumable electrode. An electrode that provides filler metal.

electrodo consumible. Un electrodo que surte el metal de relleno.

consumable insert. Preplaced filler metal that is completely fused into the root of the joint and becomes part of the weld.

inserción consumible. Metal de relleno antepuesto que se funde completamente en la raíz de la junta y se hace parte de la soldadura.

contact tube. A device that transfers current to a continuous electrode.

tubo de contacto. Un aparato que traslada corriente continua a un electrodo.

***contamination.** Any undesirable material that might enter the molten weld metal.

***contaminación.** Cualquier material indeseable que pueda ingresar en el metal de soldadura fundido.

continuous weld. A weld that extends continuously from one end of a joint to the other. Where the joint is essentially circular, it extends completely around the joint.

soldadura continua. Una soldadura que se extiende continuamente de una punta de la junta a la otra. Donde la junta es esencialmente circular, se extiende completamente alrededor de la junta.

convex fillet weld. A fillet weld with a convex face.

soldadura de filete convexa. Una soldadura de filete con una cara convexa.

convex root surface. A root surface that is convex.

raíz superficie convexa. La raíz que es convexa.

convexity. The maximum distance from the face of a convex fillet weld perpendicular to a line joining the toes.

convexidad. La distancia máxima de la cara de la soldadura convexa filete perpendicula a la linea que une los pies.

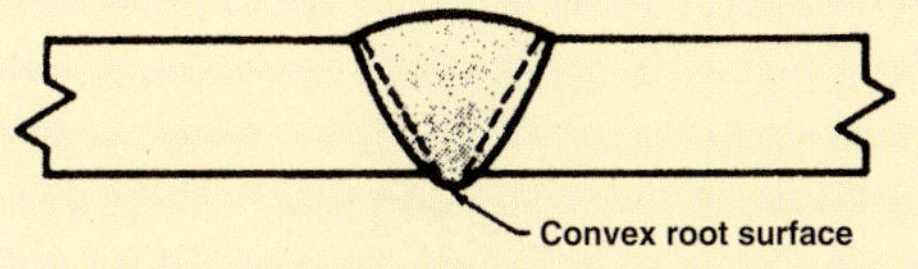

Melt -through

copper. A pinkish or peach-colored highly thermally and electrically conductive soft metal that resists corrosion.

cobre. Un metal blando de tono rosado o rosa anaranjado con gran conducción térmica y eléctrica y que resiste la corrosión.

copper alloys. An alloy primarily containing copper. The two most common copper alloys are brass, a copper tin alloy; and bronze, a copper zinc alloy.

aleaciones de cobre. Una aleación que contiene principalmente cobre. Las dos aleaciones de cobre más comunes son el latón, una aleación de cobre y estaño, y el bronce, una aleación de cobre y zinc.

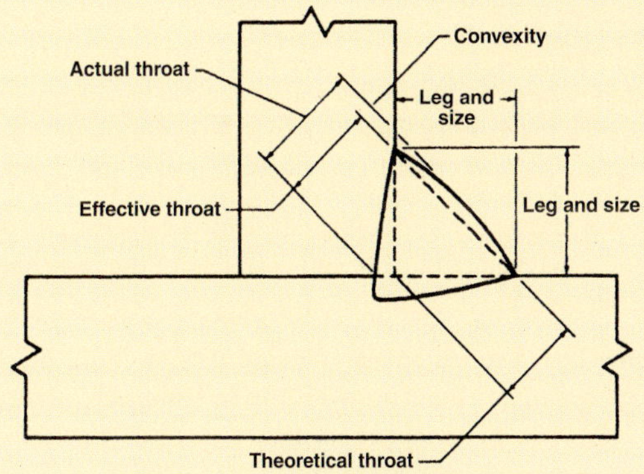

Convex fillet weld

cored solder. A solder wire or bar containing flux as a core.

soldadura de núcleo. Un alambre o barra para soldar que contiene fundente en el núcleo.

***core wire.** The wire portion of the coated electrode for shielded metal arc welding. The wire carries the welding current and adds most of the filler metal required in the finished weld. The composition of the core wire depends upon the metals to be welded.

***alambre del centro.** La porción del alambre del electrodo forrado para proteger el metal de la soldadura de arco. El alambre lleva la corriente de la soldadura y añade casi todo el metal para rellenar que es requerido para terminar la soldadura. La composición del alambre del centro depende de los metales que se van a usar para soldar.

corner joint. A joint between two members located approximately at right angles to each other.

junta de esquina. Una junta dentro de dos miembros localizados aproximadamente a ángulos rectos de unos a otros.

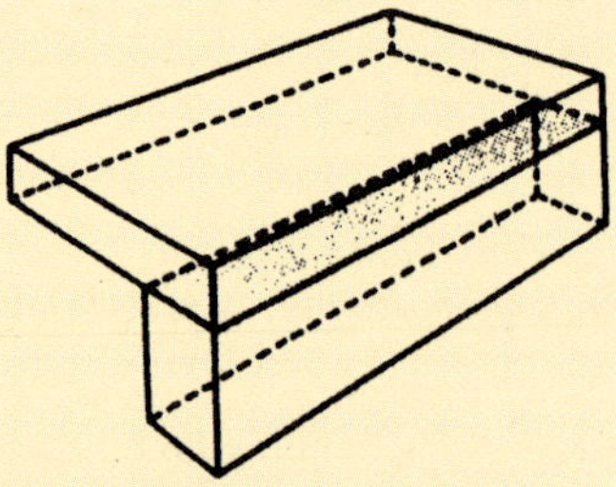

Applicable welds

Fillet	Flare Bevel-Groove
Square-Groove	Edge-Flange
V-Groove	Corner-Flange
Bevel-Groove	Spot
U-Groove	Projection
J-Groove	Seam
Flare V-Groove	Braze

Corner joint

corrosion. A chemical reaction called oxidation that occurs between oxygen atoms and a metal's atoms when metal is subjected to an electrolyte such as water.

corrosión. Una reacción química llamada oxidación que se produce entre los átomos de oxígeno y los átomos de un metal cuando éste se ve expuesto a un electrólito tal como el agua.

***corrosion resistance.** The ability of the joint to withstand chemical attack; determined by the compatibility of the base materials to the filler metal.

***resistencia a la corrosión.** La abilidad de una junta de resistir ataques químicos; determinado por la compatibilidad de los materiales bases al metal de relleno.

corrosive flux. A flux with a residue that chemically attacks the base metal. It may be composed of inorganic salts and acids, organic salts and acids, or activated rosins or resins.

fundente corrosivo. Un fundente con un residuo que ataca químicamente al metal base. Puede estar compuesto de sales y ácidos inorgánicos, sales y ácidos orgánicos, o abelinotes o resinas activados.

cosmetic pass. A weld pass made primarily to enhance appearance.

pasada cosmética. Una pasada que se le hace a la soldadura para mejorar la apariencia.

***coupling distance.** The distance to be maintained between the inner cones of the cutting flame and the surface of the metal being cut, in the range of 1/8 in. (3 mm) to 3/8 in. (10 mm).

***distancia de acoplamiento.** La distancia que debe de mantenerse entre los conos internos de la llama y la superficie del metal que se está cortando, varía de 1/8 pulgadas (3 mm) a 3/8 pulgadas (10 mm).

cover lens. A round cover plate.

lente para cubrir. Un plato redondo de vidrio para cubrir el lente obscuro.

***cover pass.** The last layer of weld beads on a multipass weld. The final bead should be uniform in width and reinforcement, not excessively wide, and free of any visual defects.

***pasada para cubrir.** La última capa de cordónes soldadura de pasadas múltiples. La pasada final debe ser uniforme en anchura y refuerzo, no excesivamente ancha, y libre de defectos visuales.

cover plate. A removable pane of colorless glass, plastic-coated glass, or plastic that covers the filter plate and protects it from weld spatter, pitting, and scratching.

plato para cubrir. Una hoja removible de vidrio claro, vidrio cubierto con plástico o plástico que cubre el plato filtrado y lo protege de salpicadura, picaduras y de que se rayen.

covered electrode. A composite filler metal electrode consisting of a core of a bare electrode or metal cored electrode to which a covering sufficient to provide a slag layer on the weld metal has been applied. The covering may contain materials providing such functions as shielding from the atmosphere, deoxidation, and arc stabilization, and can serve as a source of metallic additions to the weld.

electrodo cubierto. Un electrodo compuesto de metal para rellenar que consiste de un núcleo de un electrodo liso o electrodo con núcleo de metal el cual se le agrega cubrimiento suficiente para proveer una capa de escoria sobre el metal de la soldadura que se le aplicó. El cubierto puede contener materiales que pueden proveer funciones como protección de la atmósfera, deoxidación, y estabilización del arco, y también puede servir como fuente para añadir metales adicionales a la soldadura.

crack. A fracture-type discontinuity characterized by a sharp tip and high ratio of length and width to opening displacement.

grieta. Una desunión discontinuidada de tipo fractura caracterizada por una punta filoza y proporción alta de lo largo y de lo ancho al desplazamiento de la abertura.

crater. A depression in the weld face at the termination of a weld bead.

crater. Una depresión en la superficie de la soldadura a donde se termina el cordón de soldadura.

crater crack. A crack in the crater of a weld bead.

grieta de crater. Una grieta en el crater del cordón de soldar.

***creep.** A property of metal that allows it to be deformed under a load that is below the metal's yield point.

***fluencia.** Una propiedad de los metales que les permite deformarse bajo una carga que es inferior al límite de elasticidad.

crevice corrosion. Oxidation that occurs in the small space (crevice) between two pieces of metal as the result of moisture being trapped in the small space.

corrosión de hendiduras. Oxidación que se produce en un espacio pequeño (hendidura) entre dos piezas de metal como consecuencia de la humedad atrapada en el espacio reducido.

critical weld. A weld so important to the soundness of the weldment that its failure could result in the loss or destruction of the weldment and injury or death.

soldadura crítica. Una soldadura tan importante para la calidad del conjunto de partes soldadas, que su fracaso podría ocasionar la pérdida o destrucción de dicho conjunto, así como también lesiones o muerte.

cross-sectional sequence. The order in which the weld passes of a multiple-pass weld are made with respect to the cross section of the weld. See also **block sequence.**

secuencia del corte transversal. La orden en la cual se hacen las pasadas de la soldadura en una soldadura de pasadas múltiples hechas al respecto al corte transversal de la soldadura. Vea también **secuencia de bloque.**

crucible. A high-temperature container that holds the thermite welding mixture as it begins its thermal reaction before the molten metal is released into the mold.

crisol. Un recipiente de alta temperatura que contiene la mezcla de la soldadura con termita en el momento en que comienza su reacción térmica antes de que el metal fundido se vierta en el molde.

***crystal lattices.** See **crystalline structures.**

***redes cristalinas.** Consulte **estructuras cristalinas.**

***crystalline structures.** Orderly arrangements of atoms in a solid in a specific geometric pattern; sometimes called *lattices.*

***estructura espacial cristalina.** Un arreglo metódico de átomos en un modelo geométrico preciso. A veces se llaman *celosias.*

cup. A nonstandard term for gas nozzle.

tazón. Un término que no es la norma para de boquilla de gas.

cutting attachment. A device for converting an oxyfuel gas welding torch into an oxygen cutting torch.

equipo para cortar. Un aparato para convertir una antorcha para soldar en una antorcha para cortar con oxígeno.

cutting head. The part of a cutting machine in which a cutting torch or tip is incorporated.

cabeza de la antorcha para cortar. La parte de una maquina para cortar en donde una antorcha para cortar o una punta es incorporada.

***cutting plane line.** Lines on a drawing that represent an imaginary cut through the object. They are used to expose the details of internal parts that would not be shown clearly with hidden lines.

***línea de corte longitudinal.** Líneas en un dibujo representan un corte imaginario a través del objeto. Se utilizan para exponer los detalles de las piezas internas que no se verían claramente con líneas ocultas.

cutting tip. The part of an oxygen cutting torch from which the gases issue.

punta para cortar. Esa parte de la antorcha para cortar con oxígeno por donde salen los gases.

cutting tip, high-speed. Designed to provide higher oxygen pressure, thus allowing the torch to travel faster.

punta para cortar a alta velocidad. Diseñada para proveer presión más alta de oxígeno, asi puede caminar la antorcha más rápidamente.

cutting torch. A device used for plasma arc cutting to control the position of the electrode, to transfer current to the arc, and to direct the flow of plasma and shielding gas.

antorcha para cortar. Un aparato que se usa para cortes de arco de plasma para el control de la posición del electrodo.

***cycle time.** The period of time from starting one machine operation to starting another (in a pattern of continuous repetition).

tiempo del ciclo. El período de tiempo de cuando se empieza la operación de una maquina y cuando se empieza otra (en una norma de repetición continua).

cylinder. A portable container used for transportation and storage of a compressed gas.

cilindro. Un recipiente portátil que se usa para transportar y guardar un gas comprimido.

cylinder manifold. A multiple header for interconnection of gas sources with distribution points.

conexión de cilindros múltiple. Una tuberia con conexiones múltiples que sirve como fuente de gas con puntos de distribución.

***cylinder pressure.** The pressure at which a gas is stored in approximately 2200 pounds (15,169 kPa) per square inch (psi), and acetylene is stored at approximately 225 psi (1551 kPa).

***presión del cilindro.** La presión del cilindro en el cual un gas se guarda en aproximadamente 2200 libras (15,169 kPa) por pulgada cuadrada (psi), y el acetilino se guarda a aproximadamente 225 psi (1551 kPa).

D

defect. A discontinuity or discontinuities that by nature or accumulated effect (for example, total crack length) render a part or product unable to meet minimum applicable acceptance standards or specifications. This term designates rejectability and flaw. See also **discontinuity** and **flaw.**

defecto. Una desunión o desuniónes que por la naturaleza o efectos acumulados (por ejemplo, distancia total de una grieta) hace que una parte o producto no esté de acuerdo con las normas o especificaciones mínimas para aceptarse. Este término designado resectabilidad y falta. Vea también **discontinuidad** y **falta.**

defective weld. A weld containing one or more defects.

soldadura defectuosa. Una soldadura que contiene uno o más defectos.

***deoxidizers.** An element such as silicon that is added to a flux for the purpose of removing oxygen from the molten weld pool.

***desoxidantes.** Un elemento como por ejemplo el silicio que se agrega a un fundente con el fin de eliminar oxígeno del baño de fusión de la soldadura.

deposited metal. Filler metal that has been added during a welding operation.

metal depositado. Metal de relleno que se ha agregado durante una operación de soldadura.

deposition efficiency (arc welding). The ratio of the weight of deposited metal to the net weight of filler metal consumed, exclusive of stubs.

eficiencia de deposición. La relación del peso del metal depositado al peso neto del metal de relleno consumido, excluyendo los tacones.

deposition rate. The weight of material deposited in a unit of time. It is usually expressed as pounds per hour (lb/hr) (kilograms per hour [kg/hr]).

relación de deposición. El peso del material depositado en una unidad de tiempo. Es regularmente expresado en kilogramos por hora (kg/hora) (libras por hora [lb/hora]).

depth of fusion. The distance that fusion extends into the base metal or previous bead from the surface melted during welding.

grueso de fusión. La distancia en que la fusión se extiende dentro del metal base o del cordón anterior de la superficie que se derretió durante la soldadura.

***destructive testing.** Mechanical testing of weld specimens to measure strength and other properties. The tests are made on specimens that duplicate the material and weld procedures required for the job.

***prueba destructiva.** Pruebas mecánicas de probetas de soldadura para medir la fuerza y otras propiedades. Las pruebas se hacen en probetas que duplican el material y los procedimientos de la soldadura requeridos para el trabajo.

***dimension line.** Lines on a drawing that are drawn so that their ends touch the object being measured, or they may touch the extension line extending from the object being measured.

***línea de dimensión.** Líneas en un dibujo que se trazan para que sus extremos toquen el objeto que se está midiendo, o pueden tocar la línea de extensión que se proyecta desde el objeto que se está midiendo.

dimensioning. The measurements of an object, such as its length, width, and height; or the measurements for locating such things as parts, holes, and surfaces.

acotación. Las medidas de un objeto, tal como su longitud, ancho, y altura, o las medidas para ubicar cosas como piezas, agujeros o superficies.

dip brazing (DB). A brazing process that uses heat from a molten salt or metal bath. When a molten salt is used, the bath may act as a flux. When a molten metal is used, the bath provides the filler material.

soldadura fuerte por inmersión. Es un proceso de soldadura fuerte que usa el calor de una sal fundida o un baño de metal. Cuando se usa la sal fundida el baño puede actuar como flujo. Cuando se usa el metal fundido, el baño proporciona el metal de relleno.

dip soldering (DS). A soldering process using the heat furnished by a molten metal bath that provides the solder filler metal.

soldadura blanda de bajo punto de fusión por inmersión. Un proceso que usa calor proporcionado por un baño de metal derretido que provee el metal de relleno para soldar.

direct-current electrode negative (DCEN). The arrangement of direct-current arc welding leads on which the electrode is the negative pole and the workpiece is the positive pole of the welding arc.

corriente directa con electrodo negativo. El arreglo de los cables para soldar con la soldadura de arco donde el electrodo es polo negativo y la pieza de trabajo es polo positivo de la soldadura de arco.

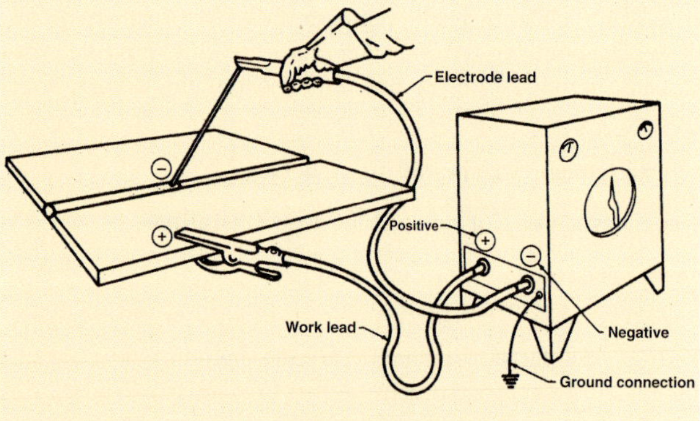

Direct-current electrode negative (straight polarity)

direct-current electrode positive (DCEP). The arrangement of direct-current arc welding leads on which the electrode is the positive pole and the workpiece is the negative pole of the welding arc.

corriente directa con el electrodo positivo. El arreglo de los cables para soldar con la soldadura de arco con el electrodo es el polo positivo y la pieza de trabajo es el polo negativo de la soldadura de arco.

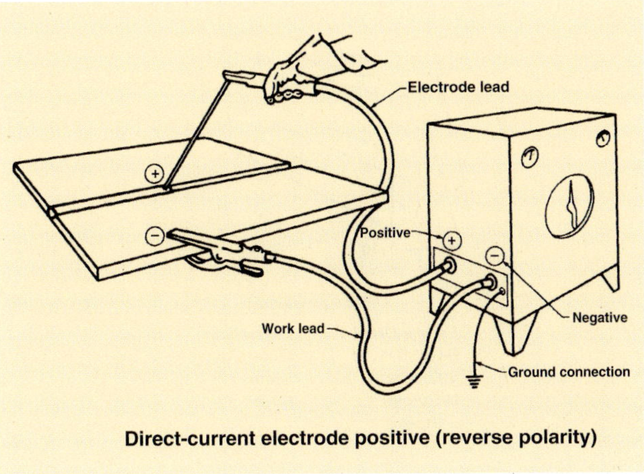

Direct-current electrode positive (reverse polarity)

discontinuity. An interruption of the typical structure of a material, such as a lack of homogeneity in its mechanical, metallurgical, or physical characteristics. A discontinuity is not necessarily a defect. See also **defect** and **flaw.**

discontinuidad. Una interrupción de la estructura típica de un material, el que falta de homogenidad en sus caracteristicas mecánicas, metalúrgicas, o fisica. Vea también **defecto** y **falta.**

***distortion.** Movement or warping of parts being welded, from the prewelding position and condition compared to the postwelding condition and position.

***deformacion.** Movimiento o torcimiento de las partes que se están soldando, comparando la posición antes de soldar a la posicion despues de soldar.

double bevel-groove weld. A type of groove weld. Refer to drawing for **groove weld.**

soldadura de ranura con doble bisel. Es un tipo de soldadura de ranura. Refiérase al dibujo para **soldadura de ranura.**

double-flare bevel-groove weld. A type of groove weld. Refer to drawing for **groove weld.**

soldadura de ranura con doble bisel acampanada. Un tipo de soldadura de ranura. Refiérase al dibujo para **soldadura de ranura.**

double-flare V-groove weld. A weld in grooves formed by two members with curved surfaces. Refer to drawing for **groove weld.**

soldadura de ranura de doble V acampanada. Una soldadura de ranura formada por dos miembros con superficies curvados. Refiérase al dibujo para **soldadura de ranura.**

double J-groove weld. A type of groove weld. Refer to drawing for **groove weld.**

soldadura de ranura de doble-J. Un tipo de soldadura de ranura. Refiérase al dibujo para **soldadura de ranura.**

double U-groove weld. A type of groove weld. Refer to drawing for **groove weld.**

soldadura de ranura de doble-U. Un tipo de soldadura de ranura. Refiérase al dibujo para **soldadura de ranura.**

double V-groove weld. A type of groove weld. Refer to drawing for **groove weld.**

soldadura de ranura de doble-V. Un tipo de soldadura de ranura. Refiérase al dibujo para **soldadura de ranura.**

downslope time. See **automatic arc welding downslope time** and **resistance welding downslope time.**

tiempo de caída del pendiente. Vea tiempo del pendiente con descenso en soldadura de arco automático y tiempo del pendiente descenso en soldadura de resistencia.

drag (thermal cutting). The offset distance between the actual and straight line exit points of the gas stream or cutting beam measured on the exit surface of the base metal.

tiro (corte termal). La distancia desalineada entre la actual y la linea recta del punto de salida del chorro de gas o el rayo de cortar medido a la salida de la superficie del metal base.

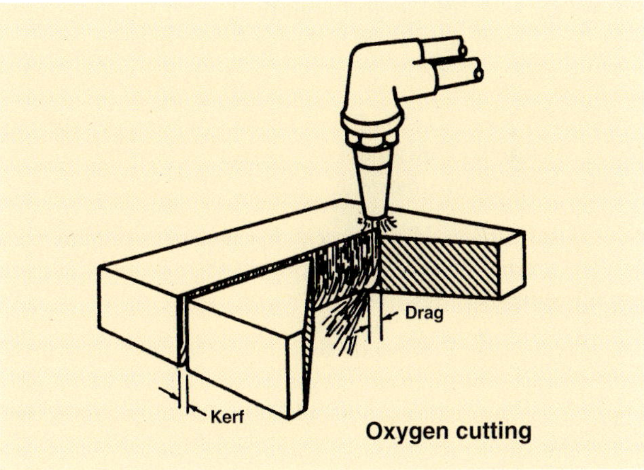

Oxygen cutting

drag angle. The travel angle when the electrode is pointing in a direction opposite to the progression of welding. This angle can also be used to partially define the position of guns, torches, rods, and beams.

ángulo del tiro. El ángulo de avance cuando el electrodo está apuntando en una dirección opuesta del progreso de la soldadura. Este ángulo también se puede usar para parcialmente definir la posición de pistolas, antorchas, varillas, y rayos.

***drag lines.** High-pressure oxygen flow during cutting forms lines on the cut faces. A correctly made cut has up and down drag lines (zero drag); any deviation from the pattern indicates a change in one of the variables affecting the cutting process; with experience the welder can interpret the drag lines to determine how to correct the cut by adjusting one or more variables.

***lineas del tiro.** La salida del oxígeno a presión elevada durante el corte forma lineas en las caras del corte. Un corte hecho correctamente tiene lineas hacia arriba y hacia abajo (zero tiro); cualquier desviación de la norma indica un cambio en uno de los variables que afectan el proceso de cortar; con experiencia el soldador puede interpretar las lineas de tiro y determinar como corregir el corte ajustando uno o más variables.

***drift.** The tendency of a system's response to gradually move away from the desired response.

***deriva.** La tendencia de la respuesta del sistema de retirarse gradualmente de la respuesta deseada.

***drooping output.** Volt-ampere characteristic of the shielded metal arc process power supply where the voltage output decreases as increasing current is required of the power supply. This characteristic provides a reasonably high voltage at a constant current.

***reducción de potencia de salida.** Característica de voltio-amperios de la alimentación de poder de un proceso de soldadura de arco protegido donde la salida del voltaje disminuye mientras un aumento de la corriente es requerida de la alimentación de poder. Está característica proporciona un voltaje razonable alto con corriente constante.

***dross.** The material expelled from the plasma arc and oxygen assist laser cutting processes which contains 40% or more of unoxidized base metal.

***escoria.** El material expulsado de los procesos del arco de plasma y corte láser asistidos con oxígeno que contiene un 40% o más de metal base sin oxidar.

***dual shield.** FCA welding process that uses both the flux core and an external shielding gas to protect the molten weld pool. See **self shielding.**

***doble protección.** Proceso de soldadura FCA que utiliza tanto el núcleo del fundente como el gas de protección externo para proteger el baño de fusión de la soldadura. Consulte **protección personal.**

***ductility.** As applied to a soldered or brazed joint, it is the ability of the joint to bend without failing.

***ductilidad.** Como aplicada a junta de soldadura fuerte o soldadura blanda, es la abilidad de la junta de doblarse sin fallar.

duty cycle. The percentage of time during an arbitrary test period that a power source or its accessories can be operated at rated output without overheating.

ciclo de trabajo. El porcentaje de tiempo durante un período a prueba arbitraria de una fuente de poder y sus accesorios que pueden operarse a la capacidad de carga de salida sin sobrecalentarse.

E

***eddy current inspection (ET).** See **nondestructive testing.**

***inspección por corrientes parásitas.** Consulte **pruebas no destructivas.**

edge joint. A joint between the edges of two or more parallel or nearly parallel members.

junta de orilla. Una junta en medio de las orillas de dos o más miembros paralelos o casi paralelos.

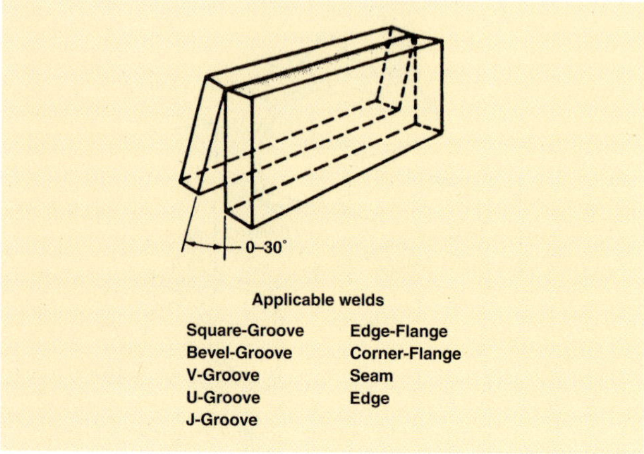

edge preparation. The surface prepared on the edge of a member for welding.

preparación de orilla. La superficie preparada en la orilla de un miembro que se va a soldar.

effective length of weld. The length of weld throughout which the correctly proportioned cross section exists. In a curved weld, it shall be measured along the axis of the weld.

distancia efectiva de soldadura. La distancia de una sección transversa correctamente proporcionada que existe por toda la soldadura. En una soldadura en curva, debe medirse por el axis de la soldadura.

effective throat. The minimum distance from the root of a weld to its face, less any reinforcement. See also **joint penetration.** Refer to drawing for **convexity.**

garganta efectiva. La distancia mínima de la raíz a la cara de una soldadura, menos el refuerzo. Vea también **penetración de junta.** Refiérase al dibujo para **convexidad.**

***elastic limit.** The maximum force that can be applied to a material or joint without causing permanent deformation or failure.

***limite elástico.** La fuerza máxima que se le puede aplicar a un material o junta sin causar deformación o falta permanente.

***elbow.** The joint that connects a robot's upper arm and forearm.

***codo.** La junta que conecta al brazo de arriba con el brazo de enfrente en un robot.

electrode. A component of the electrical circuit that terminates at the arc, molten conductive slag, or base metal.

electrodo. Un componente del circuito eléctrico que termina al arco, escoria derretida conductiva, o metal base.

***electrode angle.** The angle between the electrode and the surface of the metal; also known as the direction of travel (leading angle or trailing angle); leading angle pushes molten metal and slag ahead of the weld; trailing angle pushes the molten metal away from the leading edge of the molten weld pool toward the back, where it solidifies.

***ángulo del electrodo.** El ángulo en medio del electrodo y la superficie del metal; también conocido como la dirección de avance (apuntado hacia adelante o apuntado hacia atras); el ángulo apuntado empuja el metal derretido y la escoria enfrente de la soldadura; y el ángulo apuntado hacia atrás empuja el metal derretido lejos de la orilla delantera del charco del metal derretido hacia atrás, donde se solidifica.

***electrode classification.** Any of several systems developed to identify shielded metal arc welding electrodes. The most widely used identification system was developed by the American Welding Society (AWS). The information represented by the classification generally includes the minimum tensile strength of a good weld, the position(s) in which the electrode can be used, the type of flux coating, and the type(s) of welding currents with which the electrode can be used.

***clasificación de electrodo.** Cualquiera de los varios sistemas desarollados para identificar electrodos protegidos para soldadura de arco. El sistema de identificación que se usa mucho más fue desarrollado por la Sociedad de Soldadura Americana (AWS). La información representada por la clasificación generalmente incluye la fuerza tensible mínima de una soldadura, la posición(es) donde se puede usar el electrodo, el tipo de recubrimiento de fundente y los tipo(s) de corrientes para soldar con la cual se puede usar el electrodo.

electrode extension (GMAW, FCAW, SAW). The length of unmelted electrode extending beyond the end of the contact tube during welding.

extensión del electrodo (GMAW, FCAW, SAW). La distancia de extensión del electrodo que no está derretido más allá de la punta del tubo de contacto durante la soldadura.

electrode holder. A device used for mechanically holding and conducting current to an electrode during welding.

porta electrodo. Un aparato usado para detener mecánicamente y conducir corriente a un electrodo durante la soldadura.

electrode lead. The electrical conductor between the source of arc welding current and the electrode holder. Refer to drawing for **direct-current electrode negative.**

cable de electrodo. Un conductor eléctrico en medio de la fuente para la corriente de soldar con arco y el portaelectrodo. Refiérase al dibujo de **corriente directa con electrodo negativo.**

electrode setback. The distance the electrode is recessed behind the constricting orifice of the plasma arc torch or thermal spraying gun, measured from the outer face of the nozzle.

retroceso del electrodo. La distancia del hueco del electrodo que está detrás del orificio constringente de la antorcha de arco plasma o pistola de rocio termal, se mide de la cara de afuera a la boquilla.

***electrode tip.** The end of an electrode where the arc jumps from the electrode to the work.

***punta del electrodo.** El extremo de un electrodo en donde el arco salta del electrodo al trabajo.

electron beam cutting (EBC). A cutting process that uses the heat obtained from a concentrated beam composed primarily of high-velocity electrons, which impinge upon the workpieces to be cut; it may or may not use an externally supplied gas.

cortes a rayo de electron. Un proceso de cortar que usa calor obtenido de un rayo concentrado primeramente de electrones de alta velocidad que choca sobre la pieza de trabajo la cual se va a cortar; puede o no usar un gas surtido externamente.

electron beam welding (EBW). A welding process that produces coalescence with a concentrated beam, composed primarily of high-velocity electrons, impinging on the joint. The process is used without shielding gas and without the application of pressure.

soldadura a rayo de electron. Un proceso de soldadura la cual produce coalescencia de un rayo concentrado, compuesto primeramente de electrones de alta velocidad al chocar con la junta. Este proceso no usa gas de protección y sin la aplicación de presión.

electroslag welding electrode. A filler metal component of the welding circuit through which current is conducted from the electrode guiding member to the molten slag.

electrodo para soldadura de electroescoria. Un componente de metal de relleno del circuito para soldar por donde la corriente es conducida del miembro que guía el electrodo a la escoria derretida.

emissive electrode. A filler metal electrode consisting of a core of a bare electrode or a composite electrode to which a very light coating has been applied to produce a stable arc.

electrodo emisivo. Un electrodo de metal de relleno consistiendo de un electrodo liso o un electrodo compuesto de una capa ligera que se le aplica para producir un arco estable.

***end effector.** An actuator, a gripper, or a mechanical device attached to the wrist of a manipulator by which objects can be grasped or acted upon.

***punta que efectúa.** Un movedor, el que agarra, o un aparato mecánico fijo a la muñeca de un manipulador por el cual objectos se pueden agarrar u obrar en impulso.

entry-level welder. A person just entering the welding profession.

soldador principiante. Una persona que acaba de comenzar en la profesión de la soldadura.

***error signal.** The difference between desired response and actual response.

***señal equivocada.** La diferencia entre la respuesta deseada y la respuesta efectiva.

***etching.** The process of chemically preparing the surface of a specimen so that the metal's grain can be examined.

***decapado.** El proceso de preparar químicamente la superficie de una muestra para que se pueda examinar el grano del metal.

***eutectic composition.** The composition of an alloy that has the lowest possible melting point for that mixture of metals.

composición de tipo eutectico. La composición de un aleado que tenga el punto de fusión lo más bajo posible para esa mezcla de metales.

***exhaust pickup.** A component of a forced ventilation system that has sufficient suction to pick up fumes, ozone, and smoke from the welding area and carry the fumes, etc., outside of the area.

***recogedor de extracción.** Un componente de un sistema de ventilación forzada que tiene suficiente succión para recoger vaho, ozono, y humo de la área de soldadura y lleva al vaho, etc., a fuera de la area.

***exothermic gases.** Gasses that, when combined with oxygen (usually in a flame), release heat energy.

***gases exotérmicos.** Gases que, cuando se combinan con oxígeno (por lo general en una llama), liberan energía calórica.

***extension line.** Lines in a drawing that extend from an object which locate the points being dimensioned.

***línea de extensión.** Líneas en un dibujo que se extienden desde un objeto y que ubican los puntos a los que se intenta dar dimensión.

F

***fabrication.** An assembly whose parts may be joined by a combination of methods, including welds, bolts, screws, adhesives, etc.

***conjunto.** Un conjunto cuyas partes pueden estar unidas por una combinación de métodos que incluyen soldaduras, pernos, tornillos, adhesivos, etc.

face-bend test. A test in which the weld face is on the convex surface of a specified bend radius.

prueba de dobléz de cara. Una prueba donde la cara de la soldadura está en la superficie convexa al radio de dobléz especificado.

***face-centered cubic (FCC).** A crystalline structure that iron forms as it changes from a liquid to a solid. See also **crystalline structures.**

***cúbico de caras centradas (FCC).** Una estructura cristalina que forma el hierro a medida que cambia de estado líquido a sólido. Consulte también **estructuras cristalinas.**

face of weld. The exposed surface of a weld on the side from which welding was done.

cara de la soldadura. La superficie expuesta de una soldadura del lado de donde se hizo la soldadura.

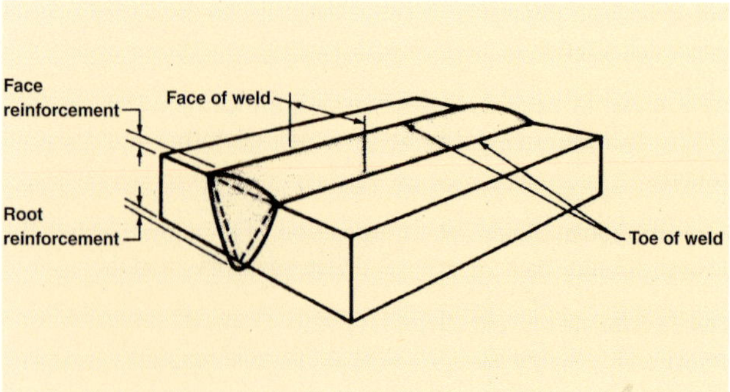

face reinforcement. Reinforcement of a weld at the side of the joint from which welding was done. See also **root reinforcement.** Refer to drawing for **face of weld.**

refuerzo de cara. Refuerzo de una soldadura en el lado de la junta de donde se hizo la soldadura. Vea también **refuerzo de raíz.** Refiérase al dibujo para **cara de la soldadura.**

face shield. A device positioned in front of the eyes and a portion of, or all of, the face, whose predominant function is protection of the eyes and face. See also **helmet.**

protector de cara sostenido a mano. Un aparato puesto en frente de los ojos y una porción, o en toda la cara, cuya función predominante es de proteger los ojos y la cara. Vea también **casco.**

***fast freezing.** A fluxing characteristic that allows it to solidify at a temperature above that of the molten weld pool so that the flux can help control the weld bead contour.

***solidificación rápida.** Una característica del fundente que permite que se solidifique a una temperatura superior a la del baño de fusión de la soldadura para que el fundente pueda ayudar a controlar el contorno del cordón de la soldadura.

***fast freezing electrode.** An electrode whose flux forms a high-temperature slag that solidifies before the weld metal solidifies, thus holding the molten metal in place. This is an advantage for vertical, horizontal, and overhead welding positions.

***electrodo de congelación rápida.** Un electrodo cuyo flujo forma una escoria a temperaturas altas que se puede solidificar antes de que el metal de soldadura se pueda solidificar, asi detiene el metal derretido en su lugar. Está es una ventaja en soldaduras de posiciones vertical, horizontal y sobrecabeza.

***fatigue resistance.** As applied to a soldered or brazed joint, it is the ability of the joint to be bent repeatedly without exceeding its elastic limit and without failure. Generally, fatigue resistance is low for most soldered and brazed joints.

***resistencia a la fatiga.** Como aplicada a una junta de soldadura con metales de bajo punto de fusión y soldadura fuerte, es la abilidad de una junta de ser doblada repetidamente sin exceder los limites elásticos y sin fracaso. Generalmente, la resistencia a la fatiga es muy baja para la mayoría de las juntas de soldadura con bajo punto de fusión y soldadura fuerte.

***faying.** See **faying surface.**

***ala de contacto.** Consulte **superficie de ala de contacto.**

faying surface. The mating surface of a member that is in contact with or in close proximity to another member to which it is to be joined.

superficie de unión. La superficie de apareamiento de un miembro del que está en contacto con otro miembro o está en proximidad cercana a otro miembro que está para ser unido.

feed rollers. A set of two or four individual rollers which, when pressed tightly against the filler wire and powered up, feed the wire through the conduit liner to the gun for GMA and FCA welding.

rodillos de alimentación. Un conjunto de dos o cuatro rodillos individuales que al ser presionados fuertemente contra el alambre de relleno y ser accionados alimentan al alambre a través del revestimiento de canal hasta la pistola, en soldaduras tipo GMA y FCA.

***ferrite.** A body-centered cubic crystalline structure that iron forms as it changes from a liquid to a solid. See also **crystalline structures.**

***ferrita.** Una estructura cristalina cúbica de cuerpo centrado que forma el hierro forma a medida que cambia de estado líquido a sólido. Consulte también **estructuras cristalinas.**

filler metals. The metals or alloys to be added in making a welded, brazed, or soldered joint.

metales de aporte. Los metales o aleados que se agregan cuando se hace una soldadura blanda o soldadura fuerte.

***filler pass.** One or more weld beads used to fill the groove with weld metal. The bead must be cleaned after each pass to prevent slag inclusions.

***pasada para rellenar.** Uno o más cordones de soldadura usados para llenar la ranura con el metal de soldadura. El cordón debe ser limpiado después de cada pasada para prevenir inclusiones de escoria.

fillet weld. A weld of approximately triangular cross section joining two surfaces approximately at right angles to each other in a lap joint, tee joint, or corner joint. Refer to drawing for **convexity.**

soldadura de filete. Una soldadura de filete de sección transversa aproximadamente triangular que une dos superficies aproximademente en ángulos rectos de uno al otro en junta de traslape, junta en- T- o junta de esquina. Refiérase al dibujo para **convexidad.**

fillet weld break test. A test in which the specimen is loaded so that the weld root is in tension.

prueba de rotura en soldadura de filete. Una prueba en donde la probeta es cargada de manera en que la tensión esté sobre la soldadura.

fillet weld leg. The distance from the joint root to the toe of the fillet weld.

pierna de soldadura filete. La distancia de la raíz de la junta al pie de la soldadura filete.

fillet weld size. For equal leg fillet welds, the leg lengths of the largest isosceles right triangle that can be inscribed within the fillet weld cross section. For unequal leg fillet welds, the leg lengths of the largest right triangle that can be inscribed within the fillet weld cross section.

tamaño de soldadura filete. Para soldaduras de filete que tienen piernas iguales, lo largo de las piernas del isósceles más grande del triángulo recto que puede ser inscribido dentro de la sección. Para soldaduras de filete con piernas desiguales, lo largo de las piernas del triángulo recto más grande puede inscribirse dentro de la sección transversal.

filter plate. An optical material that protects the eyes against excessive ultraviolet, infrared, and visible radiation.

lente filtrante. Un material óptico que protege los ojos contra ultravioleta excesiva, infrarrojo, y radiación visible.

final current. The current after downslope but prior to current shut-off.

corriente final. La corriente después del pendiente en descenso pero antes de que la corriente sea cerrada.

fisheye. A discontinuity found on the fracture surface of a weld in steel that consists of a small pore or inclusion surrounded by an approximately round, bright area.

ojo de pescado. Una discontinuidad que se encuentra en una fractura de superficie en una soldadura de acero que consiste de poros pequeños o inclusiones rodeadas de áreas aproximadamente redondas y brillantes.

fissure. A small, crack-like discontinuity with only slight separation (opening displacement) of the fracture surfaces. The prefixes *macro* and *micro* indicate relative size.

hendemiento. Una pequeña, discontinuidad como una grieta con solamente una separación (abertura desalojada) de las superficies fracturadas. El prefijo *marco* y *micro* indica el tamaño relativo.

***fixed inclined (6G) position.** For pipe welding, the pipe is fixed at a 45° angle to the work surface. The effective welding angle changes as the weld progresses around the pipe.

***posición (6G) inclinado fijo.** Para soldadura de tubo, el tubo se fija a un ángulo de 45° de la superficie del trabajo. El ángulo efectivo de la soldadura cambia cuando la soldadura progresa alrededor del tubo.

fixture. A device designed to hold parts to be joined in proper relation to each other.

fijación. Una devisa diseñada para detener partes que se van a unir en relación propia de una a la otra.

flame propagation rate. The speed at which flame travels through a mixture of gases.

cantidad de propagación de la llama. La rapidez en que la llama camina a través de una mezcla de gas.

flame spraying (FLSP). A thermal spraying process in which an oxyfuel gas flame is the source of heat for melting the surfacing material. Compressed gas may or may not be used for atomizing the propellant and surfacing material to the substrate.

rociado a llama. Un proceso de rociado termal en donde la llama del gas oxicombustible es la fuente de calor para derretir el material de revestimiento. El gas comprimido se puede o no se puede usar para automizar el propulsor y el material de revestimiento al substrato.

flange weld. A weld made on the edges of two or more joint members, at least one of which is flanged.

soldadura de reborde. Una soldadura que se hace en las orillas de dos o más miembros de junta, donde por lo menos uno tiene reborde.

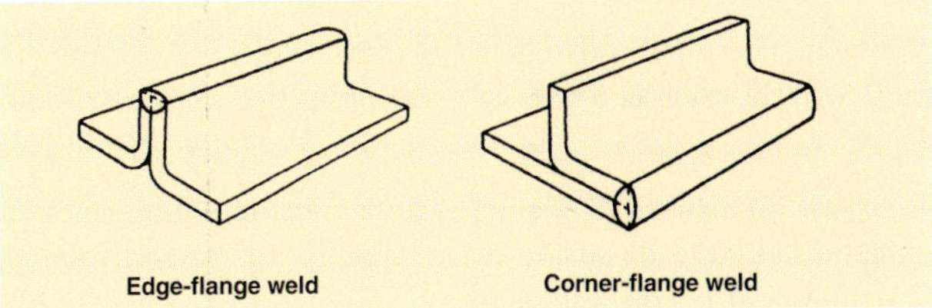

flash. The material that is expelled or squeezed out of a weld joint and that forms around the weld.

ráfaga. El material que es despedido o exprimido fuera de una junta de soldadura y se forma alrededor de la soldadura.

***flash burn.** An arc-caused burn typically on the eye that results from an extremely brief exposure to the direct ultraviolet light produced by an arc welding process.

***quemadura por fogonazo.** Una quemadura causada por el arco, por lo general en los ojos, como consecuencia de una exposición extremadamente breve a la luz ultravioleta directa producida por el proceso de soldadura por arco.

***flash glasses.** Lightly tinted safety glasses, usually a number two shade, worn by welders to protect themselves from flash burns as well as flying debris.

***lentes de seguridad.** Lentes de seguridad ligeramente coloreadas, por lo general con un tinte de nivel dos, que utilizan los soldadores para protegerse de las quemaduras por fogonazo así como también de los fragmentos que salen despedidos.

flash welding (FW). A resistance welding process that produces a weld at the faying surfaces of a butt joint by a flashing action and by the application of pressure after heating is substantially completed. The flashing action, caused by the very high current densities at small contact points between the workpieces, forcibly expels the material from the joint as the workpieces are slowly moved together. The weld is completed by a rapid upsetting of the workpieces.

soldadura de relámpago. Un proceso de soldadura de resistencia que produce una soldadura en el empalme de la superficie de una junta tope por una acción de relampagueo y por la aplicación de presión después que el calentamiento este substancialmente acabado. La acción del relampagueo, causado por densidades de corrientes altas a unos puntos de contacto pequeños en medio de las piezas de trabajo, despiden fuertemente el material de la junta cuando las piezas de trabajo se mueven despacio. La soldadura es terminada por un acortamiento rápido de las piezas de trabajo.

flashback. A recession of the flame into or in back of the mixing chamber of the oxyfuel gas torch or flame spraying gun.

llamarada de retroceso. Una recesión de la llama adentro o atrás de la cámara mezcladora de una antorcha de gas oxicombustible o pistola de rociar a llama.

flashback arrestor. A device to limit damage from a flashback by preventing propagation of the flame front beyond the point at which the arrestor is installed.

válvula de retención. Un aparato para limitar el daño de una llamarada de retroceso para prevenir la propagación del frente de la llama más allá del punto donde se instala la válvula de retención.

***flashing.** The rapid moving away of an oxyfuel torch from the molten weld pool as a way of controlling the weld's size and shape.

***rebarbar.** El apartar rápidamente un soplete que funciona con oxígeno del baño de fusión de la soldadura para controlar el tamaño y la forma de la soldadura.

flat position. The welding position used to weld from the upper side of the joint; the weld face is approximately horizontal.

posición plana. La posición de soldadura que se usa para soldar del lado de arriba de una junta; la cara de la soldadura está aproximadamente horizontal.

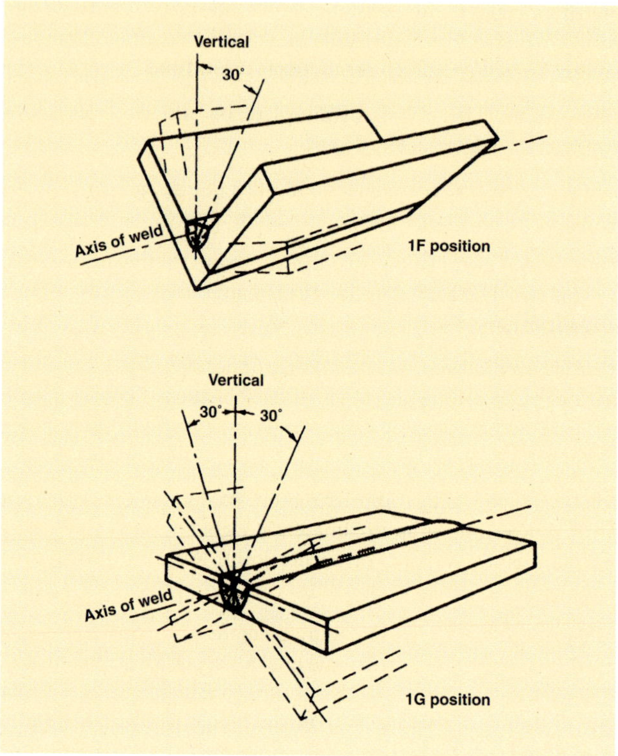

flaw. An undesirable discontinuity.

falta. Una discontinuidad indeseable.

***flow rate.** The rate at which a given volume of shielding gas is delivered to the weld zone. The units used for welding are cubic feet, inches, meters, and centimeters.

***caudal.** Velocidad a la cual llega un determinado volumen de gas protector a la zona de soldadura. Las unidades usadas para la soldadura son pies cúbicos, pulgadas, metros, y centímetros.

***flowmeter.** Device used to control the rate that a fluid flows.

***caudalímetro.** Dispositivo que se utiliza para controlar la tasa a la que fluye un fluido.

flux. A material used to hinder or prevent the formation of oxides and other undesirable substances in molten metal and on solid metal surfaces and to dissolve or otherwise facilitate the removal of such substances.

flujo. Un material que se usa para impedir o prevenir la formación de óxidos y otras substancias indeseables en el metal derretido y en las superficies del metal sólido, y para desolver o de otra manera facilitar el removimiento de dichas substancias.

***flux actions.** See **flux.**

***acciones de fundente.** Consulte **fundente.**

flux cored arc welding (FCAW). An arc welding process that uses an arc between a continuous filler metal electrode and the weld pool. The process is used with shielding gas from a flux contained within the tubular electrode, with or without additional shielding from an externally supplied gas, and without the application of pressure.

soldadura de arco con núcleo de fundente. Un proceso de soldadura de arco que usa un arco entre medio de un electrodo de metal rellenado continuo y el charco de la soldadura. El proceso es usado con gas de protección del flujo contenido dentro del electrodo tubular, y sin usarse protección adicional de abastecimiento de gas externo, y sin aplicarse presión.

flux cored electrode. A composite tubular filler metal electrode consisting of a metal sheath and a core of various powdered materials, producing an extensive slag cover on the face of a weld bead. External shielding may be required.

electrodo de núcleo de fundente. Un electrodo tubular de metal para rellenar con una compostura que consiste de una envoltura de metal y un núcleo con varios materiales de polvo, que producen un forro extensivo de escoria en la superficie del cordón de soldadura. Protección externa puede ser requerida.

flux cover. In metal bath dip brazing and dip soldering, a cover of flux over the molten filler metal bath.

tapa de fundente. En metal de baño soldadura fuerte y soldadura blanda por inmersión, una tapa de fundente sobre el baño del metal de relleno derretido.

***forced ventilation.** To remove excessive fumes, ozone, or smoke from a welding area, a ventilation system may be required to supplement natural ventilation. Where forced ventilation of the welding area is required, the rate of 200 cu ft (56 m^3) or more per welder is needed.

***ventilación forzada.** Para quitar excesivo vaho, ozono y humo de la área donde se solda, un sistema de ventilación puede ser requerido para suplementar la ventilación natural. Donde la ventilación forzada de la área de la soldadura es requerida, la rázon de 200 pies cúbicos (56 m^3) o más es requerido por cada soldador.

***forehand.** See **forehand welding.**

***directa.** Consulte **soldadura directa.**

***forehand technique.** See **forehand welding.**

***técnica de soldadura directa.** Consulte **soldadura directa.**

forehand welding. A welding technique in which the welding torch or gun is directed toward the progress of welding. See also **travel angle, work angle,** and **push angle.** Refer to drawing for **backhand welding.**

soldadura directa. Una técnica de soldar en cual la pistola o la antorcha para soldar es dirigida hacia al progreso de la soldadura. Vea también **ángulo de avance, ángulo de trabajo,** y **ángulo de empuje.** Refiérase al dibujo **soldadura en revés.**

forge welding (FOW). A solid state welding process that produces a weld by heating the workpieces to welding temperature and applying blows sufficient to cause permanent deformation at the faying surfaces.

soldadura por forjado. Un proceso de soldadura de estado sólido que produce una soldadura calentando las piezas de trabajo a una temperatura de soldadura y aplicando golpes suficientes para causar una deformación permanente en las superficies del empalme.

***frequency.** As it relates to alternating current, this refers to the rate that the current reverses direction. See **cycle.**

***frecuencia.** En relación a la corriente alterna, se refiere a la tasa en la que la corriente revierte su dirección. Consulte **ciclo.**

frogs. The large rail structures, usually castings, that form the center of a crossing or the rail crossing point of a switch.

corazónes de cruzamiento. Las grandes estructuras de rieles, normalmente piezas moldeadas, que forman el centro de un cruzamiento o el punto de cruzamiento de rieles en el sitio de convergencia.

fuel gases. Gases such as acetylene, natural gas, hydrogen, propane, stabilized methylacetylene propadiene, and other fuels normally used with oxygen in one of the oxyfuel processes and for heating.

gases combustibles. Gases como acetileno, gas natural, hidrógeno, propano, metilacetileno propodieno estabilizado, y otros combustibles normalmente usados con oxígeno en uno de los procesos de oxicombustible y para calentar.

***full face shield.** Personal protective device to protect the eyes and face from flying debris; it may be clear or tinted.

***protector facial completo.** Dispositivo de protección personal para proteger los ojos y la cara de los fragmentos que puedan salir despedidos. Puede ser transparente o coloreado.

full fillet weld. A fillet weld whose size is equal to the thickness of the thinner member joined.

soldadura de filete llena. Una soldadura de filete cuyo tamaño es igual de grueso como el miembro más delgado de la junta.

full penetration. A nonstandard term for complete joint penetration.

penetración llena. Un término fuera de la norma en vez de la penetración de junta.

furnace brazing (FB). A brazing process in which the parts to be joined are placed in a furnace heated to a suitable temperature.

soldadura fuerte en horno. Un proceso de soldadura fuerte en donde las partes que se van a unir se ponen en un horno calentado a una temperatura adecuada.

furnace soldering (FS). A soldering process in which the parts to be joined are placed in a furnace heated to a suitable temperature.

soldadura blanda en horno. Un proceso de soldadura blanda en donde las partes que se van a unir se ponen en un horno calentado a una temperatura adecuada.

fusion. The melting together of filler metal and base metal, or of base metal only, to produce a weld. See also **depth of fusion.**

fusión. El derretir el metal de relleno y el metal base juntos o el metal base solamente, para producir una soldadura. Vea también **grueso de fusión.**

fusion welding. Any welding process or method that uses fusion to complete the weld.

soldadura de fusión. Cualquier proceso de soldadura o método que usa fusión para completar la soldadura.

fusion zone. The area of base metal melted as determined on the cross section of a weld. Refer to drawing for **depth of fusion.**

zona de fusión. La área del metal base que se derritió como determinada en la sección transversa de la soldadura. Refiérase al dibujo **grueso de fusion.**

G

galling. Galling occurs between metal surfaces that move against each other under a heavy load without adequate lubrication resulting in small pieces of the surface of one part being fused to the other.

mortificante. Irritante se produce entre las superficies metálicas que se mueven uno contra el otro bajo una carga pesada sin la lubricación adecuada dando lugar a piezas pequeñas de la superficie de una parte que se fusiona con el otro.

gap. A nonstandard term used for arc length, joint clearance, and root opening.

abertura. Un término fuera de norma se usa en lugar del arco, despejo de junta, y abertura de raíz.

gas cup. A nonstandard term for gas nozzle.

tazón de gas. Un término fuera de norma en vez de boquilla de gas.

gas cylinder. A portable container used for transportation and storage of compressed gas.

cilindro de gas. Un recipiente portátil que se usa para transportación y deposito de gas comprimido.

***gas laser.** A laser in which the lasing medium is a gas.

***láser de gas.** Un láser en el que el medio de la acción láser es un gas.

gas lens. One or more fine mesh screens located in the torch nozzle to produce a stable stream of shielding gas. Primarily used for gas tungsten arc welding.

lente para gas. Uno o más cedazos de malla fina localizados en la lanza de la antorcha para producir un chorro estable de gas de protección primeramente usada para soldaduras de arco tungsteno y gas.

gas metal arc cutting (GMAC). An arc cutting process that uses a continuous consumable electrode and a shielding gas.

cortes de arco metálico con gas. Un proceso de corte con arco que usa un alambre consumible continuo y un gas de protección.

gas metal arc welding (GMAW). An arc welding process that uses an arc between a continuous filler metal electrode and the weld pool. The process is used with shielding from an externally supplied gas and without the application of pressure.

soldadura de arco metálico con gas. Un proceso de soldar con arco que usa un arco en medio de un electrodo de metal para rellenar continuo y el charco de soldadura. El proceso usa protección de un abastecedor externo de gas y sin la aplicación de presión.

gas metal arc welding–pulsed arc (GMAW-P). A gas metal arc welding process variation in which the current is pulsed.

soldadura con arco metálico con gas arco pulsado. Un proceso de soldadura de arco metálico con gas con variación en cual la corriente es de pulsación.

gas metal arc welding–short circuit arc (GMAW-S). A gas metal arc welding process variation in which the consumable electrode is deposited during repeated short circuits. Sometimes this process is referred to as MIG or CO_2 welding (nonpreferred terms).

soldadura con arco metálico con gas–arco de corto circuito. Un proceso de soldadura de arco metálico con gas variación en cual el electrodo consumible es depositado durante los cortos circuitos repetidos. A veces el proceso es referido como soldadura MIG o CO_2 (términos que no son preferidos).

gas nozzle. A device at the exit end of the torch or gun that directs shielding gas.

boquilla de gas. Un aparato a la salida de la punta de la antorcha o pistola que dirige el gas protector.

gas regulator. A device for controlling the delivery of gas at some substantially constant pressure.

regulador de gas. Un aparato para controlar la salida de gas a una presión substancialmente constante.

gas tungsten arc cutting (GTAC). An arc cutting process that uses a single tungsten electrode with gas shielding.

corte de arco con tungsteno y gas. Un proceso de corte de arco que usa un electrodo de tungsteno sencillo con gas de protección.

gas tungsten arc welding (GTAW). An arc welding process that uses an arc between a tungsten electrode (nonconsumable) and the weld pool. The process is used with shielding gas and without the application of pressure.

soldadura de arco de tungsteno con gas. Un proceso de soldadura de arco que usa un arco en medio del electrodo tungsteno (no consumible) y el charco de la soldadura. El proceso es usado con gas de protección y sin aplicación de presión.

gas tungsten arc welding-pulsed arc (GTAW-P). A gas tungsten arc welding process variation in which the current is pulsed.

soldadura de arco de tungsteno con gas-arco pulsado. Un proceso de soldadura de arco tungsteno con variación en cual la corriente es de pulsación.

***gauge (regulator).** A device mounted on a regulator to indicate the pressure of the gas passing into the gauge. A regulator is provided with two gauges—one (high-pressure gauge) indicates the pressure of the gas in the cylinder; the second gauge (low-pressure gauge) shows the pressure of the gas at the torch.

***manómetro (regulador).** Un aparato montado en un regulador para indicar la presión del gas que está pasando por el manómetro. El regulador tiene dos manómetros—uno (manómetro de alta presión) indica la presión del gas en el cilindro; el segundo manómetro (manómetro de presión baja) enseña la presión del gas en la antorcha.

***gauge pressure.** The actual pressure shown on the gauge; does not take into account atmospheric pressure.

***manómetro para presión.** La presión actual que se enseña en el manómetro; no toma en cuenta la presión atmosférica.

globular transfer. The transfer of molten metal in large drops from a consumable electrode across the arc.

traslado globular. El traslado del metal derretido en gotas grandes de un electrodo consumible a través del arco.

***GMAW.** See **gas metal arc welding.**

***GMAW.** Consulte **soldadura con arco metálico** (GMAW, por sus siglas en inglés).

***GMAW-P.** See **gas metal arc welding–pulse arc.**

***GMAW-P.** Consulte **soldadura con arco metálico–arco de pulso.**

***GMAW-S.** See **gas metal arc welding–short circuit arc.**

***GMAW-S.** Consulte **soldadura con arco metálico–arco de cortocircuito.**

***goggles.** Personal protective device to protect the eyes from flying debris; it may be clear or tinted.

***gafas de seguridad.** Dispositivo de protección personal para proteger los ojos de los fragmentos que puedan salir despedidos. Puede ser transparente o coloreado.

gouging. The forming of a bevel or groove by material removal. See also **back gouging, arc gouging,** and **oxygen gouging.**

escopleando con gubia. Formando un bisel o ranura removiendo el material. Vea también gubia trasera, gubia dura con arco, y escopleando con la gubia con oxígeno.

***grain refinement.** Is the process that occurs when larger metallic grain structures break down in size. Grain refinement can be associated with a change in temperature, mechanical working, or time.

***refinamiento de grano.** Es el proceso que se produce cuando estructuras de grano metálico más grande se reducen de tamaño. El refinamiento de grano se puede relacionar con un cambio de temperatura, trabajo mecánico o tiempo.

***graphite.** A form of carbon.

***grafito.** Un tipo de carbón.

groove. An opening or a channel in the surface of a part or between two components, that provides space to contain a weld.

ranura. Una abertura o un canal en la superficie de una parte o en medio de dos componentes, la cual provee espacio para contener una soldadura.

groove angle. The total included angle of the groove between parts to be joined by a groove weld.

ángulo de ranura. El ángulo total incluido de la ranura entre partes para unirse por una soldadura de ranura.

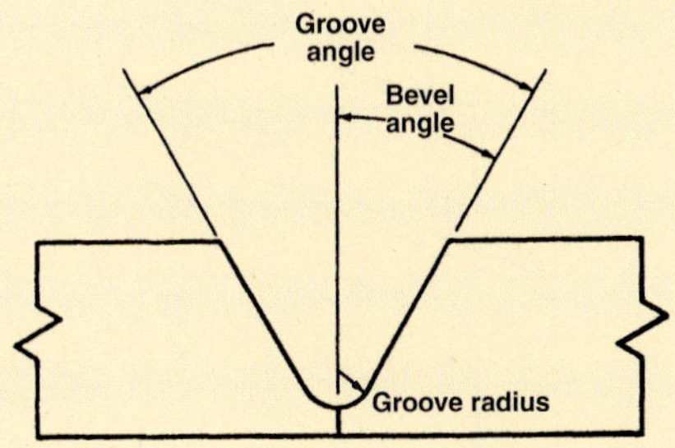

groove face. The surface of a joint member included in the groove.

cara de ranura. La superficie de un miembro de una junta incluido en la ranura.

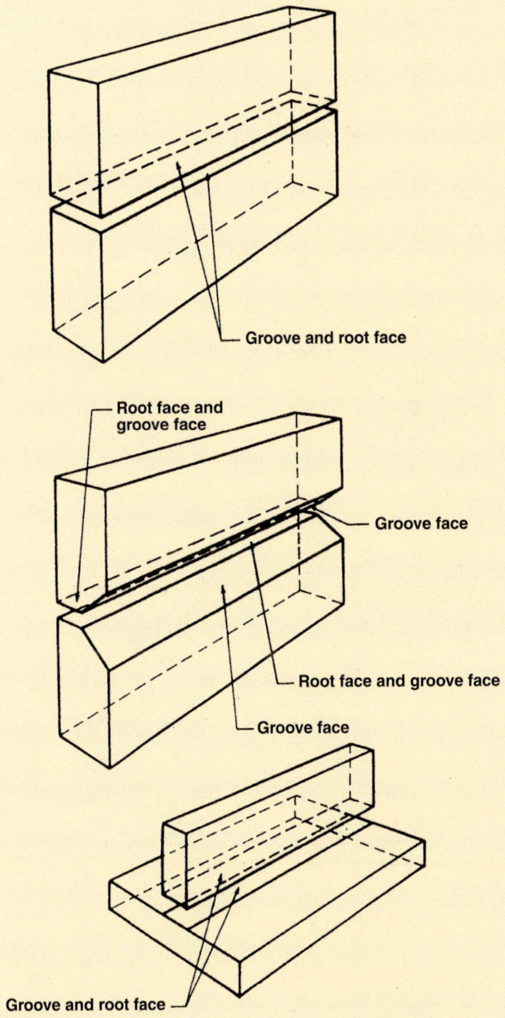

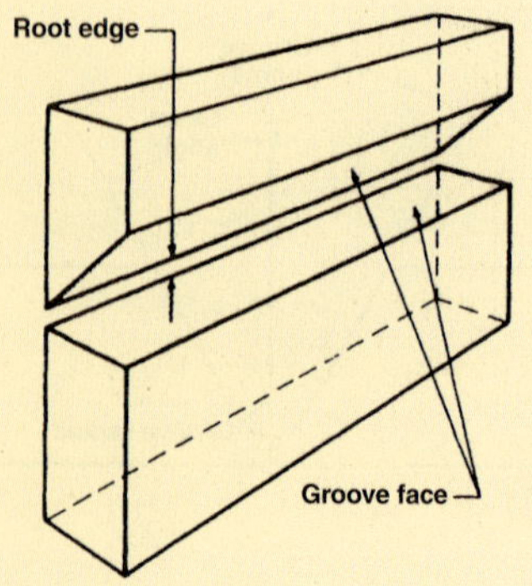

Groove face, root edge, and root face

groove radius. The radius used to form the shape of a J- or U-groove weld joint. Refer to drawings for **bevel.**

radio de ranura. La radio que se usa para formar una junta de una soldadura con una ranura de forma U o J. Refiérase a los dibujos para **bisel.**

groove weld. A weld made in the groove between two members to be joined. The standard types of groove welds are shown in the drawings.

soldadura de ranura. Una soldadura hecha en la ranura dentro de dos miembros que se unen. Los tipos normales de soldadura de ranura se ven en los dibujos.

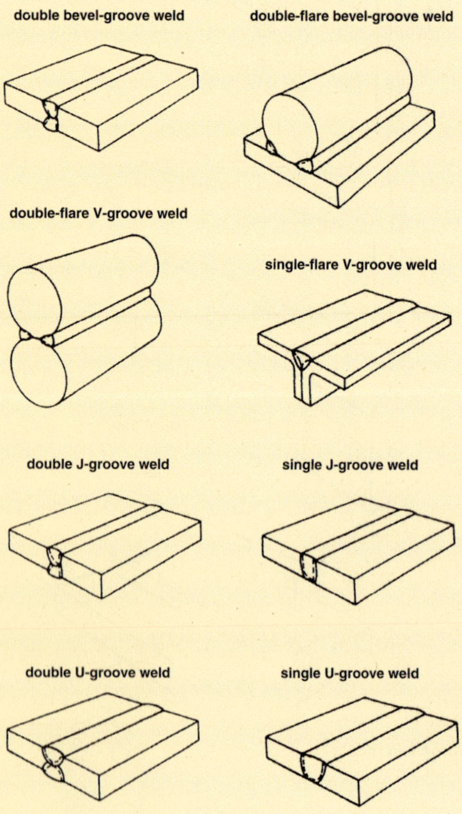

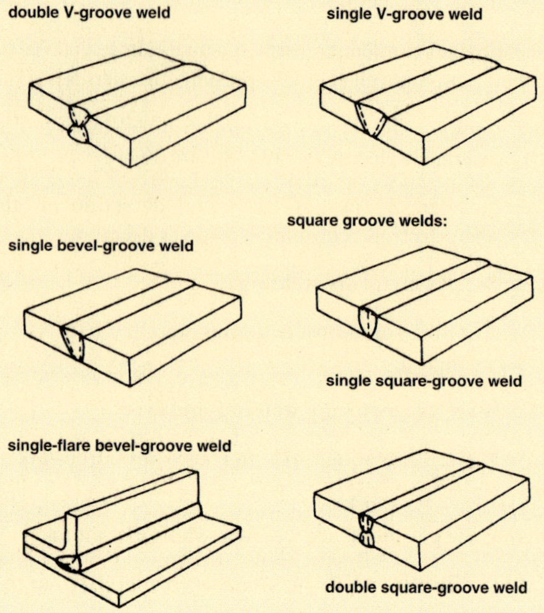

ground connection. An electrical connection of the welding machine frame to the earth for safety.

conexión a tierra. Una conexión eléctrica del marco de la máquina de soldar a la tierra para seguridad.

***ground fault circuit interpreter (GFCI).** An electrical safety device that shuts off current almost instantly if a short to ground occurs.

***intérprete de circuito de fuga a tierra (GFCI, por sus siglas en inglés).** Un dispositivo de seguridad eléctrico que corta la corriente de manera casi instantánea si se produce un cortocircuito a tierra.

ground lead. A nonstandard and incorrect term for workpiece lead.

cable de tierra. Un término fuera de norma e incorrecto que se usa en vez de cable de pieza de trabajo.

***group technology.** A system for coding parts based on similarities in geometrical shape or other characteristics of parts. The grouping of parts into families based on similarities in their production so that parts of a particular family can be processed together.

***codificador.** Un sistema para codificar partes basadas en similaridades en forma geométrica y otras características de las partes. La agrupación de partes en familias basadas en similaridades en su producción para que partes de una familia en particular puedan ser procesadas juntas.

***guided-bend specimen.** Any bend specimen that will be bend-tested in a fixture that controls the bend radii, such as the AWS bend-test fixture.

***probeta de dobléz guiada.** Cualquier probeta de dobléz en la cual se va a hacer un dobléz guiado en una máquina que controla el radio del dobléz, como la máquina de dobléz guiado del AWS.

***gun angle.** Is the relation between the welding gun, the work surface, and the direction of travel.

***ángulo del soplete.** Es la relación entre el soplete, la superficie de trabajo y la dirección de desplazamiento.

H

hardfacing. A surfacing variation in which surfacing material is deposited to reduce wear.

endurecimiento de caras. Una variación superficial donde el material superficial es depositado para reducir el desgastamiento.

heat-affected zone. The portion of the base metal whose mechanical properties or microstructure has been altered by the heat of welding, brazing, soldering, or thermal cutting.

zona afectada por el calor. La porción del metal base cuya propiedad mecánica o microestructura ha sido alterada por el calor de soldadura, soldadura fuerte, soldadura blanda, o corte termal.

***heat treatments.** Postweld heat treatment to reduce weld stresses is the most common type of heat treatment used on weldments.

***tratamientos de calor.** El tratamiento de calor posterior a la soldadura para reducir el estrés del soldeo es el tipo más común de tratamiento de calor que se utiliza en soldadura.

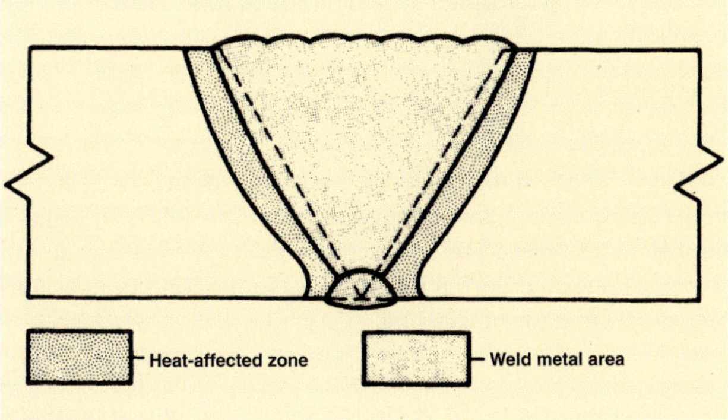

heating torch. A device for directing the heating flame produced by the controlled combustion of fuel gases.

antorcha de calentamiento. Un aparato para dirigir la llama de calentamiento que es producida por una combustión controlada de gases de combustión.

helmet. A device designed to be worn on the head to protect eyes, face, and neck from arc radiation, radiated heat, spatter, or other harmful matter expelled during arc welding, arc cutting, and thermal spraying.

casco. Un aparato diseñado para usarse sobre la cabeza para proteger ojos, cara y cuello de radiación del arco, calor radiado, salpicadura, u otra materia dañosa despedida durante la soldadura de arco, corte por arco, y rociado termal.

***hidden line.** Lines on a drawing that show the same features as object lines except that the corners, edges, and curved surfaces cannot be seen because they are hidden behind the surface of the object.

***línea escondida.** Líneas en un dibujo que muestran las mismas características que las líneas del objeto excepto que los bordes, los extremos y las superficies curvas no se pueden ver debido a que están escondidos detrás de la superficie del objeto.

high frequency. Electric current that changes polarity at a rate higher than 3 million cycles a second (3 MHz).

alta frecuencia. Corriente eléctrica que cambia de polaridad a una tasa superior a 3 millones de ciclos por segundo (3 MHz).

***high-frequency alternating current.** See **high frequency.**

***corriente alterna de alta frecuencia.** Consulte **alta frecuencia.**

horizontal fixed position (pipe welding). The position of a pipe joint in which the axis of the pipe is approximately horizontal and the pipe is not rotated during welding.

posición fija horizontal (soldadura de tubos). La posición de una junta de tubo la cual el axis del tubo es aproximadamente horizontal, y el tubo no da vueltas durante la soldadura.

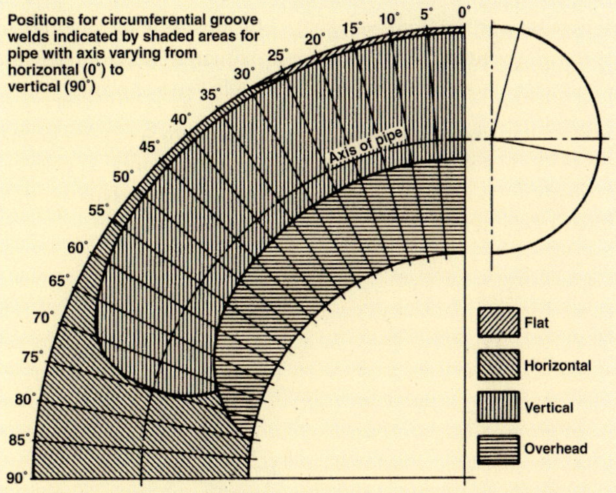

Positions of pipe welds

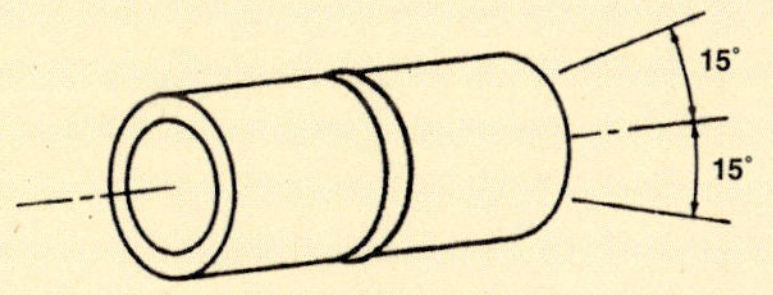

Horizontal fixed position

***horizontal fixed (5G) position weld.** For pipe welding, the pipe is fixed horizontally (cannot be rolled). The weld progresses from overhead, to vertical, to flat position around the pipe.

***soldadura de posición fija horizontal (5G).** Para soldadura de tubos, el tubo está fijo horizontalmente (no se pueder rodar). La soldadura progresa de sobre cabeza, a vertical, a la posición plana alrededor del tubo.

horizontal position (fillet weld). The position in which welding is performed on the upper side of an approximately horizontal surface and against an approximately vertical surface.

posición horizontal (soldadura de filete). La posición de la soldadura la cual es hecha en el lado de arriba de una superficie horizontal aproximadamente y junto a una superficie vertical aproximadamente.

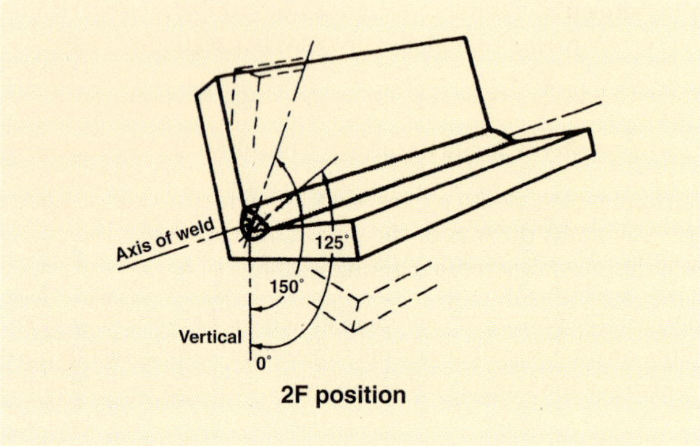

horizontal position (groove weld). The position of welding in which the weld axis lies in an approximately horizontal plane and the weld face lies in an approximately vertical plane.

posición horizontal (de ranura). La posición para soldar en la cual el axis de la soldadura está en una plana horizontal aproximadamente, y la cara de la soldadura está en una plana vertical aproximadamente.

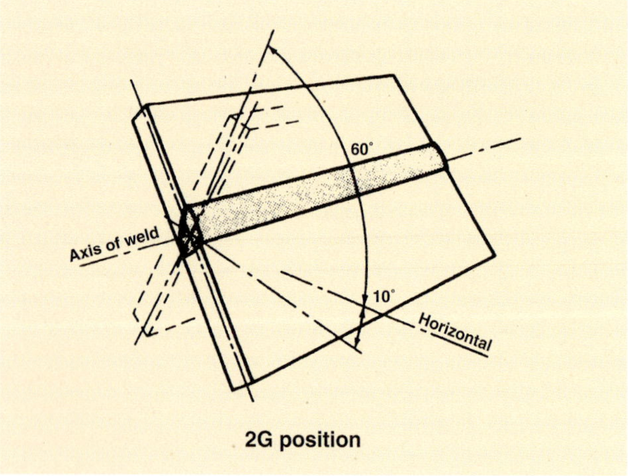

horizontal rolled position (pipe welding). The position of a pipe joint in which the axis of the pipe is approximately horizontal and welding is performed in the flat position by rotating the pipe.

posición horizontal rodada (soldadura de tubo). La posición de una junta de tubo en la cual el axis del tubo es horizontal aproximadamente, y la soldadura se hace en la posición plana con rotación del tubo.

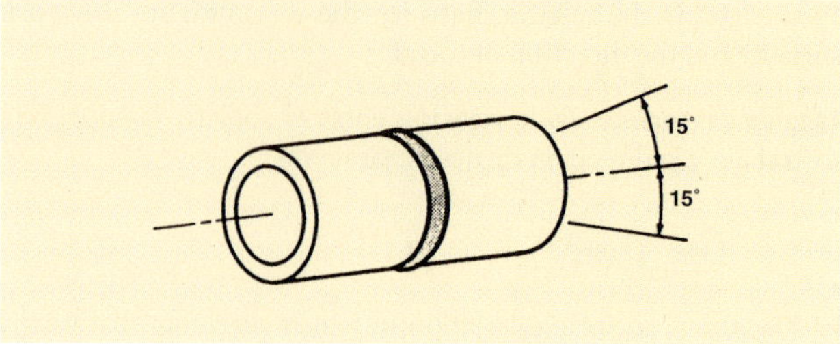

***horizontal rolled (1G) position.** For pipe welding, this position yields high-quality and high-quantity welds. Pipe to be welded is placed horizontally on the welding table in a fixture to hold it steady and permit each rolling. The weld proceeds in steps, with the pipe being rolled between each step, until the weld is complete. For plate, see **axis of a weld.**

***posición (1G) horizontal rodada.** Para soldadura de tubo, está posición produce soldaduras de alta calidad y alta cantidad. El tubo que se va a soldar se pone horizontalmente sobre la mesa de soldadura en una instalación fija que lo detiene seguro y permite cada rodadura. La soldadura procede en pasos, con el tubo siendo rodado entre cada paso, hasta que la soldadura esté completa. Para plato, vea **eje de la soldadura.**

***horizontal welds.** See **horizontal position.**

***soldaduras horizontales.** Consulte **posición horizontal.**

hot crack. A crack formed at temperatures near the completion of solidification.

grieta caliente. Una grieta formada a temperaturas cerca de la terminación de la solidificación.

***hot pass.** The welding electrode is passed over the root pass at a higher than normal amperage setting and travel rate to reshape an irregular bead and turn out trapped slag. A small amount of metal is deposited during the hot pass so the weld bead is convex, promoting easier cleaning.

***pasada caliente.** El electrodo de soldadura se pasa sobre la pasada de raíz poniendo el amperaje más alto que lo normal y proporción de avance para reformar un cordón irregular y sacar la escoria atrapada. Una cantidad pequeña de metal es depositada durante la pasada caliente para que el cordón soldado sea convexo, promoviendo más fácil la limpieza.

hot start current. A very brief current pulse at arc initiation to stabilize the arc quickly. Refer to drawing for **upslope time.**

corriente caliente para empezar. Un pulso muy breve de corriente a iniciación de arco para estabilizar el arco aprisa. Refiérase al dibujo **tiempo del pendiente en ascenso.**

hydrogen embrittlement. The delayed cracking in steel that may occur hours, days, or weeks following welding. It is a result of hydrogen atoms that dissolved in the molten weld pool during welding.

fragilidad causada por el hidrógeno. El fisuramiento retardado en el acero que puede ocurrir horas, días o semanas después de la soldadura. Es el resultado de la disolución de átomos de hidrógeno en el charco de soldadura derretido durante la soldadura.

I

inclined position. The position of a pipe joint in which the axis of the pipe is at an angle of approximately 45° to the horizontal, and the pipe is not rotated during welding.

posición inclinada. Una posición de junta de tubo en la cual el axis del tubo está a un ángulo de aproximadamente 45° a la horizontal, y no se le da vueltas al tubo durante la soldadura.

inclined position, with restriction ring. The position of a pipe joint in which the axis of the pipe is at an angle of approximately 45° to the horizontal, and a restriction ring is located near the joint. The pipe is not rotated during welding.

posición inclinada con argolla de restricción. La posición de una junta de tubo en la cual el axis del tubo está a un ángulo de aproximadamente 45° a la horizontal, y una argolla de restricción está localizada cerca de la junta. No se le da vuelta al tubo durante la soldadura.

included angle. A nonstandard term for groove angle.

ángulo incluido. Un término fuera de norma para ángulo de ranura.

inclusion. Entrapped foreign solid material, such as slag, flux, tungsten, or oxide.

inclusión. Material extraño atrapado sólido, como escoria, flujo, tungsteno, u óxido.

incomplete fusion. A weld discontinuity in which fusion did not occur between weld metal and fusion faces or adjoining weld beads.

fusión incompleta. Una discontinuidad en la soldadura en la cual no ocurrió fusión entre el metal soldado y caras de fusión o cordones soldados inmediatos.

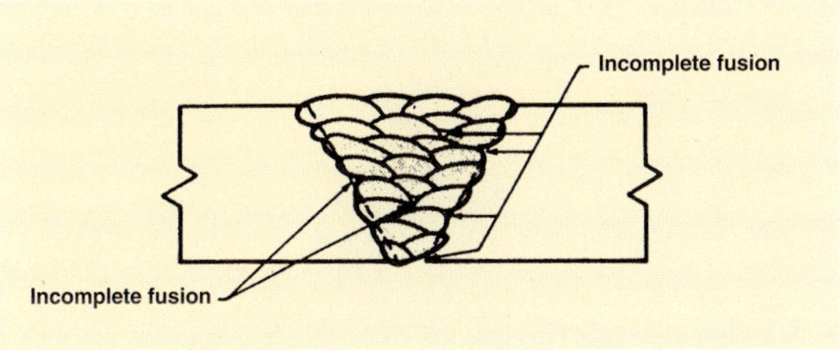

incomplete joint penetration. Joint penetration that is unintentionally less than the thickness of the weld joint.

penetración de junta incompleta. Penetración de la junta que no es intencionalmente menos de lo grueso de la junta de soldar.

***induction.** The transfer of heat obtained from the resistance of the workpieces to the flow of induced high-frequency welding current.

***inducción.** La transferencia de calor que se obtiene de la resistencia de las piezas de trabajo al flujo de corriente de soldeo de alta frecuencia.

induction brazing (IB). A brazing process that uses heat from the resistance of the workpieces to induced electric current.

soldadura fuerte por inducción. Un proceso de soldadura blanda que usa calor de la resistencia de las piezas de trabajo para inducir la corriente eléctrica.

induction soldering (IS). A soldering process in which the heat required is obtained from the resistance of the workpieces to induced electric current.

soldadura blanda por inducción. Un proceso de soldadura blanda en el cual el calor requerido es obtenido de la resistencia de las piezas de trabajo a la corriente eléctrica inducida.

inert gas. A gas that normally does not combine chemically with materials.

gas inerte. Un gas que normalmente no se combina químicamente con materiales.

***inertia welding.** A welding process in which one workpiece revolves rapidly and one is stationary. At a predetermined speed the power is cut, the rotating part is thrust against the stationary part, and frictional heating occurs. The weld bond is formed when rotation stops.

***soldadura inercial.** Un proceso en el cual una pieza de trabajo voltea rápidamente y la otra está fija. A una velocidad predeterminada se le corta la potencia, la parte que está volteando es empujada en contra de la parte fija, y el calor de fricción ocurre. La unión soldada es formada cuando se para la rotación.

***infrared light.** Light that has a wavelength longer than visible light with a wavelength ranging approximately from 1 to 430 THz.

***luz infrarroja.** Luz con una longitud de onda superior a la luz visible, con una longitud de onda que va aproximadamente de 1 a 430 THz.

infrared radiation. Electromagnetic energy with wavelengths from 770 to 12,000 nonometers.

radiación infrarrojo. Energia electromagnética con longitud de ondas de 770 a 12,000 nonometros.

***injector chamber.** One method of completely mixing the fuel gas and oxygen to form a flame. High-pressure oxygen is passed through a narrowed opening (venturi) to the mixing chamber. This action creates a vacuum, which pulls the fuel gas into the chamber and ensures thorough mixing. Used for equal gas pressures and is particularly useful for low-pressure fuel gases.

***cámara de inyector.** Un método de mezclar completamente el gas de combustión y el oxígeno para formar una llama. Oxígeno a alta presión es pasado por una abertura angosta (venturi) a la cámara de mezcla. Está acción hace un vacuo, la cual estira el gas combustible para dentro de la cámara y asegura una mezcla completa. Usada para presiones de gas que son iguales y es particularmente útil para gases combustibles de presión baja.

***inner cone.** The portion of the oxyacetylene flame closest to the welding tip. The primary combustion reaction occurs in the inner cone. The size and color of the cone serve as indicators of the type of flame (carburizing, oxidizing, neutral).

***cono interno.** La porción de la llama de oxiacetileno más cerca de la punta para soldar. La reacción de combustión principal ocurre en el cono interno. El tamaño y el calor del cono sirve como indicadores del tipo de la llama (carburante, oxidante, neutral).

***intelligent robot.** A robot that can be programmed to make performance choices contingent on sensory inputs.

***robot inteligente.** Un robot que puede ser programado para hacer preferencias contigentes en sensorios de entrada.

***interface.** A shared boundary. An interface might be a mechanical or an electrical connection between two devices; it might be a portion of computer storage accessed by two or more programs; or it might be a device for communication to or from a human operator.

***interface.** Un limite repartido. Un interface puede ser una conexión mecánica o eléctrica entre dos aparatos; puede ser una porción de deposito con accesión a dos o más programas; o puede ser un aparato para comunicarse a o con un operador humano.

***intermittent.** As it relates to a weld, it is a weld that is made for a specific length with breaks between welds.

***intermitente** En lo que se refiere a la soldadura, es una soldadura realizada para una longitud específica con cortes entre soldaduras.

interpass temperature. In a multipass weld, the temperature of the weld area between weld passes.

temperatura de pasada interna. En una soldadura de pasadas multiples, la temperatura en la área de la soldadura entre pasadas de soldaduras.

***ionized gas.** A gas that is heated to a point where it becomes conductive. See **plasma.**

***gas ionizado.** Un gas que se calienta al punto de volverse conductivo. Consulte **plasma.**

***iron.** An element. Very seldom used in its pure form. The most common element alloyed with iron is carbon.

***hierro.** Un elemento. Es muy raro que se use en forma pura. El elemento más común del aleado con hierro es carbón.

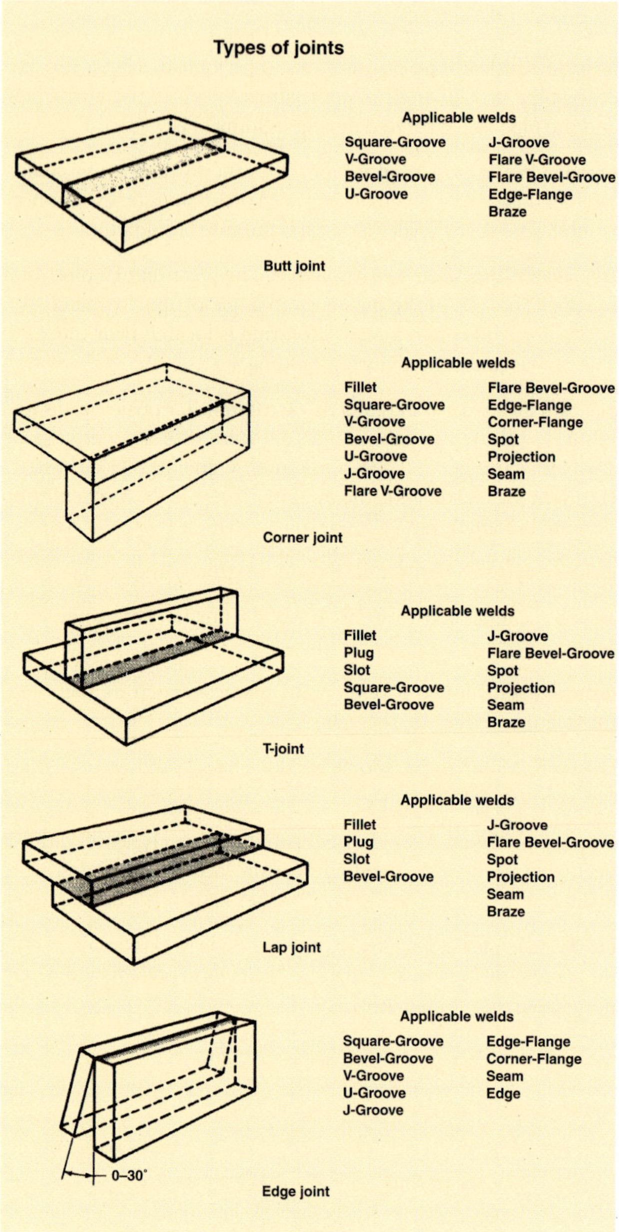

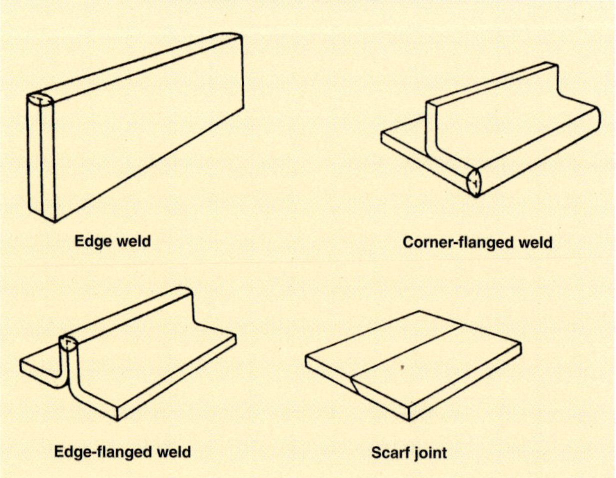

J

***J-groove.** See **joint.**

***ranura en forma de J.** Consulte **junta.**

J-groove weld. A type of groove weld.

soldadura con ranura-J. Es un tipo de soldadura de ranura.

joint. The junction of members or the edges of members that are to be joined or have been joined.

junta. El punto en que se unen dos miembros o las orillas de los miembros que están para unirse o han sido unidos.

joint buildup sequence. The order in which the weld beads of a multiple-pass weld are deposited with respect to the cross section of the joint.

secuencia de formación de una junta. La orden en la cual los cordones de soldadura en una soldadura de pasadas múltiples son depositadas con respecto a la sección transversa de la junta.

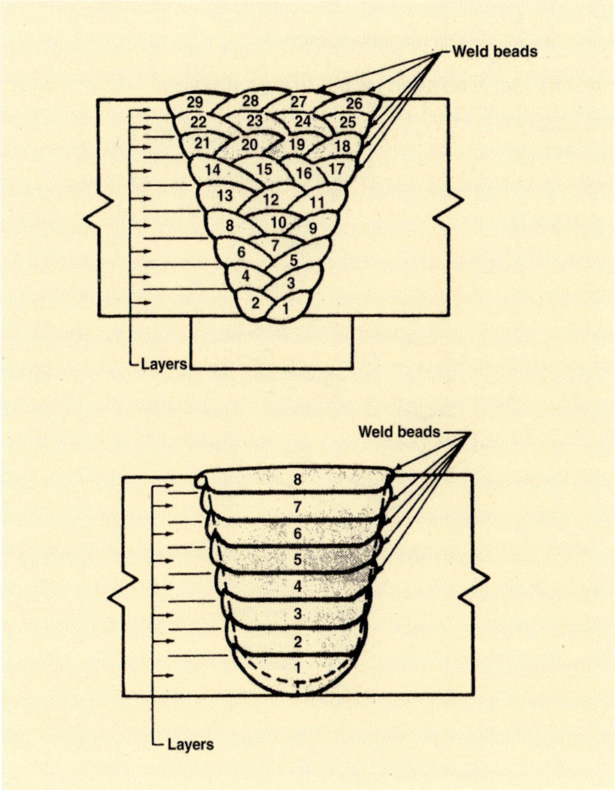

joint clearance. The distance between the faying surfaces of a joint.

despejo de junta. La distancia entre las superficies del empalme de una junta.

joint design. The joint geometry together with the required dimensions of the welded joint.

diseño de junta. La geometría de la junta junto con las dimensiones requeridas de la junta de la soldadura.

joint efficiency. The ratio of the strength of a joint to the strength of the base metal, expressed in percent.

eficiencia de junta. La razón de la fuerza de una junta a la fuerza del metal base, expresada en por ciento.

joint geometry. The shape and dimensions of a joint in cross section prior to welding.

geometría de junta. La figura y dimensión de una junta en sección transversa antes de soldarse.

joint penetration. The distance the weld metal extends from its face into a joint, exclusive of weld reinforcement.

penetración de junta. La distancia del metal soldado que se extiende de su cara hacia adentro de la junta, exclusiva de la soldadura de refuerzo.

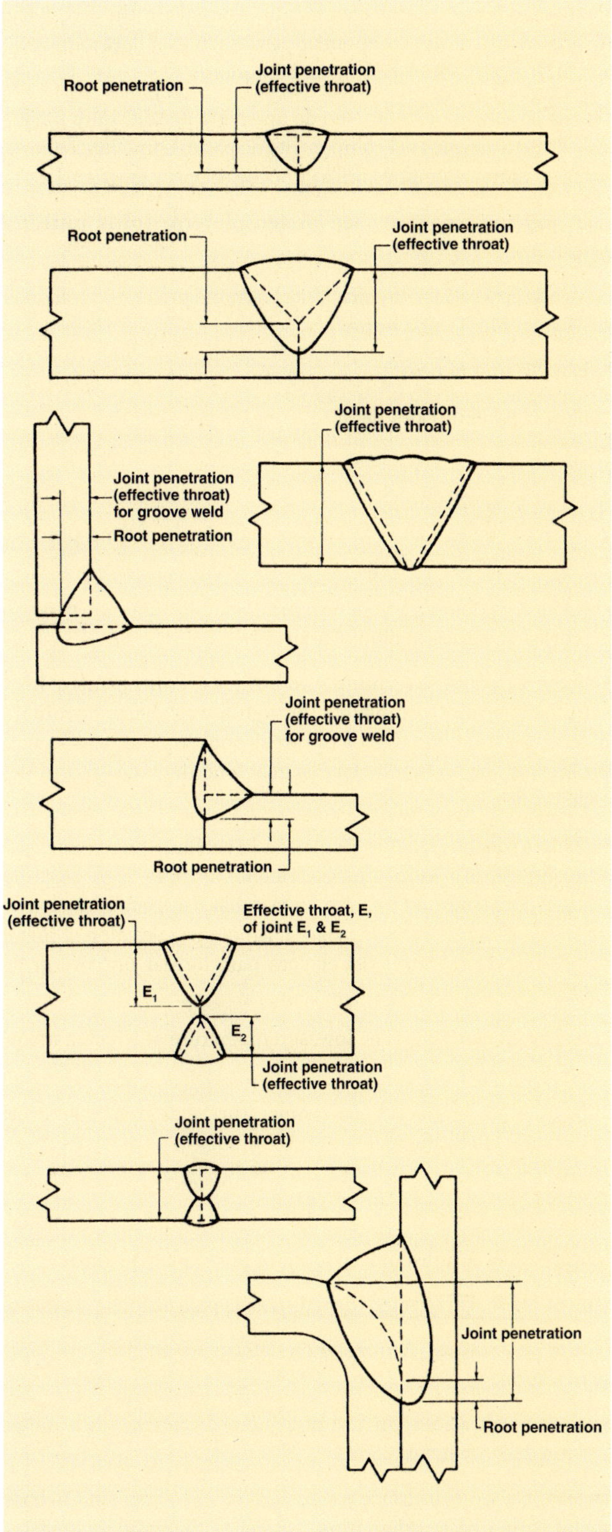

joint root. The portion of a joint to be welded where the members approach closest to each other. In cross section, the joint root may be a point, a line, or an area.

raíz de junta. Esa porción de una junta que está para soldarse donde los miembros están más cercanos uno del otro. En la sección transversa, la raíz de la junta puede ser una punta, una línea, o una área.

joint type. A weld joint classification based on the five basic arrangements of the component parts such as a butt joint, corner joint, edge joint, lap joint, and tee-joint.

tipo de junta. Una clasificación de una junta de soldadura basada en los cinco arreglos del componente de partes como junta a tope, junta en esquina, junta de orilla, junta de solape, y junta en T.

joint welding sequence. See preferred term **joint buildup sequence.**

secuencia para soldar una junta. Vea el término preferido **secuencia de formación de una junta.**

***joules.** SI unit of heat.

***julios.** Unidad de calor de SI.

K

kerf. The width of the cut produced during a cutting process. Refer to drawing for **drag.**

cortadura. La anchura del corte producido durante un proceso de cortar. Refiérase al dibujo de **tiro.**

keyhole welding. A technique in which a concentrated heat source penetrates completely through a workpiece, forming a hole at the leading edge of the weld pool. As the heat source progresses, the molten metal fills in behind the hole to form the weld bead.

soldadura con pocillo. Una técnica en la cual una fuente de calor concentrado se penetra completamente a través de la pieza de trabajo, formando un agujero en la orilla del frente del charco de la soldadura. Asi como progresa la potencia de calor, el metal derretido rellena detrás del agujero para formar un cordón de soldadura.

***kindling temperature.** The lowest temperature at which a material will burn.

***temperatura de ignición.** La temperatura más baja la cual un material se puede quemar.

L

lack of fusion. A nonstandard term for incomplete fusion.

falta de fusión. Un término fuera de norma para fusión incompleta.

lack of penetration. A nonstandard term for incomplete joint penetration.

falta de penetración. Un término fuera de norma para penetración de junta incompleta.

lamellar tear. A subsurface terrace and steplike crack in the base metal with a basic orientation parallel to the wrought surface. It is caused by stresses in the through-thickness direction of the base metals weakened by the presence of small, dispersed, planar-shaped, nonmetallic inclusions parallel to the metal surface.

rasgadura laminar. Una terraza subsuperficie y una grieta como un escalón en el metal base con una orientación paralela a la superficie forjada. Es causada por tensión en la dirección de lo grueso-continuo de los metales de base debilitados por la presencia de pequeños, dispersados, formados como plano, inclusiones no metálicas paralelas a la superficie del metal.

land. See preferred term **root face.**

hombro. Vea el término preferido **cara de raíz.**

lap joint. A joint between two overlapping members.

junta de solape. Una junta entre dos miembros traslapadas.

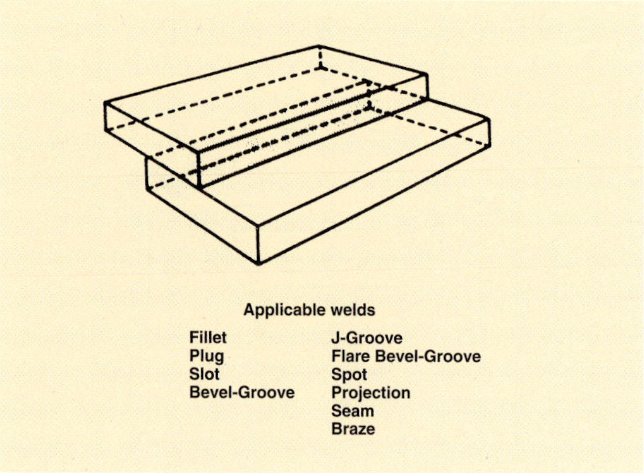

laser beam cutting (LBC). A thermal cutting process that severs metal by locally melting or vaporizing with the heat from a laser beam. The process is used with or without assist gas to aid the removal of molten and vaporized material.

cortes con rayo laser. Un proceso de cortes termal que separa al metal vaporizado o derretido localmente con el calor de un rayo laser. El proceso es usado sin gas que asiste a remover el material vaporizado o derretido.

***laser beam drilling (LBD).** A thermal cutting process used to produce holes in metal as accurately as if they had been drilled.

***perforación con rayo láser.** Un proceso de corte térmico que se utiliza para realizar orificios en el metal de manera tan precisa como si se hubieran taladrado.

laser beam welding (LBW). A welding process that produces coalescence with the heat from a laser beam impinging on the joint. The process is used without a shielding gas and without the application of pressure.

soldadura con rayo laser. Un proceso de soldar que produce coalescencia con calor de un rayo laser al golpear contra la junta.

***lasers.** Light Amplification by Stimulated Emission of Radiation is a light amplification method used to produce a narrow beam of light capable of cutting or welding metals.

***láser.** La amplificación de la luz por emisión estimulada de radiación es un método de amplificación de la luz que se utiliza para producir un haz de luz angosto capaz de cortar o soldar metales.

lattice. An orderly geometric pattern of atoms within a solid metal. The lattice structure is responsible for many of the mechanical properties of the metal.

enrejado. Es una forma geométrica bien arreglada de átomos dentro de un metal sólido. La estructura de enrejado es responsable por muchas propiedades mecánicas del metal.

layer. A stratum of weld metal or surfacing material. The layer may consist of one or more weld beads laid side by side. Refer to drawing for **joint buildup sequence.**

capa. Un estrato de metal de soldadura o material de superficie. La capa puede consistir de uno o más cordones de soldadura depositados o puestos de lado. Refiérase al dibujo **secuencia de formación de una junta.**

***leaders and arrows.** Leaders are the straight part and arrows are the pointed end that points to a part to identify it, show the location, and/or are the basis of a welding symbol.

***guías y flechas.** Las guías son las partes rectas y las flechas son los extremos puntiagudos que apuntan hacia una parte para identificarla, mostrar la ubicación y/o son la base de un símbolo de soldadura.

***leading angle.** See **electrode angle.**

***ángulo de guía.** Consulte **ángulo del electrodo.**

***leak-detecting solution.** A solution, usually soapy water, that is brushed or sprayed on the hose fittings at the regulator and torch to detect gas leaks. If a small leak exists, soap bubbles form.

***solución para descubrir escape.** Una solución, por lo regular de agua enjabonada, que se acepilla o se rocía sobre las conexiones de las mangueras y los reguladores y antorcha para detectar escape de gas. Si existe un escape pequeño, se forman burbujas de jabón.

leg of a fillet weld. See **fillet weld leg.**

pierna de soldadura filete. Vea **pierna de soldadura filete.**

lightly coated electrode. A filler metal electrode consisting of a metal wire with a light coating applied subsequent to the drawing operation, primarily for stabilizing the arc.

electrodo con recubrimiento ligero. Un electrodo de metal de aporte consistiendo de un alambre de metal con un recubrimiento ligero aplicado subsecuente a la operación del dibujo, principalmente para estabilizar el arco.

***lime-based flux.** These alkaline fluxes are commonly used on both SMA and FCA welding electrodes.

***fundente a base de óxido de calcio.** Estos fundentes alcalinos se utilizan comúnmente en los electrodos para soldadura SMA y FCA.

***line drop.** The difference between the pressure at the low-pressure gauge and the pressure at the torch; results from the resistance to gas flow offered by the hose and how it is affected by the diameter and length of the hose. The smaller the hose diameter, or the longer the hose, the greater is the line drop.

***descenso de línea.** La diferencia de la presión en el manómetro de baja presión y la presión en la antorcha; resultados de la resistencia de la corriente del gas causada por la manguera, y como es afectada por el diámetro, y lo largo de la manguera. Si el diámetro de la manguera es más chica o es más larga la manguera, más grande es el descenso.

***liquefied fuel gases.** A gas that is stored under adequate pressure so that it is a liquid.

***gases combustibles licuados.** Un gas que se almacena a la presión adecuada para mantenerlo en estado líquido.

***liquid-solid phase bonding process.** Soldering or brazing where the filler metal is melted (liquid) and the base material does not melt (solid); the phase is the state at which bonding takes place between the solid base material and liquid filler metal. There is no alloying of the base metal.

***proceso de ligación de fase líquido-sólido.** Soldando con soldadura blanda o soldadura fuerte donde el metal de relleno se derrite (líquido) y el material base no se derrite (sólido); la fase es el estado la cual el ligamento se lleva a cabo entre el material base sólido y el metal de relleno (líquido). No se mezcla con el metal base.

liquidus. The lowest temperature at which a metal or an alloy is completely liquid.

liquidus. La temperatura más baja en la cual un metal o un aleado es completamente líquido.

load. The force that a part may be subject to while it is in service.

carga. La fuerza que se ejerce sobre una parte mientras esté en servicio.

local preheating. Preheating a specific portion of a structure.

precalentamiento local. El precalentamiento de una porción especificada de un estructura.

local stress relief heat treatment. Stress relief heat treatment of a specific portion of a structure.

tratamiento de calor para relevar la tensión local. Un tratamiento de calor el cual releva la tensión de una porción especificada de una estructura.

longitudinal sequence. The order in which the weld passes of a continuous weld are made in respect to its length. See also **backstep sequence.**

secuencia longitudinal. La orden en que las pasadas de un soldadura continua son hechas en respecto a su longitud. Vea también **secuencia a la inversa.**

***low-fuming alloys.** As it relates to brazing filler rods it indicates that there is enough deoxidizers in the alloy to reduce the problem of zinc forming oxides during torch brazing.

***aleaciones de bajas emisiones.** En relación a las varillas de relleno de soldadura fuerte, indica que hay suficientes desoxidantes en la aleación para reducir el problema de que el zinc forme óxidos durante la soldadura fuerte con soplete.

M

***machine operation.** Welding operations are performed automatically under the observation and correction of the operator.

***operación de máquina.** Operaciones de soldadura son ejecutadas automáticamente bajo la observación y corrección del operador.

machine welding. Welding with equipment that performs the welding operation under the constant observation and control of a welding operator. The equipment may or may not perform the loading and unloading of the work. See also **automatic welding.**

máquina para soldadura. Soldadura con equipo que ejecutan la operación de soldadura bajo la observación constante de un operador de soldadura. El equipo pueda o no ejecutar el cargar o descargar del trabajo. Vea también **soldadura automática.**

***macro structure.** A structure large enough to be seen with the naked eye or low magnification, usually under 30 power.

***estructura macro.** Una estructura suficientemente grande que puede verse con el puro ojo o con un amplificador de aumento bajo, regularmente abajo de poder 30.

macroetch test. A test in which a specimen is prepared with a fine finish, etched, and examined under low magnification.

prueba con grabado al agua fuerte y examinado por magnificación. Una prueba en una probeta preparada con acabado fino, grabada al agua fuerte, y examinado debajo de un amplificador de aumento bajo.

***magnetic flux lines.** Parallel lines of force that always go from the north pole to the south pole in a magnet, and surround a DC current–carrying wire.

líneas magnéticas de flujo. Líneas paralelas de fuerza que siempre van del polo norte al polo sur en un magneto, y rodea un alambre que lleva corriente DC.

***malleable cast iron.** See **cast iron.**

***hierro fundido maleable.** Consulte **hierro fundido.**

***manganese.** This metallic element is an alloy added to steels to improve their strength.

***manganeso.** Este elemento metálico es una aleación que se agrega a los aceros para mejorar su resistencia.

manifold. A multiple header for interconnection of gas or fluid sources with distribution points.

conexión múltiple. Una tuberia con conexiones múltiples que sirve como fuente de gas o flúido con puntos de distribución.

***manifold system.** Used when there are a number of workstations or a high volume of gas is required. A piping system that allows several oxygen and fuel-gas cylinders to be connected to several welding stations. Normally regulators are provided at the manifold and at the stations to provide control of the oxygen and fuel-gas pressures. Safety features such as reverse flow valves, flashback arrestors, and back pressure release must be provided at the manifold.

***sistema de conexiones múltiples.** Usado cuando hay un número de estaciones de trabajo o cuando se requiere un alto volumen de gas. Un sistema de tubos que permite que se conecten varios cilindros de oxígeno y gas combustible a varias estaciones de soldadura. Normalmente se usan reguladores en el tubo de conexiones múltiples y en las estaciones para mantener el control de la presión del oxígeno y el gas combustible. Normas de seguridad como válvulas de retención, protector de agua contra retroceso de llama, y escape de presión deben usarse en el tubo de conexiones múltiples.

***manipulator.** A mechanism, usually consisting of a series of segments, joined or sliding relative to one another for the purpose of grasping and moving objects, usually in several degrees of freedom. It may be remotely controlled by a computer or by a human.

***manipulador.** Un mecanismo, que consiste regularmente de una serie de segmentos unidos o corredizos con relación del uno al otro con el propósito de que agarre y mueva objetos, por lo regular en varios grados de libertad. Puede ser controlado remotamente por una computadora o un humano.

***manual operation.** The entire welding process is manipulated by the welding operator.

***operación manual.** Todo el proceso de soldadura es manipulado por un operador de soldadura.

manual welding. Welding with the torch, gun, or electrode holder held and manipulated by hand. Variations of this term are *manual brazing, manual soldering, manual thermal cutting,* and *manual thermal spraying.*

soldadura manual. Soldando con la antorcha, pistola, porta electrodo detenido y manipulado por la mano. Variaciones de este término son *soldadura fuerte manual, soldadura blanda manual, cortes termal manual,* y *rociado termal manual.*

***MAPP®.** One manufacturer's trade name for a specific stabilized, liquefied MPS mixture. MAPP® has a distinctive odor, which makes it easy to detect; used for welding and cutting. See also **methylacetylene propadiene (MPS).**

***MAPP®.** Un nombre comercial de un fabricante para una específica estabilizada, licuada, mezcla MPS. MAPP® tiene un olor distintivo, el cual es muy fácil de descubrir; es usado para cortes y soldaduras. Vea también **metilacetileno y propadieno.**

***martensite.** A very hard and brittle solid-solution phase that is found in medium and high carbon steels.

***martensita.** Una solución sólida con un aspecto muy duro y quebradizo que se encuentra en aceros medianos y de alto carbón.

***material specification data sheet (MSDS).** A form with data regarding the properties of a particular material that is used by workers to promote the safe handling and use of the material and by emergency personnel.

***hoja de datos de seguridad del material (MSDS, por sus siglas en inglés).** Un formulario que indica las propiedades de un material específico que usan los trabajadores para promover la manipulación y el uso seguros del material, y el personal de emergencias.

***mechanical testing.** See **destructive testing.**

***prueba mecánica.** Consulte **prueba destructiva.**

meltback time. The time interval at the end of crater fill time to arc outage during which electrode feed is stopped. Arc voltage and arc length increase and current decreases to zero to prevent the electrode from freezing in the weld deposit.

tiempo de refundición. El tiempo de intervalo al fin del tiempo en que se llena el crater hasta que se apaga el arco durante el cual el alimento del electrodo se detiene. El voltaje del arco y lo largo del arco aumenta y la corriente empieza a desminuir hasta llegar a cero para prevenir la congelación del electrodo en el deposito de la soldadura.

melting range. The temperature range between solidus and liquidus.

variación de derretimiento. La variación de temperatura entre solidus y liquidus.

melting rate. The weight or length of electrode, wire, rod, or powder, melted in a unit of time.

cantidad de derretimiento. El peso o lo largo de un electrodo, alambre, varilla, o polvo derretido en una unidad de tiempo.

melt-through. Complete joint penetration for a joint welded from one side. Visible root reinforcement is produced.

derretir de un lado a otro. Una junta con penetración completa para una junta que está soldada de un lado. Refuerzo de raíz visible es producido.

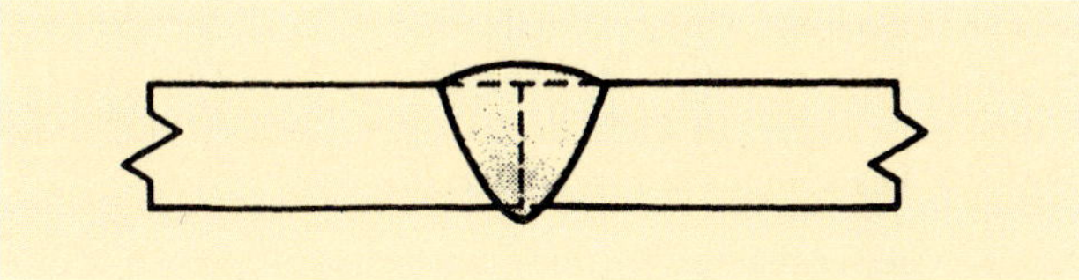

metal. An opaque, lustrous, elemental, chemical substance that is a good conductor of heat and electricity, usually malleable, ductile, and more dense than other elemental substances.

metal. Una opaca, brillante, elemental, substancia química que es una buena conductora de calor y electricidad, por lo regular es maleable, ductil, y es más densa que otras substancias elementales.

metal arc cutting (MAC). Any of a group of arc cutting processes that sever metals by melting them with the heat of an arc between a metal electrode and the base metal. See also **shielded metal arc cutting** and **gas metal arc cutting.**

cortes de metal con arco. Cualquiera de un grupo de procesos de cortes con arco que corta metales derritiéndolos con el calor de un arco entre un electrodo de metal y el metal base. Vea también **cortes de arco metálico protegido** y **cortes de arco metálico con gas.**

metal cored electrode. A composite tubular filler metal electrode consisting of a metal sheath and a core of various powdered materials, producing no more than slag islands on the face of a weld bead. External shielding may be required.

electrodo de metal de núcleo. Un electrodo de metal para rellenar tubular compuesto consistiendo de una envoltura de metal y núcleo de varios materiales en polvo, que producen nada más que islas de escoria en la cara del cordón de soldadura. Protección externa puede ser requerida.

metal electrode. A filler or nonfiller metal electrode used in arc welding or cutting, which consists of a metal wire or rod that has been manufactured by any method and that is either bare or covered with a suitable covering or coating.

electrodo de metal. Un electrodo de metal que se usa para rellenar o para no rellenar la soldadura de arco o para cortar, que consiste de un alambre de metal o varilla que ha sido fabricada por cualquier método ya sea liso o cubierto con un cubierto o revestimiento propio.

***metallurgy.** The scientific study of metals.

***metalurgia.** El estudio científico de los metales.

***methylacetylene-propadiene (MPS).** A family of fuel gases that are mixtures of two or more gases (propane, butane, butadiene, methylacetylene, and propadiene). The neutral flame temperature is approximately 5031°F (2927°C), depending upon the actual gas mixture. MPS is used for oxyfuel cutting, heating, brazing, and metallizing; rarely used for welding.

***metilacetileno y propadieno.** Una familia de gases de combustión que son mezclas de dos o más gases (propano, butano, butadiano, metilacetileno, propadieno). La temperatura de la llama natural es aproximadamente 5031°F (2927°C), dependiendo de la mezcla actual del gas. MPS es usado como gas de combustión para cortar, calentar, soldadura fuerte, y metalizar; es muy raro que se use para soldar.

***micro structure.** A structure that is visible only with high magnification or with the aid of a microscope.

***estructura micronesia.** Una estructura que es visible solamente con un amplificador de poder muy alto o con la ayuda de un microscopio.

***microcomputer.** A computer that uses a microprocessor as its basic element.

***computadora micronesia.** Una computadora que usa un procesor micronesio como su elemento básico.

microetch test. A test in which the specimen is prepared with a polished finish, etched, and examined under high magnification.

prueba con grabado al agua fuerte y examinada por un amplificador de alto poder. Una prueba en una probeta preparada con acabado fino, grabada al agua fuerte y examinado bajo un amplíficador de alto poder.

***microprocessor.** The principal processing element of a microcomputer, made as a single, integrated circuit.

***procesor micronesio.** El elemento principal de un procedimiento de una computadora micronesia, hecha con un solo circuito integrado.

mineral-based fluxes. Fluxes that use inorganic compounds such as the rutile-based flux (titanium dioxide, TiO_2). These mineral compounds do not contain hydrogen, and electrodes that use these fluxes are often referred to as low hydrogen electrodes. Less smoke is generated with this welding electrode than with cellulose-based fluxes, but a thicker slag layer is deposited on the weld. E7018 is an example of an electrode that uses this type of flux.

fundentes de base mineral. Fundentes que usan compuestos inorgánicos, como por ejemplo, el fundente a base de rutilo (bióxido de titanio, TiO_2). Estos compuestos minerales no contienen hidrógeno, y a los electrodos que usan estos fundentes se los llama con frecuencia electrodos de bajo hidrógeno. En la soldadura con electrodos se producen menos humos que en la que se realiza con fundentes celulósicos, pero se deposita una capa de escoria más gruesa en la soldadura. El F7018 es un ejemplo de un electrodo que usa este tipo de fundente.

mixing chamber. That part of a welding or cutting torch in which a fuel-gas and oxygen are mixed.

cámara mezcladora. Esa parte de una antorcha para soldar y cortar por la cual el gas combustible y el oxígeno son mezclados.

mold. A high-temperature container into which liquid metal from the thermite welding process is poured and held until it cools and hardens into the container's interior shape.

molde. Un contenedor de alta temperatura en el cual se vierte y se mantiene metal líquido del proceso de soldadura con termita hasta que éste se enfríe y se solidifique tomando la forma interior del contenedor.

molecular hydrogen. A bonded pair of hydrogen atoms (H_2). This is the configuration that all hydrogen atoms try to form.

hidrógeno molecular. Un par de átomos de hidrógeno (H_2) unidos. Ésta es la configuración que tratan de formar todos los átomos de hidrógeno.

molten weld pool. The liquid state of a weld prior to solidification as weld material.

charco de soldadura derretido. El estado líquido de una soldadura antes de solidificarse como material de soldadura.

***multipass weld.** A weld requiring more than one pass to ensure complete and satisfactory joining of the metal pieces.

***soldadura de pasadas múltiples.** Una soldadura que requiere más de una pasada para asegurar una completa y satisfactoria unión de las piezas de metal.

N

***natural ventilation.** Ventilation usually resulting from the heat generated convection currents that cause welding fumes to rise.

***ventilación natural.** La ventilación que, por lo general, se produce como consecuencia del calor que generan las corrientes de convección que hacen que se eleven los gases de la soldadura.

neutral flame. An oxyfuel gas flame that has characteristics neither oxidizing nor reducing. Refer to drawing for **cone.**

llama neutral. Una llama de gas oxicombustible que no tiene características de oxidación ni de reducción. Refiérase al dibujo para **cono.**

***noble inert gases.** See preferred term **inert gas.**

***gases inertes nobles.** Consulte el término preferido **gas inerte.**

nonconsumable electrode. An electrode that does not provide filler metal.

electrodo no consumible. Un electrodo que no provee metal de relleno.

noncorrosive flux. A soldering flux that in neither its original nor its residual form chemically attacks the base metal. It usually is composed of rosin or resin-base materials.

flujo no corrosivo. Un flujo para soldadura blanda que ni en su forma original ni en su forma restante químicamente ataca el metal base. Regularmente es compuesto de materiales de colofonia o resino de base.

***nondestructive.** See **nondestructive testing.**

***no destructiva.** Consulte **pruebas no destructivas.**

***nondestructive testing (NDT).** Methods that do not alter or damage the weld being examined; used to locate both surface and internal defects. Methods include visual inspection, penetrant inspection, magnetic particle inspection, radiographic inspection, eddy current, and ultrasonic inspection.

***pruebas no destructivas.** Métodos que no alteran ni dañan la soldadura que se está examinando. Se usa para encontrar ambos defectos internos y de superficie. Incluye métodos como inspección visual, inspección penetrante, inspección de partículas magnéticas, inspección de radiografía, y inspección ultrasónica.

nontransferred arc. An arc established between the electrode and the constricting nozzle of the plasma arc torch or thermal spraying. The workpiece is not in the electrical circuit. See also **transferred arc.**

arco no transferible. Un arco establecido entre el electrodo y la boquilla constrictiva de la antorcha del arco de plasma o pistola termal para rociar. La pieza de trabajo no está en el circuito eléctrico. Vea también **arco transferido.**

nozzle. A device that directs shielding media.

boquilla. Un aparato que dirige el medio de protección.

nugget. The weld metal joining the workpieces in spot, seam, and projection welds.

botón. El metal de soldadura que une a las piezas de trabajo en soldadura de puntos, costura, y proyección de soldaduras.

nugget size. The diameter or width of the nugget measured in the plane of the interface between the pieces joined.

tamaño del botón. El diámetro o lo ancho del botón medido en el plano del interfaze entre las piezas unidas.

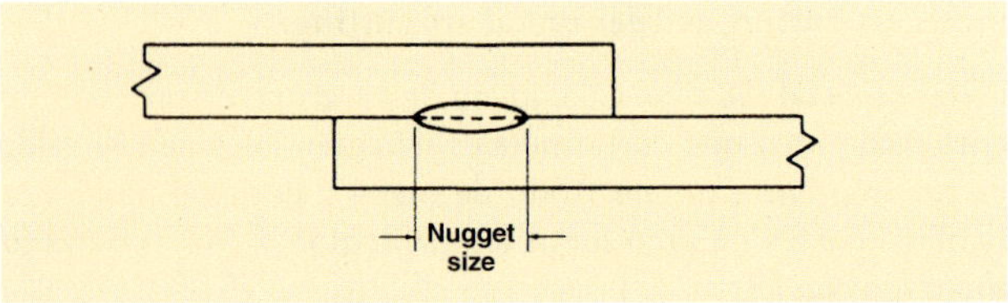

O

***object line.** Lines on a drawing that show the edge of an object, the intersection of surfaces that form corners or edges, and the extent of a curved surface, such as the sides of a cylinder.

***línea de objeto.** Líneas en un dibujo que muestran el borde de un objeto, la intersección de superficies que forman esquinas o bordes y la extensión de una superficie curvada, como por ejemplo, los laterales de un cilindro.

open circuit voltage. The voltage between the output terminals of the power source when no current is flowing to the torch or gun.

voltaje de circuito abierto. El voltaje entre los terminales de salida de una fuente de poder cuando la corriente no está corriendo a la antorcha o pistola.

open-root joint. An unwelded joint without backing or consumable insert.

junta de raíz abierta. Una junta que no está para soldarse sin respaldo o inserto consumible.

***operating voltage.** It is the actual voltage across the arc or closed circuit voltage.

***voltaje operativo.** Es el voltaje real entre el arco o voltaje de circuito cerrado.

orifice. See **constricting orifice.**

orifice. Vea **orifice de constreñimiento.**

orifice gas. The gas that is directed into the plasma arc torch or thermal spraying gun to surround the electrode. It becomes ionized in the arc to form the arc plasma and issues from the constricting orifice of the nozzle as a plasma jet.

gas para orifice. El gas que es dirigido dentro de la antorcha de plasma o la pistola de rociado termal para rodear el electrodo. Se vuelve ionizado dentro del arco para formar el arco de plasma y sale de la orifice de constreñimiento a la boquilla como chorro de plasma.

orifice throat length. The length of the constricting orifice in the plasma arc torch or thermal spraying gun.

largo de garganta del orifice. Lo largo de la orifice constreñida en la antorcha de plasma o en la pistola de plasma para rociar.

***outer envelope.** The outer boundary of the oxyacetylene flame. The secondary combustion reaction occurs in the outer envelope.

***envoltura externa.** El límite de afuera de la llama de oxiacetileno. La reacción de la combustión secundaria ocurre en la envoltura externa.

***out-of-position.** See **out-of-position welding.**

***fuera de posición.** Consulte **soldadura fuera de posición.**

***out-of-position welding.** Any welding position other than the flat position; includes vertical, horizontal, and overhead positions.

***soldadura fuera de posición.** Cualquier posición de soldadura menos la de la posición plana; incluye vertical, horizontal, y posiciones de sobrecabeza.

***outside corner joint.** See **joint.**

***junta de esquina exterior.** Consulte **junta.**

overhead position. The position in which welding is performed from the underside of the joint.

posición de sobrecabeza. La posición en la cual se hace la soldadura por el lado de abajo de la junta.

***overhead weld.** See **overhead position.**

***soldadura sobre cabeza.** Consulte **posición sobre cabeza.**

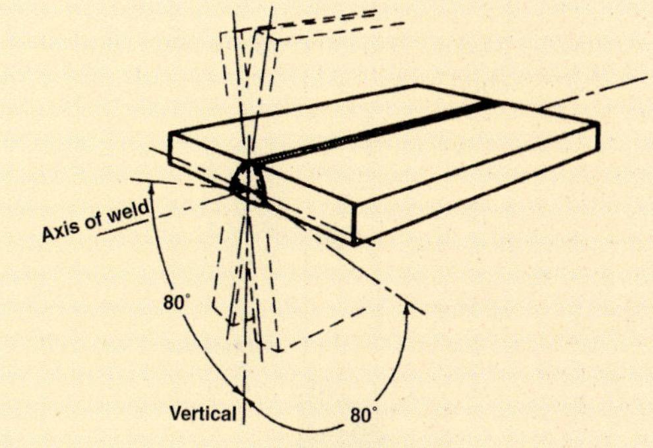

Overhead position
4G position

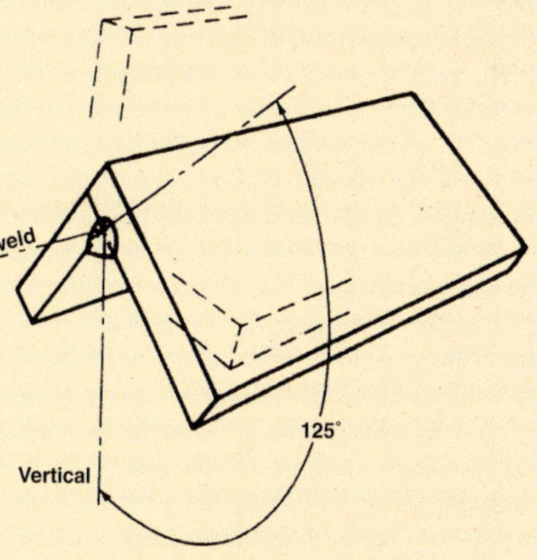

Overhead position
4F position

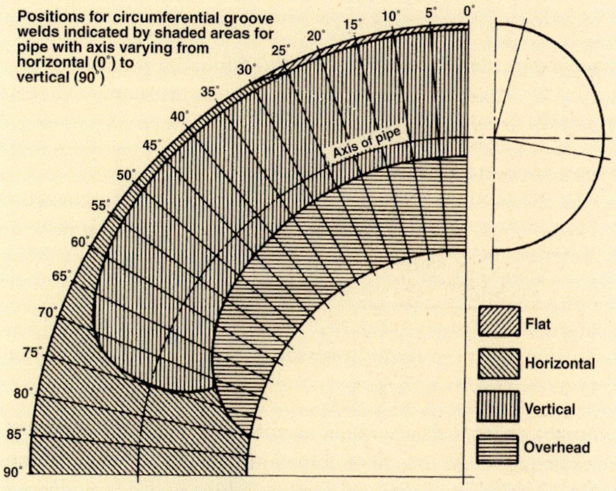

Positions of pipe welds

overlap. The protrusion of weld metal beyond the toe, face, or root of the weld; in resistance seam welding, the area in the preceding weld remelted by the succeeding weld.

traslapo. El metal de la soldadura que sobresale más allá del pie, cara, o de la raíz de una soldadura; en soldaduras de costuras por resistencia, la área de la soldadura anterior se rederrite por la soldadura subsiguiente.

***oxide layer.** A layer of oxidized metal on the surface, on steel it can be called rust.

***capa de óxido.** Una capa de metal oxidado en la superficie, en el acero se denomina herrumbre.

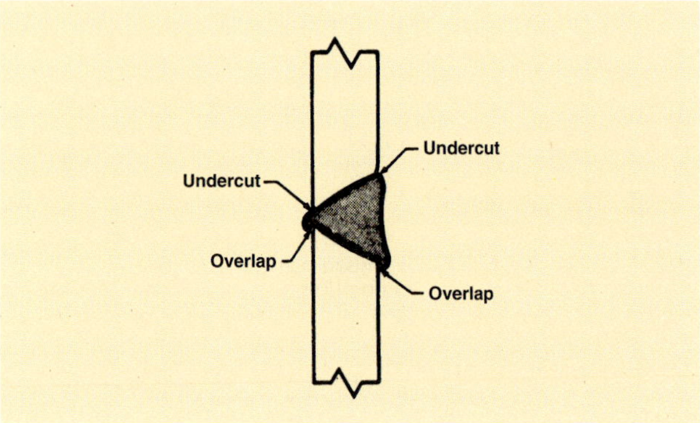

oxidizing flame. An oxyfuel gas flame in which there is an excess of oxygen, resulting in an oxygen-rich zone extending around and beyond the cone.

llama oxidante. Una llama de gas oxicombustible en la cual hay un exceso de oxígeno, resultando en una zona rica de oxígeno extendiéndose alrededor y más allá del cono.

oxyacetylene cutting (OFC-A). An oxyfuel cutting process variation that uses acetylene as the gas.

cortes con oxiacetileno. Un proceso de cortes con gas con variación que usa acetileno como gas de combustión.

***oxyacetylene flame.** A flame produced with oxygen and acetylene; it is the most popular type of flame used for fuel gas welding and cutting.

***llama oxiacetilénica.** Una llama producida con oxígeno y acetileno: Es el tipo más común de llama que se utiliza para el corte y la soldadura con gas combustible.

***oxyacetylene hand torch.** Most commonly used oxyfuel gas cutting torch; may be a cutting torch only or a combination welding and cutting torch set. On the combination set different tips can be attached to the same torch body. The torch mixes the oxygen and fuel gas and directs the mixture to the tip. The torch can be an equal pressure-type (equal pressures of oxygen and fuel gas) or an injector-type (equal pressures of high-pressure oxygen and low-pressure fuel gas).

***antorcha de mano oxiacetileno.** La antorcha de gas oxicombustible es la que se usa más frecuentemente; puede ser una antorcha para hacer cortes solamente o una combinación de un juego de antorcha para soldar y hacer cortes. En el juego de combinación diferentes puntas pueden ser conectadas al mismo mango de la antorcha. La antorcha mezcla el oxígeno y gas combustible y dirige la mezcla a la punta. La antorcha puede ser de tipo de presión igual (presiones iguales de oxígeno y gas combustible) o de tipo inyector (presiones iguales de oxígeno de alta presión y baja presión de gas combustible).

oxyacetylene welding (OAW). An oxyfuel gas welding process that uses acetylene as the fuel gas. The process is used without the application of pressure.

soldadura con oxiacetileno (OAW). Un proceso de soldadura de gas oxicombustible que usa acetileno con gas de combustión. El proceso se usa sin aplicación de presión.

***oxyfuel brazing (TB).** See **torch brazing.**

***soldadura fuerte con combustible de oxigeno (TB).** Consulte **soldadura fuerte con soplete.**

***oxyfuel flame.** A flame resulting from the combustion of oxygen mixed with a fuel gas. This intense flame is applied to two pieces of metal to cause them to melt to form weld pools. When the edges of the weld pools run together and fuse, the two pieces of metal are joined.

***llama oxicombustible.** Una llama que resulta de una combustión de oxígeno mezclado con gas combustible. Está llama intensa es aplicada a dos piezas de metal para hacer que se derritan para formar charcos de soldadura. Cuando las orillas de los charcos de soldadura se juntan y se derriten, las dos piezas de metal se unen.

***oxyfuel (OF) gas.** A general term covering all the different fuel gases such as acetylene, MAPP, natural gas, propane, hydrogen, etc., that can be used with oxygen.

***gas combustible con oxígeno (OF, por sus siglas en inglés).** Un término general que se refiere a todos los diferentes gases de combustión, tales como el acetileno, MAPP, gas natural, propano, hidrógeno, etc., que se pueden utilizar con oxígeno.

***oxyfuel gas cutting (OFC).** See **oxygen cutting.**

***corte con gas de combustible con oxígeno (OFC, por sus siglas en inglés).** Consulte **corte con oxígeno.**

oxyfuel gas cutting torch. A device used for directing the preheating flame produced by the controlled combustion of fuel gases and to direct and control the cutting oxygen.

antorcha para cortes de gas oxicombustible. Un aparato que se usa para dirigir la llama precalentada producida por la combustión controlada de los gases de combustión y para dirigir y controlar el oxígeno para cortar.

***oxyfuel gas torch.** See **oxyfuel gas welding torch.**

***soplete de gas combustible con oxígeno.** Consulte **soplete de soldadura con gas combustible con oxígeno.**

oxyfuel gas welding (OFW). A group of welding processes that produces coalescence of workpieces by heating them with an oxyfuel gas flame. The processes are used with or without the application of pressure and with or without filler metal.

soldadura con gas oxicombustible. Un grupo de procesos de soldadura que produce coalescencia de las piezas de trabajo calentándolas con una llama de gas oxicombustible. Los procesos son usados sin la aplicación de presión y con o sin el metal para rellenar.

oxyfuel gas welding torch. A device used in oxyfuel gas welding, torch brazing, and torch soldering for directing the heating flame produced by the controlled combustion of fuel gases.

antorcha para soldar con gas oxicombustible. Un aparato que se usa para soldar con gas oxicombustible, soldadura fuerte con antorcha, soldadura blanda con antorcha y para dirigir la llama calentada producida por combustión controlada de gases de combustión.

oxygen arc cutting (AOC). An oxygen cutting process that uses an arc between the workpiece and a consumable tubular electrode, through which oxygen is directed to the workpiece.

cortes de oxígeno con arco. Es un proceso de cortar con oxígeno que usa un arco entre la pieza de trabajo y un electrodo tubular consumible, por el cual el oxígeno es dirigido a la pieza de trabajo.

oxygen cutting (OC). A group of thermal cutting processes that severs or removes metal by means of the chemical reaction between oxygen and the base metal at elevated temperature. The necessary temperature is maintained by the heat from an arc, an oxyfuel gas maintained by the heat from an arc, an oxyfuel gas flame, or other sources. See also **oxyfuel gas cutting.**

cortes con oxígeno. Un grupo de procesos termales que corta y quita el metal por medio de una reacción química entre el oxígeno y el metal base a una temperatura elevada. La temperatura necesaria es mantenida por el calor del arco, un gas oxicombustible mantenido por el calor del arco, una llama de gas oxicombustible, o de otras fuentes. Vea también **gas para cortar oxicombustible.**

oxygen gouging. Thermal gouging that uses an oxygen cutting process variation to form a bevel or groove.

escopleando con la gubia con oxígeno. Gubia termal que usa un proceso de variación de corte con oxígeno para formar un bisel o ranura.

oxygen lance. A length of pipe used to convey oxygen to the point of cutting in oxygen lance cutting.

lanza de oxígeno. Un tramo de tubo usado para conducir oxígeno al punto de cortar en cortes con lanza de oxígeno.

oxygen lance cutting (LOC). An oxygen cutting process that uses oxygen supplied through a consumable lance. The preheat to start the cutting is obtained by other means.

cortes con lanza de oxígeno. Un proceso de cortar con oxígeno que usa oxígeno surtido por una lanza consumible. El precalentamiento para empezar a cortar es obtenido por otros medios.

oxyhydrogen cutting (OFC-H). An oxyfuel gas cutting process variation that uses hydrogen as the fuel gas.

cortes con oxihidrógeno. Un proceso de cortar de gas oxicombustible con variación que usa hidrógeno como gas combustible.

***oxyhydrogen flame.** A specific flame resulting from the combustion of oxygen and hydrogen; consists of primary combustion region only; used for welding and cutting.

***llama oxihidrógeno.** Una llama específica que resulta de la combustión del oxígeno e hidrógeno; consiste solamente de la región de combustión primaria; usada para cortar y soldar.

oxyhydrogen welding (OHW). An oxyfuel gas welding process that uses hydrogen as the fuel gas. The process is used without the application of pressure.

soldadura oxihidrógeno. Un proceso de soldar con gas oxicombustible que usa hidrógeno como gas de combustible. El proceso se usa sin la aplicación de presión.

oxypropane cutting (OFC-P). An oxyfuel gas cutting process variation that uses propane as the fuel gas.

cortes con oxipropano. Un proceso de cortar con gas combustible con variación que usa propano como gas combustible.

P

parent metal. See preferred term **base metal.**

metal de origen. Vea término preferido **metal base.**

***paste range.** The temperature range of soldering and brazing filler metal alloys in which the metal is partly solid and partly liquid as it is heated or cooled.

***grados de la pasta.** Los grados de la temperatura del metal para rellenar aleados para soldadura blanda o soldadura fuerte cuando se calienta o se enfria.

pattern. A replica of a part's shape that can easily be made from a thin material that is less durable, such as paper or cardboard or more durable, such as metal or wood, whose shape can be traced onto the raw stock. (See **template**)

patrón. Una réplica de la forma de una pieza hecha de un material delgado como papel, cartón, metal o madera utilizada para dibujar la forma en material crudo. (Ver **plantilla**)

***pearlite.** A two-phased lamellar iron carbon crystalline structure that forms during slow cooling.

***perlita.** Una estructura cristalina laminar de carbono y hierro de dos fases que se forma durante el enfriamiento lento.

peel test. A destructive method of inspection that mechanically separates a lap joint by peeling.

prueba por pelar. Un método de inspección destructivo de pelar que separa mecánicamente una junta de solape.

peening. The mechanical working of metals using impact blows.

martillazos (con martillo de bola). Metales que se trabajan mecánicamente con golpes de impacto.

***penetration.** The depth into the base metal (from the surface) that the weld metal extends, excluding any reinforcement.

***penetración.** La profundidad de adentro del metal base (de la superficie) que el metal de soldadura se extiende, excluyendo cualquier refuerzo.

percussion welding (PEW). A welding process that produces coalescence with an arc resulting from a rapid discharge of electrical energy. Pressure is applied percussively during or immediately following the electrical discharge.

soldadura a percusión. Un proceso de soldadura que produce coalescencia con un arco resultando de una descarga rápida de energia. Presión es aplicada a percusión durante o inmediatamente después de la descarga eléctrica.

***phantom lines.** Lines on a drawing that show an alternate position of a moving part or the extent of motion, such as the on/off position of a light switch. They can also be used as a place holder for a part that will be added later.

***líneas fantasma.** Líneas en un dibujo que muestran la posición alternada de una pieza móvil o el rango de movimiento, como por ejemplo, la posición de encendido/apagado de un interruptor de luz. También se pueden utilizar como marcadores para una parte que se agregará más adelante.

***phase diagrams.** Provide information on the crystalline constituents of metal alloys at different temperatures in three different phases: pure metal, solid solutions of two or more metals, and intermetallic compounds.

***diagramas de equilibrio.** Proporciona información en los constituyentes cristalinos de metales aleados a diferentes temperaturas en tres aspectos: metal puro, soluciones sólidas de dos o más metales, y mezclas intermetálicas.

***pick-and-place robot.** A simple robot, often with only two or three degrees of freedom, which transfers items from place to place by means of point-to-point moves. Little or no trajectory control is available. Often referred to as a "bank-bank" robot.

***robot de escoger y atar.** Un robot simple, frecuentemente con solo dos o tres grados de libertad, el cual traslada artículos de un lugar a otro por medio de movidas de punto a punto. Un poco o nada de control trayectoria es utilizado. A veces es referido como un "banco-banco" robot.

***pictorial drawings.** A type of mechanical drawing that represents an object as a picture.

***dibujos pictóricos.** Un tipo de dibujo mecánico que representa a un objeto como una pintura.

pilot arc. A low-current arc between the electrode and the constricting nozzle of the plasma arc torch to ionize the gas and facilitate the start of the welding arc.

piloto del arco. Un arco de corriente baja en medio del electrodo y la boquilla constreñida de la antorcha de arco de plasma para ionizar el gas y facilitar el arranque del arco para soldar.

***pitch.** The angular rotation of a moving body about an axis perpendicular to its direction of motion and in the same plane as its top side.

***grado de inclinación.** La rotación angular de un cuerpo en movimiento alrededor de un eje perpendicular a su dirección y en el mismo plano como el del lado de arriba.

***plain carbon steels.** See **carbon steel.**

***aceros ordinarios al carbono.** Consulte **acero al carbono.**

plasma. A gas that has been heated to an at least partially ionized condition, enabling it to conduct an electric current.

plasma. Un gas que ha sido calentado a lo menos parcialmente a una condicón ionizada permitiendo que conduzca una corriente eléctrica.

***plasma arc.** See **plasma.**

***arco de plasma.** Consulte **plasma.**

plasma arc cutting (PAC). An arc cutting process that uses a constricted arc and removes the molten metal with a high-velocity jet of ionized gas issuing from the constricting orifice.

cortes con arco de plasma. Un proceso de cortar con el arco que usa un arco constreñido y quita el metal derretido con un chorro de alta velocidad de gas ionizado que sale de la orifice constringente.

***plasma arc gouging.** See **plasma arc cutting.**

***acanalado con arco de plasma.** Consulte **corte con arco de plasma**

plasma arc welding (PAW). An arc welding process that uses a constricted arc between a nonconsumable electrode and the weld pool (transferred arc) or between the electrode and the constricting nozzle (nontransferred arc). Shielding is obtained from the ionized gas issuing from the torch, which may be supplemented by an auxiliary source of shielding gas. The process is used without the application of pressure.

soldadura con arco de plasma. Un proceso de soldadura de arco que usa un arco constreñido entre un electrodo que no se consume y el charco de la soldadura (arco transferido) o entre el electrodo y la lanza constreñida (arco no transferido). La protección es obtenida del gas ionizado que sale de la antorcha, el cual puede ser suplementado por una fuente auxiliar de gas para protección. El proceso es usado sin la aplicación de presión.

plug weld. A weld made in a circular hole in one member of a joint fusing that member to another member. A fillet-welded hole should not be construed as conforming to this definition.

soldadura de tapón. Una soldadura que se hace en un agujero circular en un miembro de una junta uniendo ese miembro con otro miembro. Un agujero de soldadura de filete no debe ser interpretado como confirmación de está definición.

***point-to-point control.** A control scheme whereby the inputs or commands specify only a limited number of points along a desired path of motion. The control system determines the intervening path segments.

***control de punto a punto.** Una esquema de control con que las entradas o las ordenes especifican solamente un número limitado de puntos a lo largo de la senda de moción deseada. El sistema de control determina el intervenio de los segmentos de la senda.

porosity. Cavity-type discontinuities formed by gas entrapment during solidification or in a thermal spray deposit.

porosidad. Un tipo de cavidad de desuniones formadas por gas atrapado durante la solidificación o en un deposito rociado termal.

***postflow.** Shielding gas that continues to flow once the GTA welding arc stops which protects both the hot tungsten and weld from atmospheric contamination.

***flujo posterior.** Gas de protección que continúa fluyendo una vez que el arco de soldadura GTA se detiene y que protege tanto al tungsteno caliente como a la soldadura de la contaminación atmosférica.

postflow time. The time interval from current shut-off to shielding gas and/or cooling water shut-off.

tiempo de poscorriente. El intervalo de tiempo de cuando se cierra la corriente a cuando se cierra el gas de protección y o cuando se cierra el agua para enfriar.

postheat current (resistance welding). The current through the welding circuit during postheat time in resistance welding.

corriente de poscalentamiento (soldadura de resistencia). La corriente que va de un lado a otro del circuito durante el tiempo de poscalentamiento en la soldadura de resistencia.

postheat time (resistance welding). The time from the end of weld heat time to the end of weld time. Refer to drawing for **downslope time.**

tiempo de poscalentamiento (soldadura de resistencia). El tiempo del fin del calor de la soldadura al tiempo al fin del tiempo de la soldadura. Refiérase al dibujo de **tiempo de cadía del pendiente.**

postheating. The application of heat to an assembly after welding, brazing, soldering, thermal spraying, or thermal cutting. See also **postweld heat treatment.**

poscalentamiento. La aplicación de calor a una asamblea después de la soldadura, soldadura fuerte, soldadura blanda, rociado termal o corte termal. Vea también **tratamiento de calor postsoldadura.**

***postpurge.** Once welding current has stopped in gas tungsten arc welding, this is the time during which the gas continues to flow to protect the molten pool and the tungsten electrode as cooling takes place to a temperature at which they will not oxidize rapidly.

***pospurgante.** Cuando la corriente de soldar se ha dentenido en la soldadura de arco gas tungsteno, este es el tiempo durante en que el gas continua a salir para proteger el charco de soldadura derretido y el electrodo de tungsteno se enfrian a una temperatura donde no se oxidan rápidamente.

***postweld finishing.** The cleaning up of a weldment after the welding is complete; it may include grinding, sandblasting, wire brushing, chipping, or painting.

***acabado posterior a la soldadura.** La limpieza de una soldadura después de completarla. Puede incluir esmerilado, arenado, cepillado con alambre, desbarbado o pintura.

postweld heat treatment. Any heat treatment subsequent to welding.

tratamiento de calor postsoldadura. Cualquier tratamiento de calor subsiguiente a la soldadura.

powder flame spraying. A thermal spraying process variation in which the material to be sprayed is in powder form. See also **flame spraying (FLSP).**

rociado de polvo con llama. Un proceso termal para rociar con variación el cual el material que está para rociar se está en forma de polvo. Vea también **rociado a llama.**

power source. An apparatus for supplying current and voltage suitable for welding, thermal cutting, or thermal spraying.

fuente de poder. Un aparato para surtir corriente y voltaje conveniente para soldar, para hacer cortes termales, o rociado termal.

***preflow.** Shielding gas that flows before the GTA welding current starts to force the air away from the arc zone which protects both the hot tungsten and weld from atmospheric contamination when welding starts.

***flujo previo.** Gas de protección que fluye antes de que comience la corriente de soldadura GTA para forzar al aire a alejarse de la zona del arco y proteger tanto al tungsteno caliente como a la soldadura de la contaminación atmosférica cuando comienza la soldadura.

preheat. The heat applied to the base metal or substrate to attain and maintain preheat temperature.

precalentamiento. El calor aplicado al metal base o substrato para obtener y mantener temperatura de precalentamiento.

***preheat flame.** Brings the temperature of the metal to be cut above its kindling point, after which the high-pressure oxygen stream causes rapid oxidation of the metal to perform the cutting.

***llama para precalentamiento.** Sube la temperatura del metal que está para cortarse a una temperatura de encendimiento, después que la corriente del oxígeno de alta presión cause una oxidación rápida del metal para hacer el corte.

***preheat holes.** The cutting tip has a central hole through which the oxygen flows. Surrounding this central hole are a number of other holes called preheat holes. The differences in the type or number of preheat holes determine the type of fuel gas to be used in the tip.

***agujeros para precalentamiento.** La boquilla para cortar tiene un agujero central por donde corre el oxígeno. Rodeando este agujero central hay un numero de otros agujeros que se llaman agujeros para precalentar. Las diferencias en el tipo o número de agujeros percalentados determina el tipo de gas combustible que se usará en la boquilla.

preheat temperature. The temperature of the base metal or substrate in the welding, brazing, soldering, thermal spraying, or thermal cutting area immediately before these operations are performed. In a multipass operation, it is also the temperature in the area immediately before the second and subsequent passes are started.

temperatura de precalentamiento. La temperatura del metal base o substrato en la soldadura, soldadura fuerte, soldadura blanda, rociado termal, o en la área de los cortes termal inmediatamente antes de que estas operaciones sean ejecutadas. En una operación multipasada, es también la temperatura en la área inmediatamente antes de empezar la segunda pasada y pasadas subsiguientes.

preheating. The application of heat to the base metal immediately before welding, brazing, soldering, thermal spraying, or cutting.

precalentamiento. La aplicación de calor al metal base inmediatamente antes de la soldadura, soldadura fuerte, soldadura blanda, rociado termal o cortes.

***prepurge.** In gas tungsten arc welding, the time during which gas flows through the torch to clear out any air in the cup or surrounding the weld zone. Prepurge time is set by the operator and is completed before the welding current is started.

***prepurgar.** En soldadura de arco de tungsteno con gas, el tiempo durante el cual el gas corre por la antorcha para quitar el aire en la boquilla o la zona de soldadura. El tiempo de prepurgar es determinado por el operador y es acabado antes de que la corriente de soldadura es empezada.

***primary combustion.** The first reaction in the chemical reaction resulting when a mixture of acetylene and oxygen is ignited. This reaction frees energy and forms carbon monoxide (CO) and free hydrogen.

***combustión primaria.** La primera reacción en una reacción química resulta cuando una mezcla de oxígeno y acetileno es encendida. Está reacción libra la energia y forma carbón monóxido (CO) e hidrógeno libre.

procedure qualification. The demonstration that welds made by a specific procedure can meet prescribed standards.

calificación de procedimiento. La demostración en que las soldaduras hechas por un procedimiento específico conformen con las normas prescribidas.

projection weld. A weld made by projection welding.

soldadura de proyección. Una soldadura hecha con soldadura de proyección.

protective atmosphere. A gas envelope surrounding the part to be brazed, welded, or thermal sprayed, with the gas composition controlled with respect to chemical composition, dew point, pressure, flow rate, etc. Examples are inert gases, combusted fuel gases, hydrogen, and vacuum.

atmósfera protectora. Una envoltura de gas que está alrededor de la parte que está para soldarse con soldadura fuerte, soldadura o rociada termal, con la composición del gas controlado con respecto a la química compuesta, punto de rocío, presión, cantidad de corriente, etc. Ejemplos son gas inerto, gases de combustión que ya están encendidos, hidrógeno, y vacuo.

***proximity sensor.** A device that senses that an object is only a short distance (e.g., a few inches or feet) away and/or measures how far away it is. Proximity sensors work on the principles of triangulation of reflected light, lapsed time for reflected sound, or intensity induced eddy currents, magnetic fields, back pressure from air jets, and others.

***sensor de proximidad.** Un aparato que siente que un objeto está solamente a una corta distancia (e.g., unas pulgadas o pies) afuera, y/o mide que tan lejos está. Sensores de proximidad trabajan en los fundamentos de triangulación de luz reflejada, tiempo lapso del sonido reflejado, o en las corrientes de Fancault inducidas con intensidad, campos magnéticos, contrapresión trasera del chorro de aire, y otras.

puddle. See preferred term **weld pool.**

charco. Vea término preferido **charco de soldadura.**

pulse start delay time. The time interval from current initiation to the beginning of current pulsation, if pulsation is used. Refer to drawing for **upslope time.**

tiempo de dilación en empezar la pulsación. El tiempo del intervalo de donde se inicia la corriente al principio de la pulsación de la corriente, si es que se use pulsación. Refiérase al dibujo de **tiempo del pendiente en ascenso.**

***pulsed-arc metal transfer.** In gas metal arc welding, pulsing the current from a level below the transition current to a level above the transition current to achieve a controlled spray transfer at lower average currents; spray transfer occurs at the higher current level.

***transferir el metal por arco pulsado.** En la soldadura de arco metálico con gas, se pulsa la corriente de un nivel más alto de la corriente de transición para lograr un traslado de rocío controlado a una corriente media baja; el traslado del rocío ocurre al nivel más alto de la corriente.

***purged.** The process of opening first one cylinder valve and then the other to replace all air in the hoses with the appropriate gas prior to welding.

***limpidor.** El proceso de abrir primero una válvula de un cilindro y luego el otro para reemplazar todo el aire en las mangueras con un gas apropiado antes de empezar a soldar.

push angle. The travel angle when the electrode is pointing in the direction of weld progression. This angle can also be used to partially define the position of guns, torches, rods, and beams.

ángulo de empuje. El ángulo de avance cuando el electrodo apunta en la dirección en que la soldadura progresa. Este ángulo también puede ser usado para parcialmente definir la posición de pistolas, antorchas, varillas, y rayos.

push weld (resistance welding). A spot or projection weld made by push welding.

soldadura de empuje (soldadura de resistencia). Una soldadura de botón o proyección hecha por soldadura de empuje.

Q

qualification. See preferred terms **welder performance qualification** and **procedure qualification.**

calificación. Vea términos preferidos **calificación de ejecución del soldador** y **calificación de procedimiento.**

***quality control (QC).** A procedure of tests set up by shops to inspect weldments as they are produced to ensure they meet the standards set up by the manufacturer.

***control de calidad (QC, por sus siglas en inglés).** Un procedimiento de pruebas que utilizan las tiendas para inspeccionar las soldaduras a medida que se producen para garantizar que cumplan las normas establecidas por el fabricante.

R

***radiographic inspection (RT).** See **nondestructive testing.**

***inspección radiográfica (RT, por sus siglas en inglés).** Consulte **prueba no destructiva.**

***raster lines.** Lines commonly known as bit map lines because the computer maps the location of every little bit (pixel) of the line.

***Líneas de barrido.** Líneas que habitualmente se conocen como líneas de mapa de bits porque la computadora mapea la ubicación de cada bit (pixel) de la línea.

reactor (arc welding). A device used in arc welding circuits for the purpose of minimizing irregularities in the flow of welding current.

reactor (soldadura de arco). Un aparato usado en los circuitos de la soldadura de arco con el propósito de reducir a lo mínimo las irregularidades en la manera que corre la corriente de soldadura de arco.

reduced section tension test. A test in which a transverse section of the weld is located in the center of the reduced section of the specimen.

prueba de tensión de sección reducida. Una prueba en la cual la sección transversa de la soldadura está ubicada en el centro de la sección reducida de la probeta.

reducing atmosphere. A chemically active protective atmosphere, which at elevated temperature will reduce metal oxides to their metallic state. (Reducing atmosphere is a relative term, and such an atmosphere may be reducing to one oxide but not to another oxide.)

atmósfera de reducción. Una atmósfera protectiva activa, la cual a una temperatura elevada reduce los óxidos del metal a sus estados metalicos. (Atmósfera de reducción es un término relativo, y cierta atmósfera puede reducir a un óxido pero no al otro óxido.)

reducing flame. An oxyfuel gas flame with an excess of fuel gas.

llama de reducción. Una llama de gas oxicombustible con un exceso de gas combustible.

regulator. A device for controlling the delivery of gas at some substantially constant pressure.

regulador. Un aparato para controlar la expedición de gas a una presión substancialmente constante.

***reinforcement.** Weld metal added to a weld that builds up the weld thickness so that the weld surface is higher than the surface of the work.

***refuerzo.** Metal de soldeo que se agrega a una soldadura para aumentar el grosor con el fin de que la superficie de soldeo sea superior a la superficie de trabajo.

residual stress. Stress present in a joint member or material that is free of external forces or thermal gradients.

fuerza residual. Fuerza presente en un miembro de una junta o material que está libre de fuerzas externas o ambulantes termales.

resistance brazing (RB). A brazing process that uses heat from the resistance to electric current flow in a circuit of which the workpieces are a part.

soldadura fuerte por resistencia. Un proceso de soldadura fuerte que usa calor de la resistencia al correr de la corriente eléctrica en un circuito en las cuales las piezas de trabajo forman parte.

resistance soldering (RS). A soldering process that uses heat from the resistance to electric current flow in a circuit of which the workpieces are a part.

soldadura blanda por resistencia. Un proceso de soldadura blanda que usa calor de la resistencia al correr de la corriente eléctrica en un circuito en las cuales las piezas de trabajo forman parte.

resistance spot welding (RSW). A resistance welding process that produces a weld at the faying surfaces of a joint by the heat obtained from resistance to the flow of welding current through the workpieces from electrodes that concentrate the welding current and pressure at the weld area.

soldadura de puntos por resistencia. Un proceso de soldar por resistencia que produce una soldadura en los empalmes de la superficie de una junta por el calor obtenido de la resistencia al correr la corriente a través de las piezas de trabajo de los electrodos que sirven para concentrar la corriente para soldar y la presión en la área de la soldadura.

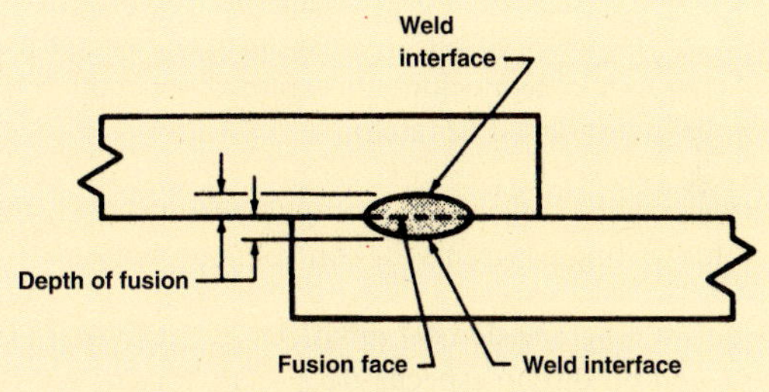

Resistance spot or seam weld

resistance welding (RW). A group of welding processes that produces coalescence of the faying surfaces with the heat obtained from resistance of the workpieces to the flow of the welding current in a circuit of which the workpieces are a part and by the application of pressure.

soldadura por resistencia. Un grupo de procesos para soldar que producen coalescencia de las superficies empalmadas con el calor obtenido de la resistencia de las piezas de trabajo al correr la corriente de soldadura en un circuito en las cuales las piezas de trabajo forman parte, y por la aplicación de presión.

resistance welding downslope time. The time during which the welding current is continuously decreasing.

tiempo del pendiente de descenso en soldadura de resistencia. El tiempo durante el cual la corriente está continuamente disminuyendo.

resistance welding electrode. The part of a resistance welding machine through which the welding current and, in most cases, force are applied directly to the workpiece. The electrode may be in the form of a rotating wheel, rotating roll, bar, cylinder, plate, clamp, chuck, or modification thereof.

electrodo para soldadura por resistencia. La parte de una máquina para soldar por resistencia por cual la corriente de soldar y, en muchos casos, la fuerza es aplicada directamente a la pieza de trabajo. El electrodo puede ser en la forma de una rueda que da vueltas, rollo rotativo, barra, cilindro, plato, empalme, calzo, o modificación de ello.

reverse polarity. The arrangement of direct-current arc welding leads with the work as the negative pole and the electrode as the positive pole of the welding arc. A synonym for direct-current electrode. Refer to drawing for **direct-current electrode positive.**

polaridad invertida. El arreglo de los cables para soldar con el arco con corriente directa con el cable de tierra como el polo negativo y el electrodo como polo positivo del arco para soldar. Un sinónimo para corriente directa electrodo. Refiérase al dibujo para **corriente directa con el electrodo positivo.**

***robot.** A reprogrammable, multifunctional manipulator designed to move material, parts, tools, or specialized devices through variable programmed motions for the performance of a variety of tasks.

***robot.** Un manipulador reprogramable, multifuncional diseñado para mover material, partes, herramienta, o aparatos especializados por medio de mociones programadas variables para la ejecución de una variedad de tareas.

***robot programming language.** A computer language especially designed for writing programs for controlling robots.

***lenguaje para programación del robot.** Un lenguaje para computadoras con un diseño especial para escribir programas para el control de los robots.

root. See preferred terms of **root of joint** and **root of weld.**

raíz. Vea las términos preferidos de **raíz de junta** y **raíz de soldadura.**

root bead. A weld bead that extends into, or includes part or all of, the joint root.

cordón de raíz. Un cordón de soldadura que se extiende adentro, o incluye parte o toda la junta de raíz.

root-bend test. A test in which the weld root is on the convex surface of a specified bend radius.

prueba de dobléz de raíz. Una prueba en la cual la raíz de la soldadura está en una superficie convexa de un radio especificado para el dobléz.

root crack. A crack in the weld or heat-affected zone occurring at the root of a weld.

grieta de raíz. Una grieta en la soldadura o en la zona afectada por el calor que ocurre en la raíz de la soldadura.

root edge. A root face of zero width. See also **root face.** Refer to drawing for **groove face.**

orilla de raíz. Una cara de raíz con una anchura de cero. Vea también **cara de raíz.** Refiérase al dibujo para **cara de ranura.**

root face. The portion of the groove face adjacent to the root of the joint. Refer to drawing for **groove face.**

cara de raíz. La porción de la cara de la ranura adyacente a la raíz de la junta. Refiérase al dibujo para **cara de ranura.**

root gap. See preferred term **root opening.**

rendija de raíz. Vea el término preferido **abertura de ráiz.**

root of joint. The portion of a joint to be welded where the members approach closest to each other. In cross section, the root of the joint may be a point, a line, or an area.

raíz de junta. Saporción de una junta que está para soldarse donde los miembros se acercan muy cerca del uno al otro. En sección transversa, la raíz de una junta puede ser una punta, una línea, o una área.

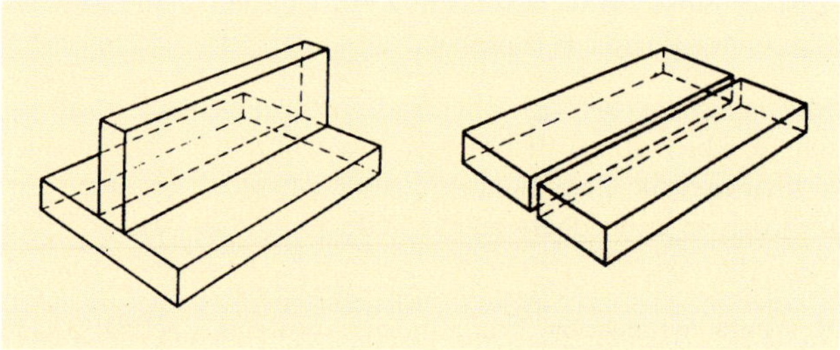

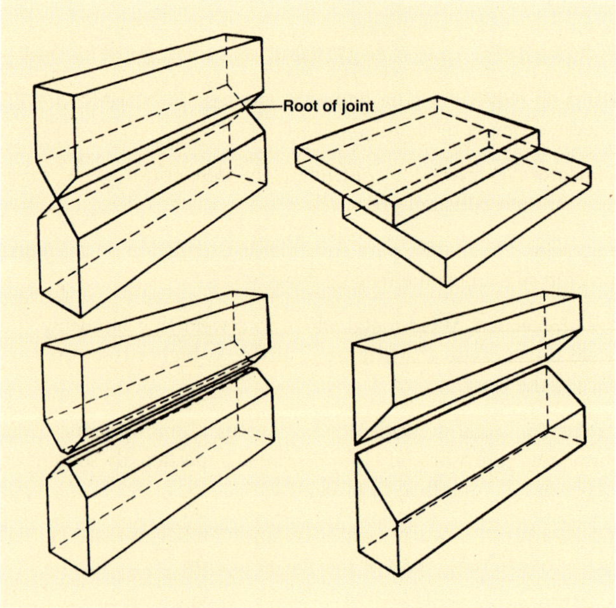

root of weld. The points, as shown in cross section, at which the back of the weld intersects the base metal surfaces.

raíz de soldadura. Las puntas, como ensena la sección transversa, donde la parte de atrás cruza con la superficie del metal base.

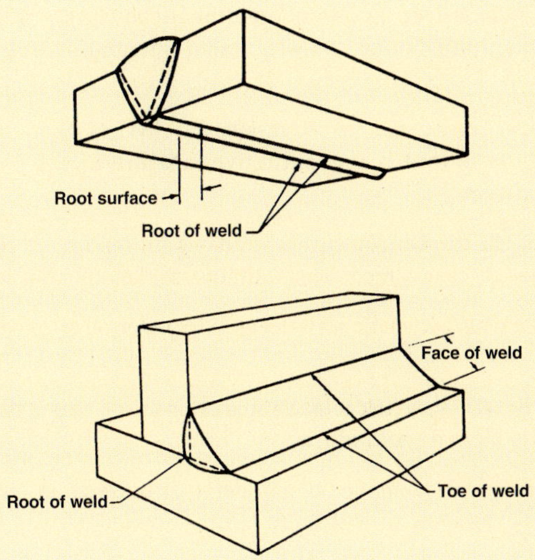

root opening. The separation between the members to be joined at the root of the joint. Refer to drawings for **bevel.**

abertura de raíz. La separación entre los miembros que están para unirse a la raíz de la junta. Refiérase al dibujo para **bisel.**

***root pass.** The first weld of a multipass weld. The root pass fuses the two pieces together and establishes the depth of weld metal penetration.

***pasada de raíz.** La primera soldadura de una soldadura de pasadas múltiples. La pasada de raíz funde las dos piezas juntas y establece la profundidad de la penetración del metal soldado.

root penetration. The distance the weld metal extends into the joint root.

penetración de raíz. La distancia que se extiende el metal de soldadura adentro de la junta de raíz.

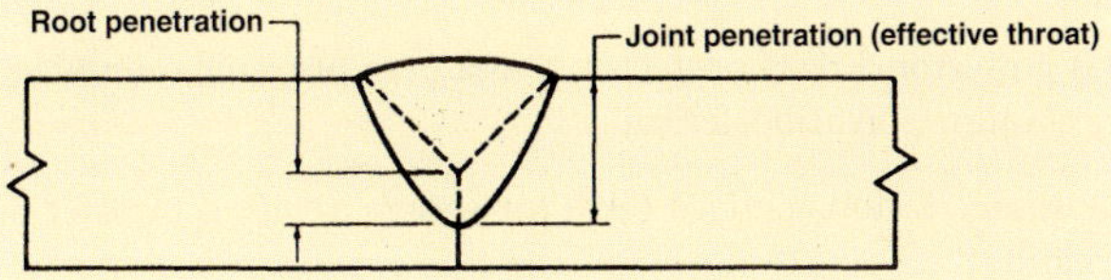

root radius. See preferred term **groove radius.**

radio de raíz. Vea el término preferido **radio de ranura.**

root reinforcement. Reinforcement of weld at the side other than that from which welding was done. Refer to drawing for **face of weld.**

refuerzo de raíz. Refuerzo de soldadura en el lado opuesto de donde se hizo la soldadura. Refiérase al dibujo para **cara de la soldadura.**

***root suck back.** The process that occurs when surface tension of the molten weld pool root surface is drawn inward causing the root surface to be concave.

***concavidad interna de la raíz.** El proceso que se produce cuando la tensión de superficie de la raíz del baño de fusión de la soldadura es atraída hacia adentro, lo que hace que la superficie de la raíz sea cóncava.

***root surface.** The exposed surface of a weld on the side other than that from which welding was done. Refer to drawings for **root of weld.**

superficie de raíz. La superficie expuesta de una soldadura en el lado opuesto de donde se hizo la soldadura. Refiérase a los dibujos para **raíz de soldadura.**

runoff weld tab. Additional material that extends beyond the end of the joint, on which the weld is terminated.

solera de carrera final de soldadura. Material adicional que se extiende más allá de donde se acaba la junta, en la cual la soldadura es terminada.

***rutile-based flux.** Acidic fluxes that produce smooth stable arcs and fast freeze slags.

***fundente con base de rutilo.** Fundentes ácidos que producen arcos estables y parejos y escorias de solidificación rápida.

S

scarf joint. A form of butt joint.

junta de echarpe. Una forma de junta a tope.

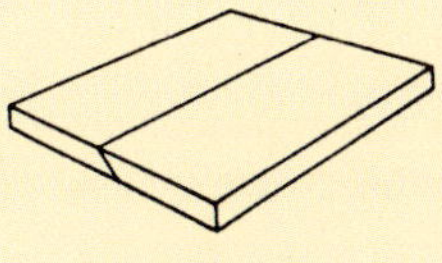
Scarf joint

***scavenger.** Elements in the flux that pick up contaminants in the molten weld pool and float them to the surface where they become part of the slag.

***limpiadores o (expulsadores).** Elementos en el flujo que levantan los contaminantes en el charco de soldadura derretida y los flotan a la superficie donde se forman parte de la escoria.

seam weld. A continuous weld made between or upon overlapping members, in which coalescence may start and occur on the faying surfaces or may have proceeded from the surface of one member. The continuous weld may consist of a single weld bead or a series of overlapping spot welds.

soldadura de costura. Una soldadura continua hecha en medio o encima de los miembros traslapados, en la cual la coalescencia puede empezar y ocurrir en la superficie del empalme, o puede haber procedido de la superficie de un miembro. La soldadura continua puede consistir de un solo cordón de soldadura o una serie de puntos traslapados en las soldaduras.

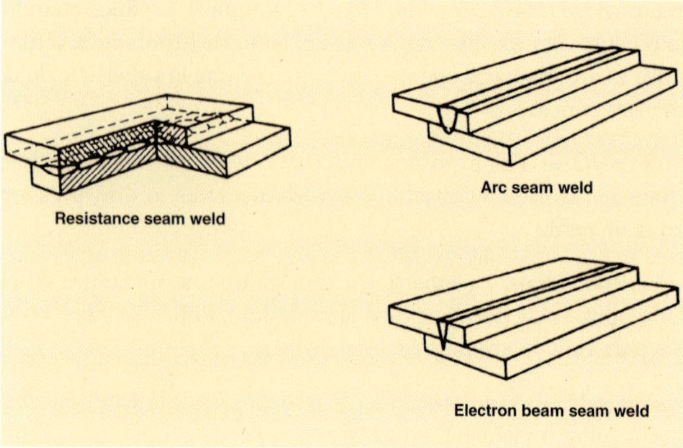

***secondary combustion.** In the combustion of acetylene and oxygen, the secondary reaction unites oxygen and the free hydrogen to form water vapor (H_2O) and liberate more heat. The carbon monoxide unites with more oxygen to form carbon dioxide (CO_2).

***combustión secundaria.** En la combustión de acetileno y oxígeno, la reacción secundaria une el oxígeno y el hidrógeno libre para formar vapor de agua (H_2O) y liberar más calor. El carbón monóxido se une con más oxígeno para formar carbón bióxido (CO_2).

***section line.** Lines on a drawing that show the surface that has been imaginarily cut away with a cutting plane line to show internal details.

***línea de sector.** Líneas de un dibujo que muestran la superficie que se ha cortado de manera imaginaria con una línea de plano de corte para mostrar los detalles internos.

***section view.** A special view of an object that shows the internal components as if the object were cut apart

***vista de corte longitudinal.** Una vista especial de un objeto que muestra los componentes internos como si el objeto estuviera cortado.

***self-shielding.** As it relates to FCA welding, it refers to electrodes that contain enough fluxing agents inside the electrode to provide complete weld protection without the need to provide a shielding gas. See **dual shield.**

***auto-protección.** En relación a la soldadura FCA, se refiere a electrodos que contienen suficientes agentes fundentes dentro del electrodo para proporcionar protección de soldadura completa sin necesidad de proveer un gas de protección. Consulte **doble protección.**

semiautomatic arc welding. Arc welding with equipment that controls only the filler metal feed. The advance of the welding is manually controlled.

soldadura de arco semiautomático. La soldadura de arco con equipo que controla solamente la alimentación del metal de relleno. El avance de la soldadura es controlado manualmente.

***semiautomatic operation.** During the welding process, the filler metal is added automatically, and all other manipulation is performed manually by the operator.

***operación semiautomática.** Durante el proceso de la soldadura, el metal de relleno es añadido automáticamente, y todas las otras manipulaciones son ejecutadas manualmente por el operador.

***sensor.** A transducer whose input is a physical phenomenon and whose output is a quantitative measure of the physical phenomenon.

***sensor.** Un transducor cuya entrada es un fenómeno físico y cuya medida es una medida cuantitativa del fenómeno físico.

sequence time (automatic arc welding). See preferred term **welding cycle.**

tiempo de secuencia (soldadura de arco automático). Vea el término preferido **ciclo de soldadura.**

***shear strength.** As applied to a soldered or brazed joint, it is the ability of the joint to withstand a force applied parallel to the joint.

***fuerza cizallada.** Asi como es aplicada a una junta de soldadura fuerte o soldadura blanda, es la habilidad de la junta de resistir una fuerza aplicada al paralelo de la junta.

shielded metal arc cutting (SMAC). An arc cutting process that uses a covered electrode.

cortes de arco métalico protegido. Un proceso de cortar con arco que usa un electrodo cubierto.

shielded metal arc welding (SMAW). An arc welding process with an arc between a covered electrode and the weld pool. The process is used with shielding from the decomposition of the electrode covering, without the application of pressure, and with filler metal from the electrode.

soldadura de arco metálico protegido. Un proceso de soldadura de arco con un arco en medio de un electrodo cubierto y el charco de soldadura. El proceso se usa con protección de descomposición del cubrimiento del electrodo sin la aplicación de presión, y con el metal de relleno del electrodo.

shielding gas. Protective gas used to prevent or reduce atmospheric contamination.

gas protector. El gas protector se usa para prevenir o reducir la contaminación atmosférica.

short arc. A nonstandard term for short-circuiting transfer arc welding.

arco corto. Un término fuera de la norma para transferir por corto circuito (soldadura de arco).

short-circuit transfer (arc welding). Metal transfer in which molten metal from a consumable electrode is deposited during repeated short circuits.

transferir por corto circuito (soldadura de arco). Transferir metal el cual el metal derretido del electrodo consumible es depositado durante repetidos cortos circuitos.

short-circuiting arc welding. A nonstandard term for short-circuiting transfer (arc welding).

soldadura de arco con corto circuito. Un término fuera de la norma para transferir por corto circuito (soldadura de arco).

shoulder. See preferred term **root face.**

hombro. Vea término preferido **cara de raíz.**

shrinkage void. A cavity-type discontinuity normally formed by shrinkage during solidification.

vacío de encogimiento. Una discontinuidad tipo cavidad normalmente formada por encogimiento durante solidificación.

side-bend test. A test in which the side of a transverse section of the weld is on the convex surface of a specified bend radius.

prueba de dobléz de lado. Una prueba en la cual el lado de una sección transversa de la soldadura está en la superficie convexa de un radio de dobléz especificado.

***silver braze.** A brazing process using an alloyed brazing rod which contains some percentage of silver. Silver is used as an alloy to promote wetting and strength.

***soldadura de plata.** Un proceso de soldadura que utiliza una varilla de soldeo de aleación que contiene cierto porcentaje de plata. La plata se utiliza como aleación para promover la fortaleza y la humidificación.

silver soldering, silver alloy brazing. Nonpreferred terms used to denote brazing with a silver-base filler metal. See preferred term **furnace brazing.**

soldadura blanda con plata, soldadura fuerte con aleación de plata. Términos no preferidos que se usan para denotar soldadura fuerte con metal para rellenar con base de plata. Vea el término preferido **soldadura fuerte en horno.**

single bevel-groove weld. A type of groove weld. Refer to drawing for **groove weld.**

soldadura de ranura de un solo bisel. Tipo de soldadura de ranura. Refiérase al dibujo para **soldadura de ranura.**

single-flare bevel-groove weld. A type of groove weld. Refer to drawing for **groove weld.**

soldadura de ranura de un solo bisel acampanado. Un tipo de soldadura de ranura. Refiérase al dibujo para **soldadura de ranura.**

single-flare V-groove weld. A type of groove weld. Refer to drawing for **groove weld.**

soldadura de ranura de una sola V acampanada. Un tipo de soldadura de ranura. Refiérase al dibujo para **soldadura de ranura.**

single J-groove weld. A type of groove weld. Refer to drawing for **groove weld.**

soldadura de ranura de una sola J. Un tipo de soldadura de ranura. Refiérase al dibujo para **soldadura de ranura.**

single-port nozzle. A constricting nozzle of the plasma arc torch that contains one orifice, located below and concentric with the electrode.

boquilla de una sola abertura. Es una boquilla constreñida de la antorcha de arco de plasma que contiene un orificio, situado debajo y concéntrico al electrodo.

single square-groove weld. A type of groove weld. Refer to drawing for **groove weld.**

soldadura de ranura de una sola escuadra. Un tipo de soldadura de ranura. Refiérase al dibujo de **soldadura de ranura.**

single U-groove weld. A type of groove weld. Refer to drawing for **groove weld.**

soldadura de ranura de una sola U. Un tipo de soldadura de ranura. Refiérase al dibujo para **soldadura de ranura.**

single V-groove weld. A type of groove weld. Refer to drawing for **groove weld.**

soldadura de ranura de una sola V. Un tipo de soldadura de ranura. Refiérase al dibujo de **soldadura de ranura.**

size of weld.

groove weld. The joint penetration (depth of bevel plus the root penetration when specified). The size of a groove weld and its effective throat are one and the same.

fillet weld. For equal leg fillet welds, the leg lengths of the largest isosceles right triangle that can be inscribed within the fillet weld cross section. Refer to drawings for concavity and convexity. For unequal leg fillet welds, the leg lengths of the largest right triangle that can be inscribed within the fillet weld cross section.

Note: When one member makes an angle with the other member greater than 105°, the leg length (size) is of less significance than the effective throat, which is the controlling factor for the strength of a weld.

flange weld. The weld metal thickness measured at the root of the weld.

tamaño de la soldadura.

soldadura de ranura. La penetración de la junta (profundidad del bisel más la penetración de la raíz cuando está especificada). El tamaño de la soldadura de ranura y la garganta efectiva son una y la misma.

soldadura filete. Para soldaduras con piernas iguales de filete, lo largo de las piernas del triángulo recto con el isosceles más grande que puede ser inscrito dentro de la sección transversa de la soldadura de filete. Refiérase al dibujo para concavidad y convexidad. Para piernas de soldadura de filete desiguales, lo largo de las piernas del triángulo recto más grande que puede ser inscrito dentro de la sección transversa de la soldadura de filete.

Nota: Cuando un miembro hace un ángulo con otro miembro más grande de 105 grados, lo largo de la pierna (tamaño) es de menor significado que la garganta efectiva, la cual es el factor de control para la fuerza de una soldadura.

soldadura de brida. Lo grueso del metal de soldadura se mide a la raíz de la soldadura.

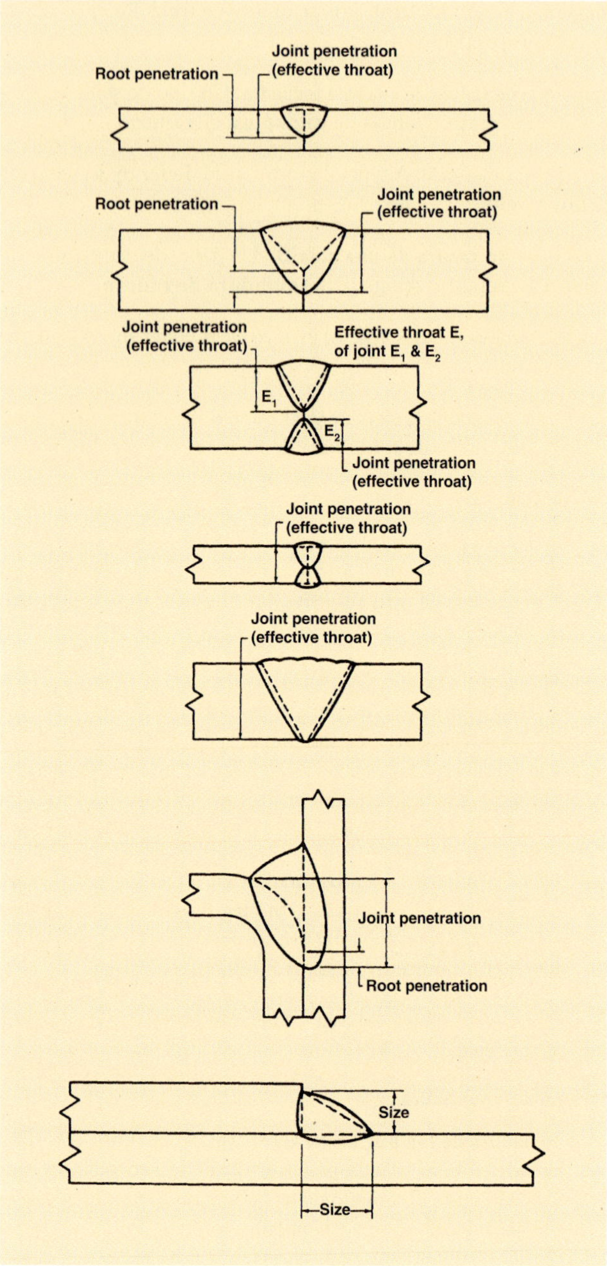

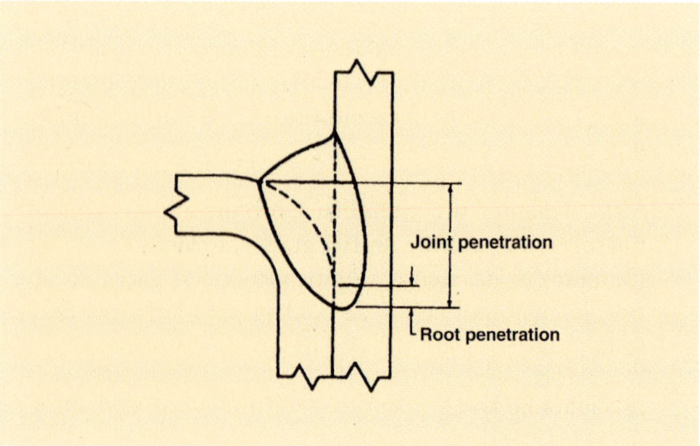

slag. A nonmetallic product resulting from the mutual dissolution of flux and nonmetallic impurities in some welding and brazing processes.

escoria. Un producto que no es metálico resultando de una disolución mutual del flujo y las impuridades no metálicas en unos procesos de soldadura y soldadura fuerte.

slag inclusion. Nonmetallic solid material entrapped in weld metal or between weld metal and base metal.

inclusion de escoria. Un material sólido que no es metálico atrapado en el metal de soldadura en medio del metal base.

slag pan. A high-temperature container that holds the slag from a thermite weld until it cools.

bandeja para escoria. Un recipiente de alta temperatura que contiene la escoria de una soldadura con termita hasta que se enfría.

***slope.** For gas metal arc welding, the volt-ampere curve of the power supply indicates that there is a slight decrease in voltage as the amperage increases; the rate of voltage decrease in the slope.

***pendiente.** Para soldadura de arco de metal con gas, la curva voltio-amperio de la fuente de poder indica que si hay un ligero decremento en voltaje cuando los amerios aumentan; la proporción del voltaje decrementa en el pendiente.

slot weld. A weld made in an elongated hole in one member of a lap or tee-joint joining that member to that portion of the surface of the other member that is exposed through the hole. The hole may be open at one end and may be partially or completely filled with weld metal. (A fillet-welded slot should not be construed as conforming to this definition.)

soldadura de ranura alargada. Una soldadura hecha en un agujero alargado en un miembro de una junta en solape o T uniendo ese miembro a esa porción de la superficie del otro miembro que está expuesto a través del agujero. El agujero puede ser abierto en una punta y puede ser parcialmente o completamente rellenado con metal de soldadura. (Una ranura alargada con soldadura de filete no debe de interpretarse como conforme a está definición.)

slugging. The act of adding a separate piece or pieces of material in a joint before or during welding that results in a welded joint not complying with design, drawing, or specification requirements.

usar trozos de metal. El acto de agregar una pieza o piezas separadas de material en una junta antes o durante la soldadura que resulta en una junta soldada que no cumple con diseño, dibujo, o las especificaciones requeridas.

solder. A filler metal used in soldering that has a liquidus not exceeding 840°F (450°C).

soldadura (material para soldar). Un metal de relleno usado para soldadura blanda que tiene un liquidus que no excede de 840°F (450°C).

soldering (S). A group of welding processes that produces coalescence of materials by heating them to the soldering temperature and by using a filler metal with a liquidus not exceeding

840°F (450°C) and below the solidus of the base metals. The filler metal is distributed between closely fitted faying surfaces of the joint by capillary action.

soldadura blanda. Un grupo de procesos de soldadura que produce coalescencia de materiales calentándolos a una temperatura de soldar y usando un metal para rellenar con un liquidus que no exceda de 840°F (450°C) y más abajo del solidus de los metales base. El metal para rellenar es distribuido en medio de las superficies empalmadas acopladas muy cerca de la junta por acción capilar.

soldering gun. An electrical soldering iron with a pistol grip and a quick heating, relatively small bit.

pistola de soldar. Un fierro eléctrico para soldar con mango de pistola, rápido para calentarse, y tiene una punta relativamente pequeña.

solid state welding (SSW). A group of welding processes that produces coalescence by the application of pressure at a welding temperature below the melting temperatures of the base metal and the filler metal.

soldadura de estado sólido. Un grupo de procesos para soldar que produce coalescencia cuando se le aplica presión a una temperatura de soldadura más baja que las temperatures que se usan para derretir el metal base y el metal de relleno.

solidus. The highest temperature at which a metal or an alloy is completely solid.

solidus. La temperatura más alta cuando un metal o una aleación está completamente sólido.

spatter. The metal particles expelled during welding and that do not form a part of the weld.

salpicadura. Las partículas de metal que se despidan cuando se está soldando y que no forman parte de la soldadura.

spatter loss. Metal lost due to spatter.

pérdida causa salpicadura. El metal perdido debido a la salpicadura.

spool. A type of filter metal package consisting of a continuous length of electrode wound on a cylinder (called the barrel), which is flanged at both ends. The flange extends below the inside diameter of the barrel and contains a spindle hole.

carrete. Un paquete de metal tipo filtro consistiendo de una extensión continua de un electrodo enrollado en un cilindro (llamado el barril), el cual tiene una brida en los dos extremos. La brida se extiende debajo del diámetro de adentro del barril y contiene un agujero huso.

spot weld. A weld made between or upon overlapping members in which coalescence may start and occur on the faying surfaces or may proceed from the surface of one member. The weld cross section (plan view) is approximately circular. See also **arc spot weld** and **resistance spot welding.**

soldadura de puntos. Una soldadura hecha en medio o sobre miembros traslapados en la cual la coalescencia puede empezar y ocurrir en las superficies empalmadas o puede continuar en la superficie de un miembro. La sección transversa (plan de vista) es aproximadamente circular. Vea también **soldadura de puntos por arco** y **soldadura de puntos por resistencia.**

spray arc. A nonstandard term for spray transfer.

arco para rociar. Un término fuera de norma para traslado rociado.

spray transfer (arc welding). Metal transfer in which molten metal from a consumable electrode is propelled axially across the arc in small droplets.

traslado rociado (soldadura de arco). Transferir el metal el cual el metal derretido de un electrodo consumible es propelado axialmente a traves del arco en gotitas pequeñas.

***square butt joint.** A joint made when two flat pieces of metal face each other with no edge preparation. See also **square groove weld.**

***junta escuadra de tope.** Una junta hecha cuando dos piezas planas de metal se enfrentan una a la otra sin preparación de orilla. Vea también **soldadura de ranura escuadra.**

square butt weld. See **butt joint.**

junta escuadra de tope. Vea **junta a tope.**

square-groove weld. A type of groove weld.

soldadura de ranura escuadra. Un tipo de soldadura de ranura.

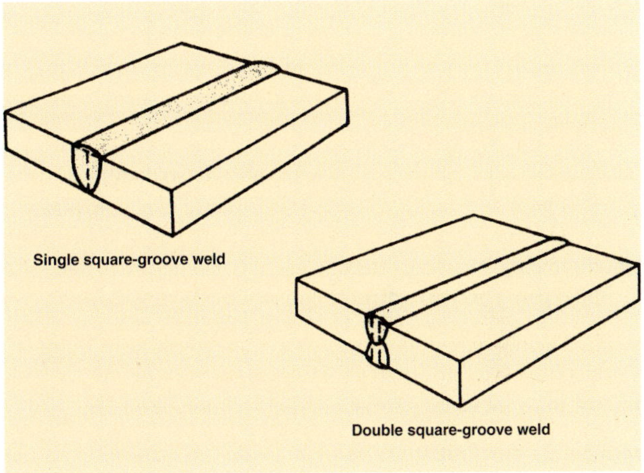

stack cutting. Thermal cutting of stacked metal plates arranged so that all the plates are severed by a single cut.

corte de metal apilado. Un corte termal de hojas de metal apilados arregladas para que todas las hojas sean cortadas por un solo corte.

staggered intermittent welds. Intermittent welds on both sides of a joint in which the weld increments on one side are alternated with respect to those on the other side.

soldadura intermitente de cadena. Soldaduras intermitentes en los dos lados de una junta en cual los incrementos de soldadura son alternados de un lado con respecto a los del otro lado.

***stainless steels.** Alloys of steel containing enough chromium so that they do not stain, corrode, or rust easily.

***aceros inoxidables.** Aleaciones de acero que contienen la cantidad de cromium necesaria para que no se manchen, no se corroyan, y no se oxiden facilmente.

standoff distance. The distance between a nozzle and the workpiece.

distancia de alejamiento. La distancia entre la boquilla y la pieza de trabajo.

starting weld tab. Additional material that extends beyond the beginning of the joint, on which the weld is started.

solera para empezar a soldar. Material adicional que se extiende más allá del principio de la junta, en donde la soldadura es empezada.

***steel.** An alloy consisting primarily of iron and carbon. The carbon content may be as high as 2.2% but is usually less than 1.5%.

***acero.** Una aleación que consiste primeramente de hierro y carbón. El contenido del carbón puede ser tan alto como 2.2% pero es regularmente menos de 1.5%.

stick electrode. A nonstandard term for covered electrode.

electrodo de varilla. Un término fuera de norma por electrodo cubierto.

stickout. See preferred term **electrode extension.**

sobresalga. Vea término preferido **extensión del electrodo.**

straight polarity. The arrangement of direct-current arc welding leads in which the work is the positive pole and the electrode is the negative pole of the welding arc. A synonym for direct-current electrode negative. Refer to drawing for **direct-current electrode negative.**

polaridad directa. El arreglo de los cables de soldadura de arco con corriente directa donde el cable de la tierra es el polo positivo y el porta electrodo es el polo negativo del arco de soldadura. Un sinónimo para corriente directa con electrodo negativo. Refiérase al dibujo para **corriente directa con electrodo negativo.**

strain. The movement or elongation of a material such as metal, wood, fiber glass, etc., that is the result of a force or stress which may or may not cause permanent deformation or failure.

tensión. El movimiento o elongación de un material como el metal, que es el resultado de una fuerza o el estrés que puede o no puede causar una deformación permanente o el fracaso.

stranded electrode. A composite filler metal electrode consisting of stranded wires that may mechanically enclose materials to improve properties, stabilize the arc, or provide shielding.

electrodo cable. Electrodo de metal para rellenar compuesto que consiste de cable de alambres que pueden encerrar materiales mecánicamente para mejorar propiedades, estabilizar el arco, o proveer protección.

stress. An internal or external force that is applied to a material or weldment.

estrés. Un interno o fuerza externa que se aplica a un material o soldadura.

***stress point.** Any point in a weld where incomplete fusion of the weld on one or both sides of the root gives rise to stress, which can result in premature cracking or failure of the weld at a load well under the expected strength of the weld.

***punto de tensión.** Cualquier punto en una soldadura donde la fusión incompleta en la soldadura en uno o en los dos lados de la raíz le aumenta la tensión, la cual puede resultar en una grieta o falta prematura en la soldadura con una carga mucho menos que la fuerza de la soldadura que se esperaba.

stress relief heat treatment. Uniform heating of a structure or a portion thereof to a sufficient temperature to relieve the major portion of the residual stresses, followed by uniform cooling.

tratamiento de calor para relevar la tensión. Calentamiento uniforme de una estructura o una porción a una temperatura suficiente para relevar la mayor porción de las tensiones restantes, seguido por enfriamiento uniforme.

stringer bead. A type of weld bead made without appreciable weaving motion. See also **weave bead.**

cordón encordador. Un tipo de cordón de soldadura sin movimiento del tejido apreciable. Vea también **cordón tejido.**

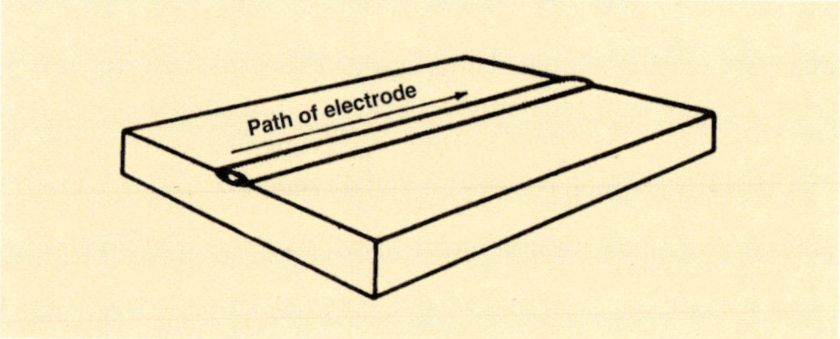

stubbing out. The process of placing a part that is roughly the correct length so it can be trimmed after assembly to the exact length or fit.

desbaste. El proceso de colocar una parte que es más o menos la longitud correcta para que pueda ser cortada después del montaje a la longitud exacta o el tamaño.

stud arc welding (SW). An arc welding process that uses an arc between a metal stud, or similar part, and the other workpiece. The process is used with or without shielding gas or flux, with or without partial shielding from a ceramic ferrule surrounding the stud, with the application of pressure after the faying surfaces are sufficiently heated, and without filler metal.

esparrago (tachón) para soldadura de arco. Un proceso de soldadura de arco que usa un arco entre un esparrago (tachón) de metal, o parte similar, y la otra pieza de trabajo. El proceso es usado con o sin flujo o protección de gas, con o sin protección parcial del casquillo cerámico que rodea el esparrago (tachón), con la aplicación de presión después que las superficies empalmadas tengan suficiente calor, y sin metal de relleno.

stud welding. A general term for the joining of a metal stud or similar part to a workpiece. Welding may be accomplished by arc, resistance, friction, or other suitable process with or without external gas shielding.

soldadura de esparrago (tachón). Un término general para la unión de esparragos (tachones) o parte similar a una pieza de trabajo. La soldadura se puede efectuar por arco, resistencia, fricción, u otro proceso conveniente con o sin gas externo para protección.

submerged arc welding (SAW). An arc welding process that uses an arc or arcs between a bare metal electrode or electrodes and the weld pool. The arc and molten metal are shielded by a blanket of granular flux on the workpieces. The process is used without pressure and with filler metal from the electrode and sometimes from a supplemental source (welding rod, flux, or metal granules).

soldadura por arco sumergido. Un proceso de soldar con arco que usa un arco o arcos entre un electrodo de metal liso o electrodos y el charco de la soldadura. El arco y el metal derretido son protegidos por una capa de flujo granular sobre la pieza de trabajo. El proceso es usado sin presión y con metal para rellenar del electrodo y a veces de una fuente suplementaria (varilla para soldar, flujo, o gránulos de metal).

substrate. Any base material to which a thermal sprayed coating or surfacing weld is applied.

substrato. Cualquier material base al cual se le aplica una capa termal o una soldadura de superficie.

suck back. See preferred term **concave root surface.**

succión del cordón de raíz. Vea el término preferido **superficie raíz concavo.**

surface preparation. The operations necessary to produce a desired or specified surface condition.

preparación de la superficie. Las operaciones necesarias para producir una deseada o una especificada condición de la superficie.

surfacing. The application by welding, brazing, or thermal spraying of a layer of material to a surface to obtain desired properties or dimensions, as opposed to making a joint. See also **buttering, cladding, coating,** and **hardfacing.**

recubrimiento superficial. La aplicación a la soldadura, a la soldadura fuerte o rociado termal de una capa de material a la superficie para obtener las deseadas propiedades o dimensiones, contrario a la hechura de una junta. Vea también **recubrimiento antes de terminar una soldadura, capa de revestimiento, revestimiento,** y **endurecimiento de caras.**

***synergic system.** Pulsed-arc metal transfer system in which the power supply and wire-feed settings are made by adjusting a single knob.

***sistema sinérgico.** Sistema de transferir metal por arco pulsado en cual la fuente de poder y los ajustes del alimentador de alambre son hechos por el ajuste de un botón solamente.

T

tab. See **runoff weld tab, starting weld tab,** and **weld tab.**

solera. Vea **solera de carrera final de soldadura, solera para empezar a soldar,** y **solera para soldar.**

tack weld. A weld made to hold parts of a weldment in proper alignment until the final welds are made.

soldadura de puntos aislados. Una soldadura hecha para detener las partes en su propio alineamiento hasta que se hagan las soldaduras finales.

***tactile sensor.** A transducer that is sensitive to touch.

***sensor táctil.** Un transducor que es sensitivo al tocar.

taps. Connections to a transformer winding that are used to vary the transformer turns ratio, thereby controlling welding voltage and current.

grifo. Conexiones al arrollamiento de un transformador que se usan para variar la proporción de vueltas del transformador, asi se puede controlar la corriente y el voltaje para soldar.

***teach.** To program a manipulator arm by guiding it through a series of points or in a motion pattern that is recorded for subsequent automatic action by the manipulator.

***enseñar.** Para programar un manipulador de brazo guiándolo por una serie de puntos o en una muestra de movimiento que está registrada para acción automática subsiguiente por el manipulador.

***teaching interface.** The mechanisms or devices by which a human operator teaches a machine.

***enseñanza de interfaze.** Los mecanismos o aparatos por los cuales un operador humano enseña a la máquina.

tee joint. A joint between two members located approximately at right angles to each other in the form of a *T.*

junta en T. Una junta en medio de dos miembros que están localizados aproximadamente a ángulos rectos de uno al otro en la forma de *T.*

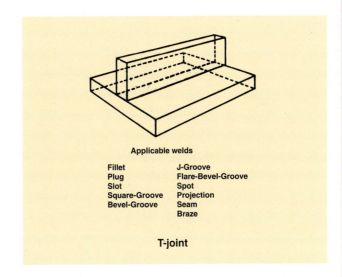

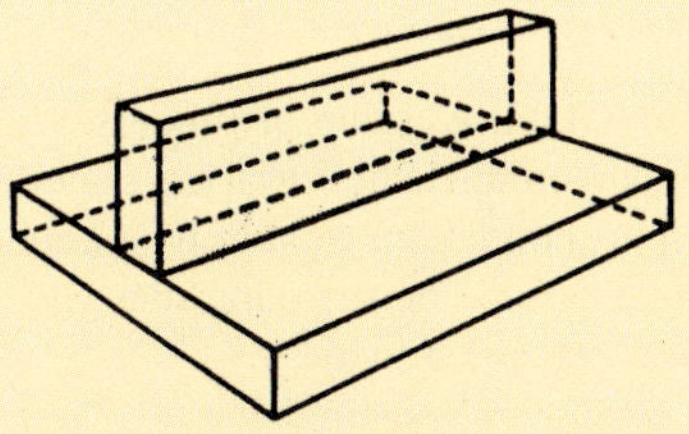

Applicable welds

Fillet	J-Groove
Plug	Flare-Bevel-Groove
Slot	Spot
Square-Groove	Projection
Bevel-Groove	Seam
	Braze

T-joint

***tempering.** Reheating hardened metal before it cools to room temperature to make it tough, not brittle.

***templar.** Recalentando un metal endurecido antes de que se enfrie a la temperatura del ambiente para hacerlo duro, no frágil.

template. A device or tool that can be traced onto the stock's surface as a guide. Templates can be specifically designed patterns or multipurpose tools, such as a circle template. (See **pattern.**)

***plantilla.** Un dispositivo o herramienta que se puede remontar a la superficie de la población como una guía. plantillas pueden ser diseñadas específicamente patrones o herramientas de propósito múltiple, como un círculo plantilla. (Ver **patrón.**)

***tensile strength.** As applied to a brazed or soldered joint, the ability of the joint to withstand being pulled apart.

***resistencia a la tensión.** Como es aplicada a una junta de soldadura fuerte o soldadura blanda, la capacidad de una junta que resista ser estirada hasta que se rompa en dos pedazos.

tension test. A test in which a specimen is loaded in tension until failure occurs. See also **reduced section tension test.**

prueba de tensión. Una prueba en la cual la probeta está cargada de tensión hasta que ocurra el fracaso. Vea también **prueba de tensión de sección reducida.**

***thermal conductivity.** A material's ability to conduct heat.

***conductividad térmica.** La capacidad de un material de conducir calor.

thermal cutting (TC). A group of cutting processes that severs or removes metal by localized melting, burning, or vaporizing of the workpiece. See also **arc cutting, electron beam cutting, laser beam cutting,** and **oxygen cutting.**

corte termal. Un grupo de procesos para cortar que desúne o quita el metal para derretir o quemar o vaporizar localmente las piezas de trabajo. Vea también **corte con arco, cortes rayo de electron, corte de rayo laser,** y **cortes de oxígeno.**

***thermal expansion.** The characteristic of a material to expand and contract with changes in temperature.

***expansión térmica.** La característica de un material de expandirse o contraerse con los cambios de temperatura.

thermal spraying (THSP). A group of processes in which finely divided metallic or nonmetallic surfacing materials are deposited in a molten or semimolten condition on a substrate to form a thermal spray deposit. The surfacing material may be in the form of powder, rod, cord, or wire.

rociado termal. Un grupo de procesos el cual los materiales metálicos o no metálicos de superficie que son depositados en una condición derretida o semiderretida sobre el substrato para formar un depósito de rociado termal. El material de superficie puede ser en forma de polvo, varilla, cordón, o alambre.

thermal stresses. Stresses in metal resulting from nonuniform temperature distributions.

tensión termal. Tensiones en el metal resultando cuando la distribución de la temperatura no está uniforme.

thermit welding (TW). A welding process that produces coalescence of metals by heating them with superheated liquid metal from a chemical reaction between a metal oxide and aluminum, with or without the application of pressure. Filler metal, when used, is obtained from the liquid metal.

soldadura termita. Un proceso de soldadura que produce coalescencia del metal calentándolos con un metal supercalentado líquido da una reacción química entre un metal óxido y el aluminio, con o sin la aplicación de presión. El metal de relleno, cuando es usado, es obtenido del metal líquido.

throat area. The area bounded by the physical parts of the secondary circuit in a resistance spot, seam, or projection welding machine. Used to determine the dimensions of a part that can be welded and determine, in part, the secondary impedance of the equipment.

área de garganta. La área limitada por las partes físicas del circuito secundario en un punto de resistencia, costura, o una máquina de soldar de proyección. Se usa para determinar las dimensiones de una parte que puede ser soldada y determinar, en parte, la impedancia secundaria del equipo.

throat depth. In a resistance spot, seam, or projection welding machine, the distance from the center line of the electrodes or plates to the nearest point of interference for flat sheets.

profundidad de garganta. En una punta de resistencia, costura, o máquina de soldar de proyección, la distancia de la línea del centro del electrodo o platinas al punto más cercano de interferencia para las hojas planas.

throat of a fillet weld.

> **actual throat.** The shortest distance from the root of weld to its face. Refer to drawing for **convexity.**
>
> **effective throat.** The minimum distance minus any reinforcement from the root of a weld to its face. Refer to drawing for **convexity.**
>
> **theoretical throat.** The distance from the beginning of the root of the joint perpendicular to the hypotenuse of the largest right triangle that can be inscribed within the fillet weld cross section. This dimension is based on the assumption that the root opening is equal to zero. Refer to drawing for **convexity.**

garganta de soldadura filete.

> **garganta actual.** La distancia más corta de la raíz de una soldadura a su cara. Refiérase al dibujo para **convexidad.**
>
> **garganta efectiva.** La distancia mínima menos cualquier refuerzo de la raíz de la soldadura a su cara. Refiérase al dibujo para **convexidad.**
>
> **garganta teórica.** La distancia de donde empieza la raíz de la junta perpendicular a la hipotenusa del triángulo recto más grande que puede ser inscrito adentro de la sección transversa de una soldadura de filete. Esta dimensión está basada en la proposición que la abertura de la raíz es igual a cero. Refiérase al dibujo para **convexidad.**

TIG welding. A nonstandard term when used for gas tungsten arc welding.

soldadura TIG. Un término fuera de norma cuando es usado por soldadura de arco de tungsteno con gas.

***titanium.** This silver-colored metal has a high strength to weight ratio which is why it is used extensively in the aerospace industry.

***titanio.** Este metal de color plateado cuenta con una gran fortaleza en proporción a su peso, motivo por el cual se lo utiliza de manera extensa en la industria aeroespacial.

toe crack. A crack in the base metal occurring at the toe of a weld.

grieta de pie. Una grieta en el metal base que ocurre al pie de la soldadura.

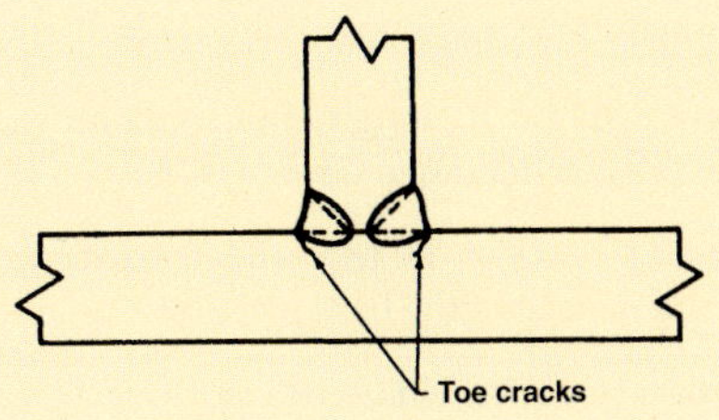

toe of weld. The junction between the face of a weld and the base metal. Refer to drawing for **face of weld.**

pie de la soldadura. La unión entre la cara de la soldadura y el metal base. Refiérase al dibujo para **cara de la soldadura.**

tolerances. The allowable deviation in accuracy or precision between the measurement specified and the part as laid out or produced.

tolerancias. Desviación permitida en la precisión entre la medida especificada y la pieza instalada o producida.

torch. See preferred terms **cutting torch** and **welding torch.**

antorcha. Vea el término preferido **antorcha para cortar** y **antorcha para soldar.**

***torch angle.** The angle between the center line of the torch and the work surface; the ideal torch angle is 45°. The torch angle affects the percentage of heat input into the metal, thus affecting the speed of melting and the size of the molten weld pool.

***ángulo de antorcha.** El ángulo en medio de la línea del centro de la antorcha y la superficie del trabajo; el ángulo ideal de la antorcha es 45°. El ángulo de la antorcha afecta el por ciento de calor que entra dentro del metal, asi afectando la rapidez de derretimiento y el tamaño del charco del metal derretido de la soldadura.

torch brazing (TB). A brazing process that uses heat from a fuel-gas flame.

soldadura fuerte con antorcha. Un proceso de soldadura fuerte que usa calor de una llama de gas combustible.

***torch manipulation.** The movement of the torch by the operator to control the weld bead characteristics.

***manipulacion de la antorcha.** El movimiento de la antorcha por el operador para el control de las características del cordón de la soldadura.

torch soldering (TS). A soldering process that uses heat from a fuel-gas flame.

soldadura blanda con antorcha. Un proceso de soldadura blanda que usa calor de una llama de gas combustible.

***trailing edge.** See **electrode angle.**

***borde posterior.** Consulte **ángulo del electrodo.**

transducer. A device that transforms one form of energy into another.

transducor. Un aparato que convierte una forma de energía a otra.

transferred arc. A plasma arc established between the electrode of the plasma arc torch and the workpiece.

arco transferido. Un arco de plasma establecido entre el electrodo de la antorcha de arco de plasma y la pieza de trabajo.

***transition current.** In gas metal arc welding, current above a critical level to permit spray transfer; the rate at which drops are transferred changes in relationship to the current. Transition current depends upon the alloy bearing welded and is proportional to the wire diameter.

***corriente de transición.** En soldadura de arco y metal con gas, corriente arriba de un nivel crítico para permitir el traslado del rociado; la proporción en la cual las gotas son transferidas cambia en relación a la corriente. La corriente de transición depende del aleado que se está soldando y es proporcional al diámetro del alambre.

transverse face bend. See **face bend.**

doblez de cara transversal. Vea **doblez de cara.**

transverse root bend. See **root bend.**

cordón de raíz transversal. Vea **cordón de raíz.**

travel angle. The angle less than 90° between the electrode axis and a line perpendicular to the weld axis, in a plane determined by the electrode axis and the weld axis. The angle can also be used to partially define the position of guns, torches, rods, and beams. See also **drag angle** and **push angle.** Refer to drawing for **backhand welding.**

ángulo de avance. El ángulo menos de 90° entre el eje del electrodo y una línea perpendicular al eje de la soldadura, en un plano determinado por el eje del electrodo y el eje de la soldadura. El ángulo también puede ser usado para parcialmente definir la posición de las pistolas, antorchas, varillas, y rayos. Vea también **ángulo del tiro** y **ángulo de empuje.** Refiérase al dibujo para **soldadura en revés.**

travel angle (pipe). The angle less than 90° between the electrode axis and a line perpendicular to the weld axis at its point of intersection with the extension of the electrode axis, in a plane determined by the electrode axis and a line tangent to the pipe surface of the same point. This angle can also be used to partially define the position of guns, torches, rods, and beams. Refer to drawing for **backhand welding.**

ángulo de avance (tubo). El ángulo menos de 90° entre el eje del electrodo y la línea perpendicular al eje a su punto de intersección con la extensión del eje del electrodo, en un plano determinado por el eje del electrodo y una línea tangente a la superficie del tubo del mismo punto. Este ángulo también puede ser usado para parcialmente definir la posición de las pistolas, varillas, y rayos. Refiérase al dibujo para **soldadura en revés.**

***tungsten.** This steel-gray metal has a melting temperature of 6170°F (3410°C) and it freely emits electrons which makes it ideal for the nonconsumable electrodes for PAC, PAW, and GTAW processes.

***tungsteno.** Este metal gris acerado tiene una temperatura de fundido de 6170°F (3410°C) y emite electrones libremente lo que lo hace ideal para los electrodos no consumibles de los procesos PAC, PAW, y GTAW.

tungsten electrode. A nonfiller metal electrode used in arc welding, arc cutting, and plasma spraying, made principally of tungsten.

electrodo de tungsteno. Un electrodo de metal que no se rellena que se usa para soldadura de arco, cortes por arco, rociado por plasma, y hecho principalmente de tungsteno.

***type A fire extinguisher.** An extinguisher used for combustible solids, such as paper, wood, and cloth. Identifying symbol is a green triangle enclosing the letter *A*.

***extinguidor para incendios tipo A.** Un extinguidor que se usa para combustibles sólidos como papel, madera, y tela. El símbolo de identificación es un triángulo verde con la letra *A* adentro.

***type B fire extinguisher.** An extinguisher used for combustible liquids, such as oil and gas. Identifying symbol is a red square enclosing the letter *B*.

***extinguidor para incendios tipo B.** Un extinguidor que se usa para liquidos combustibles, como aceite y gas. El símbolo de identificación es un cuadro rojo con la letra *B* adentro.

***type C fire extinguisher.** An extinguisher used for electrical fires. Identifying symbol is a blue circle enclosing the letter *C*.

***extinguidor para incendios tipo C.** Un extinguidor que se usa para incendios eléctricos. El símbolo de identificación es un círculo azul con la letra *C* adentro.

***type D fire extinguisher.** An extinguisher used on fires involving combustible metals, such as zinc, magnesium, and titanium. Identifying symbol is a yellow star enclosing the letter *D*.

***extinguidor para incendios tipo D.** Un extinguidor que se usa para incendios de metales combustibles, como zinc, magnesio, y titanio. El símbolo de identificación es una estrella amarilla con una letra *D* adentro.

U

***U-groove.** See **joint.**

***ranura en U.** Consulte **junta.**

U-groove weld. A type of groove weld.

soldadura de ranura en U. Un tipo de soldadura de ranura.

ultrasonic coupler (ultrasonic soldering and ultrasonic welding). Elements through which ultrasonic vibration is transmitted from the transducer to the tip.

acoplador ultrasónico (soldadura blanda ultrasónica y soldadura ultrasónica). Los elementos por los cuales la vibración ultrasónica es transmitida del transducor a la punta.

***ultrasonic inspection (UT).** See **nondestructive testing.**

***inspección ultrasónica.** Consulte **pruebas no destructivas.**

ultrasonic welding (USW). A solid state welding process that produces a weld by the local application of high-frequency vibratory energy as the workpieces are held together under pressure.

soldadura ultrasónica. Un proceso de soldadura de estado sólido que produce una soldadura por la aplicación local de energía vibratoria de alta frecuencia asi cuando las piezas de trabajo están agarradas juntas bajo presión.

***ultraviolet light.** A very short wavelength light that is above the visible light range.

***luz ultravioleta.** Una luz de longitud de onda muy corta que se encuentra por encima del rango de luz visible.

underbead crack. A crack in the heat-affected zone, generally not extending to the surface of the base metal.

grieta entre o bajo cordones. Una grieta en la zona afectada por el calor, generalmente no se extiende a la superficie del metal base.

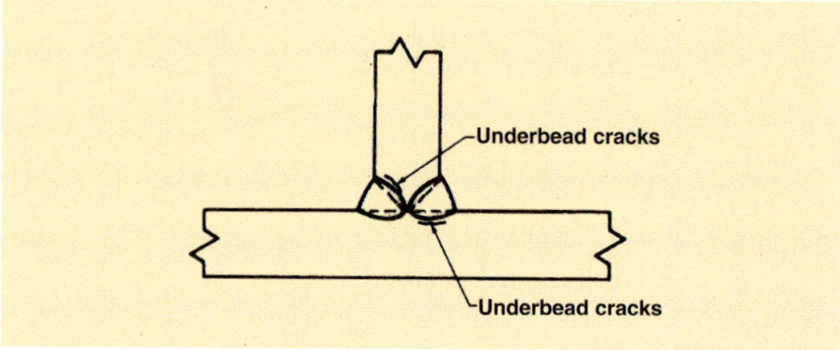

undercut. A groove melted into the base metal adjacent to the toe or root of a weld and left unfilled by weld metal. Refer to drawing for **overlap.**

socavación. Una ranura dentro del metal base adyacente al pie o raíz de la soldadura y se deja sin rellenar con el metal de soldadura. Refiérase al dibujo para **traslapo.**

underfill. A depression on the face of the weld or root surface extending below the surface of the adjacent base metal.

valle. Una depresión en la cara de la soldadura o la superficie de la raíz extendiéndose más abajo de la superficie del adyacente metal base.

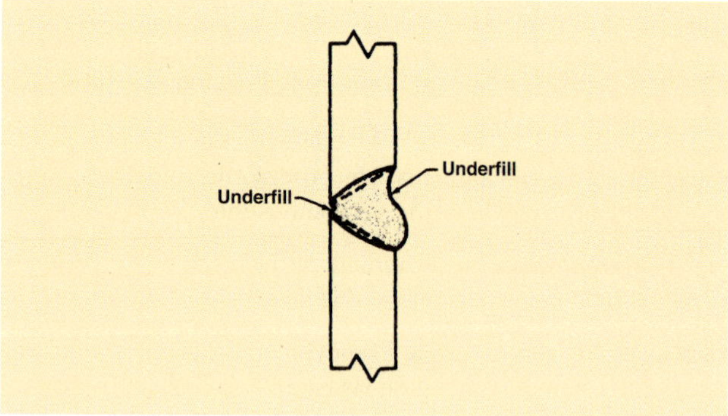

uphill. Welding with an upward progression.

soldando hacia arriba. Solando con progresión hacia arriba.

***upper arm (robot).** The portion of a jointed arm that is connected to the shoulder.

***brazo de arriba (robot).** La porción que úne al brazo que conecta con el hombro.

upset. Bulk deformation resulting from the application of pressure in welding. The upset may be measured as a percent increase in interfacial for area, a reduction in length, or a percent reduction in thickness (for lap joints).

recalada. Deformación en bulto resultado de la aplicación de presión en la soldadura. El recalado puede ser medido como un aumento en porciento en una área interfacial, una reducción en lo largo, o un porciento de reducción en lo grueso (para juntas en solape).

upset welding (UW). A resistance welding that produces coalescence over the entire area of faying surfaces or progressively along a butt joint by the heat obtained from the resistance to the flow of welding current through the area where those surfaces are in contact. Pressure is used to complete the weld.

soldadura recalada. Un proceso de soldadura por resistencia que produce coalescencia sobre toda la área de superficie empalmada o progresivamente sobre una junta a tope con el calor obtenido de la resistencia del flujo de la corriente de la soldadura por la área donde esas superficies están en contacto. Presión es usada para completar la soldadura.

upslope time (automatic arc welding). The time during which the current changes continuously from initial current valve to the welding value.

tiempo del pendiente en ascenso (soldadura automática de arco). El tiempo durante el cual la corriente cambia continuamente el valor de la corriente inicial al valor de la soldadura.

V

***vector lines.** Lines the computer draws between two points. These lines always stay crisp and sharp even when the drawing is zoomed in (magnified) hundreds of times.

***líneas de vector.** Líneas que traza la computadora entre dos puntos. Estas líneas siempre se mantienen nítidas y bien definidas aún cuando el dibujo se agrande (aumente de tamaño) cientos de veces.

vertical position. The position of welding in which the axis of the weld is approximately vertical.

posición vertical. La posición de la soldadura en la cual el eje para soldarse es aproximadamente vertical.

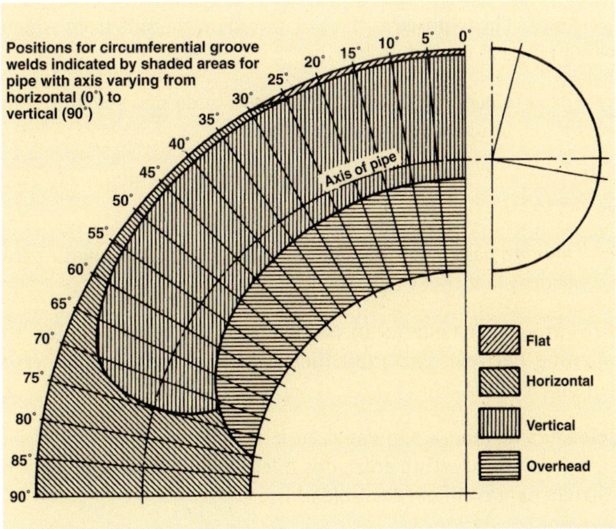

vertical position (pipe welding). The position of a pipe joint in which welding is performed in the horizontal position and the pipe may or may not be rotated.

posición vertical (soldadura de tubo). La posición de una junta de tubo en la cual la soldadura se hace en la posición horizontal y el tubo puede o no dar vueltas.

***V-groove.** See **joint.**

***ranura en V.** Consulte **junta.**

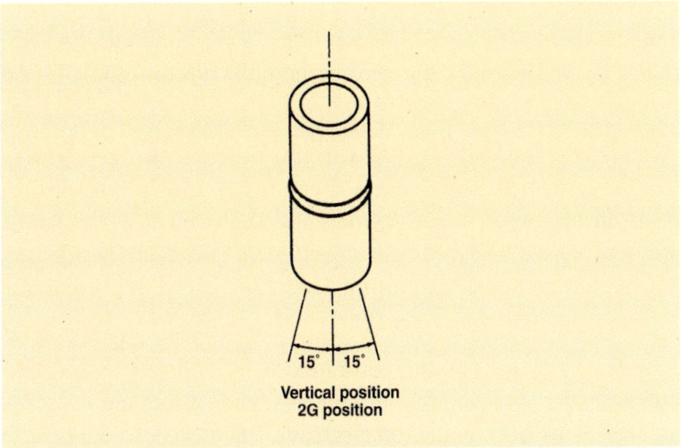

V-groove weld. A type of groove weld.

soldadura de ranura V. Un tipo de soldadura de ranura.

***visible light.** The light range that we see.

***luz visible.** El rango de luz que podemos ver.

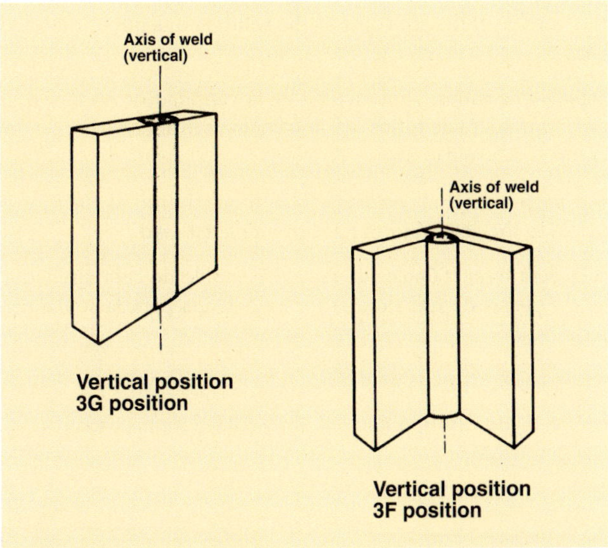

voltage range. The lower and upper limits of welding power, in volts, that can be produced by a welding machine or used with an electrode or by a process.

rango de voltaje. Los límites máximos y mínimos de poder de soldadura (en voltios) que puede tener una máquina para soldar o que pueden usarse con un electrodo o a través de un proceso.

voltage regulator. An automatic electrical control device for maintaining a constant voltage supply to the primary of a welding transformer.

regulador de voltaje. Un aparato de control eléctrico automático para mantener y proporcionar un voltaje constante a la primaria de un transformador de una soldadura.

W

***wagon tracks.** A pattern of trapped slag inclusions in the weld that show up as discontinuities in X-rays of the weld.

***huellas de carreta.** Una muestra de inclusiones de escoria atrapadas en la soldadura que enseña que hay discontinuidades en los rayos-x de la soldadura.

***water-arc plasma cutting (PAC).** In this process, nitrogen is used as the plasma gas. The plasma is created at a high temperature in an arc between the electrode and the orifice. A tap water spray applied to the plasma causes it to constrict and accelerate. This column causes the cutting action and melts a very narrow kerf in the material.

***cortes de arco-agua plasma.** En este proceso, nitrogeno es usado como el gas de plasma. La plasma es creada a una temperatura alta en un arco en medio del electrodo y la orifice. La aplicación de agua del grifo a la plasma la hace que se constriñe y acelera. Está columna causa la acción de cortar y derritir un corte que es muy estrecho en el material.

***water jet cutting.** A process using extremely high-pressure water to cut through material.

***corte por chorro de agua.** Un proceso en el que se utiliza agua a una presión extremadamente alta para cortar un material.

***wattage.** A measurement of the amount of power in the arc; the wattage of the arc controls the width and depth of the weld bead.

***número de vatios.** Una medida de la cantidad de poder en el arco; el número de vatios del arco controla lo ancho y hondo del cordón de la soldadura.

wax pattern (thermit welding). Wax molded around the workpieces to the desired form for the completed weld.

soldadura termita (molde de cera). Un molde de cera alrededor de las piezas de trabajo a la forma deseada para la soldadura terminada.

weave bead. A type of weld bead made with transverse oscillation.

cordón tejido. Un tipo de cordón de soldadura hecha con oscilación transversa.

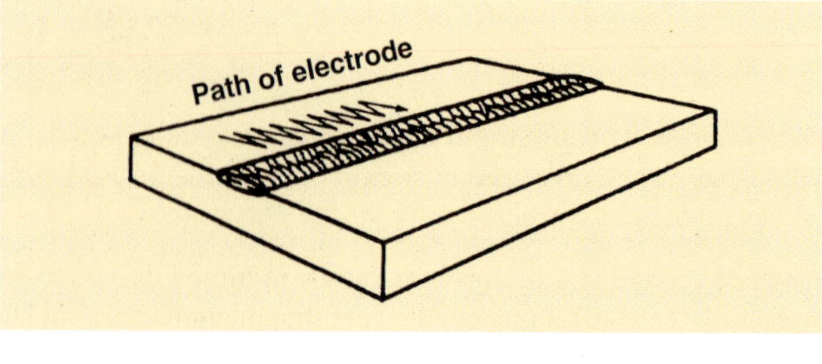

***weave pattern.** The movement of the welding electrode as the weld progresses; common weave patterns include circular, square, zigzag, stepped, C, J, T, and figure 8.

***muestra de tejido.** El movimiento del electrodo para soldar a como progresa la soldadura; las muestras de tejidos comunes incluyen circular, de cuadro, zigzag, de pasos, C, J, T, y la figura 8.

weld. A localized coalescence of metals or nonmetals produced either by heating the materials to suitable temperatures, with or without the application of pressure, or by the application of pressure alone and with or without the use of the filler material.

soldar. Una coalescencia localizada de metales o metaloides producida al calentar los materiales a una temperatura adecuada, con o sin la aplicación de presión, o por la aplicación de presión solamente y con o sin el uso del material de relleno.

weld axis. A line through the length of the weld, perpendicular to and at the geometric center of its cross section.

eje de la soldadura. Una línea a través de lo largo de la soldadura, perpendicular a y al centro geométrico de su sección transversa.

weld bead. A weld resulting from a pass. See also **stringer bead** and **weave bead.**

cordón de soldadura. Una soldadura que resulta de una pasada. Vea también **cordón encordador** y **cordón tejido.**

weld brazing. A joining method that combines resistance welding with brazing.

soldadura y soldadura fuerte. Un método de unir que combina soldadura de resistencia con soldadura fuerte.

***weld buildup.** See **buildup.**

***acumulación de soldadura.** Consulte **acumulación.**

weld crack. A crack in weld metal.

grieta en la soldadura. Una grieta en el metal de soldadura.

***weld distortion.** The deformation of the base metal as the result of welding.

***distorsión de soldadura.** La deformación del metal base como consecuencia de la soldadura.

weld face. The exposed surface of a weld on the side from which welding was done.

cara de la soldadura. La superficie expuesta de una soldadura en el lado de donde se hizo la soldadura.

weld gauge. A device designed for checking the shape and size of welds.

instrumento para medir la soldadura. Un aparato diseñado para comprobar la forma y tamaño de las soldaduras.

weld groove. A channel in the surface of a workpiece or an opening between two joint members that provides space to contain a weld.

soldadura de ranura. Un canal en la superficie de una pieza de trabajo o una abertura entre dos miembros de junta que provee espacio para contener una soldadura.

weld interface. The interface between weld metal and base metal in a fusion weld, between base metals in a solid state weld without filler metal, or between filler metal and base metal in a solid state weld with filler metal. Refer to drawing for **depth of fusion.**

interfaze de la soldadura. La interfase entre el metal de soldadura y el metal base en una soldadura de fusión, entre metales de base en una soldadura de estado sólido sin metal para rellenar, o entre metal de rellenar y metal base en una soldadura de estado sólido con metal para rellenar. Refiérase al dibujo **grueso de fusión.**

weld length. See **effective length of weld.**

largura de la soldadura. Vea **distancia efectiva de soldadura.**

weld metal. The portion of a fusion weld that has been completely melted during welding.

metal de soldadura. La porción de una soldadura de fusión que se ha derretido completamente durante la soldadura.

weld metal area. The area of the weld metal as measured on the cross section of a weld. Refer to drawing for **heat-affected zone.**

área de metal de soldadura. La área del metal de la soldadura la cual fue medida en la sección transversa de la soldadura. Refiérase al dibujo para **zona afectada por el calor.**

weld pass. A single progression of welding along a joint. The result of a pass is a weld bead or layer.

pasada de soldadura. Una progresión singular de la soldadura a lo largo de una junta. El resultado de una pasada es un cordón o una capa.

weld pass sequence. The order in which the weld passes are made. See also **longitudinal sequence** and **cross-sectional sequence.**

secuencia de pasadas de soldadura. La orden en que las pasadas de soldadura se hacen. Vea **también secuencia longitudinal** y **secuencia del corte transversal.**

weld penetration. A nonstandard term for joint penetration and root penetration.

penetracion de soldadura. Un término fuera de norma para penetración de junta y penetración de raíz.

weld pool. The localized volume of molten metal in a weld prior to its solidification as weld metal.

charco de soldadura. El volumen localizado del metal derretido en una soldadura antes de su solidificación como metal de soldadura.

weld puddle. A nonstandard term for weld pool.

charco de soldadura. Un término fuera de norma para charco de soldadura.

weld reinforcement. Weld metal in excess of the quantity required to fill a joint. See also **face reinforcement** and **root reinforcement.**

refuerzo de soldadura. Metal de soldar en exceso de la cantidad requerida para llenar una junta. Vea también **refuerzo de cara** y **refuerzo de raíz.**

weld root. The points, shown in a cross section, at which the root surface intersects the base metal surfaces.

raíz de soldadura. Los puntos, enseñados en una sección transversa, la cual la superficie de la raíz se interseca con las superficies del metal base.

weld size. See preferred term **size of weld.**

tamaño de soldadura. Vea el término preferido **tamaño de la soldadura.**

***weld spatter.** A combination of weld metal and flux material that is distributed from the end of an electrode and does not enter the molten weld.

***salpicadura de soldeo.** Una combinación de metal de soldeo y material fundente que se distribuye desde el extremo de un electrodo y no entra en el soldeo fundido.

***weld specimen.** A sample removed from a welded plate according to AWS specifications, which detail the preparation of the plate, the cutting of the plate, and the size of the specimen to be tested.

***probeta de soldadura.** Una prueba apartada del plato soldado de acuerdo con las especificaciones del AWS, las cuales detallan la preparación del plato, el corte del plato, y el tamaño de la probeta que se va a probar.

weld symbol. A graphical character connected to the welding symbol indicating the type of weld.

símbolo de soldadura. Un signo gráfico conectado al símbolo de soldadura indicando el tipo de soldadura.

weld tab. Additional material that extends beyond either end of the joint, on which the weld is started or terminated.

solera para soldar. Material adicional que se extiende más allá de cualquier punto de la junta en la cual la soldadura es empezada o terminada.

weld test. A welding performance test to a specific code or standard.

prueba de soldadura. Una prueba de ejecución de soldadura según una norma o código específico.

weld time (automatic arc welding). The time interval from the end of start time or end of upslope to beginning of crater fill time or beginning of downslope. Refer to the drawings for **upslope time.**

tiempo de soldadura (soldadura de arco automática). El intervalo de tiempo del fin del tiempo de arranque o el fin del pendiente en ascenso al principio del tiempo de llenar el crater o el principio del tiempo del pendiente en descenso. Refiérase al dibujo para **tiempo del pendiente en ascenso.**

weld timer. A device that controls only the weld time in resistance welding.

contador de tiempo para soldadura. Un aparato que controla solamente el tiempo de soldar en soldaduras de resistencia.

weld toe. The junction of the weld face and the base metal.

pie de la soldadura. La unión de la cara de la soldadura y el metal base.

weldability. The capacity of a material to be welded under the fabrication conditions imposed into a specific, suitably designed structure and to perform satisfactorily in the intended service.

soldabilidad. La capacidad de un material para soldarse bajo las condiciones de fabricación impuestas en un específico, en un diseño de estructura adecuada y para ejecutar satisfactoriamente los servicios intentados.

welder. One who performs manual or semiautomatic welding.

soldador. Uno que ejecuta soldadura manual o semiautomática.

welder certification. Written verification that a welder has produced welds meeting a prescribed standard of welder performance.

certificación del soldador. Verificación escrita de que un soldador ha producido soldaduras que cumplen con la norma prescrita de la ejecución del soldador.

welder performance qualification. The demonstration of a welder's ability to produce welds meeting prescribed standards.

calificación de ejecución del soldador. La demostración de la habilidad del soldador de producir soldaduras que cumplen con las normas prescritas.

welder registration. The act of registering a welder certification or a photostatic copy thereof.

registración del soldador. El acto de registrar una certificación del soldador o una copia fotostata de ello.

welding. A joining process that produces coalescence of materials by heating them to the welding temperature, with or without the application of pressure or by the application of pressure alone, and with or without the use of filler metal.

soldadura. Un proceso de unión que produce coalescencia de materiales calentándolos a la temperatura de soldadura, con o sin la aplicación de presión o por la aplicación de presión solamente, y con o sin el uso del metal de relleno.

welding arc. A controlled electrical discharge between the electrode and the workpiece that is formed and sustained by the establishment of a gaseous conductive medium, called an arc plasma.

arco de soldadura. Una descarga eléctrica controlada entre el electrodo y la pieza de trabajo que es formada y sostenida por el establecimiento de un medio conductivo gaseoso, llamado un arco de plasma.

welding cables. The work cable and electrode cable of an arc welding circuit. Refer to drawing for direct-current electrode positive.

cables para soldar. Los cables de pieza de trabajo y el portelectrodo de un circuito de soldadura de arco. Refierase al dibujo orriente directa con el electrodo positivo.

welding current. The current in the welding circuit during the making of a weld.

corriente para soldadura. La corriente en el circuito de soldar durante la hechura de una soldadura.

welding current (automatic arc welding). The current in the welding circuit during the making of a weld, but excluding upslope, downslope, start, and crater fill current. Refer to drawing for **upslope time.**

corriente de soldadura (soldadura de arco automático). La corriente en el circuito de soldar durante la hechura de una soldadura, pero excluyendo el pendiente en ascenso, pendiente en descenso, empiezo, y corriente par llenar el crater. Refiérase al dibujo para **tiempo del pendiente en ascenso.**

welding cycle. The complete series of events involved in the making of a weld. Refer to drawings for **downslope time** and **upslope time.**

ciclo de soldadura. Una serie completa de eventos envueltos en hacer una soldadura. Refiérase al dibujo para **tiempo de cáida del pendiente** y **tiempo del pendiente en descenso.**

welding electrode. A component of the welding circuit through which current is conducted and that terminates at the arc, molten conductive slag, or base metal. See also **arc welding electrode, bare electrode, carbon electrode, composite electrode, covered electrode, electroslag welding electrode, emissive electrode, flux cored electrode, lightly coated electrode, metal cored electrode, metal electrode, resistance welding electrode, stranded electrode,** and **tungsten electrode.**

soldadura con electrodo. Un componente del circuito de soldar por donde la corriente es conducida y que termina en el arco, en la escoria derretida conductiva, o en el metal base. Vea también **electro para soldar de arco, electrodo descubierto, electrodo de carbón, electrodo compuesto, electrodo cubierto, electrodo para soldadura de electroescoria, electrodo emisivo, electrodo de núcleo de fundente, electrodo con recubrimiento ligero, electrodo de metal de núcleo, electrodo de metal, electrodo para soldadura por resistencia, electrodo cable,** y **electrodo de tungsteno.**

welding filler metal. The metal or alloy to be added in making a weld joint that alloys with the base metal to form weld metal in a fusion welded joint.

metal de soldadura para rellenar. El metal o aleación que se va a agregar en la hechura de una junta de soldadura que se mezcla con el metal base para formar metal de soldadura en una junta de fusión de soldadura.

welding generator. A generator used for supplying current for welding.

generador para soldar. Un generador que se usa para proporcionar la corriente para la soldadura.

welding ground. A nonstandard and incorrect term for workpiece connection.

tierra de soldadura. Un término fuera de norma e incorrecto para conexión de pieza de trabajo.

welding head. The part of a welding machine in which a welding gun or torch is incorporated.

cabeza de soldar. La parte de una máquina para soldar la cual una pistola de soldadura o una antorcha se puede incorporar.

welding leads. The work lead and electrode lead of an arc welding circuit. Refer to drawing for **direct-current electrode positive.**

cables para soldar. Los cables de pieza de trabajo y el portelectrodo de un circuito de soldadura de arco. Refiérase al dibujo **corriente directa con el electrodo positivo.**

welding machine. Equipment used to perform the welding operation. For example, spot welding machine, arc welding machine, seam welding machine, etc.

máquina para soldar. El equipo que se usa para ejecutar la operación de soldadura. Por ejemplo, máquina de soldadura por puntos, máquina de soldadura de arco, máquina de soldadura de costura, etc.

welding operator. One who operates adaptive control, automatic, mechanized, or robotic welding equipment.

operador de soldadura. Uno que opera control adaptivo, automático, mecanizado, o equipo robótico para soldar.

welding position. See flat position, horizontal position, horizontal fixed position, horizontal rolled position, overhead position, and vertical position.

posición de soldadura. Vea posición plana, posición horizontal, posición fija horizontal, posición horizontal rodada, posición de sobrecabeza, y posición vertical.

POSITION OF WELDING

Flat. See flat position.
Horizontal. See horizontal position, horizontal fixed position, and horizontal rolled position.
Vertical. See vertical position.
Overhead. See overhead position.

POSITION FOR QUALIFICATION

Plate welds

Groove welds

1G. See flat position.
2G. See horizontal position.
3G. See vertical position.
4G. See overhead position

Fillet welds

1F. See flat position.
2F. See horizontal position.
3F. See vertical position.
4F. See overhead position.

Pipe welds

Groove welds

1G. See horizontal rolled position.
2G. See vertical position.
3G. See horizontal fixed position.
5G. Inclined position.
5GR. Inclined position.

Position

welding power source. An apparatus for supplying current and voltage suitable for welding. See also **welding generator, welding rectifier,** and **welding transformer.**

fuente de poder para soldar. Un aparato para surtir corriente y voltaje adecuado para soldar. Vea también **generador para soldar, rectificador para soldar,** y **transformador para soldar.**

welding procedure qualification record (WPQR). A record of welding variables used to produce an acceptable test weldment and the results of tests conducted on the weldment to qualify a welding procedure specification.

registro de calificación de procedimiento de la soldadura. Un registro de los variables usados para producir una probeta aceptable y los resultados de la prueba conducida en la probeta para calificar el procedimiento de especificación.

welding procedure specification (WPS). A document providing in detail the required variables for specific application to ensure repeatability by properly trained welders and welding operators.

calificación de procedimiento de soldadura. Un documento que provee en detalle los variables requeridos para la aplicación específica para asegurar la habilidad de repetir el procedimiento por soldadores y operadores que estén propiamente preparados.

welding process. A materials joining process that produces coalescence of materials by heating them to suitable temperatures, with or without the application of pressure or by the application of pressure alone, and with or without the use of filler metal.

proceso para soldar. Un proceso para unir materiales que produce coalescencia calentándolos a una temperatura adecuada con o sin la aplicación de presión solamente y con o sin usarse material para rellenar.

welding rectifier. A device in a welding machine for converting alternating current to direct current.

rectificador para soldar. Un aparato en una máquina para soldar para convertir la corriente alterna a corriente directa.

welding rod. A form of welding filler metal, normally packaged in straight lengths, that does not conduct the welding current.

varilla para soldar. Una forma de metal de soldadura para rellenar, normalmente empaquetada en piezas derechas, que no conduce la corriente para soldar.

welding sequence. The order of making the welds in a weldment.

orden de sucesión (para soldar). La orden de hacer las pasadas de soldar en una soldadura.

***welding speed.** The rate that the welding process moves along the weld joint, usually given in inches per minute (ipm).

***velocidad de soldadura.** La tasa a la cual avanza el proceso de soldadura a lo largo de la junta de soldeo, por lo general expresada en pulgadas por minuto (ipm, por sus siglas en inglés).

welding symbol. A graphical representation of a weld.

símbolo de soldadura. Una representación gráfica de una soldadura.

welding technique. The details of a welding procedure that are controlled by the welder or welding operator.

ejecución de soldadura. Los detalles del procedimiento que son controlados por el soldador u operador de soldadura.

welding tip. A welding torch tip designed for welding.

boquilla (punta) para soldar. Una boquilla en la antorcha de soldadura que está diseñada para soldar.

welding torch (arc). A device used in the gas tungsten and plasma arc welding processes to control the position of the electrode, to transfer current to the arc, and to direct the flow of shielding and plasma gas.

antorcha para soldar (arco). Un aparato usado en los procesos de soldadura del gas tungsteno y arco plasma para controlar la posición del electrodo, para transferir corriente al arco, y para dirigir la corriente del gas protector y gas de la plasma.

welding torch (oxyfuel gas). A device used in oxyfuel gas welding, torch brazing, and torch soldering for directing the heating flame produced by the controlled combustion of fuel gases.

antorcha para soldar (gas oxicombustible). Un aparato usado en soldadura de gas oxicombustible, soldadura blanda con antorcha y soldadura fuerte con antorcha y para dirigir la llama para calentar producida por la combustión controlada de gases de combustión.

welding transformer. A transformer used to supply current for welding. See also **reactor (arc welding).**

transformador para soldar. Un transformador que se usa para dar corriente para la soldadura. Vea también **reactor (soldadura de arco).**

welding voltage. See **arc voltage**.

voltaje para soldar. Vea **voltaje del arco**.

welding wire. A form of welding filler metal, normally packaged as coils or spools, that may or may not conduct electrical current, depending upon the filler metal and base metal in a solid state weld with filler metal.

alambre para soldar. Una forma de metal para rellenar con soldadura, normalmente empaquetado en rollos o en carretes que pueda o pueda que no conducir corriente eléctrica, dependiendo en el metal de relleno y el metal base en una soldadura que está en estado sólido con metal de relleno.

weldment. An assembly whose component parts are joined by welding.

conjunto de partes soldadas. Una asamblea cuyas partes componentes están unidas por la soldadura.

weldor. See preferred term **welder.**

soldador. Vea el término preferido **soldador.**

wetting. The phenomenon whereby a liquid filler metal or flux spreads and adheres in a thin, continuous layer on a solid base metal.

exudación. El fenómeno de que un metal para rellenar líquido o un flujo se puede desparramar y adherirse en una capa delgada, capa continua en un sólido metal base.

wire feed speed. The rate at which wire is consumed in arc cutting, thermal spraying, or welding.

velocidad de alimentador de alambre. La velocidad que el alambre es consumido en cortes de arco, rociado termal o soldadura.

work angle. The angle less than 90° between a line perpendicular to the major workpiece surface and a plane determined by the electrode axis and the weld axis. In a tee-joint or a corner joint, the line is perpendicular to the nonbutting member. This angle can also be used to partially define the position of guns, torches, rods, and beams.

ángulo de trabajo. El ángulo menos de 90° entre una línea perpendicular a la superficie de pieza de trabajo mayor y una plana determinada por el eje del electrodo y el eje de la soldadura. En una junta-T o en una junta de esquina, la línea es perpendicular a un miembro que no topa. Este ángulo puede ser usado también para parcialmente definir la posición de pistolas, antorchas, varillas y rayos.

work angle (pipe). The angle less than 90° between a line, which is perpendicular to the cylindrical pipe surface at the point of intersection of the weld axis and the extension of the electrode axis, and a plane determined by the electrode axis and a line tangent to the pipe at the same point. In a tee-joint, the line is perpendicular to the nonbutting member. This angle can also be used to partially define the position of guns, torches, rods, and beams.

ángulo de trabajo (tubo). El ángulo menos de 90° entre una línea, la cual es perpendicular a la superficie de un tubo cilíndrico al punto de intersección del eje de la soldadura y la extensión del eje del electrodo, y un plano determinado por el eje del electrodo y una línea tangente al tubo al mismo punto. En una junta-T, la línea es perpendicular a un miembro que no topa. Este ángulo puede también usarse para definir parcialmente la posición de pistolas, antorchas, varillas, y rayos.

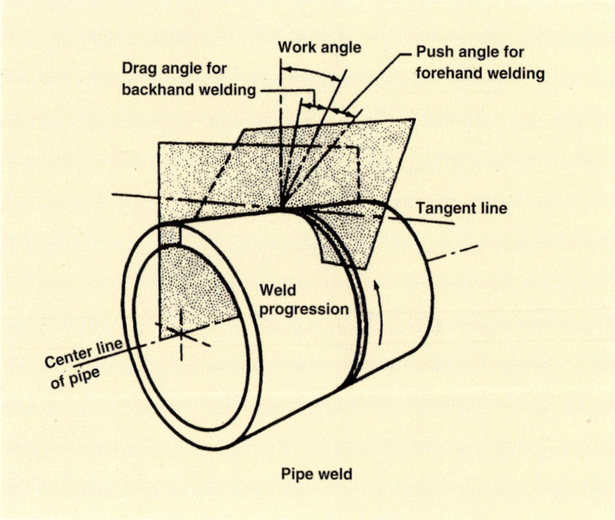

work connection. The connection of the work lead to the work. Refer to drawing for **direct-current electrode negative.**

pinza de tierra. La conexión del cable de trabajo (tierra) al trabajo. Refiérase al dibujo para **corriente directa con electrodo negativo.**

work lead. The electric conductor between the source of arc welding current and the work. Refer to drawing for **direct-current electrode negative.**

cable de tierra. Un conductor eléctrico entre la fuente de la corriente del arco y la pieza de trabajo. Refiérase al dibujo **corriente directa con electrodo negativo.**

working envelope. The set of points representing the maximum extent or reach of the robot hand or working tool in all directions.

alcance de operación. Un juego de puntos que representan la máxima extensión o alcance de la mano del robot o la herramienta del trabajo en todas las direcciones.

***working pressure.** The pressure at the low-pressure gauge, ranging from 0 to 45 psi (depending on the type of gas), used for welding and cutting.

***presión de trabajo.** La presión en el manómetro de baja presión, con escala de 0 a 45 psi (dependiendo en el tipo de gas), usado para cortar y soldar.

working range. All positions within the working envelope. The range of any variable within which the system normally operates.

extensión de trabajo. Todas las posiciones dentro del alcance del trabajo. El alcance de cualquier variable dentro del sistema que opera normalmente.

workpiece. The part that is welded, brazed, soldered, thermal cut, or thermal sprayed.

pieza de trabajo. La parte que está soldada, con soldadura fuerte, soldadura blanda, corte termal, o rociado termal.

workpiece connection. The connection of the workpiece lead to the workpiece.

conexion de pieza de trabajo (pinzas). La conexión del cable de la pieza de trabajo a la pieza de trabajo.

workpiece lead. The electrical conductor between the arc welding current source and workpiece connection.

cable de pieza de trabajo. El conductor eléctrico entre la fuente de corriente de soldadura de arco y la conexión de la pieza de trabajo.

workstation. A manufacturing unit consisting of one or more numerically controlled machine tools serviced by a robot.

estación de trabajo. Una unidad manufacturera de una o más herramienta numerada que es controlada por una máquina y abatecida por un robot.

wrist. A set of rotary joints between the arm and hand that allows the hand to be oriented to the workpiece.

muñeca. Un juego de coyunturas rotatorias entre el brazo y la mano que permite a la mano ser orientada a la pieza de trabajo.

Y

yaw. The angular displacement of a moving body about an axis that is perpendicular to the line of motion and to the top side of the body.

guiñada. El desalojamiento de un cuerpo en movimiento alrededor de un eje que está perpendicular a la línea de movimiento y al lado más alto del cuerpo.

Appendix

Appendix I

TIME SHEET					
Name:	Project Number:				
Instructor:	Project Description:		Class:		
Item	Work Performed	Date:	Starting Time	Ending Time	Hours
1					
2					
3					
4					
5					
6					
7					
8					
9					
10					
11					
12					
13					
14					
15					
16					
17					
18					
19					
20					
			Total Time on Project		

Appendix II

PARTS LIST		
Name:	Project Number:	Date:
Instructor:	Project Description:	Class:

Item	Quantity	Material Description
1		
2		
3		
4		
5		
6		
7		
8		
9		
10		
11		
12		
13		
14		
15		
16		
17		
18		
19		
20		

Appendix III

BILL OF MATERIALS				
Name:		Project Number:	Date:	
Instructor:		Project Description:	Class:	
Item	Quantity	Material Description	Unit Price	Item Cost
1				
2				
3				
4				
5				
6				
7				
8				
9				
10				
11				
12				
13				
14				
15				
16				
17				
18				
19				
20				

Get the material prices from your instructor, out of a catalog, off of the Internet, or from a local supplier.	Total Material Cost	
Calculate the labor cost by multiplying your hours from your Time Sheet times the rate of pay for welders in your area. ___________Hrs X _________Pay Rate = _________Labor Cost	Total Labor Cost	
Add up the Total Material Cost and Total Labor Cost.	Total Expenses	
To calculate Total Project Charges include a markup to cover shop expenses such as utilities, building mortgage/ rent, insurance, profit, etc. $________Total Expenses X _1._____% Markup = $________Total Project Charge*	Total Project Charge	

* Example: $25.00 Total Expense X 1.20% Markup = $30.00 Total Project Cost. The 1.20% of expenses is equal to a 20% markup. However, most shops do not show the markup as a separate item on their invoice because it is added to the Material Cost and Labor Expense.

Appendix IV

PERFORMANCE QUALIFICATION TEST RECORD

Identification No. ______________________ Weld Stamp No. ______________________

Name ______________________ Instructor ______________________

Eye Correction Required Yes ☐ No ☐ Type of Eye Correction Eye Glasses ☐
Contact Lenses ☐
Magnifiers ☐

Qualified with WPS No. ____________ Process ____________ Manual ☐ Semiautomatic ☐

Position: Flat (1G, 1F ☐), Horizontal (2G, 2F ☐), Vertical (3G, 3F ☐), Overhead (4G, 4F ☐)

Test Base Metal Specification ______________________ Thickness ______________________

AWS Filler Metal Classification ______________________ Size ______________________

Shielding Gas ______________________ Flow Rate ______________________

GUIDED-BEND TEST RESULTS

Visual Test Results Pass ☐ Fail ☐ Weld Size ______________________

Type of Bend			Result	
Face ☐	Root ☐	Side ☐	Pass ☐	Fail ☐
Face ☐	Root ☐	Side ☐	Pass ☐	Fail ☐

The above-named person is qualified for the welding process used in this test within the limits of essential variables including materials and filler metal as provided by the AWS D1.1 ________ Code. We, the undersigned, certify that the statements in this record are correct and that the test welds were prepared, welded, and tested in accordance with the requirements of the AWS D1.1 ________ Code.

Date Tested ______________ Signed by ______________________
Test Supervisor

Signed by ______________________ ______________________
Corporate Representative Title

Appendix V

Table of Conversions: U.S. Customary (Standard) Units and Metric Units (SI).

TEMPERATURE

Units

°F (each 1° change)	=	0.555°C (change)
°C (each 1° change)	=	1.8°F (change)
32°F (ice freezing)	=	0°Celsius
212°F (boiling water)	=	100°Celsius
–460°F (absolute zero)	=	0°Rankine
–273°C (absolute zero)	=	0°Kelvin

Conversions

°F to °C ______ °F – 32 = ______ × .555 = ______ °C
°C to °F ______ °C × 1.8 = ______ + 32 = ______ °F

LINEAR MEASUREMENT

Units

1 inch	=	25.4 millimeters
1 inch	=	2.54 centimeters
1 millimeter	=	0.0394 inch
1 centimeter	=	0.3937 inch
12 inches	=	1 foot
3 feet	=	1 yard
5280 feet	=	1 mile
10 millimeters	=	1 centimeter
10 centimeters	=	1 decimeter
10 decimeters	=	1 meter
1000 meters	=	1 kilometer

Conversions

in. to mm	______	in.	×	25.4	=	______	mm
in. to cm	______	in.	×	2.54	=	______	cm
ft to mm	______	ft	×	304.8	=	______	mm
ft to m	______	ft	×	0.3048	=	______	m
mm to in.	______	mm	×	0.0394	=	______	in.
cm to in.	______	cm	×	0.3937	=	______	in.
mm to ft	______	mm	×	0.00328	=	______	ft
m to ft	______	m	×	3.28	=	______	ft

AREA MEASUREMENT

Units

1 sq in.	=	0.0069 sq ft
1 sq ft	=	144 sq in.
1 sq ft	=	0.111 sq yd
1 sq yd	=	9 sq ft
1 sq in.	=	645.16 sq mm
1 sq mm	=	0.00155 sq in.
1 sq cm	=	100 sq mm
1 sq m	=	1000 sq cm

Conversions

sq in. to sq mm ______ sq in. × 645.16 = ______ sq mm
sq mm to sq in. ______ sq mm × 0.00155 = ______ sq in.

VOLUME MEASUREMENT

Units

1 cu in.	=	0.000578 cu ft
1 cu ft	=	1728 cu in.
1 cu ft	=	0.03704 cu yd
1 cu ft	=	28.32 L
1 cu ft	=	7.48 gal (U.S.)
1 gal (U.S.)	=	3.737 L
1 cu yd	=	27 cu ft
1 gal	=	0.1336 cu ft
1 cu in.	=	16.39 cu cm
1 L	=	1000 cu cm
1 L	=	61.02 cu in.
1 L	=	0.03531 cu ft
1 L	=	0.2642 gal (U.S.)
1 cu yd	=	0.769 cu m
1 cu m	=	1.3 cu yd

Conversions

cu in. to L	______	cu in.	×	0.01638	=	______	L
L to cu in.	______	L	×	61.02	=	______	cu in.
cu ft to L	______	cu ft	×	28.32	=	______	L
L to cu ft	______	L	×	0.03531	=	______	cu ft
L to gal	______	L	×	0.2642	=	______	gal
gal to L	______	gal	×	3.737	=	______	L

WEIGHT (MASS) MEASUREMENT

Units

1 oz	=	0.0625 lb
1 lb	=	16 oz
1 oz	=	28.35 g
1 g	=	0.03527 oz
1 lb	=	0.0005 ton
1 ton	=	2000 lb
1 oz	=	0.283 kg
1 lb	=	0.4535 kg
1 kg	=	35.27 oz
1 kg	=	2.205 lb
1 kg	=	1.000 g

Conversions

lb to kg	______	lb	×	0.4535	=	______	kg
kg to lb	______	kg	×	2.205	=	______	lb
oz to g	______	oz	×	0.03527	=	______	g
g to oz	______	g	×	28.35	=	______	oz

PRESSURE AND FORCE MEASUREMENTS

Units

1 psig	=	6.8948 kPa
1 kPa	=	0.145 psig
1 psig	=	0.000703 kg/sq mm
1 kg/sq mm	=	6894 psig
1 lb (force)	=	4.448 N
1 N (force)	=	0.2248 lb

Conversions

psig to kPa	______	psig	×	6.8948	=	______	kPa
kPa to psig	______	kPa	×	0.145	=	______	psig
lb to N	______	lb	×	4.448	=	______	N
N to lb	______	N	×	0.2248	=	______	psig

VELOCITY MEASUREMENTS

Units

1 in./sec	=	0.0833 ft/sec
1 ft/sec	=	12 in./sec
1 ft/min	=	720 in./sec
1 in./sec	=	0.4233 mm/sec
1 mm/sec	=	2.362 in./sec
1 cfm	=	0.4719 L/min
1 L/min	=	2.119 cfm

Conversions

ft/min to in./sec ____ ft/min × 720 = ____ in./sec
in./min to mm/sec ____ in./min × .4233 = ____ mm/sec
mm/sec. to in./min ____ mm/sec × 2.362 = ____ in./min
cfm to L/min ____ cfm × 0.4719 = ____ L/min
L/min to cfm ____ L/min × 2.119 = ____ cfm

Mild Steel

Appendix VI

FLAT BAR
BAR
HOT ROLL BAR
STRAPPING
FLATS

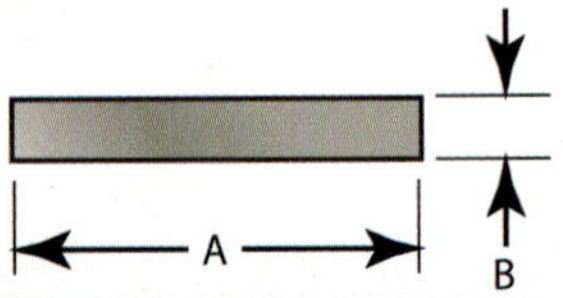

Thickness 'A' (inch)	Width 'B' (inch)	Weight per Foot (pound)*
1/8	1	0.425
	2	0.85
	3	1.275
1/4	1/2	0.425
	3/4	0.638
	1	0.85
	1 1/2	1.275
	2	1.7
	2 1/2	2.125
	3	2.55
	3 1/2	2.975
	4	3.4
	4 1/2	3.83
	5	4.25
5/16	5	5.313
	6	6.375
3/8	3/4	0.958
	1	1.275
	1 1/2	1.913
	2	2.55
	2 1/2	3.188
	3	3.83
	3 1/2	4.46
	4	5.1
	4 1/2	5.74
	5	6.375
1/2	1	1.7
	1 1/2	2.55
	2	3.4
	2 1/2	4.25
	3	5.1
	3 1/2	5.95
	4	6.8
	4 1/2	7.65
	5	8.501
5/8	2	4.25
	2 1/2	5.31
	3	6.38
	4	8.5
	5	10.63
3/4	2	5.1
	2 1/2	6.4
	3	7.65
	4	10.2
	5	12.75
1	2	6.8
	3	10.2
	3 1/2	11.9
	4	13.6
	5	17.02
1 1/4	1	4.25
	2	8.5
	3	12.75
	4	17

*Approximate Weight

Appendix VII

PLATE
MILL PLATE
HOT ROLLED PLATE

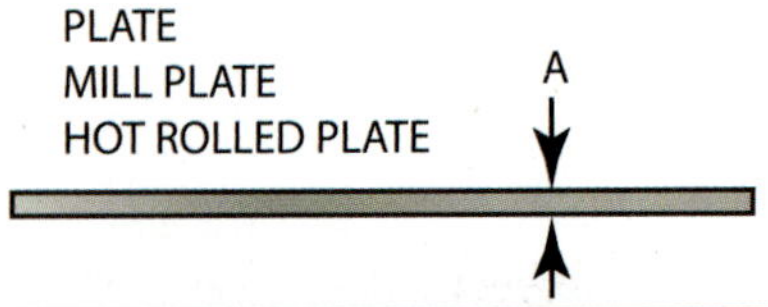

Thickness 'A' (inch)	Weight per Square Foot (lb)*
1/8	5.11
3/16	7.68
1/4	10.21
5/16	12.76
3/8	15.32
1/2	20.42
5/8	25.52
3/4	30.63
1	40.84

*Approximate Weight

Appendix VIII

PIPE

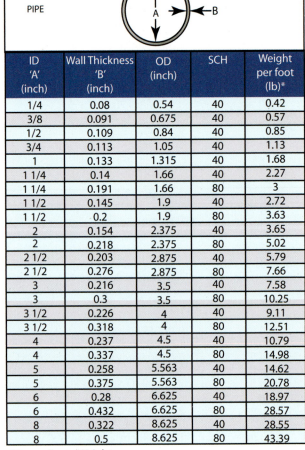

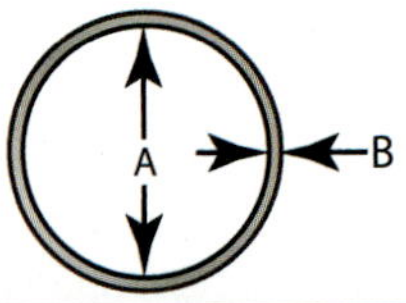

ID 'A' (inch)	Wall Thickness 'B' (inch)	OD (inch)	SCH	Weight per foot (lb)*
1/4	0.08	0.54	40	0.42
3/8	0.091	0.675	40	0.57
1/2	0.109	0.84	40	0.85
3/4	0.113	1.05	40	1.13
1	0.133	1.315	40	1.68
1 1/4	0.14	1.66	40	2.27
1 1/4	0.191	1.66	80	3
1 1/2	0.145	1.9	40	2.72
1 1/2	0.2	1.9	80	3.63
2	0.154	2.375	40	3.65
2	0.218	2.375	80	5.02
2 1/2	0.203	2.875	40	5.79
2 1/2	0.276	2.875	80	7.66
3	0.216	3.5	40	7.58
3	0.3	3.5	80	10.25
3 1/2	0.226	4	40	9.11
3 1/2	0.318	4	80	12.51
4	0.237	4.5	40	10.79
4	0.337	4.5	80	14.98
5	0.258	5.563	40	14.62
5	0.375	5.563	80	20.78
6	0.28	6.625	40	18.97
6	0.432	6.625	80	28.57
8	0.322	8.625	40	28.55
8	0.5	8.625	80	43.39

*Approximate Weight

Mild Steel (*Continued*)

Appendix IX

TUBING
ROUND TUBING

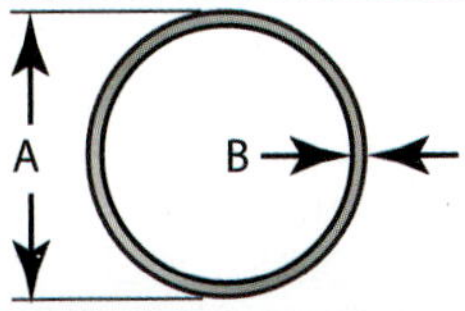

OD 'A' (inch)	Wall Thickness Gauge 'B'	Wall Thickness 'B' (inch)	Weight per Foot (lb)*
3/4	16	0.065	0.4755
3/4	14	0.083	0.59
1	16	0.065	0.6491
1	14	0.083	0.8129
1	12	0.109	1.037
1	11	0.112	1.128
1 1/4	14	0.083	1.034
1 1/4	12	0.109	1.328
1 1/4	11	0.12	1.44
1.315	14	0.083	1.09
1 1/2	14	0.083	1.26
1 1/2	12	0.109	1.619
1 1/2	11	0.125	1.769
1.66	14	0.083	1.398
1.66	12	0.112	1.76
1.66	10	0.135	2.184
1 3/4	16	0.065	1.17
1 3/4	11	0.12	2.09
1.9	14	0.083	1.611
1.9	10	0.135	2.72
2 3/8	16	0.065	1.59
2 3/8	14	0.083	2.032
2 3/8	12	0.112	2.64
2 7/8	12	0.112	3.22
2 7/8	11	0.125	3.53
2 7/8	10	0.135	3.94

*Approximate Weight

Appendix X

SQUARE TUBING
TUBING

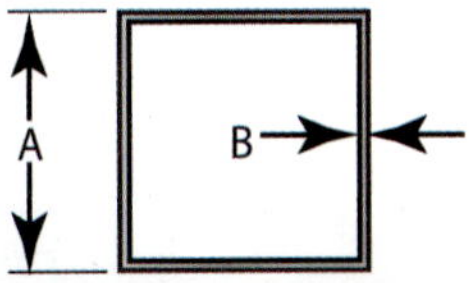

OD 'A' (inch)	Wall Thickness Gauge 'B'	Wall Thickness 'B' (inch)	Weight per Foot (lb)*
1/2	16	0.065	0.385
3/4	16	0.065	0.606
3/4	14	0.083	0.752
1	16	0.065	0.827
1	14	0.083	1.04
1	11	0.12	1.44
1 1/4	16	0.065	1.027
1 1/4	14	0.083	1.317
1 1/4	11	0.12	1.844
1 1/2	16	0.065	1.27
1 1/2	14	0.083	1.6
1 1/2	11	0.12	2.25
1 1/2	3/16	0.188	3.23
1 1/2	1/4	0.25	4.067

*Approximate Weight

Mild Steel (*Continued*)

Appendix XI

TUBING
RECTANGULAR TUBING

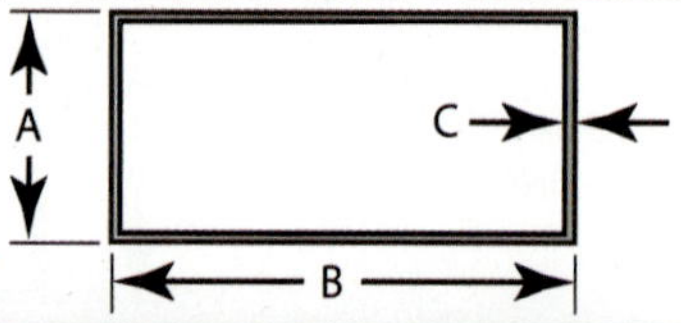

Side 'A' (inch)	Side 'B' (inch)	Wall Thickness Gauge 'C' (inch)	Wall Thickness 'C' (inch)	Weight per Foot (lb)*
1	1/2	16	0.065	0.57
1 1/2	1	11	0.12	1.84
1 1/2	1	14	0.083	1.32
1 1/2	1	16	0.065	1.027
2	1	16	0.065	1.248
2	1	14	0.083	1.6
2	1	11	0.12	2.25
2	1 1/2	11	0.12	2.66
2	1 1/2	14	0.083	2.16
2 1/2	1 1/2	14	0.083	1.84
3	1	12	0.109	2.73
3	1	14	0.083	2.45
3	1	16	0.065	1.66
3	1 1/2	11	0.12	3.63
3	1 1/2	3/16	0.188	5.07
3	1 1/2	14	0.083	2.73
3	2	1/4	0.25	7.11
3	2	3/16	0.188	5.59
3	2	11	0.12	3.88

*Approximate Weight

Appendix XII

SQUARE BAR
ROD
BAR STOCK
SQUARE STOCK

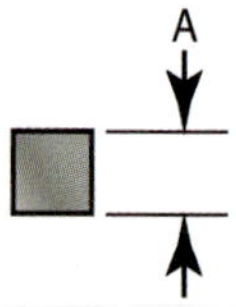

Dimension 'A' (inch)	Weight per Foot (pound)*
1/2	0.85
5/8	1.33
3/4	1.913
7/8	2.603
1	3.4

*Approximate Weight

Appendix XIII

ROUND BAR
ROD
BAR STOCK
ROUND STOCK

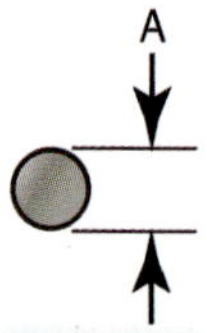

Diameter 'A' (inch)	Weight per Foot (pound)*
1/4	0.17
5/16	0.26
3/8	0.38
1/2	0.67
9/16	0.85
5/8	1.04
3/4	1.5
7/8	2.05
1	2.67

*Approximate Weight

Appendix XIV

BAR ANGLE
ANGLE IRON
ANGLE

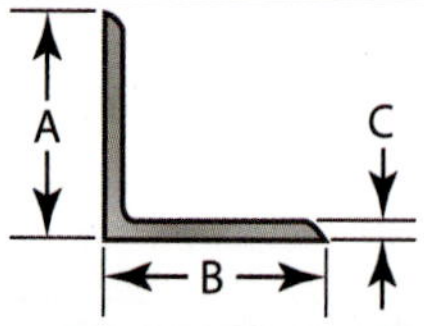

Dimension 'A' (inch)	Dimension 'B' (inch)	Thickness 'C' (inch)	Weight per Foot (pound)*
1/2	1/2	1/8	0.38
3/4	3/4	1/8	0.59
1	1	1/8	0.8
1	1	3/16	1.16
1	1	1/4	1.49
1 1/4	1 1/4	1/8	1.01
1 1/4	1 1/4	3/16	1.48
1 1/4	1 1/4	1/4	1.92
1 1/2	1 1/2	1/8	1.23
1 1/2	1 1/2	3/16	1.8
1 1/2	1 1/2	1/4	2.34
1 3/4	1 3/4	1/8	1.44
1 3/4	1 3/4	3/16	2.12
1 3/4	1 3/4	1/4	2.77
2	1 1/2	1/8	1.44
2	1 1/2	3/16	2.12
2	1 1/2	1/4	2.77
2	2	1/8	1.65
2	2	3/16	2.44
2	2	1/4	3.19
2	2	3/8	4.7

*Approximate Weight

Mild Steel (*Continued*)

Appendix XV

STANDARD CHANNEL
CHANNEL
CHANNEL IRON

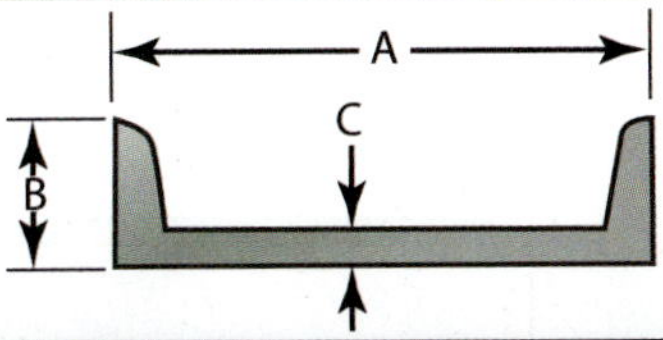

Web Width 'A' (inch)	Leg Height 'B' (inch)	Web Thickness 'C' (inch)	Weight per Foot (pound)*
3	1.355	0.13	3.5
3	1.41	0.17	4.1
3	1.498	0.258	5
3	1.596	0.356	6
4	1.56	0.14	4.5
4	1.584	0.184	5.4
4	1.647	0.247	6.25
4	1.721	0.321	7.25
5	1.75	0.19	6.7
5	1.885	0.325	9
6	1.92	0.2	8.2
6	2.034	0.314	10.5
6	2.157	0.437	13
7	2.09	0.21	9.8
8	2.26	0.22	11.5
8	2.343	0.303	13.75
8	2.527	0.487	18.75

*Approximate Weight

Appendix XVI

I-BEAM
STANDARD BEAM

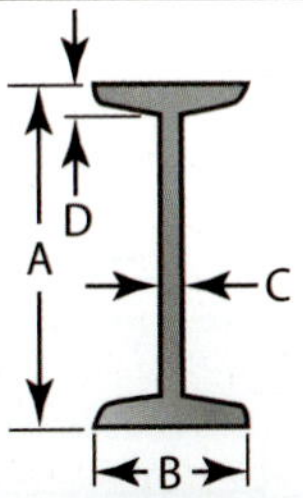

Size 'A' (inch)	Flange Width 'B' (inch)	Web Thickness 'C' (inch)	Flange Thickness 'D' (inch)	Weight per Foot (pound)*
3	2.33	0.17	0.26	5.7
3	2.51	0.35	0.26	7.5
4	2.66	0.19	0.29	7.7
4	2.80	0.33	0.29	9.5
5	3.00	0.21	0.33	10
6	1.84	0.11	0.17	4.4
6	3.33	0.23	0.36	12.5
6	3.57	0.47	0.38	17.25
8	4.00	0.27	0.43	18.4
8	4.17	0.44	0.43	23
10	2.69	0.16	0.21	9
10	4.66	0.31	0.49	25.4
10	4.94	0.59	0.49	35

*Approximate Weight

Appendix XVII

REBAR
REINFORCING ROD

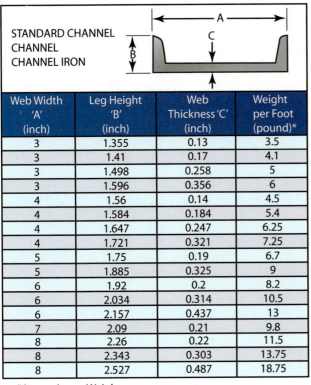

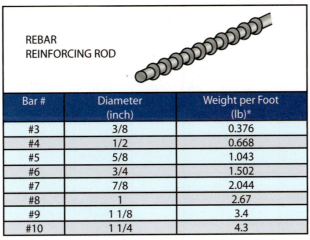

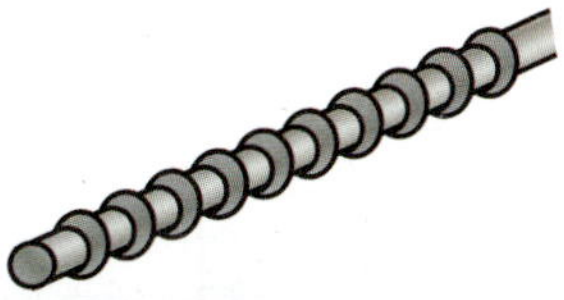

Bar #	Diameter (inch)	Weight per Foot (lb)*
#3	3/8	0.376
#4	1/2	0.668
#5	5/8	1.043
#6	3/4	1.502
#7	7/8	2.044
#8	1	2.67
#9	1 1/8	3.4
#10	1 1/4	4.3

*Approximate Weight

Mild Steel (*Continued*)

Appendix XVIII

FLAT BAR
BAR
HOT ROLL BAR
STRAPPING
FLATS

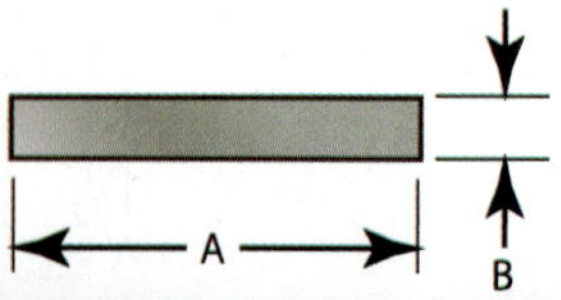

Thickness 'A' (mm)	Width 'B' (mm)	Weight per m (kg)*
3.18	25.40	0.63
	50.80	1.27
	76.20	1.90
6.35	12.70	0.63
	19.05	0.95
	25.40	1.27
	38.10	1.90
	50.80	2.53
	63.50	3.17
	76.20	3.80
	88.90	4.43
	101.60	5.06
	114.30	5.70
	127.00	6.33
7.94	127.00	7.91
	152.40	9.50
9.53	19.05	1.43
	25.40	1.90
	38.10	2.85
	50.80	3.80
	63.50	4.75
	76.20	5.70
	88.90	6.64
	101.60	7.60
	114.30	8.55
	127.00	9.50
12.70	25.40	2.53
	38.10	3.80
	50.80	5.06
	63.50	6.33
	76.20	7.60
	88.90	8.86
	101.60	10.13
	114.30	11.39
	127.00	12.66
15.88	50.80	6.33
	63.50	7.91
	76.20	9.50
	101.60	12.66
	127.00	15.83
19.05	50.80	7.60
	63.50	9.53
	76.20	11.39
	101.60	15.19
	127.00	18.99
25.40	50.80	10.13
	76.20	15.19
	88.90	17.73
	101.60	20.26
	127.00	25.35
31.75	25.40	6.33
	50.80	12.66
	76.20	18.99
	101.60	25.32

*Approximate Weight

Appendix XIX

PLATE
MILL PLATE
HOT ROLLED PLATE

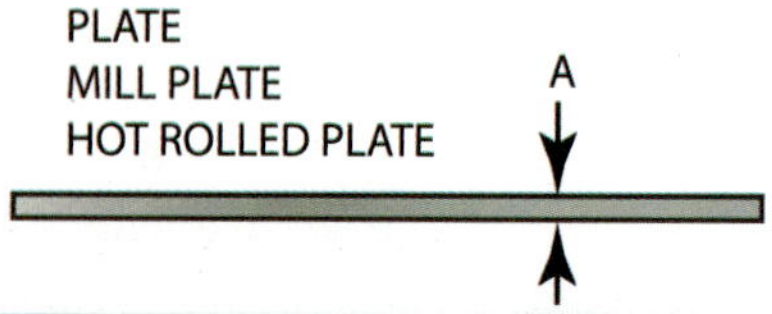

Thickness 'A' (mm)	Weight per Square m (kg)*
3.18	24.91
4.76	37.37
6.35	49.83
7.94	62.29
9.53	74.75
12.70	99.67
15.88	124.59
19.05	149.51
25.40	199.34

*Approximate Weight

Appendix XX

PIPE

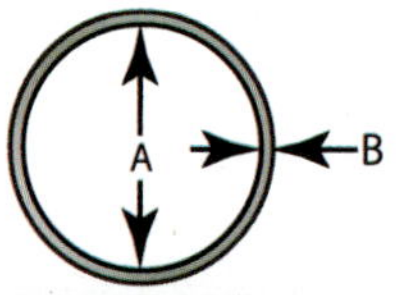

ID 'A' (mm)	Wall Thickness 'B' (mm)	OD (mm)	SCH	Weight per m (kg)*
6.35	2.03	13.72	40	0.63
9.53	2.31	17.15	40	0.85
12.70	2.77	21.34	40	1.27
19.05	2.87	26.67	40	1.68
25.40	3.38	33.40	40	2.50
31.75	3.56	42.16	40	3.38
31.75	4.85	42.16	80	4.47
38.10	3.68	48.26	40	4.05
38.10	5.08	48.26	80	5.41
50.80	3.91	60.33	40	5.44
50.80	5.54	60.33	80	7.48
63.50	5.16	73.03	40	8.62
63.50	7.01	73.03	80	11.41
76.20	5.49	88.90	40	11.29
76.20	7.62	88.90	80	15.27
88.90	5.74	101.60	40	13.57
88.90	8.08	101.60	80	18.63
101.60	6.02	114.30	40	16.07
101.60	8.56	114.30	80	22.31
127.00	6.55	141.30	40	21.78
127.00	9.53	141.30	80	30.95
152.40	7.11	168.28	40	28.26
152.40	10.97	168.28	80	42.56
203.20	8.18	219.08	40	42.53
203.20	12.70	219.08	80	64.63

*Approximate Weight

Mild Steel (*Continued*)

Appendix XXI

ROUND TUBING
TUBING

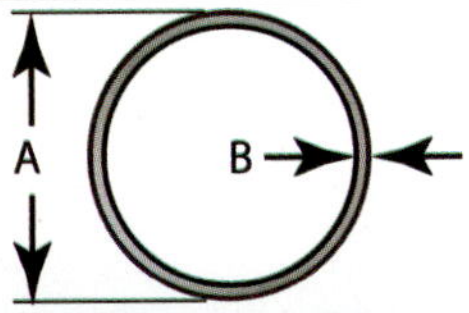

OD 'A' (mm)	Wall Thickness Gauge 'B'	Wall Thickness 'B' (mm)	Weight per m (kg)*
19.05	16	1.65	0.71
19.05	14	2.11	0.88
25.40	16	1.65	0.97
25.40	14	2.11	1.21
25.40	12	2.77	1.54
25.40	11	2.84	1.68
31.75	14	2.11	1.54
31.75	12	2.77	1.98
31.75	11	3.05	2.14
33.40	14	2.11	1.62
38.10	14	2.11	1.88
38.10	12	2.77	2.41
38.10	11	3.18	2.63
42.16	14	2.11	2.08
42.16	12	2.84	2.62
42.16	10	3.43	3.25
44.45	16	1.65	1.74
44.45	11	3.05	3.11
48.26	14	2.11	2.40
48.26	10	3.43	4.05
60.33	16	1.65	2.37
60.33	14	2.11	3.03
60.33	12	2.84	3.93
73.03	12	2.84	4.80
73.03	11	3.18	5.26
73.03	10	3.43	5.87

*Approximate Weight

Appendix XXII

SQUARE TUBING
TUBING

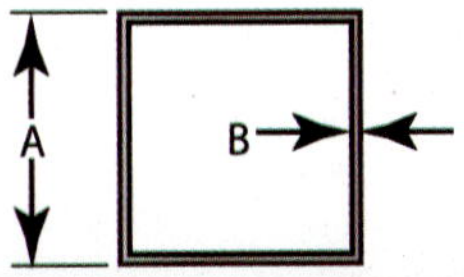

OD 'A' (mm)	Wall Thickness Gauge 'B'	Wall Thickness 'B' (mm)	Weight per m (kg)*
12.70	16	1.65	0.57
19.05	16	1.65	0.90
19.05	14	2.11	1.12
25.40	16	1.65	1.23
25.40	14	2.11	1.55
25.40	11	3.05	2.14
31.75	16	1.65	1.53
31.75	14	2.11	1.96
31.75	11	3.05	2.75
38.10	16	1.65	1.89
38.10	14	2.11	2.38
38.10	11	3.05	3.35
38.10	4.76	4.78	4.81
38.10	6.35	6.35	6.06

*Approximate Weight

Appendix XXIII

RECTANGULAR TUBING
TUBING

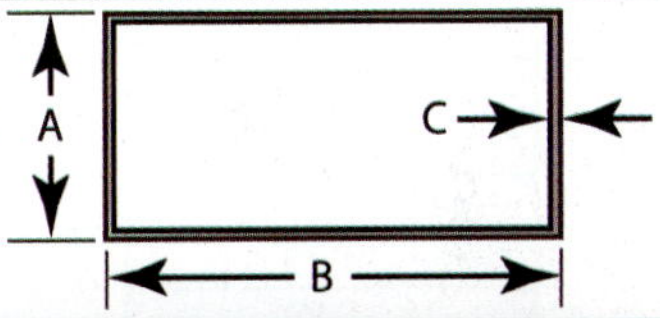

Side 'A' (mm)	Side 'B' (mm)	Wall Thickness Gauge 'C' (mm)	Wall Thickness 'C' (mm)	Weight per m (kg)*
25.4	6.35	16	1.65	0.85
38.1	25.40	11	3.05	2.74
38.1	25.40	14	2.11	1.97
38.1	25.40	16	1.65	1.53
50.8	25.40	16	1.65	1.86
50.8	25.40	14	2.11	2.38
50.8	25.40	11	3.05	3.35
50.8	38.10	11	3.05	3.96
50.8	38.10	14	2.11	3.22
63.5	38.10	14	2.11	2.74
76.2	25.40	12	2.77	4.07
76.2	25.40	14	2.11	3.65
76.2	25.40	16	1.65	2.47
76.2	38.10	11	3.05	5.41
76.2	38.10	4.76	4.78	7.55
76.2	38.10	14	2.11	4.07
76.2	50.80	6.35	6.35	10.59
76.2	50.80	4.76	4.78	8.33
76.2	50.80	11	3.05	5.78

*Approximate Weight

Mild Steel (*Continued*)

Appendix XXIV

SQUARE BAR
ROD
BAR STOCK
SQUARE STOCK

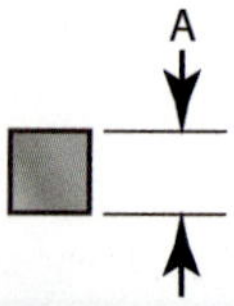

Dimension 'A' (mm)	Weight per m (k)*
12.70	1.27
15.88	1.98
19.05	2.85
22.23	3.88
25.40	5.06

*Approximate Weight

Appendix XXV

ROUND BAR
ROD
BAR STOCK
ROUND STOCK

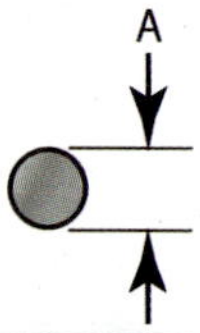

Diameter 'A' (mm)	Weight per m (kg)*
6.35	0.25
7.94	0.39
9.53	0.57
12.70	1.00
14.29	1.27
15.88	1.55
19.05	2.23
22.23	3.05
25.40	3.98

*Approximate Weight

Appendix XXVI

ANGLE IRON
BAR ANGLE
ANGLE

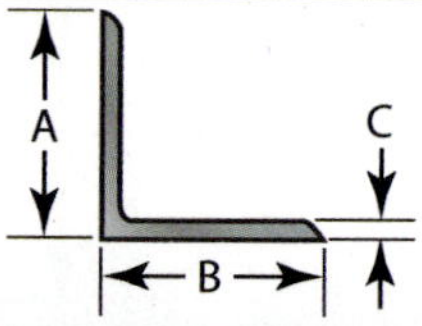

Dimension 'A' (mm)	Dimension 'B' (mm)	Thickness 'C' (mm)	Weight per m (kg)*
12.70	12.70	3.18	0.57
19.05	19.05	3.18	0.88
25.40	25.40	3.18	1.19
25.40	25.40	4.76	1.73
25.40	25.40	6.35	2.22
31.75	31.75	3.18	1.50
31.75	31.75	4.76	2.20
31.75	31.75	6.35	2.86
38.10	38.10	3.18	1.83
38.10	38.10	4.76	2.68
38.10	38.10	6.35	3.49
44.45	44.45	3.18	2.14
44.45	44.45	4.76	3.16
44.45	44.45	6.35	4.13
50.80	38.10	3.18	2.14
50.80	38.10	4.76	3.16
50.80	38.10	6.35	4.13
50.80	50.80	3.18	2.46
50.80	50.80	4.76	3.63
50.80	50.80	6.35	4.75
50.80	50.80	9.53	7.00

*Approximate Weight

Mild Steel (*Continued*)

Appendix XXVII

CHANNEL IRON
STANDARD CHANNEL
CHANNEL

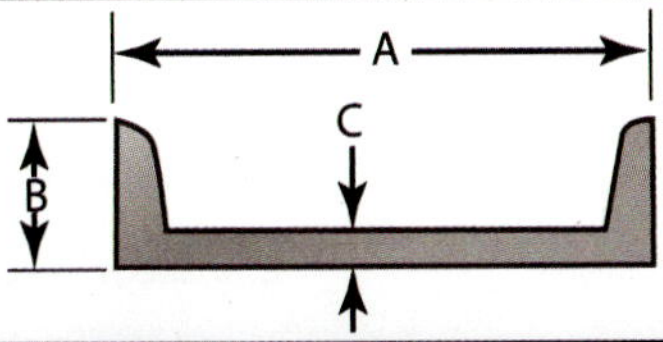

Web Width 'A' (mm)	Leg Height 'B' (mm)	Web Thickness 'C' (mm)	Weight per m (kg)*
76.2	34.42	3.30	5.21
76.2	35.81	4.32	6.11
76.2	38.05	6.55	7.45
76.2	40.54	9.04	8.94
101.6	39.62	3.56	6.70
101.6	40.23	4.67	8.04
101.6	41.83	6.27	9.31
101.6	43.71	8.15	10.80
127.0	44.45	4.83	9.98
127.0	47.88	8.26	13.41
152.4	48.77	5.08	12.21
152.4	51.66	7.98	15.64
152.4	54.79	11.10	19.36
177.8	53.09	5.33	14.60
203.2	57.40	5.59	17.13
203.2	59.51	7.70	20.48
203.2	64.19	12.37	27.93

*Approximate Weight

Appendix XXVIII

I-BEAM
STANDARD BEAM

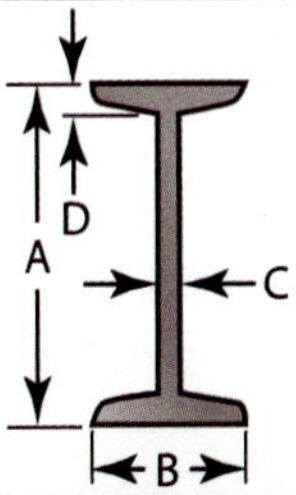

Size 'A' (mm)	Flange Width 'B' (mm)	Web Thickness 'C' (mm)	Flange Thickness 'D' (mm)	Weight per m (kg)*
76.2	59.18	4.32	6.60	8.49
76.2	63.73	8.86	6.60	11.17
101.6	67.64	4.90	7.44	11.47
101.6	71.02	8.28	7.44	14.15
127.0	76.30	5.44	8.28	14.90
152.4	46.84	2.90	4.34	6.55
152.4	84.63	5.89	9.12	18.62
152.4	90.55	11.81	9.63	25.69
203.2	101.63	6.88	10.80	27.41
203.2	105.94	11.20	10.80	34.26
254.0	68.33	3.99	5.23	13.41
254.0	118.39	7.90	12.47	37.83
254.0	125.58	15.09	12.47	52.13

*Approximate Weight

Appendix XXIX

REBAR
REINFORCING ROD

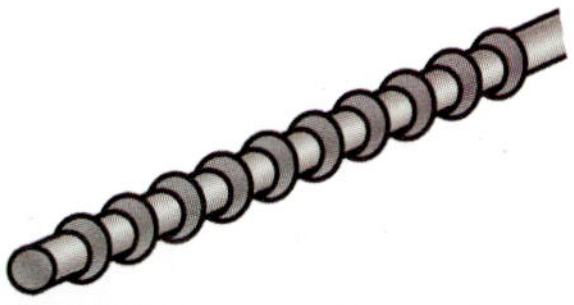

Bar #	Diameter (mm)	Weight per m (kg)*
#3	9.53	0.56
#4	12.70	0.99
#5	15.88	1.55
#6	19.05	2.24
#7	22.23	3.04
#8	25.40	3.98
#9	28.58	5.06
#10	31.75	6.40

*Approximate Weight

302, 304, and 316 Stainless Steel

Appendix XXX

FLAT BAR
BAR
STRAPPING
FLATS

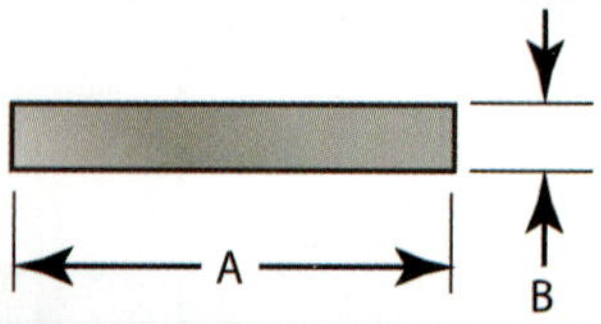

Thickness 'A' (inch)	Width 'B' (inch)	Weight per Foot (pound)*
1/8	1/2	.213
	3/4	.319
	1	.425
	1 1/2	.638
	2	.850
	2 1/2	1.06
	3	1.28
3/16	1/2	.319
	3/4	.478
	1	.638
	1 1/2	.957
	2	1.28
	2 1/2	1.59
	3	1.91
1/4	1/2	.425
	3/4	.636
	1	.850
	1 1/2	1.28
	2	1.70
	2 1/2	2.12
	3	2.55

*Approximate Weight

Appendix XXXII

FLAT BAR
BAR
STRAPPING
FLATS

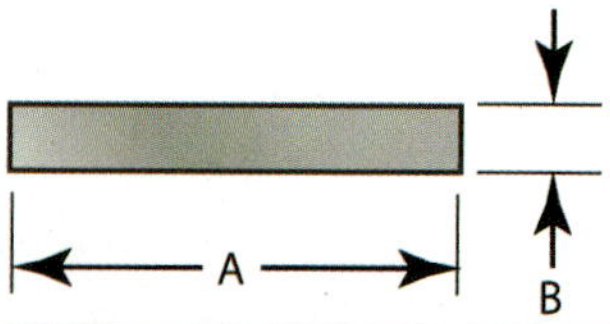

Thickness 'A' (mm)	Width 'B' (mm)	Weight per m (kg)*
3.18	12.70	0.317
	19.05	0.475
	25.40	0.633
	38.10	0.950
	50.80	1.266
	63.50	1.579
	76.20	1.907
4.76	12.70	0.475
	19.05	0.712
	25.40	0.950
	38.10	1.425
	50.80	1.907
	63.50	2.368
	76.20	2.845
6.35	12.70	0.633
	19.05	0.947
	25.40	1.266
	38.10	1.907
	50.80	2.532
	63.50	3.158
	76.20	3.798

*Approximate Weight

Appendix XXXI

PLATE

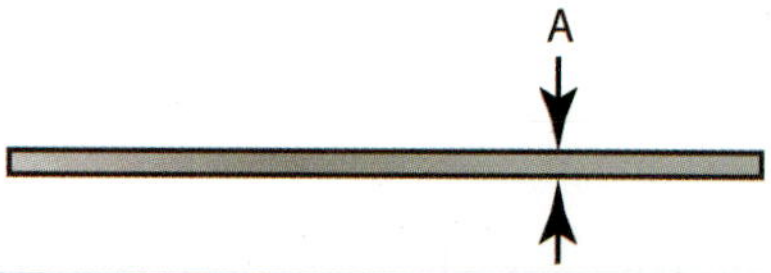

Thickness 'A' (inch)	Weight per Square Foot (lb)*
1/8	5.323
3/16	7.985
1/4	10.646
5/16	13.308
3/8	15.968
1/2	21.291
5/8	26.614
3/4	31.937
1	42.582

*Approximate Weight

Appendix XXXIII

PLATE

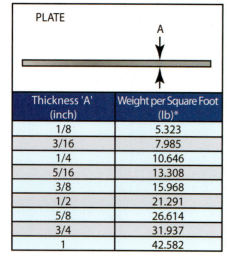

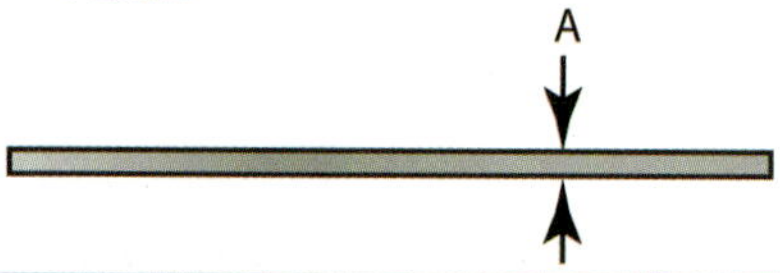

Thickness 'A' (mm)	Weight per Square m (kg)*
3.18	7.93
4.76	11.89
6.35	15.86
7.94	19.82
9.53	23.78
12.70	31.71
15.88	39.64
19.05	47.57
25.40	63.43

*Approximate Weight

1100, 6062, and 6063 Aluminum

Appendix XXXIV

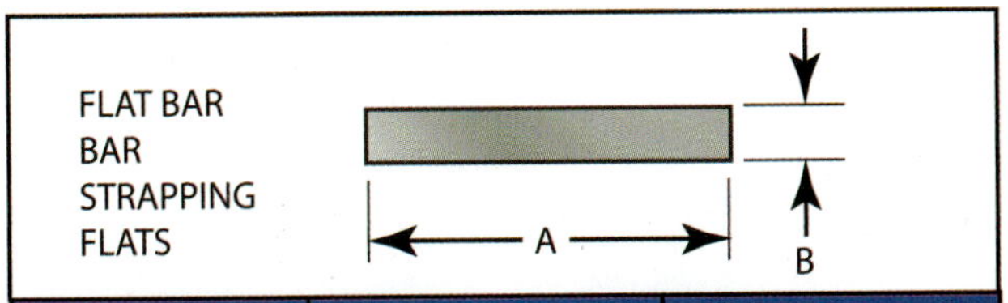

Thickness 'A' (inch)	Width 'B' (inch)	Weight per Foot (pound)*
1/8	1/2	.073
	3/4	.110
	1	.147
	1 1/2	.220
	2	.294
	2 1/2	.367
	3	.441
3/16	1/2	.110
	3/4	.165
	1	.220
	1 1/2	.330
	2	.441
	2 1/2	.551
	3	.660
1/4	1/2	.147
	3/4	.220
	1	.294
	1 1/2	.441
	2	.587
	2 1/2	.734
	3	.881

*Approximate Weight

Appendix XXXVI

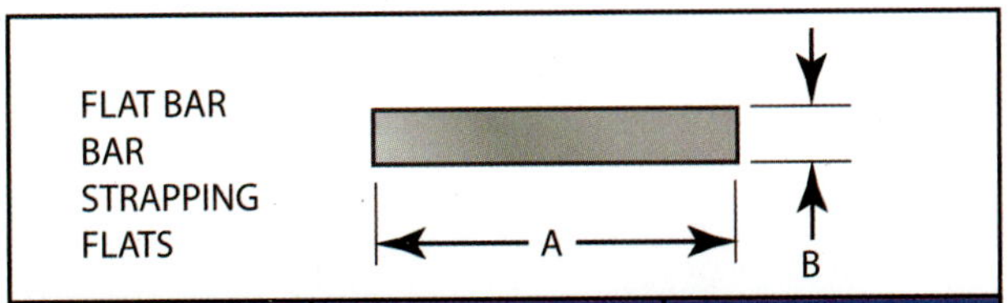

Thickness 'A' (mm)	Width 'B' (mm)	Weight per m (kg)*
3.18	12.70	0.109
	19.05	0.164
	25.40	0.219
	38.10	0.328
	50.80	0.438
	63.50	0.547
	76.20	0.657
4.76	12.70	0.164
	19.05	0.246
	25.40	0.328
	38.10	0.492
	50.80	0.657
	63.50	0.821
	76.20	0.983
6.35	12.70	0.219
	19.05	0.328
	25.40	0.438
	38.10	0.657
	50.80	0.874
	63.50	1.093
	76.20	1.312

*Approximate Weight

Appendix XXXV

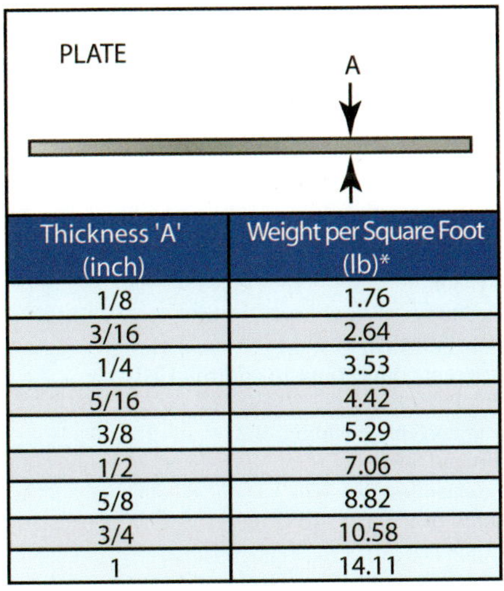

Thickness 'A' (inch)	Weight per Square Foot (lb)*
1/8	1.76
3/16	2.64
1/4	3.53
5/16	4.42
3/8	5.29
1/2	7.06
5/8	8.82
3/4	10.58
1	14.11

*Approximate Weight

Appendix XXXVII

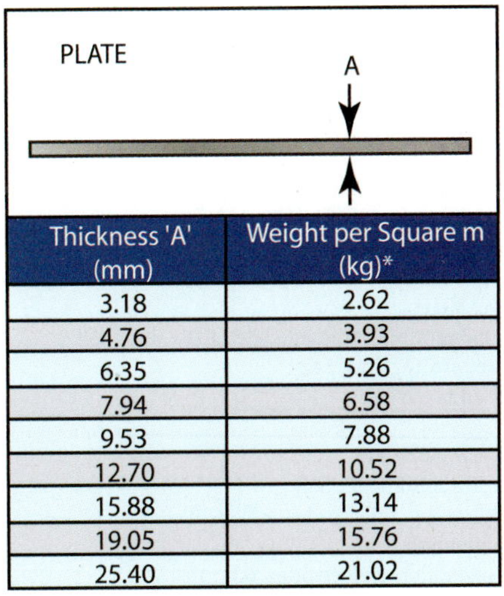

Thickness 'A' (mm)	Weight per Square m (kg)*
3.18	2.62
4.76	3.93
6.35	5.26
7.94	6.58
9.53	7.88
12.70	10.52
15.88	13.14
19.05	15.76
25.40	21.02

*Approximate Weight

Index

Note: Page numbers followed by f and t indicates figures and tables.

B

F

G

H

I

J

K

L

N

O

Q

R

T

U

V

W

X

Y

Z